SEDIMENTARY PROCESSES, ENVIRONMENTS AND BASINS

Other publications of the International Association of Sedimentologists

SPECIAL PUBLICATIONS

37 Continental Margin Sedimentation
From Sediment Transport to Sequence Stratigraphy
Edited by C.A. Nittrouer, J.A. Austin, M.E. Field, J.H. Kravitz, J.P.M. Syvitski and P.L. Wiberg
2007, 549 pages, 178 illustrations

36 Braided Rivers
Process, Deposits, Ecology and Management
Edited by G.H. Sambrook Smith, J.L. Best, C.S. Bristow and G.E. Petts
2006, 390 pages, 197 illustrations

35 Fluvial Sedimentology VII
Edited by M.D. Blum, S.B. Marriott and S.F. Leclair
2005, 589 pages, 319 illustrations

34 Clay Mineral Cements in Sandstones
Edited by R.H. Worden and S. Morad
2003, 512 pages, 246 illustrations

33 Precambrian Sedimentary Environments
A Modern Approach to Ancient Depositional Systems
Edited by W. Altermann and P.L. Corcoran
2002, 464 pages, 194 illustrations

32 Flood and Megaflood Processes and Deposits
Recent and Ancient Examples
Edited by I.P. Martini, V.R. Baker and G. Garzón
2002, 320 pages, 281 illustrations

31 Particulate Gravity Currents
Edited by W.D. McCaffrey, B.C. Kneller and J. Peakall
2001, 320 pages, 222 illustrations

30 Volcaniclastic Sedimentation in Lacustrine Settings
Edited by J.D.L. White and N.R. Riggs
2001, 312 pages, 155 illustrations

29 Quartz Cementation in Sandstones
Edited by R.H. Worden and S. Morad
2000, 352 pages, 231 illustrations

28 Fluvial Sedimentology VI
Edited by N.D. Smith and J. Rogers
1999, 328 pages, 280 illustrations

27 Palaeoweathering, Palaeosurfaces and Related Continental Deposits
Edited by M. Thiry and R. Simon Coinçon
1999, 408 pages, 238 illustrations

26 Carbonate Cementation in Sandstones
Edited by S. Morad
1998, 576 pages, 297 illustrations

25 Reefs and Carbonate Platforms in the Pacific and Indian Oceans
Edited by G.F. Camoin and P.J. Davies
1998, 336 pages, 170 illustrations

24 Tidal Signatures in Modern and Ancient Sediments
Edited by B.W. Flemming and A. Bartholomä
1995, 368 pages, 259 illustrations

23 Carbonate Mud-mounds
Their Origin and Evolution
Edited by C.L.V. Monty, D.W.J. Bosence, P.H. Bridges and B.R. Pratt
1995, 543 pages, 330 illustrations

16 Aeolian Sediments
Ancient and Modern
Edited by K. Pye and N. Lancaster
1993, 175 pages, 116 illustrations

3 The Seaward Margin of Belize Barrier and Atoll Reefs
Edited by N.P. James and R.N. Ginsburg
1980, 203 pages, 110 illustrations

1 Pelagic Sediments on Land and Under the Sea
Edited by K.J. Hsu and H.C. Jenkyns
1975, 448 pages, 200 illustrations

REPRINT SERIES

4 Sandstone Diagenesis: Recent and Ancient
Edited by S.D. Burley and R.H. Worden
2003, 648 pages, 223 illustrations

3 Deep-water Turbidite Systems
Edited by D.A.V. Stow
1992, 479 pages, 278 illustrations

2 Calcretes
Edited by V.P. Wright and M.E. Tucker
1991, 360 pages, 190 illustrations

SPECIAL PUBLICATION NUMBER 38 OF THE INTERNATIONAL ASSOCIATION OF SEDIMENTOLOGISTS

Sedimentary Processes, Environments and Basins: a Tribute to Peter Friend

EDITED BY

Gary Nichols, Ed Williams and Chris Paola

SERIES EDITOR

Ian Jarvis
School of Earth Sciences & Geography
Centre for Earth & Environmental Science Research
Kingston University
Penrhyn Road
Kingston upon Thames KT1 2EE
UK

Blackwell Publishing

BLACKWELL PUBLISHING
350 Main Street, Malden, MA 02148–5020, USA
9600 Garsington Road, Oxford OX4 2DQ, UK
550 Swanston Street, Carlton, Victoria 3053, Australia

First published 2007 by Blackwell Publishing Ltd

1 2007

Library of Congress Cataloging-in-Publication Data

Sedimentary processes, environments, and basins : a tribute to Peter Friend / edited by Gary Nichols, Ed Williams and Chris Paola.
p. cm. — (Special publication number 38 of the International Association of Sedimentologists)
Includes bibliographical references and index.
ISBN 978-1-4051-7922-5 (pbk. : alk. paper)
1. Sedimentation and deposition. 2. Environmental geology. 3. Sedimentary basins. I. Friend, P. F. II. Nichols, Gary. III. Williams, Ed, 1960– IV. Paola, C. (Chris)
QE571.S4164 2007
551.3—dc22 2007032295

A catalogue record for this title is available from the British Library.

Set in 10.5/12.5pt Palatino by Graphicraft Limited, Hong Kong
Printed and bound in Singapore by Markono Print Media Pte Ltd

The publisher's policy is to use permanent paper from mills that operate a sustainable forestry policy, and which has been manufactured from pulp processed using acid-free and elementary chlorine-free practices. Furthermore, the publisher ensures that the text paper and cover board used have met acceptable environmental accreditation standards.

For further information on
Blackwell Publishing, visit our website:
www.blackwellpublishing.com

Contents

Sedimentary processes, environments and basins – a tribute to Peter Friend: introduction

GARY NICHOLS*†, ED WILLIAMS‡ *and* CHRIS PAOLA§

**Department of Geology, Royal Holloway University of London, Egham, Surrey, TW20 0EX, UK (Email: g.nichols@gl.rhul.ac.uk)*
†University Centre on Svalbard, P.O. Box 156, Longyearbyen, N-9171, Norway
‡CRPG, B.P. 20, 15, rue Notre-Dame des Pauvres, 54501 Vandoeuvre-les-Nancy Cedex, France
§Saint Anthony Falls Laboratory, Department of Geology and Geophysics, University of Minnesota, Minneapolis, MN 55414, USA

It is one thing to be a good scientist, but the scientific community would soon be impoverished if some of those good scientists were not also able to inspire and help others. For several decades Peter Friend has been one of the leading figures in sedimentology and throughout that time he has helped scores of other people by supervising doctoral students, collaborating with colleagues, especially in developing countries, and being willing selflessly to share ideas with fellow geologists. All those who have worked with Peter know what a rich experience it is – he is not only inspirational as a scientist, but through his relaxed and friendly manner he reminds us of the pleasure both of doing good science and of doing well by people in the process. Peter's style eschews cut-throat competition and one-upmanship but rather encourages the open sharing of scholarship. The scientific community of sedimentologists has been enriched by Peter's scientific and human contribution, and this volume is a small way of saying thank you to him.

The idea of holding some form of conference 'event' started circulating soon after Peter formally retired as a full-time academic in the Department of Earth Sciences, Cambridge University, in 2001. A European meeting of the International Association of Sedimentologists seemed an appropriate forum, and the meeting being held in Coimbra, Portugal, in September 2004 was in the right place (close to the areas where Peter had worked in Spain) at the right time. The IAS Bureau, and in particular Judith McKenzie and José Pedro Calvo, provided support and encouragement, and the organizing committee of the Coimbra meeting (in particular Pedro Proença e Cunha) arranged for the first morning plenary session of the meeting to be dedicated to Peter, allowing us to invite three keynote speakers to speak on themes related to his work. We are grateful to all the Coimbra meeting organizers for allowing us to devote such a significant part of their conference to honouring Peter Friend. The contributors to the plenary session, and others presenting papers in related sessions of the meeting, were invited to contribute papers to this volume, and subsequently a more general invitation was issued to those who we thought might like to provide a manuscript.

This collection of papers is a token of thanks from a number of people who have benefited from an association with Peter, whether as doctoral students, research collaborators or just fellow scientists who have encountered him somewhere along the way.

PETER FRIEND

Academic leadership comes in many forms. In Peter's case it is a subtle blend of encouragement, enthusiasm and inspiration. The most immediate beneficiaries have been the many PhD students (over 30) who have been supervised by Peter. Some worked in areas which were core to Peter's own research interests, such as the Old Red Sandstone provinces of the North Atlantic, the Cenozoic basins of Spain and the foothills of the Himalayas, whereas others have carried out their fieldwork in exotic places as far afield as the Antarctic, Siberia and Canada, and worked on topics as varied as carbonate and evaporite sedimentology, volcaniclastics and coal basins. These doctoral students were from the United Kingdom, North America and South Asia, but there have also been researchers from other countries such as Spain and Portugal who have worked with Peter on many projects. An enthusiasm for collaboration has always been a hallmark of Peter's career, and the outcomes have been very fruitful. In some cases he has worked with researchers in other fields of earth science, such as fluid dynamics to better understand sediment transport processes in rivers (Dade & Friend, 1998; Friend & Dade, 2005), or using provenance techniques to unravel exhumation and erosion histories (White *et al.*, 2002). Peter has also collaborated with local geologists in the countries where he has carried out research, for example in Spain (Friend *et al.*, 1981; Friend & Dabrio, 1996), India (Friend & Sinha, 1993; Sinha & Friend, 1994, 1999; Sinha *et al.*, 1996) and Pakistan (Abbasi & Friend, 1989, 1993, 2000; Friend *et al.*, 2001).

International collaboration is not always easy: sometimes people feel protective about 'their' patch of geology, they do not always welcome others coming along to work in the same area, and they may be suspicious of suggestions of joint project proposals. Peter's gentle style of diplomacy seems to have allowed him to work with anyone, anywhere. Any tensions which might exist between countries do not seem to have hindered Peter working with, for example, both Pakistani and Indian colleagues during the course of his work in the Himalayan foothills, and even the rivalries which used to exist between different geology departments in Spain apparently posed few problems.

The sharing of ideas is always one of the objectives of scientific conferences, and so Peter has long been a contributor to national and international meetings. These conferences have not necessarily always been the big international jamborees, but instead the smaller, local or regional conferences, such as the annual meetings of the IAS. Every four years since 1977 an international meeting of fluvial sedimentologists has taken place, and Peter can claim to have attended more of these fluvial meetings than almost anybody else. Part of the attraction for all who attend these meetings has always been the opportunities to participate in field trips in locations like eastern Australia, South Africa, northern Spain (Fig. 1) and the Rocky Mountains. These relatively small meetings, and the field excursions associated with them, have created an international community of fluvial sedimentologists, within which Peter has long played a leading part.

Fig. 1 Oligo-Miocene alluvial-fan conglomerate body in the Ebro Basin, Spain, an area where Peter Friend has worked for many years and led field trips there as part of International Fluvial Sedimentology Conferences in 1981 and 1989.

Closer to home, in the UK geological community, Peter was one of the first to be involved with the British Sedimentological Research Group (BSRG) in the 1960s, which were the early days of modern sedimentology. At that time, the concepts of looking at sedimentary rocks in terms of processes of deposition and the recognition of facies were still relatively new, and the discipline of sedimentology has made huge advances during the course of Peter's career. Peter has continued to regularly attend the annual BSRG meetings, held at university geology departments around the British Isles, for many years. The emphasis in BSRG annual meetings has always been to provide a forum for postgraduate students and postdoctoral workers to present their work in a supportive context, and as such they strike a chord with Peter's own approach to fostering and encouraging research in sedimentology.

The 'Friends of the Devonian' is a loose association of enthusiasts of the Old Red Sandstone of the North Atlantic borderlands who have regularly held informal field meetings in Britain and Ireland. Although probably considered by many to have been dormant for a while, some of these 'friends' recently got together with others to put together a collection of papers under the editorship of Peter Friend and Brian Williams (Friend & Williams, 2000), a timely synthesis of recent work on the tectonic development and controls on depositional facies of the 'Old Red Sandstone continent'.

Fig. 2 Red beds of Devonian strata on Spitsbergen, where Peter Friend worked with Mark Moody-Stuart on the distinctive characteristics of ancient fluvial systems.

Peter's papers at conferences are typically delivered in a manner that is deceptively low-key, but they leave you thinking afterwards. Paper titles such as 'Distinctive features of some ancient river systems' (Friend, 1978) and 'Towards the field classification of alluvial architecture or sequence' (Friend, 1983) are similarly beguiling. These are landmark papers in which Peter says 'here are some issues that need to be considered' rather than providing complete answers and neat classifications. The test of these is that the aspects of fluvial sedimentology which are covered in these papers have been revisited over and over again by those who have followed after. Apart from his own presentations, Peter contributes to the conference proceedings with his ability to ask the most incisive questions in the most understated and non-confrontational way.

Many of Peter's earliest papers were on the Devonian of Spitsbergen (Fig. 2), Scotland and East Greenland, covering aspects of stratigraphy and sedimentology of the areas in which he and his colleagues and students carried out fieldwork. These thorough, detailed field studies provided the basis for new ideas about fluvial systems of the past, including the concept of downstream decrease in discharge (Friend & Moody-Stuart, 1972), and used systematic, statistical approaches to the analysis of sedimentological data (Friend *et al.*, 1970b). Studies closer to home in Norfolk in collaboration with one of the other leading figures of sedimentology, John R.L. Allen, led to a better understanding of subaqueous dune behaviour (Allen & Friend, 1976a,b). A long association with Spanish sedimentology began with fieldwork in Cenozoic fluvial deposits of the Ebro Basin leading to a much-cited paper (Friend *et al.*, 1979) which was one of the first to look at the architecture of fluvial deposits in the stratigraphic record, followed by later papers which expanded on this theme (Friend, 1983; Friend *et al.*, 1986). In particular, this work highlighted the concept of 'multistorey' sand bodies (Fig. 3), which remains a core idea and source of insight in alluvial architecture to this day. Some of the ideas on river systems which Peter had formed in Devonian and Cenozoic rocks were pursued further in the Himalayas, with studies on the Siwalik Group in Pakistan (e.g. Abbasi & Friend, 2000, Friend *et al.*, 2001) and on modern deposits of the Indo-Gangetic plain (e.g. Sinha & Friend, 1994; Fig. 4).

Fig. 3 Multistorey fluvial channel-fill sandstone bodies, Oligo-Miocene, Ebro Basin, Spain.

Fig. 4 Modern river systems, southern Himalayas.

Outside of 'pure' sedimentology, the paper of Peter Friend's which probably receives the most citations was co-authored with Gian Ori, then at the University of Bologna: 'Sedimentary basins formed and carried piggyback on active thrust sheets' (Ori & Friend, 1984). Once again, this presented a deceptively simple concept by demonstrating that basins in thrust belts can be allochthonous, and showed what features can be used to show this. Other authors have subsequently used different terminology ('thrust-top basin', 'wedge-top trough'), but the idea is essentially the same.

A look through the catalogue of Peter's publications over a period of 45 years reveals a mixture of papers which focus on documenting data (e.g. the work in East Greenland, Friend *et al.*, 1976a,b; Friend & Alexander-Marrack, 1976; Friend & Nicholson, 1976; Friend & Yeats, 1978), reviews of areas or processes (e.g. Friend, 1969, 1973, 1981, 1996) and what might be called 'ideas' papers (e.g. Friend, 1993; Stolum & Friend, 1997; Friend *et al.*, 1999). The total number of papers is in the high 80s and counting, ranging from field guides and local journals, to the most prestigious international journals. These publications have provided data, comparison between disparate areas, and ideas which have helped a couple of generations of geologists.

THIS VOLUME

The only criterion that we adopted in our invitation to contribute to this volume was that the work should be in some way related to the themes of Peter Friend's research career. Of course, this provided a huge scope because, as is apparent from the work published by Peter and his research students, there is hardly any aspect of sedimentology which would be excluded on this basis. Nevertheless, some general themes have emerged that do reflect Peter's interests, and these have formed the basis for a division of the volume into four sections.

Tectonics and sedimentation

It is easy to forget that the concept of studying sediments in their tectonic context, and looking for evidence of tectonic controls on sedimentation and stratigraphy, has not always been mainstream sedimentology. Many of Peter's papers have considered sedimentary rocks from this viewpoint, starting with some of his earliest work on the Devonian of the Isle of Arran, Scotland (Friend *et al.*, 1963), the Pyrenees (Friend *et al.*, 1996) and the Himalayas (Abbasi & Friend, 2000). These different scales of tectonic controls are also represented in the ten papers grouped under this theme in this volume.

The Spanish Pyrenees and the Himalayas provide excellent case studies of the interaction of thrust tectonics and fluvial sedimentation, themes on which Peter published a number of times (Abbasi & Friend, 1989, 2000; Friend *et al.*, 1989, 1996, 1999; Lloyd *et al.*, 1998). The Pyrenees is one of the case studies used by **Vergés** (this volume) in his review of thrust tectonics and fluvial sedimentation, which also draws on information from Iran and South America. Of the other papers that concentrate on sedimentation in compressional settings, **Brandes *et al.*** (this volume) look at the evolution of a thrust belt in Costa Rica and **Lopes & Cunha** (this volume) also use sedimentary data to unravel the development of the Algarve margin, Portugal. **Merino *et al.*** (this volume) relate cycles in Carboniferous shallow marine carbonate and clastic facies to their tectonic setting in a piggy-back basin, whereas a second paper on carbonates by **Aurell *et al.*** (this volume) considers deposition in an extensional setting. A major Neogene extensional regime, the Gulf of Corinth in Greece, is the focus of a study by **Ford *et al.*** (this volume), who provide a detailed analysis of Pleistocene fan-delta deposition. The same area is also used as a source of examples for **Leeder & Mack** (this volume) in a paper that reviews climatic and tectonic controls on erosion and incision. To complete the range of tectonic settings, there are two papers which consider deposition in strike-slip settings, one in Portugal by **Gomes *et al.*** (this volume), and the other in the Apennines by **Pascucci *et al.*** (this volume). Finally, a larger scale of tectonic control on sedimentation is tackled by **van den Belt & de Boer** (this volume) who analyse the relations between isostasy and subsidence in large evaporite basins.

Landscape evolution and provenance

The importance of the relationship between the evolution of hinterland landscape and the supply of sediment to sedimentary basins is now becoming recognized as a first-order control on the distribution of facies and the basin stratigraphy. Peter recognized this in his papers from the Pyrenees (Lloyd *et al.*, 1998) and the Himalayas (Abbasi & Friend, 1989, 2000; White *et al.*, 2002), which used provenance studies in proximal foreland basin deposits to unravel the evolution of part of the mountain belt. In this volume, one of Peter's Himalayan co-workers, **Burbank** (this volume), shows how single-crystal dating can be used to provide a record of orogenesis. On the same theme there are also case studies by **Rieser *et al.*** (this volume) and **Veiga-Pires *et al.*** (this volume), who have used radiometric dating of mica and zircons, respectively, as tools in provenance analysis. The paper by **MacDonald *et al.*** (this volume) neatly links the theme of landscape evolution with another recurring feature of Peter's career, the Old Red Sandstone: two of Peter's papers (Friend *et al.*, 1970a; Friend & Ramos, 1982) looked at the unconformity at the base of the ORS succession that has been analysed and modelled by MacDonald *et al.* (this volume).

Depositional systems

A recurring theme in the titles of the papers in this section of the volume is 'anatomy', a word which is not, of course, being used in a physiological sense, but as a term to indicate a detailed analysis of a package of sedimentary rocks. Friend *et al.* (1989) and Friend (1996) are two examples from Peter's work where a large-scale approach has been used to determine the controls on a depositional system, in these cases, fluvial facies. The development of a package of fluvial strata at a time of low relative sea level is considered by **Veiga *et al.*** (this volume) and in **Messina *et al.*** (this volume) and mixtures of different shallow marine facies are analysed in terms of deposition during rising sea level. **Leren *et al.*** (this volume) and **Janbu *et al.*** (this volume) are related case studies by a team from Bergen, Norway, working with colleagues in Turkey on aspects of clastic sedimentation in north-central Turkey.

Fluvial sedimentation

Fluvial sedimentology has always played a prominent role in Peter's research career, and he is probably more renowned for his work on ancient river systems than any other area. The five papers in this volume are concerned with very diverse themes in the subject. The study based on satellite images of western Africa by **Ori *et al.*** (this volume) demonstrates how drainage systems have been modified as a result of climate changes in the Quaternary. These modern rivers bear many similarities with the ancient systems discussed by **Nichols** (this volume), some of which are the 'terminal' systems first recognized and commented on by Friend (1978). Channel and overbank facies in a modern river system are studied by **Azevêdo *et al.*** (this volume) and experimental techniques are used by **Sheets *et al.*** (this volume) to help understand the creation of multistorey channel bodies. The last paper, by **Cuevas *et al.*** (this volume), is particularly appropriate as it is a recent study of the spectacular ribbons of fluvial channel-fill sandstone which are exposed in the southern part of the Ebro Basin. Working in the late 1970s and early 1980s with Cai Puigdefàbregas, then of the Catalan Geological Survey, and Oriol Riba and his students at the University of Barcelona, Peter drew attention to these examples of Miocene river channels which could be traced for kilometres across the landscape. These truly three-dimensional outcrops informed the classification schemes of Friend *et al.* (1979) and Friend (1983), and in the new paper by Cuevas *et al.* (this volume), the processes by which the channels are filled with sand are considered.

FINAL WORDS

Editing a collection of papers is a reminder to those of us not regularly acting as a journal editor of the essential role that reviewers play in the publication process. Asking the contributors to the volume to review other submitted contributions seemed a little incestuous to us, so in most cases we sought appropriate expertise from elsewhere. We are very grateful for the time and effort that so many people put into reading these papers, providing detailed and helpful comments, and generally helping to maintain the standards that we sought to achieve. It is only appropriate that their contribution to this volume should be formally acknowledged. The reviewers were: Lawrence

Amy, Phil Ashworth, Jaco Baas, Vic Baker, Dan Bosence, Doug Boyd, John Bridge, Bryan Cronin, Ian Davison, Tim Dooley, Cindy Ebinger, Todd Ehlers, Javier Fernandez-Suarez, Andre Freiwald, Massimiliano Ghinassi, John Graham, Stuart Hardy, Robert Hillier, Philip Hirst, James Howard, John Howell, Stuart Jones, Susan Marriott, Allard Martinius, José Martinez-Catalan, Francesco Massari, Neil Meadows, Colin North, Lesley Perg, Duncan Pirrie, Piret Plink-Bjorklund, Josep Poblet, Szczepan Porebski, Ian Reid, Alastair Robertson, Ruth Robinson, Marco Roveri, Greg Sambrook-Smith, Gary Smith, Ed Sobel, Ian Somerville, Fabrizio Storti, Esther Stouthamer, Johan ten Veen, Maurice Tucker, Jonathan Turner, David Ulicny, Steve Vincent, Jaume Vergés, Tony Watts and Moyra Wilson.

This tribute has been produced at this time because 2004 marked the 50th anniversary of when Peter started to formally study geology and this seemed an appropriate occasion to mark. To say that it marked his 'retirement' would be inaccurate: his latest project, *Southern England: Scenery and Structure*, marks a new stage in his career, as the first of a series of books that will bring geology to a wider public audience. Doubtless these books will inspire people to think about geology in the world around them in the way that Peter has inspired so many of us in our academic and other careers in sedimentology.

ACKNOWLEDGEMENTS

The contribution of Chris Paola to this work was supported by the National Center for Earth-surface Dynamics (NCED), a Science and Technology Center funded by the Office of Integrative Activities of the U.S. National Science Foundation under agreement Number EAR-0120914.

REFERENCES

Abbasi, I.A. and Friend, P.F. (1989) Uplift and evolution of the Himalayan orogenic belt, as recorded in the foredeep molasse sediments. *Z. Geomorphol. Suppl.*, **76**, 75–88.

Abbasi, I.A. and Friend, P.F. (1993) Fluvial sole structures from the Siwalik Group of Pakistan. *Geol. Bull. Univ. Peshawar*, **26**, 103–112.

Abbasi, I.A. and Friend, P.F. (2000) Exotic conglomerates of the Neogene Siwalik succession and their implications for the tectonic and topographic evolution of the Western Himalaya. In: *Tectonics of the Nanga Parbat Syntaxis and the Western Himalaya* (Eds M.A. Khan, P.J. Treloar, M.P. Searle and M.Q. Jan), pp. 455–466. Special Publication 170, Geological Society Publishing House, Bath.

Allen, J.R.L. and Friend, P.F. (1976a) Relaxation time of dunes in decelerating aqueous flows. *J. Geol. Soc. London*, **132**, 17–26.

Allen, J.R.L. and Friend, P.F. (1976b) Changes in intertidal dunes during two spring-neap cycles, Lifeboat Station Bank, Wells-next-the-Sea, Norfolk (England). *Sedimentology*, **23**, 329–346.

Dade, W.B. and Friend, P.F. (1998) Grain-size, sediment-transport regime, and channel slope in alluvial rivers. *J. Geol.*, **106**, 661–671.

Friend, P.F. (1969) Tectonic features of Old Red sedimentation in North Atlantic borders. In: *North Atlantic Geology and Continental Drift* (Ed. M. Kay), pp. 703–710. Memoir 12, American Association of Petroleum Geologists, Tulsa, OK.

Friend, P.F. (1973) Devonian stratigraphy of Greenland and Svalbard. In: *Arctic Geology* (Ed. M.G. Pitcher), pp. 467–468. Memoir 19, American Association of Petroleum Geologists, Tulsa, OK.

Friend, P.F. (1978) Distinctive features of some ancient river systems. In: *Fluvial Sedimentology* (Ed. A.D. Miall), pp. 531–542. Memoir 5, Canadian Society of Petroleum Geologists, Calgary.

Friend, P.F. (1981) Devonian sedimentary basins and deep faults of the northernmost Atlantic borderlands. In: *Geology of the North Atlantic Borderlands* (Eds J.W. Kerr and A.J. Ferguson), pp. 149–163. Memoir 7, Canadian Society of Petroleum Geologists, Calgary.

Friend, P.F. (1983) Towards the field classification of alluvial architecture or sequence. In: *Modern and Ancient Fluvial Systems* (Eds J.D. Collinson and J. Lewin), pp. 345–354. Special Publication 6, International. Association of Sedimentologists. Blackwell Scientific Publications, Oxford.

Friend, P.F. (1993) Control of river morphology by the grain-size of sediment supplied. *Sediment. Geol.*, **85**, 171–177.

Friend, P.F. (1996) The development of fluvial sedimentology in some Devonian and Tertiary basins. *Cuad. Geol. Iberica*, **21**, 55–69.

Friend, P.F. and Alexander-Marrack, P.D. (1976) Devonian sediments of East Greenland III. *Medd. Grønl.*, **206**(3), 1–121.

Friend, P.F. and Dabrio, C.J. (Eds) (1996) *Tertiary Basins of Spain: the Stratigraphic Record of Crustal Kinematics.*

World and Regional Geology, Vol. 6. Cambridge University Press, Cambridge, xvi, 400 pp.

Friend, P.F. and Dade, W.B. (2005) Transport modes and grain-size patterns in fluvial basins. In: *Fluvial Sedimentology VII* (Eds M.D. Blum, S.B. Marriott and S.F. Leclair), pp. 399–408. Special Publication 35, International. Association of Sedimentologists. Blackwell Scientific Publications, Oxford.

Friend, P.F. and Moody-Stuart, M. (1972) Sedimentation of the Wood Bay Formation (Devonian) of Spitsbergen: regional analysis of a late orogenic basin. *Nor. Polarinst. Skr.*, **157**, 77 pp.

Friend, P.F. and Nicholson, J. (1976) Devonian sediments of East Greenland V. *Medd. Grønl.*, **206**(5), 1–117.

Friend, P.F. and Ramos, A. (1982) Upper Old Red sedimentation near the unconformity at Arbroath. *Scot. J. Geol.*, **18**, 297–315.

Friend, P.F. and Sinha, R. (1993) Braiding and meandering parameters. In: *Braided Rivers* (Eds J.L. Best and C.S. Bristow), pp. 105–111. Special Publication 75, Geological Society Publishing House, Bath.

Friend, P.F. and Williams, B.P.J. (Eds) (2000) *Perspectives on the Old Red Sandstone*. Special Publication 180, Geological Society Publishing House, Bath, 623 pp.

Friend, P.F. and Yeats, A.K. (1978) Devonian sediments of East Greenland IV. *Medd. Grønl.*, **206**(4), 1–112.

Friend, P.F., Harland, W.B. and Hudson, J.D. (1963) The Old Red Sandstone and the Highland Boundary in Arran, Scotland. *Trans. Edinb. Geol. Soc.*, **19**, 363–425.

Friend, P.F., Harland, W.B. and Smith, A.G. (1970a) Reddening and fissuring associated with the Caledonian unconformity in north-west Arran. *Proc. Geol. Assoc.*, **81**, 75–85.

Friend, P.F., Alexander-Marrack, P.D. and Yeats, A.K. (1970b) Mark sensing for recording and analysis of sedimentological data. In: *Proceedings of the Symposium on Data Processing in Biology and Geology* (Ed. J.L. Cutbill), Department of Geology, University of Cambridge, 24–26 September 1969. Special Volume No. 3, The Systematics Association, pp. 1–16.

Friend, P.F., Alexander–Marrack, P.D., Nicholson, J. and Yeats, A.K. (1976a) Devonian sediments of East Greenland I. *Medd. Grønl.*, **206**(1), 1–56.

Friend, P.F., Alexander-Marrack, P.D., Nicholson, J. and Yeats, A.K. (1976b) Devonian sediments of East Greenland II. *Medd. Grønl.*, **206**(2), 1–91.

Friend, P.F., Slater, M.J. and Williams, R.C. (1979) Vertical and lateral building of river sandstone bodies, Ebro Basin, Spain. *J. Geol. Soc. London*, **136**, 39–46.

Friend, P.F., Marzo, M., Nijman, W. and Puigdefabregas, C. (1981) Fluvial sedimentology in the Tertiary South Pyrenean and Ebro Basins, Spain. In: *Field Excursion Guide, International Fluvial Conference*, Keele, September 1981, pp. 1–49.

Friend, P.F., Hirst, J.P.P. and Nichols, G.J. (1986) Sandstone-body structure and river process in the Ebro basin of Aragon, Spain. *Cuad. Geol. Iberica*, **10**, 9–30.

Friend, P.F., Hirst, J.P.P., Hogan, P.J., *et al.* (1989) *Pyrenean Tectonic Control of Oligo-Miocene River Systems, Huesca, Aragon, Spain*. Excursion Guidebook for the 4th International Conference of the International Association of Sedimentologists on Fluvial Sedimentology, 142 pp.

Friend, P.F., Lloyd, M.J., McElroy, R., Turner, J., van Gelder, A. and Vincent, S.J. (1996) Evolution of the central part of the northern Ebro margin, as indicated by its Tertiary fluvial sedimentary infill. In: *Tertiary Basins of Spain: the Stratigraphic Record of Crustal Kinematics* (Eds P.F. Friend and C.J. Dabrio), pp. 166–172. Cambridge University Press.

Friend, P.F., Jones, N.E. and Vincent, S.J. (1999) Drainage evolution in active mountain belts: extrapolation backwards from present-day Himalayan river patterns. In: *Fluvial Sedimentology VI* (Eds N.D. Smith and J. Rogers), pp. 305–313. Special Publication 28, International. Association of Sedimentologists. Blackwell Scientific Publications, Oxford.

Friend, P.F., Raza, S.M., Geehan, G. and Sheikh, K.A. (2001) Intermediate-scale architectural features of the fluvial Chinji Formation (Miocene), Siwalik Group, northern Pakistan. *J. Geol. Soc. London*, **158**, 163–178.

Lloyd, M.J., Nichols, G.J. and Friend, P.F. (1998) Oligo-Miocene alluvial-fan evolution at the southern Pyrenean Thrust Front, Spain. *J. Sediment. Res.*, **68**, 869–878.

Ori, G.G. and Friend, P.F. (1984) Sedimentary basins formed and carried piggyback on active thrust sheets. *Geology*, **12**, 475–478.

Sinha, R. and Friend, P.F. (1994) River systems and their sediment flux, Indo-Gangetic plains, Northern Bihar, India. *Sedimentology*, **41**, 825–845.

Sinha, R. and Friend, P.F. (1999) Pedogenic alteration in the overbank sediments, North Bihar Plains, India. *J. Geol. Soc. India*, **53**, 163–171.

Sinha, R., Friend, P.F. and Switzur, V.R. (1996) Radiocarbon dating and sedimentation rates in the Holocene alluvial sediments of the northern Bihar plains, India. *Geol. Mag.*, **133**, 85–90.

Stolum, H-H. and Friend, P.F. (1997) Percolation theory applied to simulated meanderbelt sandbodies. *Earth Planet. Sci. Lett.*, **153**, 265–277.

White, N.M., Pringle, M.S., Garzanti, E., *et al.* (2002) Constraints on the exhumation and erosion of the High Himalayan Slab, NW India, from foreland basin deposits. *Earth Planet. Sci. Lett.*, **195**, 29–44.

Basin-fill incision, Rio Grande and Gulf of Corinth rifts: convergent response to climatic and tectonic drivers

MICHAEL R. LEEDER* *and* GREG H. MACK†

**School of Environmental Sciences, University of East Anglia, Norwich NR4 7TJ, UK (Email: m.leeder@uea.ac.uk)*
†Department of Geological Sciences, New Mexico State University, Las Cruces, New Mexico 88003, USA

ABSTRACT

The influence of tectonics and climate on basin-fill erosion and incision in the Gulf of Corinth rift, central Greece, and Rio Grande rift, southwest USA, is examined. Overall, it is suggested that climate change controls the downstream water:sediment ratio and sediment transport capacity, via operation of the continuity equation. Tectonics, specifically the rapid growth and propagation of structures, sets up gradient contrasts and upstream-migrating changes in transport capacity via operation of the diffusivity equation. A steady-state aggradational mode operated in the southern Rio Grande rift between ~ 5 and 0.8 Ma, causing preservation of ancestral axial-channel and floodplain deposits due to relatively slow, long-term active rift subsidence. The onset of major climatic change around 0.8 Ma resulted in the axial river periodically incising to a total extent of ~ 150 m, removing about 25% by volume of previously accumulated sediment, despite continued active faulting and fault-induced subsidence. This climatic mode is interpreted to be a periodic response to positive downstream gradients in sediment transport rate during glacial and glacial-transition periods, caused by low-level external sediment sourcing and a dominance of large magnitude spring snowmelt floods from northern mountain valleys. Tectonic drivers are spectacularly demonstrated in the Gulf of Corinth, where a new theory of 'piggy-back' basin abandonment and regional uplift is proposed, as formerly active rift-margin faults are progressively dragged above the flat slab of underriding African lithosphere. Basin abandonment occurred across newly propagating faults, with erosion and basin-fill incision of up to 800 m depth, as discontinuities in drainage channel slope have migrated rapidly upstream. In both rifts, sediment relaxation time, $T_s = l^2/\kappa$, where l is a length scale and κ is sediment diffusivity, is probably short, since relevant length scales are small and diffusivities large. Thus in the Rio Grande rift, despite the great length of the river system as a whole, it is the balance between hydrological and sediment input from the many lateral tributaries that controls non-uniform transport capacity of the axial channel. In the case of the Gulf of Corinth rift, it is the high strain rate that causes diffusivity to be large in drainages cutting across rapidly vertically-growing normal faults.

Keywords Rift sedimentation, Rio Grande rift, Corinth rift, fluvial incision.

INTRODUCTION

Sedimentary basins are, by definition, sediment sinks, for which general quantitative depositional models have been developed in recent years (e.g. Paola *et al.*, 1992). Generally, tectonics creates the potential space for deposition, with changes of gradient and differential tilting allowing either deposition or erosion, depending on the circumstances. In this way, tectonics may be said to control the spatial distribution of sedimentary environments (e.g. Gawthorpe & Leeder, 2000). For example, the lateral and vertical growth of active faults and folds perturbs local or regional surface gradients. Such effects are epitomized by the workings of 'piggy-back' tectonics in propagating thrust systems (Ori & Friend, 1984), or by the lateral growth of normal faults (Leeder & Jackson,

1993; Jackson & Leeder, 1994) and thrust faults (Jackson *et al.*, 1996; Gupta, 1997). Climate on the other hand provides the water needed for vegetation and soil development, the runoff that allows hillslope sediment transport and thus the eventual sediment and water flux into a river system. Changing climate perturbs these variables, and erosion is commonplace in the Quaternary and older sedimentary record of tectonically active basins (Blum & Törnqvist, 2000). One reason for erosion is global or local sea-level fall or lake drawdown causing exposure and erosion (Posamentier & Vail, 1988). On coastal plains, river channels may override subsidence by incising in response to lowered sea level across a gradient change at the coastal plain–shelf break (Posamentier & Vail, 1988; Miall, 1991; Leeder & Stewart, 1996; Blum & Törnqvist, 2000). A second reason relevant to continental basins is that climate change can alter the balance of a river's hydrological and sediment supply variables (Dethier, 2001).

There is currently a great chasm between hydrologically based landscape process models (Kirkby & Cox, 1995; Kirkby *et al.*, 1998; Kirkby, 1999; Bogaart & van Balen, 2002; Bogaart *et al.*, 2002, 2003) and tectono-stratigraphic architectural models such as the initial development by Bridge & Leeder (1979) and more recent efforts by Mackey & Bridge (1995). Existing architectural models are kinematic rather than dynamic, and they are artificially separated from hydrologically based process models; they should in fact be a subclass driven by the hydrological budget. It is a mistake to assume that the hydrological cycle in a catchment acts kinematically: it is in fact dynamic, with both physical and thermodynamic energy transfers and transformations taking place constantly within the system. Thus catchment processes create the entire basin landscape from a number of prior conditions, rather analogous to the 'nature versus nurture' concept for individual development, whereby the genetic make-up of an individual (nature-providing) is acted upon by external circumstances (nurture-modifying). Tectonics and lithological make-up are the given catchment genes whereas hydrological variables nurture and modify the basin infill.

In this contribution, the possible reasons for Quaternary incision in the deposits of the tectonically active Rio Grande rift, southwest USA, and the Gulf of Corinth rift, central southern Greece, are explored. These two extensional rift structures occur in distinctive overall physiographic settings, the former astride the linear continental ridge of the southern Rocky Mountains (Eaton 1987), the latter as a marine Gulf that cuts the ancestral Hellenide Mountains (Zelt *et al.*, 2005). Although both rifts are tectonically active and their drainages have both suffered the vagaries of Quaternary climate change, it is proposed that climatic drivers have determined the occurrence and timing of incision in the former, whereas large-scale migration of the locus of faulting and regional uplift has determined incision in the latter.

FLUVIAL EQUILIBRIUM: SEDIMENT CONTINUITY AND DIFFUSION CONTROLS

Fluvial equilibrium

Fluvial equilibrium has a long history in geomorphology via the concepts of 'graded', 'poised', 'balanced', or 'regime' stream channels (Mackin, 1948; Lane, 1955). Significant breakthroughs in the appreciation of fluvial and landscape equilibria came with seminal papers by Schumm & Lichty (1965) and Chorley & Kennedy (1971). The stream-power approach to mechanical equilibrium introduced by Bagnold (1966) has proved fruitful in tackling the problem. This approach relates mass-transport rate to stream-power under conditions of steady, unidirectional water flow down uniformly sloping channels of constant cross-sectional area, with solid impermeable banks over beds composed of ample supplies of granular sediment. Bull (1979) and Leopold & Bull (1979) made major attempts to reconcile traditional concepts of fluvial equilibrium with Bagnold's mechanistic approach, the latter authors defining a graded equilibrium stream as '. . . one in which, over a period of years, slope, velocity, depth, width, roughness, pattern, and channel morphology delicately and mutually adjust to provide the power and efficiency necessary to transport the load supplied from the drainage basin *without aggradation or degradation of the channels*' (emphasis is ours).

Disequilibrium in the fluvial system was spectacularly demonstrated by Gilbert's (1917) early observations on aggradation in Californian rivers following the incorporation of vast volumes of tailings waste from Sierra Nevadan gold mining. Since that time, many studies have elaborated upon disequilibrium in natural channels. Particularly noteworthy is the contribution of Church & Ryder (1972) on post-glacial incision and the widespread degradation, arroyo cutting and entrenchment documented in the southwest USA from the early 1900s to about the 1970s (e.g. Hereford, 1993).

Sediment continuity and diffusion equations

Changes to sediment surface elevation over time intervals long enough to sample many disparate discharge events depend on the existence of spatial gradients in mean sediment transport rate per unit width, $\bar{i}$. This approach neglects transport unsteadiness at these longer time scales (Paola *et al.*, 1992) and enables a one-dimensional sediment continuity equation, sometimes termed the Exner Equation (see Paola, 2000), to be written as

$$\frac{\mathrm{d}\bar{h}}{\mathrm{d}t} = -\left[\frac{\partial \bar{i}}{\partial x} \cdot \frac{1}{\sigma(1-\phi)}\right] \tag{1}$$

where $\bar{h}$ is mean surface elevation measured upward (positive) or downward (negative) along a normal y axis relative to coordinates fixed below the bed (Fig. 1A), t is time, x is the downslope sediment transport direction, σ is sediment density and ϕ is deposited sediment porosity. Neglecting subsurface sediment compaction, equilibrium conditions (*sensu* Leopold & Bull, 1979) occur when $\mathrm{d}\bar{h}/\mathrm{d}t = 0$. A common climatic/hydrological reason for non-zero downstream transport gradient, $\partial\bar{i}/\partial x$, is excess water supply relative to sediment supply or vice-versa (see Lane (1955) for a pioneering exploration of this concept; also reviewed by Blum & Törnqvist (2000)).

A second reason for non-zero downstream transport gradient involves changes to gradient brought about by either tectonic tilting or base-level

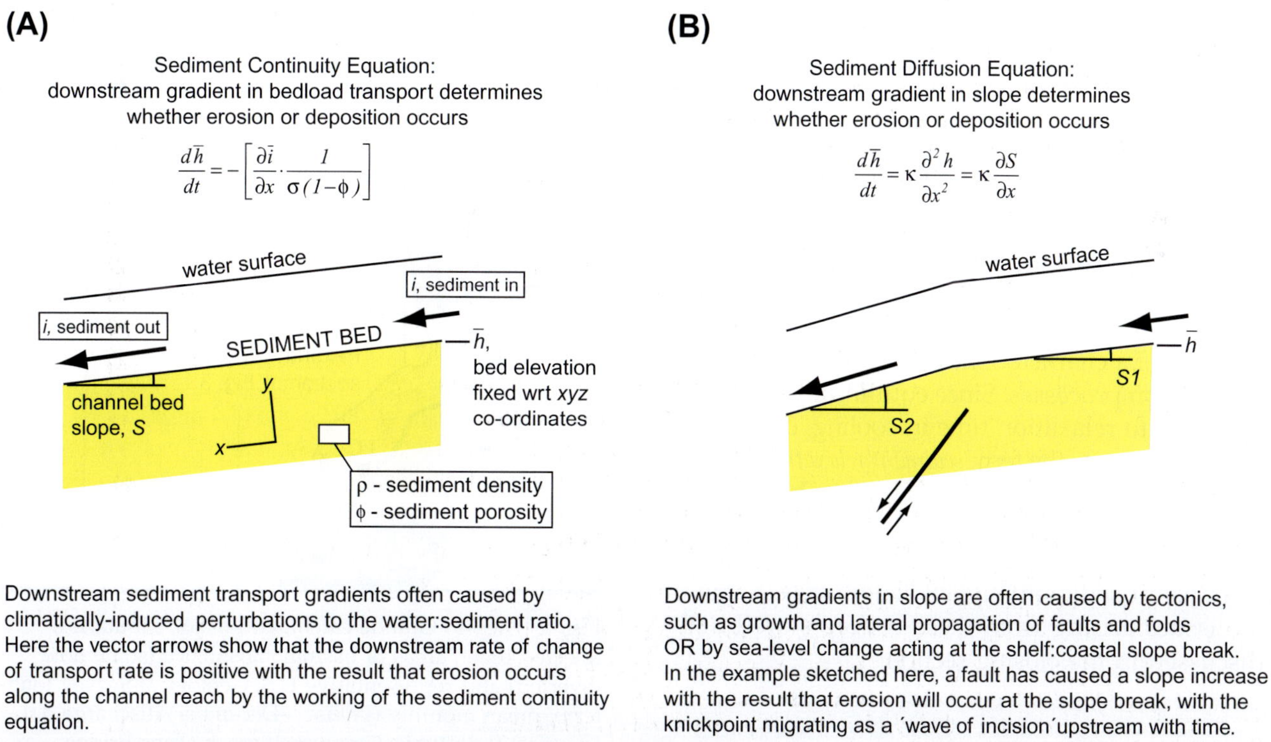

Fig. 1 Definition sketches for the sediment continuity and diffusion equations used in analysing downstream changes required to bring about deposition or erosion in river channels flowing through sedimentary basins. (A) The downstream gradient in sediment transport is positive, hence erosion occurs. (B) The downstream change of slope is positive and the diffusion equation causes erosion at the point of maximum curvature (slope change), which migrates upstream at some rate determined by both the magnitude of curvature and sediment diffusion coefficient.

fall causing a river to run over a steeper gradient reach (Fig. 1B). In both cases, the gradient curvature induces excess bed shear stress, increased stream-power and erosion downstream, the point of curvature propagating upstream as an erosional knickpoint discontinuity at a rate determined by the two-dimensional diffusion equation:

$$\frac{d\bar{h}}{dt} = \kappa \frac{\partial^2 h}{\partial x^2} = \kappa \frac{\partial S}{\partial x} \qquad (2)$$

where S is slope ($\partial h/\partial x$) and κ is a sediment diffusion coefficient (Begin, 1988; Leeder & Stewart, 1996).

Sediment relaxation time

In a notable development, Paola *et al.* (1992) introduced the concept of *equilibrium time* to the study of river deposition in sedimentary basins. This is the time required for a basin to reach equilibrium, e.g. such that the sedimentation rate balances the basin subsidence rate. It also refers to the efficiency with which a river traversing any given basin can transmit depositional or erosional 'signals', e.g. in the form of changed grain size, sediment flux or, in the cases considered below, incision. The general concept was previously used (Begin *et al.*, 1981; Begin, 1988; Paola *et al.*, 1991; Leeder & Stewart, 1996) in computing the rate of upstream propagation of fluvial incision following a base-level fall. It assumes that sediment transport is linearly proportional to local bed slope, an assumption that leads to an estimate of a characteristic diffusion coefficient for the incision processes. Since equilibrium time is an analogue to relaxation time in cooling, conducting thermal systems, the term *sediment relaxation time*, T_s, is preferred here. Generally, this is given by $T_s = l^2/\kappa$, where l is a streamwise length scale and κ is sediment diffusivity. Rivers with larger T_s take longer to transmit sediment signals lengthwise. Some of the issues arising from these concepts are briefly discussed in the closing section.

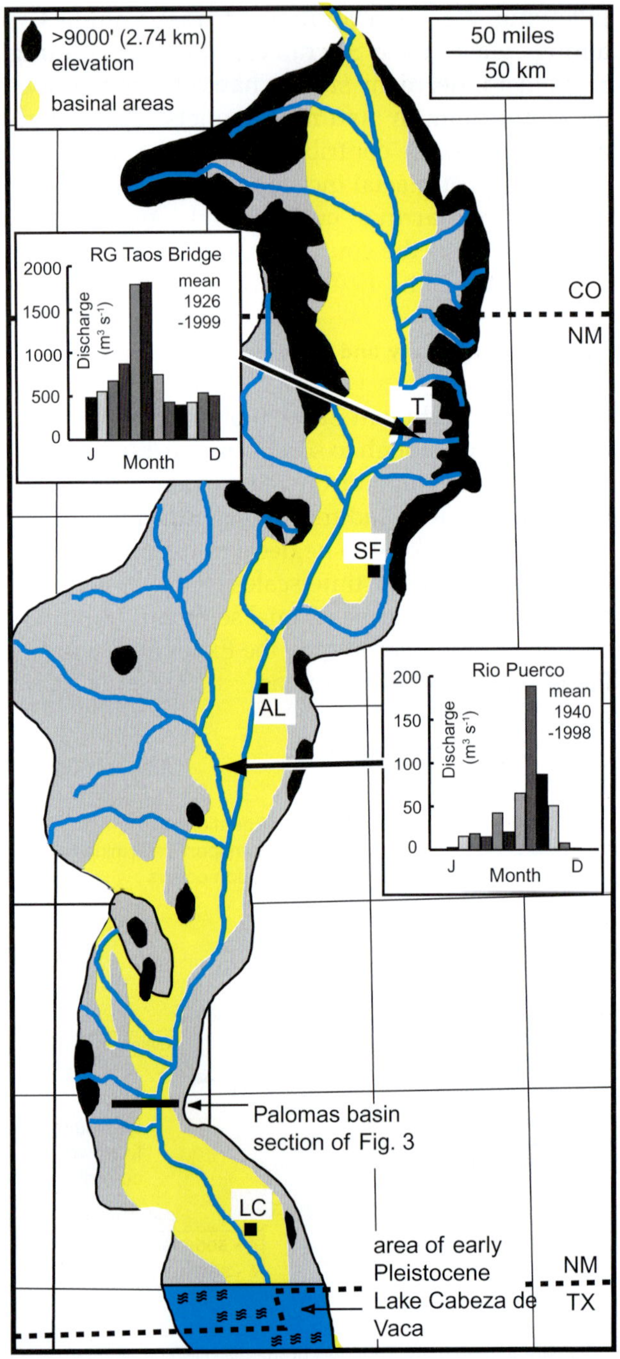

Fig. 2 The Rio Grande catchment, major tributaries, selected relief, areas of basinal subsidence (some now inverted) and representative U.S. Geological Survey long-term, mean monthly (January–December) discharge data (cumecs) for a typical snowmelt reach (Taos Bridge guaging station, note spring peak) compared with a summer monsoonal reach (Rio Puerco station, note late summer peak). CO – Colorado, NM – New Mexico, TX – Texas, T – Taos, SF – Santa Fe, AL – Albuquerque, LC – Las Cruces.

RIO GRANDE RIFT

Unique among large rivers of the world flowing through actively subsiding basins, the Rio Grande (Fig. 2) features spectacular exposures of its own past alluvial deposits in a deeply incised valley. This allows detailed comparison of both net aggradational and net degradational river activity over several million years. In southern New Mexico, alluvial sediments of the Camp Rice and Palomas Formations have long been taken to represent deposits of an ancestral Pliocene to early Pleistocene upper Rio Grande that flowed southwards through a series of adjacent and contiguous basins (Mack, 2004). The river initially emptied into ephemeral Lake Cabeza de Vaca near the Texas–New Mexico border (Fig. 2). At around 2.25 Ma, the lake was drained, presumably by headward cutting of an ancestral lower Rio Grande, allowing flow of a united Rio Grande to the Gulf of Mexico (Gustavson, 1991; Galloway *et al.*, 2000).

The aggradational phase (Gilbert to Matuyama Chrons)

In recent years, magnetostratigraphic subdivision of the Camp Rice and Palomas Formations, supported by radioisotopic dating of basalts, tuffs, and pumice beds, has revealed the presence of Gilbert, Gauss, and Matuyama Chrons and their subchrons, hence documenting some 4 Myr of ancestral Rio Grande depositional activity, stretching from early Pliocene through early Pleistocene time (Mack *et al.*, 1993, 1998, 2002; Leeder *et al.*, 1996b). Extensive three-dimensional exposures of the Camp Rice and Palomas Formations reveal river-channel and floodplain lithofacies indicative of deposition in channel-bar complexes of low-sinuosity, pebbly sand-bed channels that traversed generally dry floodplains (Mack & James, 1993; Perez-Arlucea *et al.*, 2000). Sandbody architecture is generally multistorey, with the density of sandbodies related to topographic 'funnelling' of ancestral channels through loci of maximum subsidence adjacent to active faults (Mack & Seager, 1990; Leeder *et al.*, 1996a; Mack *et al.*, 2002). In some cases, the axial river cut the toes of tributary alluvial fans, locally eroding the fans back to the border faults (Mack & Leeder, 1999; Leeder & Mack, 2001; Mack *et al.*, 2002). Despite the common presence of multistorey channel deposits, there is no evidence that ancestral Rio Grande channels ever incised older deposits more than the scour depth of one storey, about one bankfull depth (~ 3–5 m). Similarly, alluvial-fan cycles consisting of erosion, streamflood deposition, debris-flow deposition, and palaeosol development have been attributed to climatically driven changes in sediment yield within an overall aggradational regime (Mack & Leeder, 1999). The dated sedimentary records of the Palomas and Mesilla basins indicate mean aggradation rates for the Gilbert and Matuyama Chrons of around 0.03 mm yr^{-1}, with higher rates approaching 0.07 mm yr^{-1} for Gauss time (Fig. 3). For an aggrading continental basin floored by alluvial fans and an axial river, it is reasonable to assume that mean aggradation rate is roughly equivalent to mean subsidence rate.

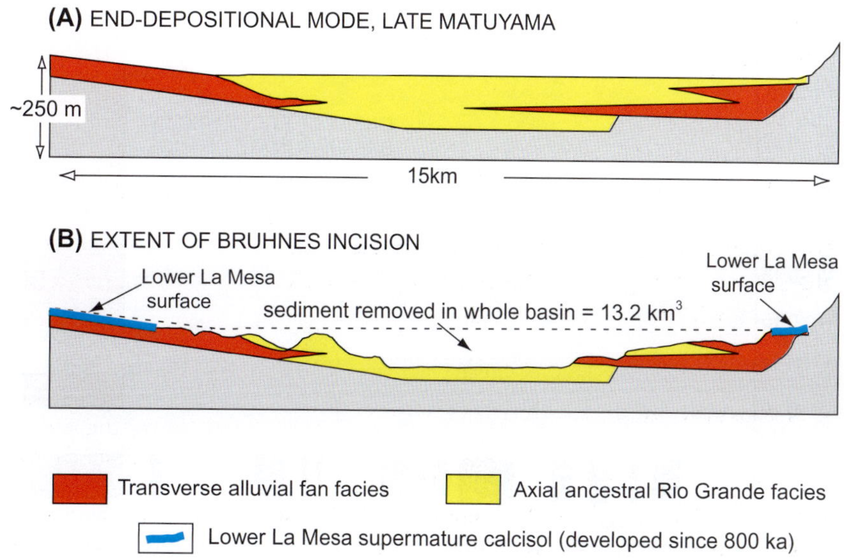

Fig. 3 Restored (A) and present-day (B) cross-sections of the Palomas basin (see Fig. 2 for location) to illustrate the extent of Bruhnes incision into the Plio-Pleistocene basin fills of the southern Rio Grande rift accumulated over the previous ~ 4 Myr (data for B from Leeder *et al.*, 1996b).

Periodic incision and erosion (Brunhes Chron)

Following its aggradational phase, the ancestral Rio Grande began a periodic process of significant incision alternating with partial backfilling that has lasted to the present day (Gile *et al.*, 1981). In the southern Rio Grande rift the onset of incision is dated near the Matuyama–Brunhes chron boundary (~ 0.8 Ma) by magnetostratigraphy and dated volcanic ashes (Kortemeier, 1982; Mack *et al.*, 1993, 1998, 2002). Although not as accurately dated, initial fluvial incision in the central and northern segments of the Rio Grande rift of New Mexico appears to have occurred at roughly the same time as in the south (Connell, 2004; Smith, 2004). In the southern Rio Grande rift the valley floor is now 100–150 m below the La Mesa geomorphological surface, which marks the level of the floodplain at 0.8 Ma (Gile *et al.*, 1981). This has occurred despite fault-displacement rates of > 0.1 mm yr^{-1}, as evidenced by the offset of middle to late Pleistocene and Holocene fan surfaces (Machette, 1987; Foley *et al.*, 1988; Seager & Mack, 2003). Although normal-fault displacement does not necessarily equate to basin subsidence, it is unlikely that basin-bounding faults created sediment accommodation space during the aggradational phase (Pliocene–early Pleistocene), but failed to do so during the subsequent degradational phase (middle Pleistocene–Holocene). The incision has resulted in the removal of up to 25% by volume (calculated for the Palomas basin) of Pliocene to lower Pleistocene alluvium. Mean rates of incision, averaged over the entire Brunhes Chron, are in the range of 0.13–1.19 mm yr^{-1}, many times that of the sediment accumulation rate in the previous aggradational regime, and comparable to other recent estimates of fluvial landscape incision in the wider western USA (Dethier, 2001). Mean late Pleistocene–Holocene subsidence rates in the southern rift are ~ 0.1 mm yr^{-1} according to the studies of Machette (1987) and Foley *et al.* (1988).

Discussion: tectonic versus climatic modes

During the interval from 5 to 0.8 Ma, the southern Rio Grande rift is interpreted to have been in a *tectonic subsidence mode*. During this time basin aggradation occurred, including the preservation of ancestral Rio Grande channel and floodplain

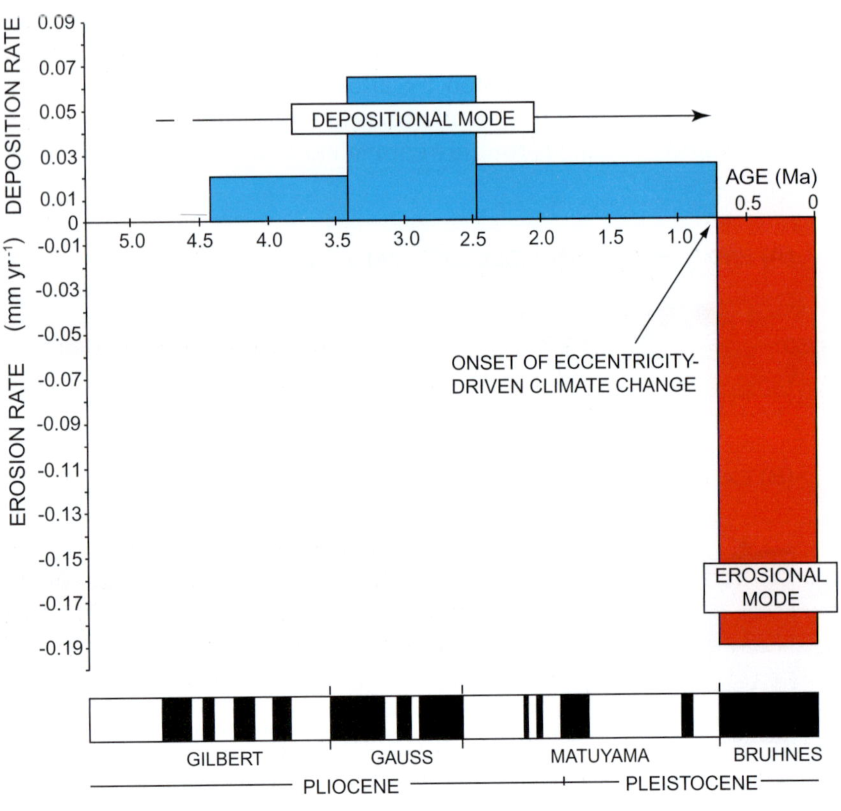

Fig. 4 Graph to show chron-averaged deposition and erosion rates derived from Plio-Pleistocene basin fill and incised valleys of southern Rio Grande rift basins (data assembled from palaeomagnetic studies by Mack *et al.*, 1993, 1998; Leeder *et al.*, 1996b; Mack & Leeder, 1999).

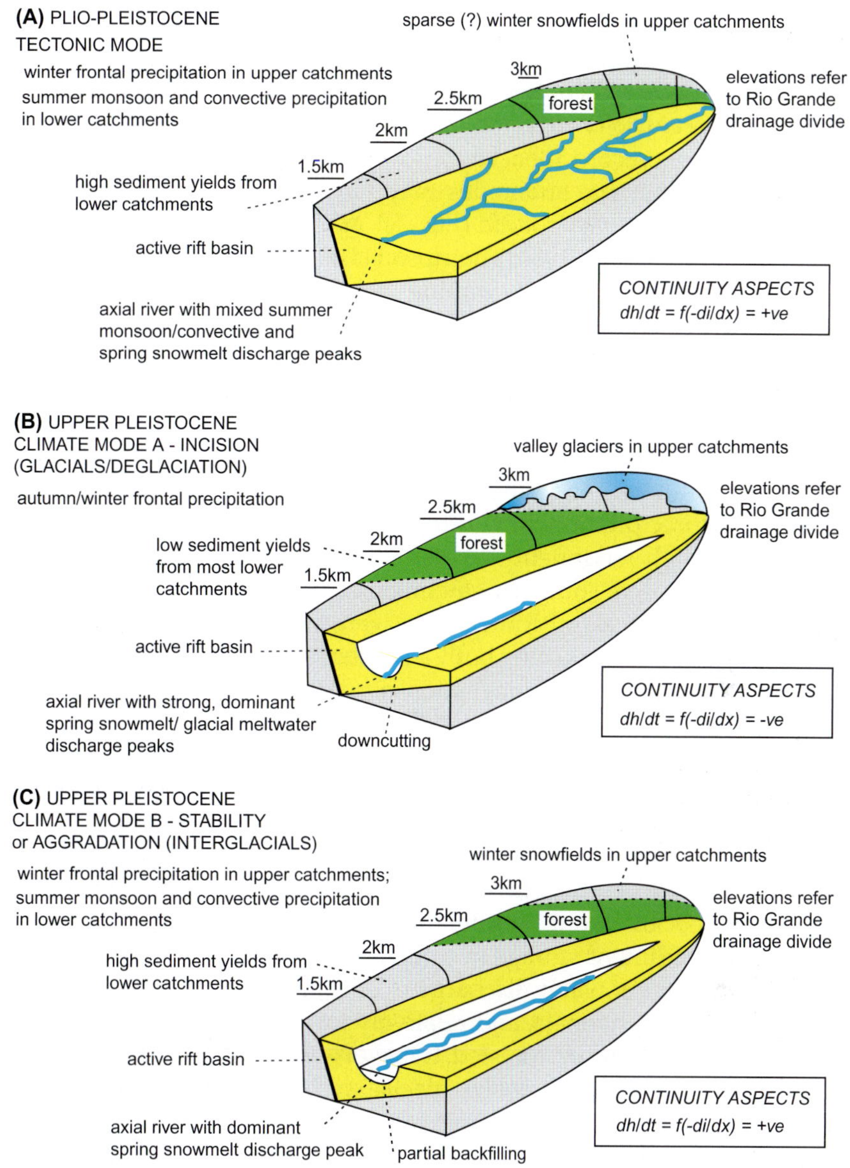

Fig. 5 Summary of Plio-Pleistocene tectonic and climatic variables affecting the whole (idealized) Rio Grande catchment.

deposits. Subsidence and sediment input from lateral sources overcame any imbalances in longitudinal gradient or transport rates to give very modest net aggradation (Figs 4 & 5A). Higher aggradation rate during Gauss time (Fig. 4) may reflect higher sediment production in transverse catchments caused by a slightly moister climatic regime, as suggested by regional stable-isotope studies of Calcisol carbonate (Smith *et al.*, 1993; Mack *et al.*, 1994). The semi-arid palaeoclimate may have produced hydrographs including, as today (Fig. 2), contributions from both winter frontal systems, particularly in mountain catchments where (?sparse) winter snowfields could accumulate, and summer monsoon with convective precipitation in southern catchments. Relatively high sediment yields would have characterized the sparsely vegetated catchments (see Fig. 5).

For the interval from 0.8 Ma to the present, it is proposed that the southern Rio Grande rift was in a *climatic mode*, during which the ancestral Rio Grande periodically incised into the deposits of the previous tectonic mode (Fig. 5B & C). The fact that the timing of onset of incision and the number and age of inset geomorphological surfaces are the same in several tectonically distinct basins in the southern Rio Grande rift supports the role of regional climatic changes rather than tectonics or changing base level as causative factors (Gile *et al.*, 1981; Seager & Mack, 2003). Moreover, incision of the ancestral Rio Grande occurred contemporaneously with the change to eccentricity-driven climate cycles in the marine isotopic record (Shackleton, 1995).

Two palaeoclimate submodes may have operated during the *climatic mode* (Fig. 5B & C; Gile *et al.*, 1981; Richmond, 1986; Thompson *et al.*, 1993; Stute *et al.*, 1995). During waning glacials and warmer interglacials, partial sediment backfilling or relative base-level stability occurred. Palaeoclimate was semi-arid overall and included contributions from winter fronts, particularly in northern catchments where generally sparse winter snowfields accumulated, and summer monsoons with convective precipitation in southern catchments. Relatively high sediment yields characterized sparsely vegetated southern catchments. Hydrographs for the ancestral Rio Grande were similar to the present day, including both a modest spring snowmelt peak and a summer monsoonal contribution. The situation was dominated by subsidence plus external sourcing, resulting in basin-floor equilibrium or aggradation. In contrast, during cooler, glacial periods, seasonal melting of snowfields in northern catchments would have seen the downstream Rio Grande hydrograph dominated by the effects of strong spring discharge peaks. Substantial sediment storage in upland transverse catchments, excess water discharge over sediment discharge in the channels, and relatively low sediment yields from more forested transverse catchments developed under cooler glacial maximum climate, and resulted in overall strong positive downstream sediment transport gradients leading to basin incision.

GULF OF CORINTH RIFT

Southern Greece, in particular the area bordering the Gulf of Corinth (Figs 6 & 7), is one of the world's most productive locations for the study of active continental extensional tectonics, chiefly on account of the high extension rates in the area (Davies *et al.*, 1997) and the exceptional clarity of onshore and offshore field evidence for extensional kinematics. The high extension rate (3–10 mm yr^{-1}, up to two orders of magnitude greater than the Rio Grande rift) and the coastal location of the rift margins lead to an abundance of kinematic markers for uplift and subsidence. A particular feature of the southern rift flank is the predominance of fluvial incision, with deep gorges (up to 800 m) cut by drainages into Quaternary sediments and Mesozoic bedrock. There remain two outstanding unresolved problems relating to the origins of this fluvial incision. The first concerns the uplift that has accompanied the evolution of the locus of maximum strain along the southern margin to the Gulf of Corinth and, more generally, over much of Peloponnisos (Kelletat *et al.*, 1976; Armijo *et al.*, 1996; Houghton *et al.*, 2003; Leeder *et al.*, 2003, 2005; McNeill & Collier, 2004). The second concerns the apparent northward migration of fault activity with time, witnessed by progressively abandoned, uplifted and incised rift sedimentary infills and inactive bounding faults (Jackson *et al.*, 1982; Ori, 1989; Ori *et al.*, 1991; Collier *et al.*, 1992; Dart *et al.*, 1994; Jackson, 1999; Goldsworthy & Jackson, 2001; Marlartre *et al.*, 2004).

Large-scale tectonics: flat-slab subduction and trench pushback

In an influential paper, Le Pichon & Angelier (1979) first proposed that rapid Aegean extension was due to southwest spatial acceleration of the Aegean part of the Anatolian plate (Figs 6 & 8A), subsequently referred to as Aegea, arising from southwards rollback of the subducting African plate (Fig. 8A). The southwestward spatial acceleration has been amply confirmed by recent satellite geodesy studies between Sterea Hellas and Peloponnisos as a ~ 10 mm yr^{-1} step-change to the south of the array of active normal faults defining the southern margin to the Gulf of Corinth (Fig. 6; Davies *et al.*, 1997; Briole *et al.*, 2000; Kahle *et al.*, 2000; McCluskey *et al.*, 2000).

Concerning the nature of Hellenic trench rollback, seismological and teleseismic studies (Spakman *et al.*, 1988; Hatzfeld *et al.*, 1989; Papazachos *et al.*, 2000; Tiberi *et al.*, 2000) have shown convincing evidence for the existence of flat-slab subduction under Peloponnisos. The most detailed recent results (Tiberi *et al.*, 2000) suggest that Peloponnisos is a 70–80 km thick block overlying flat to gently dipping Mediterranean oceanic lithosphere under the western and central Gulf of Corinth. The slab steepens eastwards and beyond under the Aegean

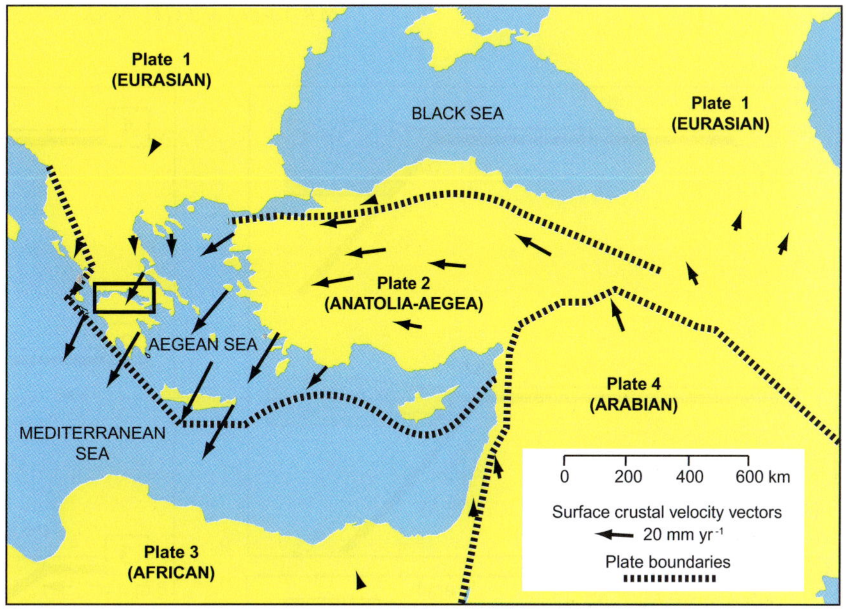

Fig. 6 Horizontal surface velocities of the plates making up the Mediterranean and Asia Minor (modified, with selected vectors redrawn after McCluskey *et al.*, 2000). Data derived from GPS satellite platforms and averaged over a few years. Note the evidence for: (i) contrast in velocity vectors between different plates; (ii) systematic east to southwest acceleration (in both magnitude and direction of velocity vectors) of the Anatolia–Aegea plate, especially the 10 mm yr^{-1} acceleration across the Gulf of Corinth strain boundary; (iii) anticlockwise spin (solid vorticity) of the Anatolia–Aegea plate. Gulf of Corinth rift is boxed.

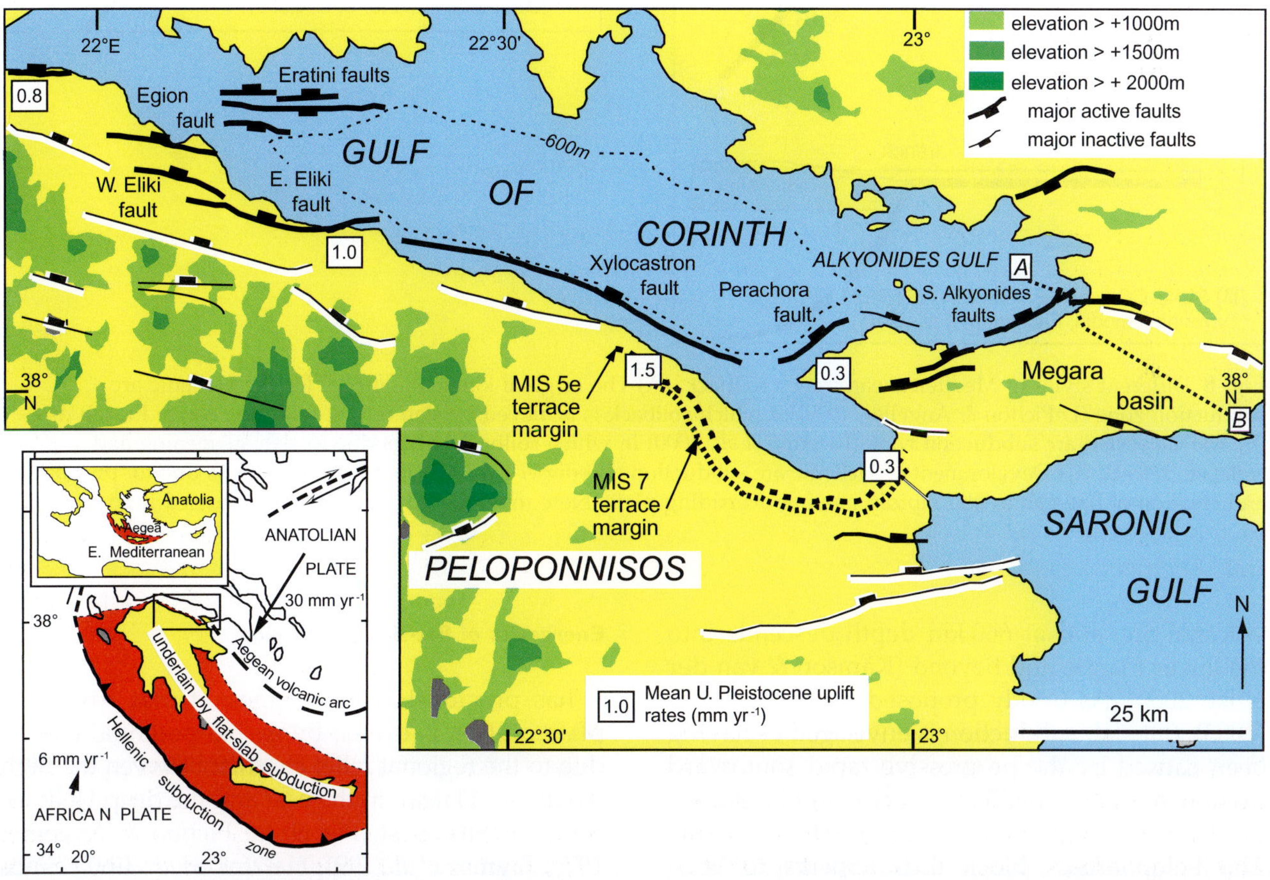

Fig. 7 General location and tectonic summary maps for Gulf of Corinth, central Greece within its Aegean context. Note the suite of abandoned faults on the southern gulf margin; these are associated with adjacent abandoned, uplifting and incising tilt-block-style synrift basins. The dashed line AB denotes the line of section of Fig. 10. Inset shows *maximum* likely extent of area underlain by flat-slab subducted plate (some 1.2 10^5 km^2). Topography and onshore and offshore faulting after Stefatos *et al.* (2002), Leeder *et al.* (2005) and McNeill *et al.* (2005).

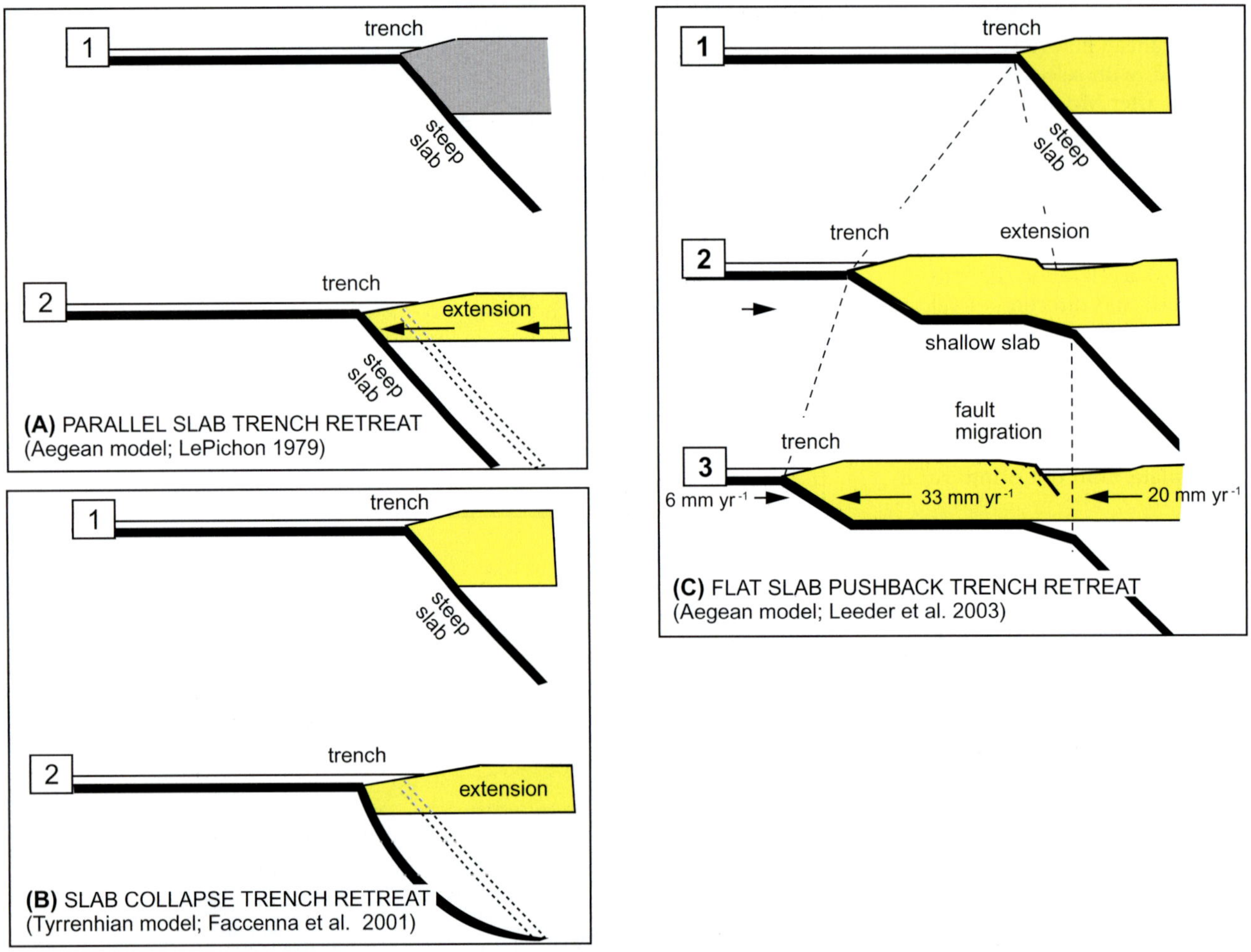

Fig. 8 Different styles of Mediterranean trench rollback. (A) The original kinematic model for the Hellenic arc subduction zone (LePichon & Angelier, 1979) in which rollback is achieved by slab retreat with no change in slab dip. (B) The Calabrian arc subduction zone (Facenna *et al.*, 2000) in which rollback occurs due to slab steepening and collapse/retreat. (C) Development of Hellenic arc subduction (Leeder *et al.*, 2003) in which rollback is due to 'pushback' and creation of flat slab by the rapidly moving overriding plate. Note moving and fixed hinges discussed in text.

volcanic arc, and at 660 km depth descends into the lower mantle and beyond (Kàrason & van der Hilst, 2000). As briefly proposed by Leeder *et al.* (2003), flat-slab subduction is envisaged as having been caused by the progressive rapid southward motion of Aegea (~ 33 mm yr^{-1}) over the slowly northward-moving (~ 6 mm yr^{-1}) African plate. The Peloponnisos block thus appears to be a south-moving, relatively rigid and aseismic lithospheric nappe riding above the flat slab. This motion has led to gradual trench *pushback* (Fig. 8C), a rather distinctive mode of slab migration in the Mediterranean context.

Energetics of flat-slab production

It has previously been suggested that Anatolian plate motion is driven by excess potential energy due to the regional relief contrast between the high Anatolian Plateau in the east and the deep Hellenic trench in the west (Fig. 9; Le Pichon & Angelier, 1979; Taymaz *et al.*, 1991; Hatzfeld *et al.*, 1997). Since the Anatolia–Aegea plate has been 'unzipped' along the bounding North Anatolian strike-slip fault, this potential energy is available to do work in driving the Anatolian plate southwest through the 'gate' between the Ionian islands and southernmost

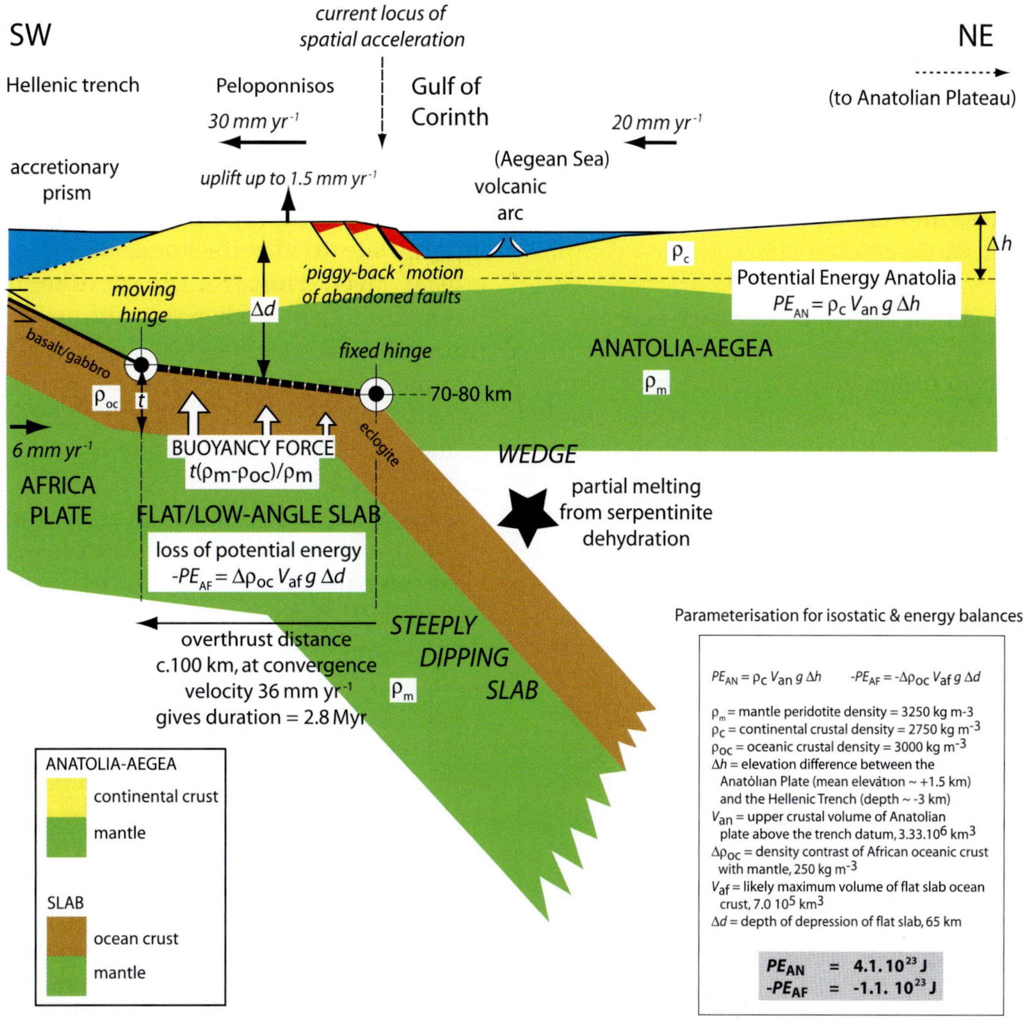

Fig. 9 General scheme for dynamics of lithospheric overthrust/flat-slab subduction of Aegea over African plate. Note the piggy-back-style southward carriage of abandoned, uplifting and incising synrift sedimentary basins over the flat slab and the concomitant northward migration of active faulting (see abandoned faults in Fig. 7). Also shown is the parameterization applied to determine the energetics of the flat-slab subduction.

Dodecanes into the Mediterranean (see analogue experiments of Hatzfeld *et al.*, 1997), overcoming frictional resistance of remaining African oceanic lithosphere at its southwestern leading edge. It is evident that this motion of southwest Aegea must dynamically load and depress the African plate. Work must therefore be done to produce the flat slab, implying that the potential energy available to Anatolia, PE_{AN}, must exceed that lost by the African flat slab, $-\Delta PE_{AF}$, as it is depressed under Peloponnisos, Crete and the southeast Aegean Sea. Assuming that the densities of African and Anatolian lithospheric mantle are similar, this involves loss of potential energy due to depression of the buoyant ocean crust of the flat slab over the minimum likely area shown in Fig. 7. With parameterization as in Fig. 9, we obtain $PE_{AN} = 4.1 \times 10^{23}$ J and $\Delta PE_{AF} = -1.1 \times 10^{23}$ J. The energetics are thus favourable, conservatively so in view of uncertainties in accurately estimating (i) the volume of flat slab involved and (ii) the extra buoyancy provided by an unknown degree of slab mantle serpentinization (see discussion below).

A further kinematic consequence of flat-slab subduction is that the rather thin, 70–80 km (Tiberi *et al.*, 2000), overriding plate may have had a certain amount of upper mantle removed during its southward translation over the flat slab. This requires mantle redistribution by three-dimensional flow as illustrated by analogue experiments on rollback subduction by Kincaid & Griffiths (2003). In the Aegean, mantle flow may have been southeastwards to the trench boundary east of Crete, contributing to steeper slab dips recorded in the Rhodes area and beyond (see Benatatos *et al.*, 2004, their fig. 8).

Origin of southern rift flank and Peloponnisos uplift

The ongoing regional uplift of the southern rift flank and Peloponnisos is manifest in the widespread occurrence of deep fluvial incision and uplifting marine Neogene to Holocene sediments (e.g. Kelletat *et al.*, 1976; Keraudren & Sorel, 1987; Seger & Alexander, 1993; Kourampas & Robertson, 2000). In particular, the southeast flank to the Corinth rift (Fig. 7) has a spectacular flight of uplifting marine terrace deposits that are well correlated with the global Quaternary sea-level curve (Keraudren & Sorel, 1987; Collier, 1991; Collier *et al.*, 1992; Armijo *et al.*, 1996; Leeder *et al.*, 2003, 2005). Uplift has been ongoing for the order of 0.6–1 Myr, based on terrace-flight elevations for steady uplift rate scenarios. Generally, rates of uplift decrease eastwards from ~ 0.7–1.5 to ~ 0.3 mm yr^{-1} (Houghton *et al.*, 2003; Leeder *et al.*, 2003, 2005; McNeill & Collier, 2004).

To explain uplift, LePichon & Angelier (1979) originally invoked sediment underplating by intrusion and thrusting from the Hellenic subduction zone and Cretan forearc, an idea taken up subsequently (e.g. Papazochos & Nolet, 1997; Knapmeyer & Harjes, 2000). However, significant sediment underplating is specifically ruled out by wide-aperture seismic refraction results (Bohnhoff *et al.*, 2001). Collier *et al.* (1992) invoked both regional uplift and local footwall uplift and noted the correlation of uplifting Peloponnisos with the extent of recently discovered flat-slab subduction under the region (Collier *et al.*, 1992, fig. 1C). Armijo *et al.* (1996) favoured only a minor role for regional uplift and developed a general footwall-uplift model for the Corinthian terraces based on a fault system located in elastic half-space undergoing co- and post-seismic deformation. Changing spatial uplift rate was related to distance of the extrapolated fault footwall from an offshore fault. The phenomenon of southerly regional backtilting was assumed to apply across the whole of the southern Gulf of Corinth rift flank, i.e. on a scale equal to the thickness of the seismogenic layer, approximately 10–15 km.

One test of the Armijo *et al.* (1996) footwall-uplift model includes prediction that terrace deposits should be progressively backtilted to the south, with the uplift decaying away over a length scale appropriate to the thickness of the seismogenic crustal layer. However, apart from the Megara basin (Fig. 10), no such systematic major southward tilting trends have been documented, indeed the regional trend of terrace elevations is equally consistent with eastwards tilting (Leeder *et al.*, 2003). A more severe test is that the modelling strategy requires 11.5 km of fault slip. Again the test is negative, only a maximum of approximately 1 km of sediment is imaged in 600 m of water above a clear pre-rift basement reflector in the offshore eastern rift (Brooks & Ferentinos, 1984), whereas deep seismic reflection data in the central gulf (Sachpazi *et al.*, 2003) reveal a maximum sediment thickness plus water depth there of < 3 km.

The idea adopted here (Fig. 9) is that regional uplift of Peloponnisos is due to deeper tectonic processes, a view also held by Moretti *et al.* (2003). Thus uplift is attributed to buoyancy of the underlying African flat slab, as augmented by local footwall uplift along the southern margin to the Gulf of Corinth. It is envisaged that the oceanic crust of the flat slab (and probably also an unknown proportion of underlying serpentinized upper mantle) initially acts as a buoyant 'sandwich' between upper mantle of the overriding plate and underthrusting slab. For subducting Mediterranean oceanic crust of thickness $h = 6$ km (Bohnhoff *et al.*, 2001) and mean density $\rho_{oc} = 3000$ kg m^{-3}, the total isostatic uplift, $h(\rho_m - \rho_{oc})/\rho_m$, expected with an upper mantle density appropriate to peridotite of $\rho_m = 3250$ kg m^{-3} is some 0.46 km. This is a minimum estimate, because of the unknown degree of additional buoyancy due to the pervasive serpentinization generally reported from oceanic upper mantle slabs (Peacock, 2001). However, the estimate is of the right order since the highest Pleistocene terraces recorded in the northern Peloponnisos are at 500–600 m above sea level. It is also possible that the coincidence may equally well be fortuitous since the calculation ignores the counter-acting tendency for progressive loss of ocean crustal buoyancy due to amphibolization and eclogization (e.g.

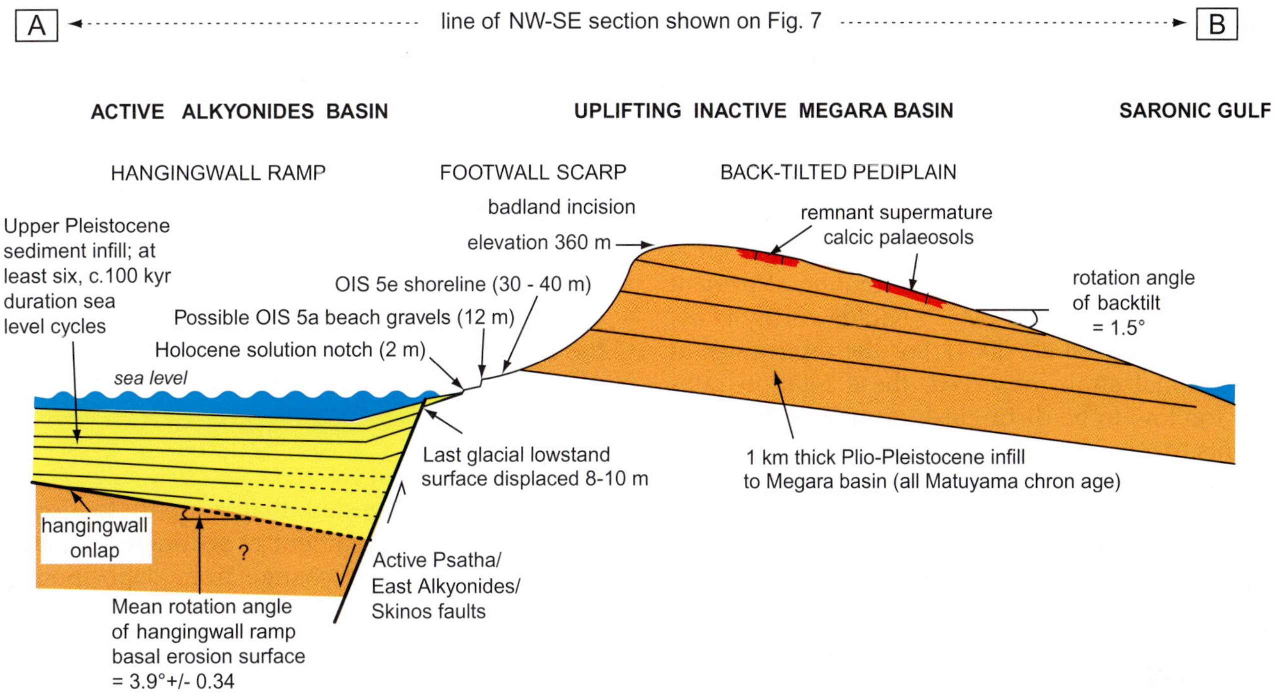

Fig. 10 Schematic section (after Leeder *et al.*, 2005) along the line A–B in Fig. 7 across the eastern part of the Gulf of Corinth to show the young, active, offshore Psatha/East Alkyonides fault bounding the Alkyonides Gulf and the abandoned, uplifting and incising Pleistocene Megara basin. Major morphological features shown are useful in parameterization of rates of uplift and subsidence associated with displacements along the active faults.

Bostock *et al.*, 2002) during subduction. Given the likely reduction in the rate of this process due to flat-slab subduction, the decrease in regional uplift eastwards from the central northern margin of Peloponnisos to the Isthmus (Leeder *et al.*, 2003) is a predictable consequence of buoyancy loss. Although the rate of loss is unknown, it must finally disappear in the eastern rift, when slab steepening occurs and additional losses of structural water occur by antigorite dehydration (Ulmer & Tromsdorff, 1995; Peacock, 2001). This process eventually facilitates melting under the Aegean volcanic arc at slab penetration depths of around 150 km.

Migration of active faulting and basin abandonment

The kinematics of the south-migrating Peloponnisos block over an underlying flat slab requires that active faults along the southern rift flank be eventually carried onto the flat slab (Fig. 9). Meanwhile, since their initiation, the faults have been rotating about horizontal axes (Proffett, 1977; Jackson *et al.*, 1982). By this combination of net southward transport and rotation it is envisaged that the normal faults are dragged piggy-back style (*sensu* Ori & Friend, 1984) onto the flat slab 'conveyor belt', whence they are disactivated and progressively uplifted. Deformation then translates northwards with respect to the moving Peloponnisos reference frame but stays above the fixed-hinge to flat-slab subduction. In this way, the fault propagation proceeds progressively northwards, as originally hypothesized by Ori (1989), and more recently by Moretti *et al.* (2003) and Malartre *et al.* (2004). The width of the zone of abandoned faults along northern Peloponnisos is in the range 10–30 km (Figs 7 & 9). For a convergence velocity between Peloponnisos and Africa of 35 mm yr^{-1}, this implies a minimum 0.29–0.86 Myr age range for initiation of the earliest abandoned faults and allows for individual fault lifetimes of 0.1–0.3 Myr for up to three generations of faults.

Finally, it should be noted that little systematic vertical axis rotation may be discerned in the process of fault abandonment in the Corinth rift, both from the evidence of Quaternary palaeomagnetic results (Duermeijer *et al.*, 1999, 2000; Hinsbergen *et al.*, 2005) and from the field observation that both 'dead' and active faults lie approximately parallel (Fig. 7; Goldsworthy & Jackson, 2001). These observations are consistent with the hypothesis put forward above, namely once hoisted piggy-back-style southwards, abandoned faults in the main rift are essentially locked into the general rigid kinematic motion of Peloponnisos, rotating at present observed rates of 3–7° Myr^{-1} (Goldsworthy *et al.*, 2002; Avallone *et al.*, 2004). For the rapid rates of extensional strain observed, these rotation rates are too low to be distinguish from field measurements. An exception might be the highly oblique (approximately 30°) trend of abandoned faults bounding the Megara basin with respect to the active coastal faults of the Alkyonides gulf in the extreme eastern rift (Figs 7 & 10; Collier *et al.*, 1992; Jackson, 1999). Here, extension is taking place above steep slab but there is no palaeomagnetic evidence from the Megara basin sediments for significant late Pleistocene vertical-axis crustal rotation (Duermeijer *et al.*, 1999, 2000; Hinsbergen *et al.*, 2005).

Sedimentary consequences of fault abandonment and uplift

Whatever the exact mode of fault abandonment and propagation, regional and footwall uplift must have acted to create slope discontinuities and knickpoints at positions where newly propagated faults were crossed by drainages issuing from antecedent valleys. This would have had the effect of eroding previously deposited sediments in the new fault footwall and led to renewed upstream propagation (according to the workings of the diffusion equation described earlier, Fig. 1B) of progressively younger knickpoints after each episode of normal faulting. In addition, upstream-migrating waves of fluvial incision (Paola *et al.*, 1991; Leeder & Stewart, 1996) would also have been produced during repeated glacial lowstand times as drainages adjusted to slope changes at former highstand shorelines. In both cases generally wetter glacial climates in central-southern Greece (Collier *et al.*, 2000) would have led to increased erosion potentials and diffusivities during such times. Over time, < 1 Myr, the net effect has been to produce spectacular gorges with 800 m of relief and the recycling of eroded earlier synrift sediment as fan deltas and submarine fans into the active rift. As noted above, footwall incision in the Megara basin has been accompanied by significant back-tilting and drainage reversal (Fig. 9). Attention has been drawn here to the development of basin-wide supermature calcrete palaeosols. These are closely analogous to the Upper and Lower La Mesa calcretes that feature as duricrusts capping the remnant pre-incision aggradational surfaces of the Rio Grande rift (see Fig. 3).

CONCLUSION

Overall, it is suggested that a trend to downstream increase in gradient of sediment transport rate accounts for the change from deposition to erosion in the upper Pleistocene of the Rio Grande rift. This change was coincident with the advent of eccentricity-driven Milankovich climatic cycles which caused significant alteration to the sediment and water discharge regimes of the river systems in the rift. In the Gulf of Corinth rift, although late Quaternary climatic changes have undoubtedly occurred, it is the combination of rapid regional uplift and extraordinarily rapid growth and propagation of faults that has largely controlled the abandonment, erosion and incision of basin sediment infills around the southern margins to the rift. This is in addition to local incision arising from knickpoint retreat during glacial age lowstand intervals.

These conclusions are relevant to the determination of sediment relaxation time, T_s, for the sedimentary systems involved. In the Rio Grande rift, and probably more generally in similar Quaternary fluvial basins, the relevant length scales used to compute T_s must be extremely small. Thus it is not so much the total length of the river channel system that should be considered, but the rapidity with which a changed discharge signal from the catchment can be transmitted downstream to cause channel degradation. Despite the great length of the whole river system, it is the balance between hydrological and sediment input from lateral and upstream tributaries that controls the non-uniform downstream sediment transporting capacity of the axial channel. Since there are very many tributaries draining catchments in the fault-bounded rift flanks, from Colorado to Texas, the relevant length scale is often small, in the order of tens of kilometres. At the same time, diffusivities during glacial-period spring snowmelt flooding are expected to have been very large, and hence overall relaxation times have generally been very short, perhaps in the order of 10^2–10^3 yr. The situation is thus analogous to the twentieth century phase of widespread arroyo-cutting in the southwest USA (e.g. Hereford, 1993) noted previously. In the case of the Gulf of Corinth rift the physical length scale from catchment to ocean is low, in the order of 30 km. This, combined with the very high strain rate (an order of magnitude greater than in the Rio Grande rift), causes diffusivity across rapidly vertically-growing normal faults to be very large and hence relaxation time to be low. It is hoped that future quantitative studies of incision rate in the two rifts, along the lines recently suggested by Carretier & Lucazeau (2005), may lead to a better understanding of the precise values of length scale, diffusion coefficient and sediment relaxation times involved.

DEDICATION AND ACKNOWLEDGEMENTS

MRL gratefully remembers early introductions to the use of sediments in basin tectonic analysis from Peter Friend, both through reading his early pioneering publications on Spitsbergen Old Red Sandstone basins and from discussions in the field on several 'Friends of the Devonian' excursions in the UK between 1970 and 1975.

We thank Cindy Ebinger and Christel Tiberi who made helpful constructive criticisms and encouraging comments on an earlier version of the paper. We also thank referees John Holbrook and Gary Smith, together with Editor Chris Paola, for their great help in making us explain our arguments in the clearest way possible.

REFERENCES

Armijo, R., Meyer, B., King, G.C.P., Rigo, A. and Papanastassiou, D. (1996) Quaternary evolution of the Corinth Rift and its implications for the late Cenozoic evolution of the Aegean. *Geophys. J. Int.*, **126**, 11–53.

Avallone, A., Briole, P., Agatza-Balodimou, A.M., *et al.* (2004) Analysis of eleven years of deformation measured by GPS in the Corinth Rift Laboratory area. *CR Acad. Sci. Paris, Geosci.*, **336**, 301–311.

Bagnold, R.A. (1966) An approach to the sediment transport problem from general physics. *US Geol. Surv. Prof. Pap.*, **422-I**.

Begin, Z.B. (1988) Application of a diffusion-erosion model to alluvial channels which degrade due to base-level lowering. *Earth Surf. Proc. Land.*, **13**, 487–500.

Begin, Z.B., Meyer, D.F. and Schumm, S.A. (1981) Development of longitudinal profiles of alluvial channels in response to base level lowering. *Earth Surf. Proc. Land.*, **6**, 49–68.

Benetatos, C., Kiratzi, A., Papazachos, C. and Karakaisis, G. (2004) Focal mechanisms of shallow and intermediate depth earthquakes along the Hellenic Arc. *J. Geodynam.*, **37**, 253–296.

Blum, M.D. and Törnqvist, T.E. (2000) Fluvial responses to climate and sea-level change: a review and look forward. *Sedimentology*, **47**(suppl. 1), 2–48.

Bogaart, P.W. and van Balen, R.T. (2002) Numerical modeling of the response of alluvial rivers to Quaternary climate change. *Global Planet. Change*, **27**, 147–163.

Bogaart, P.W., Van Balen, R.T. Vandenberghe, J. and Kasse, C. (2002) Process-based modelling of the climatic forcing of fluvial sediment flux: some examples and a discussion of optimal model complexity. In: *Sediment Flux to Basins: Causes, Controls and Consequences* (Eds S.J. Jones and L.E. Frostick), pp. 187–198. Special Publication 191, Geological Society Publishing House, Bath.

Bogaart, P.W., Van Balen, R.T. Kasse, C. and Vandenberghe, J. (2003) Process-based modelling of fluvial system response to rapid climate change II. Application to the River Maas (The Netherlands) during the Last Glacial–Interglacial Transition *Quatern. Sci. Rev.*, **22**(20), 2097–2110.

Bohnhoff, M., Makris, J., Papanikolaou, D. and Stavrakis, G. (2001) Crustal investigation of the Hellenic subduction zone using wide aperture seismic data. *Tectonophysics*, **343**, 239–262.

Bostock, M.G., Hyndman, R.D., Rondenay, S. and Peacock, S.M. (2002) An inverted continental Moho and serpentinisation of the forearc mantle. *Nature*, **417**, 536–538.

Bridge, J.S. and Leeder, M.R. (1979) A simulation model of alluvial stratigraphy. *Sedimentology*, **26**, 617–644.

Briole, P., Rigo, A., Lyon-Caen, H., *et al.* (2000) Active deformation of the Corinth rift, Greece: results from repeated Global Positioning System surveys between 1990 and 1995. *J. Geophys. Res.*, **105**, 25,605–25, 625.

Brooks, M. and Ferentinos, G. (1984) Tectonics and sedimentation in the Gulf of Corinth and the Zakinthos and Kefallinia Channels, western Greece. *Tectonophysics*, **101**, 25–54.

Bull, W.B. (1979) Threshold of critical stream power in streams. *Geol. Soc. Am. Bull.*, **90**, 453–464.

Carretier, S. and Lucazeau, F. (2005) How does alluvial sedimentation at range fronts modify the erosional dynamics of mountain catchments? *Basin Res.*, **17**, 361–381.

Chorley, R.J. and Kennedy, B.A. (1971) *Physical Geography, a Systems Approach*. Prentice Hall International.

Church, M. and Ryder, J.M. (1972) Paraglacial sedimentation: a consideration of fluvial processes conditioned by glaciation. *Geol. Soc. Am. Bull.*, **83**, 3059–3072.

Collier, R.E.Ll. (1990) Eustatic and tectonic controls upon Quaternary coastal sedimentation in the Corinth basin. *J. Geol. Soc. London*, **147**, 301–314.

Collier, R.E.Ll., Leeder, M.R., Rowe, P.J. and Atkinson, T.C. (1992) U-series dating and tectonic uplift of Upper Pleistocene marine sediments from the Corinth and Megara Basins, Greece. *Tectonics*, **11**, 1159–1167.

Collier, R.E.Ll, Leeder, M.R., Trout, M., Ferentinos, G., Lyberis, E. and Papatheodorou, G. (2000) High sediment yields and cool, wet winters: test of last glacial paleoclimates in the northern Mediterranean. *Geology*, **28**, 999–1002.

Connell, S.D. (2004) Geology of the Albuquerque basin and tectonic development of the Rio Grande rift in north-central New Mexico. *New Mex. Geol. Soc., Spec. Publ.*, **11**, 359–388.

Dart, C.J., Collier, R.E.Ll, Gawthorpe, R.L., Keller, J.V.A. and Nichols, G. (1994) Sequence stratigraphy of (?)Pliocene–Quaternary synrift, Gilbert-type fan deltas, northern Peloponnesos, Greece. *Mar. Petrol. Geol.*, **11**, 545–560.

Davies, R., England, P., Parsons, B., Billiris, H., Paradissis, D. and Veis, G. (1997) Geodetic strain of Greece in the interval 1892–1992. *J. Geophys. Res.*, **102**, 24,571–24,588.

Dethier, D.P. (2001) Pleistocene incision rates in the western United States calibrated using lava Creek B tephra. *Geology*, **29**, 783–786.

Duermeijer, C.E., Krijgsman, W., Langereis, C.G., Meulenkamp, J.E., Triantaphyllou, M.V. and Zachariasse, W.J. (1999) A Late Pleistocene clockwise rotation phase of Zakynthos (Greece) and implications for the evolution of the western Aegean arc. *Earth Planet. Sci. Lett.*, **173**, 315–331.

Duermeijer, C.E., Nyst, M., Meijer, P.Th., Langereis, C.G. and Spakman, W. (2000) Neogene evolution of the Aegean arc: paleomagnetic and geodetic evidence for a rapid and young rotation phase. *Earth Planet. Sci. Lett.*, **176**, 509–525.

Eaton, G.P. (1987) Topography and origin of the southern Rocky Mountains and Alvarado Ridge. In: *Continental Extensional Tectonics* (Eds M.P. Coward, J.F. Dewey and P.L. Hancock), pp. 355–369. Special Publication 28, Geological Society Publishing House, Bath.

Faccenna, C., Thorsten, W., Becker, F., Lucente, P., Jolive, L. and Rossetti, F. (2000) History of subduction and back-arc extension in the Central Mediterranean. *Geophys. J. Int.*, **145**, 809–820.

Foley, L.L., LaForge, R.C. and Piety, L.A. (1988) Seismotectonic study for Elephant Butte and Caballo Dams, Rio Grande Project, New Mexico. *US Bur. Recl. Seismotect. Rept.*, **88-9**, 60 pp.

Galloway, W.E., Ganey-Curry, P.E., Li, X. and Buffler, R.T. (2000) Cenozoic depositional history of the Gulf of Mexico Basin. *Am. Assoc. Petrol. Geol. Bull.*, **84**, 1743–1774.

Gawthorpe, R.L. and Leeder, M.R. (2000) Tectono-sedimentary architecture of active extensional basins. *Basin Res.*, **12**, 195–218.

Gilbert, G.K. (1917) Hydraulic-mining debris in the Sierra Nevada. *US Geol. Surv. Prof. Pap.*, **105**.

Gile, L.H., Hawley, J.W. and Grossman, R.B. (1981) *Soils and Geomorphology in the Basin and Range Area of Southern New Mexico – Guidebook to the Desert Project*. Memoir 39, New Mexico Bureau of Mineralogy and Mineral Resources, University Park, NM.

Goldsworthy, M. and Jackson, J.A. (2001) Migration of activity within normal fault systems: examples from the Quaternary of mainland Greece. *J. Struct. Geol.*, **23**, 489–506.

Goldsworthy, M., Jackson, J. and Haines, J. (2002) The continuity of active fault systems in Greece. *Geophys. Jour. Int.*, **148**, 596–618.

Gupta, S. (1997) Himalayan drainage patterns and the origin of fluvial megafans in the Ganges foreland basin. *Geology*, **25**, 11–14.

Gustavson, T.C. (1991) Arid basin depositional systems and paleosols: Fort Hancock and Camp Rice Formations (Plio-Pleistocene), Hueco Bolson, west Texas and adjacent Mexico. *Texas Bur. Econ. Geol. Rep.*, **198**, 49 pp.

Hatzfeld, D., Pedotti, G., Hatzidimitriou, P., *et al.* (1989) The Hellenic subduction beneath the Peloponnesus: first results of a microearthquake study. *Earth Planet. Sci. Lett.*, **93**, 283–291.

Hatzfeld, D., Martinod, J., Bastet, G. and Gautier, P. (1997) An analog experiment for the Aegean to describe the contribution of gravitational potential energy. *J. Geophys. Res.*, **102**, 649–659.

Hereford, R. (1993) Entrenchment and widening of the Upper San Pedro River, Arizona. *Geol. Soc. Am. Spec. Pap.*, **282**, 46 pp.

Hinsbergen, D.J.J., Langereis, C.G. and Meulenkamp, J.E. (2005) Revision of the timing, magnitude and distribution of Neogene rotations in the western Aegean region. *Tectonophysics*, **396**, 1–34.

Houghton, S.L., Roberts, G.P., Papanikolaou, I.D., McArthur J.M. and Gilmour, M.A. (2003) New ^{234}U–^{230}Th coral dates from the western Gulf of Corinth: implications for extensional tectonics. *Geophys. Res. Lett.*, **30**(19), 2013–2015.

Jackson, J.A. (1999) Fault death: a perspective from actively deforming regions. *J. Struct. Geol.*, **21**, 1003–1010.

Jackson, J.A. and Leeder, M.R. (1994) Drainage systems and the development of normal faults: an example from Pleasant Valley, Nevada. *J. Struct. Geol.*, **16**, 1041–1059.

Jackson, J.A., King, G. and Vita-Finzi, C. (1982) The neotectonics of the Aegean: an alternative view. *Earth Planet. Sci. Lett.*, **61**, 303–318.

Jackson, J.A., Norris, R. and Youngson, J. (1996) The structural development of active fault and fold systems in central Otago, New Zealand: evidence revealed by drainage patterns. *J. Struct. Geol.*, **18**, 217–234.

Kahle, H-G., Cocard, M., Peter, Y., *et al.* (2000) GPS-derived strain rate field within the boundary zones of the Eurasian, African, and Arabian Plates. *J. Geophys. Res.*, **105**, 23,353–23,370.

Kàrason, H. and van der Hilst, R. (2000) Constraints on mantle convection from seismic topography. In: *The History and Dynamics of global Plate Motion* (Eds M.R. Richards, R. Gordon and R. van der Hilst), pp. 277–288. Monograph 121, American Geophysical Union, Washington, DC.

Kelletat, D., Kowalczyk, G., Schroder, B. and Winter, K-P. (1976) A synoptic view on the neotectonic development of the Peloponnesian coastal regions. *Z. Dtsch. Geol. Ges.*, **127**, 447–465.

Keraudren, B. and Sorel, D. (1987) The terraces of Corinth (Greece) – a detailed record of eustatic sea-level variations during the last 500 000 years. *Mar. Geol.*, **77**, 99–107.

Kincaid, C. and Griffiths, R.W. (2003) Laboratory models of the thermal evolution of the mantle during rollback subduction. *Nature*, **425**, 58–62.

Kirkby, M.J. (1999) Landscape modelling at regional to continental scales. In: *Process Modelling and Landform Evolution* (Eds S. Hergarten and H.J. Neugebauer), pp. 189–203. Lecture notes in Earth Sciences, 78, Springer-Verlag, Berlin.

Kirkby, M.J. and Cox, N.J. (1995) A climatic index for soil erosion potential (CSEP) including seasonal and vegetation factors, *Catena*, **25**, 333–352, 1995.

Kirkby, M.J., Abrahart, R., McMahon, M.D., Shao, J. and Thornes, J.B. (1998) MEDALUS soil erosion models for global change. *Geomorphology*, **24**, 35–49.

Knapmeyer, M. and Harjes, H.-P. (2000) Imaging crustal discontinuities and the downgoing slab beneath western Crete. *Geophys. J. Int.*, **143**, 1–21.

Kortemeier, C.P. (1982) Occurrence of Bishop Ash near Grama, New Mexico. New Mexico. *Geology*, **4**, 22–24.

Kourampas, N. and Robertson, A.H.F. (2000) Controls on Plio-Quaternary sedimentation within an active fore-arc region: Messenia peninsula and SW Peloponnese, S. Greece. In: *Proceedings of the Third International Conference on the Geology of the Eastern Mediterranean* (Eds I. Panayides, C. Xenophontos and J. Malpas), pp. 255–285. Geological Survey Department, Nicosia, Cyprus.

Lane, E.W. (1955) The importance of fluvial morphology in hydraulic engineering. *Proc. Am. Soc. Civ. Eng.*, **81**, 1–17.

Leeder, M.R. and Jackson, J.A. (1993) The interaction between normal faulting and drainage in active extensional basins with examples from the Western United States and Greece. *Basin Res.*, **5**, 79–102.

Leeder, M.R. and Mack, G.H. (2001) Lateral erosion ('toe-cutting') of alluvial fans by axial rivers: implications for basin analysis and architecture. *J. Geol. Soc. London*, **158**, 885–893.

Leeder, M.R. and Stewart, M. (1996) Fluvial incision and sequence stratigraphy: alluvial responses to relative sea-level fall and their detection in the geological record. In: *Sequence Stratigraphy in British Geology* (Eds S.P. Hesselbo and D.N. Parkinson), pp. 47–61. Special Publication 103, Geological Society Publishing House, Bath.

Leeder, M.R., Mack, G.H., Peakall, J. and Salyards, S.L. (1996a) First quantitative test of alluvial stratigraphic models: Southern Rio Grande rift, New Mexico. *Geology*, **24**, 87–90.

Leeder, M.R., Mack, G.H. and Salyards, S.L. (1996b) Axial-transverse interactions in half-graben: Plio-Pleistocene Palomas Basin, southern Rio Grande Rift, New Mexico, USA. *Basin Res.*, **12**, 225–241.

Leeder, M.R., McNeill, L.C., Collier, R.E.Ll., *et al.* (2003) Corinth rift margin uplift: New evidence from Late Quaternary marine shorelines. *Geophys. Res. Lett.*, **30**, 1611–1614.

Leeder, M.R., Portman, C., Andrews J.E., *et al.* (2005) Normal faulting and crustal deformation, Alkyonides Gulf and Perachora peninsula, eastern Gulf of Corinth rift, Greece. *J. Geol. Soc. London*, **162**, 549–561.

Leopold, L.B. and Bull, W.B. (1979) Base-level, aggradation, and grade. *Proc. Am. Phil. Soc.*, **123**, 168–202.

Le Pichon, X. and Angelier, P. (1979) The Hellenic arc and trench system: a key to the neotectonic evolution of the eastern Mediterranean area. *Tectonophysics*, **60**, 1–42.

Machette, M.N. (1987) Preliminary assessment of Quaternary faulting near Truth or Consequences, New mexico. *U.S. Geol. Surv. Open-File Rep.*, **87-652**, 40 pp.

Mack, G.H. (2004) Middle and Late Cenozoic crustal extension, sedimentation, and volcanism in the southern Rio Grande rift, Basin and Range, and southern Transition Zone of southwestern New Mexico. *New Mex. Geol. Soc. Spec. Publ.*, **11**, 389–406.

Mack, G.H. and James, W.C. (1993) Control of basin symmetry on fluvial lithofacies, Camp Rice and Palomas Formations (Plio-Pleistocene), southern Rio Grande rift, USA. In: *Alluvial Sedimentation* (Eds M. Marzo and C. Puigdefabregas), pp. 439–449. Special Publication 17, International Association of Sedimentologists. Blackwell Scientific Publications, Oxford.

Mack, G.H. and Leeder, M.R. (1999) Climatic and tectonic controls on alluvial-fan and axial-fluvial sedimentation in the Plio-Pleistocene Palomas half-graben, southern Rio Grande rift. *J. Sed. Res.*, **69**, 635–652.

Mack, G.H. and Seager, W.R. (1990) Tectonic control on facies distribution of the Camp Rice and Palomas Formations (Plio-Pleistocene) in the southern Rio Grande rift. *Geol. Soc. Am. Bull.*, **102**, 45–53.

Mack, G.H., Salyards, S.L. and James, W.C. (1993) Magnetostratigraphy of the Plio-Pleistocene Camp Rice and Palomas Formations in the Rio Grande rift of southern New Mexico. *Am. J. Sci.*, **293**, 49–77.

Mack, G.H., Cole, D.R., James, W.C., Giordano, T.H. and Salyards, S.L. (1994) Stable oxygen and carbon isotopes of pedogenic carbonate as indicators of Plio-Pleistocene paleoclimate in the southern Rio Grande Rift, south-central New Mexico. *Am. J. Sci.*, **294**, 621–640.

Mack, G.H., Salyards, S.L., MacIntosh, W.C. and Leeder, M.R. (1998) Reversal magnetostratigraphy and radioisotope geochronology of the Plio-Pleistocene Camp Rice and Palomas Formations, southern Rio Grande rift. *New Mex. Geol. Soc., Guidebook*, **49**, 229–236.

Mack, G.H., Leeder, M. and Salyards, S.L. (2002) Temporal and spatial variability of alluvial-fan and axial-fluvial sedimentation in the Plio-Pleistocene Palomas half graben, southern Rio Grande rift, New Mexico, U.S.A. *Soc. Econ. Paleontol. Mineral. Spec. Publ.*, **73**, 165–177.

Mackey, S.D. and Bridge, J.S. (1995) Three-dimensional model of alluvial stratigraphy: theory and application. *J. Sediment. Res.*, **B65**, 7–31.

Mackin, J.H. (1948) Concept of the graded river. *Geol. Soc. Am. Bull.*, **59**, 463–512.

Malartre, F., Ford, M. and Williams, E.A. (2004) Preliminary biostratigraphy and 3D geometry of the Vouraikos Gilbert-type fan delta, Gulf of Corinth, Greece. *C.R. Geosci. Paris*, **336**, 269–280.

McCluskey, S., Balassanian, S., Barka, A., *et al.* (2000) Global positioning system constraints on plate kinematics and dynamics in the eastern Mediterranean and Caucusus. *J. Geophys. Res.*, **105**, 5695–5719.

McNeill, L.C. and Collier, R.E.Ll. (2004) Uplift and slip rates of the eastern Eliki fault segment, Gulf of Corinth, Greece, inferred from Holocene and Pleistocene terraces. *J. Geol. Soc. London*, **161**, 81–92.

McNeill, L.C., Cotterill, C.J., Henstock, T.J., *et al.* (2005) Active faulting within the offshore western Gulf of Corinth, Greece: implications for models of continental rift deformation. *Geology*, **33**, 241–244.

Miall, A.D. (1991) Stratigraphic sequences and their chronostratigraphic correlation. *J. Sediment. Petrol.*, **61**, 497–505.

Moretti, I., Sakellariou, D., Lykousis, V. and Micarelli, L. (2003) The Gulf of Corinth: an active half graben? *J. Geodynam.*, **36**, 323–340.

Ori, G.G. (1989) Geologic history of the extensional basin of the Gulf of Corinth (Miocene–Pleistocene), Greece. *Geology*, **17**, 918–921.

Ori, G.G. and Friend, P.F. (1984) Sedimentary basins formed and carried piggyback on active thrust sheets. *Geology*, **12**, 475–478.

Ori, G.G., Roveri, M. and Nichols, G. (1991) Architectural patterns in large-scale Gilbert-type delta complexes, Pleistocene, Gulf of Corinth, Greece. In: *The 3D Facies Architecture of Terrigenous Clastic Sediments and its Implication for Hydrocarbon Discovery and Recovery* (Eds A.D. Miall and N. Tyler), pp. 207–216. Concepts in Sedimentology and Paleontology, Society of Economic Paleontologists and Mineralogists, Tulsa, OK.

Paola, C., 2000, Quantitative models of sedimentary basin filling. *Sedimentology*, **47**(Suppl. 1), 121–178.

Paola, C., Heller, P.L. and Angevine, C.L. (1991) The response distance of river systems to variations in sea level. *Geol. Soc. Am. Ann. Meet. Abs. Progr.*, A170–A171.

Paola, C., Heller, P.L. and Angevine, C.L. (1992) The large-scale dynamics of grain size variation in alluvial basins. 1. Theory. *Basin Res.*, **4**, 73–90.

Papazachos, B.C. and Nolet, G. (1997) P and S deep velocity structure of the Hellenic area obtained by robust nonlinear inversion of travel times. *J. Geophys. Res.*, **102**, 8349–8367.

Papazachos, B.C., Karakostas, V.G., Papazachos, C.B. and Scordilis, E.M. (2000) The geometry of the Wadati–Benioff zone and lithospheric kinematics in the Hellenic arc. *Tectonophysics*, **319**, 275–300.

Peacock, S.M. (2001) Are the lower planes of double seismic zones caused by serpentinite dehydration in subducting oceanic mantle? *Geology*, **29**, 299–302.

Perez-Arlucea, M., Mack, G.H. and Leeder, M.R. (2000) Reconstructing the ancestral (Plio-Pleistocene) Rio Grande in its active tectonic setting, southern Rio Grande rift, New Mexico, USA. *Sedimentology*, **47**, 701–720.

Posamantier, H.W. and Vail, P.R. (1988) Eustatic controls on clastic deposition II. Sequence and systems tract models. In: *Sea-Level Changes: an Integrated Approach* (Eds C.K. Wilgus, B.S. Hastings, C.G.S.C. Kendall, H.W. Posamentier, C.A. Ross and J.C. Van Waggoner), pp. 125–154. Special Publication 42, Society of Economic Paleontologists and Mineralogists, Tulsa, OK.

Proffett, J.M. (1977) Cenozoic geology of the Yerrington district, Nevada, and implications for the nature and origin of basin and Range faulting. *Geol. Soc. Am. Bull.*, **88**, 247–266.

Richmond, G.M. (1986) Stratigraphy and correlation of glacial deposits of the Rocky Mountains, the Colorado Plateau and the ranges of the Great Basin. *Quat. Sci. Rev.*, **5**, 99–127.

Sachpazi, M., Clement, C., Laigle, M., Hirn, A. and Roussos, N. (2003) Rift structure, evolution, and earthquakes in the Gulf of Corinth, from reflection seismic images. *Earth Planet. Sci. Lett.*, **216**, 243–257.

Schumm, S.A. and Lichty, R.W. (1965) Time, space and causality in geomorphology. *Am. J. Sci.*, **263**, 110–119.

Seager, W.R. and Mack, G.H. (2003) Geology of the Caballo Mountains, New Mexico. *New Mex. Bur. Geol. Min. Res. Mem.*, **49**, 136 pp.

Seger, M. and Alexander, J. (1993) Distribution of Plio-Pleistocene and modern coarse-grained deltas south of the Gulf of Corinth, Greece. In: *Tectonics and Sedimentation* (Eds L. Frostick and R. Steel), pp. 37–48. Special Publication 20, International Association of Sedimentology. Blackwell Scientific Publications, Oxford.

Shackleton, N.J. (1995) New data on the evolution of Pliocene climatic variability. In: *Palaeoclimate and Evolution, with Emphasis on Human Origins* (Eds E.S. Vrba, G.H. Denton, T.C. Partridge and L.H. Burckle), pp. 242–248. Yale University Press, New Haven, CT.

Smith, G.A. (2004) Middle to Late Cenozoic development of the Rio Grande rift and adjacent regions in northern New Mexico. *New Mex. Geol. Soc. Spec. Publ.*, **11**, 331–358.

Smith, G.A., Wang, Y., Cerling, T.E. and Geissman, J.W. (1993) Comparison of a paleosol-carbonate isotope record to other records of Pliocene–early Pleistocene climate in the western United States. *Geology*, **21**, 691–694.

Spakman, W., Wortel, M. and Vlaar, N. (1988) The Hellenic subduction zone: a tomographic image and its geodynamic implications. *Geophys. Res. Lett.*, **15**, 60–63.

Stefatos, A., Papatheodorou, G., Ferentinos, G., Leeder, M. and Collier, R. (2002) Seismic reflection imaging of active offshore faults in the Gulf of Corinth: their seismotectonic significance. *Basin Res.*, **14**, 487–502.

Stute, M., Clark, J.F., Schlosser, P., Broecker, W.S. and Bonani, G. (1995) A 30,000 yr continental paleotemperature record derived from noble gases dissolved in groundwater from the San Juan basin, New Mexico. *Quat. Res.*, **43**, 209–220.

Taymaz, T., Jackson, J. and Mckenzie, D.P. (1991) Active tectonics of the north and central Aegean Sea. *Geophys. J. Int.*, **106**, 433–490.

Thompson, R.S., Whitlock, C., Bartlein, P.J., Harrison, S.P. and Spaulding, W.G. (1993) Climatic Changes in the Western United States since 18,000 yr B.P. In: *Global Climates since the Last Glacial Maximum* (Eds H.E. Wright *et al.*), pp. 468–513. University of Minnesota Press.

Tiberi, C. Lyon-Caen, H., Hatzfeld, D., *et al.* (2000) Crustal and upper mantle structure beneath the Corinth rift (Greece) from a teleseismic tomography study. *J. Geophys. Res.*, **105**, 28,159–28,171.

Ulmer, P. and Trommsdorff, V. (1995) Serpentinite stability to mantle depths and subduction-related magmatism. *Science*, **268**, 858–861.

Zelt, B.C., Taylor, B., Sachpazi, M. and Hirn, A. (2005) Crustal velocity and Moho structure beneath the Gulf of Corinth, Greece. *Geophys. J. Int.*, **162**, 257–268.

Drainage responses to oblique and lateral thrust ramps: a review

JAUME VERGÉS

GDL, Group of Dynamics of the Lithosphere, Institute of Earth Sciences 'Jaume Almera', CSIC, Lluís Solé i Sabarís s/n, 08028 Barcelona, Spain (Email: jverges@ija.csic.es)

ABSTRACT

The relationships between oblique or lateral ramps in fold-and-thrust belts and their impact on syntectonic fluvial drainage are analysed in this review. Both ancient and recent cases from Cenozoic belts are examined. The southern flank of the Pyrenees provides good examples to decipher the long-term effects of oblique ramps on fluvial arrangement. Recent examples from the Indus River across the front of the Himalayas in northwest Pakistan, the frontal domains of the Andes in Bolivia and the northwest Zagros Mountain Belt provide examples of the short-term interaction between oblique or lateral thrust ramps and foreland drainage systems. The interpretation of these case studies, some of them developed on top of blind thrust ramps, can facilitate the analysis of drainage distortions in active complex tectonic regions.

Oblique ramps are present either permanently or episodically at different scales in all fold-and-thrust belts. The simplest scenario is related to piggyback basins, which display an oblique ramp linked to each frontal thrust termination. This oblique ramp forms the natural outlet for confined longitudinal systems along the piggyback basin. The change in topography across the ramp constrains the position of deltaic deposits between a subaerial fluvial system deposited in its hangingwall and an open marine system deposited in its footwall. Fluvial systems can also develop either in the hangingwall or in the footwall of larger oblique ramps that grow by the tectonic inversion of earlier structures. The growth of a large oblique ramp beneath a fluvial system operates in the same way as oblique ramps related to piggyback basins. However, its larger scale causes a larger differential topographic elevation across it, accommodating fluvial deposits in its hangingwall and deep marine turbidites in its footwall. In opposition, large oblique ramps growing in front of river systems create a topographic barrier that deflects the drainage.

A complication of the interaction of oblique ramps and drainage occurs where two opposed oblique ramps form a tectonic reentrant. These tectonic reentrants form at different scales and are characterized by lower topography. The confined domains concentrate rivers flowing out from surrounding higher topographic deformed regions. Further development of deeper thrusts can uplift these reentrants, modifying the previous concentrated drainage and diverting the river courses towards regions with lower topography.

As an example, the late Miocene longitudinal fluvial system flowing along the foreland basin of the Zagros during the deposition of the lower Agha Jari Formation was shifted to the southwest in the earliest Pliocene by the uplift of the Pusht-e Kuh Arc. The present river configuration is incising through the Pusht-e Kuh Arc anticlines and flows towards the lowlands of the Dezful Embayment (tectonic reentrant), limited by two major oblique ramps along the Mountain Front Flexure. The large-scale Mountain Front Flexure confines the Tigris River towards the southwest of the front of the Pusht-e Kuh Arc.

Keywords Oblique thrusts, fluvial deposits, drainage evolution, foreland basin, piggyback basin, Pyrenees, Himalayas, Andes, Zagros.

INTRODUCTION

The interplay between fluvial deposits and basins transported on top of thrusts has been widely explored since the early 1980s when Ori & Friend (1984) described piggyback basins (also named thrust top basins; DeCelles & Giles, 1996). Among many other works, Burbank *et al.* (1996) and Friend *et al.* (1999) presented fluvial interaction with compressive structures, and fluvial interactions with extensional faults were documented by Leeder *et al.* (1996). However, little attention has been paid to oblique and lateral thrust ramps and their interaction with fluvial drainage. Oblique and lateral ramps are common features of both modern and ancient fold-and-thrust belts. These oblique ramps can be either an inherited oblique structure or a newly formed relay thrust. Inherited structures, such as a previous normal fault or basin margin, can be reactivated as a thrust during compression maintaining its oblique direction with respect to the thrust motion. A newly developed oblique thrust accommodates shortening between two separate sectors of the thrust belt. In both cases, oblique ramps connect two different segments formed by thrusts parallel to the main frontal direction of the fold-and-thrust belt. During foreland basin evolution, the sedimentary infill varies greatly, depending on several tectonic factors including the large-scale geodynamic setting and period of evolution of the thrust belt to the local geometry of the thrust system (e.g. Burbank & Raynolds, 1988; Jordan & Flemings, 1990; DeCelles *et al.*, 1991).

The interaction between present-day fluvial systems and thrust faults is well-known in active thrust belts (e.g. Formento-Trigilio *et al.*, 2003), as well as in ancient examples of fold-and-thrust systems (e.g. Burbank & Vergés, 1994; Friend *et al.*, 1996). The resultant interplay, however, will vary in response to the spatial distributions of both fluvial and thrust systems (Fig. 1). Longitudinal fluvial complexes flow roughly parallel to the thrust system, as for example along growing piggyback basins developed on top of frontal thrusts. The interaction between these longitudinal fluvial deposits and oblique ramps will depend on their relative position either in the footwall or hangingwall of oblique thrusts. Where the fluvial system is located in a footwall position and runs into the oblique ramp with a

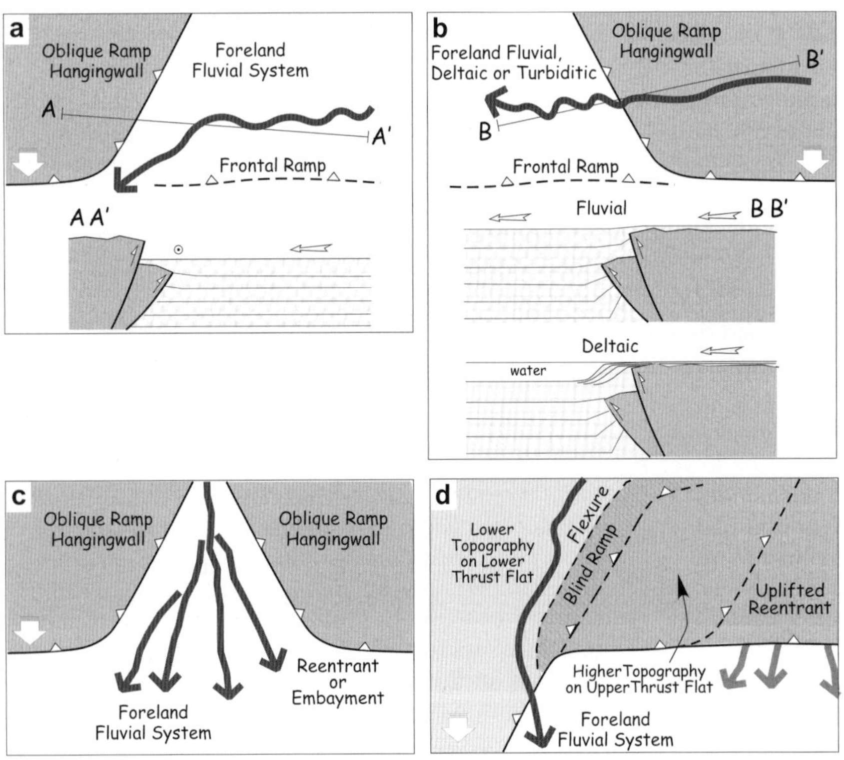

Fig. 1 Principal relations between axial depositional systems and oblique ramps in a thrust system. (a) The fluvial system flows parallel to the frontal thrusts and runs into an oblique ramp zone characterized by growing thrusts displaying a higher topography than the basin, which deflects the river towards the foreland. (b) The fluvial system flows on top of an uplifted oblique ramp and crosses it towards its footwall, characterized by either a lower topography or a marine environment. (c) Two opposed oblique ramps form a tectonic reentrant, which concentrates drainage. (d) The uplift of reentrants modifies drainage shifting it towards low topographic areas. Large white arrows indicate tectonic transport of thrusts.

higher topography, it will be diverted by the fault scarps or/and by the folded and faulted syntectonic fluvial beds near the active thrust faults (Fig. 1a). In contrast, where the fluvial system flows on top of the hangingwall of the oblique ramp and runs across it into the lower topography of its footwall, it creates a complex pattern of growth deposits in both subaerial fluvial and shallow-marine environments (Fig. 1b).

Two opposed oblique thrust ramps forming a tectonic reentrant represent a complication of the simple former scenarios (Fig. 1c). In this case, the reentrant is characterized by lower topography that constitutes a natural embayment in which rivers will flow from surrounding higher topography on both hangingwall domains of the oblique and lateral thrust ramps. The continuous shortening involving frontal and oblique segments of the thrust system can transport former reentrants on top of newly developed thrusts (Fig. 1d). In this example, a former reentrant with low topography becomes a region characterized by high topography, and thus the river system will be diverted laterally to find a better way to reach the foreland basin.

The main objective of this paper is to present a short review of the effects of oblique and lateral thrust ramps on fluvial drainage. Both ancient and recent examples from Cenozoic fold-and-thrust belts are analysed to accomplish this purpose. The study of ancient cases provides good examples of the interplay between oblique thrust ramps and synorogenic fluvial deposits. Changes in the palaeodrainage through time provided by fluvial sediments are linked to synchronous tectonic and topographic development of the oblique ramps. In most of the modern examples, however, only the very recent drainage pattern is recognizable. Changes in this drainage pattern in close proximity to the oblique or lateral thrust ramps provide information about their mutual control. This review is based on several ancient examples from the Southern Pyrenees where fluvial growth strata are broadly preserved (e.g. Vergés *et al.*, 2002), as well as on recent examples from active fold-and-thrust belts in the Himalayas of northwest Pakistan, the frontal Andes of Bolivia and the northwest Zagros of Iran. The northwest Zagros example combines late Miocene and Pliocene reconstructions with the present drainage distribution. The combination of well-recorded fossil examples to determine the long-term evolution of river palaeodrainages, and recent cases where their short-term evolution is well-documented, helps the understanding of oblique thrust ramps and drainage interactions that can be applied to other less-known fold-and-thrust-belt scenarios, either ancient or modern.

ANCIENT EXAMPLES FROM THE SOUTHERN PYRENEES

Ancient examples help to illustrate the long-term three-dimensional interaction of oblique ramps and fluvial systems, which complement the map distribution shown by modern examples. The South Pyrenean thrust belt forms an intricate system of frontal and oblique thrusts mostly inherited from late Jurassic to late Early Cretaceous rifting events followed by Late Cretaceous post-rift thermal subsidence (e.g. Puigdefàbregas & Souquet, 1986). This Cretaceous configuration was mostly reactivated and inverted during Tertiary compression, and thus determined the irregular shape of the southwards advancing thrust sheets composed of frontal and oblique ramps (Fig. 2). In the Central Pyrenees, both the South Central Unit and the Pedraforca Unit show two oblique thrust ramps bounding the thrust sheets (Martínez *et al.*, 1988; Vergés & Muñoz, 1990; Burbank *et al.*, 1992b; Nijman, 1998; Vergés, 2003). The Western Pyrenees show a large eastern oblique thrust ramp in which the Pamplona Fault corresponds to its northeast segment (Larrasoaña *et al.*, 2003; Fig. 2).

During Tertiary compression, early foreland basins were transported by thrusting and became piggyback basins (e.g. Puigdefàbregas *et al.*, 1992). These sedimentary basins, developed on top of south-directed foreland thrusts, changed from marine to continental during their infilling history, due to the change in relative basin subsidence and accumulation rates (e.g. Puigdefàbregas *et al.*, 1986; Burbank *et al.*, 1992a; Hogan & Burbank, 1996; Vergés *et al.*, 1998). The South Pyrenean foreland basin was partitioned into two different segments, both connected to the Atlantic during the early and middle Eocene: the Ripoll basin in the east (see a discussion in Nijman (1989) and Ramos *et al.* (2002), and the Tremp-Pamplona basin in the west (Puigdefàbregas, 1975; Turner, 1992; Fig. 2).

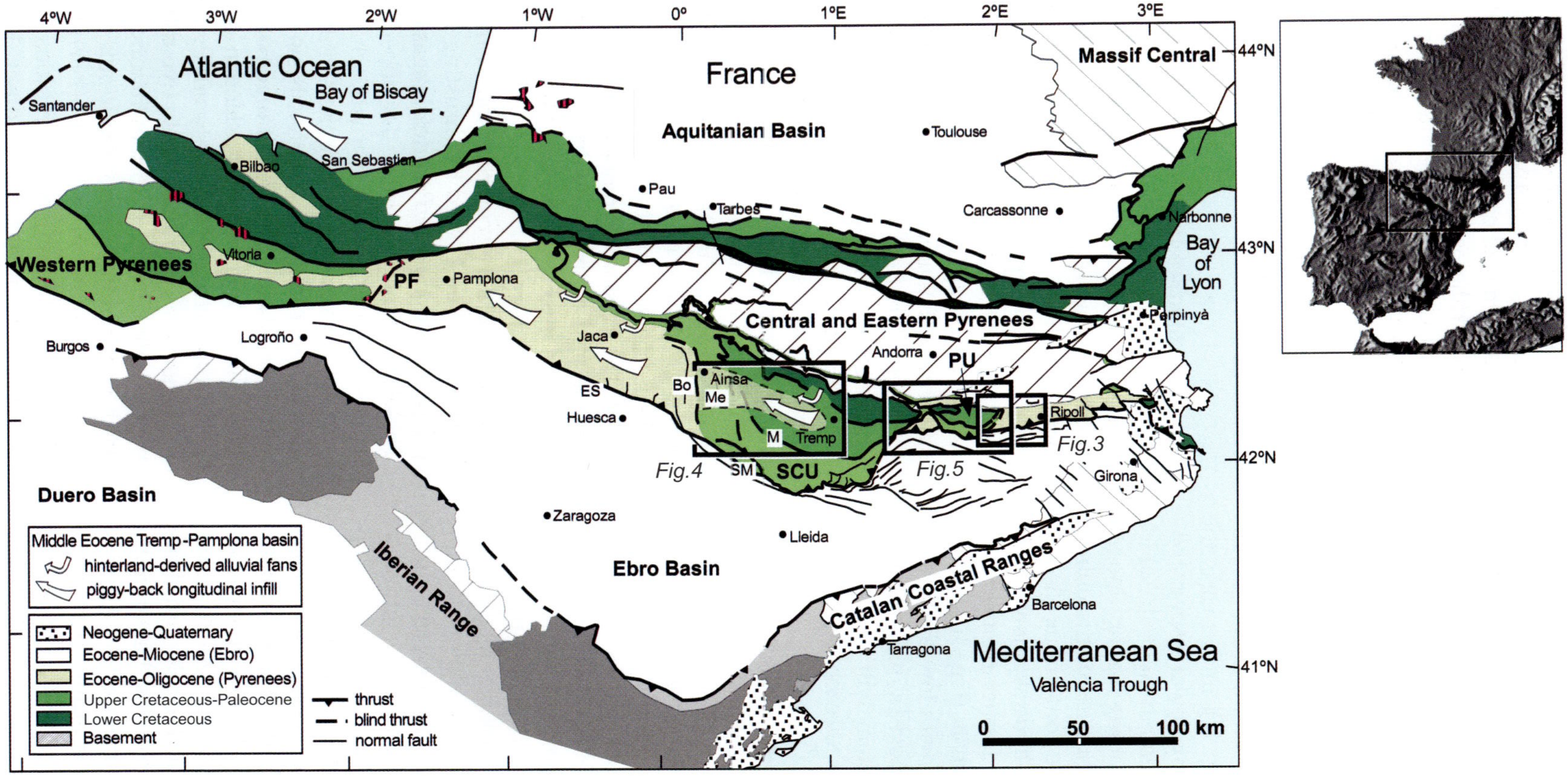

Fig. 2 Structural map of the Pyrenees showing the location of three examples of oblique ramps and Tertiary fluvial system development. The Ripoll basin example shows the interplay between a longitudinal fluvial system and a high-topography oblique ramp of the Pedraforca thrust sheet (PU), acting as a buttress (Fig. 3), as in the case illustrated in Fig. 1a. The Tremp basin example shows a fluvial system on the hangingwall of the west oblique ramp of the South Central Unit (SCU), passing through it to the foreland basin (Fig. 4), as illustrated in Fig. 1b. The Pedraforca–South Central Unit example shows the varying topography of a large reentrant evolving from low-topography to high-topography during thrust evolution (Fig. 5). PF shows the location of the Pamplona Fault. M, Me, Bo, ES and SM show the position of the Montsec thrust, Mediano anticline, Boltaña anticline, External Sierras and Serres Marginals.

Both basins were filled up by an east to west progression from fluvial to shallow-marine to deep-marine depositional sequences. The elongate Tremp-Pamplona basin is subdivided in different segments on the basis of their deposits and position with respect to underlying structures. The Tremp basin infill is continental to shallow marine and rests on top of the South Central Unit. The Ainsa basin corresponds to the transition between shallow and deep sedimentation and is located above the west oblique ramp zone of the South Central Unit. The Jaca basin corresponds to an area of distal turbiditic deposition (e.g. Labaume *et al.*, 1987). The position of these sedimentary basins is marked in the tectonic map by the Ripoll, Tremp, Ainsa, Jaca and Pamplona towns that are situated along the Paleogene Pyrenean basins (Fig. 2).

The Ripoll basin is located in the footwall of the Pedraforca thrust sheet and its eastern oblique ramp zone, whereas the Tremp basin is located in the hangingwall of the South Central Unit and its west oblique ramp zone. Three examples of interplay between fluvial and oblique thrust ramp development are shown in this section (see location in Fig. 2):

1 the middle Eocene Ripoll basin evolution on the footwall of a major oblique ramp;
2 the early–middle Eocene Tremp basin development on the hangingwall of a major oblique ramp;
3 the change from reentrant to salient of the tectonic domain delimited by the west oblique ramp of the Pedraforca unit and east oblique ramp of the South Central Unit, and its control on the fluvial development from upper middle Eocene to middle Oligocene.

The Ripoll basin: a middle Eocene example on the footwall of a major oblique ramp

The middle Eocene infill of the Ripoll basin consisted of a thick prograding sequence made up of longitudinal fluvial deposits grading into deltaic deposits and finally into open marine deposits westwards (Fig. 3). The relatively rapid westwards migration of fluvial–deltaic–marine environments along the axis of the Ripoll piggyback basin was synchronous with the activity of the Pyrenean thrust system (e.g. Puigdefàbregas *et al.*, 1986; Vergés *et al.*, 1995). The Vallfogona frontal thrust carrying the Ripoll basin started its activity in the east and propagated laterally to the west. During this time, the Pedraforca thrust sheet was emplaced along the basal thrust, and earlier thrusts were reactivated following a break-back sequence, especially well preserved along its oblique termination (Martínez *et al.*, 1988). When the west-migrating tip point of the Vallfogona thrust reached the vicinity of the Pedraforca Unit the palaeogeography of this region resulted in a complex interplay between growing frontal (Vallfogona) and oblique thrust ramps (East Pedraforca termination). These two fluvial systems interacted, with longitudinal fluvial–deltaic systems flowing west along the Ripoll basin and transverse alluvial fan systems flowing south directly from the hinterland (Fig. 3a). The break-back sequence configuration of thrusting and related thrust anticlines of the eastern oblique termination of the Pedraforca Unit sustained a significant topography during the middle Eocene period, as proved by local erosion and unconformities related to each of the thrust slices (Martínez *et al.*, 1988; Vergés *et al.*, 1994). The middle Eocene evolution of this complex region is explained using a combination of detailed tectonic and sedimentological studies (Martínez *et al.*, 1988; Vergés *et al.*, 1994; Ramos *et al.*, 2002).

The proposed sedimentary evolution for this confined region is divided into three main stages according to the analysed sedimentary succession in the southern flank of the Ripoll basin in its western termination (Ramos *et al.*, 2002). These selected stages are separated according to their thickness, as their precise duration is unknown: period T1 between 0 and 1525 m (Fig. 3a); period T2 between 1525 and 2000 m (Fig. 3b); period T3 between 2000 and 2450 m (Fig. 3c).

The T1 period is represented by three-fifths of the total sedimentary thickness and longitudinal fluvial–deltaic systems were predominant along the Ripoll basin (Fig. 3a). The position of the delta front migrated to the west synchronous with the lateral propagation of the Vallfogona thrust tip point. The westwards progression of the blind tip line of the Vallfogona thrust possibly formed a topographic step (represented by open dots in Fig. 3a), which controlled the position of the delta front through time. Locally derived Mesozoic and lower Paleogene cover rocks accumulated in alluvial fans, which formed a transverse system joining the longitudinal streams of the Ripoll basin. A

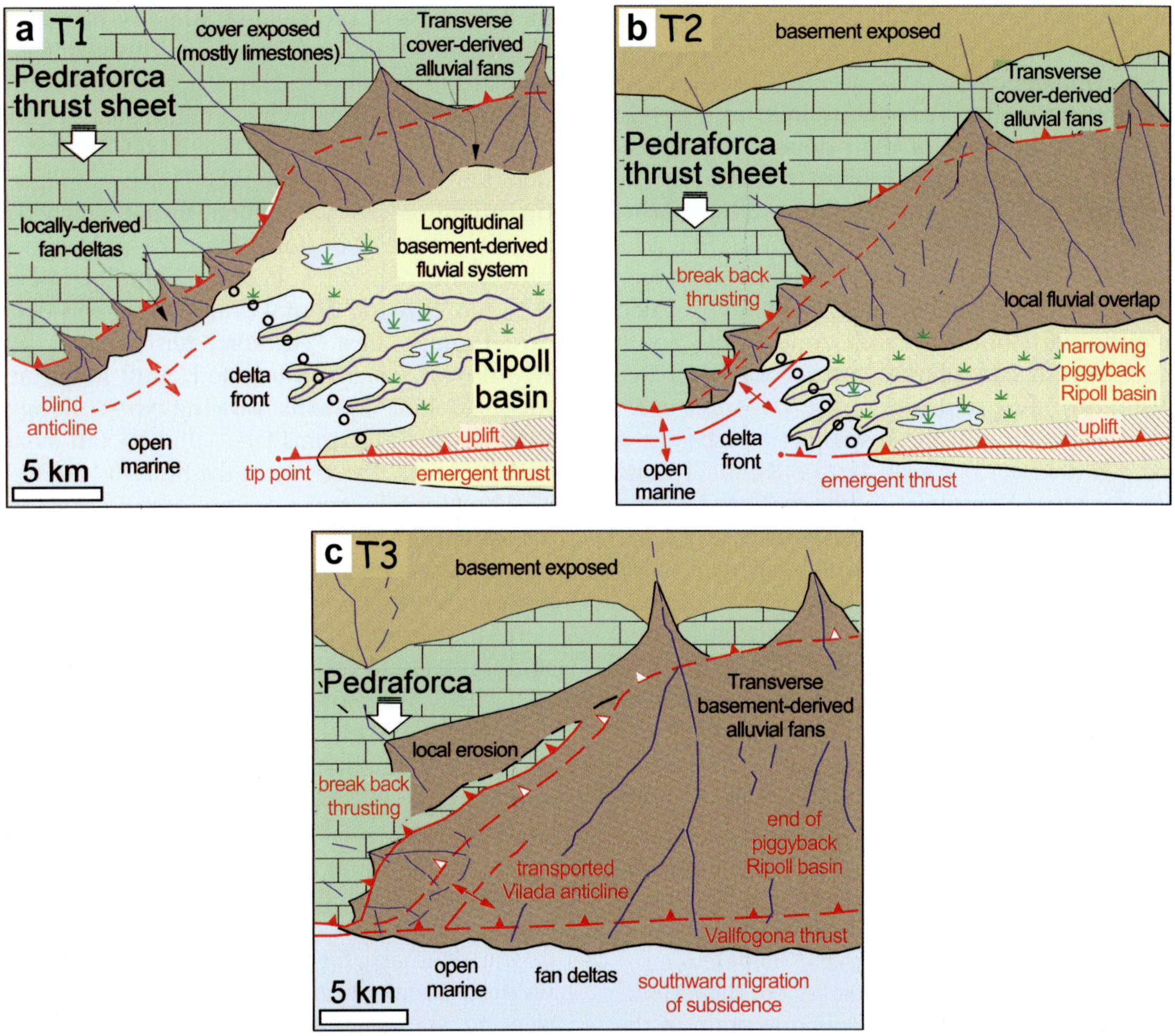

Fig. 3 Three-step evolution maps of the western segment of the Ripoll syncline in the southeast Pyrenees (a–c). The evolution from T1 to T3 times shows the interplay between longitudinal and axial fluvial and alluvial fluvial systems in the context of frontal and oblique growing thrust ramps (modified from Ramos *et al.*, 2002). Open circles show the position of the blind oblique ramp thrust linked to the west termination of the Vallfogona thrust. Thick white arrow shows the direction of the tectonic transport.

system of narrow but thick fan deltas discharged directly into the restricted bay in front of the delta mouth. The bay was confined by the growth of a blind anticline developed parallel to the oblique ramp zone of the Pedraforca thrust sheet. Although a smooth topography related to the emergence of the Vallfogona frontal thrust probably existed, there is no record of deposits derived from it in the analysed sedimentary successions (Ramos *et al.*, 2002).

The T2 period of evolution is represented by one-fifth of the total sedimentary thickness (Fig. 3b). The interplay between longitudinal fluvial sedimentation and transverse alluvial deposition was characteristic of this period. The combined effect of both the narrowing of the Ripoll piggyback basin by tectonic shortening (see discussion in Ramos *et al.*, 2002), as well as the southwards progression of the transverse alluvial fans, resulted in a restriction of the longitudinal fluvial drainage. These alluvial systems, mostly carrying Mesozoic and lower Paleogene cover-derived sediments, could temporarily overlap the longitudinal fluvial system. The break-back system of thrusting along the oblique boundary of the Pedraforca thrust sheet renewed the topographic barriers during fluvial deposition and inhibited the flow of these systems over the oblique ramp zone.

The T3 period of evolution also corresponds to one-fifth of the total sedimentary thickness (Fig. 3c). This period was characterized by the cessation of the longitudinal fluvial–deltaic system in the Ripoll piggyback basin. This was mainly due to the sudden influx of basement-derived clasts from the Pyrenean hinterland, which were deposited on large transverse alluvial fans covering the entire Ripoll basin. These large fans could finally bury, at least partially, the frontal Vallfogona thrust. The continuous growth of the Ripoll piggyback syncline was recorded by three main unconformities observed in the upper part of the sedimentary succession in both flanks of the Ripoll syncline (Vergés *et al.*, 1994; Ramos *et al.*, 2002).

The structural position of the middle Eocene fluvial longitudinal system flowing along the Ripoll piggyback basin, in the footwall of the oblique ramp zone of the Pedraforca Unit with higher topography, is similar to the first scenario in Fig. 1a. This higher topography exerted a buttress effect that diverted the fluvio-deltaic longitudinal system towards the southwest. Transverse local fans derived from the Pedraforca oblique ramps zone also could enhance the diversion effect on the longitudinal river.

The Tremp basin: a lower–middle Eocene example on the hangingwall of a major oblique ramp

The Tremp basin on top of the South Central Unit (Séguret, 1972) was geographically continuous with the turbiditic Jaca basin during lower–middle Eocene times, and continued into the Bay of Biscay in the Atlantic (Fig. 2). The western boundary of the South Central Unit was shaped by the roughly north–south trending Mediano and Boltaña anticlines (Fig. 2). The extremely good quality of the Tremp basin outcrops, as well as their continuity across the north–south anticlines into the Ainsa and Jaca basins, have permitted detailed sedimentological studies to be made of the passage from fluvial to deltaic to deep-marine deposits (Mutti *et al.*, 1972; Nijman & Nio, 1975; Puigdefàbregas, 1975; Fernández *et al.*, 2004).

The west oblique ramp zone of the South Central Unit is a complex oblique termination composed of the Montsec and Serres Marginals oblique thrust ramps and parallel related anticlines developed in front of each of these thrusts (Fig. 2). This oblique ramp zone was blind and migrated to the west-southwest through early and mid-Eocene time (Vergés, 2003). In front of the Montsec and Serres Marginals oblique thrusts, the detached Mediano anticline (Poblet *et al.*, 1998), and the fault-propagation Boltaña anticline (Fernández *et al.*, 2004), developed synchronously with the middle Eocene deposition across the west margin of the South Central Unit. In this tectonic scenario, the fluvio-deltaic Tremp basin was transported to the south as a piggyback basin on top of the Montsec thrust (e.g. Nijman, 1998). The connection between these fluvio-deltaic systems in the Tremp basin (hangingwall of the Montsec thrust) and the channelized turbidites in the Ainsa basin occurred through the oblique ramp zone (Fig. 4).

The Eocene fluvial deposits of the Montanyana Group in the Tremp basin formed a longitudinal system with west-northwest palaeoflow direction bounded to the north by a transverse alluvial fan system draining the Pyrenean Axial Zone (e.g. Nijman & Nio, 1975; Marzo *et al.*, 1988; Nijman, 1998; Vincent, 2001). The fluvial system migrated successively to the south (numbers 1 to 3 on main

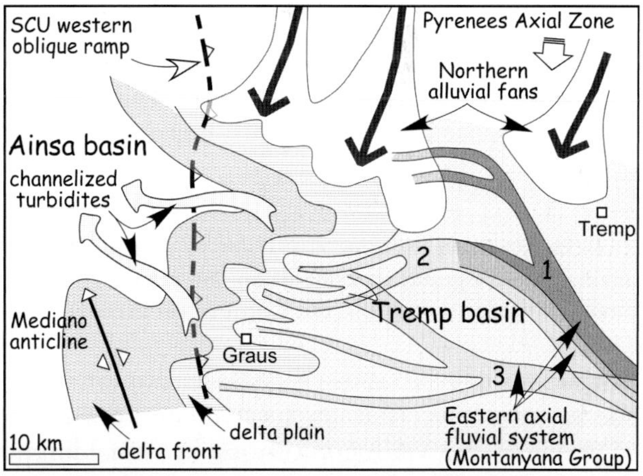

Fig. 4 Early to middle Eocene evolution of the connection zone between the Tremp basin (in the hangingwall of the west oblique ramp of the South Central Unit in the southern Pyrenees) and the Ainsa basin (in the footwall of this SCU oblique ramp). Figure modified from Nijman & Nio (1975) and Marzo *et al.* (1988). The thick white arrow represents the tectonic transport during thrusting.

fluvial channels in Fig. 4), due to the interplay between the progressive uplift in the inner parts of the thrust system and synchronous growth of the transverse alluvial systems. The fluvial systems evolved westwards into a deltaic system developed along the western edge of the South Central Unit (Nijman & Nio, 1975; Fig. 4). The delta-front facies associations overlapped the trace of the western oblique ramp of the Montsec thrust. The Ainsa basin developed to the west of the oblique ramp system and to the north of the north-plunging Mediano growth anticline (e.g. Poblet *et al.*, 1998; Fernández *et al.*, 2004).

The position of the fluvio-deltaic system on top of the Tremp piggyback basin, grading into slope deposits overlapping the west oblique thrust ramps of the Montsec thrust and grading to the west into channelized turbidites, coincides with the scenario in Fig. 1b. In this scenario, the higher topography of the hangingwall of the oblique ramp controls the position of the slope facies through time. This control of oblique ramps on the fluvial to deltaic deposition has been explored recently by numerical modelling using this particular case study (e.g. Clevis *et al.*, 2004). The concomitant growth of the Mediano anticline in the footwall of the Montsec oblique thrust ramp complicated the interrelations between tectonics and sedimentation. The northern lateral growth of the Mediano anticline pushed the turbiditic systems to the north whereas the fluvial system on the Tremp piggyback basin was pushed to the south by both uplift in the Pyrenees Axial Zone and south-migration of the transverse alluvial fans (Fig. 4).

The change from tectonic reentrant to uplifted block during thrusting and its control on the transverse fluvial development (Port del Compte reentrant)

A thick upper Eocene–Oligocene conglomeratic succession crops out in the footwall of the western segment of the Vallfogona thrust at Sant Llorenç de Morunys (Fig. 5). This continental clastic succession displays the most outstanding growth strata geometry in the world (Riba, 1976; Ford *et al.*, 1997; Suppe *et al.*, 1997). Palaeocurrent data and clast composition show a clear northern provenance, forming a transverse system of alluvial fans flowing towards the centre of the Ebro basin (Williams *et al.*, 1998). Reconstructed maps indicate that the main entrance for these conglomeratic deposits was across the region, presently uplifted above the Port del Compte thrust sheet (Williams *et al.*, 1998; Fig. 5). Two different stages of regional evolution have been illustrated during late Eocene and early Oligocene times.

During the first stage of evolution, the region has been interpreted as a tectonic reentrant (Port del Compte reentrant), bounded by the west oblique ramp of the Pedraforca thrust and by the northern segment of the Segre oblique ramps zone (Fig. 5a). The lower topography of this reentrant with respect to the surrounding oblique ramps zone acted as a collector for transverse rivers carrying basement-derived sediments from the basement thrust sheets of the Pyrenees Axial Zone to the centre of the Ebro foreland basin. Palaeoflow dispersal through the Sant Llorenç de Morunys alluvial fan was to the south and west (Williams *et al.*, 1998), whereas large-scale fluvial palaeocurrents were parallel to the growing Oliana anticline (Burbank & Vergés, 1994; Fig. 5a).

During the second stage of evolution, the Port del Compte thrust sheet was emplaced on top of the tectonic reentrant in which proximal conglomerates of the Sant Llorenç de Morunys fan deposited (Fig. 5b). As a consequence, the region occupied by the former reentrant became an uplifted region, floored by a thrust that branched to the Segre oblique ramps zone through a southeast-directed blind ramp (Vergés, 1999). This ramp was parallel to the Segre oblique ramps zone. This new palaeogeography was characterized by a high topography to the east of the Segre oblique ramps zone, including the former region of the reentrant. This higher topography to the east of the Segre oblique ramps zone significantly changed the drainage distribution of large tranverse rivers draining towards the Ebro basin. Former major Pyrenean rivers feeding the Sant Llorenç de Morunys alluvial fan were shifted to the west into the ancestral Segre River valley, which is recorded by remnants of discordant proximal conglomerates located at elevated altitude (blank patches in Fig. 5b). South-directed, locally derived alluvial fans draining the Port del Compte thrust sheet were deposited during the late stages along the active thrust front to the west of the Sant Llorenç de Morunys (Fig. 5b).

The above example is more complex than the proposed model for interactions between oblique

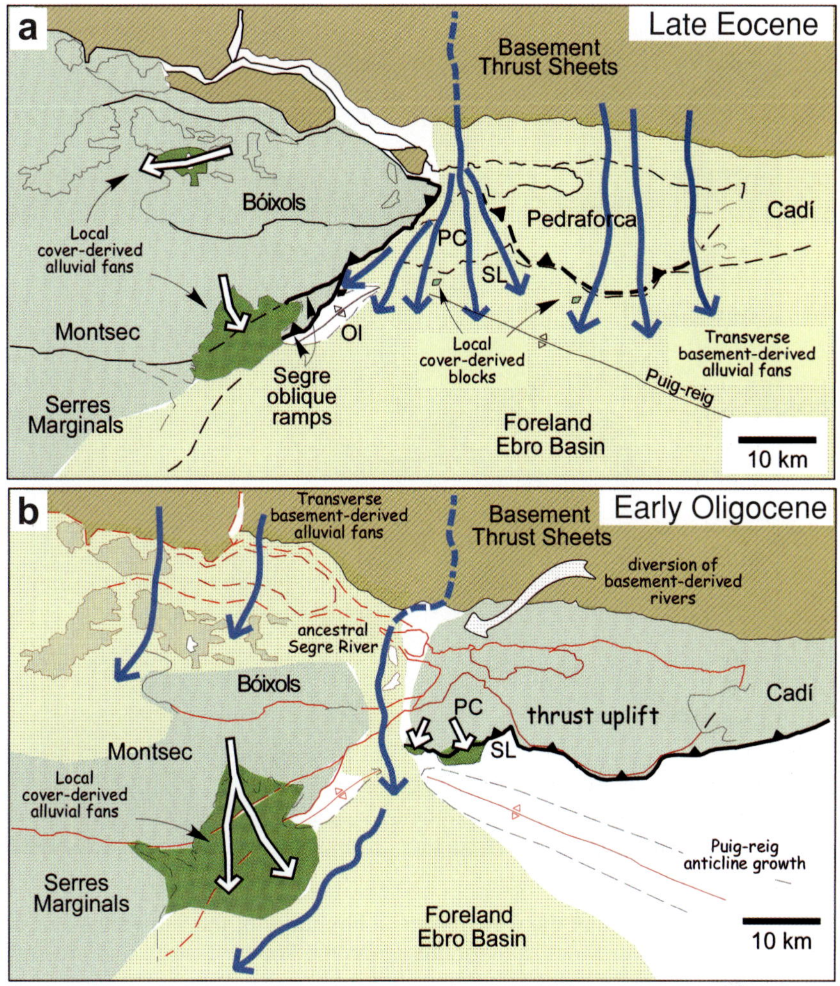

Fig. 5 Two-step evolution maps of the Pedraforca reentrant in the southern Pyrenees. (a) The first step of evolution shows the interplay between alluvial systems and oblique ramps forming a reentrant during the late Eocene (thick lines show active thrusts). This reentrant was the focus of major transverse fluvial systems infilling the closed intermountain Ebro basin. (b) During the second stage of evolution in the early Oligocene, the uplift of the reentrant by the Vallfogona thrust diverted the transverse fluvial system to the west where topography remained lower. Ol, PC and SL show the position of Oliana anticline, Port del Compte thrust sheet and Sant Llorenç de Morunys alluvial fan.

ramps and fluvial deposition illustrated in Fig. 1a. The most important elements of the example during the first stage of evolution are that the fluvial system is not longitudinal but transverse, and that there were two opposed oblique thrust ramps bounding the fluvial drainage. These two opposed oblique ramps showing higher topography formed a tectonic reentrant or embayment that collected major drainage systems flowing from the mountains to the Ebro foreland basin. During the second stage of evolution, the southeast-directed blind ramp that uplifted the entire region to the east of it formed a local flexure, parallel to the blind thrust ramp. The river system was diverted to the west of the flexure that formed the eastern side of the valley of the ancestral Segre River to the north of the Oliana anticline. Again, the high topography generated by an oblique thrust ramp, blind in this case, constrained the transverse fluvial system draining the inner parts of the mountain chain.

MODERN EXAMPLES FROM NORTHWEST HIMALAYAS, CENTRAL ANDES AND ZAGROS

Present-day examples are useful to see the short-term interplay between oblique-lateral ramps and river systems in map view. Three examples of modern rivers crossing oblique or lateral ramps are presented: the Indus River crossing the Kalabagh Fault in the Himalayas of northwest Pakistan, the drainage distribution of the tributaries of the Marmoré River along the Eva Eva piggyback basin in the front of the Andes in Bolivia, and the late Miocene to present distribution of drainage system in the northwest Zagros fold belt in Iran.

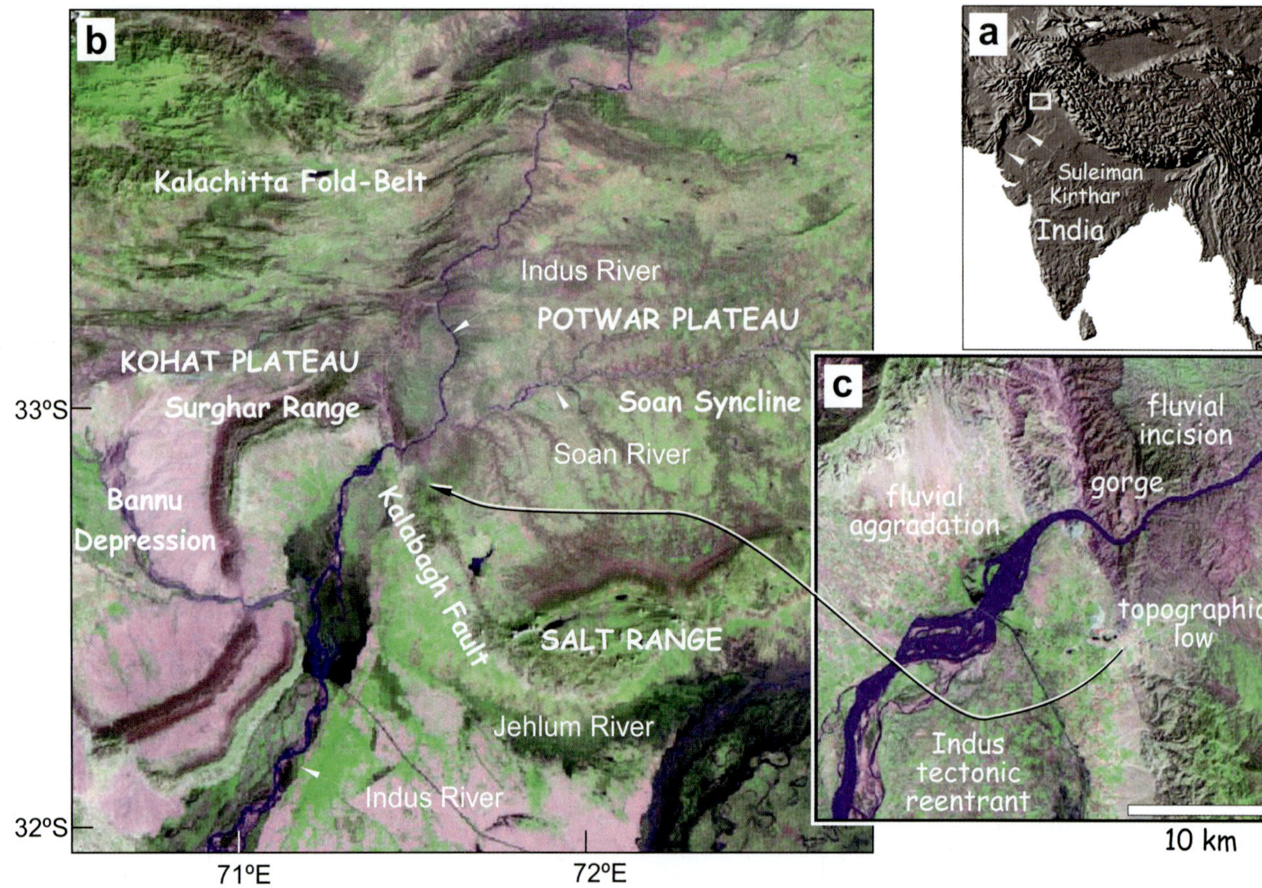

Fig. 6 Satellite mosaic of the western lateral boundary of the Salt Range and Potwar Plateau in northwest Pakistan. (a–c) The Indus River flows to the south across the Potwar Plateau and crosses to the foreland through the Kalabagh Fault. This fault connects the Salt Range and the Surghar Range. The inset (c) shows that the Indus River does not cross through the lowest topography but across the most likely south-propagating northern branch of the Kalabagh Fault. The Indus River incises into the Soan syncline deposits and aggrades in the Indus tectonic reentrant (foreland basin).

The Indus River across the Kalabagh Fault in the Himalayas of northwest Pakistan

The Indus River crosses the Kalachitta fold-belt to the north and flows along the west side of the Potwar Plateau in a north–south direction before cutting across the Kalabagh Fault (Fig. 6). The Potwar Plateau evolved as a piggyback basin starting at around 5 Ma (Burbank & Beck, 1989), or even earlier at around 10 Ma (Grelaud *et al.*, 2002), and is still active at present (Yeats & Lillie, 1991). The Potwar Plateau developed on top of a main detachment, which crops out along the frontal Salt Ranges (e.g. Butler *et al.*, 1987). The Kalabagh Fault forms the east lateral thrust ramp of the Potwar Plateau connecting the frontal thrusts along the Salt Ranges in the southeast and the Surghar Range in the northwest. This lateral fault segment limits the Indus tectonic reentrant to the east-northeast (Fig. 6). The Indus tectonic reentrant acts as a collector of waters flowing out from the more elevated regions of the Potwar and Kohat plateaux to the northeast and northwest, respectively.

The Indus river cuts and incises the Kalachitta fold-belt and the Potwar Plateau, indicating that the entire region above the detachment has been uplifted (e.g. Talling *et al.*, 1995). The Indus River joins the Soan River draining the Potwar Plateau before it crosses the Kalabagh Fault. Interestingly, the Indus River cuts through an uplifted area aligned with the fault, whereas to the south of this gorge there is a wide open gap with flat topography (Fig. 6). The fluvial process of the Indus River changes from incision along the Potwar Plateau to aggradation in the Indus tectonic reentrant, across the Kalabagh lateral fault (Fig. 6). This aggradation probably occurs because of the general flexural subsidence induced by the Himalayan thrust sheets.

Large rivers in the northwest Himalayas flow out from mountains crossing lateral and oblique ramps that limit tectonic reentrants. These tectonic syntaxes represent low topographic regions, affected by subsidence that favours fluvial aggradation. The Indus River flows out of the Himalayas through the Indus tectonic reentrant to become the major south-directed foreland river system in front of the Suleiman and Kirthar fold-belts (see their location in large-scale map of Fig. 6a).

The Indus tectonic reentrant, formed by two lateral thrust ramps linking frontal segments, corresponds to the first scenario in Fig. 1a, combined with the reentrant scenario in Fig. 1c. The Indus River flows on top of the Soan syncline and reaches the foreland basin across the Kalabagh lateral thrust ramp. The symmetric position of a lateral ramp in front of the Bannu Depression created an embayment that constrained the Indus River.

The present river system along the Eva Eva piggyback basin in the front of the Bolivian Andes

The frontal Subandean Zone to the north of the Santa Cruz bend in Bolivia shows a good example of interplay between the fluvial system in the piggyback basin and growing frontal and oblique thrust ramps (e.g. Baby *et al.*, 1995; Okaya *et al.*, 1997; Fig. 7). In

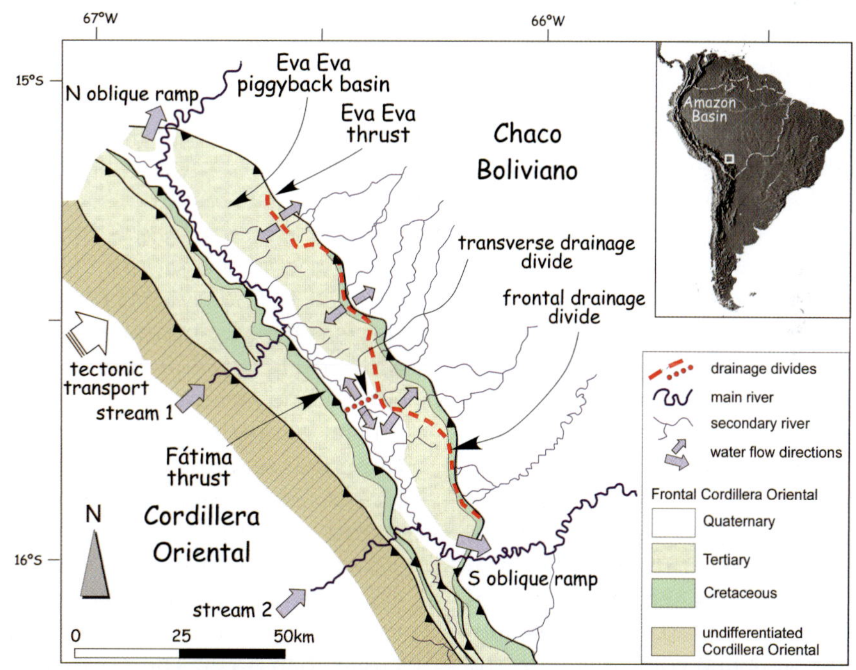

Fig. 7 Simplified tectonic map of the frontal Subandean Zone between latitudes 15°S and 17°S with river system depicted from satellite images. The map shows the two most external thrust sheets in greater detail (Fátima and Eva Eva thrusts). The Eva Eva thrust (the Subandean outer emergent thrust) carries the Eva Eva piggyback basin on top of the Chaco Boliviano foreland basin. Two main streams (labelled stream 1 and stream 2) transversely drain the Subandean Zone and become longitudinal along the Eva Eva piggyback syncline to flow out across the two oblique ramps at the north and south terminations of the Eva Eva thrust. These oblique ramps display the lowest topography along the frontal Eva Eva thrust. Newly developed rivers in the front of the thrust sheet could finally capture the longitudinal rivers and thus develop a transverse drainage system crossing the Eva Eva piggyback basin. The map of South America shows the extent of the Amazon Basin. The thin white line to the east of the study area in the white box shows the position of the Marmoré River.

this region, the most frontal tectonic unit of the Subandean Zone is formed by the Eva Eva thrust that bounds the Eva Eva piggyback basin. The infill of the basin is made up of Tertiary growth strata with typical growth geometries that are observed in both seismic lines and map pattern. The 135-km-long Eva Eva thrust emerges at the surface showing one frontal and two oblique thrust ramps at the thrust terminations. The frontal anticline is thrust related and narrow (mostly limited to the Cretaceous outcrops in Fig. 7). The anticline shows a smooth topography but enough to form topographic barriers to constrain the distribution of the fluvial drainage.

Two major streams drain the Cordillera Oriental in the hinterland of the Eva Eva basin (Fig. 7). These two rivers perpendicularly cross the major Fátima thrust and turn 90° to run parallel to the Eva Eva basin axis but with opposite directions. The transverse drainage divide between these two rivers is located in the middle of the Eva Eva basin. The outlets of both streams are located along the low topography of the presently growing oblique ramps at the terminations of the Eva Eva thrust. Out of the Subandean front, the rivers flow to the east and northeast to join the major Mamoré River that flows to the north to join the Amazon River across the plains of the Chaco Boliviano foreland basin (Fig. 7).

The frontal drainage divide separates the fluvial longitudinal system from the river system that drains the frontal anticline above the thrust, but mostly its forelimb (Fig. 7). Interestingly, in the middle part of the Eva Eva piggyback basin, this drainage divide is displaced more than 5 km reaching the Eva Eva piggyback basin to the southwest of the anticline axis. This displacement of the drainage divide is the product of frontal stream erosion. The region in which the frontal longitudinal drainage divide breached the topographical crest of the anticline coincides with the position of the transverse drainage divide separating the two axial streams in the Eva Eva piggyback basin. The concurrence of the longitudinal and transverse drainage divides, roughly in the central part of the Eva Eva thrust, probably indicates that the thrust nucleated in this central part and propagated towards both ends through time. The main longitudinal fluvial systems can potentially be defeated by transverse stream piracy in these regions, where the longitudinal drainage divide has been displaced towards the southwest into the Eva Eva piggyback basin (opposed blue arrows in Fig. 7).

Piggyback basins in frontal parts of active fold-and-thrust belts restrain fluvial systems crossing already uplifted hinterland regions of the belt. These rivers that transversely cross older thrust sheets can adjust their courses to flow parallel to the growing frontal structures, at least temporally. Most of these longitudinal fluvial systems flow out of the piggyback basin through growing oblique ramps at the terminations of frontal thrusts. This scenario resembles the proposed model in Fig. 1b. River capture is the most likely mechanism by which the fluvial arrangement of frontal active thrust sheets varies from longitudinal, along the piggyback basin, to transverse, cutting across the frontal thrust and related anticline.

The recent development of the drainage system in the Pusht-e Kuh Arc in northwest Zagros Mountains in Iran

The present configuration of the Zagros fold-and-thrust belt in Iran is composed of structural arcs and reentrants that from southeast to northwest are: the Fars Arc, the Dezful Embayment, the Pusht-e Kuh Arc and the Kirkuk Embayment (Fig. 8). The external boundary that separates arcs from reentrants is the Mountain Front Flexure (Falcon, 1961; MFF in Fig. 8). This Mountain Front Flexure has frontal and oblique segments probably matching crustal blind thrust ramps at depth. The oblique ramp that separates the Pusht-e Kuh Arc from the Dezful Embayment is called the Balarud Fault. Across this fault there are marked changes in the structural relief as well as in the topographic relief. The structural relief increases more than 2.5 km in the Pusht-e Kuh Arc but in some regions can be larger than 5 km. The mean altitude of the Dezful Embayment is around 150 m, whereas it is about 1000 m in the Pusht-e Kuh Arc, with maximum altitudes above 2700 m along the Kabir Kuh anticline.

The middle–late Miocene to present palaeogeographic reconstruction of this part of the Zagros Mountains is needed to understand its modern configuration. The lower part of the fluvial Agha Jari Formation, with a mid- to late Miocene age, was deposited in most of the Lurestan Province

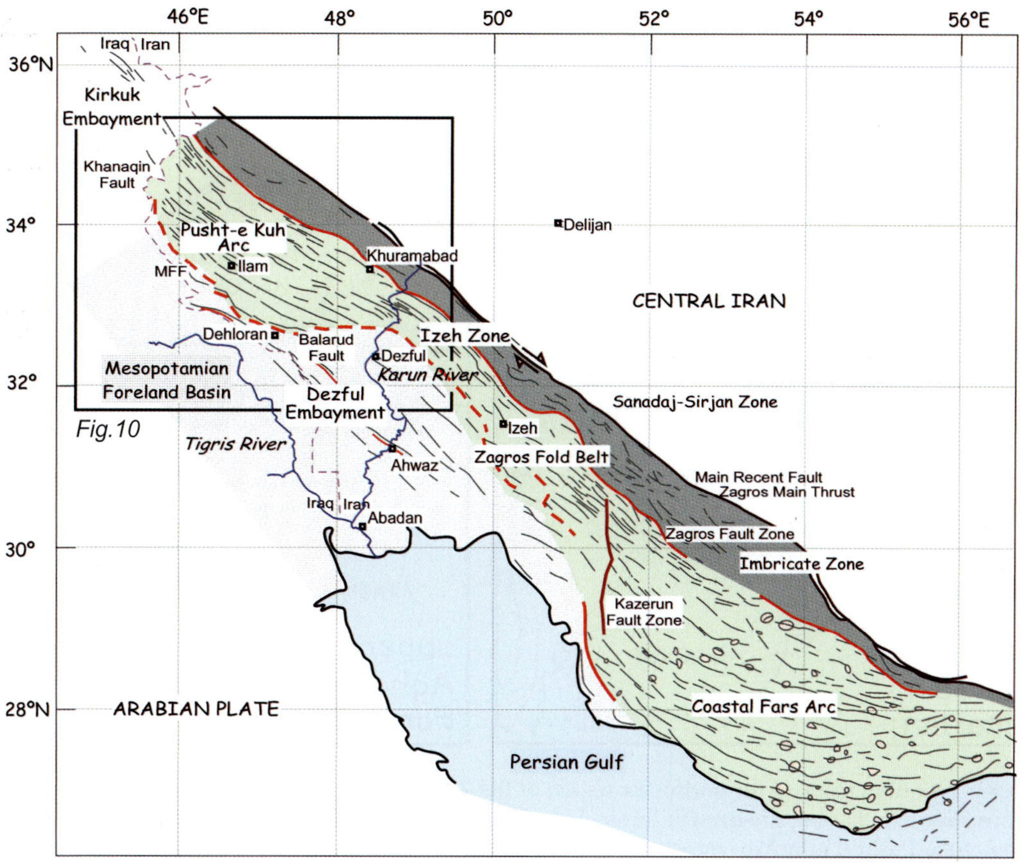

Fig. 8 Structural map of the Zagros showing the Fars and Pusht-e Kuh arcs bounding the low-topography Dezful Embayment. The Pusht-e Kuh Arc is bounded by two oblique ramps: the Khanaqin Fault to the northwest and the Balarud Fault to the southeast.

(stratigraphic region presently folded and uplifted as the Pusht-e Kuh Arc). These fluvial Agha Jari deposits constituted the major axial system during the early development of the Mesopotamian foreland basin in front of the rising Zagros fold-and-thrust belt (e.g. Elmore & Farrand, 1981; Fig. 9a). The system flowed towards the ancestral Persian Gulf, located to the northeast of its present position (e.g. Ziegler, 2001; Bahroudi & Koyi, 2004). Folding in the external belt of the Pusht-e Kuh Arc was initiated during late Tortonian times at about 7.6 Ma (Homke *et al.*, 2004). The growth in length and amplitude of Pusht-e Kuh anticlines shifted the longitudinal fluvial system towards the southwest during the deposition of the upper Agha Jari Formation (Fig. 9b). This upper Agha Jari depositional unit shows maximum thicknesses along the front of the Pusht-e Kuh Arc and Dezful Embayment, constituting the ancestral Tigris River. The uplift of the Pusht-e Kuh Arc, roughly synchronous with the deposition of the upper Agha Jari unit, took place at the beginning of the Pliocene (Homke *et al.*, 2004). The uplift of the entire tectonic arc above the Mountain Front Flexure (MFF in Fig. 9) produced the final migration of the axial rivers into the present Mesopotamian foreland basin and the onset of the recently developed drainage system in the Pusht-e Kuh Arc.

The recent large-scale tectonic structure of the northwest Zagros Mountains is characterized by uplifted arcs bounding the lowland Dezful Embayment, which constrains the present distribution of rivers draining this region (Fig. 10). Rivers flow from uplifted domains, such as the Pusht-e Kuh Arc and the Izeh Zone in the northeast of Dezful Embayment, into the Dezful Embayment tectonic reentrant. The present arrangement of the fluvial drainage in the front of

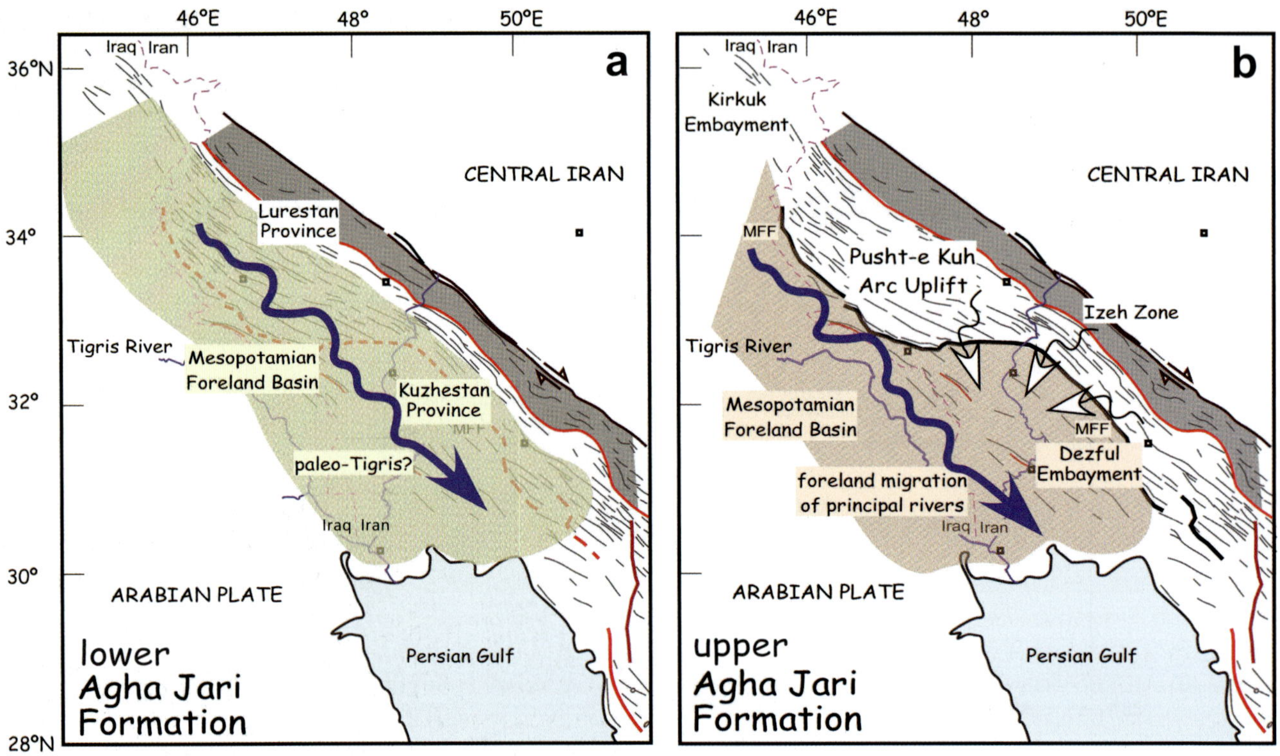

Fig. 9 Two-step evolution maps of the northwest region of the Zagros in Iran during late Miocene and Pliocene times (Agha Jari Formation). (a) During pre-growth lower Agha Jari Formation times the ancestral foreland fluvial system was located on top of the Pusht-e Kuh Arc (ancestral Tigris river?). (b) During uplift of the Pusht-e Kuh Arc by folding and thrusting the upper Agha Jari Formation fluvial system was pushed to the southwest. The Dezful Embayment became a reentrant with its low topography and bounded by oblique ramps and was the collector of fluvial networks draining the Pusht-e Kuh Arc and the Izeh Zone.

the northwest Zagros fold-and-thrust belt is the culmination of the southwest migration of the axial fluvial system during the late Miocene and Pliocene folding and thrusting. The uplift of large areas on top of the Mountain Front Flexure during the Pliocene initiated the present-day configuration of the drainage system in the Pusht-e Kuh Arc and Izeh Zone, which is characterized by rivers that flow parallel to fold trends, as well as by transverse rivers cutting through folds (Oberlander, 1985; Fig. 10a). The distribution of drainage divides in the Pusht-e Kuh Arc shows that most of this uplifted region drains towards the southeast. The 200-km-long Kabir Kuh anticline forms the southwest segment of the major drainage divide in the arc (white thick dashed line in Fig. 10a). The Seymareh River (Fig. 10c) constitutes the principal stream of the Pusht-e Kuh Arc and flows out of it across the Balarud Fault (an oblique ramp of the Mountain Front Flexure). The main drainage divide continues towards the northwest, where southeast-flowing rivers with lower base level are presently capturing northwest-flowing rivers on top of thin Quaternary gravel terraces dipping towards the Kirkuk Embayment (Fig. 10b).

The Dezful Embayment is characterized by low topography and constitutes the final destination of the rivers flowing across both the oblique ramp of the Pusht-e Kush Arc and the frontal ramp of the Izeh Zone. All of these tributaries join in a major stream (the Karun River) that flows to the south across the Abadan Plains and connects the Tigris River near the Persian Gulf.

The disposition of the Dezful Embayment and surrounding boundaries along the Mountain Front Flexure represents a large-scale tectonic reentrant like that illustrated in Fig. 1c. Large rivers cross this boundary to flow from regions with high topography in the Pusht-e Kuh Arc and Izeh Zone, to the Dezful Embayment domain that is characterized by

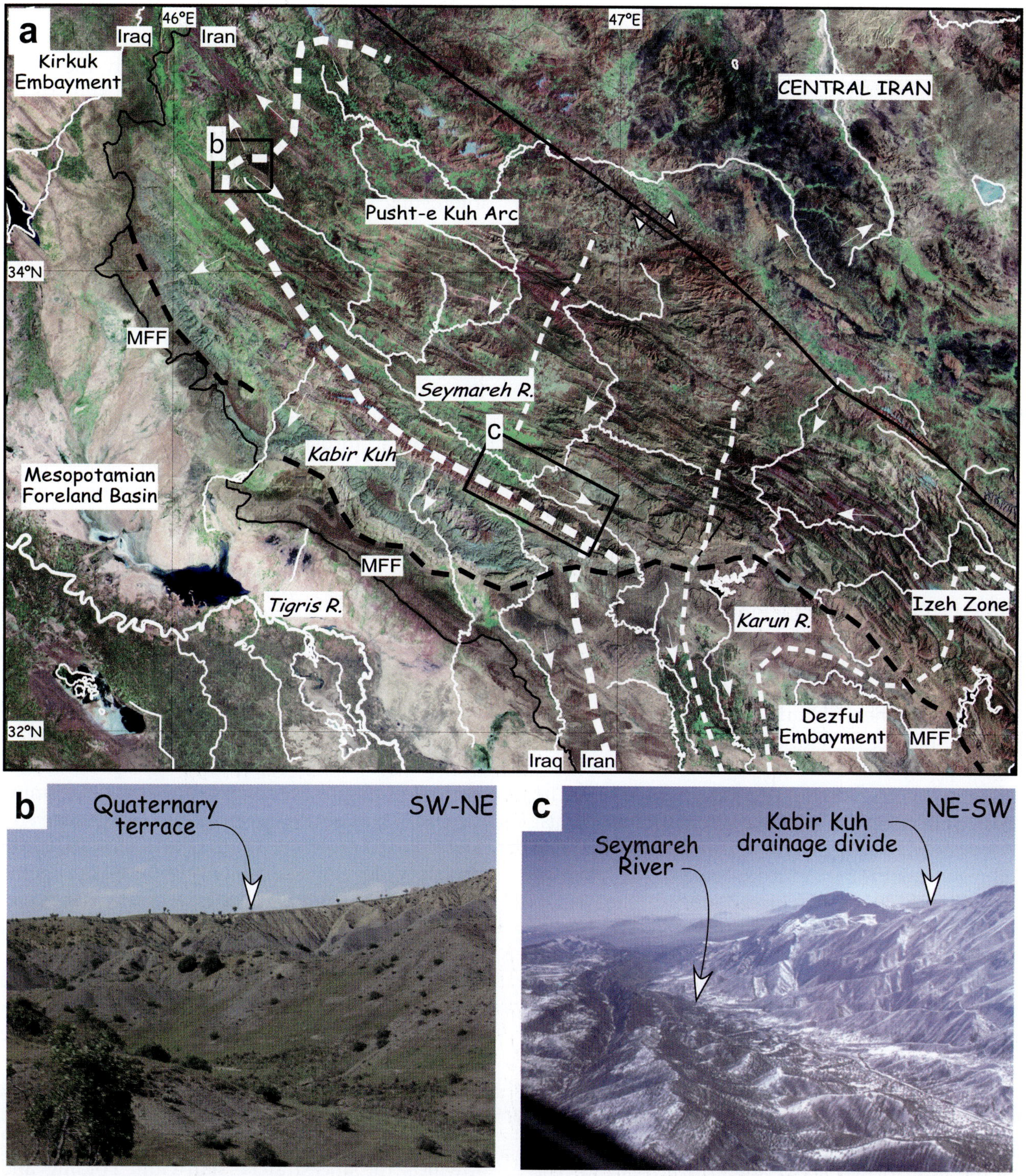

Fig. 10 Satellite mosaic of the Pusht-e Kuh Arc and Dezful Embayment in the northwest Zagros Mountains of Iran. (a) The distribution of rivers (marked in white continuous lines) and their direction of flow (white arrows) show a typical fluvial network made up of both transverse and longitudinal streams. Transverse streams cut through the tectonic structures following a NE–SW direction, which is orthogonal to folding. Longitudinal streams follow a folding-parallel direction. The 200-km-long Kabir Kuh anticline forms the southwest drainage divide (dashed thick white line), constraining the Seymareh river valley that flows towards the Dezful Embayment (c). The main water divide continues towards the northwest where this river system is capturing rivers flowing in the opposite direction (towards the Kirkuk Embayment; b).

low topography. These relatively important streams are tributaries of a major river that drains towards the Mesopotamian foreland basin and the Persian Gulf.

DISCUSSION AND CONCLUSIONS

Major oblique thrust ramps have a significant influence on drainage development in fold-belts. If these ramps are in the hangingwall of active piggyback basins they divert the fluvial and/or deltaic system because of their topographic buttress effect, as in the west termination of the Ripoll basin in contact with the east oblique ramps of the Pedraforca Unit. In contrast, where piggyback basins develop on the hangingwall of major oblique ramp zones the depositional systems grade from alluvial and fluvial in the hangingwall (Tremp basin) to channelized turbidites in the footwall (Ainsa basin). The Indus River also runs above a piggyback basin and flows out of the Himalayan ranges across a lateral ramp displaying lower topography.

Piggyback basins in front of active fold-and-thrust belts normally display one or two blind oblique ramps linked to the growing terminations of the basin. These oblique ramps are blind and piggyback basins developed in their hangingwall. These piggyback basins, if subaerial, are normally filled up by axial fluvial depositional systems. The growth of the piggyback basin constrains, at least temporally, the longitudinal drainage pattern. The natural outlet for these axial rivers to flow out of the piggyback basin corresponds to the region located near the tip point of the thrust. This region normally develops as a lateral or oblique blind ramp zone, which is characterized by lower topography given that it is only moderately involved in uplift related to thrusting, such as in the Ripoll and Eva Eva piggyback basins.

The progressive entrenchment of the river network during piggyback basin growth is associated with uplift linked to thrusting, as in the case of the Indus River above the Potwar Plateau. The change from axial to transverse drainage in a piggyback basin can be triggered by the sudden increase of transverse alluvial fan deposition that fills up the basin, such as in the case of the Ripoll basin. However, axial river capture by headwater erosion of frontal streams (linked to the emergent hangingwall of the frontal thrust) is also a potential mechanism to defeat longitudinal drainages, such as in the Eva Eva basin. Obviously, these two mechanisms together with synchronous thrust related uplift can operate simultaneously to modify the drainage system from axial to transverse.

Although not documented in this work, it is interesting to observe that the late Eocene and Oligocene infills of the Tremp and Jaca basins were subaerial (alluvial to fluvial), with longitudinal rivers confined to the north of the tectonically active External Sierras (ES in Fig. 2). The progressive infill of the Jaca basin modified the fluvial arrangement along the basin, increasing the significance of major transverse alluvial fans. This transverse drainage system finally crossed the frontal thrust of the External Sierras and became predominant during the late stages of the Ebro basin infill (e.g. Hirst & Nichols, 1986; Friend *et al.*, 1996; Nichols & Hirst, 1998; Jones, 2004).

Two opposed oblique thrusts outline a tectonic reentrant or embayment. These tectonic reentrants in the footwall of the oblique ramps constitute a morphotectonic domain with lower topography that concentrates rivers draining an active mountain belt. Examples of these reentrants at different scales are the Port del Compte reentrant in the south of the Pyrenees, the Indus reentrant in northwest frontal Himalayas and the Dezful Embayment in northwest Zagros Mountains. The Port del Compte reentrant was active during the upper Eocene and lowermost Oligocene times; the Indus reentrant probably developed in Pliocene times as did the Dezful Embayment. River processes change remarkably across these lateral or oblique thrust ramps, from incision on top of the ramp hangingwall, to aggradation on its footwall, such as in the Indus reentrant or in the Tremp basin.

The late Miocene fluvial arrangement in the foreland of the Zagros in the Lurestan Province comprised a large longitudinal river system during the deposition of the lower part of the Agha Jari Formation. This axial fluvial deposition drained towards the ancestral Persian Gulf. Folding and uplift of the Pusht-e Kuh Arc during the latest Miocene and Pliocene produced the shift of the river system to the southwest during the deposition of the upper Agha Jari Formation and part of the Bakhtyari Formation, during late Pliocene times. The uplift of the Pusht-e Kuh Arc along the Mountain

Front Flexure completed the present-day configuration of the Zagros fluvial drainage, which flows towards and merges into the structural reentrant of the Dezful Embayment. The Tigris River in the Mesopotamian foreland basin is confined to the northeast by the Mountain Front Flexure developed above a blind ramp bounding the frontal part of the Pusht-e Kuh Arc. Further uplift of the Port del Compte reentrant during the early Oligocene inverted the topography, and completely transformed the drainage installed in the low topographic domain of the tectonic reentrant by deflecting these transverse rivers towards the west.

ACKNOWLEDGEMENTS

This is a contribution of the Group of Dynamics of the Lithosphere (GDL). This paper has been possible thanks to fruitful discussions on tectonics and sedimentation with sedimentologists with whom I have collaborated over the past years. The manuscript has been greatly improved thanks to the comments of Jonathan Turner, Gary Nichols, and an anonymous reviewer. The Eva Eva example was investigated within the framework of a Repsol-YPF collaborative project. The Zagros example was documented within the framework of a collaborative research project with Norsk Hydro. Albert Martínez drafted the original map of Fig. 3. This work had the partial support of Project 2001 SGR 00339 Grup d'Estructura i Processos Litosfèrics, funded by the Comissionat per Universitats i Recerca of the Generalitat de Catalunya, Grups de Recerca Consolidats, II Pla de Recerca de Catalunya and of project Consolider-Ingenio 2010 no. CSD 2006–00041.

REFERENCES

Baby, P., Moretti, I., Guillier, B., *et al.* (1995) Petroleum system of the northern and central Bolivian sub-Andean zone. In: *Petroleum Basins of South America*, pp. 445–458. Memoir 62, American Association of Petroleum Geologists, Tulsa, OK.

Bahroudi, A. and Koyi, H.A. (2004) Tectono-sedimentary framework of the Gachsaran Formation in the Zagros foreland basin. *Mar. Petrol. Geol.*, **21**, 1295–1310.

Burbank, D.W. and Raynolds, R.G.H. (1988) Stratigraphic keys to the timing of thrusting in terrestrial foreland basins: Applications to the Northwestern Himalaya. In: *New Perspectives in Basin Analysis* (Eds K.L. Kleinspehn and C. Paola), pp. 331–351. Springer-Verlag, New York.

Burbank, D.W. and Beck, R.A. (1989) Early Pliocene uplift of the Salt Range: temporal constraints on thrust wedge development, northwest Himalaya, Pakistan. In: *Tectonics and Geophysics of the Western Himalaya* (Eds L.L. Malinconico and R.J. Lillie), pp. 113–128. Special Paper 232, Geological Society of America.

Burbank, D.W. and Vergés, J. (1994) Reconstruction of topography and related depositional systems during active thrusting. *J. Geophys. Res.*, **99**, 20,281–20,297.

Burbank, D.W., Puigdefàbregas, C. and Muñoz, J.A. (1992a) The chronology of the Eocene tectonic and stratigraphic development of the eastern Pyrenean foreland basin, northeast Spain. *Geol. Soc. Am. Bull.*, **104**, 1101–1120.

Burbank, D.W., Vergés, J., Muñoz, J.A. and Bentham, P.A. (1992b) Coeval hindward- and forward-imbricating thrusting in the central southern Pyrenees: timing and rates of shortening and deposition. *Geol. Soc. Am. Bull.*, **104**, 1–18.

Burbank, D., Meigs, A. and Brozovic, N. (1996) Interactions of growing folds and coeval depositional systems. *Basin Res.*, **8**, 199–224.

Butler, R.W.H., Coward, M.P., Harwood, G.M. and Knipe, R.J. (1987) Salt, it's control on thrust geometry, structural style and gravitational collapse along the Himalayan mountain front in the Salt Range of northern Pakistan. In: *The Dynamical Geology of Salt and Related Structures* (Eds J.J. Obrien and I. Lesche), pp. 399–418. Academic Press, Austin, TX.

Clevis, Q., de Jager, G., Nijman, W. and de Boer, P. (2004) Stratigraphic signatures of translation of thrust-sheet top basins over low-angle detachment faults. *Basin Res.*, **16**, 145–163.

DeCelles, P.G. and Giles, K.A. (1996) Foreland basin systems. *Basin Res.*, **8**, 105–123.

DeCelles, P.G., Gray, M.B., Ridgway, K.D., *et al.* (1991) Kinematic history of a foreland uplift from Paleocene synorogenic conglomerate, Beartooth Range, Wyoming and Montana. *Geol. Soc. Am. Bull.*, **103**, 1458–1475.

Elmore, R.D. and Farrand, W.R. (1981) Asphalt-bearing sediment in synorogenic Miocene-Pliocene molasse, Zagros Mountains, Iran. *Am. Assoc. Petrol. Geol. Bull.*, **65**, 1160–1165.

Falcon, N.L. (1961) Major earth-flexuring in the Zagros Mountains of south-west Iran. *Quat. J. Geol. Soc. Lond.*, **117**, 367–376.

Fernández, O., Muñoz, J.A., Arbués, P., Falivene, O. and Marzo, M. (2004) Three-dimensional reconstruction of geological surfaces: An example of growth strata and turbidite systems from the Ainsa basin (Pyrenees, Spain). *Am. Assoc. Petrol. Geol. Bull.*, **88**, 1049–1068.

Ford, M., Williams, E.A., Artoni, A., Vergés, J. and Hardy, S. (1997) Progressive evolution of a fault-related fold pair from growth strata geometries, Sant Llorenç de Morunys, SE Pyrenees. *J. Struct. Geol.*, **19**, 413–441.

Formento-Trigilio, M.L., Burbank, D.W., Nicol, A., Shulmeister, J. and Rieser, U. (2003) River response to an active fold-and-thrust belt in a convergent margin setting, North Island, New Zealand. *Geomorphology*, **49**, 125–152.

Friend, P.F., Lloyd, M.J., McElroy, R., Turner, J., Van Gelder, A. and Vincent, S.J. (1996) Evolution of the central part of the northern Ebro basin margin, as indicated by its Tertiary fluvial sedimentary infill. In: *Tertiary Basins of Spain: the Stratigraphic Record of Crustal Kinematics* (Eds P.F. Friend and C.J. Dabrio), 1st edn, *World and Regional Geology*, **6**, pp. 166–172. Cambridge University Press, Cambridge.

Friend, P.F., Jones, N.E. and Vincent, S.J. (1999) Drainage evolution in active mountain belts: extrapolation backwards from the present-day Himalayas river patterns. In: *Fluvial Sedimentology VI* (Eds N.D. Smith and J. Rogers), pp. 305–313. Special Publication 28, International. Association of Sedimentologists. Blackwell Scientific Publications, Oxford.

Grelaud, S., Sassi, W., Frizon de Lamotte, D., Jaswal, T. and Roure, F. (2002) Kinematics of eastern Salt Range and South Potwar Basin (Pakistan): a new scenario. *Mar. Petrol. Geol.*, **19**, 1127–1139.

Hirst, J.P.P. and Nichols, G.J. (1986) Thrust tectonic controls on Miocene alluvial distribution patterns, southern Pyrenees. In: *Foreland Basins* (Eds P. Allen and P. Homewood), pp. 247–258. Special Publication 8, International Association of Sedimentologists. Blackwell Scientific Publications, Oxford.

Hogan, P.J. and Burbank, D.W. (1996) Evolution of the Jaca Piggyback Basin and Emergence of External Sierra, Southern Pyrenees. In: *Tertiary Basins of Spain: the Stratigraphic Record of Crustal Kinematics* (Eds P.F. Friend and C.J. Dabrio), 1st edn, *World and Regional Geology*, **6**, pp. 153–160. Cambridge University Press, Cambridge.

Homke, S., Vergés, J., Garcés, M., Emami, H. and Karpuz, R. (2004) Magnetostratigraphy of Miocene–Pliocene Zagros foreland deposits in the front of the Push-e Kush Arc (Lurestan Province, Iran). *Earth Planet. Sci. Lett.*, **225**, 397–410.

Jones, S.J. (2004) Tectonic controls on drainage evolution and development of terminal alluvial fans, southern Pyrenees, Spain. *Terra Nova*, **16**, 121–127.

Jordan, T.E. and Flemings, P.B. (1990) From geodynamic models to basin fill – A stratigraphic perspective. In: *Quantitative Dynamic Stratigraphy* (Ed. T.A. Cross), pp. 149–163. Prentice-Hall, Englewood Cliffs, NJ.

Labaume, P., Mutti, E. and Séguret, M. (1987) Megaturbidites: a depositional model from the Eocene of the SW-Pyrenean foreland basin, Spain. *Geo-Mar Lett.*, **7**, 91–101.

Larrasoaña, J.C., Parés, J.M., Millán, H., del Valle, J. and Pueyo, E.L. (2003) Paleomagnetic, structural, and stratigraphic constraints on transverse fault kinematics during basin inversion: The Pamplona Fault (Pyrenees, north Spain). In: *Tectonics*, **22**, 1071, doi:10.1029/2002TC001446.

Leeder, M.R., Mack, G.H. and Salyards, S.L. (1996) Axial–transverse fluvial interactions in half-graben: Plio-Pleistocene Palomas Basin, southern Rio Grande Rift, New Mexico, USA. *Basin Res.*, **12**, 225–241.

Martínez, A., Vergés, J. and Muñoz, J.A. (1988) Secuencias de propagación del sistema de cabalgamientos de la terminación oriental del manto del Pedraforca y relación con los conglomerados sinorogénicos. *Acta Geol. Hisp.*, **23**, 119–128.

Marzo, M., Nijman, W. and Puigdefabregas, C. (1988) Architecture of the Castissent fluvial sheet sandstones, Eocene, South Pyrenees, Spain. *Sedimentology*, **35**, 719–738.

Mutti, E., Luterbacher, H.P., Ferrer, J. and Rosell, J. (1972) Schema stratigrafico e lineamenti di facies del paleogene marino della zona centrale sudpirenaica da Tremp (Catalogna) e Pamplona (Navarra). *Mem. Soc. Geol. Ital.*, **11**, 391–416.

Nichols, G.J. and Hirst, J.P. (1998) Alluvial fans and fluvial distributary systems, Oligo-Miocene, northern Spain: contrasting processes and products. *J. Sediment. Res.*, **68**, 879–889.

Nijman, W. (1989) Thrust sheet rotation? The South Pyrenean Tertiary basin configuration reconsidered. *Geodin. Acta*, **3**, 17–42.

Nijman, W. (1998) Cyclicity and basin axis shift in a piggyback basin: towards modellinf of the Eocene Tremp-Ager Basin, South Pyrenees, Spain. In: *Cenozoic Foreland Basins of Western Europe* (Eds A. Mascle, C. Puigdefàbregas, H.P. Luterbacher and M. Fernàndez), pp. 135–162. Special Publication 134, Geological Society Publishing House, Bath.

Nijman, W. and Nio, S.D. (1975) The Eocene Montañana delta. (Tremp-Graus Basin, provinces of Lérida and Huesca, Southern Pyrenees, N. Spain). In: *9th International Sedimentological Congress, International Association of Sedimentologists, Nice, Excursion Guidebook, 19, part B*, **2**, pp. 1–18.

Oberlander, T.M. (1985) Origin of drainage transverse to structures in orogens. In: *Tectonic Geomorphology (Proceedings of the 15th Annual Binghamton Geomorphology Symposium, September 1984)* (Eds M. Morisawa and J.T. Hack), pp. 155–182. Allen & Unwin.

Okaya, N., Tawackoli, S. and Giese, P. (1997) Area-balanced model of the late Cenozoic tectonic evolution of the Central Andean arc and back arc (lat 20°–22°S). *Geology*, **25**, 367–370.

Ori, G.G. and Friend, P.F. (1984) Sedimentary basins formed and carried piggyback on active thrust sheets. *Geology*, **12**, 475–479.

Poblet, J., Muñoz, J.A., Travé, A. and Serra-Kiel, J. (1998) Quantifying the kinematics of detachment folds using three-dimensional geometry: Application to the Mediano anticline (Pyrenees, Spain). *Geol. Soc. Am. Bull.*, **110**, 111–125.

Puigdefàbregas, C. (1975) La sedimentación molásica de la cuenca de Jaca. *Pirineos*, **104**, 1–118.

Puigdefàbregas, C. and Souquet, P. (1986) Tecto-sedimentary cycles and depositional sequences of the Mesozoic and Tertiary from the Pyrenees. *Tectonophysics*, **129**, 173–203.

Puigdefàbregas, C., Muñoz, J.A. and Marzo, M. (1986) Thrust belt development in the Eastern Pyrenees and related depositional sequences in the southern foreland basin. In: *Foreland Basins* (Eds P. Allen and P. Homewood), pp. 229–246. Special Publication 8, International Association of Sedimentologists. Blackwell Scientific Publications, Oxford.

Puigdefàbregas, C., Muñoz, J.A. and Vergés, J. (1992) Thrusting and foreland basin evolution in the Southern Pyrenees. In: *Thrust Tectonics* (Ed K.R. McClay), 1st edn, pp. 247–254. Chapman & Hall, London.

Ramos, E., Busquets, P. and Vergés, J. (2002) Interplay between longitudinal fluvial and transverse alluvial fan systems and growing thrusts in a piggyback basin (SE Pyrenees). In: *Sedimentary Geology on Growth Strata* (Eds M. Marzo, J.A. Muñoz and J. Vergés). *Sediment. Geol.*, **146**(1–2), pp. 105–131.

Riba, O. (1976) Syntectonic unconformities of the Alto Cardener, Spanish Pyrenees: A genetic interpretation. *Sediment. Geol.*, **15**, 213–233.

Séguret, M. (1972) Étude tectonique des nappes et séries décollées de la partie centrale du versant sud des Pyrénées. *Publ. Univ. Sci. Tech. Langeduc, sér. Geol. Struct.*, **2**, 155.

Suppe, J., Sàbat, F., Muñoz, J.A., Poblet, J., Roca, E. and Vergés, J. (1997) Bed-by-bed fold growth by kink-band migration: Sant Llorenç de Morunys, eastern Pyrenees. *J. Struct. Geol.*, **19**, 443–461.

Talling, P.J., Lawton, T.F., Burbank, D.W. and Hobbs, R.S. (1995) Evolution of latest Cretaceous–Eocene non-marine deposystems in the Axhandle piggyback basin of central Utah. *J. Struct. Geol.*, **107**, 297–315.

Turner, J.P. (1992) Evolving alluvial stratigraphyc and thrust front development in the West Jaca piggyback basin, Spanish Pyrenees. *J. Geol. Soc. London*, **149**, 51–63.

Vergés, J. (1999) *Estudi geològic del vessant sud del Pirineu oriental i central. Evolució cinemàtica en 3D*. Collecció Monografies tècniques 7, Institut Cartogràfic de Catalunya, 194 pp.

Vergés, J. (2003) Evolución de los sistemas de rampas oblicuas de los Pirineos meridionales: fallas del Segre y Pamplona. *Bol. Geol. Min.*, **114**, 87–101.

Vergés, J. and Muñoz, J.A. (1990) Thrust sequences in the southern central Pyrenees. *Bull. Soc. Géol. Fr.*, **8**(VI), 265–271.

Vergés, J., Martínez-Ríus, A., Domingo, F., *et al.* (1994) Mapa geológico de la hoja n. 255 de La Pobla de Lillet a escala 1/50.000 (*proyecto MAGNA*, Segunda serie, Primera edición). Memoria, 1–92. ITGE.

Vergés, J., Millán, H., Roca, E., *et al.* (1995) Eastern Pyrenees and related foreland basins: Pre-, syn- and post-collisional crustal-scale cross-sections. *Mar. Petrol. Geol.*, **12**, 903–916.

Vergés, J., Marzo, M., Santaeulària, T., *et al.* (1998) Quantified vertical motions and tectonic evolution of the SE Pyrenean foreland basin. In: *Cenozoic Foreland Basins of Western Europe* (Eds A. Mascle, C. Puigdefàbregas, H.P. Luterbacher and M. Fernàndez), pp. 107–134. Special Publication 134, Geological Society Publishing House, Bath.

Vergés, J., Marzo, M. and Muñoz, J.A. (2002) Growth strata in foreland settings. *Sediment. Geol.*, **146**, 1–9.

Vincent, S.J. (2001) The Sis palaeovalley: a record of proximal fluvial sedimentation and drainage basin development in response to Pyrenean mountain building. *Sedimentology*, **48**, 1235–1276.

Williams, E.A., Ford, M., Vergés, J. and Artoni, A. (1998) Alluvial gravel sedimentation in a contractional growth fold setting, Sant Llorenç de Morunys, southeastern Pyrenees In: *Cenozoic Foreland Basins of Western Europe* (Eds A. Mascle, C. Puigdefàbregas, H.P. Luterbacher and M. Fernàndez), pp. 69–106. Special Publication 134, Geological Society Publishing House, Bath.

Yeats, R.S. and Lillie, R.J. (1991) Contemporary tectonics of the Himlayan frontal fault system: folds, blind thrusts and the 1905 Kangra earthquake. *J. Struct. Geol.*, **13**, 215–225.

Ziegler, M.A. (2001) Late Permian to Holocene paleofacies evolution of the Arabian Plate and its hydrocarbon occurrences. *GeoArabia*, **6**, 445–504.

Stratigraphic architecture, sedimentology and structure of the Vouraikos Gilbert-type fan delta, Gulf of Corinth, Greece

MARY FORD*, EDWARD A. WILLIAMS†,
FABRICE MALARTRE‡ *and* SPERANTA-MARIA POPESCU§

**Ecole Nationale Supérieure de Géologie, CRPG, 15 Rue Notre-Dame des Pauvres, BP 20, 54501 Vandoeuvre-lès-Nancy, France (Email : mford@crpg.cnrs-nancy.fr)*
†'Geostandards and Geoanalytical Research', CRPG-CNRS, 15 rue Notre-Dame des Pauvres, BP 20, 54501 Vandoeuvre-lès-Nancy, France
‡Ecole Nationale Supérieure de Géologie, UMR 7566-G2R, Rue du Doyen Marcel Roubault, BP 40, 54501 Vandoeuvre-lès-Nancy, France
§Laboratoire Paléoenvironnements et Paléobiosphère, UMR 5125 CNRS, Université Claude Bernard – Lyon 1, 27–43 Boulevard du 11 novembre, 69622 Villeurbanne Cedex, France

ABSTRACT

In the Aegion to Kalavrita region of the Gulf of Corinth, Greece, Plio-Pleistocene syn-rift stratigraphy comprises a fluvial-dominated lower group and an upper group dominated by Gilbert-type deltas separated by an erosive unconformity. The lower group records substantial accumulation (1.3 km) of fluvial sediment across a broad area of fault-controlled grabens and half grabens, which was terminated by a marine transgression. The upper group records a great increase in accommodation space, the migration of the depocentre to the north and an increase in sediment supply. It is dominated by large gravel-rich systems that were sourced in the footwalls of active normal faults. The Vouraikos Delta is an exceptionally well-exposed Gilbert-type fan delta complex, which is >800 m thick, with a surface area of 32 km^2. It lies in the hangingwall of the Pirgaki-Mamoussia Fault and has been exhumed in the footwall of the Eastern Helike Fault. Preliminary palynological results from topset and pro-delta fine-grained facies and from lower group strata indicate that the Vouraikos Delta began forming some time before 1.1 Ma and was terminated soon after 0.7 Ma. These preliminary Early to Middle Pleistocene age estimates are coherent with published models of the uplift history of the Eastern Helike Fault. Sedimentation rates are thus estimated between 1.3 and 2 mm yr^{-1}. While the earliest delta infilled an incised palaeovalley, accommodation space was primarily tectonically controlled, first by an extensional forced fold and later by a system of major normal faults (Pirgaki-Mamoussia Fault and its splays). Several families of syn-sedimentary and late normal faults cut the delta. A listric growth fault controlled a large rollover anticline in the lowest stratigraphic package. The delta prograded (to the north-northwest) into water that reached depths of 200–600 m. Topset limestones associated with coastal conglomerate facies indicate that the delta built into a water body that was wholly or periodically marine. Internally, the Vouraikos Delta comprises five stratigraphic packages each characterized by a distinctive organization of topsets, foresets, bottomsets and pro-delta facies and bounded by major stratigraphic surfaces. These packages are tentatively correlated with regressive glacio-eustatic interglacial periods. The trajectory of the offlap break in the centre of the Vouraikos reflects early progradation-dominated behaviour followed by increasingly aggradational behaviour.

Keywords Early to Middle Pleistocene, Vouraikos Gilbert-type delta, Corinth rift, Greece, normal faults.

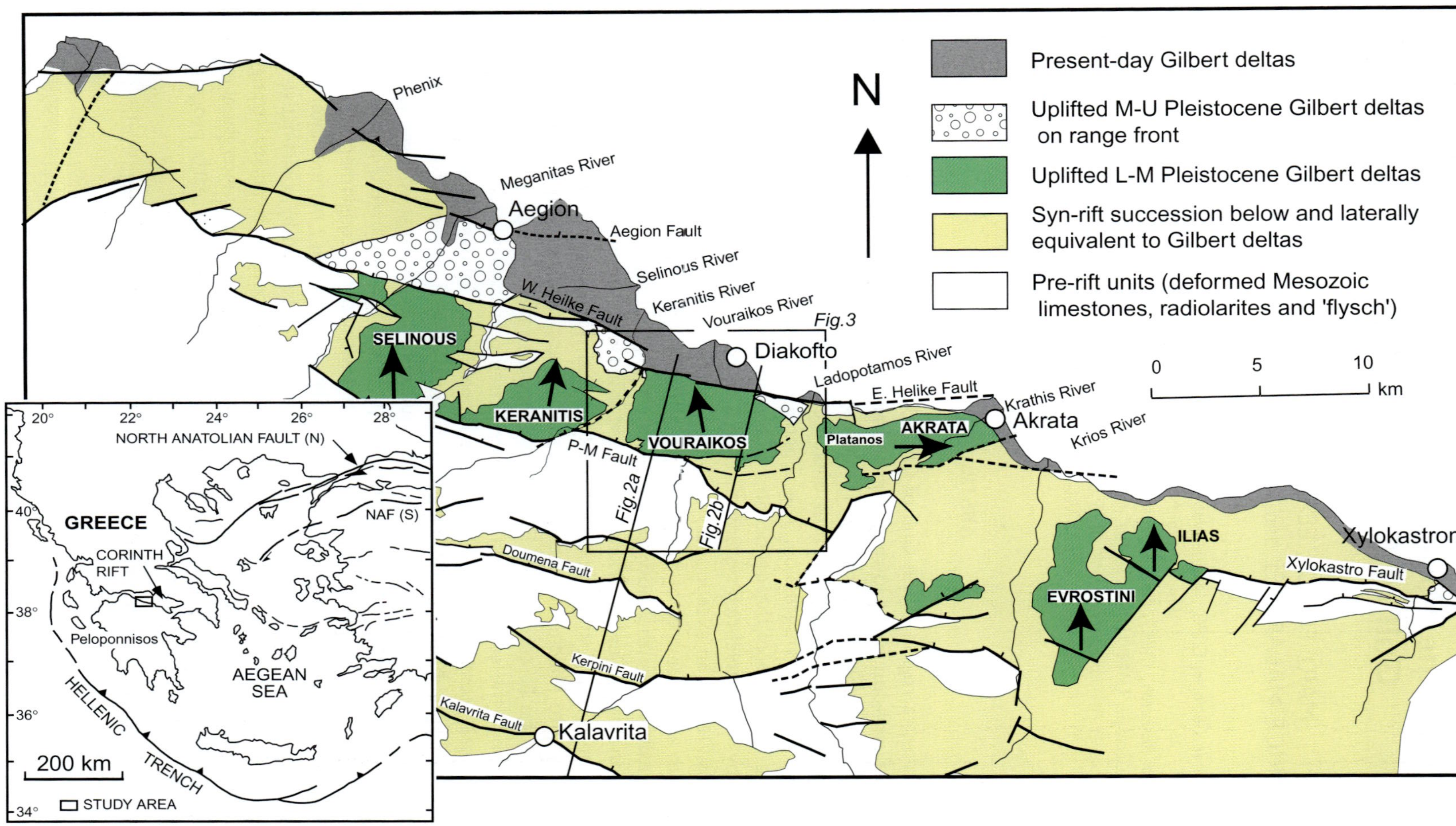

Fig. 1 Map of the southern coast of the Gulf of Corinth showing the distribution of pre-rift and syn-rift sequences. Three generations of syn-rift Gilbert-type deltas are distinguished. The principal normal faults are shown, with dip-direction and throw indicated by small ticks. The progradation directions of the principal Lower–Mid-Pleistocene Gilbert-type fan deltas are shown by large arrows. The location of the cross-sections of Fig. 2 are indicated. This map is based on Ghisetti & Vezzani (2004, 2005), Rohais *et al*. (in press) and authors' own work. Inset: Tectonic map of the Aegean region showing the Corinth rift and the location of the study area. NAF (S) is the southern branch of the North Anatolian Fault.

INTRODUCTION

Along the southern shore of the Gulf of Corinth (Fig. 1) high-relief outcrops expose some of the finest examples of ancient Gilbert-type deltas in an active extensional tectonic setting. Despite published research on several of this suite of deltas, there is little information on complete systems and some confusion about the relationships between several of the deltas. These conglomerate-rich Gilbert-type delta bodies are unusually large, varying in radius from 3 to 8 km (Fig. 1) and are up to 900 m thick. They have been uplifted to altitudes of over 800 m and are deeply incised by north-flowing rivers. While absolute and relative ages are as yet poorly constrained, various delta bodies have been attributed ages of between Pliocene and Holocene (Ori, 1989; Collier *et al.*, 1990; Ori *et al.*, 1991; Poulimenos *et al.*, 1993; Dart *et al.*, 1994). Recent provisional age estimates from palynofloras for the Vouraikos Delta give an Early Pleistocene age (Malartre *et al.*, 2004). Currently, major deltas of the same type are building out into the western Gulf, as modern rivers cannibalize their antecedent deltas (Fig. 1). The purpose of this paper is to present a detailed analysis of the Vouraikos Delta, one of the largest and best exposed of the giant Gilbert-type deltas of Corinth.

The Vouraikos Delta lies in the hangingwall of the Pirgaki-Mamoussi (PM) Fault (Figs 1–3) and has been uplifted and incised in the footwall of the Eastern Helike (or East Eliki) Fault. Three river valleys, the Keranitis, Vouraikos and Ladopotamos, provide exceptional natural sections 3 to 4 km apart and with over 700 m of incision. These sections enable the detailed sedimentological and structural study of a substantially complete delta system. Additionally, the syn-rift stratigraphy and internal structure of the whole PM Fault block is described so that delta development can be placed in the context of rift evolution. The vertical and lateral stacking pattern of the delta (its internal architecture) is interpreted in terms of sequence stratigraphy and the creation of accommodation space in order to distinguish tectonic and eustatic controls.

Detailed analyses of major cliff sections using photographic panoramas form the backbone of this work. These sections were tied together by detailed field mapping at various scales, integrating GPS technology, which forms the basis of ongoing three-dimensional database construction using GIS and gOcad. The stratigraphical architecture was established for each cliff, and key units and surfaces were correlated between cliffs. Facies associations for each stratigraphical unit were identified and detailed sedimentological analysis was carried out by logging at a scale of 1:25. Preliminary analysis of $\delta^{13}C$ and $\delta^{18}O$ isotopes was undertaken to characterize the chemical signature of critical stratigraphical horizons, and sampling for palynological analysis was carried out in order to biostratigraphically date the succession.

GROSS STRUCTURAL SETTING: THE GULF OF CORINTH

The Gulf of Corinth is an active rift that was initiated sometime in the past 5 Myr (Doutsos & Piper, 1990; Collier & Dart, 1991) in the upper plate of the Hellenic subduction zone (Fig. 1, inset). The rift is superimposed on the NNW–SSE trending Hellenide orogenic belt (Oligocene–Miocene) and is oriented 105°N. It is 120 km long, some 0.5 km wide in the west and is approximately 30 km at its widest point in the east. The basin has a maximum water depth of 900 m in the east and shallows westward to the Straits of Rion, where the water depth is only 62 m. WNW–ESE oriented north-dipping normal faults lie somewhat oblique and en échelon to the present southern coastline (Fig. 1). Active south-dipping faults, flanking the northern limit of the present graben, have recently been reported offshore (McNeill *et al.*, 2005). Seismic activity is concentrated at the western end of the basin, where geodetic measurements indicate a N–S extension rate of 1.2 cm yr^{-1} (Briole *et al.*, 2000). On the south shore, older syn-rift sediments have been uplifted and deeply incised over an area stretching south from the coast for 25–30 km (Fig. 1). Current uplift rates are estimated to be 1 to 1.5 mm yr^{-1} (De Martini *et al.*, 2004; McNeill & Collier, 2004). This unusual situation has generated superb vertical sections through the older syn-rift succession while active rifting takes place offshore.

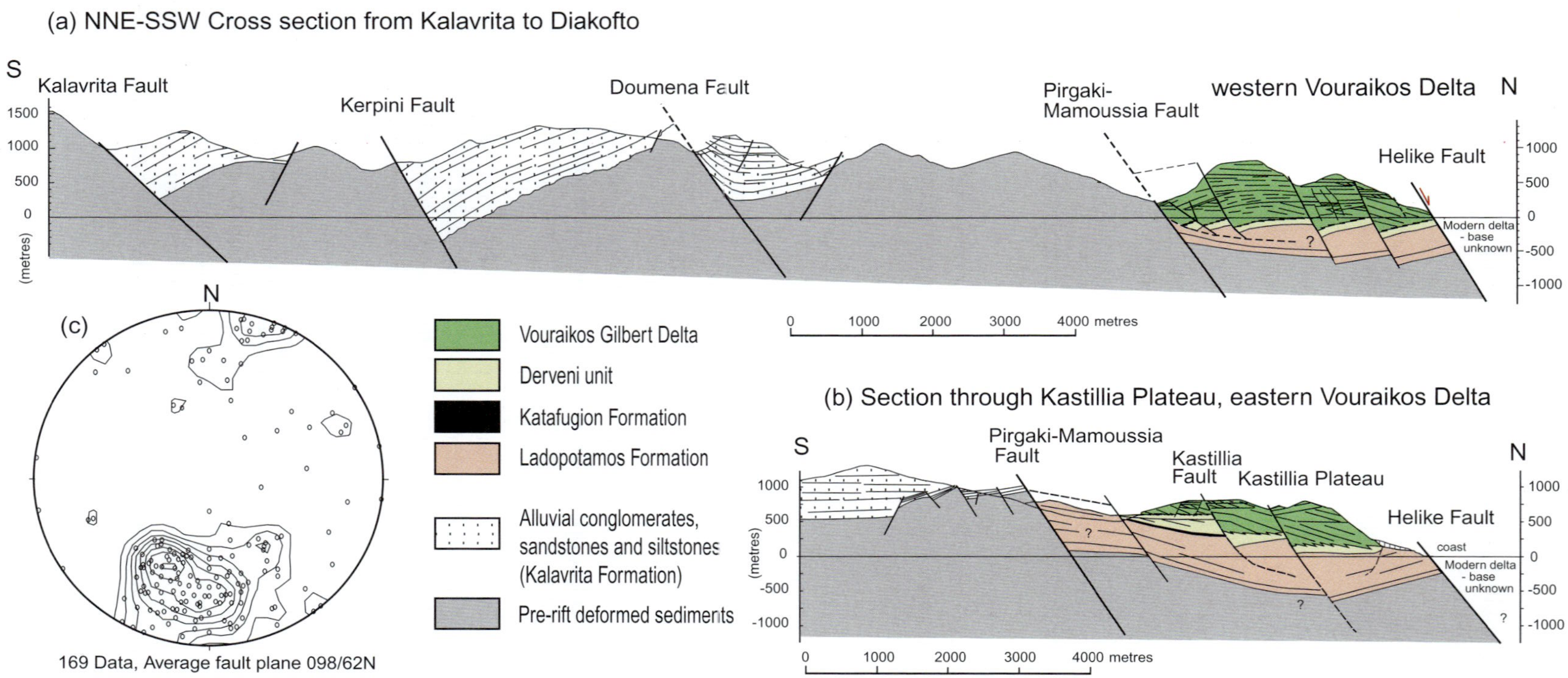

Fig. 2 Structural cross-sections through the central south coast of the Corinth Rift. (a) NNE–SSW cross-section from the Kalavrita Fault to the coast passing through the western Vouraikos Delta. (b) NNE–SSW cross-section from the Doumena Fault block to the coast passing through the eastern Vouraikos Delta. (c) Equal area stereoplot of poles to fault planes cutting the Pirgaki-Mamoussia Fault block only (data and contours), showing a dominance of north-dipping planes with an average plane of 098/62°N.

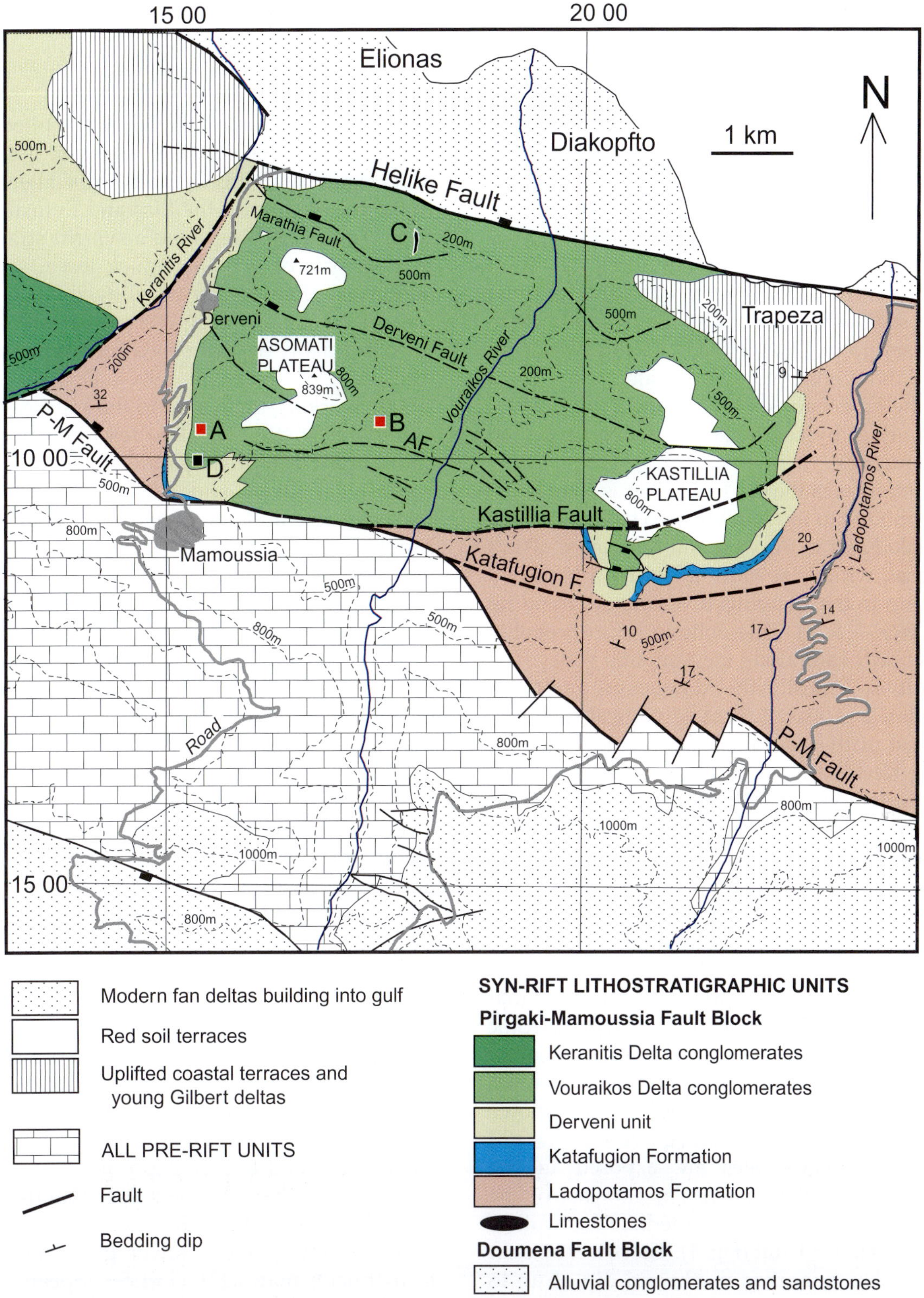

Fig. 3 Detailed geological map of the Vouraikos Delta in the Pirgaki-Mamoussia Fault block based on new mapping by the authors, revised from fig. 1 of Malartre *et al.* (2004). A is the location of heterolithic fluvial topsets (Fig. 8), B is the location of outcrops of shoreline–shallow-marine topset association (Fig. 9), C is the location of the Marathia Limestone (Fig. 11a), and D is the location of the Mamoussia Limestone (Fig. 11b). AF is the Avriyiolaka Fault. Grid coordinates are from Greek topographic base map.

LOCAL STRUCTURAL SETTING

In the Aegion–Kalavrita region (Fig. 1) major, north-dipping normal faults with a mean trend of 110° and dipping 55°N define five principal fault blocks that are between 3 and 5 km wide (Goldsworthy & Jackson, 2001; Bourlange *et al.*, 2005). From south to north these blocks are delimited by the Kalavrita, Kerpini, Doumena, Pirgaki-Mamoussi (PM) and Helike faults (Fig. 2). Each block preserves a succession of coarse alluvial syn-rift sediments, generally tilted south, with thicknesses up to 1.3 km. The PM Fault block preserves a more complex syn-rift succession, comprising heterogeneous alluvial (and other) clastic rocks overlain by conglomeratic Gilbert-type fan delta sequences.

In the study area the PM Fault block is 5 km wide, and is bounded to the south by two fault segments: the Pirgaki Fault to the west and Mamoussia Fault to the east, which are hard-linked by a breached relay ramp in the Keranitis River Valley (trending 070°N, Figs 1 & 3). This oblique ramp probably extends northward along the Keranitis Valley. The PM Fault (specifically the Mamoussia Fault) accommodated at least 1.5 km of vertical displacement. East of the Vouraikos Gorge, the east–west trending Kastillia Fault and the Katafugion Fault branch from the ESE–WNW trending PM Fault (Fig. 3) and formed important bounding faults to the Vouraikos Delta for part of its history. Farther south, the PM Fault separates the older syn-rift succession from pre-rift carbonates in its footwall (Fig. 3).

Pre-rift strata comprise Mesozoic carbonates, radiolarites and clastic turbidites ('flysch') that record multiphase deformation at low metamorphic grades related to the westward emplacement of Hellenic nappes during the Oligo-Miocene (Doutsos *et al.*, 1993). These pre-rift strata are the source rocks for the Gilbert-type delta gravels. In the PM block, pre-rift carbonates are exposed in the Selinous River Valley and just to the east of the Ladopotamos River in the immediate footwall of the Eastern Helike Fault (Fig. 1). The lower syn-rift succession has a marked northward dip of up to 30°, while the top of the syn-rift succession (i.e. the uppermost conglomeratic units *above* the Gilbert delta succession) shows a shallow (< 5°) tilt to the south.

REGIONAL SYN-RIFT STRATIGRAPHY

Published stratigraphical schemes

Two general schemes have been applied in this region (Ori, 1989; Doutsos & Piper, 1990); both differ significantly from that described here (Fig. 4). In the region around the Ilias and Evrostini deltas (Fig. 1), Ori (1989) reported a syn-rift stratigraphy that consists of a 1–3 km thick lower succession of alluvial plain–lacustrine–alluvial fan sediments (organized in a transgressive–regressive cycle), unconformably overlain by Gilbert-type fan deltas (but see Doutsos *et al.*, 1990). Doutsos & Piper (1990), working in an area to the southeast of that of Ori (1989, fig. 1), also described a two-unit stratigraphy comprising a lower unit of lacustrine and fluvial sands and silts of Middle to Late Pliocene age, and an upper unit of Quaternary conglomerates, the older part of which has been dated as 'Calabrian' (Lower Pleistocene; Doutsos & Piper, 1990, p. 815 and references therein). These conglomerates are of aerially contrasting facies, being: (i) terrestrial where they overlie basement rocks at the southern margin of the basin, and (ii) of Gilbert-type delta facies towards the north, where they are interbedded with marine, lacustrine and brackish 'marls' of locally Middle Pleistocene age.

More recently, Ghisetti & Vezzani (2005) presented a synthetic stratigraphic column for the study area, which they call the Aegion Basin, and another for the Derveni-Corinth Basin farther east. The Aegion Basin column corresponds very generally with our observations; however, we have not observed major clino-stratified conglomerates at the stratigraphic level they call 'Mid-Rift' (equivalent to our lower group). In the Evrostini-Akrata area (Fig. 1), Rohais *et al.* (in press) recognized three stratigraphic groups, which are a lower alluvial–lacustrine group, a middle group of Gilbert-type deltas and an upper group of recent small deltas and terrace deposits.

At present, there is a consensus that the onshore syn-rift stratigraphy comprises two main stratigraphical groups. The lower group of alluvial–lacustrine(?)–marine(?) clastics appears to vary rapidly in facies and thickness from west to east, so that attempts to generalize its component units across the south coast have led to some confusion. Detailed biostratigraphical dating is necessary to

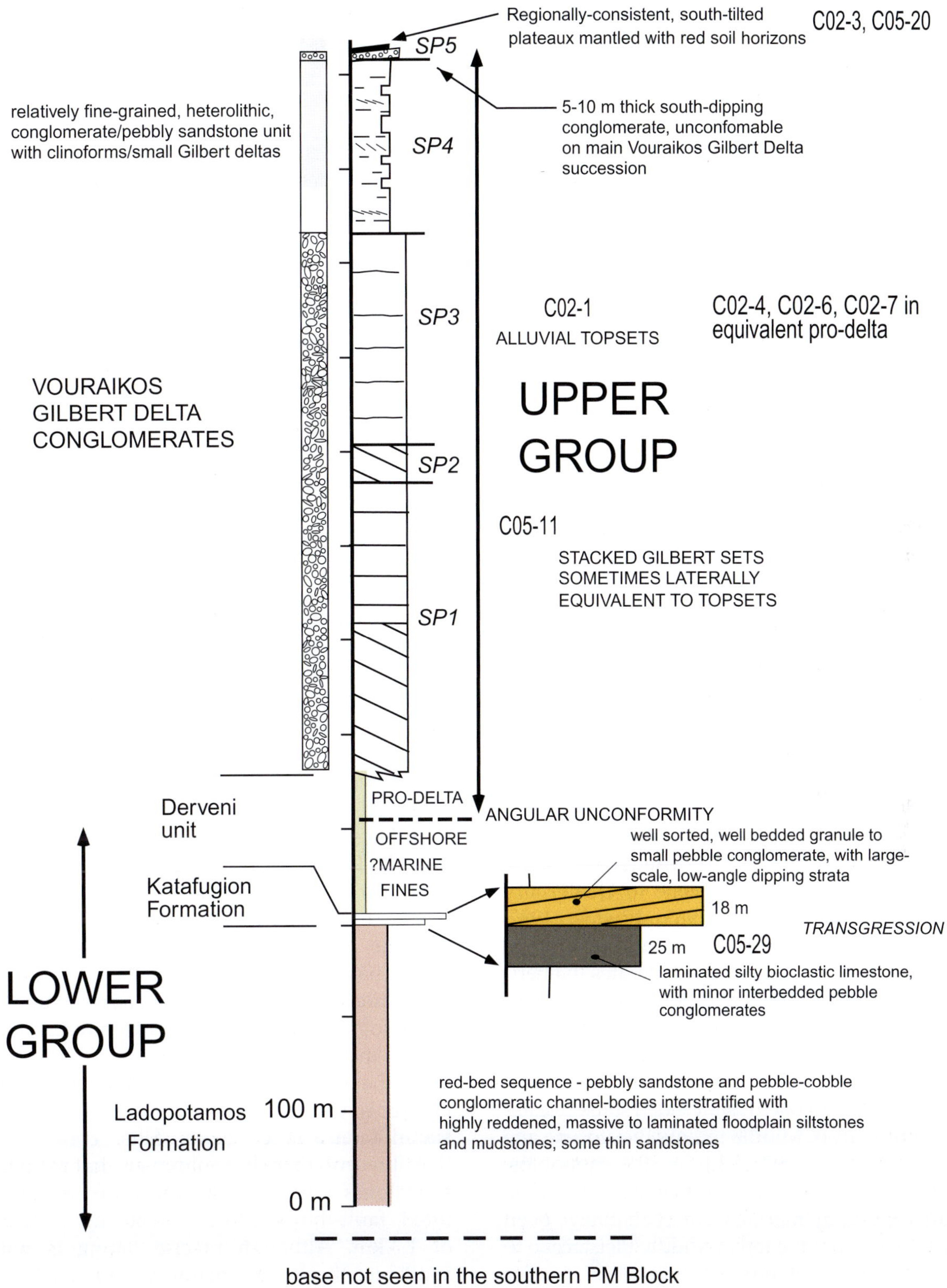

Fig. 4 Composite stratigraphical scheme for the syn-rift depositional sequence in the Pirgaki-Mamoussia Fault block between the Keranitis and Ladopotamos rivers. Note that due to poor exposure, the fine-grained offshore marine facies at the top of the lower group and the fine-grained pro-delta facies at the base of the upper group are here provisionally treated as a single mappable unit, the Derveni unit. The postulated erosive contact between the lower and upper groups lies within this unit. Pro-delta facies are equivalent to the *SP1* to *SP4* stratigraphic packages.

achieve good lateral correlation. A major unconformity probably separates the lower and upper groups, although some authors do not accept this. The upper group is dominated by the giant Gilbert-type fan deltas that are developed principally between Aegion and Xylokastron (Fig. 1). Equivalent stratigraphical levels to the east (and west?) appear to comprise thinner, smaller deltas and thick fine-grained marine clastics (e.g. Doutsos & Piper, 1990). A relatively small volume of young Gilbert-type fan deltas, fluvial deposits and terrace deposits locally unconformably overlie the main rift succession along the coastal strip, recording the late phase of surface uplift.

Stratigraphical age and dating

There is a lack of precise biostratigraphical dating of the various sedimentary successions in the Gulf of Corinth rift. This is mainly due to the dominance of conglomerates and sandstones in which biostratigraphical markers are poorly preserved. The oldest age published is Early Pliocene (Zanclean, 5.32–3.58 Ma, Papanicolaou *et al.*, 2000) from syn-rift coal-bearing rocks in the Kalavrita Basin (to the south of Vouraikos Delta), although details of the dating method are not specified. Andesites that represent the initiation of the Corinth Basin, in the east of the rift system, are dated as 4 Ma (Collier & Dart, 1991).

Along the south-eastern coast of the Gulf, brackish, lacustrine and fluvial siliciclastic sediment are dated as Middle to Late Pliocene to Quaternary (Kontopoulos & Doutsos, 1985; Frydas, 1987; Fernandez-Gonzalez *et al.*, 1994 and references therein). Thick Quaternary conglomerates (Gilbert-type deltas) overlie this lower series and in their lower levels contain mammalian fossils that have been dated as Calabrian (1.77–0.95 Ma) by Symeonidis *et al.* (1987). Intercalated marl levels within the conglomerates have yielded some calcareous nannofossils of Middle to Late Pleistocene age (Poulimenos *et al.*, 1993; Zelilidis & Kontopoulos, 1996). Gilbert-type deltas in the Xylokastron-Aegion area are capped by marine terraces that have been assigned ages from the top of Middle Pleistocene to Late Pleistocene (Keraudren & Sorel, 1987; Collier *et al.*, 1992; Dia *et al.*, 1997). Late Pleistocene to Holocene coral–algal reefal facies rocks are well developed within the eastern Gulf of Corinth (Kershaw *et al.*, 2005; Portman *et al.*, 2005).

SYN-RIFT STRATIGRAPHY IN THE PIRGAKI-MAMOUSSI BLOCK

General

The syn-rift succession of the study area (Fig. 3) is here divided into two informal stratigraphical groups (Fig. 4). The *lower group* comprises two units: the Ladopotamos and Katafugion formations. The lowest syn-rift succession within the PM block is best exposed in the Ladopotamos Valley, represented by the dominantly coarse-grained clastic Ladopotamos Formation, which dips and youngs towards the northwest to north-northwest (Fig. 3). The Ladopotamos Formation comprises at least 300 m of reddish conglomerates, sandstones and siltstones (Fig. 4) unconformably overlying Mesozoic carbonate (basement) rocks (Fig. 1). An inlier of sandstones and conglomerates, previously regarded as part of the Keranitis Delta (Ori *et al.*, 1991, fig. 2 'topset beds'; Dart *et al.*, 1994, fig. 3a, b) that crops out in the Keranitis valley (Fig. 3), is considered to be part of the Ladopotamos Formation. The Katafugion Formation (40 m thick, Fig. 5) comprises a fine white calcareous unit (20–25 m thick) overlain by a package of fine-gravel clastics (18 m thick). This is followed by poorly exposed siltstones and mudstones (included in the Derveni unit). The Katafugion Formation is observed below the southeast part of the Vouraikos Delta to the south of the Kastillia Fault (Fig. 3). To the north of this fault the fine-grained pro-delta facies of the upper group directly overlie the Ladopotamos Formation. It is suggested that the unconformity at the base of the upper group has eroded down through the Katafugion Formation to the north of the Kastillia Fault. To the south of the fault the unconformity is located within the Derveni unit.

To the south of the PM Fault (Figs 1 & 2), markedly contrasting syn-rift successions can be traced for up to 15–20 km. Here, conglomerates (locally up to small boulder grade) with minor sandstones (and red siltstones) are organized in tilted fault blocks, locally reaching thicknesses of 1.3 km. Although precise dating is not yet available, this conglomerate succession is provisionally correlated with the lower group of the PM block.

The *upper group* comprises a relatively thin succession (< 50 m) of commonly beige-coloured

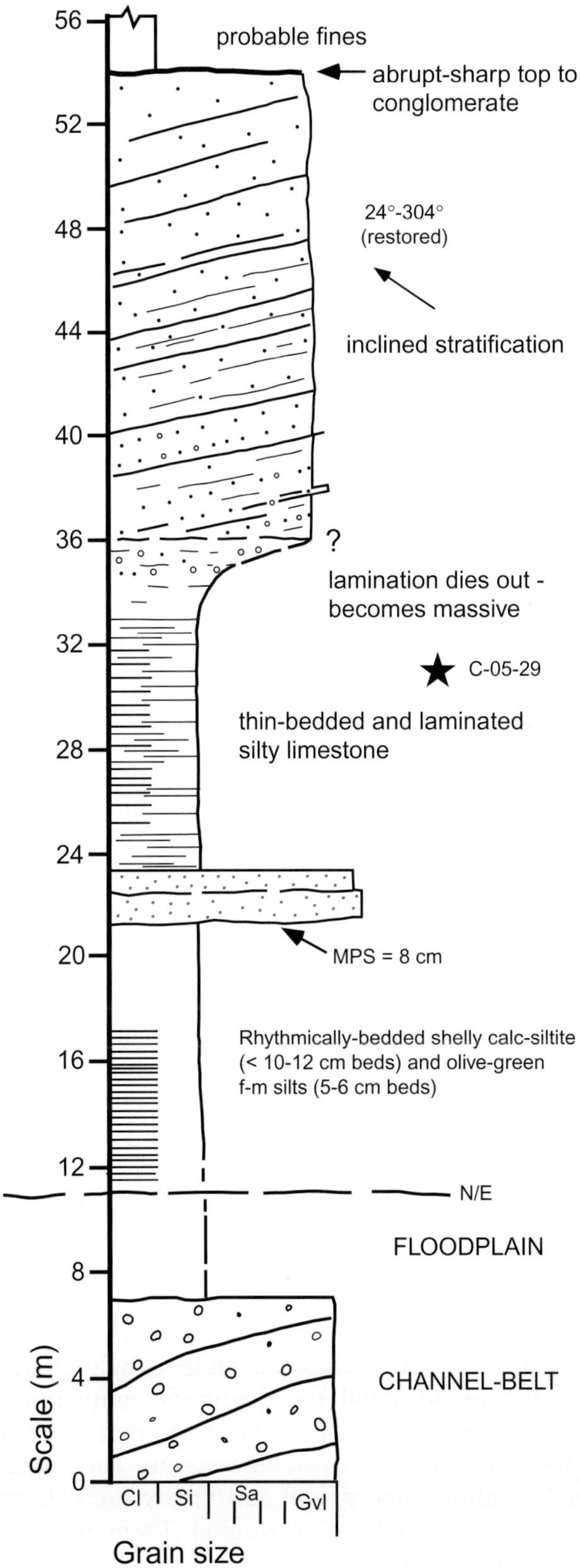

Fig. 5 Simplified graphic log of the limestone–clastic succession of the Katafugion Formation. MPS is maximum particle size.

siltstones and rarer mudstones (included in the Derveni unit; Fig. 4), overlain by a series of individual conglomeratic bodies (Gilbert-type fan deltas) that reach thicknesses of over 800 m. These are represented in the study area by the Vouraikos Delta, but also include the immediately flanking (western) Keranitis and (eastern) Plaka (or Platanos) deltas (Fig. 1). The top of the Vouraikos Delta succession is unconformably overlain by a thin (10–15 m) conglomerate unit, which is capped by recent red soils on the Asomati Plateau (Figs 3 & 4). Along the range front of the Eastern Helike Fault the upper group is incised and unconformably overlain by uplifted small Gilbert-type deltas (Fig. 3).

Lower group: Ladopotamos Formation

The Ladopotamos Formation consists of interbedded conglomerate/pebbly sandstone bodies and (minor) red siltstone/sandstone intervals. Typical conglomerate/pebbly sandstone bodies are about 2 m thick flat- and sharp-based, horizontally-bedded or more rarely cross-bedded sheets, or erosively based multi-storey bodies composed of similar sheets. Observed cross-stratification indicates north- and north-east-quadrant palaeoflows. Textures of the pebbly sandstone sheets are clast-supported, though rich in a matrix mixture of granules, sand and very small pebbles. Component clasts are extraformational, with rare instances of intraformational siltstone (pebbles and cobbles) lining basal erosive surfaces.

Coarse-grained conglomerate–sandstone bodies (5–7 m thick) observed near the top of the formation are markedly heterolithic, containing beds of medium–coarse-grained red sandstone, pebbly sandstone and small pebble to small cobble conglomerates. These show concordant channel-form structures, which incise into fine-grained red bed sequences, steep channel margins and meso- and macroscale inclined strata-sets (terminology of Bridge, 1993) with opposed dip-directions. Interbedded fine-grained sequences with the conglomerate–sandstone bodies are orange or red sandstones, small pebble conglomerates, small pebble-rich sandstones and blocky and faintly laminated red (dusky red) siltstones/mudstones. The latter contain occasional isolated small calcareous nodules and black charcoalified wood fragments of 0.75 cm size.

Interpretation

The Ladopotamos Formation is considered to be fluvial in origin, with coarse-grained probably braided river conglomerate–sandstone bodies and relatively fine-grained overbank (floodplain) siltstones and interbedded sand and gravel sheets. No lacustrine facies have been observed, as are reported farther east in this fault block (see Ori, 1989; Dart *et al.*, 1994). Moreover, the basin-wide 'fanglomerate' unit reported by Ori (1989) beneath the Gilbert-type delta sediment-bodies of the region does not occur in the study area.

Lower group: Katafugion Formation

The Katafugion Formation begins with a distinctive white-weathering flat- and parallel-laminated shelly siltstone/marly limestone (Fig. 5). Its basal contact, though not well exposed, is apparently conformable. Fresh exposures of the white-weathering facies reveal a rhythmical alternation of pale to white competent fossiliferous beds up to 10–12 cm in thickness, and olive green-coloured laminated silty to fine sandy beds that are 5–6 cm thick. The calcareous facies (typically a fine-grained, massive to faintly laminated bioclastic calcsiltite) contains distributed broken bivalve and gastropod shell material. Intact but disarticulated bivalves are concentrated on bedding planes. Much of the broken and intact bivalve material appears to be from a single genus, of small (< 1 cm) members of the suborder Pterioida. Intact small gastropods are as yet unidentified.

In thin-section, finer grained facies are finely laminated mudstones to wackestones with a high component of clay, seams of dark, fine-grained organic material and abundant calcitic microspar. Larger bioclasts include entire and broken ostracods, whereas smaller bioclasts include relatively abundant diatoms and possible coccoliths. A palynological study from the upper part of the carbonate member yielded dinoflagellate cysts. Apart from planar-horizontal lamination, these facies lack structures. Conformably interbedded in the fine-grained marly limestone facies at one interval is a bedset of 0.7–0.8 m thick small pebble conglomerate beds. Up-section within the fine calcareous unit, the horizontal lamination dies out, giving way to a massive texture that is accompanied by the appearance of occasional small pebbles.

A preliminary stable isotope analysis of shell and rock matrix from the calcareous facies was undertaken (at the CRPG, Nancy, F. Palhol, personal communication). Results for $\delta^{13}C$ were 1.71 and 1.80‰ for shell and rock material respectively, and for $\delta^{18}O$ PDB, −3.69 and −3.64‰ respectively, the determinations being consistent for the differing materials. The values for $\delta^{13}C$ are typical for a large lake or the sea with open circulation. However, the extremely negative values for $\delta^{18}O$ suggest a water body of high salinity (typical values for the eastern Mediterranean are between −0.1 and +1.0‰).

Overlying the fine calcareous unit (although an exposure gap intervenes) is a distinctive moderately sorted, well-stratified 18 m thick unit of interbedded sandy–pebbly conglomerates, coarse to very-coarse (and granule-rich) sandstones and pebbly sandstones (Fig. 5) organized in beds of 30–50 cm thickness. The conglomerates are matrix-rich, and show matrix-support of the maximum clast size population. Gravel stringers that are one to three clasts thick are composed dominantly of pebbles with occasional small cobbles. The planar stratification has a (24°) depositional dip to the northwest with respect to the underlying calcareous unit (Fig. 5). Lineations defined by grain alignment (and possible obstacle shadows) on bedding planes show a near down-dip orientation. Although not well exposed, the top of the conglomerate unit is apparently extremely abrupt and sharp (?non-erosional) where very fine-grained lithologies replace the gravel. The poorly exposed overlying succession (?70–100 m) comprises laminated, poorly consolidated yellow to red-weathering siltstones, with possible very thin fine–very fine sandstones.

Interpretation

The transitional succession of the lower group, from fluvial channel and oxidized floodplain environments (Ladopotamos Formation) to a mixed fine-grained clastic-carbonate system capped by well-stratified fine gravel and finally much finer-grained sediments (Katafugion Formation), is thought to indicate a marine transgression. The Katafugion Formation is interpreted as representing a protected subtidal (dominantly carbonate) lagoon in a microtidal regime (e.g. Anthony *et al.*, 1996), as no peritidal or obvious marsh facies were found. The restricted diversity of the mollusc–ostracod fauna plus the evidence of marine microfossils, allied to the preliminary geochemical data, suggests a saline coastal lagoon. The rare interbedded pebble gravel sheets within the lagoonal facies are likely to be storm-generated washovers or due to storm-generated flow through the barrier into the lagoon. The overlying conglomerates and sandstones suggest a high wave energy shoreface-barrier beach system (Nemec & Steel, 1984), probably at a ravinement surface. This is interpreted as part of a shoreface–shoreline retreat during transgression, leaving a seaward-dipping strata-set (Reinson, 1984). The termination of the coarse facies is thought to have occurred at a flooding surface, and the overlying fine-grained possibly offshore marine clastics locally contain thin-bedded turbidites.

The upper group

The upper group (Fig. 4) is characterized by Gilbert-type fan delta conglomerates and their laterally equivalent facies. The Vouraikos Delta is one of the largest of these uplifted deltas, covering a (preserved) surface area of 32 km^2 (Fig. 3). It is at least 800 m thick in its central region but thins considerably to the east and west (Figs 4 & 6). The delta comprises conglomerate-dominated packages of topsets, foresets and bottomsets that are grouped together as a single mapping unit (Fig. 3; Vouraikos Delta conglomerates). These are underlain by a unit (< 50 m thick) of siltstones and fine sandstones with 'floating' gravel clasts) that are

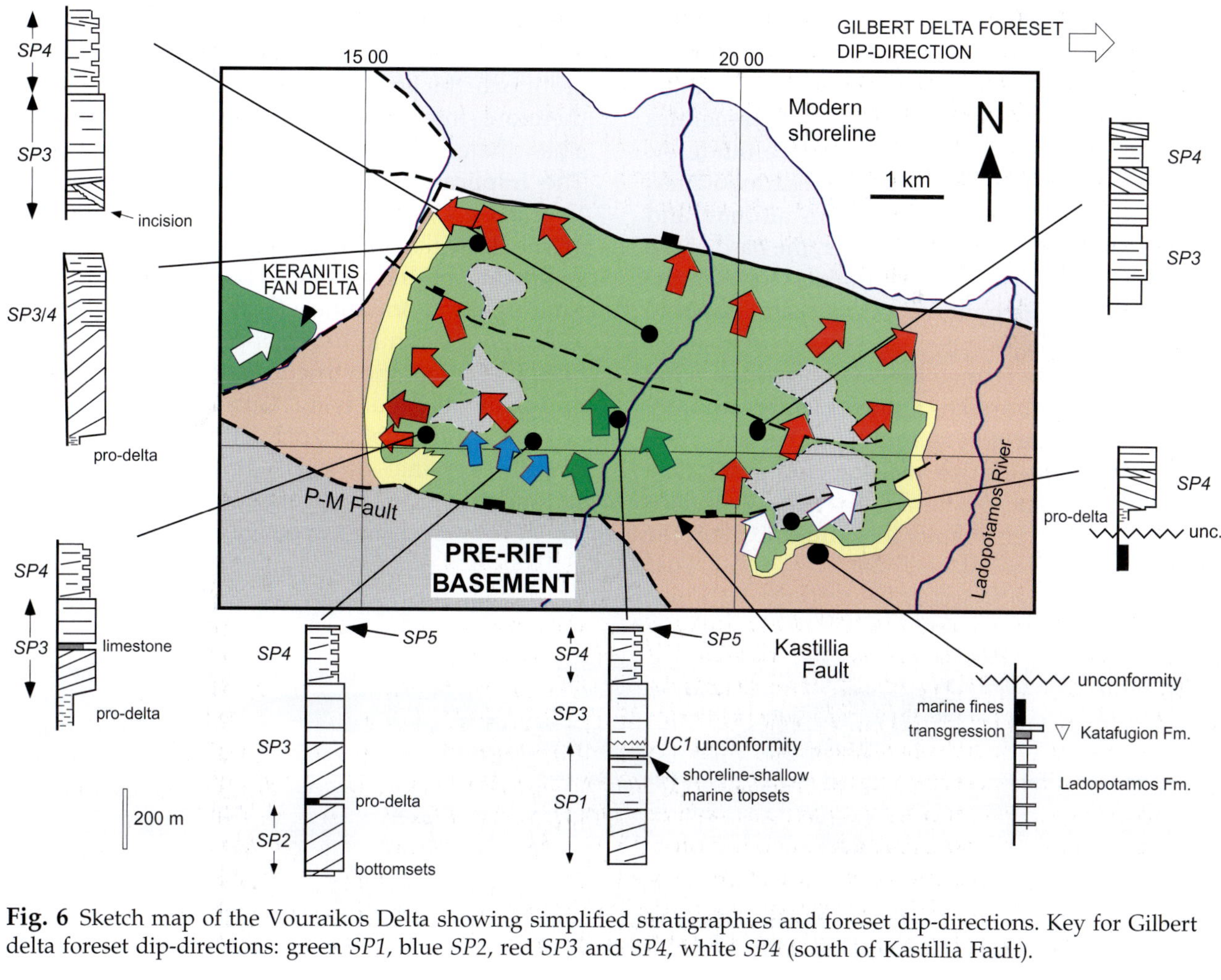

Fig. 6 Sketch map of the Vouraikos Delta showing simplified stratigraphies and foreset dip-directions. Key for Gilbert delta foreset dip-directions: green *SP1*, blue *SP2*, red *SP3* and *SP4*, white *SP4* (south of Kastillia Fault).

interpreted as proximal pro-delta facies. At several localities on the western and eastern borders of the delta, conglomeratic bottomsets are observed to pass laterally and asymptotically into these pro-delta facies. The pro-delta facies are distinguished as the Derveni unit (Fig. 3). The top of the Vouraikos Delta is characterized by two south-dipping plateaux (the Asomati to the west and the Kastillia to the east) at altitudes of between 750 and 820 m (Fig. 3). These plateaux are displaced by late secondary normal faults, and are capped by red soil deposits (Fig. 4). The detailed internal stratigraphy and sedimentology of the Vouraikos Delta are described below.

Basal contact of the upper group

The base of the Vouraikos Conglomerates can be traced on the eastern and western flanks of the delta. As the underlying pro-delta unit is consistently 20–50 m thick, it is assumed that the base of the upper group is subparallel to the base of the Vouraikos Conglomerates. In the Keranitis Valley the western base of the Vouraikos Conglomerates dips 8°N and rises southward to 600 m altitude. At the south-east edge of the delta in the Ladopotamos Valley this contact is also at 500 m altitude and locally dips 15°W (Fig. 7). However, the base of the delta is not exposed in the Vouraikos Gorge in the centre of the delta body, despite incision down to 120 m elevation.

The transverse change in elevation of the basal contact of the Vouraikos Conglomerates and, by implication, of the upper group (Fig. 7) is here interpreted as being due to incision and infill of a palaeovalley some 7–8 km wide by the Vouraikos Delta. No transverse faults that could explain the lateral change in elevation of these contacts were detected. The general cuspate base of the delta implies a relief of at least 300 m for this valley incised into the lower group. This cuspate form, however, may have been enhanced by: (i) differential subsidence below the thicker central delta succession; and (ii) by progradation of foresets across older bottomsets, thus progressively raising the delta base basinward and laterally. The palaeovalley model implies that a major regression–transgression event occurred at the boundary between the lower and upper stratigraphic groups.

In the eastern delta block (Kastillia Plateau), in the footwall block of the Kastillia Fault (Fig. 5), the character of the basal contact is different. Approximately 100 m up-section from the Katafugion Formation are conglomerates and sandstones of foreset/bottomset facies associations of the Vouraikos Delta. The thickness of the upper group (Vouraikos Conglomerates) in the footwall block of the Kastillia Fault is only around 200 m, whereas in the hangingwall the Vouraikos Conglomerates are much thicker (> 800 m; Fig. 2b). This fault is therefore interpreted as a sealed growth fault, since there is no evidence for its surface trace. The implication of this relationship is that the Vouraikos Delta conglomerates in the footwall of the Kastillia Fault are some of the youngest sediments of the system, despite overlying the Katafugion Formation (Fig. 6). Furthermore, although a definite downlapping surface has not been identified, the structural observation that the approximately flat-lying Gilbert-type delta succession (from subhorizontal topsets) structurally overlies a panel of north-dipping lower group sediments strongly suggests an angular unconformity between these respective successions.

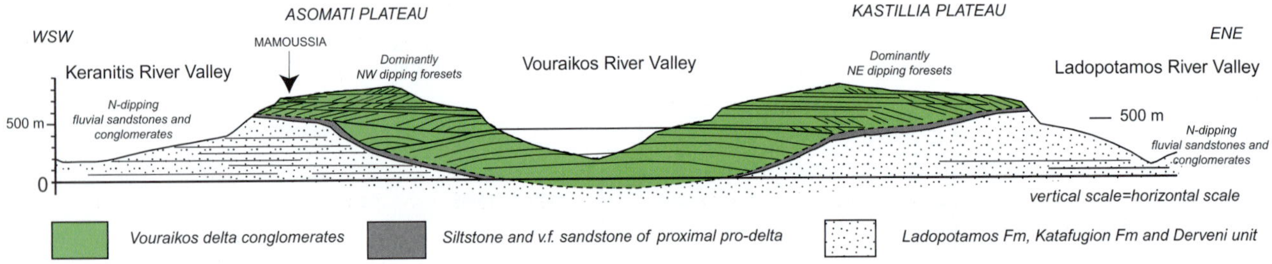

Fig. 7 Vertical ENE–WSW transverse section of the Vouraikos Delta orthogonal to its progradation direction (Fig. 6), located in the hangingwall of the Kastillia and Pirgaki-Mamoussia faults, showing incision of the delta into the lower group succession. No vertical exaggeration.

BIOSTRATIGRAPHY

Pollen content was used to date selected samples of fine-grained rocks from the lower and upper groups. The presence or absence of certain thermophilous plants that became successively extinct in this area during the past 2 Myr was used for this purpose. The chronological succession of their disappearance is relatively well known at latitude 36°–39°N in the eastern Mediterranean region based on several reference pollen successions covering the time-interval under consideration. These sections are at Crotone [Vrica Santa (Combourieu-Nebout & Vergnaud Grazzini, 1991) and Santa Lucia], Citadel of Zakynthos (Subally *et al.*, 1999), Monte San Giorgio at Caltagirone (Dubois, 2001), Tsampika in the Rhodos Island (Joannin, 2003; Cornée *et al.*, 2006), Peloponnese localities [Megalopolis (Okuda *et al.*, 2002), Phlious (Urban & Fuchs, 2005) and the Argive Plain (Jahns, 1993)] and Oeniades (Gulf of Corinth; Fouache *et al.*, 2005).

A preliminary analysis has been performed on nine samples in which pollen grains are the most abundant. Their stratigraphical positions are indicated in Fig. 4. The pollen flora is represented by 58 taxa, which are generally at the genus level for the trees and the family level for the herbs. From an ecological point of view, these taxa belong to various groups:

1 subtropical trees (Taxodiaceae, *Cathaya*);
2 warm-temperate trees (deciduous *Quercus*, *Carpinus*, *Acer*, *Ulmus*, *Zelkova*, *Castanea*, *Taxus*, *Populus*, *Salix*, *Liquidambar*, *Fraxinus*, *Ligustrum*, *Betula*, *Alnus*, *Corylus*, *Tamarix*, Cupressaceae, *Carya*);
3 Mediterranean xerophytes (evergreen *Quercus*, *Pistacia*, *Olea*, *Phillyrea*, *Cistus*, *Vitis*);
4 cool-temperate (i.e. altitudinal) trees (*Cedrus*, *Tsuga*, *Abies*, *Picea*);
5 herbs and shrubs (Poaceae, Asteraceae Asteroideae, *Centaurea*, *Artemisia*, Asteraceae Cichorioideae, Brassicaceae, *Scabiosa*, *Knautia*, *Polygonum*, *Geranium*, *Mercurialis*, *Hippophae rhamnoides*, *Convolvulus*, *Catalpa*, *Jasminum*, *Ambrosia*, Fabaceae, Cyperaceae, Caryophyllaceae, *Plantago*, Ranunculaceae, Ericaceae, *Potamogeton*, *Ephedra*, Amaranthaceae-Chenopodiaceae);
6 *Pinus* and Rosaceae, which are non-significant elements because they can be related to many different biotopes;
7 *Tricolporopollenites sibiricum*, which has an artificial species name partly based on the pollen morphology because the corresponding plant is unknown.

The prevalence of tree pollen grains versus those of herbs indicates that all the samples represent interglacial phases. Some samples yielded very scarce dinoflagellate cysts (C05-11, C02-1, C05-29), indicative of a marine influence. The presence or absence of certain thermophilous plants allow the samples to be divided into two broad age groups.

1 Early Pleistocene. Five samples (C02-4, C02-7, C02-6, C05-29, C05-4) showed a significant percentage of thermophilous tree pollen, such as Taxodiaceae (in sample C02-4 only), *Cathaya* (in samples C02-4 and C02-7 only, which also include an unidentified pollen, the so-called *Tricolporopollenites sibiricum*), *Tsuga*, *Cedrus*, *Carya*, *Liquidambar*, and *Zelkova* (Table 1). Taxodiaceae and *Cathaya* simultaneously disappeared from the area at about 1.1 Ma, whereas *Carya* and *Tsuga* persisted up to about 0.9 Ma. The extinction of *Cedrus* occurred later (around 0.7 Ma), but is still present in South Turkey and Lebanon. *Liquidambar* and *Zelkova* became rare and disappeared very recently (during the Last Glacial); they are still present in some refuge territories in Turkey. Stratigraphically, two of these samples come from the lower group, C05-29 from the Katafugion Formation and C05-4 from the alluvial succession in the footwall of the PM Fault (Kalavrita conglomerates). The three other samples, two of which contain the oldest assemblages, are from the upper group, specifically in the fine-grained beds below the western edge of the Vouraikos Delta (pro-delta facies). These pro-delta beds are estimated to belong to the SP3 package in the Vouraikos Delta (see below). These five samples are grouped into an age bracket of Early Pleistocene (1.8–0.78 Ma; Gradstein *et al.*, 2004). No distinction can be made between the lower group samples and those from the lower parts of the upper group. Hence the basal unconformity of the upper group is not detectable.
2 Middle Pleistocene. Four samples (C02-3, C05-11, C02-1, C05-20) do not contain these thermophilous plants, except some rare pollen grains of *Cedrus* (samples C02-3 and C05-11) and *Carya* (C02-3) (Table 1). In addition, they show increased percentages of *Olea* and sometimes the presence of *Vitis* (Table 1), two Mediterranean elements that developed

Table 1 Nine samples (from the lower and upper groups) were analysed for pollen content. The presence/absence of the most significant thermophilous plants used for dating in the Pleistocene of the eastern Mediterranean are shown. Values are percentages based on the total number of pollen grains counted in each sample (between 150 and 300 grains per sample). See text for further details and interpretation. Most samples are located on Figs 4 and 14

Sample	Stratigraphic position	Coordinates	Taxa (%)									
			Taxodiceae	*Cathaya*	*Tsuga*	*Cedrus*	*Carya*	*Liquidambar*	*Zelkova*	*Tricolporo-pollenites sibiricum*	*Vitis*	*Olea*
C05-20	Vouraikos topsets (SP5)	N38 09.656′ E22 12.412′	0	0	0	0	0	0	0	0	0.6	1.2
C02-3	Vouraikos topsets (SP5)	N38 09.88′ E22 09.5′	0	0	0	1	1	0	0.5	0	0.5	0.5
C02-1	Vouraikos topsets (SP3)	N38 09.69′ E22 08.56′	0	0	0	0	0	0	0	0	0	2.5
C05-11	Vouraikos topsets (SP1)	N38 09.58′ E22 10.1′	0	0	0	0.8	0	0	0	0	0	9
C02-6	Vouraikos pro-delta (SP3)	N38 10.25′ E22 08.63′	0	0	0.8	4	0	0	0.8	0	0	0.8
C02-7	Vouraikos pro-delta (SP3)	N38 10.25′ E22 08.63′	0	0.8	0.4	21	0	0	2.1	0.4	0	0
C02-4	Vouraikos pro-delta siltstone/marine clastics	N38 10.83′ E22 08.7′	1.6	0.8	0.8	5.3	1.2	0.8	0.8	0.8	0	0
C05-29	Katafugion Formation	N38 09′ E22 12.8′	0	0	6.2	2.2	1.1	6.8	0.6	0	0	0
C05-4	Kalavrita Conglomerates	N38 08.12′ E22 10.6′	0	0	1.5	0.7	0.7	2.3	0	0	0	0.7

during the Middle Pleistocene. Two samples come from within the delta itself: C05-11 comes from the upper topsets of SP1 in the centre of the delta and C02-1 comes from the western topsets of SP3. Two samples come from SP5 beds at the top of the delta (C02-3 and C05-20). These samples may be somewhat younger than those of the previous group and may belong to the Middle Pleistocene (i.e. after 0.78 Ma).

Although preliminary, these results indicate that the Vouraikos Delta was built mainly during the Early Pleistocene, beginning sometime before 1.1 Ma and probably during the beginning of the Middle Pleistocene (up to sometime after 0.6 Ma). These data indicate that the unconformity at the base of the upper group does not represent a major time gap.

VOURAIKOS DELTA SEDIMENTOLOGY

Previous work

Very little detailed stratigraphical or sedimentological research has been published on the Vouraikos Delta. Ori *et al.* (1991, fig. 2) regarded the Keranitis and Vouraikos deltas (Fig. 2) as a single system. They indicated that the sediments of the Vouraikos Delta (*sensu stricto* – as used in this paper) are composed of variously alternating sequences of topsets and foresets (their fig. 12), and the uppermost internal stratigraphical units to be 'foreset beds' (their fig. 2). Poulimenos *et al.* (1993) and Zelilidis & Kontopoulos (1996, fig. 1b and fig. 2a C–C′) regarded the western part of the Vouraikos Delta (as defined herein) to be the eastern fan delta system in their 'Egio Subbasin'. They considered the delta to lack 'toesets' (bottomsets), describing it as a 'trapezoidal' delta on the basis of its longitudinal cross-section, and attributed this form to deposition in a 'protected' or narrow basin. Dart *et al.* (1994) did not report specifically on the sediments of the Vouraikos Delta system, but they showed the Vouraikos as being a separate system from the Keranitis, as did Malartre *et al.* (2004).

Facies and facies associations

Facies associations in Gilbert-type delta systems can be defined to a first-order by their position (and attitude) in the tripartite structural division of these deltas: subhorizontal topsets, angle of repose foresets and low-angle (< 10°) bottomsets. The range of facies associations defined (see also Malartre *et al.*, 2004) further includes the markedly finer grained (subhorizontal) pro-delta and the relatively thin shoreline/coastal heterolithic types. The distinction between bottomset and pro-delta environments differs from that of several other studies. Several of these gross divisions contain distinctive subassociations, which are elaborated below.

Alluvial topset facies association

This consists of: (i) a heterolithic subassociation of sandstone, conglomerate and siltstone red-bed facies; and (ii) a volumetrically and areally dominant conglomerate-rich subassociation. Both have been observed to comprise only depositionally horizontal architectures (cf. Dart *et al.*, 1994, p. 549), and tend to occur in sequences of tens of metres thickness, although the conglomerate-rich association is consistently more thickly developed, and often reaches hundreds of metres of preserved thickness. Mean palaeoflow was towards the N–NNE.

Heterolithic subassociation. This consists of interbedded small-pebble to small-cobble conglomerates, pebbly sandstones, red or brown sandstones and reddened siltstones (Fig. 8). The small-pebble conglomerate facies are commonly horizontally stratified in sheet-like or lenticular beds, with erosive bases showing longitudinal scours. Coarser conglomerates (small cobble grade) tend to be massive, or locally cross-stratified. Normal grading has been observed in some 1 m thick beds. Subspherical carbonate nodules have been observed in red-brown siltstone facies, overlain by weakly undular-laminated carbonate (calcrete) in fine sandstone (forming a bedset > 0.7 m thick). The heterolithic subassociation is similar to 'topset facies association 2' of Dart *et al.* (1994), except that examples of their association 2 are traceable for distances up to 1 km across the fan delta top.

Conglomeratic subassociation. This is principally composed of matrix-rich (coarse sand to small pebbles) clast-supported conglomerates with modal grain sizes of medium pebble to small cobble grade. The modal grain populations are poorly sorted, and

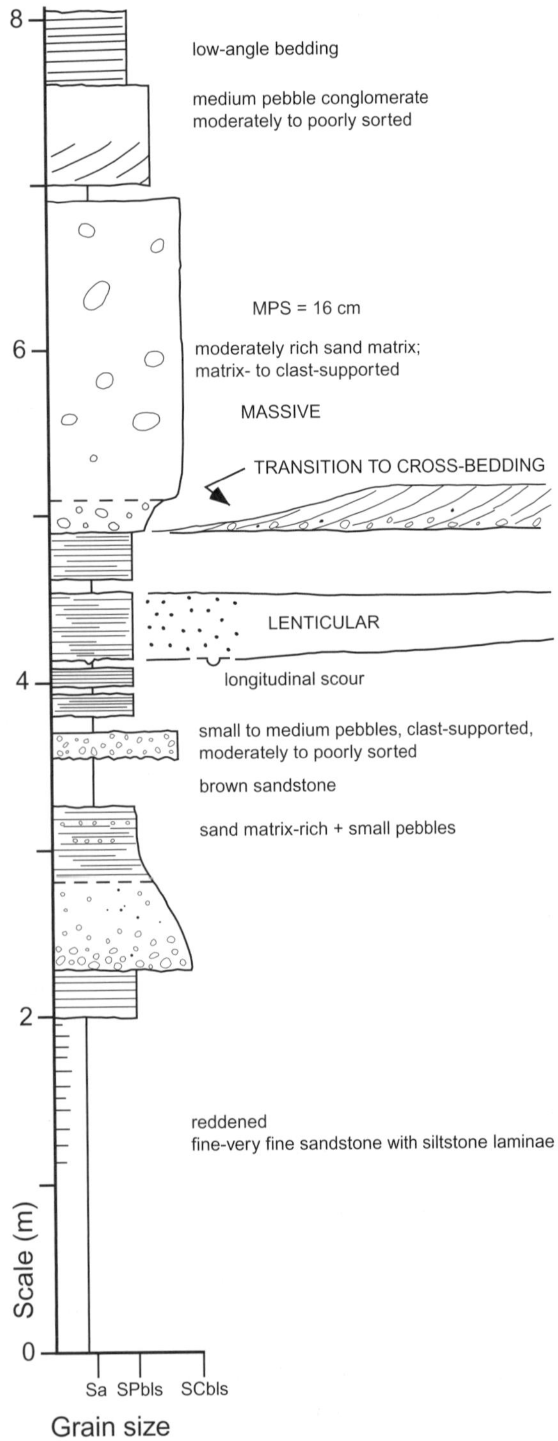

Fig. 8 Graphic log of a short section through the relatively fine-grained (heterolithic) alluvial topset facies association located on the southwest margin of the Vouraikos Delta at 675 m altitude (location A, Fig. 3), containing interbedded sandstone, conglomerate and siltstone red-bed facies. Road section to Asomati Plateau at co-ordinates 38°09′55.7″N/022°08′46.0″E.

maximum particle size is small boulder grade. Clasts are well rounded to rounded and shapes variable. Fabrics are dominantly unordered, with *a*(t)*b*(i) imbrication being subordinate. Palaeoflow data from imbrication indicated north-quadrant vectors. Bed thicknesses of the typical conglomerates range from 1 to 5 m. Finer grained conglomerate facies (e.g. small-pebble conglomerates) can have bed thicknesses of < 0.5 m. Bases of coarse facies are generally prominent irregular-erosive surfaces lacking obvious large-scale relief, but with variably developed small-scale (0.1 m) longitudinal scours. Stratification within beds is very poorly developed, with a predominance of massive (structureless) to crudely horizontally stratified beds. Stratification is enhanced occasionally by very thin (often reddened) sandstones. Large-scale cross-stratified conglomerates occur rarely, and have been observed only as solitary sets, ranging from 1.5 to 4 m in thickness. Cryptic and very low-angle planar (cross) stratification has been occasionally observed, with surfaces having (north-quadrant) dip-directions similar to associated clast imbrication palaeocurrent indicators. Grading profiles in beds are not well developed and, although a detailed study has not been carried out, there is no obvious correlation between bed thickness and maximum particle size.

This subassociation is frequently organized in cyclical sequences, metres to tens of metres thick. Topset cycles are developed from prominent basal erosional surfaces, followed by massive, crudely bedded conglomerates and capped by thinner-bedded conglomerates with development of reddened sandstones (e.g. Fig. 9). Thus, there is a tendency towards fining- and thinning-upwards cycles. Other cycles between prominent basal surfaces are neutral in terms of grain size and bed thickness.

Interpretation. The texture (particularly *a*(t)*b*(i) imbrication, clast-support and poor sorting) and bed geometry of the conglomerates (and subordinate sandstones) comprising the conglomeratic topset association are consistent with turbulent water (stream) flows (Nemec & Steel, 1984). Sheet-like geometries and the lack of channel forms suggest either deposition in wide, shallow low-sinuosity gravel rivers that occupied the subaerial fan delta top, or as unconfined high-magnitude sheet

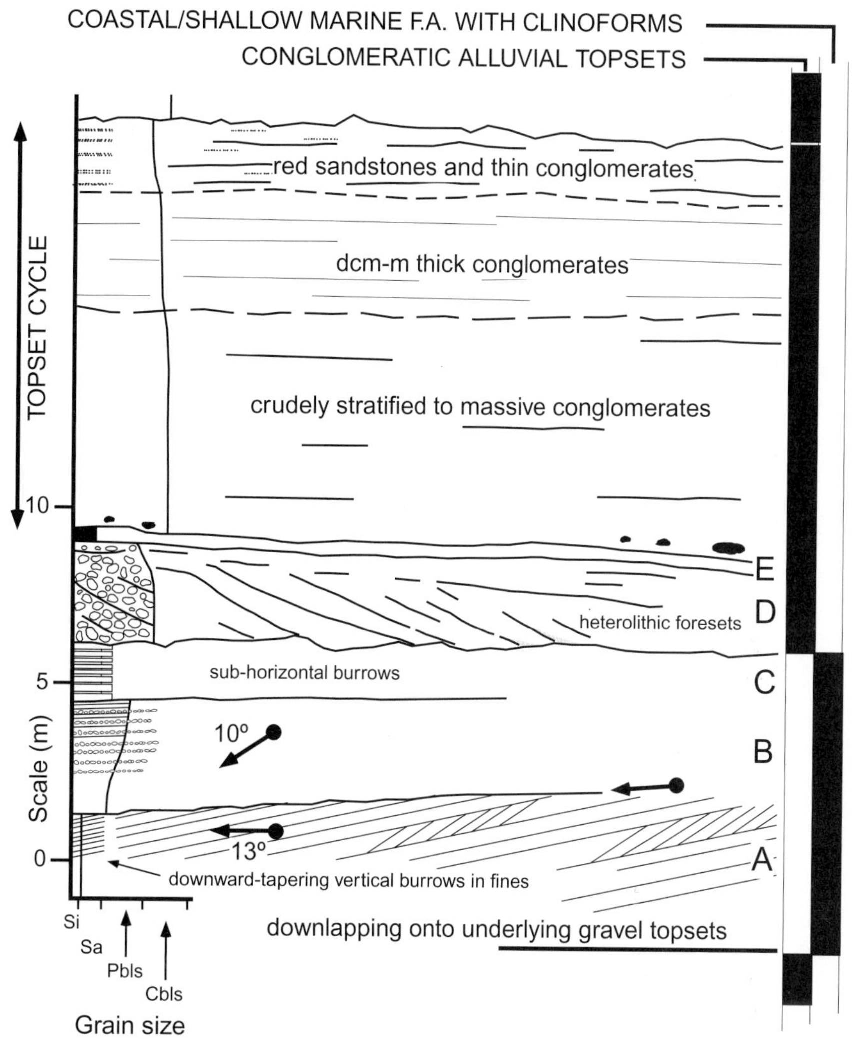

Fig. 9 Graphic log of well-stratified and sorted conglomerates and sandstones (SFA1) of the coastal–shallow-marine facies association organized in a west-dipping clinoform, and overlying flat-bedded and cross-stratified alluvial topset conglomerates (location B, Fig. 3).

floods (e.g. Young *et al.*, 2000). Occasionally preserved downstream-dipping low-angle inclined surfaces imply low-relief (?longitudinal) bars, on which gravel may have been accreted as diffuse gravel (bed load) sheets (Hein & Walker, 1977; Bridge, 2003); however, the preserved facies lack an openwork texture and well-developed grading profile typical of lower-stage plane beds (Bridge, 2003; see also Nemec & Postma, 1993).

The stacked thinning- and fining-upwards topset cycles may be a reflection of declining accommodation modulated by high-frequency fluctuations in relative sea level, where thinner bedded units correspond to progradational episodes and basal thick-bedded topsets to aggradational episodes. The heterolithic subassociation is interstratified with the gravel-dominated topset sequences, and is thought to represent subaerial, non-channelized (floodplain) environments on the delta top.

Shoreline–shallow-marine topset facies association

This lithologically and structurally diverse facies association occurs in grossly subhorizontal units and is therefore a 'topset' association. It is referred to as shoreline–shallow-marine based on evidence of distinctive textures, stratification and faunal/ichnofaunal content. It comprises four main subassociations, detailed below.

Very well sorted and stratified clinoform conglomerates (SFA1). Examples of this subassociation occur interstratified with texturally contrasting alluvial topsets (Fig. 9), in association with limestones (SFA3 below; Fig. 10c)

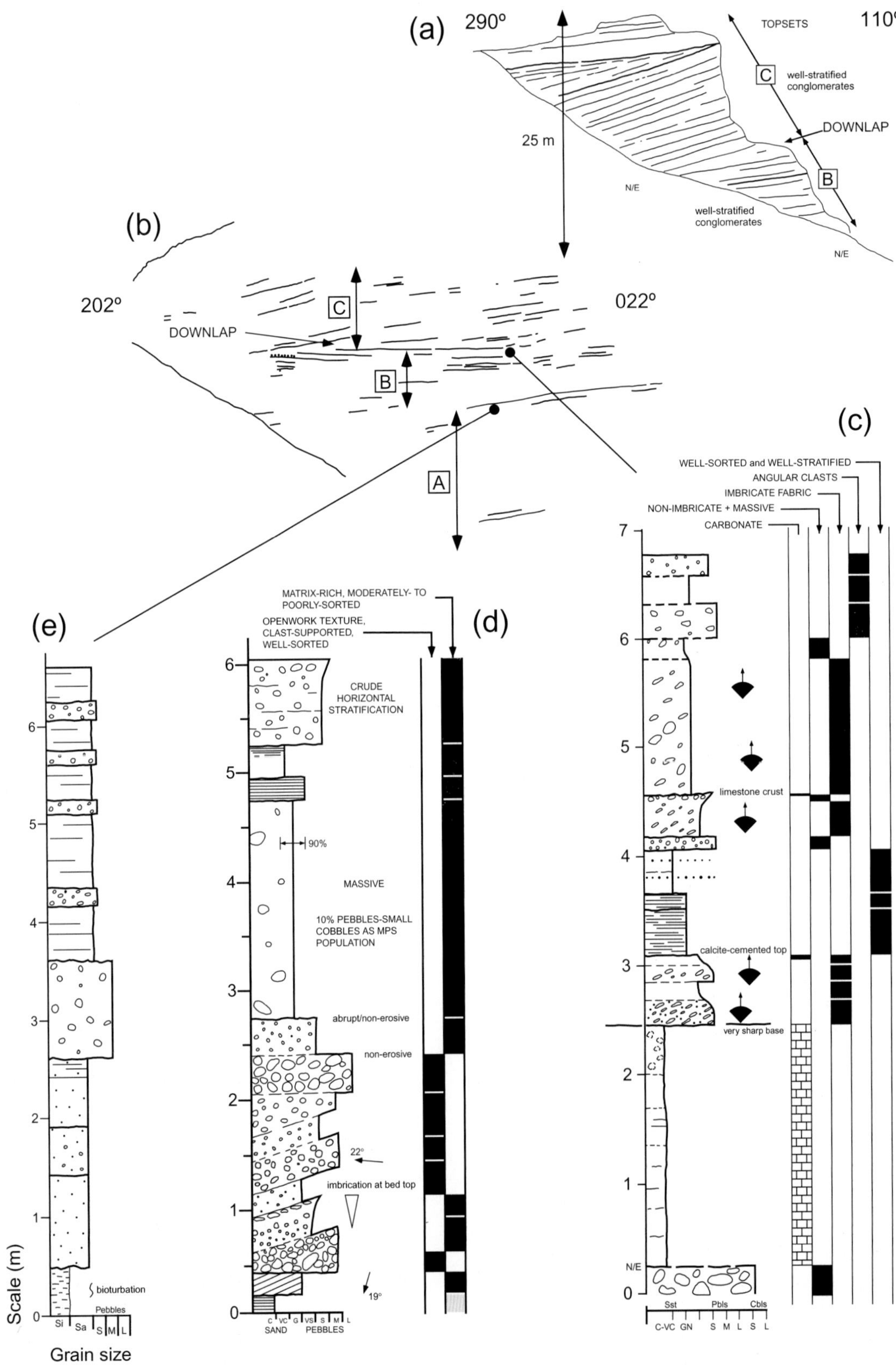

Fig. 10 (a,b) Line drawing of a valley side exposure (location C, Fig. 3) divisible into three macroscale units (A–C). Unit A contains variably sorted and structured examples of subdivision SFA1 of the shoreline–shallow-marine topset facies association (graphic logs d & e). Unit B shows an example of SFA3 (limestone) and SFA1 in erosive contact (graphic log c). Unit C comprises Gilbert delta foresets (overlain by topsets), which subtly downlap onto unit B.

and have also been observed to overlie the accretionary toplap association (SFA2, see below). This subassociation is exemplified by two detailed examples.

A well exposed 7–8 m thick sequence from the Vouraikos Gorge (Fig. 3, location B) consists of three inclined stratasets (Fig. 9, A–C). The basal unit is heterolithic, with 5 cm thick, inclined, very well sorted small-pebble to granule conglomerates interbedded with laminated red, yellow and beige fine sandstones and siltstones. There are frequent downward-tapering burrows into the tops of the fine facies from the overlying fine gravels/sandstones. Unit B coarsens upwards, commencing with subhorizontally stratified, very well sorted granule to small-pebble conglomerates (bed thicknesses 4–6 cm), which are clast-supported and normally graded. The upper two-thirds of Unit B contains evenly spaced 'stringers' of large pebbles to small cobbles. Unit C is finer grained, composed of interbedded, very well stratified small-pebble conglomerates (9–38 cm in thickness) and red, medium to coarse sandstones (5–8 cm sheets) that contain granule to small-pebble laminae. The sandy facies contain a moderate frequency of subhorizontal, tube-like burrows.

The second example of SFA1 (Fig. 3, location C) shows several diverse metre-scale successions (Fig. 10a & b) comprising:

1 well imbricated, normally-graded small-pebble conglomerates, very well sorted parallel-laminated granule conglomerates, massive angular-shaped small–medium-pebble conglomerates (Fig. 10c);
2 steeply dipping (22°) openwork small- to large-pebble conglomerates, parallel-laminated coarse sandstone to granule conglomerates and massive bimodal conglomerates (Fig. 10d);
3 bioturbated siltstones, coarse sandstones and poorly sorted small-pebble conglomerates (Fig. 10e).

Additionally, the well-imbricated and stratified succession (1, described above) occurs in conjunction with a prominent limestone bed (SFA3, below), separated by an exceptionally sharp-planar surface (Fig. 10c), and comprises a package of sediment that has low-angle dips (5–10° restored), but was probably an overall subhorizontal unit. Two conglomerate beds within the section contain calcite-cemented and limestone-encrusted bed tops (Fig. 10c). Observed clast imbrication indicated consistent north-quadrant palaeoflow.

Interpretation. SFA1 is considered to have been deposited subaqueously. The tightly packed conglomerate textures, stratified granule conglomerates and very coarse sandstones and seaward-dipping imbrication suggest moderate to high wave-energy conditions, and the lack of wave ripples in gravel sediments suggests that shallow depth conditions prevailed. The variable ichnofauna, thin carbonates, consistently directed clinoforms, and occasional cross-bedding with distinctive openwork textures again suggest wave-influenced and wave-reworked gravels in a shallow-marine shoreface to lower beachface setting (e.g. Nemec & Steel, 1984; Dabrio & Polo, 1988; Dabrio, 1990; Massari & Parea, 1990; Hart & Plint, 1995).

Accretionary toplap conglomerates/sandstones (SFA2). A suite of moderately distinctive facies is associated with accretionary toplap contacts where, rather than a single erosive contact between Gilbert foresets and flat-lying topsets, a series of areally restricted local contacts are arranged en-échelon, sometimes climbing upwards in the progradation direction (see Dart *et al.*, 1994, figs 7 & 8). Sequences arranged about accretionary toplap contacts (equivalent to the 'transition zone' of Colella, 1988; Gawthorpe & Colella, 1990) are around 5 m thick or less, and are characterized by showing moderately improved fabric development and stratification compared with alluvial topset conglomerates. The principal conglomerates' grain size varies typically from medium pebble to large cobble, and large cobble to small boulder sizes. Sorting is generally poor, with matrix-rich facies showing matrix- and local clast-support systems. Weak *a*(t)*b*(i) clast imbrication is recorded (with north-quadrant palaeoflows), but massive-unordered fabrics are the norm. Facies with matrices of coarse sand-granules to very small pebbles and outsize clasts are recorded, with variants being small- to medium-pebble conglomerates with small boulder outsize clasts. Stratification is characteristically enhanced in conglomerates of SFA2, being defined by either thin (5–10 cm thick) fine-grained conglomerates, equally thin medium–coarse (reddened) normally graded sandstones, or simply thin stratification in homogeneous gravelly sediments. Thicker interbedded sandstones within

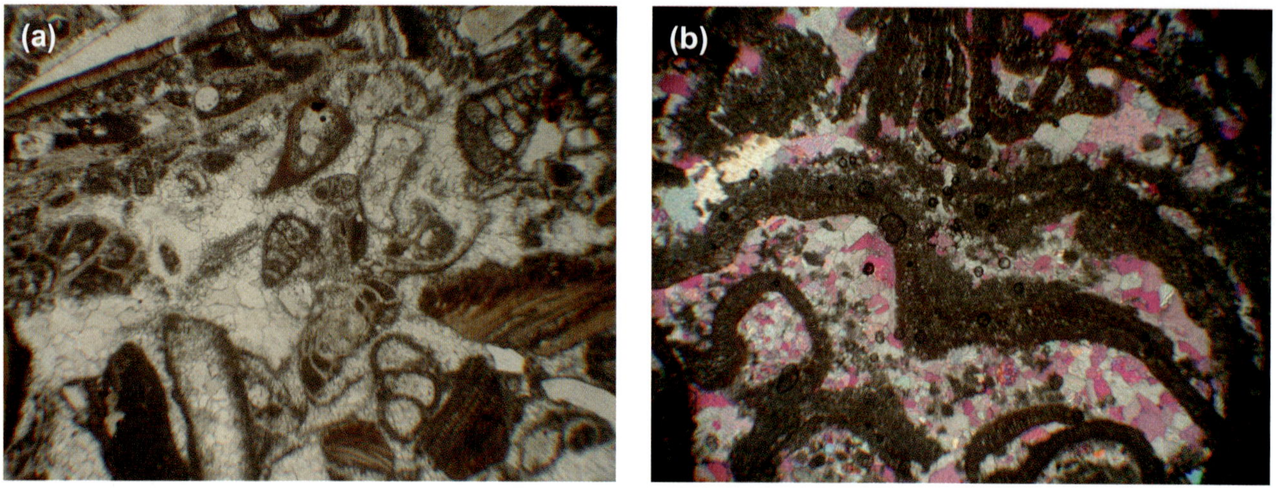

Fig. 11 (a) Photomicrograph of bioclastic grainstone carbonate facies at 38°11′02.3″N/022°10′28.1″E (Marathia Limestone). (b) Photomicrograph of algal packestone–grainstone facies from transitional section between foresets and topset facies association, at the SW extremity of the Mamoussia cliff section (location D, Fig. 3, Mamoussia Limestone). Plane-polarized light, with gypsum plate in (b). Each photomicrograph measures 5 mm across.

the accretionary toplap structure have been observed in inaccessible cliff sections, occupying the lower topset to upper foreset position. Apparently texturally distinct topset beds that overlie erosive toplap contacts have also been observed, but not directly examined. These are apparently massive and moderately to well sorted, tabular beds up to several metres thick.

Interpretation. The structural position of this subassociation alone suggests a contrasting emplacement process to alluvial topsets. The interpreted location is suggested to be a shallow subaqueous environment, which was affected by periodic flood-generated fluvial inputs of gravel (and sand) and reworking by moderate wave energy, equivalent to the 1 km wide shallow-marine topset of the modern Vouraikos Delta (Dart *et al.*, 1994, p. 549). The textural and stratification character of SFA2 contrasts with that of SFA1 where wave and current reworking of the sediment was considerably stronger. This may have been suppressed in SFA2 due to the dominance of fluvial supply during episodes of progradation and vertical accretion of the delta system. Occasionally observed distinct (single) beds erosionally overlying toplap contacts, which are apparently structureless and show closely packed textures, may represent wave-reworked gravels in a shoreline environment (see Colella, 1988).

Limestones (SFA3). Two limestone units were found in the Vouraikos Delta in this study, the Marathia and Mamoussia limestones (Figs 3, 6 & 10c). The Marathia Limestone is a bioclastic arenitic grainstone (Fig. 11a) to rudstone. It is massive to crudely horizontally stratified with a sandy texture in its lower part, and a 'rubbly' 45 cm thick top (Fig. 10c). The limestone contains low-spire gastropods, fragmentary large, thick-shelled pecten bivalves, fragmentary small bivalves, ?tubiform bryozoa encrusting/cementing clasts, ostracods, foraminifera, echinoid spines, ?rhodophytic algae and other bryozoans (Fig. 11a).

The Marathia Limestone is associated with SFA1 sediments with low-angle (7–10°) primary depositional dips (Fig. 10b & c). This limestone abruptly overlies a massive, pebble to cobble conglomerate, and is terminated by the exceptionally sharp, planar basal surface of a small-pebble conglomerate of SFA1 (Fig. 10c). The thickness of the limestone decreases southwards from 2.2 m to > 1.4 m over a distance of 30 m, where the bed terminates in a low-relief (1.4 m) palaeocliff with a steep east–west striking orientation (091/86°). Near the base of the palaeocliff there is a talus-like deposit with a northward-dipping inclined-curved fabric oriented 074/54°. However, the wall rocks of the palaeocliff have the same flat-lying bedding orientation of the main limestone, and are mixed clastic–carbonate lithologies, in which sands and granule to pebbles are organized in typically 5–30 cm thick beds with white fine calcareous matrices.

The Mamoussia Limestone is located in the extreme south-west of the Vouraikos Delta in the western part of the Mamoussia cliff (Fig. 3, location D). This is a thin bed < 2 m thick, stratigraphically located between the locally lowest Gilbert set and overlying topsets (Fig. 6). In thin section the limestone is seen to be very rich in fragmentary and complete algae (Fig. 11b). The lithology and texture is an arenitic bioclastic packestone to grainstone, containing probable peloids and extraformational clasts. Algal bioclasts have micritized envelopes. This facies, with its concentration of characteristic organisms, most closely resembles standard microfacies type 12 of Wilson (1975) and Flügel (1982).

Interpretation. The Marathia Limestone facies represents high-energy, shallow-water open-marine carbonate deposition on a flooded sector of the Vouraikos Delta top isolated from clastic input, following transgression. Similar facies from a Pliocene Gilbert-type delta setting have been described by Mortimer *et al.* (2005), where carbonate units represented marine transgression of the delta top. The Marathia palaeocliff and the contrasting mixed clastic and carbonate facies suggests an earlier phase of episodic carbonate–clastic deposition, followed by erosion and a later phase of sustained carbonate deposition lacking significant clastic input, which developed after minor fluctuations in relative sea level on the flooded delta top. The Mamoussia Limestone microfacies typifies wave-affected shelf edges (Tucker & Wright, 1990, table 1.1), which is analogous to the structural position of the unit at the seaward edge of the submerged delta topset. Carbonate sediments containing algal remains are reported in analogous shallow subaqueous (topset) settings, in (non-Gilbert) gravelly fan deltas (Ethridge & Wescott, 1984) and Gilbert-type deltas (Postma *et al.*, 1988; Young *et al.*, 2002, their facies 2b, c).

Laminated shelly siltstones and sandstones (SFA4). This rare subassociation consists of units up to 8 m thick of: (i) laminated fine-sandstone–siltstones with a uniform grain size profile; and (ii) laminated white–pale-grey siltstone (to mudstone), which contains distributed fragmentary shelly fossils, as well as monospecific bivalve assemblages. Units may be sharp-based conformably overlying stratified pebble–cobble conglomerates. This fine facies is interbedded with rare < 10–12 cm thick parallel-sided pebble conglomerates and pebbly sandstones, or interstratified with one-clast-thick layers of small pebbles.

Interpretation. This subassociation is considered to represent shallow-water back-barrier/lagoonal fines. The fine-grained laminated, weakly calcareous character plus the restricted diversity fauna point to a protected subaqueous environment at the margin of the flooded topset region of the delta. Small-pebble-clast layers and pebbly sands are interpreted to be storm washover sediments across coastal barriers or spits.

Gilbert-type delta foreset facies association

This facies association represents the largest volume of the Vouraikos fan delta. A compilation of large-scale (Gilbert) foreset orientations from the whole delta (Fig. 6) indicates a mean direction of progradation toward the north-northwest (345°, Fig. 12). The dispersion of the data suggests a convex (linguoid)-shaped sediment body. Dip values for foresets shown by the polar plot (Fig. 12; 10–35°) indicate a typical range for gravel-dominant Gilbert deltas (Nemec, 1990). It comprises well-bedded sequences of metre-scale pebble–cobble grade conglomerates, thinner bedded sand-matrix-rich very small- to small-pebble conglomerates, as well as coarse and pebbly sandstones. Notable are matrix-free openwork clast-supported pebble–cobble conglomerates (typically 10–20 cm in thickness), in units parallel to other facies.

Stratification includes types that are discordant, very low-angle planar strata and convex-up cross-stratification in single sets. Foreset stratification tends to be uniform and the texture homogeneous; subparallel, cross-cutting surfaces such as those described by Dart *et al.* (1994), or large-scale synsedimentary deformation features for the Keranitis Delta (described by Ori *et al.*, 1991), have not been observed. However, examples of interbedded fine-grained foreset intervals include mutually cross-cutting channel-form conglomerates < 2 m thick with 10–15 m wide transverse sections, interbedded

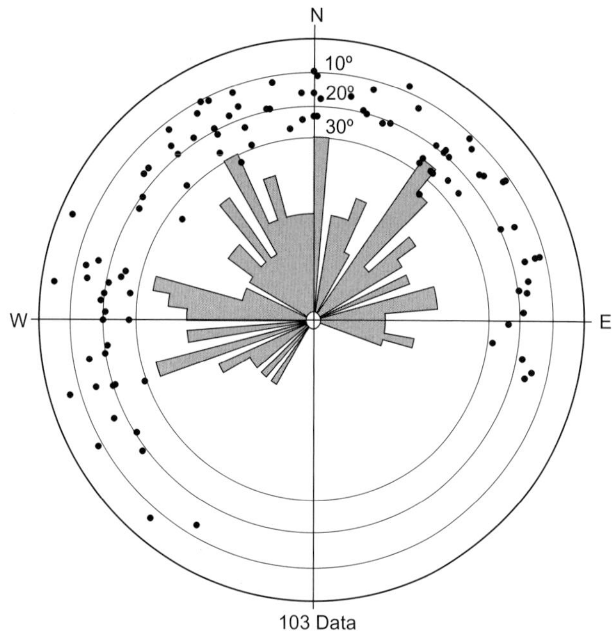

Fig. 12 Combined rose diagram and equal area polar dip-direction plot of the Vouraikos Delta foresets. Vector mean of foresets is 345° ($n = 103$). Rose diagram class interval is 3°.

with and eroding into red and brown laminated sandstones and red siltstones–mudstones. The shallow channel conglomerates are moderately to poorly sorted pebble to small cobble size with a sand matrix. Channel form bases are marked by scour structures in the form of occasional large-scale (10–15 cm thick), isolated flute marks. The channel form axes are oblique but dominantly down-dip of the foresets; interbedded associated fine-grained facies have been observed to contain small-scale down-dip verging soft sediment folds. Cliff-scale exposures of complete sets reveal very gently concave foresets, with clear reduction in dip as the bottomsets are approached.

Interpretation. Large-scale foresets represent bedload and mass-flow emplacement of gravel and sand into a standing water body (cf. Postma, 1990; Prior & Bornhold, 1990; Falk & Dorsey, 1998). The set thickness of the association indicates palaeowater depths in the range of 300–700 m. Clast-supported openwork foreset conglomerates were interpreted by Dart *et al.* (1994) as the tops of individual flow units, whereas the massive to crudely bedded, matrix-rich beds were considered as grain flows.

Bottomset facies association

This association is composed of thinly interbedded plane-laminated and rippled sandstones and conglomerates (Fig. 13a). Erosion surfaces and/or scours are common. In some places, soft-sediment deformation occurs with small-scale slump folds and/or dewatering structures (Fig. 13b). They represent the base of Gilbert foresets where low-angle slopes dip at values typically 5–10°.

Pro-delta facies association

This is an important facies association as it represents distal environments to the Gilbert-type delta, affected by basinal processes as well as sediment input from the delta. It is dominated by thinly bedded, beige and grey-green coloured, massive to finely parallel-laminated siltstones and silty sandstones with sharp bases (Fig. 13d), thin pebble conglomerates and laminated siltstones. Floating gravel clasts commonly occur in the fine (sand and silt) facies (Fig. 13c). This association is spatially related to the Gilbert-type delta bottomsets, and the two can be frequently mapped together in the field (Fig. 13e). Pro-delta and bottomsets share individual facies types (such as fine sandstones). Depositional dips are lower than those of bottomsets, and are generally undetectable.

Interpretation. These deposits can be interpreted as the products of processes ranging from suspension fallout deposits to turbidity current deposits that are *largely* beyond the influence of gravel input. They represent the deepest facies of all the associations observed in the Vouraikos fan delta system. This facies association is genetically related to the fan delta foreset–bottomset structure of the system, and results from emplacement of major increments of clastic sediment following periodic fluvial flood events. The pro-deltas of Type A feeder Gilbert-type fan deltas are affected by foreset-derived mass flows and density currents, as well as hemipelagic sedimentation (Postma, 1990; see also 'basin plain' deposits of Hwang & Chough, 2000).

VOURAIKOS DELTA ARCHITECTURE

Within the Vouraikos Delta, five internal stratigraphic packages have been defined, numbered *SP1*

Fig. 13 Field photographs of bottomset and pro-delta facies associations of the Vouraikos Delta. (a) Bottomset conglomerates and sandstones from southwest base of delta (Keranitis Valley). (b) Coarse conglomerate levels within fine-grained pro-delta facies. (c) Conglomeratic bottomsets wedging downward into fine-grained pro-delta facies eastern base of delta (Ladopotamos Valley). (d) Floating pebble in pro-delta siltstones and fine sandstones (Ladopotamos Valley). (e) Slump folds showing basinward asymmetry from southwest base of delta (Keranitis Valley). Scales: hammer shaft = 28 cm; lens cap = 6 cm.

to *SP5* (Fig. 4). A stratigraphic package is here defined as a distinct succession limited by prominent bounding surfaces. A stratigraphic package can comprise: (i) packages of topsets (representing palaeohorizontal); (ii) very large-scale foresets (at angles of repose of 20–35°); and/or (iii) multiple sets of topsets and foresets. Comparatively thin, but locally distinctive, bottomset, thin pro-delta and shallow-water coastal facies associations are occasionally found within the thicker stacked packages. There is considerable lateral variation in individual stratigraphic packages (see Fig. 6), related to: (i) thickness variations; (ii) transitions of topsets into foresets; and (iii) other variations due to intra-deltaic growth faults. Detailed correlation is further complicated by changes in structural elevation due to second-order extension faults, such as the Derveni Fault (Fig. 3). While bounding surfaces are clearly distinguishable in the proximal (topset) part of the delta, they can be lost distally, in particular within thick foreset sequences.

Despite this, good stratigraphic coherence is displayed, particularly in the southern and western parts of the delta. An analysis is presented principally of the western half of the delta (Asomati block) along two major NNE–SSW cross-sections (Fig. 14), and three east–west profiles in the Asomati block (south, centre and north). The eastern part of the delta (Kastillia block) is represented in less detail in the regional cross-section in Fig. 2b and in one natural east–west profile on the north-east side of the Kastillia block (Fig. 3).

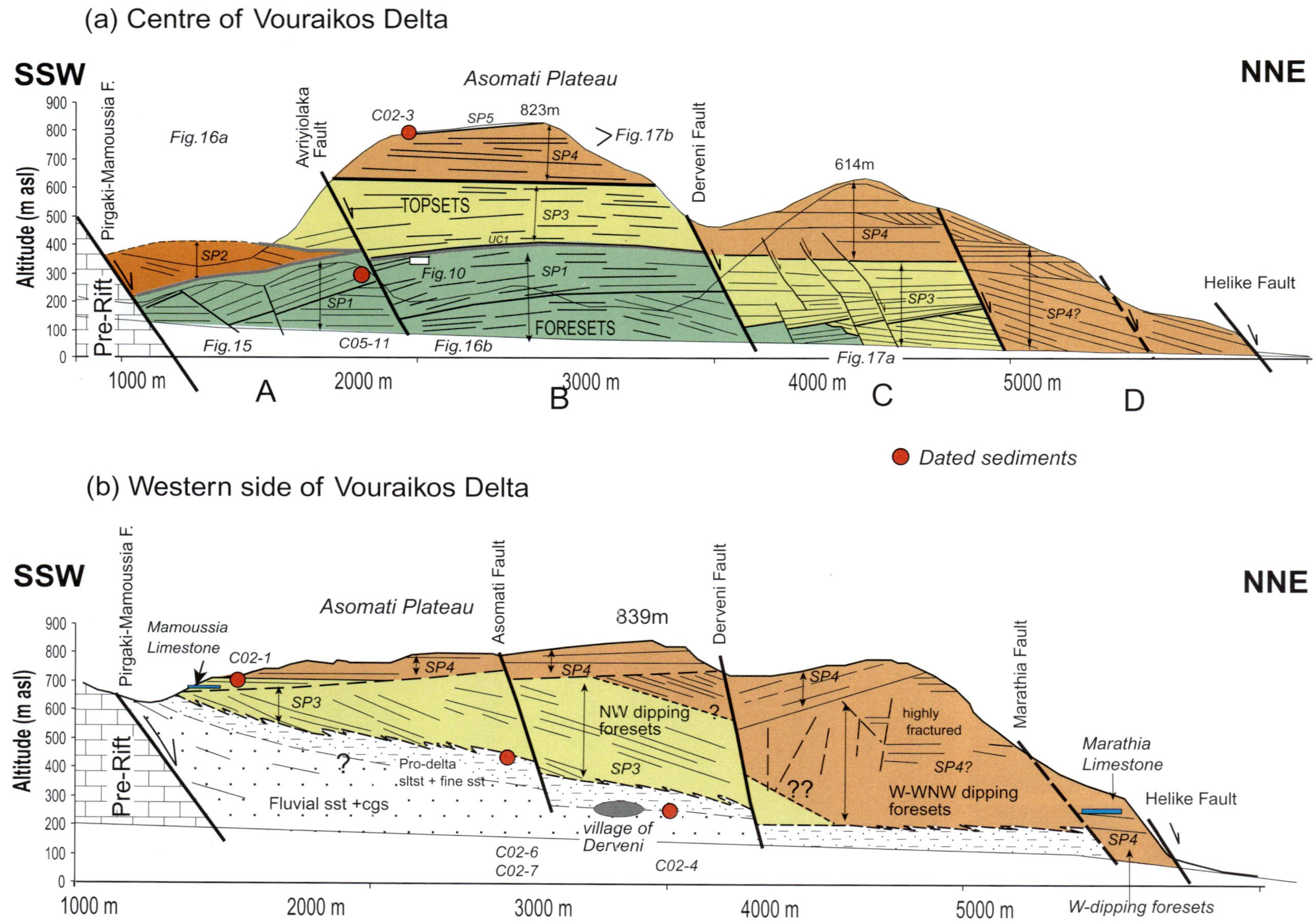

Fig. 14 Detailed SSW–NNE cross-section of (a) the western side of the Vouraikos Gorge, representing the centre of the delta and (b) east side of the Keranitis Valley, representing the western limit of the Vouraikos Delta. Circles indicate palynologically dated sample horizons (Table 1). AF is the Avriyiolaka Fault.

Centre of delta – the Vouraikos Gorge

The Vouraikos Gorge, with a relief of > 700 m, provides the most complete section through the centre of the Vouraikos Delta. For convenience, the section along the western side of the gorge is divided into four sectors demarcated by major faults and labelled A to D (Fig. 14a). The base of the delta is not exposed anywhere along this section.

The lowest stratigraphic package *SP1* is found only in the centre of the delta (sectors A, B and C in Fig. 14a). It comprises a single set of major foresets at least 200 m thick (base not seen) overlain by 200–250 m of topsets that together describe a gentle rollover anticline dissected by secondary north-dipping normal faults. At the southern extremity of the fold, in the immediate hangingwall of the PM Fault, topsets have been rotated to dip 30°S and foresets have become horizontal (Figs 14a & 15). Fold amplitude decreases upward indicating that fold activity died out during deposition of *SP3*. A simple 'chevron' construction (Verrall, 1981) suggests that the controlling listric fault soled out about 100 m below the present erosion level (not shown on Fig. 14a). At the southern end of the section, tilted foresets lie within 20 m of the PM Fault and there is no evidence of extreme basin-margin proximal facies in the southernmost *SP1* packages. This implies that the topsets equivalent to these foresets must have lain farther south. The PM Fault, hitherto regarded as the basin-bounding fault for the whole Vouraikos Delta, is therefore here interpreted as a post-*SP1* fault. The southern topsets of *SP1* were thus uplifted and eroded in the footwall of the PM Fault. On this section line, the listric fault that generated the rollover anticline

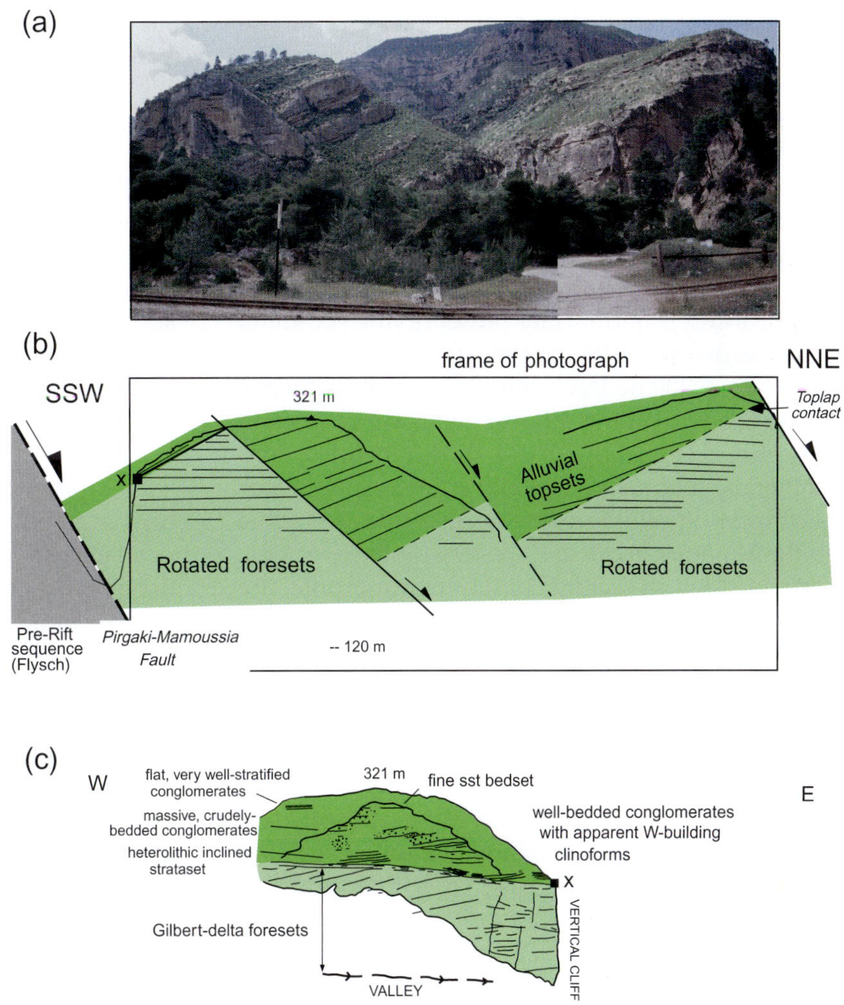

Fig. 15 Fault blocks in the Vouraikas Gorge. (a & b) View west of tilted fault blocks in *SP1* at the southern end of the gorge (located on Fig. 14a). (c) Line drawing of a cliff section at 90° to (a) viewed toward the north, showing foresets, toplap contact and overlying low-angle clinoforms. X is the common point to the two views.

was cut out by the later PM Fault, however, the Kastillia Fault identified in the eastern delta block is probably its lateral equivalent.

Across sector B, *SP1* foresets curve from horizontal to dip north, while *SP1* topsets become subhorizontal. These topsets are predominantly alluvial in character although a 12 m thick sequence of shallow-marine conglomeratic and sandy sediments occurs near the top (SFA1, Fig. 9). On the southern limb of the rollover anticline an angular unconformity of 12° marks the boundary between *SP1* and *SP3* (*UC1* on Figs 14a & 16). This unconformity disappears northward as the rollover anticline dies out. Toward the south *UC1* can be continued as the erosive boundary between *SP1* and *SP2*.

In the immediate hangingwall block of the Derveni Fault (Fig. 14a, panel C), the top of *SP1* is downthrown 300 m so that only the uppermost topsets are exposed. These south-tilted alluvial conglomerates are incised by an erosional scour with a (minimum) relief of 50 m (Fig. 17c, d) and oriented N–S. This surface is correlated with the *UC1* unconformity to the south and is overlain by a set of shallowly northwest dipping *SP3* foresets.

SP2 is an aerially restricted 200–220 m thick package of north-dipping foresets that lies between *SP1* and *SP3* in sector A (Figs 6, 7 & 14a, described below). It is found nowhere else in the delta. Northwest-dipping foresets of *SP3* abruptly overlie *SP2* foresets and locally preserve fine pro-delta facies at the *SP2*–*SP3* contact.

SP3 is a 200 m thick alluvial topset sequence showing a weakening rollover geometry up-section in sector B (Fig. 14a). These thickly-bedded conglomeratic topsets are laterally equivalent to major northwest-dipping foreset packages that are visible on the Mamoussia cliff and western profile (Fig. 14b).

Above *UC1* in sector C of the central profile, the *SP3* foresets (Fig. 14a) are cut by weakly curved faults, which have back-rotated the foresets to subhorizontal attitudes in places. The overlying SP3 alluvial conglomerate topsets thicken southward toward the Derveni Fault indicating that it was active during their deposition.

In sectors B and C (Fig. 14a), *SP3* is abruptly terminated by a sharp, apparently horizontal bedding surface that forms the base to a finer-grained, and well-bedded, sequence of conglomerates and sandstones defined as *SP4* that includes alluvial topsets and small Gilbert-type delta packages of 10–20 m thickness all building out to the northwest (Fig. 17a, b). *SP4* thickens northward across panel B from 170 m to 200 m (top exposed). The unit also thickens abruptly across the Derveni Fault to form the upper 300 m of panel C (top eroded), implying syn-*SP4* activity on this fault.

Panel D (Fig. 14a) is demarcated to the south by a weakly listric, poorly exposed, growth fault of unknown displacement. The panel is dominated by a set of large foresets dipping shallowly toward the northwest and having a minimum thickness of 400 m. Above, at least two smaller sets occur as well as at one horizon of conglomeratic topsets of unverified environment. As correlation across the fault is unclear all these strata are assigned to *SP3/4*. The Eastern Helike Fault abruptly terminates the delta to the north. At the top of the cliff in sector B (Fig. 14a) a prominent, thin (5–10 m) south-dipping conglomerate unit (*SP5*) gives a wedge-shaped aspect to *SP4*.

Western limit of delta – Keranitis Valley

This N–S cliff section (Fig. 14b, shown as a mirror image for ease of comparison) forms the eastern side of the Keranitis River valley. The profile affords a near complete section of the western limit of the Vouraikos Delta, comprising a relatively simple architecture of upper topsets, a continuous foreset succession and thin (20–50 m) underlying pro-delta sediments (Malartre *et al.*, 2004, fig. 2). The architecture of this section is markedly different from the central Vouraikos Gorge section, which lies just 3 km to the east. The basal (diachronous) 'enveloping surface' of the delta conglomerates dips markedly (8°) to the north; this is a (largely) primary dip. As the topsets dip gently south, the delta body thickens northward. Bottomsets and foresets build toward the west and west-northwest (i.e. obliquely out of the plane of the section; Fig. 6).

North-dipping sediments of the Ladopotamos Formation are exposed below the delta in the southern Keranitis Valley. Correlation of stratigraphic packages from the Vouraikos Gorge identifies the lowest stratigraphical level of the delta at the southern end of this section as *SP3* foresets and topsets (compare with Fig. 16a & b). It is estimated that somewhere to the north of Derveni Village *SP3* foresets pass up into *SP4* foresets, however, it is not possible to pinpoint this transition.

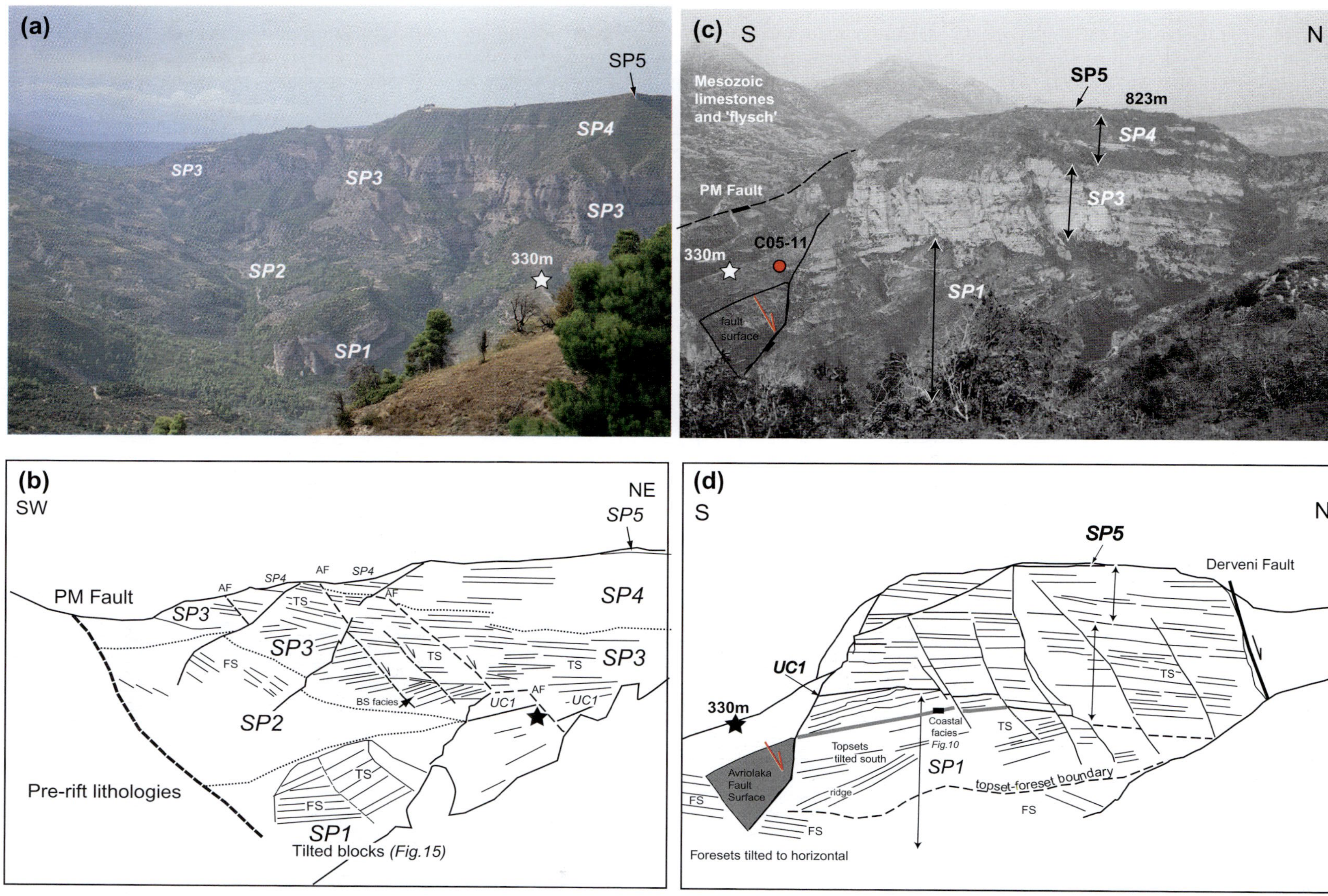

Fig. 16 Panoramas of the southern end of the Asomati Plateau and the Vouraikos Gorge. The star is a common point to the two views. (a) View westward of the southern margin of the Asomati Plateau. Relief on the section is 700 m. (b) Interpretation of (a) showing the organization of stratigraphic packages *SP1* to *SP4* in the immediate hangingwall of the Pirgaki-Mamoussia Fault. AF is the Avriyiolaka Fault. *UC1* is the unconformity between *SP1* and *SP3*. TS, topsets; BS, bottomset to pro-delta facies; FS, foresets. *SP3* foresets dip to the northwest. (c) Oblique view toward the west-southwest across the Vouraikos Gorge of the Asomati Plateau (sectors A and B of Fig. 14a). Relief in the gorge is over 700 m. (d) Interpretation of (c), showing stratigraphic packages, faults and the rollover anticline in *SP1*. The small secondary faults cutting *SP3* are not represented in Fig.14a. The heavy grey line in *SP1* topsets is the coastal facies level (Fig.10).

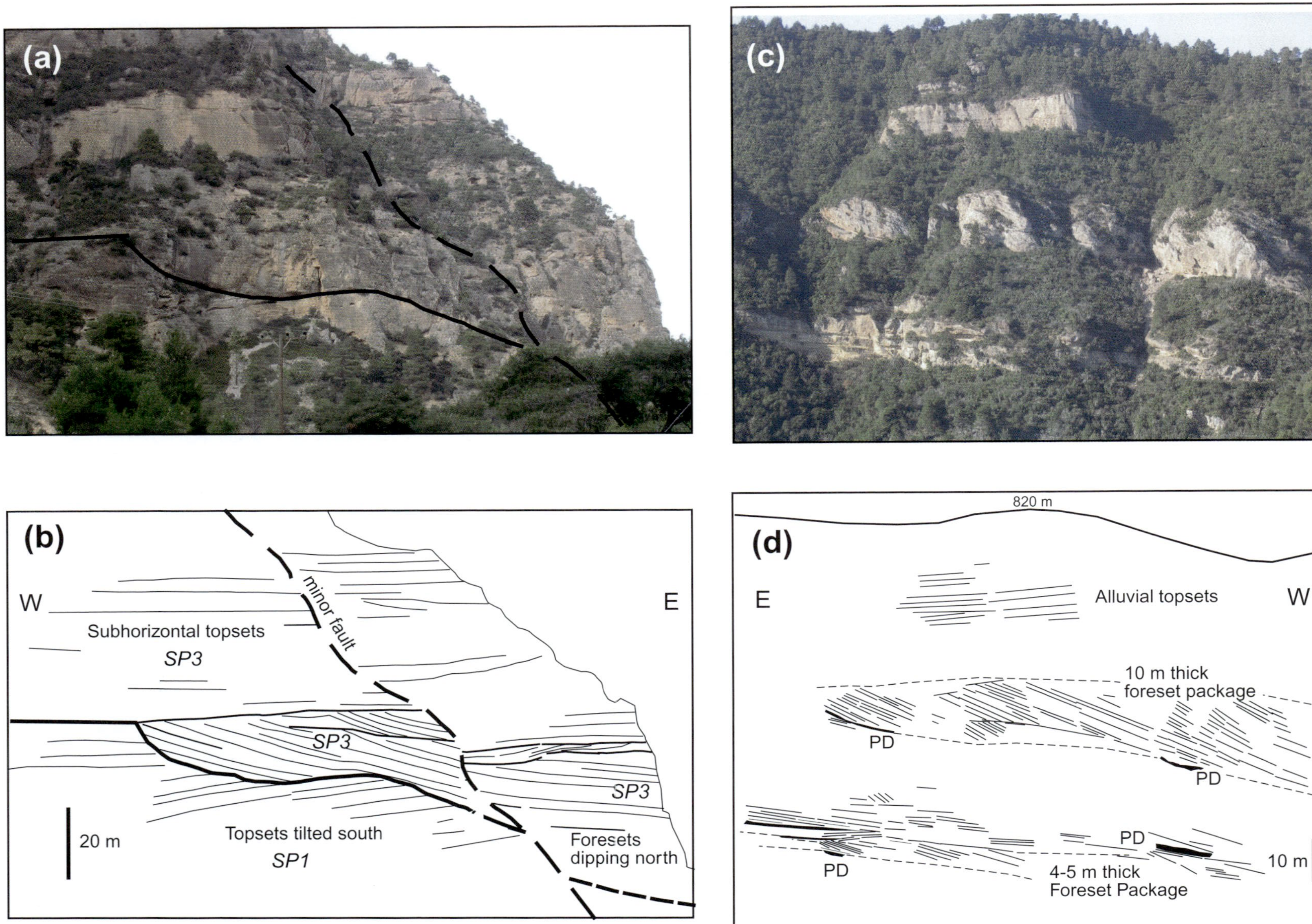

Fig. 17 (a) Photograph and (b) line drawing of major incision surface into *SP1* topsets in the central Vouraikos Gorge (panel C, Fig.14a). The incision is overlain by north-dipping foresets of *SP3*. (c) Photograph and (d) line drawing of a 110°-trending cliff in the immediate footwall of the Derveni Fault in the centre of the Asomati Plateau showing stacked Gilbert-type deltas of *SP4* building out to the northwest. PD are bottomset to pro-delta facies at the base of individual Gilbert-type deltas.

Thus the *SP3* and *SP4* topsets observed in the Vouraikos Gorge (Fig. 14a) have passed distally (toward the NW–WNW) to foresets on this profile. Thin *SP3* topsets are only observed at the southern end of the western section where they include the Mamoussia algal limestone facies (SFA3, Figs 6 & 11b). Horizontal topsets above the main succession of foresets are assigned to *SP4* because they contain small Gilbert-type deltas of 5–10 m height interspersed with alluvial topset sequences. This package can also be traced around to the central profile along the cliffs in the immediate footwall of the Derveni Fault (see Fig. 17a & b). Therefore, the major toplap surface on this section is the *SP3–SP4* boundary. *SP5* is not visible on this section line. The marked contrast in geometry and stratigraphy between this section and that in Fig. 14a is because this section represents the younger western fringe of the delta, which has overspilled the edge of the palaeovalley.

The minimum displacement on the PM Fault is considerably less on this profile than in the centre of the delta (Fig. 14a), implying that fault displacement decreases rapidly westward. The section is cut by three secondary extension faults, the Asomati, Derveni and Marathia faults. The Derveni Fault downthrows the toplap contact by 200 m to the north. Late displacement on the Derveni Fault has tilted topsets of the hangingwall block to 3°S. Between the Derveni Fault and the Helike Fault the delta conglomerates are highly fractured but comprise principally west-dipping *SP4* foresets of > 400 m height (see below).

The stratigraphic architecture of the western end of the Vouraikos Delta is quite distinct from that of the nearby eastern part of the Keranitis Delta (see Dart *et al.*, 1994, fig. 6). The bases of these deltas are separated by 400 m of altitude, while the tops are at the same level (detailed by Malartre *et al.*, 2004). This implies that they developed as independent delta systems separated by a transverse fault in the Keranitis Valley (Fig. 3).

Southwest proximal corner of delta – WSW–ENE Mamoussia section

This indented east–west cliff extends from the Vouraikos Gorge westward to just north of the village of Mamoussia (here referred to as the Mamoussia Pass, Figs 7 & 18b), a distance of approximately 2 km, and links the proximal parts of the two sections described above. An oblique view of this cliff is shown in Fig. 18a. The cliff forms a gross depositional strike section in terms of the Vouraikos Delta as a whole, but for several of the constituent stratigraphical units it forms a dip and oblique section with respect to foreset building directions. The Avriyiolaka Fault, striking E–W, obliquely cuts the indented cliff and downthrows to the north by 30–40 m, with displacement dying out rapidly to the west (fault not seen on western section, Fig. 14b). Four stratigraphical packages, *SP2* to *SP5*, can be traced between the central and western cross-sections. No cross faults (i.e. striking around N–S) are detected in this cliff, however, the base of the delta clearly rises from below 120 m in the Vouraikos Gorge westward to an altitude of 600 m at Mamoussia Pass (Fig. 18b). At the same time, the delta edifice thins from over 800 m to less than 200 m westward. These observations are interpreted to mean that the delta gradually infilled a palaeovalley of around 500 m depth as represented in Fig. 18b. A similar configuration is observed on the eastern side of the delta between the Vouraikos Gorge and the Ladopotamos Valley.

SP2 is the smallest delta package, being 1.7 km wide and around 200–220 m thick (Fig. 18). It is limited to the southwest sector of the delta and comprises principally conglomeratic foresets, although its lowest, most easterly exposures include bottomset facies. The true base to the set is not exposed but it must erosionally overlie *SP1*. In the lowest part, foresets and bottomsets build towards 041°, however, the package is dominated by foresets with a mean building direction toward 357°. Foreset and bottomset inclinations suggest that little or no rotation has occurred. *SP2* can be followed westward to within 1 km of the Mamoussia pass. In its most westerly exposure it is overlain by bottomset and pro-delta facies of *SP3*. This delta package must terminate westward because, at the same elevation 2 km further west in the Keranitis Valley, north-dipping sandstones and conglomerates of the Ladopotamos Formation crop out.

The *SP3* package can be traced across the entire length of the cliff (Fig. 18a). It is dominated by a major set (180 m thick) of foresets with a true northwest-building direction. This set and its erosional toplap contact are downthrown to the north by the extensional Avriyiolaka Fault. *SP3* foresets

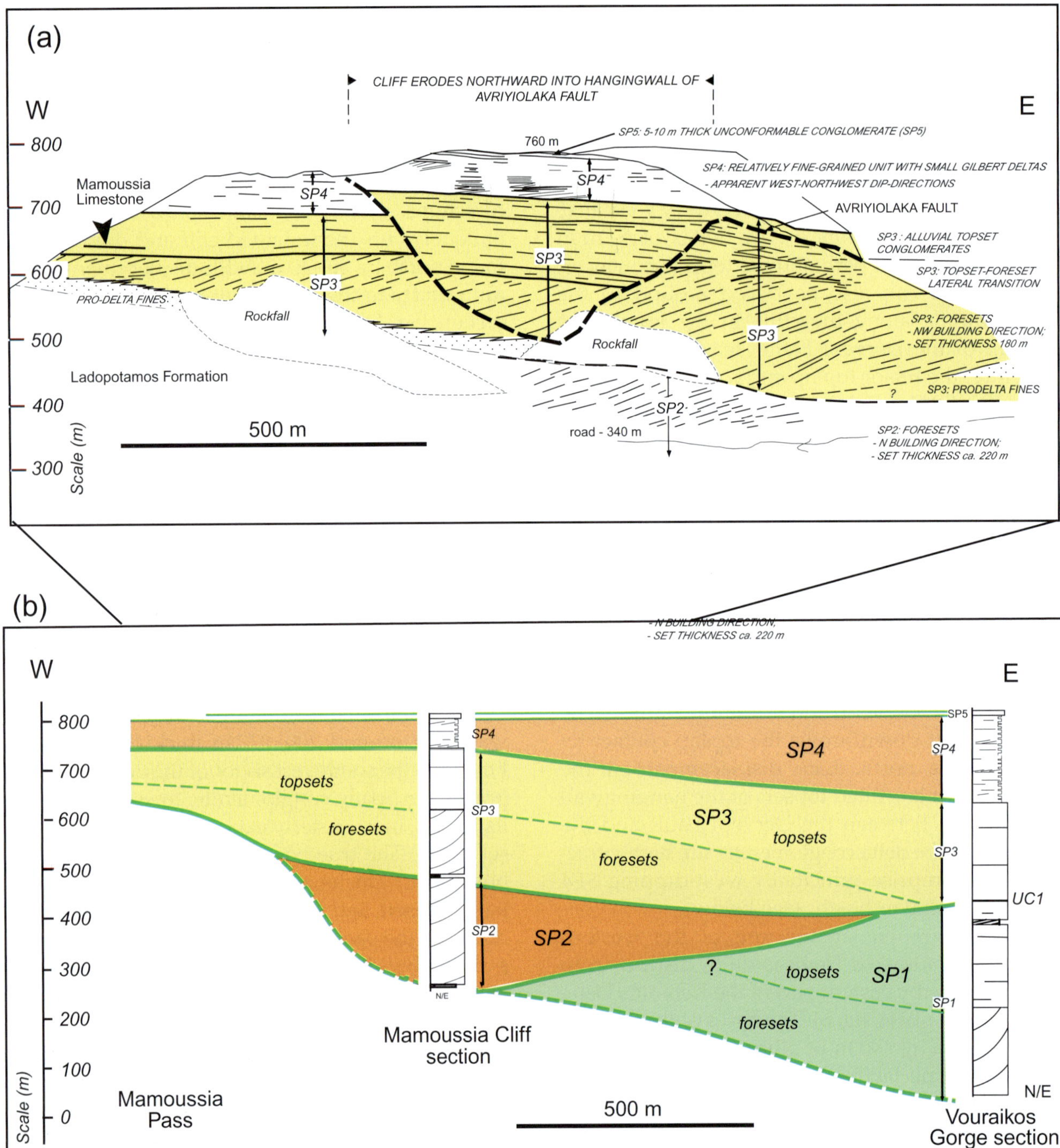

Fig. 18 (a) Field sketch of the east–west Mamoussia Cliff section representing the southwest side of the Vouraikos Delta viewed from Mamoussia village. (b) Correlation and geometry of stratigraphic packages within the proximal Vouraikos Delta from the Vouraikos Gorge (Fig. 14a) to the Keranitis Valley. The complex trace of the Avriyiolaka Fault in (a) is due to the indented cliff morphology.

pass eastward to flat-lying topset facies associations, which correlate with *SP3* alluvial topsets in the Vouraikos Gorge section (Figs 14b & 18b).

In the footwall of the Avriyiolaka Fault, fine-grained bottomset to pro-delta facies are exposed at around 500 m, marking the base of the *SP3* delta. This contact gradually rises westward in elevation to 600 m at Mamoussia Pass. *SP3* correspondingly thins rapidly to the west, where it comprises markedly curved-asymptotic foresets some 4–50 m high, passing into bottomsets and pro-delta beds (Fig. 18a).

The upper part of *SP3* (Fig. 18a) is a flat-lying sequence of horizontally stratified conglomerates of alluvial aspect. The facies are best seen in the mid-central part of the main cliff (hangingwall of the Avriyiolaka Fault), where they sharply truncate the underlying foresets. The package can be divided into three units by prominent sharp conformable bedding surfaces (Fig. 18a). The upper surface to the whole package is a very sharp, planar trace.

The uppermost major package in the profile, *SP4*, is finer grained than those units below, and contains facies showing well developed fine-scale stratification, heterolithic character, with (finer-grained) conglomerates, pebbly sandstone and sandstones. Overall it is horizontally stratified, but contains several levels comprising large-scale cross-stratification with consistent apparent inclinations to the west, which are interpreted as small-scale Gilbert-type deltas. The approximate thickness of these sets is 5–20 m. The gross horizontal stratification is conformable with the underlying *SP3* topsets. The unit is terminated by a notably continuous conglomerate bed at the top of the cliff (*SP5*), which dips to the south. *SP4* is 60–70 m thick in the Mamoussia cliff (Fig. 18a), indicating that it thins westward from 170 m at the southern end of the Vouraikos Gorge (Figs 14a & 18b).

SP5 is a 8–10 m thick grossly flat-bedded unit, comprising a variable sequence of facies that include matrix-rich, poorly sorted pebble- to cobble-grade massive conglomerates and matrix-poor, moderately sorted inclined- and flat-stratified conglomerates. Cross-bedded conglomerates occur, with sets up to 2 m thick and moderate to low-angle planar foresets. Finer-grained facies include interbedded reddish mudstones (8–10 cm bed thickness) and 12–15 cm thick bioturbated fine sandstones (with bedding parallel burrows). Highly indurated very coarse sandstone–granule facies contain sparry calcite cements and ostracod and algal bioclasts.

Northern exposures of the delta: east and west frontal profiles

The Vouraikos Delta has been cut and exhumed in the footwall of the East Helike Fault to form a range front 7 km long and 700–800 m in height. At the northwest corner of the Asomati Plateau the youngest delta foresets belonging to *SP4* are well exposed in an east–west cliff in the footwall of the secondary Marathia Fault (Fig. 19a). These frontal foresets are at least 350 m high (being cut by the Marathia Fault) and dip predominantly 23–30° to the west-northwest and west. The overlying fluvial topsets at the northwest tip of the plateau dip 10–20°S–SW (Fig. 14b). The topset–foreset transition shows that the delta front prograded toward the west. Small Gilbert delta packages are seen to build out toward the west within the *SP4* topset complex.

Similarly, the youngest northeastern frontal part of the delta is visible on the E–W range front of the Kastillia Plateau (Fig. 19b). Here, two major packages of foresets are visible; a lower north-building package below Faghia, some 250 m high; and a larger upper package of consistently northeast-dipping foresets that are at least 600 m in height. This vast (unfaulted) foreset package forms the whole northeast and eastern side of the Vouraikos Delta. These foresets are probably the equivalent of *SP4*. Thin topsets of both packages dip shallowly south on the top of the Kastillia Plateau.

At the mouth of the Vouraikos Gorge on the western side of this section, a thick sequence of south-dipping fluvial sediments occurs in front of and below the delta foresets. It is not yet clear if these strata belong to the Ladopotamos Formation and thus truly underlie the delta, or if they are younger deposits deposited along the range front during late exhumation of the delta. These deposits are incised into and overlain by the youngest Gilbert-type deltas, the tops of which form depositional terraces dipping gently toward the northeast (Fig. 19b). These young deltas have themselves been uplifted

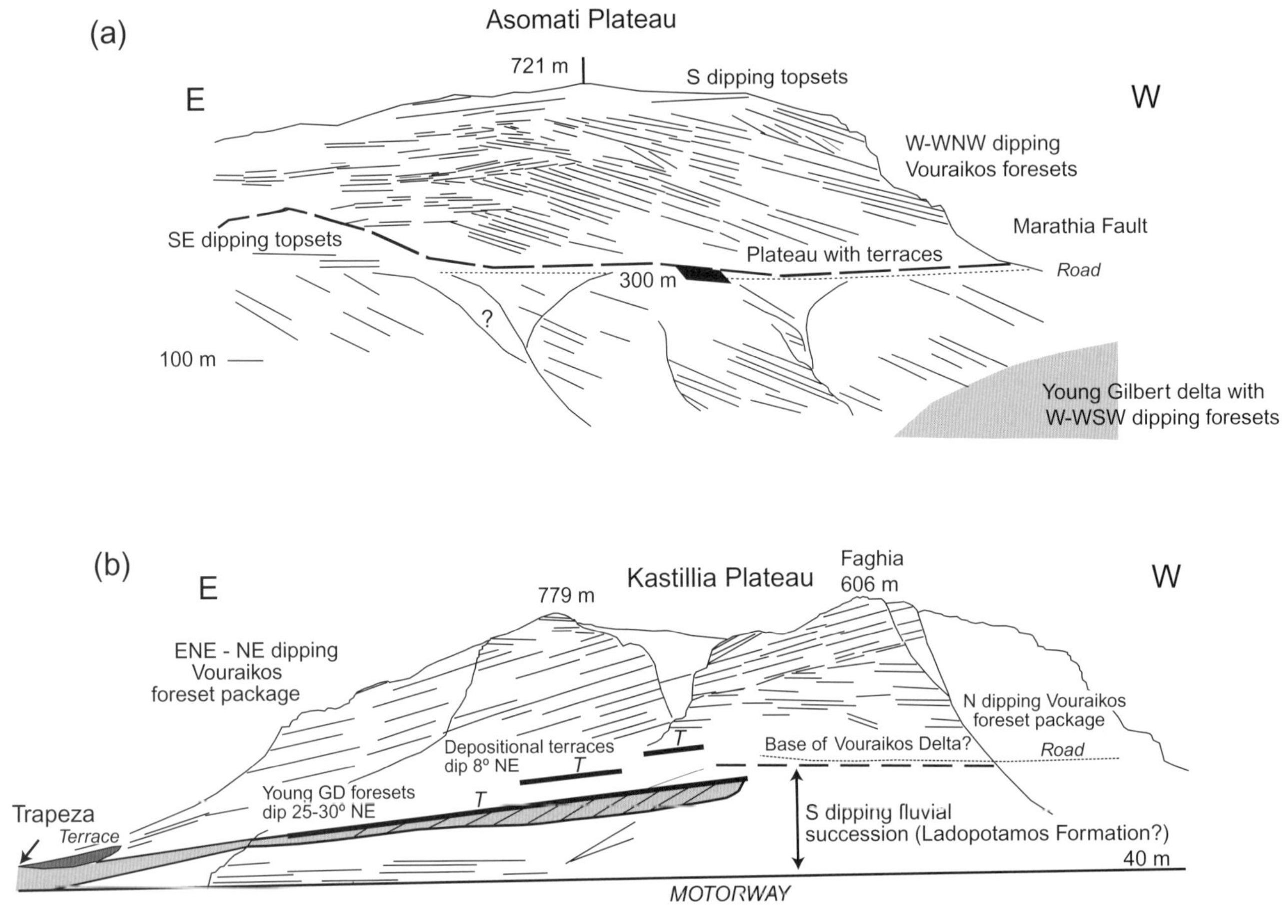

Fig. 19 Line drawing of the (a) northwest range front and (b) northeast range front of the Vouraikos Delta in the immediate footwall of the Eastern Helike Fault. Younger Gilbert deltas, deposited on the range front during its exhumation, are shown in light grey. The largest of these (100 m high) lies in the hangingwall of the Marathia Fault on the northwest range front and contains west-southwest-dipping foresets. Depositional terraces on the northeast range front dip 8°NE and are shown in heavy black lines with the letter *T* or in dark grey when dipping north.

in the footwall of the Eastern Helike Fault. Their foresets are up to 30 m high and dip predominantly to the northeast (Fig. 19b).

EVOLUTION OF THE VOURAIKOS DELTA

The data presented above are used to reconstruct the depositional history and character of the Vouraikos Delta and to identify the factors that controlled its evolution. Accommodation space was created principally by the PM normal fault system with displacement distributed on different branches at each stage of basin history (Fig. 20).

Delta deposition (800 m minimum) is estimated to have occurred during the early to mid-Pleistocene from before 1.1 Ma to after 700 ka (600–400 kyr), implying a high sedimentation rate of between 1.3 and 2 mm yr^{-1}. The delta built northward in a radial fan fault-controlled basin. The carbonate facies (in *SP3* and *SP4*) and the isotope study (Katafugion Formation) described above indicate that this basin was wholly or periodically marine. High-resolution studies on Upper Pleistocene deposits in the Gulf of Corinth indicate that the salinity of the basin fluctuated between marine and fresh water, controlled by eustatic sea-level variations (Perissoratis *et al.*, 2000; Kershaw & Guo, 2003). In addition, biostratigraphic studies on Lower and Middle Pleistocene sediments record brackish, lacustrine and marine fauna (Frydas, 1989, 1991; Fernandez-Gonzalez *et al.*, 1994).

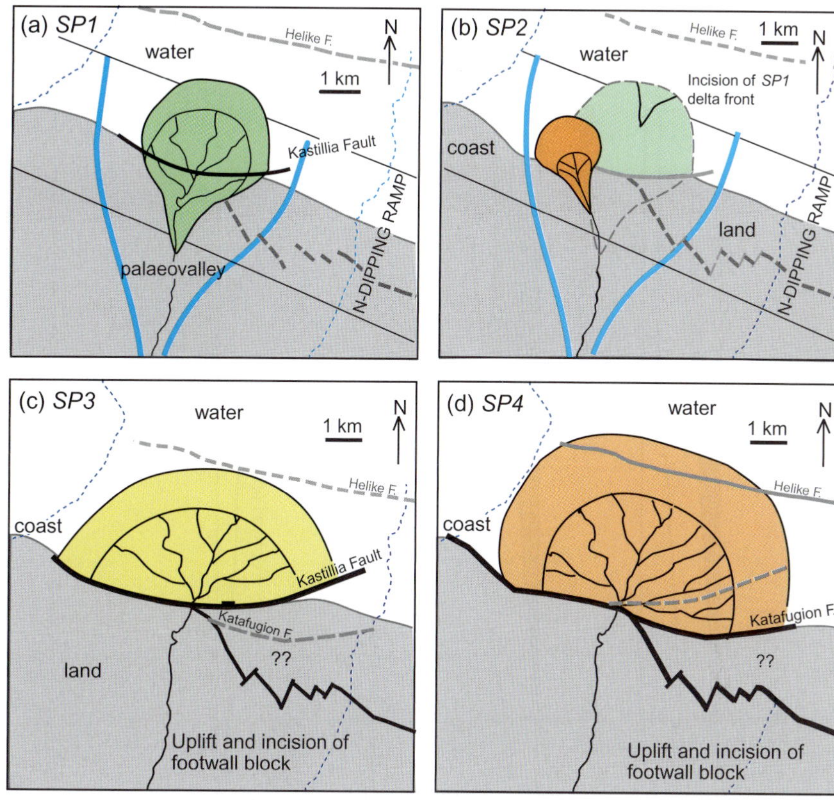

Fig. 20 Map view models for the four main stages in the evolution of the Vouraikos Delta corresponding to *SP1* to *SP4*.

Reconstruction of the western half of the delta is divided into five stages, equivalent to the five stratigraphic packages *SP1–SP5* described above. These are represented in scaled maps and in longitudinal and cross-sections (Figs 20 & 21). The cross-sections represent only the western half of the delta where the PM Fault consistently formed the basin bounding fault. It is not currently possible to define the duration of each of these stages due to lack of precise dating.

Early rifting (lower group and unconformity at base of upper group)

During the early phase of rifting (pre-1.1 Ma) the fluvial and alluvial successions of the lower group (Kalavrita conglomerates and the Ladopotamos Formation) were deposited in a series of half graben, controlled mainly by north-dipping faults spaced at between 4 and 5 km and with displacements of up to 1.5 km (Ghisetti & Vezzani, 2004, 2005; Bourlange *et al.*, 2005). The PM Fault was not active at this stage. Preliminary palaeocurrent data indicate that the main source areas lay to the south and west. Towards the end of this period a base-level rise is recorded by deposition of the Katafugion Formation.

In the early Pleistocene, a major change occurred in the tectonic and depositional dynamics of the Corinth region. The main depocentre shifted northward and became narrower and the southern area (Kalavrita to Mamoussia) became uplifted. Sediment supply increased as major rivers began to transport large volumes of coarse sediment from the southern area to newly established Gilbert-type deltas. The establishment of new Gilbert-type delta systems requires high sediment supply and the creation of significant accommodation space below base level, requiring the activity on new normal faults. The northward dip of the Ladopotamos Formation indicates that a tectonic tilting took place before the major normal fault broke surface. This tilting is interpreted as being due to forced folding above the upward propagating PM Fault. The early delta (*SP1*) was therefore deposited above an active northward-tilting ramp, in a manner similar to that described by Young *et al.* (2000) in the Gulf of Suez and

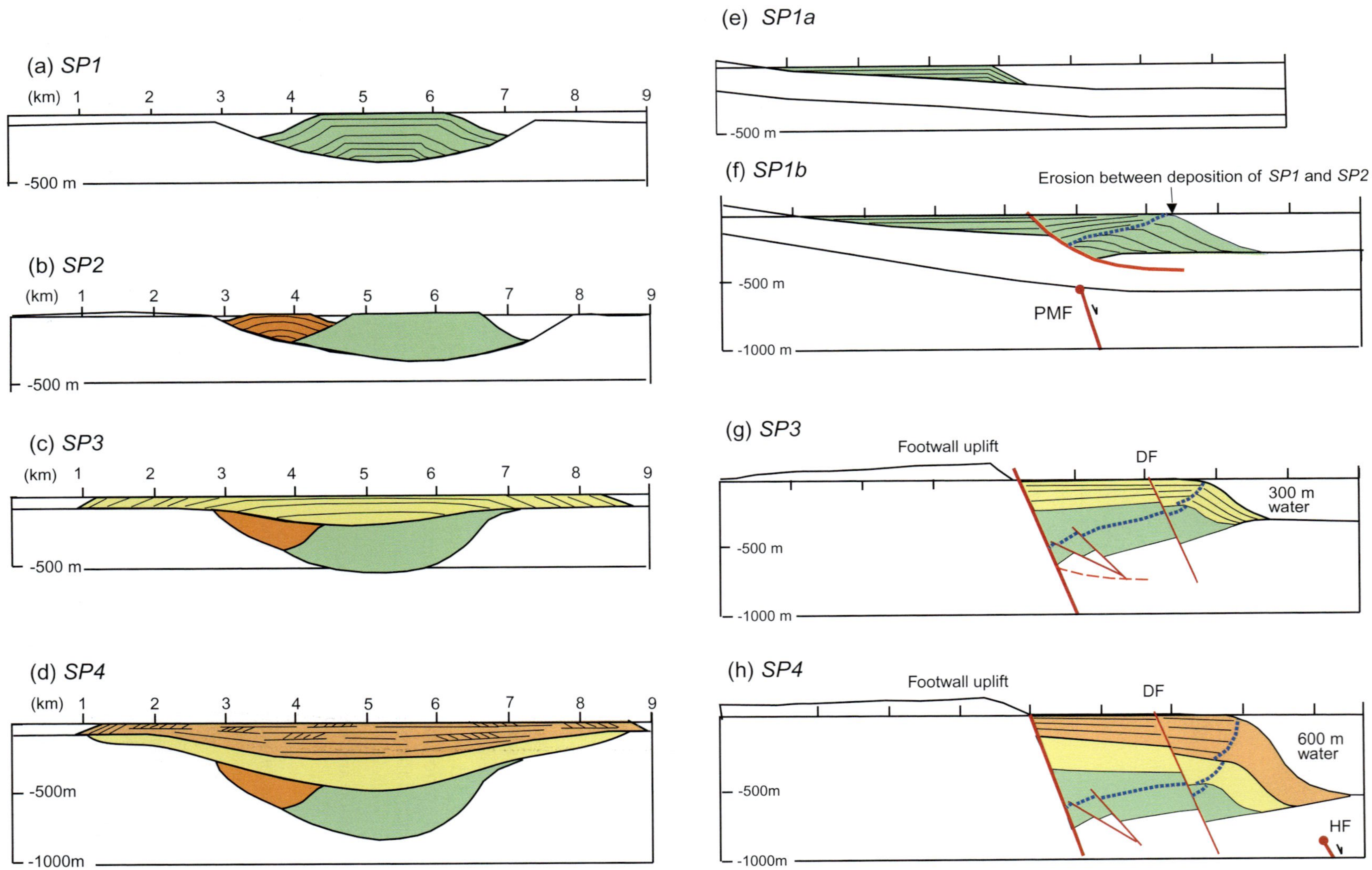

Fig. 21 Proximal longitudinal (a–d) and cross-sections through the centre of the Vouraikos Delta (e–h) representing the four stages in delta evolution corresponding to *SP1* to *SP4*, using a vertical exaggeration of two. *SP1* is represented in two cross-sections, while there is no cross-section for *SP2* as this small delta is hardly seen on the central cross-section (Fig. 14a). The offlap break is shown as a blue dashed line. The upward propagating Pirgaki-Mamoussia Fault (PMF) is shown in (f), the upward propagating Helike Fault (HF) in (h) and the Derveni Fault (DF) in (g) and (h).

as shown in the numerical models of Gawthorpe & Hardy (2002) and Ritchie *et al.* (2004a, b). Moreover, the concave erosive base of the Vouraikos Delta requires that the early delta (*SP1*) infilled a pre-existing palaeovalley of some 300 m relief. As this feature requires a major erosional (incision) event before initiation of delta deposition, a relative sea-level fall is inferred before the major relative sea-level rise.

The dramatic change in basin development between the lower and upper groups occurred sometime in the middle of the Early Pleistocene (before 1.1 Ma). The well-established change in Quaternary climate regime occurred between 0.9 and 0.6 Ma (Williams *et al.*, 1988), that is during deposition of the Vouraikos Delta. Therefore, tectonic forces must have been principally responsible for this change in basin regime.

Stage 1 of delta deposition (*SP1*)

The oldest delta package (*SP1*) was deposited in a palaeovalley incised into a gently north-dipping ramp (Figs 20a, 21a, e & f). The *SP1* delta had an estimated radius of less than 2 km. Foresets, up to 200 m high, prograded across the ramp. These foresets are overlain by alluvial topsets at an 'accretionary' toplap contact, suggesting regression and aggradation. Considerable aggradation then took place, until thin coastal facies (Fig. 9) record marine transgression across the top of the delta. Transgressive sediments over 12 m thick were deposited until terminated erosively by a return to alluvial facies *SP1* topsets. The topsets thicken southward to over 200 m across a syn-sedimentary rollover anticline generated above a listric fault that soled into a shallow décollement (Fig. 21f). This fault seems to have cut through an already well-established delta and was perhaps generated by gravitational instability on the basinward dipping ramp. Significant progradation and aggradation occurred implying rapid creation of accommodation space.

Stage 2 of delta deposition (*SP2*)

The top of *SP1* is marked by the erosional unconformity *UC1* implying a relative fall in sea level, which we correlate with the incision into the front of the *SP1* delta (Figs 14a & 17c, d). The following *SP2* package of foresets (no topsets preserved), over 200 m high, unconformably overlies the most southerly topsets of *SP1* in the southwest corner of the delta. In the map reconstruction these foresets represent the frontal part of a small northward-building delta of radius 1 km (Fig. 20c), most of which is now eroded. The delta front therefore stepped southward at the beginning of stage 2 requiring a significant relative rise in sea level. It is suggested that the *SP2* delta infilled the remaining bathymetry of the palaeovalley on the west side of the *SP2* delta (Fig. 20b). It is possible that the same phenomenon occurred in the east of the delta. While the *SP1* topsets are tilted, the *SP2* foresets do not appear to be significantly tilted, implying that the Kastillia Fault and its rollover anticline were not active during deposition of *SP2*.

Stage 3 of delta deposition (SP3)

A significant unconformity separates *SP2* from *SP3*, and pro-delta and bottomset facies associations of *SP3* are seen directly above *SP2* foresets south of the Avriyiolaka Fault (Fig. 18). North of the Avriyiolaka Fault, *SP3* was deposited directly on *SP1* topsets. *SP3* topsets in the centre of the delta (Fig. 14a) pass westward into SP3 foresets that record progradation (foresets reach heights of over 300 m) toward the NW and WNW during a relative sea-level (RSL) highstand (Figs 18 & 14b). The succeeding *SP3* topsets on the Mamoussia cliff (Fig. 18) indicate aggradation following erosional planation of the foresets probably accompanied by regression. The topsets are dominantly alluvial, although the Mamoussia Limestone may indicate a marine incursion across the (distal) delta top. The delta rapidly grew in E–W width to over 7 km and it significantly overspilled the palaeovalley (Fig. 21c). Its N–S extent, however, remained limited at just under 4 km. The significant increase in accommodation space at the *SP2* to *SP3* boundary may be explained by the emergence of the controlling normal fault. To the east of the Vouraikos Gorge displacement was distributed on two fault strands, principally on the Kastillia Fault to the north and probably on the PM Fault to the south (Fig. 20b). The Kastillia Fault formed the major bounding fault to the eastern half of the delta during Stage 3.

Stage 4 of delta deposition (*SP4*)

The *SP4* sequence is conformable on *SP3*. It is however markedly different in character, with a finer average grain size, thinner bedding and small-scale Gilbert-type deltas interspersed with alluvial topsets (Fig. 17c & d). The change occurs across a key planar surface traceable over most of the western part of the delta (Figs 14, 16 & 18), which is interpreted as a major transgressive flooding surface across the previously subaerial delta. The small delta packages record regular high-frequency relative sea-level variations right across the delta top, implying that it was regularly flooded, in marked contrast to the earlier alluvial-dominated topsets. These small delta-top foresets built out until they spilled over the delta front into the large frontal foresets. These frontal foresets can be over 600 m high, indicating a very deep basin (Fig. 19). The N–S extent of the *SP4* delta is estimated to have been at least 4.5 km (but cut by the Eastern Helike Fault), while its E–W width was over 8 km (Fig. 20d). The topsets are at least 300 m thick in the hangingwall of the Derveni Fault and are 200 m thick in its footwall, implying that this fault was active during deposition of *SP4* (Fig. 14a). At the southeast side of the delta, the Kastillia Fault was sealed by 100 m high *SP4* deltas that built north and northeast across its footwall from the Katafugion Fault (Figs 3 & 20d). It is possible that secondary point sources were active in the eastern part of the delta during this stage (Figs 20d & 21d, h),

Stage 5 of delta deposition (*SP5*)

Before the deposition of *SP5*, the Vouraikos Delta was effectively terminated during an episode when it was tilted north by 5–7° and eroded, due probably to a fault-related mechanism. This is exemplified by the *SP4* sequence having a wedge-shape, thinning southward below *SP5* (Figs 14a & 16). *SP5* is itself tilted gently south, compatible with later extensional fault block uplift and rotation. *SP5* comprises a distinctive marine-influenced (shallow-marine), dominantly conglomeratic sequence. This conglomerate is the last deposit of the Vouraikos Delta, and its planar sheet-like form represents the final approximate position of sea level prior to uplift. Following delta uplift (see below) thick red soils developed above *SP5*, which now cover the Asomati and other plateaux.

Uplift of the Vouraikos Delta

Sometime in the Middle Pleistocene the Vouraikos Delta began to be exhumed in the footwall of the newly initiated Eastern Helike Fault (EHF). During exhumation the delta was cut and tilted by a number of secondary normal faults. The EHF is still active today and its displacement history continues to be intensively studied (Koukouvelas *et al.*, 2001, 2005; Leeder *et al.*, 2003; De Martini *et al.*, 2004; McNeill & Collier, 2004; Pavlides *et al.*, 2004; McNeill *et al.*, 2005). The range front preserves a series of erosional and depositional marine terraces (see Fig. 19), which have been used to model footwall uplift rates (assuming a constant uplift rate) giving estimates of between 1 and 1.5 mm yr^{-1} (e.g. De Martini *et al.*, 2004 and references therein; McNeill & Collier, 2004). Assuming that the present-day plateau top (at around 800 m) is close to the original delta top, these rates would imply that uplift of the delta (and thus activity on the EHF) began some time between 530 and 800 ka. The biostratigraphic dates presented in this paper, bracketing the age of the Vouraikos Delta from before 1.1 Ma to sometime after 700 ka, are largely compatible with this exhumation history.

DISCUSSION

The symmetry of the gross building directions of its foresets suggests that the Vouraikos Delta had a fixed-point sediment supply from the footwall throughout its history (Type A feeder system of Postma, 1990, 1995). The approximate location of the input point, coincident with the present-day Vouraikos River, coincides with the intersection of the PM, the Katafugion and Kastillia faults, suggesting some structural control.

Although the Vouraikos was a footwall-derived delta, it has a preserved proximal profile more akin to a hangingwall delta (see e.g. Ritchie *et al.*, 2004a,b), probably due to a ramp-related steepening. This steepening during *SP1*, above the propagating footwall fault, may have achieved rapid deepening to give the high initial bathymetry modelled as being essential for the development of deltas of this architectural style (Ulicny *et al.*, 2002).

The curved structure of the foresets, and the development of prominent bottomset and pro-delta facies associations, does not support the view of Zelilidis & Kontopoulos (1996) that the Vouraikos had a simple trapezoidal dip profile, nor that it built into a laterally restricted basin. The most distal facies of the delta do not suggest this, and it is more likely that downthrow on the Eastern Helike Fault disguises a considerable part of the distal delta. Early aggradational foresets are not notably preserved, except for one case. However, frequent transitions to interstratified thick sequences of topsets record significant regressive events and aggradational episodes. Transgressions of the sub-aerial delta also occurred, with limestones indicating that the delta built into a marine basin, and associated shoreline gravels indicating significant basinal wave energy. The marine carbonates indicate that the Corinth rift basin would have been affected by Pleistocene orbitally controlled glacio-eustatic sea-level cycles.

The overall form of the Vouraikos contrasts strongly with that of the flanking Keranitis Delta, in that the latter comprises a major proximal reach composed entirely of topset facies, with relatively limited foreset progradation (e.g. Ori *et al.*, 1991, fig. 9). The Vouraikos and Keranitis are regarded as separate delta systems (cf. Ori *et al.*, 1991) of similar age that are separated by a cross-fault in the Keranitis Valley. Correlation with events recorded in the foresets of the Keranitis Delta is not straightforward; the multiple depositional sequences defined by Dart *et al.* (1994) are not characteristic of the Vouraikos. However, the large relief incision surface above Sequence 2 of Dart *et al.* (1994, fig. 6) is of similar scale to the surface identified here at the base of *SP3* (*UC1*). Possible linkage of major surfaces would suggest basin-wide changes in RSL. Sedimentation rates derived for the Keranitis Delta (1.5 mm yr^{-1}) by Dart *et al.* (1994) are similar to our estimates based on consideration of the dating and sediment thickness. Finally, an interesting contrast between the deltas occurs in the degree to which syn- and post-depositional extensional faulting affected their development. The proximal rollover anticline affecting the Vouraikos, and the suite of planar and listric syn-sedimentary faults observed, have no apparent counterparts in the Keranitis. The Keranitis also lacks the post-delta planar normal faults that disrupt the Vouraikos at a number of scales. This is probably because the Vouraikos Delta lies in the immediate footwall of the Eastern Helike Fault while the Keranitis lies farther south.

Based on the biostratigraphic dating, it is estimated that the delta was deposited within a period of ca. 0.4 to 0.6 Myr between > 1.1 Ma and 0.6–0.7 Ma. This age estimate implies that the whole Vouraikos Delta represents a third-order highstand systems tract (*sensu* Vail *et al.*, 1991). The five stratigraphic packages *SP1* to *SP5* therefore represent mainly fourth-order highstand system tracts. Each sedimentary cycle essentially comprises the regressive phase (progradation), with the vertical succession from pro-delta fines to bottomsets to foresets and finally to topsets (i.e. 'normal' regression *sensu* Posamentier *et al.*, 1992).

The development of each stratigraphic package is related to an interglacial period, which is consistent with the interglacial character of the palynological assemblages. If preserved, the lowstand deltas related to glacial periods may be situated to the north, below the present gulf. Each stratigraphic package comprises stacked fifth-order transgressive–regressive cycles, which are rarely detected in *SP1* to *SP3*. However, in *SP4* these fifth-order cycles are clearly recorded by the stacked small Gilbert deltas on the delta top.

Within the time period 1.1 to 0.6 Ma the oxygen isotope ^{18}O stages are MIS31 to MIS16. Potentially four or five major negative excursions may be recognized that could be correlated with the four stratigraphic packages *SP1* to *SP4*. This suggests that the stratigraphic packages were primarily controlled by eustasy superimposed on a high subsidence rate, in turn controlled by the PM Fault.

The bulk of this compact delta is made up of *SP1*, *SP3* and *SP4*. While *SP1* records significant progradation (> 2 km) coupled with strong aggradation (200 m) across the early ramp (Fig. 21e & f), *SP3* and *SP4* each record limited frontal progradation (< 1 km) of thick foresets coupled with strong aggradation (200–300 m) of topsets. This implies that rates of both sediment supply (S) and creation of accommodation space (A) were very high during deposition of *SP3* and *SP4* and that the ratio of S/A was about 1. The cyclic flooding of the delta top during *SP4* implies that S/A had decreased either because sediment supply was waning or because creation of accommodation space was increasing.

CONCLUSIONS

1 The syn-rift stratigraphy in the Kalavrita to Aegion region of the southern Corinth rift shows a two-phase rifting history. The coarse alluvial succession of the lower group (up to 1.3 km thick) was deposited in a series of 4–5 km wide tilted blocks. This early rift phase ended with a marine transgression, preserved in the north. An erosional unconformity marks the base of the upper group, which records a great increase in accommodation space, the migration of the depocentre to the north and an increase in sediment supply. The upper group is characterized by Gilbert-type fan deltas.

2 The Vouraikos Delta is one of several giant fault-controlled Gilbert-type fan deltas that built into the Corinth rift during the Early to Middle Pleistocene, in response to a major change in basin dynamics. It is argued that this change in basin dynamics in the Early Pleistocene was not related to climate change but was probably due to a change in large-scale regional tectonics (Fig. 1, inset).

3 A limited number of palynological dates indicate that the Vouraikos Delta was initiated sometime in the middle of the Early Pleistocene and terminated in the Middle Pleistocene, sometime after 0.7 Ma. These preliminary age estimates are consistent with published models of the uplift history on the Eastern Helike Fault. Sedimentation rates are thus estimated to have been between 1.3 and 2 mm yr^{-1}.

4 The early Vouraikos Delta (*SP1*) was constructed on a basin-dipping ramp generated by an extensional forced-fold. Internally, it was affected by a listric normal fault and its rollover anticline. Later, the Vouraikos Delta was primarily controlled by displacement on the emergent Pirgaki-Mamoussia Fault and its splays to the east (the Kastillia and Katafugion faults). Smaller planar and curved normal faults affected the delta throughout its history, and also during exhumation of the delta in the footwall of the Eastern Helike Fault.

5 The internal architecture of the conglomeratic delta records both tectonic and eustatic controls. Five stratigraphic packages (*SP1–SP5*) are separated by major surfaces. *SP1*, *SP3* and *SP4*, which make up the bulk of the delta, are each characterized by thick topsets and thick foresets, and limited bottomset and pro-delta facies. The stratigraphic packages are tentatively correlated with regressive glacio-eustatic interglacial periods. This model requires that the glacio-eustatic signal was superimposed on a relatively constant creation of accommodation space by normal faulting (by the Pirgaki-Mamoussia Fault system). While this eustatic interpretation seems quite plausible, it is not possible to eliminate the possibility that the stratigraphic packages and their bounding surfaces may have been generated wholly or partly by pulses of high and low slip-rate on the Pirgaki-Mamoussia Fault.

6 Topset limestones associated with coastal conglomerate facies indicate that the Vouraikos Gilbert-type Delta built mainly into a marine water body. Gravel-rich sediment prograded (to the north-northwest) into water that reached depths of 200–600 m.

7 The N–S radius of the 800 m thick fan delta increased only slowly (2–3.5 to 4.5 km) through time. The trajectory of the offlap break in a section through the centre of the Vouraikos Delta reflects early progradation-dominated behaviour (*SP1*), followed by increasingly aggradational behaviour during *SP3* and *SP4* deposition.

ACKNOWLEDGEMENTS

Thanks are due to F. Palhol (CRPG, Nancy) who kindly carried out stable isotope analyses. S.-M. Popescu carried out the palynological analyses. Thanks to D. Ulicny, J. ten Veen and G.J. Nichols for valuable comments on the first version of this paper. Fieldwork and dating for this project were funded by the 'Groupement de Recherche Corinthe' and by the project 'Dynamique de la Terre Interne' (DyTI) of the INSU (CNRS). MF and FM thank the following colleagues for fruitful and stimulating discussions: Nicolas Backert, Sylvain Bourlange, Rémi Eschard, Francois Guillocheau, David Jousselin, Christian Le Carlier de Veslud, Isabelle Moretti, Sébastien Rohais, and finally the students of the Nancy School of Geology (ENSG). CRPG Publication Number 1824.

REFERENCES

Anthony, E.J., Lang, J. and Oyédé, L.M. (1996) Sedimentation in a tropical, microtidal, wave-dominated coastal-plain estuary. *Sedimentology*, **43**, 665–675.

Bourlange, S., Jousselin, D., Ford, M. and Le Carlier, C. (2005) Evolution of a normal fault system on the southern flank of the Corinth Rift. *European Geophysical Union Congress, Vienna. Geophys. Res. Abstr.*, **7**, 07450, Sref-ID 107-7962/gra/EGU 05-A-07450.

Bridge, J.S. (1993) Description and interpretation of fluvial deposits: a critical perspective. *Sedimentology*, **40**, 801–810.

Bridge, J.S. (2003) *Rivers and Floodplains: Forms, Processes, and Sedimentary Record*. Blackwell Publishing, Oxford, 491 pp.

Briole, P., Rigo, A., Lyon-Caen, H., *et al.* (2000) Active deformation of the Corinth rift, Greece: Results from repeated Global Positioning System surveys between 1990 and 1995. *J. Geophys. Res.*, **105**, 25,605–25,625.

Colella, A. (1988) Pliocene–Holocene fan deltas and braid deltas in the Crati Basin, southern Italy: a consequence of varying tectonic conditions. In: *Fan Deltas: Sedimentology and Tectonic Settings* (Eds W. Nemec and R.J. Steel), pp. 51–74. Blackie and Son, Glasgow.

Collier, R.E.Ll. (1990) Eustatic and tectonic controls upon Quaternary coastal sedimentation in the Corinth Basin, Greece. *J. Geol. Soc. London*, **147**, 301–314.

Collier, R.E.Ll. and Dart C.J. (1991) Neogene to Quaternary rifting, sedimentation and uplift in the Corinth Basin, Greece. *J. Geol. Soc. London*, **148**, 1049–1065.

Collier, R.E.Ll., Leeder, M.R., Rowe, P.J. and Atkinson, T.C. (1992) Rates of tectonic uplift in the Corinth and Megara Basins, central Greece. *Tectonics*, **11**, 1159–1167.

Combourieu-Nebout, N. and Vergnaud Grazzini, C. (1991) Late Pliocene northern hemisphere glaciations: the continental and marine responses in the central Mediterranean. *Quat. Sci. Rev.*, **10**, 319–334.

Cornée, J.-J., Moissette, P., Joannin, S., *et al.* (2006) Tectonic and climatic controls on coastal sedimentation: the Late Pliocene–Middle Pleistocene of northeastern Rhodes, Greece. *Sed. Geol.*, **187**, 159–181.

Dabrio, C.J. (1990) Fan-delta facies associations in late Neogene and Quaternary basins of southeastern Spain. In: *Coarse-Grained Deltas* (Eds A. Colella and D.B. Prior), pp. 91–111. Special Publication 10, International Association of Sedimentolologists. Blackwell Scientific Publications, Oxford.

Dabrio, C.J. and Polo, M.D. (1988) Late Neogene fan deltas andassociated coral reefs in the Almanzora Basin, Almeria Province, southeastern Spain. In: *Fan Deltas: Sedimentology and Tectonic Settings* (Eds W. Nemec and R.J. Steel), pp. 354–367. Blackie and Son, Glasgow.

Dart, C.J., Collier, R.E.Ll., Gawthorpe, R.L., Keller, J.V.A. and Nichols, G. (1994) Sequence stratigraphy of (?)Pliocene–Quaternary synrift, Gilbert-type fan deltas, northern Peloponnesos, Greece. *Mar. Petrol. Geol.*, **11**, 545–560.

De Martini, P.M., Pantosti, D., Palyvos, N., Lemeille, F., McNeill, L.C. and Collier, R. (2004) Slip rates of the Aigion and Eliki Faults from uplifted terraces, Corinth Gulf, Greece. *C. R. Acad. Sci. Paris*, **336**, 325–334.

Dia, A.N., Cohen, A.S., O'Nions R.K. and Jackson J.A. (1997) Rates of uplift investigated through ^{230}Th dating in the Gulf of Corinth (Greece). *Chem. Geol.*, **138**, 171–184.

Doutsos, T. and Piper, D.J.W. (1990) Listric faulting, sedimentation, and morphological evolution of the Quaternary eastern Corinth rift, Greece: First stages of continental rifting. *Geol. Soc. Am. Bull.*, **102**, 812–829.

Doutsos, T., Kontopoulos, N., Poulimenos, G., Frydas, D. and Piper, D.J.W. (1990) Comment and Reply on 'Geologic history of the extensional basin of the Gulf of Corinth (?Miocene–Pleistocene), Greece'. *Geology*, **18**, 1256–1257.

Doutsos, T., Piper, G., Boronkay, K. and Koukouvelas, I.K. (1993) Kinematics of the central Hellenides. *Tectonics*, **12**, 936–953.

Dubois, J.-M. (2001) *Cycles Climatiques et Paramètres Orbitaux vers 1 Ma. Etude de la Coupe de Monte San Giorgio (Caltagirone, Sicile): Palynologie, Isotopes Stables, Calcimétrie*. DEA Paléontologie et Environnements Sédimentaires, Univ. C. Bernard – Lyon 1, 54 pp.

Ethridge, F.G. and Wescott, W.A. (1984) Tectonic setting, recognition and hydrocarbon reservoir potential of fan-delta deposits. In: *Sedimentology of Gravels and Conglomerates* (Eds E.H. Koster and R.J. Steel), pp. 217–235. Memoir 10, Canadian Society of Petroleum Geologists, Calgary.

Falk, P.D. and Dorsey, R.J. (1998) Rapid development of gravely high-density turbidity currents in marine Gilbert-type fan deltas, Loreto Basin, Baja California Sur, Mexico. *Sedimentology*, **45**, 331–349.

Fernandez-Gonzalez, M. Frydas, D., Guernet, C. and Mathieu, R. (1994) Foraminifères et ostracodes du Plio pléistocènes de la région de Patras (Grèce). Intérêt stratigraphique et paleogéographique. *Rev. Esp. Micropaleont.*, **26**, 89–108.

Flügel, E. (1982) *Microfacies Analysis of Limestones*. Springer-Verlag, Berlin, 633 pp.

Fouache, E., Dalongeville, R., Kunesch, S., *et al.* (2005) The environmental setting of the harbor of the classical site of Oeniades on the Acheloos Delta, Greece. *Geoarchaeol. Int. J.*, **20**, 285–302.

Frydas, D. (1987) Kalkiges Nannoplankton aus dem Neogen von NW-Peloponnes. *Neues Jb. Geol. Paläont., Monat.*, **5**, 274–286.

Frydas, D. (1989) Biostratigraphic investigations from the Neogene of the NW and W Peloponnes, Greece.

(in German). *Neues Jb. Geol. Paläont. Monat.*, **6**, 321–344.

Frydas, D. (1991) Paläoökologische und stratigraphische untersuchungen der diatomeen des Pleistozäns der N-Peloponnes, Griechenland. *Bull. Geol. Soc. Greece*, **25**, 499–513.

Gawthorpe, R.L. and Colella, A. (1990) Tectonic controls on coarse-grained delta depositional systems in rift basins. In: *Coarse-Grained Deltas* (Eds A. Colella and D.B. Prior), pp. 113–127. Special Publication 10, International Association of Sedimentolologists. Blackwell Scientific Publications, Oxford.

Gawthorpe, R.L. and Hardy, S. (2002) Extensional fault-propagation folding and base-level change as controls on growth-strata geometries. *Sediment. Geol.*, **146**, 47–56.

Ghisetti, F. and Vezzani, L. (2004) Plio-Pleistocene sedimentation and fault segmentation in the Gulf of Corinth (Greece) controlled by inherited structural fabric. *C. R. Acad. Sci. Paris*, **336**, 243–249.

Ghisetti, F. and Vezzani, L. (2005) Inherited structural controls on normal fault architecture in the Gulf of Corinth (Greece). *Tectonics*, **24**, TC4016, doi:10.1029/2004TC001696,2005.

Goldsworthy, M. and Jackson, J. (2001) Migration of activity within normal fault systems: examples from the Quaternary of mainland Greece. *J. Struct. Geol.*, **23**, 489–506.

Gradstein, F.M., Ogg, J.G. and Smith, A.G. (2004) *A Geological Time Scale 2004*. Cambridge University Press, 589 pp.

Hart, B.S. and Plint, A.G. (1995) Gravelly shoreface and beachface deposits. In: *Clastic Facies Analysis – a Tribute to the Research and Teaching of Harold G. Reading* (Ed. A.G. Plint), pp. 75–99. Special Publication 22, International Association of Sedimentologists. Blackwell Scientific Publications, Oxford.

Hein, F.J. and Walker, R.G. (1977) Bar evolution and development of stratification in the gravelly, braided, Kicking Horse River, British Colombia. *Can. J. Earth Sci.*, **14**, 562–570.

Hwang, I.G. and Chough, S.K. (2000) The Maesan fan delta, Miocene Pohang Basin, SE Korea: architecture and depositional processes of a high-gradient fan-delta-fed slope system. *Sedimentology*, **47**, 995–1010.

Jahns, S. (1993) On the Holocene history of the Argive Plain (Peloponnese, Southern Greece). *Veg. Hist. Archaeobot.*, **2**, 187–203.

Joannin, S. (2003) *Forçage Climatique des Séquences Emboîtées du Pléistocène Inférieur et Moyen de Tsampika (Ile de Rhodes, Grèce)*. DEA Paléontologie et Environnements Sédimentaires, Univ. C. Bernard – Lyon 1, 52 pp.

Keraudren, B. and Sorel, D. (1987) The terraces of Corinth (Greece). A detailed record of eustatic sea-level variations during the last 500 000 years. *Mar. Geol.*, **77**, 99–107.

Kershaw, S. and Guo, L. (2003) Pleistocene cyanobacterial mounds in the Perachora Peninsula, Gulf of Corinth, Greece: structure and applications to interpreting sea-level history and terrace sequences in an unstable tectonic setting. *Palaeogeogr. Palaeoclimatol. Palaeoecol.*, **193**, 503–514.

Kershaw, S., Guo, L. and Braga, J.C. (2005) A Holocene coral-algal reef at Mavra Litharia, Gulf of Corinth, Greece: structure, history, and applications in relative sea-level change. *Mar. Geol.*, **215**, 171–192.

Kontopoulos, N. and Doutsos, T. (1985) Sedimentology and tectonics of the Antirion area (western Greece). *Bull. Geol. Soc. Ital.*, **104**, 479–489.

Koukouvelas, I.K., Stamatopoulos, L., Katsonopoulou, D. and Pavlides, S.A. (2001) Palaeoseismological and geoarcheological investigation of the Eliki Fault, Gulf of Corinth, Greece. *J. Struct. Geol.*, **23**, 531–543.

Koukouvelas, I.K., Katsonopoulou, D., Soter, S. and Xypolias, P. (2005) Slip rates on the Helike Fault, Gulf of Corinth, Greece: new evidence from geoarchaeology. *Terra Nova*, **17**, 158–164.

Leeder, M.R., McNeill, L.C., Collier, R.E.L., *et al.* (2003) Corinth rift margin uplift: New evidence from Late Quaternary marine shorelines. *Geophys. Res. Lett.*, **30**, 13-1–13-4.

Malartre, F., Ford, M. and Williams, E.A. (2004) Preliminary biostratigraphy and 3D geometry of the Vouraikos Gilbert-type fan delta, Gulf of Corinth. *C. R. Acad. Sci. Paris*, **336**, 269–280.

Massari, F. and Parea, G.C. (1990) Wave-dominated Gilbert-type gravel deltas in the hinterland of the Gulf of Taranto (Pleistocene, southern Italy). In: *Coarse-Grained Deltas* (Eds A. Colella and D.B. Prior), pp. 311–331. Special Publication 10, International Association of Sedimentolologists. Blackwell Scientific Publications, Oxford.

McNeill, L.C. and Collier, R.E.L. (2004) Uplift and slip rates of the eastern Eliki fault segment, Gulf of Corinth, Greece, inferred from Holocene and Pleistocene terraces. *J. Geol. Soc. London*, **161**, 81–92.

McNeill, L.C., Cotterill, C.J., Henstock, T.J., *et al.* (2005) Active faulting within the offshore western gulf of Corinth, Greece: implications for models of continental rift deformation. *Geology*, **33**, 241–244.

Mortimer, E., Gupta, S. and Cowie, P. (2005) Clinoform nucleation and growth in coarse-grained deltas, Loreto Basin, Baja California Sur, Mexico: a response to episodic accelerations in fault displacement. *Basin Res.*, **17**, 337–359.

Nemec, W. (1990) Aspects of sediment movement on steep delta slopes. In: *Coarse-Grained Deltas* (Eds A. Colella and D.B. Prior), pp. 29–73. Special Publication 10, International Association of Sedimentolologists. Blackwell Scientific Publications, Oxford.

Nemec, W. and Postma, G. (1993) Quaternary alluvial fans in southwestern Crete: sedimentation processes and geomorphic evolution. In: *Alluvial Sedimentation* (Eds M. Marzo and C. Puigdefábregas), pp. 235–276. Special Publication 17, International Association of Sedimentolologists. Blackwell Scientific Publications, Oxford.

Nemec, W. and Steel, R.J. (1984) Alluvial and coastal conglomerates: their significant features and some comments on gravelly mass-flow deposits. In: *Sedimentology of Gravels and Conglomerates* (Eds E.H. Koster and R.J. Steel), pp. 1–31. Memoir 10, Canadian Society of Petroleum Geologists, Calgary.

Okuda, M., van Vugt, N., Nakagawa, T., Ikeya, M., Hayashida, A. and Setoguchi, A. (2002) Palynological evidence for the astronomical origin of lignite-detritus sequence in the Middle Pleistocene Marathousa Member, Megalopolis, SW Greece. *Earth Planet. Sci. Lett.*, **201**, 143–157.

Ori, G.G. (1989) Geologic history of the extensional basin of the Gulf of Corinth (?Miocene–Pleistocene), Greece. *Geology*, **17**, 918–921.

Ori, G.G., Roveri, M. and Nichols, G. (1991) Architectural patterns in large-scale Gilbert-type delta complexes, Pleistocene, Gulf of Corinth, Greece. In: *The Three-dimensional Facies Architecture of Terrigenous Clastic Sediments and its Implications for Hydrocarbon Discovery and Recovery* (Eds A.D. Miall and N. Tyler), pp. 207–216. Concepts in Sedimentology and Paleontology, Vol. 3, Society of Economic Paleontologists and Mineralogists, Tulsa, OK.

Papanicolaou, C., Dehmer, J. and Fowler, M. (2000) Petrological and organic geochemical characteristics of coal samples from Florina, Lava, Moscopotamos and Kalavryta coal fields. *Int. J. Coal Geol.*, **44**, 267–292.

Pavlides, S.B., Koukouvelas, I.K., Kokkalas, S., Stamatopoulos, L., Keramydas, D. and Tsodoulos, I. (2004) Late Holocene evolution of the East Eliki fault, Gulf of Corinth (Central Greece). *Quat. Int.*, **115–116**, 139–154.

Perissoratis, C., Piper, D.J.W. and Lykousis, V. (2000) Alternating marine and lacustrine sedijmentation during late Quaternary in the Gulf of Corinth Rift basin, central Greece. *Mar. Geol.*, **167**, 391–411.

Portman C., Andrews J.E., Rowe P.J., Leeder M.R. and Hoogewerff J. (2005) Submarine-spring controlled calcification and growth of large Rivularia bioherms, Late Pleistocene (MIS 5e), Gulf of Corinth, Greece. *Sedimentology*, **52**, 441–465.

Posamentier, H.W., Allen, G.P., James, D.P. and Tesson, M. (1992) Forced regressions in a sequence stratigraphic framework: concepts, examples and exploration significance. *Am. Assoc. Petrol. Geol. Bull.*, **76**, 1687–1709.

Postma, G. (1990) Depositional architecture and facies of river and fan deltas: a synthesis. In: *Coarse-Grained Deltas* (Eds A. Colella and D.B. Prior), pp. 13–27. Special Publication 10, International Association of Sedimentolologists. Blackwell Scientific Publications, Oxford.

Postma, G. (1995) Sea-level-related architectural trends in coarse-grained delta complexes. *Sediment. Geol.*, **98**, 3–12.

Postma, G., Babić, L., Zupanič, J. and Røe, S.-L. (1988) Delta-front failure and associated bottomset deformation in a marine, gravelly Gilbert-type fan delta. In: *Fan Deltas: Sedimentology and Tectonic Settings* (Eds W. Nemec and R.J. Steel), pp. 91–102. Blackie and Son, Glasgow.

Poulimenos, G., Zelilidis, A., Kontopoulos, N. and Doutsos, T. (1993) Geometry of trapezoidal fan deltas and their relationship to the extensional faulting along the south-western active margins of the Corinth rift, Greece. *Basin Res.*, **5**, 179–192.

Prior, D.B. and Bornhold, B.D. (1990) The underwater development of Holocene fan deltas. In: *Coarse-Grained Deltas* (Eds A. Colella and D.B. Prior), pp. 75–90. Special Publication 10, International Association of Sedimentolologists. Blackwell Scientific Publications, Oxford.

Reinson, G.E. (1984) Barrier island and associated strand-plain systems. In: *Facies Models* (Ed. R.G. Walker), pp. 119–140. Geological Association of Canada, St John's, Newfoundland.

Ritchie, B.D., Gawthorpe, R.L. and Hardy, S. (2004a) Three dimensional modeling of deltaic depositional sequences 1: influence of the rate and magnitude of sea level change. *J. Sed. Res.*, **74**, 202–220.

Ritchie, B.D., Gawthorpe, R.L. and Hardy, S. (2004b) Three dimensional modeling of deltaic depositional sequences 2: influence of local controls. *J. Sed. Res.*, **74**, 221–238.

Rohais, S., Eschard, R., Ford, M., Guilloucheau, F. and Moretti, I. (In press) Stratigraphic architecture of the Plio-Pleistocene infillof the Corinth rift: implications for its structural evolution. *Tectonophysics*.

Subally, D., Bilodeau, G., Tamrat, E., Ferry, S., Debard, E. and Hillaire-Marcel, C. (1999) Cyclic climatic records during the Olduvai Subchron (Uppermost Pliocene) on Zakynthos Island (Ionian Sea). *Geobios*, **32**(6), 793–803.

Symeonidis, N., Theothorou, G., Schutt, H. and Velitzelos, E. (1987) Paleontological and stratigraphic observations in the area of Achaia and Etoloakarnania (Western Greece). *Ann. Géol. Pays Hellén.*, **38**, 317–353.

Tucker, M.E. and Wright, V.P. (1990) *Carbonate Sedimentology*. Blackwell Scientific Publications, Oxford, 482 pp.

Ulicny, D., Nichols, G. and Waltham, D. (2002) Role of initial depth at basin margins in sequence architecture: field examples and computer models. *Basin Res.*, **14**, 347–360.

Urban, B. and Fuchs, M. (2005) Late Pleistocene vegetation of the basin of Phlious, NE-Peloponnese, Greece. *Rev. Palaeobot. Palynol.*, **137**, 15–29.

Vail, P.R., Audemard, F., Bowman, S.A., Eisner, P.N. and Perez-Cruz, C. (1991) The stratigraphic signatures of tectonics, eustasy and sedimentology – an overview. In: *Cycles and Events in Stratigraphy* (Eds G. Einsele, W. Ricken and A. Seilacher), pp. 617–659. Springer Verlag, Berlin.

Verrall, P. (1981) *Structural Interpretation with Applications to North Sea Problems*. Course Notes No. 3, Joint Association for Petroleum Exploration Courses, London.

Williams, D.F., Thunell, R.C., Tappa, E., Rio, D. and Raffi, I. (1988) Chronology of the Pleistocene oxygen isotope record: 0–1.88 m.y. B.P. *Palaeogeogr. Palaeoclimat. Palaeoecol.*, **64**, 221–240.

Wilson, J.L. (1975) *Carbonate Facies in Geologic History*. Springer-Verlag, Berlin, 47 pp.

Young, M.J., Gawthorpe, R.L. and Sharp, I.R. (2000) Sedimentology and sequence stratigraphy of a transfer zone coarse-grained delta, Miocene Suez Rift, Egypt. *Sedimentology*, **47**, 1081–1104.

Young, M.J., Gawthorpe, R.L. and Sharp, I.R. (2002) Architecture and evolution of syn-rift clastic depositional systems towards the tip of a major fault segment, Suez Rift, Egypt. *Basin Res.*, **14**, 1–23.

Zelilidis, A. and Kontopoulos, N. (1996) Significance of fan deltas without toe-sets within rift and piggy-back basins: examples from the Corinth graben and the Mesohellenic trough, Central Greece. *Sedimentology*, **43**, 253–262.

Anatomy of anticlines, piggy-back basins and growth strata: a case study from the Limón fold-and-thrust belt, Costa Rica

CHRISTIAN BRANDES*, ALLAN ASTORGA†, PETER BLISNIUK‡, RALF LITTKE§ *and* JUTTA WINSEMANN*

**Institut für Geologie, Leibniz Universität Hannover, Callinstr. 30, 30167 Hannover, Germany (Email: brandes@geowi.uni-hannover.de)*
†Escuela Centroamericana de Geología, San José, Costa Rica
‡Department of Geological and Environmental Sciences, Stanford University, 450 Serra Mall, Braun Hall, Stanford, CA94305-2115, USA
§Lehrstuhl für Geologie, Geochemie und Lagerstätten des Erdöls und der Kohle, RWTH Aachen, Lochnerstr. 4–20, 52056 Aachen, Germany

ABSTRACT

The Limón back-arc basin, located along the Caribbean coast of Costa Rica, is part of the Central American island-arc system. Basin evolution started in Late Cretaceous time as a response to subduction of the Farallón Plate beneath the Caribbean Plate. Today, the Limón Basin can be subdivided into northern and southern sub-basins, separated by the Trans Isthmic Fault System. Cenozoic deposits in the northern basin are nearly undeformed. The southern sub-basin, in contrast, was the site of northeast-directed folding and thrusting during the late Cainozoic. The presence of Plio-Pleistocene growth strata in seismic reflection lines from the offshore part of the South Limón Basin supports the results of previous work relating this compressive deformation to the Pliocene collision and subsequent low-angle subduction of the aseismic Cocos Ridge at the Central American subduction zone. Internally, the fold-and-thrust belt is characterized by concentric hangingwall anticlines and large southwestward-dipping thrusts. All thrusts sole into a common horizontal detachment, the position of which is probably controlled by a lithological change from shale to limestone near the base of the Middle Miocene. The geometry of growth strata in associated footwall synclines and piggy-back basins indicates that the anticlines evolved in a very steady fashion. The sediment thickness distribution in the piggy-back basins and footwall synclines varies systematically with the displacement along the thrust faults they are associated with. The greater the displacement the greater the accommodation space in the footwall syncline and the lesser the accommodation space in the piggy-back basin. Locally, thin packages of post-growth strata can be observed. In the northwestern portion of this fold-and-thrust belt, structural trends bend abruptly into a southwest–northeast orientation, thought to result from the presence of a large basement high that acted as an obstacle to the northeastward propagation of folds and thrusts.

Keywords Limón back-arc basin, fold-and-thrust belts, piggy-back basins, syn-tectonic growth strata, tectonic forward modelling, Coulomb wedge theory, Costa Rica.

INTRODUCTION

Many previous studies have shown that the subsidence history and basin-fill architecture of synorogenic foreland basins are closely linked to the tectonic evolution of the adjacent fold-and-thrust belts (e.g. Jordan, 1981, 1995; Cross, 1986; Flemings & Jordan, 1990; DeCelles & Giles, 1996). Most importantly, progressive deformation during the growth of fold-and-thrust belts leads to changes in topography and local relief, which in turn affect the rates and spatial distribution of erosion and sediment deposition. The often complex geological evolution of such regions can, therefore, be reconstructed most reliably if constraints are available not only on the geometry of folding and faulting, but also on lateral variations in the thickness of syntectonic deposits. In this study, we present new data on the tectonic and stratigraphic evolution of the South Limón Basin, NE Costa Rica, which has been transformed from an intra-oceanic back-arc basin into a retro-arc area affected by compressive deformation during the Pliocene.

The complex geology of the southern Central American island-arc has been discussed by several authors (Weyl, 1980; Astorga, 1988; Lundberg, 1991; Seyfried *et al.*, 1991; Winsemann & Seyfried, 1991; Weinberg, 1992; Winsemann, 1992; Amann, 1993; Krawinkel & Seyfried, 1994; von Huene & Flüh, 1994; Werner *et al.*, 1999; Abratis & Wörner, 2001; Gräfe *et al.*, 2002; Calvo, 2003; Krawinkel, 2003; Coates *et al.*, 2004). Many publications have focused on plate tectonic reconstructions (e.g. Pindell *et al.*, 1988; Ross & Scotese, 1988; Frisch *et al.*, 1992; Astorga, 1997; Maresch *et al.*, 2000; Meschede *et al.*, 2000). Additionally, other studies have discussed the petroleum potential of Costa Rica (Sheehan *et al.*, 1990; Barboza *et al.*, 1997; Barrientos *et al.*, 1997; Petzet, 1998; Lutz, 2002; Lutz *et al.*, 2004). Much of the recent work was also carried out in the field of marine geology/marine geophysics (Ranero & von Huene, 2000; Ranero *et al.*, 2000a, b; Barckhausen *et al.*, 2003).

The aim of the study presented here was to learn more about the driving mechanisms of fold-and-thrust belt evolution and to evaluate the interaction of the structural evolution and the deposition of growth strata. To enhance the understanding of the structural evolution of the offshore part of the Limón fold-and-thrust belt, the interpretation of seismic reflection lines is combined with forward modelling of anticlinal growth during fault-propagation and the filling of footwall synclines and piggy-back basins with syn-tectonic growth strata. The database employed includes a grid of two-dimensional onshore and offshore seismic lines, orientated parallel and perpendicular to the basin axis, and lithological information mainly derived from wells.

GEOLOGICAL SETTING

The geology of Central America is characterized by the interaction of five lithospheric plates, including the oceanic Cocos, Nazca and Caribbean Plates and the continental North and South American Plates (Fig. 1). The active tectonics of this region

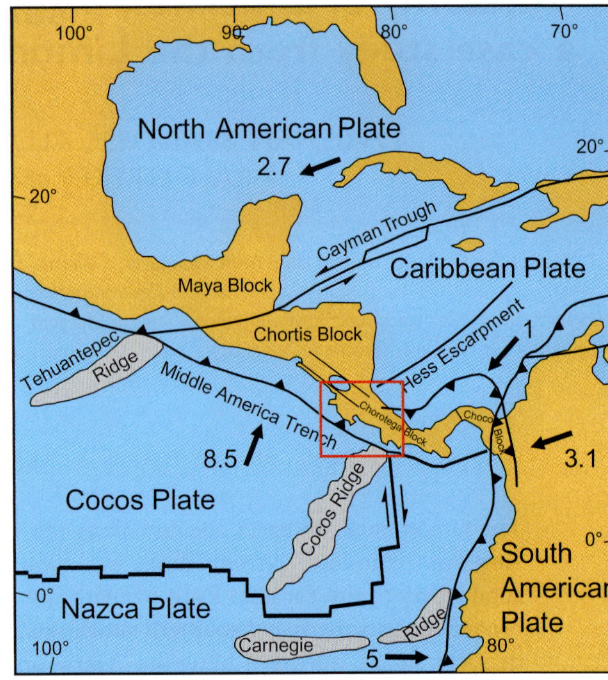

Fig. 1 Plate tectonic map of the Caribbean region (modified after Ross & Scotese, 1988; Donnelly, 1989; Meschede & Frisch, 1998). Red box shows location of detailed study area. Numbers are modern absolute plate vectors (cm yr^{-1})

are dominated by the subduction of the Cocos and Nazca Plates beneath the Caribbean Plate along the NW–SE trending Central America trench. The present-day subduction velocity off Costa Rica, relative to the Caribbean Plate, is 8.5 cm yr^{-1} (DeMets, 2001). The Cocos Plate is characterized by a large NE–SW trending aseismic ridge, the Cocos Ridge, which is interpreted to represent a hot-spot trace (e.g. Walther, 2003). The Cocos Ridge is more than 1000 km long, 250–500 km wide, rises about 2 km above the adjacent ocean floor, and has been subducted beneath southern Costa Rica since about 3.6 Ma (Collins *et al.*, 1995; Walther, 2003).

The Central American land-bridge above this subduction zone is a complex assemblage of distinct crustal blocks (Fig. 1) including, from NW to SE, the Maya, Chortis, Chorotega and Choco Blocks (Donnelly, 1989; Weinberg, 1992: Di Marco *et al.*, 1995; Campos, 2001). The Maya and Chortis Blocks have a continental basement, whereas the Chorotega and Choco Blocks comprise island-arc segments underlain by Mesozoic oceanic crust

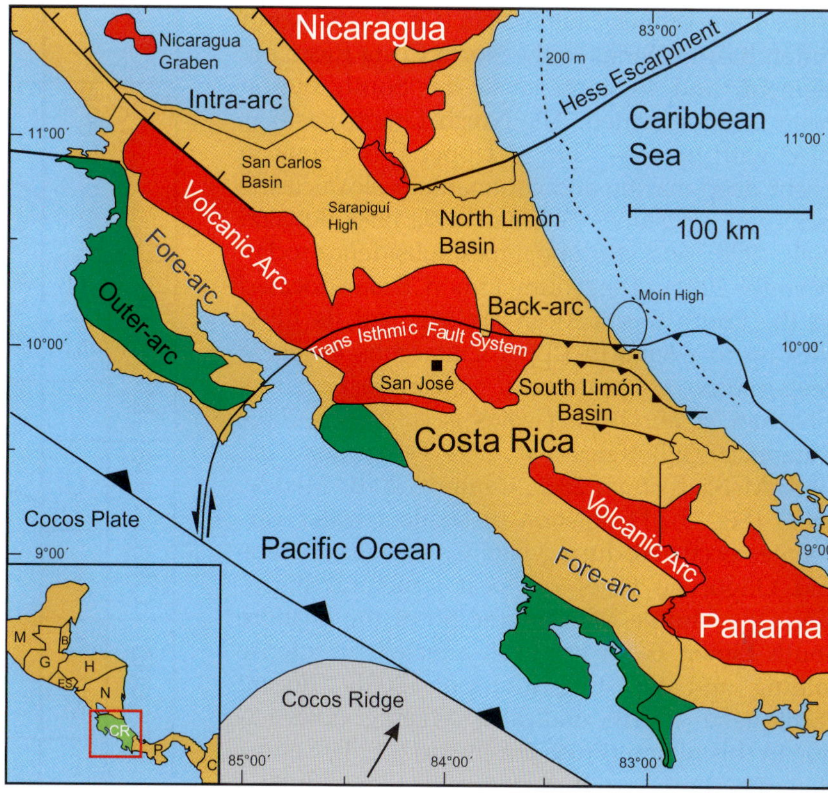

Fig. 2 Geological map of Costa Rica. The Limón Basin extends along the Caribbean coast (modified after Barboza *et al.*, 1995; Fernandez *et al.*, 1997; Campos, 2001).

(Escalante & Astorga, 1994). The Chorotega Block, which represents the Costa Rican part of the island-arc, can be subdivided into a northern and a southern arc segment (Seyfried *et al.*, 1991). The northern arc segment is bounded to the north by the Hess Escarpment and to the south by the Trans Isthmic Fault System (Fig. 2). The Hess Escarpment is a NE–SW trending bathygraphic feature in the Caribbean Sea, which separates the continental Chortis Block from the oceanic Colombia Basin (Krawinkel & Seyfried, 1994; Campos, 2001). It has been interpreted as a late Mesozoic plate boundary that acted as a strike-slip zone to compensate the movements between the Chortis and Chorotega Blocks and the Caribbean Plate (Krawinkel, 2003). The Trans Isthmic Fault System is an E–W trending active lineament with major sinistral movements (Krawinkel & Seyfried, 1994; Krawinkel, 2003). The southern Costa Rican arc segment is located south of this lineament and belongs to the Panama Microplate. Another important structural element of the Central American island-arc is the North Panama Deformed Belt, which is a typical curved fold-and-thrust belt dominating northern Panama, and extending north into southern Costa Rica. The northern edge of the deformed belt close to Puerto Limón in Costa Rica displays an abrupt bend towards the southwest (Fig. 2). The development of the fold-and-thrust belt has been controlled largely by the collision of Panama with South America since Miocene times, the forward movements of the Nazca Plate, and the oroclinal bending of the arc. The Costa Rican part of the fold-and-thrust belt is additionally affected by the low-angle subduction of the Cocos Ridge and by sediment loading (Sheehan *et al.*, 1990; Kolarsky *et al.*, 1995; Silver *et al.*, 1995).

The Limón back-arc basin is situated beneath the present-day coastal plain and continental shelf of eastern Costa Rica (Fig. 2). Its northern boundary is the Hess Escarpment; to the west and south the basin is bounded by the volcanic arc. The eastward extent is defined by the 200 m bathymetric contour line of the Caribbean Sea in the north and by the extent of the Limón fold-and-thrust belt in the south (Fig. 2). The Limón Basin can be subdivided into northern and southern sub-basins, separated by the Trans Isthmic Fault System (Fig. 2). The North Limón Basin belongs to the North Costa Rican arc segment, and in contrast to the South Limón Basin is undeformed. The North Limón Basin is filled with up to ~ 7 km of Upper Cretaceous to recent deep-marine and continental volcaniclastic rocks and limestones (Sheehan *et al.*, 1990; Bottazzi *et al.*, 1994), and still undergoes subsidence today (Mende, 2001). The South Limón Basin, located on the South Costa Rican arc segment, is filled with up to ~ 8 km of Upper Cretaceous to recent deep-marine to continental volcaniclastic rocks (Sheehan *et al.*, 1990; Coates *et al.*, 1992, 2003; Amann, 1993; Bottazzi *et al.*, 1994; Fernandez *et al.*, 1994; McNeill *et al.*, 2000; Campos, 2001; Mende, 2001) (Fig. 3). Deposition of shallow-water carbonates occurred during Late Cretaceous, Eocene and Oligocene times on local structural highs. Since the Middle Miocene the fill of the onshore South Limón Basin has been affected by intense folding and thrusting (Campos, 2001). Recent earthquake activity indicates ongoing deformation in this region (Protti & Schwartz, 1994; Suárez *et al.*, 1995).

STRATIGRAPHY OF THE SOUTH LIMON BASIN

Stratigraphic information for the South Limón Basin is largely derived from onshore outcrops and well data. The oldest sediments consist of ~ 1280-m-thick pelagic limestones and intercalated volcaniclastic rocks of Late Campanian to Maastrichtian age (Changuinola Formation, Fig. 3). The Changuinola Formation is overlain by ~ 3000 m of Palaeocene to Lower Eocene coarse-grained volcaniclastic turbidites, debris-flow deposits, lava-flows and tuffs of the Tuís Formation, representing a prograding deep-water apron-system (Mende, 2001). Early compressional deformation during Eocene to Oligocene times caused the formation of significant tectonic and topographic relief, as implied by the simultaneous deposition of 150–200 m thick shallow-water limestones of the Las Animas Formation on local structural highs (Amann, 1993; Mende, 2001), and of 700–900-m-thick hemipelagic mudstones, calcareous turbidites, and carbonate debris-flow deposits of the Senosri Formation in adjacent basin areas (Mende, 2001). During the Late Oligocene a basin-wide unconformity formed,

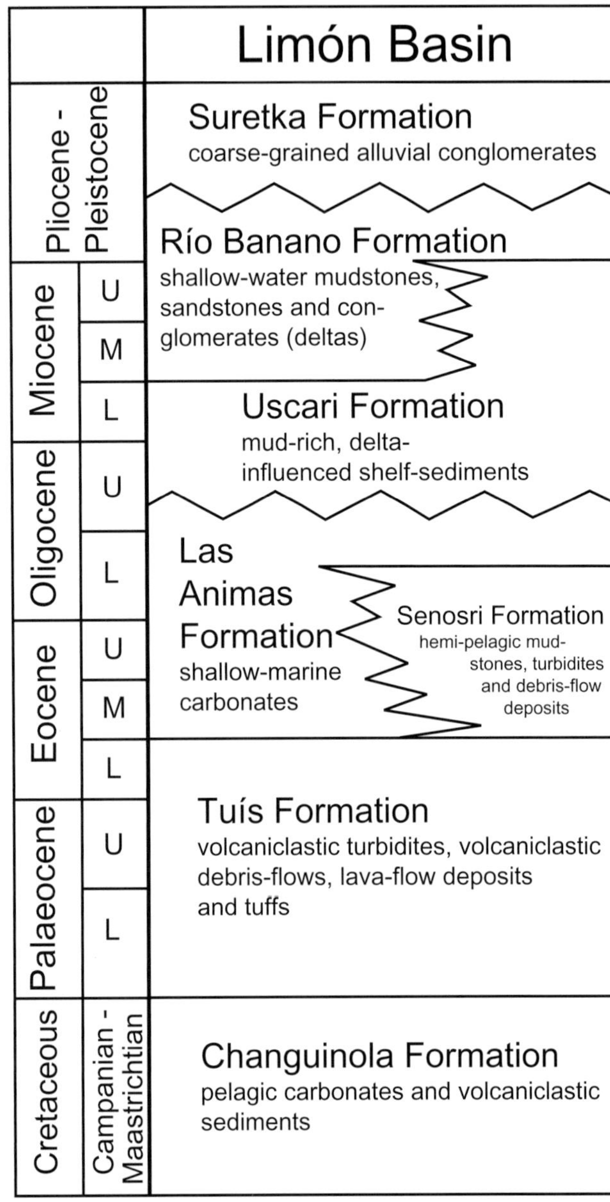

Fig. 3 Stratigraphy of the South Limón Basin (modified after Mende, 2001).

probably caused by uplift of the island-arc in combination with a major sea-level fall (Seyfried *et al.*, 1991; Amann, 1993; Krawinkel *et al.*, 2000). Subsequently, extensive carbonate ramps built above this unconformity. These carbonate ramps were overlain by ~ 2000-m-thick shallow-water volcaniclastic sediments of the Upper Oligocene to Upper Miocene Uscari Formation, interpreted as delta-influenced shelf deposits (Amann, 1993; Mende, 2001). The Uscari Formation is overlain by the shallow-water limestones and volcaniclastic rocks of the 400–1800-m-thick Río Banano Formation (Amann, 1993; Bottazzi *et al.*, 1994; Mende, 2001). During the Late Miocene to Early Pliocene, the subduction of the Cocos Ridge beneath the island-arc began (Collins *et al.*, 1995; Abratis & Wörner, 2001). This led to increased northeast-directed folding and thrusting and the development of small intramontane piggy-back basins. Subsequently shallow-marine and continental rocks of the Plio-Pleistocene Suretka Formation were deposited (Amann, 1993; Bottazzi *et al.*, 1994; Mende, 2001).

DATABASE AND METHODS

The database employed in this study includes a grid of two-dimensional seismic reflection lines acquired during onshore and offshore campaigns in the 1970s and 1980s (Fig. 4). Special emphasis was laid on the interpretation of five NE–SW-trending seismic sections, which are oriented roughly parallel to the direction of compression (lines a–e, Fig. 4). For the correlation of major reflectors between these NE–SW-trending sections, four NW–SE-trending cross-lines were used (lines f–i, Fig. 4). Stratigraphic and lithological information was derived from an onshore well located close to the NE–SW-trending seismic section (line j, Fig. 4); the well penetrates Pleistocene to Miocene sandstones, shales and limestones. The seismic interpretation was performed with the software package Kingdom Suite©. Depth conversion was performed on the basis of interval velocities. For the tectonic forward modelling the program FaultFold 4.5.4© was used, which assumes trishear kinematics to simulate the evolution of fault-propagation folds and allowed the addition of syn-tectonic growth strata during the simulation (Allmendinger, 1998). Modelling focused on the geometry of thrusts, their propagation-to-slip ratio and rates of syn-tectonic deposition of growth strata.

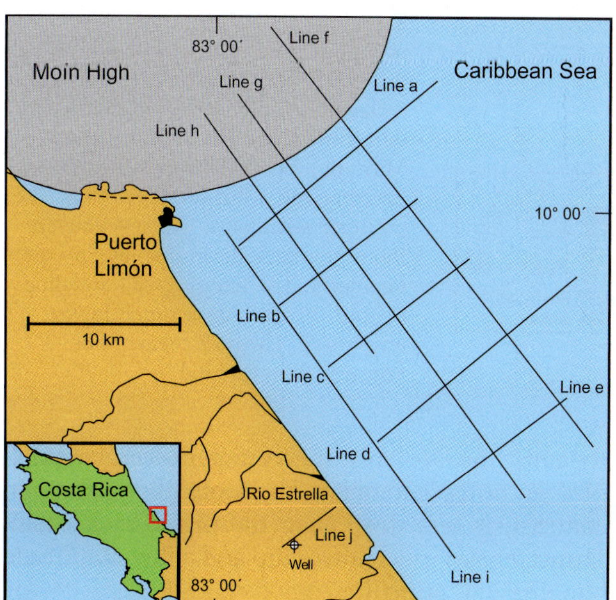

Fig. 4 Location map of seismic lines.

SEISMIC INTERPRETATION AND DEFORMATION STYLE OF THE LIMON FOLD-AND-THRUST BELT

Description

Fold and thrust architecture

The northernmost in-line (line a, Fig. 4), close to Puerto Limón, shows five thrusts (Fig. 5a), all of which sole into a subhorizontal detachment near the base of the Middle Miocene succession. The detachment is the same on all in-lines. Thrusts 2 and 4 are shorter and more planar than the others and terminate in Pliocene and Upper Miocene rocks, respectively. All other thrusts end in Pleistocene rocks. Thrust 1 shows a very pronounced listric geometry and is associated with a hangingwall anticline and a distinct piggy-back basin, filled with Plio-Pleistocene sedimentary rocks. This piggy-back basin, located on the shelf, can be traced on all in-lines. In front of thrust 1 a deep footwall syncline is present, which preserves a thick succession of syn-tectonic deposits of Pleistocene age. Like the piggy-back basin, the footwall syncline in front of thrust 1 is visible on the other in-lines.

The next in-line to the south (line b, Fig. 4) displays a low-angle listric blind thrust, which corresponds to thrust 1 on in-line a, terminating in Pleistocene rocks. Above the thrust a well-developed hangingwall anticline is present (Fig. 5b). In front of the thrust a deep footwall syncline developed, which is filled with thick syn-tectonic deposits of Plio-Pleistocene age. The offset along the thrust fault is similar to that of thrust 1 on in-line a. All other

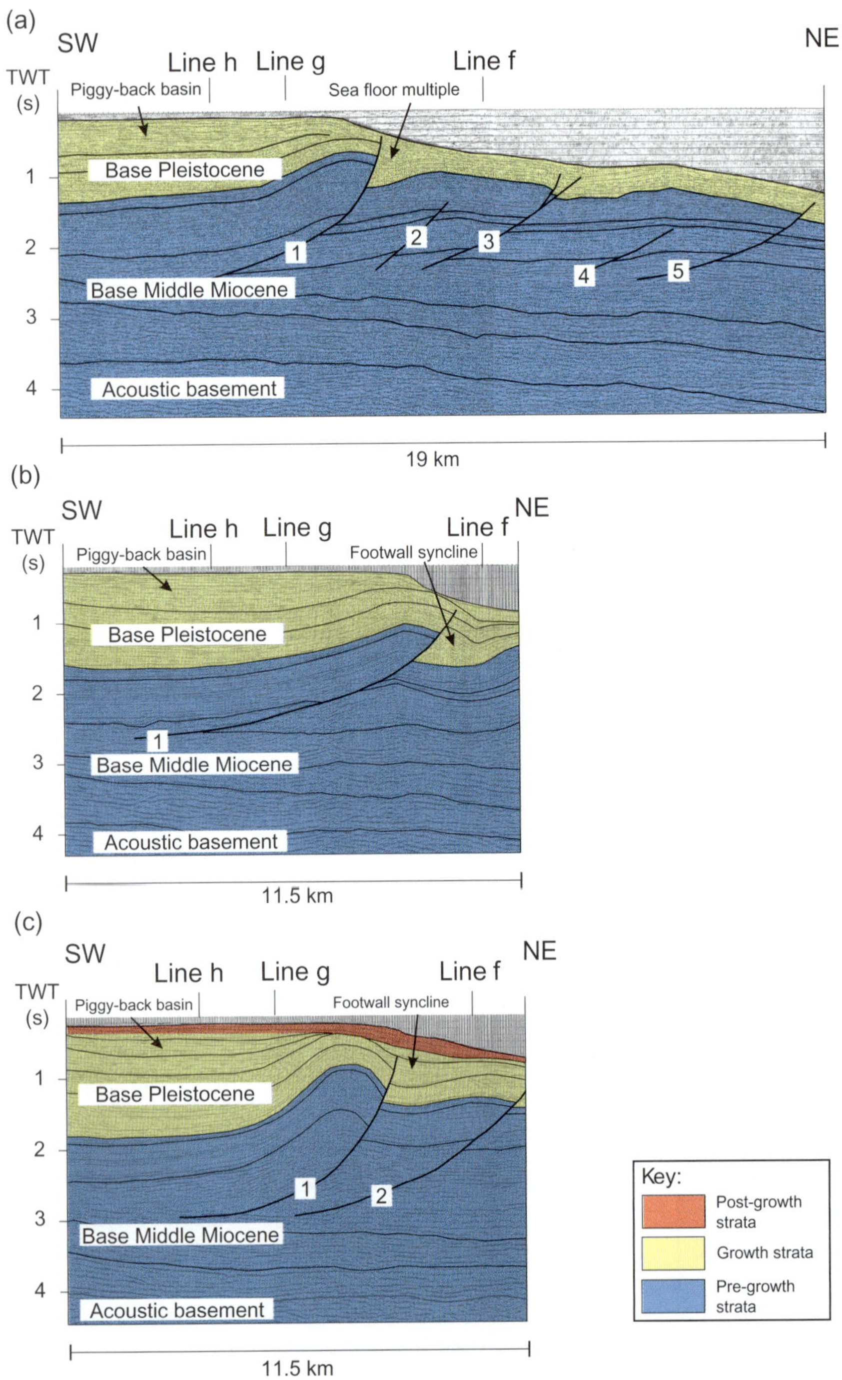

Fig. 5 (a) In-line a. The section displays five southwestward-dipping thrusts. All thrusts sole into a subhorizontal detachment (near base Middle Miocene). Thrusts 1, 3 and 5 have listric geometries. Thrust 3 has a shorter branch thrust at the tip. Behind thrust 1 a deep piggy-back basin developed, filled with Plio-Pleistocene sedimentary rocks. (b) In-line b. The seismic section shows one listric thrust. A deep footwall syncline occurs in front of the thrust. (c) In-line c. The section shows two listric thrusts that sole into a horizontal detachment (near base Middle Miocene). Behind thrust 1 a concentric hangingwall anticline developed. The younger sediments, in particular, in the piggy-back basin display an onlap geometry. The hangingwall anticline behind thrust 2 has a much lower amplitude. Thrust 2 seems to be younger than thrust 1.

thrusts on the different in-lines have smaller offsets. In-line c, located to the southeast, shows two blind listric thrusts ending in Pleistocene rocks (Fig. 5c). The significantly steeper southeastern thrust 1, interpreted to be the older one, has a concentric hangingwall anticline. The thickness of the related piggy-back basin rapidly decreases towards the anticline. Notably, the younger sediments display an onlap geometry. The footwall syncline, in contrast, is less deep than the piggy-back basin. Thrust 2 has a much lower dip and a low-amplitude hangingwall anticline.

Seismic line d (Fig. 6a) displays four thrusts with different geometries. Thrusts 1 and 2 are blind and

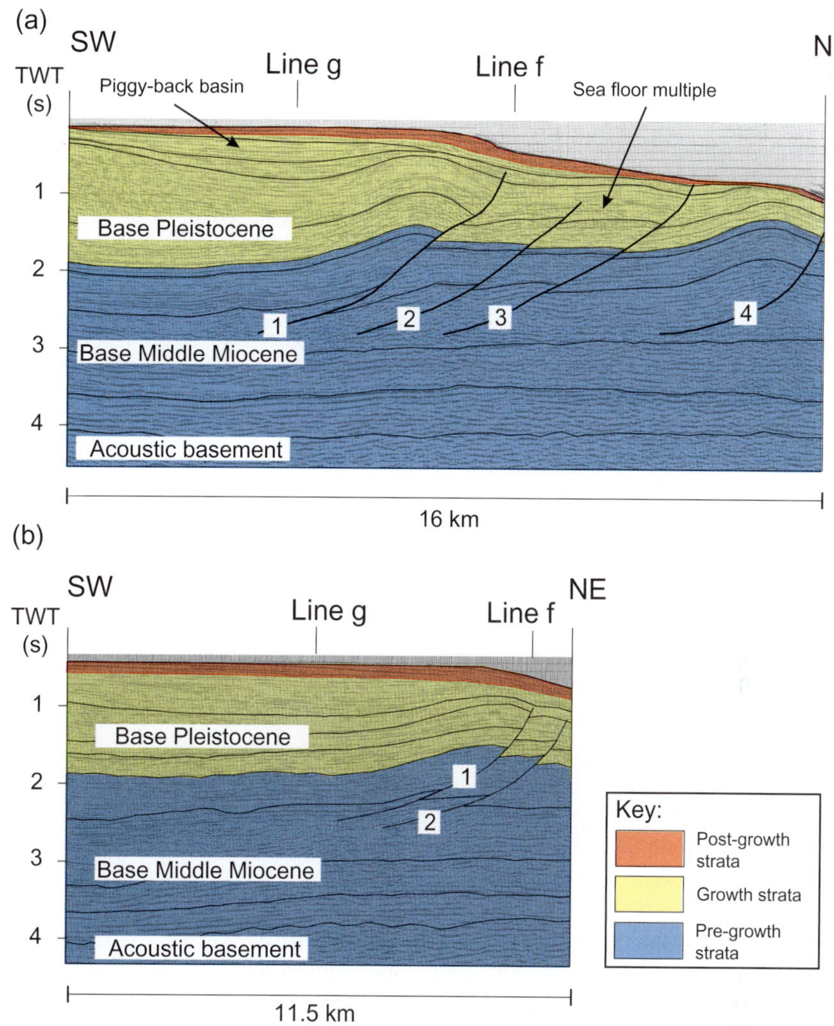

Fig. 6 (a) In-line d. The section displays four listric thrusts. Thrust 3 apparently reaches the sea floor; all other thrusts are blind. (b) In-line e. The section shows two closely spaced listric thrusts. The related hangingwall anticline has a very low amplitude.

terminate in Pleistocene deposits, whereas thrust 3 apparently reaches the sea floor. A fourth thrust at the northeastern end of the line is only partially imaged. This thrust has a very distinct listric geometry, whereas thrusts 2 and 3 appear to be much more planar. Thrust 1 has a very different geometry and displays three kinks. Hangingwall anticlines 1 and 4 are relatively pronounced compared with low-amplitude anticlines 2 and 3. All four thrusts have similar displacements. The depocentre of the piggy-back basin is close to the back-limb of the anticline. It is filled with Plio-Pleistocene sedimentary rocks. Escarpments on the sea floor are interpreted as fault scarps. The southernmost line (line e, Fig. 4) shows two closely spaced listric thrusts terminating in Pleistocene rocks. The related hangingwall anticline has a very low amplitude (Fig. 6b).

The four long NW–SE-trending cross-lines f–i (Fig. 4) display very similar structures. Again all thrusts sole into a horizontal detachment near the base of the Middle Miocene. On section f (Fig. 7a) in the very northwest, seven thrusts are present, four major listric thrusts and three smaller planar ones. Thrust 2 reaches the surface and is associated with a pronounced topographic break on the sea floor. All other thrusts are blind. The related hangingwall anticlines appear to be more asymmetric compared with most of the structures visible on the in-lines. Line g displays three large listric thrusts (Fig. 7b). Thrust 1 reaches the surface and is associated with a steep escarpment on the sea floor. Thrust 2 has no distinct hangingwall anticline. Thrust 3 is associated with a very flat anticline. Line h (Fig. 4) has three major and two minor listric thrusts (Fig. 8). Thrusts 1 and 2 are characterized

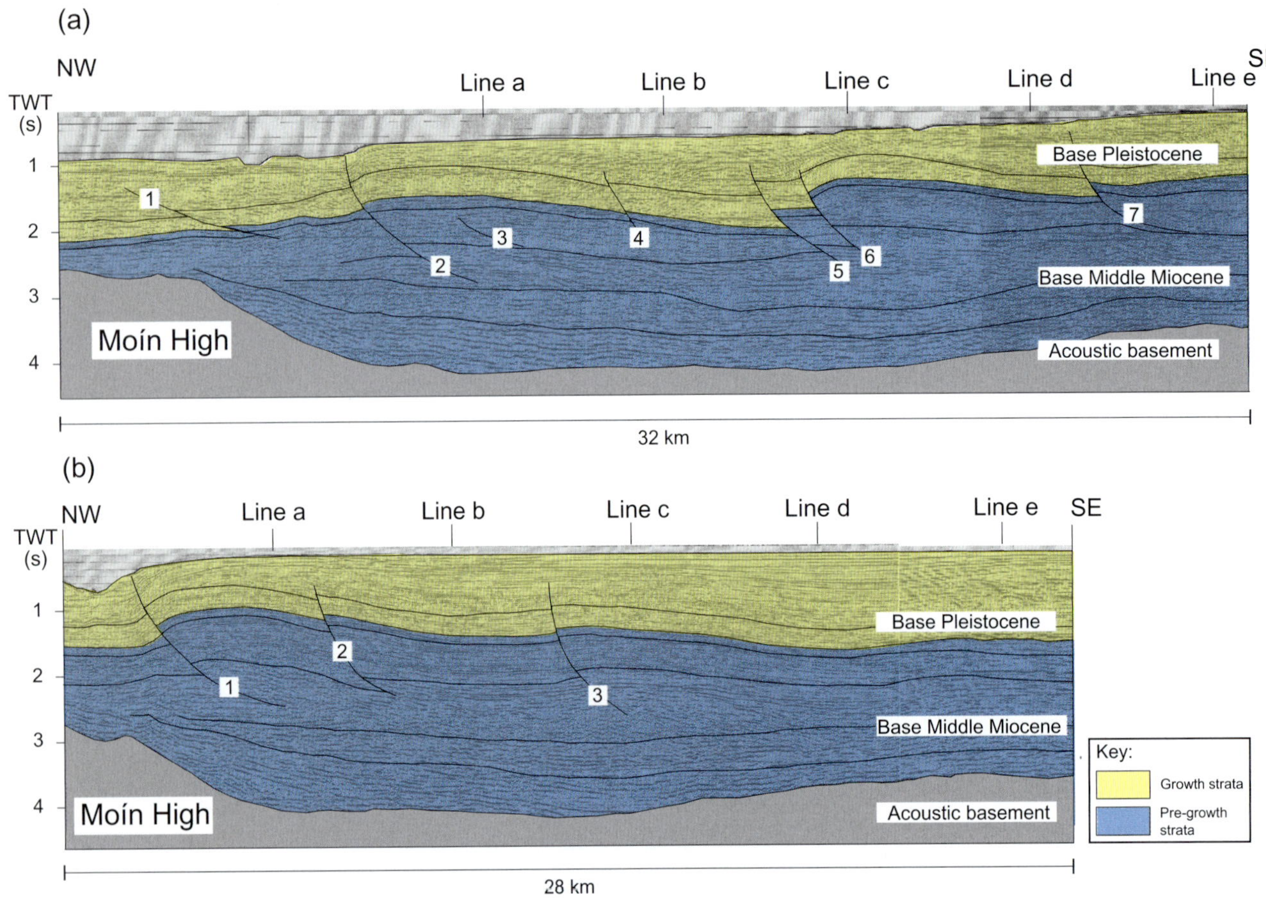

Fig. 7 (a) Cross-line f. The section displays seven thrusts. Thrust 2 reaches the surface. The tip is associated with a topographic break at the sea floor, which can be interpreted as fault scarp. The other thrusts are blind. All thrusts sole into a horizontal detachment (near base Middle Miocene). (b) Cross-line g. On the section three thrusts are present. Thrust 1 reaches the surface. The tip is associated with a topographic break at the sea floor, which can be interpreted as fault scarp. The other thrusts are blind.

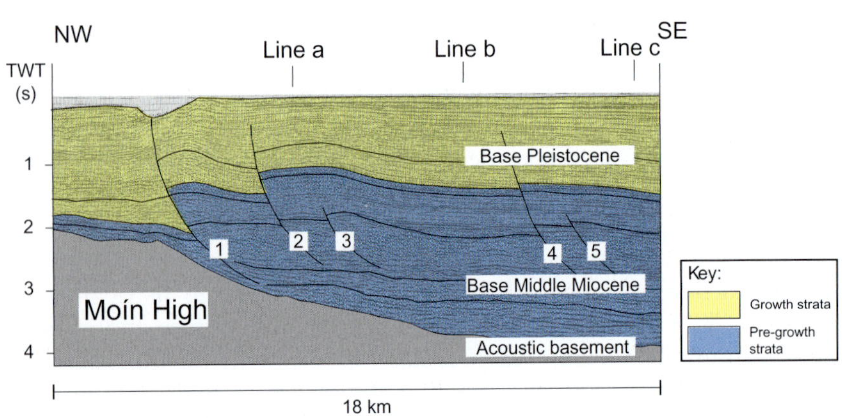

Fig. 8 Cross-line h. The section shows five listric thrusts. Thrust 1 reaches the surface. The tip is associated with a topographic break at the sea floor, which can be interpreted as fault scarp. The other thrusts are blind.

by concentric hangingwall anticlines. Piggy-back basins are present behind these anticlines. Thrust 1 reaches the sea floor. Line i (Fig. 4) is characterized by two closely spaced listric thrusts with relatively small displacements that lack pronounced hangingwall anticlines and piggy-back basins.

Growth and post-growth strata

As described above, a well-developed piggy-back basin behind thrust 1 is visible on all five in-lines allowing a three-dimensional analysis of the Pleistocene part of the fill. The basin has a lenticular shape with a steeply inclined margin close to the anticlines and a less inclined margin on the opposite side. The thickness of the Pleistocene deposits drastically decreases towards the anticlines. This demonstrates syn-tectonic filling of the piggy-back basin. The lack of onlaps in the older part of the basin shows that the uplift of the anticlines did not significantly exceed the sedimentation. Onlap features locally observed in younger Pleistocene deposits (e.g. on in-lines b and c) indicate that uplift exceeded sedimentation during that time. Late Pleistocene to Holocene post-growth strata locally drape these structures. Post-growth strata are visible on the in-lines c, d and e indicating that the tectonic activity stopped in that area. The youngest deformation is related to in-lines a and b in the north. The post-growth sediments might have been deposited from turbidity currents; a recent submarine channel visible on cross-line f (Fig. 7a) supports this assumption. The deposits next to the channel are interpreted as levees because of their wedge shape (e.g. Mutti & Normark, 1987; Klaucke *et al.*, 1998). The orientation of the channel is probably related to the nearby NE–SW-trending fault-scarp, which is visible on cross-lines f, g and h.

The reflector pattern of the piggy-back basin-fill varies from southeast towards northwest. The northern sections (in-lines a and b) show an overall horizontal reflector pattern. In the southeast on seismic lines c, d, and e, seaward inclined reflector patterns occur in the landward part of the piggy-back basin. This can be interpreted to result from prograding depositional units. Following the classification of Mitchum *et al.* (1977) the northwestern part of the basin shows an onlap fill, whereas in the southeast, locally a prograding fill is visible.

Campos (2001) described deltaic sediments in the Plio-Pleistocene deposits in the onshore part of the South Limón Basin. The prograding reflector pattern is thought to be an offshore equivalent of these deltaic systems. This pattern is very distinct on in-line d but less pronounced on in-lines c and e. This might result from a spatially limited cone-shaped sedimentary body, which progressively filled the evolving piggy-back basin. Today, the Estrella River has its mouth close to the south-western end of in-line d (Fig. 4). Due to the lower sea-level during the Pleistocene, in combination with increased sediment mobilization in the hinterland, this river may have built a small prograding sedimentary body that locally filled the piggy-back basin. A comparison of the five in-lines shows an increase in thickness of the Pleistocene deposits in the basin from northwest to southeast. This may be due to tectonically created accommodation space, but could also be related to the loading of the sediments derived from the Estrella River.

In contrast to the piggy-back basin, the fill of the footwall syncline of thrust 1 shows a decrease in thickness from northwest to southwest. The greatest sediment thickness can be observed on in-lines a and b. On in-line d the footwall syncline is less pronounced, and in front of thrust 1 on in-line e, no footwall syncline is present. This pattern correlates with the decrease in offset of thrust 1 from northwest to southeast. The evolution of the footwall syncline seems to be directly related to the displacement along the thrust fault, with higher displacement leading to a greater accommodation space.

Interpretation

The deformation style of the northeastern part of the Limón fold-and-thrust belt is characterized by mainly concentric hangingwall anticlines (on the in-lines) and planar or listric thrusts. The cross-lines often display asymmetric anticlines. Displacements along the thrusts range from 100 m to approximately 1 km. A significant feature is that all thrusts sole into the same horizontal detachment. Borehole data indicate that the position of the detachment is controlled by a lithological change from shale to limestone. Thrusts located in a more internal position within the fold-and-thrust belt are generally steeper and have greater offsets than more external

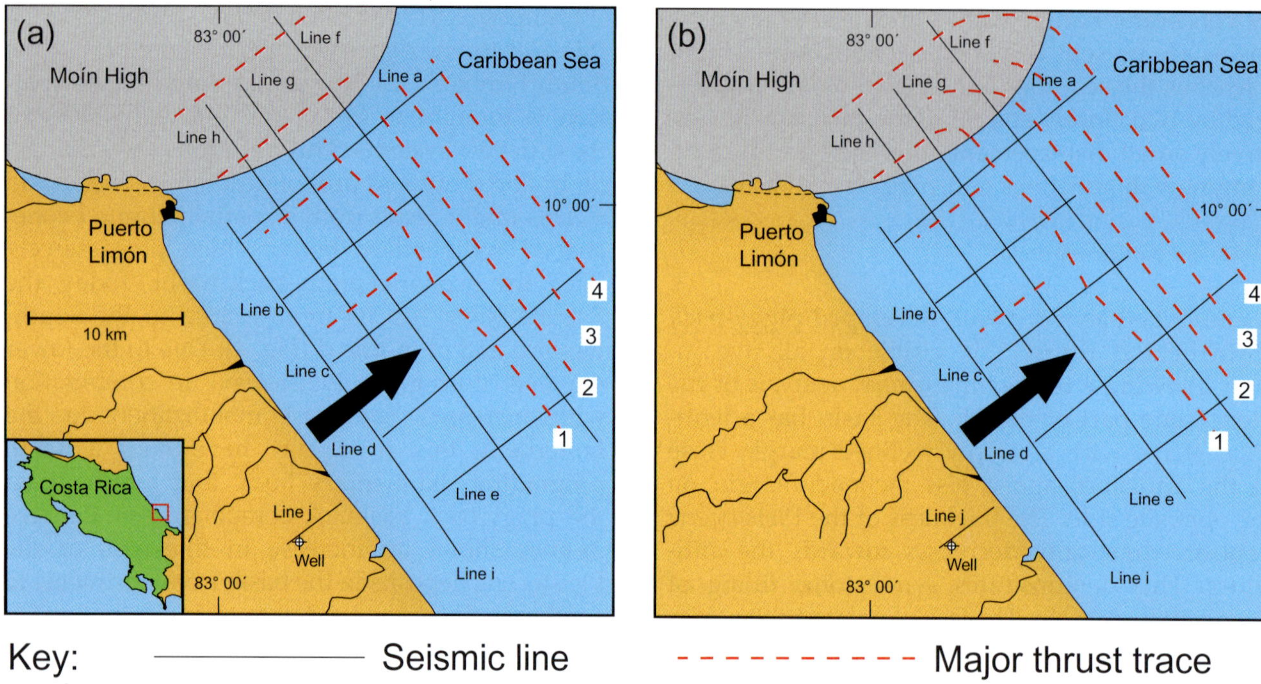

Fig. 9 The seismic data imply two possible interpretations for the structure of the study area. The numbers 1–4 refer to the thrusts visible on the in-lines. Scenario (a) has two independent thrust systems, a northeast-vergent thrust system visible on the lines a–e and a northwest-vergent one visible on the lines f–g. In the second scenario (b) only one thrust system is necessary. All thrusts bend abruptly ~ 90° in the vicinity of the Moín High.

(presumably younger) thrusts. There is no evidence for out-of-sequence thrusting.

The results described above are consistent with two possible interpretations.

1 The observed structures of the study area can be interpreted as two separate fault systems, a set of northeast-vergent faults visible on the in-lines and a set of northwest-vergent thrusts visible on the cross-lines (Fig. 9a).

2 Alternatively the observed structures can be interpreted as a single system of faults and thrusts that are characterized by a relatively abrupt bend of approximately 90° in the northern part of the study area (Fig. 9b).

Support for the first scenario is provided by similar offsets along thrust 1 on all five in-lines. Similarly, the northwest-vergent faults associated with the topographic break on the cross-lines are consistent with the assumption of a system of relatively linear northwest-vergent folds and faults. The second scenario, however, is consistent with regional structural trends in this area, including an abrupt ~ 90° bend of the Limón fold-and-thrust belt (Fernandez *et al.*, 1997), where the general trend of NW–SE-striking thrust faults swings into a NE–SW direction. Accordingly, strong support for the second scenario is provided by the presence of the Moín High (Fig. 2), a convex, mound-like antiformal structure in the northern part of the study area (Fig. 10), generally interpreted as a basement high (Barrientos *et al.*, 1997).

The internal structure is difficult to visualize from seismic sections. A strong reflector envelopes the Moín High, delineating it from the surrounding sedimentary rocks. Below this reflector, the Moín High is very weakly layered or completely structureless. Some sections, however, show a more distinct layered reflector pattern especially in the upper part of the structure. The lack of Oligocene deposits at the western flank of the structure might indicate vertical movements between Eocene and Miocene times. The Moín High is draped with Middle Miocene and younger sediments. The Moín High is therefore considered to be a type of

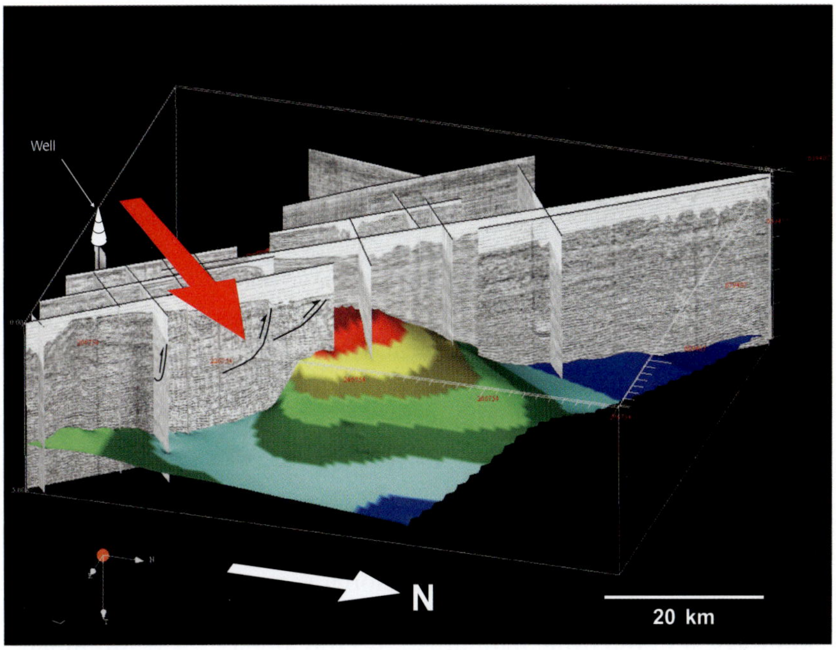

Fig. 10 Three-dimensional reconstruction of the Moín High. The Moín High is the convex mound-like structure in the centre of the figure. The deformed South Limón Basin is located to the left of the Moín High. The undeformed North Limón Basin is on the right-hand side. The depth of the seismic lines is in two-way-travel time (5 s). The central portion of the Moín High is at a depth of 1.8–1.9 s. View is towards the west. The large red arrow shows the main transport direction of the fold-and-thrust belt. The fold-and-thrust belt overthrusts the southern flank of the Moín High.

basement-cored anticline. A part of the northwestern edge of the Limón fold-and-thrust belt overthrusts the southern flank of the Moín High, implying that the abrupt bending of the fold-and-thrust belt may be an effect of interaction with this structure. Similar fold-and-thrust belt geometries are likely to be the result of the collision with such an obstacle (Marshak *et al.*, 1992). It is suggested therefore that the Moín High acted as a rigid obstacle to the propagating fold-belt, and that the northeastern edge of the fold-and-thrust belt was bent around its southern flank.

Variations in sediment thickness across the thrust faults were used to infer the age of the deformation. A few thrusts show activity in Pliocene times and there is no evidence for an earlier deformation phase in the whole study area. Most of the deformation occurred during the Pleistocene. The majority of thrusts also terminate in Pleistocene rocks, implying that deformation ceased in Pleistocene times. Only a few thrusts reach the surface, indicating recent deformation. At the tip of some of these thrusts topographic breaks, which may represent fault scarps, can be observed at the sea floor. Further evidence for continued deformation is the recent seismic activity; the Limón earthquake of 22 April 1991 was the result of thrust movements (Protti & Schwartz, 1994; Suárez *et al.*, 1995).

TECTONIC FORWARD MODELLING

The technique of tectonic forward modelling was used to quantify the controlling factors for the evolution of the fold-and-thrust belt. From an interpretation of the seismic reflection lines it is concluded that the anticline structures in the Limón Belt are fault-propagation folds, which developed in the hangingwall of the thrusts and accommodated part of the slip along the fault. Previous work on these types of folds has focused on their kinematics and the different ways to simulate their evolution (e.g. Suppe, 1983; Mitra, 1990; Suppe & Medwedeff, 1990; Mitra & Mount, 1998). It has been shown that several features of fault-propagation folding, such as the curved fold-shapes and the presence of footwall synclines, as well as systematic variations in the thickness and dip of syn-tectonic strata deposited on the anticlinal forelimbs, are difficult to explain with, for example, the parallel-kink-fold model (Allmendinger *et al.*, 2004). Trishear kinematics, which are consistent with the presence of these features, provide a way to describe the thinning of beds at the anticline hinge and thickening of beds adjacent to the syncline hinge (Erslev, 1991).

Many anticlines in the Limón fold-and-thrust belt have a curved shape, display a thinning of strata towards the anticline hinge, and distinct footwall synclines. Therefore, the trishear model seems to be appropriate for, and has been used in, this study. Most importantly, the assumption of trishear kinematics allows great flexibility in the choice of the propagation-to-slip ratio, which is one of the most important factors controlling the shape of fault-propagation folds (Hardy & Ford, 1997; Allmendinger, 1998; Allmendinger & Shaw, 2000; Allmendinger *et al.*, 2004). The forward modelling was carried out using the software FaultFold 4.5.4©, which allowed geometric simulations of fault-propagation fold growth and estimates of the amount of associated horizontal shortening in the same work-flow.

Several forward simulations were carried out for one representative section to reconstruct the structural evolution of a two-dimensional section from an undeformed to deformed state, particularly with respect to the propagation-to-slip ratio along the fault. The present-day geometry of the fold-and-thrust belt was used to calibrate the model. Forward modelling was carried out for a structurally simple seismic section with one listric thrust and a single hangingwall anticline, in order to simulate the evolution of a representative part of the fold-and-thrust belt and to learn more about the boundary conditions of the system. The section is based on seismic line b (Fig. 5b). The seismic interpretation of the section focused on the detailed assessment of the form of the thrust-fault and the shape of the related hangingwall anticline. Special emphasis was also placed on the recognition of onlap features and the evolution of the growth strata in the fore- and backlimb area of the fold. The fill of the piggy-back basin behind the anticline is very homogeneous, with only few onlap features in the youngest part of the fill. This implies that anticline growth occurred in a very steady fashion. Prior to the simulation, the interpreted seismic reflectors were depth-converted. Due to the depth conversion, the upper part of the section was vertically shortened, whereas the lower part was extended. However, these geometrical changes are only slight, and the thrust is still listric after the conversion.

The geometry of the interpreted and depth-converted line was transferred into the model. The section is 11 km long and approximately 6 km deep (Fig. 11). The detachment lies in a depth of 2800 m. As an initial condition, a dip of 5° was assumed for the ramp angle of the thrust. To choose an appropriate propagation-to-slip ratio a range from 1 to 3 in increments of 0.1 was tested. After the first two runs, the ramp angle was increased to 10°. After seven runs growth strata were added. The simulation was stopped after 12 runs at a ramp angle of 25°. Growth strata were added to enhance the comparability of the simulation output and the real cross-section. The best fit of the model and the present geometry of the hanging wall anticline was obtained with a propagation-to-slip ratio of 2.5. A higher propagation-to-slip ratio (e.g. 3) led to more asymmetric anticlines. Lower values (e.g. 1.5) created folds that were too tight. The growth strata added after run seven showed a clear thinning onto the anticline hinge, comparable with the geometry observed on the seismic lines. The model also displayed distinct thickening of beds in the footwall of the thrust, though it is not as prominent as in the footwall syncline visible on the seismic section. From the model, shortening of around 2.4 km was derived. It is remarkable that many thrusts of the Limón fold-and-thrust belt appear to be blind thrusts. It was not possible to decide whether these thrusts never reached the surface or whether they managed to break through and were only covered by sediment later. The thrust in the modelled section also occurs on the next three in-lines to the south (lines c, d and e) and on the in-line a to the north. On each section the thrust is not emergent and is therefore classified as a blind thrust (McClay, 1992).

DISCUSSION

The best-known examples of retro-arc foreland basins, in Argentina and the USA, developed on continental crust (Jordan, 1981, 1995). The South Limón Basin, in contrast, evolved on oceanic basement behind an island-arc. Similar tectonic settings are rare; examples are known from the Sea of Japan and the Sunda Arc (Protti & Schwartz, 1994; Suárez *et al.*, 1995). In general, foreland basins and their outward propagating fold-and-thrust belts form complex systems (e.g. Covey, 1986; Sinclair & Allen, 1992; DeCelles & Giles, 1996). In continental foreland basins three main processes interact: (i) thrust deformation, that builds the necessary tectonic load; (ii) sedimentary and erosional processes,

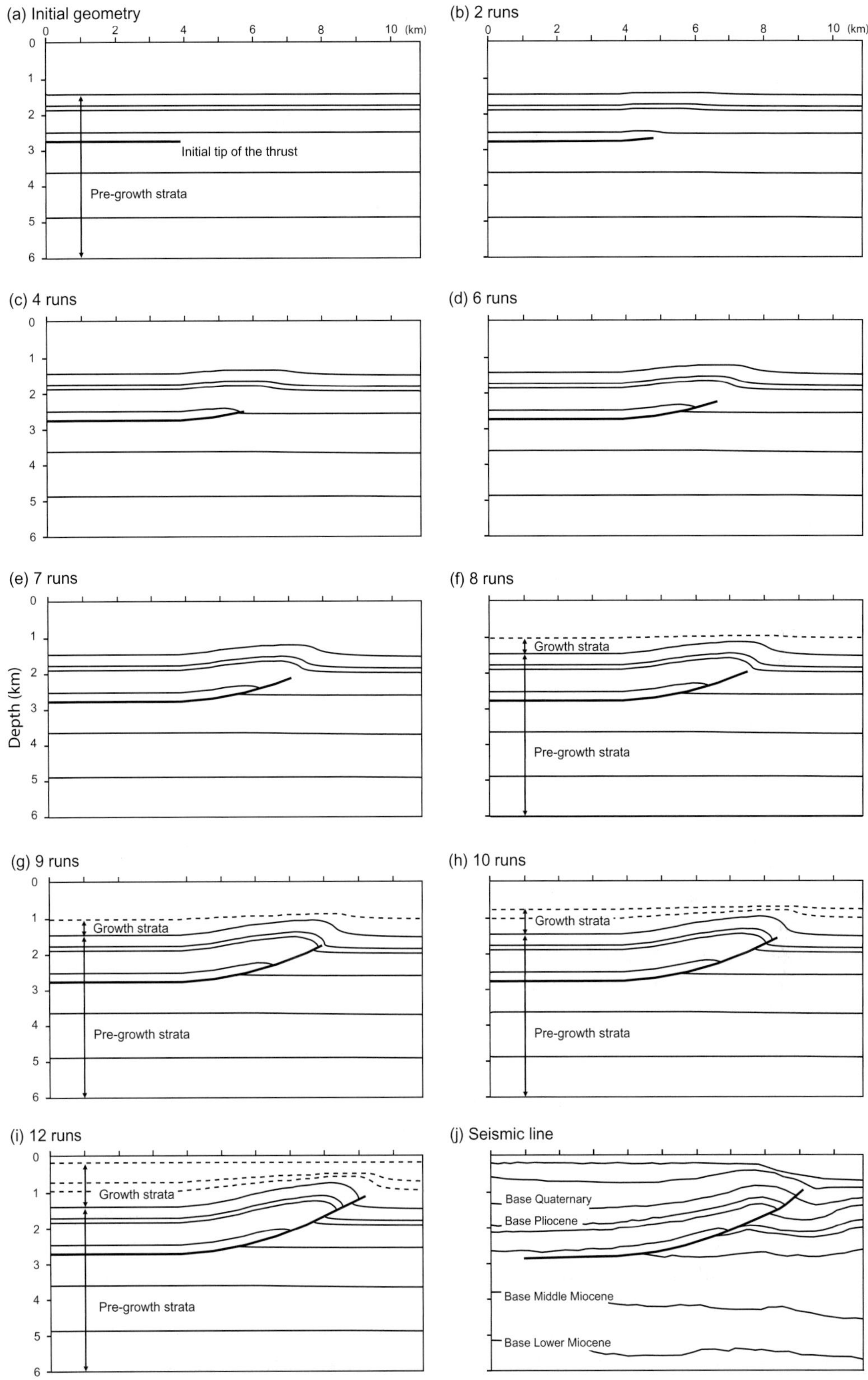

Fig. 11 Results from tectonic forward modelling. The simulation was carried out with the software FaultFold 4.5.4© by R. Allmendinger (Allmendinger, 1998; Zehnder & Allmendinger, 2000). (a) The pre-deformation state of the model. (b–i) The step-wise evolution of the thrust and the associated hangingwall anticline. Initial thrust angle was 5°. Each figure represents one run. Growth-strata were added during the simulation. (j) Seismic line b shown for comparison. The modelled section fits well to the real structure.

redistributing the load; and (iii) the flexural response of the underlying lithosphere (Sinclair *et al.*, 1991). The fill of the basin influences the evolution of the fold-and-thrust belt and *vice versa*.

Fold-and-thrust belts can be described by the Coulomb wedge theory (Davis *et al.*, 1983; Dahlen, 1984; Dahlen *et al.*, 1984; Willett, 1992), with the internal behaviour of a propagating thrust-wedge being controlled by the Mohr–Coulomb failure criterion. Models for non-cohesive and cohesive material have been developed and the effect of pore-fluid pressure has been considered. The strength of the lithosphere that underlies basin and fold-and-thrust belts is another important factor, as is the lithology of the pre-deformation strata, which can have a profound effect on the deformation style. Lithological changes can control the evolution of a basal detachment. Different studies have focused on this topic (e.g. Turrini *et al.*, 2001) and the impact of weak décollements has been discussed in the literature (e.g. Vergés *et al.*, 1992; Costa & Vendeville, 2002; Ford, 2004). For the structural development of the Limón fold-and-thrust belt the lithology of pre-deformation strata is an important factor. The detachment occurs in a position where a lithological change from shale to limestone has been observed in an onshore well. The rheological contrast between the two different lithologies therefore probably controlled the development of the detachment.

Basal friction can be another factor for the internal kinematics of fold-and-thrust belts. In general, it can be observed that with increasing basal friction the taper also increases (Davis *et al.*, 1983). Gutscher *et al.* (1996) were able to show the direct effects of variations in basal friction on the internal architecture of accretionary wedges. Low basal friction mainly leads to frontal accretion of thrust sheets. In contrast, high basal friction causes increasing underplating of weakly deformed thrusts and generates a steeper slope angle. Ford (2004) pointed out that wedges with low basal friction cannot attain a critical state and therefore the critical wedge model cannot be applied in such a case. In addition, the flexure of the lower plate influences the evolution of the wedge. For the analysis of orogenic wedges a separate analysis of the slope angle α and the angle of the base of the wedge β will be more useful than the critical taper $\alpha + \beta$ (Ford, 2004).

It is quite difficult to transfer these results to the Limón fold-and-thrust belt. Based on a depth-converted version of in-line d the surface has a slope angle α of 4° and the detachment angle β is 1.5°, implying a wedge taper of 5.5°. Following Ford (2004) natural wedges with low basal friction should have slope angles in the range of 0–1°. Therefore, the slope angle of 4° observed for the offshore parts of the Limón fold-and-thrust belt points towards high basal friction. It is clear, however, that the fold-and-thrust belt exclusively consists of repeated thrust sheets. In the study area the distances between the thrusts visible on the in-lines are about 3 km. No evidence for underplating can be found on seismic sections. This, together with the rheology at the base of the fold-belt, indicates that the detachment is characterized by low friction.

To evaluate other possible driving mechanisms, the timing of deformation provides valuable indications. Seismic interpretation shows that the deformation observed in the study area is post-Miocene in age. A few thrusts show Pliocene movements in the offshore part of the fold-belt, and most of the deformation in the study area must have taken place during the Pleistocene. Because of this timing, it is very likely that the deformation of the Limón fold-and-thrust belt is closely related to the subduction of the Cocos Ridge (Collins *et al.*, 1995). The effects of subduction of aseismic ridges have been discussed by Pilger (1981), Cross & Pilger (1982) and Cloos (1993). In general, the subduction of young and buoyant ridges leads to a decrease of the subduction angle. Following Pilger (1981), the subduction of an aseismic ridge can result in an isostatic subsidence in the back-arc area and in foreland thrust faulting.

Several authors have discussed the origin of the North Panama Deformed Belt and the influence of subduction of the Cocos Ridge on the Costa Rican part of the fold-and-thrust belt (e.g. Gardner *et al.*, 1992; Collins *et al.*, 1995; Kolarsky *et al.*, 1995; Silver *et al.*, 1995). Suárez *et al.* (1995) concluded that the Cocos Ridge does not subduct, but collides with the trench. This causes an increase in the strength of plate coupling. The resulting deformation of the upper plate is absorbed by back-arc thrusting. The subduction of the Cocos Ridge began around 3.6 Ma. At 1.6 Ma the subducted ridge should have reached the study area (Collins *et al.*, 1995).

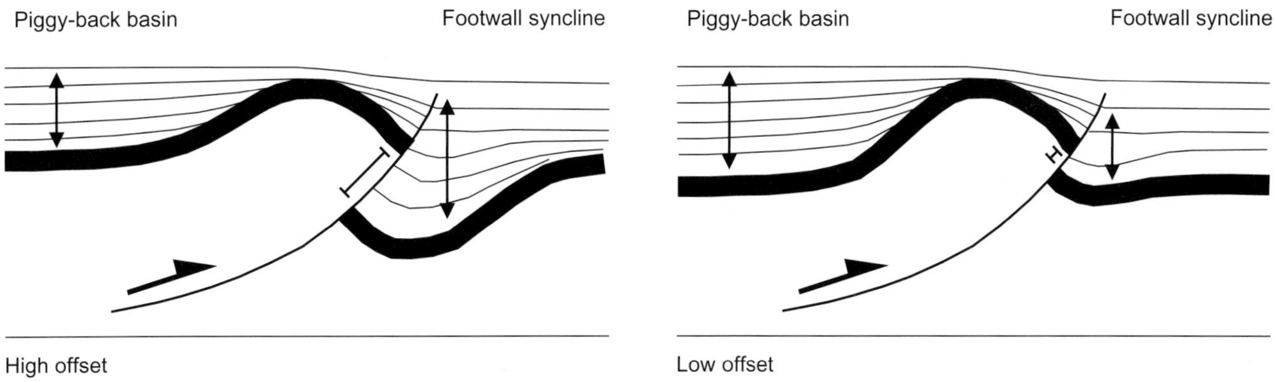

Fig. 12 Systematic relationship between fault displacement, depth of the related piggy-back basin and footwall syncline. The footwall syncline is most pronounced in front of that part of the thrust which has the highest displacement, thus creating a greater accommodation space there. The piggy-back basin, in contrast, shows the opposite behaviour: greater displacement of the thrust fault is associated with lesser sediment thickness in the piggy-back basin.

This is in good agreement with our observations, which imply Plio-Pleistocene deformation in the offshore area of the Limón fold-and-thrust belt.

The seismic lines from the South Limón Basin also provide some insight into factors controlling the internal geometry of this fold-belt–basin system. Previous studies (e.g. Silver *et al.*, 1995) have identified four principal factors controlling thrust orientations in the North Panama Deformed Belt: the collision of the Panama arc-segment with South America; the collision of the southern Costa Rica arc-segment with the Cocos Ridge; variations in sediment thickness; and variability of slope stresses. The results of the present study highlight the importance of basement morphology as an additional influence on thrust orientation.

In general, the trend of thrusts in the Limón fold-and-thrust belt is controlled by the regional stress field; compression is directed NE–SW and therefore the thrust faults generally strike NW–SE. In the area of the Moín High close to Puerto Limón, however, the whole deformed belt bends by ~ 90° (Fig. 2). The seismic lines show that, in this region, the southern flank of the Moín High is overthrust by the propagating Limón fold-and-thrust belt. It thus seems likely that the thrust faults observed on the NE–SW-directed in-lines bend by ~ 90° due to the interaction of the fold-belt with the Moín High, and correspond to those on the NW–SE running cross-lines. This hypothesis is also consistent with the results of analogue models for the collision of a fold-and-thrust belt with an obstacle (Marshak *et al.*, 1992; Marshak, 2004).

Another interesting aspect of the data from the Limón fold-and-thrust belt is that they display a systematic relationship between fault displacement, depth of the related piggy-back basin, and footwall syncline (Fig. 12). The footwall syncline is most pronounced in front of that part of the thrust which has the highest displacement, thus creating a greater accommodation space there. The piggy-back basin, in contrast, shows an opposite behaviour. Greater displacement of the thrust fault is associated with lesser sediment thickness in the piggy-back basin. This is problematic, because larger thrust displacement does not require a decrease of accommodation space there. It is interesting to note, however, that the greatest sediment thickness in the piggy-back basin occurs at the mouth of the Estrella River. This implies that the lateral thickness variation in the piggy-back basin may be influenced by locally high sediment input, leading to increased loading and subsidence in that particular area.

CONCLUSIONS

The Caribbean margin of Costa Rica displays an extensional back-arc basin (North Limón Basin) and a compressional retro-arc foreland basin (South Limón Basin), side by side. The South Limón Basin, initiated as part of a larger back-arc basin during the Late Cretaceous, was later transformed into a retro-arc foreland basin that has been affected by large-scale folding and thrusting since the Pliocene. The late Cenozoic shortening occurred due to: (i) the collision of the Panama arc with the South American continent; and (ii) the low-angle subduction of the Cocos Plate in the area of southern Costa Rica (this subduction geometry is caused by the young and buoyant crust and the influence of the Cocos Ridge). From the interpretation of seismic reflection lines, it is concluded that deformation in the offshore parts of the Limón fold-and-thrust belt started during the Pliocene, and that the main deformation occurred during the Pleistocene. The seismic sections studied contained no unambiguous evidence for young and active deformation, but such evidence is independently provided by seismic activity in this region (Protti & Schwartz, 1994; Suárez *et al.*, 1995).

The results from the Limón fold-and-thrust belt study provide several insights regarding the controlling factors of fold-and-thrust belts in general (Fig. 13). In summary, the main factors for the spatial and temporal evolution of deformed belts are:

1 the regional geodynamic framework that largely determines the regional stress field;
2 the lithology of the pre-deformation strata, which controls the location of the regional detachment and the basal friction;
3 the propagation-to-slip ratio of the thrust faults, which controls the shape of hangingwall anticlines;
4 the interaction of the fold-and-thrust belt with basement structures.

The architecture of the growth-strata is directly related to the structural evolution of the fold-and-thrust belt. Piggy-back basins behind and footwall synclines in front of the hangingwall anticlines catch the sediments on their way downslope. The evolution of the footwall syncline seems to be directly related to the displacement along the thrust fault, with higher displacement leading to a greater accommodation space. The greatest sediment thickness in the piggy-back basin occurs at the mouth of the Estrella River. The lateral thickness variations in the piggy-back basin may be

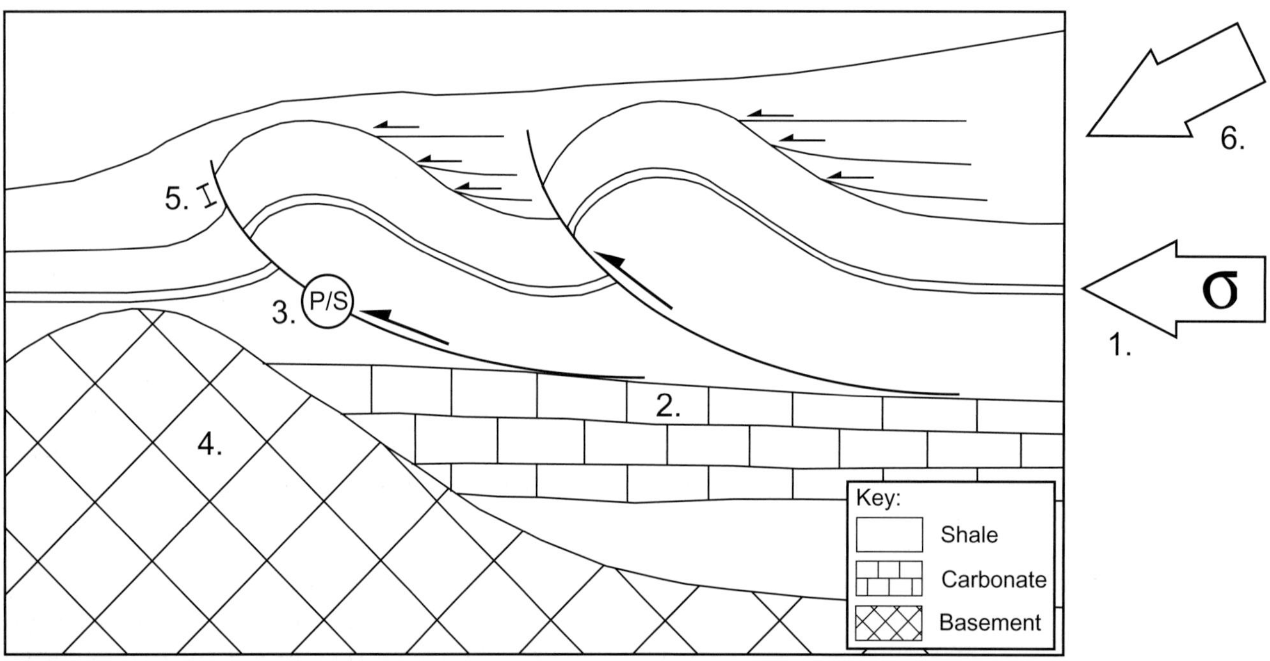

Fig. 13 Controlling factors for the fold-and-thrust belt. (1) Orientation of the regional stress field (controls the large-scale geometry of the fold-belt). (2) Lithology of the basin-fill (controls the position of the detachment and the basal friction). (3) The propogation-to-slip ratio along the thrust faults (controls the shape of the anticlines). (4) Interaction with basement structures (controls the outline of the fold-belt). (5) Displacement along the thrust faults (controls the accommodation space for the growth strata). (6) Sediment input (controls the thickness distribution of the growth strata).

influenced by locally high sediment input leading to increased loading and subsidence in that particular area.

ACKNOWLEDGEMENTS

We would like to thank the Costa Rican Ministry of Environment and Energy (MINAE) for providing the database. We are indebted especially to Alvaro Aguilar and Gustavo Segura for logistical help. Financial support from the German Research Foundation (DFG) Project Wi 1844/6-1 and a graduate scholarship from the University of Hannover are greatly acknowledged. Seismic Micro Technology Inc. is gratefully thanked for the sponsoring of Kingdom Suite©. We are indebted to Gary Nichols and Edward Williams for the possibility to participate in this special publication. The constructive comments of Edward Williams, Fabrizio Storti and an anonymous reviewer are greatly acknowledged and helped to improve the manuscript. Many thanks to Ulrich Asprion, Stefan Back, Franz Binot, Lolita Campos, Christoph Gaedicke, Stefan Ladage, Andreas Mende and Imke Struß for discussion and help with seismic interpretation.

REFERENCES

Abratis, M. and Wörner, G. (2001) Ridge collision, slab-window formation, and the flux of Pacific asthenosphere into the Caribbean realm. *Geology*, **29**, 127–130.

Allmendinger, R.W. (1998) Inverse and forward numerical modeling of trishear fault-propagation folds. *Tectonics*, **17**, 620–656.

Allmendinger, R.W. and Shaw, J.H. (2000) Estimation of fault propagation distance from fold shape: Implications for earthquake hazard assessment. *Geology*, **28**, 1099–1102.

Allmendinger, R.W., Zapata, T., Manceda, R. and Dzelalija, F. (2004) Trishear kinematic modeling of structures, with examples from the Neuquén Basin, Argentina. In: *Thrust Tectonics and Hydrocarbon Systems* (Ed. K.R. McClay), pp. 356–371. Memoir 82, American Association of Petroleum Geologists, Tulsa, OK.

Amann, H. (1993) Randmarine und terrestrische Ablagerungsräume des neogenen Inselbogensystems in Costa Rica (Mittelamerika). *Profil*, **4**, 161 pp.

Astorga, A. (1988) Geodinámica de las cuencas del Cretácico Superior – Paleógeno de la región 'forearc' del Sur de Nicaragua y Norte de Costa Rica. *Rev. Geol. Amér. Central*, **9**, 1–40.

Astorga, A. (1997) El puente-istmo de América Central y la evolución de la Placa caribe (con éfasis en el Mesozoico). *Profil*, **12**, 201 pp.

Barboza, G., Barrientos, J. and Astorga, A. (1995) Tectonic evolution and sequence stratigraphy of the central Pacific margin of Costa Rica. *Rev. Geol. Amér. Central.*, **18**, 43–63.

Barboza, G., Fernández, A., Barrientos, J. and Bottazzi, G. (1997) Costa Rica: Petroleum geology of the Caribbean margin. *Lead. Edge*, **16**, 1787–1794.

Barckhausen, U., Ranero, C.R., von Huene, R., Cande, S.C. and Roeser, H.A. (2003) Revised tectonic boundaries in the Cocos Plate off Costa Rica: Implications for the segmentation of the convergent margin and for plate tectonic models. *J. Geophys. Res.*, **106**, 19207–19220.

Barrientos, J., Bottazzi, G., Fernández, A. and Barboza, G. (1997) Costa Rican data synthesis indicates oil, gas potential. *Oil Gas J.*, **95**, 76–80.

Bottazzi, G., Fernandez, A. and Barboza, G. (1994) Sedimentología e historia tectono-sedimentaria de la cuenca Limón Sur. In: *Geology of an Evolving Island Arc, The Isthmus of Southern Nicaragua, Costa Rica and Western Panamá* (Eds H. Seyfried and W. Hellmann). *Profil*, **7**, 351–389.

Calvo, C. (2003) Provenance of plutonic detritus in cover sandstones of Nicoya Complex, Costa Rica: Cretaceous unroofing history of a Mesozoic ophiolite sequence. *Geol. Soc. Am. Bull.*, **115**, 832–844.

Campos, L. (2001) Geology and basins history of middle Costa Rica: an intraoceanic island arc in the convergence between the Caribbean and the central pacific plates. *Tübinger Geowiss. Arb., Reihe A*, **62**, 138 pp.

Cloos, M. (1993) Lithospheric buoyancy and collisional orogenesis: Subduction of oceanic plateaus, continental margins, island-arcs, spreading ridges, and seamounts. *Geol. Soc. Am. Bull.*, **105**, 715–737.

Coates, A.G., Jackson, J.B.C., Collins, L.S., *et al.* (1992) Closure of the Isthmus of Panama: The near-shore marine record of Costa Rica and western Panama. *Geol. Soc. Am. Bull.*, **104**, 814–828.

Coates, A.G., Aubry, M-P., Berggren, W.A., Collins, L.S. and Kunk, M. (2003) Early Neogene history of the Central American arc from Bocas del Toro, western Panama. *Geol. Soc. Am. Bull.*, **115**, 271–287.

Coates, A.G., Collins, L.S., Aubry, M-P. and Berggren, W.A. (2004) The geology of the Darien, Panama, and the late Miocene-Pliocene collision of the Panama arc with northwestern South America. *Geol. Soc. Am. Bull.*, **116**, 1327–1344.

Collins, L.S., Coates, A.G., Jackson, J.B.C. and Obando, J.A. (1995) Timing and rates of emergence of the Limón and Bocas del Toro basins: Caribbean effects of Cocos Ridge subduction? In: *Geologic and Tectonic Development of the Caribbean Plate Boundary in Southern Central America* (Ed. P. Mann), pp. 263–289. Special Paper 295, Geological Society of America, Boulder, CO.

Costa, E. and Vendeville, B.C. (2002) Experimental insights on the geometry and kinematics of fold-and-thrust belts above weak, viscous evaporitic décollement. *J. Struct. Geol.*, **24**, 1729–1739.

Covey, M. (1986) The evolution of foreland basins to steady state: evidence from the western Taiwan foreland basin. In: *Foreland Basins* (Eds P.A. Allen and P. Homewood), pp. 77–90. Special Publication 8, International Association of Sedimentologists. Blackwell Scientific Publications, Oxford.

Cross, T.A. (1986) Tectonic controls on foreland basin subsidence and Laramide style deformation, western United States. In: *Foreland Basins* (Eds P.A. Allen and P. Homewood), pp. 15–39. Special Publication 8, International Association of Sedimentologists. Blackwell Scientific Publications, Oxford.

Cross, T.A. and Pilger, R.H. (1982) Controls of subduction geometry, location of magmatic arcs, and tectonics of arc and back-arc regions. *Geol. Soc. Am. Bull.*, **93**, 545–562.

Dahlen, F.A. (1984) Noncohesive critical Coulomb wedges: An exact solution. *J. Geophys. Res.*, **89**, 10125–10133.

Dahlen, F.A., Suppe, J. and Davis, D. (1984) Mechanics of fold-and-thrust belts and accretionary wedges: cohesive Coulomb theory. *J. Geophys. Res.*, **89**, 10087–10101.

Davis, D., Suppe, J. and Dahlen, F.A. (1983) Mechanics of fold-and-thrust belts and accretionary wedges. *J. Geophys. Res.*, **88**, 1153–1172.

DeCelles, P.G. and Giles, K.A. (1996) Foreland basin systems. *Basin Res.*, **8**, 105–123.

DeMets, C. (2001) A new estimate for present-day Cocos-Caribbean plate motion: Implications for slip along the Central American volcanic arc. *Geophys. Res. Lett.*, **28**, 4043–4046.

Di Marco, G., Baumgartner, P.O. and Channell, J.E.T. (1995) Late Cretaceous-early Tertiary paleomagnetic data and a revised tectonostratigraphic subdivision of Costa Rica and western Panama. In: *Geologic and Tectonic Development of the Caribbean Plate Boundary in Southern Central America* (Ed. P. Mann), pp. 1–27. Special Paper 295, Geological Society of America, Boulder, CO.

Donnelly, T.W. (1989) Geologic history of the Caribbean and Central America. In: *The Geology of North America – An overview* (Eds A.W. Bally and A.R. Palmer), pp. 299–321. Geological Society of America, Boulder, CO.

Erslev, E.A. (1991) Trishear fault-propagation folding. *Geology*, **19**, 617–620.

Escalante, G. and Astorga, A. (1994) Geología del este de Costa Rica y el norte de Panama. *Rev. Geol. Amér. Central. Vol. esp. Terremoto de Limón*, 1–14.

Fernandez, J.A., Bottazzi, G., Barboza, G. and Astorga, A. (1994) Tectónica y estratigrafia de la Cuenca Limón Sur. *Rev. Geol. Amér. Central. Vol. esp. Terremoto de Limón*, 15–28.

Fernandez, J., Alvaro, A., Guillermo, B., *et al.* (1997) *Mapa Geológico de Costa Rica*. Ministerio del Ambiente y Energía, Costa Rica.

Flemings, P.B. and Jordan, T.E. (1990) Stratigraphic modeling of foreland basins: Interpreting thrust deformation and lithosphere rheology. *Geology*, **18**, 430–434.

Ford, M. (2004) Depositional wedge tops: interaction between low basal friction external orogenic wedges and flexural foreland basins. *Basin Res.*, **16**, 361–375.

Frisch, W., Meschede, M. and Sick, M. (1992) Origin of the Central American ophiolites: Evidence from paleomagnetic results. *Geol. Soc. Am. Bull.*, **104**, 1301–1314.

Gardner, T.W., Verdonck, D., Pinter, N.M., *et al.* (1992) Quaternary uplift astride the aseismic Cocos Ridge, Pacific coast, Costa Rica. *Geol. Soc. Am. Bull.*, **104**, 219–232.

Gräfe, K., Frisch, W., Villa, I.M. and Meschede, M. (2002) Geodynamic evolution of southern Costa Rica related to low-angle subduction of the Cocos Ridge: constraints from thermochronology. *Tectonophysics*, **348**, 187–204.

Gutscher, M.-A., Kukowski, N., Malavieille, J. and Lallemand, S. (1996) Cyclic behaviour of thrust wedges: Insights from high basal friction sandbox experiments. *Geology*, **24**, 135–138.

Hardy, S. and Ford, M. (1997) Numerical modelling of trishear fault-propagation folding and associated growth strata. *Tectonics*, **16**, 841–854.

Jordan, T.E. (1981) Thrust load and foreland basin evolution, Cretaceous, western United States. *Am. Assoc. Petrol. Geol. Bull.*, **65**, 2506–2520.

Jordan, T.E. (1995) Retroarc foreland and related basins. In: *Tectonics of Sedimentary Basins* (Eds C.J. Busby and R.V. Ingersoll), pp. 331–362. Blackwell Science, Oxford.

Klaucke, I., Hesse, R. and Ryan, W.B.F. (1998) Seismic stratigraphy of the northwest Atlantic mid-ocean channel: growth pattern of a mid ocean channel-levee complex. *Mar. Petrol. Geol.*, **15**, 575–585.

Kolarsky, R.A., Mann, P. and Montero, W. (1995) Island arc response to shallow subduction of the Cocos

Ridge, Costa Rica. In: *Geologic and Tectonic Development of the Caribbean Plate Boundary in Southern Central America* (Ed. P. Mann), pp. 235–262. Special Paper 295, Geological Society of America, Boulder, CO.

Krawinkel, J.J. (2003) Struktur und Kinematik am konvergenten Plattenrand der südlichen Zentralamerikanischen Landbrücke (Zentral- und Süd-Costa Rica, West-Panamá). *Profil*, **20**, 36 pp.

Krawinkel, J. and Seyfried, H. (1994) A review of plate-tectonic processes involved in the formation of the southwestern edge of the Caribbean Plate. In: *Geology of an Evolving Island Arc, The Isthmus of Southern Nicaragua, Costa Rica and Western Panamá* (Eds H. Seyfried and W. Hellmann). *Profil*, **7**, 47–61.

Krawinkel, H., Seyfried, H., Calvo, C. and Astorga, A. (2000) Origin and inversion of sedimentary basins in southern Central America. *Z. Angew. Geol.*, **SH 1**, 71–77.

Lundberg, N. (1991) Detrital record of the early Central American magmatic arc: Petrography of intraoceanic forearc sandstones, Nicoya Peninsula, Costa Rica. *Geol. Soc. Am. Bull.*, **103**, 905–915.

Lutz, R. (2002) Numerische Simulation der Kohlenwasserstoffgenese an der Subduktionszone vor Costa Rica. *Diss. RWTH Aachen*, 120 pp.

Lutz, R., Littke, R., Gerling, P. and Bönnemann, C. (2004) 2D numerical modelling of hydrocarbon generation in subducted sediments at the active continental margin of Costa Rica. *Mar. Petrol. Geol.*, **21**, 753–766.

Maresch, W.V., Stöckhart, B., Baumann, A., *et al.* (2000) Crustal history and plate tectonic development in the southern Caribbean. *Z. Angew. Geol.*, **SH 1**, 283–290.

Marshak, S. (2004) Salients, recesses, arcs, oroclines, and syntaxes – a review of ideas concerning the formation of map-view curves in fold-thrust belts. In: *Thrust Tectonics and Hydrocarbon Systems* (Ed. K.R. McClay), pp. 131–156. Memoir 82, American Association of Petroleum Geologists, Tulsa, OK.

Marshak, S. Wilkerson, M.S. and Hsui, A.T. (1992) Generation of curved fold-thrust belts: Insight from simple physical and analytical models. In: *Thrust Tectonics* (Ed. K.R. McClay), pp. 83–92. Chapman and Hall, London.

McClay, K.R. (1992) Glossary of thrust tectonic terms. In: *Thrust Tectonics* (Ed. K.R. McClay), pp. 419–433. Chapman and Hall, London.

McNeill, D.F., Coates, A.G., Budd, A.F. and Borne, P.F. (2000) Integrated paleontologic and paleomagnetic stratigraphy of the Upper Neogene deposits around Limon, Costa Rica: A coastal emergence record of the Central American Isthmus. *Geol. Soc. Am. Bull.*, **112**, 963–981.

Mende, A. (2001) Sedimente und Architektur der Forearc- und Backarc-Becken von Südost-Costa Rica und Nordwest-Panamá. *Profil*, **19**, 130 pp.

Meschede, M. and Frisch, W. (1998) A plate tectonic model for the Mesozoic and Early Cenozoic history of the Caribbean Plate. *Tectonophysics*, **296**, 269–291.

Meschede, M., Frisch, W., Chinchilla Chavez, A.L., López Saborio, A. and Calvo, C. (2000) The plate tectonic evolution of the Caribbean Plate in the Mesozoic and Early Cenozoic. *Z. Angew. Geol.*, **SH 1**, 275–281.

Mitchum. R.M., Vail, P.R. and Sangree, J.B. (1977) Seismic stratigraphy and global changes in sea-level, Part 6: stratigraphic interpretation of seismic reflection patterns in depositional sequences. In: *Seismic Stratigraphy – Applications to Hydrocarbon Exploration* (Ed. C.E. Payton), pp. 117–133. Memoir 26, American Association of Petroleum Geologists, Tulsa, OK.

Mitra, S. (1990) Fault-propagation folds: geometry, kinematic evolution, and hydrocarbon traps. *Am. Assoc. Petrol. Geol. Bull.*, **74**, 921–945.

Mitra, S. and Mount, V.S. (1998) Foreland basement involved structures. *Am. Assoc. Petrol. Geol. Bull.*, **82**, 70–109.

Mutti, E. and Normark, W.R. (1987) Comparing examples of modern and ancient turbidite systems: problems and concepts. In: *Marine Clastic Sedimentation* (Eds J.K. Leggett and G.G. Zuffa), pp. 1–38, Graham and Trotman, London.

Petzet, G.A. (1998) Costa Rica awards blocks on Caribbean coast. *Oil Gas J.*, **95**, 76–80.

Pilger, R.H. (1981) Plate reconstructions, aseismic ridges, and low-angle subduction beneath the Andes. *Geol. Soc. Am. Bull.*, **92**, 448–456.

Pindell, J.L., Cande, S.C., Pitman, W.C., *et al.* (1988) A plate-kinematic framework for models of Caribbean evolution. *Tectonophysics*, **155**, 121–138.

Protti, M. and Schwartz, S.Y. (1994) Mechanics of back arc deformation in Costa Rica: Evidence from an aftershock study of the April 22, 1991, Valle de la Estrella, Costa Rica, earthquake ($M_w = 7.7$). *Tectonics*, **13**, 1093–1107.

Ranero, C.R. and von Huene, R. (2000) Subduction erosion along the Middle America convergent margin. *Nature*, **404**, 748–752

Ranero, C.R., von Huene, R., Flueh, E., Duarte, M., Baca, D. and McIntosh, K. (2000a) A cross section of the convergent Pacific margin of Nicaragua. *Tectonics*, **19**, 335–357.

Ranero, C.R., von Huene, R., Flueh, E.R., *et al.* (2000b) Lower plate control on subduction erosion processes along the middle America convergent margin. *Z. Angew. Geol.*, **SH 1**, 291–296.

Ross, M.I. and Scotese, C.R. (1988) A hierachical tectonic model of the Gulf of Mexico and the Caribbean region. *Tectonophysics*, **155**, 139–168.

Seyfried, H., Astorga, A., Amann, H., *et al.* (1991) Anatomy of an evolving island arc: tectonic and eustatic control in the south Central American forearc area. In: *Sea-level Changes at Active Plate Margins: Processes and Products* (Ed. D.I.M. MacDonald), pp. 273–292. Special Publication 12, International Association of Sedimentologists. Blackwell Scientific Publications, Oxford.

Sheehan, C.A., Penfield, G.T. and Morales, E. (1990) Costa Rica geologic basins lure wildcatters. *Oil Gas J.*, **88**, 74–79.

Silver, E.A., Galewsky, J. and McIntosh, K.D. (1995) Variation in structure, style, and driving mechanism of adjoining segments of the North Panama deformed belt. In: *Geologic and Tectonic Development of the Caribbean Plate Boundary in Southern Central America* (Ed. P. Mann), pp. 225–234. Special Paper 295, Geological Society of America, Boulder, CO.

Sinclair, H.D. and Allen, P.A. (1992) Vertical versus horizontal motions in the Alpine orogenic wedge: stratigraphic response in the foreland basin. *Basin Res.*, **4**, 215–232.

Sinclair, H.D., Coakley, B.J., Allen, P.A. and Watts, A.B. (1991) Simulation of foreland basin stratigraphy using a diffusion model of mountain belt uplift and erosion: an example from the Central Alps, Switzerland. *Tectonics*, **10**, 599–620.

Suárez, G., Pardo, M., Domínguez, J., *et al.* (1995) The Limón, Costa Rica earthquake of April 22, 1991: Back arc thrusting and collisional tectonics in a subduction environment. *Tectonics*, **14**, 518–530.

Suppe, J. (1983) Geometry and kinematics of fault-bend folding. *Am. J. Sci.*, **283**, 684–721.

Suppe, J. and Medwedeff, D.A. (1990) Geometry and kinematics of fault-propagation folding. *Eclogae Geol. Helv.*, **83**, 409–454.

Turrini, C., Ravaglia, A.S. and Perotti, C.R. (2001) Compressional structures in a multilayered mechanical stratigraphy: Insights from sandbox modelling with three-dimensional variations in basal geometry and friction. In: *Tectonic Modeling: a Volume in Honor of Hans Ramberg* (Eds H.A. Koyi and N.S. Mancktelow), pp. 153–178. Special Paper 193, Geological Society of America, Boulder, CO.

Vergés, J., Muñoz, J.A. and Martínez, A. (1992) South Pyrenean fold and thrust belt: The role of foreland evaporitic levels in thrust geometry. In: *Thrust Tectonics* (Ed. K.R. McClay), pp. 255–263. Chapman and Hall, London.

Von Huene, R. and Flüh, E. (1994) A review of marine geophysical studies along the Middle America Trench off Costa Rica and the problematic seaward terminus of continental crust. In: *Geology of an Evolving Island Arc, The Isthmus of Southern Nicaragua, Costa Rica and Western Panamá* (Eds H. Seyfried and W. Hellmann). *Profil*, **7**, 143–159.

Walther, C.H.E. (2003) The crustal structure of the Cocos ridge of Costa Rica. *J. Geophys. Res.*, **108**, 1–21.

Weinberg, R.F. (1992) Neotectonic development of western Nicaragua. *Tectonics*, **11**, 1010–1017.

Werner, R., Hoernle, K., van den Bogaard, P., Ranero, C. and von Huene, R. (1999) Drowned 14-m.y.-old Galápagos archipelago off the coast of Costa Rica: Implications for tectonic and evolutionary models. *Geology*, **27**, 499–502.

Weyl, R. (1980) *Geology of Central America*. Borntraeger, Berlin, 371 pp.

Willett, S.D. (1992) Dynamic and kinematic growth and change of a Coulomb wedge. In: *Thrust Tectonics* (Ed. K.R. McClay), pp. 19–31. Chapman and Hall, London.

Winsemann, J. (1992) Tiefwasser-Sedimentationsprozesse und -produkte in den Forearc-Becken des mittelamerikanischen Inselbogensystems: eine sequenzstratigraphische Analyse. *Profil*, **2**, 218 pp.

Winsemann, J. and Seyfried, H. (1991) Response of deep-water forearc systems to sea-level changes, tectonic activity and volcaniclastic input in central America. In: *Sea-level Changes at Active Plate Margins: Processes and Products* (Ed. D.I.M. MacDonald), pp. 217–240. Special Publication 12, International Association of Sedimentologists. Blackwell Scientific Publications, Oxford.

Zehnder, A.T. and Allmendinger, R.W. (2000) Velocity field for the trishear model. *J. Struct. Geol.*, **22**, 1009–1014.

Tectono-sedimentary phases of the latest Cretaceous and Cenozoic compressive evolution of the Algarve margin (southern Portugal)

FERNANDO C. LOPES*† *and* P. P. CUNHA*

**Centro de Geofisica, Department of Earth Sciences, Faculty of Sciences and Technology, Universidade de Coimbra, Largo Marquês de Pombal, 3000-272 Coimbra, Portugal (Email: fcarlos@dct.uc.pt)*

†IMAR – Instituto de Mar, Department of Earth Sciences, Faculty of Sciences and Technology, Universidade de Coimbra, Av. Dr. Dias da Silva, 3000-134 Coimbra, Portugal

ABSTRACT

The latest Cretaceous and Cenozoic tectono-sedimentary evolution of the central and eastern Algarve margin (southwestern Iberia) is reconstructed as a series of structural maps and three-dimensional diagrams based on multichannel seismic reflection data. Six seismic stratigraphic units, bounded by unconformities related to tectonic events during the African–Eurasian convergence, have been identified. Several episodes of major regional change in palaeogeography and tectonic setting are distinguished: they occurred in the Campanian, Lutetian, Oligocene–Aquitanian transition, middle Tortonian, Messinian–Zanclean transition and Zanclean–Piacenzian transition. These changes were induced by geodynamic events primarily related to the relative motions of the African and Eurasian plates. The Late Cretaceous and Cenozoic in the Algarve margin were dominated by compressional deformation. Triggered by the regional tectonics that affected the basement, Upper Triassic–Hettangian evaporites played an important role in tectono-sedimentary evolution by localizing both extensional and thrust detachments and generating both salt structures and salt-withdrawal sub-basins. During middle Eocene and Oligocene times, coeval development of compressive structures and normal fault systems in the eastern Algarve domain is interpreted as resulting from gravity gliding due to a general tilt of the margin. The increasing effects of the African–Eurasian convergent plate boundary zone resulted in the uplift of some areas, overprinted by an increasingly general subsidence of the domains studied.

Keywords Cenozoic, Algarve margin, Gulf of Cadiz, Iberia, Europe, tectonics.

INTRODUCTION

Differential motions between tectonic plates create intense deformation along their boundaries. Interaction between the African/Arabian and Eurasian plates has generated a broad collision zone comprising the Himalayan–Alpine chains, running from southeast Asia to southwest Europe. In the case of Iberia, located at the western end of this zone of convergence, the progressive opening of the North Atlantic Ocean has been the most important control in the complex pattern of differential motion between Iberia, Eurasia and Africa (e.g. Ziegler, 1988). After a long period in the Mesozoic, during which extension was the dominant mode of deformation, the Late Cretaceous to present-day has been a period of compression in the Iberian peninsula. The major compressive tectonic intervals can be related to the Pyrenean collision, opening of the western part of the Mediterranean basin, and collision in the Betics. According to Andeweg & Cloetingh (2001), Iberia has been dominated by compressive regimes with the maximum horizontal compressive stress (Sh_{max}) ranging between northeast and northwest; the dominant stress regimes range from uniaxial compression to transpression.

The main aim of this paper is to characterize the latest Cretaceous and Cenozoic tectono-sedimentary phases of the Algarve margin, a region in a critical location for the study of evolving plate boundaries, showing sedimentary evolution and salt tectonics in a compressional setting. A secondary aim is to discuss the tectono-stratigraphic interrelationships between the coeval development of the study area and the adjacent domains, which is important in understanding the large-scale tectonic processes that caused the plate deformation and in placing its evolution in a broader tectonic context. This integration of data improves the interpretation of the regional geodynamic evolution (Gulf of Cadiz), which is relevant to the understanding of the Azores–Gibraltar plate boundary.

GEOLOGICAL SETTING

The Algarve margin, in southwestern Iberia, is situated on the northern border of the Gulf of Cadiz (Fig. 1) at the eastern end of the Azores–Gibraltar fracture zone (AGFZ), a diffuse transpressional plate boundary between the Iberian and African plates (Sartori *et al.*, 1994). Its complex geodynamic evolution, particularly during the latest Cretaceous and Cenozoic, has resulted from the convergence between Africa and Iberia along the eastern segment of the AGFZ (Dewey *et al.*, 1989; Srivastava *et al.*, 1990a, b) and the westward migration of the front of the Gibraltar Arc (e.g. Ribeiro *et al.*, 1990; Sanz de Galdeano, 1990; Gràcia *et al.*, 2003). During Neogene compressional phases, concentric wedges of fold and thrust belts and large allochthonous masses were emplaced in the Gulf of Cadiz (Campo de Gibraltar, External Betics and Guadalquivir Allochthon; e.g. Flinch *et al.*, 1996; Gràcia *et al.*, 2003), from the southeast (pre-early Langhian) towards the northwest (late Tortonian). Large gravitational accumulations and submarine landslides formed the 'Giant Chaotic Body' identified in the outer part of the Gulf of Cadiz (e.g. Bonnin *et al.*, 1975; Lajat *et al.*, 1975; Auzende *et al.*, 1981; Malod, 1982; Flinch *et al.*, 1996; Maestro *et al.*, 2003). The present-day geodynamics in the region of the Gulf of Cadiz, Gibraltar Arc and westernmost Alboran Sea, where the relative convergence between Iberia and Africa is only 4 mm yr^{-1}, are compatible with an active east-dipping subduction zone beneath the Gibraltar Arc (Gutscher *et al.*, 2002).

The stratigraphic record of the Algarve basin, both onshore and offshore, spans from Upper Triassic to Quaternary times, with several unconformity-bounded sequences (Terrinha, 1998; Lopes & Cunha, 2000; Lopes, 2002). This record can be briefly summarized as follows. Triassic to Lower Jurassic units are 500 m thick onshore. The Triassic red fluvial siliciclastics are capped by Hettangian evaporites and volcanics, followed by Sinemurian to Toarcian dolomites and marly limestones. The Middle Jurassic succession is 960 m thick and comprises bioclastic limestones that change upwards to marls and limestones, whereas the Upper Jurassic consists of 1000 m of dolomites and limestones. Lower Cretaceous strata are 900 m thick, comprising limestones, dolomites, sandstones and clays, but Upper Cretaceous to Paleocene sediments are not widely developed. Paleogene sediments have been reported from offshore wells but are not known onshore. The 675-m-thick upper Campanian(?) to middle Eocene succession comprises dolomites and some limestones. The middle Eocene to Oligocene succession is 200 m thick, comprising micritic limestones and minor dolomites. Probable Aquitanian to lower Tortonian deposits could be 100 m thick and are mainly limestones that are overlain by fine sandstones. The 1000-m-thick upper Tortonian to Quaternary succession comprises siltstones and sandstones.

The basement consists of metasediments and some igneous rocks, belonging to the South Portuguese Zone of the Variscan External Belt. Basement-related movements may have controlled a significant part of the structural deformation of the Algarve basin, under the changing stress field; pre-Tertiary structures played a major role in the later deformation (Ribeiro *et al.*, 1979). The Cenozoic tectonic style was thin-skinned, both onshore and offshore; Hettangian evaporites acted as a detachment layer during the extensional and compressional stages (Ribeiro *et al.*, 1990; Terrinha, 1998; Lopes, 2002).

METHODS

The present study is based on the interpretation of a 1974 Chevron and Challenger multichannel seismic reflection (MCS) survey, consisting of a grid of seismic profiles covering an area of about 125 × 100 km,

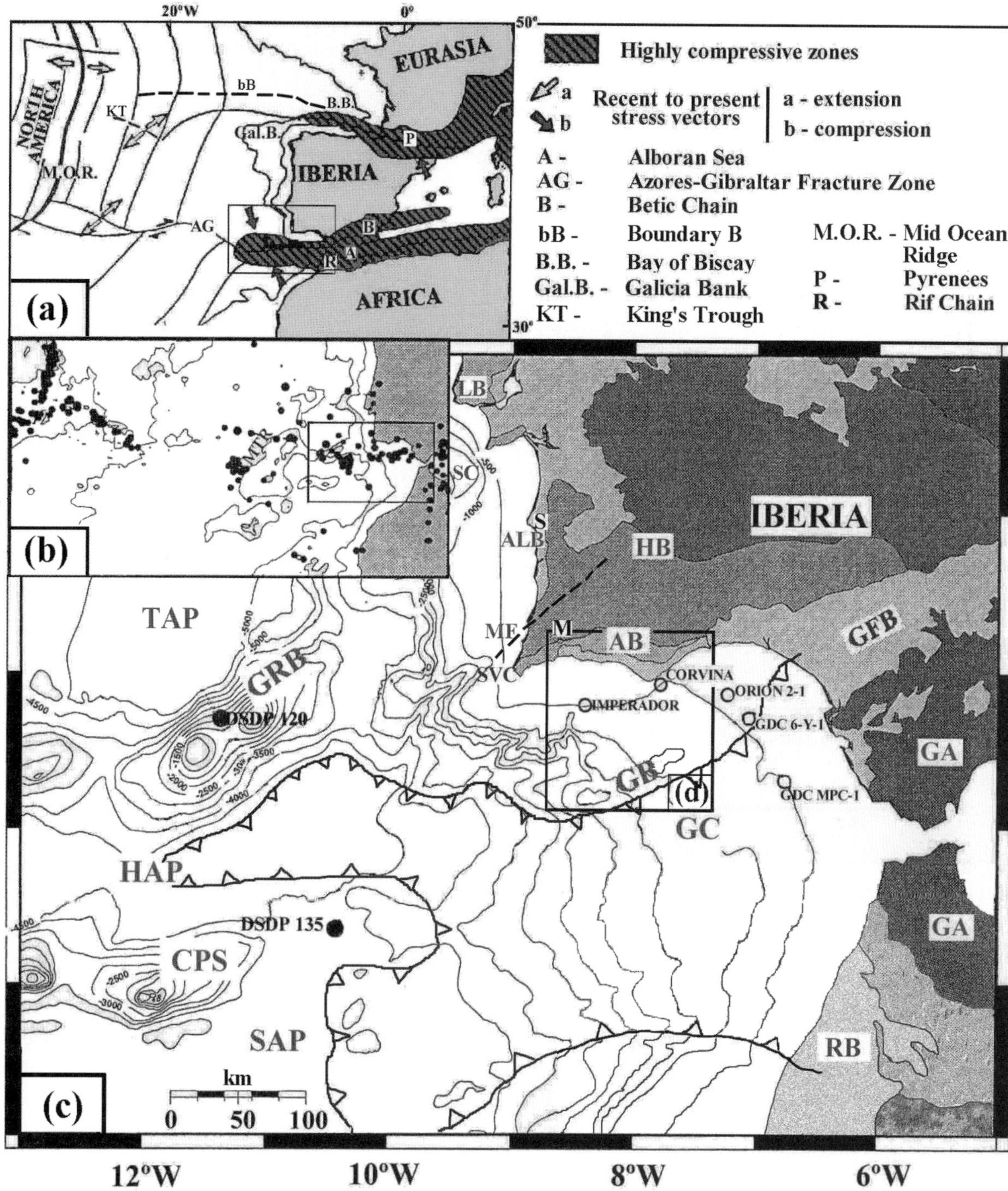

Fig. 1 (a) Present-day stress field at the periphery of the Iberian microplate (adapted from Olivet, 1996). (b) Location of the earthquakes with M > 3 in the Azores–Alboran region; MTR: Madeira-Tore Ridge (adapted from Buforn *et al.*, 1988). (c) Geological setting and simplified bathymetry of the Gulf of Cadiz and surrounding areas (adapted from Le Gall *et al.*, 1997; Tortella *et al.*, 1997). AB, Algarve basin; ALB, Alentejo basin; CPS, Coral Patch Seamont; GA, Gibraltar Arc; GB, Guadalquivir Bank; GC, Gulf of Cadiz; GFB, Guadalquivir foreland basin; GRB, Gorringe Bank; HAP, Horseshoe Abyssal Plain; HB, Variscan Basement; LB, Lusitanian basin; M, Monchique; MF, Messejana fault; RB, Raarb basin; SAP, Seine Abyssal Plain; S, Sines; SC, Setúbal canyon; SVC, São Vicente canyon; TAP, Tagus Abyssal Plain; filled circles, DSDP sites; open circles, exploration wells; line with open triangles, 'Giant body' boundary. (d) Study area.

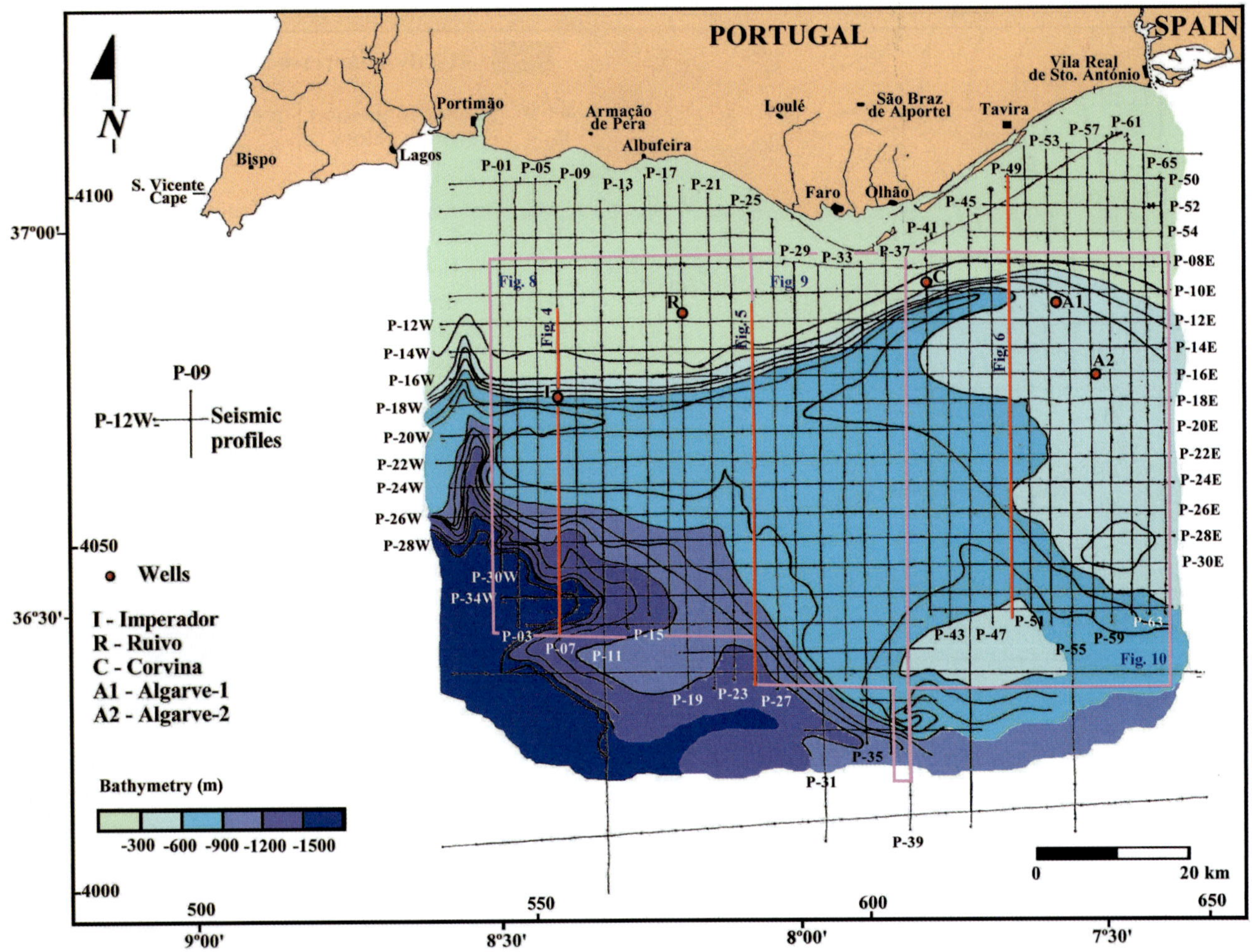

Fig. 2 Simplified bathymetric chart of the study area. The grid of the multichannel seismic (MCS) profiles and location of the five exploration wells are shown. The bold red lines indicate the location of the seismic profiles shown in Figs 4–6. The pink boxes represent the areas displayed in Figs 8, 9 & 10.

in the central and the eastern sectors of the Algarve margin (longitudes 8°30′W and 7°30′W; latitudes 36°10′N and 37°00′N). The seismic profiles are tied to five oil exploration wells (Imperador-1, 1976; Ruivo-1, 1975; Corvina-1, 1976; Algarve-1, 1982; Algarve-2, 1982) drilled as deep as 3 km, in this part of the Algarve margin (Figs 2 & 3).

The offshore uppermost Cretaceous to recent seismic units (labelled B–G), bounded by unconformities (labelled as reflectors *H6–H1*), previously identified and characterized by Lopes & Cunha (2000) and Lopes (2002), support the establishment of tectono-sedimentary phases presented here (Fig. 3). It is not possible to show all the seismic data used for this study, so only three representative lines and interpretations are presented (Figs 4–6) in order to validate the interpretation of the seismic data.

The ages of the seismic units have been interpreted on the basis of:

1 biostratigraphic data from the oil exploration well reports;
2 the intersection between the Portuguese seismic grid and an adjacent Spanish MCS profile interpreted by Maldonado *et al.* (1999), allowing the correlation of the Cenozoic seismic units recognized in both margins;
3 the presence of the Guadalquivir Allochthonous front, dated as middle to late Tortonian in the adjacent area (e.g. Gràcia *et al.*, 2003);
4 correlation with unconformities dated in adjacent Portuguese basins (Cunha, 1992a, b; Pais *et al.*, 2000;

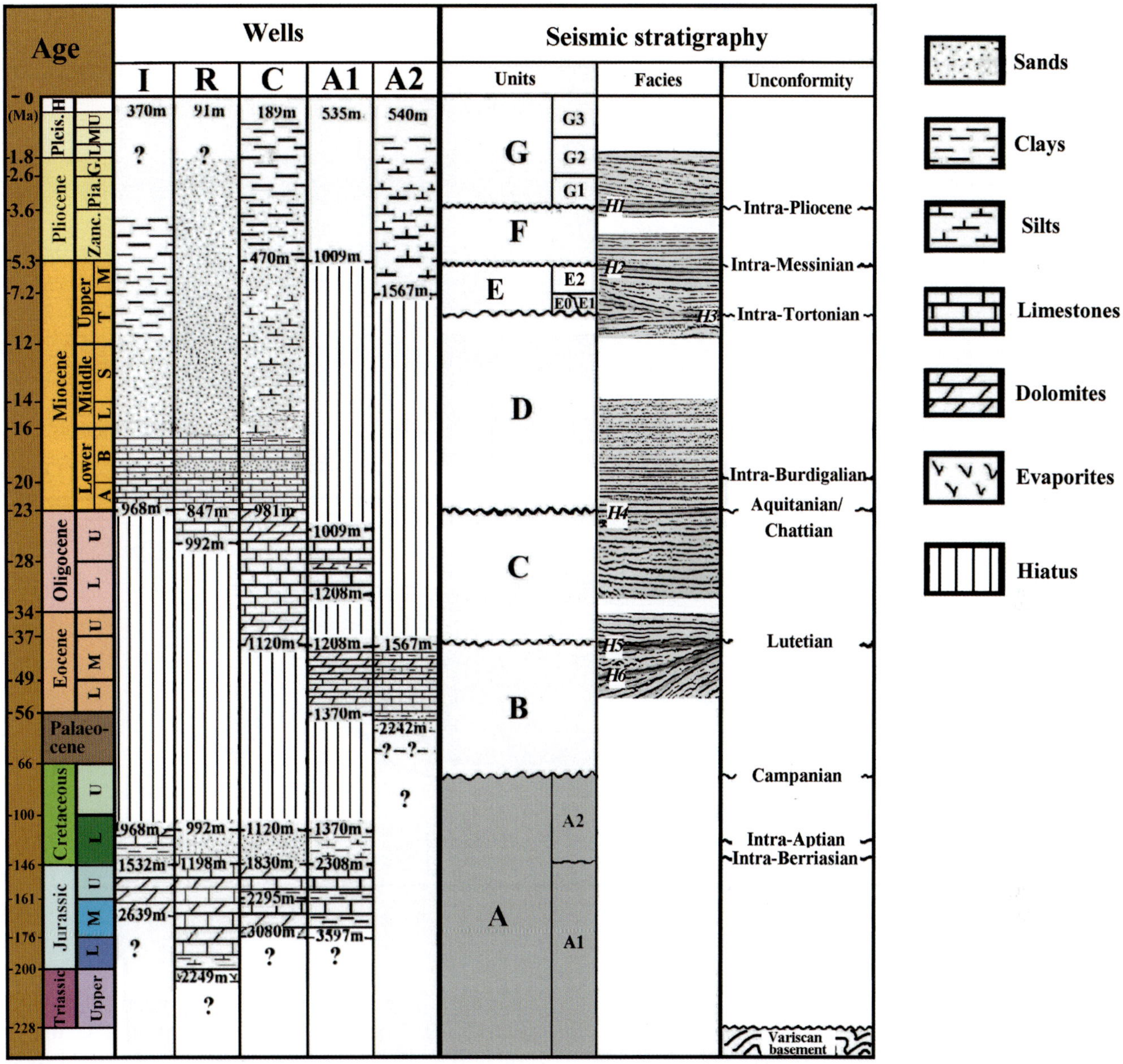

Fig. 3 Seismic stratigraphy, main unconformities and wells in the Algarve offshore. Wells: I, Imperador-1; R, Ruivo-1; C, Corvina-1; A1, Algarve-1; A2, Algarve-2. Chronological time-scale from Gradstein *et al.* (2004).

Alves *et al.*, 2003) and related to the tectonic events that affected Iberia.

STRUCTURAL FRAMEWORK

Four major fault zones, roughly transverse to the Azores–Gibraltar Fracture Zone, segment the Algarve margin.

1 The Messejana fault zone, striking N60°E, crops out onshore and offshore. Its recent activity is indicated by the São Vicente submarine canyon (Fig. 1) and seismic activity.

2 The Portimão–Monchique fault zone (PMFZ), striking N–S and also identified onshore, is about 70 km long offshore (Fig. 7). It is well documented in the E–W seismic reflection profiles, the westernmost ends of which intersect this fault. Its recent activity

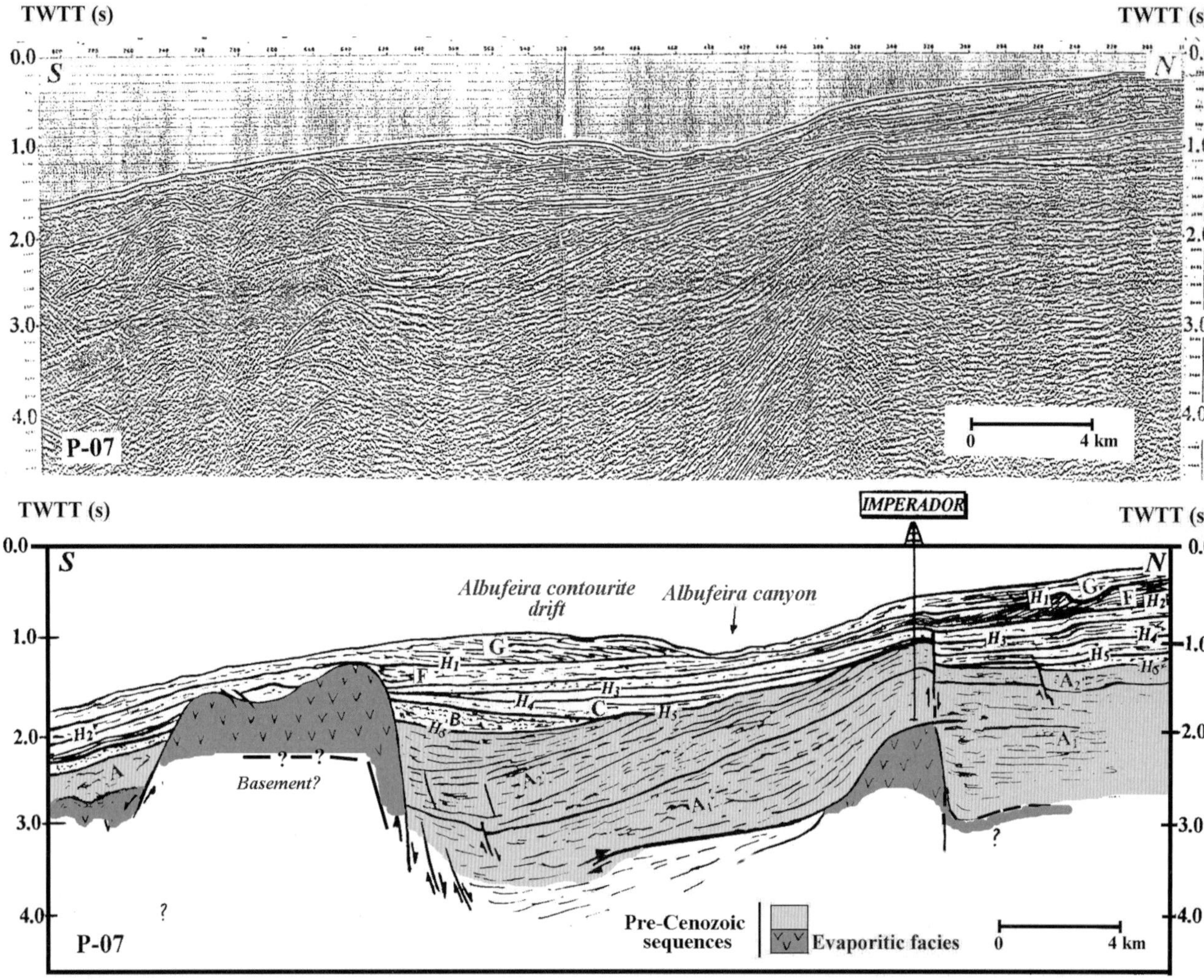

Fig. 4 P-07 seismic profile and interpretation (see Fig. 2 for location).

is indicated by the Portimão submarine canyon and seismic activity. According to Terrinha (1998) and Terrinha *et al.* (1999), this structure is a segment of an intermittent late Variscan dextral vertical fault that was reactivated as a main transfer fault during tectonic extension and tectonic inversion of the Algarve Basin and as a dextral strike-slip fault during the Late Cretaceous rotation of Iberia. As a consequence of the NW–SE middle Tortonian compressive event, PMFZ became a sinistral strike-slip fault.

3 The Albufeira fault zone (ALFZ) strikes approximately N–S and appears to be a segmented listric extensional fault involving three main fault segments with opposite polarities. Its activity was diachronous along-strike, with younger fault displacements in its southernmost segment. Here, there is evidence for important extensional displacements along the eastern margin of an easterly facing half-graben filled with syntectonic sequences ranging from unit C up to unit F (Fig. 8). The central segment is marked by a 2–3-km-thick elongate salt-body intrusion.

4 The São Marcos–Quarteira fault (SMQF) zone strikes N40°W and also crops out onshore; it is 70 km long offshore and coincides with the Diogo Cão deep. According to Terrinha (1998), this is an inherited Variscan thrust reactivated as a major dextral transtensional fault during Mesozoic extension. In the eastern area of the basin, the downthrow of the eastern block allowed deposition of sediments more than twice as thick as the western equivalent. During tectonic inversion, the São Marcos–Quarteira fault zone was reactivated mainly as a dextral strike-slip fault. The SMQF zone is thought to be a transfer fault of the offshore southward verging E–W to ENE–WSW thrust front.

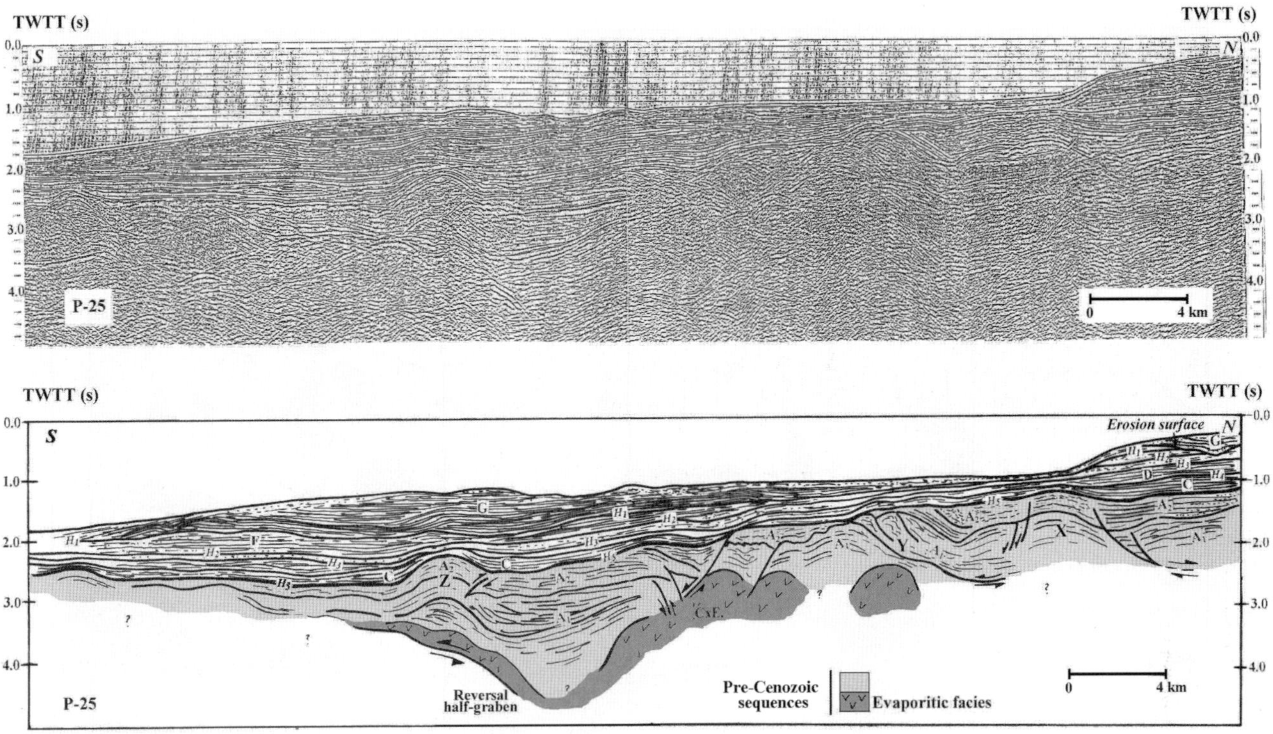

Fig. 5 P-25 seismic profile and interpretation (see Fig. 2 for location).

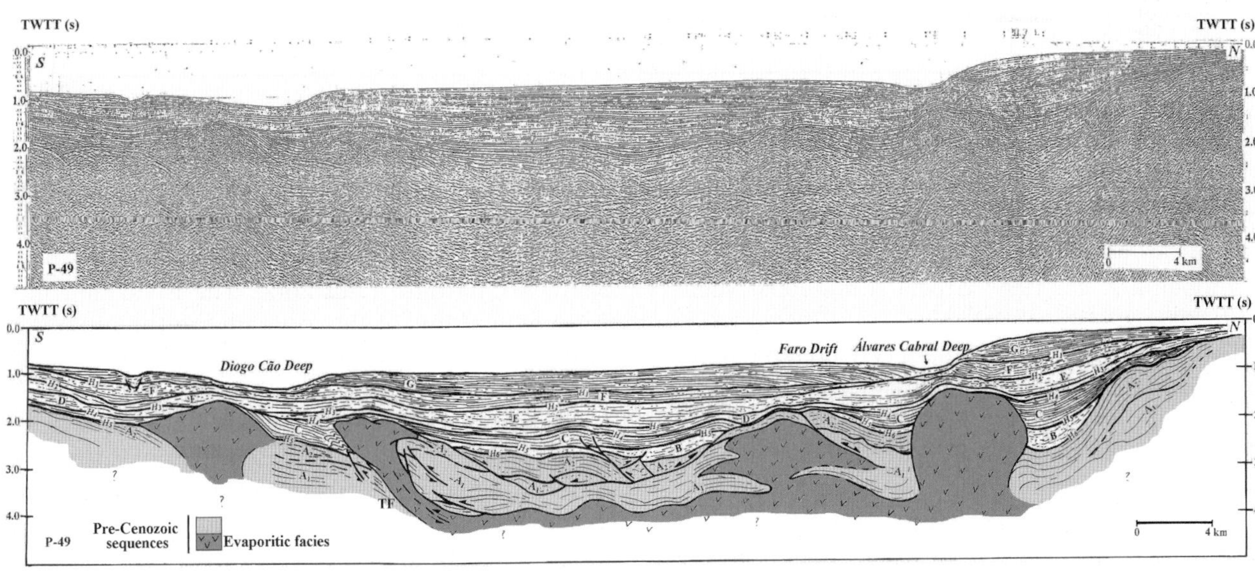

Fig. 6 P-49 seismic profile and interpretation (see Fig. 2 for location). A thin-skinned syn-unit C gravitational gliding and the later inversion of the structures are dominant.

The latter three fault zones (b–d, above) define the three tectonic domains of the study area (Fig. 7), all bounded to the south by the N70°E-trending Guadalquivir Bank, a morphotectonic high located on the middle continental slope of the Atlantic Southern Iberian margin, 100 km south of Faro (Portugal). The Guadalquivir Bank is the offshore continuation of the Iberian Variscan Massif

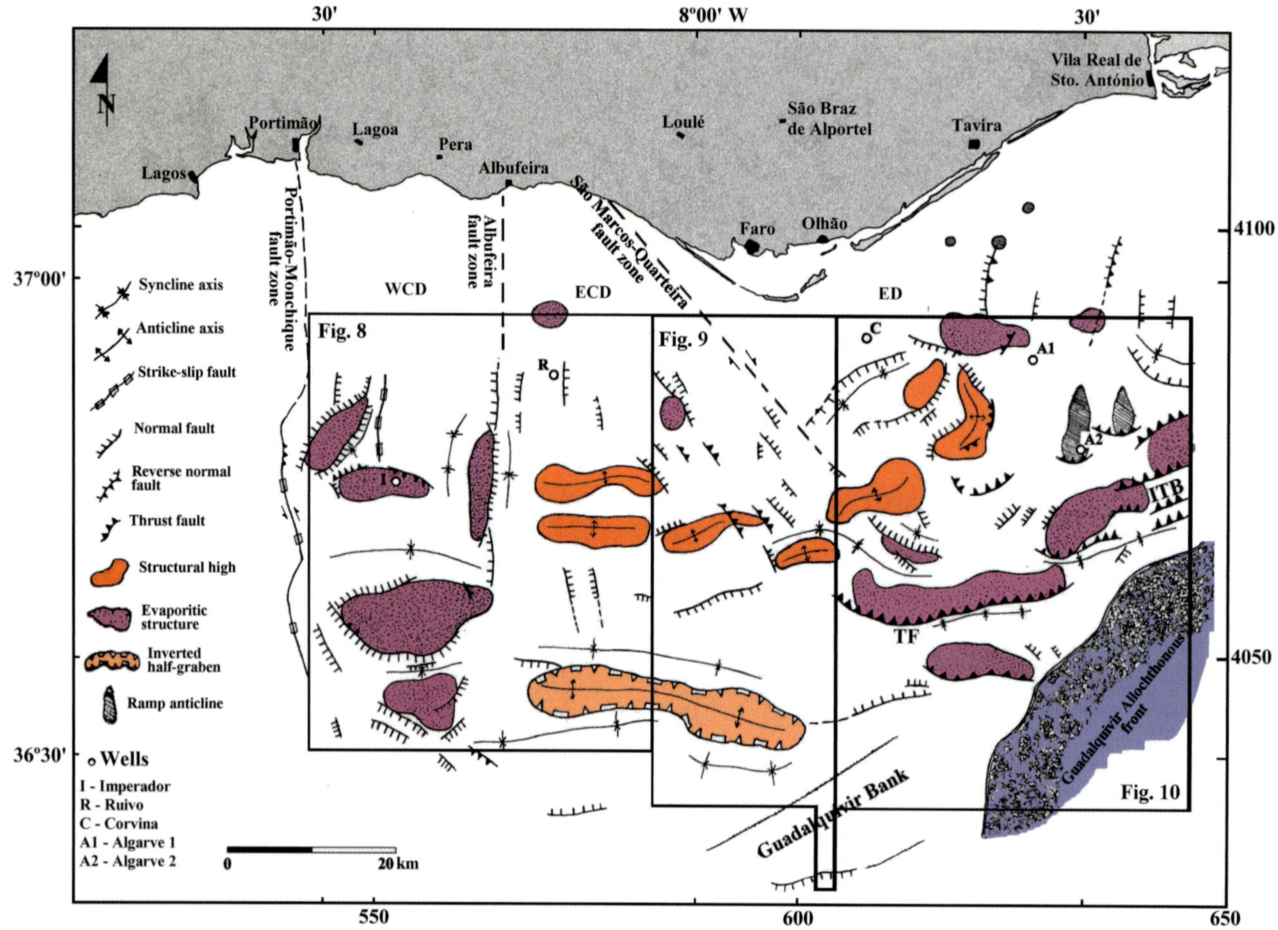

Fig. 7 Summary map of the main Cenozoic tectonic structures. Boxes show the areas covered by Figs 8–10: WCD, Western Central Domain; ECD, Eastern Central Domain; ED, Eastern Domain; TF, thrust front; ITB, imbricated thrust belt.

(Dañobeitia *et al.*, 1999; Gràcia *et al.*, 2003; Vegas *et al.*, 2004).

The Western Central Domain (WCD) is a narrow (25 km wide) N–S-trending domain, about 1500 km^2 in area, limited to the west by the Portimão–Monchique fault zone and to the east by the Albufeira fault zone. It includes predominant N–S- and E–W-trending structures and, secondarily, NW–SE and N40°E structures. The main morphotectonic features are four evaporitic walls associated with N–S (central segment of the ALFZ), E–W and N40°E lineaments respectively (Figs 4, 7 & 8).

The Eastern Central Domain (ECD) is a triangular area (1300 km^2) bounded to the west by the Albufeira fault zone and to the east by the São Marcos–Quarteira fault zone. The main morphotectonic features of this domain are three parallel antiforms with E–W- to ENE–WSW-trending axes (Figs 5 & 7–9).

The Eastern Domain (ED) is an irregular-shaped area (1800 km^2), tectonically more complex than the others, that corresponds to a structural depression dominated by three main lineaments (Figs 7 & 10): a WSW–ENE 20-km-long thrust front, verging to the south, located north of the Guadalquivir Bank (near latitude 36°38′N), which involves salt slices at depth (Fig. 6); N60°E, southeasterly dipping listric normal faults, located close to the upper slope, and a 20-km-wide zone of imbricate thrust faults verging to the south, located at the southeast margin of the domain; NNE–SSW reverse faults,

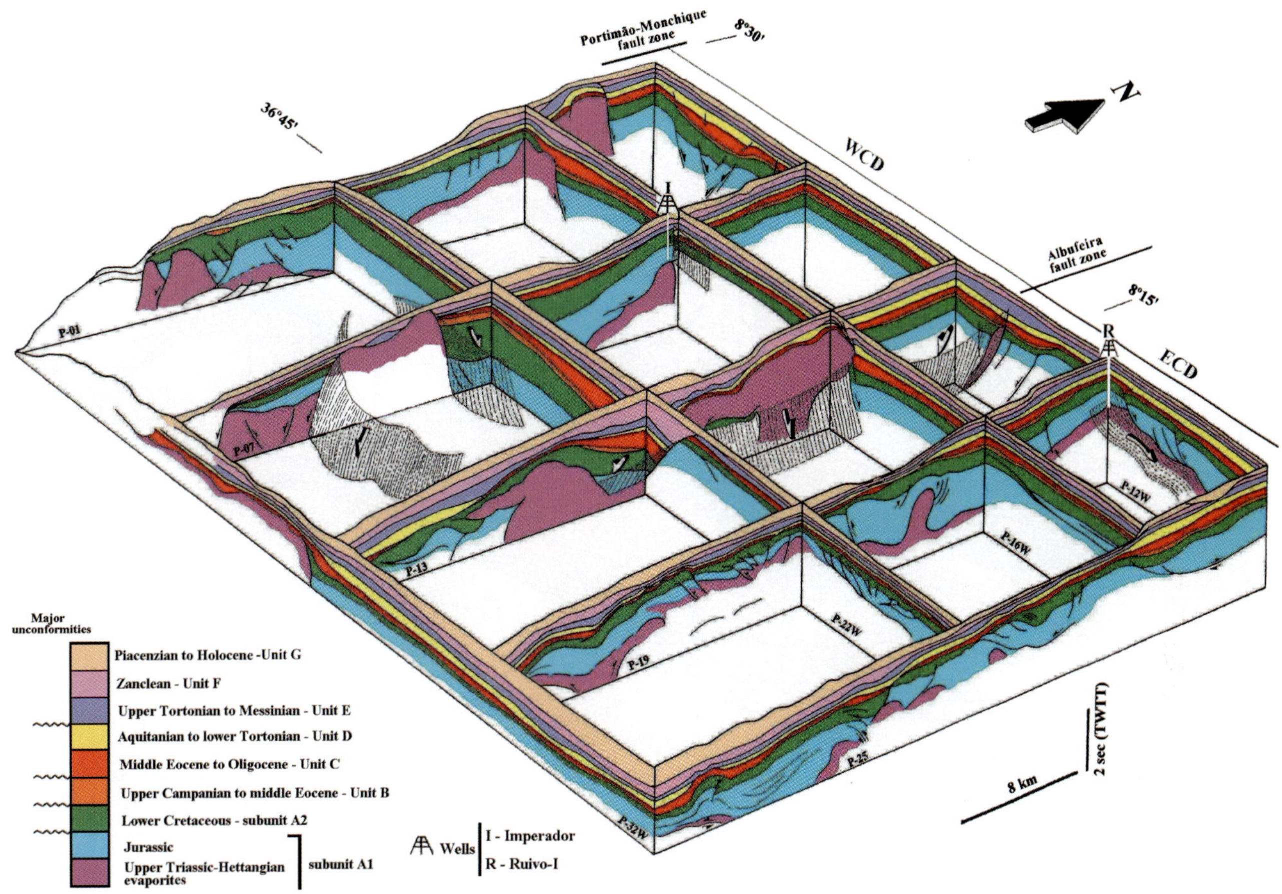

Fig. 8 Three-dimensional diagram of the Western Central Domain (WCD) and part of the Eastern Central Domain (ECD), showing the seismic units and their geometrical relationship to tectonic structures.

verging to the west, located southeast of Tavira. These reverse faults resulted from the inversion of previous extensional structures. The Eastern Domain is also dominated by the Guadalquivir Allochthonous front, located in the southeastern extremity of the study area. This 50-km-wide front has a wedge-shaped geometry, with decreasing thickness northwards and westwards (Figs 7 & 10).

TECTONO-SEDIMENTARY PHASES OF THE ALGARVE MARGIN

Evaluation of the tectono-stratigraphic interrelations makes it possible to infer episodes of major change that simultaneously affected the adjacent parts of the convergent plate boundary zone in the past 80 Myr. The following sections characterize the six tectono-sedimentary phases documented in the Algarve margin (Fig. 11).

Late Campanian to middle Eocene tectono-sedimentary phase

In the Algarve margin, the late Campanian to middle Eocene phase started with the deposition of marls and sandstones, followed by marine grey dolomites intercalated with marly limestones and micritic limestones. This is documented by well data (Fig. 3) and corresponds to seismic unit B. This unit is better represented in the Eastern Domain where it can reach more than 0.4 s TWTT equivalent thickness (Figs 6, 10 & 12). In some areas unit B is only preserved in E–W-trending synclines; some later erosion, prior to deposition of unit C, may have occurred.

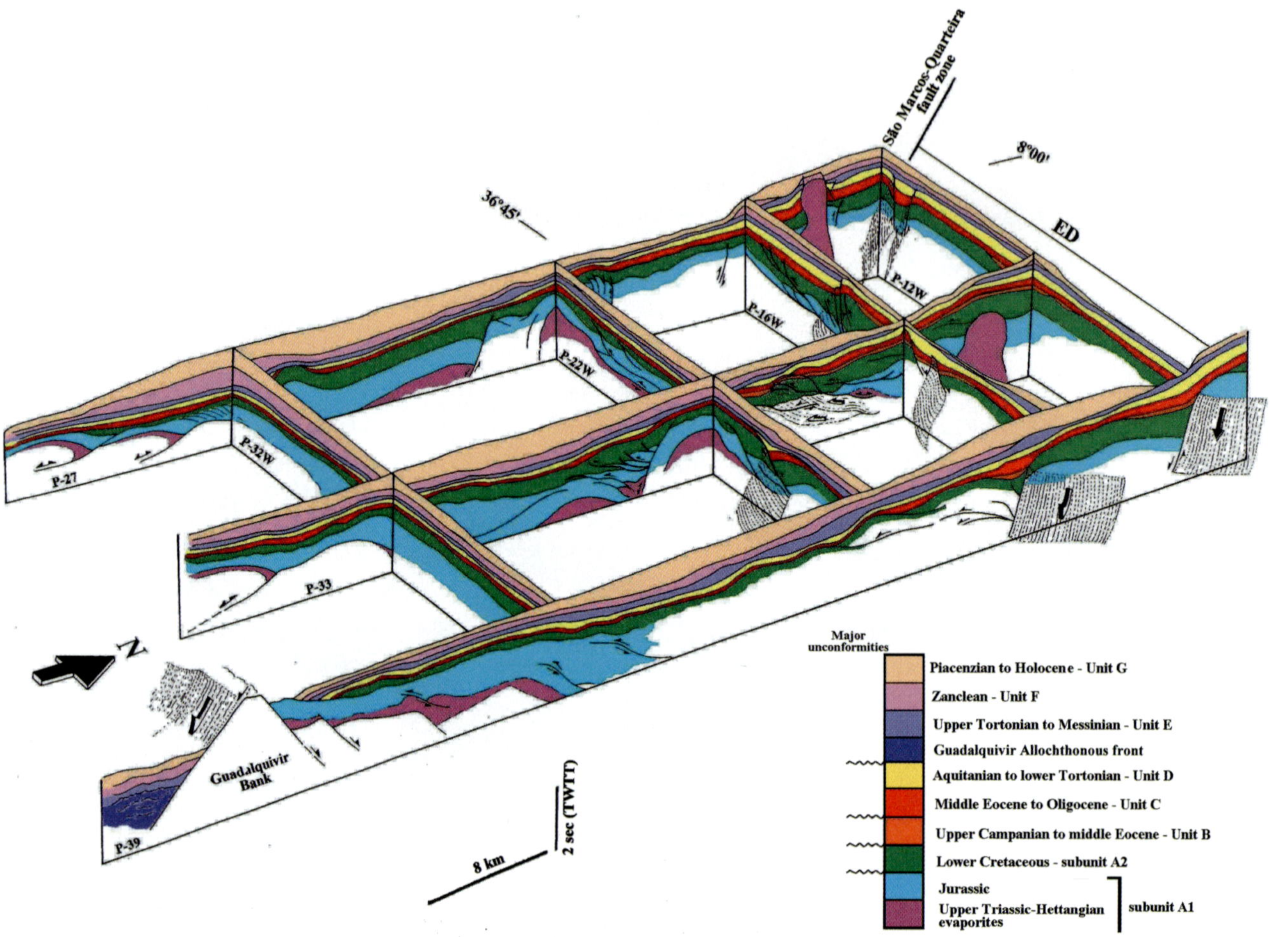

Fig. 9 Three-dimensional diagram of the eastern part of the Central Domain and the boundary with the Eastern Domain (ED), showing the seismic units and their geometrical relationship to tectonic structures.

Middle Eocene to Oligocene tectono-sedimentary phase

In the Western Central Domain of the Algarve margin, an important angular unconformity truncates the folded pre-unit-C deposits, testifying to a major tectonic event (Fig. 4). The middle Eocene to Oligocene phase was characterized by the deposition of micritic limestones (seismic unit C; Fig. 3), suggesting that a carbonate platform developed over the entire margin. Although the thickness of unit C is variable, values of more than 0.6 s TWTT are found in half-grabens and foredeep basins mainly at the eastern Algarve margin (Fig. 6).

Seismic data show that the middle Eocene to Oligocene evolution of the Algarve margin was marked by intense and widespread halokinesis (Figs 6, 8–10 & 13). Salt withdrawal from interdiapiric areas and transfer into growing salt pillows or salt walls resulted in the formation of salt-withdrawal sub-basins. A salt-/fault-controlled (thin-skinned) subsidence influenced the thickness and the lateral distribution of unit C. In the Western Central Domain, the southern part of the Albufeira fault zone was active during this phase. In the northern sector of the Eastern Central Domain, a NE–SW flexural sub-basin was active (Figs 8 & 13). In the Eastern Domain gravity gliding of the cover was associated with uplift and tilting of the northern sector of the margin, enhanced by tectonic inversion of the basement. The resultant glide tectonics formed an area under tension upslope and an area under compression downslope. The extensional sector was characterized by the development of a N60°E-striking listric normal fault system. Half-grabens were developed in the

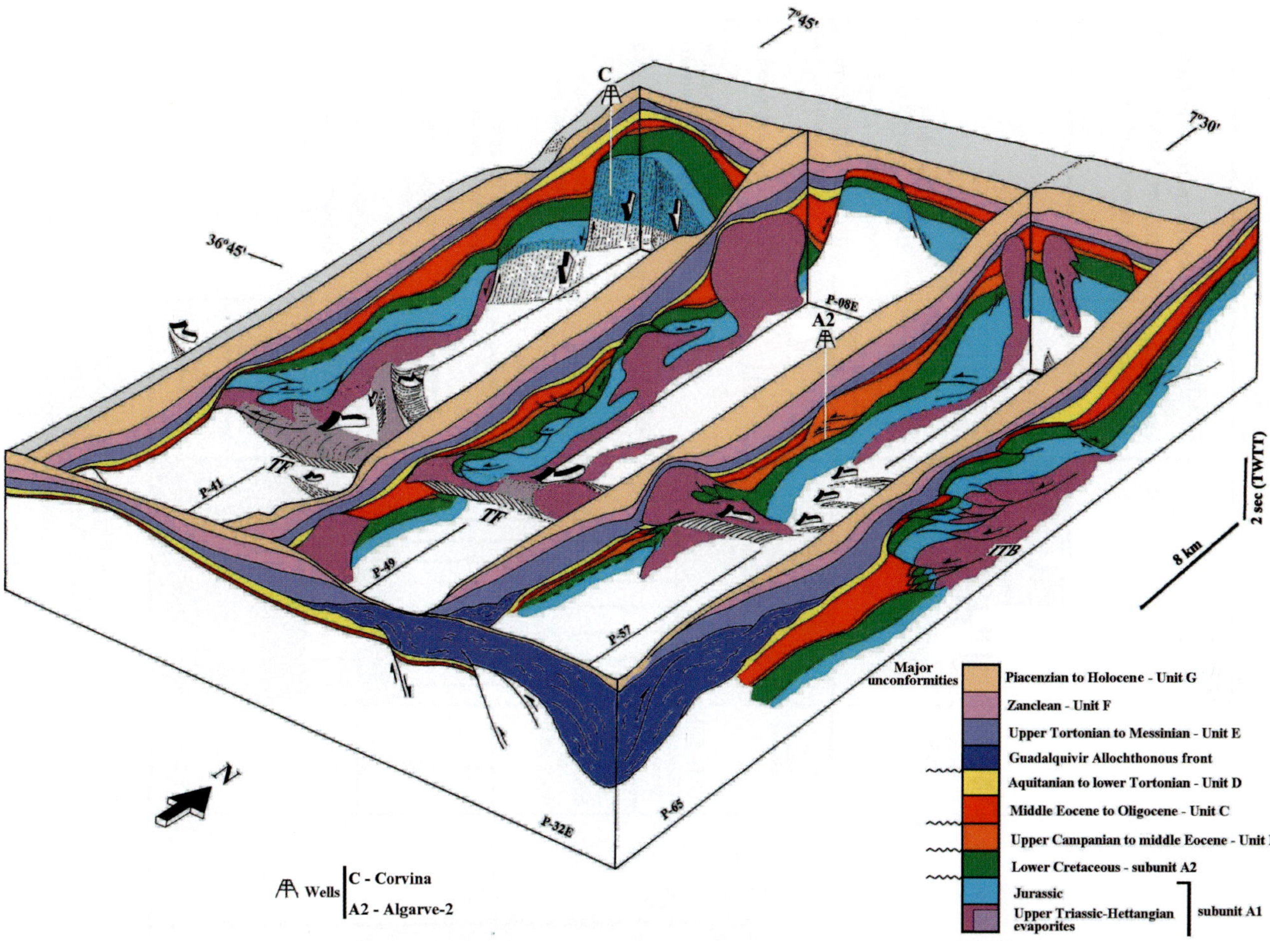

Fig. 10 Three-dimensional diagram of the eastern part of the Eastern Domain (ED), showing the seismic units and their geometrical relationship to tectonic structures: TF, thrust front; ITB, imbricated thrust belt.

northwestward tilted hanging-wall blocks. The contractional sector was characterized by the development of salt anticlines and turtle structures and the ENE–WSW 20-km-wide thin-skinned imbricate thrust front. The frontal contractional structures were controlled by basinward salt pinch-out (Letouzey *et al.*, 1996). During this time, the NNE–SSW lineament was a steep westerly-dipping extensional fault system.

Aquitanian to early Tortonian tectono-sedimentary phase

The Aquitanian(?) to lower Tortonian sequence (seismic unit D) comprises marine carbonates and later siliciclastics. Unit D is widespread and exhibits variable seismic facies across the study area, reaching more than 0.25 s TWTT in thickness (Figs 4–6 & 14). The first deposits of unit D, mainly corresponding to marine platform carbonates, reached the modern onshore (Lagos–Portimão Formation). The upper part of unit D, represented in the onshore by the Tortonian 'Fine Sands and Sandstones' (Pais *et al.*, 2000; Fig. 11), indicates that marine environments were replaced by transitional ones and the deposits became carbonate–siliciclastic.

During this phase, regional halokinesis decreased. In the Eastern Central Domain new N–S normal faults and E–W antiforms were developed (Fig. 14). Two W–E- to ENE–WSW-trending subbasins appeared north and south of the meridional antiform. At the end of this stage, the northeastern sector of the Eastern Domain was subjected to major uplift and southward tilting; the inversion of the NNE–SSW-striking fault set, the attenuation

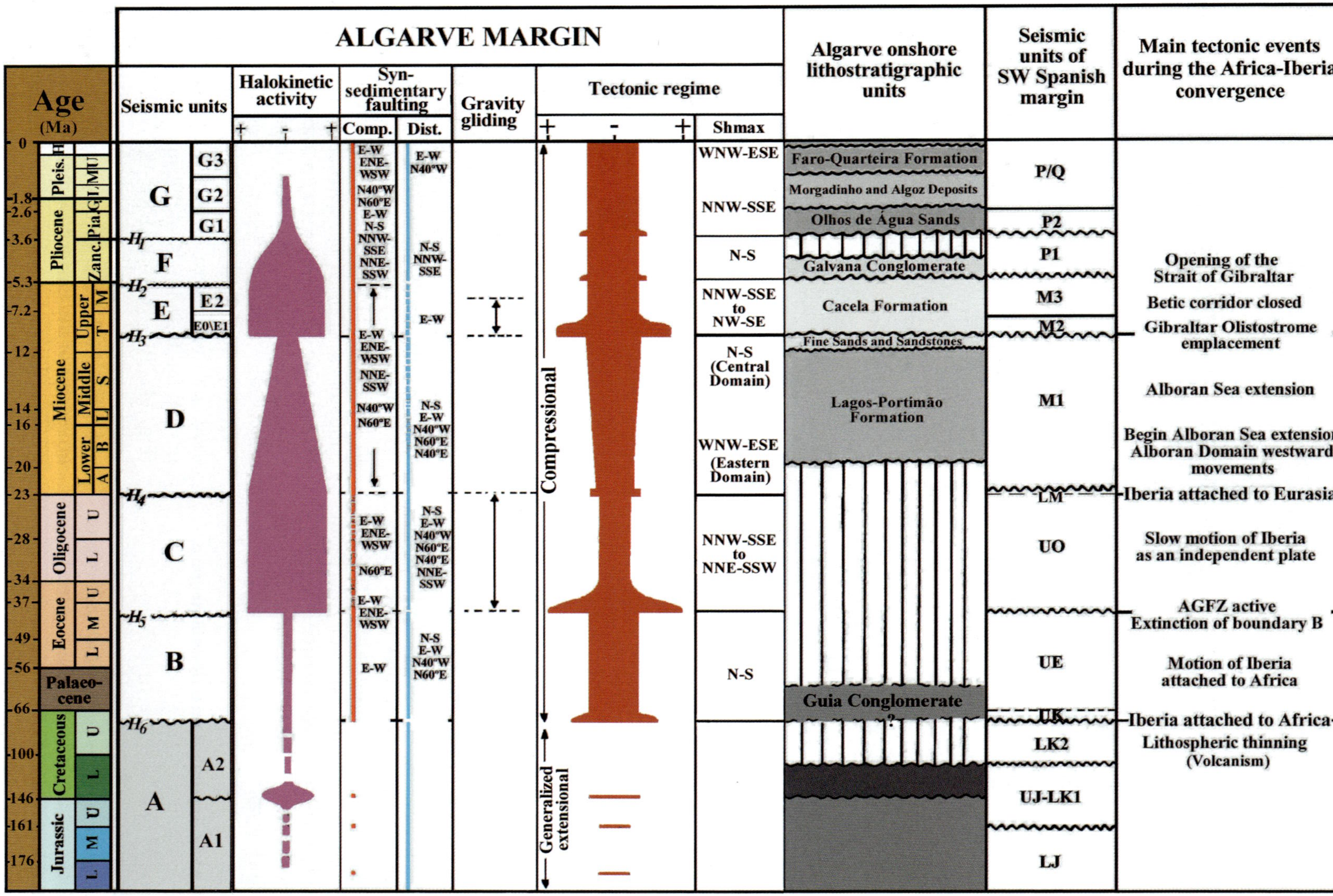

Fig. 11 Synthesis of the tectono-sedimentary stages of the Algarve margin. Cenozoic seismic units and bounding unconformities are correlated to the onshore lithostratigraphic units (Pais *et al.*, 2000) and to the seismic units of southwest Spanish margin (Maldonado *et al.*, 1999). The tectonic events and relative motion of Iberia and Africa are correlated with the development of the seismic units. Chronological time-scale from Gradstein *et al.* (2004).

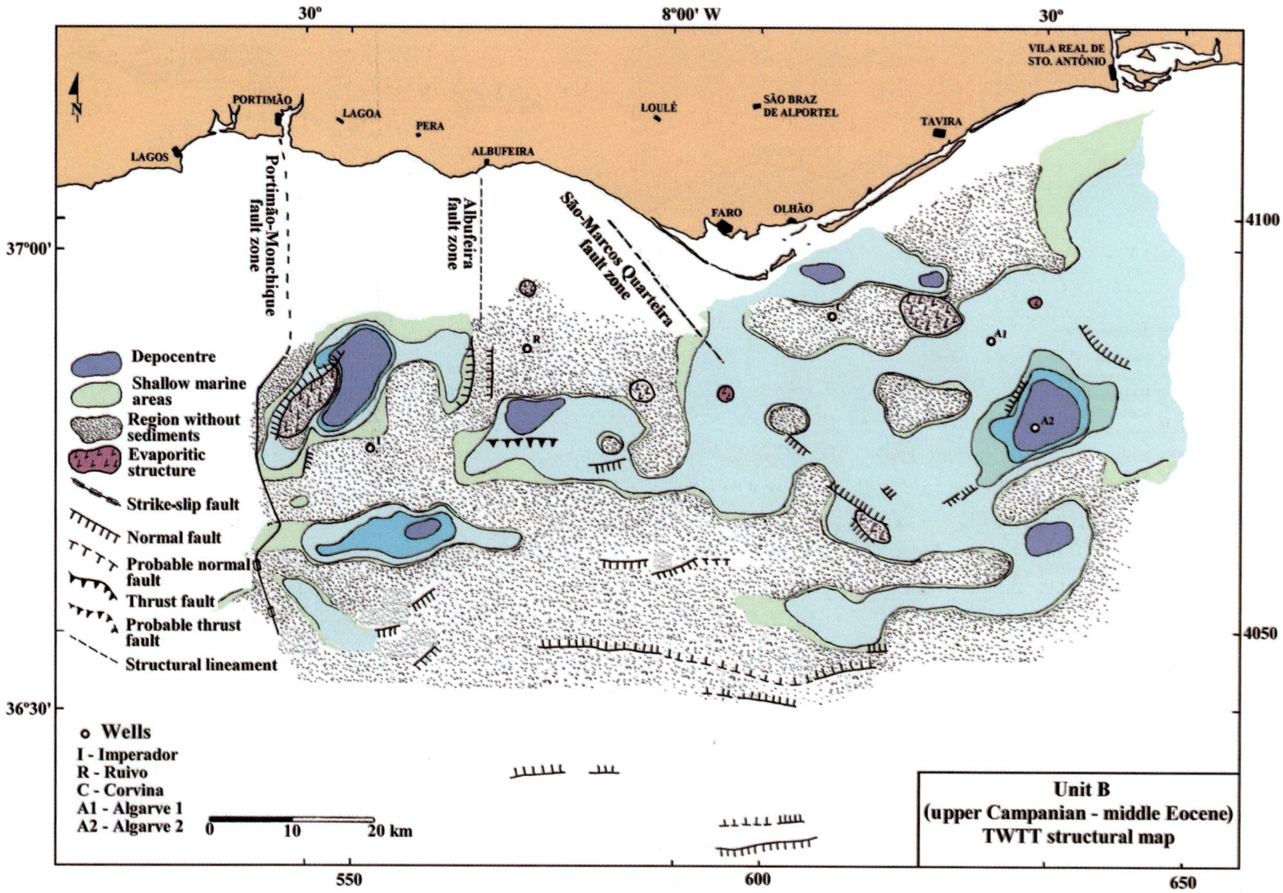

Fig. 12 Unit B (upper Campanian to middle Eocene) TWTT structural map. Areas with no data are represented in white.

of syn-sedimentary folding and the activity in the E–W to ENE–WSW thrust front all occurred.

Late Tortonian to Messinian tectono-sedimentary phase

This phase was marked by significant uplift and southward tilting of the northeastern sector of the Eastern Domain (Figs 14 & 15), leading to the erosion of unit D over a 15-km-wide N–S-trending zone located between Tavira and Vila Real de Santo António and extending southwards to the Algarve-1 and Algarve-2 well sites. In this region, southward gravitational sliding occurred, leading to the formation of ramp anticlines downslope (Figs 10 & 15). A subsiding N60°E-trending central sub-basin was developed, with northeastward migration. General siliciclastic sedimentation (seismic unit E) began with the arrival of the Guadalquivir Allochthonous front in the southeastern Algarve margin during the late Tortonian. Close to the Guadalquivir Allochthonous front and in some small depressions on the top of this chaotic body, detrital deposits accumulated, grading northwards into pelagic deposits.

During this phase, a generalized NNW–SSE compressional regime induced the tectonic inversion of most previous structures and reactivation of the ENE–WSW thin-skinned thrust faults (Figs 6, 10 & 15). Widespread halokinesis also occurred, with reactivation of the previous salt structures that pierced their cover. In the Western Central Domain, the southern part of the Albufeira fault zone was characterized by quiescence during the deposition of unit E, which has the same thickness on both sides of the fault (Fig. 15). Uplift is documented in some sections of the Portimão–Monchique fault zone and at the

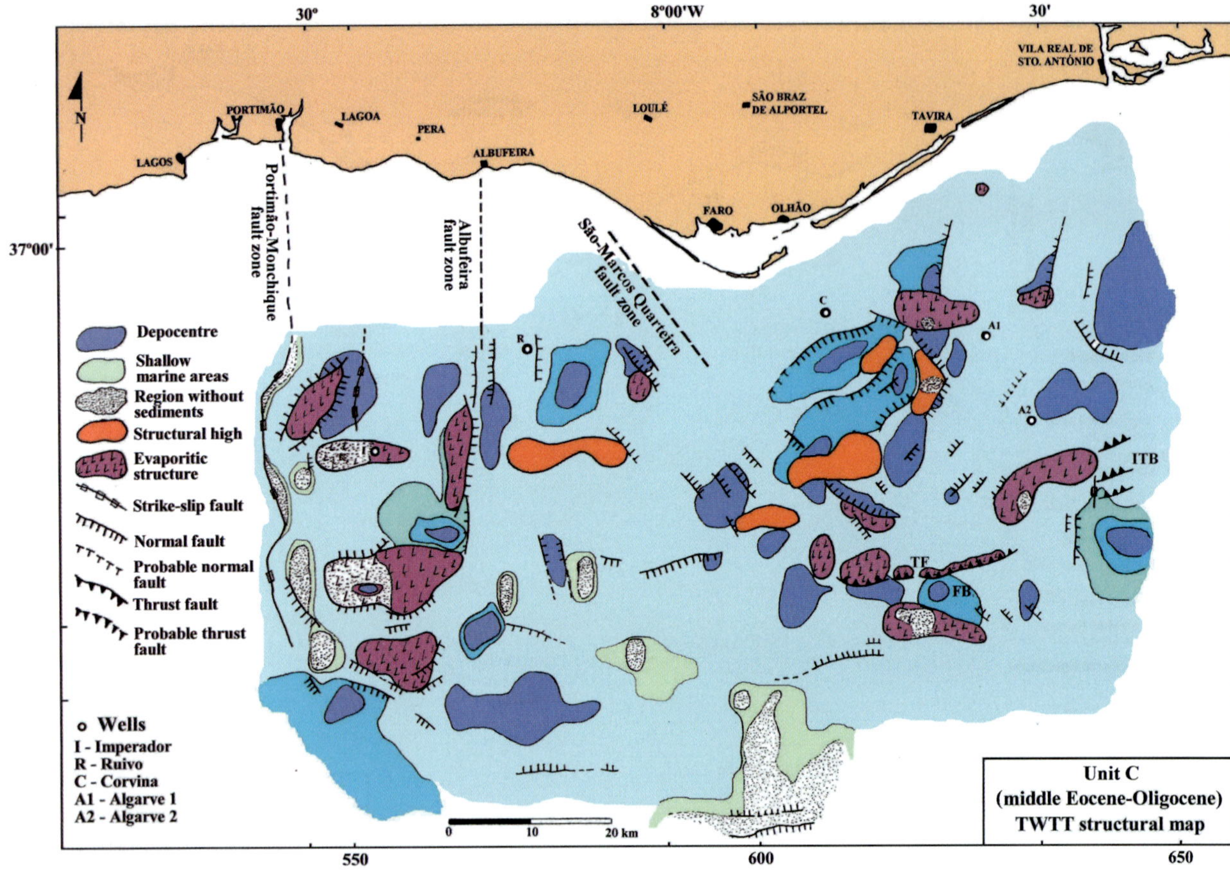

Fig. 13 Unit C (middle Eocene to Oligocene) TWTT structural map. TF, thrust front; FB, foredeep basin; ITB, imbricated thrust belt. Areas with no data are represented in white.

northeastern end of the Eastern Domain (Figs 4, 8 & 15). In the Eastern Central Domain the anticlines were active.

Zanclean tectono-sedimentary phase

Data from the wells (Imperador, Ruivo and Corvina) indicate that the lithologies corresponding to seismic unit F consist of upper to lower bathyal mudstones and sandstones with interbedded sandy mudstones (Fig. 3). The thickness of these deposits is variable and was controlled by the underlying fault/salt structures (Figs 4–6, 8–10 & 16). Values of more than 0.6 s TWTT are documented in half-grabens, particularly in the western Algarve margin.

During the Zanclean, in all the study area, the depressions underlying unit F were filled. In the Western Central Domain, the southern part of the Albufeira fault zone was reactivated. In the Central Eastern Domain, the southern anticline became inactive and its northern and southern boundary sub-basins became a single, rapidly subsiding N60°E-trending depocentre. In the Eastern Domain, strong subsidence occurred in a N60°E-trending depocentre located northwestwards of the Guadalquivir Bank. Decreasing halokinesis is documented, with a more localized diapirism, forming small rim synclines.

Piacenzian to Holocene tectono-sedimentary phase

Seismic unit G comprises hemipelagic silts and sands, turbiditic sands and current-drift sands. Basinwards, this phase was characterized by rapid subsidence along a roughly N60°E-trending axis, where a considerable thickness was accumulated (more than 0.7 s TWTT) (Figs 4–6, 8–10 & 17).

During the Piacenzian to Holocene phase the present-day Gulf of Cadiz marine current regime

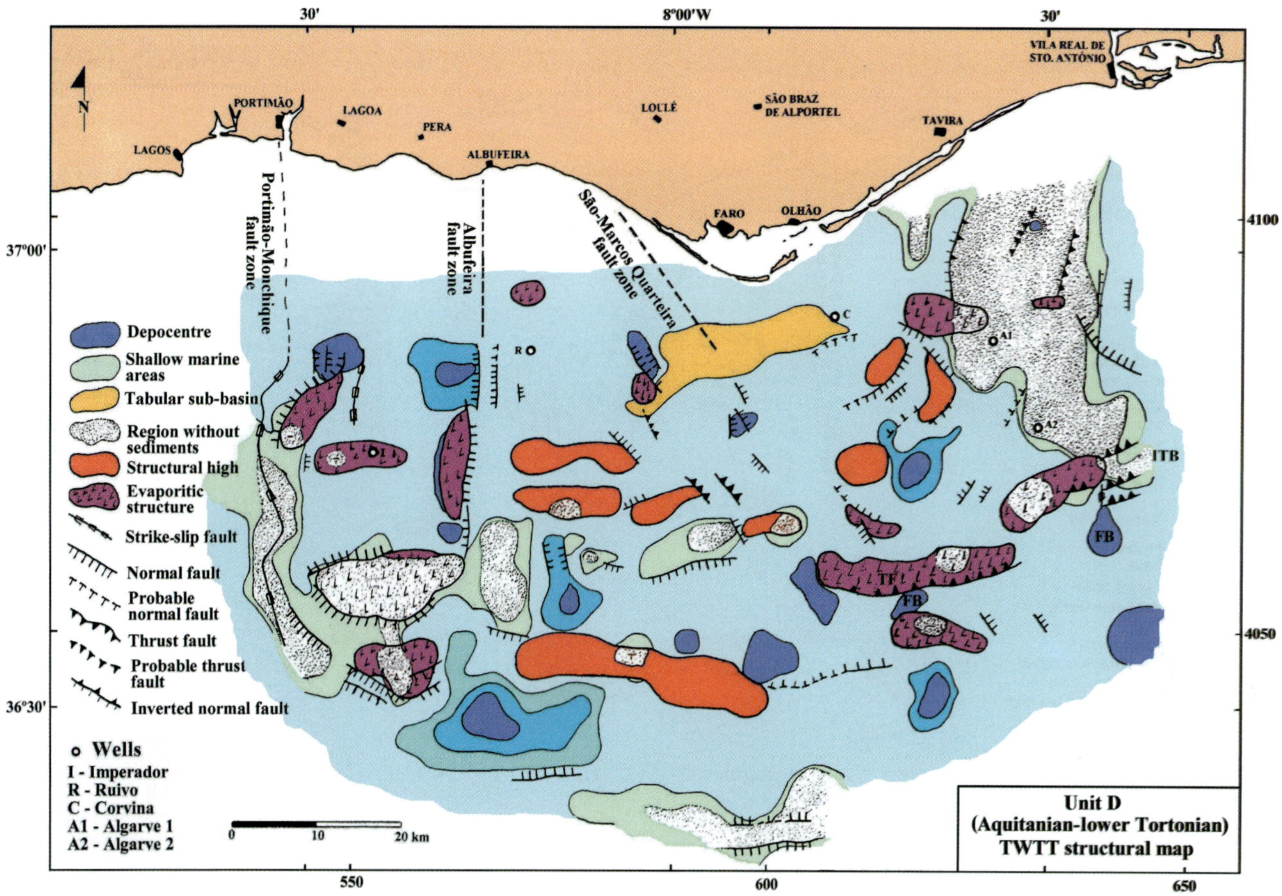

Fig. 14 Unit D (Aquitanian to lower Tortonian) TWTT structural map. TF, thrust front; FB, foredeep basin; ITB, imbricated thrust belt. Areas with no data are represented in white.

became established (e.g. Mougenot & Vanney, 1982). The Portimão and Albufeira canyons and the Álvares Cabral and Diogo Cão deeps were developed and the north-northwestwards progradational contourite drifts of Albufeira and Faro established their present-day positions (e.g. Nelson *et al.*, 1999) (Figs 4, 6 & 17). The halokinesis seems to decrease (onshore, some salt structures such as the Loulé Diapir were still active in the Quaternary; Terrinha *et al.*, 1990) and the previous depressions were progressively filled. The orientation and type of the syn-sedimentary faults suggest the development of a stress field with Sh_{max} oriented NNW–SSE, but also WNW–ESE. Significant present-day seismicity is dominantly offshore (Cabral, 1995), mainly related to the Portimão–Monchique and São Marcos–Quarteira fault zones, the ENE–WSW thrust front, NNE–SSW reverse faults and the Guadalquivir Bank (Lopes, 2002).

SYNTHESIS OF THE REGIONAL GEODYNAMIC EVOLUTION

Late Cretaceous to Lutetian

At Chron 34 (Santonian, 84 Ma), Iberia was attached to the African plate and the plate boundary with Eurasia was then located in the Bay of Biscay (boundary B; Srivastava *et al.*, 1990a, b). The new geodynamic setting caused north–south convergence (Dewey *et al.*, 1973, 1989; Argus *et al.*, 1989). This resulted in inversion of the northern margin of Iberia, developing into northward subduction/underthrusting of the plate (starting in the Campanian; Puigdefàbregas & Souquet, 1986) and creating the Pyrenees. In mainland Portugal, a compressive episode occurred in the middle Campanian (around 80 Ma; e.g. Mougenot, 1981, 1989), with Sh_{max} oriented north–south, leading to

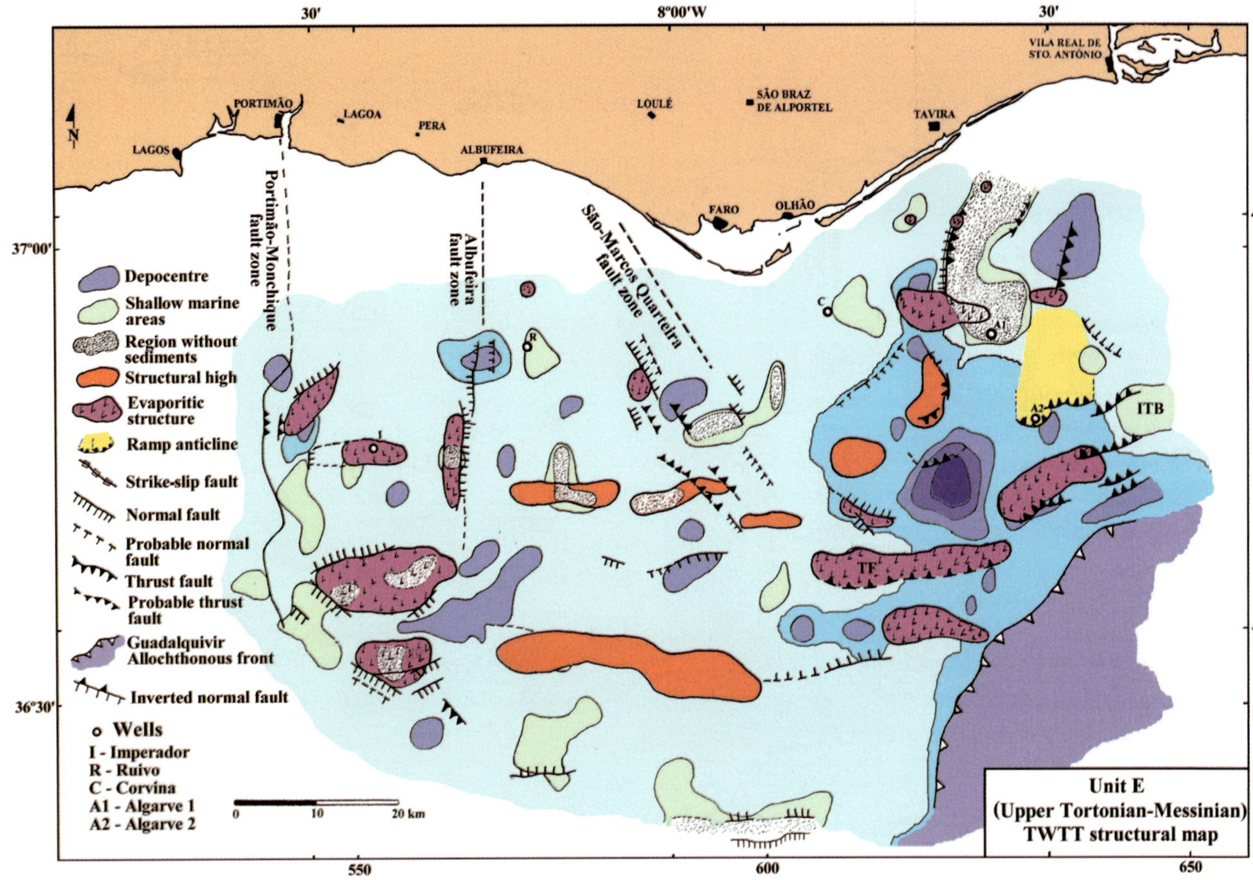

Fig. 15 Unit E (upper Tortonian to Messinian) TWTT structural map. TF, thrust front; ITB, imbricated thrust belt. Areas with no data are represented in white.

the intrusion of the Sintra, Sines and Monchique alkaline plutons, probably along a deep-seated dextral strike-slip fault (e.g. Ribeiro *et al.*, 1979; Abranches & Canilho, 1981; Terrinha, 1998; Gomes & Pereira, 2004). Significant volcanic activity, diapirism and faulting also occurred in central Portugal (Cunha & Pena dos Reis, 1995; Pinheiro *et al.*, 1996). In the Algarve margin, this event is recorded by unconformity H6 (Figs 3 & 11). The upper Campanian to middle Eocene sequence was deposited irregularly, with significant facies variations, as documented by unit B in the Algarve margin (Lopes, 2002) and the unit UK-UE in the southwest Spanish margin (Maldonado *et al.*, 1999).

Lutetian to Chattian

At the start of the Lutetian an important event occurred – the inception of rifting in western Europe, initiating basins of the European Cenozoic Rift System (Sissingh, 2001). Compression related to the Pyrenean collision was transmitted into the central part of the Iberian mainland: NNE–SSW compression and perpendicular extension generated the Portuguese Tertiary basins (Mondego, Lower Tejo and Sado basins) and a large number of basins in Spain were created (Lutetian compressive phase), filled by arkose sediments resulting from the erosion of the Hesperic Massif (Variscan basement). Despite the belief that deformation decreased southward, away from the active boundary, the southwestern border of Iberia (Gulf of Cadiz) was affected by a compressive event (Fig. 11) that resulted in:

1 The cessation of movement along boundary B, and the jumping of the plate boundary to the region of King's Trough, extending eastward along the

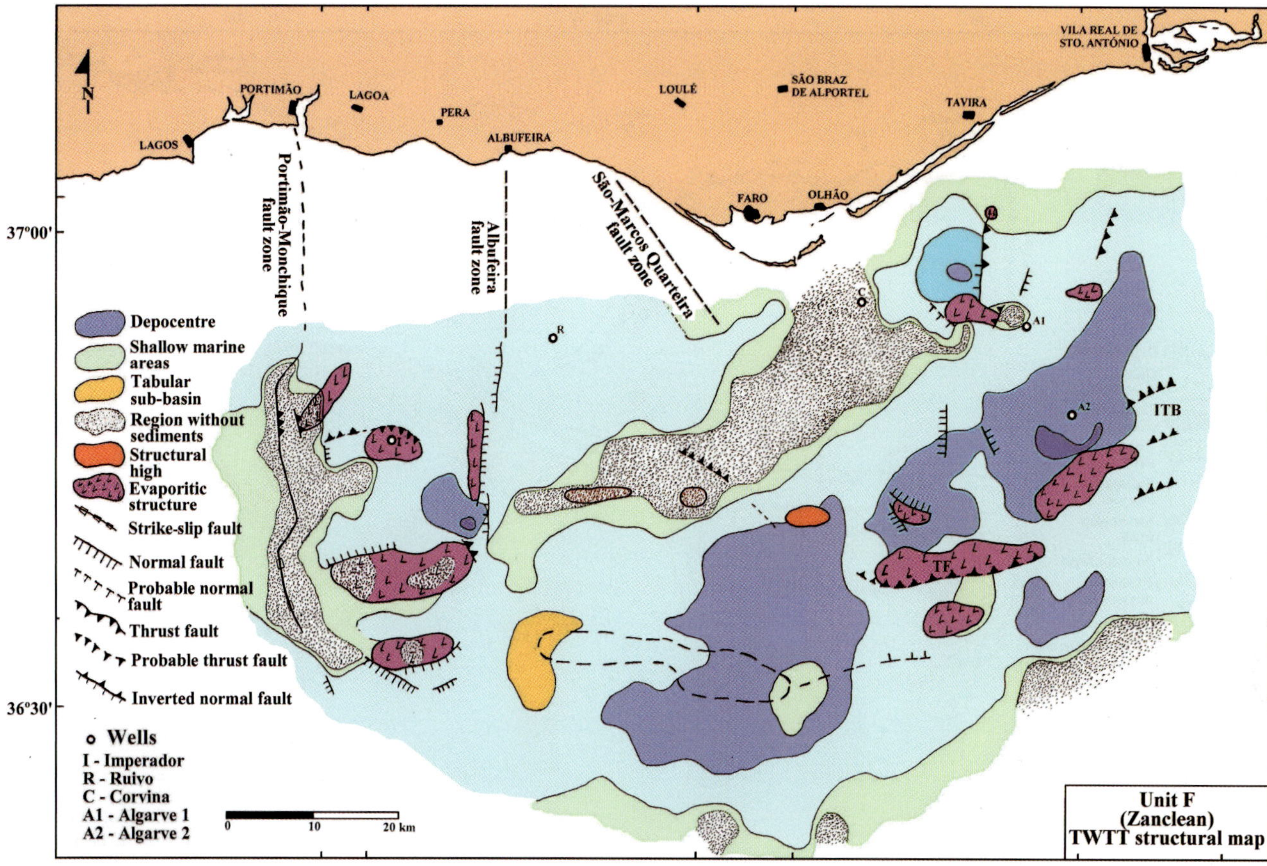

Fig. 16 Unit F (Zanclean) TWTT structural map. TF, thrust front; ITB, imbricated thrust belt. Areas with no data are represented in white.

Azores–Biscay to the North Spanish Trough and Pyrenees (Srivastava *et al.*, 1990a, b; Roest & Srivastava, 1991); the movement was extensional in the King's Trough and compressive along the Pyrenees (Fig. 1a).

2 The reactivation of the Azores–Gibraltar fracture zone, constituting again a plate boundary between Africa and Iberia (Chron 18, 42 Ma; Srivastava *et al.*, 1990a, b). Until the amalgamation of Iberia with Eurasia along the Pyrenean suture, Iberia moved as an independent plate from 42 to 24 Ma (Roest & Srivastava, 1991).

According to Maldonado *et al.* (1999) the transpressive movement between Iberia and Africa along the Gulf of Cadiz started at this time, with probable subduction of thinned Tethyan crust towards the south.

In the Algarve margin, this major compressive episode provoked strong tectonic inversion (uplift, folding, thrusting) and the generation of the important H5 unconformity (Fig. 11). The northern sector and the Guadalquivir Bank emerged. Westward, this intense instability was recorded by a very thin or absent sedimentary record (Hayes *et al.*, 1972) and by important uplift and amplification of the Gorringe Bank (Le Gall *et al.*, 1997).

After this intense compressive episode, the structures identified as active suggest that the tectonic regime became NNW–SSE to NNE–SSW moderately compressive, until the end of the Oligocene. In the southern border of Iberia, along a corridor that linked the central Atlantic and Mediterranean basins, a vast carbonate platform developed, with deposition of unit E1 in the Alentejo margin (Alves *et al.*, 2003), unit C in the Algarve margin (Lopes, 2002) and of unit UO-LM in the southwest Spanish margin (Maldonado *et al.*, 1999). In the Algarve margin, intense halokinesis occurred; in the Eastern Domain, gravitational

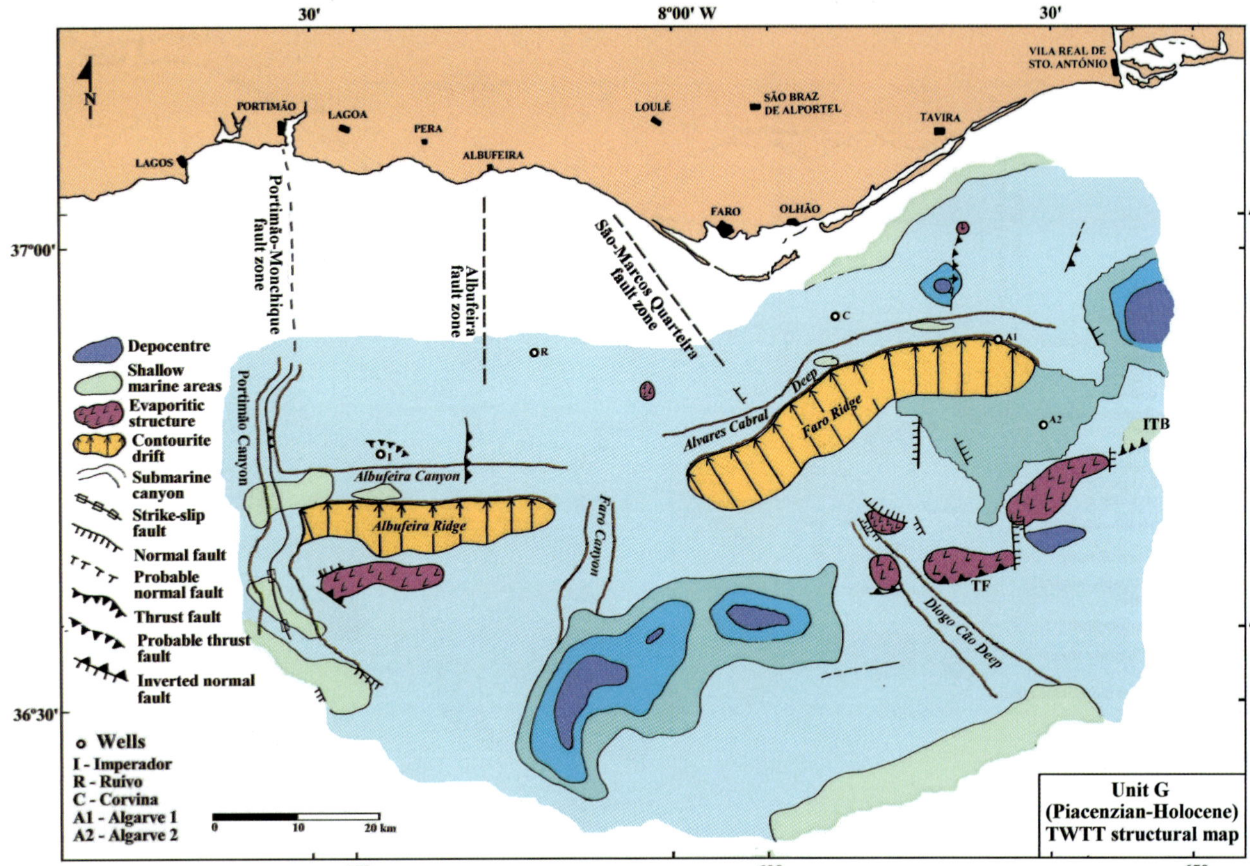

Fig. 17 Unit G (Piacenzian to Holocene) TWTT structural map. TF, thrust front; ITB, imbricated thrust belt. Areas with no data are represented in white.

extension and concomitant compression in the distal and deeper parts of the basin were due to a general tilt of the margin.

During the Chattian to Aquitanian, in the Mediterranean area, the Algerian–Provençal Basin was developed (Sanz de Galdeano, 1990), acting as a back-arc basin relative to the subduction of Africa under the South Sardinian Domain or Alkapeca (Bouillin *et al.*, 1986), located between Africa and Eurasia (Fig. 18a).

Aquitanian to middle Tortonian

At the Chattian–Aquitanian transition (anomaly 6c; around 23 Ma), the plate boundary along the King's Trough–Azores–Bay of Biscay–Pyrenees became extinct and Iberia was integrated with the Eurasian plate. The plate boundary became located along the Azores–Gibraltar fracture zone (Srivastava *et al.*, 1990a, b; Roest & Srivastava, 1991; Fig. 11).

At this time, a widespread change to marine conditions in the western domains of the Peri-Tethyan platforms was probably related to the counterclockwise rotation of the Corsica–Sardinia block (Meulenkamp & Sissingh, 2003). A major sedimentary break at the Chattian–Miocene boundary is recognized in the Iberian Tertiary basins (Cunha, 1992a, b; Calvo *et al.*, 1993; Alves *et al.*, 2003). In the Gulf of Cadiz, the Chattian–Aquitanian boundary was also marked by an important regional unconformity (H4 in the Algarve margin; Lopes, 2002), followed by the deposition, respectively, of unit D (Algarve margin) and unit M1 (southwest Spanish margin; Fig. 11). By this time, the opening of the Algerian–Provençal Basin became accentuated, provoking the fragmentation of the South Sardinian Domain (Fig. 18a; Sanz de

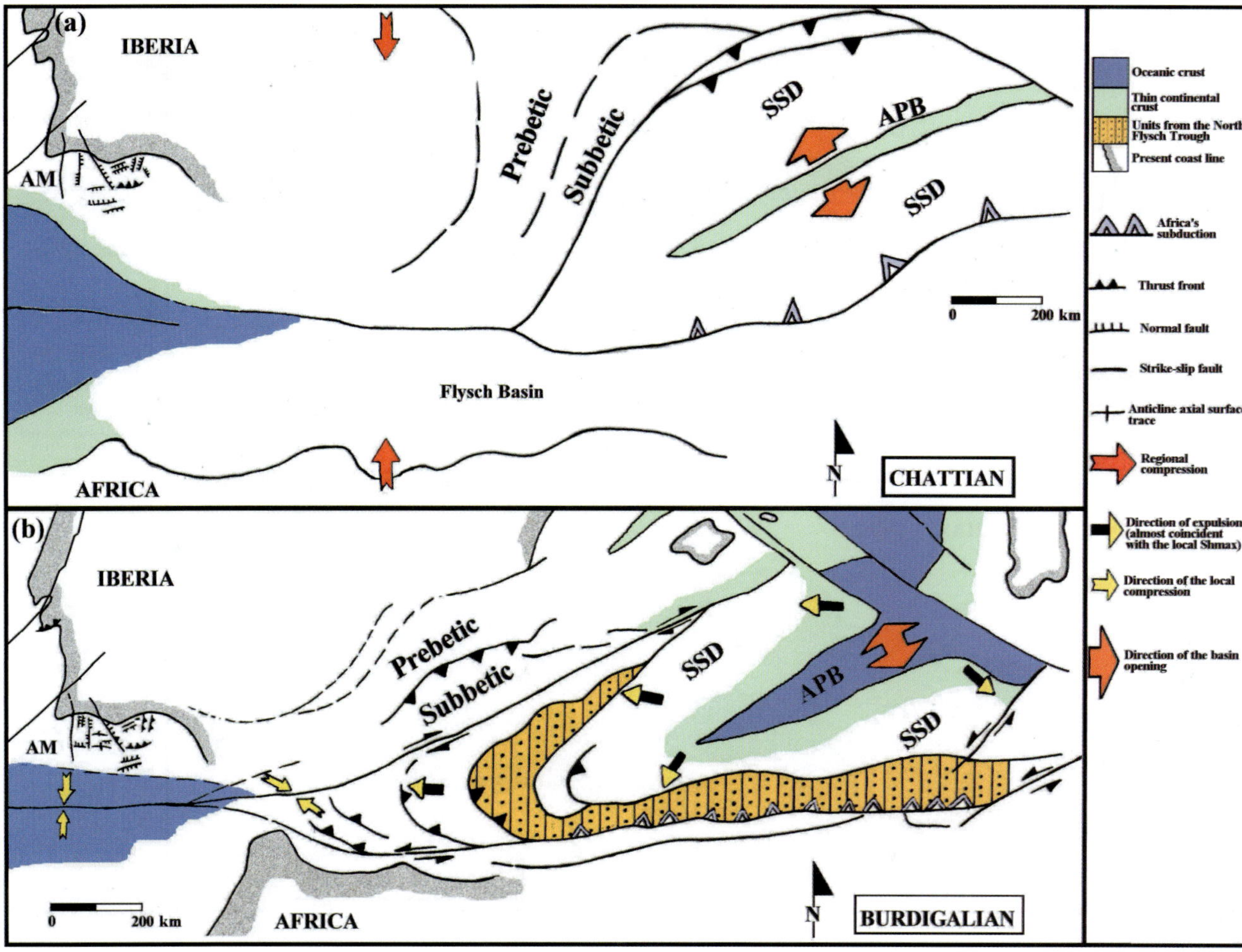

Fig. 18 Synthetic reconstruction of the geodynamic evolution of the westernmost alpine Mediterranean area, during the (a) Chattian and (b) Burdigalian (modified from Sanz de Galdeano, 1990, 2000; Sanz de Galdeano & Vera, 1991; Sanz de Galdeano & Rodríguez-Fernández, 1996). AM, Algarve margin; APB, Algerian-Provençal basin; SSD, South Sardinian Domain.

Galdeano, 1990; Sanz de Galdeano & Vera, 1991) and the expulsion towards the west-southwest of one of its constituents, the Alboran Domain (Andrieux *et al.*, 1971; Durand-Delga & Fontboté, 1980).

The South Sardinian Domain expulsion reached its climax during the Burdigalian (Hermes, 1985), reflected by significant compressive effects in the sedimentary cover of the South Iberian and North African continental margins, leading to the formation of the Rift and Betic External Zones (Fig. 18b). The Sub-Betic Zone was compressed by the western movement of the Internal Zones and the North Betic Strait appeared, linking the Atlantic to the Mediterranean Sea (Sanz de Galdeano & Vera, 1991). In the most active sector of this chain-front basin (Betic trough), with migration towards the north-northwest, large volumes of allochthonous masses were deposited. According to Sanz de Galdeano & Rodríguez-Fernández (1996), the main displacement of the Internal Zones ended in the early Langhian. In consequence of the Internal Zones' emplacement, progressive lithospheric delamination of the African plate provoked the extensional collapse of the Alboran Sea (Platt & Vissiers, 1989; Maldonado *et al.*, 1999).

Late Tortonian to Holocene

A fourth episode of major regional change in palaeogeography and tectonic setting occurred in

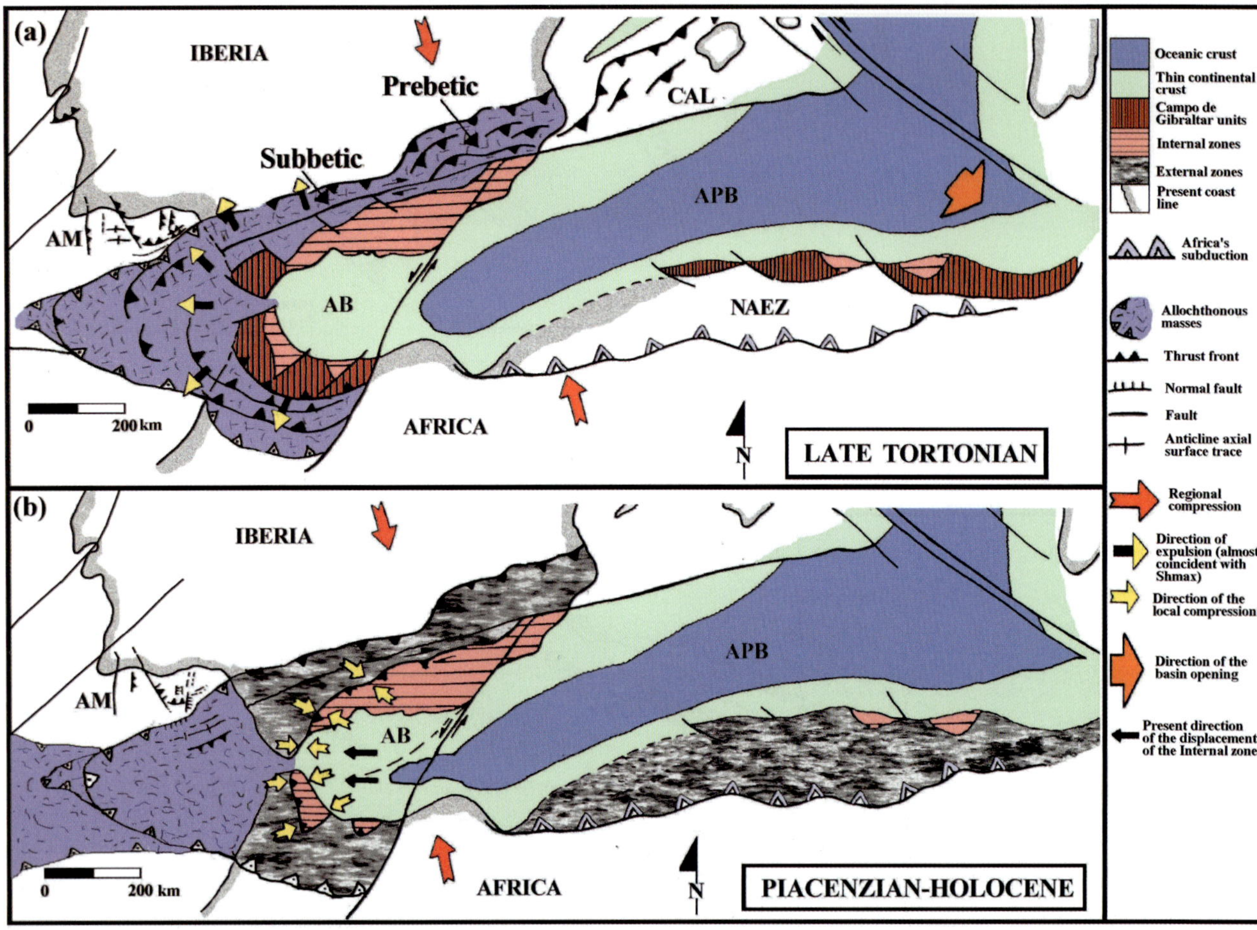

Fig. 19 Synthetic reconstruction of the geodynamic evolution of the westernmost alpine Mediterranean area, during the (a) late Tortonian and (b) Piacenzian to Holocene (modified from Sanz de Galdeano, 1990, 2000; Sanz de Galdeano & Vera, 1991; Sanz de Galdeano & Rodríguez-Fernández, 1996). AB, Alboran basin; AM, Algarve margin; APB, Algerian–Provençal basin; CAL, Cadiz–Alicante line; NAEZ, North African External Zones.

the Tortonian, around 9–8 Ma, affecting the majority of the domains of the African–Eurasian convergent plate boundary zone. The resultant modifications included enhanced uplift and emergence of large parts of western and central Europe in association with the end of sedimentation in the northern Alpine foreland (Meulenkamp & Sissingh, 2003). It coincided with a change in the direction of convergence from north-northwest to northwest between Africa and Eurasia and led to the development of the Gibraltar Arc. Inversion tectonics became active in the interior of the Iberian plate in the Spanish Central System (Vicente *et al.*, 1996) and in the Portuguese Central Range (Ribeiro *et al.*, 1990).

The middle Tortonian highy compressive event, with Sh_{max} oriented NNW–SSE, affected the whole Gulf of Cadiz and Betic areas (Sanz de Galdeano, 1990; Sanz de Galdeano & Vera, 1991; Sanz de Galdeano & Rodríguez-Fernández, 1996; Maldonado *et al.*, 1999) and is recognized by an important unconformity (H3 reflector, in the Algarve margin; unconformity BFU, in the southwest Spanish margin).

This event coincided with the last major radial expulsion of the External Zones (Prebetics and Flysch Basin, coeval with the stretching of the Internal Zones); the North Betic Strait became restricted to the western sector of the Betic trough (Sanz de Galdeano & Vera, 1991; Fig. 19a) and most of the Betic Neogene basins were developed (Sanz de Galdeano & Vera, 1992). It led to the emplacement, in the Southern Iberian margin and in the central Gulf of Cadiz, of an accretionary prism (Guadalquivir Allochthonous; Gràcia *et al.*, 2003)

(Figs 10 & 19a) where, because of imbricating thrusts with low-angle detachments, Mesozoic and Cenozoic fragments of the Betic margin were included (Bonnin *et al.*, 1975; Lajat *et al.*, 1975; Malod, 1982; Sanz de Galdeano, 1990; Flinch *et al.*, 1996; Maldonado *et al.*, 1999; Maestro *et al.*, 2003). Large masses of Triassic–Hettangian evaporites, tectonically incorporated, were responsible for the halokinesis in the central part of the gulf since the Messinian (Flinch *et al.*, 1996). Gravity processes were largely responsible for the migration of the allochthonous mass towards the Horseshoe and Seine abyssal plains (Gràcia *et al.*, 2003; Fig. 1).

The Tortonian event was also marked by the arrival of the Guadalquivir Allochthonous front to the southeast Algarve margin; its progression towards the interior may have been inhibited by the Guadalquivir Bank. Intense halokinesis and inversion tectonics were also recorded. In the southwest Iberian margin, this phase of intense instability was responsible for some vertical development of the Gorringe Bank (Le Gall *et al.*, 1997).

During the late Tortonian and Messinian, the widespread compressive regime led to the emergence of a great part of the Betic Range, coeval with an important sea-level fall (Haq *et al.*, 1987). The straits between the Betics and Rif were closed (Sanz de Galdeano, 1990), which led to the 'Mediterranean salinity crisis' (e.g. Maldonado & Nelson, 1999). In the west sector of the Betic trough, clockwise rotation of the depocentres was coeval with the development of the Guadalquivir Basin (Sierro *et al.*, 1996; Fig. 1). In the far eastern Algarve basin, onshore (Cachão, 1995) and offshore, an increase in subsidence and a migration towards the northeast of the depocentre were recorded. The sedimentary units that can be related to this tectonic phase are units B, C and D in the Guadalquivir Basin (Sierro *et al.*, 1992a, b, 1996), units M2–M3 in the southwest Spanish margin (Maldonado *et al.*, 1999) and unit E in the Algarve margin. According to Alves *et al.* (2003), the third Cenozoic deformation event affecting the Alentejo margin relates to late Tortonian–Zanclean tectonics and is responsible for the initiation of the modern Setúbal and São Vicente submarine canyons (Fig. 1).

Zanclean

By the late Messinian, Sh_{max} became oriented roughly N–S (Phillip, 1987, in Maldonado & Nelson, 1999), a transtensional regime became dominant in the Betic range and a connection between the Atlantic and Mediterranean through the Gibraltar Strait was opened. This opening, coeval with a significant increase in subsidence (Maldonado & Nelson, 1999; Maldonado *et al.*, 1999), allowed the establishment in the Zanclean of the marine hydrodynamic setting that has continued to the present (Malod, 1982). North of the Gibraltar axis, sedimentation was controlled by halokinesis coeval with high subsidence; south of the Gibraltar axis, sedimentation continued in the same style as in the latest Miocene.

During the Zanclean, coeval with a eustatic sea-level high (Haq *et al.*, 1987), the Gulf of Cadiz was dominated by the incursion of saline Mediterranean water and the sedimentary regime was characterized by the formation of deposits of deep-currents and contourites (Nelson *et al.*, 1993; Maldonado & Nelson, 1999). Ongoing compressive strike-slip activity of N20–40°E-trending faults is documented in the eastern Betics (Andeweg & Cloetingh, 2001). In the southwestern Iberian border, the southwest Spanish margin unit P1 (Maldonado *et al.*, 1999) and the Algarve margin unit F were deposited (Fig. 11). In the Algarve margin, the old depocentres underwent progressive infill. A N–S to NNW–SSE oriented Sh_{max} is suggested by the strike and type of syn-sedimentary faults.

In mainland Portugal, during the late Tortonian to Zanclean, endorheic alluvial fans developed along active NNE–SSW indent-linked strike-slip faults and NE–SW reverse faults (Cunha, 1992a, b; Cunha *et al.*, 2000), controlled by intense NNW–SSE crustal shortening (Ribeiro *et al.*, 1990); exorheic drainage systems were developed only at the transition to the more humid conditions of the Piacenzian.

Piacenzian to Holocene

The Piacenzian, Gelasian and the Quaternary (Fig. 19b) are represented by units P2 and Q/P in the southwest Spanish margin and by unit G in the Algarve margin. The spatial distribution of these deposits was controlled by a complex interplay between sea-level changes, sediment supply and variation in the speed of marine currents (Nelson *et al.*, 1993, 1999; Rodero *et al.*, 1999).

Present-day seismicity indicates that the majority of the tectonic structures are still active (Cabral,

1995), controlled by complex dextral slip along the crustal segment between the Gorringe Bank and the Gualdalquivir basin (Maestro *et al.*, 1998). A high Sh_{max}, acting obliquely to the western Portuguese continental margin, is interpreted by Ribeiro *et al.* (1996) as reactivating this passive margin, with the nucleation of a subduction zone in the Gorringe Bank (Fig. 1), propagating northward along the base of the continental slope.

CONCLUSIONS

The position of Iberia, located at a critical point on active plate boundaries throughout the Late Cretaceous and Cenozoic, provides a setting in which the relationship between the changing plate boundary conditions and the tectono-sedimentary processes is relatively direct. Several tectonically controlled breaks in deposition, induced by the increasing effects of African–Eurasian convergence, occurred during the regional differentiation in basin development and depositional setting in the Algarve margin; their timing is similar to those identified in adjacent areas.

Six tectono-sedimentary phases have been reported: (i) late Campanian to middle Eocene, (ii) middle Eocene to Oligocene, (iii) Aquitanian to early Tortonian, (iv) late Tortonian to Messinian, (v) Zanclean and (vi) Piacenzian to Holocene. Their sedimentary character changed through time, from carbonate to siliciclastic, and they are widely involved in the polyphase structures of the different tectonic domains. The increase in the siliciclastic content may be related to the concomitant growth of the land mass, reflecting the impact of a large-scale, tectonically induced inversion process.

Evaporitic structures occur mainly in the Western Central and Eastern Domains of the Algarve margin, related to major structural lineaments. Thin and thick-skinned thrusts, orientated E–W to ENE–WSW, and N60°E imbricate thrusts are concentrated in the Eastern Domain and they generally exhibit a southward or southeastward vergence. This tectonic signature is attributed to the proximity of the Betic Orogen and the Guadalquivir Allochthonous front, and to the São Marcos–Quarteira fault zone that acts as a buttress fault to the westward propagation of the compression of the Gibraltar Arc.

An important role in the tectono-sedimentary evolution was played by Triassic–Hettangian evaporites, which acted as a major detachment during the extensional and compressional stages and generated both salt structures and salt-withdrawal sub-basins. From the wedge-shaped geometry of the sedimentary packages in the salt-withdrawal sub-basins between the salt structures, major halokinetic activity is likely to have occurred during the middle Eocene to Oligocene and the late Tortonian to Messinian; halokinesis was limited during the Aquitanian to earlier Tortonian and later decreased. During the middle Eocene to Oligocene phase, widespread halokinesis was generated by a moderate compressional reactivation of basement-related structures. The progressive basement graben inversion in the Eastern Domain, with uplift and tilting of the northern sector of the margin, led to the gravity gliding of the sedimentary cover above a salt detachment layer. Folds and the thrust front were generated downslope coeval with extension upslope. Southeastward, N60°E-trending imbricated thrust faults were induced by the basement contraction.

A generalized subsidence increased during the Cenozoic. The Paleogene was characterized by fault/salt control and flexure, leading to the formation of numerous and widespread depocentres. Since the middle Tortonian, the structural control exerted by the northern border of the basin and by the Guadalquivir Bank (in the south) was probably caused by the NW–SE to NNW–SSE compressive regime. This allowed the development of a strongly subsiding N60°E-trending basin, with increasing flexure of the margin; large subsidence in the Guadalquivir Basin, located northeastward along this axis, was coeval.

In summary, the main compressive structures were: the E–W to ENE–WSW thrust front; N60°E imbricate thrusts; E–W anticlines; NNE–SSW reverse faults; N40°W thrusts. Normal fault systems were also identified, with development of half-grabens oriented N–S to NNE–SSW; N40°E; N60°E; E–W; N40°W. The coeval development of compressive structures and normal fault systems is considered a consequence of:

1 Paleogene horizontal migration of evaporites and the development of a gravity gliding structural style controlled by the inversion of the basement structures related to the convergence of Africa and Eurasia along the Azores–Gibraltar fracture zone (transpressive regime);

2 Neogene horizontal migration of evaporites into the rising salt structures, convergence of Africa and Eurasia along the AGFZ (transpressive regime), and the westward migration of the Gibraltar Arc, causing a radial trajectory of Sh_{max} around it.

The interpretation of the tectono-sedimentary features of the Algarve margin contributes to the understanding of the geodynamic evolution of the Gulf of Cadiz, primarily controlled by the enhanced coupling of the African and Iberian plates. A generalized compressive tectonic regime can be recognized, with two highly compressional phases that occurred in the Lutetian and in the Tortonian.

ACKNOWLEDGEMENTS

This study is part of the research projects POCTI/CTA/38659/2001 and DESIRE, of the Fundação para a Ciência e a Tecnologia and co-founded by the FEDER. The work has also been supported by the Centro de Geociências da Universidade de Coimbra, the Centro de Geofísica da Universidade de Coimbra and the Centro de Universidade de Lisboa. The authors wish to thank the Portuguese Núcleo para a Pesquisa e Prospecção de Petróleo for the permission to use seismic profiles and well data. We are grateful to B. Le Gall (Institut Univisité Européen de La Mer, France) for help in the interpretation of seismic profiles, Claudia Magno for perusing the English version and Peter F. Friend (University Cambridge) for reviewing an early version of the manuscript. Edward Williams (editor) and two anonymous reviewers made helpful comments and suggestions at a later stage.

REFERENCES

Abranches, M.C.B. and Canilho, M.H. (1981) Estudos de geocronologia e geologia isotópica, pelo método do Rubídio-Estrôncio, dos três maciços Mesozóicos portugueses: Sintra, Sines e Monchique. *Bol. Soc. Geol. Portugal*, **22**, 385–390.

Algarve-1 (1982) Final Well Report. *Arquivo Gabinete para a Prospecção e Exploração de Petróleo, n° 21967 – off/Exxon*. Esso, Prospecção e Produção, S.A.R.L.

Algarve-2 (1982) Final Well Report. *Arquivo Gabinete para a Prospecção e Exploração de Petróleo, n° 21973 – off/Exxon*. Esso, Prospecção e Produção, S.A.R.L.

Alves, T.M., Gawthorpe, R.L., Hunt, D.W. and Monteiro, J.H. (2003) Cenozoic tectono-sedimentary evolution of the western Iberian margin. *Mar. Geol.*, **195**, 75–108.

Andeweg, B. and Cloetingh, S. (2001) Evidence for an active sinistral shear zone in the western Alboran region. *Terra Nova*, **13**, 44–50.

Andrieux, J., Fontboté, J.M. and Mattauer, M. (1971) Sur un modèle explicatif de l'Arc de Gibraltar. *Earth Planet. Sci. Lett.*, **12**, 191–198.

Argus, D.F., Gordon, R.G., Demets, C. and Stein, S. (1989) Closure of the Africa–Eurasia–North America plate motion circuit and tectonics of the Gloria Fault. *J. Geophys. Res.*, **94**, 5585–5602.

Auzende, J.M., Olivet, J-L. and Pastouret, L. (1981) Implications structurales et paléogéographiques de la présence de Messinien à l'Ouest de Gibraltar. *Mar. Geol.*, **43**, 9–18.

Bonnin, J., Olivet, J.L. and Auzend, J.M. (1975) Structure en nappe à l'Ouest de Gibraltar. *C. R. Acad. Sci. Paris*, **280**, 559–562.

Bouillin, J.P., Durand-Delga, M. and Olivier, Ph. (1986) Betic–Rifian and Tyrrhenian Arcs: distintive features, genesis and development stages. In: *The Origin of Arcs* (Ed. F.C. Wezel), pp. 281–304. Elsevier, Amsterdam.

Buforn, E., Udías, A. and Colombás, M.A. (1988) Seismicity, source mechanisms and tectonics of the Azores–Gibraltar plate boundary. *Tectonophysics*, **152**, 89–118.

Cabral, J. (1995) Neotectónica de Portugal continental. *Mem. Inst. Geol. Min. Portugal*, **31**, 265 pp.

Cachão, M. (1995) *Utilização de nanofósseis calcários em biostratigrafia, paleoceanografia e paleoecologia. Aplicações ao Neogénico do Algarve (Portugal) e do Mediterrâneo Ocidental (ODP 653) e à problemática do Coccolithus pelagicus*. Unpublished PhD thesis, Universidade de Lisboa, 356 pp.

Calvo, J.P., Daams, R., Morales, J., *et al.* (1993) Up-to-date Spanish continental Neogene synthesis and paleoclimatic interpretation. *Rev. Soc. Geol. Esp.*, **6**, 29–40.

Corvina-1 (1976) Final Well Report. *Arquivo Gabinete para a Prospecção e Exploração de Petróleo, n° 21301 – off/Challenger*. Challenger Portugal Inc.

Cunha, P.P. (1992a) *Estratigrafia e sedimentologia dos depósitos do Cretácico Superior e Terciário de Portugal Central, a leste de Coimbra*. Unpublished PhD thesis, Universidade de Coimbra, 262 pp.

Cunha, P.P. (1992b) Establishment of unconformity-bounded sequences in the Cenozoic record of the western Iberian margin and synthesis of the tectonic and sedimentary evolution in central Portugal during

Neogene. *First Congress R.C.A.N.S. – 'Atlantic general events during Neogene' (Abstracts)*, Lisbon, pp. 33–35.

Cunha, P.P. and Pena dos Reis, R. (1995) Cretaceous sedimentary and tectonic evolution of the northern sector of the Lusitanian Basin. *Cretaceous Res.*, **16**, 155–170.

Cunha, P.P., Pimentel, N. and Pereira, D.I. (2000) Assinatura tectono-sedimentar do auge da compressão bética em Portugal: a descontinuidade sedimentar Valesiano terminal-Turoliano. *Ciênc. Terra (UNL)*, **14**, 61–72.

Dañobeitia, J.J., Bartolomé, R., Checa, A., Maldonado, A. and Slootweg, A.P. (1999) An interpretation of a prominent magnetic anomaly near the boundary between the Eurasian and African plates (Gulf of Cadiz, SW margin of Iberia). *Mar. Geol.*, **155**, 45–62.

Dewey, J.F., Pitman III, W., Ryan, W. and Bonnin, J. (1973) Plate tectonics and the evolution of the Alpine system. *Geol. Soc. Am. Bull.*, **84**, 3137–3180.

Dewey, J.F., Helman, M.L., Turco, E., Hutton, D.H.W. and Knott, S.D. (1989) Kinematics of the western Mediterranean. In: *Alpine Tectonics* (Eds M.P. Coward, D. Dietrich and R.G. Park), pp. 265–283. Special Publication 45, Geological Society, London.

Durand-Delga, M. and Fontboté, J.M. (1980) Le cadre structural de la Méditerranée Occidentale. *Bur. Rech. Géol. Min. Mem.*, **115**, 67–85.

Flinch, J.F., Bally, A.W. and Wu, S. (1996) Emplacement of a passive-margin evaporitic allochthon in the Betic Cordillera of Spain. *Geology*, **24**, 67–70.

Gomes, C.S.R. and Pereira, L.C.G. (2004) Paleomagnetismo do Maciço de Monchique (Sul de Portugal): implicações tectónicas. *Cadernos Lab. Xeológico de Laxe*, **29**, 291–297.

Gràcia, E., Dañobeitia, J., Vergés, J., Bartolomé, R. and Córdoba, D. (2003) Crustal architecture and tectonic evolution of the Gulf of Cadiz (SW Iberian margin) at the convergence of the Eurasian and African plates. *Tectonics*, **22**, 1–19.

Gradstein, F.M., Ogg, J.G. and Smith, A.G. (2004) *A Geologic Time Scale 2004*. Cambridge University Press, Cambridge, 589 pp.

Gutscher, M.-A., Malod, J., Rehauult, J.-P., *et al.* (2002) Evidence for active subduction beneath Gibraltar. *Geology*, **30**, 1071–1074.

Haq, B.U., Hardenbol, J. and Vail, P.R. (1987) Chronology of fluctuating sea levels since the Triassic. *Science*, **235**, 1156–1166.

Hayes, D.E., Pimm, A.C., Beckmann, J.P., *et al.* (1972) Site 135. In: *Initial Reports of the Deep-Sea Drilling Project*, JOIDES (Ed. A.C. Pimm), **14**, 15–48.

Hermes, J.J. (1985) Algunos aspectos de la estrutura de la Zona Subbética (Cordilleras Béticas, España Meridional). *Estud. Geol. Madrid*, **41**, 157–176.

Imperador-1 (1976) Final Well Site Report. *Arquivo Gabinete para a Prospecção e Exploração de Petróleo, n° 21413 – off/Chevron*. Chevron Oil Company of Portugal.

Lajat, D., Biju-Duval, B., Gonnard, R., Letouzey, J. and Winnock, E. (1975) Prolongement dans l'Atlantique de la partie externe de l'Arc bético–rifain. *Bull. Soc. Géol. Fr.*, **7**, 481–485.

Le Gall, B., Piqué, A., Réhault, J.P., Specht, M. and Malod, J. (1997) Structure et mise en place d'une ride océanique dans un contexte de limite de plaques convergentes: le Banc de Gorringe (SW Ibéria). *C. R. Acad. Sci. Paris*, **325**, 853–860.

Letouzey, J., Colletta, B., Vially, R. and Chermette, J.C. (1996) Evolution of salt-related structures in compressional settings. In: Salt Tectonics: a Global Perspective (Eds M.P.A. Jackson, D.G. Roberts and S. Snelson), pp. 41–60. Memoir 65, American Association of Petroleum Geologists, Tulsa, OK.

Lopes, F.C. (2002) *Análise tectono-sedimentar do Cenozóico da Margem Algarvia*. Unpublished PhD thesis, Universidade de Coimbra, 593 pp.

Lopes, F.C. and Cunha, P.P. (2000) Estratigrafia sísmica do Cenozóico na Plataforma Continental Algarvia: interpretação do controle tectónico da sedimentação. *Ciênc. Terra (UNL)*, **14**, 257–276.

Maestro, A., Somoza, L., Díaz-del-Rio, V., *et al.* (1998) Neotectónica transpresiva en la plataforma continental Suribética Atlantica. *Geogaceta*, **24**, 203–206.

Maestro, A., Somoza, L., Medialdea, T., *et al.* (2003) Large-scale slope failure involving Triassic and middle Miocene salt and shale in the Gulf of Cadiz (Atlantic Iberian Margin). *Terra Nova*, **15**, 380–391.

Maldonado, A. and Nelson, C.H. (1999) Interaction of tectonic and depositional processes that control the evolution of the Iberian Gulf of Cadiz margin. *Mar. Geol.*, **155**, 217–242.

Maldonado, A., Somoza, L. and Pallarés, L. (1999) The Betic orogen and the Iberian–African boundary in the Gulf of Cadiz: geological evolution (central North Atlantic). *Mar. Geol.*, **155**, 9–43.

Malod, J.A. (1982) *Comparaison de l'évolution des marges continentales au Nord et au Sud de la Péninsule Ibérique*. Thése d'Etat, Mém. Sc. Terre, Univ. Paris VI, 235 pp.

Meulenkamp, J.E. and Sissingh, W. (2003) Tertiary palaeogeography and tectonostratigraphic evolution of the Northern and Southern Peri-Tethys platforms and the intermediate domains of the African–Eurasian convergent plate boundary zone. *Palaeogeog. Palaeoclimatol. Palaeoecol.*, **196**, 209–228.

Mougenot, D. (1981) Une phase de compression au Crétacé terminal à l'Ouest du Portugal: quelques arguments. *Bol. Soc. Geol. Portugal*, **22**, 233–239.

Mougenot, D. (1989) *Geologia da Margem Portuguesa*. Instituto Hidrográfico, Lisboa, 259 pp.

Mougenot, D. and Vanney, J.R. (1982) Les rides de contourites plio-quaternaires de la pente continentale sud-portugaise. *Bull. Inst. Géol. Bassin Aquitaine*, **31**, 131–139.

Nelson, C.H., Baraza, J. and Maldonado, A. (1993) Mediterranean undercurrent sandy contourites, Gulf of Cadiz, Spain. *Sed. Geol.*, **82**, 103–131.

Nelson, C.H., Baraza, J., Maldonado, A., Rodero, J., Escutia, C. and Barber, J. (1999) Influence of the Atlantic inflow and Mediterranean outflow currents on Late Quaternary sedimentary facies of Gulf of Cadiz continental margin. *Mar. Geol.*, **155**, 99–129.

Olivet, J.L. (1996) La Cinématique de la Plaque Ibérique. *Bull. Centres Rech. Explor. – Prod. Elf Aquitaine*, **20**, 131–195.

Pais, J., Legoinha, P., Elderfield, H., Sousa, L. and Estevens, M. (2000) The Neogene of Algarve (Portugal). *Ciênc. Terra (UNL)*, **14**, 277–288.

Pinheiro, L.M., Wilson, R.C.L., Pena dos Reis, R., Whitmarsh, R.B. and Ribeiro, A. (1996) The western Iberia margin: A geophysical and geological overview. In: *Proceedings of the Ocean Drilling Program, Scientific Results, Volume 149* (Eds R.B. Whitmarsh, D.S. Sawer, A. Klaus and D.G. Masson), pp. 3–23. Ocean Drilling Program, College Station, TX.

Platt, J.P. and Vissers, R.L.M. (1989) Extensional collapse of thickened continental lithosphere: A working hypothesis for the Alboran Sea and Gibraltar arc. *Geology*, **17**, 540–543.

Puigdefàbregas, C. and Souquet, P. (1986) Tectosedimentary cycles and depositional sequences of the Mesozoic and Tertiary from the Pyrenees. *Tectonophysics*, **129**, 173–203.

Ribeiro, A., Antunes, M.T., Ferreira, M.P., *et al.* (1979) *Introduction à la Géologie Générale du Portugal.* Serviços Geológicos de Portugal, Lisboa, 114 pp.

Ribeiro, A., Kullberg, M.C., Kullberg, M.C., Manuppella, G. and Phipps, S. (1990) A review of Alpine tectonics in Portugal: Foreland detachment in basement and cover rocks. *Tectonophysics*, **184**, 357–366.

Ribeiro, A., Cabral, J., Baptista, R. and Matias, L. (1996) Stress pattern in Portugal mainland and the adjacent Atlantic region, West Iberia. *Tectonics*, **15**, 641–659.

Rodero, J., Pallarés, L. and Maldonado, A. (1999) Late Quaternary seismic facies of the Gulf of Cadiz Spanish margin: depositional processes influenced by sea-level change and tectonic controls (central North Atlantic). *Mar. Geol.*, **155**, 131–156.

Roest, W.R. and Srivastava, S.P. (1991) Kinematics of plate boundaries between Eurasia, Iberia and Africa in the North Atlantic from the Late Cretaceous to the present. *Geology*, **19**, 613–616.

Ruivo-1 (1975) Evaluation Report. *Arquivo Gabinete para a Prospecção e Exploração de Petróleo, n° 21350 – off/ Chevron.* Chevron Oil Company of Portugal.

Sanz de Galdeano, C. (1990) Geologic evolution of Betic Cordilleras in the Western Mediterranean, Miocene to the present. *Tectonophysics*, **172**, 107–119.

Sanz de Galdeano, C. (2000) Evolution of the Iberia during the Cenozoic with special emphasis on the formation of the Betic cordillera and its relation with the western Mediterranean. *Ciênc. Terra (UNL)*, **14**, 9–24.

Sanz de Galdeano, C. and Rodríguez-Fernández, J. (1996) Neogene palaeogeography of the Betic Cordillera: an attempt at reconstruction. In: *Tertiary Basins of Spain. The Stratigraphic Record of Crustal Kinematics* (Eds P.F. Friend and C.J. Dabrio), pp. 323–329. Cambridge University Press.

Sanz de Galdeano, C. and Vera, J.A. (1991) Una propuesta de clasificación de las cuencas neogénicas béticas. *Acta Geol. Hisp.*, **26**, 205–227.

Sanz de Galdeano, C. and Vera, J.A. (1992) Stratigraphic record and palaeogeographical context of the Neogene basins in the Betic Cordillera, Spain. *Basin Res.*, **4**, 21–36.

Sartori, R., Torelli, L., Zitellini, N., Peis, D. and Lodolo, E. (1994) Eastern segment of the Azores–Gibraltar line (central-eastern Atlantic): an ocean plate boundary with diffuse compressional deformation. *Geology*, **22**, 555–558.

Sierro, F.J., González Delgado, J.A., Dabrio, C., Flores, J.A. and Civis, J. (1992a) Excursion C (Spanish part): The Neogene of the Western Guadalquivir Basin (SW Spain). *Ciênc. Terra (UNL)*, **11**, 73–97.

Sierro, F.J., González Delgado, J.A., Dabrio, C., Flores, J.A. and Civis, J. (1992b) The Neogene of the Guadalquivir Basin (SW Spain). In: *Guias de las Excursiones Geológicas.* III Congresso Geológico de España y VIII Congresso Latinoamericano de Geología, 180–236.

Sierro, F.J., González Delgado, J.A., Dabrio, C., Flores, J.A. and Civis, J. (1996) Late Neogene depositional sequences in the foreland basin of Guadalquivir (SW Spain). In: *Tertiary Basins of Spain. The Stratigraphic Record of Crustal Kinematics.* (Eds P.F. Friend and C.J. Dabrio), pp. 339–345. Cambridge University Press.

Sissingh, W. (2001) Tectonostratigraphy of the west Alpine foreland: correlation of Tertiary sedimentary sequences, changes in eustatic sea-level and stress regimes. *Tectonophysics*, **333**, 361–400.

Srivastava, S.P., Schouten, H., Roest, W.R., *et al.* (1990a) Iberian plate kinematics: a jumping plate boundary between Eurasia and Africa. *Nature*, **344**, 756–759.

Srivastava, S.P., Roest, W.R., Kovacs, L.C., *et al.* (1990b) Motion of the Iberia since the Late Jurassic: Results from detailed aeromagnetic measurements in the Newfoundland Basin. *Tectonophysics*, **184**, 229–260.

Terrinha, P. (1998) *Structural Geology and Tectonic Evolution of the Algarve Basin, South Portugal*. Unpublished PhD thesis, Imperial College, London, 430 pp.

Terrinha, P., Coward, M.P. and Ribeiro, A. (1990) Salt tectonics in the Algarve basin: the Loulé diapir. *Comun. Serv. Geol. Portugal*, **76**, 33–40.

Terrinha, P., Dias, R.P., Ribeiro, A. and Cabral, J. (1999) The Portimão Fault, Algarve Basin, South Portugal. *Comun. Inst. Geol. Min.*, **86**, 107–120.

Tortella, D., Torne, M. and Pérez-Estaún, A. (1997) Geodynamic evolution of the eastern segment of the Azores–Gibraltar Zone: The Gorringe Bank and the Gulf of Cadiz Region. *Mar. Geophys. Res.*, **19**, 211–230.

Vegas, R., Medialdea, T., Muñoz, M., Díaz del Río, V. and Somoza, L. (2004) Nature and tectonic setting of the Guadalquivir Bank (Gulf of Cadiz, SW Iberian Peninsula). *Rev. Soc. Geol. Esp.*, **17**, 49–60.

Vicente, G., Gonzalez-Casado, J.M., Muñoz, A., Giner, J.L. and Rodriguez-Pascoa, M.A. (1996) Structure and Tertiary evolution of the Madrid Basin. In: *Tertiary Basins of Spain. The Stratigraphic Record of Crustal Kinematics* (Eds P.F. Friend and C.J. Dabrio), pp. 263–267. Cambridge University Press.

Ziegler, P.A. (1988) *Evolution of the Arctic–North Atlantic and the Western Tethys*. Memoir 43, American Association of Petroleum Geologists, Tulsa, OK, 198 pp.

Late Cenozoic basin opening in relation to major strike-slip faulting along the Porto–Coimbra–Tomar fault zone (northern Portugal)

ALBERTO GOMES*, HELDER I. CHAMINÉ†, JOSÉ TEIXEIRA‡, PAULO E. FONSECA§, LUÍS C. GAMA PEREIRA¶, ARY PINTO de JESUS**, AUGUSTO PÉREZ ALBERTÍ††, MARIA A. ARAÚJO*, ALEXANDRA COELHO‡, ANTÓNIO SOARES de ANDRADE‡ *and* FERNANDO T. ROCHA‡

**Departamento de Geografia, FLUP, Universidade do Porto, Via Panorâmica s/n, 4150-565 Porto, Portugal (Email: atgomes@netcabo.pt)*
†Laboratório de Cartografia e Geologia Aplicada, Departamento de Engenharia Geotécnica, Instituto Superior de Engenharia do Porto, ISEP, Rua Dr. A. Bernadino de Almeida, 431, 4200-072 Porto, Portugal; and Centro de Minerais Industriais e Argilas, Universidade de Aveiro, Portugal
‡Departamento de Geociências, Universidade de Aveiro, Campus de Santiago 3810-193 Aveiro, Portugal
§Departamento de Geologia, Universidade de Lisboa (LATTEX), Campo Grande, Edifício C6, 1749-016 Lisboa, Portugal
¶Departamento de Ciências da Terra, Universidade de Coimbra (GMSG), Largo Marquês de Pombal; 3000-272 Coimbra, Portugal
***Departamento de Geologia, Universidade do Porto (GIPEGO), Rua do Campo Alegre, 4169-007 Porto, Portugal*
††Departamento de Xeografía, Universidade de Santiago de Compostela, Faculdad de Xeografía e Historia, Campus Universitario Norte, 15782 Universidade de Santiago de Compostela, Spain

ABSTRACT

Northern Portugal is located in a tectonically complex area affected by major strike-slip zones, namely the north-northwest-trending Porto–Coimbra–Tomar fault zone and the north-northeast-trending Verin–Régua–Penacova sinistral strike-slip fault. Within this region, the sector between Albergaria-a-Velha and Águeda is crucial since it is highly affected by large-scale strike-slip faults and extensional deformation events. Late Cenozoic tectonics in northern Iberia resulted from the collision of the Africa and Eurasia plates, especially in the eastern segment of the Azores–Gibraltar plate boundary. The continued plate indentation originated the movement of major strike-slip faults in the Iberian Massif. The movement on these faults, accompanying the regional stress-field during the early Miocene, initiated the formation of incipient Cenozoic pull-apart basins.

The Albergaria-a-Velha–Águeda fault segment has been studied in an attempt to clarify the dynamic relationship between this active fault zone and the evolving landscape. Three geomorphological sectors were identified in the Albergaria-a-Velha region: (i) a littoral platform consisting of a polygenic erosion surface overlain by late Cenozoic alluvial–fluvial sequences; (ii) a tectonically controlled basin (Valongo do Vouga basin) located between hillslopes of two river valleys and normal faults with N–S orientation, where late Cenozoic subsidence is suggested by an influx of alluvial sandy conglomerates; and (iii) a domain of inner elevations of wide metapelitic landforms. Reactivation of the prevailing north-northwest-striking Upper Proterozoic/Palaeozoic basement is a regionally important control on the orientation and kinematics of late Cenozoic faults. Thus, the opening and development of these basins was influenced by the intersection of the north-northwest-trending dextral faults with north-northeast-trending sinistral faults associated with north–south shortening and east–west extension.

Keywords Basement, Iberian Massif, Porto–Coimbra–Tomar fault zone, relief, strike-slip basins.

INTRODUCTION

The tectonosedimentary architecture and mechanical and seismic properties of large fault systems can be better understood if structural geometries within the fault zone are characterized (e.g. Sylvester, 1988; Davison, 1994; Woodcock & Schubert, 1994; Richard *et al.*, 1995; Stone *et al.*, 1997; Davis, 2000; Friend *et al.*, 2000; Lagarde *et al.*, 2000). In general, large-displacement faults produce wide deformation zones either by extension or by compression. Wide damaged zones develop a complex internal geometry, which may influence the intrinsic behaviour of major faults (Cunningham *et al.*, 2003). These structures comprise upward-diverging faults, typically cutting antiformal push-ups or synformal pull-aparts, that normally form an anastomosed network of faults, where strike-slip faulting is one of the most important deformation mechanisms (Cabral, 1989; Woodcock & Schubert, 1994; Dooley & McClay, 1997; Rahe *et al.*, 1998). Out-of-sequence thrust faults are commonly found in many orogenic belts (McKerrow *et al.*, 1977; Morley, 1988; Friend *et al.*, 2000; Little & Mortimer, 2001; Gutiérrez-Alonso *et al.*, 2004). Over geological time-scales, however, fault systems frequently undergo modifications of their pattern style in response to variations in their regional stress field (Andeweg & Cloetingh, 2001; Ribeiro, 2002; Arjannikova *et al.*, 2004).

The complexity of sedimentary basins associated with strike-slip fault systems is almost as great as that observed for all other types of basins (Badham, 1982; Sylvester 1988; Wood *et al.*, 1994; Woodcook & Schubert, 1994; Richard *et al.*, 1995; Wakabayashi *et al.*, 2004). Furthermore, strike-slip fault systems within continental crust are likely to experience alternating periods of extension and compression as slip directions adjust along major crustal faults (Crowell, 1974; Ingersoll, 1988). The occurrence of offsets and bifurcations in strike-slip fault systems can lead to the formation of either transtensional or transpressional areas (Mann *et al.*, 1983; Woodcook & Schubert, 1994; Basile & Brun, 1999; Wakabayashi *et al.*, 2004). The shape of pull-aparts varies with their progressive development from spindle shape, through 'lazy-S' or 'lazy-Z' and rhomb shapes, to complex multi-rhomb shapes (Mann *et al.*, 1983).

The basement of intra and/or interplate settings consists mainly of exposures of highly deformed crystalline rocks, often with a smooth topography (Cabral, 1989; Ribeiro *et al.*, 1996; Bonnet *et al.*, 2000; Cunningham *et al.*, 2003). According to recent data from studies on surface processes and topographic relief of the formation of basement rocks in a collisional framework, the gravity collapse during the orogenic late stage of evolution is an important mechanism by which the elevation of mountain chains is strongly reduced (Burbank & Anderson, 2000; Summerfield, 2000).

Basement relief cannot be regarded merely as the result of long-term erosional activity (Stone *et al.*, 1997; Bonnet *et al.*, 2000). The study of the landforms and the adjacent depocentres generated by active and persistent tectonic processes may, consequently, provide insights into fault-generated mountain fronts and large-scale relief development related to tectonic uplift of the basement (e.g. Silva *et al.*, 1993, 2003; Bonnet *et al.*, 2000; Eusden *et al.*, 2000; Tippet & Hovius, 2000). Recent studies of relief development in crystalline basement relate mainly to the recognition and reconstruction of old palaeosurfaces in relation to the geological record of ancient kinematic processes (Bonnet *et al.*, 2000; Summerfield, 2000). Large-displacement faults often produce wide zones of deformation that commonly have complex internal geometries, which in turn may lead to significant modifications of the properties of the discontinuities and the sedimentary deposition.

Tectonics drives geodynamic background processes that, over time, directly shape surface topography. The effects of tectonics on topography occur over a large range of temporal and spatial scales. Surface topography in active deformation zones also incorporates the effects of processes such as climate, lithology and vegetation. The relationships between tectonics and relief formed by drainage network development in active zones depend directly on the role of erosion (displayed through isostatic responses and climate change) on the control of large-scale tectonic uplift (Bonnet *et al.*, 2000; Schumm *et al.*, 2000; Tippet & Hovius, 2000).

The Porto–Coimbra–Tomar (PCT) fault zone is an almost linear narrow belt with a north-northwest trend comprised within the crystalline polymetamorphosed belt of the Iberian Variscides (e.g. Lefort & Ribeiro, 1980; Ribeiro *et al.*, 1990a, 1996). During pre-Mesozoic times (Late Proterozoic to Palaeozoic; Beetsma, 1995; Chaminé *et al.*, 2003b) this fault system was a major dextral imbricated thrust zone (Gama Pereira, 1987; Dias & Ribeiro,

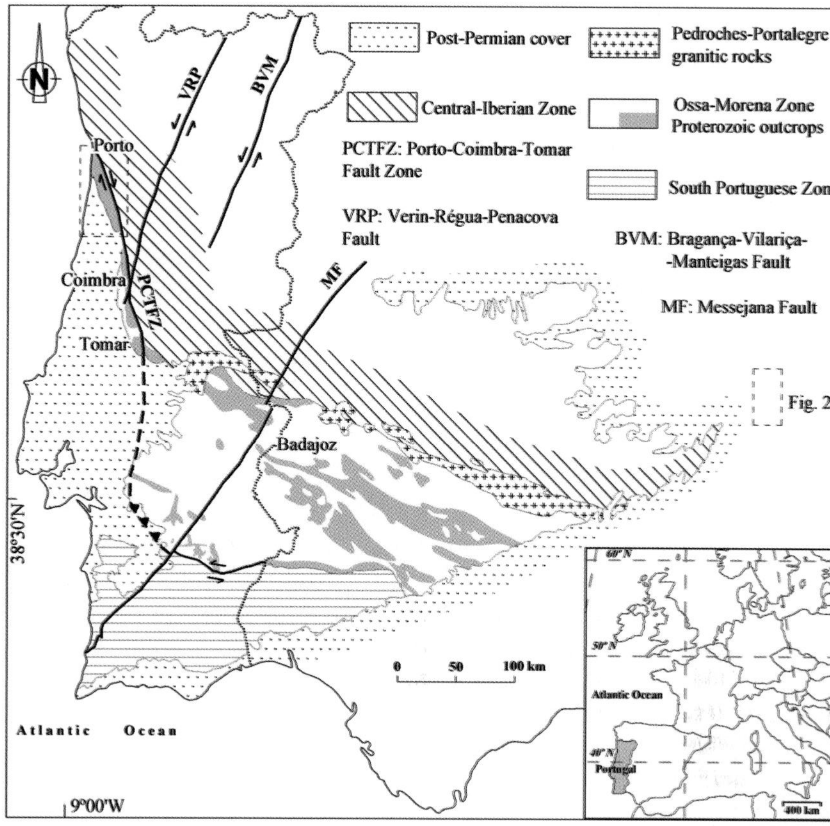

Fig. 1 Geotectonic setting of Iberian Massif with the location of the Porto–Coimbra–Tomar fault zone. (Adapted from Ribeiro *et al.*, 1990a.)

1993; Chaminé, 2000). Moreover, the PCT fault zone (Fig. 1) is part of the Porto–Tomar–Ferreira do Alentejo major shear zone (Chaminé, 2000; Ribeiro *et al.*, 2003; Chaminé *et al.*, 2003a,b), and is enclosed within the Western Iberian Line (WIL; Chaminé *et al.*, 2003a,b). The WIL delineates a north-northwest-trending tectonic corridor more than 520 km long, from Tomar (Portugal) to Finisterre (Galicia, Spain). This westernmost deformation corridor is characterized by out-of-sequence thrusting with affinity to the Ossa-Morena Zone (Chaminé *et al.*, 2003a,b). Mainly it comprises dextral strike-slip parallel overthrusts and dip-slip faults, as well as normal faults originating from transtensional basins forming within the inner part of the major shear zone. These configurations typically formed releasing bend structures during the Variscan Orogeny. This scenario is most probably responsible for the scattering of several imbricated metapelitic and blastomylonitic slices of Late Proterozoic/Palaeozoic tectonostratigraphic units (Chaminé *et al.*, 2003a,b, 2004, 2007).

The present study analyses the late Cenozoic landscape development and the structural evolution of a collapsed transpressive system along the Albergaria-a-Velha–Águeda segment (Valongo do Vouga basin) of the PCT fault zone. The purpose of this assessment is to achieve a better understanding of the geometry of near-surface strike- and dip-slip structures and their relationships to the evolving surface landscape. The observations of the internal structure of this large fault zone reported here provide new insights into the geometry and kinematics of basin evolution. This work outlines, initially, the geotectonic and geomorphological settings of the PCT fault zone, and then describes detailed field observations of the Albergaria-a-Velha–Águeda fault segment in an attempt to clarify the dynamic relationship between the principal displacement zone and the evolving landscape.

REGIONAL GEOLOGICAL SETTING

The Porto–Coimbra–Tomar fault zone was formed as a large lineament of complex accretionary thrusts during Variscan times. It comprises autochthonous and parautochthonous tectonostratigraphic units

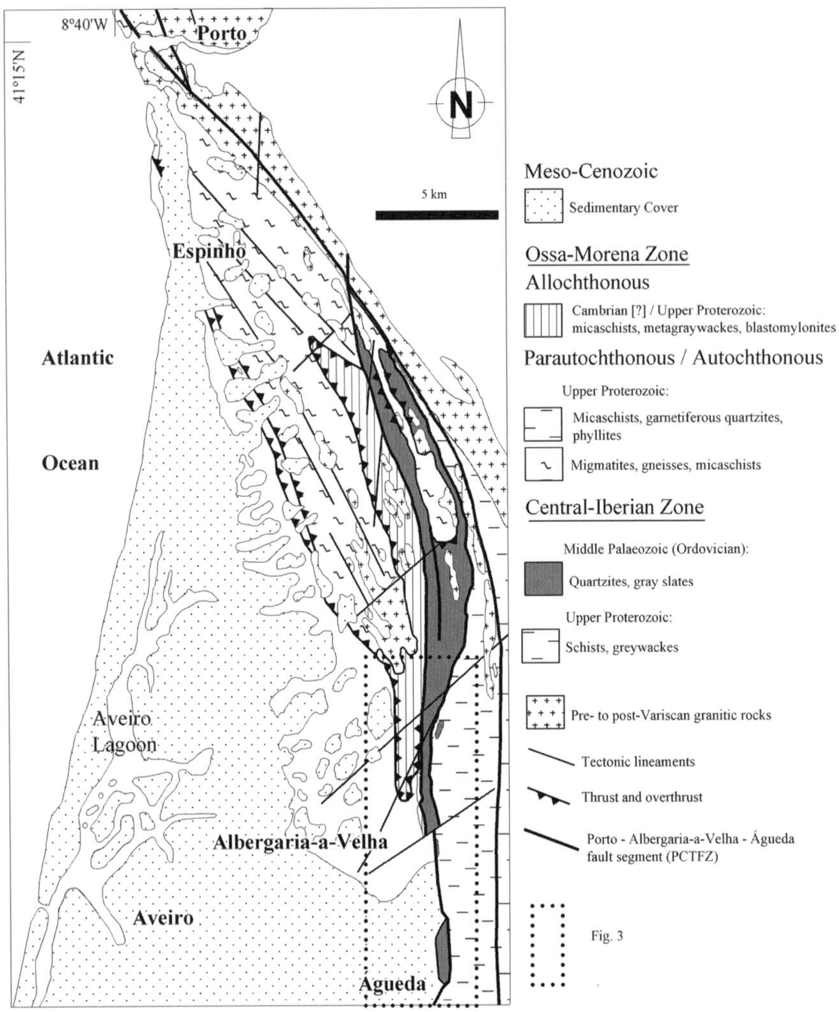

Fig. 2 Regional geotectonic framework of the Porto-Albergaria-a-Velha metamorphic belt, Ossa-Morena Zone, western Portugal. (Adapted from Chaminé, 2000.)

(Fig. 2) of low- to high-grade metamorphic rocks, as well as allochthonous units of medium- to high-grade metamorphic rocks, assumed to be of Late Proterozoic age (Gama Pereira, 1987; Beetsma, 1995; Chaminé *et al.*, 2003a,b; and references therein).

The general features for the region suggest two main regional tectonometamorphic stages of Variscan deformation (Severo Gonçalves, 1974; Gama Pereira, 1987; Chaminé, 2000) sometimes overprinting an earlier Cadomian migmatite sequence (Gama Pereira, 1987; Dias & Ribeiro, 1993; Chaminé, 2000). The first Variscan stage produced important folding and thrusts, as well as the dominant regional cleavage. The second regional stage (related to Central Iberian Zone Variscan-D_3; Dias & Ribeiro, 1995), also associated with megashear zones, produced a typical C–S shear deformation fabric and a non-coplanar cleavage schistosity with mylonitic or blastomylonitic foliation and crenulation. The metamorphic recrystallization coincided with the first Variscan stage, and continued in the second stage, when the major event of deformation resulted in metamorphic blastesis and metasomatism (Severo Gonçalves, 1974; Gama Pereira, 1987; Chaminé, 2000). Two major fault branches of the S. João-de-Ver thin skin thrust sheet (Chaminé, 2000), in a N–S direction, dominate the Albergaria-a-Velha–Águeda sector. During late Cenozoic times, a sinistral strike-slip faulting was associated with transtensional kinematics triggered by the post-orogenic collapse of the structure along the ancient Porto–Coimbra–Tomar thrust planes. These processes generated a multitude of discrete ENE–WSW, NNE–SSW to NE–SW regional fault systems (e.g. Açores-Carvalhal fault, Vouga River fault) with the generation of several pull-apart basins.

Basement rocks of the Albergaria-a-Velha–Águeda sector generally comprise subgreenschist- to amphibolite-grade metasedimentary rocks, as well as metavolcanic and blastomylonitic rocks (Chaminé *et al.*, 2003a). The Upper Proterozoic substratum rocks (Beetsma, 1995) comprise monotonous greenschists, phyllites, slates and staurolite–garnet schists, which generally dip northeast and strike 25°NW; these bedrocks dominate the structure of the northern Águeda region. These metapelitic rocks are internally folded and foliated reflecting at least two phases of Variscan regional ductile deformation. Cutting the basement are various granitoids and blastomylonitic rocks of early–late Palaeozoic age (Chaminé *et al.*, 1998). Unconformably overlying the polymetamorphic basement and infilling the pull-apart basins are Upper Devonian/Lower Carboniferous fossiliferous black shales (Chaminé *et al.*, 2003b).

Triassic coarse clastic sediments (red conglomerate–sandstone deposits) are also found in the region. The basement in the region is largely covered by post-Miocene continental sedimentary deposits (Soares de Carvalho, 1946a; Palain, 1976; Telles Antunes *et al.*, 1979). The tectonostratigraphy of the Porto–Albergaria-a-Velha–Águeda sector is summarized in Table 1.

METHODS

Structural and morphotectonic mapping of the Albergaria-a-Velha–Águeda segment on the Porto–Coimbra–Tomar fault zone, during late Cenozoic times, was the first goal. Based upon fault kinematics identified in the field and theoretically expected fault geometries, an attempt was made to reconstruct the original structural system. The landforms were also mapped in order to integrate fault kinematics and landform generation over time; relief formation was then used to help understand the development of the brittle fault system.

Structural geomorphology and landform maps were created by using a combination of aerial photograph interpretation, standard field mapping and digital terrain models. Portuguese Aerial Mapping photos at 1/33,000 and 1/15,000 scales, obtained from the National Army Geographical Institute, were used for photo-interpretation of tectonic network lineaments and as base maps for compilation and fieldwork. Bedrock structural geology of the Albergaria-a-Velha–Águeda region has been taken from the sketch geological map of Águeda presented by Soares de Carvalho (1946a) and further updated, while for the Albergaria-a-Velha sector, information was taken from Severo Gonçalves (1974) and Chaminé (2000). Figure 3 presents the new geological map synthesis achieved for the region under study after the fieldwork campaigns. This geological map was used as a basis for further refinements in order to better understand the overall lithological and structural framework.

Detailed information on the pre-Mesozoic evolution of the PCT basement rocks is beyond the scope of this paper and the main results have been published in Chaminé *et al.* (2003a,b, 2004, 2007) and Fernández *et al.* (2003).

MORPHOLOGY AND FAULT ARCHITECTURE

The segment studied (Albergaria-a-Velha–Águeda area, northwest Portugal) consists of an asymmetric fault system 25 km long and 5 km wide, bounded to the south by the Águeda River and to the northeast by the Caima River, which narrows northwards to a single fault trace. Between the bounding faults of the structure there are several subsidiary fault scarps that initially formed an imbricate set of footwall propagating thrust faults. Thus, inside the corridors of the PCT dextral strike-slip zone, major strain partitioning boundaries were defined (Dias & Ribeiro, 1993). These strike-slip partitioning corridors separate predominantly pre-Mesozoic regional-scale oblique strike-slip faults from thrust-dominated structures (Chaminé, 2000). The pattern of late Cenozoic faulting indicates north-northwest-driven compressional stresses from the Africa–Eurasia collision occurring in the eastern segment of the Azores–Gibraltar plate boundary (Cabral, 1995; Ribeiro *et al.*, 1996; Andeweg *et al.*, 1999; Cloetingh *et al.*, 2002; Ribeiro, 2002). Consequently, Alpine stress trajectories of the Atlantic margin clearly changed from the post-Miocene to the present-day (e.g. Cabral, 1989; Ribeiro *et al.*, 1990b; Andeweg, 2002; Cloetingh *et al.*, 2002; Jabaloy *et al.*, 2002). These

Table 1 Summary of stratigraphic and tectonometamorphic features of the Porto–Albergaria-a-Velha fault system

Province	Zone	Stratigraphy	Tectonometamorphic events	Timing (Chaminé *et al.*, 2003a,b)	Orogeny
Porto–Albergaria-a-Velha–Águeda platform		Sedimentary cover	Diagenesis, rifting process, tectonic inversion	Early Triassic–Quaternary	Alpine
					Late Variscan
Metasedimentary basement	Ossa–Morena	Allochthonous			
		Black shales bearing metacarbonates	Very low-grade metamorphism, organic-rich rocks	Early Carboniferous (Namurian), Late Devonian (Givetian/Frasnian)	
		Parautochthonous/autochthonous Micaschists, garnetiferous quartzites, phyllites; migmatites, gneisses	Middle- to high-grade metamorphism, folding in higher-grade areas, thrusting Early deformation, low- to high-grade metamorphism, peak metamorphism (*c.* 311 Ma), folding in higher-grade areas; post-metamorphism deformation, cross-folding, shear zones, thrusting, extensional cleavage	Cambrian[?]–Late Proterozoic	Pre- and late-Variscan; Cadomian[?]
	Central Iberian	Parautochthonous/autochthonous Armorican quartzites, fossiliferous grey slates	Low-grade metamorphism (greenschist facies), folding, Variscan structures, pre- to syn-peak metamorphism, shear zones, fabric development	Ordovician	Variscan and Cadomian[?]
		Schists, greywackes		Late Proterozoic	
Granitic rocks		Lavadores granite	Post-tectonic granite	298 ± 12 Ma	Late Variscan
		Oliveira de Azeméis–Feira–Lourosela granitic belt	Granitic antiform structure	320 ± 3 Ma; 379 ± 12 Ma; 421 ± 4 Ma; 419 ± 4 Ma	Variscan and pre-Variscan
		Ossela–Milheirós de Poiares blastomylonitic belt	Granitic synform structure		
		Foz do Douro Complex	Shear zones, mylonitic fabric	575 ± 5 Ma; 607 ± 17 Ma	Cadomian

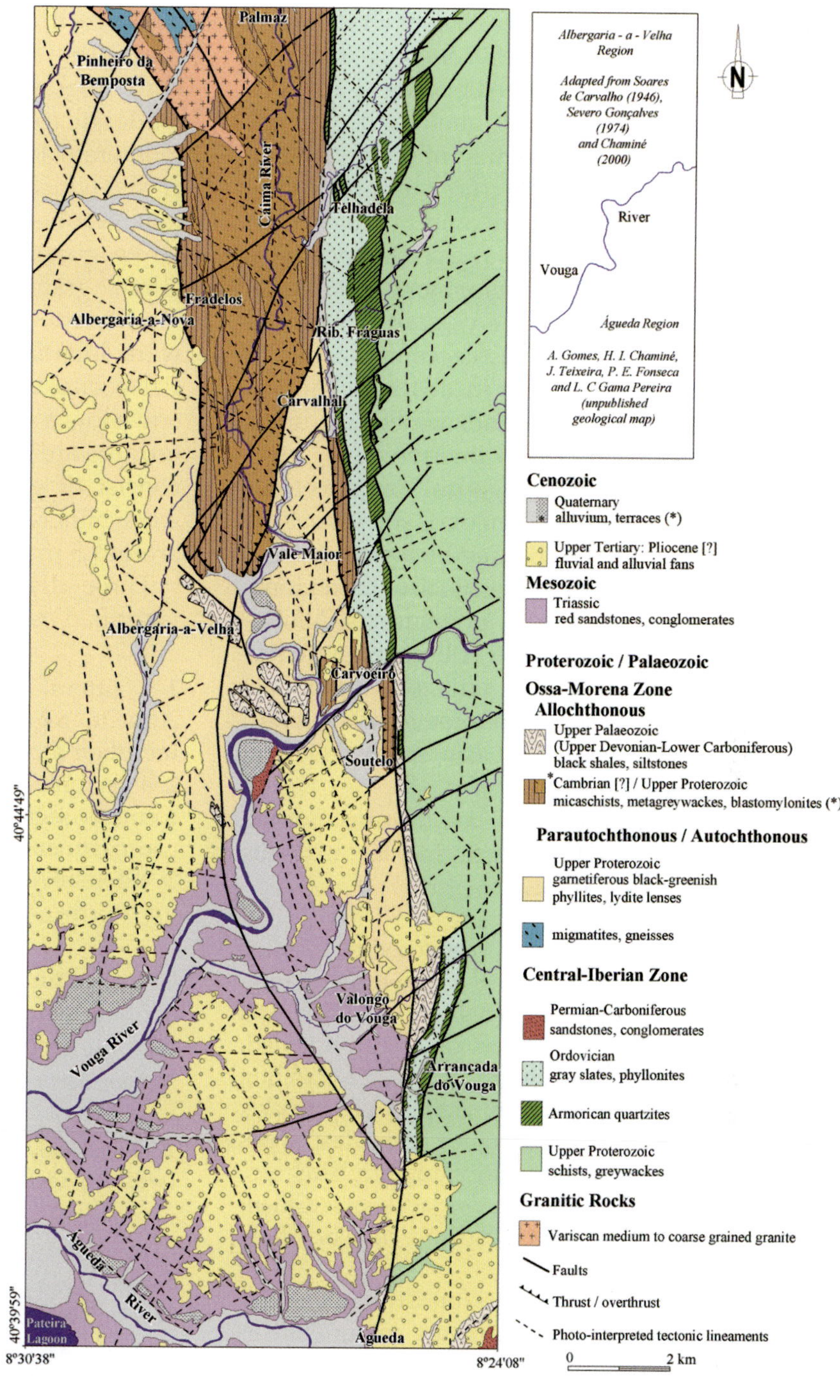

Fig. 3 Geological map of the Albergaria-a-Velha–Águeda region (northwest Portugal).

features suggest a recent change of the regional strain regime by a reversal of kinematics inducing tectonic activity on the north-northeast strike-slip faults. The recent geotectonic evolution of northwest Iberia has been dominated by north-northeast-trending fault zones reactivating inherited crustal structures in a sinistral strike-slip regime, namely the Verin–Régua–Penacova fault and the Bragança–Vilariça–Manteigas fault (Cabral, 1989, 1995; Brum Ferreira, 1991; Ribeiro, 2002). Nevertheless, total displacement as evidenced by the regional geomorphological framework is actually the result of the periodic reactivation of different PCT fault segments since the end of the Variscan Orogeny until the present-day (Ribeiro *et al.*, 1990b; Cabral, 1995).

The morphotectonic and geological surveys allowed an original outline map of the crystalline basement relief to be reproduced (Fig. 3). This new structural geology and geomorphology mapping of the northern Águeda region demonstrates that the scale of relief development is strongly associated with the existence of scarps inherited from the PCT principal displacement zone. In addition, relief development in this segment was mainly controlled by late Cenozoic tectonic uplift during fault reactivation and interference.

The study area, the Valongo do Vouga tectonic basin, occupies a NNW–SSE narrow strip along the PCT fault zone, which is limited by a major range-bounding fault and an uplift block. The regional block steps show two main orientations, NNW–SSE to N–S and NE–SW. The Albergaria-a-Velha–Águeda segment of the PCT fault zone is the most prominent scarp of this morphology, with a length of 35 km and a height that can reach 200 m in some localities.

Filling the tectonic basin are unconsolidated Miocene–Pliocene alluvial continental deposits that are outlined as follows.

1 The base of the sequence is composed of massive boulder conglomerate beds with a grey-brown sandy matrix. The bed thickness ranges from 4 to 5 m. The sorting is very poor with clasts randomly oriented. The clasts have a maximum clast size ranging from 40 to 70 cm and are dominated by white quartz (83% of the total) and quartzite; both clast types are mainly subrounded. Individually, beds are matrix-supported conglomerates lacking sedimentary structures, a characteristic feature of debris-flow deposits.
2 In the middle of the sequence, massive grey–white silty–muddy beds occur, which are interpreted as alluvial plain deposits.
3 The top of the sequence comprises massive coarse sandy beds with a red silty matrix. Outcrops include some pebble or conglomerate lenses, with rounded clasts showing a red patina (iron oxides). The clasts have been reworked, and are dominated by quartzite (64%), white quartz (33%) and metagreywacke (3%). In addition, within these beds there are sets of trough cross-stratified sandstone facies that were interpreted as representing stream channel deposits.

The early Quaternary fluvial terraces are characterized by massive clast-supported coarse sands with a grey–white silty matrix. The sorting is moderate to poor. The deposits include some conglomerate lenses with elongated and imbricated clasts, which are also rounded. The clasts have a maximum size of 20 cm and are composed of white quartz (37%), slate (20%), metagreywacke (16%), granite (14%), gneiss (9%) and quartzite (4%). Metre-scale sets of cross-stratified sandstone were identified in the stream channel deposits. Palaeocurrent directions are towards N225°E ± 10° and show a small amount of dispersion. The fluvial terraces are covered by colluvium deposits, which consist of grey–brown very fine silty clay with some mineral grains (quartz, mica), estimated to be around 5 m thick.

Three distinct morphostructural sectors were identified in this strike-slip fault segment (Fig. 4A):

1 a littoral platform in the Albergaria-a-Velha–Águeda region (western sector) subdivided by a meridian tectonic corridor;
2 a tectonic basin in the Valongo do Vouga area (central sector), comprising two morphotectonic compartments;
3 a domain of smooth stepped elevations to the interior (northeastern sector).

Their main morphotectonic characteristics, described below, bear the signature of the major tectonic structures occurring in the Albergaria-a-Velha–Águeda fault segment.

Albergaria-a-Velha littoral platform (western sector)

The Albergaria-a-Velha littoral platform corresponds to a planar surface gently dipping to the west of the Caima and Vouga river valleys (Soares de Carvalho, 1946a,b, 1949). The area is dominantly composed of greenschist-grade basement rocks. In the study sector, the platform shows an elevation varying between 60 and 200 m (on its eastern side) and ends against the S. João-de-Ver thrust sheet, which is marked by small elongated hills, of N–S orientation (Fig. 4), and extends from Fradelos to Senhora do Socorro (216 m). This planation surface is interrupted by a meridian graben of flat-topped hills which are 20 m to 40 m

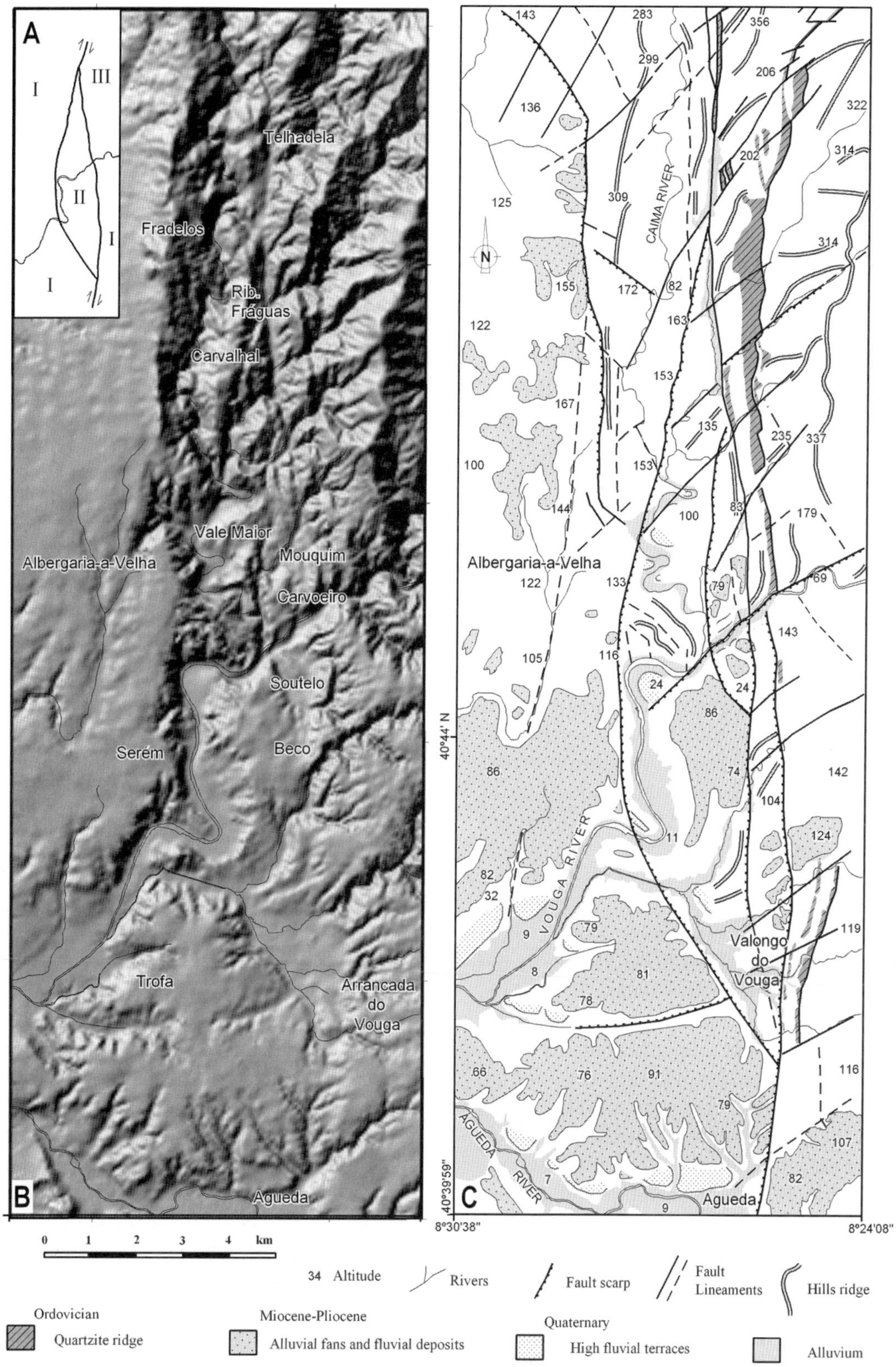

Fig. 4 Morphotectonic map of the Albergaria-a-Velha–Águeda region. (A) Morphostructural sectors: Albergaria-a-Velha–Águeda littoral plataform (I), Valongo do Vouga tectonic basin (II), inner elevations domain (III). (B) Digital elevation model of the area studied, obtained from digitization of elevation contour lines of the 1:25,000 scale map and generated by kriging. Ground resolution is 50 m. Shadowed relief map of the digital terrain model, artificially illuminated from the west. (C) Morphotectonic interpretation.

below the littoral platform top. According to Brum Ferreira (1980), the platform is a polygenic erosion surface overlain by late Cenozoic alluvial–fluvial sequences (Cunha *et al.*, 2005). The down-thrown hills are separated from the tectonic basin of Valongo do Vouga and, particularly from the PCT fault zone (*sensu stricto*), by deeply incised river valleys which are under regional tectonic control (Fig. 4).

Valongo do Vouga tectonic basin (central sector)

The Valongo do Vouga basin corresponds to a tectonic corridor located between the western hillslopes of the Caima and Vouga river valleys and the normal faults, of N–S orientation, extending from the Telhadela to Águeda regions (Fig. 4). It is characterized by the presence of steep hills to the west, whereas a more gentle topography is observed to the east (near Albergaria-a-Velha platform). The Valongo do Vouga basin is composed of two major morphotectonic compartments (Fig. 5): (i) the Carvalhal compartment north of Vouga river and (ii) the Soutelo compartment to the south.

The Carvalhal compartment is a major uplift block with a southeast tilt due to northwest-directed thrusting along northwest inner thrust fronts. Along this block, the main frontal fault scarp is remarkably linear (Figs 5 & 6). The main fault zone is made up of a distinct zone of soft black–greenish gouge 5–8 m wide. This compartment is bounded, on the east, by the PCT strike-slip fault with a normal component, on the west by the eastern branch of the S. João-de-Ver thrust sheet and, on the south, by the Vouga River normal fault (Fig. 5). At this location, the Vouga River normal fault is up to 4.5 km long and contains many subparallel fracture surfaces striking on average N45°E ± 5° and dipping vertically or steeply to the southeast. The area is dominantly composed of metapelitic substratum rocks of greenschist–amphibolite grade. Where the Vouga River crosses the Valongo do Vouga tectonic basin, three levels of Quaternary fluvial terraces can be distinguished. In the vicinity of Carvoeiro (Fig. 5), several terraces are preserved on the right bank of the valley. To the north of the site lies the highest terrace, made of unconsolidated alluvial deposits (conglomerates and sands), 40 m above the present-day riverbed. A NE–SW normal fault scarp passing along the Vouga valley is responsible for a vertical displacement of 8–10 m between the highest fluvial terraces near Carvoeiro. The NNE–SSW sinistral component is clearly observed on the Digital Terrain Model (DTM) where several rivers show deflection and a typical offset corner in a sinistral sense (Fig. 4A). In the Carvalhal compartment the morphology is characterized by the occurrence of deeply incised valleys that down-cut the relief, localizing steep sloped hill tops.

According to Soares de Carvalho (1946a), the Soutelo compartment is down-thrown with a gentle southeast tilt due to the northwest-directed thrusting with rotation (Fig. 6). The Soutelo depression shows a mean altitude of 24 m and is filled with Miocene to Quaternary sediments. The mean thickness of these deposits is around 30 m, their origin being alluvial or colluvial. The Lower Triassic red sandstones crop out mainly in the western part of the compartment. Morphostructural highs, aligned in a N–S direction and reaching 104 m, comprise metasedimentary rocks cropping out in the eastern part of the Soutelo block. The linear trend of the scarp fault is indicative of the near-vertical dip angle of this segment included in the Albergaria-a-Velha–Águeda fault system (Figs 5 & 6).

Inner elevations domain (northeastern sector)

The inner elevations domain is a major uplift block positioned to the northeast of the Valongo do Vouga basin (Fig. 4). It is composed of several steep hills (reaching 600 m height). The western front of this group of hills is marked by a major active fault system that is clearly visible on aerial photographs and the DTM. This fault is part of the regional PCT system (Chaminé *et al.*, 2003a), and in this study is referred to as the Albergaria-a-Velha–Águeda fault. The presence of wide blocks of metapelitic rocks at high elevations, bounded by deep valleys, more incised than those of the Valongo do Vouga basin, is the most prominent feature. Near the main fault, the relief is dominated by elongated N–S residual resistant quartzitic ridges, with altitudes reaching 400 m. The inner elevations contrast with the smooth surfaces to the west (200 m), and define chiefly regional tectonic lineaments of N–S to NNW–SSE preferential trends.

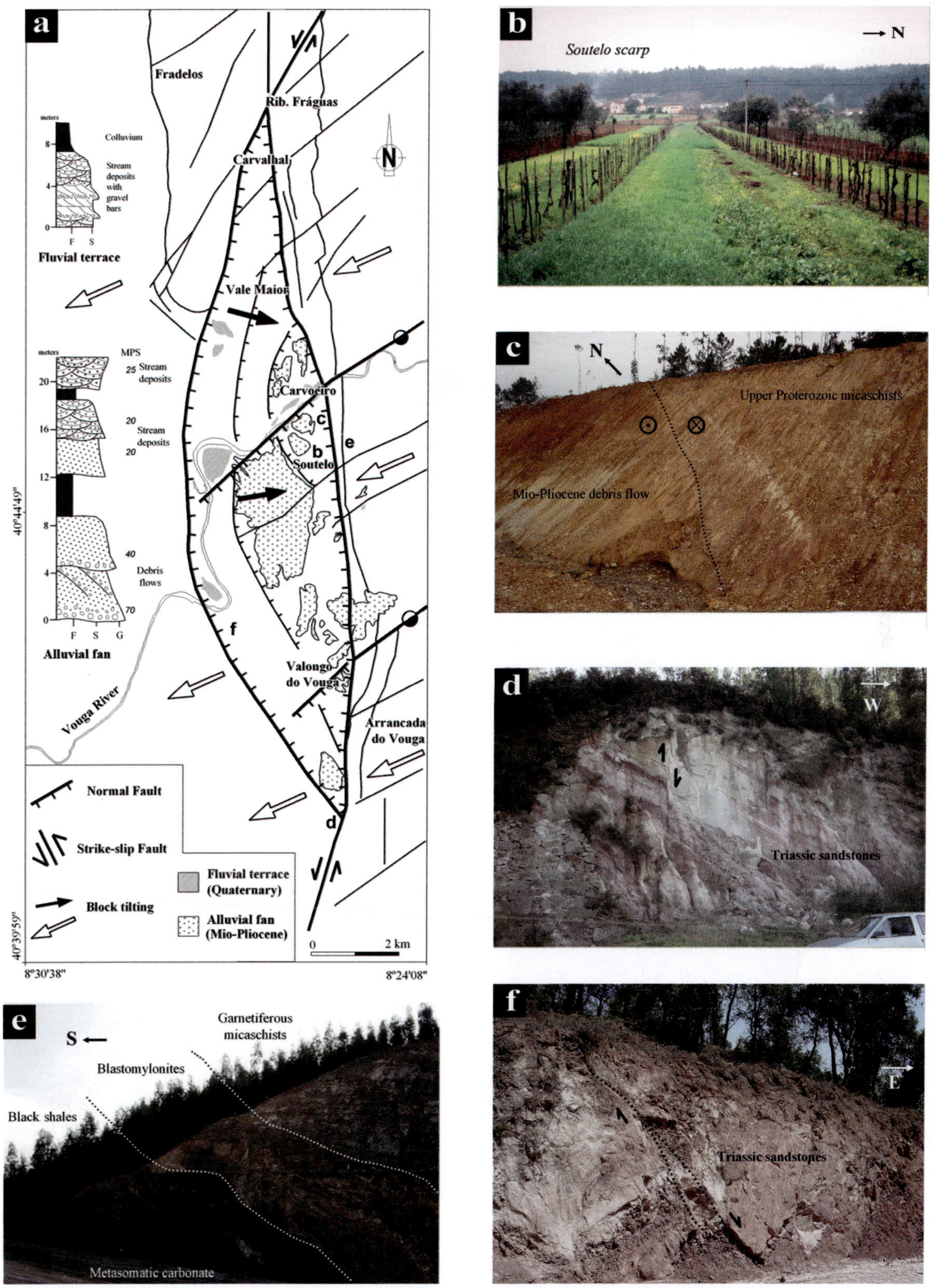

Fig. 5 (a) Morphostructural interpretation of the Valongo do Vouga tectonic basin. (b) View of the Soutelo flat-floor depression and the terminating frontal scarp to the west. (c) Detailed view of a normal fault affecting Mio-Pliocene debris-flow deposits in the Soutelo compartment in the east. (d) Normal fault of N–S orientation affecting Triassic sandstones near Águeda. (e) View west of old brittle frontal thrust zone (Porto–Coimbra–Tomar fault zone) cutting Upper Palaeozoic black shales bearing metasomatic carbonates, near Soutelo. (f) Normal fault of NW–SE direction affecting Triassic sandstones near the Vouga River.

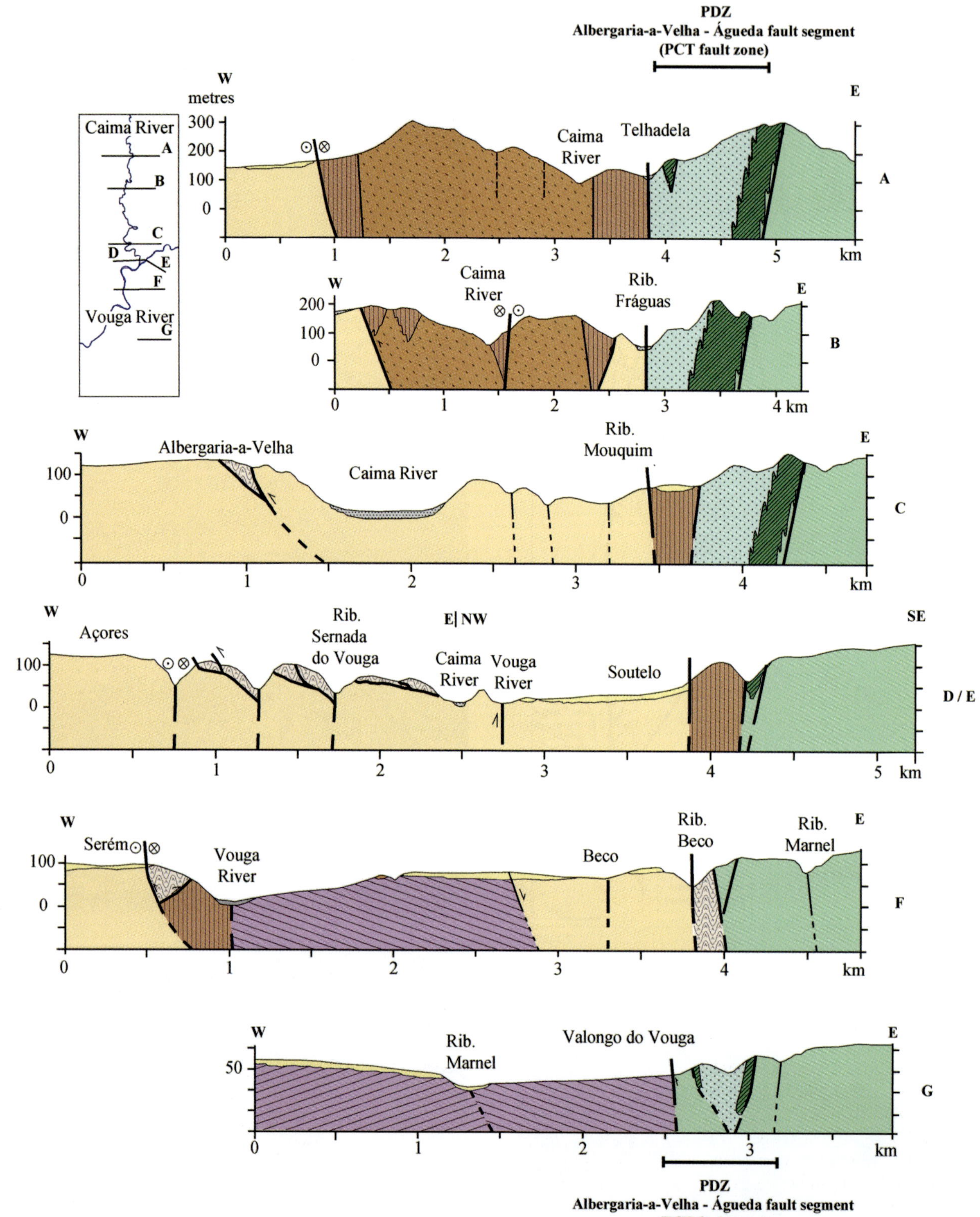

Fig. 6 Cross-sections through the Valongo do Vouga tectonic basin. Cross-sections are constructed perpendicular to the Porto–Albergaria-a-Velha–Águeda fault system (see Fig. 3 for location and explanation).

A MORPHOTECTONIC MODEL: INTERPRETATION AND DISCUSSION

The data presented here for the Valongo do Vouga basin provide new insights to the Porto–Coimbra–Tomar major strike-slip fault zone. In addition to improving knowledge of the evolving landforms and deposits developed in this region, they clarify the understanding of the geodynamic evolution of the Albergaria-a-Velha–Águeda fault segment. The complex geometry originating from a stretched zone encompassed between two tectonostratigraphic Iberian megadomains (Ossa-Morena and Central Iberian Zones) with a long polyphase geotectonic history of activity (Gama Pereira, 1987; Chaminé, 2000; Chaminé *et al.*, 2003a,b) explains the complex general pattern of the pre-Mesozoic substratum, the so-called Porto–Albergaria-a-Velha Proterozoic metamorphic belt (Chaminé, 2000).

Given the geological and geomorphological complexity of the Valongo do Vouga tectonic basin we focused the present work on the following aspects: (i) the occurrence of rigid crustal anisotropy in mechanically weak pre-Mesozoic metasedimentary and blastomylonitic basement rocks; (ii) late Cenozoic movements occurring along the north-northeast-trending fault segments displaying both a normal and a left-lateral component, thus indicating a transtensional regime; (iii) the morphotectonic compartments (Carvalhal and Soutelo blocks) of the Valongo do Vouga basin, which show major fault scarps with southeast tilting; these faults bound the northwest sides of uplifted/downthrown blocks. By contrast, the fault architecture on other morphostructural sectors is essentially large west-tilted thrust blocks.

Altogether, the data suggest that the morphology of the Valongo do Vouga basin has been formed by late Cenozoic displacement on offset segments of the Porto–Coimbra–Tomar fault system due to sinistral movement of the north-northeast-trending faults (Fig. 7). Originally, the Valongo do Vouga basin had a flat-rhombic shape with its long axis oriented in a north direction, and bounded by arc-shaped sidewall faults linking two offset segments of this major N–S-trending fault system. A sigmoid structure then succeeded as the surface expression of a fault-bounded frame developed along the margin of the basin. The terraces that formed along the southern basin sidewall fault system are interpreted as corresponding to down-faulted blocks. The structure may be formed either before, or simultaneously with NE–SW to E–W faults. The basin sidewall faults change from curved in the middle of the basin, to steep and planar toward the corners, where they are connected

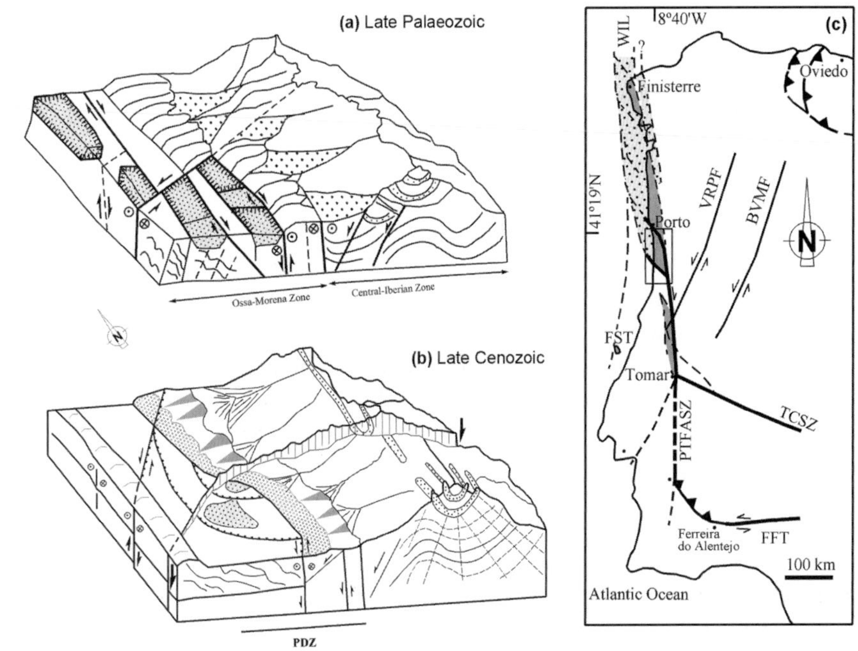

Fig. 7 (a and b) Tentative three-dimensional model block diagrams and (c) structural sketch map illustrating the relief development related to pre-Mesozoic substratum legacy of the late Cenozoic infill basins. (a) Early basin formation fault-block and growth fault sequence development due to dextral displacement along the Porto–Albergaria-a-Velha fault system; (b) illustration of the present rhomboidal basin architecture (PDZ: principal displacement zone). Shear zones and terranes: WIL, Western Iberian Line; PTFASZ, Porto–Tomar–Ferreira do Alentejo dextral major shear zone; FFT, Ferreira do Alentejo–Ficalho thrust; TCSZ, Tomar–Córdoba sinistral shear zone; FST, Farilhões suspect terrane; VRPF, Verin–Régua–Penacova Fault; BVMF, Bragança–Vilariça–Manteigas Fault.

to the principal displacement zones of the PCT fault system. The basin floor consists dominantly of pre-Permian metamorphic rocks and Lower Triassic sedimentary cover (Figs 3 & 6), surrounded by side-wall fault scarps of up to 200 m high. Likewise, elongated small depressions are defined on the secondary valley floors (e.g. at Telhadela, Ribeira-de-Fráguas, Vale Maior, Mouquim and Soutelo). The tendency for a local subsidence pattern in this corridor is marked by the deposition and preservation of Miocene–Pliocene alluvial fans and the conservation of the main Quaternary fluvial terraces of the Vouga, Caima and Águeda rivers.

Another relevant aspect of the tectonic control on the drainage network is the emergence of a lithological resistance barrier (e.g. the Armorican quartzite) along a NE–SW regional fault system as a result of bedrock incision. In fact, the spatial organization of the regional drainage network is strongly controlled by the large-scale topography, namely by the location of high elevation domains of the Armorican quartzite relief. The main rivers (e.g. the Caima and Vouga rivers) flow along the principal displacement zone, with a north to south trend, before reaching the Atlantic Ocean. The river valleys have moderate depths, ranging from a minimum of 150 to a maximum of 200 m.

CONCLUSIONS

The Porto–Albergaria-a-Velha–Águeda metamorphic belt, which is at least 65 km long, is segmented into several morphotectonic compartments. One of these compartments is the Albergaria-a-Velha–Águeda fault system, which reaches about 35 km long and is the focus of this study. The major faults bounding the Valongo do Vouga basin are inherited ancient discrete structures that have been reactivated in a transtensional tectonic regime since, at least, late Cenozoic times.

An evolutionary model for the Valongo do Vouga tectonic basin may thus be constructed, using the comparative analysis of structural mapping and geomorphological data. A tentative approach has been developed by linking the structural Variscan substratum evolution of the Albergaria-a-Velha–Águeda segment of the Porto–Coimbra–Tomar fault system to its dynamic landscape elements in an active tectonic setting. Indeed, the basement structure with uplifted and deformed Triassic deposits, caused by post-Variscan kinematics, in the southwestern part of the Valongo do Vouga region, defines the late evolution of the segmented fault system. The formation of the early northern segment consisted of uplifted/downthrown fault blocks with evidence of brittle deformation, due to late Cenozoic displacement along the Albergaria-a-Velha–Águeda fault system and sinistral north-northeast-trending related system (e.g. Verin–Régua–Penacova fault). Furthermore, these events were closely followed by the deposition and displacement of the sedimentary succession on the fault-block. Finally, the formation of the present rhomboidal basin architecture may have triggered the emergence of the NNE–SSW to NE–SW conjugated trending faults.

ACKNOWLEDGEMENTS

We gratefully acknowledge Peter Friend (Cambridge), Gaspar Soares de Carvalho (Braga), Christian Palain (Nancy) and João Cabral (Lisbon) for stimulating discussions and for their assistance in fieldwork. This research was partially supported by TBA (FCT-POCTI/CTA/38659/2001), Geodyn (POCTI-ISFL-5-32/Lisbon), GROUNDURBAN (POCTI/CTE-GIN/59081/2004) and FCT-SFRH/BPD/3641/2000 (Aveiro) research grants. We acknowledge James Howard, Tim Dooley and Gary Nichols for constructive reviews that helped to improve the clarity of the manuscript.

REFERENCES

Andeweg, B. (2002) *Cenozoic tectonic evolution of the Iberian Peninsula, the effects and causes of changing stress fields*. Unpublished PhD thesis, Vrije Universiteit, Amsterdam, 178 pp.

Andeweg, B. and Cloetingh, S. (2001) Evidence for an active sinistral shear zone in the Western Alboran region. *Terra Nova*, **13**, 44–50.

Andeweg, B., Vicente, G., Giner, J. and Muñoz Martin, A. (1999) Local stress fields and intraplate deformation of Iberia: variations in spatial and temporal interplay of regional stress sources. *Tectonophysics*, **305**, 153–164.

Arjannikova, A., Larroque C., Ritz, J.-F., *et al.* (2004) Geometry and kinematics of recent deformation in the Mondy–Tunka area (south-westernmost Baikal rift zone, Mongolia–Siberia). *Terra Nova*, **16**, 265–272.

Badham, J.P.N. (1982) Strike-slip orogens: an explanation for the Hercynides. *J. Geol. Soc. London*, **139**, 493–504.

Basile, C. and Brun, J.P. (1999) Transtensional faulting patterns ranging from pull-apart basins to transform continental margins: an experimental investigation. *J. Struct. Geol.*, **21**, 23–27.

Beetsma, J.J. (1995) *The late Proterozoic/Paleozoic and Hercynian crustal evolution of the Iberian Massif, N Portugal, as traced by geochemistry and Sr-Nd-Pb isotope systematics of pre-Hercynian terrigenous sediments and Hercynian granitoids*. Unpublished PhD thesis, Vrije Universiteit, Amsterdam, 223 pp.

Bonnet, S., Guillocheau, F., Brun, J.P. and Driessche, J. (2000) Large-scale relief development related to Quaternary tectonic uplift of the Proterozoic-Paleozoic basement: The Armorican Massif, NW France. *J. Geophys. Res.*, **105**(B8), 19,273–19,288.

Brum Ferreira, A. (1980) Surfaces d'aplanissement et tectonique récente dans le Nord de la Beira (Portugal). *Rev. Géol. Dynam. Géog. Phys.*, **22**(1), 51–62.

Brum Ferreira, A. (1991) Neotectonics in northern Portugal: a geomorphological approach. *Z. Geomorphol.*, **82**, 73–85.

Burbank, D.W. and Anderson, R.S. (2000) *Tectonic Geomorphology*. Blackwell Science, Oxford, 274 pp.

Cabral, J. (1989) An example of intraplate neotectonic activity, Vilariça basin, Northeast Portugal. *Tectonics*, **8**(2), 285–303.

Cabral, J. (1995) Neotectónica em Portugal Continental. *Mem. Inst. Geól. Min. Lisboa*, **31**, 1–256.

Chaminé, H.I. (2000) *Estratigrafia e estrutura da faixa metamórfica de Espinho-Albergaria-a-Velha (Zona de Ossa-Morena): implicações geodinâmicas*. Unpublished PhD thesis, Universidade do Porto, 497 pp.

Chaminé, H.I., Leterrier, J., Fonseca, P.E., Ribeiro, A. and Lemos de Sousa, M.J. (1998) Geocronologia U/Pb em zircões e monazites de rochas ortoderivadas do sector Espinho-Albergaria-a-Velha (Zona de Ossa Morena, NW de Portugal). *Com. Inst. Geol. Min. Lisboa*, **84**(1), B115–B118.

Chaminé, H.I., Gama Pereira, L.C., Fonseca, P.E., Noronha, F. and Lemos de Sousa, M.J. (2003a) Tectonoestratigrafia da faixa de cisalhamento de Porto–Albergaria-a-Velha–Coimbra–Tomar, entre as Zonas Centro-Ibérica e de Ossa-Morena (Maciço Ibérico, W de Portugal). *Cad. Labor. Xeol. Laxe Coruña*, **28**, 37–78.

Chaminé, H.I., Gama Pereira, L.C., Fonseca P.E., *et al.* (2003b) Tectonostratigraphy of middle and upper Palaeozoic black shales from the Porto–Tomar–Ferreira do Alentejo shear zone (W Portugal): new perspectives on the Iberian Massif. *Geobios*, **36**(6), 649–663.

Chaminé, H.I., Gomes A., Teixeira J., *et al.* (2004) Geologia, geomorfologia e estratigrafia dos domínios estruturais de Carvoeiro-Caldas de S. Jorge e de Arrancada do Vouga-Águeda (faixa de cisalhamento de Porto-Tomar, NW de Portugal): implicações paleogeográficas. *Cad. Labor. Xeol. Laxe Coruña*, **29**, 299–330.

Chaminé, H.I., Fonseca, P.E., Pinto de Jesus, A., *et al.* (2007) Tectonostratigraphic imbrications along strike-slip major shear zones: an example from the early Carboniferous of SW European Variscides (Ossa-Morena Zone, Portugal). In: *Proceedings of the XVth ICCP (Utrecht, 2003)* (Ed. T.E. Wong), pp. 405–416. Special Volume of the Royal Netherlands Academy of Arts and Sciences, Edita KNAW, Amsterdam.

Cloetingh S., Burov E., Beekman F., *et al.* (2002) Lithospheric folding in Iberia. *Tectonics*, **21**(5), 5.1–5.25.

Crowell, J.C. (1974) Sedimentation along the San Andreas fault. *Spec. Publ. Soc. Econ. Paleontol. Mineral.*, **19**, 292–303.

Cunha, P., Martins, A., Daveau, S. and Friend, P. (2005) Tectonics control of the Tejo river fluvial incision during the Late Cenozoic in Ródão, Central Portugal (Atlantic Iberian Border). *Geomorphology*, **64**, 271–298.

Cunningham, D., Dijkstra, A., Howard, J., Quarles, A. and Bardarch, G. (2003) Active intraplate strike-slip faulting and transpressional uplift in the Mongolian Altai. In: *Intraplate Strike-slip Deformation Belts* (Eds F. Storti, R.E. Holdsworth and F. Salvini), pp. 65–87. Special Publication 210, Geological Society Publishing House, Bath.

Davis, G.H., Bump, A.P., García, P.E. and Ahlgren, S.G. (2000) Conjugate Riedel deformation band shear zones. *J. Struct. Geol.*, **22**(2), 169–190.

Davison, I. (1994) Linked fault systems: extensional, strike-slip and contractional. In: *Continental Deformation* (Ed. P.L. Hancock), pp. 121–142. Pergamon Press, Oxford.

Dias, R. and Ribeiro, A. (1993) Porto–Tomar shear zone, a major structure since the beginning of the variscan orogeny. *Com. Inst. Geol. Min. Lisboa*, **79**, 31–40.

Dias, R. and Ribeiro, A. (1995) The Ibero–Armorican arc: a collision effect against an irregular continent?. *Tectonophysics*, **246**, 113–128.

Dooley, T., and McClay, K. (1997) Analog modeling of pull-apart basins. *Am. Assoc. Petrol. Geol. Bull.*, **81**(11), 1804–1826

Eusden, J.D., Pettinga, J.R. and Campbell, J.K. (2000) Structural evolution and landscape development of a collapsed transpressive duplex on the Hope Fault, North Canterbury, New Zealand. *NZ J. Geol. Geophys.*, **43**, 391–404.

Fernández, F.J., Chaminé, H.I., Fonseca, P.E., *et al.* (2003) HT-fabrics in a garnet-bearing quartzite from Western Portugal: geodynamic implications for the Iberian Variscan Belt. *Terra Nova*, **15**(2), 96–103.

Friend, P.F., Williams, B.P.J., Ford, M. & Williams, E.A. (2000) Kinematics and dynamics of Old Red Sandstone basins. In: *New Perspectives on the Old Red Sandstone* (Eds P.F. Friend and B.P.J. Williams), pp. 29–60. Special Publication 180, Geological Society Publishing House, Bath.

Gama Pereira, L.C. (1987) *Tipologia e evolução da sutura entre a Zona Centro Ibérica e a Zona Ossa Morena no sector entre Alvaiázere e Figueiró dos Vinhos (Portugal Central).* Unpublished PhD thesis, Universidade de Coimbra, 331 pp.

Gutiérrez-Alonso, G., Fernández-Suárez, J. and Weil, A.B. (2004) Orocline triggered lithospheric delamination. In: *Orogenic Curvature, Integrating Paleomagnetic and Structural Analyses* (Eds. A.J. Sussman and A.B. Weil). *Geol. Soc. Am. Spec. Pap.*, **383**, 121–130.

Ingersoll, R.V. (1988) Tectonics of sedimentary basins. *Geol. Soc. Am. Bull.*, **100**, 1704–1719.

Jabaloy, A., Galindo-Zaldívar, J. and González-Lodeiro, F. (2002) Palaeostress evolution of the Iberian Peninsula (Late Carboniferous to present-day). *Tectonophysics*, **357**, 159–186.

Lagarde, J.L., Baize, S., Amorese, D., Delcaillau, B., Font, M. and Volant, P. (2000) Active tectonics, seismicity and geomorphology with special reference to Normandy (France). *J. Quatern. Sci.*, **15**, 745–758.

Lefort, J.P. and Ribeiro, A. (1980) La faille Porto–Badajoz–Cordoue a-t-elle contrôllé l'evolution de l'océan paléozoique sud-armoricain? *Bull. Soc. Géol. Fr.*, **22**(3), 455–462.

Little, T.A. and Mortimer, N. (2001) Rotation of ductile fabrics across the Alpine Fault and Cenozoic bending of the New Zealand orocline. *J. Geol. Soc. London*, **158**, 745–756.

Mann, P., Hempton, M.R., Bradley, D.C. and Burke, K. (1983) Development of pull-apart basins. *J. Geol.*, **91**, 529–554.

McKerrow, W.S., Leggett, J.K. and Eales, M.H. (1977) Imbricate thrust model of the Southern Uplands of Scotland. *Nature*, **267**, 237–239.

Morley, C.K. (1988) Out of sequence thrusts. *Tectonics*, **7**, 539–561.

Palain, C. (1976) Une série détritique terrigène les 'Grés de Silves': Trias et Lias inférieur du Portugal. *Mem. Inst. Geol. Min. Lisboa*, **25**, 1–377.

Rahe, B., Ferrill, D.A. and Morris, A.P. (1998) Physical analog modelling of pull-apart basin evolution. *Tectonophysics*, **285**, 21–40.

Ribeiro, A. (2002) *Soft Plate and Impact Tectonics*. Springer-Verlag, Berlin, 324 pp.

Ribeiro, A., Quesada, C. and Dallmeyer, R.D. (1990a) Geodynamic evolution of the Iberian Massif. In: *Pre-Mesozoic Geology of Iberia* (Eds. R.D. Dallmeyer and E. Martínez-García), pp. 397–410. Springer-Verlag, Berlin.

Ribeiro, A., Kullberg, M.C., Kullberg, J.C., Manuppella, G. and Phipps, S. (1990b) A review of Alpine tectonics in Portugal: foreland detachment in basement and cover rocks. *Tectonophysics*, 184, 357–366.

Ribeiro, A., Cabral, J., Baptista, R. and Matias, L. (1996) Stress pattern in Portugal Mainland and the adjacent Atlantic region, West Iberia. *Tectonics*, **15**(2), 641–659.

Ribeiro, A., Marcos, A., Pereira, E., *et al.* (2003) 3-D strain distribution in the Ibero-Armorican Arc: a review. *Ciênc. Terra (UNL)*, **5**, D62–D63. (CD-Rom)

Richard, P.D., Naylor, M.A. and Koopman, A. (1995) Experimental models of strike-slip tectonics. *Petrol. Geosci.*, 1(1), 71–80.

Schumm, S., Dumont, J. and Holdbrook, J. (2000) *Active Tectonics and Alluvial Rivers*. Cambridge University Press, 276 pp.

Severo Gonçalves, L. (1974) *Geologie und petrologie des gebietes von Oliveira de Azeméis und Albergaria-a-Velha (Portugal).* Unpublished PhD thesis, Freien Universität, Berlin, 261 pp.

Silva, P.G., Goy, J.L., Somoza, L., Zazo, C. and Bardají, T. (1993) Landscape response to strike-slip faulting linked to collisional settings: Quaternary tectonics and basin formation in the Eastern Betics, Southeast Spain. *Tectonophysics*, **224**, 289–303.

Silva, P.G., Goy, J.L., Zazo, C. and Bardají, T. (2003) Fault-generated mountain fronts in southeast Spain: geomorphologic assessment of tectonic and seismic activity. *Geomorphology*, **50**, 203–225.

Soares de Carvalho, G. (1946a) *As formações geológicas mais antigas da Orla Mesozóica Ocidental de Portugal.* Unpublished PhD thesis, Universidade de Coimbra, 126 pp.

Soares de Carvalho, G. (1946b) Subsídios para o estudo das formações geológicas do Distrito de Aveiro. O Alto da Pedra Aguda e uma memória de Carlos Ribeiro. *Mem. Notícias Coimbra*, **15**(1), 5–15.

Soares de Carvalho, G. (1949) Subsídios para o estudo das formações geológicas do Distrito de Aveiro. Depósitos de sopé no Concelho de Oliveira de Azeméis. *Arquiv. Distr. Aveiro*, **15**, 5–10.

Stone, P., Kimbell, G.S. and Henney, P.J. (1997) Basement control on the location of strike-slip shear in the Southern Uplands of Scotland. *J. Geol. Soc. London*, **154**, 141–144.

Summerfield, M. (2000) *Geomorphology and Global Tectonics*. Wiley, Chichester, 367 pp.

Sylvester, A.G. (1988) Strike-slip faults. *Geol. Soc. Am. Bull.*, **100**, 1666–1703.

Telles Antunes, M., Ferreira, M.P., Rocha, R.B., Soares, A.F. and Zbyszewski, G. (1979) Essai de reconstitution paléogéographique par cycles orogéniques: Le cycle Alpin. In: *Introduction à la géologie générale du Portugal* (Eds A. Ribeiro, M. Telles Artunes, M.P. Ferreira, *et al.*), pp. 45–89. Serviços Geológicos de Portugal, Lisboa.

Tippet, J.M. and Hovius, N. (2000) Geodynamic process in the Southern Alps, New Zealand. In: *Geomorphology and Global Tectonics* (Ed. M.A. Summerfield), pp. 109–134. Wiley, Chichester.

Wakabayashi, J., Hengesh, J.V. and Sawyer, T.L. (2004) Four-dimensional transform fault process: progressive evolution of step-overs and bends. *Tectonophysics*, **392**(1–4), 279–301.

Wood, R.A., Pettinga, J.R., Bannister, S., Lamarche, G. and McMorran, T.J. (1994) The structure of the Hanmer strike-slip basin, Hope fault, New Zealand. *Geol. Soc. Am. Bull.*, **106**, 1459–1473.

Woodcock, N.H. and Schubert, C. (1994) Continental strike-slip tectonics. In: *Continental Deformation* (Ed. P.L. Hancock), pp. 251–263. Pergamon Press, Oxford.

Effects of transverse structural lineaments on the Neogene–Quaternary basins of Tuscany (inner Northern Apennines, Italy)

VINCENZO PASCUCCI*, I. PETER MARTINI†, MARIO SAGRI‡ *and* FABIO SANDRELLI§

**Istituto di Scienze Geologico-Mineralogiche, Università di Sassari, Corso Angioy, 10, 07100 Sassari, Italy (Email: pascucci@unisi.it)*
†Land Resources Science, University of Guelph, Ontario, N1G 2WI, Canada
‡Dipartimento di Scienze della Terra, Università di Firenze, Via La Pira 4, 50121 Firenze, Italy
§Dipartimento di Scienze della Terra, Università di Siena, Via Laterina 8, 53100 Sienna, Italy

ABSTRACT

Several mountain arcs formed in the Mediterranean area during the Alpine orogeny, among them the Northern Apennines. They show diachronous development with the outer thrust front prograding eastward, and being progressively replaced to the west by relatively thin, post-orogenic extensional or transtensional basins. The outer and inner parts of the orogen are linked together through a series of transverse structural lineaments. Segments of such lineaments have through time acted as transcurrent faults, lateral ramps of thrusts, strike- and oblique-slip faults, and normal faults. The main lines of evidence about transverse lineaments of the Northern Apennines are reviewed, and their effects on some Neogene–Quaternary basins of the inner part of the orogen in Tuscany and the northern Tyrrhenian Sea shelf are assessed. New information from recently released commercial seismic profiles and from surface sedimentological studies has made it possible to confirm that the post-orogenic basins formed in tectonic depressions delimited by major, quiescent substrate thrusts. The depressions were longitudinally separated into basins by the transverse lineaments. The stratigraphy of the basins in each tectonic depression is similar; in most cases, initial narrow syn-rift sedimentation was followed by extensive post-rift successions due to thermal subsidence. However, the thickness and distribution of their sedimentary sequences vary according to different subsidence (or uplift) and extension that have occurred along each side of the same transverse lineament, or in blocks delimited by different lineaments. Furthermore, portions of the lineaments, such as those of the Livorno–Sillaro, may have temporarily acted as strike-slip faults (late Miocene–Pliocene, in this case), and equivalent substrate highs and parts of the same basin may have been shifted left-laterally for about 15–20 km. A further effect of the transverse lineaments on basin sedimentation has been the development of major alluvial fans at relay ramps developed near the intersection of lineaments and quasi-orthogonal, listric boundary faults.

Keywords Neogene basins, transverse lineaments, seismic stratigraphy, Northern Apennines, transfer fault, transfer zone, alluvial fan, Tuscany.

INTRODUCTION

The problem and objectives of the paper

The Northern Apennines is one of the Alpine arcs that developed in the Mediterranean area (Boccaletti & Guazzone, 1974; Wezel, 1986, 1988). One of the characteristics of these arcs is that at the same time they show compression in the frontal, outer zone, and mostly extension in their back, inner zone (Royden *et al.*, 1982; Royden, 1988). Furthermore, the progradation of the arcs has occurred with differential movements along structural lineaments oriented perpendicular (transverse) to the tectonic front. Some of these transverse lineaments are old features, perhaps even relict from the pre-orogenic

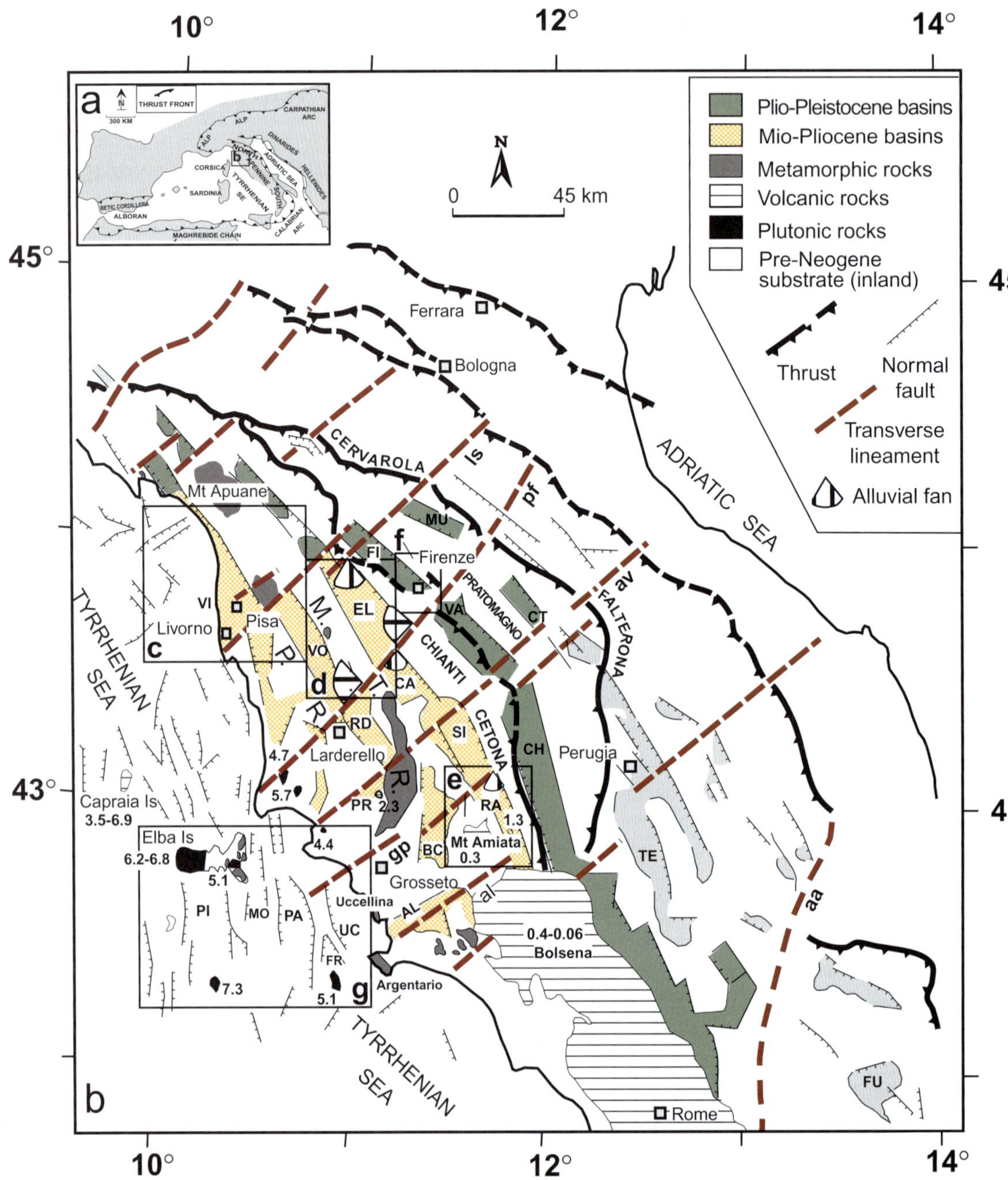

Fig. 1 Generalized structural maps of the area. (a) Schematic map showing the mountain chains of the Mediterranean region and geological subdivisions of the Northern Apennines. (b) Neogene–Quaternary basins of the Northern Apennines. (Basins: AL, Albegna; BC, Baccinello; CA, Casino; CH, Chiana; CT, Casentino. EL, Elsa; FI, Firenze; FR, Formiche; FU, Fucino; MO, Montecristo; MU, Mugello; PA, Punta Ala; PI, Pianosa; RA, Radicofani; RD, Radicondoli; SI, Siena; TE, Tiberino; UC, Uccellina; VA, Valdarno; VI, Viareggio; VO, Volterra. Transverse lineaments: aa, Olevano–Antrodoco; al, Albegna; av, Arbia–Marecchia; gp, Grosseto–Pienza; ls, Livorno–Sillaro; pf, Piombino–Faenza.) MTR, Middle Tuscany Ridge; PR, Perityrrhenian Ridge; 3.5, radiometric age of igneous rocks in Ma. Tuscany is a province enclosed approximately between the transverse lineaments just northwest of Apuane Mountains and southeast of Argentario, the Cervarola–Falterona thrust to the east, and the Tyrrhenian Sea coast to the west. (c) Location of Fig. 6. (d) Location of Fig. 9. (e) Location of Fig. 13. (f) Location of Fig. 5b. (g) Location of Fig. 17.

parent oceanic basin. They were active in a punctuated fashion throughout the evolution of the mountain chain, segments of them acting at times as transcurrent faults, during other times as normal faults. The lineaments of the Northern Apennines have been recognized since 1935 (Sacco, 1935; Signorini, 1935), and have been studied in detail by many authors. A brief synthesis of the main evidence of the influence of these lineaments on the regional geology will be presented here. The main objectives of this paper are, however, to present new evidence on these lineaments derived from the analysis of seismic profiles, and a brief analysis of the tectono-sedimentary evolution of selected basins of the inner Northern Apennines in Tuscany (Fig. 1).

Methods of study

A brief literature review has been augmented by the analysis of geological maps, wells drilled for hydrocarbons and geothermal prospecting, and numerous industrial seismic profiles of selected Neogene–Quaternary basins (Viareggio (VI), Elsa (EL), Siena (SI), Radicofani (RA) and in the northern Tyrrhenian Sea shelf (PI, MO, PA, UC, FR; Fig. 1). The offshore seismic data were acquired using an airgun source with shot spacing interval of 26 m; a Vibroseis source with a shot spacing interval of 40 m was used inshore. The seismic data were recorded with 24 channels. The procedure reported by Mariani & Prato (1988) was used for data processing.

The Northern Apennines

The Northern Apennines is a complex mountain chain that has developed by the interaction between Adria (a promontory of the Africa plate) and the Europe plate (Fig. 2a). The Adria promontory of the Africa plate protruded into the Ligurian Piedmont oceanic basin, a narrow western arm of the Jurassic Tethys. The Apennines are characterized by imbricate fold-thrust belts accreted eastward on the Adria microplate in response to the westward-dipping subduction zone (Fig. 2b).

The Northern Apennines is composed of deformed sedimentary successions belonging to different domains: the ophiolitic-bearing Ligurides derived from the Ligurian Piedmont oceanic basin, the Subligurides deposited adjacent to the Adria continental crust, and the Tuscan and Umbrian units formed on the Adria continental margin (Fig. 2b). The Ligurides are composed of lower Jurassic to Eocene rocks (ophiolites, radiolarites, pelagic carbonates, shales and turbidites). The Subligurides include shales, pelagic limestones, and turbidite Eocene to Oligocene deposits. Here the Ligurides and Subligurides are combined and referred to as 'Ligurides'. The Tuscan and Umbrian units consist primarily of Mesozoic carbonates, radiolarites, shales, and thick Cenozoic turbidites Macigno, Cervarola, Falterona (Tuscan units) and Marnoso arenacea (Umbrian unit) (Vai, 2001).

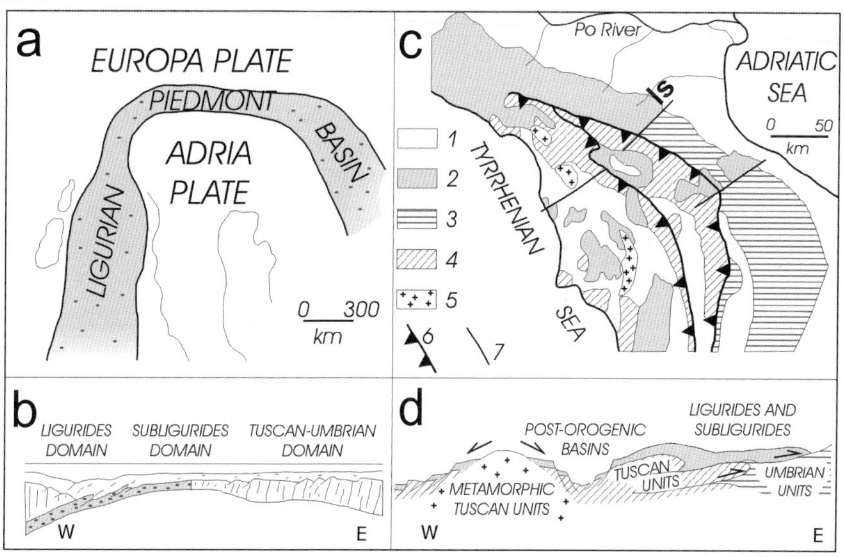

Fig. 2 Structural features of the central-north area of Italy. (a) Paleogeographic map of the Ligurian-Piedmont oceanic basin. (b) Cross-section showing original sedimentary domains of various units of the Northern Apennines. (c) General structural map of the Northern Apennines with major structures and distribution of the tectono-sedimentary units (1, Miocene to Quaternary deposits; 2, Ligurides; 3, Umbrian units; 4, non-metamorphic Tuscan units; 5, Metamorphic Tuscan Unit). (d) Schematic cross-section showing the relations among the tectono-sedimentary units of the Northern Apennines. (From Sagri *et al.*, 2004.)

The Ligurian Piedmont oceanic basin started closing during the late Cretaceous, and the Ligurides began to be deformed and thrust eastward. From the Oligocene onward, the Adria continental margin has been involved in a continent-to-continent collision (Patacca *et al.*, 1990). During this collision, imbricate thrust structures developed and the lowermost Tuscan units underwent low-grade metamorphism (Metamorphic Tuscan Unit; Fig. 2c). From the Miocene, the imbricate thrust belt advanced eastward, and the Ligurides overrode the thrust pile as a nappe (Fig. 2d). The resulting major structural features are the Middle Tuscany Ridge (MTR) uplift, the Chianti–Cetona Ridge (CCR), and the Cervarola–Falterona thrust (Fig. 1). The still active thrust front of the orogen lies further to the east under the Adriatic Sea and the Po River plain (Figs 1 & 2; Castellarin, 2001).

After the main early Miocene compressional phases, extensional basins 10–40 km long, 15–20 km wide developed in the inner, western part of the Northern Apennines (Fig. 1). For the most part, these basins are now bounded by normal faults, many of Plio-Pleistocene formation or reactivation. They are separated longitudinally from each other by transverse structural lineaments and are filled with up to 3 km of upper Miocene to Quaternary deposits (Sequence 1 to Sequence 6; Fig. 3). The basins west of the Chianti–Cetona Ridge have developed on a thin (20–25 km) continental crust, whereas those to the east are on thicker crust (about 35 km; Giese, 1981; Nicolich, 1987) (Fig. 1). Furthermore, the basins west of the Middle Tuscany Ridge contain the whole upper Miocene–Pleistocene succession; those between the Middle Tuscany Ridge and the Chianti–Cetona Ridge have similar stratigraphy except that they lack upper Messinian evaporite facies. The basins east of the Chianti–Cetona Ridge contain mainly continental Pliocene–Pleistocene deposits (Figs 1 & 4; Bossio *et al.*, 1993).

The stratigraphy of the main basins has been determined on the basis of field and seismic information. Six major unconformity-bounded units (Sequence 1 to Sequence 6) have been recognized as follows (Fig. 4).

- Sequence 1 is present only locally, and is characterized by shallow-marine sandstones with occasional marlstones, capped by conglomerates. It is considered to be Serravalian to early Tortonian in age.
- Sequence 2 mostly consists of conglomerates, sandstones and clays. The lower part is known as 'series lignitifera' because it contains thin layers of lignite. Thin gypsum layers, marine clays and local patch-reefs carbonates characterize its top part. It developed during the late Tortonian to early Messinian.
- Sequence 3 has some marine gypsum layers in the lower part, but it is mostly composed of lacustrine to brackish clays with some layers of sandstone and conglomerate. This interval is called 'lago mare' and represents part of the so-called 'Messinian salinity crisis' (Hsu *et al.*, 1973; Roveri *et al.*, 2003). Such a crisis resulted from the drying out of most of the Mediterranean Sea area with development of local hypersaline basins, followed by a widespread inundation of lacustrine–brackish water (the 'lago mare'). It is considered to be late Messinian in age.
- Early Pliocene Sequence 4 is mostly composed of marine clays with a few interstratified conglomeratic layers. Clay dominates in the centre of the basins, with the coarser clastics occurring on the margins.
- Middle Pliocene Sequence 5 is composed of marine clays; thin sandstone units occur at its base and top, where there also are local, thin biocalcarenites.
- Sequence 6 represents Pleistocene deposition in shallow-marine settings near the Tyrrhenian Sea and fluvio-lacustrine settings inland.

TRANSVERSE LINEAMENTS AND THEIR EFFECTS ON BASINS

Transverse lineaments of the Northern Apennines (that is, oriented perpendicular to the crest of the orogen in a NE–SW direction, also called 'antiapenninic-oriented') have been variously interpreted as transfer zones, transfer faults, lateral ramps of thrusts, strike- and oblique-slip faults, and normal faults. Some of them may have actually acted as all these fault types at different times during the evolution of the orogen. The lineaments have been recognized through morphological signature, on aerial photographs and satellite images (Boccaletti *et al.*, 1977; Bemporad *et al.*, 1986). In the stratigraphic record they are recognized by analysing the different thickness in the sedimentary succession of two adjacent areas (Bortolotti, 1966; Liotta, 1991), and in the field from structural characteristics, observing fault-plane slickensides, and horizontal displacement of analogous geological units as depicted on maps.

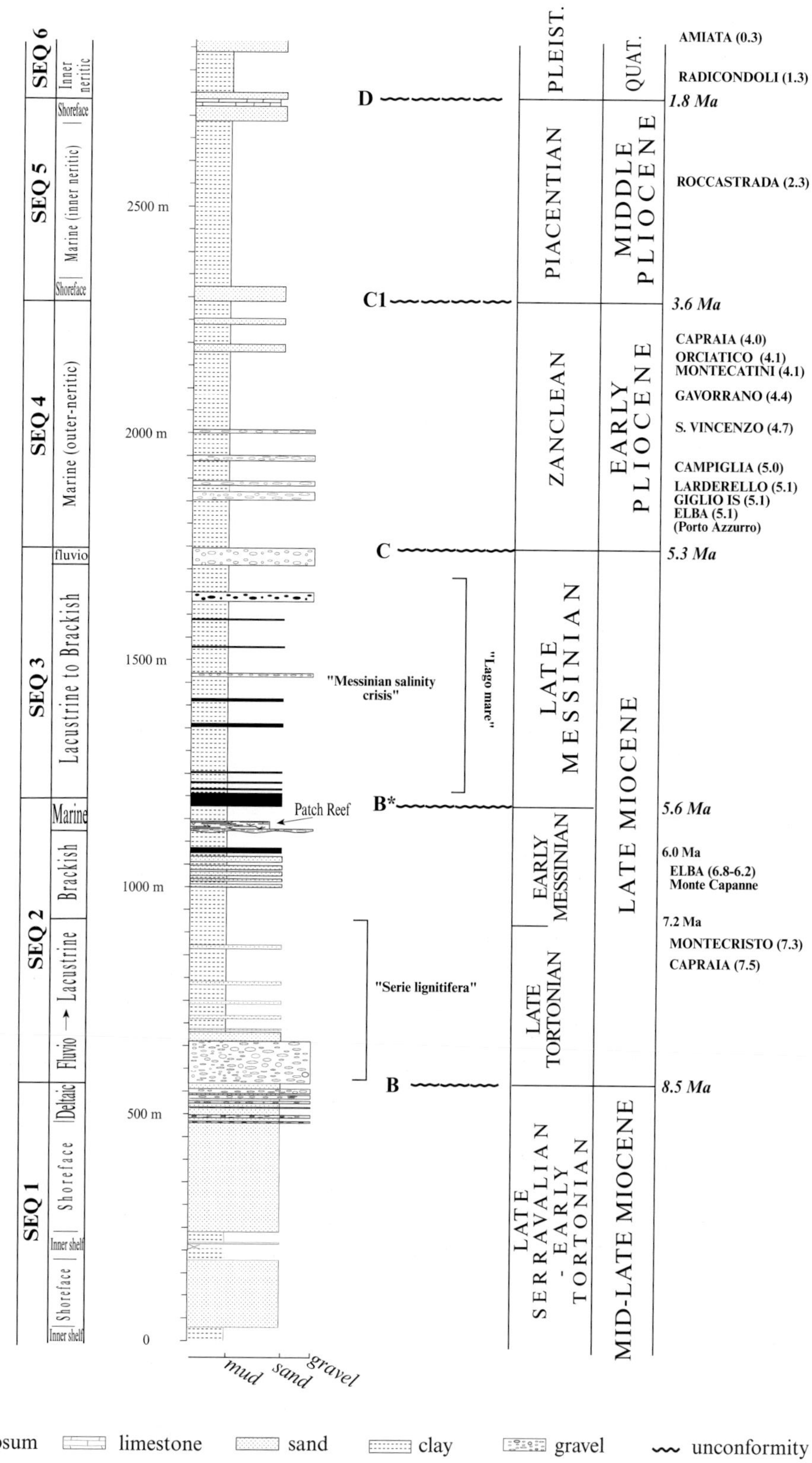

Fig. 3 Stratigraphic type column for Miocene to Pleistocene basin fills of Tuscany and the northern Tyrrhenian Sea. Representative thickness derives from the Volterra Basin. Dates for biostratigraphic boundaries and the facies interpretations are according to Bossio *et al.* (1997), Pascucci *et al.* (1999), Martini *et al.* (2001) and Sagri *et al.*, (2004). Dates (Ma) for volcanic events (right column) are from Serri *et al.* (2001). Seq = sequence.

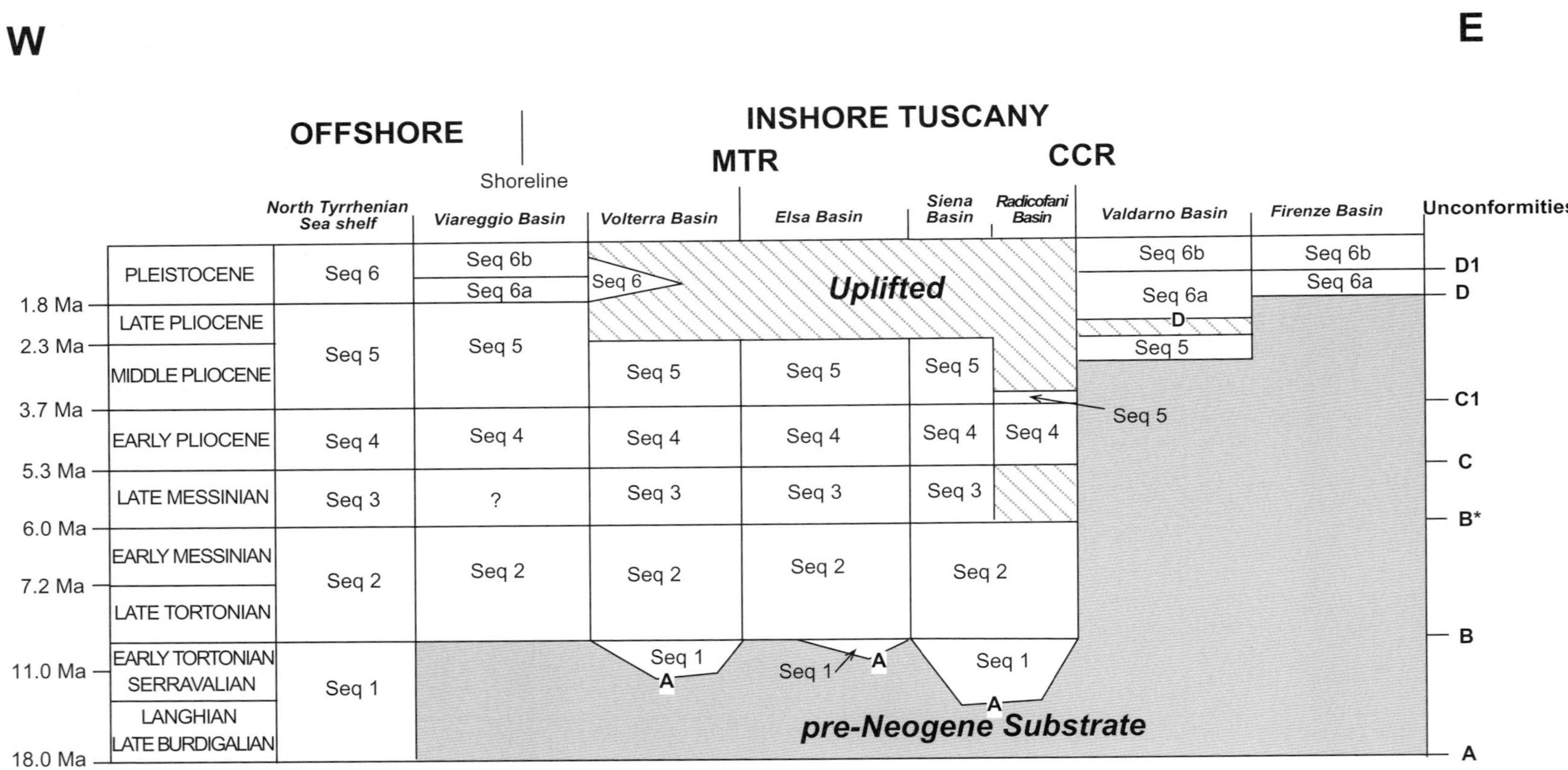

Fig. 4 Sedimentary sequences (Seq 1 to Seq 6) and their bounding unconformities (A, B, B*, C, C1, D, D1) in selected Neogene–Quaternary basins of Tuscany (MTR, Middle Tuscany Ridge; CCR, Chianti–Cetona Ridge). Seq = sequence.

For each transverse lineament the general, most important information attesting to its existence is presented first. Second, major tectono-stratigraphic characteristics of the selected basins it crosses will be reconstructed from seismic and borehole information to determine what influence, if any, the transverse lineament had on basin development. The lineaments dealt with in this paper are, from north to south (Fig. 1):

1 Livorno–Sillaro (ls) with the Viareggio (VI) and Elsa (EL) basins;
2 Piombino–Faenza (pf) with brief notes on the Firenze Basin (FI);
3 Arbia–Marecchia (av) with the analysis of the basins of the northern Tyrrhenian Sea shelf;
4 Grosseto–Pienza (gp) with the Siena (SI) and Radicofani (RA) basins;
5 Albegna (al) with a brief note on the Albegna Basin (AL), and the offshore basins of the northern Tyrrhenian Sea shelf.

Livorno–Sillaro lineament

Bortolotti (1966) first defined the Livorno–Sillaro lineament. It consists of two major segments, one shifted in respect to the other within the Firenze Basin (FI, Fig. 1). This, as with all other lineaments, is not a single entity, but consists of a series of subparallel structural–geomorphological features, in places occurring en échelon over a corridor up to 10–20 km wide.

Major lines of evidence for this lineament and its effects on the geology are the following, from east to west.

1 On the eastern outer flank of the Apennines the lineament separates the Marnoso arenacea (Miocene, Umbrian Units) to the southeast from the Ligurides (Jurassic–Eocene) in the northwest (Fig. 5a). This is the result of a NW downthrow that formed a depression receiving the Ligurides (Barchi *et al.*, 2001).
2 In Tuscany, the Mesozoic stratigraphy of the Tuscan Unit is significantly different, and it has a different thickness north and south of the lineament. The tectonic deformation features of the pre-Neogene rocks are more intense and include considerable tectonic pile-ups north of the lineament, whereas such deformation features are subdued to the south (Bortolotti, 1966).
3 The lineament delimits to the northwest the Neogene–Quaternary Mugello (MU) (Ghelardoni, 1965) and Firenze basins (Capecchi *et al.*, 1975; Fig. 1).
4 The Neogene–Quaternary basins of central western Tuscany are more extensively developed south of the lineament (Fig. 1).
5 Large alluvial fans mark the intersection between the boundaries of Neogene–Quaternary basins and the transverse lineament (Benvenuti & Degli Innocenti, 2001; Fig. 1).
6 Across the main Apennines chain there is a concentration of earthquakes along the lineament (Bortolotti, 1966). Earthquakes have also occurred during the past 100 yr near Livorno, which suggest still active fault movements (Ghelardoni, 1965; Cantini *et al.*, 2001).

New relevant information has been obtained for this transverse lineament from the analysis of seismic profiles and hydrocarbon exploratory wells of the Neogene–Quaternary Viareggio and Elsa basins (Fig. 1).

Viareggio Basin

Setting. The Viareggio Basin (VI, Fig. 1) has an offshore and an inshore part and is subdivided into a northern and a southern basin (Fig. 6; Mariani & Prato, 1988; Argnani *et al.*, 1997; Pascucci, 2006). The southern basin is centred in the Arno River mouth area and is oriented NW–SE. It is about 20 km wide and 25 km long. It is bordered by the Pisani Mountains to the northeast, by the Meloria–Maestra shoal to the southwest, and by the Livornesi Mountains to the southeast (Fig. 6).

Stratigraphy. The basin is filled with up to 2500 m of Neogene–Quaternary deposits (Sequence 2 to Sequence 6, Fig. 4), mainly sand and clay, resting unconformably on the Oligocene to lower Miocene Macigno sandstone, the uppermost part of the Tuscan Units. The Ligurides are not present at this locality. They are present, however, in the offshore Maria 1 well to the west, where about 1800 m have been penetrated, and they form the onshore Livornesi Mountains to the south (Fig. 6).

The Neogene to Quaternary succession commences with 300 m of marine clay and sandstone considered to be Messinian by Mariani & Prato (1988), overlain by Pliocene sequences. The Pliocene succession can be subdivided into two sequences,

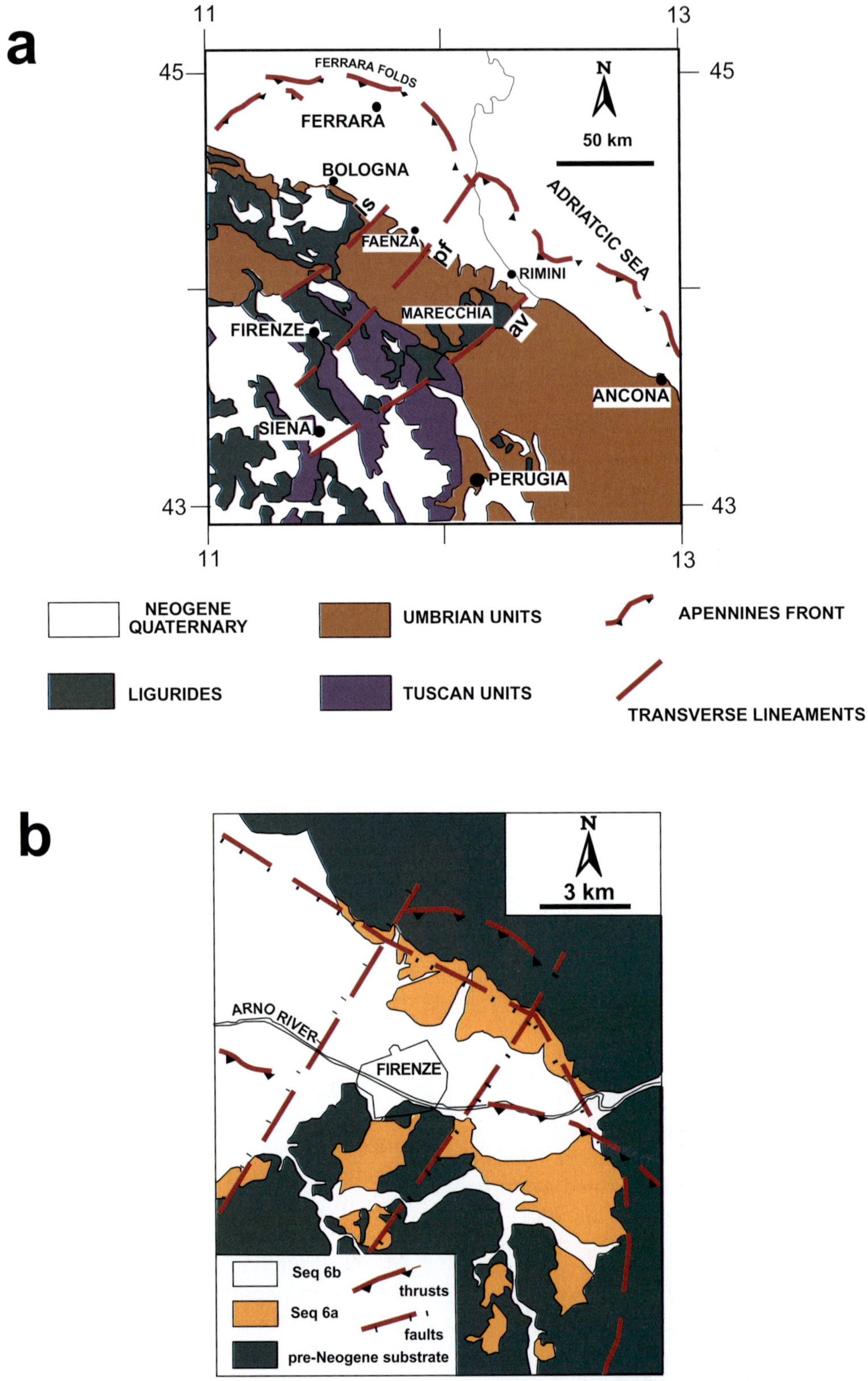

Fig. 5 Schematic geological maps of the Northern Apennines. (a) Central-eastern part of the mountain chain (after Vai, 2001; Castellarin, 2001). (b) Southestern part of the Firenze Basin (after Capecchi *et al.*, 1975; Boccaletti *et al.*, 1982).

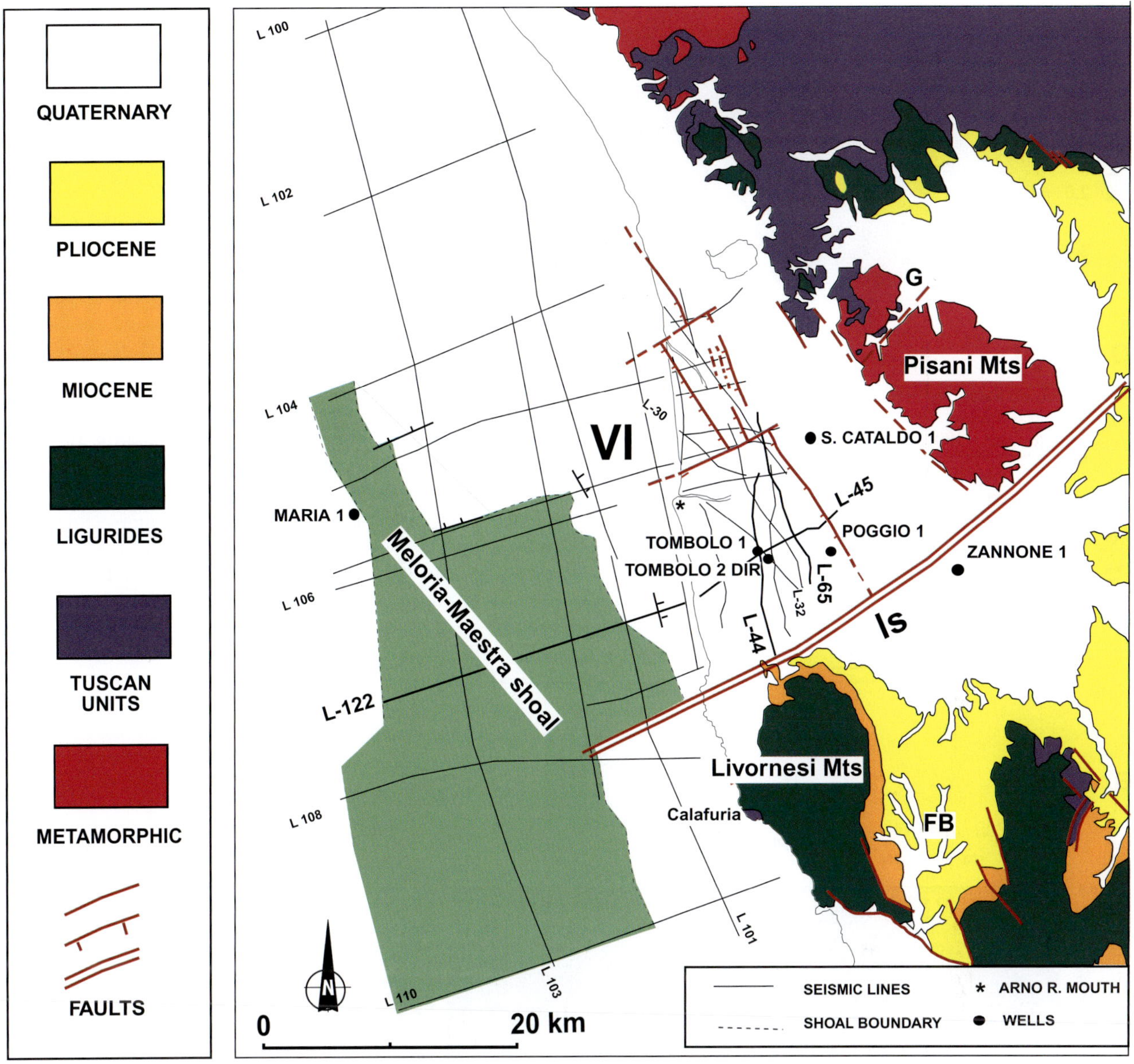

Fig. 6 Generalized geological map and location of seismic profiles and wells of the Viareggio Basin. Offshore data are from seismic prolfiles and Pascucci (2006); inland data are from Boccaletti *et al.* (1982). G, Guappero line; ls, Livorno–Sillaro; FB: Fine Basin; VI: Viareggio. In bold are the presented seismic lines.

Sequences 4 and 5, on the basis of differences in seismic facies and a seismically definable unconformity (Fig. 7). The topmost Quaternary deposits consist of 700 m of open marine (Sequence 6a) to littoral (Sequence 6b) clay and sand. Inclined, well-defined reflectors present in sequence Sequence 6a indicate a seaward prograding sandy mouth-bar related to a palaeo-Arno delta.

Basin geometry and interpretation. In NE–SW seismic profiles, the basin fill shows a triangle-shaped geometry consistent with a half-graben model in the lower part (primarily Sequence 2, Sequence 4 and lower part of Sequence 5), with a southwestward-dipping listric master fault that flattens at a depth of 2.5 s (TWT), about 3.5 km. The upper part the basin has a wide bowl-shaped geometry typical of

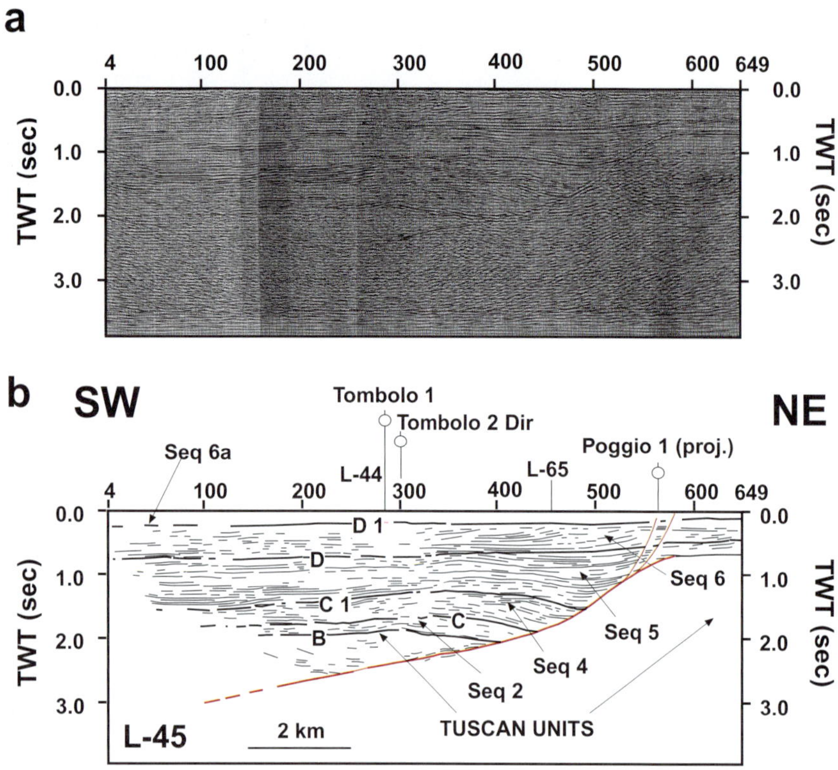

Fig. 7 SW–NE seismic profiles (L-45) across the Viareggio Basin. (a) Original seismic profile. (b) Interpreted line drawing, with location of wells and intersecting seismic profiles L-44 and L-65 (see Fig. 6 for location). B, C, C1, D, D1 are unconformities; Seq = sequence; proj = projected; in red on the northeast side is the listric master fault.

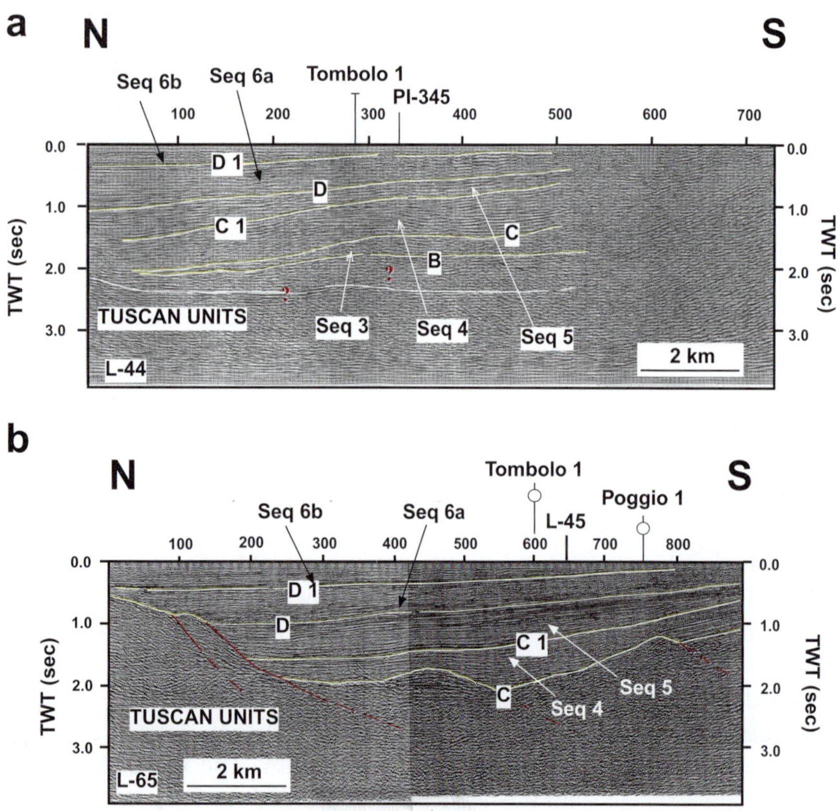

Fig. 8 Interpreted seismic profiles from the Viareggio Basin, with location of wells and intersecting seismic profiles Pl-345 and L-45. (a) Southwestern portion of profile L-44. Note the ill-defined seismic response to the south. (b) Northwestern portion of profile L-65 (see Fig. 6 for location). Note the faulted substrate high to the north. B, C, C1, D, D1 are unconformities; Seq = sequence; in red are faults.

post-rift sedimentation (upper part of Sequence 5 and Sequence 6, Fig. 7).

The basin geometry in a NW–SE direction is relatively well defined by two seismic profiles: profile L-44 from the centre to the southeastern end; and profile L-65 from the centre to the northwestern end (Figs 6 & 8).

Along profile L-44 the Neogene–Quaternary units are characterized by well-defined, quasi-continuous reflectors in the centre of the basin; the seismic response becomes chaotic and the units are ill defined toward the southeastern end (Fig. 8a). The substrate is, however, still recognizable; it shallows rapidly, and eventually crops out to the southeast on the Livornesi Mountains. We interpret the chaotic seismic pattern as an intensely faulted zone associated with the Livorno–Sillaro transverse lineament.

To the northwest, profile L-65 shows again well-defined, subparallel reflectors in the centre of the basin (Figs 6 & 8b). The reflectors terminate sharply against a faulted substrate high. A similar termination of the reflectors occurs to the northwest along the adjacent, parallel line L-32, whereas this substrate high is not recognized along the parallel line L-30 farther to the southwest where a depocentre exists instead (Fig. 6). This geometry records the existence of an indentation in the substrate and a seaward shift of the previously defined master fault of the northeastern flank of the basin.

The structural map reconstructed utilizing the available seismic and geological information shows the following (Fig. 6).

1 The basin is delimited by faults on three sides: a southwest-dipping, listric master fault to the northeast; and inferred, left-lateral transverse (antiapenninic-oriented) faults to the northwest and southeast. Offshore, the deposits onlap onto the substrate without evidence of major fault dislocations. Faults were mainly active during the early Pliocene.

2 The transverse fault at the northwest end of the basin can be extended inland into the antiapenninic-oriented Guappero fault (G) mapped in the Pisani Mountains (Ghelardoni, 1965). This fault parallels the Livorno–Sillaro (ls) lineament.

The Elsa Basin

As previously indicated, the transverse lineaments are in reality zones of subparallel structures (faults and folds) locally occurring in an en échelon fashion. One such secondary fault is reported to verge out of the main Livorno–Sillaro track in a northeast direction, into the Elsa Basin (Ghelardoni *et al.*, 1968; Cantini *et al.*, 2001).

Setting. The Elsa Basin (EL) is oriented NW–SE and is bordered by two elongated ridges (the Albano Mount to the northeast, and the partially exposed Middle Tuscany Ridge to the southwest), and by the Chianti Ridge to the southeast and the Cerbaia hills to the northwest (Fig. 9). The basin is 40 km

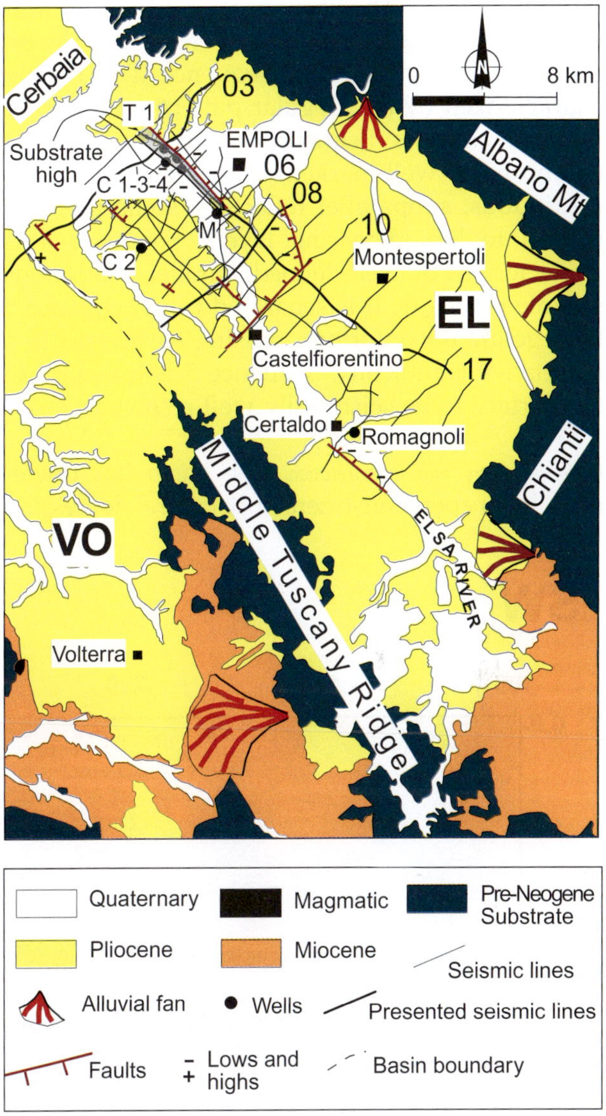

Fig. 9 Generalized geological map of the Elsa (EL) and Volterra (VO) basins, and location of seismic profiles and wells. C 1-3-4 are the locations of the Certaldo 1, 3, 4 wells; T1 is the Tolomei 1 well.

long and 20 km wide. It is reasonably well exposed, numerous seismic profiles are available, and it has been penetrated by several deep exploratory wells.

Stratigraphy. The basin is filled with up to 1000 m of upper Miocene and 1000 m of Pliocene deposits (Sequence 2 to Sequence 5, Fig. 4). The upper Miocene sand and clay lacustrine Sequences 2 and 3 are overlain by marine Pliocene sand and clay, and, locally, by conglomeratic alluvial fan deposits (Sequences 4 and 5) (Benvenuti & Degli Innocenti, 2001). The sequences and their bounding unconformities are readily recognizable on seismic profiles (Fig. 10). Sequence 2 is mostly characterized by an irregular pattern with poorly defined reflectors. Sequence 3 to the northeast is composed of well-marked, closely spaced, quasi-continuous, subparallel reflectors, some possibly related to lignite-bearing layers. Sequence 4 is characterized by well-defined, parallel, continuous seismic reflectors. It is separated from Sequence 5 by a gentle unconformity passing basinward to a correlative conformity. Sequence 5 is too close to the surface to be seismically well-resolved.

Basin geometry and interpretation. The basin has developed in an apenninic-oriented structural depression longitudinally subdivided into two parts by a pre-Neogene antiapenninic-oriented fault (Figs 9 & 11). The southeast part of the basin is shallower and is filled with 800 m of upper Miocene–Pliocene deposits (Sequences 3 and 4, and the exposed lowermost part of Sequence 5; Fig. 11, shotpoints 550 to 768). The northwest part of the basin is deeper and structurally more complex (Fig. 11, shotpoints ~100–550). An apenninic-oriented bounding fault was locally active during the late Tortonian to early Messinian (Sequence 2), generating a half-graben (between profiles L-10 and L-08; Figs 9 & 10). Between profiles L-06 and L-03 a narrow, southeastward-plunging high, which is delimited on one side by a northeastward-dipping normal fault, subdivides the basin longitudinally into two (Figs 9 & 12). The Pliocene deposits (Sequences 4 & 5) blanket the area and rest everywhere unconformably over the Miocene deposits or on the pre-Neogene substrate rocks.

The analysis of the Elsa Basin reveals three major features.

1 The depression where the basin is located has been strongly influenced by the uplift of the adjacent Mid-Tuscany Ridge (MTR, Fig. 1).
2 The antiapenninic-oriented fault that separates the Elsa Basin into two parts may represent the eastern

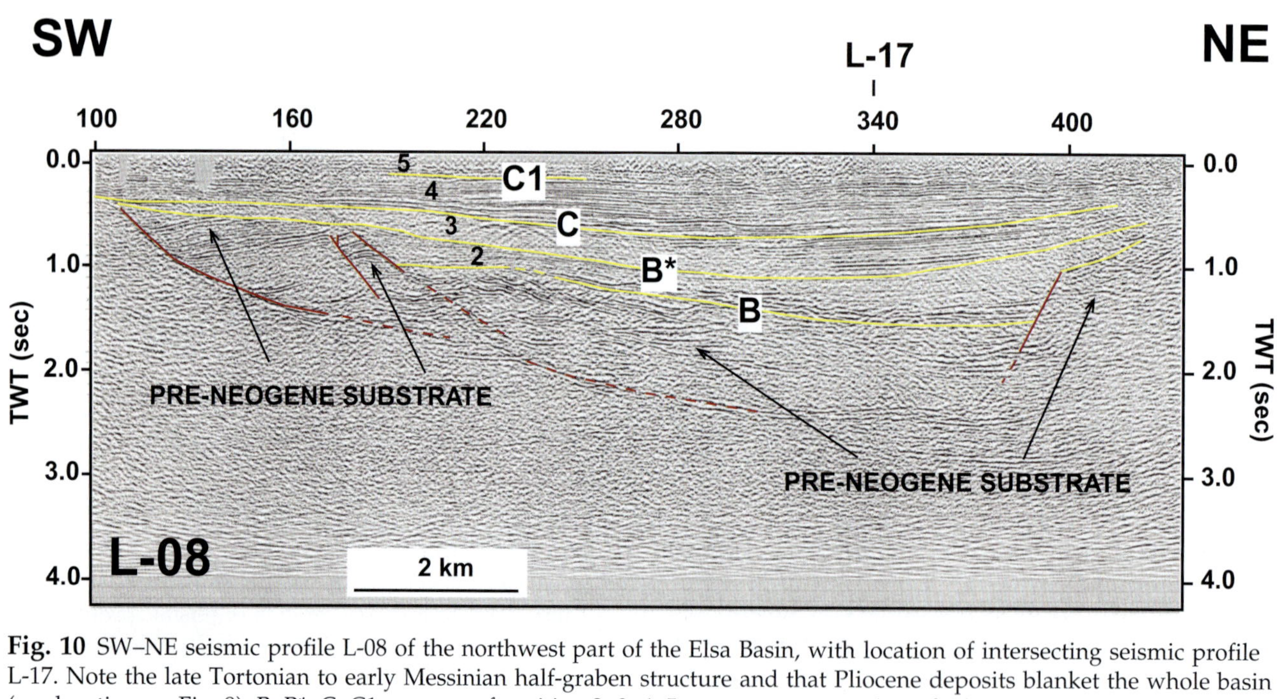

Fig. 10 SW–NE seismic profile L-08 of the northwest part of the Elsa Basin, with location of intersecting seismic profile L-17. Note the late Tortonian to early Messinian half-graben structure and that Pliocene deposits blanket the whole basin (see location on Fig. 9). B, B*, C, C1 are unconformities; 2, 3, 4, 5 = sequences; in red are faults.

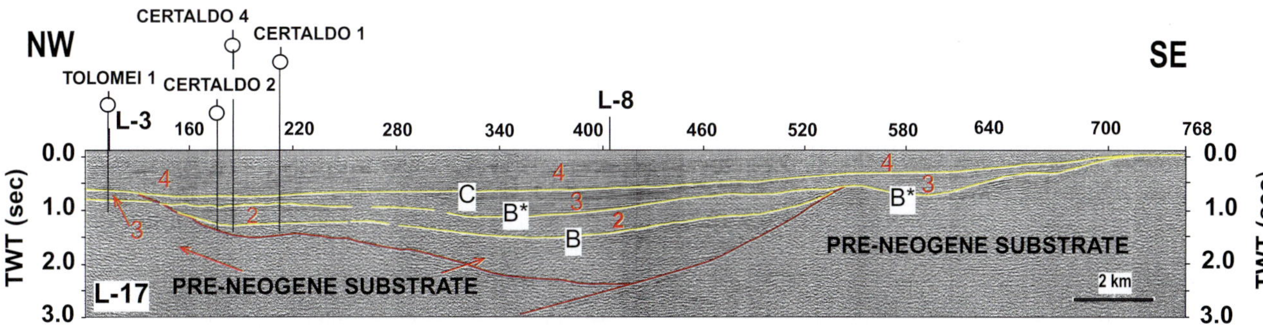

Fig. 11 NW–SE seismic profile L-17 of the Elsa Basin with location of wells and intersecting seismic profiles L-3 and L-8. Note the pre-Neogene antiapenninic-oriented fault (see location on Fig. 9). B, B*, C are unconformities; 2, 3, 4 = sequences.

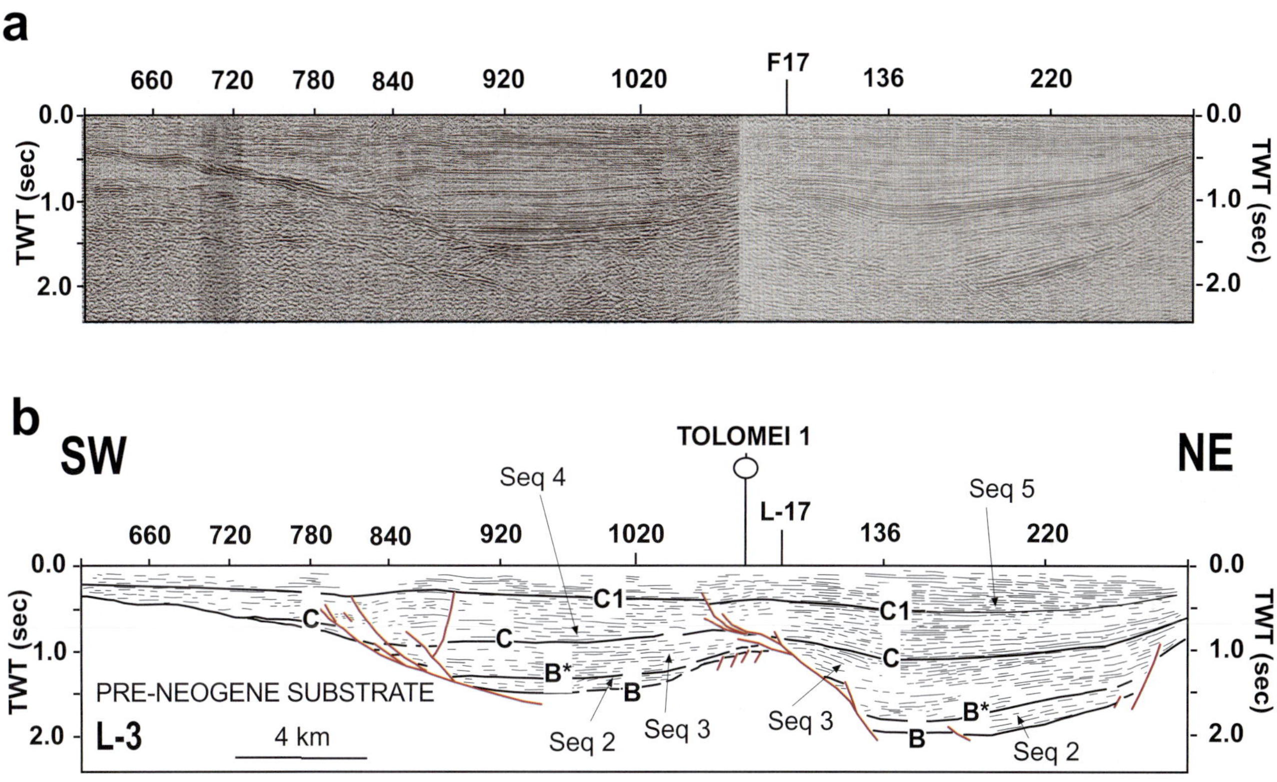

Fig. 12 SW–NE seismic profiles L-03 of the northernmost part of the Elsa Basin with location of well and intersecting seismic profile L-17. (a) Original seismic profile. (b) Interpreted line drawing. Note the complex intrabasinal structure (see location on Fig. 9). B, B*, C, C1 are unconformities; Seq = sequence; in red are faults.

flank of the complex-faulted low area associated with the Livorno–Sillaro lineament.

3 The complex intrabasinal structure shown on profile L-3 (Fig. 12) remains unexplained, and it is not possible at this stage to associate it with transverse lineament activities.

Piombino–Faenza lineament

The Piombino–Faenza is another major lineament that cuts across the Northern Apennines (pf, Fig. 1). In the Adriatic Sea it marks the southernmost extension of the so-called 'Ferrara Folds' (part of the thrust orogen front; Fig. 5a; Castellarin, 2001; Vai, 2001). In the same area, a major deepening of the base of the Pliocene basin occurs from a depth of 3000 m just north of the lineament to about 9000 m south of it (Castellarin, 2001). In the inner part of the Northern Apennines, the lineament marks the southeast border of the Neogene–Quaternary Mugello (MU), Firenze (FI), Elsa (EL) and Volterra (VO) basins, and the northwest border of the Valdarno (VA) and Radicondoli (RD) basins (Fig. 1). In the Firenze Basin (FI) the lineament delimits a substrate high and contributes to a northeastward shift of the nose of a major substrate thrust fault (Figs 1 & 5). Evidence of the influence of this lineament on sedimentation are the large conglomeratic alluvial fans that developed at the intersection between the lineament and the master faults in basins such as the Elsa (EL; Canuti *et al.*, 1966), Casino (CA; Bossio *et al.*, 2002) and Volterra (VO; Martini *et al.*, 1995) (Fig. 1). The lineament also cuts through the Larderello geothermal area and extends offshore into Elba Island (Fig. 1). In the Larderello area, the effect of this lineament on sedimentation is recorded in a late Messinian antiapenninic-oriented palaeovalley filled with Elba-Island-sourced igneous rocks (Pascucci *et al.*, 2006b).

Arbia–Marecchia lineament

Segments of this lineament are relatively well defined from the Adriatic Sea across the Apennines crest to the Siena Basin (Fig. 1). The lineament is less well defined farther west but it can still be extended into the north Tyrrhenian Sea shelf. In the outer part of the Northern Apennines, it cuts the front of the Apennines thrusts under the Adriatic Sea (Argnani, 1998). It delimits the Marecchia valley where a large, isolated body of Ligurides occurs (Fig. 5a; Barchi *et al.*, 2001).

Along the axis of the mountain chain the lineament marks the change in direction of two main thrusts (Chianti–Cetona and Cervarola–Falterona) from NW–SE to N–S (Fig. 1). Liotta (1991) conducted a detailed study in the central-western basins of Tuscany and recognized major differences in Messinian–Pleistocene stratigraphy between the north and south side of the lineament.

The Arbia–Marecchia lineament continues into the northern Tyrrhenian Sea shelf, where it delimits the southeastern termination of the substrate high of the Elba Island igneous and metamorphic complex, as indicated by the Bouger residual anomalies magnetic map (Bartole *et al.*, 1991). Indication of this seaward extension of the lineament can also be observed on seismic profiles of the northern Tyrrhenian Sea shelf (see below: Northern Tyrrhenian Sea shelf).

Grosseto–Pienza lineament

The Grosseto–Pienza is a secondary lineament recognizable from the Chianti–Cetona Ridge to the Tyrrhenian Sea. It marks the boundary between the Siena Basin and the Radicofani Basin. This lineament also separates the Larderello plutonic area to the northwest from the Amiata Mount volcanic zone to the southeast, and cuts through the Triassic metamorphic complex separating the main body of the MTR from that of the Uccellina Mountains (Fig. 1). It can be extended into the northern Tyrrhenian Sea shelf where it has influenced the development of the complex system of basins (see below: Northern Tyrrhenian Sea shelf).

Evidence of the effects of this lineament can be readily found in the Siena–Radicofani area.

Siena and Radicofani basins

Setting. The so-called 'Siena–Radicofani basin' is a NW–SE oriented depression 50 km long and 15 km wide, bordered by the Chianti–Cetona ridge to the east, and the Montalcino–Amiata high to the west (Figs 1 & 13). It has well-exposed rocks, it is covered by numerous seismic profiles, and is drilled in the southeast part by four stratigraphic wells (Fig. 13). The Siena–Radicofani depression can be subdivided longitudinally into three parts: the Siena Basin (SI) proper to the northwest; the Pienza area in the central part; and the Radicofani Basin (RA) proper to the southeast (Pascucci *et al.*, 2006a) (Figs 1 & 13).

1 The *Siena Basin* is filled primarily with lower to middle Pliocene clay in the centre, and sand and conglomerate at the margins (Sequences 4 and 5, up to 800 m thick). Upper Miocene deposits (continental facies of Sequence 2 and Sequence 3, up to 800 m thick) are recognized on SW–NE oriented seismic profiles (not shown here) only in a narrow, local sub-basin to the northwestern corner. The upper

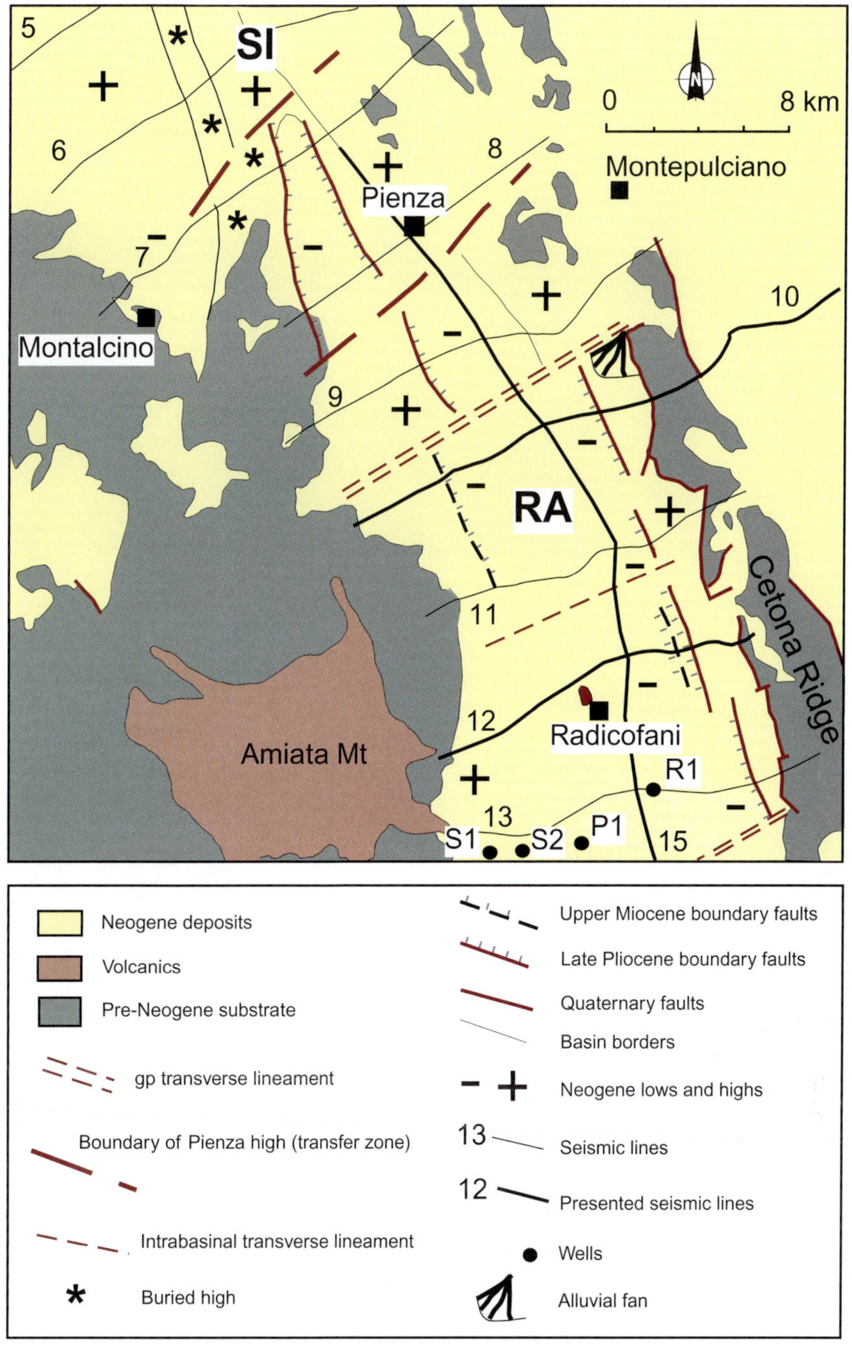

Fig. 13 Generalized structural map and location of seismic profiles and wells of the Radicofani Basin. SI, southernmost part of Siena Basin; RA, Radicofani Basin.

Miocene sub-basin has a triangular-shaped geometry and is interpreted as a half-graben. The overlying Pliocene deposits, however, show an overall wide bowl-shaped geometry with a depocentre located in the middle of the basin. This is interpreted as a post-rift depositional setting.

2 The *Pienza area* is a substrate high covered by thin (up to 200 m) middle Pliocene deposits. It separates the Siena from the Radicofani basins (Figs 1 & 13). To the west it is bordered by a shallow (400 m of sedimentary fill) Pliocene basin (low area west of Pienza; Fig. 13) bounded by normal faults reactivated during the Quaternary.

3 The *Radicofani Basin* contains middle Miocene marine deposits (Sequence 1, approximately 400 m thick) overlain by upper Miocene fluvio-lacustrine

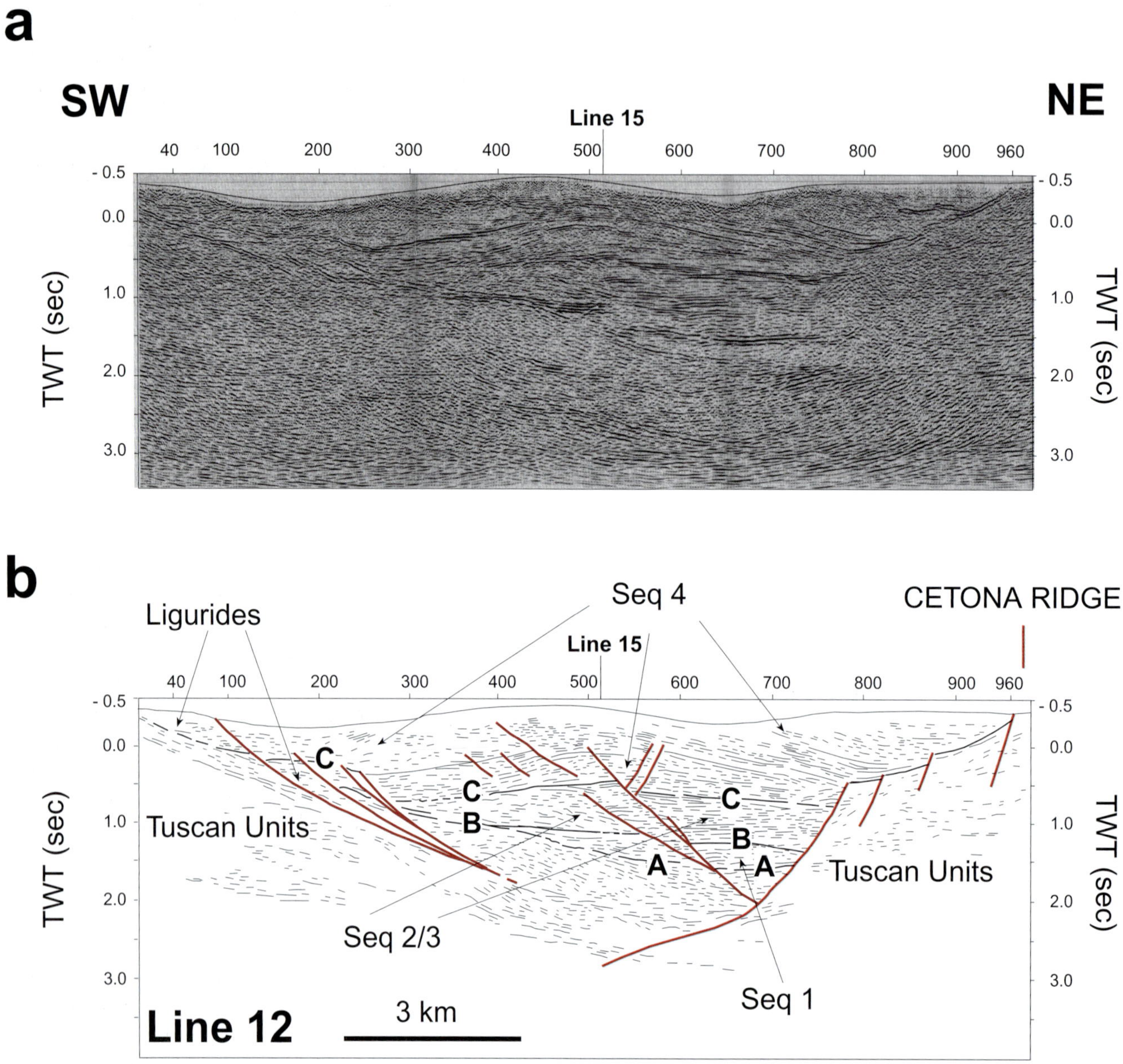

Fig. 14 Seismic profile, Line 12, of the southern part of the Radicofani Basin. (a) Original profile, with location of wells and intersecting seismic profile L-15. (b) Interpreted line drawing. Note the Messinian half-graben and the Pliocene open anticline (see Fig. 13 for location). L-15 = intersecting line; A, B, C are unconformities; Seq = sequence; in red are faults.

clastics (Sequences 2 and 3, about 900 m thick), and by thick lower Pliocene marine clay with locally interbedded sandstones and conglomerates (Sequence 4, about 1200 m thick) (Figs 4, 13 & 14; Liotta, 1996). A thin veneer of middle Pliocene carbonates and sandstones occurs along the southeast margin. Several conglomeratic alluvial fans have developed along the eastern flank of the basin during the early Pliocene. In particular, a large fan-delta complex (600 m thick) has developed to the northeast near the intersection between the Grosseto–Pienza transverse lineament and the apenninic-oriented border fault of the basin (Figs 1 & 13; Costantini & Dringoli, 2003; Pascucci *et al.*, 2006a). Numerous indentations associated with closely spaced, antiapenninic-oriented faults have also been mapped along the eastern border of the basin (Fig. 13). The large volcanic edifice of the Amiata Mount bounds the basin to the west. A volcanic neck

(Radicofani) occurs in the central southern part of the basin (Fig. 13).

The subsurface structural map, based on seismic data, shows that the Radicofani Basin can be divided longitudinally in two parts: the upper Miocene succession that developed in two sub-basins with opposite-dipping boundary faults, and the lower Pliocene succession (Fig. 13).

In the southern part, NE–SW seismic profiles show that the upper Miocene succession (Sequences 2 and 3) has a triangular-shaped geometry characteristic of a half-graben with the master fault to the east (Fig. 14). The overlying Pliocene succession has a complex structure because it is involved in an open anticline. The origin of the anticline is debated, but probably it is related to the emplacement of the Radicofani neck and Amiata Mount volcano, which has deformed the Pliocene deposits (Acocella *et al.*, 2002; Pascucci *et al.*, 2006a). Nevertheless, the Pliocene deposits are shown to onlap on the western side of the basin, whereas they are affected by a suite of normal faults to the east (Fig. 14; Liotta, 1996). In the northern part of the basin, a NE–SW seismic profile shows a Miocene half-graben with master faults to the west, and a Pliocene basin with the down-faulted portion of a major alluvial fan to the east (Fig. 15, Sequences 2 and 3; Pascucci *et al.*, 2006a).

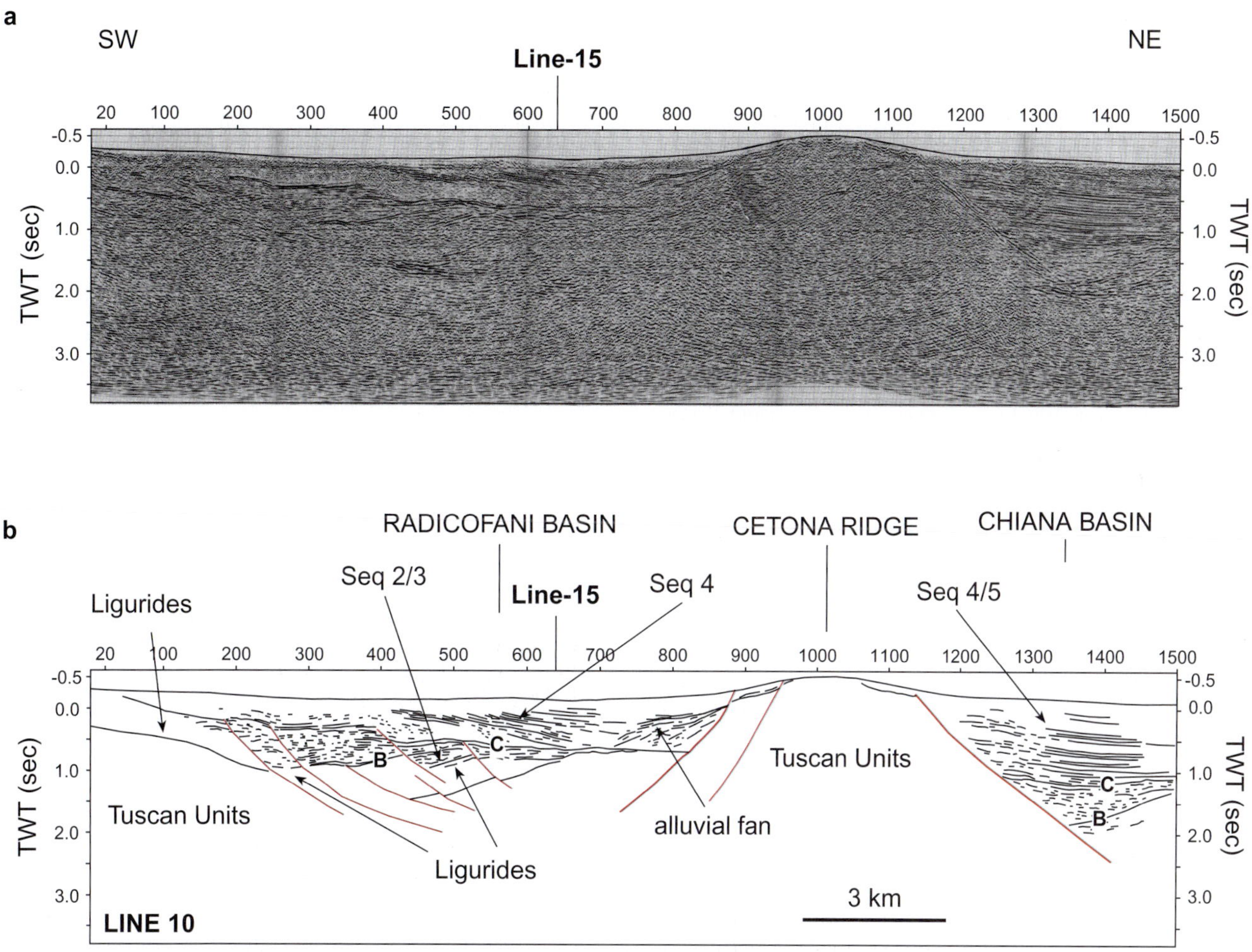

Fig. 15 Seismic profile, Line 10, of the northern part of the Radicofani Basin, with location of intersecting seismic profile L-15. (a) Original profile. (b) Interpreted line drawing. Note the Messinian half-graben with the master fault to the west (see Fig. 13 for location). L-15 = intersecting line; B, C are unconformities; Seq = sequence; in red are faults.

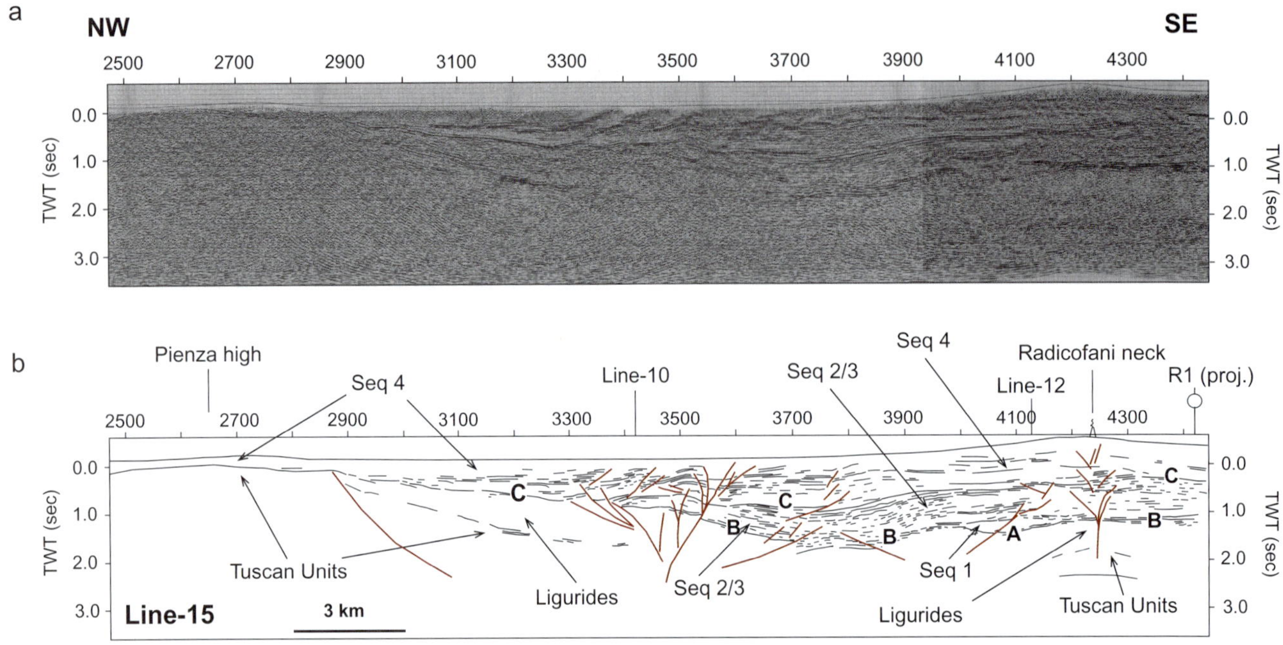

Fig. 16 NW–SE seismic profile, Line 15, of the Radicofani Basin, with location of intersecting seismic profiles L-10 and L-12. (a) Original profile. (b) Interpreted line drawing. Note the positive flower structures (see Fig. 13 for location). L-10, L-12 = intersecting lines; A, B, C are unconformities; R1 = Radicofani 1 well; Seq = sequence; in red are faults.

A longitudinal, NW–SE oriented seismic profile shows several characteristic features of the Radicofani Basin, as follows (Fig. 16).

1 The Pliocene deposits overextend the underlying Miocene sequence and onlap on the substrate Pienza high, with possible local deformation by post-lower Pliocene faults.

2 Five kilometres to the southeast of Pienza (shot points 1850–2050), a complex reflector pattern is present, which is interpreted as a positive flower structure that may be associated with the Grosseto–Pienza transverse lineament (Fig. 16). Other researchers have, however, interpreted this same pattern as a kinematically complex set of southeastward-directed thrust faults and northward-directed backthrusts (Bonini & Sani, 2002).

3 The southeast part of the profile runs near the Radicofani volcanic neck. There is a doming in the reflectors, a local lack of seismic response, and the distribution and tilting of some of the stronger reflectors suggest the presence of several faults. All this is probably associated with the emplacement of magma in the nearby Radicofani volcanic complex.

The Grosseto–Pienza lineament can be extended into the north Tyrrhenian Sea shelf (see below: Northern Tyrrhenian Sea shelf).

Albegna lineament

The Albegna lineament is relatively well defined only on the western side of the Northern Apennines (Fig. 1). It probably delimits to the southeast the Chianti–Cetona Ridge and the Radicofani Basin. It separates the Amiata Mount volcano area from the larger volcanic complex of Bolsena. It crosses longitudinally the Albegna Basin where marked differences occur in the Miocene deposits between the northern and southern sides of the lineament (Bossio *et al.*, 2004). The Albegna Basin is oriented NE–SW. It has been variously interpreted as a graben or as a pull-apart basin associated with a dextral strike-slip fault (Boccaletti *et al.*, 1977).

Northern Tyrrhenian Sea shelf

Setting. Three transverse lineaments (av, gp and al) converge into the northern Tyrrhenian Sea shelf area (Figs 1 & 17). The area is characterized by several N–S and NW–SE trending Neogene basins (Pianosa, Montecristo, Cerboli, Punta Ala, Formiche and Uccellina) separated by substrate highs (Elba–Pianosa Ridge, Montecristo, Montecalamita and the composite GFR Ridge) (Fig. 17). Pre-Neogene

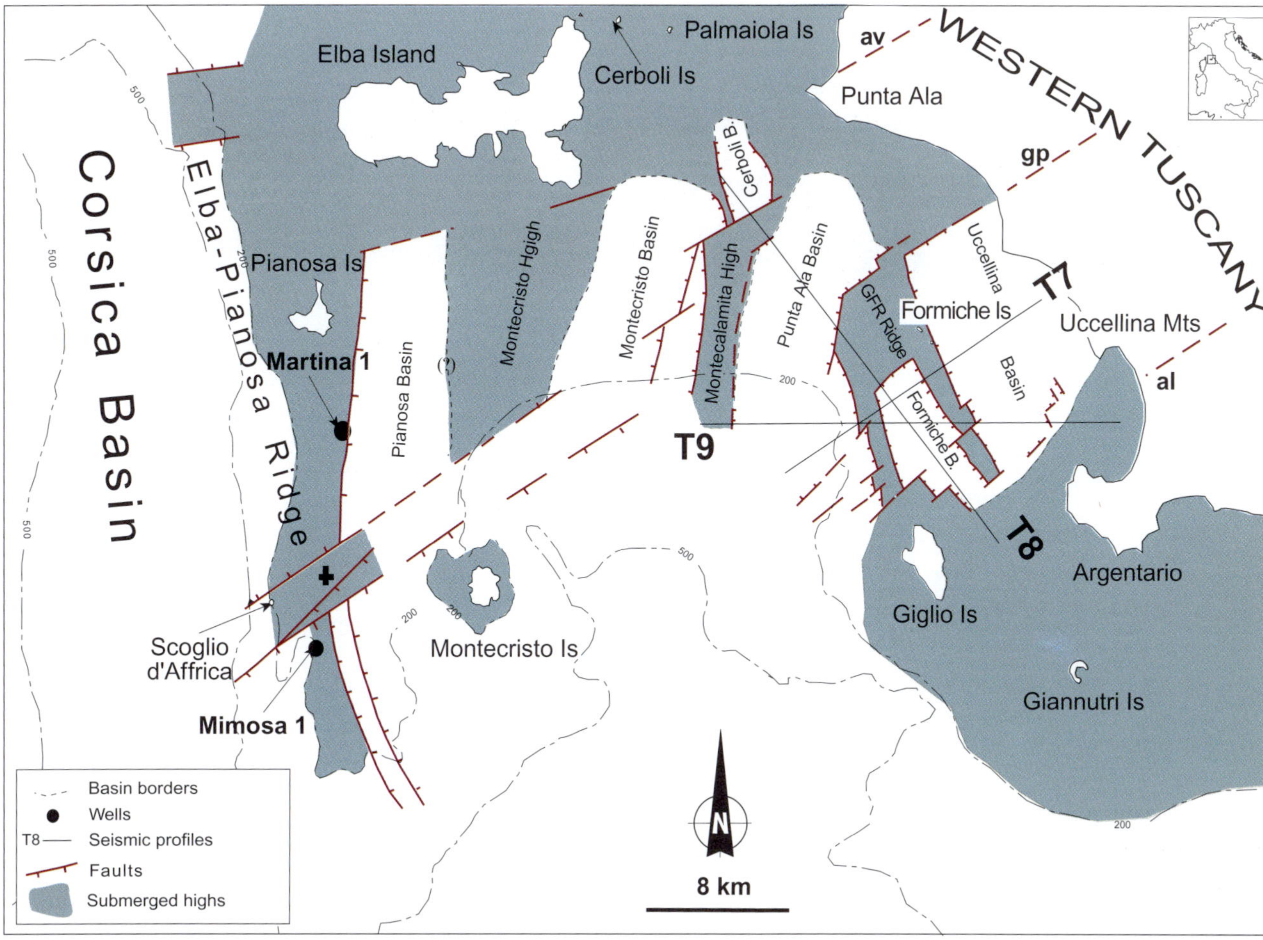

Fig. 17 Generalized structural map and location of seismic profiles and wells (Martina 1 and Mimosa 1) of the northern Tyrrhenian Sea shelf: al, Albegna; av, Arbia–Marecchia; gp, Grosseto–Pienza; Is, Island; Mts, Mountains.

substrate is exposed on the mainland and islands. Direct geological information on the Neogene deposits can be obtained from inshore basins and from two wells drilled offshore along the Elba–Pianosa Ridge (Fig. 17; Pascucci *et al.*, 1999; Cornamusini *et al.*, 2002).

Stratigraphy. All middle Pliocene (lower Tortonian) to Pleistocene sequences observed in the inshore basins (Fig. 4) are present, and they are readily recognizable in seismic profiles. Well-defined, continuous subhorizontal reflectors characterize every sequence except Sequence 2 (Figs 18 & 19). The sequences are delimited by onlap and downlap surfaces which are well defined at the basin margin; they become correlative conformities towards the centre. Sequence 2 is recognizable because of the characteristic seismic response to its topmost gypsum layers. The seismic response consists of concave downward reflectors probably associated with an increase of seismic velocity in the gypsum, and discontinuity of the layers (Figs 18b & 19).

Basin geometry and interpretation. The Neogene–Quaternary basins have developed on a thrust substrate locally dissected by normal faults (Bartole *et al.*, 1991; Bartole, 1995). The basins have a triangular geometry in the lower part, changing into a wide bowl and blanket-shaped geometries in the upper (Pascucci *et al.*, 1999). The triangle shape of the upper Tortonian–Messinian sedimentary fill (Sequence 2 and possibly Sequence 3) of the Formiche Basin, for instance, is interpreted as a half-graben fill with the master fault dipping to the east (Figs 18a & 19). Mostly undeformed, post-rift,

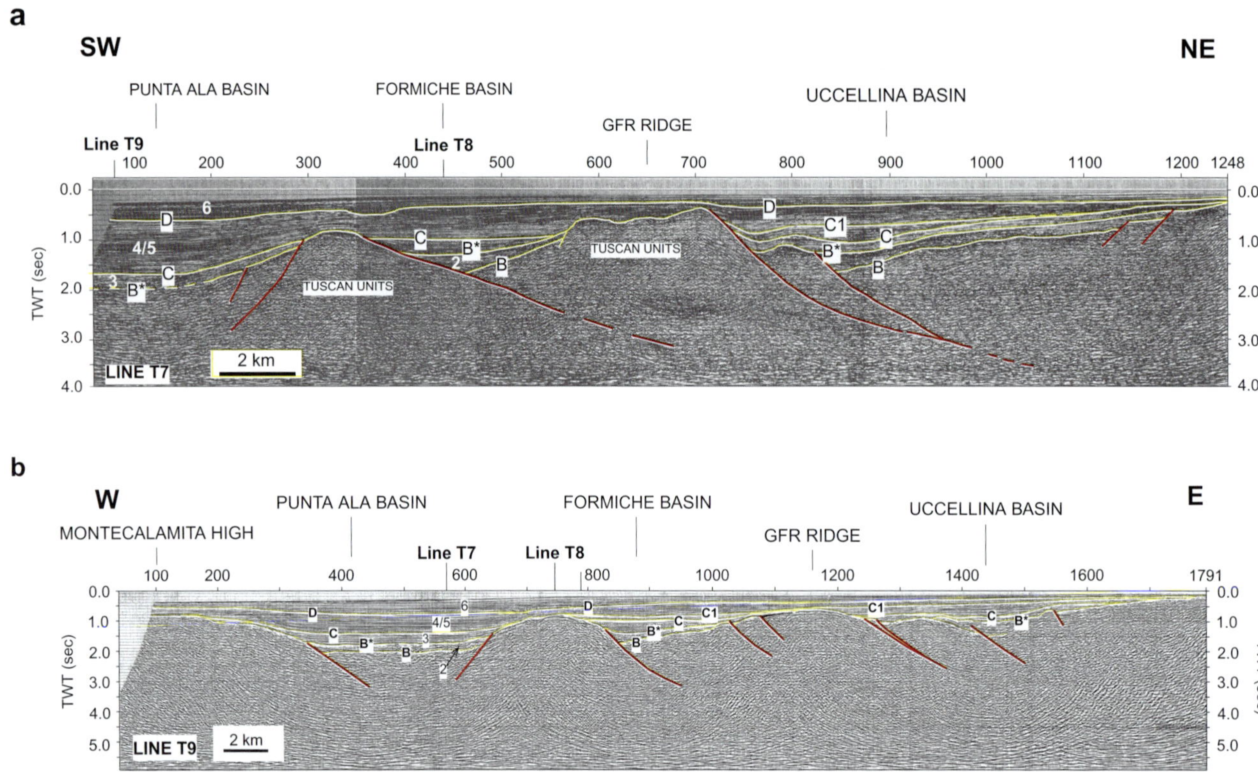

Fig. 18 Seismic profiles of the northern Tyrrhenian Sea shelf. (a) SW–NE, interpreted seismic profile T7 crossing the Formiche and Uccellina basins. (b) W–E, interpreted profile T9 crossing the Punta Ala, Formiche basins and the southernmost corner of the Uccellina Basin. Note that the graben or half-graben basins are in the lower part, and the change into wide bowl and blanket-shaped structures is in the upper part (see Fig. 17 for location). B, B*, C, C1, D are unconformities; 2, 3, 4, 5, 6 = Sequences 2, 3, 4, 5 and 6; in red are faults.

Pliocene–Quaternary deposits overlie this. A similar eastward dipping boundary fault system exists in the Uccellina Basin. The Punta Ala Basin, similar to the Montecristo and the Pianosa, is not well delimited to the south toward the open Tyrrhenian Sea. The available seismic profiles indicate that it may have been a rift during the late Miocene (Fig. 18b).

There is no strong direct evidence that the basins of the north Tyrrhenian Sea shelf have been affected by major strike- or oblique-slip movements associated with transverse lineaments mapped onshore, except for possible occasional flower structures (Bartole, 1995). However, the en échelon distribution of the basins and the antiapenninic-oriented indentations of some margins indicate that the basins have been affected, and some are terminated by the seaward extensions of the transverse lineaments recognized inland, in particular the Arbia–Marecchia (av) and the Grosseto–Pienza (gp) lineaments. The Arbia–Marecchia lineament has affected primarily the Cerboli Basin and the northern part of the Montecristo and Pianosa basins; the Grosseto–Pienza lineament has affected mainly the Uccelina and Formiche basins (Fig. 17). The two lineaments may have converged farther offshore and affected the Elba–Pianosa Ridge in the area of the Scoglio d'Africa high.

DISCUSSION

The transverse lineaments of the Northern Apennines have different origins, importance, and times of formation and of reactivation. Nonetheless, they have acted through time as a means of linking the outer part of the developing orogen where thrust imbrications were and are still active, and the inner part where post-orogenic basins have and are developing.

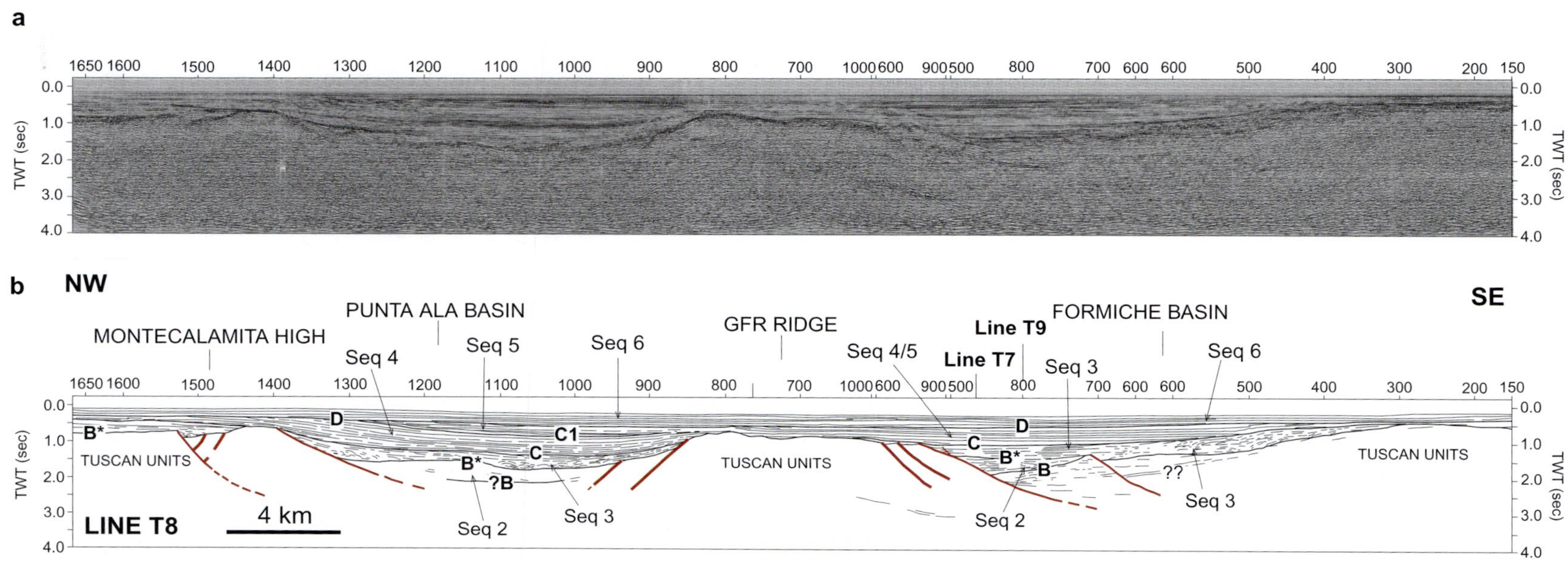

Fig. 19 NW–SE seismic profile T8 of the northern Tyrrhenian Sea shelf: (a) Original profile. (b) Interpreted line drawing. Note the triangle shape of the upper Tortonian–Messinian sedimentary fill (see Fig. 17 for location). Lines T7 and T9 are intersecting lines; B, B*, C, C1, D are unconformities; Seq = sequence; in red are faults.

The transverse lineaments are locally well defined by clear fault traces. In other places along the same or in other lineaments, they appear at the surface as diffuse deformation zones best related to transfer zones (Sorgi *et al.*, 1998). They may be closely associated to the extensional basins they intersect such as in Brazil (Milani & Davison, 1988), or they may represent weakness zones that have acted in a different manner and variously reactivated at different times.

The Miocene to Quaternary geological effects of these lineaments can be followed both through space, at the present time, from the Adriatic Sea to the Tyrrhenian Sea, and through time at the same locality in western Tuscany and the northern Tyrrhenian Sea shelf.

1 In the first case, active thrust fronts in the Adriatic Sea side are locally separated (cut through or lateral thrust ramps) by the major transverse lineaments. In this outer part of the orogen, the transverse lineaments are acting as transfer faults in an overall compressive regime (Liotta, 1991; McClay, 1992). Farther to the west, on the Tyrrhenian Sea side of the orogen, the transverse lineaments delimit laterally the basins within the NW–SE oriented structural depressions associated with the fronts of major pre-Neogene thrusts. Except for some segments, the lineaments serve primarily as weakness corridors, separating zones (basins) with different rates of subsidence and/or different throw and dip of apenninic-oriented master faults. In this respect, therefore, they mostly act as transfer zones (Liotta, 1991).

2 In the second case, the effect through time of the lineaments on the development of the inner zone of the Northern Apennines (Tuscany and northern Tyrrhenian Sea shelf) was initially that of active elements cutting through pre-Tortonian thrusts. After the inner part of the orogen changed from a predominantly compressive to a predominantly extensive system, the major lineaments continued to act as dividers between differently evolving blocks. What this evolution has been and continues to be is a matter of conjecture (Migliorini, 1949; Merla, 1952; Ghelardoni, 1965; Bortolotti, 1966; Boccaletti *et al.*, 1982; Fazzini & Gelmini, 1982; Liotta, 1991; Martini & Sagri, 1993; Carmignani *et al.*, 1994; Bonini & Sani, 2002; Pascucci *et al.*, 2006a). The fact remains that beyond basic similarities due to similar origin and overall tectonic system, the blocks have significant differences due to the Neogene to Quaternary evolution. Among others, major differences are associated with the very rapid uplift (average Neogene–Quaternary exhumation rates of 0.4 mm yr^{-1}; Ballestrieri *et al.*, 2003) of the Apuane Mountains (mostly metamorphic rocks) north of the Livorno–Sillaro (ls) lineament, the large intrusion to shallow depth of acid anatectic magma plutons in the Larderello area (Lavecchia, 1988; Serri *et al.*, 2001) between the Piombino–Faenza (pf) and Arbia–Marecchia (av) lineaments, and the large volcanic systems of Amiata Mount between the Grosseto–Pienza (gp) and the Albegna (al) lineaments (Fig. 20). These features have affected the development and preservation of the Neogene–Quaternary basins.

Furthermore, west of the Chianti–Cetona Ridge the lineaments separate more elevated blocks from adjacent relatively more subdued ones. These are the Apuane Mountains high area north of the Livorno–Sillaro (ls), the generally lower area between the Livorno–Sillaro (ls) and the Piombino–Faenza (pf) lineaments, the relatively higher block between the Piombino–Faenza and the Grosseto–Pienza (gp), and the relative low between the Grosseto–Pienza and the Albegna (al) lineaments (Figs 1 & 20). Some of these relations had probably evolved before the Neogene, whilst others are more recent. Indeed, considerable, variable change in elevation has occurred throughout the Northern Apennines from the Pliocene to the Present as mapped in the neotectonic map by Bartolini *et al.* (1982).

The transverse lineaments had several functions. Fazzini & Gelmini (1982), for instance, suggested that the blocks bounded by the major transverse lineaments shifted northeastward during the development of the mountain chain, but at different rates at different times, generating various degrees of extension and subsidence or uplift at either side of transfer faults/zones. Furthermore the lineaments did not act synchronously nor in the same manner along the whole length, rather the various segments behaved independently. This is well illustrated by the Livorno–Sillaro lineament. This lineament is indeed a major one, probably associated with transcurrent faults of the original Jurassic ocean basin where the substrate rocks of the Northern Apennines were formed (Fazzini & Gelmini, 1982). It probably cuts through the whole crust (Royden *et al.*, 1987). It was active during the eastward translation of the Apennines area, and separated constrained blocks to the north from more freely migrating blocks to

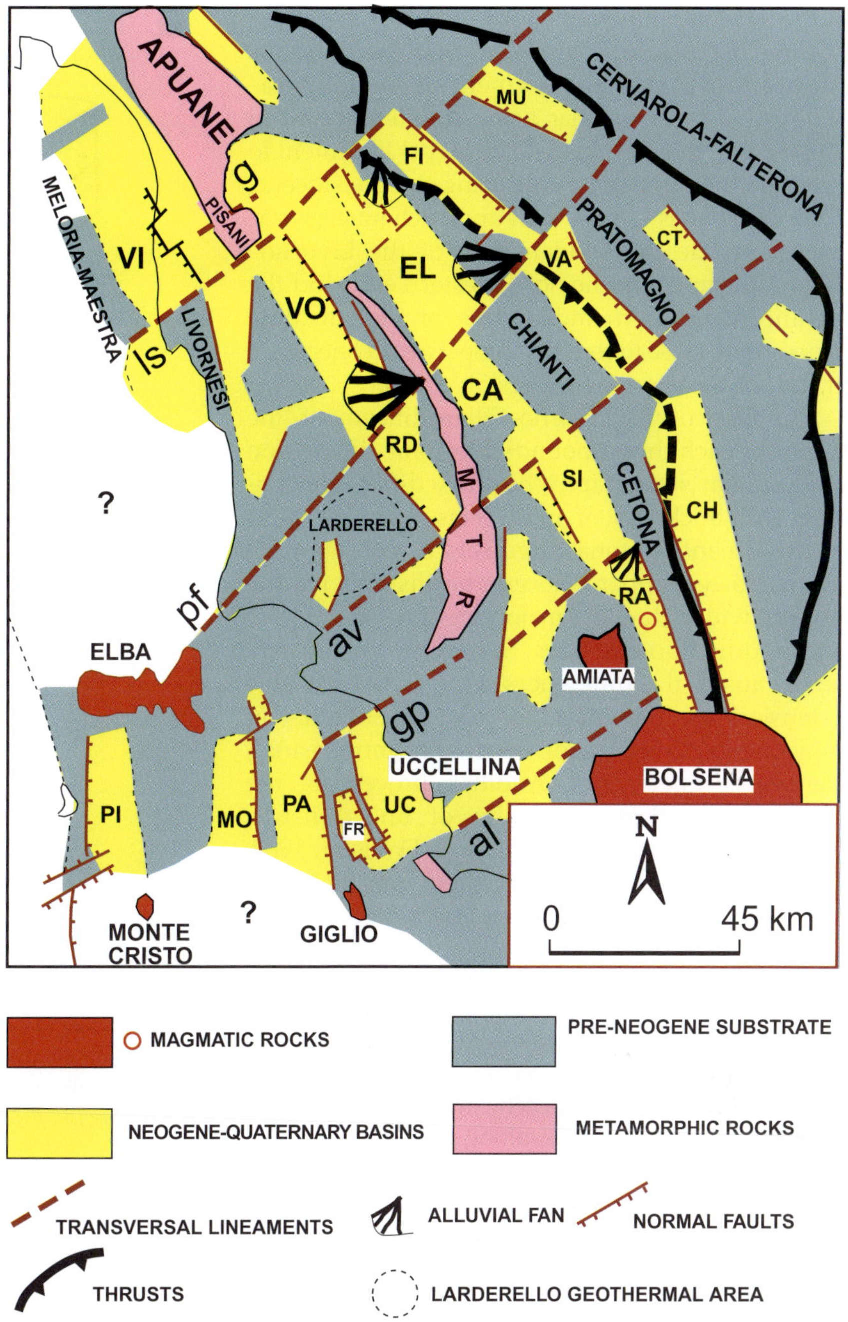

Fig. 20 Generalized structural map of Tuscany showing principal thrusts and Neogene–Quaternary basins, transverse lineaments and major intrabasinal faults.

the south (Fig. 20; Bortolotti, 1966). Throughout its existence, starting in the Jurassic–Cretaceous, it has delimited areas to the north and to the south experiencing differential subsidence and uplift as demonstrated by different stratigraphy and structural deformation (Bortolotti, 1966).

In Tuscany, one old feature reactivated during the Neogene is the left-lateral movement along the western segment of the Livorno–Sillaro lineament, which most likely shifted the Apuane Mountains and Pisani Mountains southwestward in respect of the Middle Tuscany Ridge, and the Meloria–Maestra shoal in respect of the Livornesi Mountains (Figs 1, 6 & 20). If this were correct it is also likely that the Viareggio and Volterra basins that contain similar thick Pliocene successions (Pascucci *et al.*, 1999; Pascucci, 2006) are equivalent, but shifted one with respect to the other. This would imply that part of the left-lateral movement of the Livorno–Sillaro lineament has also occurred during the Pliocene. Farther to the east, the segment of the Livorno–Sillaro (ls) lineament has acted differently and in unison with the Piombino–Faenza (pf) lineament to generate the Firenze Basin during the Pleistocene. This basin has some pull-apart characteristics. These include the sudden shift of the Livorno–Sillaro lineament at the northwestern end, and the significant northern indentation of the substrate thrust-front, close to the Piombino–Faenza lineament at the southeastern end (Figs 5b & 20).

Southeast of the Livorno–Sillaro lineament, the basins of each thrust-bounded tectonic depression have similar stratigraphy, but they differ in extension and thickness of infill (Figs 1 & 20). West of the Chianti–Cetona ridge this is true for both the narrower, upper Miocene basins and for the usually wider, due to a major transgression, lower and middle Pliocene ones. Several processes may have caused the differences. One is the variable extension at different times due to oblique-slip movements along major or minor (intrabasinal) transverse lineaments. Another process is the difference in throw and slip of linked listric, boundary normal faults (Milani & Davison, 1988; Davison, 1994). Examples of these can be found in the Viareggio and in the Radicofani basins, as follows.

1 The indentation of the boundary faults of the Viareggio Basin along the seaward propagation of the Guappero fault (G) is interpretable as being due to either oblique-slip Pliocene movements and/or changing dip and throw along master fault segments (Fig. 21a; Milani & Davison, 1988).
2 In the Radicofani Basin the reversal of the boundary faults in the Miocene sub-basins recognized from seismic profiles records a transfer of depocentre with the master faults of the sub-basins shallowing toward an intrabasinal transfer zone (Figs 13 & 21b; Rosendahl *et al.*, 1986; Schlische, 1995).
3 The positive flower structure recognized along the longitudinal seismic profile in the Radicofani Basin suggests post-early Pliocene strike-slip movement along the Grosseto–Pienza lineament (Figs 13 & 16). Structural activity along this transverse lineament and the apenninic-oriented boundary fault of the basin is also indicated by the large, lower Pliocene alluvial fan-delta that has developed at the relay ramp near their intersection (Figs 13 & 15).

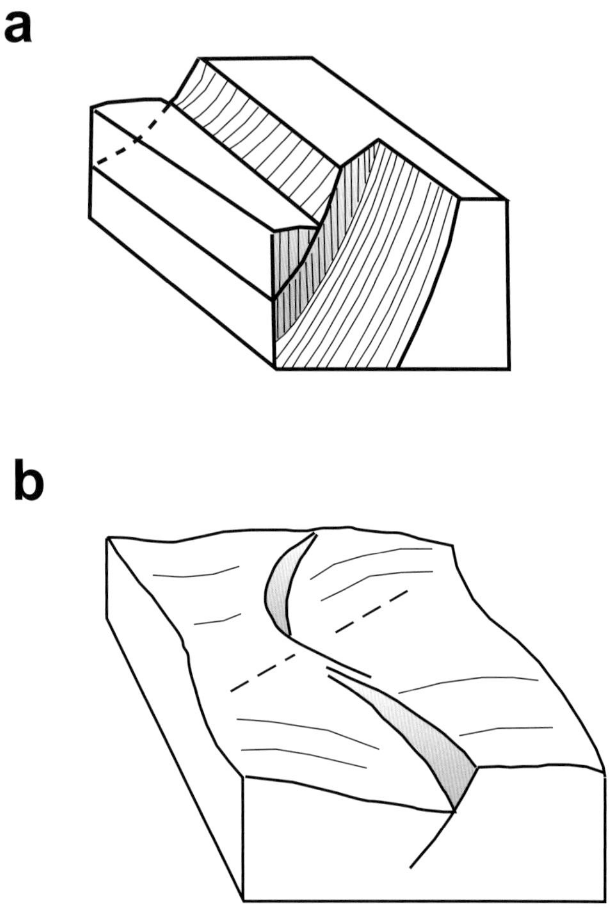

Fig. 21 Schematic features of transfer faults/zones: (a) Change in throw and dip of listric normal fault either side of the transfer fault (after Milani & Davison, 1988). (b) Linkage of two half-grabens with decreasing throw of master faults toward an intrabasinal transfer zone (after Rosendahl *et al.*, 1986; Schlische, 1995).

Transverse lineaments similar to those of the Northern Apennines occur in most other orogens, and some have acted throughout the evolution of the complex mountain systems, such as in the Carpathian–Pannonian system (Royden *et al.*, 1982, Royden & Horváth, 1988), the Caledonian orogeny in Scotland (Watson, 1984) and in Sub-Andean basins (Jacques, 2003). In contrast to these other examples, the transverse lineaments of the Northern Apennines can be seen today cutting through thrust faults that are active in the outer part of the orogen and are essentially quiescent in the inner part.

CONCLUSIONS

Much has been written about the transverse lineaments of the Northern Apennines. Geological and geophysical evidence attest to their existence and various activities. New subsurface information is presented here for selected Neogene to Quaternary basins of Tuscany affected by the lineaments. This information derives from the analysis of newly released, commercial seismic profiles.

1 The study of seismic profiles of Neogene–Quaternary basins has allowed a tectono-stratigraphic analysis of several basins, and indicates the following.

(a) Complex (chaotic seismic response, intensely faulted) structures developed near major lineaments, such as the Livorno–Sillaro in the Viareggio Basin.

(b) Secondary, Neogene, antiapenninic-oriented faults are associated with the lineaments; for example, in the Livorno–Sillaro system in the Viareggio Basin and in the Elsa Basin they occur over a corridor 10–20 km wide.

(c) A secondary, intrabasinal transfer zone with reversal of the dip orientation of the master faults is documented in the Miocene of the Radicofani Basin.

(d) A few possible positive flower structures have been identified, such as the one cutting through the lower Pliocene deposits of the Radicofani Basin, which indicates strike-slip movement along a segment of the Grosseto–Pienza lineament.

(e) Several transverse lineaments can be extended into the shelf of the northern Tyrrhenian Sea where they delimit various Neogene–Quaternary basins.

2 Pre-latest Miocene tectonic activity (thrusts and uplifts) generated major structural depressions in Tuscany. The major transverse lineaments determined the formation and development of the upper Miocene to Quaternary basins by doing the following:

(a) dissecting the structural depression longitudinally;

(b) acting as oblique-slip faults accommodating different overall amount and rate of shifts in different blocks, at different times;

(c) acting as transfer faults or, at times along some segments, transfer zones accommodating different throws and dips of apenninic-oriented boundary faults of the evolving basins;

(d) fostering development of different type and thickness of sedimentary successions on opposite sides of the lineaments, and causing the formation of particular sedimentary edifices such as large alluvial fans.

Alluvial fans occur frequently in the Neogene–Quaternary basins of the inner part of the Northern Apennines. However, major fans are preferentially located at relay ramps near the intersection of the transverse (antiapenninic-oriented) lineaments and apenninic-oriented boundary faults.

3 Structural lineaments transverse to the main orogen trend are not unique to the Apennines. They are common features of most mountain chains. Major lineaments that formed as transform faults in the original oceans, such as the Livorno–Sillaro (ls) and Olevano–Antrodoco (aa) in the Northern Apennines), were reactivated throughout the ages and guided the translation and rotation of the various lithospheric blocks during the mountain building process. Within each block, secondary, shallower lineaments guided thrusts and some were the result of differential displacements of thrust edifices or of extensional blocks.

ACKNOWLEDGEMENTS

We thank A. Lazzarotto for encouraging and supporting this research, and L. Casini for critically reading an early version of the manuscript. Dr M. Roveri and Dr I. Davison are kindly acknowledged for the constructive final reviews of the paper. S. Pizzolante helped with the analysis of some seismic lines of the Elsa Basin. Special thanks go to F. Benelli, L. Burbi and S. Merlini of ENI-AGIP Division for their suggestions and permission to publish seismic data. The Italian Consiglio Nazionale della Ricerca (CNR), MIUR (COFIN 2003, Me.La. project, to VP and FS), the Universities of Siena, Sassari, Firenze and the Natural Science and Engineering Research Council of Canada (NSERC, Grant 0GP0007371 to IPM) financed the work.

REFERENCES

Acocella, V., Pascucci, V. and Dominici, G. (2002) Basin deformation due to laccolith emplacement at Radicofani (Southern Tuscany, Italy). *Boll. Soc. Geol. It.*, **Vol. Spec. 1**, 749–756.

Argnani, A. (1998) Structural elements of the adriatic foreland and their relationships with the front of the apennine fold-and-thrust belt. *Mem. Soc. Geol. It.*, **52**, 647–654.

Argnani, A., Bernini, M., Di Dio, G.M., *et al.* (1997). Stratigraphic record of crustal-scale tectonics in the Quaternary of the Northern Apennines (Italy). *Quaternario*, **10**, 595–602.

Ballestrieri, M.L., Bermet, M., Brandon, M.T., *et al.* (2003) Pliocene and Pleistocene exhumation and uplift of two key areas of the Northern Apennines. *Quat. Int.*, **101–102**, 67–73.

Barchi, M., Landuzzi, A., Minelli, G. and Pialli, G. (2001) Outer Northern Apennines. In: *Anatomy of an Orogen – The Apennines and Adjacent Mediterranean Basins* (Eds G.B. Vai and I.P. Martini), pp. 215–254. Kluwer Academic Publisher, Dordrecht.

Bartole, R. (1995) The North Tyrrhenian–Northern Apennines post-collisional system: constraints for a geodynamic model. *Terra Nova*, **7**, 7–30.

Bartole, R., Torelli, L., Mattei, G., *et al.* (1991) Assetto stratigrafico del Tirreno Settentrionale: stato dell'arte. *Stud. Geol. Camerti*, **Vol. Spec. 1**, 115–140.

Bartolini, C., Bernini, M., Carloni, G.C., *et al.* (1982) Carta neotettonica dell'Appennino Settentrionale. Note illustrative. *Boll. Soc. Geol. It.*, **101**, 523–549.

Bemporad, S., Conedera, C., Dainelli, P., *et al.* (1986) Landsat imagery: a valuable tool for regional and structural geology. *Mem. Soc. Geol. It.*, **31**, 287–298.

Benvenuti, M. and Degli Innocenti, D. (2001) The Pliocene deposits in the central-eastern Valdelsa Basin (Florence, Italy) revised through facies analysis and unconformity-bounded stratigraphic units. *Riv. It. Paleontol. Stratigr.*, **107**, 265–286

Boccaletti, M. and Guazzone, G. (1974) Remnant arcs and marginal basins in the Cainozoic development of the Mediterranean. *Nature*, **25**, 18–21.

Boccaletti, M., Coli, M. and Napoleone, G. (1977) Nuovi allineamenti strutturali da immagini Landsat e rapporti con l'attivita' sismica negli Appennini. *Boll. Soc. Geol. It.*, **96**, 79–694.

Boccaletti, M., Decandia, F.A., Gasperi, G., *et al.* (1982) Carta strutturale dell'Appennino Settentrionale. *CNR-Consiglio Nazionale delle Ricerche, Pubblicazione*, **429–1982**, Pisa.

Bonini, M. and Sani, F. (2002) Extension and compression in the Northern Apennines (Italy) hinterland: Evidence from the late Miocene-Pliocene Siena-Radicofani Basin and relations with basement structures. *Tectonics*, **21**, 1–35.

Bortolotti, V. (1966) La tettonica trasversale dell'Appennnino 1. La Linea Livorno Sillaro. *Boll. Soc. Geol. It.*, **85**, 529–540.

Bossio, A., Costantini, A., Lazzarotto, A., *et al.* (1993) Rassegna delle conoscenze sulla stratigrafia del neoautoctono toscano. *Mem. Soc. Geol. It.*, **49**, 17–98.

Bossio, A., Foresi, L.M., Mazzei, R., *et al.* (1997) Allostratigraphy and seismic stratigraphy of the Miocene sediments of the Spicchiaola-Pomarance area, southern side of the Volterra basin (Tuscany, Italy). *Riv. It. Paleontol. Statigr.*, **103**, 357–368.

Bossio, A., Mazzei, R., Salvatorini, G. and Sandrelli, F. (2002) Geologia del'area compresa tra Siena e Poggibonsi ('Bacino del Casino'). *Atti Soc. Toscana Sci. Nat. Mem. Ser. A*, **107**, 69–85.

Bossio, A., Foresi, L.M., Mazzei, R., *et al.* (2004) Geology and Stratigraphy of the southern sector of the Neogene Albegna River basin (Grosseto, Tuscany, Italy). *Geol. Romana*, **37** (2003–2004), 165–173.

Cantini, P., Testa, G., Zanchetta, G. and Cavallini, R. (2001) The Plio-Pleistocene evolution of extensional tectonics in northern Tuscany, as constrained by new gravimetric data from the Montecarlo Basin (lower Arno Valley, Italy). *Tectonophysics*, **330**, 25–43.

Canuti, P., Pranzini, G. and Sestini, G. (1966) Provenienza ed ambiente di sedimentazione dei cittolami del Pliocene di San Casciano (Firenze). *Mem. Soc. Geol. It.*, **5**, 340–364.

Capecchi F., Guazzone, G. and Pranzini, G. (1975) Il bacino lacustre di Firenze-Prato-Pistoia. Geologia del sottosuolo e ricostruzione evolutiva. *Boll. Soc. Geol. It.*, **94**, 637–660.

Carmignani, L., Decandia, F.A., Fantozzi, P.L., *et al.* (1994) Tertiary extensional tectonics in Tuscany (Northern Apennines, Italy). *Tectonophysics*, **238**, 295–315.

Castellarin, A. (2001) Alps-Apennines and Po Plain-frontal Apennines relations. In: *Anatomy of an Orogen – The Apennines and Adjacent Mediterranean Basins* (Eds G.B. Vai and I.P. Martini), pp. 177–196. Kluwer Academic Publisher, Dordrecht.

Cornamusini, G., Lazzarotto, A., Merlini, S. and Pascucci, V. (2002) Eocene–Miocene evolution of the north Tyrrhenian Sea. *Boll. Soc. Geol. It.*, **Vol. Spec. 1**, 769–787.

Costantini, A. and Dringoli, R. (2003) Le rocce raccontano. Nascita del territorio tra Chianciano e Sartiano. In: *Musei Senesi. Quaderni Scientifico-Naturalistici*, Vol. 2 (Ed. Museo di Storia Naturale dell'Accademia dei Fisiocritici), pp. 1–40. Protagon Editori Toscani, Siena.

Davison, I. (1994) Linked fault systems; extension, strike-slip and contractional. In: *Continental Deformation* (Ed. P.L. Hancock), pp. 121–142. Pergamon Press, New York.

Fazzini, P. and Gelmini, R. (1982) Tettonica trasversale nell'Appennino settentrionale. *Mem. Soc. Geol. It.*, **10** (1984), 299–311.

Ghelardoni, R. (1965) Osservazioni sulla tettonica trasversale dell'Appennino settentrionale. *Boll. Soc. Geol. Ital.*, **84**, 277–290.

Ghelardoni, R., Giannini, E. and Nardi, R. (1968) Ricostruzione paleogeogafica dei bacini neogenici e quaternari nella bassa Valle dell'Arno sulla base dei sondaggi e dei rilievi sismici. *Mem. Soc. Geol. It.*, **7**, 91–106.

Giese, P. (1981) Intramontane basins and crustal structure. In: *Sedimentary Basins of Mediterranean Margins* (Ed. F.C. Wezel), pp. 55–61. CNR Italian Project of Oceanography, Technoprint, Bologna

Hsu, K.J., Ryan, W.F.B. and Cita, M.B. (1973) Late Miocene desiccation of the Mediterranean. *Nature*, **242**, 240–244.

Jacques, J.M. (2003) A tectnostrastigraphic synthesis of the Sub-Andean basins: inferences on the position of South American intraplate accommodation zones and their control on South Atlantic opening. *J. Geol. Soc. London*, **160**, 703–717.

Lavecchia, G. (1988) The Tyrrhenian-Apennines system: structural setting and seismotectogenesis. *Tectonophysics*, **147**, 263–296.

Liotta, D. (1991) The Arbia-Val Marecchia line, Northern Apennines. *Eclogae Geol. Helv.*, **84**, 413–430.

Liotta, D. (1996) Analisi del settore centro-meridionale del Bacino pliocenico di Radicofani (Toscana meridionale). *Boll. Soc. Geol. Ital.*, **115**, 115–143.

Mariani, M. and Prato, R. (1988) I bacini neogenici costieri del margine tirrenico: approccio sismico-stratigrafico. *Mem. Soc. Geol. It.*, **41**, 519–531.

Martini, I.P. and Sagri, M. (1993) Tectono-sedimentary characteristics of late Miocene–Quaternary extensional basins of the Northern Apennines, Italy. *Earth Sci. Rev.*, **34**, 197–233.

Martini, I.P., Pascucci, V. and Sandrelli, F. (1995) Late Miocene paleogeography of the Monte Soldano area, southeastern part of Volterra basin, Tuscany, Italy. *Riv. Ital. Paleontol. Strat.*, **101**, 381–388.

Martini, I.P, Sagri, M. and Colella, A. (2001) Neogene-Quaternary basins of the inner Apennines and Calabrian Arc. In: *Anatomy of an Orogen – The Apennines and Adjacent Mediterranean Basins* (Eds G.B. Vai and I.P. Martini), pp. 375–400. Kluwer Academic Publisher, Dordrecht.

McClay, K.R. (1992) Glossary of thrust tectonic terms. In: *Thrust Tectonics* (Ed. K.R. McClay), pp. 419–433. Chapman & Hall, London.

Merla, G. (1952) Geologia dell'Appennino settentrionale. *Bol. Soc. Geol. It.*, **70**, 95–382.

Migliorini, C.I. (1949) I cunei composti nell'orogenesi. *Bol. Soc. Geol. It.*, **67**, 29–142.

Milani, E.J. and Davison, I. (1988) Basement control and transfer tectonics in the Recôcavo–Tucano–Jatobá rift, Northeast Brazil. *Tectonophysics*, **154**, 41–70.

Nicolich, R. (1987) Crustal structures from seismic studie in the frame f the European geotraverse (southern segment) and CROP Projects. In: *The Lithosphere in Italy* (Eds A. Boriani, M. Bonafede, G.B. Piccardo and G.B. Vai). Advances in Earth Science Research. *Atti Conveg. Lincei*, **80**, 41–60

Pascucci, V. (2006) Neogene evolution of the Viareggio Basin, Northern Tuscany (Italy). *GeoActa*, **1**, 12–24, (2005).

Pascucci, V., Merlini, S. and Martini, I.P. (1999) Seismic Stratigraphy of the Miocene-Pleistocene Sedimentary Basins of the Northern Tyrrhenian Sea and Western Tuscany (Italy). *Basin Res.*, **11**, 337–356.

Pascucci, V., Costantini, A., Martini, I.P. and Dringoli, R. (2006a) Tectono-sedimentary analysis of a complex, extensional, Neogene basin formed on thrust-faulted, Northern Apennines hinterland: Radicofani Basin, Italy. *Sediment. Geol.*, **1–2**, 71–97.

Pascucci, V., Gibling, M.R. and Sandrelli, F. (2006b) Valley formation and filling in reponse to Neogene magmatic doming of Elba island, Tuscany, Italy. In: *Incised Valleys: Dynamic and Processes* (Eds R. Dalrymple, R. Tillman and D. Leckie), pp. 327–343. Special Publication 85, Society of Economic Paleontologists and Mineralogists, Tulsa, OK.

Patacca, E., Sartori, R. and Scandone, P. (1990) Tyrrhenian basin and Apennines arcs: kinematic relations since late Tortonian times. *Mem. Soc. Geol. It.*, **45**, 435–451.

Rosendahl, B.R., Reynolds, D.J., Lober, P.M., *et al.* (1986) Structural Expression of rifting: lessons form Lake Tanganyika, Africa. In: *Sedimentation in the African Rifts* (Eds L.E. Frostick, R.W Renaut, I. Reid and J.J. Tiercelin), pp. 29–43. Special Publication 25, Geological Society, London.

Roveri, M., Manzi, V., Ricci Lucchi, F. and Rogledi, S. (2003) Sedimentary and tectonic evolution of the Vena del Gesso basin (Northern Apennines, Italy): Implications for the onset of the Messinian salinity crisis. *Geol. Soc. Am. Bull.*, **115**, 387–405.

Royden, L.H. (1988) Late Cenozoic tectonics of the Pannonian Basin system. In: *The Pannonnian Basin – a Study in Basin Evolution* (Eds L.H. Royden and F. Horwáth), pp. 27–48. Memoir 45, American Association of Petroleum Geologists, Tulsa, OK.

Royden, L.H. and Horváth, F. (Eds) (1988) *The Pannonnian Basin – a Study in Basin Evolution*. Memoir 45, American Association of Petroleum Geologists, Tulsa, OK.

Royden, L.H., Horváth, F. and Burchfiel, B.C. (1982) Transform faulting, extension and subduction in the Carpathian Pannonian region. *Bull. Geol. Soc. Am.*, **93**, 717–725.

Royden, L.H., Patacca, E., and Scandone, P. (1987) Segmentation and configuration of subducted lithosphere in Italy: and important control on thrust belt and foredeep basin evolution. *Geology*, **15**, 714–717.

Sacco, F. (1935) Le direttrici tettoniche trasversali nell'Appennino settentrionale. *Atti R. Acc. Lincei*, **2**, 371–375.

Sagri, M., Martini, I.P. and Pascucci, V. (2004) Sedimentology and tectonic evolution of selected Neogene-Quaternary basins of the Apennines (Italy). Field Trip Guide Book P15. In: *Field Trip Guide Books, 32nd International Geological Conference, Florence 20–28 Agosto 2004* (Eds L. Guerrieri, I. Rischia and L. Serva). *Memorie Descrittive della Carta Geologica d'Italia*, from P14 to P36, **63**(4), 1–46. APAT, Roma.

Schlische, W. (1995) Geometry and origin of fault-related folds in extensional settings. *Am. Assoc. Petrol. Geol. Bull.*, **79**, 1661–1678.

Serri, G., Innocenti, F. and Manetti, P. (2001) Magmatism from Mesozoic to Present: petrogenesis, time-space distribution and geodynamic implications. In: *Anatomy of an Orogen – The Apennines and Adjacent Mediterranean Basins* (Eds G.B. Vai and I.P. Martini), pp. 77–104. Kluwer Academic Publisher, Dordrecht.

Signorini, R. (1935) Linee tettoniche trasversali nell'Appennino settentrionale. *Rendiconti R. Accad. Naz. Lincei*, **21**, 42–45.

Sorgi, C., Deffontaines, B., Hippolyte, J.C. and Cadet, J.P. (1998) An integrated analysis of transverse structures in the northern Apennines, Italy. *Geomorphology*, **25**, 193–206.

Vai, G.B. (2001) Structure and statigraphy: an overview. In: *Anatomy of an Orogen – The Apennines and Adjacent Mediterranean Basins* (Eds G.B. Vai and I.P. Martini), pp. 15–31. Kluwer Academic Publisher, Dordrect.

Watson, J. (1984) The ending of the Caledonian orogeny in Scotland. *J. Geol. Soc. London*, **141**, 193–214.

Wezel, F.C. (Ed.) (1986) *The Origin of Arcs*. Elsevier, Amsterdam, 567 pp.

Wezel, F.C. (Ed.) (1988) The origin and evolution of arcs. *Tectonophysics* (Spec. Issue), **146**, 384 pp.

Facies architecture and cyclicity of an Upper Carboniferous carbonate ramp developed in a Variscan piggy-back basin (Cantabrian Mountains, northwest Spain)

OSCAR MERINO-TOMÉ*, JUAN R. BAHAMONDE†, LUIS P. FERNÁNDEZ† *and* JUAN R. COLMENERO*

**Department of Geology, Salamanca University, 37008 Salamanca, Spain (Email: bolanos@usal.es)*
†Department of Geology, Oviedo University, 33005 Oviedo, Spain

ABSTRACT

The latest Moscovian (late Myachkovsky) to Gzhelian succession in the northern sector of the Picos de Europa Province (Cantabrian Mountains, northwest Spain) was deposited in a rapidly subsiding *piggy-back* basin. The succession has been subdivided into 11 mappable depositional sequences (3rd–4th order), which have in turn been grouped into two *sequence sets*, and they are formed of several higher order cycles (4th–5th order). This work focuses on the study of a carbonate ramp system (Puentellés Formation, northern proximal deposits of Sequences 8–10), where the cyclicity is best developed. Each of the three depositional sequences forming the Puentellés Formation comprises a fining upward lower part, which consists of several higher order (metric to decimetric) cycles with similar internal organization, and an upper part mainly arranged in metric to decametric shallowing upward parasequences. The results of this study suggest that tectonic activity controlled the configuration and long-term development of the latest Moscovian (late Myachkovsky) to Gzhelian basins, being responsible for the large-scale stratigraphical architecture (sequence sets). Eustasy, on the other hand, was the driving mechanism for the higher frequency architecture (3rd–5th-order cyclicity).

The subsidence curves are similar for both sequence sets, showing a maximum subsidence rate in the beginning that gradually decreased, reflecting the tectonic load during two major phases of thrust-sheet emplacement. The unconformity between both sequence sets records an intense phase of erosion linked to uplift of the northern sector of the Picos de Europa Province due to the onset of the emplacement of the Picos de Europa thrust sheets. The tectonic deformation was also responsible for the angular unconformities and the syntectonic unconformities that bound the depositional sequences, but the unconformities in themselves were a consequence of eustatic sea-level falls. The lower part of each sequence in the Puentellés Formation records late lowstand clastic sedimentation in flood-dominated deltas and fan-deltas, which co-existed with a reduced carbonate production in a narrow shallow-water ramp. The minor unconformities that bound the 4th–5th-order cycles are only present in the lower part of the 3rd-order sequences, and are interpreted to have developed when minor tectonic uplift was enhanced during lowstand stages. The upper part, mainly composed of autochthonous carbonates (including microbial and algal boundstones, among others), represents the abandonment of the previous clastic systems and the encroachment and aggradation of the carbonate ramps during rising and highstand stages.

The development of the Puentellés carbonate ramp was exceptional in that, normally, terrigenous influx typical of active tectonic regimes adversely affects carbonate production, preventing the development of carbonate depositional systems. In this case, the clastic supply was mainly calcareous, due to the composition of the source area, and as a consequence, almost no terrigenous clay was shed into the basin to inhibit carbonate production.

Keywords Carbonate ramps, piggy-back basins, Carboniferous, Cantabrian Mountains.

INTRODUCTION

Sedimentation in tectonically active settings is largely controlled by the structural grain and the timing and rate of emplacement of structures, although a eustatic imprint is commonly also recognized (Crumeyrolle *et al.*, 1991; Luterbacher *et al.*, 1991; Déramond *et al.*, 1993; Berástegui *et al.*, 1998; Nijman, 1998; Zweigel *et al.*, 1998). In the Variscan Cantabrian Zone (northwest Spain), the upper Moscovian–Gzhelian synorogenic succession that crops out in the northern sector of the Picos de Europa Province comprises 11 depositional sequences (numbered 1–11), each of them constituted by higher order, metric to decametric, cycles. The integration of biostratigraphical data, a large number of measured sections and detailed field maps has enabled the documentation of the interplay between eustasy and tectonics on the development of the depositional sequences. This paper is focused on the calcareous Puentellés Formation (proximal deposits of Sequences 8–10) because shallow-water carbonates represent more sensitive sea-level indicators in the geological record (Kendall & Schlager, 1981).

Carbonate ramps in marine foreland basins commonly form linear belts located on the distal side of the foreland margin (peripheral 'bulge') and typically display a ramp-like profile (Burchette & Wright, 1992; Dorobek, 1995). In the shallow-water proximal areas of the late Kasimovian (Dorogomilovsky) to Gzhelian piggy-back basin of the northern sector of the Picos de Europa Province, high rates of carbonate production characterized transgressive and highstand situations. However, carbonates were also formed during late lowstand stages, co-existing with high rates of clastic input.

The aims of this work are twofold. First, to describe the lithofacies and facies associations forming this carbonate ramp-like system. Second, to study the different orders of cyclicity and construct a model for their origin, as a response to the interplay between tectonics and glacioeustasy.

REGIONAL GEOLOGICAL SETTING

The Cantabrian Zone (the most external part in the north of the Iberian Massif, Fig. 1A) displays a thick Palaeozoic succession deformed by thin-skinned tectonics during the Variscan Orogeny. At least in the Bashkirian–Moscovian period, the Cantabrian Zone constituted a wide (a few 100 km) marine foreland basin mostly filled by thick clastic wedges, which crop out in the Fold and Nappe, Central Asturian Coalfield, central and southern sectors of Ponga Nappe, and Pisuerga–Carrión provinces (Fig. 1A; see Colmenero *et al.*, 2002). Nevertheless, in some weakly subsiding distal parts of this basin, an extensive carbonate platform was developed (the Picos de Europa Province and the NW sector of Ponga Nappe Province; e.g. Bahamonde *et al.*, 1997, 2000; Kenter *et al.*, 2003). This carbonate system (Valdeteja and Picos de Europa formations) nucleated on a laterally extensive carbonate unit (Serpukhovian Barcaliente Formation) composed of thinly bedded and laminated, dark lime mudstones. The Valdeteja and Picos de Europa platform consists of a wide range of deposits from shallow-shelf skeletal limestones to upper slope micritic boundstones and lower slope breccias and calciturbidites.

Later during the Variscan Orogeny, the evolution of the Picos de Europa Province and northwest sector of the Ponga Nappe can be described in terms of two stages of development. During the first stage (late Myachkovsky–Khamovnichesky), the advance of the orogenic front affected the northern areas of the carbonate platform (Ponga–Cuera unit, see Fig. 1B for location), which emplaced southwards as a set of E–W-trending, imbricate thrust sheets (Marquínez, 1989). This led to the generation of a highly subsiding basin in the northern part of Picos de Europa Province, on the frontal part of the thrust wedge (wedge-top depozone of DeCelles & Giles, 1996) that was filled by fan deltas, deep-water depositional systems and rare carbonate ramps. Coevally, carbonate sedimentation, with several drowning episodes, continued in the south and central parts of the Picos de Europa Province. Here subsidence rates were much lower, except at the end of this first stage (late Krevyakinsky–Khamovnichesky), when this southern area began to be affected by thrusting. During the second stage (Dorogomilovsky to Gzhelian), the southward advance of the orogenic front fully affected the whole Picos de Europa Province, which gave rise to the individualization of several small marine piggy-back basins that

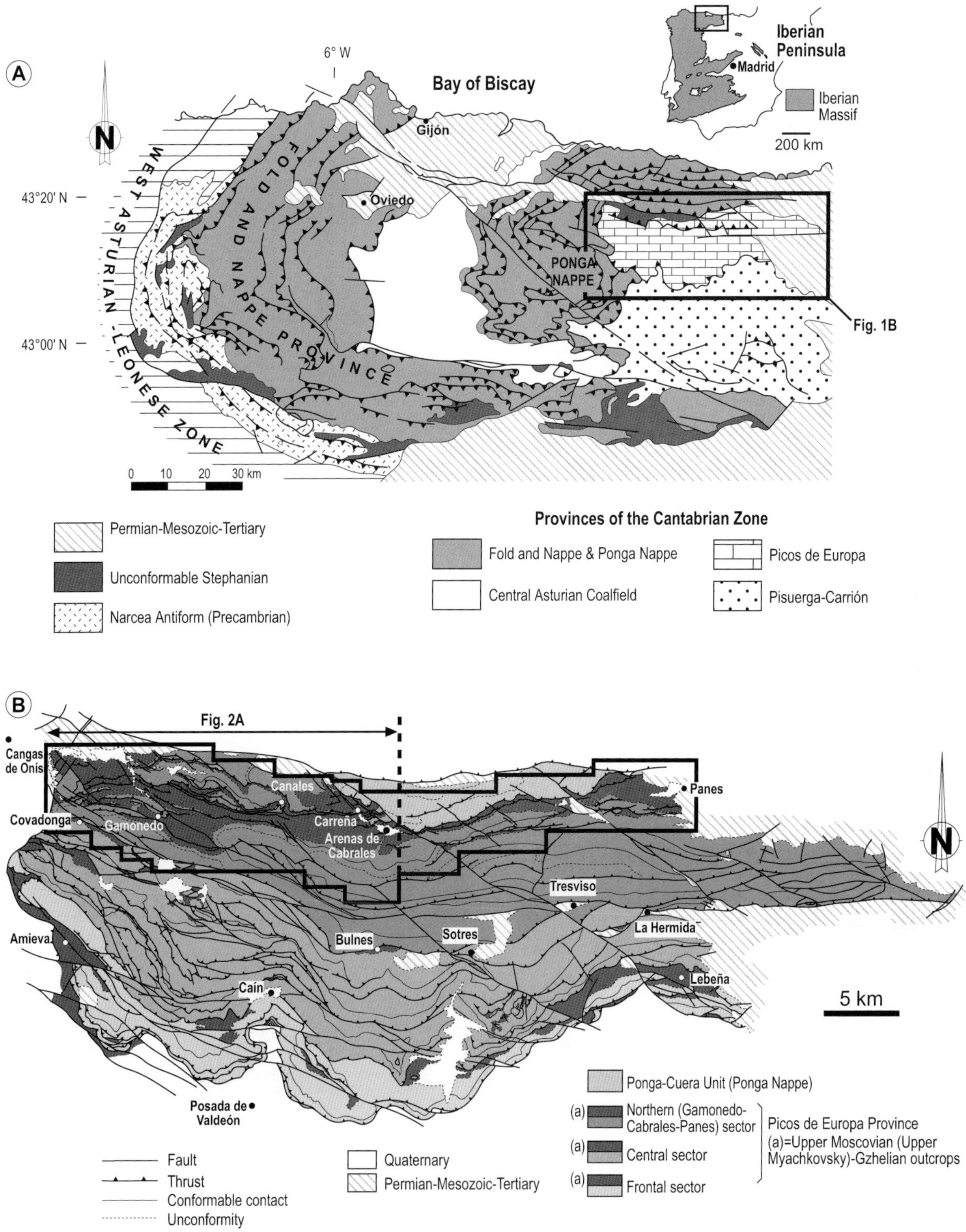

Fig. 1 (A) Geological sketch map of the Cantabrian Zone showing the tectonostratigraphic provinces (based on Julivert, 1971; Pérez-Estaún *et al.*, 1988) with the location of the Picos de Europa Province marked by a box. (B) Simplified geological map of the Picos de Europa Province showing the three sectors differentiated by Merino-Tomé (2004) and the distribution of the upper Moscovian (upper Myachkovian) to Gzhelian outcrops. Boxed area indicates the area of study, showing also the location of Fig. 2A.

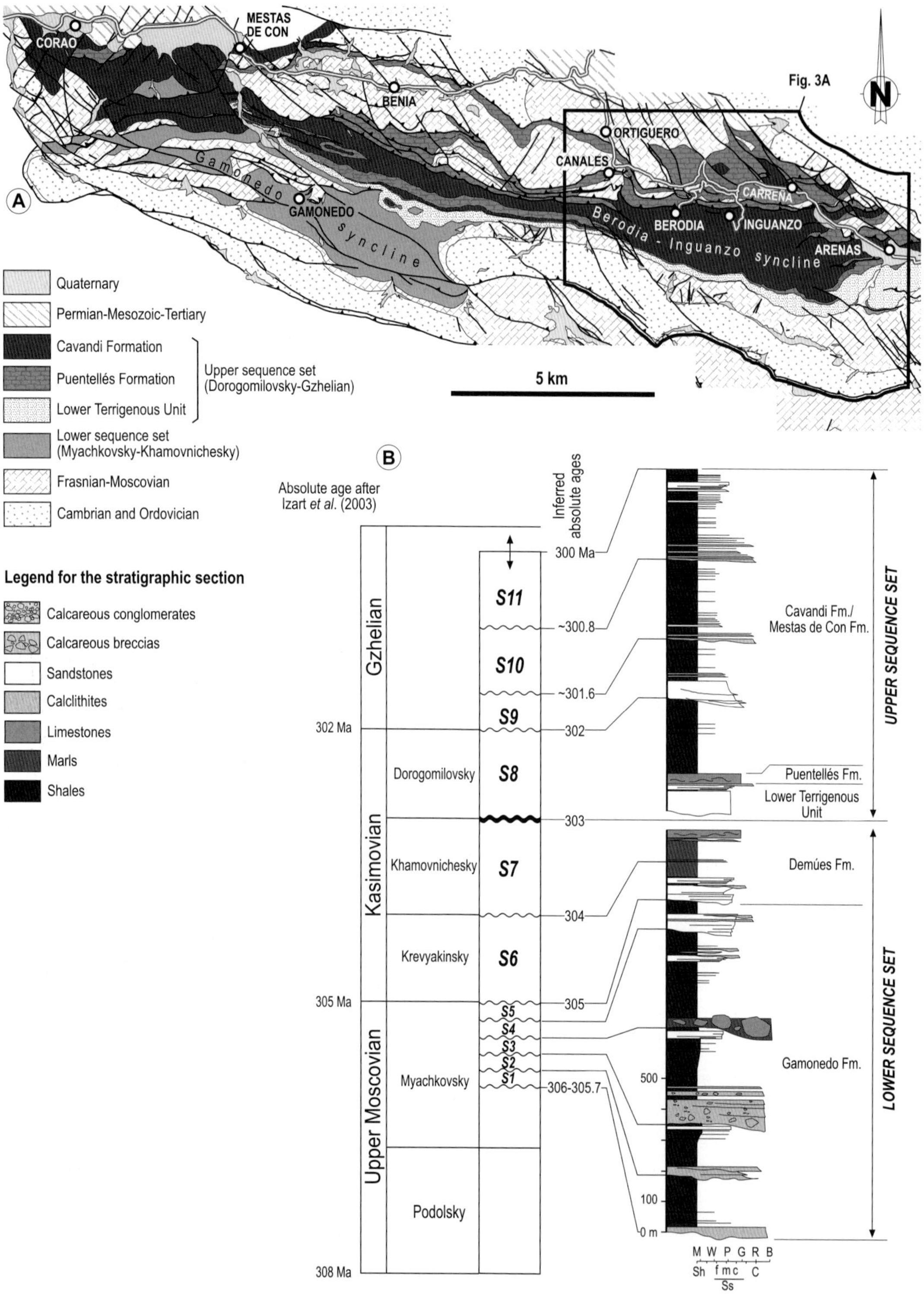

Fig. 2 (A) Simplified geological map of the central western part of the northern sector of the Picos de Europa Province showing the outcrops of the Puentellés Formation (see Fig. 1B for location) and the location of Fig. 3A (boxed area). (B) Synthetic stratigraphy of the synorogenic deposits in the northern sector of the Picos de Europa Province showing the sequences (S1–S11) and sequence sets established by Merino-Tomé (2004) and their correspondence with the lithostratigraphic formations of Martínez García & Villa (1998).

were fed by deltaic systems from the north and west. This tectono-sedimentary evolution resulted in a different stratigraphy for the uppermost Moscovian (upper Myachkovsky) to Gzhelian successions across the Picos de Europa Province, which, consequently, has been subdivided into the northern, central and southern sectors (Merino-Tomé, 2004; see also Marquínez, 1989; Fig. 1B).

The northern sector of the Picos de Europa Province is the area studied in this paper. It represents the transition zone between the Ponga Nappe Province (Sierra del Cuera area) and the Picos de Europa Province. In this sector, the synorogenic succession records the development of two successive piggy-back basins, related to the emplacement of the Ponga Nappe and the Picos de Europa thrust sheets (Merino-Tomé, 2004). Several features indicate the piggy-back character of these basins. The most important are:

1 these successions overlie a previously deformed substratum;
2 their basal deposits, being of shallow-water and coastal origin, overlie deep-water basinal and slope deposits related to the Bashkiran–Moscovian carbonate shelf;
3 the abundance of syntectonic unconformities, which indicates that deposition was coeval with tectonic deformation (Merino-Tomé, 2004).

The succession crops out in two E–W-trending synclines forming a band between the Gamonedo and Panes localities in the so-called Gamonedo–Cabrales Basin and its eastern extension (Fig. 2A). Classically, it was subdivided into five formations: the upper Myachkovsky–Gamonedo Formation, the Krevyakinsky–Khamovnichesky Demúes Formation, the Dorogomilovsky–Lower Gzhelian Puentellés Formation and the Gzhelian Cavandi and Mestas de Con formations (Martínez-García, 1981; Martínez-García & Villa, 1998, Fig. 2B).

METHODS

The stratigraphical architecture of the synorogenic succession studied is complex and characterized by many unconformities, which have allowed the definition of 11 mappable depositional sequences (S1–S11), grouped into two sequence sets (Fig. 2B) that are equivalent to the sequence sets of Déramond *et al.* (1993). This sequence stratigraphic framework has been tied to the stablished lithostratigraphic units. In this sense, the proximal parts of three sequences (Sequences 8–10) form the Puentellés Formation and are termed Puentellés I, II and III (Figs 3 & 4). The new stratigraphy is based on detailed geological mapping at 1:25,000 scale (Fig. 3), the logging of numerous stratigraphical sections and biostratigraphical studies. The biostratigraphical data involve foraminifera, mostly a fusulinoidean fauna, and are derived from the work of previous authors (van Ginkel, 1971; Sánchez de Posada *et al.*, 1993, 1996, 1999; Villa, 1995; Villa & van Ginkel, 1999) and sampling during this study.

Subsidence analysis was performed for two synthetic sections, which are located in the Gamonedo and Berodia–Inguanzo synclines and are representative of the lower and upper sequence sets, respectively (Fig. 5). The thickness of decompacted units was calculated by the *backstripping* computer program of Allen & Allen (1990). Porosity reduction with depth was estimated using the methodology of Sclater & Christie (1980), Schmoker & Halley (1982) and Vergés *et al.* (1998). The density values of the different types of sediment (Fig. 5) were taken from Allen & Allen (1990), and the porosity from Magara (1980), Bond & Kominz (1984), Stam *et al.* (1987) and Vergés *et al.* (1998).

UPPER MOSCOVIAN–GZHELIAN STRATIGRAPHY OF THE NORTHERN SECTOR OF PICOS DE EUROPA PROVINCE

The 11 depositional sequences recognized here range between 100 and 380 m in thickness and had estimated durations of ~ 0.14–0.2 Myr in the case of the 4th-order sequences and of 0.4–1 Myr in the case of the 3rd-order sequences (Fig. 2B). Each sequence set is more than 1000 m in thickness and had a duration of ~ 3 Myr (Fig. 2B). The lower sequence set comprises Sequences 1–7 (upper Myachkovsky to upper Khamovnichesky) and mainly crops out in the southwestern sector. The upper sequence set comprises Sequences 8–11 (Dorogomilovsky to Gzhelian) and has greater lateral extent than the lower one, overlying the lower sequence set in the western outcrops (Fig. 2A).

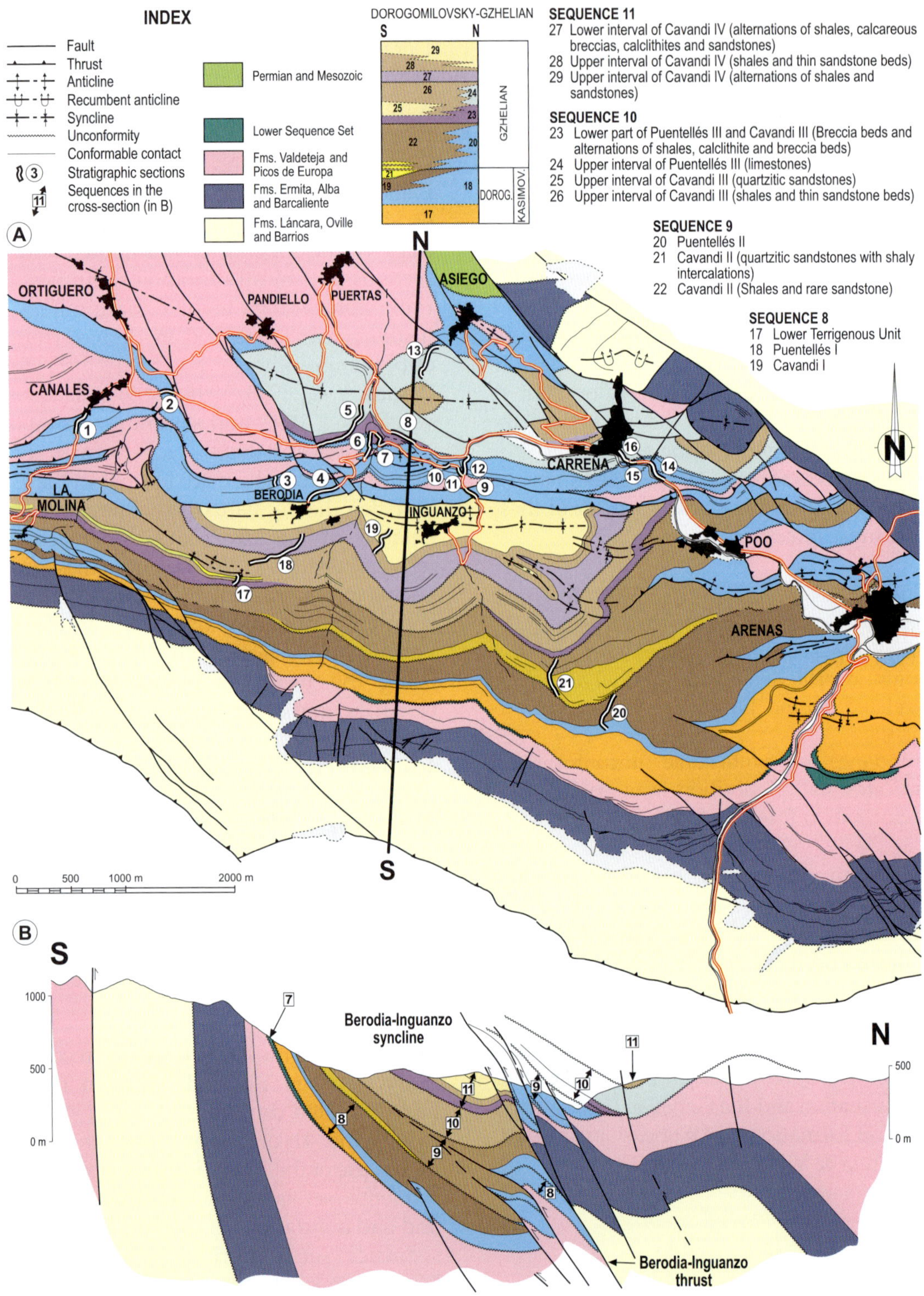

Fig. 3 (A) Detailed geological map of the central part of the northern sector of the Picos de Europa Province (see Fig. 2A for location) showing the location of the stratigraphic sections of the upper sequence set discussed in this paper. The location of the cross-section shown in (B) is also included. (B) North–south cross-section showing the northwards replacement of the thick clastic deposits of the Cavandi Formation, present in the Berodia–Inguanzo syncline, by the Puentellés Formation. Arrows mark the vertical extent of the upper sequence-set sequences.

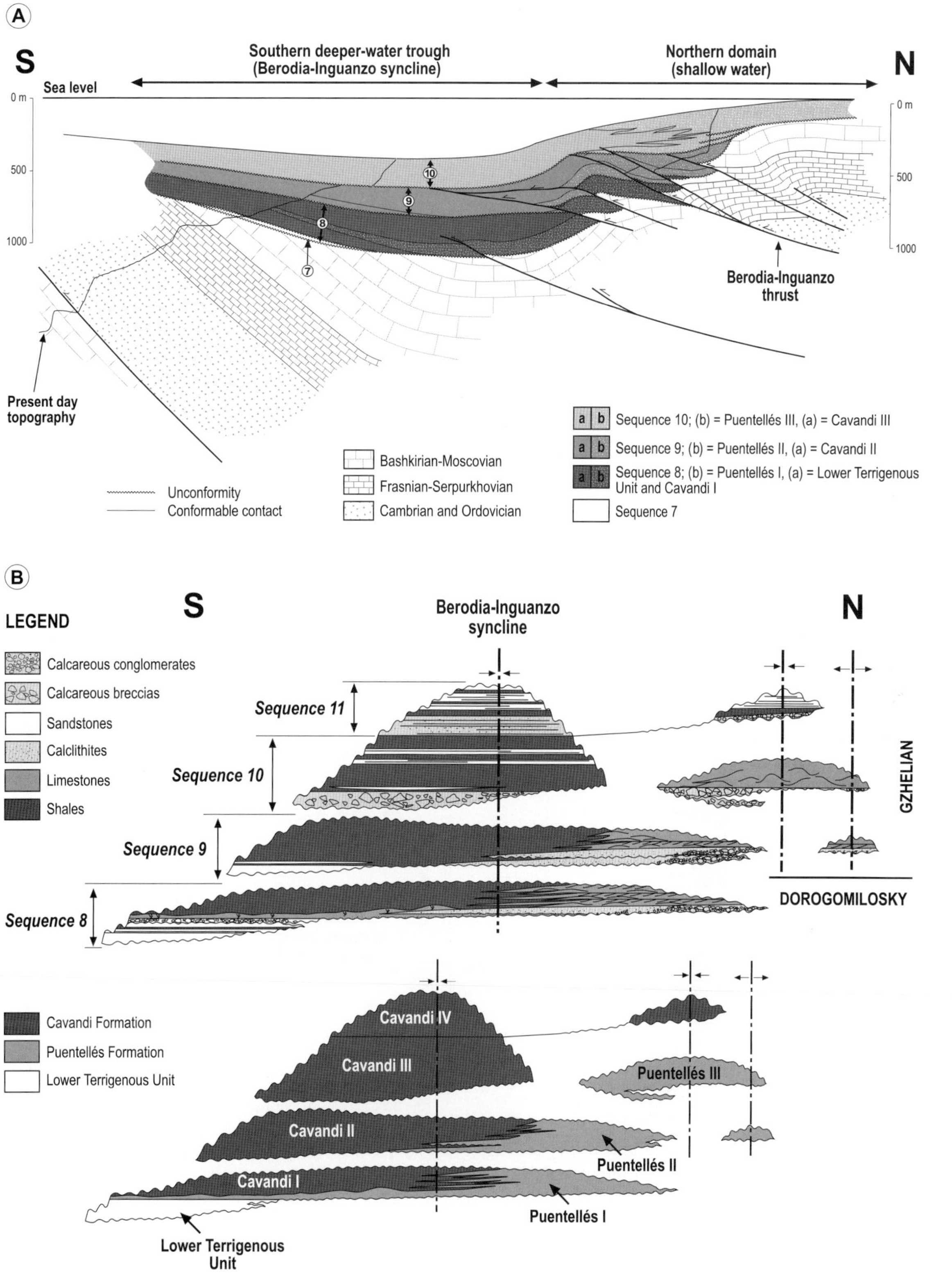

Fig. 4 (A) Aproximate palinspastic restoration of the N–S cross-section in Fig. 3 during the deposition of Sequence 10. The thickness of the mostly shaly successions of Sequences 8, 9 and 10 in the Berodia–Inguanzo syncline corresponds to their present-day thickness, and compaction effect has not been taken into account. (B) Chronostratigraphic diagram showing the sequence architecture of the upper sequence set along the N–S cross-section of Fig. 3B (cross-section has been palinspastically restored). The proximal calcareous deposits, mainly present to the north of the Berodia–Inguanzo syncline, represent the Puentellés Formation. The thick distal clastic succession of shales, sandstones, calclithites and calcareous breccias in the Berodia–Inguanzo syncline to the south forms the Cavandi Formation.

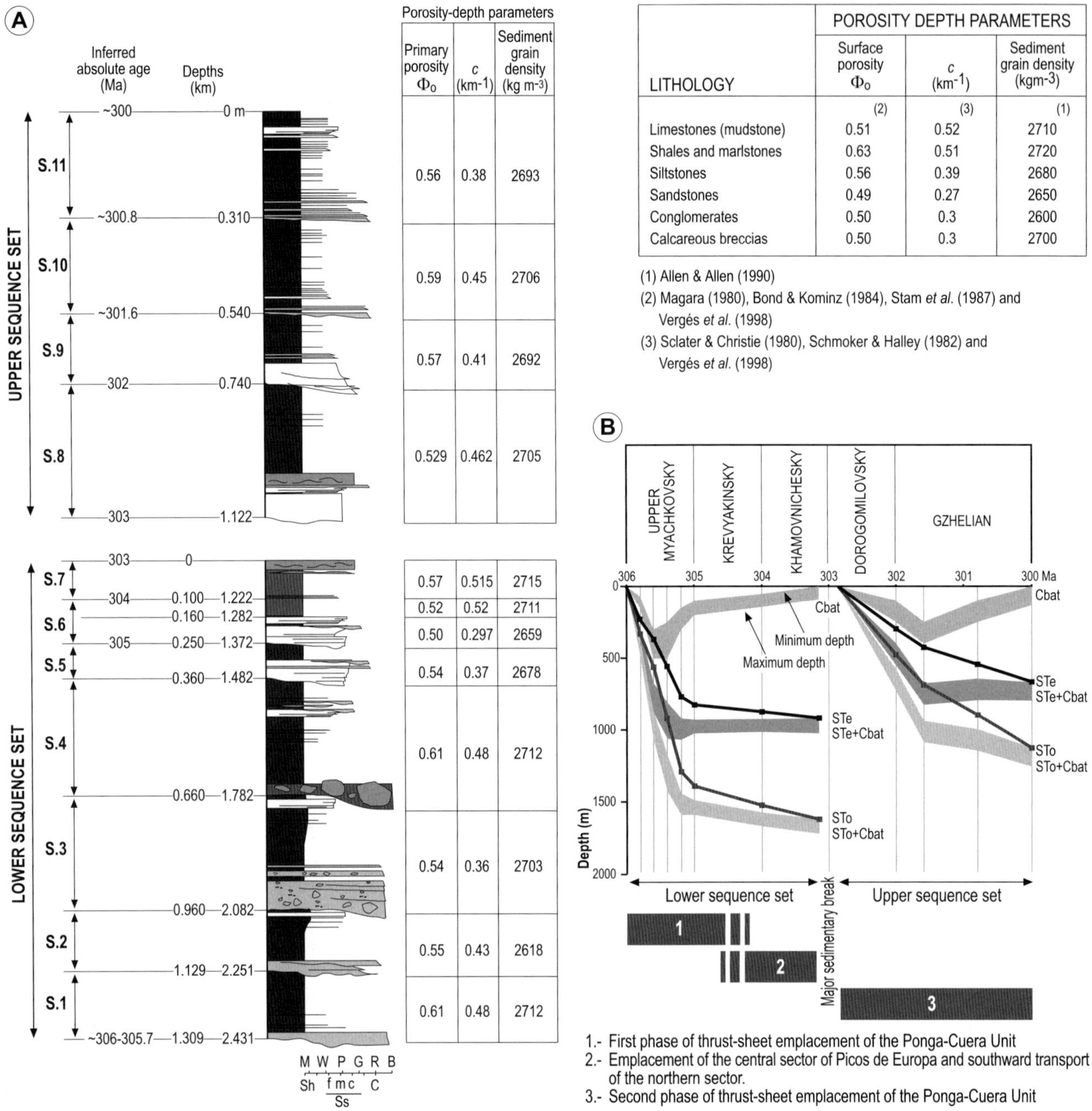

LITHOLOGY	POROSITY DEPTH PARAMETERS Surface porosity Φo	c (km-1)	Sediment grain density (kgm-3)
	(2)	(3)	(1)
Limestones (mudstone)	0.51	0.52	2710
Shales and marlstones	0.63	0.51	2720
Siltstones	0.56	0.39	2680
Sandstones	0.49	0.27	2650
Conglomerates	0.50	0.3	2600
Calcareous breccias	0.50	0.3	2700

(1) Allen & Allen (1990)
(2) Magara (1980), Bond & Kominz (1984), Stam *et al.* (1987) and Vergés *et al.* (1998)
(3) Sclater & Christie (1980), Schmoker & Halley (1982) and Vergés *et al.* (1998)

Fig. 5 Subsidence curves for the lower and upper sequence sets. (A) Representative sections of the lower and upper sequence sets (Gamonedo and Berodia–Inguanzo sections, respectively) used for the analysis. The inferred absolute ages and the porosity depth parameters, which were estimated on the basis of the values for single lithologies (see text for details), are also shown. (B) Palaeobathymetric (Cbat), tectonic subsidence (STe) and total subsidence (STo) curves. STe + Cbat and STo + Cbat are the result of combining palaeobathymetric and tectonic or total subsidence curves, respectively.

These two sequence sets record the two stages of geodynamic evolution in the northern sector of the Picos de Europa Province described above.

At least during the sedimentation of the upper sequence set, the basin consisted of a northern and relatively shallow-water proximal domain, where fault-propagation folds developed in relation with lateral and frontal ramps of blind thrusts, passing to the south into a distal and deeper-water trough (Fig. 4A). In the northern proximal domain, the depositional sequences are usually bound by unconformities, commonly with a syntectonic character, involving subaerial exposure and karstification. They consist of fan-deltaic deposits overlain by carbonate-ramp sediments (Puentellés Formation; see Merino-Tomé *et al.*, 2001; Villa & Bahamonde, 2001). The lithofacies distribution of the Puentellés Formation follows a ramp-like pattern (possibly distally steepened), deepening to the south and passing into a deep-water setting with terrigenous turbidite sedimentation (Cavandi Formation). In this southern distal domain, the Cavandi Formation almost lacks unconformities (Fig. 4B). Although the Cavandi turbidites were also fed from the north, the main terrigenous supply came from the west and northwest and the palaeocurrent patterns show an eastwards sediment dispersal, lateral to the carbonate ramp and following the structural trend of the elongate basin.

The Puentellés Formation

This unit comprises the proximal deposits of Sequences 8–10 (Puentellés I–III respectively; see Fig. 4). Puentellés I unconformably overlies the Moscovian Picos de Europa Formation in the northeastern outcrops, whereas towards the south it conformably overlies terrigenous deposits that form the base of sequence 8, and will not be dealt with here. Puentellés II and III unconformably overlie the previous sequence (Puentellés I and II, respectively), although in their northernmost outcrops each of them overlies the Picos de Europa Formation. Each sequence can be described in terms of two parts, a lower and an upper part, with different lithology and cyclical internal organization (Figs 6 & 7).

The lower part is composed of alternations of clastic deposits (calcareous breccias and conglomerates, graded and laminated pebbly quartz arenites and calclithites) and autochthonous carbonates (marls, skeletal wackestones and algal bafflestones). This lower part displays an overall fining-upward trend and consists of several higher-order cycles (4th–5th order) bound by minor unconformities that display karstic features in the northernmost outcrops (Merino-Tomé *et al.*, 2001). These minor sequences are metric to decametric in thickness and show an internal organization that is similar to the 3rd-order sequences.

The upper part of each sequence is entirely composed of autochthonous carbonates including micritic (microbial) boundstones, phylloid algal and *Anthracoporella* bafflestones, dark pseudonodular mudstones and skeletal wackestones, and skeletal to ooidal grain- to packstones (Merino-Tomé *et al.*, 2001). These deposits are mainly arranged in metric to decametric 4th–5th-order shallowing-upward cycles, bound by marine flooding surfaces, and are comparable to parasequences as defined by Mitchum (1977) and Mitchum *et al.* (1977). These shallowing-upward cycles are similar to those described in carbonate ramps by Elrick & Read (1991), Burchette & Wright (1992) and Proust *et al.* (1998).

Puentellés I is up to 225 m thick in the easternmost outcrops (near the Panes locality; see Fig. 1B). There, the lower part of the sequence is up to 112 m thick and thins towards the north. In the central part, in the vicinity of Carreña (Fig. 3A), it is composed of 2-m-scale, fining-upwards, minor sequences bound by conspicuous syntectonic unconformities (Section 14, Fig. 7). The upper part of the sequence reaches a thickness of 125 m in the outcrops east of Arenas (Invernales de La Nava section, Fig. 7), where five shallowing-upward cycles made of microbial boundstones and dark pseudonodular skeletal mud- to wackestones passing upwards into algal bafflestones and skeletal grain- to packstones on top can be recognized. To the south, Puentellés I thins (Fig. 4) and is mainly composed of micritic limestones and marls.

Puentellés II displays the same cyclical arrangement as Puentellés I. Its lower part is well exposed in the central part of the outcrop belt (near the localities of Canales and Carreña, see Figs 1B & 3A). It reaches a maximum thickness of 60 m near Carreña (Sections 14–16, Fig. 7), and it is composed of three fining-upwards minor sequences (Fig. 8B). The boundstone-dominated upper part reaches a minimum thickness of 70 m near Canales, and consists of several shallowing-upward cycles (Sections 1 and 2, Fig. 7).

Puentellés III only crops out in the northernmost thrust sheets (in the surroundings of Ortiguero, Asiego, Berodia and Carreña localities, Fig. 3A). The lower part of the sequence is up to 60 m thick (Sections 5–8, Fig. 9), wedging out rapidly to the north, and is composed of clast-supported calcareous breccias, with clasts derived from Puentellés I and II, and of minor sandstones, calclithites, marls

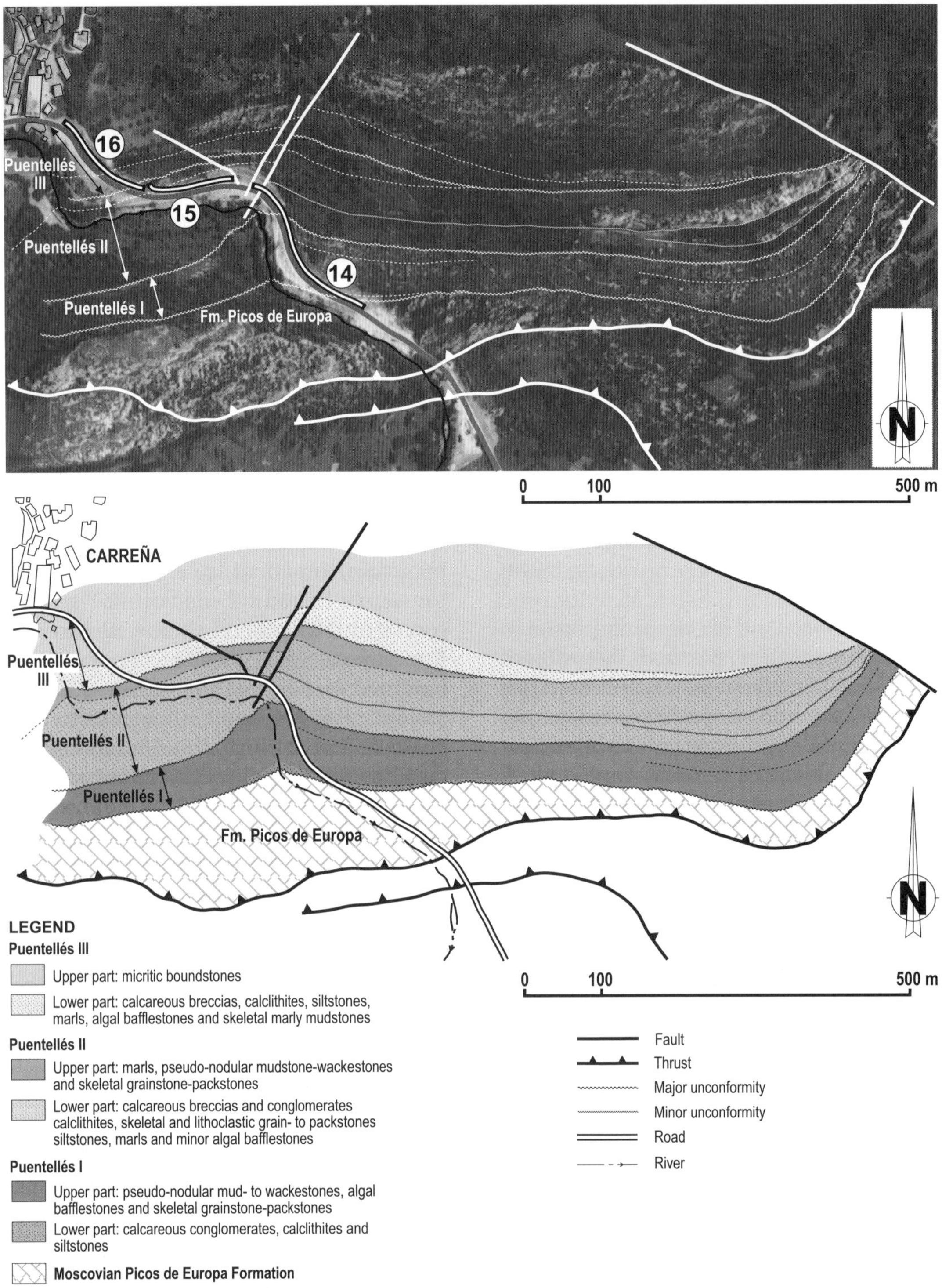

Fig. 6 Vertical aerial photograph and reconstructed geological sketch map of the Puentellés Formation outcrops near Carreña on the Carreña-Poo road (see Fig. 3A for location) showing the major and minor unconformities and the resulting sequence architecture (see text for details). The locations of the stratigraphic logs 14–16 in Fig. 7 are also depicted.

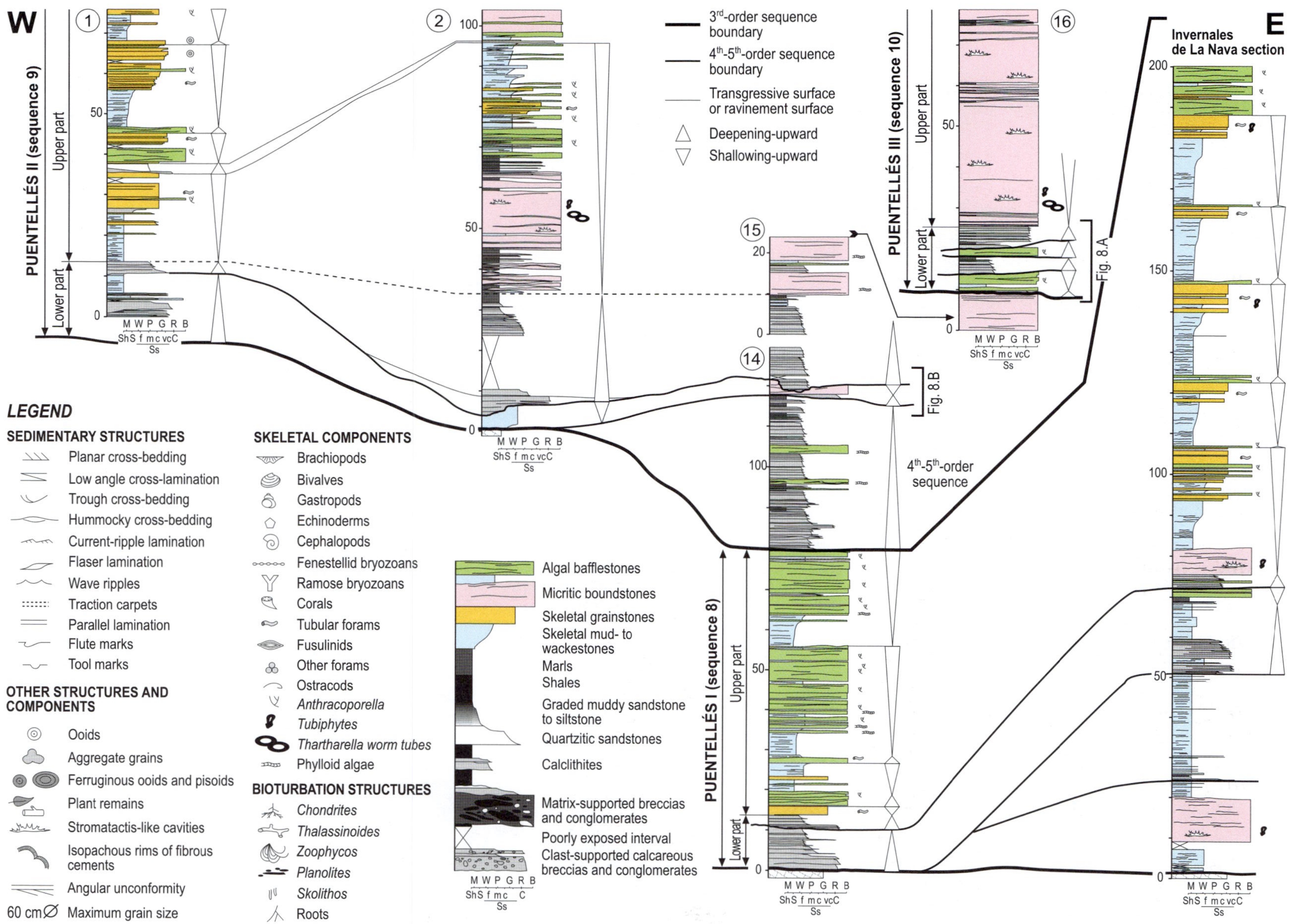

Fig. 7 Stratigraphic logs of the Puentellés Formation showing its internal arrangement into 3rd-order sequences (Puentellés I–III) and 4th–5th-order sequences and parasequences. Notice the vertical arrangement of facies in the three sequences (see text for details) and the location of the photographs of Fig. 8. Sections oriented W–E are numbered as in Fig. 3A except Invernales de La Nava (see Fig. 1B). Legend applies also to Figs 9, 11, 12, 13 & 15.

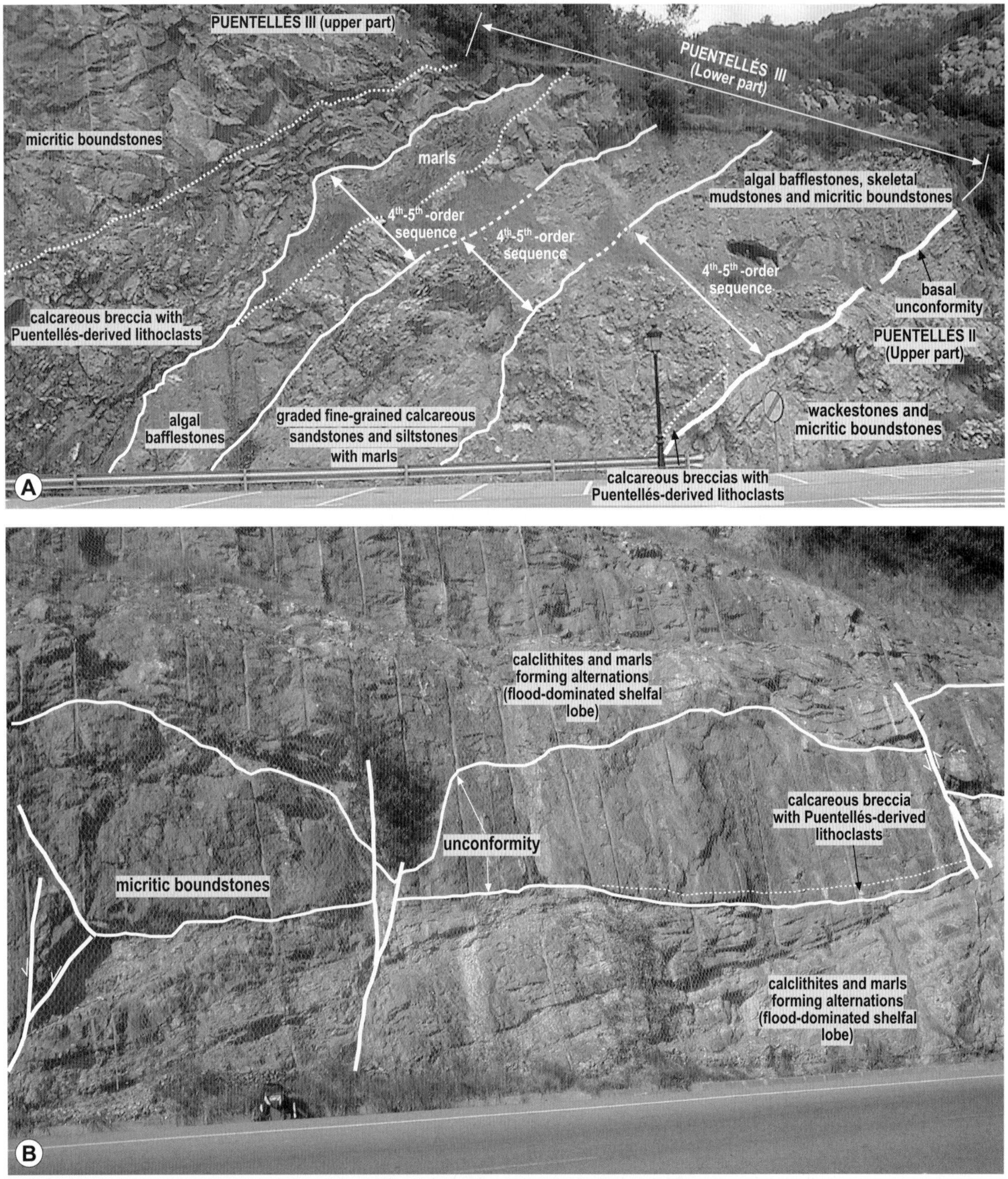

Fig. 8 Field aspect of the Puentellés Formation in the Carreña outcrops (Sections 16 and 14, see Fig. 7 for location). (A) Lower part of Puentellés III showing the basal unconformity and other minor unconformities defining several 4th–5th-order sequences (white arrows). (B) Close-up view of two minor unconformities in the lower part of Puentellés II, which bound a thin, almost completely eroded, 4th–5th-order sequence. In this example, the clastic calcareous deposits that usually form the lower part of these cycles are almost absent, only forming a thin discontinuous interval of calcareous breccias.

and autochthonous carbonates. These lithologies are organized in four fining-upward minor sequences bound by erosional unconformities with karstic features (Figs 8A, 9 & 10). The upper part reaches 150 m in thickness in the Asiego section (Section 13, Fig. 9) and consists of two parts. The lower is dominated by microbial boundstone (~ 80 m thick), and the upper (~ 70 m thick) is made of nodular

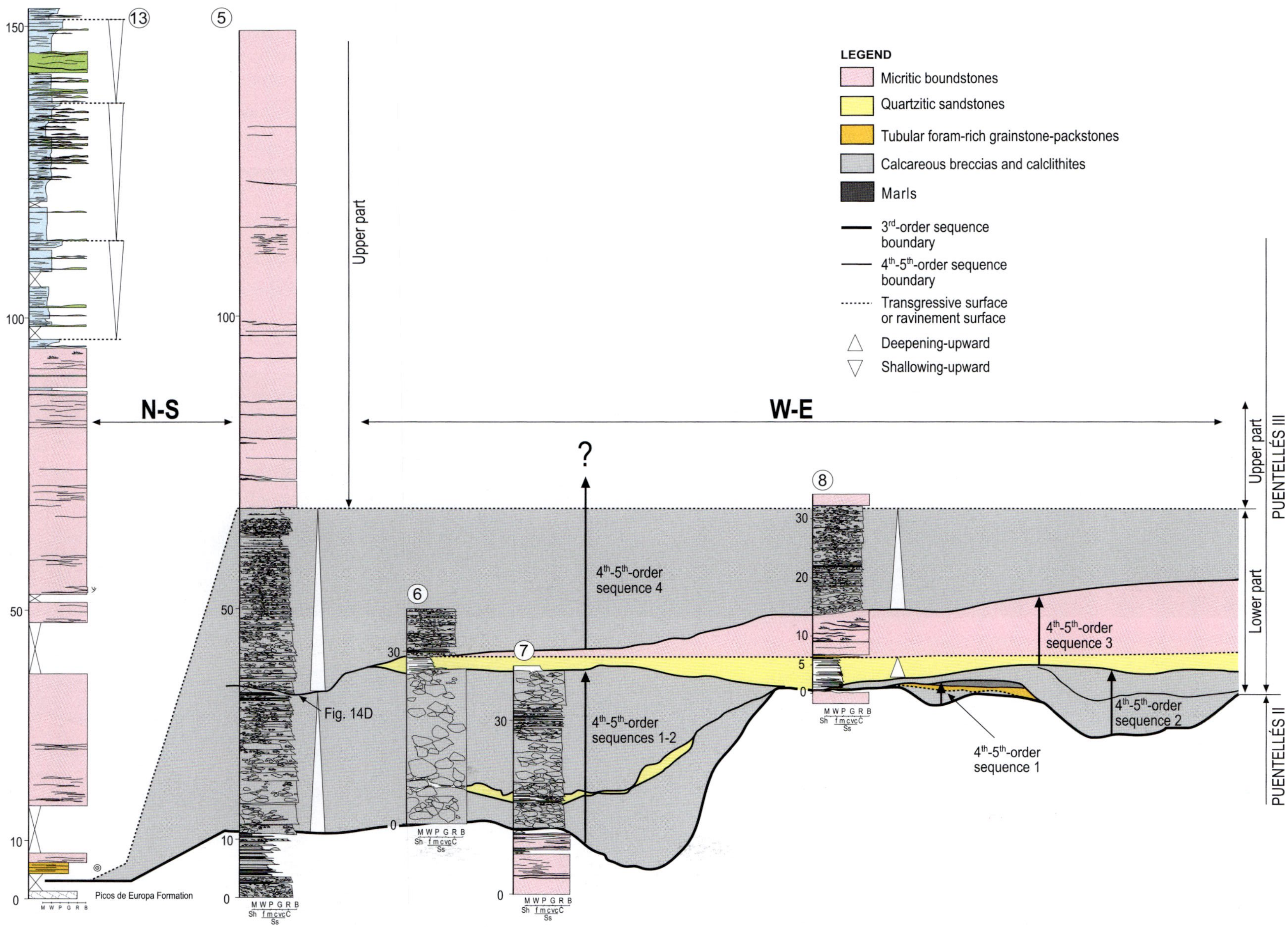

Fig. 9 Stratigraphic cross-section of the lower part of Puentellés III in the Ortiguero–Asiego–Berodia area. Notice the vertical arrangement of facies in the minor sequences (see text for details). The location of the photograph of Fig. 14D is also shown. Sections oriented N–S are numbered as in Figs 3A & 10A. See Fig. 7 for a key to symbols.

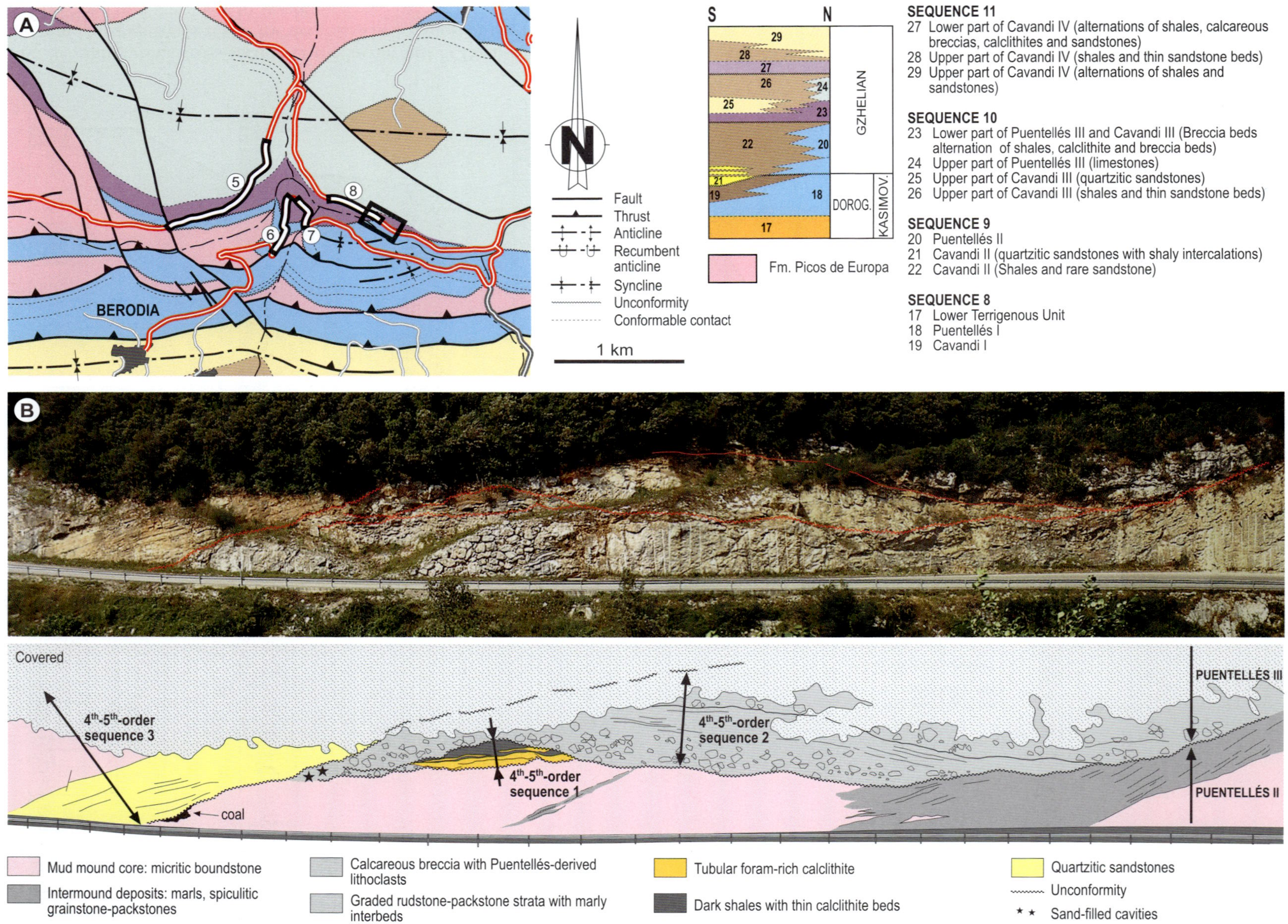

Fig. 10 (A) Detailed geological map of the Ortiguero–Asiego–Berodia area showing the location of the sections 5–8 depicted in Fig. 9 and of the photo-mosaic shown in B (boxed area). (B) Photo-mosaic and interpretative diagram showing the contact between Puentellés II and III. Three angular unconformities can be recognized bounding three 4th–5th-order sequences (1–3).

mud- to wackestones, oncolitic packstones and some algal-rich layers, and displays a poorly defined cyclical organization.

The Cavandi Formation

This unit comprises the distal deposits of Sequences 8–10 (Cavandi I–III, respectively) and Sequence 11 (Cavandi IV) (see Figs 3 & 4). In the Berodia–Inguanzo syncline, Cavandi I is a 150–200 m thick interval of dark shales with thin-bedded sandy turbidites, representing a prodeltaic mudstone wedge (Fig. 11; Type III turbidite systems of Mutti 1985). Cavandi II, III and IV have a similar internal organization displaying an overall fining-upward trend, formed of metre-scale fining-upward cycles (Fig. 11). Two parts can be distinguished in each of these three units. The lower part, up to some 100 m thick, consists of clastic deposits, including one megaturbidite, with Puentellés-derived lithoclasts and quartzitic sandstones. This lower interval is interpreted as being deposited in fan deltas that evolved eastwards into sand-rich, elongate turbidite systems (Types I and II of Mutti, 1985). The upper interval, up to some 150 m thick, is made of dark shales with subordinate sandy and silty turbidites. In the case of Cavandi IV, the upper part displays a coarsening-upward trend and the sandstones at its top show planar and wave-ripple lamination, being arranged in minor fining- and coarsening-upward cycles. The upper interval is considered to represent thick prodeltaic mudstone wedges (Type III turbidite systems of Mutti, 1985), which in the case of Cavandi IV evolves upwards into shelfal deposits.

FACIES ASSOCIATIONS OF THE PUENTELLÉS FORMATION

The deposits of the Puentellés Formation have been grouped into a number of facies, which correspond to clastic sediments, mainly of calcareous composition (Fig. 12), and to autochthonous carbonates (Fig. 13). The clastic deposits are present in the lower part of the sequences where they alternate with autochthonous carbonates, which, in turn, form the upper part of the sequences.

The clastic facies of Fig. 12 have been grouped into six main clastic facies associations, namely: karstic deposits and regoliths, braided-channel fills, estuarine-channel fills, mouth-bar deposits, flood-dominated shelfal lobes and deltaic debris cones. They belong to flood-dominated deltaic and fan-deltaic systems and associated shelfal lobes. These types of sedimentary systems have been reported in tectonically active settings characterized by small- to medium-sized and high-gradient fluvial systems with high-elevation catchment areas located close to a marine basin (Mulder & Syvitsky, 1995; Mutti *et al.*, 1996, 2003).

Karstic deposits and regoliths

This association is found overlying a calcareous substratum and comprises: (i) poorly sorted breccias with calcareous clasts derived from the underlying units, and calcareous and ferruginous cements (Facies 1, Fig. 14A); (ii) reddish ferruginous and argillaceous massive accumulations and crusts including iron-rich glaebules and nodules (Facies 2); and (iii) cavity- and fracture-fills made of calcareous to quartzitic sandstones, pebbles and cobbles (Facies 3; Fig. 14B). These deposits have been interpreted as regoliths and karstic breccias filling cavities (Esteban & Klappa, 1983; Cooper & Keller, 2001), lateritic crusts developed under subtropical climatic conditions (Duchaufour, 1982; Tardy, 1993), and sand-filled dykes (Kerans & Donalson, 1988; Cooper & Keller, 2001), respectively.

Braided-channel fills

This association comprises moderately to poorly sorted calcareous conglomerates organized into amalgamated, massive to normally graded, rarely cross-stratified, lenticular beds with erosive bases (Facies 5). These beds form metre-scale lenticular bodies, which fine upwards and overlie the depositional sequence boundaries.

Estuarine-channel fills

These comprise: (i) lenticular bodies of medium- to fine-grained litharenites to quartz arenites, which display trough and sigmoidal cross-bedding (Facies 9 and 11 respectively); (ii) lenticular and wedge-shaped beds and bedsets of fine-grained litharenites with planar cross-lamination and common mud-drapes (Facies 10); and (iii) mudstones and fine-grained sandstones showing lenticular to flaser bedding and current- to wave-ripple

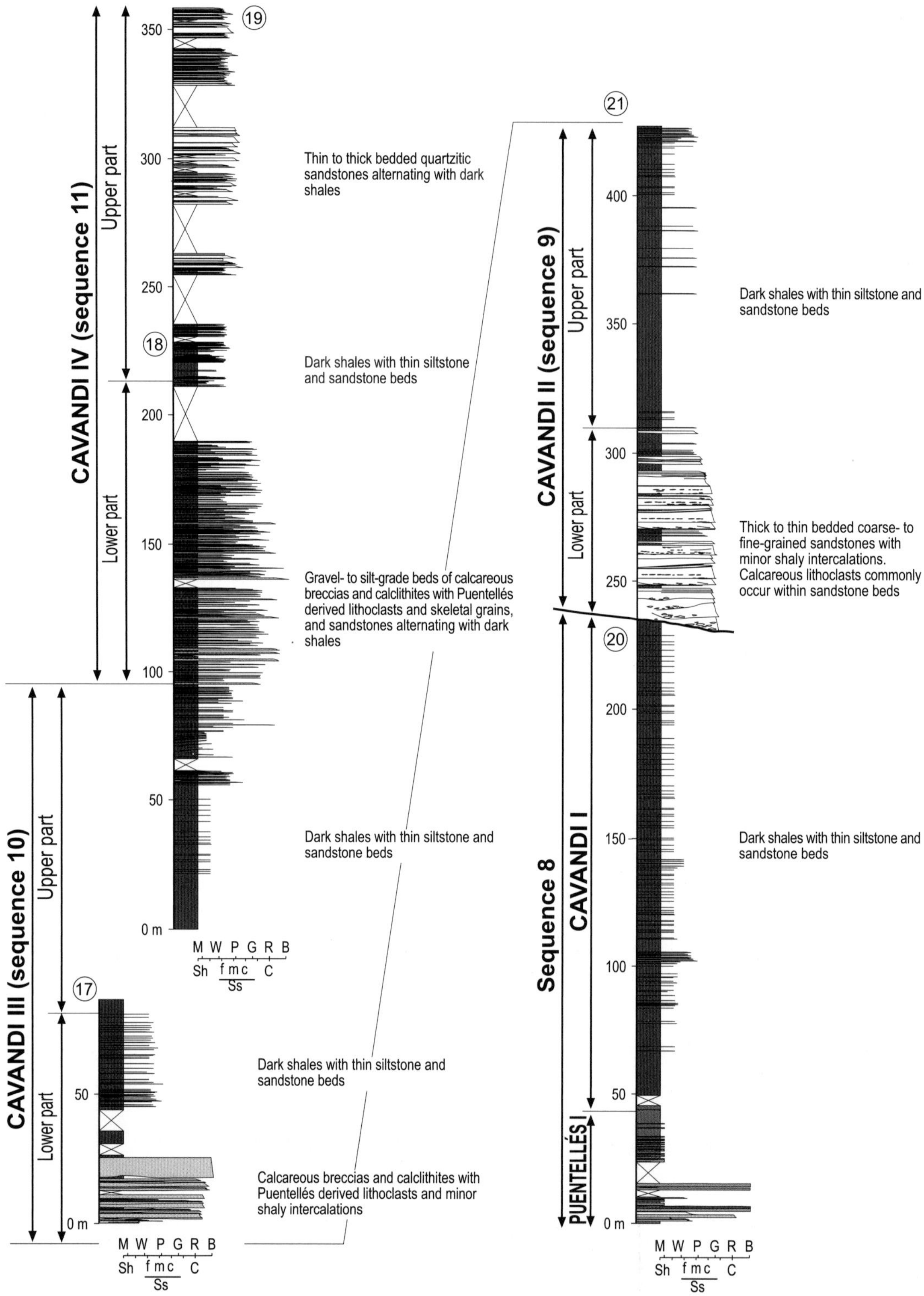

Fig. 11 Stratigraphic logs of the Cavandi Formation showing the internal arrangement of the Cavandi I–IV 3rd-order sequences. Notice the vertical arrangement of facies in the three sequences (see text for details). Sections are numbered as in Fig. 3A. See Fig. 7 for a key to symbols.

Clastic (terrigenous and calcareous) facies

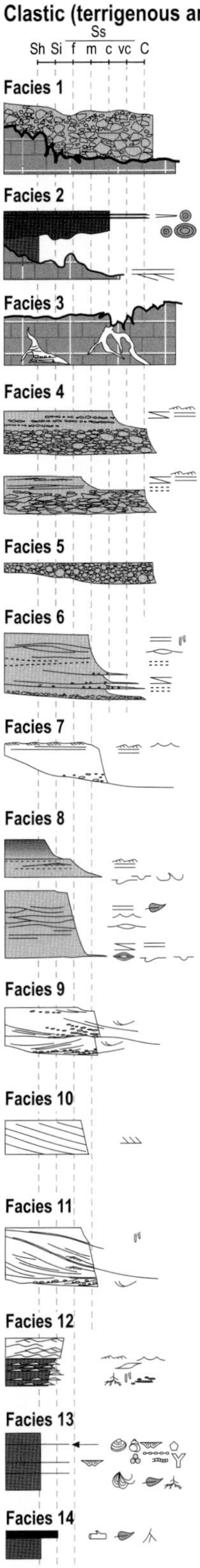

Facies	Lithology	Geometry	Sedimentary structures: Inorganic	Sedimentary structures: Biogenic	Interpretation
Facies 1	Poorly sorted breccias with angular calcareous clasts derived from the underlying units. Calcareous to ferruginous cement	Discontinuous beds with irregular and sharp to gradational base	None (massive beds). Subordinately normal grading		Karstic breccias and calcareous regoliths
Facies 2	Dense, dark to reddish (Fe-rich) crusts with massive appearance. Locally Fe glaebules and nodules can be observed	Tabular to discontinuous beds. Thickness < 5 m			Ferruginous crusts developed on the top of lateritic weathering profiles
Facies 3	Calcareous to quartzitic sandstones with calcareous and quartzitic pebbles and cobbles	Cavity- and fracture-fill masses and dykes			Infills of karstic cavities developed in calcareous units underlying unconformities
Facies 4	Pebble to boulder clast-supported calcareous breccias and conglomerates, in some cases with skeletal grains, passing upwards into sandstones and calclithites	Decimetric to metric (<3.5 m) tabular or lenticular beds with erosive bases and sharp to gradational tops	Massive to normal grading. Some beds with a basal inversely graded division		Bipartite hyperconcentrated and concentrated density flows (Mulder & Alexander, 2001). High-density gravelly turbidity currents (Lowe, 1982)
Facies 5	Pebble to cobble, well to moderately rounded and sorted calcareous conglomerates	Decimetric to metric thick lenticular beds with erosive bases	Massive to normal grading and locally trough cross-bedding. Locally a(t)b(i) imbrication		Conglomeratic lags and longitudinal to 3D bars in fluvial channels
Facies 6	Very coarse- to medium-grained calclithites to quartzitic sandstones with scattered pebbles and cobbles towards the base	Decimetric to metric (<1.3 m) tabular to lenticular beds with erosive bases	Low-angle cross-bedding, parallel and undulose lamination and, locally, hummocky cross-bedding on top	Burrows on top	Sediment-laden waning flows
Facies 7	Coarse- to fine-grained quartzites, sublitharenites and calclithites with scattered skeletal grains	Decimetric (<1.2 m) lenticular beds with erosive bases	Massive to normal grading, occasionally parallel and ripple cross-lamination on top	Burrows and locally roots	Channelized sand-laden waning currents
Facies 8	Medium- to fine-grained quartzites to calclithites and siltstones passing into mudstones. Common skeletal grains	Centimetric to decimetric (<1m) tabular beds with erosive bases. Tops are sharp (planar or undulating) or gradational into mudstones	Sole (flutes, grooves and crescents) marks and load casts. Beds with a massive to graded (either normally or inversely) lower division, and an upper division with parallel lamination, hummocky cross-bedding and current- or wave-ripple lamination	*Planolites*, *Chondrites* and other burrows on top	Turbidity currents (*sensu* Mulder & Alexander, 2001) generated at river mouths (hyperpycnal flows) and by storms
Facies 9	Coarse- to fine-grained quartzites, sublitharenites and calclithites with scattered pebbles and cobbles	Decimetric to metric (<3m) lenticular beds and bedsets with erosive bases	Medium-scale trough cross-bedding	Locally vertical burrows	3D sandy dunes within channels
Facies 10	Medium- to fine-grained quartzites, sublitharenites and calclithites with scattered skeletal grains	Decimetric tabular and wedge-shaped beds and bedsets	Medium- to small-scale tabular cross-bedding. Mud-drapes and reactivation surfaces may be present	*Thalassinoides*, *Skolithos*, and horizontal burrows	2D sandy dunes
Facies 11	Medium- to fine-grained quartzites, sublitharenites and calclithites with thin mudstone and siltstone intervals	Metric (< 2.5 m) lenticular beds with erosive bases and sigmoids separated by shale intervals	Sigmoidal cross-bedding, parallel and ripple cross-lamination. Mud drapes and mud partings occur in each sigmoid	*Skolithos*	Sand bars in tidally influenced channels
Facies 12	Fine-grained sandstones, siltstones and mudstones	Lenticular to flaser bedding	Parallel and current- or wave-ripple cross-lamination	*Chondrites*, *Planolites* and *Zoophycos*	Fine grained clastics accumulated offshore or in onshore tidal dominated settings
Facies 13	Dark marly shales and marls with marine fauna (bivalves, gastropods, brachiopods, crinoids, bryozoans, algae,...) and plant fragments	Decimetric to metric tabular intervals	Massive to laminated	*Zoophycos*, *Planolites* and *Chondrites*	Low-energy offshore environments
Facies 14	Impure coals	Decimetric (<0.85 m) tabular to lenticular beds	Massive		Peat accumulations

Fig. 12 Main clastic facies recognized in the Puentellés Formation. See Fig. 7 for a key to symbols.

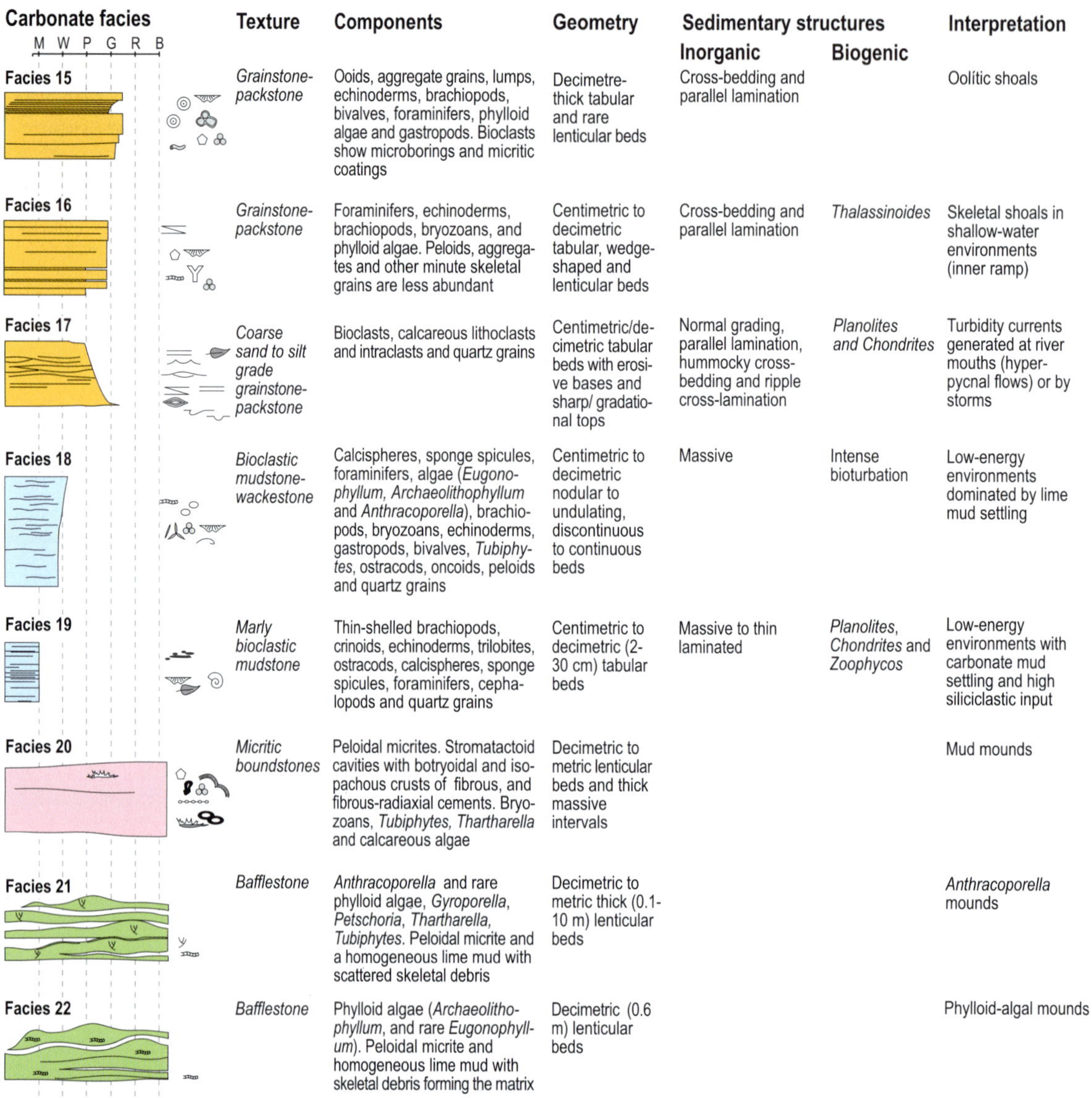

Fig. 13 Main autochthonous carbonate facies recognized in the Puentellés Formation. See Fig. 7 for a key to symbols.

lamination (Facies 12). These facies form fining-upward sequences (Fig. 15A), filling metre-scale erosional features (channels) of the depositional sequence boundaries. These are interpreted as estuarine to tidal channel-fill deposits, developed probably during transgressions, similar to the examples described by Devine (1991), Shanley *et al.* (1992), Ulicný (1999) and Mellere *et al.* (2002).

Mouth-bar deposits

This association comprises coarsening-upward sequences (Fig. 15B), which are up to several tens of metres in thickness and are made up of: (i) shales to marls (Facies 13); (ii) siltstones and fine-grained calclithites and litharenites forming tabular beds with parallel and current ripple-cross lamination (Facies 12 and 8); (iii) coarse- to fine-grained calclithites and litharenites in tabular and wedge-shaped amalgamated beds, which are cross-stratified (Facies 10); and (iv) coarse- to fine-grained calclithites to quartz arenites, locally with scattered pebbles, forming lenticular beds which are massive or trough cross-stratified (Facies 7 and 9, respectively). These deposits are similar to the delta mouth-bar deposits described by Coleman & Wright (1975) and Reading & Collinson (1996).

Fig. 14 Photographs showing some of the clastic facies and facies associations of the Puentellés Formation. (A) Karst features in the contact between the Moscovian Picos de Europa Formation and the overlying Puentellés II. *In situ* breccias with red-stained matrix and clasts derived from the underlying Picos de Europa Formation (Facies 1) constitute the basal deposits of the Puentellés II. Sandstone dykes (Facies 3) cutting both the Puentellés II breccias and the underlying Picos de Europa limestones indicate that, at least, two different karstification episodes occurred. (B) Karstic cavity below the basal unconformity of the Puentellés III sequence filled with laminated quartzitic sandstone (Facies 3), which was probably originated by sand infiltration from an overlying channel fill deposit. (C) Example of shelfal-lobe facies association formed of stacked tabular beds of graded fine-sand to silt grade calclithites (Facies 8) in the lower part of Puentellés II in Section 14. (D) Example of deltaic debris-cone deposits consisting of clast-supported calcareous breccias (Facies 4) in the lower part of Puentellés III (Section 5, see Fig. 9 for location).

Flood-dominated shelfal lobes

These form laterally continuous bodies that fine upwards and are composed of tabular beds (Figs 14C & 15C). These bodies are correlatable between stratigraphic sections several kilometres apart and consist of: (i) calcareous breccias and conglomerates forming amalgamated, massive to graded beds (Facies 4); (ii) pebbly coarse-grained to fine-grained calclithites in beds with a lower massive to graded division overlain by an upper division with parallel lamination and low-angle to hummocky cross-bedding (Facies 6); (iii) fine-grained calclithites to calcisiltites forming graded beds with parallel and ripple cross-lamination (Facies 8); and (iv) marls and marly mudstones (Facies 13 and 19). These bodies are comparable to the shelfal-lobe deposits laid down by hyperpycnal flows related to flood-dominated deltas and fan-deltas (Mulder & Syvitsky, 1995; Mutti *et al.*, 1996, 2000, 2003; Mulder *et al.*, 1998, 2003; Mulder & Alexander, 2001).

Deltaic debris cones

This association forms metric to decametric thick fining-upward cycles (Fig. 15D) composed of: (i) clast- to matrix-supported calcareous breccias, calclithites and rud- to packstones in tabular to

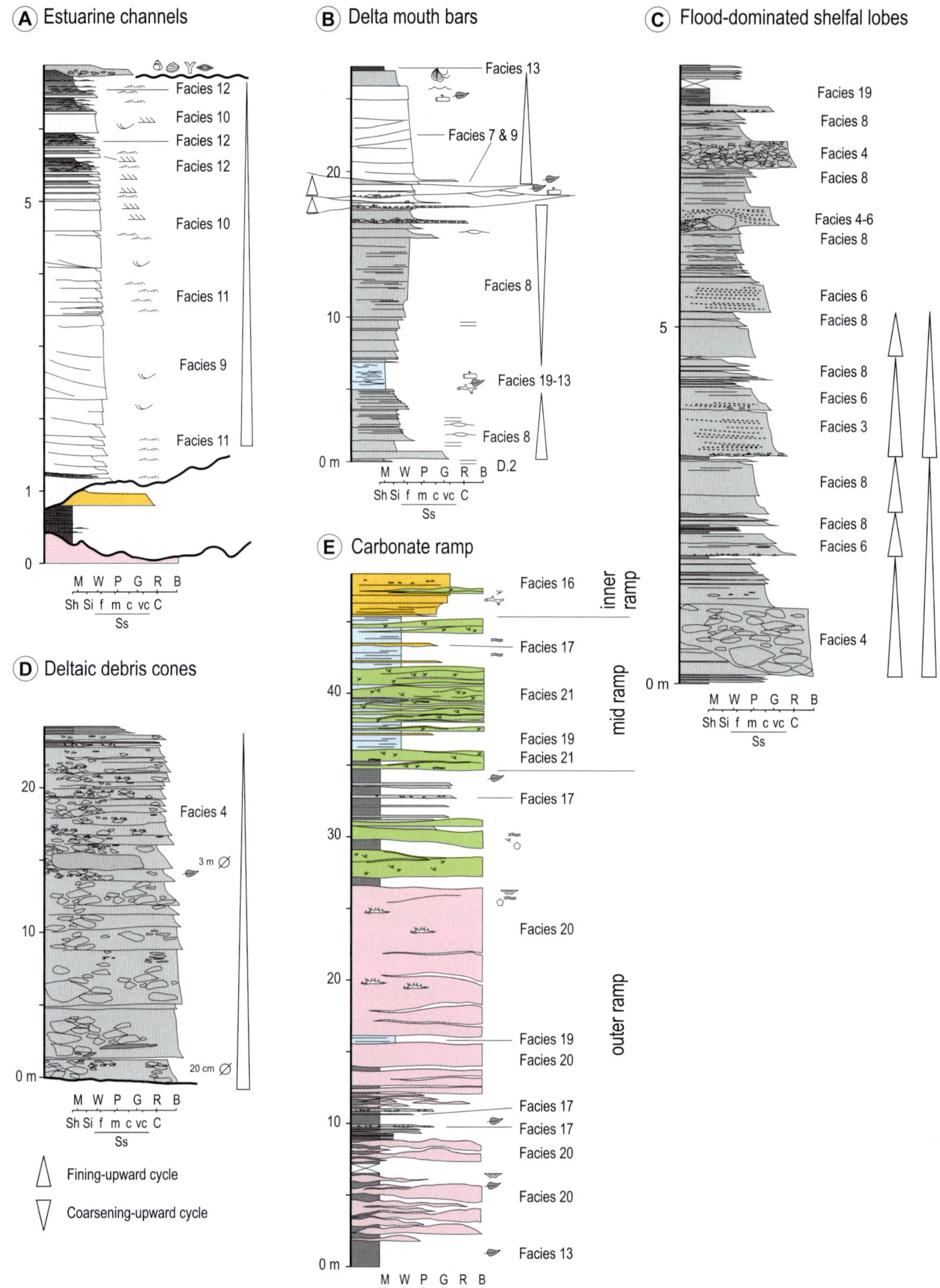

Fig. 15 Detailed stratigraphic sections showing the facies associations described in the Puentellés Formation (see Fig. 7 for a key to symbols). (A) Estuarine channel-fill deposits (lower part of the Puentellés III sequence, Section 8, see Fig. 3A for location). (B) Delta mouth-bar deposits (lower part of Puentellés-II near Panes locality, see Fig. 1B for location). (C) Stacked tabular beds of calcareous breccias and calclithites of shelfal lobes fed from flood-dominated fan-delta systems (lower part of Puentellés II, lower part of Section 5, Fig. 9). (D) Calcareous clast-supported breccia beds forming a fining-upward cycle interpreted as delta debris-cones (lower part of Puentellés III, Section 5, Fig. 9). (E) Carbonate ramp deposits in Puentellés II (Section 2, Fig. 7). Notice the large-scale shallowing-upward trend from outer to inner ramp facies associations. The outer ramp deposits of this example are dominated by microbial boundstones (Facies 20).

lenticular, massive to graded beds with erosive bases (Facies 4 and 8); and (ii) grey marls (Facies 13). These deposits occur forming local accumulations in the northern active border of the basin (lower part of Puentellés III; see Figs 9 & 14D) and are comparable to the debris cones of Postma (1990). They were laid down by dense to dilute gravity flows in a steep coastal setting during the early stages of deltaic development.

Carbonate ramp deposits

Autochthonous carbonate deposits comprise a wide spectrum of facies and microfacies (Fig. 13), which are interpreted to have been deposited in a carbonate ramp. This carbonate ramp was probably of the distally steepened type and developed in a narrow and shallow-water northern domain, which was connected northwards with clastic coastal depositional systems. It passed southwards into a deeper-water area with turbidite sedimentation.

In general terms, these facies and microfacies can be grouped into three facies associations, inner, mid- and outer ramp, following the model of Burchette & Wright (1992; Fig. 15E).

Inner-ramp deposits

These comprise skeletal, less commonly ooidal, pack- to grainstones (Facies 16 and 15, respectively; Fig. 16A), and minor nodular mud- to wackestones (Facies 18; Fig. 16B) and algal bafflestones (phylloid algae and *Anthracoporella*, Facies 22 and 21, respectively). The skeletal limestones contain a diverse biota, being usually rich in foraminifers. In some cases, they also include *Tubiphytes*, oncoids and lumps. Inner-ramp deposits correspond to skeletal and ooidal shoals, beaches and tidal channels (see Burchette & Wright, 1992). The mud-rich deposits and the bafflestones were deposited in protected areas between sand banks, and also in distal environments. Inner-ramp deposits form intervals up to 13 m thick in the uppermost part of the shallowing-upward cycles that occur in the upper part of the sequences (Fig. 15E). This association also forms the basal deposits of sequences in those areas where the lower clastic part was not deposited.

Mid-ramp deposits

This association forms intervals up to 20 m thick that comprise nodular mud to-wackestones (Facies 18) with minor intercalations of algal bafflestones (Facies 21 and 22; Fig. 16C) and coarse to fine-grained, graded beds of skeletal grainstones, locally displaying hummocky cross-stratification (Facies 17; Fig. 15E). These features are characteristic of mid-ramp environments of storm-dominated carbonate ramps (e.g. Aigner, 1985; Wright, 1986; Faulkner, 1988; Somerville & Strogen, 1992). *Anthracoporella* mounds formed in subtidal quiet-water environments located in the photic zone below the active wave base (Krainer, 1995; Samankassou, 1998), thus pertaining to mid-ramp settings.

Outer-ramp deposits

This association usually forms packages up to 20 m thick composed of nodular marly mudstones, spiculites (Facies 18 and 19, respectively; Figs 16B & D), and marls and marly-shales (Facies 13). Thin graded beds of skeletal packstones with parallel and current-ripple lamination (Facies 17) are subordinate. Similar deposits have been described by numerous authors (e.g. Wilson, 1969; Read, 1980; Aigner, 1985; Faulkner, 1988; Burchette & Wright, 1992) and have been interpreted as being deposited below storm wave-base (outer ramp), where the settling of lime mud, transported from shallower-water areas, is the main sedimentary process. In some sections (Fig. 15E), however, this facies association is dominated by amalgamated sheet-like (or flat lens-shaped) units composed mainly of mud and microspar with a clotted peloidal texture, containing fenestellid and ramose bryozoans, *Tubiphytes*, *Terebetella*-like worm tubes, foraminifers (mainly *Tuberitina* and calcitornellids) and rare calcareous algae (Facies 20). Irregular cavities filled with a thin isopachous crust of early marine cement post-dated by homogeneous to laminated internal sediment (stromatactis-like cavities) are a characteristic feature of these deposits (Fig. 16E). This type of deposit has been interpreted as a microbial boundstone (e.g. Pickard, 1996; Riding, 2000), forming mud mounds, which dominantly occur in the distal parts of carbonate ramps (Lees & Miller, 1995; Jeffery & Stanton, 1996, Wendt *et al.*, 2001).

ARCHITECTURE AND COMPOSITION OF THE SEQUENCES OF THE PUENTELLÉS FORMATION

From the cyclical arrangement displayed by the Puentellés Formation, either on the 3rd-order cycle

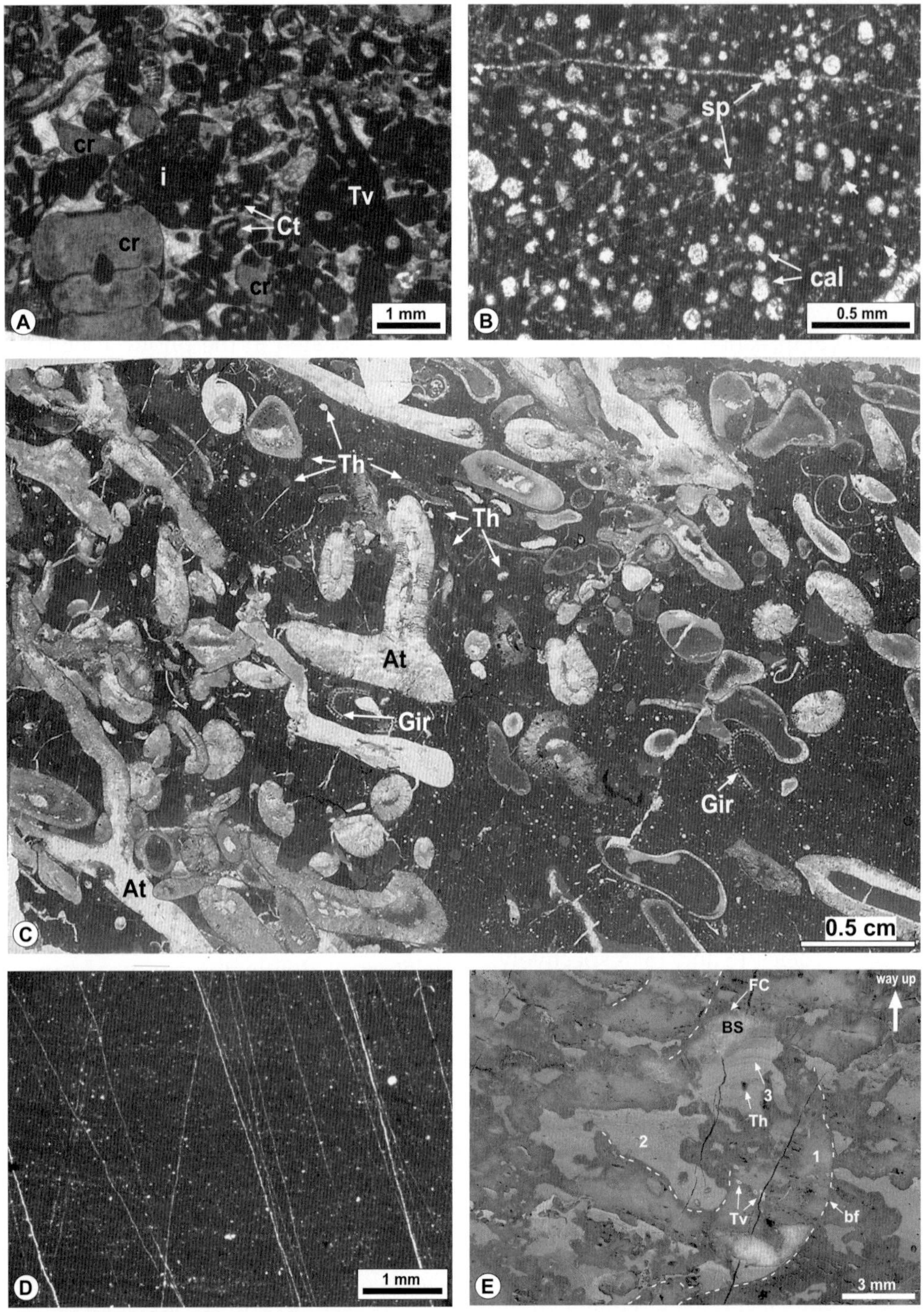

Fig. 16 Photographs showing the autochthonous carbonate facies of the Puentellés Formation. (A) Photomicrograph of the skeletal, tubular-foraminifer-rich grainstones (Facies 16) that form the upper part of the 4th–5th-order shallowing-upward cycles in the upper part of Puentellés I and II. Abundant tubular foraminifers (mainly calcitornellids, black skeletal grains, Ct), *Tubiphytes* (Tv), echinoderms (cr) and intraclasts (i) are the most conspicuous grains in this example. Most of the grains show thin micritic envelopes. (B) Calcisphere-rich mud- to wackestone (Facies 18) typical of the mid- and outer-ramp facies associations (cal, calcispheres; sp, sponge spicule). (C) Photomicrograph of *Anthracoporella* bafflestone (Facies 21). Large branches of *Anthracoporella* (At), scattered *Gyroporella* thalli (Gir) and rare *Thartharella* or *Terebetella* worm tubes (Th) are embedded in a homogeneous dark micrite with scattered skeletal grains. (D) Photomicrograph of a marly mudstone (Facies 19) with a poorly developed parallel lamination marked by silt-sized grains. (E) Polished slab of micritic boundstone (Facies 20) made of clotted peloidal micrite (1) with irregular solution cavities filled first with thin fibrous cement crusts (FC), then with homogeneous (2) and laminated (3) muddy internal sediment, and finally with blocky spar cement (BS). *Tubiphytes* (Tv), *Thartharella–Terebetella*-like worm tubes (Th), and fenestellid bryozoans (dashed lines labelled with bf) are also present.

or at 4th and 5th-order scales, four stages can be recognized in the development of the sequences. These are described below in terms of a theoretical relative sea-level curve.

Rapid sea-level fall (falling stage) and early lowstand

This is recorded by lateritic soils and karstic fills that are present on the basal unconformities. These features reflect the emergence of the northern sector of the basin and the ensuing erosion and karstification of the previous carbonate deposits. In the subsiding trough, this stage is recorded by submarine erosional surfaces overlain by carbonate-rich and siliciclastic flood-dominated deltas and fan deltas. These pass eastwards into sand-rich turbidite systems (Type I and II turbidites of Mutti, 1985) and minor slope aprons, belonging to the Cavandi Formation (Fig. 17A).

Late lowstand

In the northern domain, deltaic and fan-deltaic shelfal lobes (*sensu* Mutti *et al.*, 1996, 2000, 2003), fed from the north and alternating with autochthonous carbonates, accumulated in a narrow (10–15 km) ramp-like shelf, in inner to outer ramp environments (lower part of Puentellés I and II; Fig. 17B$_1$). In Puentellés III, minor slope aprons formed along the northern border of the subsiding trough (Fig. 17B$_2$). The 3rd-order late lowstand deposits are punctuated by numerous syntectonic unconformities that bound the minor (4th–5th order) cycles (see above). In the northernmost outcrops, these deposits pinch out and only scarce conglomeratic alluvial channel-fills, which represent the feeder systems of the fan-deltas and deltas, are preserved locally. In the distal subsiding trough, the turbidite systems were progressively abandoned.

Rising sea level

During this stage, the clastic systems were abandoned and mid- and outer-ramp carbonate deposits accumulated marking the onset of widespread carbonate ramp deposition. These carbonate deposits overlie the late lowstand deposits, although, northwards, they overlie the basal unconformity of each sequence (Fig. 17C). This is interpreted to record a rapid rise of sea level and northwards retreat of the coastline. At this time, the sediment supply was very low in the subsiding distal trough, where mainly mud accumulated, with minor amounts of silt and sand derived from the west.

Highstand

During this stage, carbonate ramps continued developing with a dominant aggradational style, probably due to significant rates of subsidence (Fig. 17D). In the southern trough, thick prodeltaic mudstone wedges and related basinal deposits were deposited offshore of deltaic systems that, located in the western end of the basin, are not now preserved.

DISCUSSION

Sequence stratigraphy deals with the stratigraphic response to the interaction between sediment supply and accommodation space. In foreland basins, these two factors are the result of the interplay of several geological processes. The rates of tectonic uplift and denudation in the orogenic wedge and the climate control the sediment flux, whereas the accommodation space is governed by the rheology of the lithosphere, the tectonic loading, the sediment compaction and the eustatic sea-level changes (e.g. Posamentier & James, 1993; Mascle & Puigdefàbregas, 1998). The cyclic behaviour of some of these processes results in the fill of foreland basins being organized into cycles of several orders. In this study, three different orders of cyclicity have been recognized, giving rise to sequence sets (2nd order), sequences (3rd and 4th order) and minor cycles (4th and 5th order) (Fig. 2B). As discussed in detail below, the tectonic activity in the orogenic wedge is considered to be responsible for the basin configuration and for the major break separating the two sequence sets. This break could be correlated with the so-called Asturian unconformity and related breaks that can be recognized in other parts of the Cantabrian Zone (see Colmenero *et al.*, 2002 and references therein). In contrast, eustatic sea-level fluctuations are interpreted to have controlled the higher-order (3rd–5th order) hierarchical arrangement of the succession of the Puentellés Formation (e.g. the 3rd-order sequences Puentellés I, II and III). A

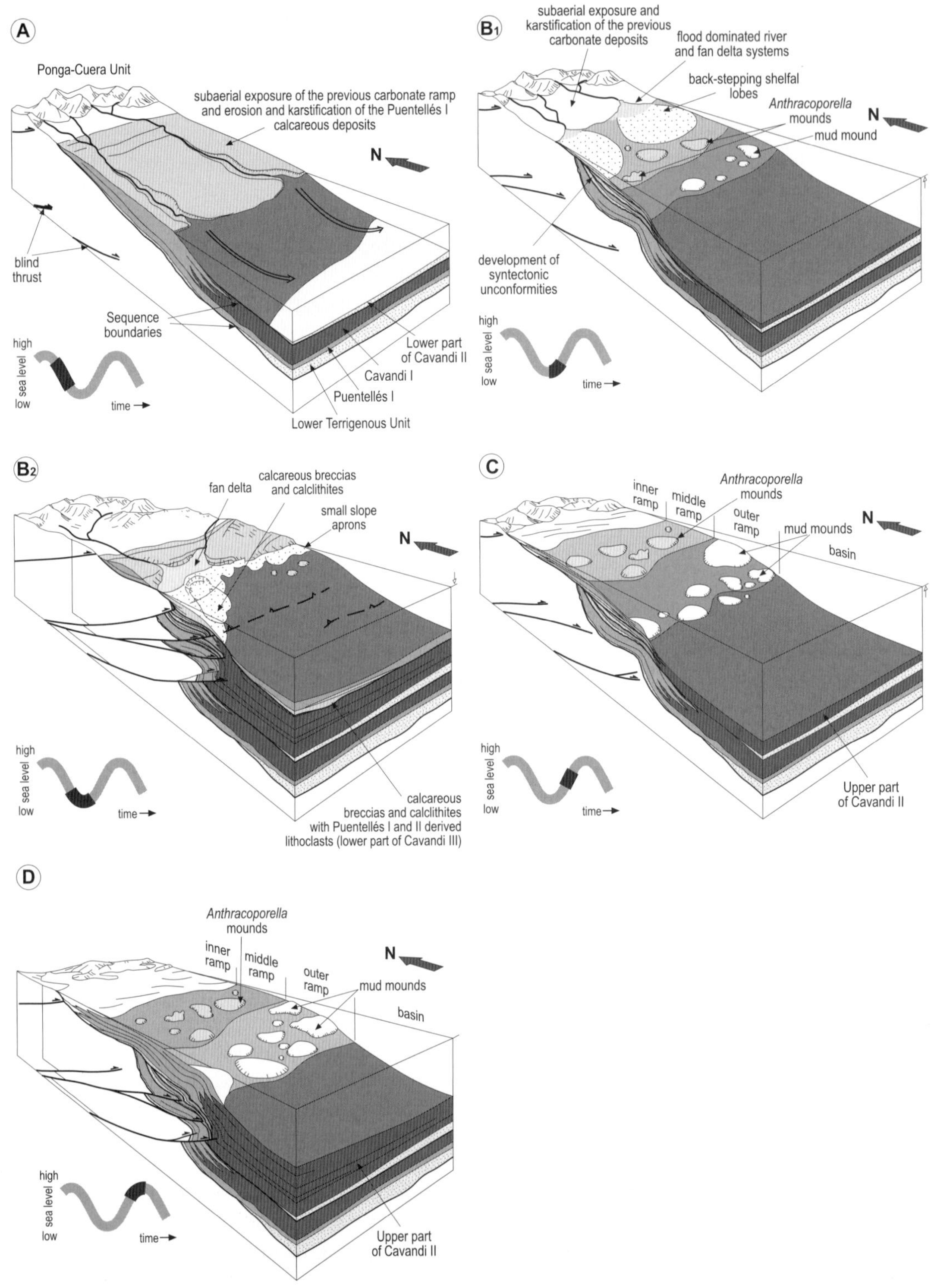

Fig. 17 Block diagrams showing the depositional model inferred for the northern sector of the Picos de Europa Province along a complete cycle of sea-level oscillation: (A) rapid sea-level fall and early lowstand; (B_1 & B_2) late lowstand; (C) rising sea level; and (D) highstand. Diagrams B_1 and B_2 differ in that the latter, corresponding to the Puentellés III sequence, reflects a slightly different basin configuration with a steeper proximal–distal slope due to increased tectonic activity.

eustatic overprint has also been well documented in the Alpine foreland basins of the Pyrenees (Luterbacher *et al.*, 1991; Déramond *et al.*, 1993; Nijman, 1998), Betics (Berástegui *et al.*, 1998) and Alps (Crumeyrolle *et al.*, 1991; Zweigel *et al.*, 1998).

Tectonic control

The tectonic structures affecting the succession and the numerous unconformities, some of them with a syntectonic character (Fig. 18), demonstrate that tectonic movements took place during sedimentation in the latest Moscovian (late Myachkovsky) to Gzhelian in the northern sector of the Picos de Europa Province.

The angular and syntectonic unconformities present along the more active northern border of the basin (see Figs 6 & 18) display geometric patterns that are similar to those described from the active flanks of fault-propagation folds related to blind thrusts (Arbués & Berástegui, 1996; Ford *et al.*, 1996; Den Bezemer *et al.*, 1998) and that fit some of the numerical models of growth strata developed above active monoclines (e.g. Patton, 2004). In the present case, mapping of tectonic structures suggests that fault-propagation folds are related to both frontal and lateral ramps of the Ponga Nappe thrust sheets.

The cross-cutting relationships between the tectonic structures, the infill of the main sedimentary depocentres and the timing of formation of the growth structures point to two main phases of deformation and thrust development (see Fig. 18). During the first phase, the Gamonedo syncline (see Fig. 2A for location) was formed. This is demonstrated by the fact that these synclines display an upper Myachkovsky–Khamovnichesky syntectonic fill, corresponding to the lower sequence set. During the second phase, the Berodia–Inguazo syncline (Fig. 2A) was formed and syntectonically filled by a Dorogomilovsky to Gzhelian succession, corresponding to the upper sequence set.

The subsidence curves, constructed for each sequence set, show a concave-upward morphology (see Fig. 5C), consisting of two parts. The initial part displays a steep slope that points to high rates of subsidence. The second part, which forms the remainder of the curve, is more gentle, suggesting a strong reduction in the subsidence rate. The maximum subsidence rate at the beginning of each sequence set would reflect the flexural downwarping of the lithosphere due to the tectonic load during the emplacement of thrust sheets from the north. The diminished subsidence rate recorded in the second part of the lower sequence set is interpreted to result from the onset of emplacement of the central sector of the Picos de Europa Province during the late Krevyakinsky–Khamovnichesky, and the concomitant transport southward of the northern sector, which became transformed into a piggy-back basin. This resulted in the uplift of the northern sector and the formation of the unconformity between the two sequence sets at the Khamovnichesky–Dorogomilovsky boundary. This evolution is also detectable southwards (basinwards), in the central and southern sectors of the Picos de Europa Province, where deep-water olistostrome-type sediments unconformably overlie older deposits.

Eustatic control

The unconformities that bound the three 3rd-order depositional sequences forming the Puentellés Formation (and similarly the other upper Myachovsky–Gzhelian unconformities in the northern sector of the Picos de Europa Province) are interpreted to be a consequence of eustatic sea-level falls. At least during the Namurian to Permian period, changes in continental ice volume in high latitudes of Gondwana gave rise to a prolonged (though fluctuating) episode of glaciation that started in the Namurian and collapsed in the early Sakmarian (Veevers & Powell, 1987). These glaciations resulted in high-amplitude sea-level fluctuations, which are recorded as transgressive–regressive, 2nd- to 5th-order cycles in the stratigraphic successions (Heckel, 1977; Bush & Rollins, 1984; Ross & Ross, 1985, 1988; Heckel, 1986; Veevers & Powell, 1987; Klein & Willard, 1989; Crowley & Baum, 1991; Maynard & Leeder, 1992; Soreghan, 1994; Soreghan & Giles, 1999; Izart *et al.*, 2003).

The duration of the Kasimovian and Gzhelian 3rd-order transgressive–regressive cycles has been estimated in the range 0.8–1.5 Myr in the Russian Platform, Donets Basin and Carnic Alps (Izart *et al.*, 2003). These cycles were generated by sea-level fluctuations ranging from 100 to 200 m (Ross &

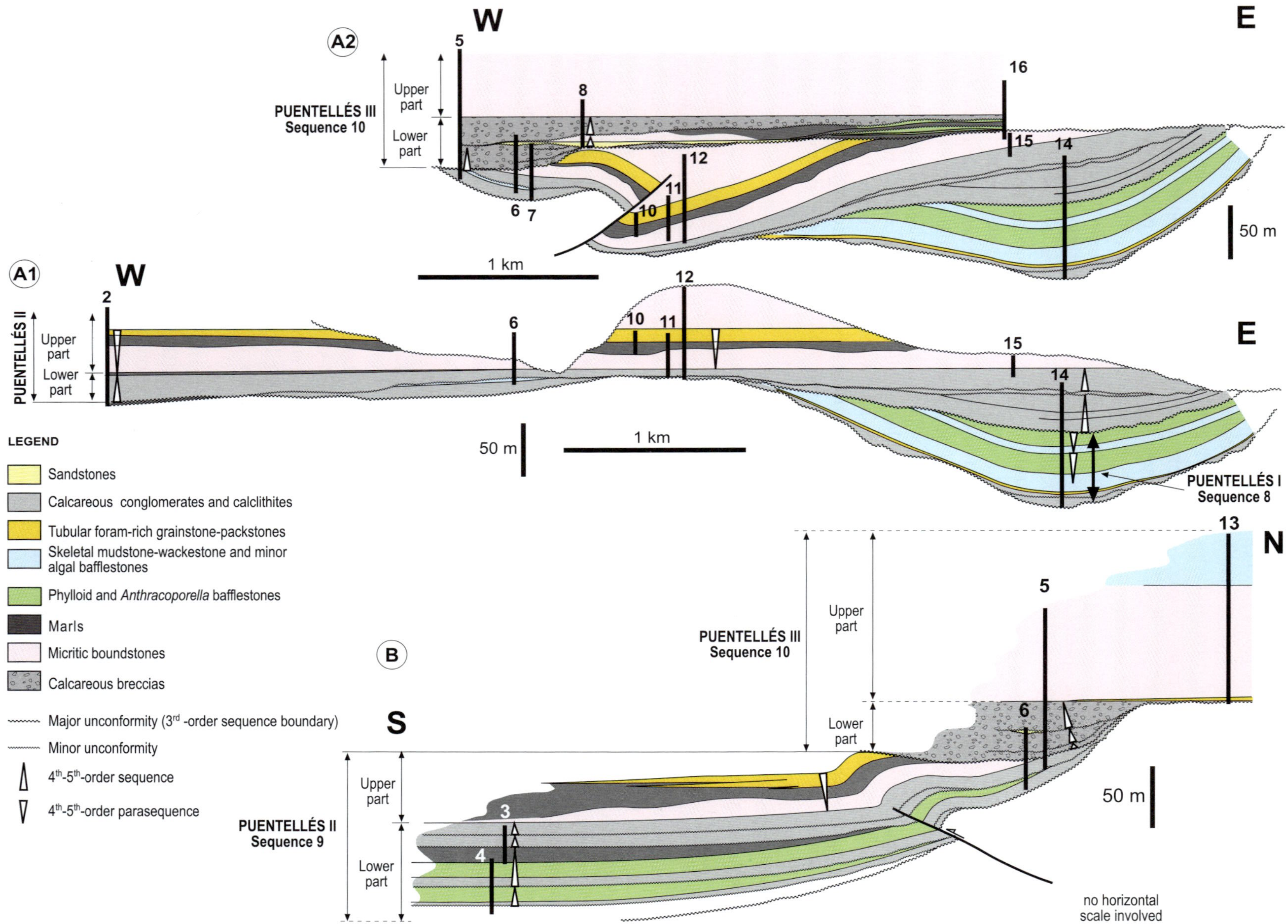

Fig. 18 Restored structural cross-sections of the Puentellés Formation in the central part of the basin (see Fig. 3A for location) showing the syntectonic character of the unconformities. Notice the fanning pattern of both the unconformities and stratigraphic packages, the 'loosening-upwards' character of synclines (right part of A1–A2) and the post-dating of thrusts by the younger deposits (left part of A2 and right part of B). The A2 cross-section depicts a later evolutionary stage of the eastern half (right half) of the A1 cross-section after the sedimentation of the Puentellés III sequence.

Ross, 1985). The 4th–5th-order cycles appear to be related to sea-level oscillations of 42 to 100 m or even more (Table 1), with periodicities comparable to the Milankovitch cycles of eccentricity (400 and 100 kyr), obliquity (100 kyr) and precession (20 kyr) (Bush & Rollins, 1984; Heckel, 1986; Veevers & Powell, 1987; Maynard & Leeder, 1992; Soreghan, 1994; Soreghan & Giles, 1999; Smith & Read, 2000; Izart *et al.*, 2003). The high magnitude of these sea-level changes supports the interpretation that they must have left an imprint in the sedimentary record of the synorogenic basin studied, especially in the carbonates of the Puentellés Formation, as has occurred in other Upper Carboniferous successions (Mid-continent USA, Russian Platform and Donets Basin, see references above).

The latest Moscovian to Gzhelian 3rd-order cycles seem to have been approximately synchronous along the palaeo-Tethys domain (East Europe and North Africa) and the American Midcontinent (Ross & Ross, 1985, 1988; Izart *et al.*, 2003). The duration of these 3rd-order late Moscovian (upper Myachkovsky) to Gzhelian cycles (0.8–1.5 Myr) is similar to that of the 3rd-order sequences forming the Puentellés Formation (0.4–1 Myr, see Fig. 2B). In addition, the basal unconformities of Sequences 6, 7, 8 (Puentellés I) and 9 (Puentellés II) can be roughly correlated with the maximum eustatic sea-level falls of the 3rd-order cycles recorded in the USA, Russian Platform, Donets Basin and Carnic Alps (Ross & Ross, 1985, 1988; Izart *et al.*, 2003). On other hand, the onset of carbonate sedimentation in Sequence 9 (Puentellés II) coincides with the arrival of cosmopolitan fusulinoidean faunas in the basin. Villa & Ueno (2002) and Villa *et al.* (2003) interpreted this event as the result of a significant transgression that occurred at the beginning of Gzhelian times and that is also recorded in the previously cited areas.

The duration inferred for the higher order (4th–5th) transgressive–regressive cycles (see above and Fig. 2B) makes these compatible with Milankovich obliquity and eccentricity cycles (40 ka and 100–400 ka respectively; De Boer & Smith, 1994). The fact that the boundaries of 3rd- to 5th-order sequences correspond to angular unconformities in the northernmost proximal areas of the basin suggests that they have a composite origin. These boundaries would reflect the interplay between tectonics and eustatic sea-level oscillations (Fig. 17). The angular relationships and the geometric stratal patterns observed (Fig. 18) would have been generated by the progressive tilting and folding of the syntectonic deposits by active tectonic structures (fault-propagation and recumbent folds related to blind thrusts), while the unconformities themselves would correspond to eustatic sea-level falls. In a similar way, Patton (2004) showed in numerical models that changes in the accommodation space due to base-level falls during the growth of active folds can generate unconformities.

Sequence versus parasequence development

The 4th–5th-order cycles have a different expression depending on their location within 3rd-order sequences. In the lower part of 3rd-order sequences, they are recorded by high-frequency 4th–5th-order sequences. In turn, the 4th–5th-order cycles in the upper part of 3rd-order sequences mainly correspond to parasequences, although in some of these cycles the capping inner-ramp deposits display a rather sharp base that could represent a regressive surface marked by subtle submarine erosion and basinward shift of facies as occurs during forced regressions.

This different development of high-frequency 4th-order cycles was described by Van Wagoner *et al.* (1990), who differentiated type-A and type-B 4th-order cycles, and is shown in Fig. 19. During the 3rd-order falling stages, long-term sea-level fall would reinforce the lower amplitude 4th–5th-order sea-level falls, resulting in the generation of 4th–5th-order sequence boundaries. As a consequence, the effects of the tectonic deformation would be enhanced during the 3rd-order lowstands in the northern domain of the basin, favouring the development of syntectonic unconformities, such as has been described by Castelltort *et al.* (2003). On the other hand, the 3rd-order rising stages would amplify the lower magnitude 4th–5th-order sea-level rises, resulting in the generation of 4th–5th-order flooding events bounding parasequences. Nevertheless, due to the relatively high amplitude of 4th–5th-order sea-level fluctuations during the Late Carboniferous (Table 1), the parasequences in the upper part of the 3rd-order sequences display a rapid shallowing-upward trend and usually contain sharp-based shoreface deposits in their upper part.

Table 1 Estimated periodicity and amplitude of 3rd and 4th–5th-order sea-level fluctuations during Late Carboniferous to Early Permian time

Authors	Basin	Duration of sea-level cycles			Amplitude
		3rd order	4th order	5th–6th order	
Heckel (1977)	Midcontinent (Middle–Upper Pennsylvanian)		400 kyr 129–216 kyr*		100 m
Bush & Rollins (1984)	Northern Appalachian Basin (Missourian)	0.9–1.5 Myr	400–450 kyr	100–225 kyr	No estimation
Ross & Ross (1985)	Russian Platform and Midcontinent (Carboniferous and Permian)	1.2–4 Myr Kasim.-Gzhel. 0.5–0.8 Myr			100–200 m
Heckel (1986)	Midcontinent (Pennsylvanian)		235–393 kyr 129–216 kyr*	40–120 kyr 24–65 kyr*	No estimation
Adlis *et al.* (1988)	Midland Basin, Texas (Virgilian)			High-frequency cycles	At least 70 m
Crowley & Baum (1991)			Milankovitch orbital forced glacioeustatic sea-level cycles		60 ± 15 m
Maynard & Leeder (1992)	Pennine Basin, Kansas, Midcontinent, Central Appalachian Basin (Westphalian)			100–120 kyr	Minimum mean values of 42 m
Soreghan (1994)	Pedregosa and Orogrande Basins (Upper Pennsylvanian)		478–347 kyr 230–250 kyr		No estimation
Klein (1994)	Midcontinent (Pennsylvanian)				86 m using the Gerhar (1991) model. 96.4 m using the Heckel (1977) model
Chesnut (1994)	Central Appalachian Basin (Lower and Middle Pennsylvanian)	2.5 Myr 1.45 Myr	400 kyr 244 kyr		No estimation
Rasbury *et al.* (1998)	New Mexico and Texas (Upper Pennsylvanian–Lower Permian)			**143 ± 64 kyr**	No estimation
Saller *et al.* (1999)	West Texas (Middle Pennsylvanian–Lower Permian)			160 kyr	No estimation
Soreghan & Giles (1999)	Orogrande Basin (Southern New Mexico)			High-frequency sequences	Minimum value of 80 m Likely in excess of 100 m
Olszewski & Patzkowskyr (2003)	Midcontinent (Pennsylvanian–Permian)		400 kyr	< 100 kyr	No estimation
Izart *et al.* (2003)	Moscow, Donets basin and Carnic Alps (Kasimovian and Gzhelian)	Gzhelian: **0.8–1.5 Myr** Kasimovian: **0.8–1 Myr**	Gzhelian: **375–666 kyr** Kasimovian: **428–375 kyr**	Gzhelian: **254–200 kyr** Kasimovian: **115–300 kyr**	No estimation

Underlined entries indicate cycle duration recalculated by the present authors using the time-scale of Gradstein *et al.* (2004).
Bold entries indicate cycle duration obtained applying time-scales similar to that of Gradstein *et al.* (2004).
* Cycle duration recalculated by Klein (1990) applying the Carboniferous time-scale of Lippolt *et al.* (1986) and Hess & Lippolt (1986), which is similar for the Late Carboniferous to that of Gradstein *et al.* (2004).

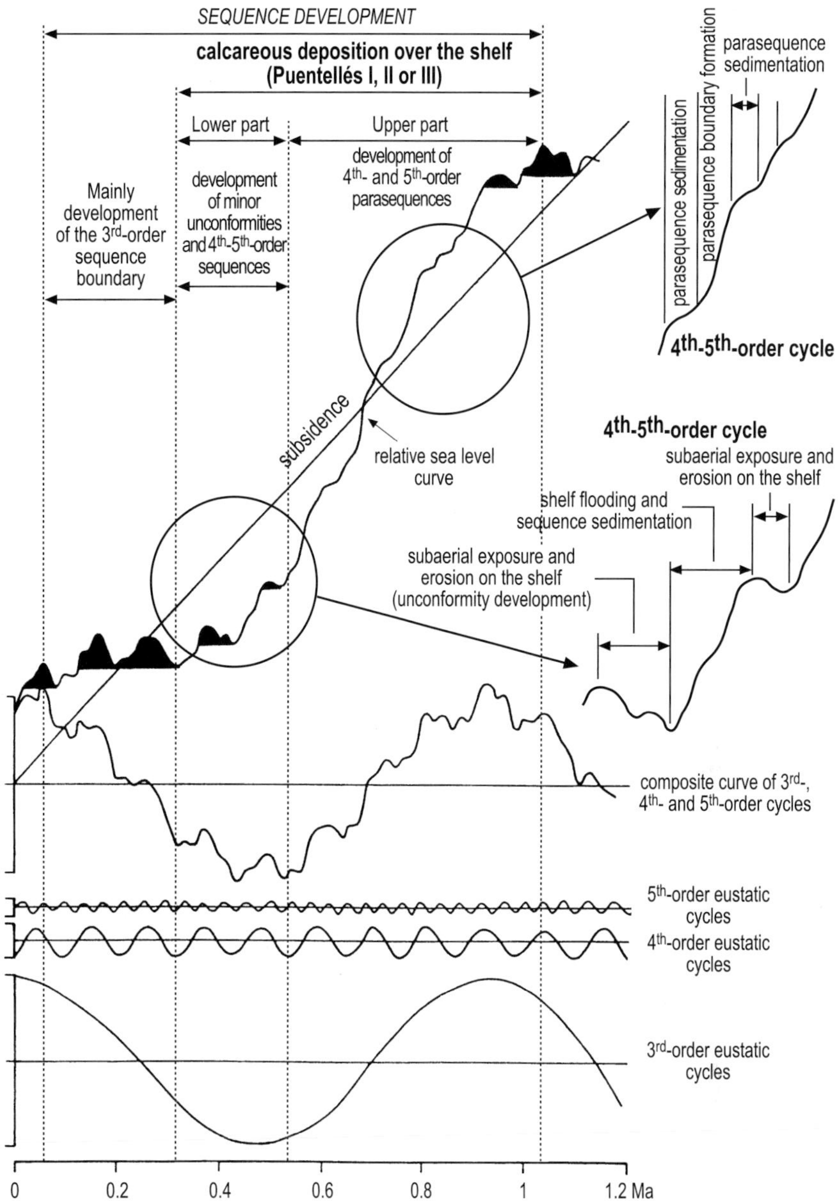

Fig. 19 Conceptual model explaining the expression of the 4th–5th-order cycles as sequences or as parasequences depending on their location within 3rd-order cycles (adapted from Van Wagoner *et al.*, 1990).

The Puentellés Formation carbonate ramp

An interesting point that is worthy of discussion concerns the development of a carbonate ramp (Puentellés Formation) within a synorogenic clastic wedge deposited in a piggy-back basin. Marine carbonates in foreland basins typically display a ramp-like profile and preferentially develop on the foreland margin (Burchette & Wright, 1992; Dorobek, 1995), with exceptional examples occurring in the synorogenic clastic wedge in relation to blind-thrust or salt-diapir highs (Luterbacher *et al.* (1991) and Purser (1973) respectively). To our knowledge, such a thick carbonate ramp succession as in the Puentellés Formation, present in an orogen-attached synorogenic clastic wedge (Fig. 17), is an almost unique example in the geological record (cf. Sanders & Höfling, 2000).

The sedimentary regime in foreland basins, piggy-back basins included, involves huge amounts of clastic sediment shed into the basin. The prevailing terrigenous nature of the orogen-derived detritus leads to high amounts of suspended clay in the sea water, which inhibits carbonate production except during major transgressions.

Tectonic deformation in the study area resulted in a Dorogomilovsky–Gzhelian basin configuration with a northern shallow-water domain, where carbonate sedimentation took place (Puentellés Formation), and a southern trough with deep-water clastic sediments (Cavandi Formation; Fig. 17). High terrigenous input from deltaic and fan-deltaic systems in the west was funnelled into the southern trough and almost no terrigenous material reached the northern domain, especially during transgressive and highstand times. In addition, most of the clastic sediments that were shed into the northern domain of the basin from the northern hinterland had a calcareous composition. This is because this source area (Sierra del Cuera) is almost exclusively made of Carboniferous limestones (Barcaliente, Valdeteja and Picos de Europa formations). As a consequence, little clay was produced and transported into the basin to inhibit carbonate production. Indeed, the concomitant carbonate enrichment of the sea water is interpreted to have favoured the maintenance of high rates of carbonate production in the basin, particularly since the region was located in a subtropical palaeolatitude (Torsvik & Cocks, 2004). This indicates that, under certain conditions, piggy-back basins can be suitable settings for prolific carbonate production and development of narrow carbonate ramps with marked proximal–distal facies changes.

CONCLUSIONS

The latest Moscovian (late Myachkovsky) to Gzhelian succession in the northern sector of the Picos de Europa Province was deposited in a rapidly subsiding piggy-back basin located in front of the Ponga Nappe thrust sheets, where two successive phases of thrust emplacement are recorded. The Dorogomilovsky–Gzhelian synorogenic basin was structured into two domains: a northern shallow-water domain with carbonate clastic and carbonate ramp sedimentation (Puentellés Formation); and a southern deep-water trough with mainly turbidite sedimentation (Cavandi Formation).

The succession in this piggy-back basin has been subdivided into 11 mappable depositional sequences (3rd–4th order) consisting of several higher order cycles (4th–5th order), bound by angular unconformities, which in some cases correspond to syntectonic unconformities. The sequences have in turn been grouped into two *sequence sets* (2nd-order sequences) recording the two phases of thrust emplacement.

The proximal deposits of Sequences 8–10 (upper sequence set) constitute the Puentellés Formation. Each of these 3rd-order sequences comprises a fining-upward lower part, composed of several 4th–5th-order high-frequency sequences with a similar internal organization, and an upper part arranged in metre to decametre-scale shallowing-upward parasequences.

The integration of biostratigraphical and lithostratigraphic data and field mapping suggests that, although thrust-sheet emplacement was responsible for the basin development and configuration, eustasy controlled the higher order (3rd to 5th order) cyclical arrangement of the succession.

The lower part of each 3rd-order sequence in the Puentellés Formation records late lowstand clastic sedimentation in flood-dominated deltas and fan-deltas, which co-existed with carbonate deposition in the protected proximal parts of the carbonate ramp. The upper part of each 3rd-order sequence, mainly composed of autochthonous carbonates, represents the abandonment of the previous clastic systems and the encroachment and aggradation of the carbonate ramps during rising and highstand stages.

The Puentellés Formation demonstrates that, under certain conditions, piggy-back basins can be suitable settings for prolific carbonate production and the development of narrow carbonate ramps with marked proximal–distal facies changes. In this case, basin configuration and the predominant carbonate lithologies of the northern source area seem to have been the controlling factors.

ACKNOWLEDGEMENTS

Financial support by Dirección General de Investigación Científica y Técnica of Spain (Project MCT-00-BTE-0580) is acknowledged. We thank Elisa Villa for providing biostratigraphical data and Nemesio Heredia for stimulating discussions in the field. Many thanks to I.D. Somerville, M.E. Tucker and J. Vergés and the editor E.A. Williams for providing critical reviews and comments that helped to significantly improve this paper.

REFERENCES

Adlis, D.S., Grossman, E.L., Yancey, T.E. and McLerran, R.D. (1988) Isotope stratigraphy and paleodepths changes of Pennsylvanian cyclical sedimentary deposits. *Palaios*, **3**, 487–506.

Aigner, T. (1985) *Storm Depositional Systems. Dynamic Stratigraphy in Modern and Ancient Shallow-marine Sequences*. In: *Lecture Notes in Earth Sciences*, Vol. 3 (Eds G.M. Friedman, H.J. Neugebauer and A. Seilacher), 174 pp. Springer-Verlag, Berlin.

Allen, P.A. and Allen, J.R. (1990) *Basin Analysis: Principles and Applications*. Blackwell Scientific Publications, Oxford, 45 pp.

Arbués, P., Pi, E. and Berástegui, X. (1996) Relaciones entre la evolución sedimentaria del Grupo de Areny y el cabalgamiento de Boixols (Campaniense terminal-Maastrichtiense del Pirineo Meridional-Central). *Geogaceta*, **20**, 446–449.

Bahamonde, J.R., Colmenero, J.R. and Vera, C. (1997) Growth and demise of late Carboniferous carbonate platforms in the eastern Cantabrian Zone, Asturias, northwestern Spain. *Sediment. Geol.*, **110**, 99–122.

Bahamonde, J.R., Vera, C. and Colmenero, J.R. (2000) A steep-fronted Carboniferous carbonate platform: clinoformal geometry and lithofacies (Picos de Europa Region, NW Spain). *Sedimentology*, **47**, 645–664.

Berástegui, X., Banks, C.J., Puig, C., Taberner, C., Waltham, D. and Fernández, M. (1998) Lateral diapiric emplacement of Triassic evaporites at the southern margin of Guadalquivir Basin, Spain. In: *Cenozoic Foreland Basins of Western Europe* (Eds A. Mascle, C. Puigdefàbregas, H.P. Luterbacher and M. Fernández), pp. 49–68. Special Publication 134, Geological Society Publishing House, Bath.

Bond, G.C. and Kominz, M.A. (1984) Construction of tectonic subsidence curves for the Early Paleozoic miogeocline, southern Canadian Rocky Mountains: implications for subsidence mechanisms, age of breakup, and crustal thinning. *Geol. Soc. Am. Bull.*, **95**, 155–173.

Burchette, T.P. and Wright, V.P. (1992) Carbonate ramp depositional systems. *Sediment. Geol.*, **79**, 3–57.

Bush, R.M. and Rollins, H.B. (1984) Correlation of Carboniferous strata using hierarchy of transgressive-regressive units. *Geology*, **12**, 471–474.

Castelltort, S., Guillocheau, F., Robin, C., *et al.* (2003) Fold control on the stratigraphic record: a quantified sequence stratigraphic study of the Pico del Aguila anticline in the south-western Pyrenees (Spain). *Basin Res.*, **15**, 527–551.

Chesnut, D.R. (1994) Eustatic and tectonic control of deposition of the Lower and Middle Pennsylvanian strata of the central Appalachian Basin. In: *Tectonic and Eustatic Controls on Sedimentary Cycles* (Eds J.M. Denninson and F.R. Ettensohn), pp. 51–64. *Concepts in Sedimentology and Palaeontology* 4, Society of Economic Paleontologists and Mineralogists, Tulsa, OK.

Coleman, J.M. and Wright, L.D. (1975) Modern river deltas: variability of processes and sand bodies. In: *Deltas, Models for Exploration* (Ed. M.L. Broussard), pp. 99–149. Houston Geological Society, Houston, TX.

Colmenero, J.R., Fernández, L.P., Moreno, C., *et al.* (2002) Carboniferous. In: *The Geology of Spain* (Eds W. Gibbons and M.T. Moreno), pp. 93–116. Geological Society Publishing House, Bath.

Cooper, J.D. and Keller, M. (2001) Paleokarst in the Ordovician of the southern Great Basin, USA: implications for sea-level history. *Sedimentology*, **48**, 855–873.

Crowley, T.J. and Baum, S.K. (1991) Estimating Carboniferous sea-level fluctutations from Gondwana ice extent. *Geology*, **19**, 975–977.

Crumeyrolle, P., Rubino, J.-L. and Clauzon, G. (1991) Miocene depositional sequences within a tectonically controlled transgressive-regressive cycle. In: *Sedimentation, Tectonics and Eustasy: Sea-Level Changes at Active Margins* (Ed. D.I.M. McDonald), pp. 373–390. Special Publication 12, International Association of Sedimentologists. Blackwell Scientific Publications, Oxford.

De Boer, P.L. and Smith, D.G. (1994) Orbital forcing and cyclic sequences. In: *Orbital Forcing and Cyclic Sequences* (Eds P.L. de Boer and D.G. Smith), pp. 1–14. Special Publication 19, International Association of Sedimentologists. Blackwell Scientific Publications, Oxford.

DeCelles, P.G. and Giles, K.A. (1996) Foreland basin systems. *Basin Res.*, **8**, 105–123.

Den Bezemer, T., Kooi, H., Podladchikov, Y. and Cloetingh, S. (1998) Numerical modelling of growth strata and grain-size distributions associated with fault-bend folding. In: *Cenozoic Foreland Basins of Western Europe* (Eds A. Mascle, C. Puigdefàbregas, H.P. Luterbacher and M. Fernández), pp. 381–401. Special Publication 134, Geological Society Publishing House, Bath.

Déramond, J., Souquet, P., Fondecave-Wallez, M.-J. and Specht, M. (1993) Relationships between thrust tectonics and sequence stratigraphy surfaces in foredeeps: model and examples from Pyrenees (Cretaceous-Eocene, France, Spain). In: *Tectonics and Seismic Sequence Stratigraphy* (Eds G.D. Williams and A. Dobb), pp. 193–219. Special Publication 71, Geological Society Publishing House, Bath.

Devine, P.E. (1991) Transgressive origin of channeled estuarine deposits in the Point Lookout Sandstone, Northwestern New Mexico: A model for Upper Cretaceous, cyclic regressive parasequences of U.S. Western Interior. *Am. Assoc. Petrol. Geol. Bull.*, **75**, 1039–1063.

Dorobek, S.L. (1995) Synorogenic carbonate platforms and reefs in foreland basins: controls on stratigraphic evolution and platform/reef morphology. In: *Stratigraphic Evolution of Foreland Basin.* (Eds S.L. Dorobek and G.M. Ross), pp. 127–147. Special Publication 52, Society of Economic Paleontologists and Mineralogists, Tulsa, OK.

Duchaufour, P. (1982) *Pedology*. George Allen & Unwin, London, 449 pp.

Elrick, M. and Read, J.F. (1991) Cyclic ramp-to-basin carbonate deposits, Lower Mississippian, Wyoming and Montana: a combined field and computer modeling study. *J. Sediment. Petrol.*, **61**, 1194–1224.

Esteban, M. and Klappa, C.F. (1983) Subaerial exposure environment. In: *Carbonate Depositional Environments* (Eds P.A. Scholle, D.G. Bebout and C.H. Moore), pp. 1–54. Memoir 33, American Association of Petroleum Geologists, Tulsa, OK.

Faulkner, T.J. (1988) The Shipway Limestone of Gower: sedimentation on a storm-dominated early Carboniferous ramp. *Geol. J.*, **23**, 85–100.

Ford, M., Williams, E.A., Artoni, A., Vergés, J. and Hardy, S. (1996) Progressive evolution of a fault-related fold pair from growth strata geometries, Sant Llorenç de Morunys, SE Pyrenees. *J. Struct. Geol.*, **19**, 413–441.

Gerhar, L.C. (1991) Stratigraphy in the modern generation: Lawrence. *Kansas Geol. Surv. Open File Rep.*, **91–24**, 13 pp.

Gradstein, F.M., Ogg, J.G., Smith, A.G., Bleeker, W. and Lourens, L.J. (2004) A new geologic time scale, with special reference to Precambrian and Neogene. *Episodes*, **2**, 83–100.

Heckel, P.H. (1977) Origin of phosphatic black shale facies in Pennsylvanian cyclothems of Mid-Continent North America. *Am. Assoc. Petrol. Geol. Bull.*, **61**, 1045–1068.

Heckel, P.H. (1986) Sea-level curve for Pennsylvanian eustatic marine transgressive-regressive depositional cycles along Midcontinent outcrop belt, North America. *Geology*, **14**, 330–334.

Hess, J.C. and Lippolt, H.J. (1986) ^{40}Ar/^{39}Ar ages of tonstein and tuff sanidines: new calibration points for the improvement of the Upper Carboniferous time scale. *Isotope Geosci.*, **59**, 143–154.

Izart, A., Stephenson, R., Battista Vai, G., *et al.* (2003) Sequence stratigraphy and correlation of late Carboniferous and Permian in the CIS, Europe, Tethyan area, North Africa, Arabia, China, Gondwanaland and the USA. *Palaeogeogr. Palaeoclimatol. Palaeoecol.*, **196**, 59–84.

Jeffery, D.L. and Stanton, R.J. Jr (1996) Autochthonous outer ramp sedimentation: the Alamogordo Member of the Lake Valley Formation, New Mexico. *Facies*, **35**, 9–28.

Julivert, M. (1971) Décollement tectonics in the Variscan Cordillera of the northwest Spain. *Am. J. Sci.*, **270**, 1–29.

Kendall, C.G. St.C. and Schlager, W. (1981) Carbonates and relative changes in sea level. *Mar. Geol.*, **44**, 181–212.

Kenter, J.A.M., Hoeflaken, F., Bahamonde, J.R., Bracco Gartner, G.L., Keim, L. and Besems, R.E. (2003) Anatomy and lithofacies of an intact and seismic-scale Carboniferous carbonate platform (Asturias, NW Spain): analogs of hydrocarbon reservoirs in the Pricaspian basin (Kazakhstan). In: *Paleozoic Carbonates of the Commonwealth of Independent States (CIS): Subsurface Reservoirs and Outcrop Analogs* (Eds W.G. Zempolich and H.E. Cook), pp. 181–203. Special Publication 74, Society of Economic Paleontologists and Mineralogists, Tulsa, OK.

Kerans, C. and Donalson, J.A. (1988) Proterozoic paleokarst profile, Dismal Lakes Group, N.W.T. Canada. In: *Paleokarsts* (Eds N.P. James and P.W. Choquette), pp. 167–183. Springer-Verlag, New York.

Klein, G.deV. (1990) Pennsylvanian time scales and periods. *Geology*, **18**, 455–457.

Klein, G.deV. (1994) Depth determination and quantitative distinction of the influence of tectonic subsidence and climate on changing sea level during deposition of Midcontinent Pennsylvanian cyclothems. In: *Tectonic and Eustatic Controls on Sedimentary Cycles* (Eds J.M. Denninson and F.R. Ettensohn), pp. 35–50. Concepts in Sedimentology and Palaeontology, Vol. 4, Society of Economic Paleontologists and Mineralogists, Tulsa, OK.

Klein, G.deV. and Willard, D.A. (1989) Origin of the Pennnsylvanian coal-bearing cyclothems of North America. *Geology*, **17**, 152–155.

Krainer, K. (1995) *Anthracoporella* mounds in the Late Carboniferous Auernig Group, Carnic Alps (Austria). *Facies*, **32**, 195–214.

Lees, A. and Miller, J. (1995) Waulsortian banks. In: *Carbonate Mud Mounds. Their Origin and Evolution* (Eds C.L.V. Monty and D.W.J. Bosence), pp. 191–272. Special Publication 23, International Association of Sedimentologists. Blackwell Scientific Publications, Oxford.

Lippolt, H.J., Hess, J.C. and Burger, K. (1986) Isotopische alter von pyroklastischen sanidinen aus Kaolin-Kohlentonstein als korrelationsmarken fur das mitteluеroaische Oberkarbon. *Fortschr. Geol. Rheinl. Westfalen*, **32**, 119–150.

Lowe, D.R. (1982) Sedimentary gravity flows: II Depositional models with special reference to the deposits of high-density turbidity currents. *J. Sediment. Geol.*, **52**, 279–297.

Luterbacher, H.P., Eichenseer, H., Betzler, C.H. and van den Hurk, A.M. (1991) Carbonate-siliciclastic depositional systems in the Paleogene of the South Pyrenean foreland basin: a sequence-stratigraphic approach. In: *Sedimentation, Tectonic and Eustasy*. (Ed. D.I.M. Macdonald), pp. 391–407. Special Publication 12, International Association of Sedimentologists. Blackwell Scientific Publications, Oxford.

Magara, K. (1980) Comparison of porosity-depth relationships of shale and sandstone. *J. Petrol. Geol.*, **3**, 175–185.

Marquínez, J. (1989) Mapa Geológico de la Región del Cuera y Picos de Europa (Cordillera Cantábrica, NW de España). *Trab. Geol. Univ. Oviedo*, **18**, 137–144.

Martínez García, E. (1981) El Paleozoico de la Zona Cantábrica oriental (Noroeste de España). *Trab. Geol. Univ. Oviedo*, **11**, 95–127.

Martínez García, E. and Villa, E. (1998) El desarrollo estratigráfico de las unidades alóctonas del área de Gamonedo-Cabrales (Picos de Europa, Asturias, NW de España). *Geogaceta*, **24**, 219–222.

Mascle, A. and Puigdefàbregas, C. (1998) Tectonics and sedimentation in foreland basins: results from the Integrated Basin Studies project. In: *Cenozoic Foreland Basins of Western Europe* (Eds A. Mascle, C. Puigdefàbregas, H.P. Luterbacher and M. Fernández), pp. 1–28. Special Publication 134, Geological Society Publishing House, Bath.

Maynard, J.R. and Leeder, M.R. (1992) On the periodicity and magnitude of Late Carboniferous glacio-eustatic sea-level changes. *J. Geol. Soc. London*, **149**, 303–311.

Mellere, D., Plink-Björklund, P. and Steel, R. (2002) Anatomy of shelf-edge deltas at the edge of a prograding Eocene shelf margin, Spitsbergen. *Sedimentology*, **49**, 1181–1206.

Merino-Tomé, O.A. (2004) *Estratigrafía, Sedimentología y evolución tectono-sedimentaria de las sucesiones estefanienses en la Región de Picos de Europa*. Unpublished PhD thesis, Universidad de Salamanca, Vol. I, 295 pp. and Vol. II, 109 pp.

Merino-Tomé, O.A., Colmenero, J.R., Bahamonde, J.R. and Fernández, L.P. (2001) Estratigrafía y Sedimentología de la sucesión Estefaniense del sector nororiental de la Region de Picos de Europa (Zona Cantábrica). *Stud. Geol. Salmant.*, **37**, 25–90.

Mitchum, R.M. Jr. (1977) Seismic stratigraphy and global changes of sea level. Part 11: Glossary of terms used in seismic stratigraphy. In: *Seismic Stratigraphy – Applications to Hydrocarbon Exploration* (Ed. C.E. Payton), pp. 205–212. Memoir 26, American Association of Petroleum Geologists, Tulsa, OK.

Mitchum, R.M. Jr., Vail, P.R. and Thomson III, S. (1977) Seismic stratigraphy and global changes of sea level. Part 2: The depositional sequence as a basic unit for stratigraphic analysis. In: *Seismic Stratigraphy – Applications to Hydrocarbon Exploration* (Ed. C.E. Payton), pp. 53–62. Memoir 26, American Association of Petroleum Geologists, Tulsa, OK.

Mulder, T. and Alexander, J. (2001) The physical character of subaqueous sedimentary density flows and their deposits. *Sedimentology*, **48**, 269–300.

Mulder, T. and Syvitski, J.M.P. (1995) Turbidity currents generated at river mouths during exceptional discharges to the world oceans. *J. Geol.*, **103**, 285–299.

Mulder, T., Syvitsky, J.P.M. and Skene, K.I. (1998) Modelling of erosion and deposition by turbidity currents generated at river mouths. *J. Sediment. Res.*, **68**, 124–137.

Mulder, T., Syvitsky, J.P.M., Migeon, S., Faugères, J-C. and Savoye, B. (2003) Marine hyperpycnal flows: initiation, behavior and related deposits. A review. *Mar. Petrol. Geol.*, **20**, 861–882.

Mutti, E. (1985) Turbidite systems and their relations to depositional sequences. In: *Provenance of Arenites* (Ed. G.G. Zuffa), pp. 65–93. NATO-ASI Series, Series C: Mathematical and Physical Sciences, Vol. 148, Reidel, Amsterdam.

Mutti, E., Davoli, G., Tinterri, R. and Zavala, C. (1996) The importance of ancient fluviodeltaic systems dominated by catastrophic flooding in tectonically active basins. *Mem. Sci. Geol. Padova*, **48**, 233–291.

Mutti, E., Tinterri, R., di Biase, L., Fava, L., Mavilla, N., Angella, S. and Calabrese, L. (2000) Delta-front facies associations of ancient fluvio-deltaic systems. *Rev. Soc. Geol. Esp.*, **13**, 165–190.

Mutti, E., Tinterri, R., Benevelli, G., di Base, D. and Cavanna, G. (2003) Deltaic, mixed and turbidite sedimentation of ancient foreland basins. *Mar. Petrol. Geol.*, 20, 733–755.

Nijman, W. (1998) Cyclicity and basin axis shift in a piggy-back basin: towards modelling of the Eocene Tremp-Ager Basin, South Pyrenees, Spain. In: *Cenozoic Foreland Basins of Western Europe* (Eds A. Mascle, C. Puigdefàbregas, H.P. Luterbacher and M. Fernández), pp. 135–162. Special Publication 134, Geological Society Publishing House, Bath.

Olszewski, T.D. and Patzkowsky, M.E. (2003) From cyclothems to sequences: the record of eustasy and climate on an icehouse epeiric platform (Pennsylvanian-Permian, North American Mid-continent). *J. Sediment. Res.*, **73**, 15–30.

Patton, T.L. (2004) Numerical models of growth-sediment development above an active monocline. *Basin Res.*, **16**, 25–39.

Pérez-Estaún, A., Bastida, F., Alonso, J.L., *et al.* (1988) A thin-skinned tectonics model for an arcuate fold and thrust belt: The Cantabrian Zone (Variscan Ibero-Armorican Arc). *Tectonics*, **7**, 517–537.

Pickard, N.A.H. (1996) Evidence for microbial influence on the development of Lower Carboniferous buildups. In: *Recent Advances in Lower Carboniferous Geology* (Eds P. Strogen, I.D. Somerville and G.Ll. Jones), pp. 65–82. Special Publication 107, Geological Society Publishing House, Bath.

Posamentier, H.W. and James, D.P. (1993) An overview of sequence-stratigraphic concepts: uses and abuses. In: *Sequence Stratigraphy and Facies Associations* (Eds H.W. Posamentier, C.P. Summerhayes, B.U. Haq and G.P. Allen), pp. 3–18. Special Publication 18, International Association of Sedimentologists. Blackwell Scientific Publications, Oxford.

Postma, G. (1990) Depositional architecture and facies of rivers and fan deltas: a synthesis. In: *Coarse-grained Deltas* (Eds A. Colella and D.B. Prior), pp. 13–27. Special Publication 10, International Association of Sedimentologists. Blackwell Scientific Publications, Oxford.

Proust, J.N., Chuvashov, B.I., Vennin, E. and Boisseau, T. (1998) Carbonate platform downing in a foreland setting: The mid-Carboniferous platform in western Urals (Russia). *J. Sediment. Res.*, **68**, 1175–1118.

Purser, B.H. (1973) Sedimentation around bathymetric highs in the southern Persian Gulf. In: *The Persian Gulf – Holocene Carbonate Sedimentation and Diagenesis in a Shallow Water Epicontinental Sea* (Ed. B.H. Purser), pp. 1–9. Springer-Verlag, New York.

Rasbury, E.T., Hanson, G.N., Meyers, W.J., Holt, W.E., Goldstein, R.H. and Saller, A.H. (1998) U–Pb dates paleosols: Constraints on late Paleozoic cycle durations and boundary ages. *Geology*, **26**, 403–406.

Read, J.F. (1980) Carbonate ramp to basin transitions and foreland basin evolution, Middle Ordovician, Virginia Appalachians. *Am. Assoc. Petrol. Geol. Bull.*, **64**, 1575–1612.

Reading, H.G. and Collinson, J.D. (1996) Clastic Coasts. In: *Sedimentary Environments: Processes, Facies and Stratigraphy*, 3rd edn (Ed. H.G. Reading), pp. 154–231. Blackwell Science, Oxford.

Riding, R. (2000) Microbial carbonates: the geological record of calcified bacterial-algal mats and biofilms. *Sedimentology*, **47**(Suppl. 1), 179–214.

Ross, C.A. and Ross, J.R.P. (1985) Late Paleozoic depositional sequences are synchronous and worldwide. *Geology*, 13, 194–197.

Ross, C.A. and Ross, J.R.P. (1988) Late Paleozoic transgressive-regressive deposition. In: *Sealevel Changes: an Integrated Approach* (Eds C.K. Wilgus, B.S. Hasting, C.G. Kendall, H.W. Posamentier, C.A. Ross and J.C. Van Wagoner), pp. 227–247. Special Publication 42. Society of Economic Paleontologists and Mineralogists, Tulsa, OK.

Saller, A.H., Dickson, J.A.D. and Matsuda, F. (1999) Evolution and distribution of porosity associated with subaerial exposure in Upper Paleozoic platform limestones, West Texas. *Am. Assoc. Petrol. Geol. Bull.*, **83**, 1835–1854.

Samankassou, E. (1998) Skeletal framework mounds of dasycladacean alga *Anthracoporella*, Upper Paleozoic, Carnic Alps, Austria. *Palaios*, **13**, 297–300.

Sánchez de Posada, L.C., Martínez Chacón, M.L., Méndez, C., Menéndez-Álvarez, J.R., Truyols, J. and Villa, E. (1993) El Carbonífero de las regiones de Picos de Europa y Manto del Ponga (Zona Cantábrica, N de España): fauna y bioestratigrafía. *Rev. Esp.Paleontol.*, no. extraordinario, 89–108.

Sánchez de Posada, L.C., Martínez Chacón, M.L., Méndez, C., *et al.* (1996) El Carbonífero marino del ámbito Astur-Leonés (Zona Cantábrica): síntesis paleontológica. *Rev. Esp.Paleontol.*, no. extraordinario, 82–96.

Sánchez de Posada, L.C., Villa, E., Martínez Chacón, M.L., Rodríguez, R.M., Rodríguez, S. and Coquel, R. (1999) Contenido paleontológico y edad de la sucesión de Demúes (Carbonífero, Zona Cantábrica). *Trab. Geol., Univ. Oviedo*, **21**, 339–352.

Sanders, D. and Höfling, R. (2000) Carbonate deposition in mixed siliciclastic-carbonate environments on top of an orogenic wedge (Late Cretaceous, Northern Calcareous Alps, Austria). *Sediment. Geol.*, 137, 127–146.

Schmoker, J.G. and Halley, R.B. (1982) Carbonate porosity versus depth: a predictable relation for South Florida. *Am. Assoc. Petrol. Geol. Bull.*, **66**, 2561–2570.

Sclater, J.G. and Christie, P.A.F. (1980) Continental stretching: an explanation of the post-mid-Cretaceous subsidence of the Central North Sea Basin. *J. Geophys. Res.*, **85**, 3711–3739.

Shanley, K.W., McCabe, P.J. and Hettinger, R.D. (1992) Tidal influence in Cretaceous fluvial strata from Utah, USA: a key to sequence stratigraphic interpretation. *Sedimentology*, **39**, 905–930.

Smith, L.B. and Read, J.F. (2000) Rapid onset of late Paleozoic glaciation on Gondwana: evidence from the Upper Mississippian strata of the Midcontinent of United States. *Geology*, **28**, 279–282.

Somerville, I.D. and Strogen, P. (1992) Ramp sedimentation in the Dinantian limestones of the Shannon Trough, Co. Limerick, Ireland. *Sediment. Geol.*, **79**, 59–75.

Soreghan, G.S. (1994) Stratigraphic responses to geologic processes: Late Pennsylvanian eustasy and tectonics in the Pedregosa and Orogrande basins, Ancestral Rocky Mountains. *Geol. Soc. Am. Bull.*, **106**, 1195–1211.

Soreghan, G.S. and Giles, K.A. (1999) Amplitudes of Late Pennsylvanian glacioeustasy. *Geology*, **27**, 255–258.

Stam, B., Gradstein, F.M., Lloyd, P. and Gillis, D. (1987) Algorithms for porosity and subsidence history. *Comput. Geosci.*, **13**, 317–349.

Tardy, Y. (1993) *Pétrologie des latérites et des sóls tropicaux*. Masson, Paris, 459 pp.

Torsvik, T.H. and Cocks, L.R.M. (2004) Earth geography from 400 to 250 Ma: a palaeomagnetic faunal and facies review. *J. Geol. Soc. London*, **161**, 555–572.

Ulicný, D. (1999) Sequence stratigraphy of the Dakota Formation (Cenomanian), Southern Utah: interplay of eustasy and tectonics in a foreland basin. *Sedimentology*, **46**, 807–836.

Van Ginkel, A.C. (1971) Fusulinids from uppermost Myachkovian and Kasimovian strata of northwestern Spain. *Leidse Geol. Meded*, **47**, 115–161.

Van Wagoner, J.C., Mitchum, R.M., Campion, K.M. and Rahmanian, V.D. (1990) *Siliciclastic Sequence Stratigraphy in Well Logs, Cores, and Outcrops*. Methods in Exploration Series, Vol. 7, American Association of Petroleum Geologists, Tulsa, OK, 55 pp.

Veevers, J.J. and Powell, C.M. (1987) Late Paleozoic glacial episodes in Gondwanaland reflected in transgressive-regressive depositional sequences in Euramerica. *Geol. Soc. Am. Bull.*, **28**, 475–487.

Vergés, J., Marzo, M., Santaeularia, T., *et al.* (1998) Quantified vertical motions and tectonic evolution of the SE Pyrenean foreland basin. In: *Cenozoic Foreland Basins of Western Europe* (Eds A. Mascle, C. Puigdefàbregas, H.P. Luterbacher and M. Fernández), pp. 107–134. Special Publication 134, Geological Society Publishing House, Bath.

Villa, E. (1995) *Fusulináceous carboníferos del este de Asturias (N de España)*. Bioestratigraphie du Paleozoique, Vol. 13. Université Claude Bernard-Lyon I, Lyon, 261 pp.

Villa, E. and Bahamonde, J.R. (2001) Accumulations of Ferganites (Fusulinacea) in shallow turbidite deposits from the Carboniferous of Spain. *J. Foramin. Res.*, **31**, 173–190.

Villa, E. and Ueno, K. (2002) Characteristics and paleogeographic affinities of the early Gzhelian fusulinoideans from the Cantabrian Zone (NW Spain). *J. Foramin. Res.*, **32**, 135–154.

Villa, E. and van Ginkel, A.C. (1999) First record of Gzhelian fusulinids from the Carboniferous of northern Spain. *Rev. Esp. Paleontol.*, Vol. Hom. Prof. Truyols, 205–216.

Villa, E., Merino-Tomé, O., Bahamonde, J.R. and Ueno, K. (2003) Fusulinoideans from the Puentellés Formation (Upper Carboniferous, NW Spain): discussion on phylogeny, paleoecology and paleobiogeography. *Riv. Ital. Paleontol. Stratigr., Milano*, **109**, 241–253.

Wendt, J., Kaufmann, B. and Belka, Z. (2001) An exhumed Paleozoic underwater scenary: the Visean mud mounds of the eastern Anti-Atlas (Morocco). *Sediment. Geol.*, **145**, 215–233.

Wilson, J.L. (1969) Microfacies and sedimentary structures in deeper-water lime mudstones. In: *Depositional Environments in Carbonate Rocks* (Ed. G.H. Friedman), pp. 4–19. Special Publication 14, Society of Economic Paleontologists and Mineralogists, Tulsa, OK.

Wright, V.P. (1986) Facies sequences on a carbonate ramp: the Carboniferous Limestone of South Wales. *J. Sediment. Petrol.*, **33**, 221–241.

Zweigel, J. Aigner, T. and Luterbacher, H. (1998) Eustatic versus tectonic controls on Alpine foreland basin fill: sequence stratigraphy and subsidence analysis in the SE German Molasse. In: *Cenozoic Foreland Basins of Western Europe* (Eds A. Mascle, C. Puigdefàbregas, H.P. Luterbacher and M. Fernández), pp. 299–323. Special Publication 134, Geological Society Publishing House, Bath.

Peritidal carbonate–evaporite sedimentation coeval to normal fault segmentation during the Triassic–Jurassic transition, Iberian Chain

MARC AURELL*, BEATRIZ BÁDENAS*, ANTONIO M. CASAS* *and* RAMÓN SALAS†

*Departamento Ciencias de la Tierra, Universidad Zaragoza, 50.009 Zaragoza, Spain (Email: maurell@unizar.es)
†Departament GPPG, Universitat Barcelona, 08.028 Barcelona, Spain

ABSTRACT

Around the Triassic–Jurassic transition, a major tectonic rifting phase affected the northern part of the Iberian Basin (northeast Spain). Extensive normal faulting resulted in basin segmentation, as reflected by the rapid thickness and facies variation of the syn-rift, shallow-marine to supratidal carbonate–evaporite units. The upper Rhaetian–Hettangian syn-rift units exposed around the locality of Morata de Jalón (northeast Spain) provide good exposures that enable a precise sedimentological and structural analysis. The study of these units was achieved through geological mapping of an area of 10 km^2, facies characterization, correlation of seven selected logs (vertical thickness from 40 m to 135 m), and measurement of normal faults, fractures and slump folds. The orientation of the major syn-sedimentary normal faults (NW–SE to NNW–SSE) suggests an origin through reactivation of late Variscan faults located within the Palaeozoic basement. However, the orientation of the newly formed faults and joints indicates a main N–S extension direction. The tectono-sedimentary evolution of the studied basin can be summarized in two stages:

1 at the end of the Triassic, a subsiding salina formed in the downthrown blocks of two normal faults – salina deposits were sourced from marine waters and, also probably, from the weathering of previously deposited evaporite-rich units;
2 during the Hettangian, tectonic reactivation combined with the long-term regional sea-level rise resulted in the formation of a tidal flat complex.

These graded, in the areas of greater subsidence, into shallow-marine carbonate platform deposits. Syn-sedimentary fracturing affected the early lithified carbonates, favouring the formation of collapse-breccias in the tidal-flat environment, and limestone rudites (sedimentary breccias and conglomerates) in the subtidal domain, formed and partly transported as submarine debris flows.

Keywords Extensional basins, shallow-marine carbonates, carbonate breccias, evaporites, normal faults.

INTRODUCTION

During Mesozoic times, the western part of Iberia formed an uplifted massif surrounded by intracratonic basins. The evolution and the amount of accommodation created in these basins were controlled by discontinuous extensional tectonic activity, which was mainly concentrated in the two rifting episodes in the Triassic and in the latest Jurassic to Early Cretaceous. More stable tectonic periods, characterized by broad and homogeneous thermal subsidence, favoured the formation of wide marine epicontinental platforms (Salas & Casas, 1993; Salas *et al.*, 2001).

A Late Triassic to Early Jurassic extensional tectonic event can be considered the last pulse of

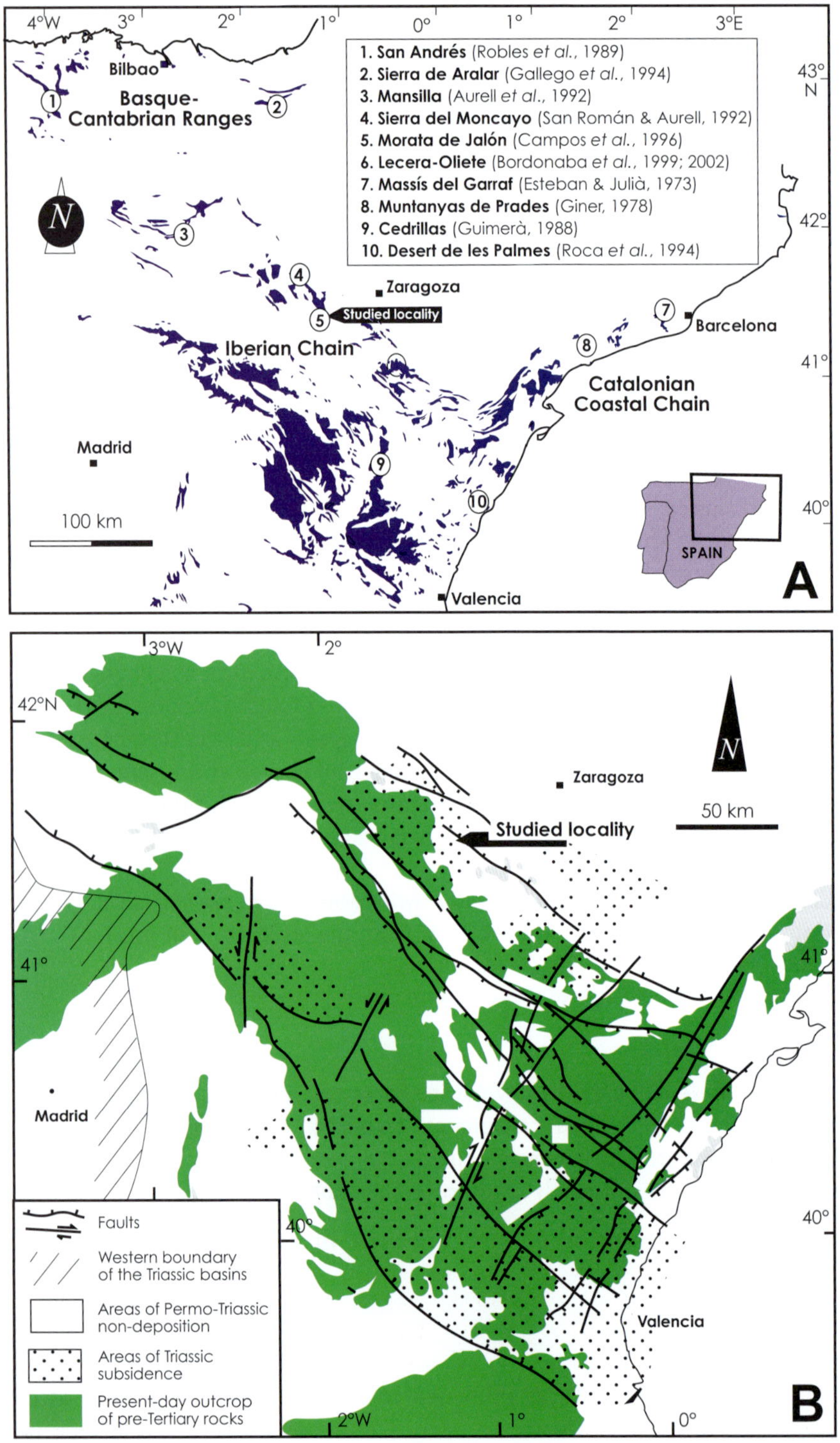

Fig. 1 (A) Distribution of the Jurassic outcrops (blue) in northeast Spain, indicating areas (numbered 1–10) where extensional tectonic activity developed around the Triassic–Jurassic transition has been documented. The locality studied in this work (5, Morata de Jalón) was first described by Campos *et al.* (1996). (B) An outline geological map of the area of study.

the Triassic rift episode. This tectonic event has been studied in several localities across northeast Spain (see Fig. 1 and references therein). The sedimentary units that formed coeval to this tectonic activity locally show thickness variation and facies changes (shallow-marine carbonates, breccias of different origin and evaporites), reflecting tilting of blocks due to normal faulting, as has been documented in different areas (Fig. 1), such as the Sierra del Moncayo (San Román & Aurell, 1992), the Desert de les Palmes (Roca *et al.*, 1994) and the Lecera–Oliete area (Bordonaba *et al.*, 1999; Bordonaba & Aurell, 2002). In some localities, the subaereal erosion of the uplifted blocks resulted in the formation of a prominent angular unconformity, and the syn-rift and post-rift units may lie over older, Triassic or even Palaeozoic, rocks (Riba *et al.*, 1971; Esteban & Juliá, 1973; Robles *et al.*, 1989; San Román & Aurell, 1992; Roca *et al.*, 1994).

In addition to the localities previously documented in the northern part of the Iberian Chain, the Morata de Jalón area (Fig. 1) is of special relevance in understanding the tectonics and sedimentary events occurring around the Triassic–Jurassic transition. This is because:

1 the excellence of the outcrops allows a precise reconstruction of thickness and facies changes of the syn-rift units;
2 the coexistence in an area of a few square kilometres of the main peritidal to shallow-marine facies developed around the Triassic–Jurassic transition in the Iberian basin;
3 the presence of a number of mappable syn-sedimentary normal faults, which have not been significantly inverted during Alpine compression.

The sedimentological and structural analysis of the uppermost Triassic to lowermost Jurassic units reported in this work (Morata de Jalón, northeast Spain) has resulted in the characterization of a number of detrital, evaporitic and carbonate facies developed in terrestrial, peritidal to shallow-marine environments. This set of facies developed coevally with intense extensional tectonic activity, as documented by the existence of syn-sedimentary normal faults of different scales controlling thickness and facies variations, as well as the formation of depositional slopes containing decametre-scale olistoliths and slumps.

The case study provides some key arguments that unravel the origin of a wide spectrum of sedimentary, diagenetic and tectono-sedimentary carbonate breccias developed in the shallow-subtidal to supratidal environments. In addition, the reconstruction model provides a tool to further understand the complex geological history occurring around the Triassic–Jurassic boundary, not only in the Iberian basin, but also in other Alpine–Mediterranean areas. The fragmentation and subsidence of the carbonate platforms in the whole Alpine–Mediterranean region began with the Triassic–Jurassic transition (e.g. Bernoulli & Jenkyns, 1974; Fütchbauer & Ricchter, 1983). Similar to the model reconstructed in the presented case study, Cozzi & Hardie (2003) have reported on normal faulting prior to and at the Triassic–Jurassic boundary in the Carnian Prealps, which was related to a major rifting phase for the northwestern Tethys starting at the middle Norian.

GEOLOGICAL SETTING

Palaeogeographical remarks and stratigraphy

During Triassic times, three successive carbonate epeiric platforms covered large areas of northeast Iberia: the lower two developed during the Middle Triassic (lower and upper Muschelkalk facies), the upper one during the Late Triassic (upper Norian to lower Rhaetian Imón–Isabena formations; Calvet *et al.*, 1990; Arnal *et al.*, 2002). Major tectonic activity occurred around the Triassic–Jurassic transition, causing the break-up of the wide Late Triassic epeiric carbonate platform and the formation of a prominent unconformity below the syn-rift sequence (i.e. the Cortes de Tajuña Formation) in the northern part of the Iberian basin.

The unconformity located below the Cortes de Tajuña Formation was regarded as the lower boundary of the so-called Lower Jurassic Cycle (Fig. 2; Aurell *et al.*, 2003). However, its age cannot be precisely established. The Rhaetian–Hettangian boundary has been placed in the lower part of this unit on the basis of scarce benthic fossils and palynomorph associations (Pérez-López *et al.*, 1996; Barrón *et al.*, 2001; Arnal *et al.*, 2002). In the northwestern Iberian Chain, the Hettangian–Sinemurian boundary is assumed to be located

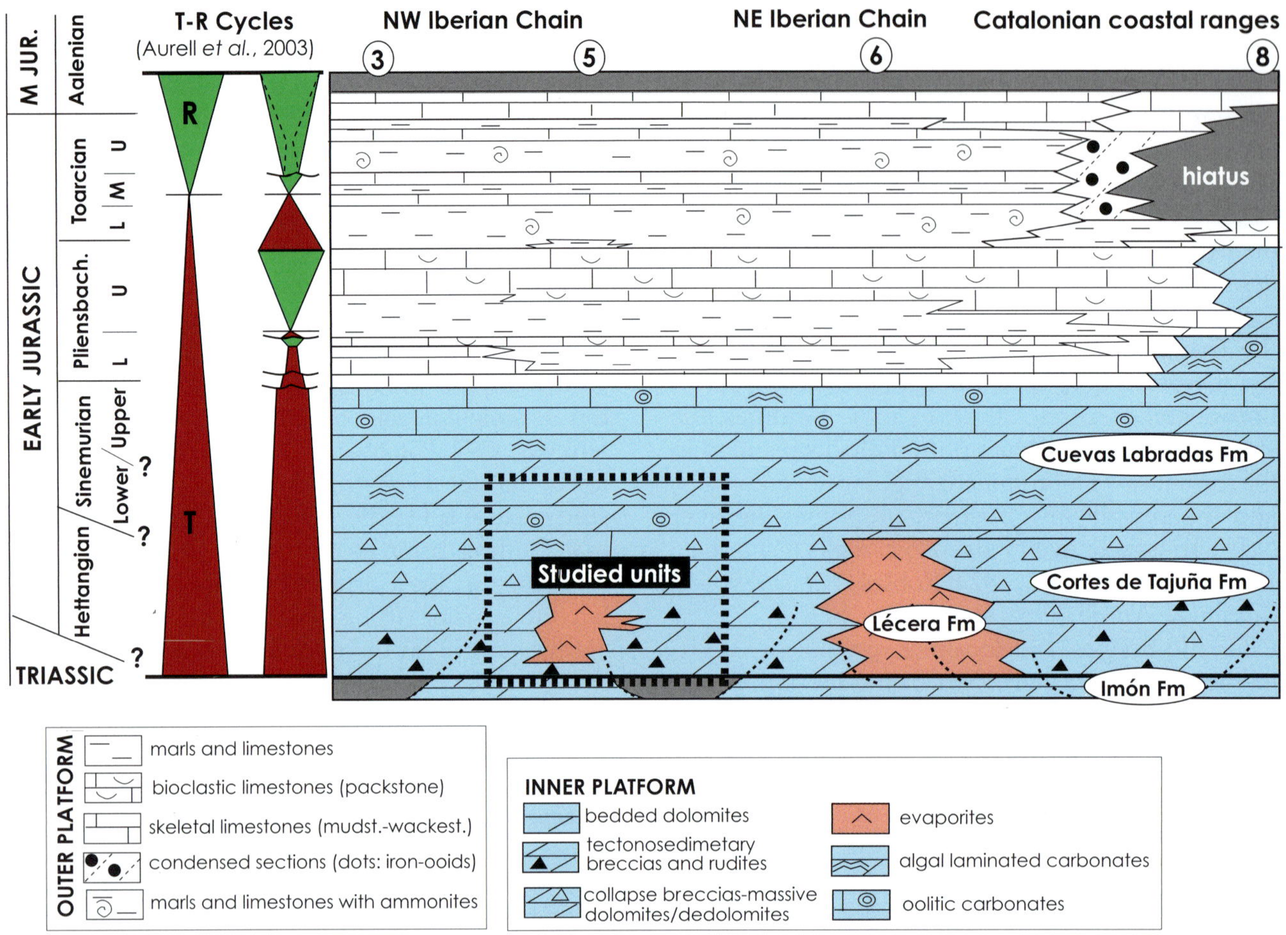

Fig. 2 Main facies distribution, transgressive–regressive (T–R) cycles and lithostratigraphy of the Lower Jurassic rocks found in the northern Iberian Chain. The numbers in the upper part of the figure correspond to the reference localities indicated in Fig. 1A.

around the middle or in the upper part of the Cortes de Tajuña Formation (e.g. Comas-Rengifo & Yébenes, 1988). This formation gradually changes upwards into the well-bedded Sinemurian carbonates of the Cuevas Labradas Formation. As a whole, the Cortes de Tajuña Formation and the Cuevas Labradas Formation define a transgressive hemicycle (long-term evolution from peritidal to shallow subtidal), bounded at the top by the major flooding event of the latest Sinemurian (Aurell *et al.*, 2003). This flooding event resulted in hemipelagic outer platform conditions all across the northern part of the Iberian basin (Fig. 2).

Overall structure of the Morata de Jalón area

The present-day structure around the Morata de Jalón area (Fig. 3) is the result of successive deformational episodes, comprising part of the overall tectonic evolution of the Iberian Chain. This evolution can be summarized as a Mesozoic extensional stage with two episodes of rifting, followed by a Tertiary compressional stage with basement uplift and tectonic inversion (see e.g. Guimerà & Alvaro, 1990; Salas & Casas, 1993; Casas *et al.*, 2000). The Mesozoic evolution of the Iberian basin was strongly conditioned by the structural framework resulting from the late Variscan fracturing (Permian, Arthaud & Matte, 1975), with basement faults oriented NW–SE and NE–SW.

In the Morata de Jalón area, the Tertiary compressional stage resulted in the formation of a faulted, antiform-like wide fold that can be followed 5 km along trend, limited to the south by the Río Grío Fault, which shows a right-lateral strike-slip combined with reverse movement (Fig. 3; Campos

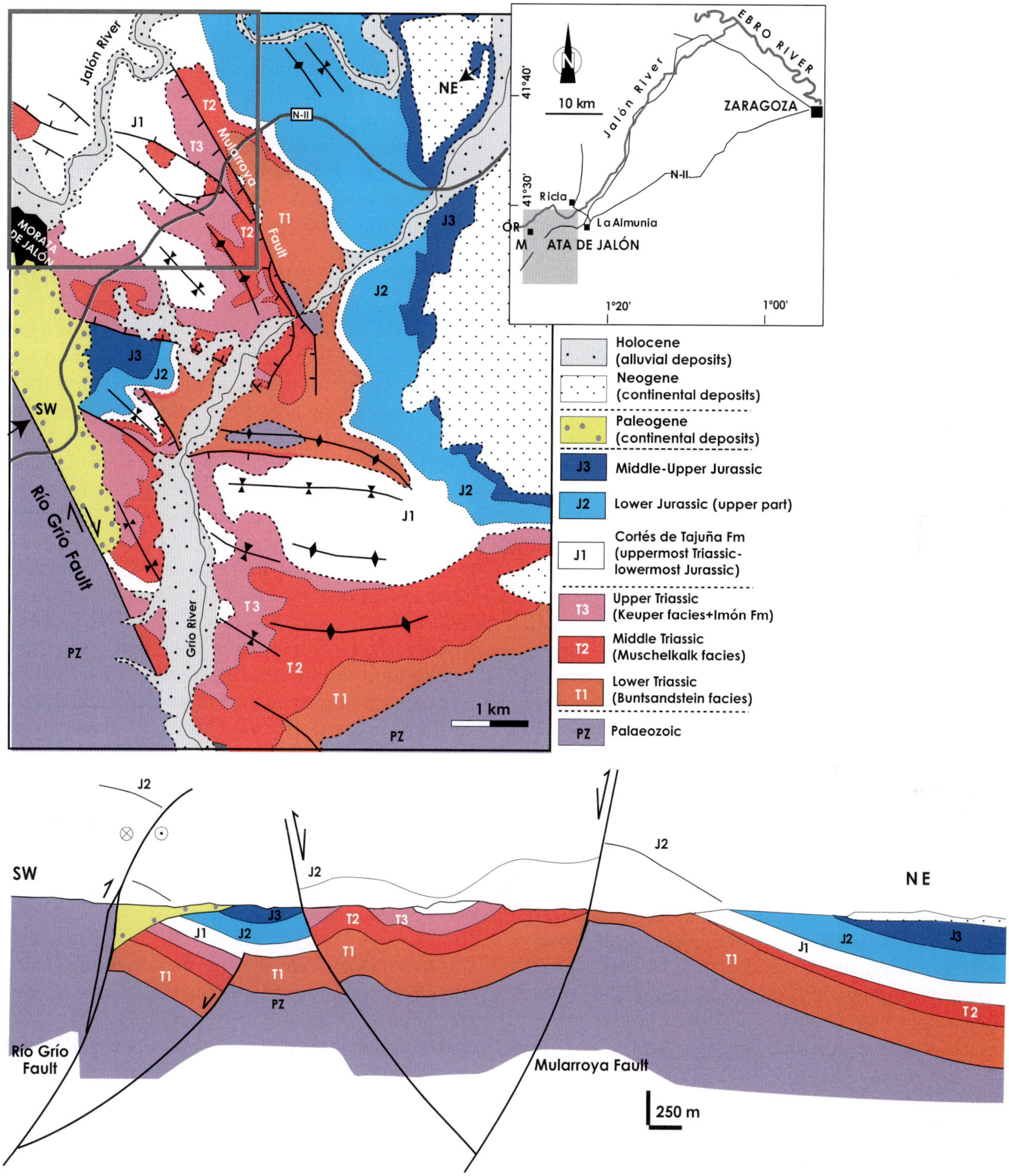

Fig. 3 Geographical and geological location of the outcrops studied northeast of Morata de Jalón (boxed area). The NE–SW cross-section is modified from Campos *et al.* (1996).

et al., 1996). In the southwestern, uplifted block of this fault, the Mesozoic sequence is completely eroded. The Tertiary activity of this fault, inherited from the late Variscan fracturing episode, is evidenced by syn-tectonic Paleogene deposits cropping out in its foot-wall (northeast of the fault). Most of the E–W trending folds located in the southern part of the studied area can be related to compressional steps or jogs located along the Río Grío Fault. To the northeast of the study area, the Neogene deposits of the Ebro basin unconformably cover the Mesozoic units, precluding direct observation of the structure. Subsurface data (San Román, 1994; Cortés, 2004) point to a gradual deepening of the Mesozoic units toward the north, folded with an E–W trend and cut by NNE–SSW strike-slip faults (Cortés & Casas, 1996).

The Mesozoic extensional episodes in the Morata de Jalón area resulted in the formation of several NW–SE normal faults, most of them dipping to the southwest. The syn-sedimentary origin of these faults can be constrained by the existence of deposits of Late Triassic to Early Jurassic age (Campos *et al.*, 1996). Furthermore, a regional unconformity between the Lower–Middle Triassic Buntsandstein and Muschelkalk facies and the syn-tectonic Cortes de Tajuña Formation can also be distinguished (Fig. 3). The latest Jurassic to Early Cretaceous extensional stage, the most important in the western and eastern sectors of the Iberian Chain (Salas & Casas, 1993), is represented by less than 100 m of continental deposits, cropping out southeast of the study area. Some of the Mesozoic extensional faults were reactivated as reverse faults during the Tertiary compression (Fig. 3), although for the most part the original extensional movement on the faults was not completely recovered. Tertiary compression formed a wide antiform in the area, rather than inversion of the normal faults (see cross-section in Fig. 3).

FACIES ANALYSIS

Methods and results

The data presented in this work were collected from fieldwork over an area of around 10 km^2. The geological map (Fig. 4) shows four main NW–SE trending sets of normal faults. For the most part, the movement along these faults is thought to have occurred after the deposition of the Upper Triassic Imón Formation. In the uplifted blocks, the previously deposited Triassic units were partly eroded and the syn-rift unit locally overlies Middle Triassic units (Muschelkalk facies). In the downthrown blocks, most of the Triassic was preserved from erosion, and the syn-rift unit generally rests conformably over the Imón Formation.

Fieldwork provided data for the reconstruction of the facies and thickness distribution of the syn-rift Cortes de Tajuña Formation from the measurements of seven selected sedimentary logs (Fig. 5). The facies characterization was completed with the aid of thin sections. The vertical and lateral distribution of the main facies types and their relationship to the active normal faults are shown in two transects, correlating the four logs located near the A-2 Highway and the Jalón River (transects A and B respectively, in Fig. 5). The proposed correlation is well constrained by the lateral continuity of the outcrops and by the existence of some marker beds that can be followed across the area studied (i.e. the *slump marker beds* and the *laminated marker beds* 1 and 2; see Fig. 5). The upper surface of the section was 'hung' from the well-bedded limestones (the *laminated marker beds* 2) that show the transition from the studied syn-rift unit (Cortes de Tajuña Formation) to the overlying Sinemurian(?) Cuevas Labradas Formation (Fig. 2). In the syn-rift sequence, four main facies associations were defined: well-bedded carbonates, massive carbonates, sulphate evaporites and detrital rocks.

Well-bedded carbonates

This facies association has in common the existence of well-defined plane beds forming decimetre- to metre-thick strata, frequently forming thinning-upward successions (Fig. 6A). Three different facies associations have been recognized: mud-supported limestones, grain-supported limestones and algal-laminated limestones–dolostones. The mud-supported limestones are the more common facies, whereas most of the grain-supported facies are found around Section B3 (Fig. 5).

Mud-supported limestones

These are bioclastic lime mudstones with scattered whole bivalves and bioclasts (bivalves, gastropods,

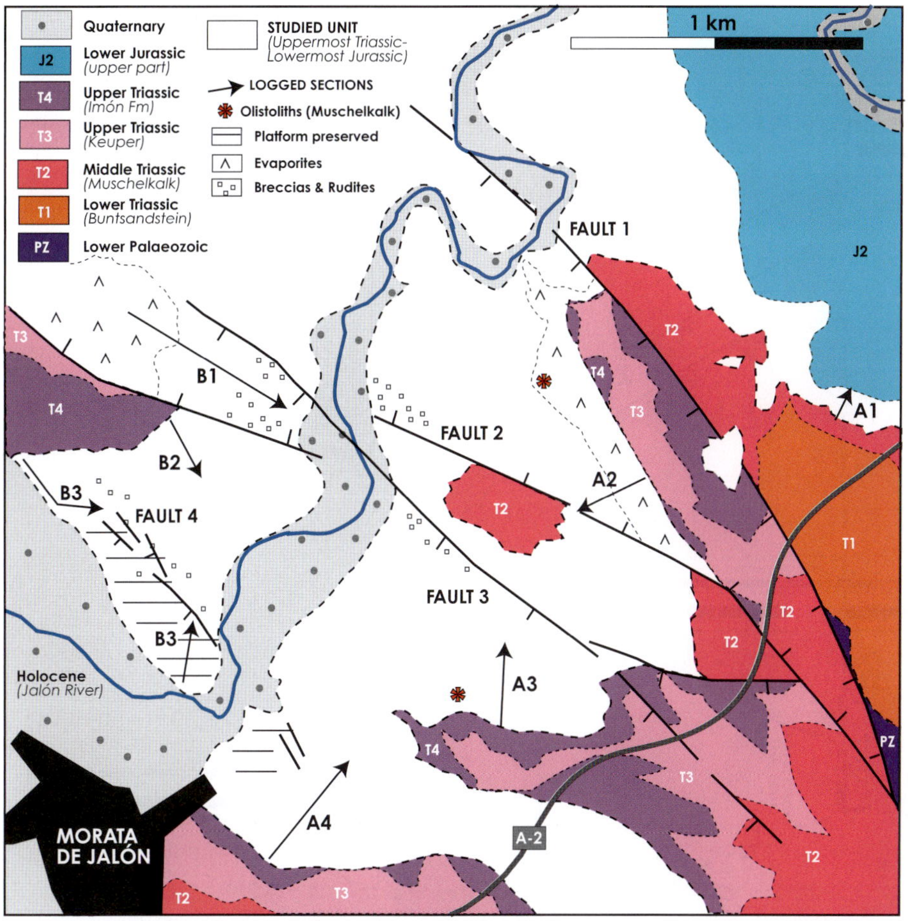

Fig. 4 Geological map northeast of Morata de Jalón, indicating the distribution of the main facies recognized in the units studied. The syn-sedimentary normal Faults 1–4 controlled the facies and thickness distribution in the unit studied. The locations of the sections logged near the A-2 Highway (A1–A4) and near the Jalón River (B1–B4) are also indicated.

brachiopods, sponge spicules, echinoderms). They occasionally show millimetre- to centimetre-thick graded levels of ooids and peloids with erosive bases and planar lamination that are interpreted as storm levels (i.e. tempestites). Bioturbation is scarce, although millimetre burrows filled with ooids and peloids may occur. The facies was deposited in a shallow subtidal platform: the abundance of carbonate mud and the fossil diversity indicate low-energy and normal salinity depositional conditions; the peloidal and oolitic tempestite levels correspond to resedimented material derived from the high-energy facies belts.

Grain-supported limestones

This facies is composed of packstones to grainstones with variable proportions of peloids, ooids, oncoids and bioclasts. Planar and cross-lamination and burrows are occasionally present. Based on the main components, four subfacies have been differentiated:

Oolitic packstones–grainstones. Dominated by ooids with peloidal and bioclastic nuclei and well-developed cortical layers (micritic and fine sparitic laminae with concentric crystal structure). Peloids (1–2 mm in diameter) and micritic and peloidal intraclasts are also present in variable proportion. The fossil content is scarce and is represented by scattered bivalves, echinoderms, gastropods, brachiopods, foraminifera (textularids, miliolids), dasycladacean algae and *Favreina*. This facies was deposited in active oolitic shoals. The ooids are similar to those interpreted by Strasser (1986) as having been deposited in relatively high-energy, non-restricted

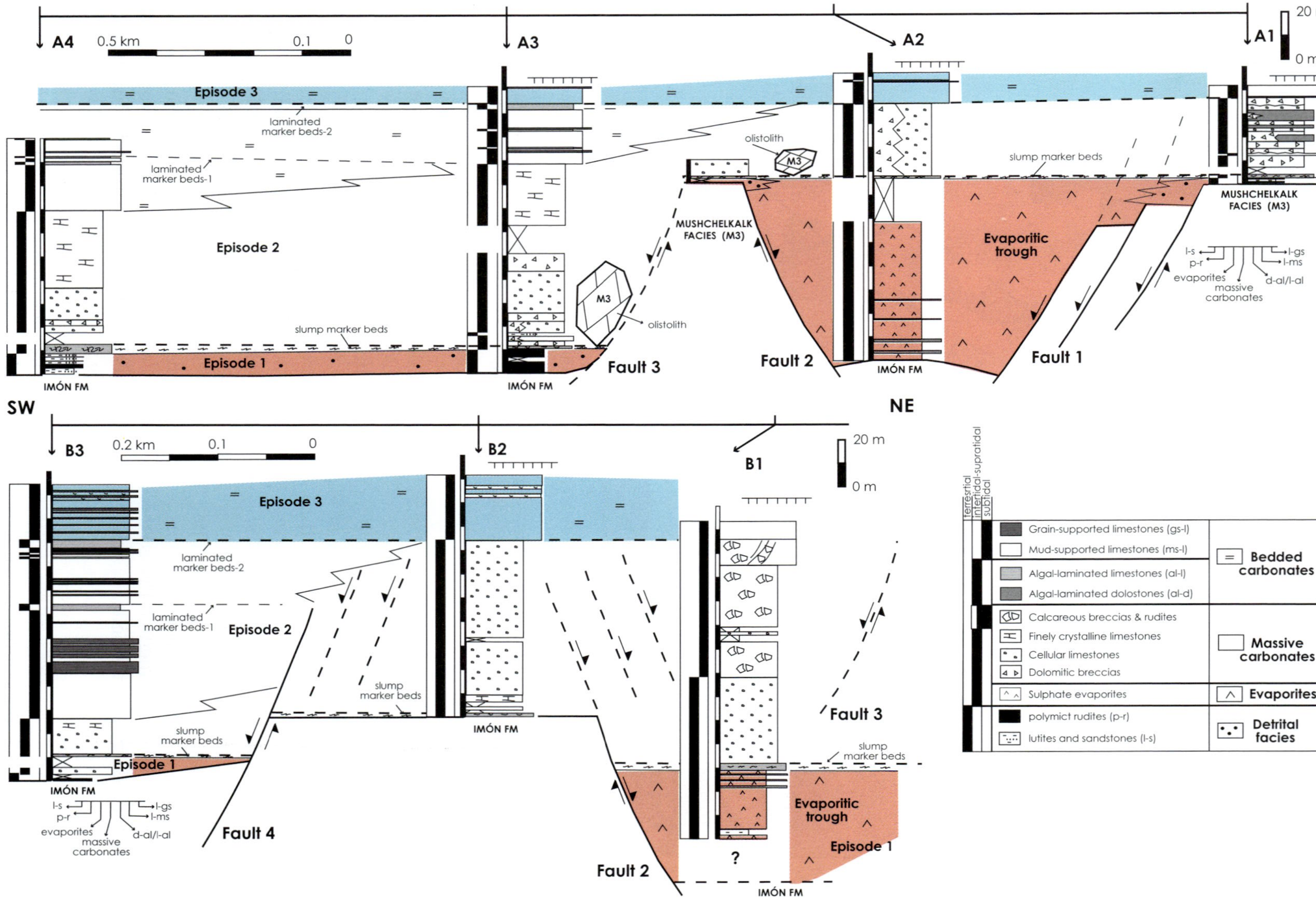

Fig. 5 Synthetic representation of the seven sections logged around the A-2 Highway (A1–A4) and near the Jalón River (B1–B4). Notice the change of the horizontal scale between transects A and B. The correlation was based on the lateral tracing in the field of the slump and laminated marker beds.

Fig. 6 Lithofacies. (A) Well-bedded limestones exposed in the middle part of Section B3, which is 15 m high. (B) Laminated carbonates affected by small-scale normal faults and fractures, in the lower part of Section A1. Pen is 13 cm long. (C) Massive to dark grey–white laminated gypsum exposed in a quarry from the lower part of Section B1. Hammer shaft is 30 cm long. (D) Poorly sorted dolomitic breccia in the middle part of Section A1. Pen is 15 cm long.

marine environments. The high fossil diversity also indicates normal salinity.

Burrowed oncolitic–intraclastic packstones. Made up of poorly sorted ooids–oncoids that are frequently aggregated. They display micritic-intraclastic cores and well-developed cortical layers (micritic fine laminae and intercalated sparitic laminae with concentric crystal structure). Micritic intraclasts and micritized gastropods and bivalves are also abundant. Echinoderms, corals and foraminifera (miliolids, textularids) are present in lower proportions. On the basis of the aggregates and micritized bioclasts, deposition in relatively low-energy conditions in inactive or stabilized oolitic shoals is inferred (cf. Tucker & Wright, 1990).

Peloidal packstones–grainstones. Mainly composed of well-sorted and well-rounded peloids, most of them originating from organic activity (*pellets*). However, the presence of micritic intraclasts and partially micritized bioclasts suggests a non-faecal origin for some of the peloids. Ooids are present in variable proportion. Scattered bioclasts of bivalves, gastropods, brachiopods (occasionally forming graded, sharp-based and continuous levels, interpreted as tempestites), echinoderms, foraminifera (textularids, miliolids), ostracods and dasycladacean algae are recognized. The fossil diversity and the presence of high-energy ooids indicate the formation of shoals in relatively open and high-energy marine conditions (Strasser, 1986; Tucker & Wright, 1990).

Oolitic and bioclastic wackestones–packstones. Mainly made up of poorly sorted ooids with diverse nuclei (peloids, bioclasts, fragmented ooids, intraclasts) and variably developed cortical layers (2–3 thick sparitic laminae with radial crystal structure). The bioclastic fraction is mainly composed of ostracods, gastropods and bivalves. Foraminifera (miliolids, textularids), dasycladacean algae and echinoderms, as well as scattered micritic intraclasts of various sizes, peloids and quartz silt are also present. The morphology of the ooids indicates relatively restricted, low-energy marine environments (Strasser, 1986). The predominance of ostracods, gastropods and bivalves indicates fluctuations in the salinity in a semi-restricted lagoon.

Algal laminated limestones and dolostones

These are made up of submillimetre- to millimetre-thick microbial and micropeloidal laminae with frequent fenestrae. The observed structures include tepees, mud cracks, flat pebbles (desiccation breccias) and lenticular evaporite moulds generally filled with calcite (evaporite pseudomorphs). The presence of algal lamination and desiccation features indicates low-energy depositional environments between the intertidal to supratidal zones. The inference of pseudomorphs of evaporite indicates arid climatic conditions. The facies includes a particular level with abundant slumps (see Fig. 9C), which was used as a marker bed throughout the area studied (*slump marker beds*).

Massive carbonates

The common feature of this group of facies is the absence of well-defined bedding, resulting in a massive appearance in the field (Fig. 6C). A wide spectrum of carbonate facies have been grouped into massive to poorly bedded dolomitic breccias, cellular limestones, finely crystalline limestones, and calcareous breccias and rudites.

Dolomitic breccias

The clast and matrix-supported breccias are formed by poorly sorted, centimetre- to decimetre-sized dolomitic clasts, and calcareous matrix (Fig. 6B & D). The clasts are angular to poorly rounded and show frequent algal lamination and evaporite pseudomorphs. Occasionally, the clasts show deformation features (which can be traced following the algal lamination), indicating that the sediment was poorly lithified during the process of formation of the breccia. The breccias change laterally and vertically to mud-supported and algal-laminated dolostones at the outcrop scale. The lateral change may be gradual or controlled by the presence of small fractures. The lower boundary of the breccia levels corresponds to irregular surfaces (often with dessication-crack cavities) and in some cases to highly irregular karstic surfaces.

The origin of this breccia has been traditionally related to the collapse of carbonate layers formed under peritidal conditions after dissolution of evaporites (i.e. collapse breccia: e.g. Morillo & Meléndez, 1979; San Román & Aurell, 1992). The time of formation of this breccia, either during early burial and/or during late diagenesis, is open to discussion (see Bordonaba & Aurell, 2002; Ortí & Salvany, 2004). In the outcrops studied, a model of formation during the early stages of burial and cementation is further supported by the presence of deformed soft clasts and by the relationship between the breccias and the parent facies. The process of brecciation could also be controlled and favoured by the development of small-scale extensional faults and fractures in the early cemented carbonates (Fig. 6B).

Cellular limestones

These facies correspond to grey–reddish coarsely crystalline limestones with abundant open secondary porosity of variable size (from micropores to centimetre-sized pores). Lateral changes between the cellular limestones and dolomitic breccias have been observed within a stratigraphic level at outcrop scale. The cellular limestones show occasional intercalated decimetre-thick levels of preserved algal-laminated dolostones. This evidence, along with the presence of pseudomorphs of dolomitic crystals (rhombic and zoned calcite crystals) and preserved dolomitic clasts with algal lamination and evaporite pseudomorphs, supports their origin from dolomite dissolution and/or replacement (calcitization) of previous dolomitic breccias and algal dolostones, generating the secondary porosity. This dedolomitization process is thought to have occurred by meteoric water circulation, most probably during the latest stages of diagenesis.

Finely crystalline limestones

This facies represents a gradual transition between the cellular limestones and the mud-supported limestone facies described above. They show evidence of a texture predominantly formed by a mosaic of sparitic crystals without relict grains (sparstones, *sensu* Wright, 1992). The occasional presence of pseudomorphs of dolomitic crystals and preserved centimetre-thick levels with algal lamination indicate that they may correspond to the diagenetic alteration of carbonate mud-supported sediments deposited in peritidal environments.

Limestone breccias and rudites

The facies is best seen in the upper part of Section B1 and its lateral extent is controlled by faults (Figs 4 & 5). Similar facies also crop out in the area located between Sections B2 and B3, concentrated in the damage zone of several normal faults dipping to the southwest (Fig. 7B). The facies is composed of centimetre- to decimetre-sized (occasionally metric) calcareous and dolomitic clasts, surrounded by calcareous cement (Fig. 7A). Locally, a poorly developed mudstone matrix is present. Lime mudstone clasts are ubiquitous; grain-supported (peloidal–oolitic) limestone clasts and dolomitic clasts (frequently algal laminated and with fenestrae and evaporite pseudomorphs) are also present. Grain-supported breccias with angular to subangular poorly sorted clasts are common. Decimetre- to metre-thick beds of grain-supported and mud-supported conglomerates (subangular to subrounded, well-sorted centimetre-sized clasts) are intercalated within these breccias. They show lobe and sheet-like geometry with irregular bottom surfaces, pinching laterally out into breccias.

The composition of the clasts indicates that the parent sediment corresponded to the subtidal to supratidal well-bedded and massive carbonate facies described above. The genetic relationship between breccias and preserved (autochthonous) platform facies is also evidenced by the lateral change at the same stratigraphic level between preserved limestones, fractured limestones cut by cemented fractures with decimetre–metre-scale vertical extent and subperpendicular to the bedding (see Fig. 9A), and breccias. It is suggested that debris fall (sliding?) from submarine fracturing produced in the semilithified or lithified autochthonous limestones was the origin of a significant part of these rudites and breccias. This hypothesis will be further discussed below, after the presentation of the structural data.

Sulphate evaporites

This group of facies forms decimetre- to metre-thick levels of predominantly laminated to massive grey-white gypsum, with intercalations of decimetre-thick units of red–orange gypsiferous claystones. Centimetre- to decimetre-thick levels of grey dolostones and limestones are locally intercalated in these evaporitic successions. These carbonates show a locally defined sequence from massive to algal-laminated levels. Fenestrae, tepees and mud cracks may appear in the algal-laminated levels. Alternations of millimetre–centimetre-thick carbonate and evaporitic laminae are also recognized.

Different workers focusing on the Upper Triassic–Lower Jurassic evaporite deposits of the Iberian Chain have inferred different marine-influenced depositional environments: subsiding sabkhas (Ortí, 1987) or an arid tidal flat complex

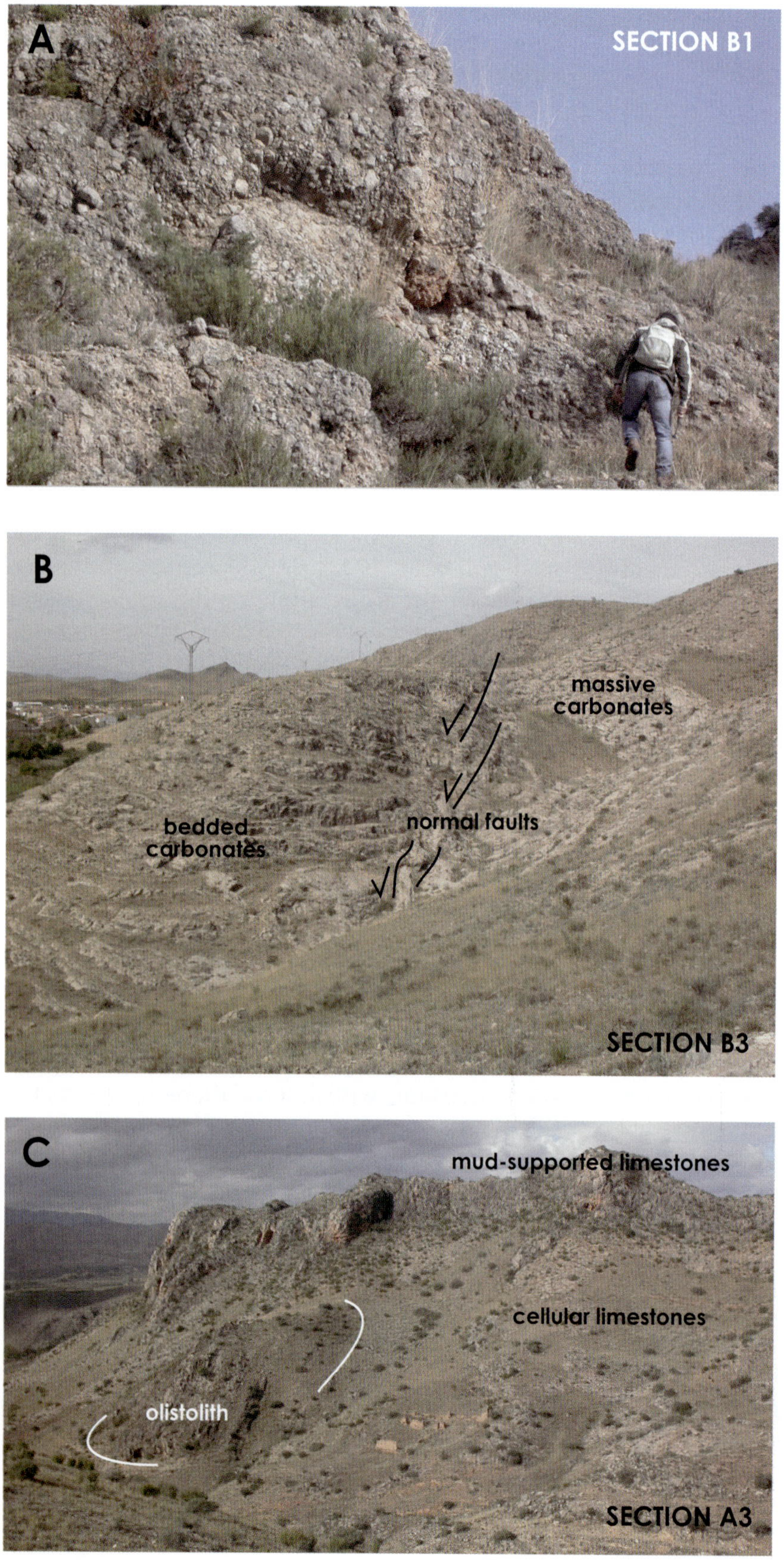

Fig. 7 Study sections. (A) Massive units of clast-supported calcareous rudites interpreted as submarine mass flows, observed in the upper part of Section B1. (B) Fault 4 zone with well-bedded limestones (tilted carbonate platform facies) preserved in the downthrown block, and massive carbonates (breccias and rudites) in the area of intensive faulting and cracking. The total thickness of bedded carbonates exposed to the left of the fault zone is approximately 20 m. (C) General view of Section A3, showing a decametre-scale block (olistolith eroded from the Middle Triassic Muschelkalk facies). The vertical height of the block is approximately 40 m.

(Bordonaba & Aurell, 2002). A recent borehole study of these Ca-sulphates by Ortí & Salvany (2004) indicates that sedimentation mainly occurred in a subsiding coastal basin of salina or lagoon type. A marine supply for the sulphate has been demonstrated by geochemical data in age-equivalent deposits found in other areas of the Iberian basin (Utrilla *et al.*, 1992; Ortí *et al.*, 1996). In the study area, additional supply from the weathering of the Late Triassic evaporite-rich units (i.e. the Keuper facies) also can be considered as a likely source.

Detrital facies

This facies association is thin and locally present in the lower part of some of the sections studied (Fig. 5). It is formed by a wide spectrum of detrital facies including reddish claystones with intercalated calcareous sandstones and polymict conglomerates and breccias. Most of the conglomerates and breccias are grain-supported and are formed by subangular to subrounded centimetre-sized carbonate clasts, included in a muddy to sandy matrix. Some of these rudites were derived from the erosion of the older Triassic units exposed in the uplifted blocks of the faults. A distinct example of this facies is the existence of olistoliths (up to 40 m thick) derived from the erosion of the Muschelkalk facies (Fig. 7C). Red sandstone clasts probably derived from the erosion of the Lower Triassic units are also found locally. There is no evidence pointing to marine influence during the deposition of these detrital facies, and most of them are interpreted as having been deposited in terrestrial environments.

STRUCTURAL ANALYSIS

Structural analysis is a useful tool to determine the relationships between tectonics and sedimentation and the activity of faults in each tectonic stage. However, it must be taken into account that the Iberian Chain underwent two Mesozoic stages of rifting followed by thermal subsidence (Salas & Casas, 1993) and later inversion. In the Triassic sediments it is not easy to a priori differentiate between Triassic and Cretaceous faulting. This difficulty becomes greater when we also consider that the Early Cretaceous extensional stage strongly affected the Iberian Chain and that the features related to the Late Triassic–Early Jurassic episode are of relatively minor importance (total thickness up to 200 m) when compared with the whole basinal history (total thickness of about 1200 m in the northern sector of the Iberian Chain).

Nevertheless, there are some features that allow us to infer the presence of active faults during the Late Triassic to Early Jurassic, comprising: (i) structures cutting the lower part of the series and unconformably covered by the upper layers; (ii) thickness changes in the stratigraphic units cut by faults; (iii) structures linked to a particular stratigraphic level, which cannot be followed in the upper or lower levels, indicating a structural topographic control for these structures within the basin (i.e. slump folds).

Normal faults, fractures and slump folds

In the study area the activity of the main syn-sedimentary faults was deduced from facies and stratigraphic analysis (see faults 1–4 in Figs 4 & 5). Photogeological analysis of lineaments (Fig. 8) shows a dominance of the NW–SE fracture set, together with a N–S to NNW–SSE set. Fractures observed in aerial photographs are frequently localized on the NW–SE lines coincident with the mapped traces of the normal faults, except for a group of NNW–SSE lineaments located north of the Mularroya Fault. In the southwestern part of the area studied, 800 m north of Morata de Jalón, the photogeological lineaments coincide with normal faults with top to the south movement, cutting across the preserved platform limestones, and probably contemporaneous with deposition of this unit.

Fractures at the outcrop scale are very pervasive, with spacing of 5–10 cm in most of the calcareous units (Fig. 9A). In some places, the transition between fractured limestone and angular breccia is gradual, clasts within the breccia being polyhedra bounded by first-generation fractures. This may suggest a tectonic, syn-sedimentary origin for some of the brecciated units (see interpretation of the breccia and rudite facies and discussion below). It is commonly accepted (e.g. Cozzi, 2000, and references therein) that tensile fractures develop readily in extensional regimes and are found frequently in incipient extensional regimes.

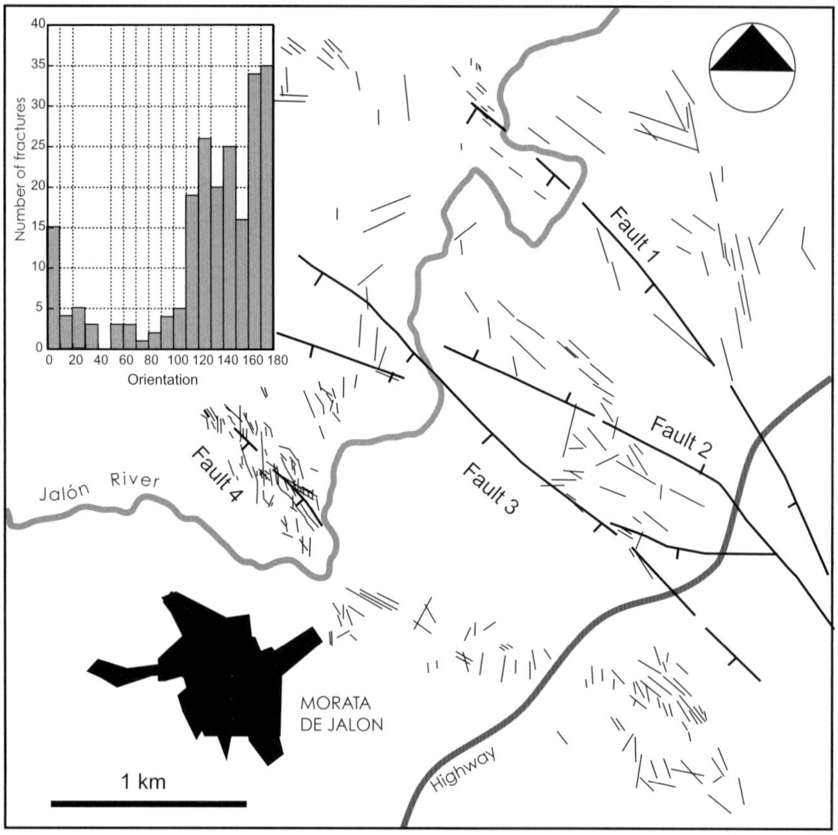

Fig. 8 Photogeological lineaments in the Morata de Jalón area. The main mapped faults (obtained from facies and thickness analysis of stratigraphic units) are also shown. Inset shows the statistical analysis of lineament orientation (number of fractures versus strike). The significance of photogeological lineaments and their correspondence with faults was checked by identifying many of them on the field. See Fig. 3 for location of map area.

Orientations of fractures at the outcrop scale show an E–W maximum for their orientation (Fig. 9B), with subvertical dips, and secondary N–S and NW–SE maxima. It is remarkable that only the two secondary maxima coincide with orientation of fractures recorded from the photogeological analysis (Fig. 8), and that the E–W set is unrecorded. This apparent contradiction can, however, be easily explained by the switching of σ_2 and σ_3 axes common in extensional regimes (Simón *et al.*, 1988; Cozzi, 2000). This also supports the synsedimentary extensional regime proposed for the studied period.

Structural analysis of the slump folds (Fig. 9C) indicates a dominant N–S direction for most of the slump fold axes, with a strong scattering within the average bedding plane and opposite vergences between some of the folds. This scattering is consistent with arcuate axes of the slump folds (Fig. 9E), changing in trend from nearly parallel to perpendicular to the strike of the slope, with apparently opposite vergences along trend. The transport direction obtained from these data according to Hansen's (1971) method is east-southeast, and is therefore parallel to the main faults limiting the subsidence area (Fig. 9D). This implies tilting of the platform, located in the hanging wall of normal faults, associated with the differential along-strike displacement of the normal faults (Fig. 9E).

Interpretation

Analysis of deformational structures in the Morata de Jalón area point to an extensional, immediately post-sedimentary origin for most of the faults and fractures found in Upper Triassic to Lower Jurassic rocks. The major faults bounding the main blocks and limiting the subsiding areas show a NW–SE direction and are responsible for most thickness changes seen in the sedimentary units. The extension direction during this stage was probably N–S, since the dominant set of fractures can be attributed to extensional joints, perpendicular to the main horizontal extension axes. However, the lack of unequivocally dated structures at the outcrop

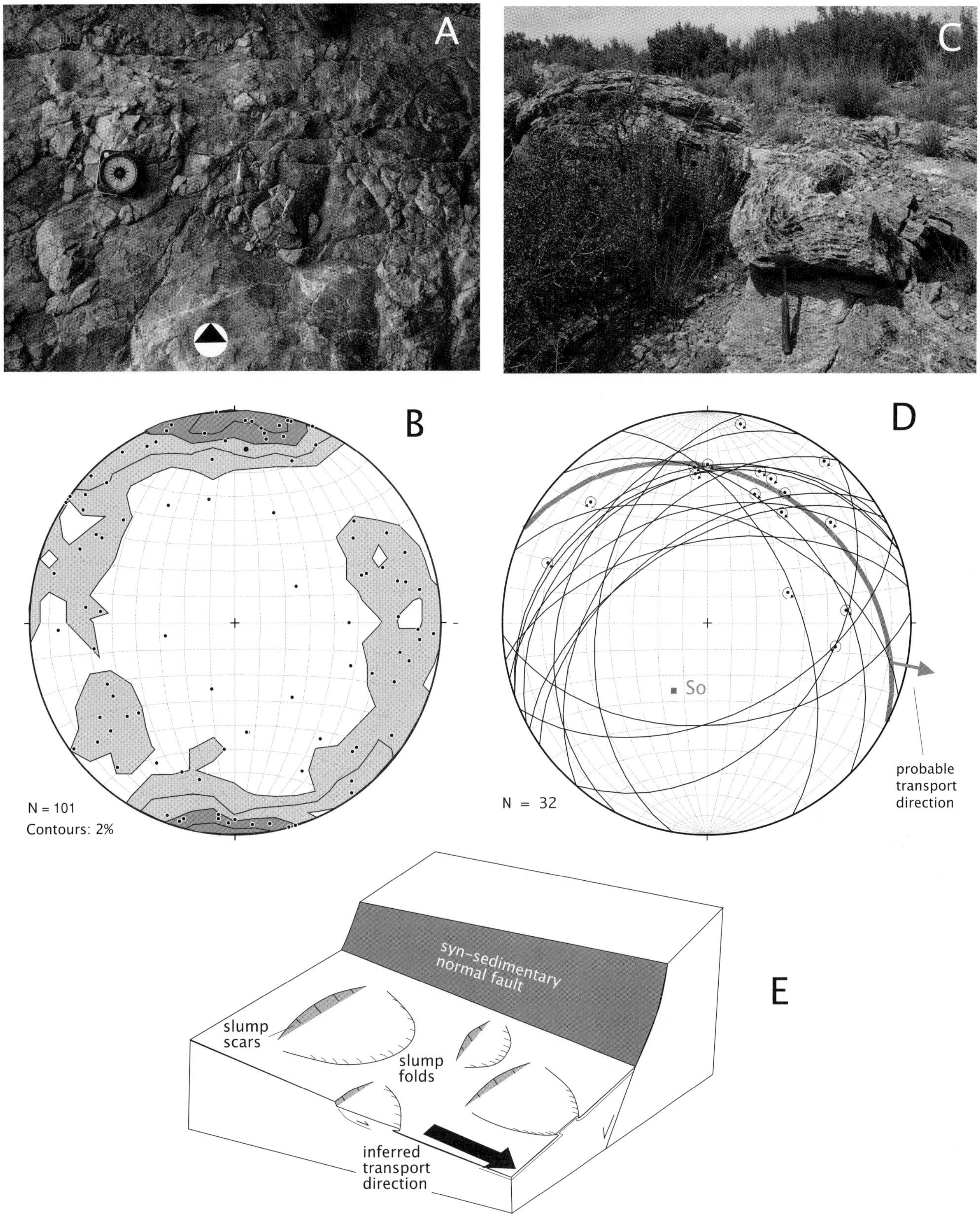

Fig. 9 Structures at the outcrop scale found in the Morata area. (A) Fractures and faults with centimetric spacing (Section B1, upper part). The compass for scale is 10 cm across. Arrow points north. The main set is oriented E–W. (B) Stereoplot (Schmidt net, lower hemisphere) showing the orientation of fractures at the outcrop scale measured throughout the area studied. (C) Slump fold in the lower laminated marker bed (*slump marker beds*, Section A1, lower part). The hammer shaft is 30 cm long. (D) Stereoplot (Schmidt net, lower hemisphere) showing the orientation of slump folds in this level, analysed using Hansen's (1971) method. The transport direction obtained is indicated. The average bedding orientation ('So' and great circle) is also shown. (E) Three-dimensional sketch showing the relationship between slump folds, the sediment slope and the main normal fault limiting the basin to the north.

scale precludes the establishment of the orientation of stress axes.

The difference in orientation of fractures at the outcrop scale and at the map scale can be related to the mechanisms involved in fracturing and basin formation during the Late Triassic in this area. Hectometre- to kilometre-scale faults are probably controlled by the reactivation of ancient, late Variscan faults located within the Palaeozoic basement, that show a similar orientation in the central part of the Iberian Chain (e.g. Alvaro *et al.*, 1979). Moreover, these two directions (NW–SE and NNW–SSE) are the typical late Variscan trends in this area (Cortés & Casas, 1996). In comparison, E–W faults and joints (absolute maximum in Fig. 9B) are probably newly formed during the Late Triassic to Early Jurassic extensional stage and only controlled by the extensional stress. Further support for this hypothesis comes from the pervasive character of the fracturing and the consistent perpendicular relationships between fractures and bedding. The N–S faults may have controlled some of the depositional depth changes within the basin studied, segmenting them obliquely to the main normal faults. This can explain the transport direction obtained from asymmetric folds in the slumped level, indicating an eastward palaeoslope, at least locally, within the basin. An alternative explanation would be the presence of relay ramp-type structures between normal faults.

TECTONO-SEDIMENTARY EVOLUTION

The structural and sedimentological data presented above can be interpreted as the result of two well-defined episodes of tectono-sedimentary evolution (i.e. Episodes 1 and 2 in Fig. 5). The two sedimentary models presented in Fig. 10 show the reconstruction of the studied portion of the basin at the end of Episodes 1 and 2. The onset of Episode 3 (indicated also in Fig. 5) represents the transition over the entire study area from the massive carbonates of the Cortes de Tajuña Formation to the well-bedded carbonates of the Cuevas Labradas Formation. As a whole, these two units were deposited during the Hettangian–Sinemurian long-term transgressive event, observed not only in the Morata de Jalón areas, but also at the basin scale (San Román & Aurell, 1992; Aurell *et al.*, 2003; Fig. 2).

Episode 1: formation of an evaporitic trough

The first stage of active tectonic extension involved the formation and main movement of Faults 1 and 2. A rapidly subsiding trough was formed in the graben between these two faults (Fig. 10A). The graben was filled by an 80 m thick succession dominated by sulphate evaporites (see Section A2 in Fig. 5). In this stage of evolution, sedimentary breccias were found only locally near the shoulders of the graben, and consist mainly of dolomitic blocks and gravel-size clasts, most of them derived from the erosion of older Triassic units.

Sedimentary breccias are common in a variety of geological environments. Marine calcareous or dolomitic breccias are usually associated with steep escarpments and slopes linked to faults in graben and half-graben basins (e.g. Leeder & Gawthorpe, 1987; Carulli *et al.*, 1998; Della Pierre *et al.*, 2002). In symmetric graben-type basins the distribution of breccia-type deposits can be more widespread throughout the basin because fault scarps appear on both margins. However, the restricted extension of the breccias during this first stage of evolution indicates that the elevation created by the fault scarps was never significant, probably due to the rapid evaporitc deposition that was able to compensate for the accommodation created.

Age-equivalent evaporitic units are also found locally elsewhere in the northern Iberian Chain. Ortí & Salvany (2004) documented the sedimentary evolution of an evaporitic unit from borehole analysis in the Lecera-Oliete area (number 6 in Fig. 1A). Sedimentation mainly occurred in a subsiding coastal basin of salina or lagoon type, in which the water depth was progressively reduced due to infilling. This interpretation was based on the vertical evolution of the restricted lagoon–subtidal carbonate deposits, salinas (subaqueous), gypsum deposits (preserved as anhydrites) and sabkha (subareal) anhydrite deposits. This subsiding evaporite environment was clearly controlled by fault activity (Bordonaba & Aurell, 2002). In areas located north of Morata de Jalón (Sierra del Moncayo, number 4 in Fig. 1), San Román & Aurell (1992) also illustrated the preferential

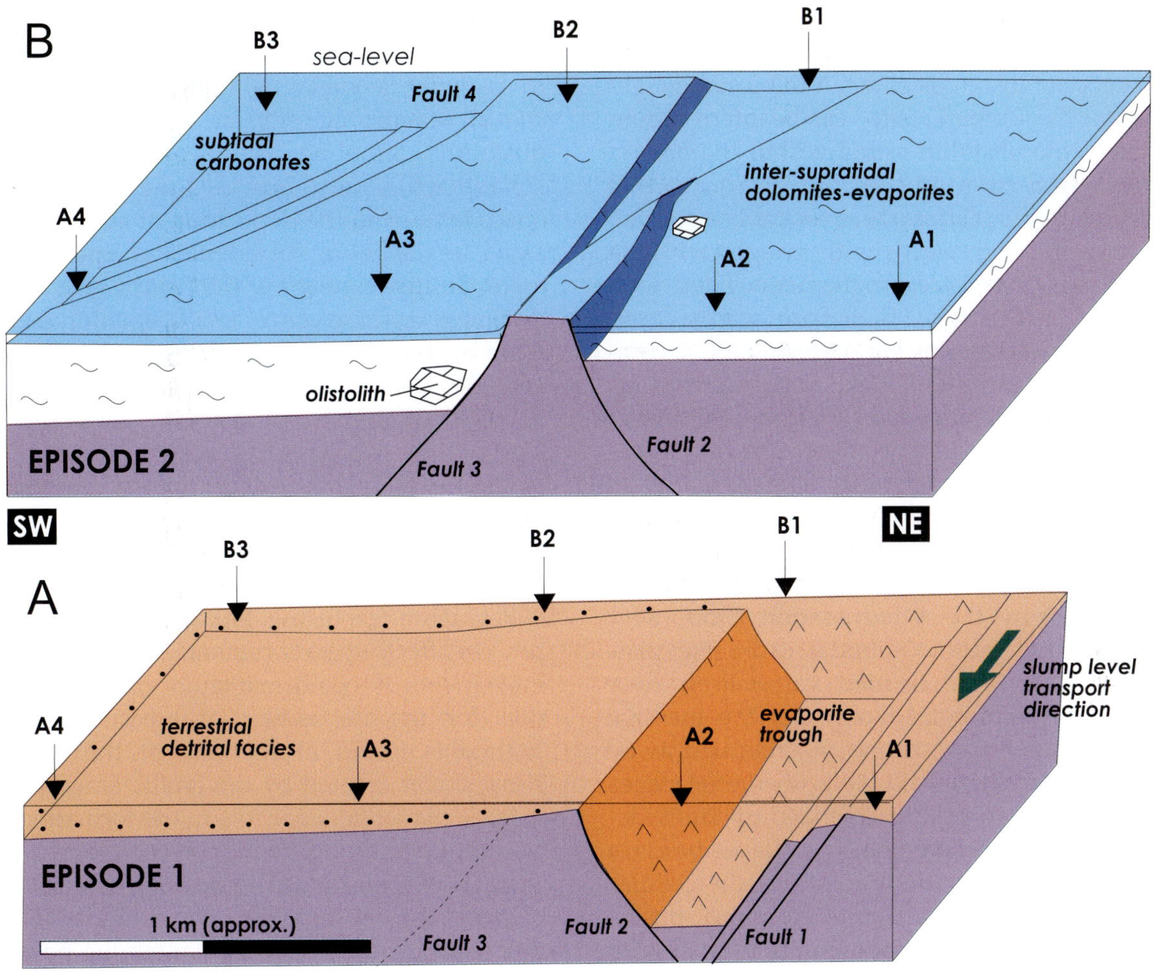

Fig. 10 Block diagrams showing the reconstruction of the basin. (A) At the end of Episode 1 (latest Rhaetian?). (B) At the end of Episode 2 (Hettangian). The locations of the study sections (A1–A4; B1–B3) are indicated.

accumulation of the uppermost Triassic to lowermost Jurassic evaporites in the downthrown blocks of normal faults.

Terrestrial detrital facies were sparsely represented on the uplifted blocks located to the north and south of faults 1 and 2 respectively. The erosive gap observed below these detrital facies (see, for instance, the area located northeast of fault 1 in Fig. 4) indicates significant weathering and erosion of the uplifted areas. The erosion and dissolution of the evaporites and claystones of the Upper Triassic Keuper facies in the graben shoulders may have provided the source of the Ca-sulphates and the fine detrital sediments accumulated in the evaporite trough.

After this first stage of erosion and terrestrial to coastal-evaporite sedimentation, a relative sea-level rise created a carbonate tidal-flat environment. The sedimentation during this brief stage of tectonic quiescence resulted in the formation of a 1–3 m thick laminated bed, which can be recognized throughout the study area with no significant thickness variation. After its partial lithification, the laminated bed was broken and slumped at the onset of the tectonic Episode 2.

Episode 2: carbonate platform and calcareous breccias and rudites

The onset of Episode 2 is marked by normal fault reactivation, including major activity of the newly formed Faults 3 and 4 (Fig. 10B). The uplifted areas were eroded, as indicated by the existence of Muschelkalk olistoliths that accumulated at the toe of Faults 2 and 3. Some hints on the basinal history during this episode can be obtained from the presence of these olistoliths. A minimum height of 40 m for the scarp of Fault 3 can be inferred from the size of the olistolith found within the breccia sediments in Section A3 (Fig. 5). This means that the rupture of the platform was probably a sudden event, creating high scarps with troughs within a starved basin that were progressively filled with sediments. Furthermore, high fault scarps could have been the triggering mechanisms for decompressional fracturing of the exposed Triassic units or the opening of previous, closely spaced joints, thus favouring the formation of extensive sedimentary breccia at the toe of the slopes created by the fault scarps.

The filling of the sedimentary basin during Episode 2 shows a continuous relative sea-level rise, as evidenced by the retrogradation (from west to east) of the shallow subtidal carbonate platform facies (i.e. the bedded carbonates) over the intertidal–supratidal facies (i.e. the massive carbonates, see Fig. 5). In the areas located away from the scarps of the main faults, the initial stages of sedimentation were characterized by the widespread presence of the dolomitic collapse breccias or their equivalent altered facies (i.e. the cellular limestones, see Fig. 5). Similar to other basins showing extensive brecciation due to evaporite dissolution (e.g. Eliassen & Talbot, 2005), the time of formation of the collapse breccia in the Iberian basin is open to discussion (e.g. Bordonaba & Aurell, 2002; Ortí & Salvany, 2004). In the Morata de Jalón area, aspects such as the existence of deformed soft clasts and the relationship between the breccias and the parent facies support the genesis of the breccias during the early stages of cementation. In this case, water circulation and evaporite dissolution could be clearly favoured by the development of small-scale extensional faults and fractures during this episode of extensional tectonic activity.

Other factors could also have contributed to the formation of breccia deposits at this stage. Seismically induced soft-sediment deformation is recognized as a process able to form breccias, associated with progressive bed break-ups, sedimentary dykes and asymmetric folding in soft beds (Onash & Kahle, 2002). However, in the Morata de Jalón area the widespread occurrence of breccias and their distribution along the stratigraphic filling of the basin point to a rather more regional process linked to regional tectonics, although the influence of coeval seismic activity in the emplacement of olistoliths cannot be dismissed.

Fault activity also provided the accommodation space favouring the deposition of subtidal successions in the more subsident and open platform areas. Southwestward of the Fault 4 area a notable thickness increase coeval to a deepening evolution from intersupratidal facies (cellular limestones) to subtidal facies is observed (see Section B3 in Figs 5 & 10B). The predominance of subtidal-derived clasts (lime mudstones and grain-supported peloidal–oolitic limestones) in the breccias and rudites found in the downthrown areas of Faults 2 and 3 also reflects the existence of a subtidal environment in this subsiding area.

Pervasive fracturing linked to the extensional process affecting early cemented beds, followed by movements of the normal faults, offers an explanation for the deposition of breccia and rudite sediments found at the foot of the submarine steep slopes created by the faults. Fractures seem to have played a major role in the formation of the calcareous and dolomitic breccias and rudites, since all the intermediate geometries can be seen between the end-members consisting of fractured dolostones and the breccias containing polyhedral clasts. Some of these intermediate members include dolostones with open fractures limiting small microlithons and fault-bounded clasts within a calcareous matrix (Fig. 6B)

Fault activity has been interpreted as a triggering mechanism for the origin of similar matrix-poor, shallow-platform-derived breccias (e.g. Wilson, 1999; see compilation by Drzewiecki & Simó, 2002). However, in our case study, tectonic extension created a zone of high fracture density, instead of a definite normal fault plane, and maintained a steep and unstable slope on the platform. The presence within the breccias of interbedded peritidal facies (represented by the secondary facies of cellular limestones), and the absence of a slip surface between the breccias and the overlying autochthonous limestones, supports this interpretation. However, depositional slopes could have been created across the fracture-fault zone, as is indicated by the presence of interbedded sorted (rudite) levels, which may be interpreted as debris-flow deposits. The observed relationship between fractured limestones, breccias and mass flows has some similarities with those described by Fütchbauer & Ricchter (1983).

The end of this episode is marked by the presence of poorly bedded micritic and algal-laminated mudstones throughout the study area, indicating the gradual establishment of a shallow carbonate platform during a stage of broad and homogeneous subsidence, which characterizes the Sinemurian over most of the Iberian Basin (e.g. San Román & Aurell, 1992; Aurell *et al.*, 2003; see Fig. 2).

CONCLUSIONS

The sedimentological and structural analysis of the Upper Triassic to Lower Jurassic carbonate and evaporite units cropping out around the Morata de Jalón area (northeast Spain) illustrates in detail a significant stage in the evolution of the northern part of the Iberian Basin. The model presented in this work provides further support to previous interpretations of the long-term tectono-sedimentary evolution of the Iberian Basin during the latest Triassic to early Jurassic, proposed by Aurell & San Román (1992) and Aurell *et al.* (2003). The alternative interpretation of Gómez & Goy (2005), which neglected the latest Triassic unconformity and proposed continuous sedimentation during a late Norian to early Sinemurian transgressive–regressive cycle (with a regressive trend for the Cortes de Tajuña Formation), is not consistent with the data reported here.

In the northern Iberian basin, a first stage of tectonic activity, thought to have begun at the end of the Triassic, involved the formation of strongly subsiding evaporite-rich troughs in a graben. A coastal salina for evaporite precipitation with episodic marine influence was filled by Ca-sulphates. A possibly additional source was the weathering of ancient evaporite-rich units (i.e., the Keuper facies), exposed on the horsts. The reported data give support to the suggestion that the late Rhaetian to Hettangian evaporites accumulated only locally in subsiding areas of the Iberian basin (e.g. San Román & Aurell, 1992; Bordonaba & Aurell, 2002).

A second stage of tectonic extensional activity, most probably occurring at the beginning of the Jurassic (Hettangian), combined with the Hettangian–Sinemurian long-term regional sea-level rise (Aurell *et al.*, 2003), led to the development of a carbonate tidal-flat environment (with episodic precipitation of evaporites) that graded in areas of greater subsidence to a shallow-water carbonate platform. Two main types of carbonate breccias (and rudites) were formed:

1 in the tidal-flat environment, the formation of a dissolution-collapse breccia was favoured by the syn-sedimentary cracking and fracturing affecting the early lithified carbonates;
2 in the subtidal environment, carbonate breccias and rudites were derived in part by submarine debris flows sourced by the incipiently lithified and fractured shallow-marine carbonate facies developed adjacent to syn-sedimentary faults.

The orientation of the faults and joints newly formed during the latest Triassic to earliest Jurassic extensional stage indicates a main N–S extension direction. However, a NNW–SSE to N–S faulting direction appears both at the macro- and mesostructural scale. This orientation was probably controlled by the existence of late Variscan faults located within the Palaeozoic basement. The reactivation of some of the late Variscan faults also occurred during the latest Jurassic to Early Cretaceous rifting cycle. However, the other main faulting direction active during this rifting cycle (NE–SW; Capote *et al.*, 2002) is not represented in the area studied. Although regional variations can be invoked to explain these changes, a temporal change in the extensional field between the Early Jurassic and the Early Cretaceous could be an explanation for the different extension direction found.

ACKNOWLEDGEMENTS

Financial support was provided by MCT, Spain (Projects BTE2002-04453 and BTE2002-04168) and by the Aragon Government (Financiación de Grupos Consolidados). We are grateful to the two reviewers, Dan Bosence and Moyra Wilson, for their constructive comments. The paper was also improved as a result of suggestions by Gary Nichols.

REFERENCES

Alvaro, M., Capote, R. and Vegas, R. (1979) Un modelo de evolución geotectónica para la Cadena Celtibérica. *Acta Geol. Hisp. Hom. a Lluis Solé Sabaris*, **14**, 172–177.

Arnal, I., Calvet, F., Márquez, L., Márquez-Aliaga, A. and Solé de Porta, N. (2002) La plataforma carbonatada epírica (Formaciones Imón e Isábena) del Triásico superior del noreste de la Península Ibérica. *Acta Geol. Hisp.*, **37**, 299–328.

Arthaud, F. and Matte, P. (1975) Les décrochements tardi-herciniens du Sud-Ouest de l'Europe: Geómetrie et essai de reconstruction des conditions de la deformation. *Tectonophysics*, **25**, 139–171.

Aurell, M., Meléndez, A., San Román, J., *et al.* (1992) Tectónica sinsedimentaria distensiva en el límite Triásico-Jurásico en la Cordillera Ibérica. *III Congr. Geol. Esp. (Salamanca) Actas*, **1**, 50–54.

Aurell, M., Robles, S., Bádenas, B., *et al.* (2003) Transgressive/regressive cycles and Jurassic palaeogeography of northeast Iberia. *Sediment. Geol.*, **162**, 239–271.

Barrón, E., Gómez, J.J. and Goy, A. (2001) Dataciones con palinomorfos en los materiales del tránsito Triásico-Jurásico de Poza de la Sal (Burgos). *Publ. Sem. Paleont. Univ. Zaragoza*, **5**, 46–55.

Bernoulli, D. and Jenkyns, H.C. (1974) Alpine, Mediterranean and central Atlantic Mesozoic facies in relation to the early evolution of the Tethys. In: *Modern and Ancient Geosynclinal Sedimentation* (Eds R.H. Dott and R.H. Shaver), pp. 129–160. Special Publication 19, Society of Economic Paleontologists and Mineralogists, Tulsa, OK.

Bordonaba, A.P., Aurell, M. and Casas, A. (1999) Control tectónico y distribución de las facies en el tránsito Triásico-Jurásico en el sector de Oliete (Teruel). *Geogaceta*, **25**, 43–46.

Bordonaba, A.P. and Aurell, M. (2002) Variación lateral de facies en el Jurásico basal de la Cordillera Ibérica Central: orígen diagenético temprano y tectónica sinsedimentaria. *Acta Geol. Hisp.*, **37**, 335–368.

Calvet, F., Tucker. M.E. and Henton. J.M. (1990) Middle Triassic carbonate ramp systems in Catalan Basin, northeast Spain: facies, systems tracts, sequences and controls. In: *Carbonate Platforms: Facies, Sequences and Evolution* (Eds M.E. Tucker, J.L. Wilson, P.D. Crevello, J.R. Sarg and J.F. Read), pp. 79–108. Special Publication **9**, International Association of Sedimentologists. Blackwell Scientific Publications, Oxford.

Campos, S., Aurell, M. and Casas, A. (1996) Origen de las brechas de la base del Jurásico en Morata de Jalón (Zaragoza). *Geogaceta*, **20**(4), 887–890.

Capote, R., Muñoz, J.A., Simón, J.L., Liesa, C.L. and Arlegui, L.E. (2002) Alpine Tectonics I: the Alpine System north of the Betic cordillera. In: *The Geology of Spain* (Eds W. Gibbons and T. Moreno), pp. 367–400. Geological Society Publishing House, Bath.

Carulli, G.B., Cozzi, A., Longo, G., Ponton, M. and Podda, F. (1998) Evidence of synsedimentary tectonic activity during the Norian–Lias (Carnian Prealps, northern Italy). *Mem. Soc. Geol. It.*, **53**, 403–415.

Casas, A.M., Cortés, A.L. and Maestro, A. (2000) Intraplate deformation and basin formation during the Tertiary within the Northern Iberian plate: origin and evolution of the Almazán Basin. *Tectonics*, **19**, 258–289.

Comas-Rengifo, M.J. and Yébenes, A. (1988) El Lías al Sur de la Sierra de Urbión (Castrovido, Burgos). *III Coloquio de Estratigrafía y Paleogeografía del Jurásico de España, libro guía de las excursiones, Ciencias de la Tierra (Instituto de Estudios Riojanos)*, **11**, 149–166.

Cortés, A.L. (2004) Geometría del subsuelo de la Cuenca del Ebro en el Campo de Belchite. *Geo-Temas*, **6**(3), 225–228.

Cortés, A.L. and Casas, A.M. (1996) Deformación alpina de zócalo y cobertera en el borde norte de la Cordillera Ibérica (Cubeta de Azuara-Sierra de Herrera). *Rev. Soc. Geol. Esp.*, **9**, 51–66.

Cozzi, A. (2001) Synsedimentary tensional features in Upper Triassic shallow-water platform carbonates of the Carnian Prealps (northern Italy) and their importance as palaeostress indicators. *Basin Res.*, **12**, 133–146.

Cozzi, A. and Hardie, L.H. (2003) Third-order depositional sequences controlled by synsedimentary extensional tectonics: evidence from Upper Triassic carbonates of the Carnian Prealps (NE Italy). *Terra Nova*, **15**, 40–45.

Della Pierre, F., Clari, P., Cavagna, S. and Bicchi, E. (2002) The Parona chaotic complex: a puzzling record of the Messinian (Late Miocene) events in Monferrato (NW Italy). *Sediment. Geol.*, **152**, 289–311.

Drzewiecki, P.A. and Simó, T. (2001) Depositional processes, triggering mechanisms, and sediment composition of carbonate gravity flow deposits: examples from the Late Cretaceous of the south-central Pyrennes, Spain. *Sediment. Geol.*, **146**, 155–189.

Eliassen, A. and Talbot, M.R. (2005) Solution-collapse breccias of the Minkinfjellet and Wordiekammen Formations, Central spitsbergen, Svalbard: a large gypsum paleokarst system. *Sedimentology*, **52**, 775–794.

Esteban, M. and Juliá, R. (1973) Discordancias erosivas intrajurásicas en las Catalánides. *Acta Geol. Hisp.*, **8**, 153–157.

Fütchbauer, H. and Ricchter, D.K. (1983) Relations between submarine fissures, internal breccias and mass flows during Triassic and earlier rifting periods. *Geol. Rundsch.*, **72**, 53–66.

Gallego, M.R., Aurell M., Badenas, B., Fontana, B. and Meléndez, G. (1994) Origen de las brechas de la base del Jurásico de Leitza (Cordillera Vasco Cantábrica-Oriental, Navarra). *Geogaceta*, **15**, 26–29.

Giner, J. (1978) Origen y significado de las brechas del Lías de la Mesa de Prades (Tarragona). *Estud. Geol.*, **34**, 529–533.

Gómez, J.J. and Goy, A. (2005) Late Triassic and Early Jurassic palaeogeographic evolution and depositional cycles of the Western Tethys Iberian platform system (Eastern Spain). *Palaeogeogr. Palaeoclimatol. Palaeoecol.*, **222**, 77–94.

Guimerà, J. (1988) *Estudi estructural de l'enllaç entre la serralada Iberica i la serralada costanera catalana.* Unpublished PhD thesis, Univ. Barcelona, 2 vols, 599 pp.

Guimerà, J. and Alvaro, M. (1990) Structure et évolution de la compression alpine dans la Chaîne Ibérique de la Chaîne Côtière Catalane (Espagne). *Bull. Soc. Géol. France*, **8**, 339–348.

Hansen, E. (1971) *Strain Facies.* Springer-Verlag, New York.

Leeder, M.R. and Gawthorpe, R.L. (1987) Sedimentary models for extensional tilt-block/half-graben basins. In: *Continental Extensional Tectonics* (Eds M.P. Coward, J.F. Dewey and P.L. Hancock), pp. 139–152. Special Publication 28, Geological Society, London.

Morillo, M.J. and Meléndez, F. (1979) El Jurásico de la Alcarria-La Mancha. *Cuad. Geol. Univ. Granada*, **10**, 149–166.

Onash, C.M. and Kahle, C.F. (2002) Seismically induced soft-sediment deformation in some Silurian carbonates, eastern U.S. Midcontinent. In: *Ancient Seismites* (Eds F.R. Ettensoh, N. Rast and C.E. Brett). *Geol. Soc. Am. Spec. Pap.*, **359**, 165–176.

Ortí, F. (1987) Aspectos sedimentológicos de las evaporitas del Triásico y del Liásico inferior en el este de la Península Ibérica. *Cuad. Geol. Ibérica*, **11**, 837–858.

Orti, F. and Salvany, J.M. (2004) Coastal saline evaporites of the Triassic–Liassic boundary in the Iberian Península: the Alacón borehole. *Geol. Acta*, **2**, 291–304.

Orti, F., García Veigas, J., Rosell, L., Jurado, M.J. and Utrilla, R. (1996) Formaciones Salinas de las cuencas triásicas en la Península Ibérica: caracterización petrológica y geoquímica. *Cuad. Geol. Ibérica*, **20**, 13–35.

Pérez-López, A., Sole de Porta, N. and Orti, F. (1996) Facies carbonato-evaporíticas del Trias Superior y tránsito al Lías en el Levante español: nuevas precisines estratigráficas. *Cuad. Geol. Ibérica*, **20**, 245–269.

Riba, O., Maldonado, A., Puigdefabregas, C., Quirantes, J. and Villena, J. (1971) *Mapa Geológico e España, Escala 1: 200000, nº 32 (Zaragoza).* I.G.M.E. ed., 33 pp.

Robles, S., Pujalte, V. and Valles, J.C. (1989) Sistemas sedimentarios del Júrasico de la parte occidental de la Cuenca Vasco-Cantábrica. *Cuad. Geol. Ibérica*, **13**, 185–198.

Roca, E., Guimerà, J. and Salas, R. (1994) Mesozoic extensional tectonics in the southeast Iberian Chain. *Geol. Mag.*, **131**, 155–168.

Salas, R. and Casas, A. (1993) Mesozoic extensional tectonics, stratigraphy and crustal evolution during the Alpine cycle of the eastern Iberian basin. *Tectonophysics*, **228**, 33–55.

Salas, R., Guimerà, J., Mas, R., Martín-Closas, C., Meléndez, A. and Alonso, A. (2001) Evolution of the Mesozoic Central Iberian Rift System and its Cenozoic inversion (Iberian Chain). In: *Peritethyan Rift/Wrench Basins and Passive Margins* (Eds W. Cavazza, A. Robertson and P.A. Ziegler). Peritethyan Memoirs Vol. 6. *Mém. Mus. Natn. Hist. Nat.*, **186**, 145–185.

San Román, J. (1994) *Estudio hidrogeológico del interfluvio Queiles-Jalón (Zaragoza).* Unpublished PhD Thesis, Univ. de Zaragoza, 231 pp.

San Román, J. and Aurell, M. (1992) Palaeogeographical significance of the Triassic-Jurassic unconformity in the north Iberian basin (Sierra del Moncayo, Spain). *Palaeogeogr. Palaeoclimatol. Palaeoecol.*, **99**, 101–117.

Simón, J.L., Serón, F. and Casas, A.M. (1988) Stress deflection and fracture development in a multidirectional extension regime. Mathematical and experimental Approach with field examples. *Ann. Tect.*, **2**, 21–32.

Strasser, A. (1986) Ooids in Purbeck limestones (lowermost Cretaceous) of the Swiss and French Jura. *Sedimentology*, **33**, 711–727.

Tucker, M.E. and Wright, V.P. (1990) *Carbonate Sedimentology.* Blackwell Scientific Publications, Oxford, 482 pp.

Utrilla, R., Pierre, C., Orti, F. and Pueyo, J.J. (1992) Oxygen and sulfur isotope compositions as indicators of the origin of Mesozoic and Cenozoic evaporites from Spain. *Chemical Geology (Isotope Geoscience Section)*, **102**, 229–244.

Wilson, M.E.J. (1999) Prerift and synrift sedimentation during early fault segmentation of a Tertiary carbonate platform, Indonesia. *Mar. Petrol. Geol.*, **16**, 825–848.

Wright, V.P. (1992) A revised classification of limestones. *Sediment. Geol.*, **76**, 177–185.

A shallow-basin model for 'saline giants' based on isostasy-driven subsidence

FRANK J.G. VAN DEN BELT *and* POPPE L. DE BOER

University of Utrecht, Faculty of Earth Sciences, PO Box 80021, 3508 TA Utrecht, The Netherlands (Email: fjgvandenbelt@geo.uu.nl; pdeboer@geo.uu.nl)

ABSTRACT

The common assumption that 'saline giants' must have formed in deep basins and that their thickness reflects initial basin depth ignores the principle of isostasy. Due to the high density of anhydrite and high precipitation rates for evaporite minerals, isostatic compensation is much more important in evaporite than in non-evaporite settings. The main implication is that evaporite precipitation drives subsidence rather than the other way round, and that thick evaporite deposits require an initial basin depth much less than their final thickness. Once initiated, evaporite precipitation and consequent isostatic subsidence is a self-sustaining process that can result in kilometre-scale evaporite stratigraphy. Rapid isostatic compensation is facilitated by thin, fractured crust in extensional basins, which explains the typical occurrence of saline giants in such settings. It is shown that a shallow-basin origin in combination with rapid isostatic compensation can well explain the extreme thickness of saline giants as well as the commonly associated shallow-water sedimentary facies. Although there is no reason to exclude the possibility of a basin-wide dropdown of a few thousand metres as proposed for some saline giants, a desiccated deep basin is certainly not a requirement. An initially shallow basin that rapidly deepens by isostatic adjustment in response to the precipitation of evaporites eliminates the need for deep-basin desiccation, gigantic waterfalls, and repeated opening and closure of a connection to the world ocean, and makes the extreme thickness of saline giants less enigmatic.

Keywords Saline giants, isostasy, evaporites, halite, anhydrite, Zechstein, Messinian.

INTRODUCTION

A number of evaporite successions are characterized by extraordinary thickness and are therefore commonly referred to as 'saline giants'. They are up to 4 km thick and typically consist of a number of stacked, thinning-upward evaporite cycles (Table 1). For example, the carbonate–evaporite succession from the Permian Zechstein reaches a thickness of 2 km (Taylor, 1998); individual halite bodies are up to 600 m thick (Sannemann *et al.*, 1978) and anhydrite bodies are up to 280 m (Van der Baan, 1990). The major Messinian evaporite succession in the western Mediterranean was estimated to be 2–3 km thick (Hsü *et al.*, 1973) and is 2 km in the eastern Mediterranean (Tay *et al.*, 2002). According to Krijgsman *et al.* (1999) these Mediterranean evaporites were deposited in no more than 0.6 Myr.

In the absence of recent analogues, developing models for saline giants has proven speculative. In the late 19th century Ochsenius (1877) developed a depositional model based on evaporite precipitation in a restricted lagoonal environment. Hsü *et al.* (1973, 1977) felt it could not explain the new data from the Mediterranean, which they interpreted as deposits formed by precipitation from shallow-water salt lakes that occupied the deepest parts of kilometres deep, desiccated basins (Fig. 1). The model is known as the deep-basin shallow-water model and is often used in explaining thick halite deposits (e.g. Sonnenfeld, 1984; Warren, 1999).

The formation of Zechstein halite bodies has also been attributed to deep-basin shallow-water

Table 1 Summary of stratigraphic, facies and thickness data for various Palaeozoic to Cenozoic saline giants

Basin	Age	Setting	Associated facies	Number of cycles	Total thickness (m)	Average cycle thickness (m)	Halite thickness (m)
Dead Sea (Al-Zoubi *et al.*, 2002; Neev & Emery, 1967)	Pleistocene	Transtension	Lacustrine, alluvial	6	1500	250	375
Western Mediterranean Basin (Blanc, 2000; Dercourt *et al.*, 1986; Hsü *et al.*, 1973)	Miocene	Various	Various	nd	2000	nd	nd
Eastern Mediterranean Basin (Blanc, 2000; Tay *et al.*, 2002)	Miocene	Various	nd	nd	3500	nd	nd
Red Sea (Sonnenfeld, 1984; Orszag-Sperber *et al.*, 1998)	Miocene	Extension	nd	nd	3000–4000	nd	nd
Khorat Basin (Anderson *et al.*, 1972; El Tabakh *et al.*, 1999)	Cretaceous	Extension	Red beds	3	1100	350	350
Cuanza Basin (Siesser, 1978)	Cretaceous	Extension	Shallow marine	nd	1500	nd	nd
Gulf of Mexico Basin (Reed, 1994)	Jurassic	Extension	Red beds, volcanics	nd	4000	nd	nd
Southern Permian Basin (Zechstein) (Sannemann *et al.*, 1978; Van der Baan, 1990; Ziegler, 1990)	Permian	Extension, strike slip	Aeolian, shallow marine, starved basin	4	2000	500	600
East European Basin (Ural) (Northrup & Snyder-Walter, 2000; Zharkov, 1984)	Permian	Transtension	Aeolian, shallow marine	6	2500	420	500
Precaspian Basin (Volozh *et al.*, 2003)	Permian	Thin crust	Deep marine?	nd	4000	nd	nd
Delaware Basin (Anderson *et al.*, 1972)	Permian	Transtension	Starved basin	2	1100	550	400
Paradox Basin (Catacosimos *et al.*, 1990; Williams-Stroud, 1994; Zharkov, 1984)	Carboniferous	Transtension	Shallow marine	5–7	2000	300	270
Michigan Basin (Cercone, 1988; Stevenson & Baars, 1986; Zharkov, 1984)	Silurian	Extension	Shallow marine	5	1000	200	400
East Siberian Basin (Zharkov, 1984)	Cambrian	Extension	Shallow marine	12	2500	210	nd

nd, no data available.

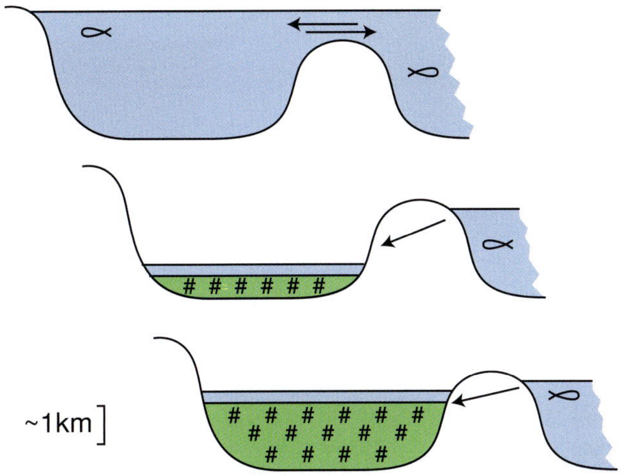

Fig. 1 Deep-basin shallow-water model developed for saline giants. (After Kendall, 1992.) The model does not take into account any syn-depositional isostatic compensation due to evaporite loading.

deposition, although of different order (e.g. Tucker, 1991). Here the estimate of maximum basin depth equals the thickness of the thickest halite body (approximately 600 m) (Tucker, 1991; Warren, 1999). Estimated basin depth before evaporite deposition has been calculated in a similar way in, for example, the Delaware Basin and the Paradox Basin (Anderson *et al.*, 1972; Williams-Stroud, 1994).

For the Zechstein (Southern Permian Basin) abundant drilling has shown that the thick evaporite succession consists of at least four major cycles, the thickest basal cycle being more than 600 m thick locally (Sannemann *et al.*, 1978; Van der Baan, 1990; Tucker, 1991; Taylor, 1998). These cycles are composed of a marginal carbonate wedge, an anhydrite platform and an onlapping halite body (Fig. 2). It is widely accepted that at the termination of each cycle, halite had filled the basin approximately to sea level, and that after continued tectonic subsidence the deposition of a subsequent evaporite cycle started (Van der Baan, 1990; Tucker, 1991; Taylor, 1998; Warren, 1999). Such an internal architecture, with anhydrite pre-dating halite, is common in evaporite basins (Sonnenfeld, 1984; Warren, 2000).

Despite the wide acceptance of a deep-basin origin of halite bodies, a number of aspects of their formation have not been adequately explained. Following Nesteroff (1973), Sonnenfeld (1985) argued against a deep-basin shallow-water origin for the Messinian evaporites, giving a long list of arguments among which was the unexpected occurrence of tidal sediments. Recently, the deep-basin shallow-water origin of Messinian evaporites has been challenged by Hardy & Lowenstein (2004) and Manzi *et al.* (2005).

Although a *shallow-basin* shallow-water model well explains the occurrence of mainly shallow-water depositional structures, the model is qualified as 'unlikely in most tectonic environments' by Kendall (1992) because it requires subsidence and deposition to be in equilibrium during the deposition of kilometre-scale evaporite successions. In the discussion about the depth of such basins prior to the formation of saline giants, the role of isostasy on basin evolution and stratigraphic development is commonly not appraised. Here, the focus will be on isostatic compensation as a mechanism that can explain how thick evaporite sequences can form

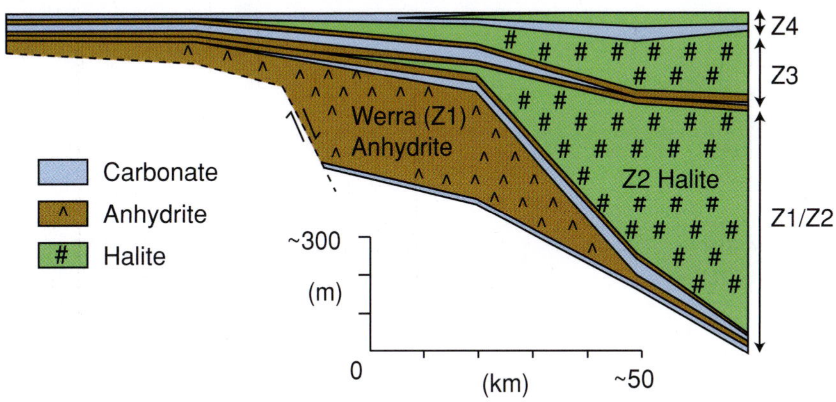

Fig. 2 Stratigraphy and cyclical character of the Zechstein evaporites. (Modified from Visser, 1956.) The Zechstein 1 halite from the original figure is not represented here, as it did not precipitate in the main basin (e.g. Van der Baan, 1990).

in shallow-water basins under long-term gradual subsidence.

ISOSTASY

Isostatic compensation is the response of the lithosphere to a change of overburden by flexure or elastic rebound to achieve regional equilibrium (e.g. Watts, 2001). Such corrections are accommodated by lateral displacement of more ductile, high-density asthenosphere beneath the flexing plate. That such corrections may be implemented rapidly is shown by the fast response to polar deglaciation, where unloading has been 90% compensated by glacial rebound during the 10 kyr of the Holocene (Watts, 2001).

It has been demonstrated that the deposition of a thick siliciclastic wedge at a basin edge causes a strong isostatic response (Watts, 2001), and this should be even more pronounced for an anhydrite wedge due to the higher density of anhydrite (Table 2). Hence, evaporite deposits such as from the Zechstein or the Miocene Mediterranean, which are 2–3 km thick and occupy basins many hundreds of kilometres across, must have created much of their own accommodation space by means of loading. It is therefore expected that the mechanism of isostatic subsidence during salt precipitation explains, at least partly, the great *apparent basin depth* of many evaporite basins (Fig. 3). A factor that is expected to facilitate isostatic correction during salt precipitation is the condition of the basement of many saline giants, which consist of thin, fragmented crust due to rifting or post-orogenic collapse (Table 1) (Burke, 1975; Stanley, 1986; Volozh *et al.*, 2003).

Table 2 Rock, mineral and water densities relevant to this study (Valyashko, 1972; Schumann, 1987; Watts, 2001)

Constituent	Density (g cm^{-1})
Quartz	2.65
Calcite	2.8–2.9
Sediment (30% water)*	~ 2.2
Halite	2.1–2.2
Gypsum	2.2–2.4
Anhydrite	2.9–3.0
Water	1.0
Sea water	1.03
Asthenosphere	3.3

*Calculated value.

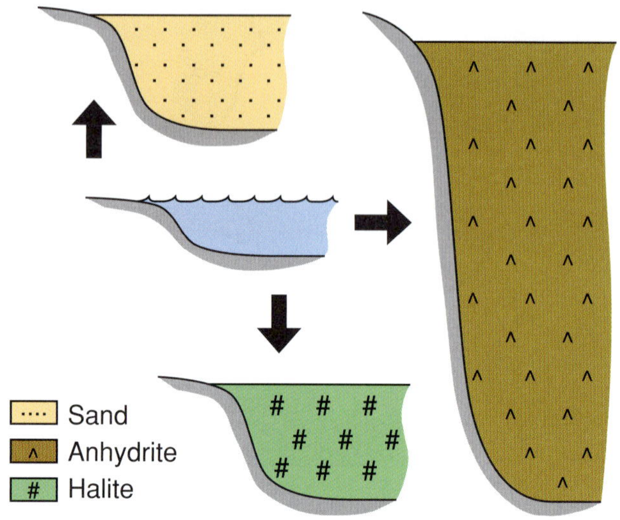

Fig. 3 Isostatic response to sediment loading. Due to a higher density, anhydrite precipitation causes a high degree of isostatic compensation, allowing the formation of thick successions in shallow basins.

Several authors have acknowledged the loading effect on the crust of thick salt deposits (Norman & Chase, 1986; Diegel *et al.*, 1995; Van Wees *et al.*, 2000), but they have not considered this to be a syn-depositional phenomenon. An advanced analysis of isostatic compensation in relation to evaporite basin evolution was published by Norman & Chase (1986). They applied the 'Lake Bonneville' principle of Gilbert (1890), who showed that the Late Pleistocene desiccation of the present Great Salt Lake caused a 40 m uplift of the lake-shore deposits. Norman & Chase (1986) demonstrated that desiccation of the Mediterranean must have resulted in large-scale uplift of the basin floor as well as its margins. Besides that they argued that the Messinian Mediterranean was much shallower than now due to isostatic compensation in response to salt loading. Fabbri & Curzi (1979) invoked an isostasy model to calculate the depth of deposition for the lower Messinian evaporites in the Tyrrhenian Sea, and concluded that they had been deposited in a shallow rather than a deep basin. On the other hand, Ryan (1976) performed a quantitative reconstruction incorporating the effect of loading, and concluded that the Mediterranean Sea was locally more than 2.5 km deep (Balearic Basin).

HALITE

The deep-basin theory that was developed for saline giants requires that the unusually steep basin margins as they are observed now in the subsurface (Warren, 1999) were already in place before the onset of evaporite precipitation (Fig. 1). If the basin margins were indeed as steep prior to halite deposition as after, the marginal successions within such basins should be characterized by abundant clastic deposits. However, evaporite cycles are typified by an absence of clastic interbeds except for anhydrite breccias, while such deposits may be common in underlying or overlying formations (e.g. Sonnenfeld, 1984). It is assumed, therefore, that the tectonic component of total subsidence in evaporite basins is low.

The implications of isostatic compensation during the precipitation of evaporites may be assessed by making simple calculations based on the Airy isostasy model (Fig. 4). It was not the intention here to perform a state-of-the-art basin-scale modelling study. Instead, it has been explored how the incorporation of isostatic compensation may help to develop an alternative model that explains the large-scale subsidence history of salt basins, as well as their sedimentary development.

The calculations are based on two assumptions. First, it is assumed that isostatic adjustment of the lithosphere takes place *during* deposition. Note that the Late Permian, which was characterized by evaporite formation worldwide, lasted approximately 10 Myr. Krijgsman *et al.* (1999) have demonstrated that the Messinian salinity crisis lasted only 600 kyr: a short period for the precipitation of 2–3 km of evaporites. This should, however, be sufficient for isostatic compensation, as it operates on an even shorter time-scale of 10 kyr (Watts, 2001). Second, it is assumed that deposition occurs in a large basin (e.g. 300 × 1500 km for the Southern Permian 'Zechstein' Basin (Ziegler, 1990)), such that the flexural wavelength of the lithosphere is significantly smaller than the scale of the basin. For these conditions, the maximum thickness of the evaporite columns was determined, assuming that salt precipitation occurred under continuous isostatic compensation.

Rates of precipitation of halite are of the order of 10–150 mm yr^{-1} (Schreiber & Hsü, 1980; Sonnenfeld, 1984 and references therein), which is

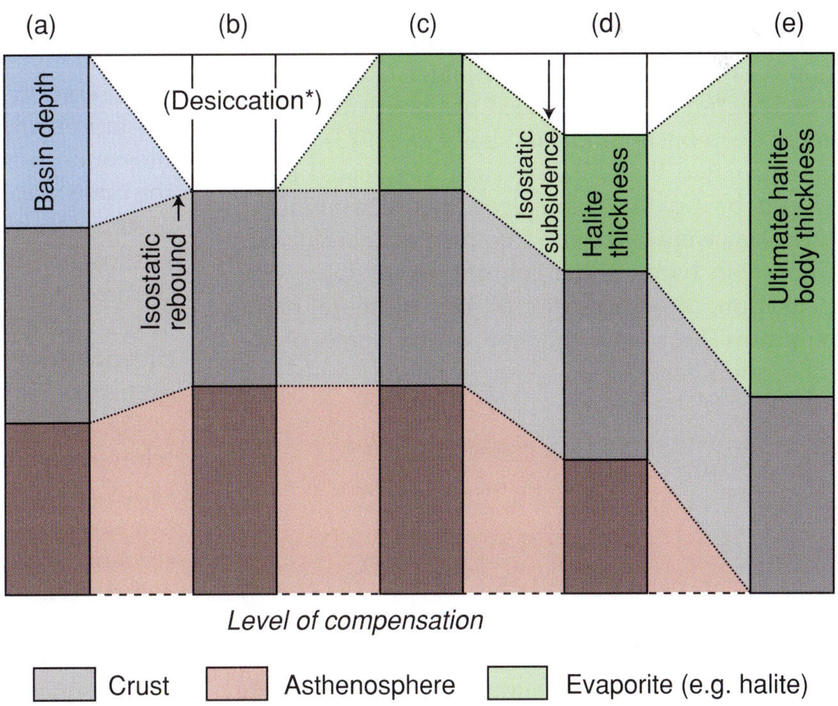

Fig. 4 An Airy isostasy model for basin drawdown and evaporite precipitation. (a) Water-filled basin (isostatic equilibrium). (b) Uplift due to basin desiccation. (c) Halite precipitation. (d) Subsidence due to halite precipitation. (e) Maximum halite-accumulation potential for a 'stage a' basin (isostatic equilibrium). *According to the deep-basin, shallow-water model, the basin desiccates causing isostatic rebound; according to the shallow-basin shallow-water model, the basin remains water-filled.

up to three orders of magnitude greater than subsidence for extensional basins with average rates up to a few millimetres per year (Einsele, 1992, and references therein). Precipitation rates for gypsum and anhydrite are of the order of 1–10 mm yr^{-1} (Sonnenfeld, 1984), thus of the same order as subsidence rates of extensional basins. It is concluded that the tectonic component of overall subsidence during halite precipitation can be ignored, whereas it is important during gypsum/anhydrite precipitation. Hence subsidence of a halite-accumulating basin is likely to be entirely controlled by loading due to halite precipitation.

Balancing the columns in Fig. 4 for a case of a deep basin that dries out demonstrates that the amount of uplift due to desiccation is a function of the initial basin depth (D_{basin}):

$$\text{Uplift} = D_{basin} \times \left(\frac{\rho_{water}}{\rho_{asthenosphere}}\right)$$

$$= D_{basin} \times \left(\frac{1.0}{3.3}\right) = 0.3 \times D_{basin}$$

The density values (ρ) used in the equations are presented in Table 2. As a density range applies to halite and anhydrite, the mean density has been used here. Hence the results vary slightly if lower or higher density values are used.

From the above equation, it follows that the depth of a desiccated basin equates to 70% of the initial depth of a water-filled basin. For the desiccated deep-basin model of Hsü *et al.* (1973), it may be calculated that a 2.0 km deep desiccated basin would be up to 2.9 km deep before drawdown if isostasy were taken into account. If that basin were filled with halite under continuous isostatic compensation, the thickness of the ultimate halite column (Th_{halite}) is a function of the depth of the desiccated basin:

$$Th_{halite} = D_{basin} \times \left(\frac{\rho_{asthenosphere} - \rho_{air}}{\rho_{asthenosphere} - \rho_{halite}}\right)$$

$$= D_{basin} \times \left(\frac{3.3}{1.15}\right) = 2.9 \times D_{basin}$$

This equation predicts that a 2.0 km deep desiccated basin is filled with a maximum of 5.8 km of halite if precipitation takes place under a condition of rapid isostatic adjustment. On the other hand, a desiccated basin only 690 m deep would be sufficient to accommodate a 2.0 km thick halite sequence if halite precipitation occurred under rapid isostatic compensation.

The deep-basin shallow-water model of Hsü *et al.* (1973) implies that the filling with halite of a 2 km deep desiccated basin is followed by up to 2 km of subsidence to regain isostatic equilibrium. Note that a shallow basin and a deep basin both allow the formation of a 2 km thick evaporite succession (Fig. 5). However, it is felt that the shallow-basin model is more generally applicable and less restrictive where tectonic and geographical conditions are concerned. For example, it accounts for the occurrence of shallow-water sediments (early stage) as well as deeper-water sediments (late stage), without repetitive kilometre-scale marine desiccation and refilling.

The above calculations show that there is a simple alternative to the deep-basin shallow-water evaporite model, which explains the thickness of saline giants, as well as the occurrence of shallow-water sedimentary structures. The main implication of isostatic compensation in evaporite-basin evolution is that evaporite precipitation drives subsidence instead of the other way round, and that thick halite deposits as they are observed in the rock record require an initial basin depth much less than their eventual thickness.

A halite-deposition model, which explains the formation of saline giants under the condition of isostatic compensation, is shown in Fig. 6. First the connection of a shallow water-filled basin with the open ocean becomes restricted such that much of the oceanic inflow evaporates and that little outflow of dense brines occurs. This restricted outflow is attributed to a progressive narrowing of a straight that, for example, may be controlled by anhydrite precipitation along the margins of a graben.

The precipitation of halite is a rapid process allowing halite to rapidly fill a shallow basin. The rapid deposition of halite causes disturbance of the isostasy balance, thereby forcing a subsidence reaction of nearly 50% of the thickness of the halite column (Fig. 4). This newly created accommodation space may consequently be filled with halite, again causing a subsidence reaction. As long as the basin receives ocean water, which is to be expected if no tectonic events occur, the process can continue

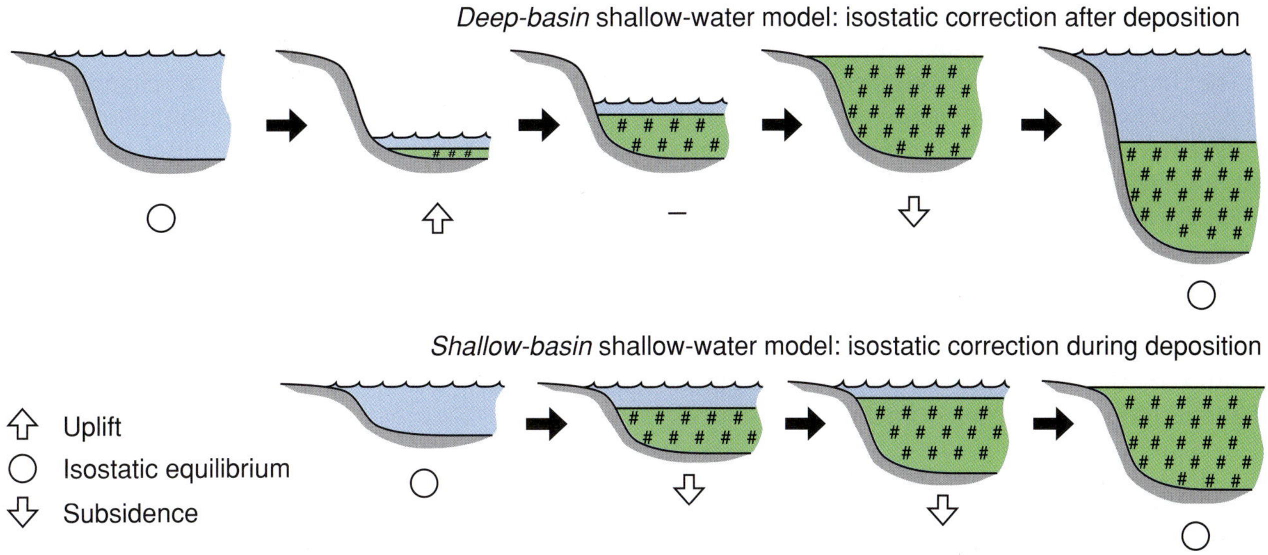

Fig. 5 Comparison of deep-basin and shallow-basin models. The deep-basin shallow-water model for saline giants is based on isostatic compensation *after* salt precipitation. The shallow-basin shallow-water model for saline giants is based on isostatic compensation *during* salt precipitation. Note that the latter model is characterized by an initially shallow basin, whereas the former model is characterized by an initially deep basin, which after filling with halite is subjected to a phase of isostatic subsidence.

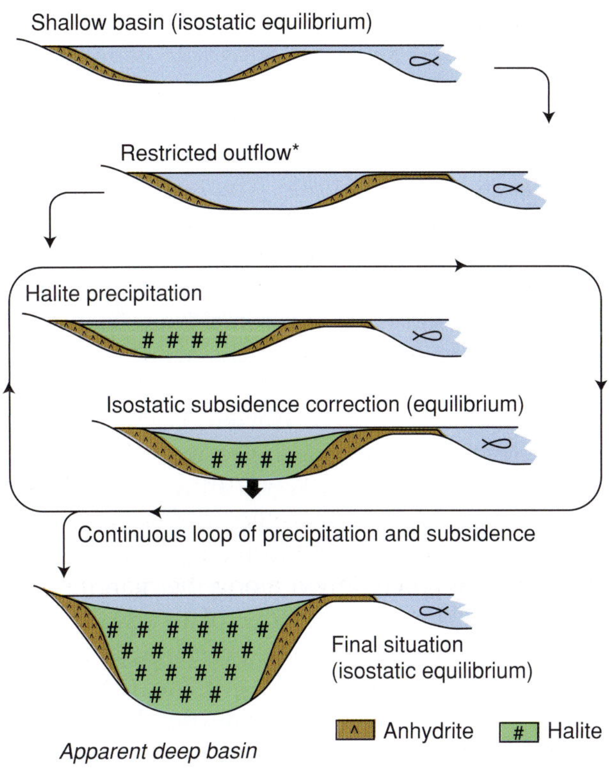

Fig. 6 Model for halite precipitation under continuous isostatic compensation: rapid precipitation of halite in a shallow basin causes isostatic subsidence, thereby resulting in an apparent deep-basin structure. *Restricted outflow may be controlled by anhydrite-platform progradation into a narrow strait (e.g. rift), connecting the evaporite basin with the world ocean.

until subsidence approaches zero. By that time a halite column of up to three times the desiccated basin depth or twice the water-filled basin depth will have been accommodated. Note that this model requires continuous oceanic inflow and restricted outflow, whereas the deep-basin shallow-water model is based on repeated phases of complete isolation from the world's oceans (Fig. 1).

For a water-filled basin the ultimate halite thickness is a function of the initial water depth (D_{basin}):

$$Th_{\text{halite}} = D_{\text{basin}} \times \left(\frac{\rho_{\text{asthenosphere}} - \rho_{\text{water}}}{\rho_{\text{asthenosphere}} - \rho_{\text{halite}}} \right)$$

$$= D_{\text{basin}} \times \left(\frac{2.3}{1.2} \right) = 2.1 \times D_{\text{basin}}$$

The above equation suggests that a halite body such as the thick Zechstein-2 halite (600 m) may form in a basin with an initial water depth of 285 m. A 2 km thick evaporite succession representing a single precipitation event may be accommodated within a 950 m deep water-filled basin. In the case of two evaporite units separated by a period with tectonic subsidence, an average basin depth of 425 m is sufficient to accommodate 2 km of halite in two phases. Note that many saline giants consist of four or more anhydrite–halite cycles (Table 1).

Hence, the average basin depth is then reduced to a few hundred metres or less.

The derived depths are within the depth range of current desiccated continental depressions such as Death Valley, California (−85 m), the Dead Sea rift, Jordan (−411 m), the Qattara Depression, Egypt (−134 m) and the Danakil Depression in the Afar Triangle, Ethiopia (−116 m). The flooded evaporite-precipitating Gulf of Karabokhaz, Turkmenistan is currently 35 m below global sea level. Hence, these depressions which are characterized by thinned and fractured crust may well host future saline giants if connected to the marine domain.

ANHYDRITE

Evaporation of marine-sourced brines causes calcium sulphate ($CaSO_4$) to precipitate well before the halite saturation point is reached (Hardy, 1967). Consequently major halite bodies are found in association with $CaSO_4$ precipitates. Major anhydrite bodies have been shown to be basin-margin wedges, and the bulk of these bodies have precipitated in shallow coastal sabkha environments (Sonnenfeld, 1984). Evaporation has the greatest net effect in shallow water and thus coastal platforms act as evaporite traps. Primary formation of anhydrite is inhibited by chemical boundary conditions, but primary gypsum may be directly converted into anhydrite under high temperature and/or high brine salinity, conditions commonly observed in coastal sabkha environments (Hardy, 1967).

The stability of either of the two $CaSO_4$ minerals is important with respect to isostasy, since their density values are markedly different (Table 2). The density of anhydrite is much higher than that of porous sediment, thus a change from non-evaporite to anhydrite deposition has a major effect on isostatic balance and subsidence. The density of gypsum is approximately equal to the density of porous sediment, so that both have a similar effect on the isostasy balance in terms of density. Gypsum precipitation also may result in accelerated isostatic subsidence because of the common high precipitation rate. Based on constraints of brine concentration and temperature, it is assumed (cf. Tucker, 1991; Warren, 1999) that anhydrite generally precipitates on the platform (high net evaporation) while gypsum (selenite) precipitates on the platform slope (low net evaporation).

In an aggradational-platform situation where tectonic subsidence, sedimentation and isostatic subsidence are in equilibrium, a change from non-evaporite to anhydrite precipitation would approximately cause tripling of total subsidence according to the equation:

$$Th_{\text{anhydrite}} = Th_{\text{sediment}} \times \left(\frac{\rho_{\text{asthenosphere}} - \rho_{\text{sediment}}}{\rho_{\text{asthenosphere}} - \rho_{\text{anhydrite}}} \right)$$

$$= Th_{\text{sediment}} \times \left(\frac{1.10}{0.35} \right) = 3.1 \times Th_{\text{sediment}}$$

The long-term laterally equivalent deposition of coastal-sabkha anhydrite and inland sabkha clay under continuous isostatic compensation therefore would result in a rapid basinward thickening rock column, where the anhydrite column is up to three times thicker (Fig. 7). Locally such differential subsidence may be facilitated by passive (non-tectonic) fault movement, as has been observed on seismic cross-sections for Zechstein anhydrite bodies (Van der Baan, 1990).

The proposed model for anhydrite deposition is illustrated in Fig. 7. As long as some tectonic subsidence occurs and fresh seawater is supplied, aggradational anhydrite precipitation along the basin margin can continue. The rate of anhydrite precipitation is expected to be higher on the platform than on the platform slope, so a progressive steepening of platform clinoforms is predicted. This may result in mass movement, as observed in the Zechstein and other basins where slumped anhydrite and anhydrite turbidites occur (Van der Baan, 1990; Tucker, 1991; Warren, 1999).

The thickest anhydrite body in the Zechstein is the up to 280 m thick Werra anhydrite (Van der Baan, 1990). Most of its relief is filled with the 600 m of halite belonging to the Zechstein-2 cycle (Fig. 2). The precipitation of a thick anhydrite wedge results in the formation of an equally deep adjacent basin, which may be characterized by the precipitation of pelagic gypsum (varves) as observed in the Zechstein (Van der Baan, 1990) or in the Delaware Basin, Texas (Sonnenfeld, 1984; Van der Baan, 1990), and is the later site of halite precipitation. Hence, the prolonged basin-margin precipitation of anhydrite initiated under shallow-water, lagoonal conditions may contribute to the formation of very thick halite bodies (Fig. 7).

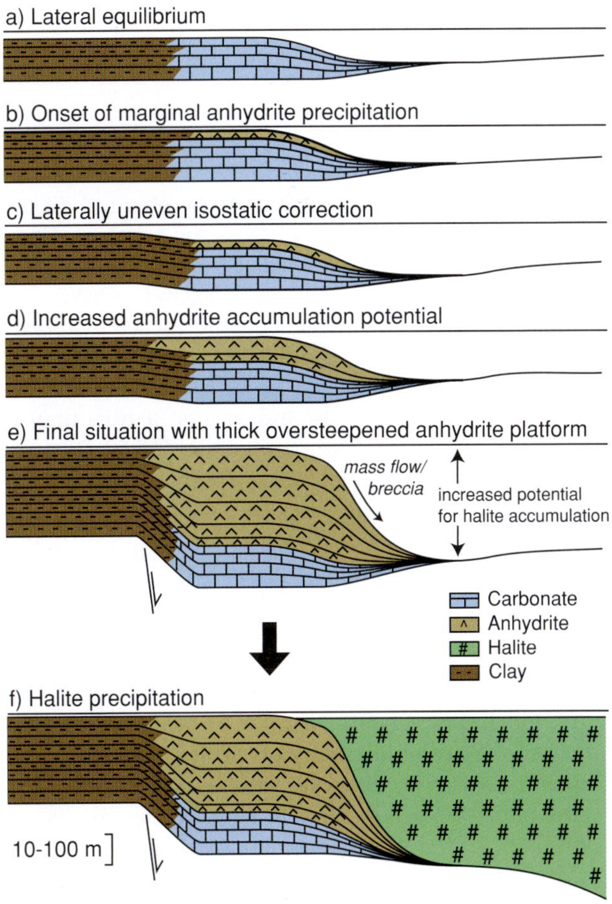

Fig. 7 Anhydrite precipitation model on a platform margin. The high density of anhydrite causes accelerated isostatic subsidence, thus allowing the accommodation of a thick anhydrite platform at the site of an initially shallow basin. On the other hand, anhydrite loading may result in the formation of a deep adjacent basin, which allows the consequent accumulation of a thick halite body.

DISCUSSION

Isostatic compensation during evaporite deposition is expected to have a major influence on evaporite-basin evolution, due to the high density values of the minerals involved and the high deposition rates of evaporite minerals. Saline giants typically formed in (post-orogenic) rifting-dominated areas (Table 1). Late-Carboniferous to Permian evaporites formed on thin, fragmented crust that developed during the collapse of Hercynian mountain chains (Stanley, 1986; Volozh *et al.*, 2003). Triassic to earliest Cretaceous evaporites formed in rift-basins that were transgressed by the sea during the break-up of Pangea (Burke, 1975).

A relatively weak and thin crust, dissected by faults, thus seems to characterize sites of major evaporite formation. Such conditions would have allowed rapid isostatic adjustment that favoured thick evaporite accumulations. The location of major evaporite bodies at the downthrown sides of major faults (Van der Baan, 1990) suggests that reactivation of existing faults may have allowed a quick isostatic response locally.

The application of the isostasy principle predicts that kilometres-deep continental depressions are not a prerequisite for the formation of saline giants, but that relatively shallow basins located on a weak crust, such as the Dead Sea or the Danakil and Qattara depressions in the northernmost East African rift, offer favourable conditions. Brine and influx modelling by Tucker & Cann (1986) has shown that deep-brine basins are not required for the formation of thick evaporite series, and that 'for most geological examples it is possible to postulate a shallow-basin origin in which the basin is continuously replenished by new influx', i.e. that many saline giants could have formed in basins of only a few hundred metres deep that were replenished by normal sea water. Their model requires that evaporite deposition is balanced by constant basin subsidence, a condition that is met in a shallow-basin model by the isostatic effect of salt loading.

The shallow basin concept requires a shallow depth of deposition of sediments underlying saline giants. The evaporites from the Zechstein conformably overlie aeolian sands, playa shales and evaporites from the Rotliegend Group, the basal Zechstein Coppershale and a thin unit of basal Zechstein ramp carbonates (Taylor, 1998). The Rotliegend itself developed on thin crust after orogenic decay. Similarly, other saline giants appear to be associated with shallow-marine deposits as well as arid terrestrial siliciclastics (Table 1).

Due to rapid precipitation during phases of strong evaporation, isostatic subsidence (loading) outpaced tectonic subsidence. Most evaporite successions are characterized by a number of anhydrite–halite cycles separated by relatively thin carbonate sequences. To allow such repetitive phases of evaporation, it is necessary that halite precipitation phases of short duration are followed by longer periods of carbonate or anhydrite deposition to allow the (tectonic) formation of accommodation space for a new phase of halite precipitation.

Considering the similarity in thickness and extent of the Mediterranean Messinian halite bodies and other saline giants (Table 1), the obvious question is whether their genesis was similar. A deep-basin model has been advocated for the Messinian evaporites based on assumed evidence of pre- and post-Messinian deep-marine deposition (Hsü *et al.*, 1973; Cita, 2001) and on the assumption that the Mediterranean Sea was already deep before the Messinian period. It has been demonstrated here that isostatic compensation of salt precipitating from continuously inflowing marine water in arid climate zones may allow relatively shallow basins to develop into saline giants. Hence this model could provide a simple alternative for the deep-basin theory. Note that the model eliminates the need for repeated opening and closure of the oceanic connection, deep-basin desiccation and gigantic waterfalls.

Future saline giants may form in presentday arid-region continental depressions such as the Dead Sea and the Qattara and Danakil depressions of the East African rift. Once such depressions tens to some few hundred metres deep are connected to the world ocean by a relatively shallow strait, the formation of evaporites is expected to cause gradual isostatic subsidence and thus allow the deposition of evaporites much thicker than the present-day depth of these depressions. The size of many of these areas is comparatively small, however continued rifting and subsequent flooding with sea water may result in a rapid increase of surface area. For example, continued widening and collapse of the main East African rift, now only below sea level in the northernmost Afar Triangle, could result in an evaporite basin as large as the Southern Permian Basin (Zechstein).

ACKNOWLEDGEMENTS

We thank G. Nichols, C.J. Ebinger (Royal Holloway University of London) and A.B. Watts (University of Oxford) for their reviews. Earlier versions of the manuscript have benefited from discussions with various collegues at the University of Utrecht and the Netherlands Institute of Applied Geoscience, Utrecht.

REFERENCES

Al-Zoubi, A., Shulman, H. and Ben-Avraham, Z. (2002) Seismic reflection profiles across the southern Dead Sea basin. *Tectonophysics*, **346**, 61–69.

Anderson, R.Y., Dean, W.A., Kirkland, D.W. and Snider, H.I. (1972) Permian Castile varved evaporite sequence, West Texas and New Mexico. *Geol. Soc. Am. Bull.*, **83**, 59–86.

Blanc, P.L. (2000) Of sills and straits: a quantitative assessment of the Messinian Salinity Crisis. *Deep-Sea Res. Part I-Oceanogr. Res. Pap.*, **47**, 1429–1460.

Burke, K. (1975) Atlantic evaporites formed by evaporation of water spilled from the Pacific, Tethyan and Southern oceans. *Geology*, **3**, 613–616.

Catacosimos, P.A., Daniel., P.A. and Harrison, W.B. (1990) Structure, stratigraphy and petroleum geology of the Michigan Basin. In: *Interior Cratonic Basins* (Eds M.W. Leighton, D.R. Kolata, D.F. Oltz and J.J. Eidel), pp. 561–601. Memoir 51, American Association of Petroleum Geologists, Tulsa, OK.

Cercone, K.R. (1988) Evaporative sea-level drawdown in the Silurian Michigan Basin. *Geology*, **16**, 387–390.

Cita, M.B. (2001) The Messinian salinity crisis in the Mediterranean. In: *Paradoxes in Geology* (Eds U. Briegel and W. Xiao), pp. 353–360. Elsevier Science, Amsterdam.

Dercourt, J., Zonenshain, L.P. and Ricou, L.E. (1986) Geological evolution of the Tethys belt from the Atlantic to the Pamirs since the Lias. *Tectonophysics*, **123**, 241–315.

Diegel, F.A., Schuster, D.C., Karlo, J.F., Schoup, R.C. and Tauvers, P.R. (1995) Cenozoic structural evolution and tectonostratigraphic framework of the northern Gulf Coast continental margin. In: *Salt Tectonics: a Global Perspective* (Ed. M.P.A. Jackson, D.G. Roberts and S. Nelson), pp. 109–151. Memoir 65, American Association of Petroleum Geologists, Tulsa, OK.

Einsele, G. (1992) *Sedimentary Basins: Evolution, Facies and Sediment Budget*. Springer-Verlag, Berlin, 628 pp.

El Tabakh, M., Utha-Aroon, C. and Schreiber, B.C. (1999) Sedimentology of the Cretaceous Maha Sarakham evaporites in the Khorat Plateau of northeastern Thailand. *Sediment. Geol.*, **123**, 31–62.

Fabbri, A. and Curzi, P. (1979) The Messinian of the Tyrrhenian Sea: seismic evidence and dynamic implications. *Giorn. Geol.*, **43**, 215–248.

Gilbert, G.K. (1890) *Lake Bonneville*. *U. S. Geol. Surv. Mem.*, **1**, 438 pp.

Hardy, L.A. (1967) The gypsum–anhydrite equilibrium at one atmosphere pressure. *Am. Mineral.*, **52**, 171–200.

Hardy, L.A. and Lowenstein, T.K. (2004) Did the Mediterranean Sea dry out during the Miocene? A reassessment of evaporite evidence from DSDP legs 13 and 42A cores. *J. Sediment. Res.*, **74**, 453–461.

Hsü, K.J., Ryan, W.B.F. and Cita, M.B. (1973) Late Miocene desiccation of the Mediterranean. *Nature*, **242**, 240–244.

Hsü, K.J., Montadert, L., Bernoulli, D., Cita, M.B., Erickson, A., Garrison, R.E., Kidd, R.B., Melieres, F., Mueller, C. and Wright, R. (1977) History of the Mediterranean salinity crisis. *Nature*, **267**, 399–403.

Kendall, A.C. (1992) Evaporites. In: *Facies Models; Response to Sea Level Change*. (Eds R.G. Walker and N.P. James), pp. 375–409. Geological Association of Canada, St Johns, Newfoundland.

Krijgsman, W., Hilgen, F.J., Raffi, I., Sierro, F.J. and Wilson, D.S. (1999) Chronology, causes and progression of the Messinian salinity crisis. *Nature*, **400**, 652–655.

Manzi, V., Lugli, S., Ricci Lucchi, F. and Roveri, M. (2005) Deep-water clastic evaporites deposition in the Messinian Adriatic foredeep (northern Apennines, Italy): did the Mediterranean ever dry out? *Sedimentology*, **52**, 875–902.

Neev, D. and Emery, K.O. (1967) *The Dead Sea Depositional Processes and Environments of Evaporites*. Bulletin 41, Geological Survey of Israel, Jerusalem.

Nesteroff, W.D. (1973) Mineralogy, petrography, distribution, and origin of the Messinian Mediterranean evaporites. In: *Inititial Reports: Deep Sea Drilling Project, Leg 13*, Vol. 13-2 (Eds W.B.F. Ryan, K.J. Hsu and M.B. Cita), pp. 673–694. Texas A & M University, College Station, TX.

Norman, S.E. and Chase, C.G. (1986) Uplift of the shores of the western Mediterranean due to Messinian desiccation and flexural isostacy. *Nature*, **322**, 450–451.

Northrup, C.J. and Snyder-Walter, S. (2000) Transpressional and transtensional tectonics of the southern Ural Orogen. *AAPG International Conference and Exhibition 15–18 October, Bali*, Abstract, pp. 107–108.

Ochsenius, C. (1877) *Die Bildung der Steinsalzlager und Ihrer Mutterlaugensalze*. Pfeffer, Halle.

Orszag-Sperber, F., Harwood, G., Kendall, A. and Purser, B.H. (1998) A review of the evaporites of the Red Sea–Gulf of Suez rift. In: *Sedimentation and Tectonics of Rift Basins: Red Sea–Gulf of Aden* (Eds B.H. Purser and D.W.J. Bosence), pp. 409–426. Chapman & Hall, London.

Reed, J.M. (1994) Probable Cretaceous-to-Recent rifting in the Gulf of Mexico Basin. *J. Petrol. Geol.*, **17**, 429–444.

Ryan, W.B.F. (1976) Quantitative evaluation of depth of western Mediterranean before, during and after Late Miocene salinity crisis. *Sedimentology*, **23**, 791–813.

Sannemann, D., Zimdars, J. and Plein, E. (1978) Der basale Zechstein (A2–T1) zwischen Weser und Ems. *Z. Deut. Geol. Ges.*, **129**, 33–69.

Schreiber, B.C. and Hsü, K.J. (1980) Evaporites. In: *Developments in Petroleum Geology*, Vol. 2 (Ed. G.D. Hobson), pp. 87–138. Elsevier Science, Amsterdam.

Schumann, W. (1987) *Stenen en Mineralen*. Thieme, Baarn, 381 pp.

Siesser, W.G. (1978) Leg 40 results in relation to continental shelf and onshore geology. In: *Initial Reports of the Deep Sea Drilling Project*, Vol. 40 (Eds H.M. Bolli and W.B.F. Ryan), pp. 965–979. Texas A & M University, College Station, TX.

Sonnenfeld, P. (1984) *Brines and Evaporites*. Academic Press, Orlando, 613 pp.

Sonnenfeld, P. (1985) Models of Upper Miocene evaporite genesis in the Mediterranean region. In: *Geological Evolution of the Mediterranean Basin* (Eds D.J. Stanley and F.-C. Wezel), pp. 323–346. Springer-Verlag, New York.

Stanley, M.S. (1986) *Earth and Life through Time*. W.H. Freeman, New York, 619 pp.

Stevenson, G.M. and Baars, D.L. (1986) The paradox; a pull-apart basin of Pennsylvanian age. In: *Paleotectonics & Sedimentation* (Ed. J.A. Peterson), pp. 513–539. Memoir 41, American Association of Petroleum Geologists, Tulsa, OK.

Tay, P.L., Lonergan, L., Warner, M. and Jones, K.A. (2002) Seismic investigation of thick evaporite deposits on the central and inner unit of the Mediterranean Ridge accretionary complex. *Mar. Geol.*, **186**, 167–194.

Taylor, J.C.M. (1998) Upper Permian, Zechstein. In: *Petroleum Geology of the North Sea; Basic Concepts and Recent Advances* (Ed. K.W. Glennie), pp. 174–210. Blackwell Science, Oxford.

Tucker, M.E. (1991) Sequence stratigraphy of carbonate evaporite basins – models and application to the Upper Permian (Zechstein) of northeast England and adjoining North Sea. *J. Geol. Soc. London*, **148**, 1019–1036.

Tucker, R.M. and Cann, J.R. (1986) A model to estimate the depositional brine depth of ancient halite rocks: Implications for ancient subaqueous evaporite depositional environments. *Sedimentology*, **33**, 401–412.

Valyashko, M.G. (1972) Playa lakes – a necessary stage in the development of a salt bearing basin. In: *Geology of Saline Deposits* (Ed. G. Richter-Bernberg), pp. 41–51. UNESCO, Paris.

Van der Baan, D. (1990) Zechstein reservoirs in The Netherlands. In: *Classic Petroleum Provinces* (Ed. J. Brooks), pp. 379–398. Special Publication 50, The Geological Society, London.

Van Wees, J.-D., Stephenson, R.A., Ziegler, P.A., *et al.* (2000) On the origin of the Southern Permian Basin, Central Europe. *Mar. Petrol. Geol.*, **17**, 43–59.

Visser, W.A. (1956) The Upper Permian in the Netherlands. *Leidse Geol. Mededel.*, 20.

Volozh, Y.A., Antipov, M.P., Brunet, M.-F., Garagash, I.A., Lobkovskii, L.I. and Cadet, J.-P. (2003) Pre-Mesozoic geodynamics of the Precaspian Basin (Kazakhstan). *Sediment. Geol.*, **156**, 35–58.

Warren, J. (1999) *Evaporites; their Evolution and Economics*. Blackwell Science, Oxford, 438 pp.

Warren, J. (2000) Dolomite: occurrence, evolution and economically important associations. *Earth Sci. Rev.*, **52**, 1–81.

Watts, A.B. (2001) *Isostasy and Flexure of the Lithosphere*. Cambridge University Press, Cambridge, 458 pp.

Williams-Stroud, S. (1994) The evolution of an inland sea of marine origin to a non-marine saline lake: the Pennsylvanian Paradox Salt. In: *Sedimentology and Geochemistry of Modern and Ancient Saline Lakes* (Eds R.W. Renaut and W.M. Last), pp. 293–306. Special Publication 50, Society of Economic Paleontologists and Mineralogists, Tulsa, OK.

Zharkov, M.A. (1984) *Paleozoic Salt Bearing Formations of the World*. Springer-Verlag, Berlin.

Ziegler, P.A. (1990) *Geological Atlas of Western and Central Europe*. Shell International Petroleum Maatsch., Den Haag, 239 pp.

Single-crystal dating and the detrital record of orogenesis

DOUGLAS W. BURBANK*, IAN D. BREWER†, EDWARD R. SOBEL‡
and MICHAEL E. BULLEN†

**Department of Earth Science, University of California, Santa Barbara, CA 93106, USA (Email: burbank@crustal.ucsb.edu)*
†Department of Geosciences, Pennsylvania State University, University Park, PA 16802, USA
‡Institut fuer Geowissenschaften, Universitaet Potsdam, Postfach 601553, 14415 Potsdam, Germany

ABSTRACT

Single-crystal dating of detrital mineral grains confers a remarkable ability to reconstruct cooling histories of orogens and to place limits on the timing, magnitude, and spatial variations of erosion. Numerous grains from a detrital sample are typically dated, and the statistical variability between populations of ages in different samples provides keys to variations in cooling histories and exhumation rates within the hinterland. Given that detrital samples comprise minerals drawn from an entire catchment, they offer an integrated perspective that is almost always unattainable with bedrock samples. Moreover, because detrital ages are preserved within stratigraphic successions, the evolution of populations of cooling ages through time and across an orogen can be reconstructed from the sedimentary record. When combined with a known hinterland 'stratigraphy' of bedrock cooling ages, studies of detrital ages in modern river systems demonstrate the fidelity of the detrital signal, and reveal both the power and limitations of detrital single-crystal dating in sedimentary basins. Low-temperature thermochronometers can be sensitive to variations in hinterland erosion of as little as 1–2 km. Although recognized previously from a theoretical viewpoint, the impact exerted on modern detrital ages by the interplay between erosion rates and lithology within tributary catchments has only recently been documented and provides a basis for refining orogenic histories using detrital ages. Documentation of the downstream evolution of detrital ages emphasizes that the distribution of ages that reaches the mouth of a river may bear little resemblance to age distributions in the headwaters. Similarly, because lithological concentrations of minerals used for single-crystal dating can vary by many fold within the hinterland, rapidly eroding tributary catchments do not necessarily dominate populations of detrital ages. An ability to exploit detrital ages to place limits on kinematic rates within collisional orogens as a function of cooling rates provides a potent new analytical tool. If uncertainties regarding kinematic geometries, related particle pathways through orogens and steady-state assumptions can be reduced, detrital ages in both modern rivers and the recent stratigraphical record can serve to reconstruct rates of deformation and erosion and to test the viability of proposed models of orogenic evolution.

Keywords Detrital ages, single-crystal dating, methodologies, erosion rates, controls on detrital record, Himalaya, Tien Shan.

INTRODUCTION

Cooling histories of orogens represent responses to tectonic denudation, such as extensional faulting (Davis, 1988), and to erosion by geomorphological processes. As erosion removes rock at the Earth's surface, rock at depth moves toward the surface and cools. As individual minerals in these rocks cool below their radiometric 'closure' temperature and retain the products of radiometric decay, they begin to record the time since cooling below that critical temperature. Minerals with high closure temperatures may not be affected by cooling events that affect minerals with sensitivity to lower temperatures. For example, U–Pb ages on zircons commonly represent crystallization ages of rocks, whereas $^{39}Ar/^{40}Ar$ ages on hornblende or muscovite and fission-track ages on zircon record cooling of a rock below ~ 525°C, 350°C and ~ 250°C, respectively (McDougall & Harrison, 1988; Yamada *et al.*, 1995).

It can be difficult to determine whether extensional faulting, waning magmatic processes, or geomorphic erosion has caused the cooling recorded by various thermochronometers. Cooling of minerals through higher temperatures (300–600°C) typically occurs at depths > 10 km, so field evidence of extensional faulting may be removed during subsequent erosion. Consequently, the driving mechanism for cooling may remain ambiguous. For minerals with lower closure temperatures (< 150°C), however, geological evidence for extensional faulting, if it has occurred, is likely to be preserved. For low-temperature thermochronology, therefore, absence of evidence for extensional faulting indicates that cooling is primarily or entirely due to erosion by surface processes.

Irrespective of the crustal depth at which thermochronologically relevant cooling occurs, upon reaching the surface, the sediment derived from these rocks typically retains the age information about when they cooled through their respective closure temperatures. Hence, the distribution of cooling ages in detrital sediment contains a record of the cooling history of the rocks from which the sediment was eroded (Garver & Brandon, 1994).

The advent of single-crystal dating has allowed precise determination of individual cooling ages. Initial success in the 1980s utilized fission-track dating of detrital zircon (Cerveny *et al.*, 1988), but now single-crystal dating is routinely done with $^{39}Ar/^{40}Ar$, U–Pb, and other methods (Gehrels & Kapp, 1998; Brewer *et al.*, 2003; Wobus *et al.*, 2003 Ruhl & Hodges, 2005). Thus, it is now possible to examine hundreds of individual grain ages from a sediment sample, either modern or ancient. It is even possible (though rarely done) to date minerals from the same sample with different methods, such as by combining fission-track with U–Pb ages (Carter & Moss, 1999; Reiners *et al.*, 2004).

Single-crystal dating provides a new perspective on reconstructing the cooling/erosional history of an orogen. Two specific applications have become common. Samples collected sequentially within a stratigraphic section can be analysed to yield a step-by-step reconstruction of changes in the suite of mineral cooling ages emerging from an orogen (e.g. Carter & Moss, 1999; White *et al.*, 2002; Reiners *et al.*, 2004). Analysis of detrital sediments in modern streams yields a broad sampling of the realm of cooling ages presently exposed within a tributary catchment (e.g. Bernet *et al.*, 2004a). As opposed to the single cooling age typically derived from an individual bedrock sample, detrital ages display an unparalleled attribute: they represent a collection of bedrock cooling ages from throughout a catchment. As such, detrital samples can provide a potent synthesis of information on catchment-wide cooling ages.

Occasionally, the stratigraphic and modern-stream approaches have been combined. For example, one of the first single-crystal age studies (Cerveny *et al.*, 1988) looked at the age distributions of detrital zircon both in the modern Indus River in Pakistan, as well as in Indus foreland strata that dated back to ~ 14 Ma (Fig. 1a). This study identified very young detrital ages (~ 1 Ma) in the modern river and concluded that, because these ages indicated cooling rates of ~ 200°C Myr^{-1}, they represented very rapid erosion (≥ 3 mm yr^{-1}) somewhere in the hinterland. Moreover, within the stratigraphic sequence, when the depositional ages and the cooling ages were compared, Cerveny *et al.* (1988) showed that zircons with similarly young cooling ages (1–2 Myr) had been deposited in the foreland throughout the past 14 Myr (Fig. 1b). The only known modern source for cooling ages this young was the Nanga Parbat–Haramosh massif (Zeitler, 1985), and Cerveny *et al.* (1988) reached the important conclusion that uplifts similar to Nanga Parbat must have persisted in the northwestern Himalaya since at least middle Miocene times. This pioneering study demonstrated the potential of detrital ages to reveal a time series of cooling histories and of reconstructed erosion rates that had previously been inaccessible.

The present study focuses on concepts related to detrital mineral ages, on tests of the assumptions that underpin interpretations of cooling ages, and on some new applications of detrital cooling ages to tectonic problems. After reviewing how bedrock and sedimentary cooling ages are generated and common ways of interpreting detrital age data, we ask:

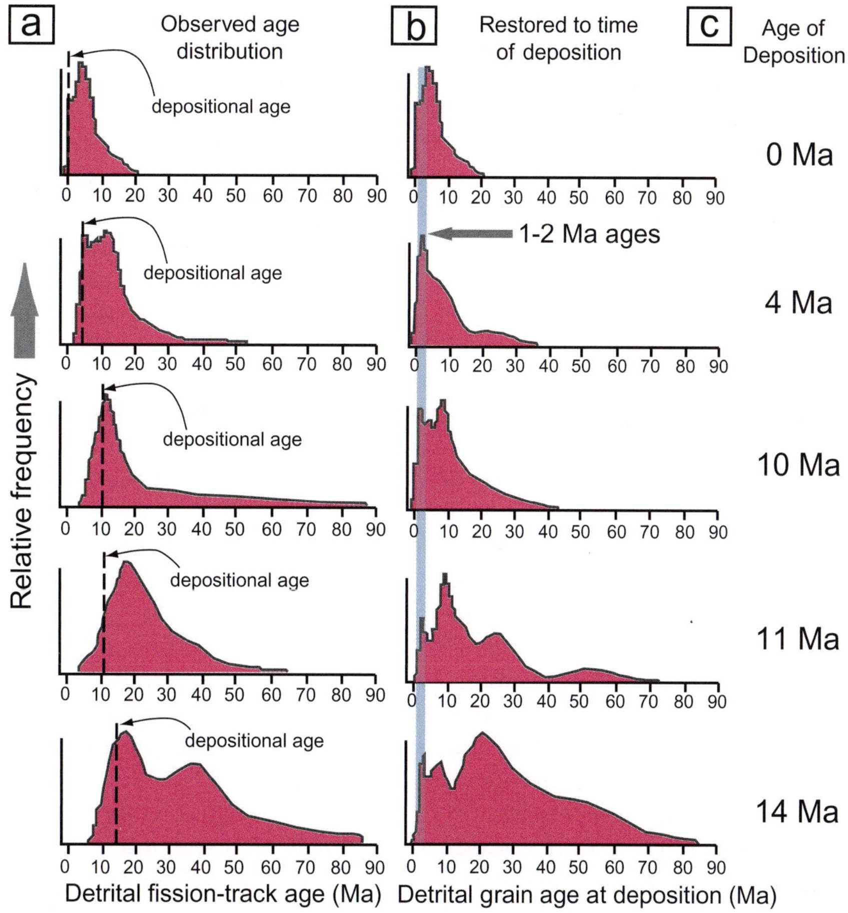

Fig. 1 Single-crystal fission-track age populations of detrital zircons in the northwest Himalaya (modified from Cerveny *et al.*, 1988). (a) Observed ages in the modern Indus River (0 Ma) and at stratigraphic horizons of known age extending back to 14 Ma. The stratigraphic levels were dated using magnetostratigraphy (Johnson *et al.*, 1985) and have uncertainties of ~ 0.5 Myr. Note that for all sites other than the modern site, some fission-track ages are younger than the time of deposition. These samples have never been heated sufficiently to anneal fission tracks following deposition (which reset the ages), so these 'too-young' ages are likely to result from the statistics of counting small numbers of spontaneous fission tracks, and due to the relatively large uncertainties (typically ~ 10%) that characterize fission-track dating. (b) Detrital ages restored back to the time of deposition. Restoration is accomplished both by subtracting the depositional age from each detrital age (c) and through accounting for the statistical uncertainties inherent in populations of fission-track dates (see Cerveny *et al.* (1988) for a more thorough description). Notably, young (1–2 Myr) ages are present in each of the restored population of ages. These young ages require cooling at rates of > 100°C Myr^{-1}, and they indicate that rapid erosion was occurring somewhere in the Indus catchment throughout the past 14 Myr.

1 To what extent do cooling ages in sediments match model predictions of age distributions and frequencies?
2 How does the detrital cooling-age signal evolve as it passes through an orogen from headwater regions to an adjacent basin?
3 To what extent do hinterland variations in lithology or erosion rates control the contributions of cooling ages from each tributary?
4 How can distributions of detrital mineral ages be used to place viable constraints on tectonic rates?

Through comparison of observed bedrock cooling ages in the northern Tien Shan with nearby ancient and modern sediments, it is shown that: (i) modern detrital ages are matched by a combination of a known bedrock age stratigraphy, basin relief and basin hypsometry; and (ii) along-strike differences in modern detrital age distributions correlate with changes in the timing and magnitude of uplift and dissection of the range. By tracking cooling-age distributions in modern sediment along a river that traverses the Himalaya of central Nepal, we explore how variations in rates of bedrock erosion, lithology and basin size are convolved to create the trunk-stream detrital age signal. Finally, cooling ages in a collisional orogen are predicted using a simple thermo-mechanical model and then the observed detrital cooling ages are used to place limits on the rates of deformation within the orogen.

CONCEPTS OF BEDROCK COOLING AGES

All minerals used in thermochronological studies have critical temperatures at which they begin to record time. In reality, this critical temperature is a range of values that can be compositionally dependent, but conceptually this temperature can be considered as a discrete value: the 'closure' temperature (Dodson, 1979). For most radiometric approaches, such as the ^{39}Ar/^{40}Ar or U–Pb systems, the ratio of parent radiometric nuclides to their daughter products is used to define a cooling age. At temperatures higher than the closure temperature, radiometric daughter products are lost by rapid diffusion through the crystal lattice. Once cooled below the closure temperature, the mineral lattice retains the daughter nuclides and the radiometric clock starts.

In fission-track dating, rather than producing daughter nuclides that are subsequently measured, fissioning of uranium creates fragments that tear in opposite directions through the mineral lattice for ~ 8 μm in each direction in apatites and ~ 5.5 mm in zircon (Carter, 1999). At high temperatures, the lattice gradually restores its original geometry and anneals the damage zone. At temperatures less than the 'annealing' temperature, repair of the lattice damage is sufficiently slow that fission tracks are preserved. The two minerals apatite and zircon, most commonly used in fission-track dating, have nominal closure temperatures (below which tracks are preserved) of 110°C and ~ 250°C, respectively (Naeser, 1979; Yamada *et al.*, 1995). The closure temperature is sensitive to the rate of cooling, such that apatite crystals that cool at > 100°C Myr^{-1} have closure temperatures of ~ 140°C (Dodson, 1979). The closure temperature is also sensitive to composition: chlorapatites have closure temperatures that are as much as ~ 50°C higher than the more commonly occurring fluorapatites (Ketcham *et al.*, 1999).

Annealing does not stop abruptly at a given temperature, so a 'partial annealing zone' (PAZ) exists between the nominal closure temperature and temperatures < 60°C at which fission tracks become essentially permanent. Thus, a structure of ages is predicted to exist in the subsurface (Fig. 2). The trends of ages in the rock above the PAZ should reflect the previously cooling history. If these ages show little variation with depth, they indicate that this part of the rock column cooled rapidly during the penultimate episode of erosion and uplift, whereas a steady and large downward decrease in ages would indicate a sustained interval of slow erosion. For typical geothermal gradients, the PAZ is 2–3 km thick. At its top (Fig. 2a), a downward trajectory of decreasing ages begins. The rate of decrease of ages with depth in the PAZ depends on the time since the last major cooling episode (compare Fig. 2a & b): the rate increases with greater elapsed time since the penultimate cooling event. The base of the PAZ is marked by a 'kink' below which all ages are zero, because temperatures greater or equal to the closure temperature have been encountered. If this column of rock is suddenly subjected to rapid rock uplift and erosion in response to tectonic events, the former PAZ will be raised toward the surface (Fig. 2c), and the kink formerly at its base should be preserved during uplift (Fitzgerald *et al.*, 1995). If the amount of erosion exceeds 4–5 km, then the rocks at the base of the former PAZ are likely to be exposed at the surface. The ages just below the kink are commonly interpreted to indicate the time that accelerated erosion began (Fig. 2c), whereas the thickness of the zone of nearly uniform ages beneath the kink provides a minimum limit on the amount of erosion during that event.

For a particular mineral, the cooling ages observed at the surface of a mountain belt are a function of the rate of erosion and the depth of the relevant closure isotherm. Most commonly, cooling ages have been interpreted using a geothermal gradient that is assumed to be both vertically and spatially uniform (Zeitler, 1985; Tippett & Kamp, 1993; Fitzgerald *et al.*, 1995; Blythe *et al.*, 2002). In this case, the depth of a given isotherm will be everywhere the same beneath the mean surface elevation. By applying a known or assumed geothermal

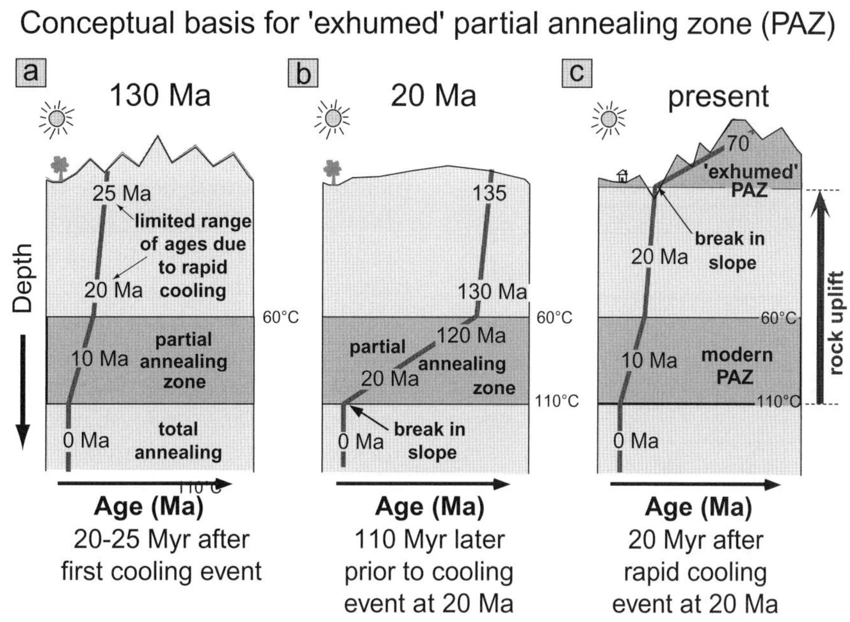

Fig. 2 Patterns of fission-track ages and the partial annealing zone (PAZ) versus depth at different times that span an interval of rapid erosion. (a) Predicted pattern of ages at 130 Ma, which illustrates conditions 20 Myr after a still-earlier major cooling event. Rapid cooling and extensive erosion could be interpreted for that event because the ages above the PAZ are very similar, such that only a small change in age versus depth occurs. Note that ages within the PAZ gradually decrease to 0 at its base where temperatures > 110°C are first encountered. (b) Pattern of ages at 20 Ma following a long interval (110 Myr) of quiescence. Note the greatly increased range of ages within the PAZ compared with (a). (c) Ages at present. The former PAZ (now 'exhumed') has been raised to the surface. The ages at the kink at the base of the PAZ indicate the time (20 Ma) of the uplift/erosion event. Note that the gradient of ages in the PAZ is now identical to that in the first panel which also depicts a time 20 Myr after a major uplift/erosion event.

gradient to define that depth (z), cooling ages (t) can be readily converted into erosion rates (dz/dt).

Such an analysis is a simplification of the more typical situation in a mountain range, in which ridges act like radiator fins and affect the pattern of cooling and position of isotherms in the subsurface (Stüwe *et al.*, 1994; Mancktelow & Grasemann, 1997; Stüwe & Hintermüller, 2000). Isotherms are warped upward beneath ridges and are more widely spaced than they are beneath valleys. The amount of warping is a function of the topographic relief and the rate of erosion (Fig. 3). This compression of isotherms toward the surface and especially beneath valleys can be understood in the context of rock that is advected toward the erosional surface: rapid erosion causes rapid advection which in turn causes hotter rocks from depth to be brought more quickly toward the surface, thereby increasing the near-surface geothermal gradient. The greater the topographic relief, the more that isotherms are deflected upward beneath the ridges. The greater the erosion rate, the greater the compression of the isotherms beneath the valleys.

One key conclusion from analyses of isotherms in the context of variable topographic relief and erosion rates is that, at any point in the landscape, the geothermal gradient is not uniform (Stüwe *et al.*, 1994; Mancktelow & Grasemann, 1997; Stüwe & Hintermüller, 2000). As well as having higher gradients under valleys than beneath ridges and higher gradients when erosion is rapid, the gradient is also higher near the surface than at depth (Fig. 3). As a consequence, application of an assumed uniform geothermal gradient is compromised, especially for minerals with low closure temperatures and in regions of rapid erosion and high relief. Thus, the effects of topography are particularly significant for fission-track or [U–Th]/He dating of apatite with their sensitivity to the 110°C and 70°C isotherms, respectively. For minerals with higher closure temperatures, the assumption of a uniform geotherm is less problematic. For

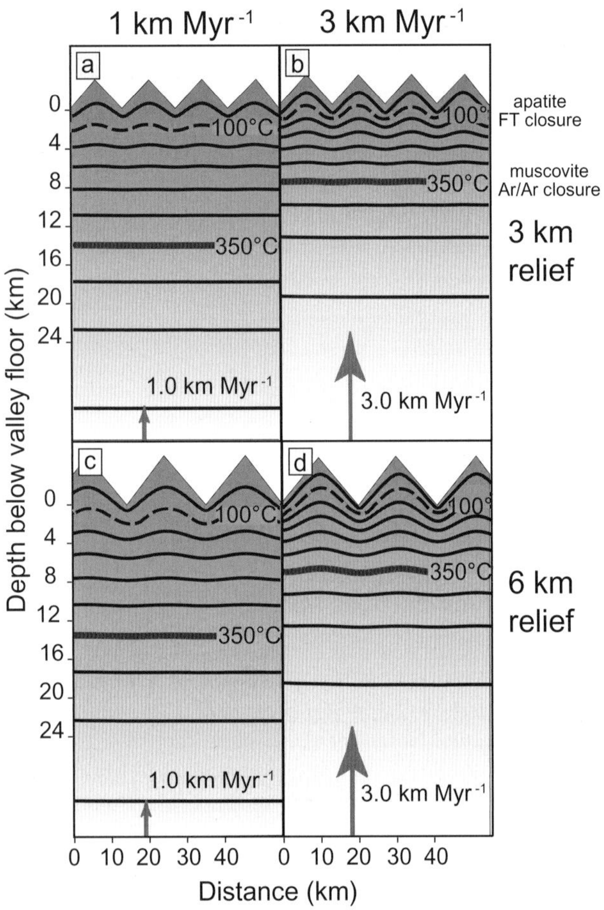

Fig. 3 Numerical modelling prediction of thermal structure of continental crust, given specified erosion rate, topographic relief and wavelength, and hillslope angle. The thermal structure is equivalent to a thermal steady state and occurs before 20 Myr in model runs. Relief production is instantaneous at the start of the model run, with a steady-state landscape that contains 30° slopes, which simulate threshold conditions for landsliding. The depth and deflection of individual isotherms depend on topographic relief and the rate of erosion, with the lower temperature isotherms being most affected. High relief and rapid erosion rates cause the maximum amount of deflection of the isotherms beneath the peaks, as well as the maximum compression of isotherms beneath valleys. (a & b) Erosion rates of 1.0 km Myr^{-1} and 3.0 km Myr^{-1} are imposed on a landscape with 3 km of relief. (c & d) Relief is increased to 6 km with the same erosion rates. Note that the 350°C closure isotherm for $^{40}Ar/^{39}Ar$ in muscovite is predicted to be essentially flat, except when erosion is ≥ 3 km Myr^{-1} and topographic relief is ≥ 6 km, in which case the depth of the 350°C isotherm varies by ~ 200 m. (Modified from Brewer, 2001.)

example, for muscovite that is dated using $^{39}Ar/^{40}Ar$ techniques, the relevant closure temperature is ~ 350°C. As long as topographic relief remains < 6 km and erosion rates are ≤ 3 mm yr^{-1}, numerical modelling suggests that the 350°C isotherm remains nearly horizontal beneath the mean topography (Brewer *et al.*, 2003), although this isotherm will have been advected closer to the surface when erosion rates are high (Fig. 3).

In addition to topographic controls on isotherms, faulting can also have important thermal effects (Ehlers & Farley, 2003; Bollinger *et al.*, 2004; Jamieson *et al.*, 2004). Underthrusting of colder rock beneath a hangingwall refrigerates the hangingwall and perturbs the thermal structure. In extensional settings, tilting of the footwall during faulting will reorient isotherms (Ehlers *et al.*, 2003). The net result of faulting in active orogens is that cooling ages become less directly linked to the rate of vertical erosion at any particular site on the surface. Additional variability in isotherms results from spatial differences in radioactive heat production and from variations in subsurface fluid flow and thermal conductivity, which promote heat advection that can be largely independent of the rate of rock advection.

CONCEPTS OF DETRITAL COOLING AGES

Sediments eroded from orogens are commonly preserved in the foreland basin or in large delta complexes of major river systems (e.g. the Indus or Bengal fans). Detrital cooling ages extracted from such sediment should be a representation of cooling ages within the river's catchment (Stock & Montgomery, 1996; Bernet *et al.*, 2004b). The ways in which this signal is produced can be more readily conceptualized if it is considered how tributary areas of the main river contribute cooling ages to the trunk river as it flows from the hinterland to the site of deposition. It is helpful to target tributary areas at spatial scales for which erosion and cooling rates within a catchment are nearly uniform.

Given a prediction of bedrock cooling ages in a tributary catchment, conceptualization of how the detrital age signal should develop within an orogen is straightforward (Stock & Montgomery, 1996). Every tributary to a trunk stream should

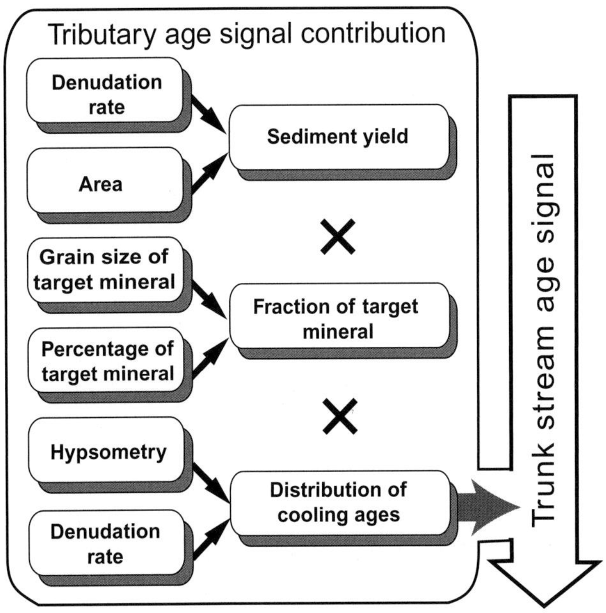

Fig. 4 Parameters controlling the contribution of cooling ages from an individual tributary to a trunk-stream cooling-age signal. Either the trunk-stream or the foreland-basin signal can be modelled as a specified mix of several such tributaries. The tributary area and the average denudation rate define the total sediment yield. The abundance of the mineral to be dated and the size distribution of the target mineral determine the fraction of grains in the total sample that could be dated. The denudation rate (assumed to have persisted long enough to create steady-state thermal conditions) and the hypsometry combine to determine the range and abundance of cooling ages. The product of the total sediment yield, the fraction of the target mineral and the frequency distribution of cooling ages define the tributary's contribution to the distribution of individual grain ages in the trunk stream.

contribute a sediment volume proportional to its catchment area and the average erosion rate within it (Fig. 4). If erosion has persisted sufficiently long to achieve a thermal steady state (Willett & Brandon, 2002), the range of cooling ages should be a predictable function of the basin relief and erosion rate (Fig. 5a). If the distribution of the mineral that is targeted for dating is uniform within the catchment, the distribution of cooling ages (Fig. 5b) will be a direct function of the erosion rates and catchment hypsometry (the distribution of area versus altitude; Brewer *et al.*, 2003). Finally, to calculate how a tributary's flux of cooling ages will affect the cooling-age distribution of the trunk stream, the fraction of the target mineral, including its abundance in the size fraction being dated, must be compared between the tributary sediment and that of the trunk stream (Fig. 4). If the target mineral for dating is in low abundance, e.g. zircon in limestone, then even a large and rapidly eroding catchment will have a minimal impact on the cooling ages in the trunk stream (Spiegel *et al.*, 2004). Within this conceptual framework, it is possible to work progressively downstream and model how the trunk stream signal will evolve with the addition of material from tributaries with varying characteristics.

The direct prediction of the distribution of detrital ages (Fig. 5) that results from combining an erosion rate and a catchment hypsometry (Brewer *et al.*, 2003) means that observed detrital age distributions can, in theory, be used to infer combinations of hypsometry and erosion rates within the tributary catchment. For example, if a hypsometry with topography distributed as a simple Gaussian function is assumed, then straightforward predictions of the effects of changes in erosion rates and relief can be made (Fig. 6). As erosion rates increase, the mean of the detrital ages becomes increasingly young. As relief increases, the breadth of detrital ages increases, due to the age difference between valley bottoms and ridge crests (Figs 5 & 6). Such changes provide a theoretical basis for using cooling-age distributions to test hypotheses, such as those related to increased relief production (Small & Anderson, 1998) and/or enhanced erosion rates during Late Cenozoic times (Zhang *et al.*, 2001). Successful testing with this approach, however, requires high-resolution dating and a stratigraphic section in a basin that has had a stable tributary catchment over the period of interest. Recent experimental and numerical studies, however, suggest that drainage divides will migrate over time (Hasbargen & Paola, 2000, 2002; Pelletier, 2004), and relative changes in catchment location and geology due to divide migration must be small relative to overall catchment size for the catchment to be considered stable. Without an extensive knowledge of the evolution of the hinterland that permits fingerprinting of distinct source areas (Spiegel *et al.*, 2004), assessment of the dominance and persistence of source areas is rarely possible.

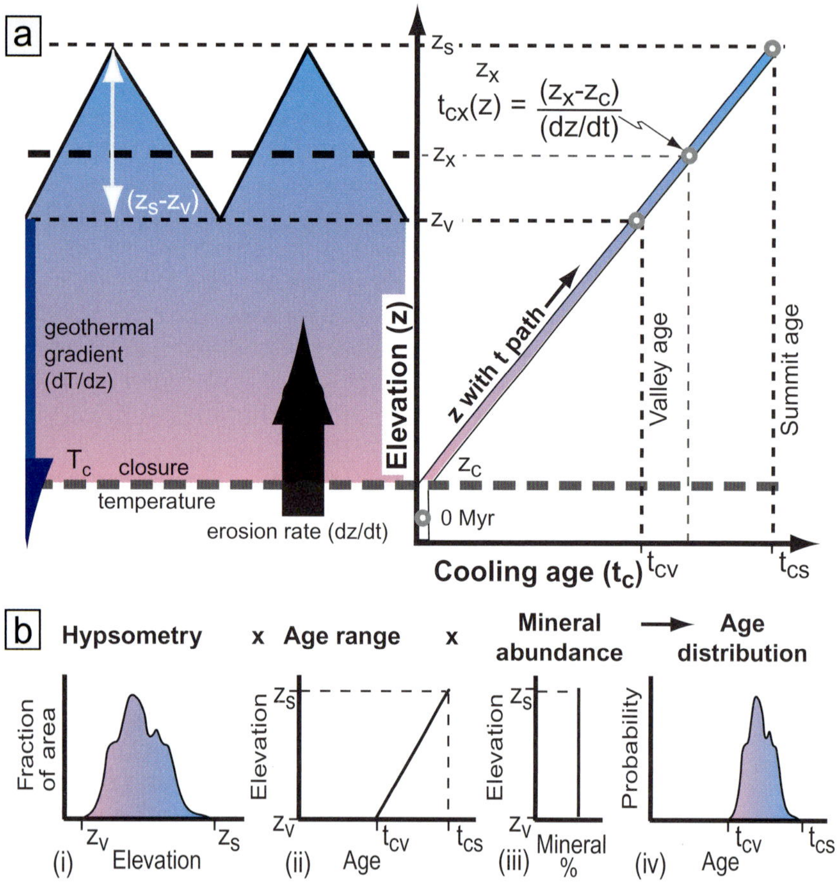

Fig. 5 Construction of a 'theoretical' distribution of bedrock cooling ages for an individual catchment. (a) A cooling age (t_c) is calculated from the depth (z_c) of the closure temperature (T_c) for which z_c results from a thermal model and the erosion rate (dz/dt). The difference between summit elevation (z_s) and valley elevation (z_v) creates a difference between summit cooling ages (t_{cs}) and the valley cooling ages (t_{cv}). The cooling age (t_{cx}) of a sample 'x' derived from elevation z_x can be calculated using the equation shown. (b) The frequency distribution of cooling ages is governed by the combination of the age range (t_{cv} to t_{cs}) and the altitude-dependent frequency of the target mineral (here assumed to be uniform) with the hypsometry of the catchment. The direct correspondence of the hypsometry and the distribution of cooling ages results from the uniform distribution of the target mineral throughout the catchment. (Modified from Brewer *et al.*, 2003.)

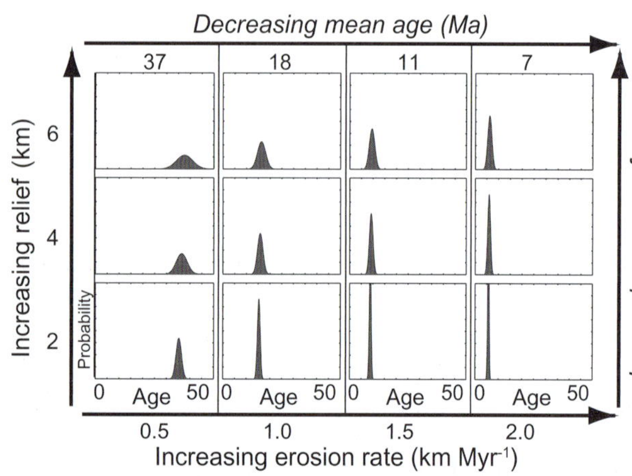

Fig. 6 (*left*) Theoretical effects of variations in uplift rate and relief on cooling-age distributions for a source-area catchment with a Gaussian-distributed hypsometry. The depth to the 350°C isotherm is modelled as a function of erosion rate and topographic relief (Stüwe *et al.*, 1994). The scale of each inset theoretical probability density function plot is the same, with the x axis ranging from 0 to 50 Myr and probability on the y axis. The area under the curve in each plot is normalized to one. According to these models, changes in relief or erosion rate should be manifested by changes in the central age and the spread of ages found within a catchment. For any given relief, slower erosion rates produce a broader range of cooling ages. Similarly, for a given erosion rate, greater relief yields a broader range of ages. (Modified from Brewer *et al.*, 2003.)

Thus, although changes in the range and distribution of cooling ages are predictable based on variations in relief and erosion rate (Fig. 6), these concepts can be practically applied only in unusual circumstances in which the source area can be shown to be stable through time.

OROGENIC EVOLUTION, LAG TIMES AND APPLICATIONS OF DETRITAL AGES

The preceding discussion indicates that, prior to an interval of accelerated rock uplift and erosion, a 'stratigraphy' of cooling ages for low-temperature

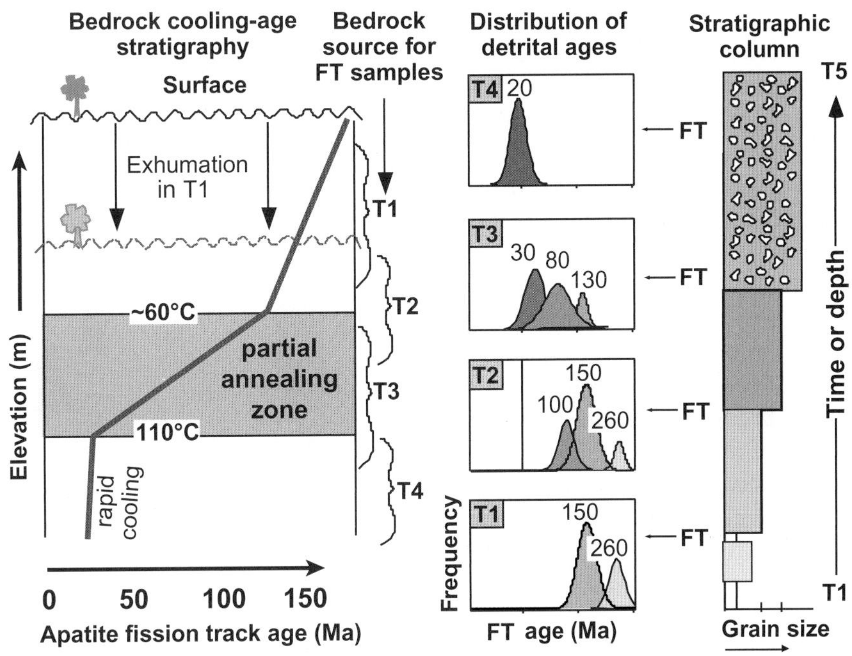

Fig. 7 Model for detrital mineral populations resulting from progressive unroofing (at times T1 to T4) through a crustal column that preserves an exhumed apatite partial annealing zone (PAZ) in which ages span from 25 to 125 Ma. Apatite is assumed to be uniformly distributed in the crustal column. The stratigraphic column on the right depicts a coarsening upward sequence in which 'FT' indicates the stratigraphic level at which four detrital fission-track samples are analysed. The primary component age distributions (Brandon, 1992) are illustrated for each detrital sample (centre column) with the central age of each component shown in 'million years ago'. Progressive erosion of a crustal column with a given age–elevation succession is predicted to produce an inverse age stratigraphy in an adjacent basin. Time-step T1 represents erosion only through the region of older cooling ages that sits above the PAZ in the crustal column. Consequently, the detrital age distribution contains only older age populations. For successively higher and younger stratigraphic levels, erosion in the hinterland has progressed down into the (now exhumed) PAZ. The youngest component age-peak reflects the deepest level of erosion at any given time. When Tertiary ages first appear (stage T3), erosion has progressed deeply into the PAZ. For the youngest sample (T4), only reset samples with Tertiary ages are predicted, suggesting erosion entirely through the PAZ.

thermochronometers commonly exists within a vertical column of slowly eroding bedrock (Stock & Montgomery, 1996). The youngest ages (0 Ma) occur at depths where the temperature exceeds the closure temperature, while the oldest ages occur near the surface, where they reflect previous thermal events (Fig. 2). Accelerated erosion that slices progressively deeper through this rock column would be expected to yield an inverted age stratigraphy in an adjacent basin (Fig. 7). Consequently, detrital ages from a stratigraphic section should be expected to record the progressive unroofing of the hinterland (Brown, 1991; Gallagher *et al.*, 1998; Carter & Moss, 1999). Most significantly, in the context of fission-track dating, the beginning of unroofing of the partial annealing zone (PAZ) should be clearly evidenced by the abrupt appearance of younger detrital ages. Such an event indicates that erosion in the hinterland has proceeded significantly below a depth that was formerly at ~ 60°C (top of the PAZ: Fig. 2), typically at 2–3 km initial depth. As detrital ages more closely approach the depositional age, it is likely that the rocks previously situated beneath the PAZ (at depths exceeding 3.5–5 km) are now being eroded. Under these conditions and assuming the occurrence of apatite is approximately uniform in the source area, the detrital record within a well-dated basin stratigraphy provides detailed insights on the progressive erosion of the hinterland (Fig. 7).

The 'lag time' is defined as the time it takes for a mineral to pass from its closure temperature at some depth below the surface to its deposition in a sedimentary basin (Cerveny *et al.*, 1988; Garver & Brandon, 1994). In a mountain belt, this lag time encompasses two suites of processes: those responsible for bringing the mineral to the surface from some depth-dependent temperature; and those responsible for transporting the mineral from the orogen to a sedimentary basin (Ruiz *et al.*, 2004). In many active orogens, the transport time from when the mineral first reaches the surface to when it is deposited in a basin is considered to be negligible. In rapidly eroding mountains (> 0.5 mm yr^{-1}), such an assumption can be broadly validated simply by comparing volumes of stored sediments, or of potential storage within the mountain catchments, with the amount of sediment produced by persistent erosion. Typically, only a tiny fraction of the overall volume of sediments eroded could be stored for more than a few hundred thousand years. If the transport time can be argued to be negligible, then the lag time represents the time required for a mineral to pass from the closure isotherm to the surface and becomes a proxy for the rate of erosion. Therefore, for the record of unroofing in a nearby basin, the lag time represents the difference between the cooling age and the depositional age. As the detrital age approaches progressively closer to the depositional age, i.e. lag times shorten, the reconstructed rate of cooling of the source area becomes increasingly rapid, as does the correlative rate of hinterland erosion.

Given this framework, lag times can be used to assess the erosional state of an orogen (Fig. 8). Much current debate revolves around the question of whether orogens can attain a steady-state condition (Willett & Brandon, 2002), and if they do so, how rapidly and by what processes does this occur. For example, orogenic steady state has been defined in terms of exhumational steady state (in which cooling ages at a given position in the orogen remain constant through time) or thermal steady state (in which the thermal structure with respect to the surface is invariant; Willett & Brandon, 2002). In either case, the overall distribution of cooling ages at the surface of a steady-state orogen is predicted to remain constant through time. If the sediment transport time is either negligible or predictable, such that it can be subtracted from the cooling age, then a steady-state orogen should yield consistent lag times.

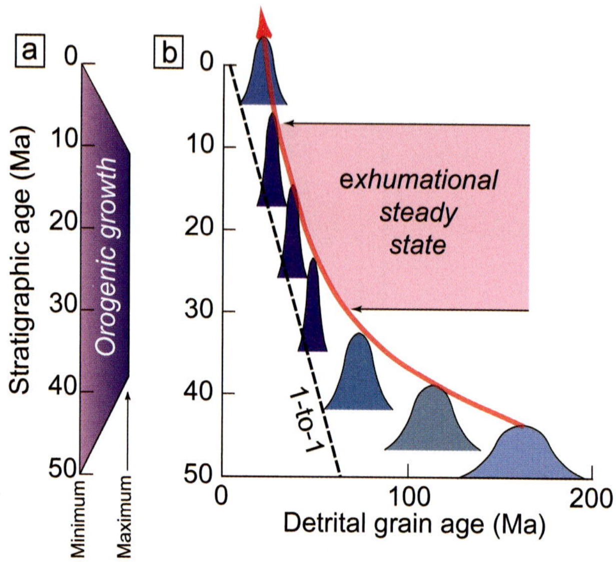

Fig. 8 Relationships among orogenic growth, populations of detrital cooling ages and lag times. (a) Orogenic growth during constructional (40–50 Ma), steady-state (10–40 Ma) and destructional (0–10 Ma) phases. (b) Theoretical sequence of populations of detrital cooling ages represented by Gaussian distributions of ages. The lag time represents the difference between the peak of the cooling age distributions and the stratigraphic age. The 1-to-1 line occurs where the detrital age and the stratigraphic age are equal. Detrital age populations that fall on this line represent a lag time of 0 Myr. During constructional phases, lag times decrease as erosion proceeds increasingly deeply into the hinterland. Lag times remain constant during exhumational steady state and increase again during the destructional phase as orogenesis and erosion rates wane.

Although a spectrum of cooling ages from a given orogen is typically measured in a single sedimentary sample, the distribution of ages is commonly deconvolved into a small number of component populations (Brandon, 1992) that, when summed together, approximate the observed age distribution. The time difference between the youngest component peak of the detrital ages and the depositional age is used to define the lag time (Garver & Brandon, 1994).

Consider the growth and decay of an orogen over 50 Myr (Fig. 8). During its initial growth, the detrital ages are old, but the lag time becomes progressively younger. During steady state, a constant lag time is displayed, and this interval of the shortest lag times equates with the most rapid erosion. During a waning stage of orogenesis, lag times should increase, but not as rapidly as they decreased in the initial growth stages, because in the post-orogenic stage, many minerals that have recently passed through their closure temperature will be exhumed.

Irrespective of considerations of orogenic steady state, the abrupt appearance of a suite of younger detrital ages within a stratigraphic sequence could be interpreted to define or slightly post-date the onset of deformation in the hinterland (as long as the change has not resulted from capture of a new source area). Detrital ages can also be used as a provenance tracer with respect to tectonic reconstructions: exposure of a new source area with a distinctive suite of cooling ages, even very old ones (Carter & Moss, 1999), can provide a readily discernable detrital signal to a nearby basin. Finally, detrital ages can be used to place useful limits on poorly dated continental strata. For example, the occurrence of detrital grains with cooling ages younger than the previously assigned depositional age can force upward revisions in the depositional age (Najman *et al.*, 2001).

TESTING THE FIDELITY AND SENSITIVITY OF DETRITAL AGES

Each of the interpretive strategies or applications described above relies on the assumption that the detrital signal in a basin provides a faithful representation of the age distribution in the hinterland source area (Garver *et al.*, 1999). For fission-track dating in deep sedimentary basins, this assumption is commonly violated due to burial heating that partially or fully resets fission-track ages, especially for apatite grains (Green *et al.*, 1989). Zircon fission-track ages are less susceptible to resetting due to their higher closure temperature, but given this higher temperature they are also relatively insensitive to exhumation that is $< 5-8$ km. Hence, when applying fission-track dating to detrital samples, the likely thermal history of the depositional basin from which samples are collected and the magnitude of hinterland exhumation should dictate whether apatite or zircon is dated.

Even in the absence of reset ages, the correspondence between observed detrital ages and bedrock ages in their source area is rarely assessed. For example, muscovite is a typical mineral exploited for single-crystal detrital dating, and yet the platy nature of muscovite's mineral habit would seem to make it particularly susceptible to comminution during transport. Consequently, any downstream changes in populations of detrital muscovite ages might be an artefact of the loss of grains from farther upstream, rather than an indication of downstream changes in erosion rates and cooling ages. In this case, downstream changes in populations of ages would provide scant insights on variable erosion rates. On the other hand, if muscovite were to travel primarily in a river's wash load, it could experience only minor comminution through grain-to-grain collisions, such that observed detrital ages would be independent of transport distance. One way to test whether comminution is likely to have modified age distributions is to compare detrital age populations of minerals with different susceptibilities to comminution, but collected from the same site. For example, micas are highly susceptible to comminution, whereas zircon is resistant. If, using dating methods with comparable closure temperatures, the distribution of ages from muscovite and zircon are similar at the same site, this would argue against transport distance or comminution as an important control on observed ages.

When 55 $^{40}Ar/^{39}Ar$ muscovite ages (closure temperature: ~ 350°C) and 70 zircon fission-track ages (closure temperature: ~ 250°C) are compared from the same site in a Himalayan river ~ 150 km below its headwaters, the populations of ages resemble each other, although the primary peak in the zircon age population is shifted 1–2 Myr younger in comparison to the muscovite ages (Fig. 9). In fact, such a shift is expected, given the lower closure temperature of zircon fission-track ages. This example suggests, therefore, that the detrital age signal of muscovite can yield a reliable proxy for the distribution of cooling ages in the entire upstream catchment and that comminution during transport creates only minor perturbations.

Detrital age studies assume that meaningful errors can be assigned to dates for individual grains. Most of the detrital studies published to date have utilized zircon fission-track, $^{40}Ar/^{39}Ar$ and U–Pb dating of single crystals, because the uncertainty on any single-crystal age is commonly small. The U and Th content of apatite, however, is typically an order of magnitude less than that of

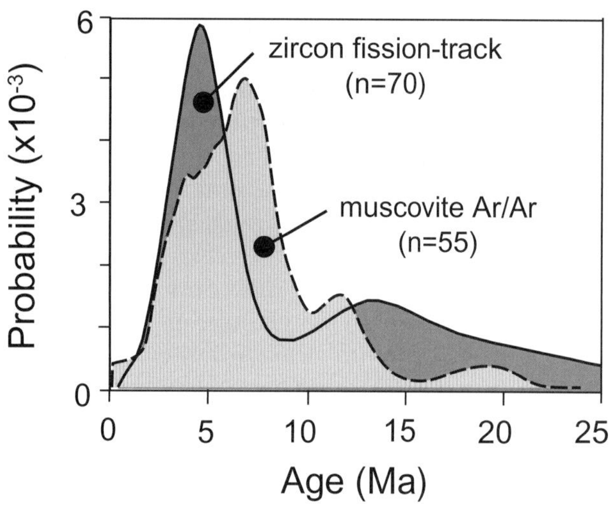

Fig. 9 Comparison of the age distribution of a resistant mineral (zircon) and a mineral susceptible to comminution (muscovite), to assess whether downstream transport causes selective loss of non-resistant minerals. In this example, zircon and muscovite were separated from detrital sand samples collected at the same site and were dated by fission-track (closure temperature ~ 240°C) and $^{40}Ar/^{39}Ar$ (closure temperature ~ 350°C) methodologies, respectively. The basic patterns of detrital ages are in agreement, but the zircon ages are shifted ~ 2 Myr younger. This shift is expected, given the lower closure temperature of zircon. The overall similarity of the age spectra, including the 2-Myr shift, suggests that, despite having been collected over 150 km from the headwaters, the muscovite ages have not experienced significant attrition. This lends support to arguments that micas travel largely in the wash load where little loss occurs due to comminution.

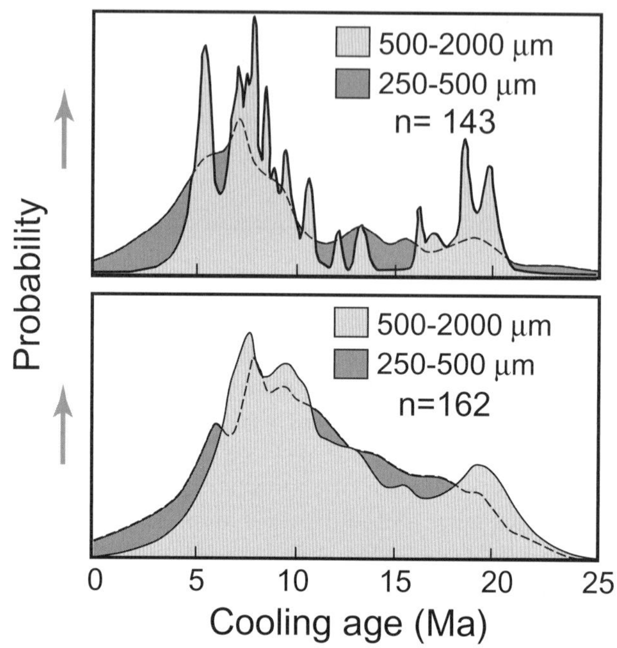

Fig. 10 Comparisons of detrital muscovite $^{40}Ar/^{39}Ar$ ages for different size fractions in the same sample at two different sites in the Marsyandi catchment, central Nepal. The composite probability curve is constructed by summing up individual grain-age measurements by assigning each age a Gaussian 'kernel' with an area equal to unity, a width that is defined by the uncertainty on the age, and a most probable age that equals the measured age. The summed probability of all grains is then re-normalized to unity. Overall, the distributions of ages are independent of grain size, suggesting a similar spatial distribution and susceptibility to erosion for both size ranges in each catchment. The less peaked nature of the probability curve for the smaller grain sizes in the upper sample is due primarily to the higher analytical uncertainties associated with small grains with young cooling ages. (Modified from Ruhl & Hodges, 2005.)

zircon. As a consequence, fewer fission tracks are generated, and even the best of apatite fission-track bedrock samples will have a distribution of grain ages that is dispersed around the 'true' cooling age. The implication of this inherent age uncertainty for detrital studies is that it is difficult to assign an uncertainty to individual grain ages beyond the uncertainty based on the counting statistics, and yet the expected uncertainty would be considerably larger. As a consequence, the approach most commonly used for interpreting detrital fission-track data is to determine the statistically significant component populations that can be arithmetically combined to yield the observed distribution of detrital ages (Brandon, 1992; Brandon & Vance, 1992).

Detrital ages of individual crystals are also typically assumed to be independent of grain size. As a consequence, grain sizes that are most amenable to a given dating approach are commonly analysed. For example, because recently cooled muscovite grains contain only small quantities of radiogenic argon, large detrital grains (> 500 μm) are typically dated in order to minimize the error on each age. Comparisons of single-grain age distributions (Ruhl & Hodges, 2005) can show similar age distributions for different size fractions at a given site (Fig. 10) that validate the assumed size independence of ages. In these situations, the distribution of grain sizes in contributing areas with different cooling histories must remain fairly uniform. On the other hand, if all of the fine-grained muscovite occurred in a rapidly eroding area, whereas the coarse-grained muscovite derived from a slowly eroding domain, then analysis of only the coarse fraction would miss the young, rapidly cooled grains. Thus, if schists and granites experienced different erosion rates, analysis of coarser muscovite grains might only reflect the cooling history of the granite.

The distribution of detrital ages in a river is assumed to be unchanging at annual time-scales and homogeneous at spatial scales of tens of metres. It is commonly assumed that sampling in the spring or autumn would not change the observed age distribution. Similarly, grab samples from different positions on a gravel bar are assumed to yield comparable detrital ages. Such assumptions are infrequently tested, yet one can envision circumstances that would create instability in the detrital age signal: large landslides in a given catchment; different, seasonally dependent erosion mechanisms or sediment transport events in different catchments (glacial meltwater versus winter rainstorms); preferential storage of some sediments; or different grain sizes from different catchments that are hydraulically sorted into different positions on a bar. For example, Amidon *et al.* (2005a) found > 50-fold differences in zircon abundance for different size fractions at the same site. In stratigraphic studies of detrital minerals, the cost and labour of dating individual grains typically precludes testing of duplicate samples separated laterally by tens of metres or vertically by centimetres. Nonetheless, such tests are needed to verify the stability (or arbitrariness) of the detrital signal (see Ruhl & Hodges (2005) for examples of time-varying distributions).

Dated grains from detrital samples are assumed to provide a reliable portrait of the population of grain ages. Only a limited number (40–100) of grains are typically dated, however. For a simple distribution of grain ages, such as when a single dominant age peak is present, 40 dates are sufficient to capture its essential characteristics. For distributions with multiple peaks, 100 dates may only begin to mimic the true complexity of the age distribution (Brewer *et al.*, 2003; Vermeesch, 2004). Although dating larger numbers of grains is perhaps the most reliable means to address this issue, numerical smoothing of age distributions improves the match between 'daughter' and 'parent' distributions, especially for complex age distributions (Amidon *et al.*, 2005b). The technique of extracting component age peaks from the observed distributions also serves as a smoothing function (Brandon, 1996, 2002) that reduces mismatches.

Despite a common desire to extract as much detailed information as possible from detrital age data, the sensitivity of detrital ages to small changes in the magnitude or rate of erosion has rarely been well assessed. To evaluate this sensitivity, apatite fission-track ages from the Kyrgyz Range in the northern Tien Shan provide a test case. Here the distribution of bedrock cooling ages is particularly well documented (Sobel *et al.*, 2006) and differences of 1–2 km in differential erosion along the length of the range can be discerned. Two previous studies by Bullen *et al.* (2001, 2003) in the Ala Archa catchment (central Kyrgyz Range) have demonstrated the presence of an exhumed partial annealing zone and of an underlying zone of reset ages where, throughout a zone > 1 km thick, all of the apatite ages are ~ 11 Ma (Fig. 11a–d). The kink in the age–elevation trend at the base of the exhumed partial annealing zone has been interpreted to represent the time at which rock uplift and erosion accelerated (see Fig. 2 for the conceptual background) in the Kyrgyz Range following a 100-Myr-long interval of quiescence and slow erosion (Bullen *et al.*, 2003). Along the sampled bedrock transect (Fig. 11d), the oldest ages are ~ 20 Ma (~ 4000 m elevation) and lie well within the partial annealing zone (Fig. 11d). The catchment containing this transect, however, extends deeper within the range, where, given a southward tilt of the range that is imposed by the north-vergent thrusts beneath it, still older cooling ages are expected to be prevalent.

With this known 'stratigraphy' of cooling ages, it becomes possible to assess whether a catchment yields ages of a predictable range and abundance (Stock & Montgomery, 1996). Following the logic of Brewer *et al.* (2003), it would be predicted that the relative frequency of ages should result directly from convolving the hypsometry of the catchment (Fig. 5) with the cooling age 'stratigraphy'. To test this concept, detrital apatites were dated in a sample of sand from the modern Ala Archa River, which drains the catchment that encompasses the bedrock transect. The youngest age component of the detrital sample displays a peak of 17 ± 2 Ma (Fig. 11e) and is highly consistent with the detrital ages that would be expected, given the observed altitudinal distribution of cooling ages (Fig. 11d).

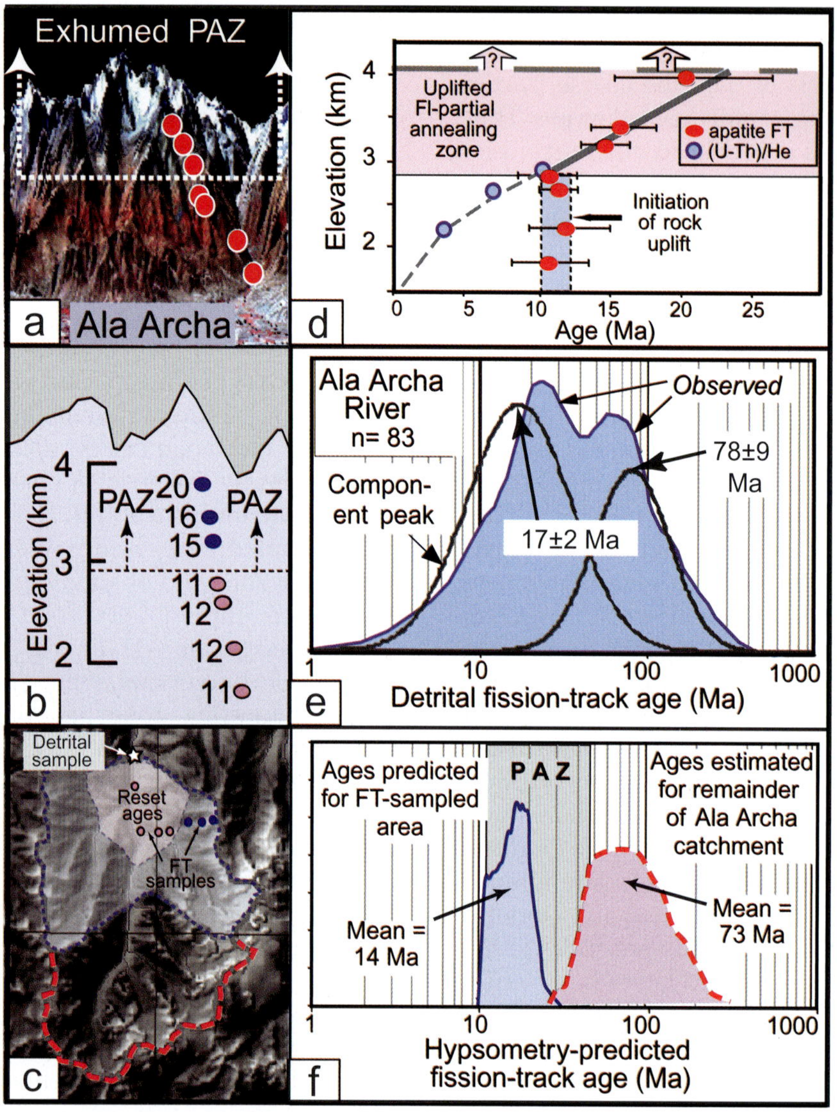

Fig. 11 Bedrock and detrital apatite fission-track ages from Ala Archa, central Krygyz Range, Tien Shan. (a) Vertically exaggerated DEM draped with a satellite image and showing location of fission-track samples. Dark circles lie beneath the exhumed partial annealing zone (PAZ). (b) Cooling ages (Ma) in their relative vertical positions. (c) Digital elevation model showing the location of the bedrock samples (coloured dots), detrital sample (star) and the outline (blue dashed line) of the northern part of the Ala Archa drainage containing the FT samples, and the southern part of the catchment (red dashed line). (d) Age–elevation profile depicting the base of the PAZ and indicating an age of 11 Ma for the beginning of rapid cooling and rock uplift. (e) Detrital FT data from modern sand in Ala Archa River, showing two component populations that account for most of the observed distribution. The mean of the younger population appears consistent with the observed FT relief section. (f) Computed distributions of FT ages for the Ala Archa catchment based on the areas outlined in (c). The younger population results from convolving the hypsometry with the predicted age-elevation trend (d) and provides a close match to the observed 17 ± 2 Ma population. The older population is calculated assuming the PAZ is tilted southward at ~ 12° and that the age-elevation trend in the PAZ remains similar to the observed trend (d). The shaded 'PAZ' area indicates all samples derived from within the PAZ.

When the altitude-dependent cooling ages are combined with the hypsometry of the northern Ala Archa sub-basin from which the samples were collected (Fig. 11c), the resulting distribution of ages yields an average age of 14 Ma (Fig. 11f) and provides a rather good match to the younger component peak in the observed detrital ages (Fig. 11e).

Not unexpectedly (given the young bedrock cooling ages: Fig. 11d), these predicted ages fail to match the older observed ages with a component age peak at ~ 80 Ma, which are probably derived from the more southerly part of the catchment. To mimic this older age component, it is suggested that the range has been tilted southward at ~ 12°, such that the base of the partial annealing zone is located below 1600 m altitude in the southern part of the catchment. To obtain a satisfactory match to the older (80 Ma) observed age peak (Fig. 11e), a much larger range of bedrock ages is required. Therefore, above the PAZ, a strong altitude dependency is required to produce ages ranging from Tertiary to Palaeozoic. Although the overall match to the observed ages is inexact, this hypsometric approach mimics the observed data and provides some insight into the age distribution in the bedrock. In particular, in order to produce the observed ages, it requires that unsampled, southern parts of the catchment have much older cooling ages than those observed in the fission-track vertical transect. Overall, this analysis suggests that the detrital ages provide a high-fidelity record of the bedrock cooling ages in the tributary catchment.

Along the trend of the Kyrgyz Range, two additional vertical relief sections of apatite ages have been analysed (Sobel *et al.*, 2006). Together, the three sections span about 60 km laterally (Fig. 12b) and display consistent trends of young ages at the lowest altitudes and older ages higher up (Fig. 12a). From east to west, however, the altitude of the top and base of the partial annealing zone varies significantly. For example, at 2800 m in the east at Shamsi, a cooling age of ~ 17 Ma occurs within the PAZ (Fig. 12a), whereas at the same altitude farther west, fully reset cooling ages of ~ 7 Ma and ~ 11 Ma occur beneath the PAZ in the Issyk Ata and Ala Archa sections. At 3800 m, samples from all three sections lie within the PAZ, but also show a strong east–west gradient, ranging from ~ 70 Ma in the east to ~ 18 Ma in the west at Ala Archa. The height of the base of the PAZ is greatest in the Issyk Ata drainage (Fig. 12a), thereby suggesting that more erosion and rock uplift have occurred here. The oldest ages occur at Shamsi, where the onset of rapid cooling (~ 7 Ma) appears later than at the other sites and the total magnitude of erosion appears to be the least.

If detrital samples faithfully reflect the cooling ages in their source rock, these along-strike fission-track profiles should provide a test to discern the sensitivity of modern samples to the bedrock ages in their source areas. Indeed, the contrast between samples from the modern Shamsi and Ala Archa Rivers (Fig. 12c) is striking. In Shamsi, only a very small representation exists of ages < 17 Ma, whereas the dominant component peaks are ~ 70 Ma and ~ 150 Ma, more than 50 Myr older than the dominant peak at Ala Archa. In both cases, the detrital ages are highly consistent with the observed bedrock cooling ages (Fig. 12a). Moreover, the contrast in the primary age component suggests that, under appropriate circumstances, detrital fission-track ages can resolve differences on the order of 1–2 km in the magnitude of erosion between different sections.

When combined with geological data (Bullen *et al.*, 2003), the known bedrock-cooling ages from the Kyrgyz Range also underpin a reconstruction of the sequential dissection of the range that is based on detrital fission-track ages. Prior to deformation beginning at ~ 11 Ma, the region was characterized by rocks with Mesozoic and late Palaeozoic cooling ages that extended from the surface to the top of the PAZ, yielding a vertical succession of ages (cf. Fig. 7). After rock uplift commenced in the late Middle Miocene, accelerated erosion created an enhanced sediment flux to the nearby Chu basin, the Cenozoic foreland basin that bounds the northern margin of the Kyrgyz Range.

Based on a magnetostratigraphic section that provides time control from ~ 9 Ma to 3 Ma (Bullen *et al.*, 2001), four different stratigraphic horizons, ranging from 8.5 Ma to ~ 1.5 Ma, were analysed for detrital apatite fission-track ages (Fig. 13). Even though the oldest stratigraphic sample was collected from strata that post-dated the initiation of uplift by some 2–3 Myr, the youngest component of its detrital ages is centred at ~ 165 Ma and contains no post-Mesozoic ages. This absence indicates that erosion at this time had not progressed into the former partial annealing zone (Fig. 7). Within

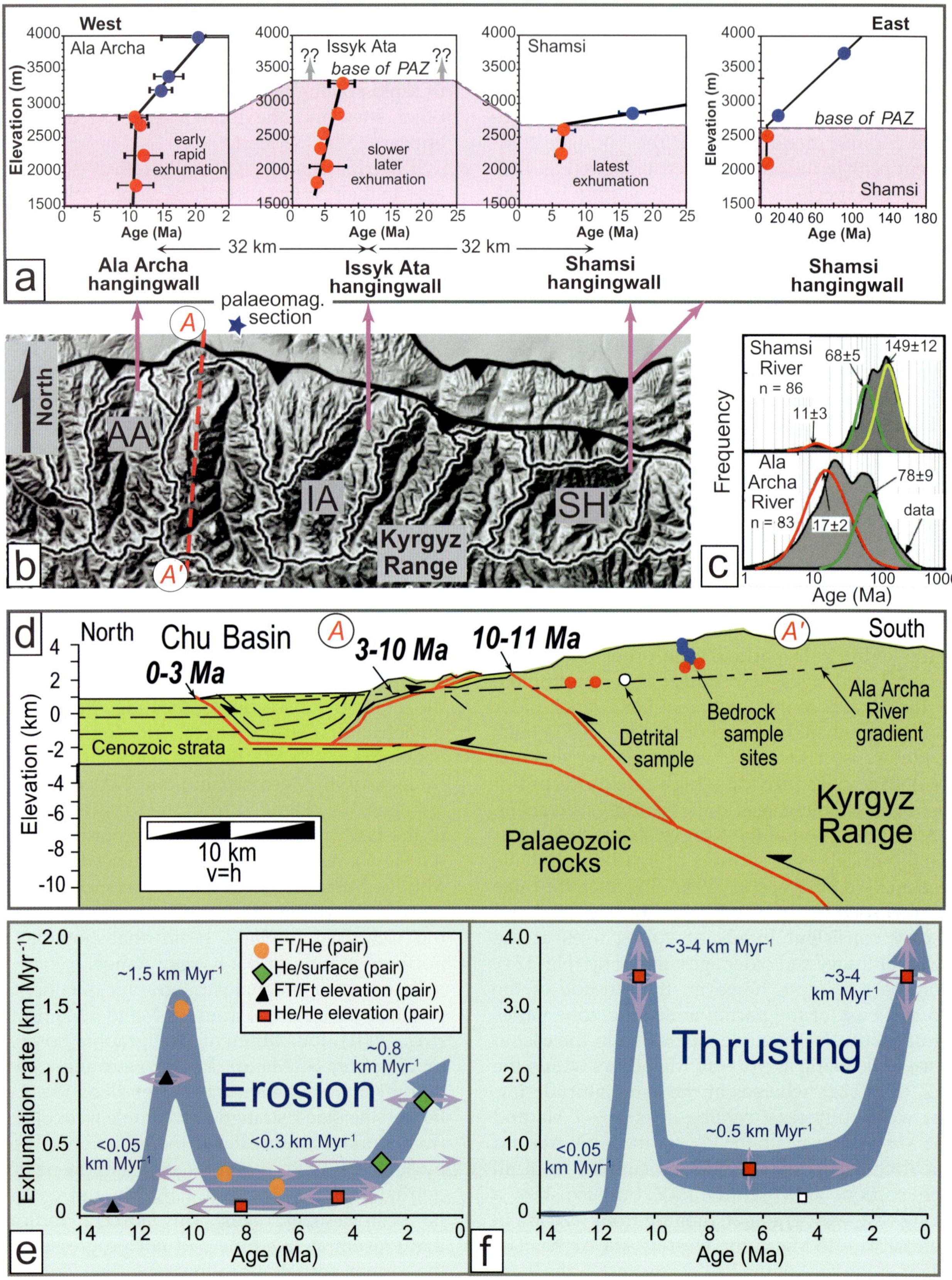
West
East
Ala Archa
Issyk Ata
Shamsi
Elevation (m)
Age (Ma)
base of PAZ
??
early rapid exhumation
slower later exhumation
latest exhumation
32 km
Ala Archa hangingwall
Issyk Ata hangingwall
Shamsi hangingwall
palaeomag. section
North
AA
IA
SH
Kyrgyz Range
Shamsi River n = 86
68±5
149±12
11±3
Ala Archa River n = 83
78±9
17±2
data
Frequency
North
Chu Basin
South
0-3 Ma
3-10 Ma
10-11 Ma
Cenozoic strata
Elevation (km)
10 km v=h
Detrital sample
Bedrock sample sites
Ala Archa River gradient
Kyrgyz Range
Palaeozoic rocks
Exhumation rate (km Myr-1)
FT/He (pair)
He/surface (pair)
FT/Ft elevation (pair)
He/He elevation (pair)
~1.5 km Myr-1
~0.8 km Myr-1
Erosion
<0.05 km Myr-1
<0.3 km Myr-1
~3-4 km Myr-1
~3-4 km Myr-1
Thrusting
~0.5 km Myr-1
a
b
c
d
e
f

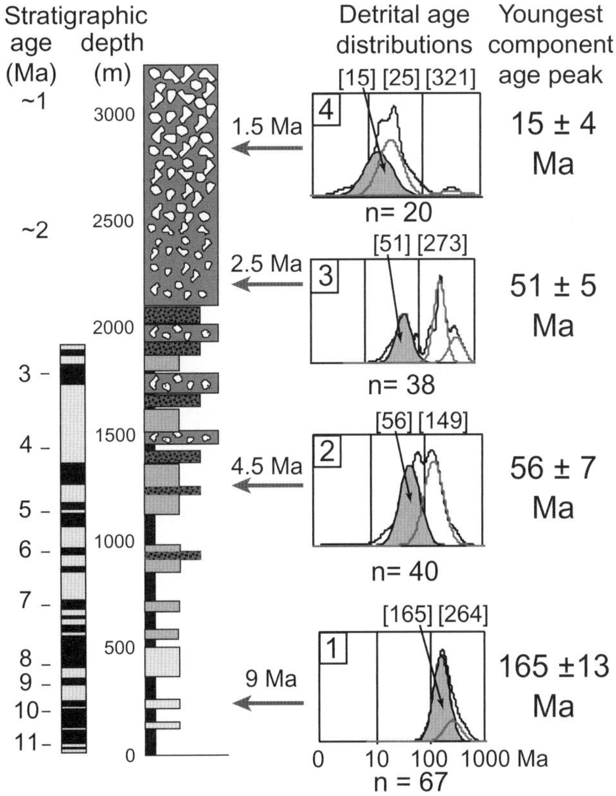

Fig. 13 (*left*) Detrital fission-track data at four stratigraphic horizons in the Chu basin. Section lies in the foreland of the Kyrgyz Range, offset 15 km east from the outlet of the Ala Archa catchment (star, Fig. 12b). Stratal time control derives from magnetostratigraphy (Bullen *et al.*, 2001). Grain-age populations, or peaks (labelled in Ma), were statistically separated using the binomial peak-fitting routine of Brandon (1996). Youngest peaks at each level (shaded grey) decrease in age up-section from 165 Ma to 56 Ma, then 51 Ma and finally to 15 Ma at 9, 4.5, 2.5 and 1.5 Ma, respectively. The large change in detrital ages from 9 to 4.5 Ma suggests that erosion progressed into the partial annealing zone (PAZ) during this interval. The very small change in ages between 4.5 and 2.5 Ma suggests that hinterland erosion was slow during this interval and remained within the PAZ, whereas the young ages from 1.5 Ma indicate erosion has proceeded well into the zone of reset ages. This youngest age peak (~ 15 Ma) is indistinguishable from the analogous peak in the modern Ala Archa River (17 Ma: Fig. 11f), suggesting exhumational steady state. The small change in detrital ages between 4.5 and 2.5 Ma indicates very limited erosion during this interval and is synchronous with slower rates of erosion and shortening in the hinterland (Fig. 12e, f).

the next 4 Myr, an abrupt change in detrital ages is manifested (Fig. 13), as the youngest component peak drops to ~ 55 Ma. Given the absence of significant thermal events affecting this region in Cenozoic times, the presence of Cenozoic detrital ages clearly indicates that the PAZ had been breached by 4.5 Ma. Over the next 2 Myr, the youngest component age peak remains almost constant (50–55 Ma). The persistence of a peak of nearly the same age can be interpreted to indicate that the rate of erosion decreased, because no significantly younger ages from greater depths in the nearby hinterland were being introduced to the

Fig. 12 (*opposite*) Bedrock and detrital cooling ages from multiple, along-strike sites in the Kyrgyz Range, northern Kyrgyzstan. (a) Vertical apatite fission-track sections from the central (Ala Archa) to the eastern (Shamsi) Kyrgyz Range. All three sections contain some completely reset cooling ages near the base, but pronounced differences occur in ages within and adjacent to the PAZ at a given elevation, and the base of the PAZ occurs at different elevations, thereby indicating variable amounts of rock uplift and erosion. Note two depictions of the Shamsi data (right) with different age scales. (b) Shaded relief map of the Kyrgyz Range, showing catchments where bedrock and modern rivers were sampled. Star marks location of magnetostratigraphic section. AA, Ala Archa; IA, Issyk Ata; SH, Shamsi. (Modified after Sobel *et al.*, 2006.) (c) Modern detrital apatite fission-track ages from the Shamsi River (top) and Ala Archa (bottom) clearly capture the differences in the cooling age stratigraphy in the catchments from which they were derived. (From Bullen *et al.*, 2001.) (d) Simplified geological cross-section of the central Kyrgyz Range showing major forethrusts and backthrusts, plus locations of vertical-relief bedrock samples and detrital fission-track sample. (From Bullen *et al.*, 2003.) The interval of primary displacement on each major fault is indicated. Location of the central part of the cross-section (A-A′) is shown in (b). (e) Summary of punctuated rates of erosion deduced from bedrock cooling ages in the central Kyrgyz Range near Ala Archa. Various combinations of data are used to define rates, such as the elevation difference between two samples divided by the difference in their ages. The fast-slow-fast pattern of erosion is consistent with the changes in populations of detrital ages from these same time intervals in the stratigraphic record (Fig. 13). (Modified from Bullen *et al.*, 2003.) (f) Variations in shortening rates through time deduced from structural cross-section at Ala Archa and ages assigned to faulting episodes. Note the overall temporal concurrence with the erosion rate changes (e). (Modified after Bullen *et al.*, 2003.)

foreland. This is consistent with the record of hinterland erosion as deduced from (U–Th)/He bedrock cooling ages (closure temperature ~ 70°C: Farley, 2000; Fig. 11) and structural analysis (Fig. 12d) that shows a threefold decrease in the rate of shortening, erosion and rock uplift between 8 Ma and 2 or 3 Ma (Fig. 12e, f), when compared with deformation and erosion between 10 and 12 Ma (Bullen *et al.*, 2003). The youngest detrital sample at 1.5 Ma shows another abrupt decrease in the age of its youngest component peak. In fact, its ~ 15 Ma age is indistinguishable from the 17-Ma peak seen in the modern Ala Archa River (Figs 11 & 13). These young ages indicate that erosion had progressed through the exhumed PAZ and into the zone of reset ages that is presently exposed at elevations < 2800 m (Fig. 11d). This inference is also consistent with the (U–Th)/He dates, which require erosion to exhume rocks from depths of ~ 3 km in the past 3 Myr (Fig. 11). In sum, over the 8-Myr-long stratigraphic record of detrital ages, the youngest component age peak decreased by ~ 150 Myr, and hinterland incision (as inferred from the pre-erosion hinterland age stratigraphy) totalled ~ 5–6 km. The decreasing lag time (Fig. 8) clearly indicates that the range had not attained an exhumational steady state until at least Pleistocene times.

This succession of detrital age samples aptly illustrates progressive hinterland unroofing. Such an analysis of detrital ages is clearly strengthened by the documented cooling-age stratigraphy of the bedrock in the hinterland (Figs 8 & 11) that permits quantification of the approximate magnitude of erosion at each time step recorded by the foreland-basin samples. Even in the absence of documentation of the hinterland age stratigraphy, robust inferences can be drawn on the magnitude of erosion from the component ages in each depositional level in the basin. For example, the initial appearance of sediments that were eroded from the PAZ is clearly evident, and changes in relative rates of erosion can be estimated from the persistence or changes in component ages from one depositional level to the next (Fig. 8).

EVOLUTION OF DETRITAL AGES THROUGH AN ACTIVE OROGEN

The detrital minerals that are preserved within basinal strata contain a final amalgam of ages that emerged from the hinterland. The way in which that detrital signal is created within the hinterland, however, typically remains unknown and, thereby, leaves unresolved questions. What combination of tectonic, lithological and erosional controls determines the suite of minerals and cooling ages that are transported out of the hinterland? How do spatial variations in erosion rates or lithology in the hinterland affect the downstream evolution of the suite of detrital ages? Although few studies have examined the downstream evolution of detrital cooling ages within a hinterland (Bernet *et al.*, 2004a; Brewer *et al.*, 2006), the expectation is that regions within the hinterland with rapid erosion rates, high abundances of the target mineral, and large areas will dominate the detrital age signal.

To test these expectations and to document the evolution of the detrital age signal through an active orogen, over 400 muscovite grains have been analysed from 11 sites along the Marsyandi River in central Nepal (Brewer *et al.*, 2006). This catchment extends from the southern edge of the Tibetan Plateau to the Gangetic foreland (Fig. 14a). Along its course, the river flows across both the South Tibetan Detachment fault, a down-to-the-north normal fault, and the Main Central Thrust fault, a major south-vergent thrust fault (Hodges, 2000). These two faults bound the Greater Himalaya, which contains about one-third of the Marsyandi catchment. Another third lies in the Tethyan strata of Palaeozoic age associated with the Tibetan Plateau, and the remainder lies within Lesser Himalaya, south of the Main Central Thrust. If all three subcatchments had similar hypsometries and distributions of muscovite and were eroding at the same rate, no downstream changes in detrital ages would be anticipated.

Quite expectedly, however, the detrital age signal changes dramatically downstream (Fig. 14b & c). The sample highest in the catchment has over 80% of its source area in Tethyan rocks (Site 11, Fig. 14a). Its detrital ages display a strong peak centred at ~ 14 Ma. Only 20 km farther downsteam (Site 10, Fig. 14b), the 14-Ma peak has nearly disappeared and has been replaced by a 17–18-Ma peak. Such a change leads to an apparent contradiction: the dominance of the older, 17–18-Ma peak suggests that it derives from a region that is supplying more sediment to the trunk stream, whereas the fact that the dominant age is becoming older downstream

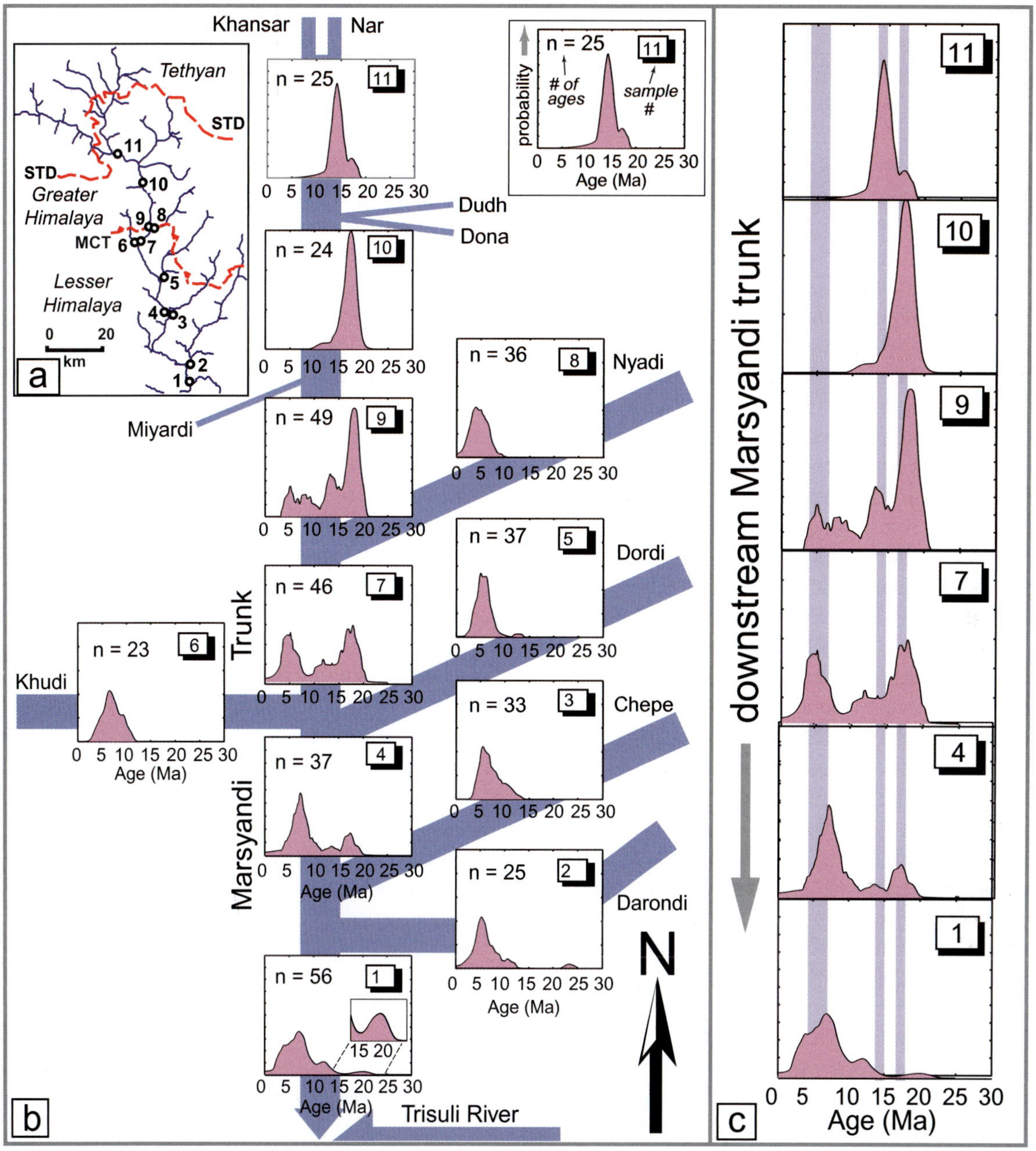

Fig. 14 Detrital muscovite $^{40}Ar/^{39}Ar$ ages along the Marsyandi River, central Nepal. (a) With headwaters in the Tibetan Plateau, the Marsyandi traverses the Greater and Lesser Himalaya. Numbered sample locations for both trunk stream and major tributaries are shown. STD, South Tibetan Detachment (down-to-the-north normal fault); MCT, Main Central Thrust. (b) Topological map of the Marsyandi drainage showing probability density functions of detrital muscovite ages at each sample site. Ages are smoothed with a 2-Myr scrolling window. Downstream changes in age distribution reflect contributions due to catchment erosion rate, size and lithology. Note that the northern sites have older cooling ages, southern tributaries show young (< 10 Ma) ages, and the sample at the mouth (Site 1) contains few of the older ages that are abundant in the headwaters. These data suggest that an influx of younger ages from tributaries in the lower part of the catchment overwhelms the older ages that characterize the headwater region of the catchment. (c) Downstream changes in cooling ages along the main stem of the Marsyandi River. Shaded bars indicate the approximate age range for prominent detrital age peaks that vary in significance along the Marsyandi's course. (Modified after Brewer *et al.*, 2006.)

implies slower erosion rates. These observations can be reconciled by the fact that muscovite is 5–10 times more abundant in the tributaries emerging within the Greater Himalaya (Brewer *et al.*, 2006). The resultant flux, despite somewhat slower erosion rates (Fig. 6), could cause the Greater Himalayan signal to overwhelm that from the upstream mica-poor Tethyan regions (Fig. 14c).

As the Marsyandi traverses the Greater and then the Lesser Himalaya, two significant trends emerge within the detrital age data. First, the 17–18-Ma peak which is so dominant high within the Himalaya has almost disappeared when the Marsyandi debouches into the Trisuli River: a trans-Himalayan river with a considerably larger catchment. Second, a peak with ages concentrated in the 5–8-Ma range becomes increasingly important downstream. The reason for the emergence of the 5–8-Ma peak becomes clear when the age signal from the major tributaries is examined: all of them are dominated by the same 5–8-Ma peak and contain almost no ages older than ~ 12 Ma (Fig. 14b). These younger ages indicate more rapid erosion in these tributaries, and their muscovite fraction is also another three to ten times higher than in the Greater Himalayan tributaries that feed Site 10 (Brewer *et al.*, 2006).

Several clear conclusions can be drawn from this analysis of modern detrital age samples. First, the detrital signal that is delivered from a mountain range to an adjacent basin is commonly transformed along its passage through the mountains. If the foreland basin were closer to the upland area (in this case, the Tibetan Plateau), a very different detrital age spectrum would be present. Second, shifts toward younger detrital ages typically reflect higher rates of erosion for catchments contributing the younger ages (Fig. 6). Third, at the scale of tributary catchments, the abundance of the target mineral (in this case, muscovite) can change by an order of magnitude or more across a few tens of kilometres. Interpretations offered by studies that assume the distribution of a target mineral is uniform (Bernet *et al.*, 2004a) may need modification when the actual distribution of the target mineral is known.

The available detrital age data make it possible to estimate erosion rates in each of the tributary catchments (Brewer *et al.*, 2006), given several assumptions and observations. It is assumed that: (i) within a given tributary, erosion rates and muscovite distributions are approximately uniform; (ii) there is no significant relief on the 350°C isotherm – an assumption consistent with topographic relief of ≤ 6 km and erosion rates ≤ 3 mm yr^{-1} (Fig. 3); (iii) rock-particle trajectories are vertical. If true, then the altitude dependence of the distribution of ages can be predicted for any erosion rate (Fig. 5). Combining these ages with the observed catchment hypsometry yields a prediction of detrital ages (Fig. 5). The optimal erosion rate for each tributary catchment is calculated by minimizing the mismatch between the predicted and observed distributions of detrital ages (Brewer *et al.*, 2003).

The resultant map of variations in predicted erosion rates at the catchment scale depicts regional trends across the Himalaya (Fig. 15). The calculated rates vary from 2.3 mm yr^{-1} to 0.9 mm yr^{-1} with higher rates along the southern flank of the Himalaya than along the northern flank. The highest rates occur in the Nyadi to Darondi catchments, each of which straddles the Main Central Thrust. Strikingly, those catchments that include the highest Himalayan topography (e.g. Dona, Dudh and Khansar) have significantly lower erosion rates. These spatial variations indicate that, at the present time, the most rapid erosion is displaced well south of the Himalayan crest. Such a position coincides spatially both with the swath of highest monsoonal precipitation (Burbank *et al.*, 2003) and with the zone adjacent to and immediately south of the Main Central Thrust, where young bedrock-cooling ages (Harrison *et al.*, 1998), steepened stream gradients (Wobus *et al.*, 2003) and brittle faulting (Hodges *et al.*, 2004) suggest active deformation.

DETRITAL AGES AND COLLISIONAL TECTONICS

Most studies that attempt to convert cooling ages into erosion rates consider only the vertical transport of rocks toward the surface (e.g. Stüwe *et al.*, 1994; Fitzgerald *et al.*, 1995; Garver *et al.*, 1999). Such vertical kinematics were used in the previous estimates of Himalayan erosion rates (Figs 3, 6 & 15) described above (Brewer *et al.*, 2006). Yet, in most convergent orogens rock primarily advects laterally, not vertically, such that convergence is commonly five to ten times greater than vertical rock uplift (Willett, 1999; Stüwe & Hintermüller, 2000; Batt & Brandon, 2002). Rocks with different thermal

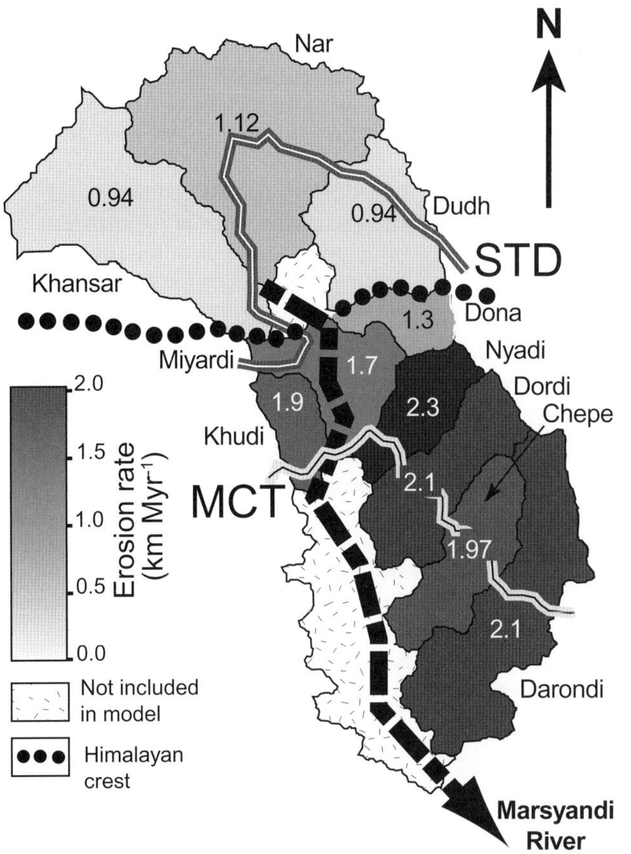

Fig. 15 Spatial variation in erosion rates at the catchment scale for the Marsyandi River. Erosion rates are taken from the results of modelling the detrital cooling-age probability density distributions for individual tributaries. The stippled areas indicate zones not included in the calculations and the dashed black line indicates the approximate path of the trunk stream. The Himalayan crest is indicated by the black dots. Highest erosion rates are predicted along the southern flank of the Himalaya, where catchments straddle the Main Central Thrust (MCT). Regionally, rates are predicted to vary by ~ 2–2.5 fold. STD, South Tibetan Detachment. (After Brewer *et al.*, 2006.)

conductivity and radioactive heat production on one flank of an orogen are typically thrust over or under rocks on the other flank. In such conditions, isotherms are not just warped by topography, but are strongly perturbed (Fig. 16) by the relatively cold, underthrust slab (Koons, 1989; Harrison *et al.*, 1997; Beaumont *et al.*, 2001; Jamieson *et al.*, 2002). On their way to the orogen's surface in response to erosion, rock particles move obliquely through this thermal field (Stüwe & Hintermüller, 2000; Batt & Brandon, 2002). Although their cooling ages when they reach the surface still reflect the transport time since crossing the appropriate closure isotherm, their path toward the surface is now oblique and longer than when only the vertical component of motion is considered (Stüwe & Hintermüller, 2000). Hence, lag times now integrate the complete horizontal and vertical travel history of rocks with respect to a warped closure isotherm (Fig. 16).

As a consequence of the above, cooling ages at the surface depend on multiple factors (Ehlers & Farley, 2003; Brewer & Burbank, 2006): the depth of the critical closure isotherm; the particle path from the closure isotherm to the surface; the rate of motion along that particle path toward the surface; and the topographic relief at the surface, which continues to produce differences in ages between valleys and ridges (Fig. 16). If the cooling ages that emerge at the surface can be modelled successfully in this complex thermal and kinematic regime, and if sediment transport times to the basin are short, then detrital cooling ages can be used to gain insight on orogenic dynamics and erosion rates (Brewer & Burbank, 2006).

Although the overall plate convergence rate across an orogen may be known through geodetic or geological data, the way in which that convergence is accommodated is commonly poorly known. For example, with two plates colliding, one of the two could be imagined as passive and unchanging, whereas the other plate would either subduct beneath or override the 'stationary' plate. Although such end-member models rarely apply, the actual amount of convergence that is partitioned into each plate is typically difficult to assess. Nonetheless, that partitioning and associated erosion will define the particle pathways and related thermal histories within an orogen. Here, a model (Brewer & Burbank, 2006) is described that predicts orogenic cooling ages and the resultant detrital ages in a catchment as a function of the partitioning of convergence and erosion rates. Through comparisons of the model predictions with observed detrital ages (Brewer *et al.*, 2006), it is possible to assess various partitioning and erosion scenarios and obtain new insights on orogenic kinematics.

The Himalayan orogen has been examined in this manner, using a simplified numerical model relevant to $^{39}Ar/^{40}Ar$ cooling ages of muscovite. At the coarsest scale, the orogen is defined by an underthrusting Indian plate and an overthrusting

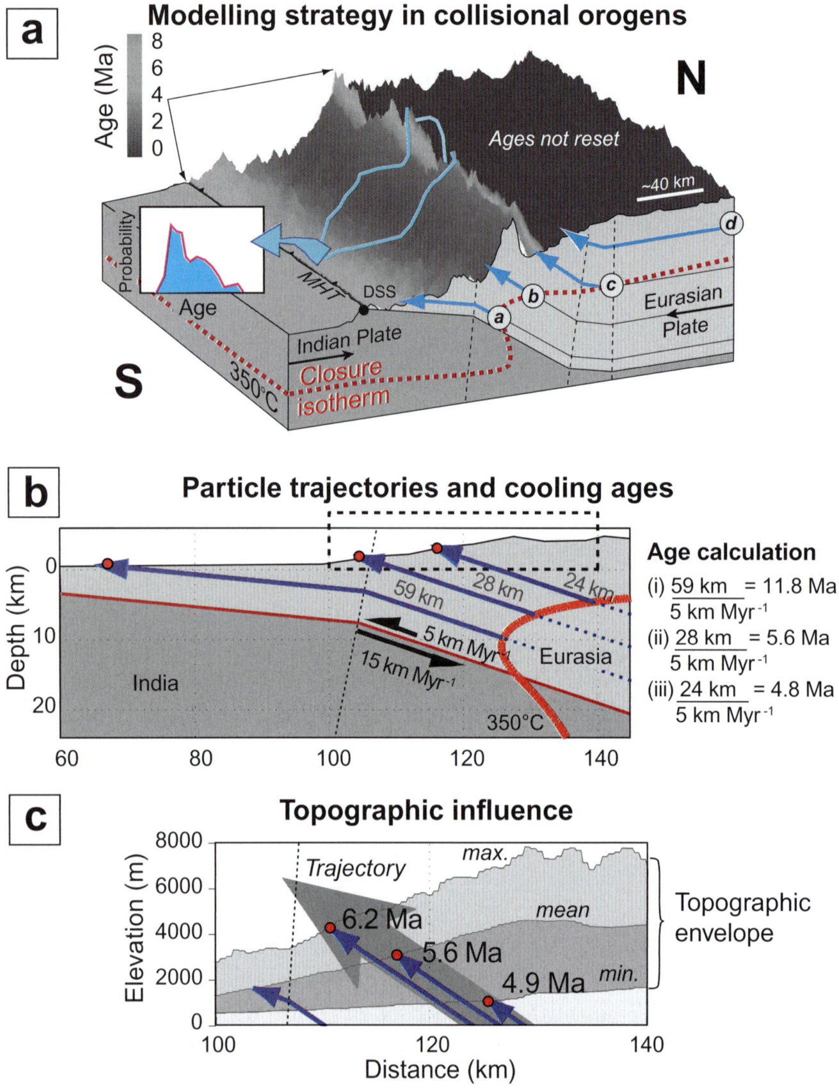

Fig. 16 Conceptual basis for combining thermal and kinematic models to predict a detrital signal in a convergent orogen in which lateral advection rates are high. (a) Given a simplified ramp-flat geometry and known convergence rate, particle velocities and depth to the muscovite closure temperature (350°C) are calculated, assuming thermal and topographic steady state. In a digital elevation model (DEM), an age is calculated for each point based on its position with respect to the closure isotherm and particle pathways to the surface. Ages 'collected' from a catchment in the DEM define the age-population distribution for a detrital sample. Among the particle trajectories (*a–d*), trajectory *b* will have the youngest age because the 350°C isotherm is closest to the surface along this trajectory, while trajectory *d* will have old or un-reset ages. (b) Examples of surface age calculations based on three trajectories and an overthrusting rate of 5 km Myr^{-1}. (c) Particles following the same trajectory in two dimensions commonly travel different distances to the surface due to along-strike changes in topography in a three-dimensional landscape. The predicted effects of maximum, minimum and mean topography on cooling ages are illustrated. (After Brewer & Burbank, 2006.)

Asian plate. (Note that the 'Asian' plate actually consists of accreted Indian plate rocks in the area of interest within the Himalaya.) The overall modelling strategy requires specifying a kinematic geometry that defines how particles move through the thermal structure of the orogen (Fig. 16a). A critical assumption is that the topography is in steady state at the time-scales of interest (10^6–10^7 Myr: Willett, 1999): this dictates that the surface erodes at exactly the same rate as rocks are moving along particle pathways toward the surface. Hence, the rate of overthrusting and the rate of erosion are equal, and the rate of cooling is inextricably linked to them, because the thermal structure varies with erosion rate and particle pathway beneath a fixed surface topography. Thermal attributes in terms of heat production and conductivity are assumed and a two-dimensional kinematic geometry is defined that dictates particle paths. Subsequently, the position of the closure isotherm is solved by assuming a thermal steady state: a condition that requires a few million years to achieve (Brewer *et al.*, 2003). The rate of rock movement along the pathway, the erosion rate and the cooling ages at the surface (Fig. 16b) all depend on how convergence is partitioned into the overriding or underthrusting plate (Brewer & Burbank, 2006).

To extrapolate to a pseudo-three-dimensional model of bedrock cooling ages, it is assumed that the kinematic geometry and thermal characteristics are homogeneous along strike in the orogen, and then the time it takes (i.e. the cooling age) for a particle

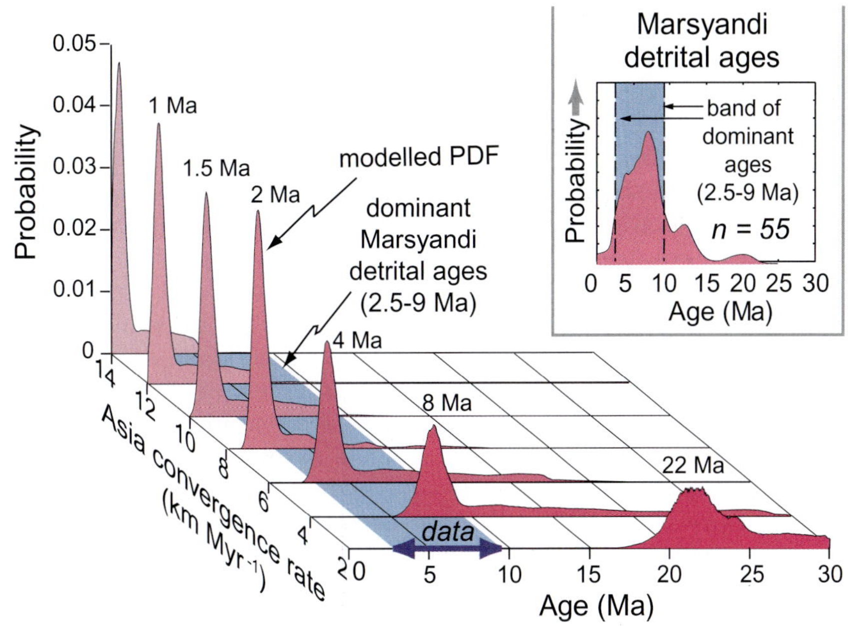

Fig. 17 Modelled detrital cooling-age signals resulting from partitioning the relative convergence rate between India and Asia. Topographic steady state is assumed, such that the erosion and overthrusting rates are equal in the 'Asian' plate. Probability density functions (PDFs) of the predicted age distribution represent the reset age signals from a topographic swath across the Nepalese Himalaya centred on the Marsyandi catchment. The central age for each PDF is shown by the peak in the PDF. The Asian convergence (or overthrusting) rate varies between 2 and 14 km Myr^{-1}, while total convergence rate (20 mm yr^{-1}) and all else remain constant. The inset shows the observed detrital ages from the most downstream site from the Marsyandi River (Brewer *et al.*, 2006) that define the dominant range of ages (blue band) against which the model data are compared. Convergence (or erosion) rates of 4–6 km Myr^{-1} for Asia provide the best match to the bulk of the observed data. Faster overthrusting ($\geq$ 8 km Myr^{-1}) is predicted to yield ages that are too young, whereas slower overthrusting ($\leq$ 2 km Myr^{-1}) yields ages that are too old. See Fig. 16 for the schematic framework of the age and overthrusting calculations.

to travel from the closure isotherm to the surface along a discrete trajectory to each point on the surface (Fig. 16c) is calculated. These predicted bedrock-cooling ages are then 'extracted' from a catchment in a digital elevation model of the orogen (Fig. 16a) and compared with observed cooling ages from the same catchment. Comparisons of observed ages with those predicted for various scenarios for partitioning overthrusting and underthrusting set clear limits on which scenarios are viable.

This modelling approach, the details of which are described in Brewer & Burbank (2006), has been applied to the Marsyandi catchment in central Nepal. A simple fault-bend fold kinematic model is employed in which the geometry reflects that inferred from deep seismic profiling, modern seismicity and surface geology (Schelling, 1992; Pandey *et al.*, 1995; Nelson *et al.*, 1996; Nábelek *et al.*, 2005). Across the Himalaya, the Indian subcontinent is converging with southern Tibet at a geodetic rate of ~ 20 mm yr^{-1} (Bilham *et al.*, 1997; Wang *et al.*, 2001). The goal is to use the observed detrital ages and the numerical model to determine how much of this convergence is partitioned into 'Asian' overthrusting and how much is partitioned into 'Indian' underthrusting.

Given the assumption of topographic steady state, the more convergence that is partitioned into Asian overthrusting, the younger the cooling ages are expected to be within the Himalaya, because erosion rates and overthrusting rates are equivalent (Fig. 17). As extracted from the digital Marsyandi catchment, and based on different overthrusting rates, the predicted cooling ages exhibit dominant age peaks that range from 0.5 Ma to 22 Ma for Asian overthrusting, and erosion rates of 14 mm yr^{-1} to 2 mm yr^{-1}, respectively (Fig. 17). The observed detrital data from the most downstream sample from the Marsyandi (Figs 14 & 17) are dominated by ages between 4 and 9 Ma. Hence, both rapid (8–12 mm yr^{-1}) and slow (2 mm yr^{-1}) overthrusting and erosion can be eliminated. The best match to the data clearly occurs for an overthrusting and erosion rate of between 4 and 6 mm yr^{-1} (Brewer & Burbank, 2006).

Despite many assumptions and simplifications, comparison of the observed detrital age data and model predictions makes it possible to gain useful insights on the Himalayan collision. Previous models that hypothesized overthrusting rates of ~ 10 mm yr^{-1} (Harrison *et al.*, 1997) clearly would produce much younger ages than those observed at the surface if the overthrusting rates were analysed within the thermal and kinematic context of the current model. When constrained by detrital ages and assuming a steady-state topography, modelling suggests that only about 25% of the Indo-Tibetan convergence occurs as overthrusting (Fig. 17). Further exploration of orogenic parameters could be accomplished by varying the kinematic geometry or by relaxing some of the model assumptions, such as that regarding steady-state topography. Each such change in the model would yield different cooling ages at the surface. Whereas the temporal resolution afforded by the observed suite of cooling ages is typically limited to a few million years, such precision is sufficient to distinguish among several model predictions.

DISCUSSION

The use of the stratigraphic record to interpret tectonic histories is a core goal of many stratigraphic studies. Advances in geochemical techniques now facilitate analysis of individual minerals. Single-crystal dating of detrital minerals confers a remarkable ability to utilize ages to reconstruct cooling histories of orogens and to place limits on the timing, magnitude and spatial variations of erosion. As opposed to bedrock cooling ages, which are obtained from single outcrops, detrital samples have a tremendous advantage: they comprise minerals drawn from the entire catchment. Thus, detrital samples offer an integrated perspective that is almost always unattainable at the outcrop scale. Moreover, detrital ages are preserved within stratigraphic successions, such that the evolution of cooling ages through time and across an orogen can be reconstructed from the sedimentary record (e.g. Cerveny *et al.*, 1988; Bullen *et al.*, 2001; White *et al.*, 2002).

Despite a burgeoning suite of applications of single-crystal dating to stratigraphic problems, many concepts that underpin interpretations of the data remain poorly explored: the fidelity of the sedimentary detrital age signal with respect to the bedrock ages from which it was derived; the effects of spatial and temporal variations in erosion on detrital ages; the ability of detrital ages to record subtle variations in erosion; the influence of variable source-area lithology and grain sizes on detrital ages. In this study, an attempt has been made to quantify several key controls on detrital ages and to examine some new applications of detrital ages to tectonic interpretations.

When a research objective is to define the initiation of major deformation, low-temperature dating approaches, such as apatite fission track or [U-Th]/He (Stockli *et al.*, 2000; Ehlers & Farley, 2003), are typically preferable because the shallow depth of the closure isotherm renders it particularly sensitive to surface cooling due to erosion. Not only do these dating approaches utilize a low-temperature thermochronometer that varies spatially at shallow depths due to warping of the closure isotherm beneath topographic relief (Stüwe *et al.*, 1994), but the existence of a partial annealing zone for fission-track dating and a partial retention zone for [U–Th]/He dating (Wolf *et al.*, 1996) generates distinctive changes in cooling ages within this thermal layer. Low-temperature dating of detrital grains is particularly well suited for dating of nascent uplifts and for slowly eroding ranges (Naeser *et al.*, 1983; Sobel & Dumitru, 1997; Blythe, 2002). In rapidly eroding ranges (≥ 2–4 mm yr^{-1}), the paucity in fission tracks in most apatite crystals increases analytical uncertainties and limits the age resolution. For example, along the Marsyandi River within the Greater Himalaya, apatite fission-track bedrock cooling ages are very young (~ 0.5 Ma) and have large uncertainties (20–50%; Burbank *et al.*, 2003) that preclude detailed analyses of detrital apatite grains. Dating of apatite by [U–Th]/He utilizes an even lower closure temperature (~ 70°C) and presents similar analytical challenges in rapidly eroded sediments.

The closure isotherm (~ 350°C) for $^{39}Ar/^{40}Ar$ dates on muscovite remains nearly insensitive to topography, as long as erosion rates and topographic relief are less than 3 mm yr^{-1} and 6 km, respectively (Fig. 3). As a consequence, cooling-age variations are easy to predict as a function of altitude. On the other hand, interpretations of cooling ages as recorders of erosion rates are complicated by typical detrital muscovite ages that range into millions of years (Carrapa *et al.*, 2003, 2004). Changes in erosion rates, thermal regimes and kinematic pathways are likely to occur during the extended time it takes a rock to transit from the closure isotherm to the surface, and it commonly requires several million years to attain a new thermal equilibrium (Brewer *et al.*, 2003).

Studies from the Kyrgyz Range in the Tien Shan suggest that detrital fission-track ages can provide a faithful sampling of the distribution of cooling ages within a given catchment. When erosion rates are relatively slow (0.1–1 mm yr^{-1}), as seen in the Kyrgyz Range, and the kinematic geometry is simple (Fig. 12d), detrital ages can be sensitive to differences of as little as 1–2 km in the magnitude of erosion. Slow erosion rates underpin this sensitivity, because they allow incremental incision through a stratigraphy of bedrock cooling ages (Stock & Montgomery, 1996), thereby producing discernible variations in the detrital ages derived from an eroding range. Under such circumstances, spatial differences in detrital ages along a range can record differential incision (Bullen *et al.*, 2001). It is important to note that these Kyrgyz examples are drawn from simple catchments draining a single range, such that variations in detrital ages can be linked directly to erosion in the nearby mountains. In typical foreland basins, however, multiple sources feed sediment into the basin, and the detrital age signal is expected to be a complex integration of those sources. To minimize the uncertainty in source areas in pre-Quaternary orogens, samples should be drawn from sections where provenance studies suggest a consistent source area and palaeocurrent directions indicate transverse flow. Rivers with this orientation are more likely to drain a simple orogenic catchment, whereas longitudinal rivers almost always drain a broad suite of hinterland catchments (Burbank, 1992).

Analyses in active orogens of detrital ages in modern rivers, such as the Marsyandi River in the Nepalese Himalaya (Fig. 14), clearly demonstrate a striking downstream evolution of the suite of ages. This evolution results from convolving the sediment fluxes from each tributary catchment, each of those fluxes being dependent on catchment size, erosion rate and abundance of the mineral targeted for dating. Two important aspects that are related to the interpretation of detrital ages in stratigraphic successions emerge from the analysis of the modern Marsyandi sediment. First, major parts of a given catchment may be poorly represented in the detrital signal preserved in a sedimentary basin. Immediately upstream of the junction of the Marsyandi and Trisuli Rivers, for example, almost no detrital ages from the northern 50% of the catchment are represented among the 56 grains dated. Rapid erosion within more downstream tributaries apparently overwhelms the contributions from upstream. Second, spatial variations in the abundance of the mineral targeted for dating can exert a major control on the detrital age signal that emerges from a range. In the Marsyandi catchment, a >100-fold variation is observed in muscovite abundances among catchment areas varying from 10^2 to 10^3 km^2. Zircon, another mineral commonly used for detrital single-crystal dating, has up to eightfold differences in abundance in the Greater and Lesser Himalaya, and is almost absent from a few regions (Amidon *et al.*, 2005a). Quantitative analyses of detrital ages that assume uniform mineral distribution can be biased by ignoring the actual variability in mineral abundance. Although abundances of commonly occurring minerals such as muscovite can be determined through point counting, determining the grain frequency of zircon or apatite, which occur in trace amounts, is more tedious. Amidon *et al.* (2005b) have recently developed a methodology based on grain shape and volume that generates an estimate of zircon abundance within the size fraction being dated with a precision of ±10%: a resolution better than that typically achieved with point counting of sparse minerals (Van der Plas & Tobi, 1965; Brewer *et al.*, 2006).

In analyses such as the attempt to calculate erosion rates within the Marsyandi (Fig. 15), a trade-off exists between spatial resolution and research investment. Clearly, by dividing the Marsyandi catchment into tributary catchments, a more highly resolved image of spatial variations in erosion rates is attainable than if the entire catchment were modelled as a uniformly eroding entity. The tributary catchments themselves, however, are typically

> 200 km^2 and would be expected to have variations in erosion rates within them. Further subdivision of each catchment, additional detrital mineral dates and measurement of variations in muscovite abundance (the target mineral for dating) would be required to refine the erosion-rate estimates.

In collisional orogens, the lateral component of particle pathways should not be ignored, because ages observed at the surface depend on the oblique trajectories of rocks through the orogen. In addition to the altitudinal dependence of ages that pertains when rocks move only vertically toward the surface, a spatial dependence will exist that reflects the intersection of the closure isotherm with various particle trajectories. Thus, even with uniform erosion across an orogen, cooling ages may differ by several fold when rocks from proximal and distal sites are compared. If an appropriate kinematic and thermal model can be developed, detrital ages derived from the orogen can be used to test large-scale kinematic parameters, such as average rates of overthrusting.

As single-crystal dating of detrital minerals becomes more efficient and less expensive, many of the assumptions that underpin current interpretations can be more systematically tested. In addition, it will be possible to quantify modern fluvial systems much more thoroughly, in order to understand how detrital ages evolve within a variably eroding and lithological heterogeneous mountain range. Finally, combining observed detrital ages with improved numerical models should permit further exploration of the dynamics of orogens and the diversity of erosion styles and rates.

CONCLUSIONS

Whereas many methodological improvements can be envisioned to render analyses of detrital ages more robust, the advent of single-crystal dating has opened a new era in the analysis of both stratigraphic successions and geomorphic systems. Studies of modern systems demonstrate how the detrital signal is generated, and reveal both the power and limitations of single-crystal dating in sedimentary basins. Although recognized previously from a theoretical viewpoint, the impact exerted on modern detrital ages by the interplay between erosion rates and lithology within numerous tributaries has only recently been documented, and provides a basis for refining orogenic histories using detrital ages. These studies of modern rivers provide both insights and cautions with respect to the interpretation of detrital ages within the stratigraphic record: detrital ages can record subtle variations in the history of erosion and rock uplift, but rapidly eroding areas can overwhelm slowly eroding ones and lithological variability can introduce strong biases in detrital ages. An ability to exploit detrital ages to constrain kinematic rates within collisional orogens provides a potent new analytical tool. If uncertainties regarding kinematic geometries and related particle pathways through orogens can be reduced, detrital ages in both modern rivers and the recent stratigraphic record can serve to reconstruct rates of deformation and erosion and to test the viability of proposed models of orogenic evolution.

ACKNOWLEDGEMENTS

This research was primarily supported by grants from the National Science Foundation (EAR-9627865, 9614765, 9909647) and NASA (NAG5-9039; 10520; 13758). We are grateful for helpful discussions with M. Brandon, W. Amidon and R. Slingerland, and for the contributions of Mike Oskin and Bodo Bookhagen with respect to the digital elevation model analysis. We thank John Garver for his assistance with the fission-track analyses. Incisive reviews by C. Paola and T. Ehlers greatly improved this manuscript.

REFERENCES

Amidon, W.H., Burbank, D.W. and Gehrels, G.E. (2005a) Construction of foreland mineral populations: insights from mixing of U–Pb zircon ages in Himalayan rivers. *Basin Res.*, **17**, 463–485.

Amidon, W.H., Burbank, D.W. and Gehrels, G.E. (2005b) U–Pb zircon ages as a sediment mixing tracer in the Nepal Himalaya. *Earth Planet. Sci. Lett.*, **235**, 244–260.

Batt, G.E. and Brandon, M.T. (2002) Lateral thinking: 2-D interpretation of thermochronology in convergent orogenic settings. *Tectonophysics*, **349**, 185–201.

Beaumont, C., Jamieson, R.A., Nguyen, M.H. and Lee, B. (2001) Himalayan tectonics explained by extrusion of a low-viscosity crustal channel coupled to focused surface denudation. *Nature*, **414**, 738–742.

Bernet, M., Brandon, M.T., Garver, J.I. and Molitor, B.R. (2004a) Downstream changes of Alpine zircon fission-track ages in the Rhône and Rhine Rivers. *J. Sediment. Res.*, **74**, 82–94.

Bernet, M., Brandon, M.T., Garver, J.I. and Molitor, B.R. (2004b) Fundamentals of detrital zircon fission-track analysis for provenance and exhumation studies with examples from the European Alps. In: *Detrital Thermochronology–Provenance Analysis, Exhumation, and Landscape Evolution of Mountain Belts* (Eds M. Bernet and C. Spiegel). *Geol. Soc. Am. Spec. Pap.*, **378**, 25–36.

Bilham, R., Larson, K. and Freymueller, J. (1997) GPS measurements of the present-day convergence across the Nepal Himalaya. *Nature*, **386**, 61–63.

Blythe, A.E., House, M.A. and Spotila, J. (2002) Low-temperature thermochronology of the San Gabriel and San Bernardino Mountains, southern California: constraining structural evolution. In: *Contributions to Crustal Evolution of the Southwestern United States* (Ed. A.P. Barth). *Geol. Soc. Am. Spec. Pap.*, **365**, 231–250.

Bollinger, L., Avouac, J.P., Beyssac, O., *et al.* (2004) Thermal structure and exhumation history of the Lesser Himalaya in central Nepal. *Tectonics*, **23**, TC5015, doi:10.1029/2003TC001564.

Brandon, M.T. (1992) Decomposition of fission-track grain-age distributions. *Am. J. Sci.*, **292**, 535–564.

Brandon, M.T. (1996) Probability density plot for fission-track grain-age samples. *Radiat. Meas.*, **26**, 663–676.

Brandon, M.T. (2002) Decomposition of mixed grain-age distributions using BINOMFIT. *On Track*, **24**, 13–18.

Brandon, M.T. and Vance, J.A. (1992) Fission-track ages of detrital zircon grains: implications for the tectonic evolution of the Cenozoic Olympic subduction complex. *Am. J. Sci.*, **292**, 565–636.

Brewer, I.D. (2001) *Detrital-mineral thermochronology: investigations of orogenic denudation in the Himalaya of central Nepal*. Unpublished PhD thesis, Pennsylvania State University, 201 pp.

Brewer, I.D., Burbank, D.W. and Hodges, K.V. (2003) Modelling detrital cooling-age populations: insights from two Himalayan catchments. *Basin Res.*, **15**, 305–320.

Brewer, I.A., Burbank, D.W. and Hodges, K.V. (2006) Downstream development of a detrital cooling-age signal: Insights from 40Ar/39Ar muscovite thermochronology in the Nepalese Himalaya. *Geol. Soc. Am. Spec. Pap.*, **398**, 321–338.

Brewer, I.D. and Burbank, D.W. (2006) Thermal and kinematic modeling of bedrock and detrital cooling ages in the Central Himalaya. *J. Geophys. Res.*, **III**, B09409, doi: 09410.01029/02004JB0003304.

Brown, R.W. (1991) Backstacking apatite fission-track 'stratigraphy': a method for resolving the erosional and isostatic components of tectonic uplift histories. *Geology*, **19**, 74–77.

Bullen, M.E., Burbank, D.W., Abdrakhmatov, K.Y. and Garver, J. (2001) Late Cenozoic tectonic evolution of the northwestern Tien Shan: constraints from magnetostratigraphy, detrital fission track, and basin analysis. *Geol. Soc. Am. Bull.*, **113**, 1544–1559.

Bullen, M.E., Burbank, D.W. and Garver, J.I. (2003) Building the northern Tien Shan: integrated thermal, structural, and topographic constraints. *J. Geol.*, **111**, 149–165.

Burbank, D.W. (1992) Causes of recent Himalayan uplift deduced from deposited patterns in the Ganges basin. *Nature*, **357**, 680–682.

Burbank, D.W., Blythe, A.E., Putkonen, J., *et al.* (2003) Decoupling of erosion and precipitation in the Himalayas. *Nature*, **426**, 652–655.

Carrapa, B., Wijbrans, J. and Bertotti, G. (2003) Episodic exhumation in the Western Alps. *Geology*, **31**, 601–604.

Carrapa, B., Wijbrans, J. and Bertotti, G. (2004) Detecting provenance variations and cooling patterns within the western Alpine orogen through $^{40}Ar/^{39}Ar$ geochronology on detrital sediments: The Tertiary Piedmont Basin, northwest Italy. In: *Detrital Thermochronology–Provenance Analysis, Exhumation, and Landscape Evolution of Mountain Belts* (Eds M. Bernet and C. Spiegel). *Geol. Soc. Am. Spec. Pap.*, **378**, 67–103.

Carter, A. (1999) Present status and future avenues of source region discrimination and characterization using fission track analysis. *Sediment. Geol.*, **124**, 31–45.

Carter, A. and Moss, S.J. (1999) Combined detrital-zircon fission-track and U–Pb dating: a new approach to understanding hinterland evolution. *Geology*, **27**, 235–238.

Cerveny, P.F., Naeser, N.D., Zeitler, P.K., Naeser, C.W. and Johnson, N.M. (1988) History of uplift and relief of the Himalaya during the past 18 million years; evidence from sandstones of the Siwalik Group. In: *New Perspectives in Basin Analysis* (Eds K.L. Kleinspehn and C. Paola), pp. 43–61. Springer-Verlag, New York.

Davis, G.A. (1988) Rapid upward transport of mid-crustal mylonitic gneisses in the footwall of a Miocene detachment fault, Whipple Mountains, southeastern California. *Int. J. Earth Sci.*, **77**, 191–209.

Dodson, M.H. (1979) Theory of cooling ages. In: *Lectures in Isotope Geology* (Eds E. Jaeger and C.J. Hunziker), pp. 194–202. Springer-Verlag, New York.

Ehlers, T.A. and Farley, K.A. (2003) Apatite (U–Th)/He thermochronometry: methods and applications to problems in tectonic and surface processes. *Earth Planet. Sci. Lett.*, **206**, 1–14.

Ehlers, T.A., Willett, S.D., Armstrong, P.A. and Chapman, D.S. (2003) Exhumation of the central

Wasatch Mountains, Utah: 2. Thermokinematic model of exhumation, erosion, and thermochronometer interpretation. *J. Geophys. Res.*, **108**, 2173, doi:10.1029/2001JB001723.

Farley, K.A. 2000. Helium diffusion from apatite; general behavior as illustrated by Durango fluorapatite. *J. Geophys. Res.*, **105**, 2903–2914.

Fitzgerald, P.G., Sorkhabi, R.B., Redfield, T.F. and Stump, E. (1995) Uplift and denudation of the central Alaska Range: a case study in the use of apatite fission track thermochronology to determine absolute uplift parameters. *J. Geophys. Res.*, **100**, 20,175–20,191.

Gallagher, K., Brown, R. and Johnson, C. (1998) Fission track analysis and its applications to geological problems. *Ann. Rev. Earth Planet. Sci.*, **26**, 519–572.

Garver, J.I. and Brandon, M.T. (1994) Erosional exhumation of the British Columbia Coast Ranges as determined from fission-track ages of detrital zircon from the Tofino basin, Olympic Peninsula, Washington. *Geol. Soc. Amer. Bull.*, **106**, 1398–1412.

Garver, J.I., Brandon, M.T., Roden-Tice, M. and Kamp, P.J.J. (1999) Exhumation history of orogenic highlands determined by detrital fission-track thermochronology. In: *Exhumation Processes; Normal Faulting, Ductile Flow and Erosion* (Eds U. Ring, M.T. Brandon, G.S. Lister and S. Willett), pp. 283–304. Special Publication 154, Geological Society Publishing House, Bath.

Gehrels, G.E. and Kapp, P.A. (1998) Detrital zircon geochronology and regional correlation of metasedimentary rocks in the Coast Mountains, southeastern Alaska. *Can. J. Earth Sci.*, **35**, 269–279.

Green, P.F., Duddy, I.R., Gleadow, A.J.W. and Lovering, J.F. (1989) Apatite fission track analysis as a paleotemperature indicator for hydrocarbon exploration. In: *Thermal History of Sedimentary Basins: Methods and Case Histories* (Eds N.D. Naeser and T.H. McCulloh), pp. 181–195. Springer-Verlag, New York.

Harrison, T.M., Ryerson, F.J., Le Fort, P., Yin, A., Lovera, O.M. and Catlos, E.J. (1997) A late Miocene–Pliocene origin for the central Himalayan inverted metamorphism. *Earth Planet. Sci. Lett.*, **146**, E1–E7.

Harrison, T.M., Grove, M., Lovera, O.M. and Catlos, E.J. (1998) A model for the origin of Himalayan anatexis and inverted metamorphism. *J. Geophys. Res.*, **103**, 27,017–27,032.

Hasbargen, L. and Paola, C. (2000) Landscape instability in an experimental drainage basin. *Geology*, **28**, 1067–1070.

Hasbargen, L. and Paola, C. (2002) How predictable is local erosion rate in erosional landscapes? In: *Prediction in Geomorphology* (Eds P.R. Wilcock and R.M. Iverson), pp. 231–240. Monograph **135**, American Geophysical Union, Washington, DC.

Hodges, K.V. (2000) Tectonics of the Himalaya and southern Tibet from two perspectives. *Geol. Soc. Am. Bull.*, **112**, 324–350.

Hodges, K., Wobus, C., Ruhl, K., Schildgen, T. and Whipple, K. (2004) Quaternary deformation, river steepening, and heavy precipitation at the front of the Higher Himalayan ranges. *Earth Planet. Sci. Lett.*, **220**, 379–389.

Jamieson, R.A., Beaumont, C., Nguyen, M.H. and Lee, B. (2002) Interaction of metamorphism, deformation and exhumation in large convergent orogens. *J. Metamorph. Geol.*, **20**, 9–24.

Jamieson, R.A., Beaumont, C., Medvedev, S. and Nguyen, M.H. (2004) Crustal channel flows: 2. Numerical models with implications for metamorphism in the Himalayan-Tibetan orogen. *J. Geophys. Res.*, **109**, B06407, doi:10.1029/2003JB002811.

Johnson, N.M., Stix, J., Tauxe, L., Cerveny, P.F. and Tahirkheli, R.A.K. (1985) Paleomagnetic chronology, fluvial processes, and tectonic implications of the Siwalik deposits near Chinji Village, Pakistan. *J. Geol.*, **93**, 27–40.

Ketcham, R.A., Donelick, R.A. and Carlson, W.D. (1999) Variability of apatite fission-track annealing kinetics: III. Extrapolation to geologic time scales. *Amer. Mineral.*, **84**, 1235–1255.

Koons, P.O. (1989) The topographic evolution of collisional mountain belts: a numerical look at the Southern Alps, New Zealand. *Am. J. Sci.*, **289**, 1041–1069.

Mancktelow, N.S. and Grasemann, B. (1997) Time-dependent effects of heat advection and topography on cooling histories during erosion. *Tectonophysics*, **270**, 167–195.

McDougall, I. and Harrison, T.M. (1988) *Geochronology and Thermochronology by the $^{40}Ar/^{39}Ar$ Method*. Oxford University Press, New York, 212 pp.

Nábelek, J.L., Vergne, J., Hetenyi, G. and Team, H.-C. (2005) Project Hi-CLIMB: a synoptic view of the Himalayan collision zone and Southern Tibet. *Eos Trans, AGU Fall Meet. Suppl.*, **86**, T52A-02.

Naeser, C.W. (1979) Fission-track dating and geological annealing of fission tracks. In: *Lectures in Isotope Geology* (Eds E. Jäger and J.C. Hunziker), pp. 154–169. Springer-Verlag, New York.

Naeser, C.W., Bryant, B., Crittenden, M.D., Jr. and Sorensen, M.L. (1983) Fission-track ages of apatite in the Wasatch Mountains, Utah: an uplift study. *Geol. Soc Am. Mem.*, **157**, 29–36.

Najman, Y., Pringle, M., Godin, L. and Oliver, G. (2001) Dating of the oldest continental sediments from the Himalayan foreland basin. *Nature*, **410**, 194–197.

Nelson, K.D., Zhao, W., Brown, L.D., *et al.* (1996) Partially molten middle crust beneath southern Tibet;

synthesis of Project INDEPTH result. *Science*, **274**, 1684–1688.

Pandey, M.R., Tandukar, R.P., Avouac, J.P., Lavé, J. and Massot, J.P. (1995) Interseismic strain accumulation on the Himalayan crustal ramp (Nepal). *Geophys. Res. Lett.*, **22**, 751–754.

Pelletier, J.D. (2004) Persistent drainage migration in a numerical landform evolution model. *Geophys. Res. Lett.*, **31**, doi:10.1029/2004GL020802.

Reiners, P.W., Spell, T.L., Nicolescu, S. and Zanetti, K.A. (2004) Zircon (U–Th)/He thermochronometry: He diffusion and comparisons with ^{40}Ar/^{39}Ar dating. *Geochem. Cosmochim. Acta*, **68**, 1857–1887.

Ruhl, K.W. and Hodges, K.V. (2005) The use of detrital mineral cooling ages to evaluate steady-state assumptions in active orogens: an example from the central Nepalese Himalaya. *Tectonics*, **24**, TC4015, 10.1029/2004TC001712.

Ruiz, G.M.H., Seward, D. and Winkler, W. (2004) Detrital thermochronology – a new perspective on hinterland tectonics, an example from the Andean Amazon Basin, Ecuador. *Basin Res.*, **16**, 413–430.

Schelling, D. (1992) The tectonostratigraphy and structure of the eastern Nepal Himalaya. *Tectonics*, **11**, 925–943.

Small, E.E. and Anderson, R.A. (1998) Pleistocene relief production in Laramide mountain ranges, western United States. *Geology*, **26**, 123–136.

Sobel, E.R. and Dumitru, T.A. (1997) Thrusting and exhumation around the margins of the western Tarim basin during the Indian-Asia collision. *J. Geophys. Res.*, **102**, 5043–5063.

Sobel, E.R., Oskin, M., Burbank, D. and Mikolaichuk, A. (2006) Exhumation of basement-cored uplifts: Example of the Kyrgyz Range quantified with apatite fission-track thermochronology. *Tectonics*, **25**, TC2008, 10,1029/2005TC001809.

Spiegel, C., Siebel, W., Kuhlemann, J. and Frisch, W. (2004) Toward a comprehensive provenance analysis: a multi-method approach and its implications for the evolution of the Central Alps. In: *Detrital Thermochronology – Provenance Analysis, Exhumation, and Landscape Evolution of Mountain Belts* (Ed. M. Bernet and C. Spiegel). *Geol. Soc. Am. Spec. Pap.* **378**, 37–50.

Stock, J.D. and Montgomery, D.R. (1996) Estimating palaeorelief from detrital mineral age ranges. *Basin Res.*, **8**, 317–328.

Stockli, D., Farley, K.A. and Dumitru, T.A. (2000) Calibration of the apatite (U–Th)/He thermochronometer on an exhumed fault block,White Mountains, California. *Geology*, **28**, 983–986.

Stüwe, K. and Hintermüller, M. (2000) Topography and isotherms revisited: The influence of laterally migrating drainage divides. *Earth Planet. Sci. Lett.*, **184**, 287–303.

Stüwe, K., White, L. and Brown, R. (1994) The influence of eroding topography on steady-state isotherms: application to fission track analysis. *Earth Planet. Sci. Lett.*, **124**, 63–74.

Tippett, J.M. and Kamp, P.J.J. (1993) Fission track analysis of the Late Cenozoic vertical kinematics of continental Pacific crust, South Island, New Zealand. *J. Geophys. Res.*, **98**, 16,119–16,148.

Van der Plas, L. and Tobi, A.C. (1965) A chart for judging the reliability of point counting results. *Am. J. Sci*, **263**, 87–90.

Vermeesch, P. (2004) How many grains are needed for a provenance study? *Earth Planet. Sci. Lett.*, **224**, 441–451.

Wang, Q., Zhang, P.-Z., Freymueller, J.T., *et al.* (2001) Present-day crustal deformation in China constrained by Global Positioning System measurements. *Science*, **294**, 574–577.

White, N.M., Pringle, M., Garzanti, E., *et al.* (2002) Constraints on the exhumation and erosion of the High Himalayan Slab, NW India, from foreland basin deposits. *Earth Planet. Sci. Lett.*, **195**, 29–44.

Willett, S.D. (1999) Orogeny and orography: the effects of erosion on the structure of mountain belts. *J. Geophys. Res.*, **104**, 28,957–28,982.

Willett, S.D. and Brandon, M.T. (2002) On steady states in mountain belts. *Geology*, **30**, 175–178.

Wobus, C.W., Hodges, K.V. and Whipple, K.X. (2003) Has focused denudation sustained active thrusting at the Himalayan topographic front? *Geology*, **31**, 861–864.

Wolf, R.A., Farley, K.A. and Silver, L.T. (1996) Helium diffusion and low temperature thermochronometry of apatitie. *Geochim. Cosmochim. Acta*, **60**, 4231–4240.

Yamada, R., Tagami, T., Nishimura, S. and Ito, H. (1995) Annealing kinetics of fission tracks in zircon: an experimental study. *Chem. Geol.*, **122**, 249–258.

Zeitler, P.K. (1985) Cooling history of the NW Himalaya, Pakistan. *Tectonics*, **4**, 127–151.

Zhang, P., Molnar, P. and Downs, W.R. (2001) Increased sedimentation rates and grain sizes 2–4 Myr ago due to the influence of climate change on erosion rates. *Nature*, **410**, 891–897.

Modelling and comparing the Caledonian and Permo-Triassic erosion surfaces with present-day topography across Highland Scotland: implications for landscape inheritance

DAVID MACDONALD*, BARRY ARCHER*, SELMA MURRAY†, KIEREN SMITH* *and* ANN BATES*[1]

*Department of Geology & Petroleum Geology, University of Aberdeen, Meston Building, King's College, Aberdeen AB24 3UE, UK (Email: dmacdonald@abdn.ac.uk)

†PGL, Ternan House, North Deeside Road, Banchory AB31 5YR, UK*

ABSTRACT

The Caledonian Orogeny marks a starting point for the evolution of the Scottish Highlands. There is debate as to the level of erosion that the Highlands have experienced since the Devonian and the extent to which the Highland landscape reflects Permo-Triassic rather than Caledonian events. Data on the position and elevation of the Caledonian and Permo-Triassic unconformities have been used to create topographic models of both surfaces. A variety of computer mapping packages have been used that allow the interpreter to control many of the mapping parameters, creating models of surfaces that honour the data and maintain realistic surface trends. The effects of these parameters have been tested in a series of sensitivity experiments. The modelled Caledonian erosion surface has proved to be a good indicator of the present-day surface, suggesting that the Highlands are an exhumed landscape. This indicates that there has been limited denudation of basement rocks since the end of the Devonian. The model of the Permo-Triassic erosion surface has lower altitude and less relief than the model of the Caledonian surface, suggesting onlap onto a positive Highland block. Palaeomagnetic results showing Permo-Triassic reddening and fissuring of Highland basement rocks are interpreted as reflecting re-occupation of an older surface.

Keywords Landscape evolution, landscape inheritance, Old Red Sandstone, Permo-Triassic, Scotland, topographic modelling, unconformity.

INTRODUCTION

'We have an irregular body of land, raised above the level of the ocean' (Hutton, 1788)

Hutton identified the base of the Old Red Sandstone (ORS) in Scotland as a subaerial unconformity, post-dating a major episode of folding. 'Hutton's Unconformity' is not just the classic localities at Siccar Point and Cock of Arran, but is a record of an end-Caledonian landscape that is widespread in the Midland Valley of Scotland and across the eastern part of the Highlands (Fig. 1). One of the key questions in Scottish palaeogeography has been the extent to which this surface controlled development of later landscapes from the end of the Caledonian Orogeny to the present day. This has been addressed by palaeogeography, geomorphology, thermochronology and study of the erosion surfaces themselves. This paper is a contribution to that study, using surface modelling of the Caledonian unconformity to emphasize the longevity of the landscape established across the eastern Highlands and its influence on later erosion surfaces.

There is also a wider context to this work: the extent to which older erosion surfaces control younger landscapes. There are many anecdotal examples of this (e.g. Collinson *et al.*, 1989; Johnstone & Mykura,

[1]Present address: IHS Energy, Enterprise House, Cirencester Road, Tetbury GL8 8LD, UK.

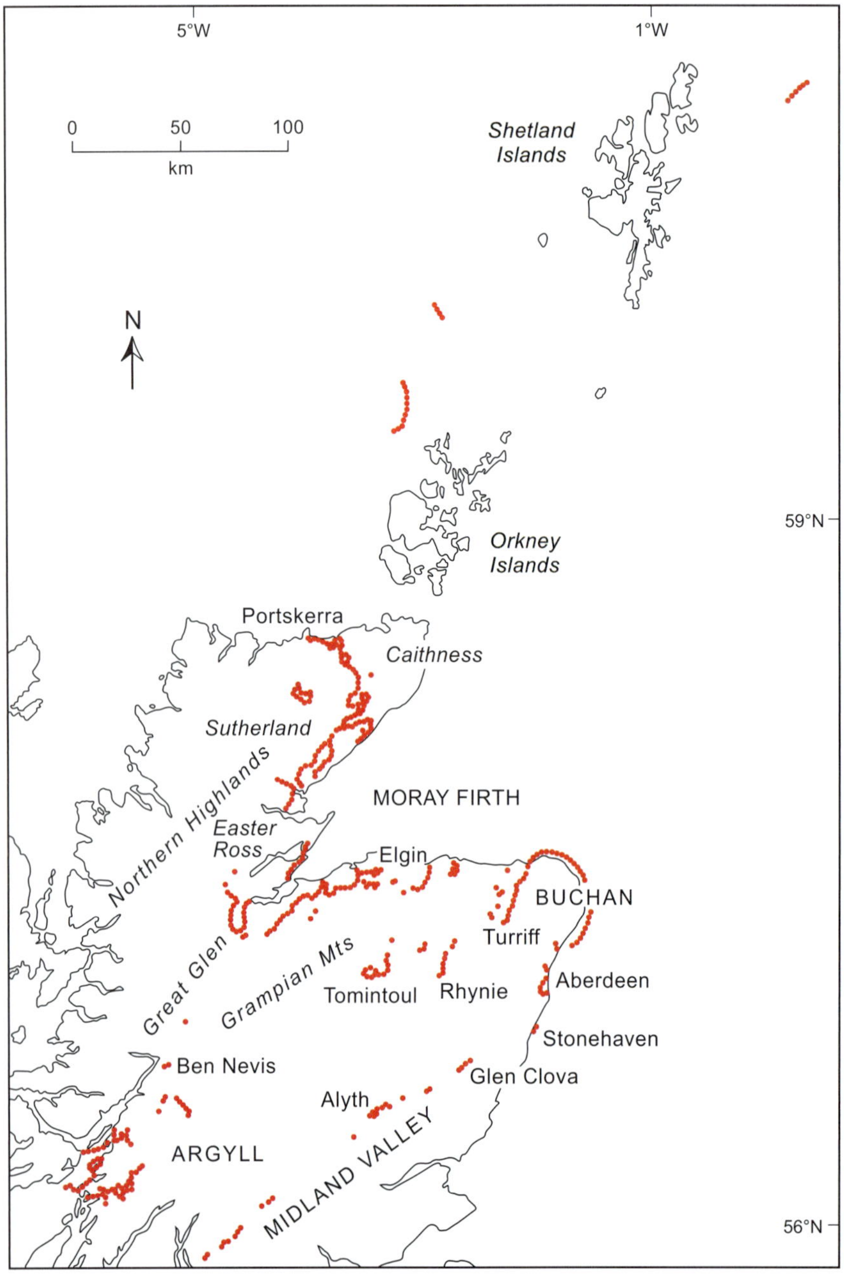

Fig. 1 Map of northern Scotland, showing localities where the Caledonian unconformity is present, both onshore and in adjacent offshore areas (as proved by drilling). The area of this diagram represents the project area. Key Highland outliers are named.

1989), but this paper is the first attempt to test and quantify these conceptual models.

Nomenclature

The surface that separates metamorphic and igneous rocks associated with the Caledonian Orogeny and overlying post-orogenic red beds is easy to define, but hard to date. The underlying rocks mostly went through peak metamorphism in the Late Silurian (Strachan *et al.*, 2002), while overlying strata can be anything from Late Silurian to Middle Devonian (an age range of as much as 35 Myr). 'Hutton's Unconformity' has specific meaning for the two classic localities, while 'Devonian unconformity' is inappropriate, given the possibility of Silurian deposits above the unconformity. Following the suggestion of Friend *et al.* (1970), this surface is referred to here as the *Caledonian unconformity*. The surface that separates the New Red

Sandstone from older strata is termed the *Permo-Triassic unconformity*.

Palaeogeography

The Caledonian Orogeny involved ductile deformation and tectonic thickening associated with the closure of the Iapetus Ocean (Strachan *et al.*, 2002; Fig. 2A). This led to isostatic uplift and erosion, which removed 25–30 km of overburden in the period 500–410 Ma, during the final stages of the Caledonian Orogeny (Watson, 1984). The land surface at the end of this period provides a valuable reference plane (Watson, 1985).

All palaeogeographical reconstructions of the post-Caledonian period, from Wills (1951) to Cope *et al.* (1992), and the various authors in Trewin (2002), have shown that the bones of the geography of Scotland were established at the end of the Caledonian Orogeny, some time between the Late Silurian and the Middle Devonian. The main elements of Scotland are easily recognizable: a large embayment in the Moray Firth, the Buchan peninsula, the line of the Great Glen and the Midland Valley (Fig. 2B). These elements are defined by the margins of major basins (Bluck, 1978), which persist into the Carboniferous (Besly, 1998). Most authors also tend to show the Highlands as a persistent emergent high, although this commonly reflects a lack of depositional remnants, rather than direct evidence.

The one truly problematic period is the Late Cretaceous. Here, a combination of no clastic input attributable to a Scottish landmass and some enigmatic deposits in Buchan and the Western Isles has led to the suggestion that the Highlands may have been at least partially inundated. However, even during this period, most authors believe that parts of the Northern Highlands and the Grampians were emergent (Wills, 1951; Cope *et al.*, 1992).

Geomorphology

Bremner (1943), in his study of the origin of the Scottish river system, noted the presence of outliers of ORS in valleys in the eastern Highlands, and stated that 'The present features [of the drainage] are in large measure due to the resuscitation of the old floor of schists . . .', i.e. the Caledonian unconformity. Hall (1991) proposed that the Caledonian Orogeny marked the starting point for the evolution of the Scottish Highlands, suggesting that the Scottish Highlands have been a relatively stable positive feature since Devonian time, based on the proximity of the basal Devonian surface to the present-day surface, the preservation of Devonian roof rocks over Caledonian granites, and the fact that exhumed or partially exhumed Devonian landforms are an important element in eastern parts of the Highlands. There is an apparent conflict between this view of the Highlands representing an exhumed Devonian landscape, and research which emphasizes the extensive, near-flat erosion planes, such as the 2000 ft (610 m) surface (George, 1955). The relationship between the principal surfaces has been discussed by Hall & Bishop (2002).

Thermochronology

There is an ongoing debate as to the amount of post-Devonian erosion across the Highlands, which is linked to both the regional palaeogeography and to the disagreement on the inheritance of geomorphological features discussed above. Hall (1991) postulated modest (< 1–2 km) denudation since the Devonian. However, this raises questions over the source of the thick Mesozoic and Tertiary sediments in surrounding sedimentary basins (see e.g. Watson, 1985; Evans, 1997). Thomson *et al.* (1999) used apatite fission track analysis to argue that minimum erosion values for the Northern Highlands are approximately 3 km. In addition, they suggested that a significant amount of this erosion occurred at the end of the Carboniferous, with around 2 km removed during Variscan uplift. This question has been revisited by Hall & Bishop (2002) who point out that there are inconsistencies between a deep denudation scenario and the geological evidence. They suggested that earlier apatite fission track thermochronology (AFTT) had overestimated the depth of former cover rocks and concluded that a modest denudation scenario was more probable.

The Caledonian unconformity

The numerous outliers of Old Red Sandstone across the Eastern and Northern Highlands (Blackbourn, 1981; Johnstone & Mykura, 1989; Stephenson & Gould, 1995) provide an opportunity to study the

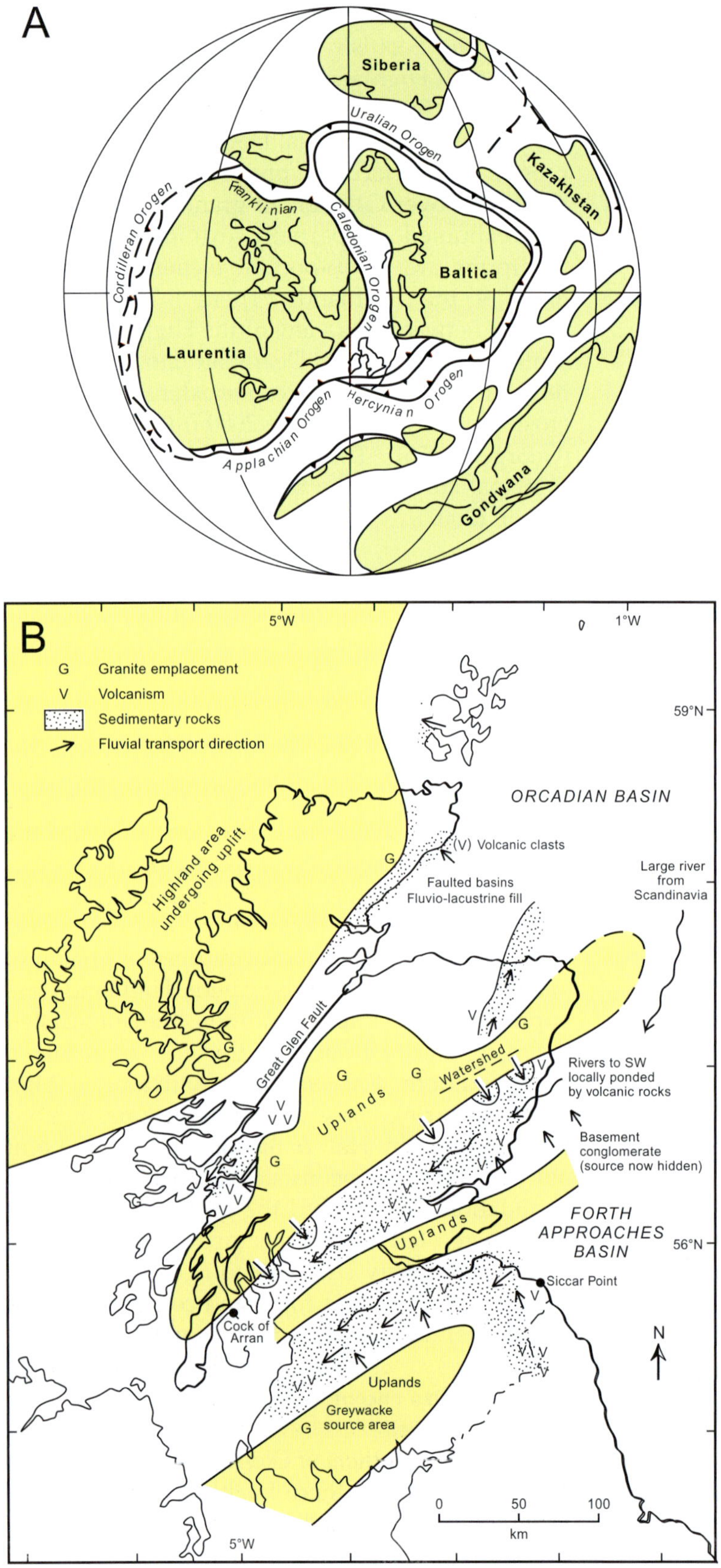

Fig. 2 (A) Palaeogeography of the Late Silurian–Devonian supercontinent formed during collision of Laurentia and Baltica; the Caledonian orogenic belt formed as a suture between them (Bluck, 1990). (Redrawn after Ziegler, 1988.) (B) Major palaeogeographic elements of Scotland in Early Devonian times. (After Trewin & Thirwall, 2002.)

surface that underlies them. Since the outliers are of non-marine Old Red Sandstone facies, their basal unconformity can be identified as the Devonian or pre-Devonian land surface (Watson, 1985). Almost all of the Highland Devonian outcrop is not at high altitude, and is not positive relief, suggesting preservation of the lower parts of an original irregular unconformity surface. There is also more subtle evidence of the widespread nature of the Caledonian unconformity in the form of reddening of underlying units. This phenomenon was noted by Friend *et al.* (1963) in northeast Arran, where Dalradian rocks were reddened and fissured below an obvious unconformity. Friend *et al.* (1970) used these observations to argue that the Caledonian unconformity must be close above the present-day land surface across a wide area of northwest Arran where reddened, fractured Dalradian was present.

Parnell *et al.* (2000) noted that reddening of the Dalradian is not restricted to Arran, but can be found across most of Argyll, Islay and Bute and also in Northern Ireland. They noted a close association between reddened Dalradian and dolomitic breccia veins, which they ascribed to Carboniferous extension. They suggested that the red colour was associated with haematite precipitation, dated as Late Permian to Early Triassic by palaeomagnetic means. Elmore *et al.* (2003) also found evidence of Permo-Triassic haematite precipitation along the Moine Thrust, well outside the known limits of ORS deposition. These results suggest that Permo-Triassic erosion is also regionally important and may have modified or deepened the end-Caledonian erosion.

There are several other lines of evidence for the presence of the Caledonian surface at no great height above the present-day Grampians.

1 Late Caledonian (Silurian) granitoids and attendant mineralization show features indicating a shallow level of emplacement (no more than 4 km): vuggy textures, intrusion breccias, associated porphyritic rocks and fluid inclusion evidence from quartz veins (Dr C.M. Rice, personal communication, 2005).
2 Silcrete palaeosols, plausibly interpreted as Devonian in age, have been found in upper Glen Clova (Goodman *et al.*, 1990).
3 Unpublished field data from the senior author shows extensive reddened fracture networks at intermediate levels in upper Deeside and Donside.

Rationale for this paper

In this paper, three-dimensional modelling techniques are used to reconstruct the original extent of the Caledonian unconformity. This model is compared with a similar model of the Permo-Triassic surface to assess whether the latter surface was inherited. Both are then compared with the present-day land surface to quantify the extent to which the modern Highland landscape is controlled by older surfaces.

METHODOLOGY

Surfer models

An initial series of models of the Caledonian unconformity surface was constructed using an abbreviated dataset in the gridding and contouring package *Surfer® Version 6 for Windows* using a kriging algorithm to model the surface. These models were used as a first-pass assessment of the extent to which the shape of the Caledonian unconformity resembles the present-day surface.

These models were based on sampling 10 km by 10 km squares based on an arbitrary origin at the southwest corner of UK National Grid 100 km square NR. For this phase, the project area included all of the Scottish Highlands, including Kintyre and Arran. British Geological Survey (BGS) maps of this region (Table 1; see also Fig. 3) were examined for the presence of the Caledonian unconformity. Within each square where it occurs, the highest and lowest elevations were identified, and a visual estimate made of the modal elevation. Three *XYZ* files were thus created: maximum, minimum and modal elevations of the Caledonian unconformity.

Detailed data collection

For the second phase of the project, data were collected on all of the mapped unconformity surfaces in Highland Scotland and the shallow offshore. In order to eliminate 'island effects' the elongate area of Kintyre was excluded from the project area, which was defined as 56–62°N and 3°E–10°W. This area was deliberately defined with a large extent in order to incorporate subsurface data from offshore oil wells in the future.

Table 1 List of British Geological Survey (BGS) map sheets used for digitization of the data used in the modelling of the Caledonian and Permo-Triassic unconformities

Scale	BGS sheet name	Sheet number
1:250,000	Argyll	
	Moray-Buchan	
1:50,000	Ardnamurchan	52
	Loch Torridon	81
	Ballater	65E
	Skye Broadford	71W
	Glenbuchat	75E
	Glenlivet	75W
	Alford	76W
	Glenfiddich	85E
	Orkney	
	Peterhead	
1:63,360	Mull	44
	Gairloch	91
	Elgin	95
	Ullapool	101
	Lewis & Harris (N)	105
	Tongue	105
	Rhum	(Special Sheet)

Data were gathered by digitizing unconformities directly from BGS maps (Table 1) into ArcView. The resulting lines were draped on a digital elevation model (DEM) of Scotland cropped from the USGS GTopo 30 DEM of the world. The DEM was also used as the comparative present-day surface in all difference models. In a second stage of sampling, marine outcrop data from around the Scottish coast and the Orkney and Shetland islands were added; these were draped on bathymetry taken from the General Bathymetric Chart of the Oceans (*GEBCO Digital Atlas – Centenary Edition*).

The resulting lines were sampled at a variety of length intervals between 100 and 1000 m in order to provide *XYZ* files of the elevation of the Caledonian unconformity. The different sampling lengths were used in sensitivity tests. Data collected were Universal Transverse Mercator co-ordinates (spheroid WGS84) and height above sea level in metres.

Sensitivity tests and volumetrics

Sensitivity tests were carried out to evaluate the best spacing of sampling points on the lines representing the present-day outcrop of the unconformities. This is important as the Caledonian unconformity locally has relief of tens to hundreds of metres, but

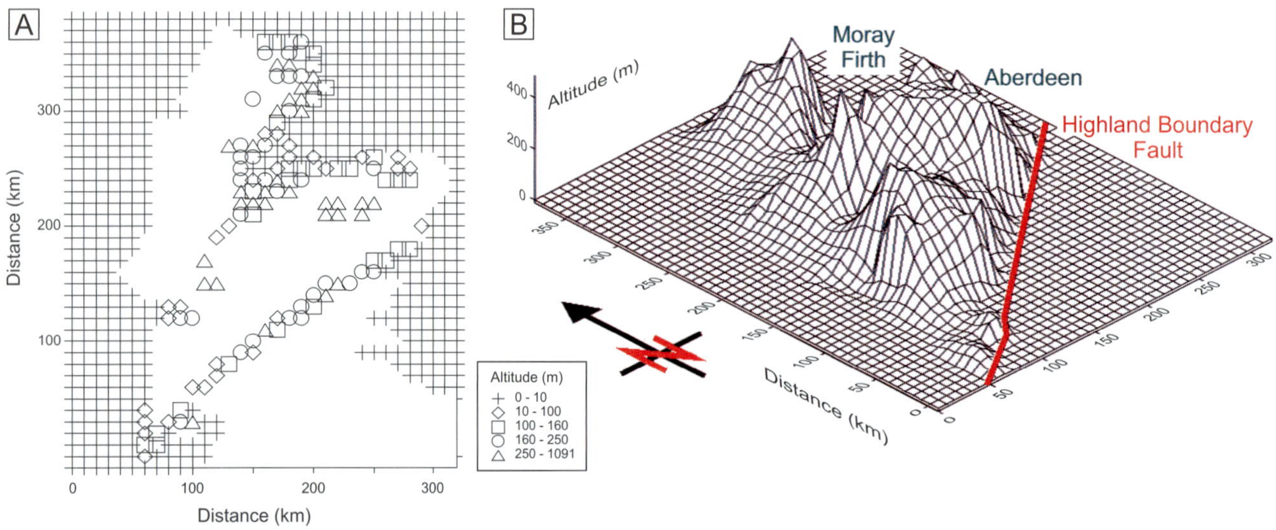

Fig. 3 Surfer™ model of the Caledonian erosion surface across the Scottish Highlands. (A) Classed post diagram (arbitrary geographical co-ordinates in 10 km squares from an origin southwest of Scotland) showing distribution and altitude of data points used to construct the model. (B) View from the southwest across the Surfer™ model of the Caledonian unconformity.

outcrops can be spaced as much as 50 km apart. A small spacing might capture the local variability but unduly weight the gridding. After a number of trials, a spacing of 500 m was chosen, resulting in 1535 points on the Caledonian unconformity. For the Permo-Triassic unconformity, which has a more restricted outcrop, it was necessary to vary the spacing from 100 to 1000 m in order to provide a balanced data set of 372 points.

The closeness of the modelled Caledonian unconformity to the present-day surface was modelled using the RMS modelling package within the GeoQuest suite of seismic interpretation software. This was also used for further sensitivity tests of the gridding algorithm and on edge effects, primarily towards the west, where ORS outcrop disappears.

Modelling in *ZMap Plus*

The main models used to compare the Caledonian and Permo-Triassic erosion surfaces were created using the Landmark mapping package *ZMap Plus*. This program allows good control over the parameters that control how model surfaces will be created.

Basemap setup

Basemaps for both data sets were created prior to surface modelling. The dimensions of the basemap were specified using the maximum and minimum longitude/latitude coordinates creating an Area of Interest (AOI) within the project area. The basemap was then set up with the appropriate projection system: Transverse Mercator with reference spheroid WGS 84, UTM zone 30, and map scale 1:1,000,000. The *X* and *Y* data were plotted on the basemap; these become control points that hold the *Z* value (elevation).

Gridding

The control points are defined by the *X*, *Y* and *Z* values measured from the maps. In order to be able to contour these points, *ZMap Plus* creates a grid. The grid is made up of grid nodes: both the spacing of the nodes and the algorithm that extrapolates between nodes is controlled by the interpreter. Both the Caledonian and Permo-Triassic erosion surface models were created using a least squares algorithm, which assigns grid node values by fitting a weighted planar least squares fit to the data in a circular area around the grid node (Fig. 4). This algorithm tends to pass a smooth surface through the data with no sharp peaks. It was chosen largely based on its ability to create a meaningful trend and honour the individual data points across the map.

A large search radius was required for both surfaces to overcome a problem of data clusters, which is a consequence of the natural position of outcrops across the Scottish Highlands. A large search radius allows the grid nodes to produce values based on control points across the map rather than be biased totally to the area around them. This allows the trend of the surface to be extrapolated by the gridding process across data-void areas. The search radius was set differently for both the maps. The Caledonian erosion surface

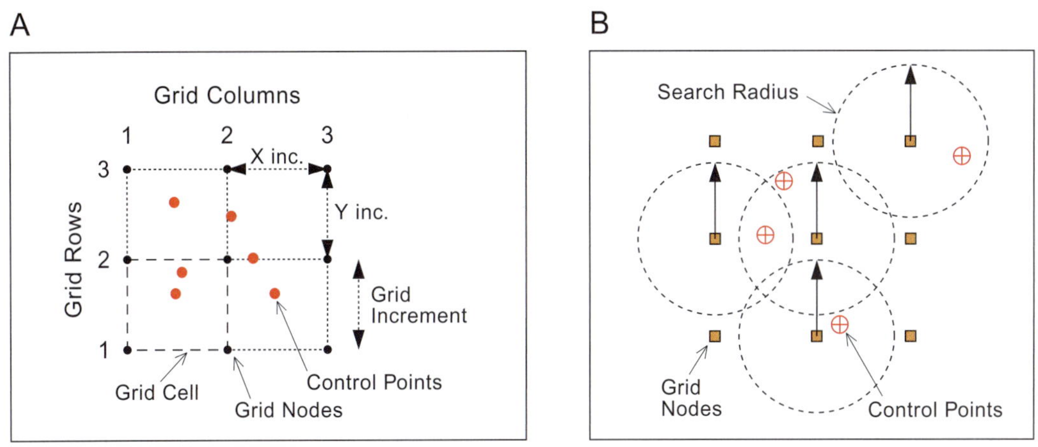

Fig. 4 (A) Gridding terminology used in the text. (B) Search radius around the grid nodes.

Table 2 Starting, intermediate and final increments used for refining the grids on which the models were based

Grid parameter	Caledonian grid	Permian-Triassic grid
Starting increment	8000	4800
Four intermediate	4000	2400
increments	2000	1200
	1000	600
	500	300
Final increment	250	150

search radius was 250,000 arbitrary units and the Permo-Triassic erosion surface was set at a larger 300,000 arbitrary units. The difference was required in order to accommodate the smaller Permo-Triassic data set and the larger void between data clusters.

Additionally, the gridding process can be controlled in terms of its grid increment (Fig. 4 & Table 2). The grid is initially coarse in order to capture the trend of the surface, and then is made progressively finer to honour the data. For both surface maps the grid went through six refinements; the grid is halved six times to reach a final grid increment. The two surfaces produced were given different final grid increments; this was a product of the data spacing. The Permo-Triassic surface has fewer data points than the Caledonian surface, but these are more closely spaced, and therefore required a smaller grid increment to honour the data more accurately. When the Permo-Triassic surface was gridded with the same final increment as the Caledonian surface (250) as a test, this led to exaggerated smoothing and closely clustered data not being honoured.

The gridded surface was filtered using an algorithm to remove noise (unjustifiable surface variations) while retaining the best fit at the data locations. The filtering algorithm filters the grid on each refinement; this is typically four to six times for each refinement. Grid flexing is repeated until the amount of change from one pass to the next is smaller than 0.25, then the next refinement of the grid is allowed to begin. A biharmonic filtering algorithm was used on both surfaces; this algorithm varied smoothly from point to point and continued trends beyond the data-rich areas into void areas of the map.

Contouring

The gridded surface was then contoured and colour filled, highlighting surface trends at 100 m spacing.

Subtracting surfaces

A dual grid operation was undertaken on the two surfaces; the Caledonian grid was subtracted from the Permo-Triassic grid. This resulted in the creation of a new grid that effectively was an isopach map representing the difference between the surfaces. The new grid assumed a grid increment of 250 from the Caledonian surface grid; adopting the larger increment of the two grids was the default and advised by the product manual. The operation places the two grids on top of each other, but grid nodes do not line up, as the grid cells are different sizes. The program overlays the new 250 grid and performs the subtraction calculation by making a best fit.

RESULTS

Limitations and assumptions

There are a number of limitations to a study of this type that must be borne in mind when discussing the results. First, and most importantly, both surfaces are composite, formed over several millions to tens of millions of years. The Caledonian surface in particular is a record of a long period of erosion, being overlain at different places by Silurian, and Lower, Middle and Upper Devonian deposits. It could span as much as 35 Myr (Trewin & Thirlwall, 2002). The time period represented by the Permo-Triassic surface is probably shorter: no more than 15 Myr (Glennie, 2002).

The second problem is the variable geometry of the Caledonian surface. At some localities, especially in the Northern Highlands, the surface represents the passive infilling of relatively short-wavelength topography: hills tens of metres high, spaced hundreds of metres apart (Donovan, 1973, 1975). The

Fig. 5 The Caledonian unconformity south of Rubha Bhrà, Portskerra (NC877666). Fluvial Middle ORS facies overlie Moine gneisses which form buried hills up to 20 m high and spaced 100–200 m apart. Where the ORS cover has been removed, the Moine hills remain as exhumed topography.

spectacular exposures at Portskerra are typical of this style of unconformity (Fig. 5). In other places, including the Brora Outlier and the eastern Grampians, the unconformity has a much longer wavelength: palaeohills are hundreds of metres high and kilometres apart. Areas such as Tom an-t Suidhe Mor in the Tomintoul Outlier illustrate this (Fig. 6). Farther east, the Rhynie and Turriff outliers occupy partially exhumed half grabens, and the topographic relief on the surface must be several hundred metres. The variable relief and wavelength of the Caledonian surface make the modelling much more difficult, and it is impossible to capture the detail that must exist in the data-poor areas.

Third, in some areas, boundaries originally interpreted as the Caledonian unconformity have been reinterpreted as faults. The largest reduction in the inferred outcrop of the unconformity is in the Rhynie Outlier, where Rice & Ashcroft (2004) have reinterpreted much of the eastern boundary, interpreted as an unconformity on BGS Sheet 76W, as being faulted. This problem is not regarded as critical, as the faulting is minor, and the outcrop of the basal conglomerate is a reasonable proxy for the Caledonian surface.

The fourth problem stems from the isolated nature of the outcrop of the Permo-Triassic surface, far less extensive than that of the Caledonian surface. Permian–Triassic exposure across the Scottish Highlands is isolated and predominantly found onshore to the west of Scotland, although small outcrops are exposed onshore on the north coast, south of the Moray Firth, to the west of Elgin; and around Golspie on the northwest coast of the Moray Firth. The Permo-Triassic surface comes to the seabed in the Moray Firth, around Moray–Buchan, and offshore Orkney. In the east of Scotland the Permo-Triassic sediments are unconformable on the Old Red Sandstone. In western Scotland, they generally lie unconformably on pre-Devonian basement rocks of Lewisian and Torridonian age.

Surfer models

Figure 3B shows a typical Surfer model of the shape of the Caledonian surface, using the modal elevation dataset. This model clearly shows the shape of Highland Scotland, and provided encouragement to go ahead with the more detailed work. In this view, the shape of the Moray Firth and Great Glen can be seen clearly, while the model correctly predicts the high topography of east Sutherland, Easter Ross and the central Grampians.

The model provides a clear indication of the extent to which the Caledonian surface controls

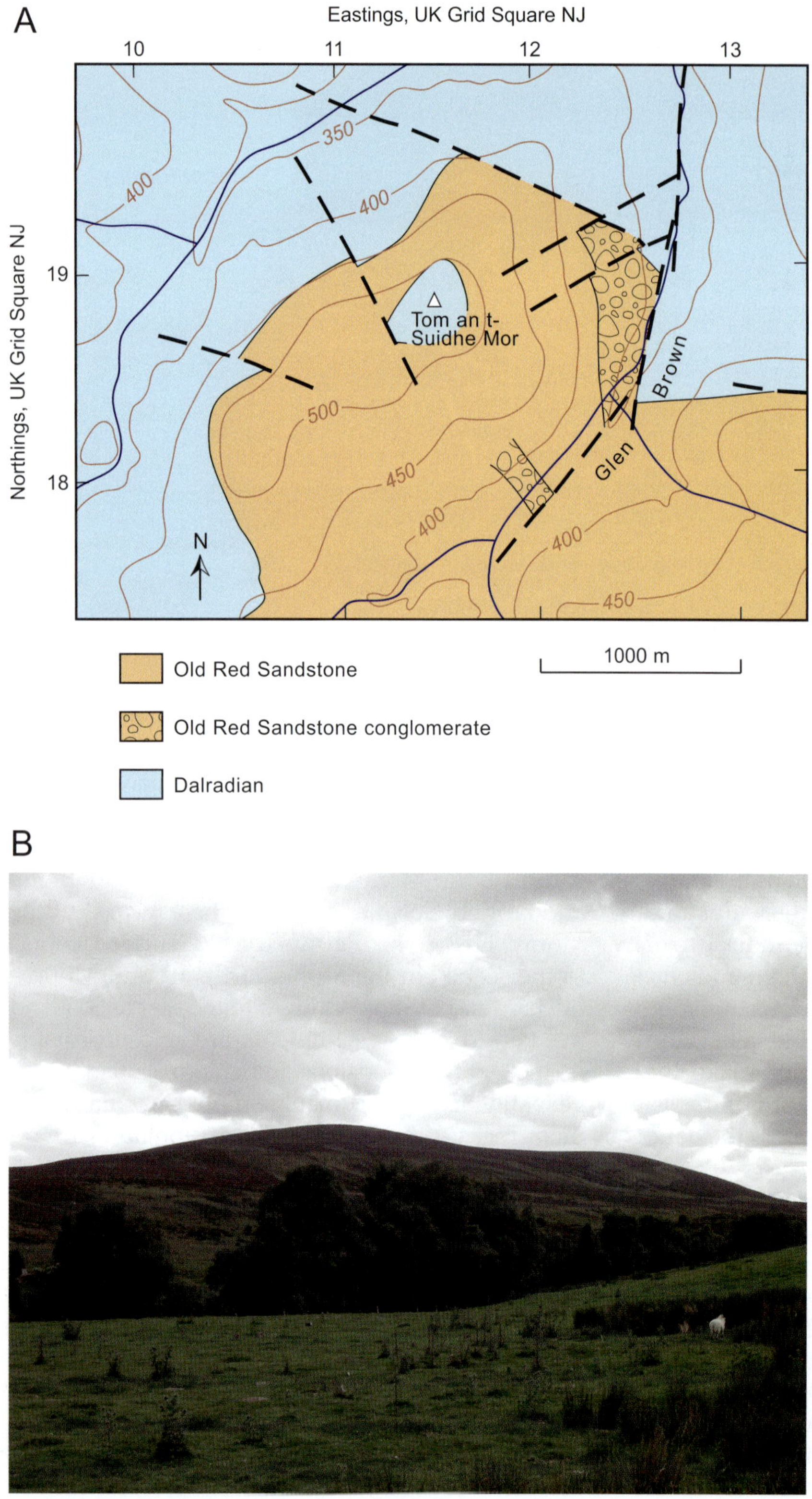

Fig. 6 (A) Geological sketch map redrawn from a section of British Geological Survey 1:50,000 Sheet 75W (Glenlivet), showing the close relationship between Caledonian and present-day topography. The hill of Tom an-t Suidhe Mor has a peak of Dalradian rocks, but slopes mantled by ORS deposits. The map also shows that coarse facies of the ORS are restricted to the modern valleys. (B) View of Tom an-t Suidhe Mor from the east, showing the moderate topography that is typical of the eastern Grampians; the Dalradian-cored peak is just left of centre.

present-day topography. The model is less successful at predicting the absolute topography, with maximum values of only a little more than 400 m, about one-third of the true value. This arises from the fact that the Devonian outliers are generally at intermediate altitudes, and the model cannot project the surface effectively into areas without control points. This is one of the reasons for doing the detailed modelling in *ZMap Plus*, where there is more control over the gridding parameters.

The model also fails to predict the existence of high topography in Northwest Scotland, reverting to zero values in areas without ORS outcrop.

Sensitivity tests

Figure 7 shows the results of a sensitivity experiment used to test the gridding parameters. Here an RMS model of the present-day land surface of Highland Scotland (based on the GTopo 30 dataset) has been intersected with an early model of the Caledonian surface; if the Caledonian surface lies below the present-day surface, the negative difference is a prediction of areas where Devonian rocks should be present in outcrop. In the test illustrated in Fig. 7, it can clearly be seen that there are relatively large regions where the modelled Caledonian surface is below the present-day surface. This model overestimates the extent of small outliers (mostly inland), and underestimates the size of the large outliers (mostly coastal). In this case, it is clear that the model has not been allowed to generate enough curvature of the surface, or to create enough high topography. This is partly due to the fact that the ORS outliers are at intermediate altitudes.

There are two ways that this problem can be tackled. First, the modelled Caledonian surface could be constrained not to cut the present-day surface. However, since the outliers exist, the two surfaces clearly intersect. Hence, repeat tests with different increments and search radii have to be undertaken, until a surface is produced that approximates to reality.

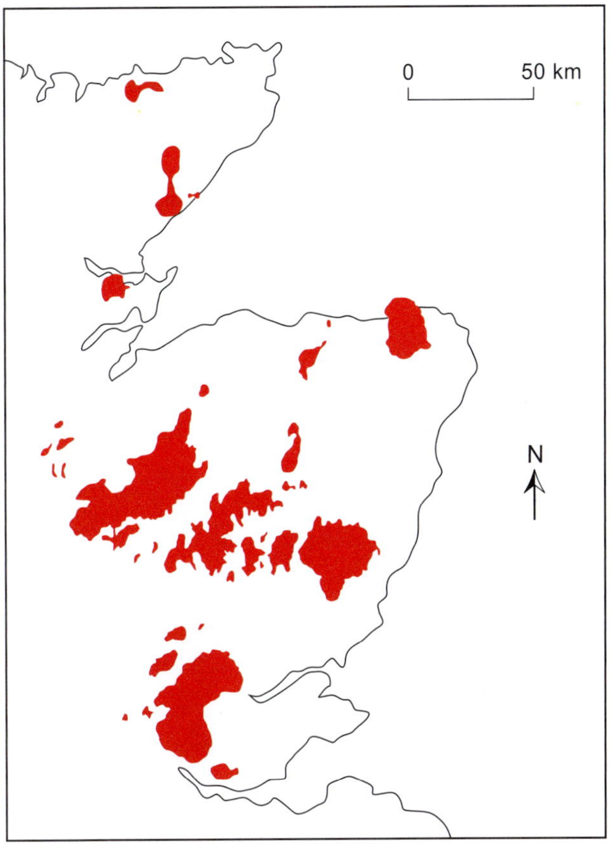

Fig. 7 Sensitivity test run in RMS to test the gridding parameters. In this test, the red areas are regions where the modelled Caledonian surface is below the present-day surface, i.e. predicted ORS and younger outcrop. This model overestimates the extent of small outliers, and underestimates the size of the large outliers. See text for discussion.

ZMap Plus models

The modelled Caledonian and Permo-Triassic surfaces are presented as colour-filled contour maps of the gridded surface (Figs 8 & 9). The land surface above the present-day mean sea level is shown in green and that below sea level in blue.

The modelled Caledonian erosion surface (Fig. 8) is a composite surface representing a blurred view of the landscape prior to the deposition of the Old Red Sandstone. There is a clear correlation of this surface to the present-day Scottish Highlands. The shape of the Moray–Buchan coast and the main ridge of high ground running northeast from Ben Nevis are well displayed. The model correctly predicts the presence of the Orcadian Basin, the West of Shetland Basin and the Forth Approaches Basin.

The Permo-Triassic erosion surface (Fig. 9) has been mapped with a similar colour scheme to the

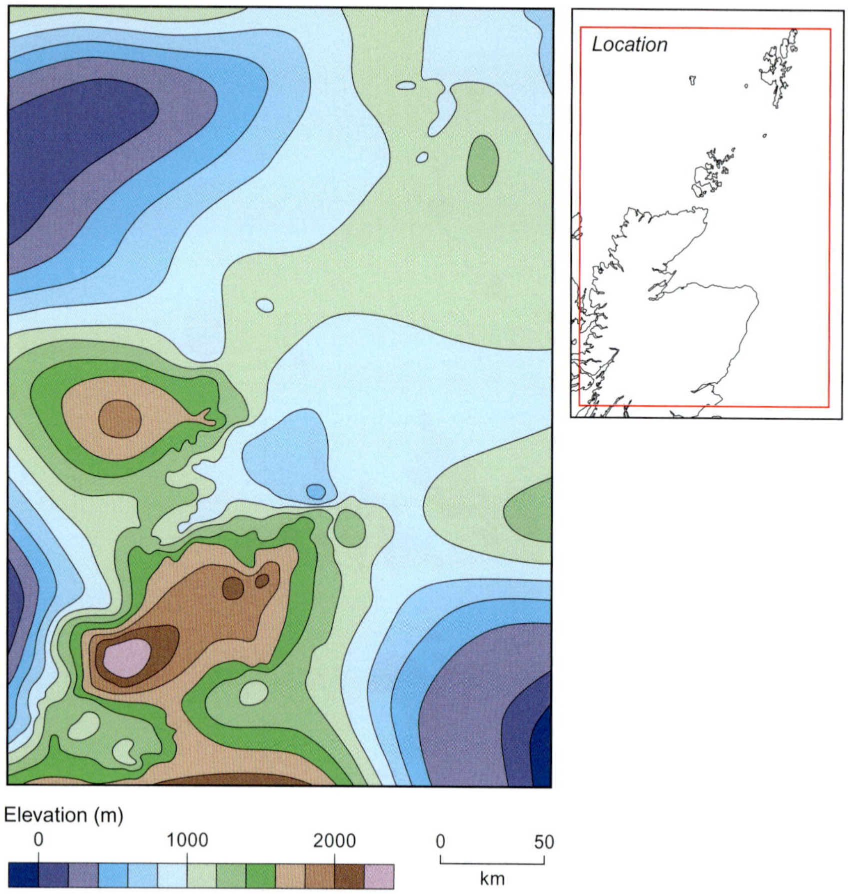

Fig. 8 Modelled Caledonian erosion surface (*ZMap Plus*), clearly showing the Scottish Highlands and surrounding basins. Elevations in metres above an arbitrary datum; the reference map of northern Scotland shows the area covered by the model (red box).

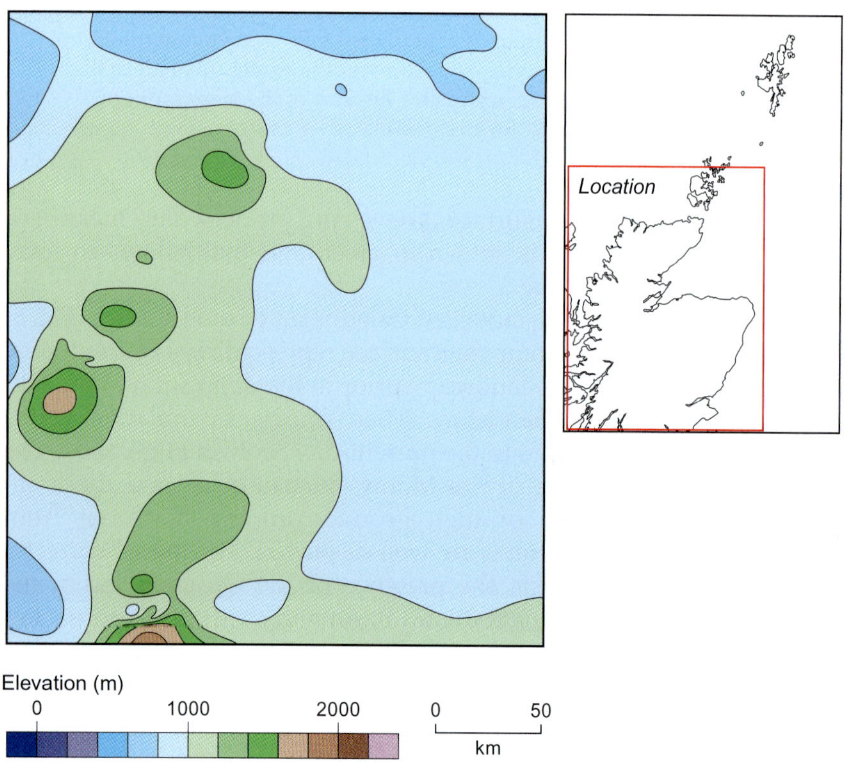

Fig. 9 Modelled Permo-Triassic erosion surface (*ZMap Plus*). Note the subdued relief compared to that of the Caledonian surface (Fig. 8).

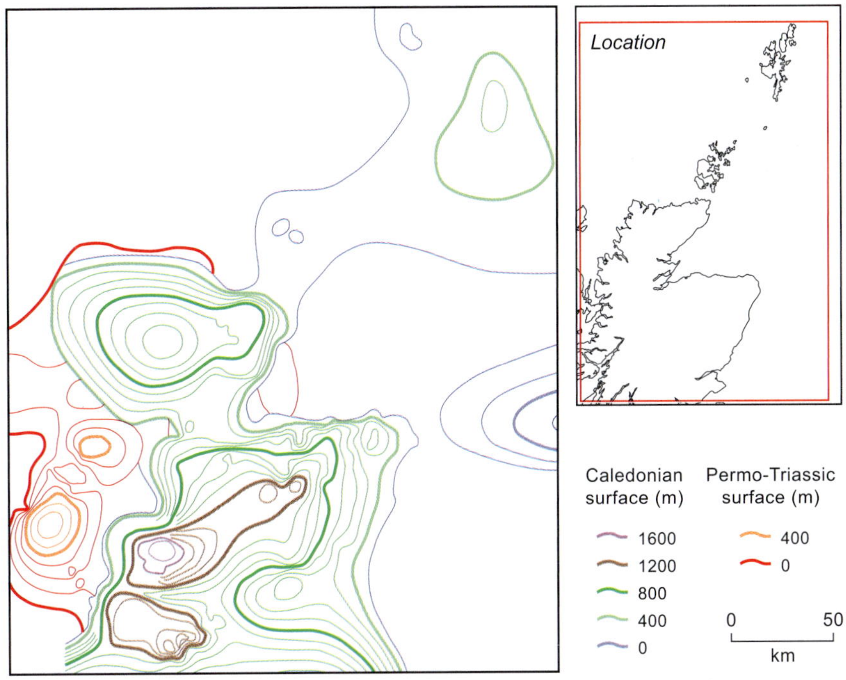

Fig. 10 Relationship of the Permo-Triassic surface to the Caledonian surface. Red and orange contours show areas where the Permo-Triassic surface is above the Caledonian surface, while the blue-purple spectrum shows where the Caledonian surface is above the modelled level of the Permo-Triassic surface. Contours are at 100 m intervals; surface minimum set at 0 for both sets of contours. Some of the intermediate contours have been omitted for clarity in western Scotland.

Caledonian surface shown in Fig. 8. The Permo-Triassic surface is notably flatter and lower than the Caledonian surface; what high elevations there are, lie in the west. This is a natural artefact of the generally isolated onshore outcrop of Permo-Triassic across the Scottish Highlands. This surface does not define any recognizable topography.

Relationships between modelled surfaces

The modelled Permo-Triassic erosion surface is clearly lower than the modelled Caledonian surface. Figure 10 shows the relationship between them. The Caledonian surface defines a prominent high area (Highland Scotland), surrounded by lower-lying areas where it lies below the Permo-Triassic surface. This suggests that the Scottish Highlands have been a significant positive feature since the Devonian, and the Permo-Triassic surface onlaps the Caledonian surface. The lower marginal areas are offshore basins with significant preserved Devonian (and possibly Carboniferous) sediment. These areas are defined in Fig. 11.

Relationship between the modelled surface and the present-day surface

The model presented in Fig. 12 predicts that the modelled Caledonian surface lies very close above the present-day land surface (within a few tens of metres). The difference in volume (i.e. the space above the present-day surface and below the modelled Caledonian surface) is about 3000 km^3. This means that there has been very little net erosion below the Caledonian surface. Across the map area, this would equate to a layer about 50 m thick. It further suggests that post-Devonian denudation has been of sediments rather than Highland basement. The implications of this are profound for modelling the sediment provenance of post-Devonian sedimentary rocks in offshore areas around northern Britain.

DISCUSSION

This paper touches on two major topics:

1 the morphology of the Caledonian unconformity and its relationship to the present land surface;
2 the relationship between the Caledonian and Permo-Triassic unconformities.

The research presented here relies on the fact that the ORS sediments are everywhere non-marine, implying that their basal unconformity is a Devonian or pre-Devonian land surface (Watson, 1985). There are few parts of the world where

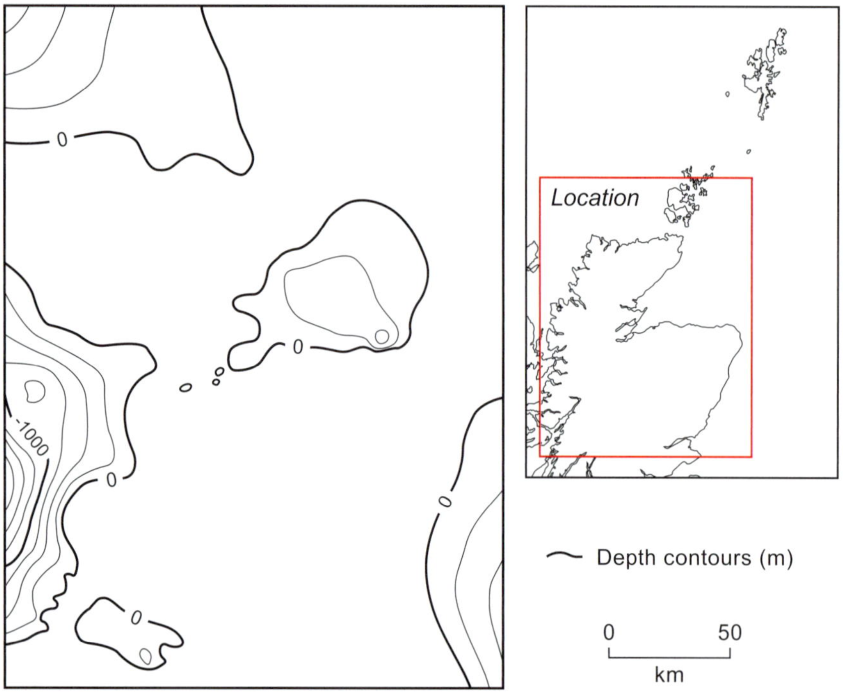

Fig. 11 Devonian–Carboniferous basins predicted by subtracting the Caledonian erosion surface from the Permo-Triassic surface.

exposures of a major unconformity are extensive enough to carry out the sort of work outlined here, but in a qualitative sense, control of recent topography by an older landscape unconformity is well-known. For instance, the present landscape of the Lewisian outcrop of Northwest Scotland is clearly exhumed from beneath the sub-Torridonian unconformity (Johnstone & Mykura, 1989). A similar relationship between Proterozoic palaeovalleys and present topography has been described from North Greenland by Collinson *et al.* (1989).

The modelled Caledonian erosion surface is a good predictor of the present-day landscape of the Scottish Highlands (Figs 3 & 8). Although this modelled surface probably represents a smoothed-off surface, which has lost some of the topographic 'noise', it does suggest that the present-day Scottish Highlands were largely shaped immediately prior to and during the deposition of the Old Red Sandstone. In particular, a significant positive relief trends NE–SW into Buchan. This represents the preservation of the Devonian watershed separating the Orcadian Basin and Midland Valley, as suggested by Watson (1985).

It is also evident that the present-day surface across the Highlands not only mimics the shape of the Caledonian surface, but lies in close proximity to it (Fig. 12). This suggests that most of the Mesozoic and Tertiary denudation of Scotland was of sedimentary material. This makes a modest denudation scenario (1–2 km), as proposed by Hall & Bishop (2002), much more plausible than the deep (3 km) scenario of Thomson *et al.* (1999). This is in interesting contrast to the situation on the other side of the North Sea, where there is deep Cenozoic denudation of southern Norway (Huuse, 2002).

Since there clearly has been some denudation of Scotland, this raises the question of what has been removed. Thomson *et al.* (1999) provided evidence for partial Carboniferous cover across the Scottish Highlands. There are two small outcrops of Upper Carboniferous deposits in the Southwest Highlands, while the glacial erratics of the Outer Hebrides have been shown to contain Carboniferous sediments. This appears to show that the Scottish Highlands were at least partially covered with Carboniferous sediments, with these sediments being reworked and subsequently deposited into the surrounding evolving Mesozoic basins.

This leads to the last strand of the discussion: the Permo-Triassic surface. This onlaps the Devonian surface (Fig. 10), which would imply that most Carboniferous cover would have been removed in

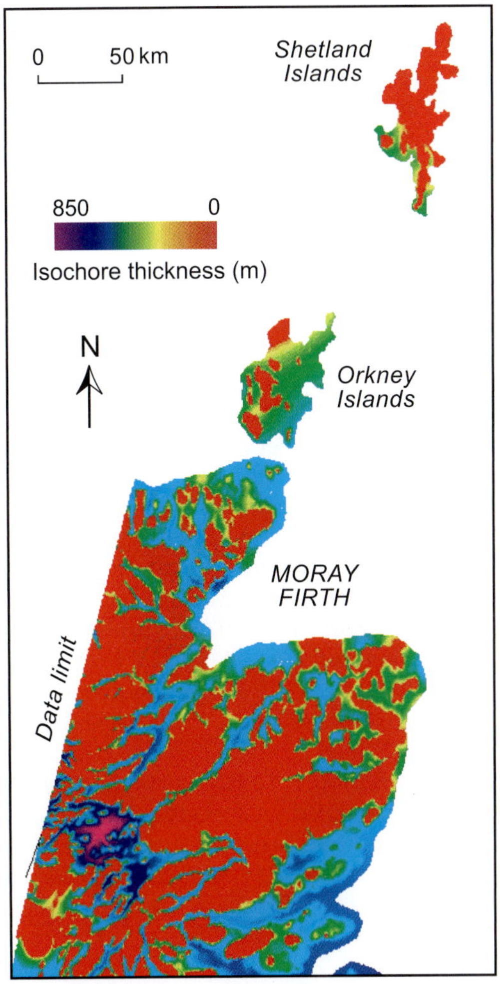

Fig. 12 Difference between the modelled Caledonian surface and the present-day land surface (which mostly lies below the Caledonian surface). It is notable that over much of the eastern Highlands, the two surfaces are predicted to lie within a few tens of metres of each other.

the Permian. The difference map in Fig. 11 represents the net gain of sediment between the Devonian and the Permian. The thicknesses of sediment predicted by this model also correspond to the suggestion by Hall (1991) that erosion over the Scottish Highlands has been less than 1–2 km. Figure 13 shows the conceptual relationship between the Devonian and Permo-Triassic surfaces. Notwithstanding the palaeomagnetic data of Parnell *et al.* (2000), the Caledonian surface appears to be the dominant control on Highland landscape development; Friend *et al.* (1970) were right.

CONCLUSIONS

Data on the position and elevation of the Caledonian and Permo-Triassic unconformities have been used to create topographic models of both surfaces. By comparing these with each other and with the present-day land surface, the following conclusions can be drawn:

1 The modelled Caledonian erosion surface has proven to be a good indicator of the present-day surface, suggesting that the Highlands are an exhumed landscape. This suggests that there has been limited denudation of basement rocks since the end of the Devonian.

2 The model of the Permo-Triassic erosion surface has lower altitude and less relief than the model of the Caledonian surface, suggesting onlap onto a positive Highland block. Palaeomagnetic results suggesting Permo-Triassic reddening and fissuring of Highland basement rocks are interpreted as reflecting reoccupation of an older surface.

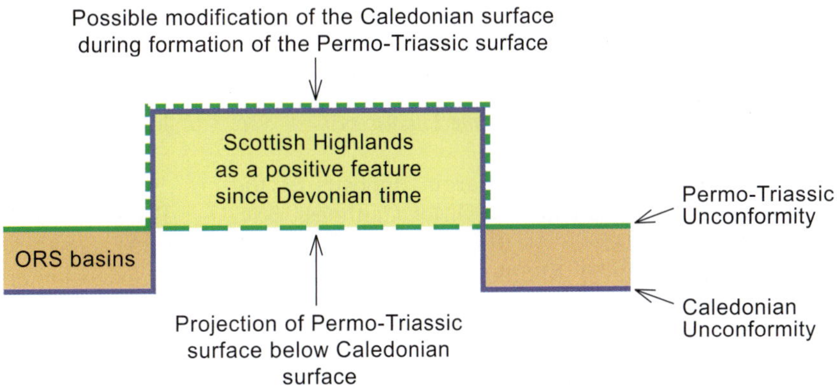

Fig. 13 Conceptual sketch illustrating the relationship between the Caledonian and Permo-Triassic surfaces across Highland Scotland.

ACKNOWLEDGEMENTS

The work for this paper was sponsored in part by NERC grant NER/T/S/2000/01367 to MacDonald and partly from the University of Aberdeen ASTARTE consortium (Anadarko, Enterprise Oil, and Sakhalin Energy Investment Company). We are indebted to Henry Allen of PGL for his interest and support and to Paul Bishop, Alan Crane, Adrian Hall, Adrian Hartley, Clive Rice and Nigel Trewin for useful discussion. We also gratefully acknowledge thorough reviews by Doug Boyd and Neil Meadows. The senior author is indebted to Peter Friend for inspiring a lifelong love of the relationship between geology and landscape.

REFERENCES

Besly, B.M. (1998) Carboniferous. In: *Petroleum Geology of the North Sea: Basic Concepts and Recent Advances* (Ed. K.W. Glennie), pp. 104–136. Blackwell Science, Oxford.

Blackbourn, G.A. (1981) Correlation of Old Red Sandstone (Devonian) outliers in the Northern Highlands of Scotland. *Geol. Mag.*, **118**, 409–414.

Bluck, B.J. (1978) Sedimentation in a late orogenic basin: the Old Red Sandstone of the Midland Valley of Scotland. In: *Crustal Evolution in Northwestern Britain and Adjacent Regions* (Eds D.R. Bowes and B.E. Leake). *Geol. J. Spec. Issue*, **10**, 249–278.

Bluck, B.J. (1990) Terrane provenance and amalgamation: examples from the Caledonides. *Phil. Trans Roy. Soc. London*, **A331**, 599–609.

Bremner, A. (1943) The origin of the Scottish river system, Parts I-III. *Scot. Geogr. Mag.*, **58–59**, 15–20, 54–59, 99–103.

Collinson, J.D., Bevins, R.E. and Clemmensen, L.B. (1989) Post-glacial mass-flow and associated deposits preserved in palaeovalleys: the Late Precambrian Morænesø Formation, North Greenland. *Medd. Grønl. Geosci.*, **21**, 3–26.

Cope, J.C.W., Ingham, J.K. and Rawson, P.F. (eds) (1992) *Atlas of Palaeogeography and Lithofacies. Geol. Soc. London Mem.*, **13**, 153 pp.

Donovan, R.N. (1973) Basin margin deposits of the Middle Old Red Sandstone at Dirlot, Caithness. *Scot. J. Geol.*, **9**, 203–211.

Donovan, R.N. (1975) Middle Devonian limestones developed at the margin of the Orcadian Basin, Caithness. *J. Geol. Soc. London*, **131**, 489–510.

Elmore, R.D., Blumstein, R., Engel, M. and Parnell, J. (2003) Palaeomagnetic dating of fluid flow events along the Moine Thrust Fault, Scotland. *J. Geochem. Expl.*, **78–79**, 45–49.

Evans, D.J. (1997) Estimates of the eroded overburden and the Permian-Quaternary subsidence history of the west of Orkney. *Scot. J. Geol.*, **33**, 169–182.

Friend, P.F., Harland, W.B. and Hudson, J.D. (1963) The Old Red Sandstone and the highland boundary in Arran, Scotland. *Trans. Edinb. Geol. Soc.*, **19**, 363–425.

Friend, P.F., Harland, W.B. and Smith, A.G. (1970) Reddening and fissuring associated with the Caledonian unconformity in north-west Arran. *Proc. Geol. Assoc.*, **81**, 75–85.

George, T.N. (1955) British Tertiary landscape evolution. *Sci. Prog.*, **43**, 291–307.

Glennie, K.W. (2002) Permian and Triassic. In: *The Geology of Scotland*, 4th edn (Ed. N.H. Trewin), pp. 301–321. Geological Society, London.

Goodman, S., Leslie, A.G., Ashcroft, W.A. and Crane, A. (1990) *The Geology of the Central Part of Sheet 65E (Ballater); Contribution to the Memoir.* Unpublished Technical Report WA/90/59, British Geological Survey, Edinburgh.

Hall, A.M. (1991) Pre-Quaternary landscape evolution in the Scottish Highlands. *Trans. Roy. Soc. Edinb. Earth Sci.*, **82**, 1–26.

Hall, A.M. and Bishop, P. (2002) Scotland's denudational history: an integrated view of erosion and sedimentation at an uplifted passive margin. In: *Exhumation of the North Atlantic Margin: Timing, Mechanisms and Implications for Petroleum Exploration* (Eds A.G. Doré, J.A. Cartwright, M.S. Stoker, J.P. Turner and N. White), pp. 271–290. Special Publication 196, Geological Society Publishing House, Bath.

Hutton, J., 1788. Theory of the Earth; or an investigation of the laws observable in the composition, dissolution, and restoration of land upon the globe. *Trans. Roy. Soc. Edinb.*, 1.

Huuse, M. (2002) Cenozoic uplift and denudation of southern Norway: insights from the North sea Basin. In: *Exhumation of the North Atlantic Margin: Timing, Mechanisms and Implications for Petroleum Exploration* (Eds A.G. Doré, J.A. Cartwright, M.S. Stoker, J.P. Turner and N. White), pp. 209–233. Special Publication 196, Geological Society Publishing House, Bath.

Johnstone, G.S. and Mykura, W. (1989) *The Northern Highlands of Scotland*, 4th edn. British Regional Geology, British Geological Survey, HMSO, 219 pp.

Parnell, J., Baron, M., Davidson, M., Elmore, D. and Engel, M. (2000) Dolomitic breccia veins as evidence for extension and fluid flow in the Dalradian of Argyll. *Geol. Mag.*, **137**, 447–462.

Rice, C.M. and Ashcroft, W.A. (2004) The geology of the northern half of the Rhynie Basin, Aberdeenshire, Scotland. *Trans. Roy. Soc. Edinb. Earth Sci.*, **94**, 299–308.

Stephenson, D. and Gould, D. (1995) *The Grampian Highlands*, 4th edn. British Regional Geology, British Geological Survey, HMSO, 261 pp.

Strachan, R.A., Smith, M., Harris, A.L. and Fettes, D.J. (2002) The Northern Highlands and Grampian terranes. In: *The Geology of Scotland*, 4th edn (Ed. N.H. Trewin), pp. 81–147. Geological Society, London.

Thomson, K., Underhill, J.R., Green, P.F., Bray, R.J. and Gibson, H.J. (1999) Evidence from apatite fission track analysis for the post-Devonian burial and exhumation history of the northern Highlands, Scotland. *Mar. Petrol. Geol.*, **16**, 27–39.

Trewin, N.H. (ed.) (2002) *The Geology of Scotland*. The Geological Society, London.

Trewin, N.H. and Thirlwall, M.F. (2002) The Old Red Sandstone. In: The Northern Highlands and Grampian terranes. In: *The Geology of Scotland*, 4th edn (Ed. N.H. Trewin), pp. 213–249. Geological Society, London.

Watson, J.V. (1984) The ending of the Caledonian Orogeny in Scotland. *J. Geol. Soc. London*, **141**, 193–214.

Watson, J.V. (1985) Northern Scotland as an Atlantic-North Sea divide. *J. Geol. Soc. London*, **142**, 221–243.

Wills, L.J. (1951) *A Palaeogeographic Atlas of the British Isles and Adjacent Parts of Europe*. Blackie and Sons, London.

Ziegler, P.A. (1988) Laurussia – The Old Red Continent. In: *Devonian of the World* (Eds N.J. McMillan, A.F. Embry and D.J. Glass). *Can. Soc. Petrol. Geol. Mem.*, **14**, 15–48.

$^{40}Ar/^{39}Ar$ dating of detrital white mica as a complementary tool for provenance analysis: a case study from the Cenozoic Qaidam Basin (China)

ANDREA B. RIESER*[1], FRANZ NEUBAUER*, YONGJIANG LIU†, JOHANN GENSER*, ROBERT HANDLER*, XIAO-HONG GE† *and* GERTRUDE FRIEDL*

**Department of Geography and Geology, Division General Geology and Geodynamics, University of Salzburg, Hellbrunnerstr. 34, 5020 Salzburg, Austria (Email: andrea.rieser@nagra.ch)*
†College of Earth Sciences, Jilin University, Jianshe Str. 2199, 130061 Changchun, China

ABSTRACT

When classic petrographic analysis of the modal composition of sandstones yields no distinction between different source regions, $^{40}Ar/^{39}Ar$ dating of detrital white mica can provide vital information on the age of a source area and thus link the sediments to a specific provenance in the hinterland. This approach is exemplified by a case study of the intramontane Qaidam Basin (western China). While the geology of the surrounding mountains of the Qaidam Basin shows considerable lithological variation and the basin's palaeoclimate changed from semi-arid to arid, modal analysis of sandstones from two sections in the northwestern basin, as well as a section on the eastern margin, yielded no significant spatial or temporal differences. All sandstones, most of them classified as lithic wackes with matrix/cement contents between 14 and 39%, plot mainly in the recycled orogenic field of Dickinson's ternary discrimination diagrams for a tectonic environment. The sandstones are quartz dominated, with quartz contents of 33–65% and relative high contents of feldspar and lithic grains. On the other hand, $^{40}Ar/^{39}Ar$ total-fusion age data obtained from detrital white mica of between 123 and 546 Ma yielded three age clusters (120–180, 220–280, 350–450 Ma) that could be assigned to certain provenance areas within the early Palaeozoic and Permian basement in the Altyn and Qimantagh mountains. This contrasts with the Lulehe section in the east of the basin, where exclusively Permian ages between 250 and 279 Ma were found. This significant difference in age distribution, and thus provenance, could not be deduced from sandstone composition. The results of this study show how $^{40}Ar/^{39}Ar$ thermochronology can complement classic point-count analysis.

Keywords Tibetan plateau, Qaidam Basin, provenance, $^{40}Ar/^{39}Ar$ age dating, Inner Asian orogens, sandstone composition.

INTRODUCTION

Classic provenance analysis on sandstones does not always yield clear indications of clastic sources. In such cases, other methods or a combination of methods can provide better constraints on sediment provenance. Clastic sediments reveal information on continental and oceanic source regions that have been eroded or metamorphosed as a result of subsequent tectonic events. The provenance and geodynamic development of sandstone-rich basin-fill successions can be determined by a variety of methods including: petrographic analysis; whole rock and mineral chemistry; and radiometric dating. Sandstone composition mostly depends on the nature of the source area. However, climate,

[1]Present address: Nagra, Hardstrasse 73, 5430 Wettingen, Switzerland.

weathering, erosion, relief, length of transport, sedimentation, burial and diagenesis are further factors that can alter the original source signature (Pettijohn *et al.*, 1987; Johnsson, 1993; Fralick & Kronberg, 1997). Proportions of detrital framework minerals can be used to classify siliciclastic rocks on a volumetric basis and the provenance can be interpreted in terms of tectonic setting (e.g. Crook, 1974; Schwab, 1975; Dickinson & Suczek, 1979; Dickinson *et al.*, 1983; Dickinson, 1985).

$^{40}Ar/^{39}Ar$ thermochronology and other low-temperature thermochronometers (e.g. fission-track analysis on zircon, Rb–Sr on biotite) are helpful tools in evaluating erosion rates over geological time and allowing exhumation rates to be inferred (e.g. Willett & Brandon, 2002; Brewer *et al.*, 2003). Furthermore, $^{40}Ar/^{39}Ar$ dating of detrital minerals such as feldspar, mica and amphibole allows large-scale palaeogeographical relationships and the tectonothermal evolution of orogens and their denudation to be monitored (e.g. Copeland & Harrison, 1990; Najman *et al.*, 1997). Different closure temperatures (500°C for amphibole, 350°C for white mica, 300°C for biotite, 200°C for K-feldspar) and the slow cooling rates typical for regional-metamorphic areas make the various minerals suitable for different applications, depending on the geological setting. While feldspars are well suited for modelling cooling histories, amphiboles and mica usually yield better plateau ages, provided they are not overprinted and reset. For this study, white mica was selected because of its abundance in the basin. Although mica usually has low chemical and mechanical resistance, it is well preserved in the Qaidam Basin due to the short transport distances as suspended load. The $^{40}Ar/^{39}Ar$ method has the advantage that it can be applied to both sedimentary rocks and the surrounding basement, allowing straightforward comparison of the results. Furthermore, single-grain ages can be verified by using the $^{40}Ar/^{39}Ar$ step-wise heating method. Both methods, modal analysis and $^{40}Ar/^{39}Ar$ geochronology, require certain prerequisites: suitable material for analysis, i.e. fresh medium-grained sandstones and mica-bearing sandstones must be available; and hinterland geology should be well-known, in terms of discernible mineralogy and/or different formation ages. This is the case for the basement rocks surrounding the Qaidam Basin, for which a number of $^{40}Ar/^{39}Ar$ white mica ages have been published.

In this paper, a case study from the Qaidam Basin is presented, in which it is shown that a $^{40}Ar/^{39}Ar$ geochronological analysis of detrital white mica from Cenozoic formations has proved a more satisfactory method in revealing different provenance areas, after the classic point-count method failed to identify the sediment sources.

GEOLOGICAL SETTING

The present-day active Qaidam Basin of western China is situated at the northeasternmost margin of the Tibetan plateau (Fig. 1A & B), at altitudes between 2800 and 3500 m. It is considered to be part of the convergent systems at the northern margin of the Tibetan plateau (e.g. Meyer *et al.*, 1998), with deformation still going on today. The Qaidam block was amalgamated with the North China block in Late Devonian time. However, it was only after the Triassic Indo-Sinian orogenesis that a proper basin history began. The Qaidam block was overridden from the north and the south by the Qilian and Kunlun orogenic belts, respectively (Xia *et al.*, 2001), forming a flexural basin (Meyer *et al.*, 1998). Situated between the sinistral Altyn Tagh and Central Kunlun faults, the rhomb-shaped Qaidam Basin was considered to have formed in response to oblique compression (Métivier *et al.*, 1998; Meyer *et al.*, 1998), which is supported by recent compressive and left-lateral transpressive earthquake focal mechanisms in the region (Meyer *et al.*, 1998; Bedrosian *et al.*, 2001). Distributed shortening by crustal thickening has been the dominant deformation mechanism in the basin area since mid-Miocene time (Yin & Harrison, 2000; Tapponnier *et al.*, 2001b; Yue *et al.*, 2003; Ritts *et al.*, 2004). Part of the thickening appears to have been due to thrusting of the Kunlun (Songpan–Ganzi) terrane northwards upon the Qaidam block.

The Qaidam Basin, with a surface area of 120,000 km^2, has an unusually thick Mesozoic to Cenozoic sedimentary sequence of more than 10 km. Cenozoic sediments are exclusively terrestrial, and since at least the Early Oligocene (Huo, 1990) a 6–8 km thick fluvial-lacustrine succession, comprising mainly sandstones and siltstones, has

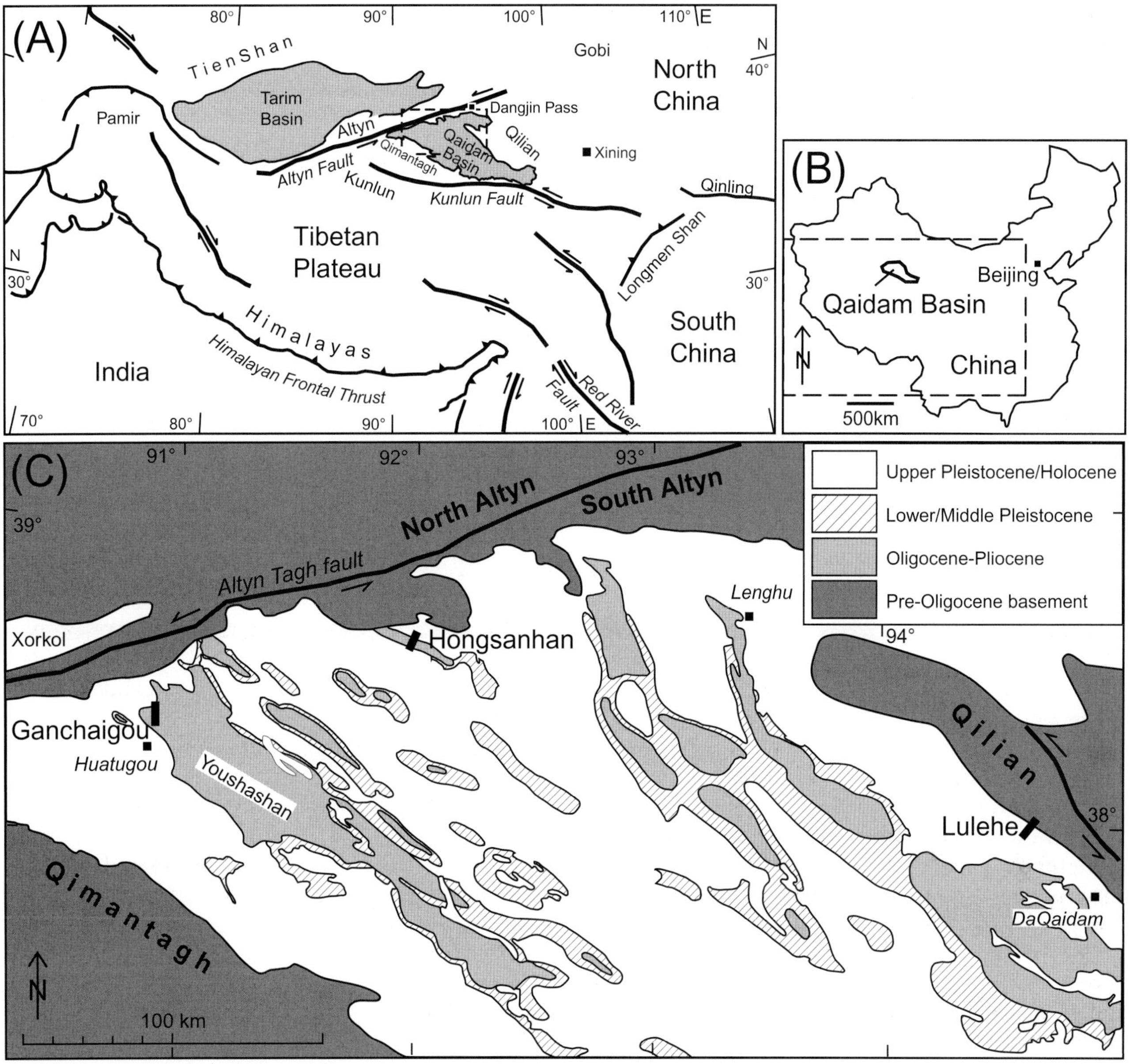

Fig. 1 (A) Simplified tectonic sketch map of the Himalaya–Tibetan region showing the Altyn Tagh and Kunlun faults bounding the Qaidam Basin to the north and south, respectively (redrawn and modified from Tapponnier *et al.*, 2001b). (B) Position of the Qaidam Basin within China. (C) Simplified geological map of the northwestern Qaidam Basin showing the outlines of Pliocene fold structures. Bars indicate the location of the Ganchaigou, Hongsanhan and Lulehe sections.

accumulated due to internal drainage (e.g. Liu *et al.*, 1998; Shi *et al.*, 2001).

Despite lithostratigraphic correlations, magnetostratigraphy, seismic stratigraphy and microfossils from lake sediments (Sun *et al.*, 1999, 2005; Xia *et al.*, 2001), it has remained difficult to develop a comprehensive chronostratigraphy for the Qaidam Basin. Nevertheless, distinct seismic reflectors (T-layers), which can be traced from marginal wells right across the basin centre, have allowed a detailed correlation of all seven Cenozoic formations (Table 1) to be made (Huang *et al.*, 1997; Xia *et al.*, 2001). The lake sediments can be divided into near-shore and deep-water

Table 1 Stratigraphy of the Qaidam Basin fill. Ages are based on Gradstein *et al.* (2004). Numbers 1–7 refer to the various formations as used in some of the figures

Age (Ma)	Epoch	Formation		Environment
	Holocene	Qigequan	7	Alluvial–fluvial
	Pleistocene			
1.81	Pliocene	Shizigou	6	Fluvial–lacustrine
5.33	Miocene (Messinian)			
7.25	Miocene (Tortonian–Langhian)	Shangyoushashan	5	Lacustrine–fluvial
15.97	Lower Miocene	Xiayoushashan	4	Fluvial–lacustrine
23.03	Oligocene (Chattian)			
	Oligocene (Chattian–Rupelian)	Shangganchaigou	3	Fluvial–lacustrine
33.9	Eocene (Priabonian)			
37.2	Eocene (Bartonian–Ypresian)	Xiaganchaigou	2	Fluvial
	Eocene (Ypresian)	Lulehe	1	Fluvial

palaeoenvironments (< 50 m), depending on facies (cf. Figs 2 & 3). In the western and central Qaidam Basin these sediments are exposed in Pliocene and Pleistocene fold structures (e.g. Song & Wang, 1993; Meyer *et al.*, 1998). Alluvial fan sediments (conglomerates and breccias) along the margins, reaching 20–30 km into the basin, directly link to source regions in the adjacent hinterland.

The present climate in the Qaidam region is arid. However, from Oligocene to Quaternary time a palaeolake existed, which migrated from the western part of the basin to the east (Liu *et al.*, 1998). The lake reached its maximum extent during a semi-arid interval in the Miocene (Wang *et al.*, 1999). It shrank dramatically in Pliocene and Pleistocene times, when folding started and the driest climatic conditions occurred (Wang *et al.*, 1999), and subsequently abundant evaporites were formed. This is similar to today's climate, when evaporites form in small saline lakes in the southeastern basin. During the past two million years, annual precipitation was in the same order as today, when the 25 mm yr^{-1} (< 50 mm yr^{-1}; Lehmkuhl & Haselein, 2000) precipitation greatly exceeded the annual potential evaporation of about 3000 mm yr^{-1} (e.g. Wang *et al.*, 1999; Duan & Hu, 2001).

The Altyn Mountains in the north, the Kunlun Mountains/Qimantagh to the southwest, and the Qilian Mountains in the east have confined the Qaidam Basin completely since the Oligocene. The Altyn Mountains include the still active Altyn Tagh Fault, one of the longest (1600 km, with 350–400 km sinistral Cenozoic offset) continental strike-slip faults in Asia (Tapponnier and Molnar, 1977; Wittlinger *et al.*, 1998; Bendick *et al.*, 2000; Tapponnier *et al.*, 2001a; Yue *et al.*, 2003). Lithological units known from the east-southeast-striking Qilian Mountains can nowadays be found in the Xorkol region on the northern side of the Altyn Fault (Fig. 1C), a ~ 200 km westward offset. This implies that the present-day hinterland is different from that of the Paleogene (Yue *et al.*, 2004, and references therein).

The following units are exposed from north to south in the North Altyn Mountains (Sobel & Arnaud, 1999): (i) Early Proterozoic medium-grade metamorphic basement rocks, including paragneiss and orthogneiss; (ii) Early Palaeozoic ophiolites with metamorphic ages ranging from 500 to 440 Ma. To the south of the Altyn Tagh Fault, medium-grade schists with an Early Palaeozoic metamorphic age (440–360 Ma) are exposed (Sobel & Arnaud, 1999;

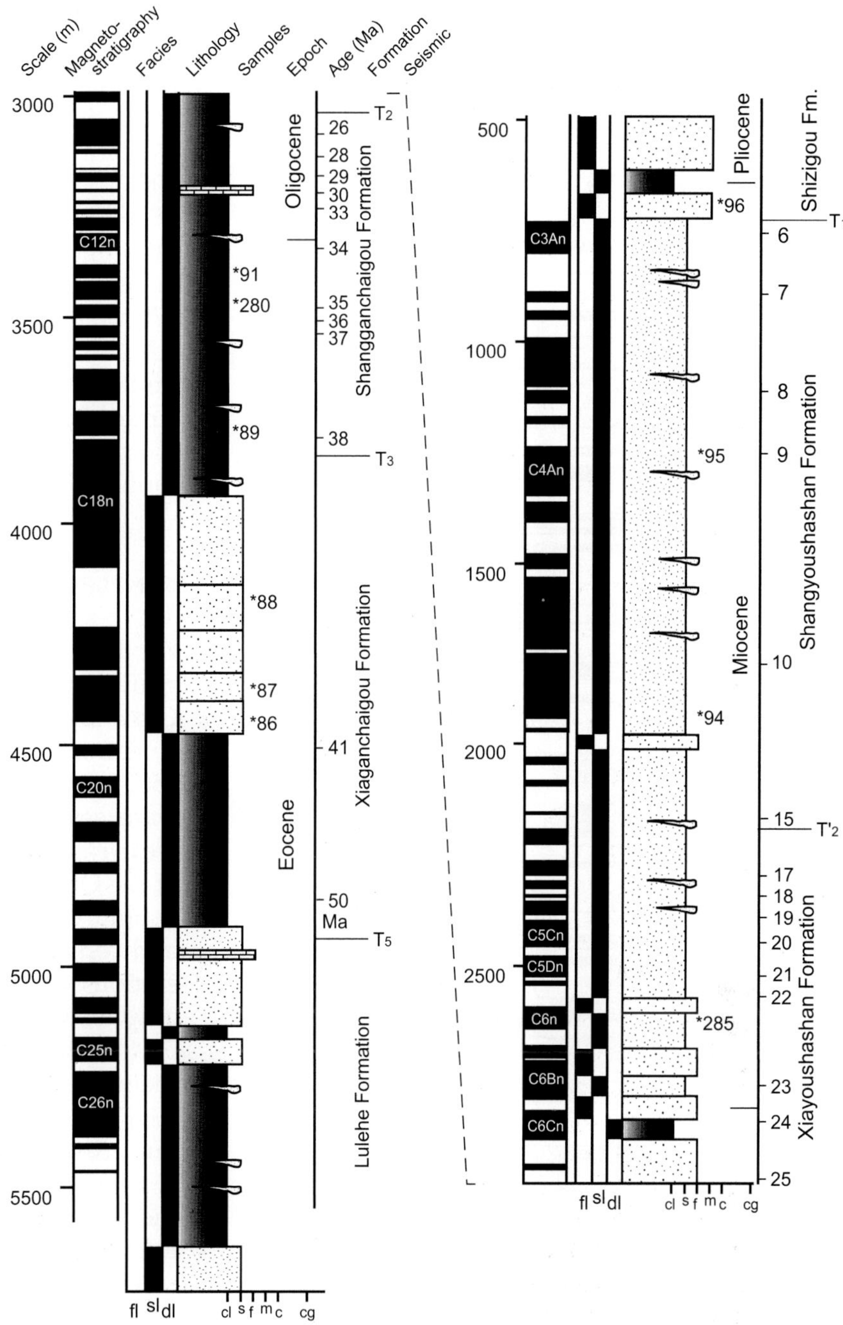

Fig. 2 Simplified stratigraphical column of the Ganchaigou section shown together with the magnetostratigraphy (both based on Yang *et al.*, 1992). Stars indicate sample locations and T_1 to T_5 seismic reflectors defining formation boundaries. Abbreviations: fl, fluvial facies; sl, shallow lake facies; dl, deep lake facies; cl, clay; s, silt; f, fine sand; m, medium-grained sand; c, coarse sand; cg, gravel.

Delville *et al.*, 2001; Sobel *et al.*, 2001; Liu *et al.*, 2003), which are intruded by Late Palaeozoic and Permian granites (Gehrels *et al.*, 2003b, and references therein). The metamorphic rocks of the South Altyn Mountains are locally overprinted by: (i) low-grade shear zones of Jurassic age that formed fine-grained white mica with ages ranging from 180 to 160 Ma; and (ii) by an event at 30–25 Ma (Liu *et al.*, 2003), observed near the Dangjin Pass (Fig. 1A).

The geological map of the Qaidam Basin (Wang & Zhang, 1999) shows that the Qimantagh is composed of low-grade phyllites, Ordovician to Carboniferous flysch and limestone successions. These units are intruded by granitoids of uncertain age. The Qilian Mountains represent a mid-Palaeozoic suture zone with exposed metamorphic and plutonic basement rocks between the North China block and the basement of the

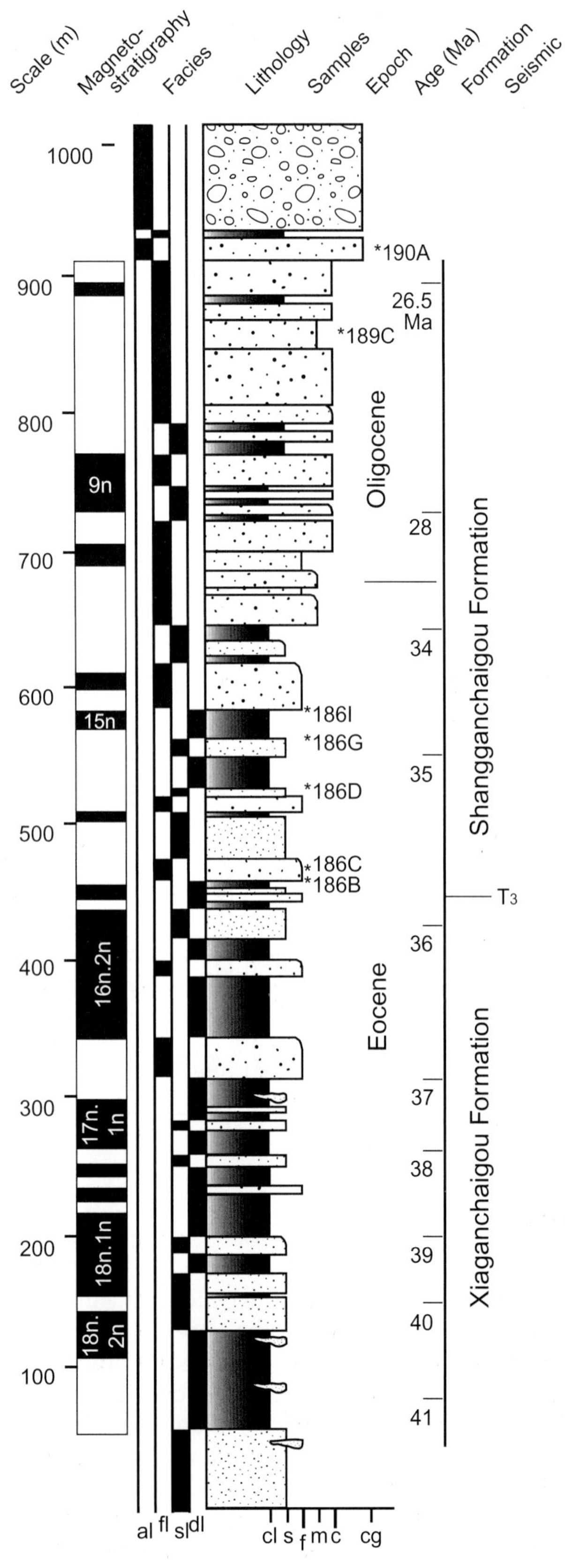

Qilian Mountains and Qaidam (Zhang *et al.*, 1984). Granitic bodies of both Silurian to earliest Devonian age and Permian to earliest Triassic age occur in both the South Qilian and northern Qaidam regions (Gehrels *et al.*, 2003b; Yue *et al.*, 2003). The southern zone includes a wide zone of the South Qilian metamorphic belt rocks, and ultrahigh-pressure eclogites and associated gneisses of the Qaidam belt with $^{40}Ar/^{39}Ar$ ages of 470 Ma (Yang *et al.*, 2001a,b; Song *et al.*, 2003). The eclogite belt is separated from the main Qilian units by a 350 km long and 2 km wide sinistral strike-slip shear belt (Xu *et al.*, 2002). The metamorphic basement is intruded by largely undeformed Palaeozoic and Jurassic granites.

SAMPLING

For this study, 27 sandstone samples with a suitable grain-size for modal analysis by the Dickinson–Gazzi method (Dickinson, 1985), and 18 sandstones containing abundant detrital white mica (125–350 μm) for $^{40}Ar/^{39}Ar$ total-fusion geochronology, were selected from three sections at the north-western, north-central and northeastern margins of the Qaidam Basin (Fig. 1C). Along the northern margin, cross-cut anticlines offer good access to long and continuous sections. Additionally, two of the sections chosen are also dated by magnetostratigraphy, which provides better age control. For the 5000 m thick Ganchaigou section (Fig. 2) an older magnetostratigraphy is available (Yang *et al.*, 1992), while the 1100 m thick Hongsanhan section (Fig. 3) has been measured more recently (Sun *et al.*, 2005). Due to different interpretations of the magnetic reversal pattern by the respective authors, the formation boundaries in these two sections are not situated at the same chrons. Both

Fig. 3 (*left*) Simplified stratigraphical column (based on Ma, personal communication) of the Hongsanhan area shown together with the magnetostratigraphy (Sun *et al.*, 2005). Stars indicate sample locations and T_3 the seismic reflector defining the boundary between the Xiaganchaigou and Shangganchaigou Formations. Abbreviations: al, alluvial fan facies; fl, fluvial facies; sl, shallow lake facies; dl, deep lake facies; cl, clay; s, silt; f, fine sand; m, medium-grained sand; c, coarse sand; cg, gravel.

sections show an overall coarsening-upward trend, reflecting progradation of coarse alluvial fan sediments over fine-grained deep lake facies, as a result of lake withdrawal. This is in agreement with facies maps from the northwestern basin (Internal Report, 2005), which are based on well log data. Coarsening in the Hongsanhan section started earlier because of the more marginal setting and larger effect by uplift in the Altyn Mountains. Well-known hinterland geology and availability of age information of the Altyn Mountains in the north facilitate discussion of age data.

The Ganchaigou section, near the northwestern margin of the basin, is a N–S striking valley incised into the NW–SE trending Ganchaigou anticline, offering access to a continuous section from Eocene (Xiaganchaigou Formation) to Pliocene (Shizigou Formation) strata (Fig. 2). From its geographical setting (i.e. in the corner of two merging mountain ranges), likely source areas for the Ganchaigou area are both the Altyn Mountains and the Qimantagh. The Hongsanhan anticline is located at the northern basin margin and perpendicularly cross-cut by three valleys. The middle valley, the Hongsanhan Third High Peak Valley, exposes Eocene (Xiaganchaigou Formation) to Oligocene (Shangganchaigou Formation) strata (Fig. 3). Samples for age-dating are from the Hongsanhan First High Peak Valley, which is located a few kilometres to the west and extends into the Lulehe Formation. Based on the close proximity, samples from the Hongsanhan area are likely to have had a source within the Altyn Mountains. The third section, the Lulehe section, lies at the eastern margin of the basin about 30 km north of DaQaidam (Fig. 1C). Along the southwestern margin of the Qilian Mountains, the Lulehe section offers a well-exposed sequence of Eocene (Lulehe Formation) to Pliocene (Shizigou Formation) strata, characterized by fine- to medium-grained reddish and greenish sandstones.

METHODS

Modal framework analysis

By counting 300 to 500 framework mineral grains in a sandstone thin-section and determining the mineralogy, a range of tectonic settings for source areas can be distinguished (e.g. Dickinson & Suczek, 1979; Dickinson, 1985). Modal analysis of framework grains of the size range 0.063–2 mm involved the counting of the following types: monocrystalline quartz (Q_m), polycrystalline quartz (Q_p), plagioclase (P) and K-feldspar (K1), constituting feldspar (F) and including microcline (M), lithic sedimentary and metasedimentary clasts (L_s) and lithic volcanic clasts (L_v) (following Dickinson & Suczek, 1979; Dickinson, 1985). The L_s and L_v clasts constitute together the lithic clastics (L) and, together with Q_p, the total lithic clastics (L_t); Q_m and Q_p together make total quartz (Q_t). Furthermore, detrital white mica (Ms), biotite (Bt) and carbonates (C), including monocrystalline, polycrystalline and biogenic carbonate, have been distinguished. Ooids, opaques, heavy minerals, chlorite and amphiboles were combined in a separate category (others, O).

The distorting effects of grain size on provenance identification in quartzo-feldspathic rocks have been treated carefully by following methods outlined by Ingersoll *et al.* (1984). In addition to the framework constituents described above, cement and matrix were counted. As matrix/cement make up about 20–30% of all the counts, 500 points per thin-section have been counted in this study. Within the three distinct profiles for each formation with multiple samples, an average value was calculated. The averaged data points in Fig. 4 contain data from two to six individual thin-sections per sampled section. Although such low numbers are not fully representative, when compared with the larger data-set in Rieser *et al.* (2005) and Table 2 they display a similar compositional pattern.

$^{40}Ar/^{39}Ar$ mineral dating

Using the $^{40}Ar/^{39}Ar$ method, whole-rock or mineral samples can be dated by determining the ratio of radiogenic ^{40}Ar to neutron-induced ^{39}Ar. Given the constant atmospheric ratio of $^{40}Ar/^{36}Ar$ (= 295.5) and reference sample minerals of known ages, the target minerals can be dated after additional corrections for isotopic interference effects.

Sandstone samples for mineral dating were mechanically crushed and sieved. In this study the fraction with a grain-size of 125–350 μm was used. The whole fraction was cleaned in an ultrasonic bath with deionized water and alcohol.

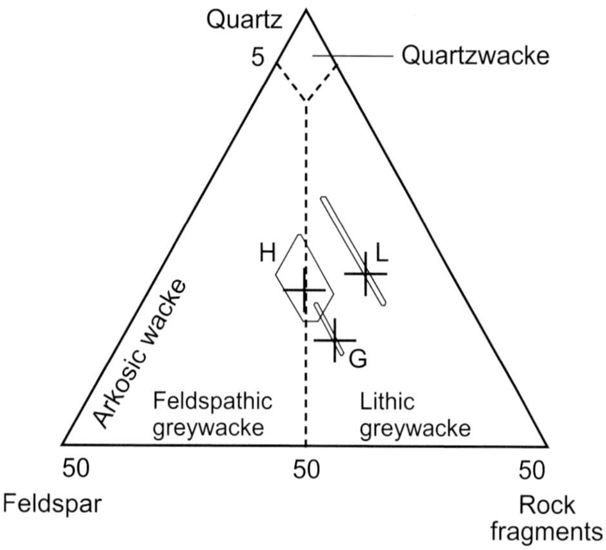

Fig. 4 Wacke classification after Pettijohn *et al.* (1987). Mean values for each section are given with their respective error polygons: G, Ganchaigou; H, Hongsanhan; L, Lulehe. Note that only the upper part of the triangle is shown. Numbers are mineral percentages.

Optically clean mica grains without inclusions and alteration rims were handpicked under a binocular microscope and, as concentrates, wrapped in aluminium foil. These capsules were packed into sealed quartz vials and irradiated in the MTA KFKI reactor in Debrecen, Hungary for 16 h. In between the samples, DRA1 sanidine with a known $^{40}Ar/^{39}Ar$ plateau age of 25.03 ± 0.05 Ma (Wijbrans *et al.*, 1995) was packed to monitor variations in the neutron-flux along the length of the irradiation assembly.

For measurement, each irradiated grain was packed into a one-way aluminium sample-holder, which was then placed into an UHV Ar-extraction line equipped with a combined MERCHANTEK^{TM} UV/IR laser-ablation facility and a VG-ISOTECH^{TM} NG3600 mass spectrometer at the ARGONAUT laboratory at Salzburg University, Austria. Using a defocused (~ 1.5 mm diameter) 2 W CO_2-IR laser operating in Tem_{00} mode at wavelengths between 10.57 and 10.63 μm, detrital grains were heated until fusion. The laser was controlled by a personal computer, while the laser position on the sample was monitored through a double-vacuum window on the sample chamber via a video camera in the optical axis of the laser beam. Gas was cleaned on both a hot and a cold Zr–Al SAES getter. Gas admittance and pumping of the mass spectrometer and the Ar-extraction line were also computer-controlled using pneumatic valves. Measurement was performed on an axial electron multiplier in static mode. A Hall-probe controlled peak-jumping and stability of the magnet.

For each increment, the intensities of all Ar isotopes (^{36}Ar, ^{37}Ar, ^{38}Ar, ^{39}Ar and ^{40}Ar) were measured, from which the baseline readings on mass 35.5 were automatically subtracted. Peak intensities were back-extrapolated over 16 measured intensities to the time of gas admittance, either by a linear or curved fit. Intensities were automatically corrected for system blanks, background, post-irradiation decay of ^{37}Ar and interfering isotopes. Correction factors for interfering isotopes have been calculated from 10 analyses of two Ca-glass samples and 22 analyses of two pure K-glass samples ($^{36}Ar/^{37}Ar_{(Ca)} = 0.00026025$, $^{39}Ar/^{37}Ar_{(Ca)} = 0.00065014$ and $^{40}Ar/^{39}Ar_{(K)} = 0.015466$). The calculation of ages was carried out using ISOPLOT/EX (Ludwig, 2001).

RESULTS

The raw data on sandstone composition from the three sections are given in Table 2 (abstracted from Rieser *et al.*, 2005). The sandstones of the northwestern Qaidam Basin are relatively immature. As they contain a high proportion of micritic matrix or fine cement (14–39%) they are classified as lithic wackes (Fig. 4; Pettijohn *et al.*, 1987). Cements usually consist of calcite and in a few cases anhydrite. The sandstones comprise a high proportion of unstable lithic fragments and feldspar, but are quartz-rich (33–65%) and plot in the field of recycled orogenic provenance with a continental block provenance (Fig. 5). Mica concentrations range between 0 and 5%, and up to 11% in the Lulehe section. From a consideration of the main framework constituents, hinterland petrology seems to show no major changes between the sections analysed (Fig. 5).

$^{40}Ar/^{39}Ar$ isotopic and age data are given in Table 3. $^{40}Ar/^{39}Ar$ total-fusion ages of single white mica grains yielded more discernible results (Fig. 6). Each point in the plot represents the age of a single grain, with the age given on the *y* axis and single grains plotted side by side with increasing

Table 2 Sandstone composition data (%) from three sections in the Qaidam Basin. Samples are in stratigraphic order

Section	Sample	Q_m	Q_p	P	K1	M	L_v	L_s	Bt	Ms	C	O	MC	Total	Q_t	K	F	L	L_t	Formation
Ganchaigou	QA-287A	31	2	4	11	3	1	10	0	0	7	3	28	100	33	14	14	11	13	Qigequan
	QA-96	33	4	6	6	3	1	12	0	2	2	2	31	100	37	8	8	13	17	Shizigou
	QA-94	34	2	5	5	1	0	19	0	1	0	1	30	100	37	7	7	19	21	Shangyoushashan
	QA-285A	48	0	5	7	2	0	3	1	1	0	3	31	100	48	9	9	3	3	Xiayoushashan
	QA-91	44	2	7	10	1	0	17	0	1	2	2	16	100	46	11	11	17	19	Shangganchaigou
	QA-89	45	4	7	7	1	0	15	0	1	2	3	14	100	49	8	8	15	19	Shangganchaigou
	QA-280A	41	0	8	5	2	0	2	0	1	1	1	39	100	41	7	7	2	2	Shangganchaigou
	QA-88	45	1	5	6	3	0	11	0	1	4	6*	18	100	46	9	9	11	13	Xiaganchaigou
	QA-87	34	2	8	9	3	0	20	0	1	3	1	21	100	36	11	11	20	22	Xiaganchaigou
	QA-86	53	1	9	9	2	0	10	0	1	0	3	12	100	54	11	11	10	11	Xiaganchaigou
Hongsanhan	QA-190A	52	0	5	6	3	0	4	3	2	5	2	18	100	53	9	9	4	5	Xiayoushashan
	QA-189C	50	1	3	8	2	0	5	0	1	2	2	26	100	51	11	11	5	6	Shangganchaigou
	QA-186I	47	2	5	7	3	0	9	0	1	1	1	24	100	49	10	10	9	11	Shangganchaigou
	QA-186G	37	1	6	15	1	0	11	1	1	5	6	16	100	38	16	16	11	12	Shangganchaigou
	QA-186D	39	1	7	10	4	0	7	0	0	10	0	22	100	40	14	14	7	8	Shangganchaigou
	QA-186C	43	0	7	9	4	0	6	0	0	7	1	23	100	43	12	12	6	6	Shangganchaigou
	QA-186B	36	2	14	8	5	1	11	1	1	1	4	18	100	38	13	13	12	14	Shangganchaigou
	QA-130B	43	2	8	6	2	0	15	0	0	0	2	21	100	45	8	8	15	17	Xiaganchaigou
	QA-133B	45	1	15	6	4	0	10	1	1	0	2	14	100	47	10	10	10	12	Xiaganchaigou
	QA-133C	48	1	14	6	3	0	9	1	2	0	3	15	100	49	8	8	9	10	Xiaganchaigou
	QA-133A	37	2	5	6	3	0	10	0	0	0	2	34	100	40	9	9	10	13	Lulehe
	QA-132A	50	4	5	4	0	0	8	0	1	0	2	26	100	54	5	5	8	11	Lulehe
Lulehe	QA-239A	54	0	2	3	1	0	5	6	5	3	3	17	100	55	4	4	5	6	Shizigou
	QA-239B	43	2	2	2	1	0	25	2	1	10	3	8	100	45	4	4	25	27	Shizigou
	QA-238A	63	1	3	4	2	0	11	4	4	0	4	2	99	65	6	6	11	12	Xiayoushashan
	QA-238B	55	0	5	5	1	0	7	2	4	1	5	15	100	55	6	6	7	7	Xiayoushashan
	QA-232A	32	3	0	8	1	1	13	0	0	4	1	38	100	34	9	9	13	16	Lulehe

Q_m, monocrystalline quartz; Q_p, polycrystalline quartz; P, plagioclase; K1, K-feldspar; M, microcline; L_v, lithic volcanic clasts; L_s, lithic sedimentary clasts; Bt, biotite; Ms, muscovite; C, carbonates; MC, matrix/cement; O, others; $Q_t = Q_m + Q_p$; $K = K1 + M$; $F = P + K1 + M$; $L = L_v + L_s$; $L_t = Q_p + L_s + L_v$.

*5% of total framework constituents are ooids.

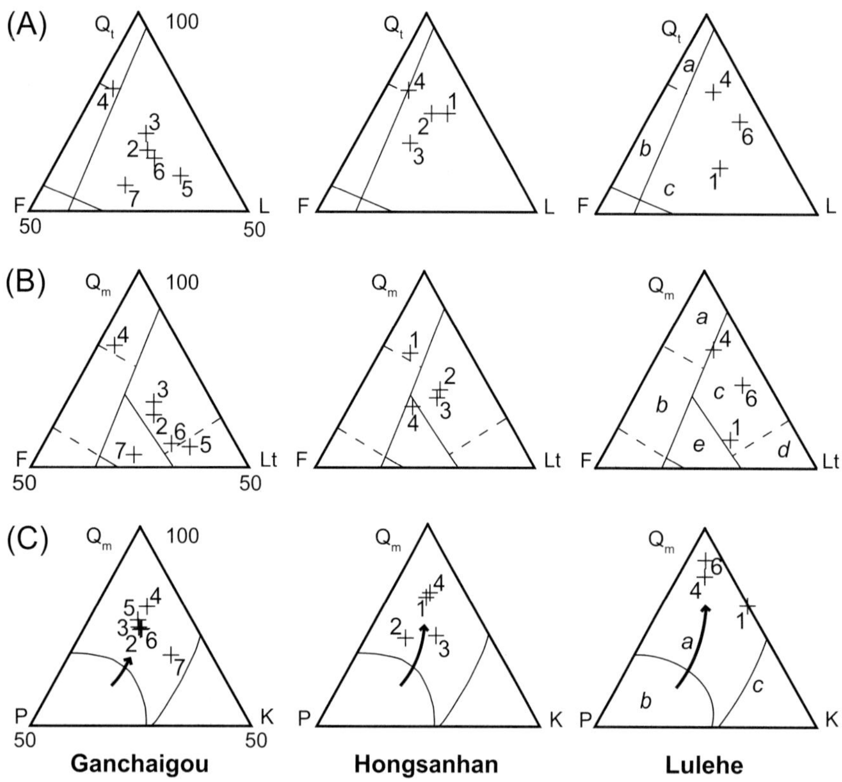

Fig. 5 Ternary discrimination diagrams (Dickinson, 1985) of sandstones from the Ganchaigou, Hongsanhan and Lulehe sections. Mean values of two to six samples per formation are shown with a number, indicating the formation: 7, Qigequan; 6, Shizigou; 5, Shangyoushashan; 4, Xiayoushashan; 3, Shangganchaigou; 2, Xiaganchaigou; 1, Lulehe. Note that for all triangles only the upper half is shown. For mineral abbreviations see Table 2 and text. (A) Framework-grain assemblage Q_t–F–L. Fields represent: a, craton interior; b, transitional continental; c, recycled orogenic. (B) Framework-grain assemblage Q_m–F–L_t. Fields represent: a, craton interior; b, transitional continental; c, quartzose recycled; d, transitional recycled; e, mixed. (C) Framework mineral grains Q_m–P–K. Arrow in a indicates increasing maturity/stability from continental block provenances towards the Q_m–pole; b, circum-Pacific volcanoplutonic suites; c, limit of detrital modes.

age on the *x* axis. Most measurements yielded 2σ uncertainties (Table 3) in the range of 1–2% of the apparent mineral age (not shown in the figure). Samples from the lower part of the Ganchaigou section yielded a wide age-distribution, with ages between 122.5 and 542.5 Ma. However, two groups (220–280 Ma and 350–450 Ma) dominate, while in the upper part of the section the age group of 350–450 Ma clearly prevails (highlighted in Fig. 6), representing 80% of the dated grains (Fig. 7). The Hongsanhan samples yielded ages similar to the Ganchaigou samples, with a wide range in ages from 210.3 ± 2.1 up to 515.0 ± 4.6 Ma. Again, the group of 350–450 Ma old grains is best represented. The Lulehe section yielded solely Permian ages from 250.2 to 279.4 Ma (Reiser *et al.*, 2006) as highlighted by the grey bars in Fig. 6. Figure 7 summarizes the results as formation mean percentages. Figure 7A shows the similarity for the sandstone compositions across all formations and sections. In Fig. 7B the change between the upper and lower Ganchaigou section and the completely different age distribution in the Lulehe section become clearly visible.

DISCUSSION

The main conclusion of point-counting analyses is that, although variable climatic conditions have been reconstructed for the Cenozoic (Wang *et al.*, 1999) and the Altyn Tagh Fault has offset northern units several hundreds of kilometres to the west, there are no significant differences in petrographical composition revealed through time. However, the small shift in provenance fields from a continental block source to a more recycled orogen source (Fig. 5) may have followed the major slip of the Altyn Tagh Fault in the Oligocene (Yue *et al.*, 2001) and the Oligocene–Neogene onset of exhumation in the South Altyn Mountains and the Qimantagh (Jolivet *et al.*, 2001). Based on the hinterland geology, it should be expected that the sandstones would plot into the recycled orogen field (Q_t–F–L), but with a closer affinity to the magmatic arc setting and less mature compositions (Q_m–P–K). Sandstones plotting within the recycled orogen field are typical of foreland basins where texturally and compositionally immature

Table 3 $^{40}Ar/^{39}Ar$ isotopic and age results from total-fusion analyses on single grains of detrital white mica from the Ganchaigou (samples G), Hongsanhan (samples H) and Lulehe (samples L) sections. Samples within any section are in stratigraphic order

Formation	Sample details	$^{36}Ar/^{39}Ar$ measured	$^{36}Ar/^{39}Ar$ 1-sigma absolute	$^{37}Ar/^{39}Ar$ corrected	$^{37}Ar/^{39}Ar$ 1-sigma absolute	$^{40}Ar/^{39}Ar$ measured	$^{40}Ar/^{39}Ar$ 1-sigma absolute	%^{40}Ar*	Age (Ma)	± (Myr) 1-sigma absolute
Qigequan	Sample G7	0.0068	0.0001	1.2423	0.0001	16.910	0.031	88.0	425.8	3.9
	(QA-97D-01)	0.0152	0.0003	0.6767	0.0002	16.292	0.077	72.4	345.3	3.8
	with $J = 0.0179$	0.0155	0.0003	2.8715	0.0003	18.143	0.103	74.7	391.5	4.4
	±0.00018	0.0187	0.0003	4.2280	0.0003	17.607	0.103	68.6	352.8	4.2
		0.0070	0.0001	0.4383	0.0001	13.884	0.027	85.0	345.4	3.2
		0.0023	0.0001	0.4268	0.0001	15.657	0.024	95.6	427.9	3.9
		0.0041	0.0001	0.1908	0.0001	14.746	0.023	91.7	390.8	3.6
		0.0008	0.0001	2.6837	0.0001	15.983	0.021	98.6	447.8	4.0
		0.0016	0.0001	2.7425	0.0000	12.598	0.018	96.4	354.3	3.3
		0.0045	0.0001	14.2511	0.0002	14.065	0.043	90.5	369.8	3.5
		0.0015	0.0001	5.9917	0.0001	12.797	0.015	96.6	360.3	3.3
		0.0073	0.0008	37.3962	0.0007	15.935	0.227	86.4	397.1	6.9
		0.0101	0.0002	16.7971	0.0002	19.123	0.064	84.3	457.1	4.4
		0.0052	0.0002	12.5412	0.0003	15.277	0.074	90.0	396.5	4.1
		0.0053	0.0001	3.3114	0.0001	14.073	0.019	89.0	364.4	3.4
Shizigou	Sample G6	0.0063	0.0001	0.4175	0.0001	15.413	0.037	87.9	385.0	3.7
	(QA-97C-01	0.0427	0.0008	2.2191	0.0005	19.226	0.244	34.3	197.6	7.2
	+ QA-96-01)	0.0201	0.0004	0.9438	0.0003	18.340	0.119	67.6	355.1	4.5
	with $J = 0.01758$	0.0190	0.0003	0.8052	0.0003	18.751	0.090	70.0	374.1	4.2
	± 0.00018	0.0080	0.0001	0.4299	0.0001	16.516	0.035	85.6	400.1	3.8
	$J = 0.0168$	0.0144	0.0002	0.8521	0.0002	17.158	0.070	75.2	368.3	3.9
	± 0.00017	0.0092	0.0001	0.5459	0.0001	14.724	0.043	81.6	345.3	3.4
		0.0164	0.0003	0.8272	0.0003	18.046	0.082	73.1	376.1	4.1
		0.0156	0.0002	0.8037	0.0002	20.120	0.068	77.1	434.5	4.3
		0.0045	0.0004	3.8332	0.0005	16.215	0.132	91.8	419.0	5.1
		0.0059	0.0005	9.9781	0.0005	15.933	0.148	89.1	401.5	5.3
		0.0020	0.0001	2.4513	0.0001	13.701	0.022	95.7	373.8	3.5
		0.0026	0.0002	10.4196	0.0002	14.535	0.070	94.7	390.7	4.0
		0.0026	0.0003	11.2074	0.0002	16.282	0.076	95.4	435.0	4.4
		0.0014	0.0002	11.8819	0.0002	14.470	0.050	97.1	397.8	3.9
		0.0012	0.0001	3.8558	0.0001	11.672	0.018	97.0	327.0	3.1
		0.0131	0.0002	0.6298	0.0002	22.454	0.074	82.7	489.6	4.7
		0.0054	0.0003	0.3856	0.0003	16.990	0.075	90.6	414.7	4.2

Table 3 *(cont'd)*

Formation	Sample details	$^{36}Ar/^{39}Ar$ measured	$^{36}Ar/^{39}Ar$ 1-sigma absolute	$^{37}Ar/^{39}Ar$ corrected	$^{37}Ar/^{39}Ar$ 1-sigma absolute	$^{40}Ar/^{39}Ar$ measured	$^{40}Ar/^{39}Ar$ 1-sigma absolute	$\%^{40}Ar^*$	Age (Ma)	± (Myr) 1-sigma absolute
		0.0154	0.0003	0.8462	0.0003	20.190	0.099	77.5	420.5	4.5
		0.0019	0.0002	7.3426	0.0002	16.465	0.056	96.6	426.8	4.1
		0.0004	0.0002	10.6453	0.0002	14.219	0.065	99.2	383.1	3.8
		0.0017	0.0002	0.5755	0.0001	13.916	0.045	96.4	366.1	3.5
		0.0026	0.0002	7.9609	0.0002	14.002	0.066	94.5	361.8	3.7
		0.0000	0.0002	1.5463	0.0002	17.860	0.072	99.9	472.8	4.5
		0.0012	0.0002	0.6585	0.0003	15.171	0.067	97.7	400.9	4.0
		0.0008	0.0002	9.3007	0.0002	17.734	0.063	98.7	464.5	4.4
		0.0019	0.0003	17.7065	0.0003	14.071	0.097	96.1	368.8	4.1
		0.0034	0.0006	2.6213	0.0006	16.748	0.172	93.9	422.7	5.6
		0.0012	0.0004	9.5047	0.0004	15.110	0.113	97.6	399.0	4.5
Shangyoushashan	Sample G5	0.0051	0.0001	0.2288	0.0001	14.626	0.026	89.6	369.5	3.3
	(QA-95-01	0.0046	0.0001	0.2194	0.0001	17.547	0.038	92.2	446.3	4.0
	+ QA-94-01)	0.0003	0.0001	0.0359	0.0001	18.292	0.032	99.4	494.7	4.3
	with $J = 0.01736$	0.0046	0.0001	0.2311	0.0001	18.433	0.028	92.6	467.8	4.1
	$\pm$ 0.00017	0.0045	0.0001	0.2469	0.0001	16.399	0.029	91.9	419.0	3.7
	$J = 0.01713$	0.0106	0.0001	0.3605	0.0002	16.701	0.039	81.3	381.5	3.5
	$\pm$ 0.00017	0.0043	0.0044	35.4199	0.0036	21.496	1.321	94.1	542.5	31.0
		0.0029	0.0002	13.5510	0.0002	16.125	0.074	94.8	424.1	4.1
		0.0018	0.0001	5.2534	0.0001	13.069	0.029	96.0	355.2	3.2
		0.0014	0.0002	10.6766	0.0002	14.029	0.063	97.0	382.3	3.7
		0.0020	0.0002	4.8393	0.0001	14.201	0.056	95.8	382.1	3.7
		0.0005	0.0001	2.8161	0.0001	15.339	0.021	99.0	421.6	3.7
		0.0018	0.0002	7.5896	0.0002	12.431	0.059	95.6	338.0	3.4
		0.0007	0.0000	1.8527	0.0000	12.219	0.010	98.2	341.0	3.1
		0.0079	0.0001	0.3311	0.0001	11.240	0.027	79.1	255.4	2.5
		0.0101	0.0001	0.4340	0.0001	9.110	0.040	67.3	179.7	2.0
		0.0061	0.0001	0.2954	0.0001	14.027	0.017	87.2	342.9	3.1
		0.0092	0.0001	0.4953	0.0001	9.976	0.041	72.6	210.6	2.3
		0.0142	0.0003	0.7280	0.0002	10.863	0.079	61.3	194.4	2.9
		0.0082	0.0001	0.4350	0.0001	13.891	0.032	82.5	323.1	3.1
		0.0077	0.0001	0.4325	0.0001	8.442	0.035	73.2	181.0	2.0
		0.0093	0.0001	0.5071	0.0001	10.607	0.039	74.0	227.1	2.4
		0.0008	0.0000	2.2789	0.0000	12.013	0.006	97.9	330.8	3.0

		0.0025	0.0001	1.8063	0.0001	13.672	0.028	94.5	360.4	3.3
		0.0017	0.0000	2.2059	0.0000	13.156	0.014	96.2	353.6	3.2
		0.0018	0.0001	5.8155	0.0001	13.269	0.027	96.1	355.9	3.3
		0.0010	0.0001	8.2469	0.0001	7.919	0.039	96.1	220.7	2.3
		0.0011	0.0001	3.6969	0.0001	12.170	0.023	97.4	333.0	3.1
		0.0014	0.0001	6.4410	0.0001	10.136	0.028	95.8	277.2	2.7
Shangganchaigou	Sample G3	0.0044	0.0001	0.2446	0.0001	13.230	0.020	90.2	331.3	3.1
	(QA-89-01)	0.0284	0.0007	1.5289	0.0005	22.086	0.211	62.0	375.6	6.2
	with $J = 0.01691$	0.0152	0.0003	0.7388	0.0003	22.441	0.088	80.0	477.7	4.7
	± 0.00017	0.0284	0.0006	1.8425	0.0009	43.724	0.210	80.8	844.8	7.9
		0.0043	0.0007	4.2927	0.0006	14.577	0.201	91.3	365.6	6.0
		0.0012	0.0022	11.8021	0.0021	17.187	0.655	98.0	451.8	16.1
		0.0017	0.0006	3.2227	0.0006	15.584	0.182	96.7	409.1	5.8
		0.0052	0.0006	13.7468	0.0006	20.437	0.194	92.4	499.8	6.3
		0.0040	0.0005	6.1371	0.0005	9.152	0.134	87.0	227.4	4.2
		0.0100	0.0012	5.1522	0.0013	11.431	0.363	74.3	241.6	10.0
		0.0006	0.0007	27.8801	0.0006	13.562	0.203	98.6	367.4	6.0
		0.0022	0.0008	6.7764	0.0008	14.683	0.251	95.5	383.5	7.1
		0.0020	0.0013	3.1669	0.0013	18.796	0.382	96.8	483.6	9.9
		0.0001	0.0003	1.9524	0.0003	7.933	0.077	99.8	226.2	3.0
		0.0034	0.0012	32.3143	0.0012	16.552	0.355	93.9	420.4	9.4
		0.0022	0.0003	17.5452	0.0003	8.630	0.098	92.5	228.0	3.4
Xiaganchaigou	Sample G2	0.0179	0.0003	0.8917	0.0005	19.574	0.100	73.0	383.2	4.3
	(QA-86-01)	0.0391	0.0008	1.9309	0.0009	23.259	0.226	50.3	319.9	6.4
	with $J = 0.01659$	0.0225	0.0006	1.6574	0.0004	18.392	0.170	63.9	320.9	5.2
	± 0.00017	0.0220	0.0004	1.3174	0.0005	18.085	0.111	64.1	316.8	4.1
		0.0217	0.0007	1.9602	0.0008	18.975	0.200	66.2	341.1	5.9
		0.0161	0.0005	1.0210	0.0004	9.009	0.137	47.2	122.5	4.0
		0.0244	0.0010	1.3801	0.0009	15.710	0.283	54.2	237.9	7.8
		0.0008	0.0000	2.2789	0.0000	12.013	0.006	97.9	321.2	3.0
		0.0042	0.0007	38.5238	0.0007	15.761	0.216	92.1	388.9	6.3
		0.0040	0.0016	91.1020	0.0015	18.123	0.469	93.5	446.4	11.7
		0.0036	0.0012	68.2944	0.0013	9.283	0.366	88.4	229.8	9.9
		0.0024	0.0006	49.6778	0.0007	17.150	0.182	95.8	434.6	5.8
		0.0006	0.0009	114.0900	0.0011	16.148	0.272	99.0	423.9	7.5
		0.0003	0.0011	51.2572	0.0011	16.907	0.330	99.5	443.7	8.7
		0.0004	0.0016	29.5104	0.0018	16.511	0.486	99.3	433.7	12.1
		0.0003	0.0009	34.5053	0.0010	17.748	0.270	99.5	463.2	7.5
		0.0001	0.0001	1.1236	0.0001	15.379	0.031	99.9	409.1	3.8
		0.0043	0.0012	17.0406	0.0010	15.663	0.342	92.0	386.2	9.0

Table 3 *(cont'd)*

Formation	Sample details	$^{36}Ar/^{39}Ar$ measured	$^{36}Ar/^{39}Ar$ 1-sigma absolute	$^{37}Ar/^{39}Ar$ corrected	$^{37}Ar/^{39}Ar$ 1-sigma absolute	$^{40}Ar/^{39}Ar$ measured	$^{40}Ar/^{39}Ar$ 1-sigma absolute	$\%^{40}Ar^*$	Age (Ma)	± (Myr) 1-sigma absolute
Xiaganchaigou	Sample H2	0.00048	0.00003	0.00052	0.00002	10.411	0.008	98.6	315.2	3.0
	(QA-133C-01)	0.00028	0.00002	0.00113	0.00002	15.044	0.007	99.4	442.9	4.0
	with $J = 0.01862$	0.00323	0.00020	0.02698	0.00019	11.591	0.059	91.8	325.6	3.5
	± 0.00019	0.00109	0.00006	0.00819	0.00006	9.074	0.017	96.4	271.9	2.6
		0.00038	0.00007	0.00043	0.00007	12.625	0.020	99.1	377.4	3.5
		0.00039	0.00005	0.00007	0.00005	17.873	0.016	99.3	515.0	4.6
		0.00026	0.00006	0.00010	0.00005	6.732	0.018	98.8	210.3	2.1
		0.00029	0.00007	0.00064	0.00008	8.700	0.022	99.0	268.0	2.6
		0.02734	0.00013	0.00229	0.00007	17.903	0.039	54.9	302.6	3.0
		0.00030	0.00006	0.00197	0.00006	16.772	0.017	99.5	487.6	4.4
		0.00167	0.00018	0.01721	0.00019	18.524	0.054	97.3	521.9	4.8
Lulehe	Sample H1	0.00018	0.00004	0.00004	0.00003	16.392	0.011	99.7	479.2	4.3
	(QA-133A-01	0.00108	0.00009	0.00102	0.00007	11.251	0.025	97.2	334.3	3.2
	+ QA-132A-01)	0.00065	0.00006	0.00099	0.00005	12.353	0.017	98.5	368.3	3.4
	with $J = 0.01862$	0.00461	0.00009	0.00267	0.00007	11.856	0.026	88.5	322.0	3.1
	± 0.00019	0.00299	0.00006	0.00298	0.00004	12.526	0.017	92.9	354.0	3.3
	$J = 0.01855$	0.02466	0.00016	0.00153	0.00009	18.095	0.047	59.7	330.8	3.4
	± 0.00019	0.00304	0.00014	0.00097	0.00013	12.060	0.042	92.5	340.6	3.4
		0.00130	0.00004	0.00005	0.00004	11.276	0.012	96.6	333.2	3.1
		0.00019	0.00006	0.00026	0.00005	12.386	0.018	99.6	372.9	3.5
		0.00071	0.00008	0.00015	0.00007	12.478	0.023	98.3	371.2	3.5
		0.03214	0.00532	0.05867	0.00550	23.780	1.582	60.1	424.0	42.0
		0.00031	0.00007	0.00082	0.00007	14.528	0.020	99.4	428.1	3.9
		0.00171	0.00006	0.00099	0.00005	13.443	0.019	96.2	388.1	3.6
		0.00521	0.00023	0.00498	0.00017	15.679	0.067	90.2	420.3	4.2
		0.00073	0.00006	0.00237	0.00005	10.733	0.018	98.0	321.5	3.1
		0.00020	0.00016	0.00200	0.00013	16.373	0.046	99.6	476.9	4.5
		0.00030	0.00006	0.00042	0.00005	14.787	0.018	99.4	435.0	4.0
		0.00071	0.00009	0.00019	0.00007	15.006	0.026	98.6	437.6	4.0
		0.00101	0.00009	0.00054	0.00007	15.298	0.027	98.0	442.9	4.1
		0.00008	0.00012	0.00005	0.00011	10.012	0.037	99.8	306.7	3.1
		0.00064	0.00010	0.00004	0.00008	14.432	0.029	98.7	422.9	3.9

Shizigou	Sample L6	0.00024	0.00002	0.00002	0.00002	8.114	0.007	99.1	257.8	2.4
	(QA-239A-02)	0.00656	0.00355	0.00058	0.00295	14.259	1.051	86.4	381.2	29.5
	with $J = 0.01910$	0.00079	0.00008	0.00037	0.00005	8.028	0.023	97.1	250.4	2.4
	± 0.00019	0.00018	0.00004	0.00016	0.00004	8.454	0.012	99.4	268.5	2.5
		0.00264	0.00035	0.00350	0.00032	9.459	0.103	91.8	276.7	4.0
		0.00016	0.00004	0.00043	0.00003	8.383	0.011	99.4	266.5	2.5
		0.00027	0.00007	0.00043	0.00007	8.353	0.022	99.0	264.7	2.5
		0.00085	0.00008	0.00007	0.00007	8.794	0.024	97.1	272.7	2.6
		0.00042	0.00003	0.00037	0.00003	8.151	0.010	98.5	257.4	2.4
		0.00003	0.00006	0.00008	0.00005	8.023	0.018	99.9	256.9	2.4
		0.00019	0.00014	0.00197	0.00013	8.668	0.042	99.3	274.7	2.8
uppermost	Sample L5	0.00002	0.00005	0.00008	0.00005	8.579	0.016	99.9	273.9	2.6
Shangyoushashan	(QA-240A-02)	0.00001	0.00005	0.00002	0.00005	7.992	0.015	100.0	256.4	2.4
	with $J = 0.01912$	0.00023	0.00006	0.00010	0.00006	7.923	0.019	99.1	252.5	2.4
	± 0.00019	0.00000	0.00007	0.00000	0.00006	8.545	0.020	100.0	273.0	2.6
		0.00003	0.00006	0.00003	0.00005	8.483	0.017	99.9	270.9	2.6
		0.00008	0.00005	0.00007	0.00005	8.549	0.015	99.7	272.4	2.6
		0.00066	0.00007	0.00105	0.00006	8.564	0.021	97.7	267.8	2.6
		0.00045	0.00003	0.00023	0.00003	8.201	0.010	98.4	258.8	2.4
		0.00029	0.00016	0.00031	0.00014	8.712	0.047	99.0	275.4	2.9
		0.00034	0.00007	0.00019	0.00006	8.287	0.020	98.8	262.4	2.5
		0.00013	0.00003	0.00005	0.00003	8.539	0.008	99.5	271.7	2.5
Xiayoushashan	Sample L4	0.000344	0.000043	0.000058	0.000035	8.487	0.013	98.8	267.6	2.5
	(QA-238A-02	0.000345	0.000044	0.000099	0.000042	8.469	0.013	98.8	267.1	2.5
	+ QA-237A-02	0.000061	0.000036	0.000051	0.000030	8.301	0.011	99.8	264.6	2.5
	+ QA-234A-02)	0.000090	0.000051	0.000631	0.000044	8.542	0.015	99.7	271.5	2.6
	with $J = 0.01907$	0.000018	0.000053	0.001105	0.000044	8.491	0.016	99.9	270.6	2.5
	$J = 0.01904$	0.001129	0.000045	0.001335	0.000039	8.547	0.013	96.1	262.5	2.5
	$J = 0.01901$	0.000261	0.000011	0.000373	0.000011	8.364	0.003	99.1	264.7	2.5
	all ± 0.00019	0.000484	0.000030	0.000110	0.000026	8.451	0.009	98.3	265.3	2.5
		0.015528	0.001013	0.017977	0.000895	11.349	0.300	59.6	218.8	9.4
		0.000570	0.000042	0.000946	0.000038	8.496	0.012	98.0	265.9	2.5
		0.000127	0.000186	0.000055	0.000202	8.420	0.055	99.6	267.1	3.0
		0.000066	0.000074	0.000103	0.000094	8.475	0.022	99.8	269.3	2.6
		0.000337	0.000079	0.000000	0.000080	8.439	0.023	98.8	265.9	2.6

Table 3 *(cont'd)*

Formation	Sample details	$^{36}Ar/^{39}Ar$ measured	$^{36}Ar/^{39}Ar$ 1-sigma absolute	$^{37}Ar/^{39}Ar$ corrected	$^{37}Ar/^{39}Ar$ 1-sigma absolute	$^{40}Ar/^{39}Ar$ measured	$^{40}Ar/^{39}Ar$ 1-sigma absolute	$\%^{40}Ar^*$	Age (Ma)	± (Myr) 1-sigma absolute
		0.003576	0.000368	0.000563	0.000423	9.504	0.109	88.9	269.1	4.1
		0.000296	0.000051	0.000028	0.000058	8.242	0.015	98.9	260.4	2.5
		0.000854	0.000085	0.002706	0.000081	8.485	0.025	97.0	262.7	2.6
		0.000561	0.000044	0.000940	0.000041	8.268	0.013	98.0	258.8	2.4
		0.000007	0.000087	0.000147	0.000086	8.290	0.026	100.0	264.3	2.6
		0.000069	0.000130	0.001946	0.000136	8.462	0.039	99.8	268.9	2.7
		0.000220	0.000098	0.001167	0.000098	8.521	0.029	99.2	269.3	2.6
		0.000708	0.000148	0.000153	0.000136	8.401	0.044	97.5	261.5	2.8
		0.00039	0.00003	0.00001	0.00002	8.114	0.009	98.6	255.4	2.4
		0.00039	0.00005	0.00014	0.00004	8.391	0.015	98.6	263.6	2.5
		0.00065	0.00004	0.00004	0.00003	8.409	0.011	97.7	261.9	2.5
		0.00124	0.00009	0.00507	0.00008	8.645	0.028	95.7	263.7	2.6
		0.00010	0.00004	0.00068	0.00004	8.140	0.012	99.7	258.7	2.4
		0.00006	0.00002	0.00046	0.00002	8.230	0.007	99.8	261.7	2.4
		0.00120	0.00012	0.00477	0.00010	8.690	0.036	95.9	265.3	2.7
		0.00097	0.00003	0.00032	0.00002	8.624	0.008	96.7	265.4	2.5
		0.00391	0.00033	0.00696	0.00030	9.841	0.098	88.3	275.7	3.9
		0.00000	0.00005	0.00140	0.00004	8.285	0.015	100.0	263.8	2.5
		0.00009	0.00003	0.00031	0.00002	8.349	0.008	99.7	265.0	2.5
Shangganchaigou	Sample L3	0.00021	0.00011	0.00009	0.00008	8.645	0.031	99.3	272.0	2.7
	(QA-233A-02)	0.00006	0.00004	0.00000	0.00003	8.350	0.011	99.8	264.7	2.5
	with $J = 0.01896$	0.00470	0.00023	0.00019	0.00019	10.164	0.068	86.3	277.7	3.3
	± 0.00019	0.00011	0.00044	0.00103	0.00036	9.539	0.130	99.7	298.9	4.7
		0.00105	0.00014	0.00059	0.00013	9.062	0.040	96.6	277.0	2.8
		0.02377	0.00008	0.00005	0.00004	16.400	0.024	57.2	295.2	2.8
		0.00017	0.00007	0.00017	0.00006	8.436	0.020	99.4	266.2	2.5
		0.02184	0.00017	0.00013	0.00013	16.187	0.050	60.1	305.5	3.2
		0.00022	0.00007	0.00003	0.00007	8.339	0.022	99.2	262.9	2.5
		0.00111	0.00007	0.00064	0.00005	8.746	0.019	96.3	267.2	2.6
		0.00159	0.00007	0.00137	0.00006	8.960	0.021	94.8	269.3	2.6

Xiaganchaigou	Sample L2	0.00210	0.00012	0.00248	0.00012	9.054	0.035	93.2	265.8	2.7
	(QA-231A-02)	0.00132	0.00009	0.00383	0.00009	8.537	0.026	95.4	257.4	2.5
	with $J = 0.01882$	0.00074	0.00007	0.00010	0.00007	8.656	0.020	97.5	265.9	2.6
	± 0.00019	0.00097	0.00007	0.00008	0.00006	8.548	0.019	96.6	260.7	2.5
		0.00075	0.00003	0.00000	0.00003	8.535	0.010	97.4	262.3	2.5
		0.00014	0.00002	0.00000	0.00002	8.358	0.007	99.5	262.3	2.5
		0.00002	0.00004	0.00000	0.00004	8.341	0.011	99.9	262.8	2.5
		0.00270	0.00024	0.00006	0.00021	9.694	0.071	91.8	279.3	3.3
		0.00084	0.00016	0.00328	0.00014	8.821	0.048	97.2	269.8	2.9
		0.00218	0.00019	0.00334	0.00016	8.548	0.057	92.5	250.2	2.9
		0.00044	0.00006	0.00108	0.00005	8.509	0.018	98.5	264.1	2.5
Lulehe	Sample L1	0.00237	0.00004	0.00005	0.00003	15.516	0.012	95.5	445.8	4.0
	(QA-232A-02)	0.00055	0.00003	0.00010	0.00002	8.651	0.008	98.1	268.7	2.5
	with $J = 0.01892$	0.00061	0.00007	0.00039	0.00006	8.623	0.021	97.9	267.4	2.6
	± 0.00019	0.00032	0.00006	0.00001	0.00005	8.535	0.017	98.9	267.3	2.5
		0.00878	0.00045	0.00203	0.00035	12.217	0.132	78.8	301.8	4.7
		0.00128	0.00006	0.00071	0.00005	8.983	0.018	95.8	272.1	2.6
		0.00051	0.00009	0.00095	0.00008	8.683	0.026	98.2	269.9	2.6
		0.00054	0.00011	0.00161	0.00009	9.014	0.033	98.2	279.4	2.8
		0.00035	0.00008	0.00128	0.00007	8.797	0.024	98.8	274.7	2.7
		0.00049	0.00014	0.00346	0.00012	8.911	0.043	98.4	276.8	2.9
		0.00164	0.00019	0.00699	0.00017	8.993	0.056	94.6	269.3	3.0

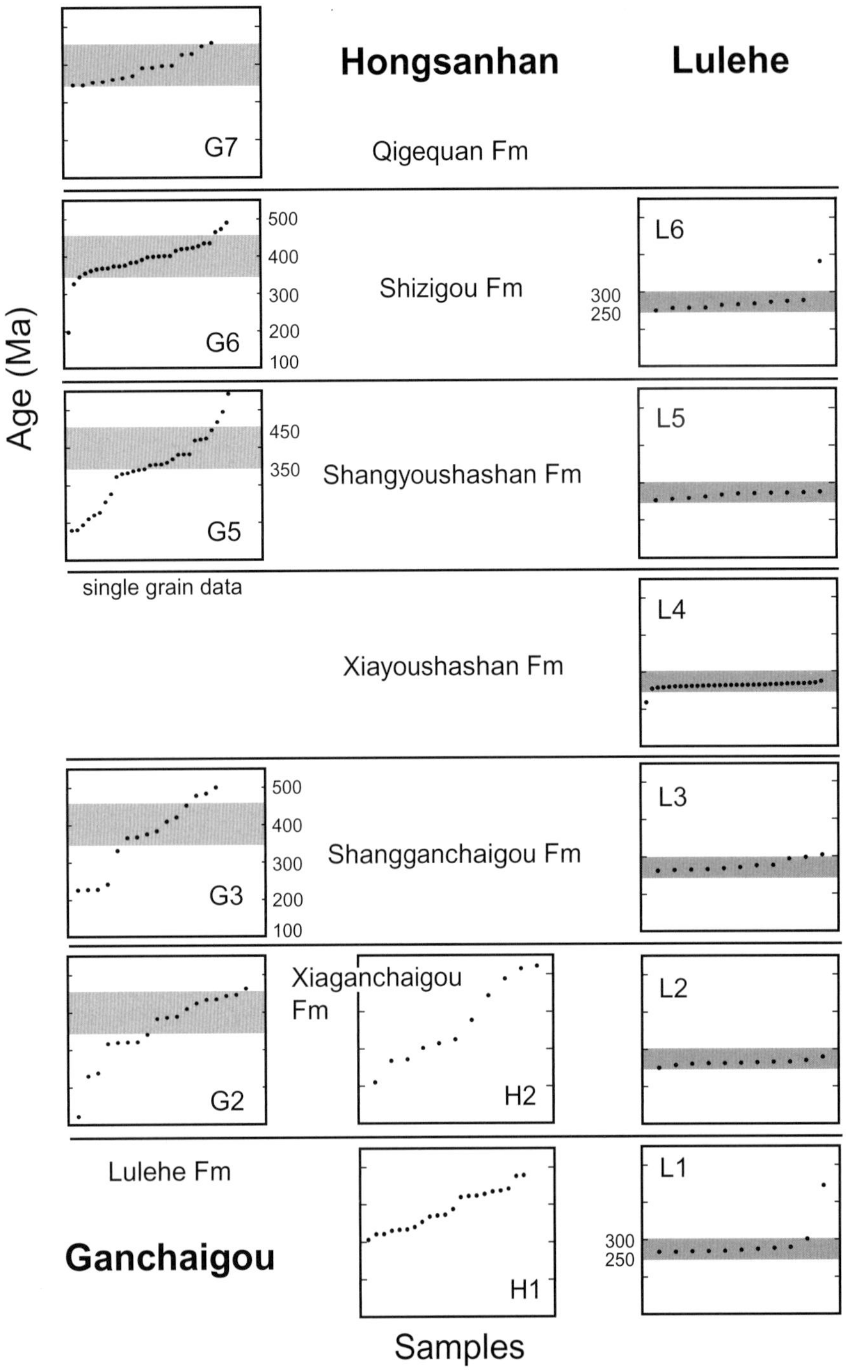

Fig. 6 Detrital white mica $^{40}Ar/^{39}Ar$ total-fusion ages from the Ganchaigou (G2-7), Hongsanhan (H1-2) and Lulehe (L1-6) sections. Each point represents a single white mica grain age. Shaded bars in the Ganchaigou section highlight the 350–450 Ma interval and in the Lulehe section the 250–300 Ma interval. No error bars are shown, as usually 2σ uncertainties are in the range of 1–2% of the age, which cannot be displayed at this scale.

sediments are derived from nearby uplifting ranges (e.g. Robinson *et al.*, 2003).

A high mica content is considered to be a reliable indicator to characterize molasse-type basins (foreland basins and intramontane basins) that formed during orogeny (Mader & Neubauer, 2004). During humid climatic phases, a decrease in feldspar and mica content should be expected (Johnsson, 1993), but this was not observed in the Qaidam Basin data. On the contrary, relatively high concentrations of both mineral groups have been observed and mica concentrations continuously increase with time. This can be explained by short transport distances from the catchment area

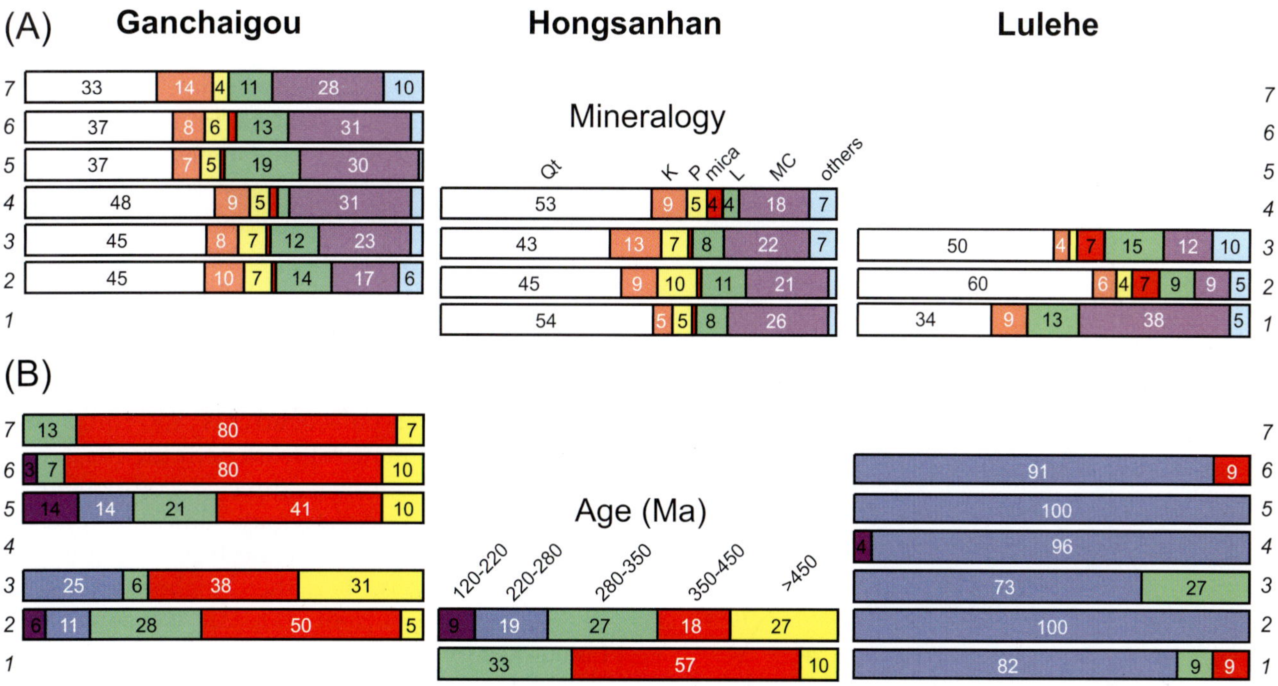

Fig. 7 Mean percentage of framework constituents and mineral ages per formation in the Ganchaigou, Hongsanhan and Lulehe sections. Numbers indicate the formations: 7, Qigequan; 6, Shizigou; 5, Shangyoushashan; 4, Xiayoushashan; 3, Shangganchaigou; 2, Xiaganchaigou; 1, Lulehe. (A) Percentage of framework constituents (for abbreviations, see Table 2) in the sandstones. (B) Percentage of single white mica grains with ages of 120–220 Ma, 220–280 Ma, 280–350 Ma, 350–450 Ma and > 450 Ma.

to the sink within the Qaidam Basin. Rapid burial and cementation isolated water from the chemically unstable grains after deposition. The discrimination diagrams (Fig. 5) and the summary diagram (Fig. 7a) show no difference between the three sections analysed that could be explained by different source lithologies, which would have been expected on the basis of the source region geology (Wang & Zhang, 1999).

Hanson's (1999) directional palaeocurrent data (Fig. 8) provide so far the only direct indication of the source location, beside general considerations regarding the distribution of sample locations in relation to facies zones (i.e. large alluvial fans along the basin margins reaching far into the basin and lake sediments in the basin centre). The general dominance of the 350–450 Ma age group suggests, at first sight, that most material has been shed from the Altyn Mountains in the north, where ages of this range are well-documented in the metamorphic and granitic basement (Fig. 9). In particular, this age group dominates in basement rocks of the Xorkol area (Jolivet *et al.*, 1999; Sobel and Arnaud, 1999; Sobel *et al.*, 2001; Gehrels *et al.*, 2003a), where they have been interpreted as cooling ages after peak metamorphic conditions or after magmatic intrusions. Figure 9 summarizes published $^{40}Ar/^{39}Ar$ ages from both granitic and metamorphic basement rocks around the Qaidam Basin. There are many small magmatic bodies in the surrounding hinterland of the Qaidam Basin most of which are undated. However, it can be assumed that they are mid-Palaeozoic, Mesozoic or Jurassic–Cretaceous in age, as some of them are dated. Thus, parts of the North Altyn Mountains that were offset westwards will possibly be of the same age, but neither an age signal nor a compositional change could be observed that directly documents faulting.

The two predominant age groups in the Ganchaigou section (220–280 and 350–450 Ma) seem to reflect the change observed in palaeocurrent directions (Figs 8 & 10). For the Oligocene, Hanson (1999) found a polymodal palaeocurrent distribution with a main southwest-directed component, but also northeast- and southeast-directed components, indicating general uplift of the

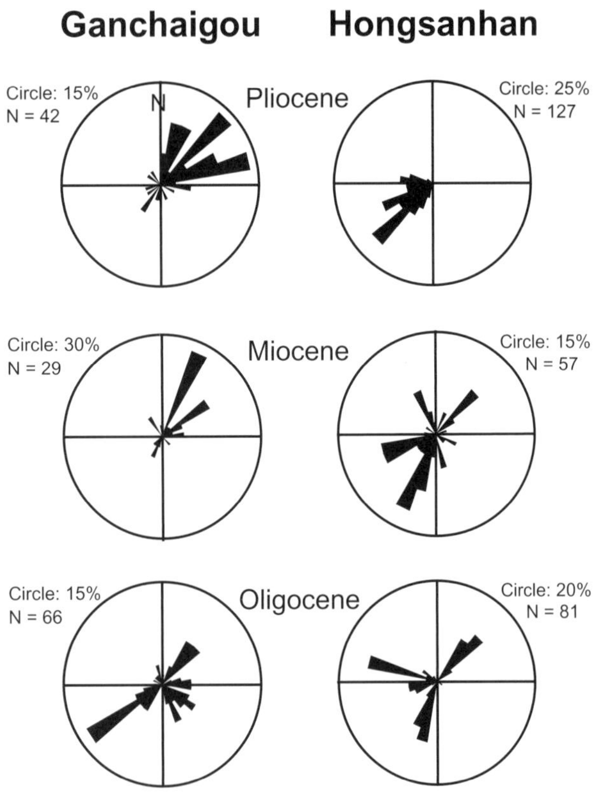

Fig. 8 Palaeocurrent data (rose diagrams) from the Ganchaigou and Hongsanhan sections. (Redrawn from Hanson, 1999.)

northwestern corner of the Qaidam Basin. For the Miocene and Pliocene, almost exclusively northeastward palaeocurrents have been reported (Fig. 8). The fact that palaeocurrents show a change from NE–SW divergent directions during the Oligocene to a unidirectional northeast distribution in the Miocene implies that in the Qimantagh also, Early to Mid-Palaeozoic magmatic and metamorphic rocks are present, which formed the source for the Late Neogene sedimentary deposits in the Ganchaigou section. An alternative interpretation would be that the notheast transport was due to the growth and surface uplift of the Youshashan anticline (Fig. 1C), which started to form during Late Miocene and Pliocene times.

The proportion of mica with ages younger than 200 Ma remained relatively small. Such micas were found only in the Oligocene formations of the Ganchaigou section, whereas corresponding micas are known only in basement rocks in limited areas along and immediately north of the Altyn Tagh Fault to the north of the northeastern edge of the basin, where Neubauer *et al.* (2004) reported earliest Triassic $^{40}Ar/^{39}Ar$ total-fusion single-grain white mica ages of 200 Ma from widespread Jurassic and Cretaceous sandstones (South Altyn Mountains). The young ages require another source closer to the Ganchaigou section, possibly offset along the Altyn Tagh Fault.

The Hongsanhan samples, which obtained their detrital material largely from the north (Fig. 10), show a very similar age distribution to the Paleogene Ganchaigou samples. Both sections show polymodal palaeocurrents and thus probable mixing of material from both the north and the south. The northeast-directed component in the Hongsanhan section may indicate recycling of basin sediments, when the basin margin was slightly depressed. For the Lulehe section no Cenozoic palaeocurrent data are available. The Lulehe section most probably received its material from a nearby area in the Qilian Mountains. However, no basement rocks with dominant Permian ages are reported from the south and middle Qilian Mountains (Fig. 9) beside Triassic–Jurassic sandstones, which could have yielded a small proportion of Permian ages, but have been dated only in the Altyn Mountains so far (Neubauer *et al.*, 2004). On the other hand, metamorphic basement rocks with Early and Late Palaeozoic ages ranging from 360 to 470 Ma seem to be widespread.

The Himalayas, the Tibetan plateau and the adjacent mountains north of the plateau with an average elevation of 4000–5000 m form the most outstanding present-day topographic feature resulting from the India–Asia collision (Molnar & Tapponnier, 1975; Allègre *et al.*, 1984; Yin & Nie, 1996). The data presented allow discussion of the linkages between sedimentation and tectonic relationships in this special geological setting: a closed basin with mountains and faults along its margins and changing climatic conditions. The data from the Qaidam Basin show that differences between distinct sources can be identified using $^{40}Ar/^{39}Ar$ single-grain dating of white mica, although framework constituents of sandstones do not show significant compositional variations through time.

The lack of mineralogical differences within sandstones of the Cenozoic formations implies that over a larger part of the Altyn Mountains basement lithologies must have been very similar

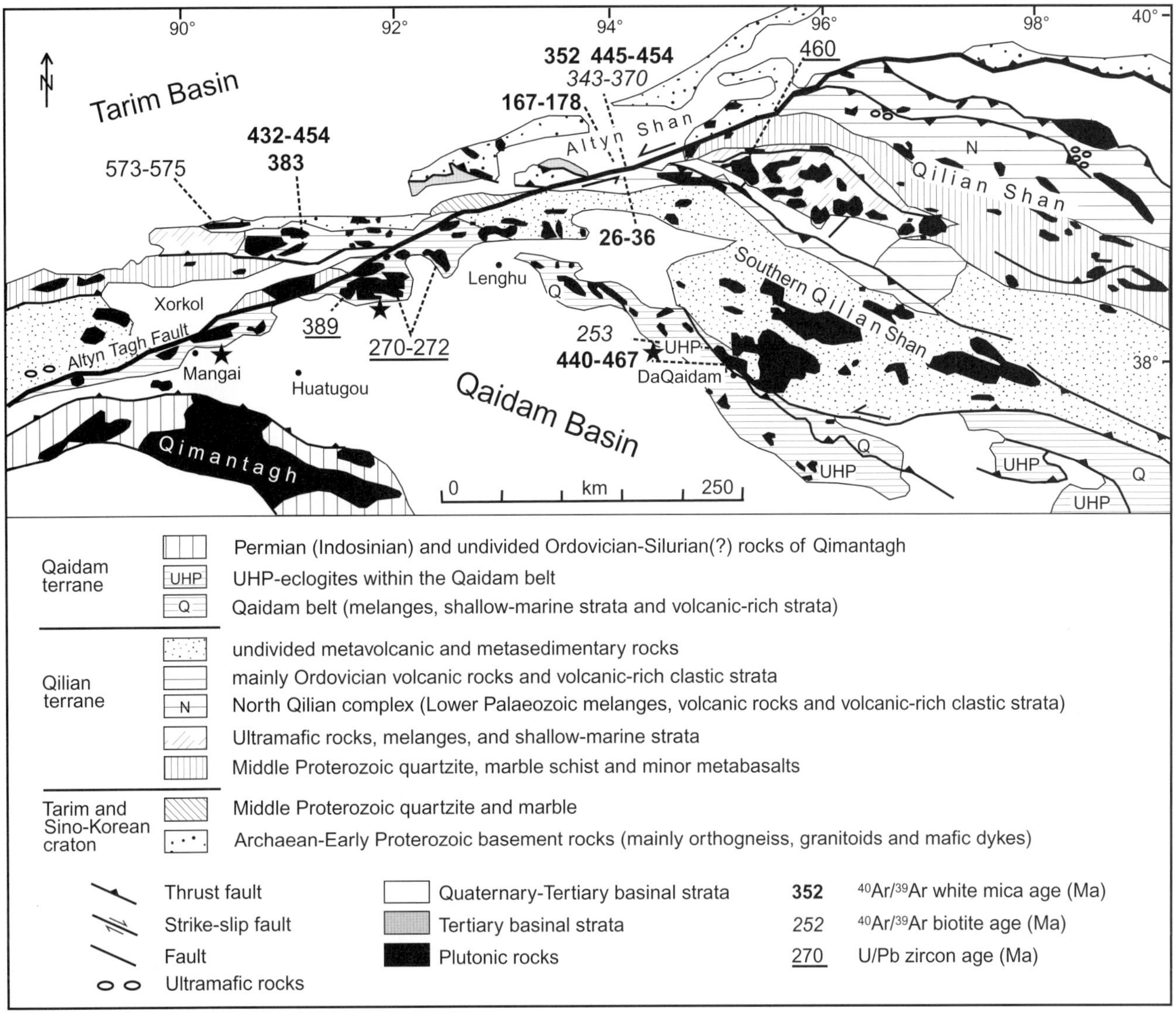

Fig. 9 Geological map of the basement rocks surrounding the Qaidam Basin showing age data (Ma) of relevant magmatic, metamorphic and sedimentary units. Stars indicate the locations of the three sections sampled: Ganchaigou, Hongsanhan, Lulehe. See text for details of data and sources. (Redrawn and simplified from Gehrels *et al.*, 2003a.)

on both the northern and southern sides of the Altyn Tagh Fault. The fact that strike-slip faults with large Cenozoic offsets bound the basin, implies that the source regions for dated sediments have not remained constant with time. This is particularly important for evaluating the former relationship between the Palaeozoic belts in the north (North Altyn Mountains) and the Qaidam Basin.

Age groups from Eocene to Oligocene formations of the Ganchaigou and the Hongsanhan sections show single-grain $^{40}Ar/^{39}Ar$ ages close to 450–500 Ma or even higher, which are not known from the South Altyn Mountains, but from distinct regions within the North Altyn Mountains. We tentatively suggest an origin for these mica grains from the latter region. This matches the observation that the South Altyn Mountains were uplifted during Early Miocene times (e.g. Jolivet *et al.*, 2001) successively forming a barrier that interrupted direct drainage to the Qaidam Basin.

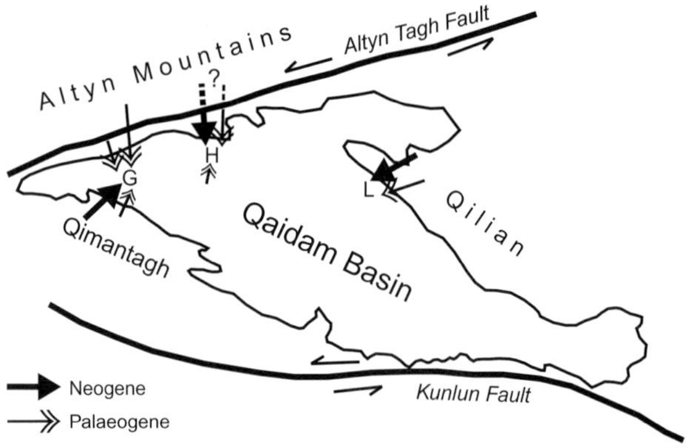

Fig. 10 Simplified map of the Qaidam Basin with arrows showing the Paleogene and Neogene sediment transport directions. Small Paleogene arrowheads represent less important directions. Capital letters mark the respective sections: G, Ganchaigou; H, Hongsanhan; L, Lulehe.

CONCLUSIONS

Although the geology of the surrounding mountains of the Qaidam Basin shows large variations, modal analysis of sandstones from three sections in the northwestern and eastern basin yielded no significant differences in source-area signature. However, $^{40}Ar/^{39}Ar$ total-fusion data yielded several age clusters that can be assigned to certain provenance areas within the Early Palaeozoic and Indosinian basement complexes in the Altyn and Kunlun Mountains. The lack of 350–450 Ma ages, well-known in the Altyn Mountains, indicates a different provenance for the Lulehe sandstones, although modal analysis shows similar petrographical compositions.

It is postulated that under certain circumstances (i.e. in the case of Qaidam basement rocks with significantly different magmatic and metamorphic ages), $^{40}Ar/^{39}Ar$ dating of detrital white mica can provide critical insights into sediment composition and origin, and link the sediments to specific source areas in the hinterland.

ACKNOWLEDGEMENTS

We thank Ma Lixiang for the lithostratigraphical profile from the Hongsanhan section and Andrew Hanson for providing unpublished data from his PhD thesis. We acknowledge continuous support for fieldwork in the Qaidam Basin by both the NSF of China (Grant Nr. 40272099/40572135) and the Qinghai Oil Company. Ed Sobel, Duncan Pirrie and Ed Williams are thanked for their constructive reviews.

REFERENCES

Allègre, C.J., Courtillot, V., Tapponnier, P., *et al.* (1984) Structure and evolution of the Himalaya–Tibet orogenic belt. *Nature*, **307**, 17–22.

Bedrosian, P.A., Unsworth, M.J. and Wang, F. (2001) Structure of the Altyn Tagh Fault and Daxue Shan from magnetotelluric surveys: implications for faulting associated with the rise of the Tibetan Plateau. *Tectonics*, **20**, 474–486.

Bendick, R., Bilham, R., Freymueller, J., Larson, K. and Yin, G. (2000) Geodetic evidence for a low slip rate in the Altyn Tagh fault system. *Nature*, **404**, 69–72.

Brewer, I.D., Burbank, D.W. and Hodges, K. (2003) Modelling detrital cooling-age populations: insights from two Himalayan catchments. *Basin Res.*, **15**, 305–320.

Copeland, P. and Harrison, M. (1990) Episodic rapid uplift in the Himalaya revealed by $^{40}Ar/^{39}Ar$ analysis of detrital K-feldspar and muscovite, Bengal fan. *Geology*, **18**, 354–357.

Crook, K.A.W. (1974) Lithogenesis and geotectonics: The significance of compositional variations in flysch arenites (graywackes). In: *Modern and Ancient Geosynclinal Sedimentation* (Eds R.H. Dott and R.H. Shaver), pp. 304–310. Special Publication 19, Society of Economic Paleontologists and Mineralogists, Tulsa, OK.

Delville, N., Arnaud, N., Montel, J.-M., *et al.* (2001) Paleozoic to Cenozoic deformation along the Altyn-Tagh fault in the Altun Shan massif area, eastern Qilian Shan, northeastern Tibet, China. In: *Paleozoic and Mesozoic Tectonic Evolution of Central Asia: From Continental Assembly to Intracontinental Deformation* (Eds M.S. Hendrix and A.M. Davis), pp. 269–292. Memoir 194, Geological Society of America, Boulder, CO.

Dickinson, W.R. (1985) Interpreting provenance relations from detrital modes of sandstones. In: *Provenance of Arenites* (Ed. G.G. Zuffa), pp. 333–361. D. Reidel, Dordrecht.

Dickinson, W.R. and Suczek, C.A. (1979) Plate tectonics and sandstone compositions. *Am. Assoc. Petrol. Geol. Bull.*, **63**, 2164–2182.

Dickinson, W.R., Beard, L.S., Brakenridge, G.R., *et al.* (1983) Provenance of North American Phanerozoic sandstones in relation to tectonic setting. *Geol. Soc. Am. Bull.*, **94**, 222–235.

Duan, Z. and Hu, W. (2001) The accumulation of potash in a continental basin: The example of the Qarhan Saline Lake, Qaidam Basin, West China. *Eur. J. Mineral.*, **13**, 1223–1233.

Fralick, P.W. and Kronberg, B.I. (1997) Geochemical discrimination of clastic sedimentary rock sources. *Sediment. Geol.*, **113**, 111–124.

Gehrels, G.E., Yin, A. and Wang, X.-F. (2003a) Detrital-zircon geochronology of the northeastern Tibetan plateau. *Geol. Soc. Am. Bull.*, **115**, 881–896.

Gehrels, G.E., Yin, A. and Wang, X.-F. (2003b) Magmatic history of the northeastern Tibetan Plateau. *J. Geophys. Res.*, **108**, doi:10.1029/2002JB001876.

Gradstein, F.M., Ogg, J.G. and Smith, A. (2004) *A Geologic Time Scale 2004*. Cambridge University Press.

Hanson, A.D. (1999) *Organic geochemistry and petroleum geology, tectonics and basin analysis of southern Tarim and northern Qaidam basins, northwest China*. Unpublished PhD thesis, Stanford University, Stanford, CA, 388 pp.

Huang, Q., Huang, H. and Ma, Y. (1997) *Geology of Qaidam basin and its Petroleum Prediction*. Geological Publishing House, Beijing, 158 pp.

Huo, G.M. (1990) *Petroleum Geology of China: Oil Fields in Qinghai and Xizang*, 14. Chinese Petroleum Industry Press, Beijing, 483 pp.

Ingersoll, R.V., Bullard, T.F., Ford, R.L., Grimm, J.P., Pickle, J.D. and Sares, S.W. (1984) The effect of grain size on detrital modes: a test of the Gazzi-Dickinson point-counting method. *J. Sediment. Petrol.*, **54**, 103–106.

Internal Report (2005) *Tectonic evolution of the Altyn strike-slip fault, the frontal nappes of the Kunlun Mountains, and their influences and controls to the structural evolution of the Qaidam Basin as well as the prediction of potential oil and gas reservoirs in the western Qaidam Basin*. Unpublished Internal Project Report, Jilin University, Qinghai Oil Company, Salzburg University, Dunhuang, China.

Johnsson, M.J. (1993) The system controlling the composition of clastic sediments. In: *Processes Controlling the Composition of Clastic Sediments* (Eds M.J. Johnsson and A. Basu), pp. 1–19. Special Paper 284, Geological Society of America, Boulder, CO.

Jolivet, M., Roger, F., Arnaud, N., Brunel, M., Tapponier, P. and Seward, D. (1999) Histoire de l'exhumation de l'Altun Shan: indications sur l'âge de la subduction du bloc du Tarim sous le système de l'Altun Thag (Nord Tibet). *C. R. Acad. Sci. Paris*, **329**, 749–755.

Jolivet, M., Brunel, M., Seward, D., *et al.* (2001) Mesozoic and Cenozoic tectonics of the northern edge of the Tibetan plateau: fission-track constraints. *Tectonophysics*, **343**, 111–134.

Lehmkuhl, F. and Haselein, F. (2000) Quaternary paleoenvironmental change on the Tibetan Plateau and adjacent areas (Western China and Western Mongolia). *Quatern. Int.*, **65/66**, 121–145.

Liu, Y., Genser, J., Ge, X., *et al.* (2003) $^{40}Ar/^{39}Ar$ age evidence for Altyn fault tectonic activities in western China. *Chinese Sci. Bull.*, **48**, 2024–2030.

Liu, Z., Wang, Y., Ye, C., Li, X. and Li, Q. (1998) Magnetostratigraphy and sedimentologically derived geochronology of the Quaternary lacustrine deposits of a 3000 m thick sequence in the central Qaidam basin, western China. *Palaeogeogr. Palaeoclimatol. Palaeoecol.*, **140**, 459–473.

Ludwig, K.R. (2001) *Isoplot/Ex – A Geochronological Toolkit for Microsoft Excel*. Special Publication 1a, Berkeley Geochronological Center.

Mader, D. and Neubauer, F. (2004) Provenance of Palaeozoic sandstones from the Carnic Alps (Austria): petrographic and geochemical indicators. *Int. J. Earth Sci.*, **93**, 262–281.

Meyer, B., Tapponnier, P., Bourjot, L., Métivier, F., Gaudemer, Y., Peltzer, G., Shunmin, G. and Zhitai, C. (1998) Crustal thickening in Gansu-Qinghai, lithosperic mantle subduction, and oblique, strike-slip controlled growth of the Tibet plateau. *Geophys. J. Int.*, **135**, 1–47.

Molnar, P. and Tapponnier, P. (1975) Cenozoic tectonics of Asia: effects of a continental collision. *Science*, **189**, 419–425.

Métivier, F., Gaudemer, Y., Tapponnier, P. and Meyer, B. (1998) Northeastward growth of the Tibet plateau deduced from balanced reconstruction of two depositional areas: The Qaidam and Hexi Corridor basins, China. *Tectonics*, **17**, 823–842.

Najman, Y., Pringle, M., Johnson, M.R.W., Robertson, A.H.F. and Wijbrans, J.R. (1997) Laser $^{40}Ar/^{39}Ar$ dating of single detrital muscovite grains from early foreland-basin sedimentary deposits in India: Implications for early Himalayan evolution. *Geology*, **25**, 535–538.

Neubauer, F., Liu, Y., Genser, J., Handler, R., Friedl, G. and Ge, X. (2004) $^{40}Ar/^{39}Ar$ detrital white mica ages of Palaeozoic and Mesozoic sandstones from Qilian and Altyn Mountains, China: constraints on accretion history and Palaeozoic palaeogeographic relationships. *Geophys. Res. Abstr.*, **6**, EGU04-A-03510.

Pettijohn, F.J., Potter, P.E. and Siever, R. (1987) *Sand and Sandstone*. Springer-Verlag, Berlin, 553 pp.

Rieser, A.B., Neubauer, F., Liu, Y. and Ge, X. (2005) Sandstone provenance of north-western sectors of the intracontinental Cenozoic Qaidam Basin, western China: tectonic vs. climatic control. *Sediment. Geol.*, **177**, 1–18.

Rieser, A.B., Liu, Y., Genser, J., Neubauer, F., Handler, R. and Ge, X.-H. (2006). Uniform Permian $^{40}Ar/^{39}Ar$

detrital mica ages in the eastern Qaidam Basin (NW China): where is the source? *Terra Nova*, **18**, 79–87.

Ritts, B.D., Yue, Y. and Graham, S.A. (2004) Oligocene-Miocene Tectonics and Sedimentation along the Altyn Tagh Fault, Northern Tibetan Plateau: analysis of the Xorkol, Subei, and Aksay Basins. *J. Geol.*, **112**, 207–229.

Robinson, D.M., Dupont-Nivet, G., Gehrels, G.E. and Zhang, Y. (2003) The Tula uplift, northwestern China: evidence for regional tectonism of the northern Tibetan Plateau during late Mesozoic–early Cenozoic time. *Geol. Soc. Am. Bull.*, **115**, 35–47.

Schwab, F.L. (1975) Framework mineralogy and chemical composition of continental margin-type sandstones. *Geology*, **3**, 487–490.

Shi, Y., Yu, G., Liu, X., Li, B. and Yao, T. (2001) Reconstruction of the 30–40 ka BP enhanced Indian monsoon climate based on geological records from the Tibetan Plateau. *Palaeogeogr. Palaeoclimatol. Palaeoecol.*, **169**, 69–83.

Sobel, E.R. and Arnaud, N. (1999) A possible middle Paleozoic suture in the Altyn Tagh, NW China. *Tectonics*, **18**, 64–74.

Sobel, E.R., Arnaud, N., Jolivet, M., Ritts, B.D. and Brunel, M. (2001) Jurassic to Cenozoic exhumation history of the Altyn Tagh range, NW China constrained by $^{40}Ar/^{39}Ar$ and apatite fission track thermochronology. In: *Paleozoic and Mesozoic Tectonic Evolution of Central and Eastern Asia: from Continental Assembly to Intracontinental Deformation* (Eds M.S. Hendrix and G.A. Davis), pp. 1–15. Memoir 194, Geological Society of America, Boulder, CA.

Song, S., Yang, J., Liou, J.G., Wu, C., Shi, R. and Xu, Z. (2003) Petrology, geochemistry and isotopic ages of eclogites from the Dulan UHPM Terrane, the North Qaidam, NW China. *Lithos*, **70**, 195–211.

Song, T. and Wang, X. (1993) Structural styles and stratigraphic patterns of syndepositional faults in a contradictional setting: Examples from Qaidam Basin, NW China. *Am. Assoc. Petrol. Geol. Bull.*, **77**, 102–117.

Sun, Z., Feng, X., Li, D., Yang, F., Qu, Y. and Wang, H. (1999) Cenozoic ostracoda and palaeoenvironments of the northeastern Tarim Basin, western China. *Palaeogeogr. Palaeoclimatol. Palaeoecol.*, **148**, 37–50.

Sun, Z., Yang, Z., Pei, J., *et al.* (2005) Magnetostratigraphy of Paleogene sediments from northern Qaidam Basin, China: implications for tectonic uplift and block rotation in northern Tibetan plateau. *Earth Planet. Sci. Lett.*, **237**, 635–646.

Tapponnier, P. and Molnar, P. (1977) Active faulting and tectonics in China. *J. Geophys. Res.*, **82**, 2905–2930.

Tapponnier, P., Ryerson, F.J., Van der Woerd, J., Mériaux, A.-S. and Lasserre, C. (2001a) Long-term slip rates and characteristic slip: keys to active fault behaviour and earthquake hazard. *C. R. Acad. Sci. Paris*, **333**, 483–494.

Tapponnier, P., Xu, Z., Roger, F., *et al.* (2001b) Oblique stepwise rise and growth of the Tibet Plateau. *Science*, **294**, 1671–1677.

Wang, J., Wang, J.Y., Liu, Z.C., Li, J.Q. and Xi, P. (1999) Cenozoic environmental evolution of the Qaidam Basin and its implications for the uplift of the Tibetan Plateau and the drying of central Asia. *Palaeogeogr. Palaeoclimatol. Palaeoecol.*, **152**, 37–47.

Wang, Y. and Zhang, M. (1999) *Geological map of Qaidam Basin 1:500,000.* Jin Zhijun and Zhang Bingshan edn, Qinghai Oil Company and China Oil University, Beijing.

Wijbrans, J.R., Pringle, M.S., Koopers, A.A.P. and Schveers, R. (1995) Argon geochronology of small samples using the Vulkaan argon laserprobe. *Proc. K. Ned. Akad. Wet.*, **98**, 185–218.

Willett, S.D. and Brandon, M.T. (2002) On steady states in mountain belts. *Geology*, **30**, 175–178.

Wittlinger, G., Tapponnier, P., Poupinet, G., *et al.* (1998) Tomographic evidence for localized lithospheric shear along the Altyn Tagh fault. *Science*, **282**, 74–76.

Xia, W., Zhang, N., Yuan, X., Fan, L. and Zhang, B. (2001) Cenozoic Qaidam basin, China: A stronger tectonic inversed, extensional rifted basin. *Am. Assoc. Petrol. Geol. Bull.*, **85**, 715–736.

Xu, Z., Li, H., Chen, W., Wu, C., Yang, J., Jin, X. and Chen, F. (2002) A large ductile sinistral strike-slip shear zone and its movement timing in the South Qilian Mountains, Western China. *Acta Geol. Sinica*, **76**, 183–193.

Yang, F., Ma, Z., Xu, T. and Ye, S. (1992) A Tertiary paleomagnetic stratigraphic profile in Qaidam Basin. *Acta Petrol. Sinica*, **13**, 97–101.

Yang, J., Xu, Z., Song, S., Zhang, J., Wu, C., Shi, R., Li, H. and Brunel, M. (2001a) Discovery of coesite in the North Qaidam Early Palaeozoic ultrahigh pressure (UHP) metamorphic belt, NW China. *C.R. Acad. Sci. Paris*, **333**, 719–724.

Yang, J., Xu, Z., Zhang, J., Chu, C.-Y., Zhang, R. and Liou, J.G. (2001b) Tectonic significance of early Paleozoic high-pressure rocks in Altun-Qaidam-Qilian Mountains, northwest China. In: *Paleozoic and Mesozoic Tectonic Evolution of Central and Eastern Asia: from Continental Assembly to Intracontinental Deformation* (Eds M.S. Hendrix and G.A. Davis), pp. 151–170. Memoir 194, Geological Society of America, Boulder, CA.

Yin, A. and Harrison, T.M. (2000) Geologic evolution of the Himalayan-Tibetan orogen. *Annu. Rev. Earth Planet. Sci.*, **28**, 211–280.

Yin, A. and Nie, S.Y. (1996) Phanerozoic palinspastic reconstruction of China and neighbouring areas. In: *Tectonic Evolution of Asia* (Eds A. Yin and T.M. Harrison), pp. 442–485. Cambridge University Press, Cambridge.

Yue, Y., Ritts, B.D. and Graham, S.A. (2001) Initiation and long-term slip history of the Altyn Tagh fault. *Int. Geol. Rev.*, **43**, 1087–1093.

Yue, Y., Ritts, B.D., Graham, S.A., Wooden, J.L., Gehrels, G.E. and Zhang, Z. (2003) Slowing extrusion tectonics: lowered estimate of post-Early Miocene slip rate for the Altyn Tagh fault. *Earth Planet. Sci. Lett.*, **217**, 111–122.

Yue, Y., Ritts, B.D., Hanson, A.D. and Graham, S.A. (2004) Sedimentary evidence against large strike-slip translation on the Northern Altyn tagh fault, NW China. *Earth Planet. Sci. Lett.*, **228**, 311–323.

Zhang, Z.M., Liou, J.G. and Coleman, R.G. (1984) An outline of the plate tectonics of China. *Geol. Soc. Am. Bull.*, **95**, 295–312.

Provenance of Quaternary sands in the Algarve (Portugal) revealed by U–Pb ages of detrital zircon

CRISTINA VEIGA-PIRES*, DELMINDA MOURA*, BRUNO RODRIGUES*, NUNO MACHADO†,‡, LEIF CAMPO† *and* ANTONY SIMONETTI†

**FCMA – CIMA, Universidade do Algarve, Campus de Gambelas, 8000-117 Faro, Portugal (Email: cvpires@ualg.pt)*
†Centre de recherche en géochimie et géodynamique – GEOTOP-UQAM-McGILL, CP 8888, Succ. Centre-Ville, Montréal, Quebec H3C 3P8, Canada
‡Département des Sciences de la Terre et de l'Atmosphère, UQAM, CP 8888, Succ. Centre-Ville, Montréal, Quebec H3C 3P8, Canada

ABSTRACT

The application of U–Pb dating by laser ablation–inductively coupled plasma–mass spectrometry (LA–ICP–MS) to determine the sources of detrital zircon in Neogene and Quaternary sands of the Algarve (southern Portugal) revealed the presence of three age groups: Palaeozoic to Neoproterozoic (200–800 Ma), Palaeoproterozoic (ca. 1700–2100 Ma) and Neo- to Meso-Archaean (2600–3200 Ma). The results suggest that at least some of the detrital grains were derived from pre-existing formations from the Southern Portuguese Zone of the Variscan orogen (namely Palaeozoic metasediments) and Mesozoic sedimentary rocks. The older sources from which zircons probably derived originally seem to be metamorphic rocks cropping out northeast of the Southern Portuguese Zone. Previous work shows that these older rocks from the Variscan Ossa–Morena and Central Iberian Zones contain inherited zircons with the same ages as those obtained in the present study from sediments. In parallel, the results also show a consistency in the source ages of the detrital zircons from Lower Pliocene to Holocene sediments, thus contributing to the discussion of known changes to the river drainage network in the Algarve region during the Pliocene based on published field observations.

Keywords Detrital zircons, Neogene, Quaternary, U–Pb ages, river drainage network, tectonics.

INTRODUCTION

Present-day river drainage networks are strongly influenced by climatic factors and changes in climate through the Holocene. However, in the Algarve region (southern Portugal), the present drainage network characteristics have been defined by the geomorphological and tectonic settings established during Pliocene–Pleistocene time. Drainage networks are modified, therefore, by changes in two main factors: climate and tectonics. In the Algarve region, a relatively important change in the river drainage network occurred between Early and Late Pleistocene time, as recorded by the change in the fluvial system from deeply incised channels to braided rivers migrating laterally on top of older formations (Cabral, 1993; Moura & Boski, 1994, 1997). However, the most recent tectonic event in the Algarve, which corresponds to the uplift of the northern part of the region and was due to the NW–SE compression linked to the convergence of tectonic plates, seems to have occurred between the Early and Late Pliocene (Cabral, 1993; Moura & Boski, 1999).

The cause and timing of the observed reorganization of the drainage network in the Algarve region are still under discussion. The above interpretations are based on field observations and stratigraphic evidence, lacking absolute dating to constrain timing. The present work aims, therefore, to highlight and quantify the contribution of geomorphological changes to drainage network modifications through numerical dating of detrital minerals, which allows the characterization and identification of sediment sources since the Pliocene. Accordingly, the present study focuses on zircons from the heavy mineral assemblages in Pliocene and Pleistocene sandstone formations, as well as in sands from modern beaches.

The mineralogical composition of beach and fluvial sands results from several factors, the most important of which are the composition of the source rocks and mixing by hydrodynamic processes. Since the pioneering work of Trask (1952), who used the mineral augite as a tracer, the study of the heavy mineral assemblages of beach sands has been used to determine sediment sources and to characterize sediment transport and mixing. This is important not only in order to reconstruct the evolution of river drainage networks but also to understand ongoing and evolving coastal processes. However, specific mineral tracers are seldom found in most beach sediments and physicochemical characteristics of heavy mineral assemblages are often difficult to relate to specific sources. In contrast, the crystallization age of single zircon grains, a common constituent of heavy mineral assemblages, is a direct indication of the age of the source rock.

Although traditionally used for precise dating of geological events such as magmatism and metamorphism (e.g. Krogh, 1993), U–Pb dating of zircon has been shown to be a valuable tool in studies of sediment provenance (Machado & Gauthier, 1996; Machado *et al.*, 1996; Fernandéz-Suárez *et al.*, 1999, 2002; Sircombe, 1999). Zircon dating has not been used often in sedimentary studies, in part because the method most commonly used, isotope dilution–thermal ionization–mass spectrometry, although more precise, is expensive and time consuming. However, it is likely that the recent development of affordable and relatively fast U–Pb dating methods based on laser ablation (e.g. Horn *et al.*, 2000; Machado & Simonetti, 2001) will widen their application in unravelling sedimentary processes. The results reported here represent the first attempt at using U–Pb ages of zircon from Pliocene–Pleistocene formations and Holocene beach sands from the central Algarve in order to determine their provenance.

GEOLOGICAL SETTING

Located in the western part of the Iberian Peninsula, Portugal is bordered to the west and south by the Atlantic Ocean and to the north and east by Spain. The western part of the Iberian Peninsula is mostly represented by the Iberian Massif, also known as the Hesperic Massif, which forms the most continuous portion of the European Variscan orogen (Fig. 1). The Iberian Variscan belt is divided into five structural zones (Fig. 1), the: (i) Cantabrian Zone (CZ); (ii) West Asturian–Leonese Zone (WALZ); (iii) Central Iberian Zone (CIZ); (iv) Ossa–Morena Zone (OMZ); and (v) South Portuguese Zone (SPZ). Isolated during the Pangean fragmentation, the Iberian block also shows some remnant ophiolitic units identified as oceanic exotic terranes (Terrinha, 1998; Fig. 1).

The Algarve, 5019 km^2 in area, is Portugal's southernmost region and is underlain by Variscan basement rocks of the South Portuguese Zone (Fig. 1). This zone is overthrust by the Ossa–Morena Zone (Fig. 1), in which Lower Palaeozoic and Upper Proterozoic formations are widespread (Terrinha, 1998), in contrast to the South Portuguese Zone (SPZ) where the oldest rocks consist essentially of Upper Devonian shales and greywackes. Acidic and mafic volcanic rocks of Mississipian (Tournaisian and Early Visean) age are also present in the southeastern sector of the SPZ. These rocks are of economic interest due to their high content of pyrite and other sulphides, which makes this region one of the most important in the world for massive sulphide ore deposits. The South Portuguese Zone is also characterized by low-grade regional metamorphism decreasing from northeast to southwest and containing diverse mineral assemblages (Oliveira, 1990).

Sedimentary basins located at the periphery of the Iberian Massif (Fig. 1) contain Mesozoic successions first represented by Triassic sandstones, which overlie Palaeozoic metasediments. The genesis of the Algarve Basin is attributed to the opening of Tethys and the Atlantic Ocean (Terrinha & Ribeiro, 1995; Andeweg, 2002). Carbonate sediments deposited during the Jurassic Period now form an extensive E–W orientated zone, locally called the Barrocal (Fig. 2), that comprises limestones and dolomites. During the Cretaceous, sedimentation alternated between carbonate and terrigenous depending on sea-level fluctuations (Terrinha & Ribeiro, 1995; Andeweg, 2002). The Late Cretaceous was marked by the intrusion of a nephelinic syenite subvolcanic massif, the Monchique Massif

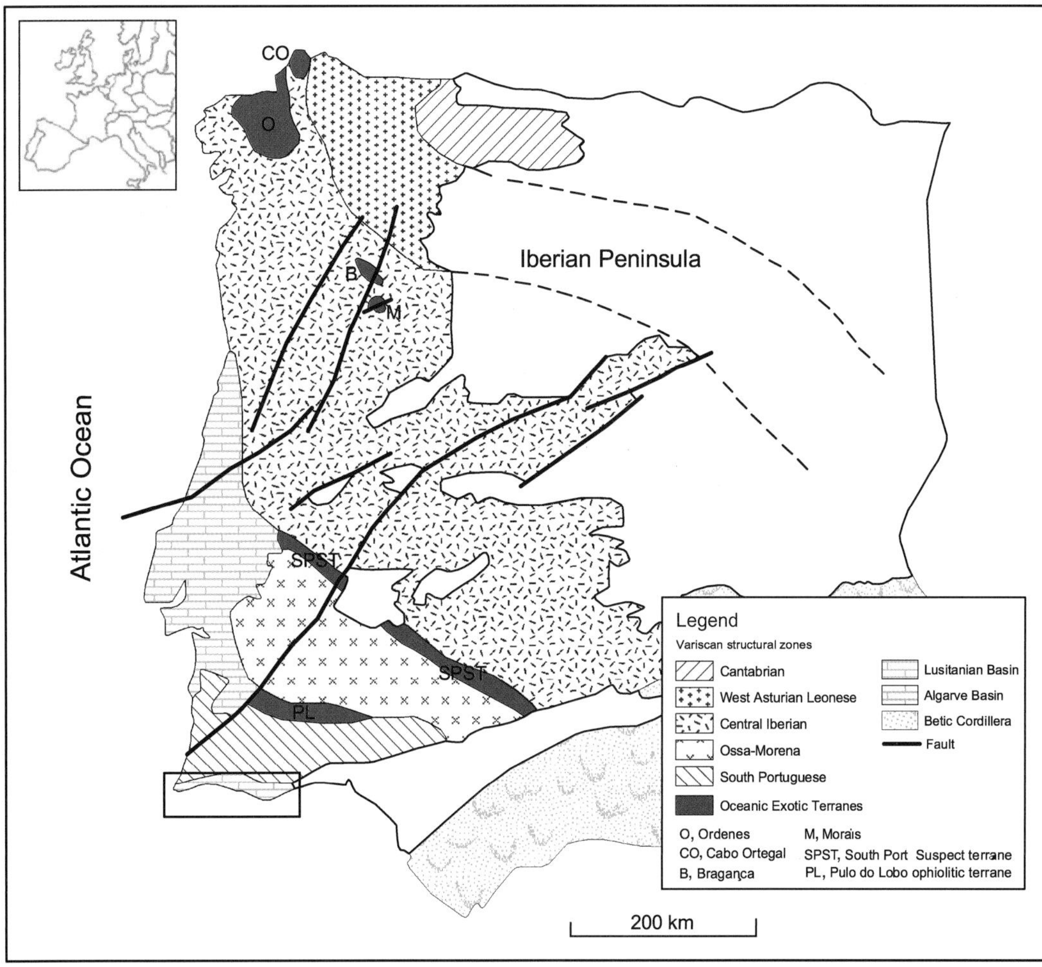

Fig. 1 Variscan structural zones of the Iberian Massif, as well as exotic terranes and main tectonic structures of the Iberian Peninsula. The South Portuguese Zone includes several domains such as the Baixo Alentejo Flysch group (Upper Visean to Namurian), the Pyritic belt (Upper Famennian to Mid-Visean), the Southwest Portuguese Flysch Group (Famennian to Lower Tournaisian) and the Pulo do Lobo terrane (PL; Tournaisian to Lower Devonian). The rectangle represents the area of Fig. 2.

(Rock, 1983). During the Cenozoic, sedimentation in the Algarve was first carbonate-dominated, represented by the Upper Miocene *Olhos de Água* calcarenites (unit F) and *Cacela* (unit E) Formation (Figs 2 & 3). Subsequently, during Early and Late Pliocene times, fluvio-marine and shallow continental shelf sedimentation took place, as recorded by the *Falésia* (unit D) and *Quarteira* (unit C) sandstones (Figs 2 & 3). Finally, the sedimentation became fluvial (units B and A) during the Pleistocene (Figs 2 & 3; Moura & Boski, 1999). The oldest formations are typically located in the north of the region and the youngest in the south. Both Neogene and Quaternary formations show maximum lateral extent along the present-day coastline, and they dip and thicken to the south-southeast (Figs 1, 2 & 4).

METHODS

In general, the Plio-Pleistocene outcrops are neither continuous nor easy to reach (Figs 2 & 4). Accordingly, whenever possible, several samples of the same sedimentary units (Table 1) were taken from locations where outcrops had previously been

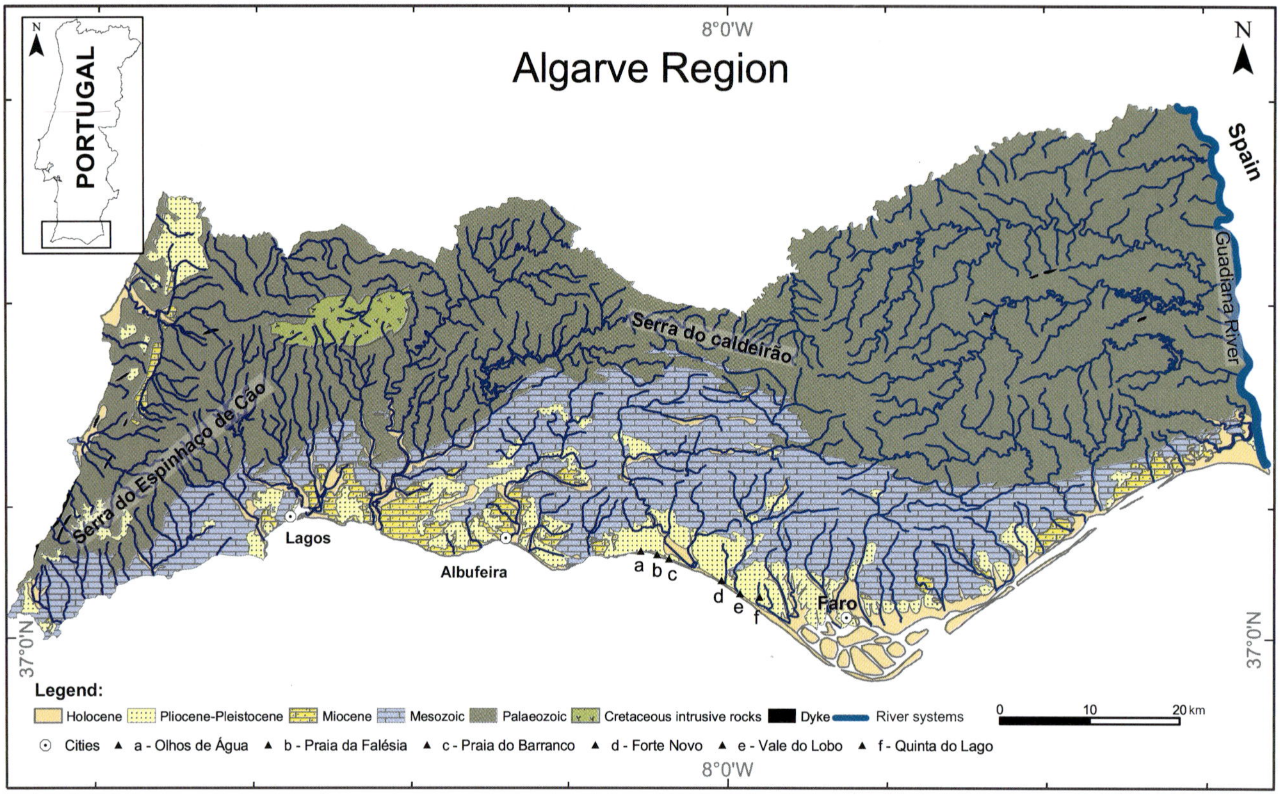

Fig. 2 Simplified geological map of the Algarve with the present-day river drainage network and the sampling sites (a–f).

identified (Moura & Boski, 1994, 1999). Samples consisted of poorly consolidated sandstones, which allowed them to be treated as if they were loose sediments. Heavy minerals were separated using a Wilfley table, heavy liquids and a Frantz isodynamic magnetic separator. Zircon was the only mineral found that is appropriate for U–Pb dating. Only zircon grains devoid of fractures, inclusions, alteration and other imperfections were selected for analysis. The grains analysed were representative of the chromatic and morphological types present in each sample. The zircons were analysed by laser ablation–multicollector–inductively coupled plasma–mass spectrometry (LA–MC–ICP–MS) at the GEOTOP-UQAM-McGILL Research Centre, Montreal (Canada).

Selected grains were mounted in epoxy known to be devoid of lead and uranium from previous analyses (Machado & Simonetti, 2001). Fragments of an in-house reference sample zircon (UQ-Z8) were also added at this stage to the sample mount. Samples were manually polished and cleaned with distilled water in an ultrasonic bath, sub-boiling 6.2 mol L^{-1} HCl and sub-boiling H_2O, and left to dry under a class 100 clean hood. The results were obtained with an excimer laser coupled to a multicollector mass spectrometer (Micromass Isoprobe) with an ICP source and a hexapole collision cell. Data were acquired in static, multicollection mode using six Faraday collectors, in the only possible configuration that allowed the large mass spread between ^{204}Pb and ^{238}U to be encompassed.

After determining the frequency and energy of the laser, as well as the beam diameter (Machado & Simonetti, 2001), the UQ-Z8 reference sample was ablated and the argon nebulizer gas flow rate adjusted to obtain a mean ^{206}Pb/^{238}U as close as possible to that of the reference sample (Machado & Gauthier, 1996). The 2σ precisions obtained for

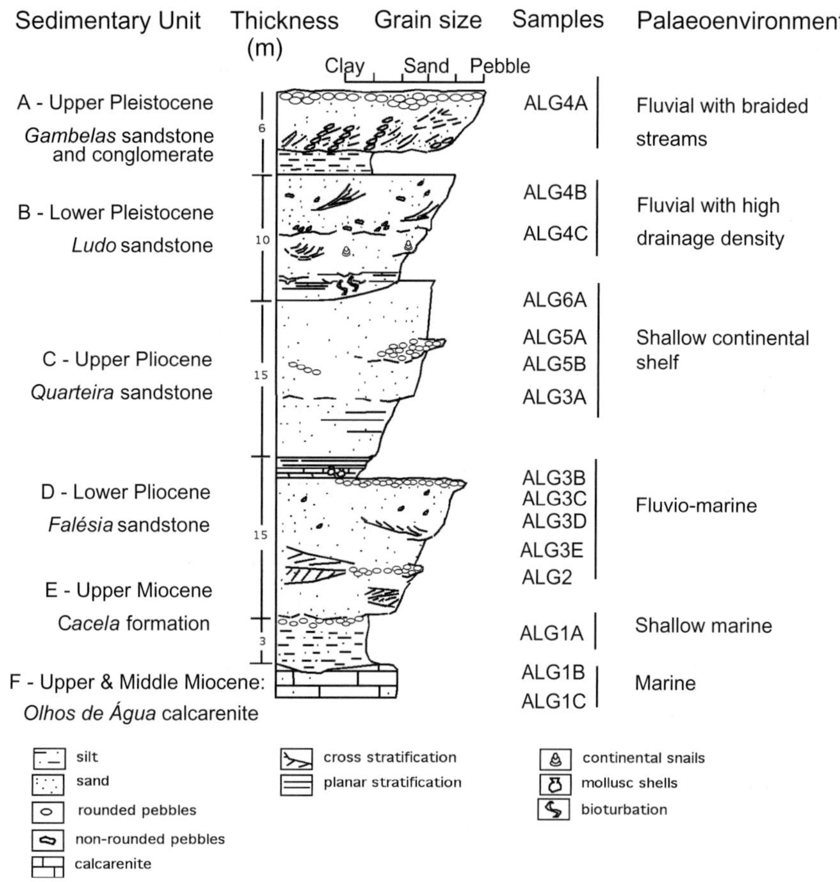

Fig. 3 Lithostratigraphic column from Miocene to Pleistocene. Sedimentary units are referred to by capital letters (A–F). (Adapted from Moura, 1998.)

$^{207}Pb/^{206}Pb$ and $^{238}U/^{206}Pb$ varied between 0.1 and 1.3%, respectively. All analyses were corrected for U fractionation relative to a $^{238}U/^{235}U$ ratio of 137.88. Age calculations were performed using Isoplot/Ex Version 2 (Ludwig, 2000).

RESULTS

All samples contained similar heavy mineral suites, dominated by staurolite, andalusite and sillimanite, typical of intermediate- to high-temperature metamorphism. Other minerals present in minor amounts were zircon, epidote and rutile. The majority of zircon grains were colourless and a minor proportion was light brown to pink. Two distinct types of zircon were present in all samples: a predominant group of rounded grains; and a lesser number of euhedral ones. The latter included both perfect crystals and crystals with percussion marks and slightly abraded edges. The lack of intermediate types is noteworthy.

The sample numbers and their stratigraphic position are indicated in Table 1 and the isotopic data in Table 2 and Figs 5–7. The $^{207}Pb/^{206}Pb$ ages of 103 zircon grains range between 199 Ma and 3.18 Ga, and can be classified into three groups: Neoproterozoic–Palaeozoic, Palaeoproterozoic and Archaean. Zircons with ages younger than 800 Ma were the most abundant, followed by those with ages around 2 Ga (Fig. 6) and finally by those in the 2.6–2.9 Ga range. The younger group could be subdivided, but it is possible that the observed gaps (Fig. 6) are due to sampling bias. Excepting the fact that the Archaean zircons are rounded, no correlation was found between zircon types and ages or between ages and stratigraphic position. This suggests that the same source or sources with identical ages have been available since the Pliocene.

Fig. 4 Photographs of the sampling locations. (a) Olhos de Água to Falésia Beach; the double arrow shows the specific location of Falésia sampling. (b) Falésia sampling location. (c) Quinta do Lago sampling outcrop. (d) Vale do Lobo sampling location. Capital letters refer to the sedimentary units in Fig. 3.

Table 1 Sample location and identification

Site	Location	Holocene beach (n = 54)	Pleistocene		Pliocene		Miocene (n = 1)
			Upper (n = 4)	Lower (n = 0)	Upper (n = 27)	Lower (n = 17)	
1	Olhos de Água						ALG1A*, ALG1B, ALG1C
2	Falésia					ALG2*	
3	Barranco	ALG3F*			ALG3A	ALG3B*, ALG3C*, ALG3D, ALG3E	
4	Quinta do Lago		ALG4A*	ALG4B, ALG4C			
5	Vale do Lobo	ALG5C*			ALG5A*, ALG5B*		
6	Forte Novo	ALG6B*			ALG6A		

n, total number of dated zircons.
* Samples analysed.

Table 2 Isotopic ratios and U–Pb ages

Sample-grain number	Isotopic ratios						Ages (Ma)		
	$^{206}Pb/^{238}U$	±1σ%	$^{207}Pb/^{235}U$	±1σ%	$^{207}Pb/^{206}Pb$	±1σ%	$^{206}Pb/^{238}U$	$^{207}Pb/^{235}U$	$^{207}Pb/^{206}Pb$
2A-1	0.4147	1.09	6.743	1.16	0.12608	0.10	2236	2078	2044
2A-2	0.4122	0.83	6.835	0.98	0.12226	0.05	2225	2090	1990
2A-3	0.3867	0.92	6.582	1.33	0.12908	0.38	2107	2057	2085
2A-4	0.0994	1.93	0.855	2.86	0.07417	3.58	611	628	1046
2A-5	0.1081	0.88	0.926	1.47	0.06310	0.58	662	665	712
2A-6	0.0984	0.65	0.854	0.71	0.06358	0.34	605	627	728
2A-7	0.1338	0.95	1.218	1.04	0.07100	0.35	810	809	957
2A-8	0.0604	0.69	0.473	1.64	0.05813	1.70	378	394	535
2A-9	0.1032	0.69	0.872	0.47	0.06127	0.32	633	637	649
2A-10	0.0969	0.44	0.844	0.75	0.06419	0.29	596	621	748
2A-22	0.0644	0.31	0.823	3.11	0.08061	3.62	402	610	1212
3B-2	0.0912	0.67	0.752	1.59	0.05982	0.94	563	569	597
3B-5	0.0770	0.38	0.593	0.98	0.05956	0.80	478	473	588
3B-6	0.0424	0.35	0.356	0.85	0.06153	1.07	268	310	658
3B-7	0.0452	0.57	0.328	1.08	0.05266	0.91	285	288	314
3B-8	0.0453	0.93	0.333	1.61	0.05298	1.73	286	292	328
3C-2	0.3205	0.91	4.915	2.35	0.12614	0.49	1792	1805	2045
3C-3	0.1004	0.35	0.747	4.69	0.06035	2.95	616	566	616
3C-4	0.1143	0.85	1.066	1.14	0.06840	0.27	698	737	881
3C-5	0.0852	0.58	0.702	0.74	0.06282	0.52	527	540	702
3C-6	0.0444	0.60	0.309	1.02	0.05299	0.83	280	273	328
3C-7	0.0452	0.64	0.336	0.98	0.05455	0.92	285	294	394
3C-8	0.3284	0.65	5.532	2.30	0.12165	0.44	1831	1906	1981
3C-9	0.0504	0.67	0.368	0.94	0.05299	0.71	317	318	328
3C-10	0.0933	0.71	0.788	0.71	0.06117	0.51	575	590	645
3C-11	0.0518	0.64	0.420	1.49	0.05893	1.35	326	356	565
3F-1	0.1000	0.72	0.869	0.79	0.06302	0.27	614	635	709
3F-3	0.1355	0.25	1.449	1.81	0.08048	1.28	819	909	1209
3F-4	0.0971	0.75	0.805	0.62	0.06052	0.25	597	599	622
3F-5	0.1010	0.76	0.853	0.79	0.06155	0.14	620	626	658
3F-6	0.1127	0.67	1.000	0.77	0.06421	0.20	688	704	749
3F-7	0.0549	0.56	0.395	1.23	0.05345	1.12	344	338	348
3F-8	0.0567	0.38	0.446	0.76	0.05748	0.64	355	375	510
3F-9	0.0555	0.36	0.416	1.44	0.05427	1.35	348	353	382
3F-10	0.0564	0.30	0.476	0.76	0.06292	0.77	353	395	706
3F-11	0.0535	0.20	0.453	0.74	0.06259	0.75	336	379	694
3F-12	0.3708	0.71	6.581	0.87	0.13007	0.05	2033	2057	2099
4A-1	0.0540	0.63	0.403	0.54	0.05472	0.37	339	344	401
4A-2	0.0892	0.57	0.805	2.83	0.06560	2.15	551	600	794
4A-4	0.3379	0.38	5.610	0.47	0.12306	0.08	1877	1918	2001
4A-5	0.0922	0.85	0.780	1.44	0.06196	1.14	569	585	673
5A-1	0.3578	0.55	6.597	1.12	0.12578	0.12	1972	2059	2040
5A-2	0.0525	0.71	0.395	1.07	0.05580	0.76	330	338	445
5A-3	0.0520	0.84	0.385	0.84	0.05412	0.54	327	331	376
5A-4	0.0482	0.59	0.384	0.74	0.05801	0.95	304	330	530
5A-5	0.0484	0.63	0.339	0.92	0.05257	1.02	305	296	310
5A-6	0.3693	0.80	6.339	1.10	0.13026	0.15	2026	2024	2101
5A-7	0.0926	0.72	0.735	1.48	0.05909	1.03	571	560	570
5A-8	0.1045	0.82	0.876	0.93	0.06128	0.31	641	639	649

Table 2 *(cont'd)*

Sample-grain number	Isotopic ratios						Ages (Ma)		
	$^{206}Pb/^{238}U$	±1σ%	$^{207}Pb/^{235}U$	±1σ%	$^{207}Pb/^{206}Pb$	±1σ%	$^{206}Pb/^{238}U$	$^{207}Pb/^{235}U$	$^{207}Pb/^{206}Pb$
5A-9	0.4776	1.25	13.101	1.51	0.20103	0.20	2517	2687	2835
5A-10	0.0682	0.32	0.858	1.90	0.10002	1.47	425	629	1625
5B-1	0.0849	0.62	0.695	1.39	0.06286	1.10	525	536	703
5B-2	0.0605	0.79	0.478	8.74	0.06450	7.23	379	397	758
5B-3	0.0468	0.69	0.354	0.96	0.05474	0.73	295	307	402
5B-4	0.1045	0.46	0.915	0.44	0.06554	0.42	641	660	792
5B-5	0.0485	0.50	0.357	1.04	0.05491	0.80	305	310	408
5B-6	0.0819	0.81	0.678	0.82	0.05994	0.25	507	526	601
5C-1	0.0453	0.87	0.344	1.64	0.05483	1.02	286	300	405
5C-2	0.0103	0.68	0.071	2.07	0.05009	1.78	66	70	199
5C-3	0.0506	0.67	0.399	1.27	0.05590	1.21	318	341	449
5C-4	0.0481	0.52	0.383	1.33	0.05881	1.16	303	329	560
5C-5	0.0499	0.54	0.432	2.02	0.06248	1.88	314	365	690
5C-6	0.0455	0.61	0.355	1.90	0.05771	2.00	287	309	519
5C-9	0.3425	0.65	5.709	1.11	0.12760	0.12	1899	1933	2065
6B-1	0.0588	0.75	0.500	1.29	0.06316	1.24	368	412	714
6B-2	0.0467	0.85	0.351	0.85	0.05464	0.66	294	305	397
6B-3	0.0496	0.72	0.373	1.31	0.05508	1.44	312	322	416
6B-4	0.0488	0.71	0.347	1.60	0.05194	1.25	307	302	283
6B-5	0.0454	0.62	0.341	1.32	0.05510	1.09	286	298	416
6B-6	0.0467	0.74	0.354	1.05	0.05473	0.49	294	308	401
6B-7	0.0940	0.74	0.806	0.99	0.06320	0.47	579	600	715
6B-8	0.0454	0.45	0.350	0.74	0.05563	0.51	286	305	438
6B-9	0.0506	0.23	0.469	1.34	0.06762	1.17	318	391	857
6B-10	0.0502	0.41	0.417	1.86	0.06060	2.11	316	354	625
6B-11	0.0480	0.50	0.364	0.67	0.05443	0.37	302	315	389
6B-12	0.0450	0.40	0.332	1.63	0.05500	1.41	284	291	412
6B-13	0.0466	0.23	0.340	2.15	0.05360	1.75	294	297	354
6B-14	0.0734	0.73	0.586	0.67	0.05952	0.35	457	468	586
6B-15	0.0456	0.28	0.337	1.56	0.05506	1.64	287	295	414
6B-16	0.0843	0.70	0.651	1.18	0.05758	1.08	522	509	514
6B-17	0.0960	0.59	0.829	0.59	0.06293	0.20	591	613	706
6B-18	0.0735	0.63	0.601	0.59	0.05933	0.21	457	478	579
6B-19	0.0795	0.66	0.616	0.76	0.05790	0.60	493	487	526
6B-20	0.0495	0.66	0.374	1.33	0.05533	1.32	312	322	425
6B-21	0.0852	0.73	0.741	3.59	0.06497	2.97	527	563	773
6B-22	0.0946	0.46	0.737	2.93	0.06075	2.21	583	561	630
6B-23	0.0897	0.57	0.721	3.59	0.06072	2.75	554	551	629
6B-24	0.0949	0.75	0.788	0.76	0.06097	0.44	584	590	638
6B-25	0.3416	0.92	5.966	1.21	0.12868	0.07	1894	1971	2080
6B-26	0.3254	0.67	5.525	0.69	0.12355	0.05	1816	1904	2008
6B-27	0.3442	0.92	6.112	1.26	0.13197	0.23	1907	1992	2124
6B-28	0.3295	0.61	5.613	0.96	0.12581	0.18	1836	1918	2040
6B-29	0.5065	0.76	14.166	1.44	0.20365	0.09	2642	2761	2856
6B-30	0.4637	0.62	11.470	2.07	0.18424	0.20	2456	2562	2691
6B-31	0.5306	0.54	15.515	0.98	0.21546	0.04	2744	2847	2947
6B-32	0.3454	0.58	6.169	0.56	0.12843	0.15	1913	2000	2077
6B-33	0.0120	0.39	0.100	4.25	0.06007	4.26	77	97	606
6B-34	0.0106	0.72	0.089	3.00	0.06103	3.12	68	86	640

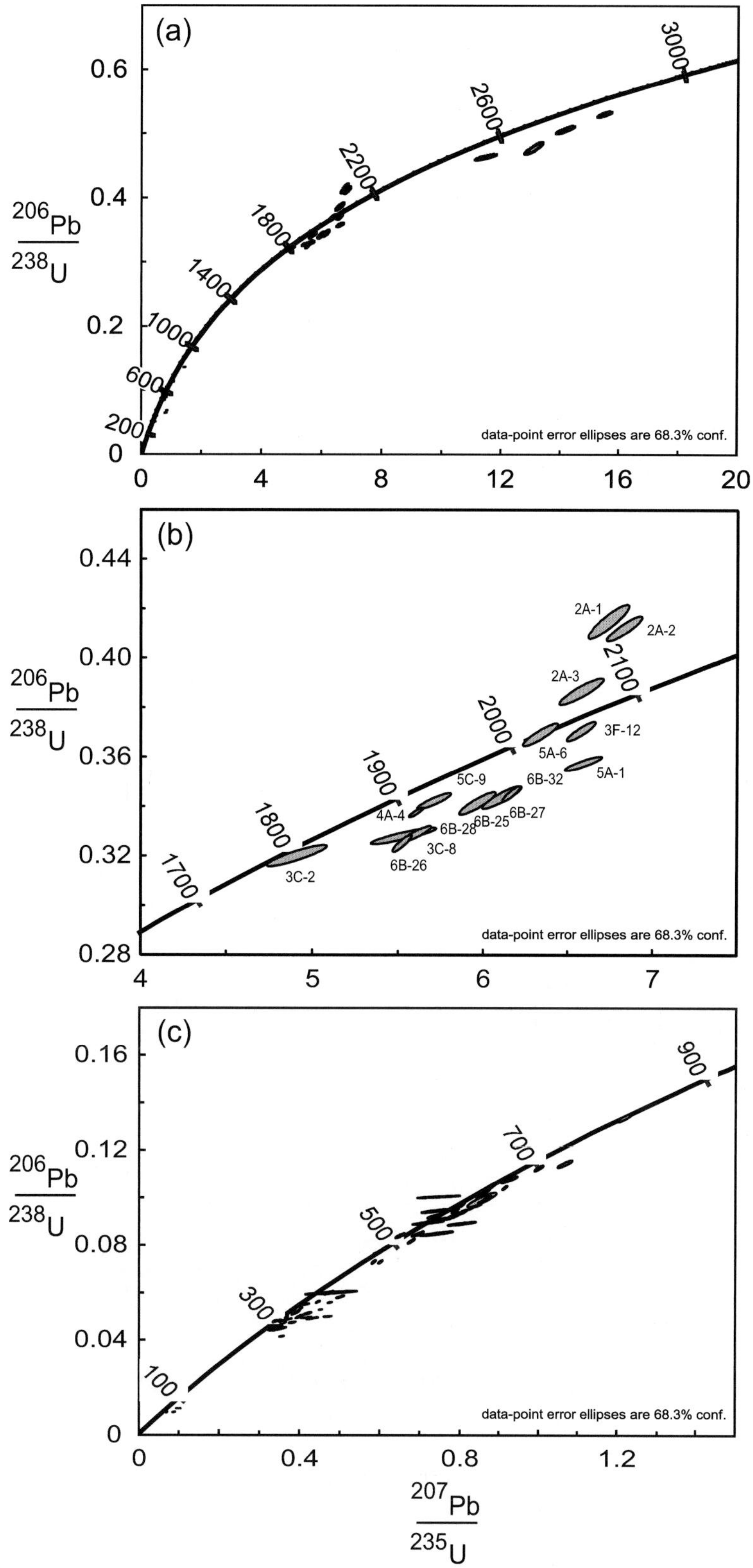

Fig. 5 Concordia diagrams for the detrital zircons analysed. (a) All grains. (b) Grains of Palaeoproterozoic age. The grain numbers are keyed to those in Table 2. (c) Neoproterozoic and younger zircons.

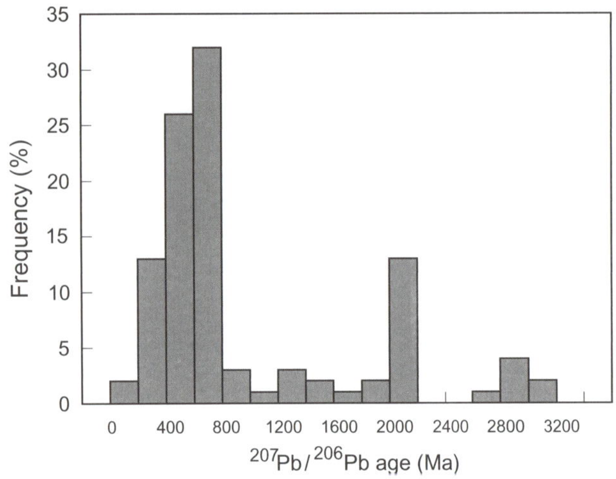

Fig. 6 Histogram illustrating the general age pattern of the detrital zircons.

DISCUSSION

The latest tectonic event in the South Iberian Peninsula occurred in the Late Pliocene due to tectonic plate movements that accompanied the formation of the Betics (Cabral, 1993). Along the southwestern Iberian margin, NW–SE compression and perpendicular NE–SW extension occurred, resulting in the uplift of the northern part of the Algarve region (Dias & Cabral, 1997; Andeweg, 2002). This tectonic event was probably responsible for changes in the drainage network, such as the change in the drainage path of the Guadiana River (Fig. 2), which started draining for the first time towards the south during Late Pliocene times (Martinez Del Olmo *et al.*, 1984; Cabral, 1993; Hurtado *et al.*, 1993; Vidal *et al.*, 1993;

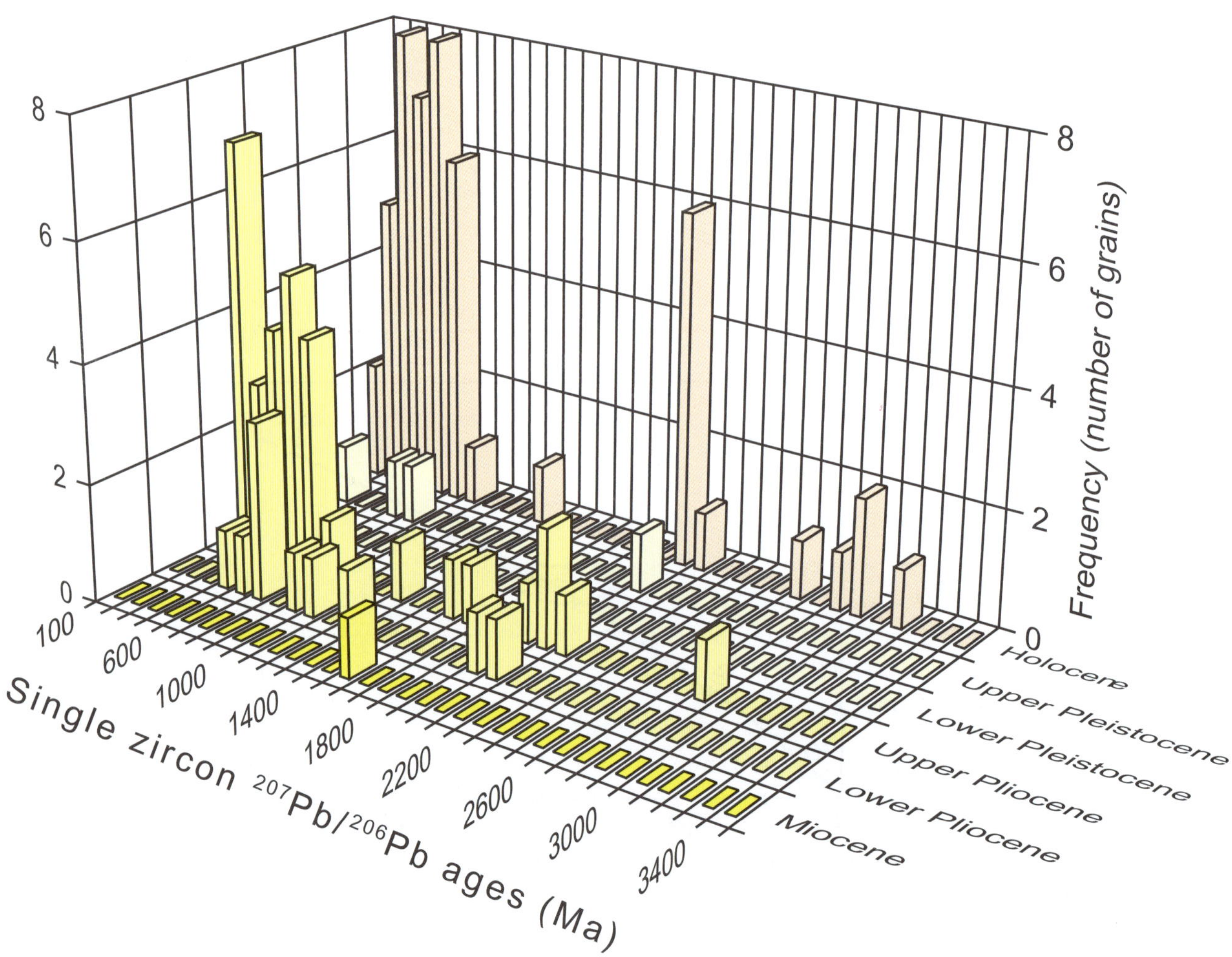

Fig. 7 Histogram of $^{207}Pb/^{206}Pb$ ages from single zircon grains plotted in relation to the age of the sample formation.

Hurtado & Vidal, 1994; Moura & Boski, 1997; Andeweg, 2002).

The Quaternary sandy fluvial facies, which cover Pliocene feldspathic sands, have a geometry and structure that point to very different fluvial regimes during the Early and Late Quaternary. The existence of euhedral and subrounded grains, evenly distributed in all samples, indicates at least two different sources. High grain-roundness can be due to either long transport distance, long residence time in shallow and/or high-energy marine waters, or to several cycles of transport and deposition. On the other hand, euhedral grains can be associated with short distance of transport if entrained as individual clasts and, hence, linked to a closer source, or they can lack abrasion marks due to their inclusion in other minerals, in which case, their sedimentary history can be difficult to reconstruct.

Nevertheless, detrital zircons from Late Pliocene formations to present-day beaches show the same three groups based on $^{207}Pb/^{206}Pb$ ages (Figs 5–7). The first and most abundant group is associated with Palaeozoic to Neoproterozoic ages (200–800 Ma); the second group is associated with Palaeoproterozoic ages (1700–2100 Ma); and the third and less common group with Archaean ages (> 2600 Ma). Zircons displaying Archaean to Lower Palaeozoic ages are superficially difficult to explain, since Archaean rocks are unknown in the entire Iberian Peninsula, and Neoproterozoic to Palaeozoic rocks are unknown in the South Portuguese Zone (Fig. 1). However, Archaean zircons are reported as being incorporated in Neoproterozoic and Palaeozoic rocks from Iberia, namely in the Ossa–Morena and Cantabrian Zones (Fig. 1; de la Rosa *et al.*, 2002; Fernández-Suárez *et al.*, 2000, 2002). These Archaean detrital zircons are probably derived from the West African craton and parts of the old craton remobilized by the Pan-African orogeny in northern Africa (Fernández-Suárez *et al.*, 2002; Zeck *et al.*, 2004). Nevertheless, to explain the wide range of ages obtained from detrital zircons in the present study, and their morphological and colour variations, three hypotheses can be formulated.

First, detrital zircons in Plio-Pleistocene formations could come from the erosion of northern Algarve Palaeozoic schists and greywackes, considering that these formations already contain inherited zircons from older formations. In this case, the Serra do Caldeirão region (Fig. 2) would be a possible Palaeozoic source since it is included in the Baixo Alentejo Flysch Domain of the SPZ, which is formed of Carboniferous terrigenous sediments of syn-orogenic character derived from the OMZ and probably also from the CIZ (Fig. 1; Oliveira, 1990).

Second, if it is considered that zircons come directly from Precambrian and Palaeozoic formations, then the drainage network must have been oriented from northeast to southwest. Indeed, the domains where such older rocks exist are located northeast of the Algarve and correspond to the formations of the Ossa–Morena and Central Iberian Zones (Fig. 1). This hypothesis could corroborate the change in the river drainage network orientation from NE–SW to the present-day NW–SE, documented in the Algarve and Andalusia (Martinez del Olmo *et al.*, 1984; Hurtado & Vidal, 1994; Moura & Boski, 1997). Such a change could have been due to a late Alpine orogenic phase, which caused the general uplift of southwestern Iberia during the Late Pliocene (Cabral, 1993; Andeweg, 2002).

Third, the detrital zircons and the associated metamorphic minerals could have been derived from both previous sources. With the existing data and the actual knowledge of the South Portuguese Zone, more specifically the Baixo Alentejo Flysch Domain, it is difficult to preferentially support one or another hypothesis.

In any case, if the roundness and altered colour of some individual grains are taken into consideration, then the provenance for the Pliocene drainage network could have been pre-existing Cretaceous and Triassic clastic formations (Andeweg, 2002). Accordingly, zircons would have undergone several sedimentary cycles, the first of which must have had a source area northeast of the South Portuguese Zone. This would explain the observed sparse distribution and remains of Cretaceous clastic formations associated with the main E–W oriented Jurassic fold belt (Moura, 1998).

At the present day, the main rivers in the Algarve are oriented NW–SE or N–S (Fig. 2). Their drainage basin source areas are either Palaeozoic shales and greywackes or Mesozoic carbonate rocks (Fig. 2). Sedimentological evidence, such as palaeocurrent patterns and facies distributions, points to a drastic change of the drainage network in the south Iberian Peninsula after the Late Pliocene (Gouvêa, 1938; Feio, 1946; Vidal *et al.*, 1993; Moura & Boski, 1997). Assuming that the southeastwards-directed drainage pattern was initiated in Plio-Pleistocene times, detrital zircon should be derived from a unique source.

The oldest rocks cropping out in the southern Alentejo–northern Algarve region are Devonian flysch sequences of the South Portuguese Zone of the Iberian Massif, deformed and metamorphosed during the Variscan orogeny (Oliveira, 1990; Terrinha, 1998). However, the lithostratigraphy of southern Portugal reveals the absence of significant magmatic and metamorphic events that could produce zircon-bearing rocks. Rather, it shows that detrital zircon from the Palaeozoic flysch units could have been through sedimentary cycles during the Triassic, Cretaceous and Cenozoic, and underwent further recycling in the Holocene beaches. Apparently, most detrital zircons therefore could be derived from this Palaeozoic flysch, which can be found in the Serra do Caldeirão (Fig. 2).

However, a striking observation is that the most frequent heavy minerals found in both the Holocene beaches and the Plio-Pleistocene rocks are staurolite and andalusite typical of medium-grade schists. These minerals are absent in the Palaeozoic flysch sequences of southern Alentejo, which are characterized by low-grade metamorphism (Oliveira, 1990), implying that these sequences are themselves derived from older metamorphic rocks. The source(s) of the sediments typical of the South Portuguese Zone are still a long-standing problem, but their age characterization by the method reported here is underway.

It is also of interest to note that with the exception of Archaean zircons that are well rounded, the younger ones are either euhedral to little rounded or very well rounded. This bimodality suggests that zircon grains display two abrasion histories, whereby rounded grains probably underwent multiple sedimentary cycles, whereas the euhedral ones could have been liberated from the source rock in the Holocene. The processes leading to rounding of detrital grains are not well understood, but if the conclusion of Kuenen (1959) that grains are not significantly rounded during transport is accepted, it is appropriate to suggest that zircon underwent rounding in beach settings. This would imply the recycling of coastal sediments.

The current study did not directly assess evidence for multiple sedimentary cycles, but a project with such an objective has been initiated. It is possible, however, to identify three main orogenic episodes: at 300–450 Ma, 450–800 Ma and 1.8–2.2 Ga, corresponding to Variscan tectonometamorphic events in Europe, and to Pan-African and Eburnean events in Gondwana. The interval 2.6–3.18 Ga is too poorly defined to be ascribed to a specific event, but several events in this age range are known in Gondwana (e.g. Machado *et al.*, 1996). These Archaean ages obtained on several zircon grains open a vast new domain to study on sediment recycling and palaeogeographical reconstructions.

Nevertheless, reworking of previous clastic formations, probably of Cretaceous and Triassic age, seem to be responsible for the existing Pliocene and Pleistocene formations, just as at the present time, when Holocene beach and fluvial sediments are mainly the result of the erosion of Pliocene and Pleistocene formations. Moreover, even the difference in drainage network characteristics observed between the Lower and Upper Pleistocene (Fig. 3) cannot be explained by a change in the drainage orientation. Indeed, based on the obtained single detrital zircon ages, the detrital sources do not seem to differ from the Pliocene to the present.

CONCLUSIONS

This work illustrates the application of U–Pb dating of individual detrital zircon grains to investigate the sources of Plio-Pleistocene detrital formations and of Holocene sands of the Algarve region. The results suggest that these units comprise both newly liberated and recycled detritus derived mainly from the pre-existing clastic formations or extensive Palaeozoic flysch sequences present in the South Portuguese Zone. Besides indicating that most zircons crystallized during Variscan, Pan-African and Eburnean tectonometamorphic events, these results also corroborate the existence of several sedimentary cycles through the Iberian Massif. However, the expected quantitative support for the timing of the river drainage network reorganization, previously documented in the Southern Iberian Peninsula during the Late Pliocene and linked to a late Alpine orogenic phase, has not been achieved. This is because single zircon ages from Lower Pliocene through Holocene sediments reveal the same source ages, ranging from the Palaeozoic to Archaean.

ACKNOWLEDGEMENTS

This work benefited from financial support from a Natural Sciences and Engineering Research Council of Canada (NSERC) grant to N. Machado and from individual financial support from the Portuguese Foundation for Science and Technology (FCT), through the CIMA centre pluri-annual grant, to D. Moura and C. Veiga-Pires. The Micromass Isoprobe instrument and the LambdaPhysik-Merchantek-New Wave laser system were financed through NSERC with contributions from FCAR (Québec) and Fondation UQAM. The laboratory is maintained in part with a NSERC MFA grant. R. Lapointe is thanked for providing essential technical support. We also acknowledge the precious help from Carlos Loureiro who kindly edited some of the figures. Finally, this work has benefited from the constructive and detailed reviews from Javier Fernández-Suárez, José R. Martínez Catalán and from the editor Edward A. Williams.

REFERENCES

Andeweg, B. (2002) *Cenozoic tectonic evolution of the Iberian Península: causes and effects of changing stress fields.* Unpublished PhD thesis, Vrije Universiteit, Amsterdam, 178 pp.

Cabral, J. (1993) *Neotectónica de Portugal Continental.* Unpublished PhD thesis, Universidade de Lisboa, 265 pp.

De la Rosa, J.D., Jenner, G.A. and Castro, A. (2002) A study of inherited zircons in granitoid rocks from the South Portuguese and Ossa-Morena Zones, Iberian Massif: support for the exotic origin of the South Portuguese Zone. *Tectonophysics*, **352**, 245–256.

Dias R.P. and Cabral, J. (1997) Plio-Quaternary crustal vertical movements in Southern Portugal – Algarve. In: *Actas da IV Reunião do Quaternário Ibérico* (Ed. J. Rodriguez Vidal), pp. 61–68. A.E.Q.U.A. – G.T.P.E.Q., Huelva.

Feio, M. (1946) Os terraços do Guadiana a Jusante do Ardila. *Comun. Inst. Geol. Min.*, **27**, 1–84.

Fernández-Suárez, J., Gutierrez, A.G., Jenner, G.A. and Tubrett, M.N. (1999) Crustal sources in lower Palaeozoic rocks from NW Iberia: insights from laser abalation U–Pb ages of detrital zicons. *J. Geol. Soc. London*, **156**, 1065–1068.

Fernández-Suárez, J., Gutierrez-Alonzo, G., Jenner, G.A. and Tubrett, M.N. (2000) New ideas on the Proteorozoic-Early Paleozoic evolution of NW Iberia: insights from U–Pb detrital zircon ages. *Precambrian Res.*, **102**, 185–206.

Fernández-Suárez, J., Gutiérrez Alonso, G. and Jeffries, T.E. (2002) The importance of along-margin terrane transport in northern Gondwana: insights from detrital zircon parentage in Neoproterozoic rocks from Iberia and Brittany. *Earth Planet. Sci. Lett.*, **204**, 75–88.

Gouvêa, A.M. (1938) *Algarve (aspectos fisiográficos).* Publicação do Instituto para a Alta Cultura, 145 pp.

Horn, I., Rudnick, R.L. and McDonough, W.F. (2000) Precise elemental and isotope determination by simultaneous solution nebulization and laser ablation–ICP–MS: application to U–Pb geochronology. *Chem. Geol.*, **164**, 281–301.

Hurtado, E.F. and Vidal, J.R. (1994) Rasgos morfotectónicos del interfluvio costero Guadiana-Guadalquivir (Golffo de Cádiz). In: *Geomorfologia en España* (Eds J. Arnaez, J.M. Garcia Ruiz and A. Gomez Villar), pp. 13–19. Sociedad Española de Geomorfologia, Logroño.

Hurtado, E.F., Rodriguez Vidal, J. and Cáceres Puro, L.M. (1993) Nuevos dados sobre compresión pleistocena en el extremo oriental del Algarve (Portugal). *Geogaceta*, **14**, 123–125.

Krogh, T.K. (1993) High precision U–Pb ages for granulite metamorphism and deformation in the archean Kapuskasing structural zone, Ontario: impications for structure and development of the lower crust. *Earth Planet. Sci. Lett.*, **119**, 1–18.

Kuenen, P.H. (1959) Experimental abrasion, 3: fluviatile action on sand. *Am. J. Sci.*, **257**, 172–190.

Ludwig, K.R. (2000) *Isoplot/Ex version 2.3. A Geochronological Toolkit for Microsoft Excel.* Special Publication 1a, Berkeley Geochronological Center, 54 pp.

Machado, N. and Gauthier, G. (1996) Determination of Pb-207/Pb-206 ages on zircon and monazite by laser-ablation ICPMS and application to a study of sedimentary provenance and metamorphism in southeastern Brazil. *Geochim. Cosmochim. Acta*, **60**, 5063–5073.

Machado, N. and Simonetti, A. (2001) U–Pb dating and Hf isotopic composition of zircon by laser ablation-

MC-ICP-MS. In: *Laser Ablation–ICPMS in the Earth Sciences – Principles and Applications* (Ed. P. Sylvester), pp. 121–146. Short Course Series 29, Mineralogical Association of Canada, St John's, Newfoundland.

Machado, N., Schrank, A., Noce, C.M. and Gauthier, G. (1996) Ages of detrital zircon from Archean–Paleoproterozoic sequences: Implications for Greenstone belt setting and evolution of a Transamazonian foreland basin in Quadrilatero Ferrifero, southeast Brazil. *Earth Planet. Sci. Lett.*, **141**, 259–276.

Martinez del Olmo, W., Mallo, J.G., Leret, J., Onãte, A.S. and Alba, J.S. (1984) Modelo tectosedimentario del Bajo Guadalqivir. In: *I Congresso Espanhol de Geologia*, Vol. I, pp. 199–213.

Moura, D. (1998) *Litostratigrafia do Neogénico terminal e Plistocénico na Baia Central-Algrave, Evolução paleoambiental.* Unpublished PhD thesis, Universidade do Algarve, Faro, 255 pp.

Moura, D. and Boski, T. (1994) Ludo Formation – a new lithostratigraphy unit in Quaternary of central Algarve. *Gaia*, **9**, 95–98.

Moura, D. and Boski, T. (1997) Sistema fluvial do Plistocénico Superior na Bacia Algarvia (portugal). *Cuatern. Ibérico*, AEQUA 69–76.

Moura, D. and Boski, T. (1999) Pliocene and Pleistocene lithostratigrafic units in Algarve. *Comun. Inst. Geol. Min.*, **86**, 85–106.

Oliveira, J.T. (1990) Stratigraphy and synsedimentary tectonism. In: *Pre-Mesozoic of Iberia* (Eds R.D. Dallmeyer and M. Garcia), pp. 334–347. Springer-Verlag, Berlin.

Rock, N.M.S. (1983) Alguns aspectos geológicos, petrológicos e geoquímicos do complexo eruptivo de Monchique. *Comum. Ser. Geol. Portugal*, **69**, 325–372.

Sircombe, K.N. (1999) Tracing provenance through the isotope ages of littoral and sedimentary detrital zircon, eastern Australia. *Sediment. Geol.*, **124**, 47–67.

Terrinha, P.A.G. (1998) *Structural geology and tectonic evolution of the Algarve Basin, south portugal*. Unpublished PhD thesis, University of London, 430 pp.

Terrinha, P. and Ribeiro A. (1995) *Tectonics of the Algarve Basin, South Portugal*. Memórias 4, Faculdade de Ciências, Museu e Laboratório Mineralógico e Geológico, Porto, pp. 321–325.

Trask, P.D. (1952) *Sources of Beach Sand at Santa Barbara, California, as Indicated by Mineral Grain Studies*. Technical Memo 28, Beach Erosion Board, U.S. Army Corp of Engineers.

Vidal, J.R., Cáceres, L.M. and Ramirez, A.R. (1993) Modelo evolutivo de la red fluvial cuaternaria en el suroeste de la Peninsula Iberica. In: *3ª Reunião do Quaternário Ibérico*, pp. 93–96, Coimbra.

Zeck, H.P., Wingate, M.T.D., Pooley, G.D. and Ugidos, J.M. (2004) A sequence of Pan-African and Hercynian events recorded in zircons from an orthogneiss from the Hercynian belt of the Western Central Iberia – an ion microprobe U–Pb study. *J. Petrol.*, **45**, 1613–1629.

Anatomy of a fluvial lowstand wedge: the Avilé Member of the Agrio Formation (Hauterivian) in central Neuquén Basin (northwest Neuquén Province), Argentina

GONZALO D. VEIGA*, LUIS A. SPALLETTI* *and* STEPHEN S. FLINT†

**Centro de Investigaciones Geológicas, Universidad Nacional de La Plata-CONICET, Calle 1 #644, B1900TAC La Plata, Argentina (Email veiga@cig.museo.unlp.edu.ar)*

†Stratigraphy Group, Department of Earth and Ocean Sciences, University of Liverpool, 4 Brownlow Street, Liverpool, UK

ABSTRACT

The Hauterivian (Lower Cretaceous) Avilé Member of the Agrio Formation constitutes a non-marine lowstand wedge dominated by fluvial and aeolian deposits that sharply overlie deep-marine, ammonite-bearing shales of the Lower Member of the Agrio Formation in the central part of the Neuquén Basin. Detailed sedimentological logging at 12 localities allowed the identification of 11 sedimentary bodies that record the evolution of fluvial environments through this lowstand wedge. Channel units identified include complex sheets and ribbons as well as simple ribbons developed under contrasting accommodation/supply conditions. Small-scale sandy and heterolithic channels are related to fine-grained floodplain/lacustrine deposits, together with small-scale bars and sandstone lobes indicating overbank splays. In addition, large-scale lacustrine bars are present, associated with complex ribbons, suggesting the development of distributary systems that fed relatively deep water bodies. Locally, aeolian reworking of fluvial channels and aeolian deposits (dunes and sandsheets) occurs.

Regional and vertical changes in fluvial style were recorded within this lowstand wedge. The up-dip area is characterized by a relatively small thickness and is almost completely dominated by the superimposition of complex sandstone sheets. Towards the north of the study area, in a down-dip position, the unit studied shows a much greater thickness and a high proportion of fine-grained floodplain deposits. However, the intercalation of bedload dominated and mixed-load, high-sinuosity fluvial intervals is recorded. This alternation represents contrasting accommodation/sediment supply conditions, associated either with climatic fluctuations or with oscillations in fluvial base-level that could be related to eustatic changes due to orbital processes. Although the vertical evolution in the upstream sector is obscured by reduced accommodation, in the downstream area the increase in the proportion of fine-grained facies and the gradual change to a mixed-load fluvial system reflect a gradual increase in accommodation (relative to coarse-grained sediment supply) that could be associated with an overall (low frequency) transgressive trend developed after the relative sea-level fall that produced the onset of non-marine accumulation in the central part of the basin.

Keywords Argentina, Neuquén Basin, Hauterivian, Cretaceous, fluvial deposits, sediment architecture, sequence stratigraphy.

INTRODUCTION

One of the most striking features of the Mesozoic fill of the Neuquén Basin in west central Argentina is the development of several non-marine successions that sharply overlie shallow- and even deep-marine deposits across major erosional surfaces (Legarreta, 2002). These deposits have been interpreted as second-order lowstand wedges developed in response to major relative sea-level falls

(Legarreta & Gulisano, 1989; Legarreta & Uliana, 1991). Even when these wedges have similar lithological characteristics and genesis, they differ significantly in facies development and distribution. Therefore, it is not easy to define a common vertical evolution or the external controlling factors for these deposits, and detailed facies and architectural analysis is needed to improve the understanding of these lowstand wedges.

The Hauterivian (Lower Cretaceous) Avilé Member of the Agrio Formation developed in response to one of these major relative sea-level falls, and is dominated by fluvial and aeolian deposits that are both underlain and overlain by deep-marine, ammonite-bearing shales of the Lower and Upper Members of the Agrio Formation in the central part of the basin. This paper aims to provide a more general characterization of these non-marine deposits and to describe how they relate to the marine deposits of the Agrio Formation. The data are then used to address the evolution of the basin. The Avilé Member is also a very important hydrocarbon reservoir in the Neuquén Basin and the added understanding of its character in outcrop may improve subsurface exploration and production.

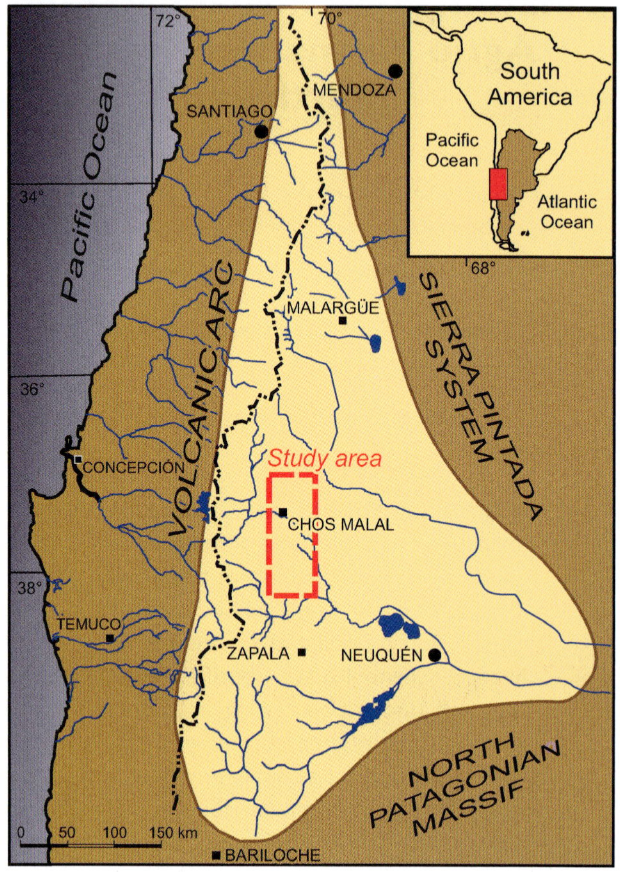

Fig. 1 Geological setting of the Neuquén Basin and location of the study area.

GEOLOGICAL SETTING AND PREVIOUS WORK

The Neuquén Basin, located in west-central Argentina (Fig. 1), is a large triangular-shaped depocentre that was active between Late Triassic and Early Tertiary time (Legarreta & Gulisano, 1989). It evolved as a back-arc basin on the active southwestern margin of Gondwana. The main characteristics of its sedimentary record were controlled by a combination of eustatic oscillations and a complex tectonic history (Vergani *et al.*, 1995), related both to the dynamics of the proto-Andean active margin and intraplate activity related to the break-up of the Gondwana supercontinent.

Since Late Valanginian times, the central part of the Neuquén Basin was dominated by the accumulation of deep-marine deposits of the Agrio Formation (Fig. 2). This unit is characterized by a thick succession of dark shales accumulated in the deep portions of a ramp environment. However, this deep-marine succession is punctuated by the development of fluvial and aeolian deposits that constitute the Avilé Member (also known as the 'Avilé Sandstone', as defined by Weaver, 1931). This unit is widely distributed throughout the central part of the basin and has a variable thickness, ranging from only a few metres at the southern margin of the depositional area, up to 140 m in the northern part of the study area. The unit developed on top of a major erosion surface and is overlain by deep-marine deposits of the Upper Member of the Agrio Formation (Fig. 2) across a major transgressive surface. The Avilé Member represents one of the most important hydrocarbon reservoirs in the subsurface of the eastern part of the Neuquén Basin.

The Avilé Member has been previously described in terms of its sequence stratigraphic significance (Legarreta & Gulisano, 1989; Veiga & Vergani, 1993) and was interpreted as a Lowstand Systems Tract of a 3rd-order sequence (Legarreta

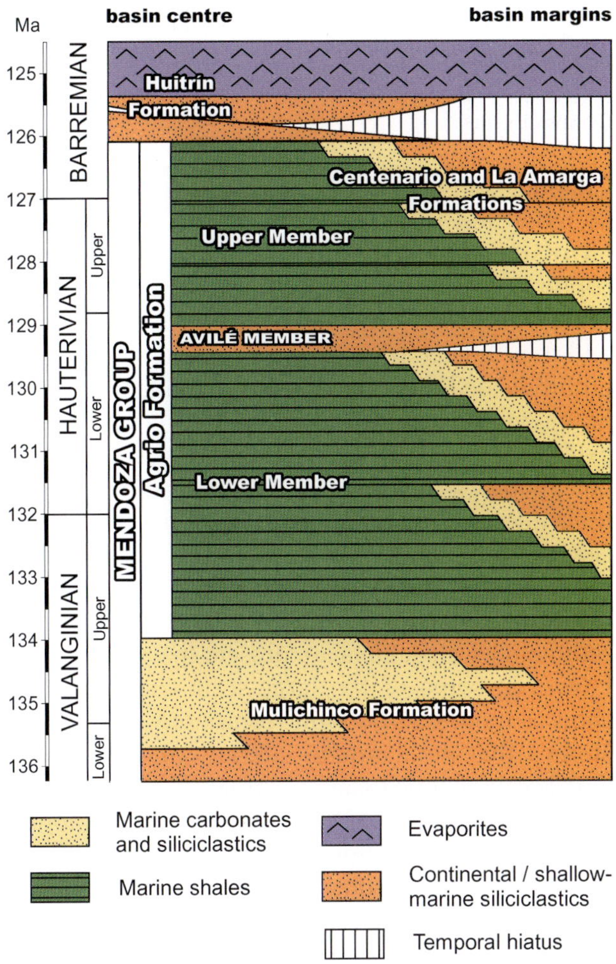

Fig. 2 Chronostratigraphic chart for the Lower Cretaceous in the central Neuquén Basin. (Modified from Veiga *et al.*, 2002.)

& Uliana, 1991). Many authors consider that the Avilé Member represents a complete desiccation event in the Neuquén Basin (Rossi, 2001; Legarreta, 2002) but the presence of marginal marine deposits towards the north (Mendoza Province) suggests that at least a restricted marine connection might have existed during part of the accumulation of the unit (Sagasti, 2002). Also, the nature of the interfingering between fluvial and aeolian deposits in marginal areas suggests an overall transgressive stacking (Veiga *et al.*, 2002), in contrast to the progradational organization expected in solely lowstand deposits.

The excellent biostratigraphic control on the accumulation of the Agrio sequence reveals that the Avilé Member lies between black shales containing ammonites of the *Weavericeras vacaensis* Zone and is capped again by black shales of the *Spitidiscus ricardii* Zone. This, combined with the presence of relevant calcareous nanofossils (*Cruciellipsis cruvilleri*), implies that the Avilé Sandstone belongs to the uppermost Lower Hauterivian and that its accumulation could have spanned 0.5 Myr (Aguirre-Urreta & Rawson, 1997; Aguirre-Urreta *et al.*, 1999).

STUDY AREA AND METHODS

The study area is located in northern Neuquén Province in the western part of the Neuquén Basin (Fig. 1). This area, known as the Agrio Fold and Thrust Belt (Ramos, 1978), is characterized by strong Andean (Cenozoic) deformation that resulted in multiple N–S oriented anticline structures (with narrow synclines) that expose the Cretaceous succession. The characteristics of the Avilé Member differ considerably between the eastern and western sectors. In the eastern sector, the Avilé Sandstone is dominated by thick packages of aeolian deposits that intercalate with minor fluvial units (Veiga *et al.*, 2001, 2002). This study focuses on the western sector, where most of the unit is represented by fluvial deposits and where aeolian intercalations are subordinate and less than 2 m thick.

The Avilé Member has been studied in 12 localities (Fig. 3), where detailed sedimentary logging was undertaken (Fig. 4). On a detailed outcrop scale, lithosomes were defined in terms of geometry and lateral facies variations. Facies and facies boundaries were mapped on photomosaics of selected outcrops. On a broader scale, and when outcrops were good enough, key depositional surfaces and rock packages were correlated and traced between logged sections, in order to clarify their stratigraphic and sedimentological importance and to define a general evolutionary depositional model.

FACIES ASSOCIATIONS/SEDIMENTARY UNITS

Eleven different sedimentary units (lithosomes) have been identified in the Avilé Member of the Agrio Formation in the study area (Table 1 & Fig. 5). These deposits have been grouped in terms of their external geometry into channel units (with an

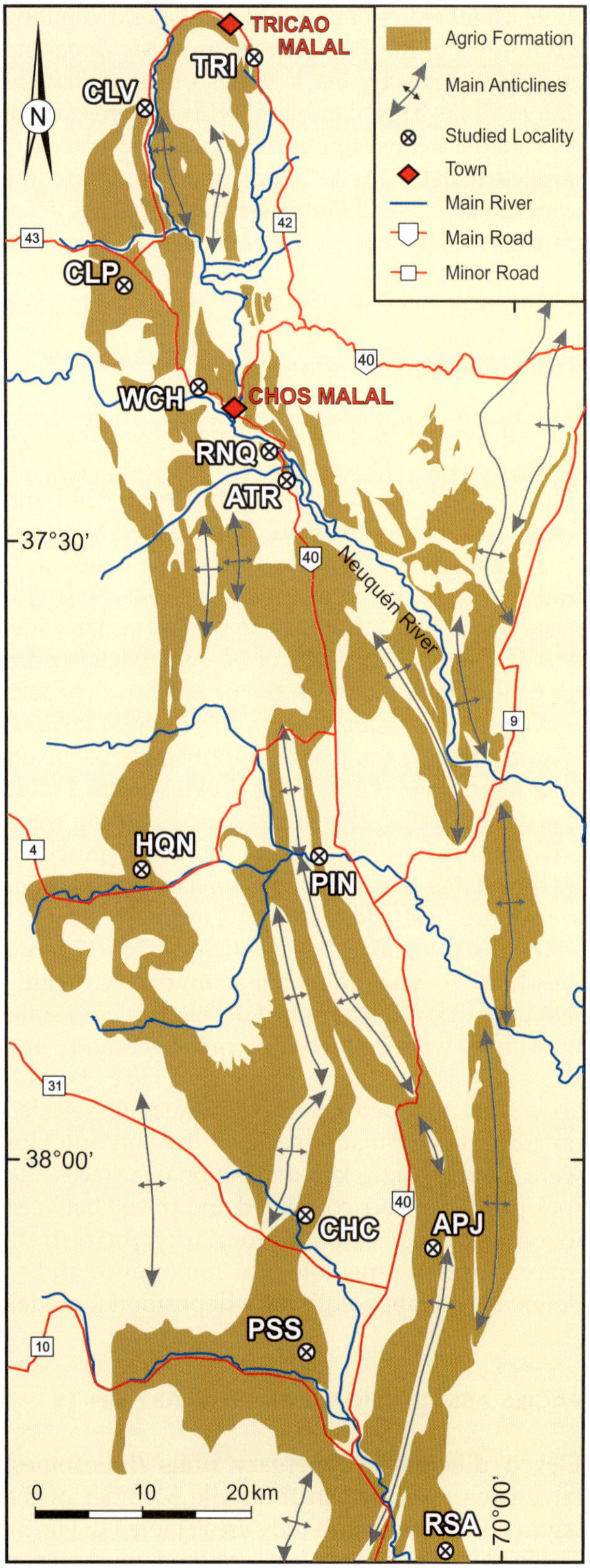

Fig. 3 (*left*) Map of the study area with outcrops of the Agrio Formation and location of the studied localities. APJ, Puesto Jara; ATR, Truquico Creek; CHC, Coihueco Creek; CLP, Cerro de la Parva; CLV, Currileuvu Creek; HQN, Huaiquillan Community; PIN, Pichi Neuquén Creek; PSS, Pampa del Salado South; RNQ, Neuquén River; RSA, Salado River; TRI, Tricao Malal; WCH, Chos Malal West.

erosive and concave lower boundary and an overall lenticular geometry) and non-channel units.

Channel units

Channel units are very common within the Avilé Member and five different types were identified, all of them representing different accumulation conditions in a fluvial/lacustrine setting. Channel units comprise large- and small-scale channels and their fill can be from exclusively sandy to heterolithic, the latter characterized by the alternation of fine-grained sandstone/mudstone couplets. The external geometry of these deposits is also variable, with a wide range of W/D ratios. In terms of their internal organization, most of the large-scale channels are filled mainly by trough cross-bedded sandstone. A small proportion show large-scale inclined strata (*sensu* Bridge, 1993), where different sedimentary structures are present in each stratum. However, some of the large-scale channels may show a more complex fill, composed of different storeys that amalgamate laterally and vertically. Small-scale channels can be simple, filled exclusively with cross-bedded and cross-laminated sandstones; however, some of them may be internally structureless, suggesting some degree of post-depositional modification. Due to the great variability shown by channel units, they were classified in terms of their dimensions, external geometry and internal architecture according to the terms proposed by Friend *et al.* (1979; Fig. 5 & Table 1).

Large-scale complex sheets

These units are the most common channel deposits of the Avilé Member in the study area and they are ubiquitous in every locality studied. They comprise thick (up to 10 m) sandstone bodies with a conspicuous erosive lower boundary. Their

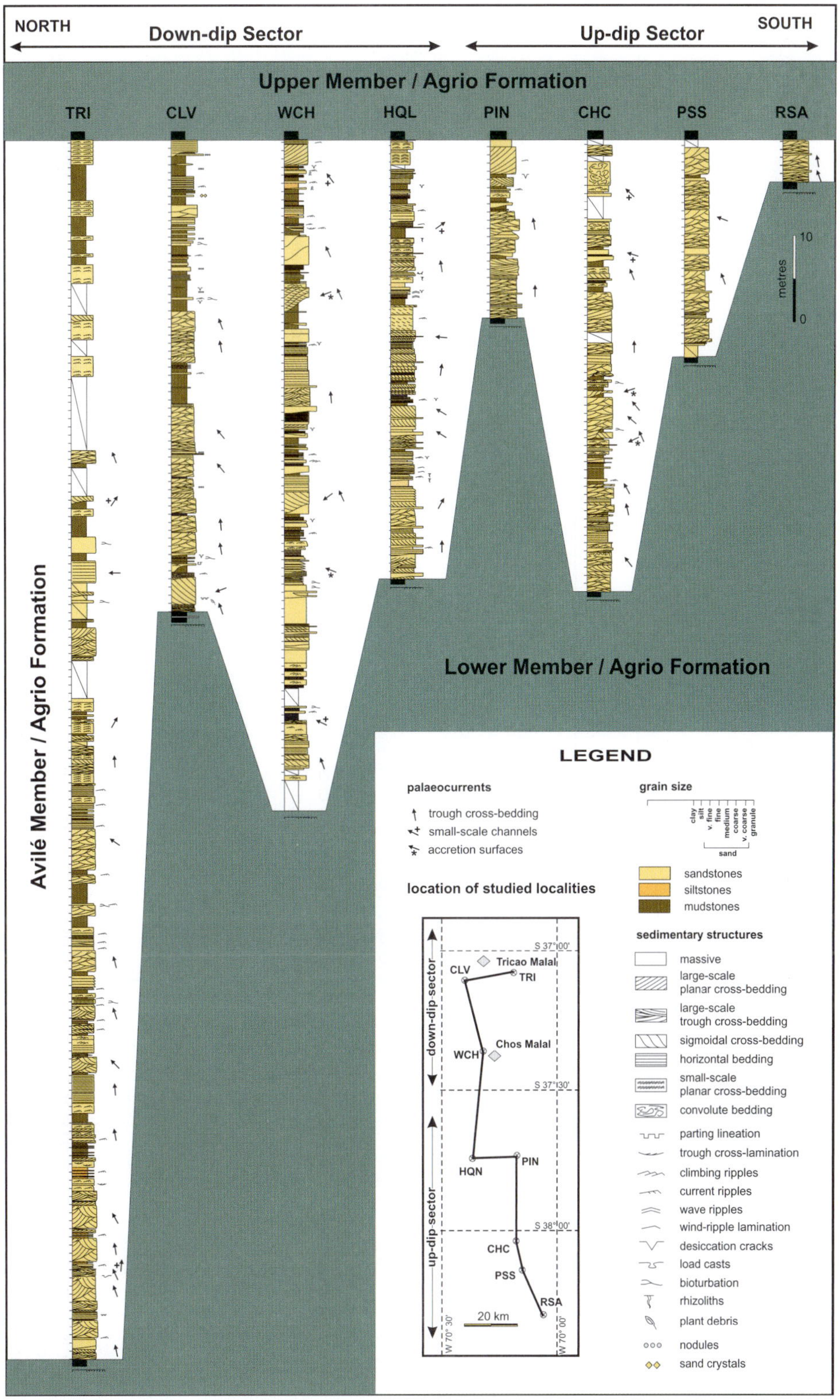

Fig. 4 Vertical and lateral distribution of sedimentary units in the Avilé Member. No horizontal scale. Localities as in Fig. 3. Inset: location of logs.

Table 1 Sedimentary units identified in the Avilé Member

Group	Sedimentary unit	Lithology	Sedimentary structures	Geometry	Dimensions	Bounding surfaces	Interpretation
Channel units	Large-scale complex sheets	Very coarse- to very fine-grained sandstones. Rip-up clast conglomerates	Large-scale trough cross-bedding, horizontal and planar cross-bedding; bioturbation and soft-sediment deformation. Bioturbation	Tabular at outcrop scale. Lenticular storeys	1–3 m thick; up to 100 m wide	Base: horizontal and erosional. Top: sharp and horizontal	Fluvial channel complexes. Sandy braided fluvial system?
	Large-scale complex ribbons	Coarse- to very fine-grained sandstones. Rip-up clast conglomerates	Large-scale inclined strata. Large-scale trough cross-bedding. Parallel stratification. Bioturbation	Lenticular	2–5 m thick; up to 20 m wide	Base: concave-up and erosional. Top: sharp and horizontal	Distributary channel complexes
	Large-scale simple ribbons	Coarse- to very fine-grained sandstones. Rip-up clast conglomerates	Large-scale inclined strata. Trough and planar cross-bedding. Current ripples and cross-lamination. Bioturbation in the upper and lower boundaries	Lenticular	1–3 m thick; up to 10 m wide	Base: concave-up and erosional. Top: horizontal. Interfingers with fine-grained deposits	Meandering channels
	Small-scale heterolithic ribbons	Fine-grained sandstone–mudstone cuplets	Imbricate coset of inclined sets. Sandstones massive or laminated. Massive mudstones	Lenticular	Up to 2 m thick; tens of metres wide	Base: concave-up and erosional. Top: sharp and horizontal	Small distributaries
	Small-scale ribbons	Coarse- to fine-grained sandstones, heterolithics, mudstones	Mainly massive. Current ripples and small-scale trough cross-bedding (sandstones), wavy and horizontal lamination. Bioturbation	Lenticular	Up to 0.9 m thick; tens of metres wide	Base: concave-up and erosional. Top: sharp and horizontal	Crevasse channels

Non-channel units	Fine-grained background	Mudstones, fine-grained sandstones and heterolithic intervals	Mudstones massive, horizontal lamination. Desiccation cracks and rhizoliths. Sandstones massive, current ripples, climbing ripples and horizontal lamination. Bioturbation	Tabular	Centimetres to more than 2 m thick	Base and top: horizontal, sharp to transitional	Floodplain, background lacustrine sedimentation. Distal lacustrine mouth bars
	Fining upward tabular sandstones	Fine- to medium-grained sandstones in a fining upward succession. Isolated small rip-up clasts	Massive, small-scale cross lamination, asymmetrical subcritical climbing ripples and horizontal lamination	Tabular	0.2–0.6 m thick. Vertical stacking in sequences up to 1.5 m thick with no vertical grain-size trend	Base: erosive and horizontal. Top: sharp or transitional to fine-grained units	Unconfined floods
	Lobes	Fine- to medium-grained sandstones	Horizontal lamination, subcritical climbing ripples and current ripples. Massive, with bioturbation and convolute bedding	Lenticular	< 1 m thick and tens of metres wide	Base: sharp and horizontal. Top: sharp, convex up	Crevasse splays
	Small-scale bars	Fine- to medium-grained sandstones and mudstones in a coarsening upward succession	Small-scale trough cross-bedding, current ripples and horizontal lamination. Rootlets and desiccation cracks	Tabular	Small coarsening upward sequences 0.5–1 m thick	Base: transitional from fine-grained sediments. Top: horizontal and sharp	Levee, crevasse splays, lacustrine mouth bars
	Large-scale bars	Medium- to coarse-grained sandstones. Abundant rip-up clasts	Large-scale cross-bedding. Locally, trough and low-angle cross-bedding	Tabular. Wedge shaped	2–5 m thick. Tens of metres wide	Base: sharp and horizontal. Top: sharp and inclined	Large-scale mouth bars
	Aeolian dunes and sandsheets	Fine- to medium-grained, well-sorted sandstones	Grainflow/grainfall laminae. Wind-ripple lamination. Horizontal lamination. Planar, tangential cross-stratification	Tabular	0.4–3 m thick	Base: horizontal and sharp. Top: sharp and horizontal	Aeolian dunes and sandsheets

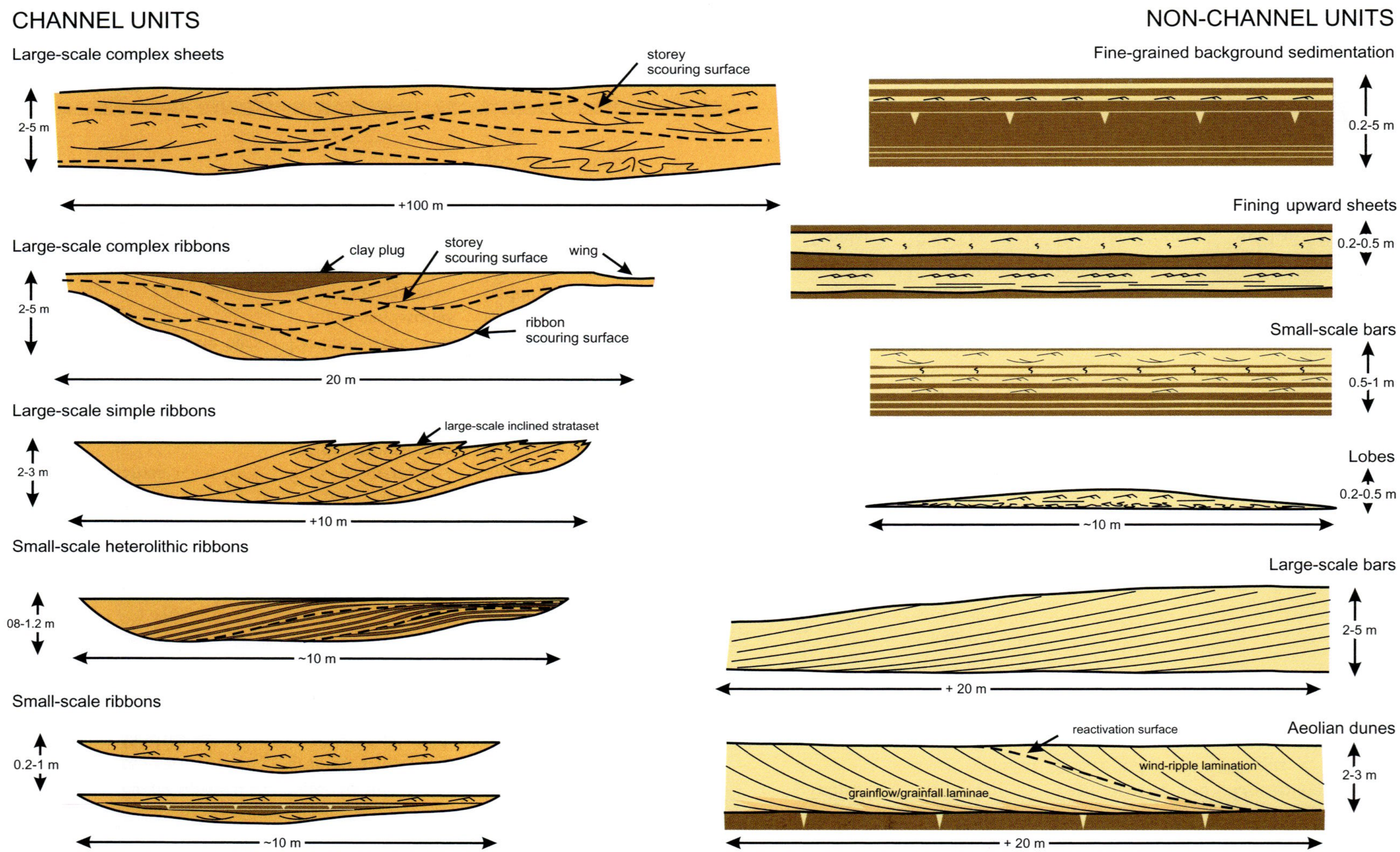

Fig. 5 Sedimentary units identified for the Avilé Member in the study area. For brief description see Table 1; symbols as in Fig. 4.

Fig. 6 Stacking of large-scale complex sheets at the base of the Avilé Member at Coihueco Creek (CHC). Note the horizontal nature of the lower surface of the sheet (black arrow) and the concave geometry of the storey scouring surfaces (white arrows). Person for scale.

external geometry is tabular on outcrop scale (> 100 m), although they are internally composed of lenticular storeys up to 3 m thick, each of them with an erosive and concave lower boundary (Fig. 6). Storeys are composed of coarse- to fine-grained sandstones with abundant rip-up mudclasts up to 10 cm in diameter. Intraformational clasts are concentrated at the basal portions of these units, related to the scouring in the lower boundary. Exceptionally, granule-conglomerate layers may be present at the bases of these channels. Some of the storeys show a conspicuous fining upward succession with intraformational conglomerates at the base, coarse- to medium-grained sandstones comprising the bulk of the unit, and fine- to very fine-grained sandstones towards the top.

Internally, lenticular storeys are composed almost exclusively of large-scale trough cross-bedded sets up to 1 m thick. Palaeocurrent direction measured on these cross-bedded sets shows a unimodal distribution with a main flow direction towards the northwest (328°; Fig. 4). Less common is the presence of layers up to 60 cm with horizontal stratification and parting lineation and isolated planar cross-bedded sets. Soft-sediment deformation structures and bioturbation (horizontal tubes) are also frequent in these units and can include abundant flutes at the base. Towards the tops of these units, together with the decrease in grain size, current ripples and cross-lamination may occur.

These units are interpreted as fluvial channel deposits where bedload was primarily transported as three-dimensional dunes at the bottom of the channels and without the development of major cross-channel or marginal bars. The amount of lateral amalgamation of these units suggests the development of non-fixed channels (promoted by the absence of cohesive banks?) that wandered across an alluvial plain, probably with a pattern of multiple shallow channels and without the preservation of fine-grained floodplain facies due to continuous lateral reworking. These units are very common and may amalgamate vertically (Fig. 6), suggesting that they might represent recurring periods of increased sediment supply but under a low aggradation rate of the alluvial plain. They also typify the complete thickness of the Avilé in the southern sector, where the unit is thinnest.

Large-scale complex ribbons

These sandstone units are characterized by a lenticular geometry (ribbons *sensu* Friend *et al.*, 1979; *W/D* ratios between 4 and 10) and by a complex internal organization, defined by the vertical (and less common lateral) amalgamation of individual channel units (Fig. 7). These complex ribbons usually erode into thick packages of cross-stratified sandstones (large-scale bars) or may be intercalated within heterolithic intervals (floodplain/lacustrine

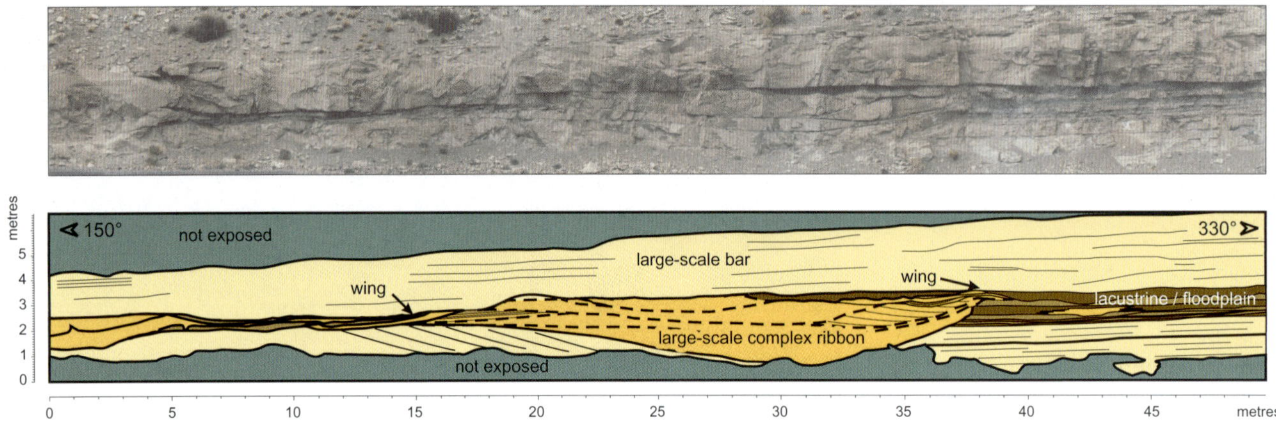

Fig. 7 Architectural element analysis panel for the Neuquén River (RNQ) locality, showing the relationship between fine-grained floodplain deposits, large-scale complex ribbons and large-scale bars. Note the development of wings on both margins of the complex ribbon.

deposits) (Fig. 8). They may reach 5 m thickness, with lateral extents of less than 20 m. Individual storeys are less than 2 m thick and 10 m wide. If the individual thickness of each storey is considered, a thinning upward succession can be recognized. The thickness of the basal storey can be almost the same as the depth of the basal scour surface. Storeys are composed of medium- to coarse-grained sandstones with abundant rip-up clasts mantling the storey scouring surface. The upper section of individual channels is usually eroded, although lenticular mudstones can be found at the tops of these complexes. Internally each individual unit may show large-scale inclined surfaces roughly perpendicular to the main channel orientation or trough cross-bedded sets. Individual channels (especially the upper ones) can be traced laterally beyond the limit of maximum scour into adjacent mudstones and fine-grained sandstones defining 'wings' (*sensu* Friend *et al.*, 1979; Fig. 7).

The organizational characteristics of these complex ribbons suggest that they were deposited under very different conditions to the large-scale complex sheets, even when some of their internal features may resemble particular storeys within complex sheets. The vertical stacking of channel units, the lack of lateral migration of the main basal erosion surface, and the fact that this surface seems to have been developed during the initial stages of development of these ribbons indicate an important initial incision (up to 5 m) where fixed channel complexes were developed. No aggradation of fine-grained deposits took place lateral to these channels until the last stages where wings are developed. These complexes might have been filled with solitary channels with some minor degree of lateral migration (and the development of point bars), or by relatively straight channels with lateral bars and a meandering thalweg. In each case, the mobility of the channels was always restricted to the basal main scouring surface of the ribbon. Coeval fine-grained deposits, if present lateral to these channels, were not preserved due to subsequent fluvial erosion. However, lenticular mudstone layers may be present as thin clay plugs related to channel abandonment, especially in the upper storeys.

Large-scale simple ribbons

These channel units are characterized by single lenticular bodies, up to 3 m thick, with low W/D ratios, composed of coarse- to medium-grained sandstone encased in fine-grained floodplain deposits. They show a conspicuous erosive lower boundary and a sharp, flat top. They have fairly steep flanks with vertical steps that define a well-developed lenticular geometry.

The fills include large-scale inclined surfaces (*sensu* Bridge, 1993) that dip between 10° and 20° defining inclined sets up to 40 cm thick. The orientation of these inclined surfaces is almost perpendicular to the orientation of the ribbon (which is similar to the orientation of other channel units

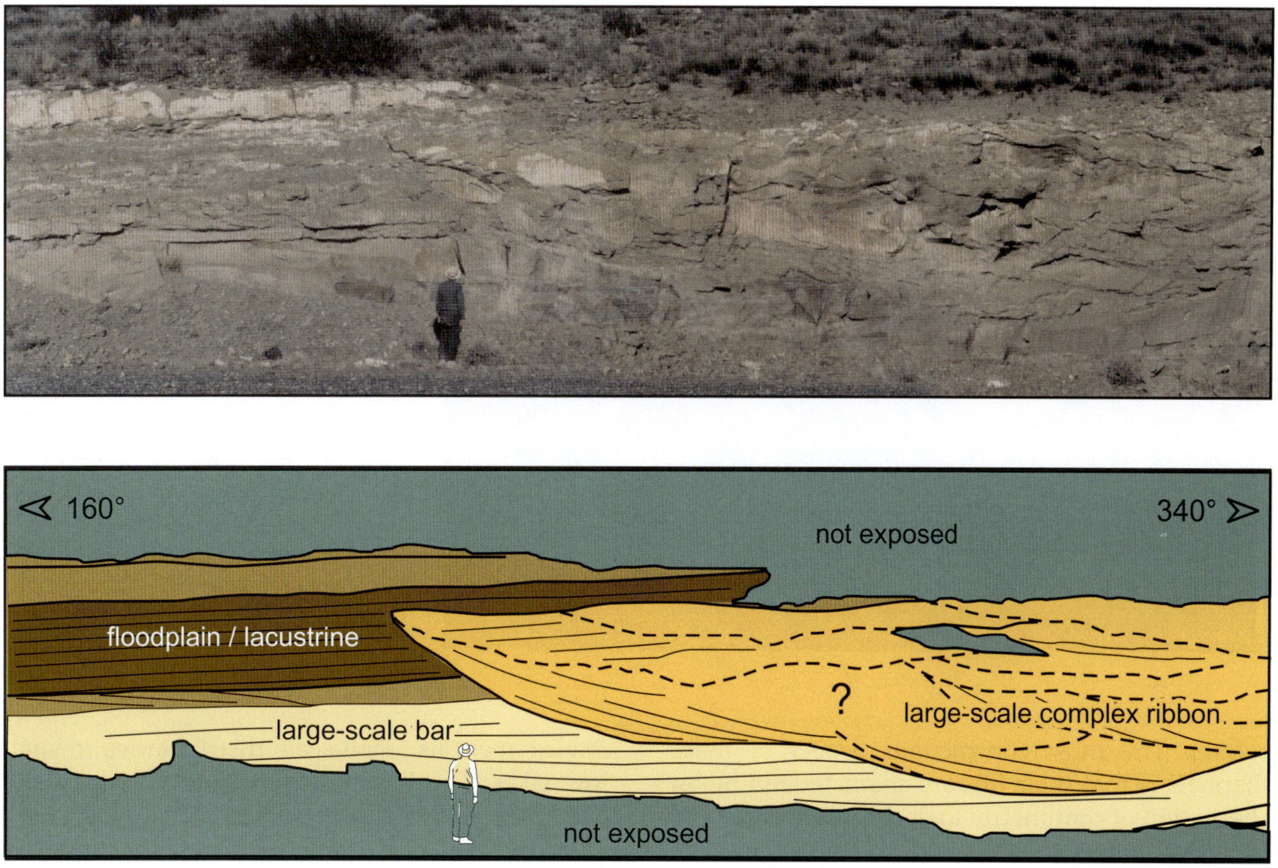

Fig. 8 Architectural element analysis panel for the Truquico Creek (ATR) locality, showing the relationship between fine-grained floodplain deposits, large-scale complex ribbons and large-scale bars. Note the complex internal organization of the ribbon and its lateral relationship with fine-grained floodplain/lacustrine facies. Person for scale.

– mainly northwest), dipping indistinctively to the southwest or northeast. The anatomy of these inclined sets is very well preserved and presents a complex pattern. Each set is characterized by abundant rip-up clasts at the base, and may show a lamination parallel to the inclined surfaces or planar cross-stratification that dips in the opposite direction to that of the large-scale surfaces. These structures may grade upwards in the set to small-scale cross-lamination with preserved current ripples at the tops. In the upper portion of the inclined stratasets, these may interfinger with fine-grained deposits (Fig. 9). In this upper part of the inclined stratasets abundant bioturbation was recorded.

The presence of large-scale inclined surfaces within these channel deposits oriented almost perpendicular with respect to the main orientation of the channel suggests the development of lateral accretion structures in the convex margins of relatively highly sinuous channels. The fact that these deposits are represented by isolated ribbons within floodplain deposits suggests a single-channel pattern which, when combined with the amount of lateral accretion structures, may suggest the development of meandering channels that flowed across muddy floodplains.

Small-scale heterolithic ribbons

These units are also characterized by an external lenticular geometry and by a concave, erosive lower boundary. However, they are smaller scale features in comparison to the main fluvial channels, having thicknesses of around 1 m and tens of metres lateral extent, always isolated within fine-grained floodplain/lacustrine deposits (see description

Fig. 9 Large-scale simple ribbon at Chos Malal West (WCH). Note the development of large-scale inclined surfaces (black arrows) defining inclined stratasets and the interdigitation with fine-grained deposits towards the top of each inclined strataset (white arrows).

below). The most conspicuous aspect of these channels is that their fills are characterized by an alternation of centimetre- to decimetre-scale layers of fine-grained sandstone and mudstone.

Using the descriptive nomenclature proposed by Thomas *et al.* (1987), the sedimentary fill of these units can be defined as an imbricate coset of several sets of inclined coarse-to-fine heterolithic couplets. Heterolithic sets have slightly erosive concave basal surfaces and convex-up tops (Fig. 10). Each member of the couplet ranges from 20 to 100 mm thick. The fine member of the couplet is composed of continuous layers of siltstones and mudstones with no internal structures. Fine-grained sandstones with horizontal lamination or ripple cross-lamination characterize the coarse member of these couplets.

Erosive lenticular channels showing inclined heterolithic stratification (IHS) have been interpreted as point-bar deposits accumulated in high-sinuosity channels (Thomas *et al.*, 1987; Plint & Browne, 1994). The fact that these channels are encased in fine-grained floodplain deposits, are smaller and have a completely different fill style compared with the main fluvial channels to which they are vertically and laterally related suggests that these deposits may represent high-sinuosity channels developed in a low-gradient floodplain environment as small-scale distributaries to shallow lakes. The inferred low gradient, potentially combined with local stabilizing factors (e.g. vegetated banks), might have increased the sinuosity of these channels, promoting a local increase in the rate of suspended load aggradation (Plint & Browne, 1994). The fact that the degree of lateral migration of these channels is reduced, and that they show a ribbon external geometry developed under low-gradient conditions, may suggest that these are relatively short-lived channels.

Small-scale simple/complex ribbons

These units are also characterized by a lenticular geometry and by an erosive basal surface, but thicknesses range from 0.3 to 1 m and lateral extents are in the order of 10 m or less (W/D ratios between 8 and 16). They may cap coarsening upward successions on top of fine-grained floodplain sandstones or can be found on top of large-scale complex sheets (Fig. 11). They are also isolated between floodplain mudstones and sandstones (Fig. 12). In these cases, the orientation of these units can diverge up to 60° from the orientation of the main channel units (Fig. 4). These units are composed of coarse- to fine-grained sandstones and no vertical or lateral variation in grain size is observed,

Fig. 10 Small-scale heterolithic channel at Chos Malal West (WCH). Note the concave and erosive lower boundary (black arrow) and the convex geometry of the inclined set boundaries (white arrows). Person for scale.

Fig. 11 Small-scale channel on top of a large-scale complex sheet at Tricao Malal (TRI). Note the heterolithic nature of the fill and the development of desiccation cracks (c) in the fine-grained intercalations. Hammer is 40 cm long.

except for the presence of isolated rip-up mudclasts up to 20 mm diameter in the bases of the channels. Sedimentary structures include small-scale trough cross-stratification or current ripples, but the units are more commonly massive with some degree of bioturbation. However, some of these ribbons contain a more complex fill of alternating 0.1 to 0.2 m thick layers of fine-grained sandstones with current ripples or low-angle and trough cross-stratification, and massive to horizontally laminated mudstone beds with occasional desiccation cracks (Fig. 11). These layers are usually horizontal and onlap towards the margins.

These small-scale lenticular bodies, with erosive bases, relatively low *W/D* ratio and closely related to fine-grained floodplain deposits, are interpreted as crevasse channels developed in a fine-grained floodplain due to episodic floods that cut through the channel banks into the adjacent, lower relief areas (Clemente & Pérez-Arlucea, 1993; Mjøs *et al.*, 1993; Plint & Browne, 1994). Those channels spatially related to large-scale complex ribbons may reflect the development of cross-bar chutes. While most of these units may represent responses to single flood events, those with a more complex fill may represent multi-event processes. In these cases,

Fig. 12 Fine-grained floodplain deposits at Chos Malal West (WCH). Note the intercalation of small-scale channel (CH) and coarsening-upward succession related to small-scale bars (white arrow). c, desiccation cracks. Person for scale.

an initial flood may be responsible for the development of the erosional relief and the subsequent, more passive fill may be related to less energetic floods or waning periods where mudstones are accumulated by settling from suspension, with subsequent subaerial exposure and development of desiccation cracks.

Non-channel units

Non-channel units are characterized by a wide range of grain sizes and external geometries (Table 1 & Fig. 5). They are most likely to show sharp contacts, but they may also show transitional boundaries. Most of these units are related to the development of subaqueous bars and background sedimentation in a floodplain/lacustrine environment. Aeolian bedforms and originally fluvial deposits reworked by wind activity were also identified.

Background fine-grained sedimentation

These fine-grained units are composed of dark grey to green mudstones intercalated with very fine- to medium-grained sandstones, and range from exclusively muddy deposits up to 2 m thick, to heterolithic intervals in which sandstone and mudstone are present in the same proportions (Fig. 11). In a few cases only, sandstones dominate with thin (millimetre) mudstone intercalations. However, these heterolithic intervals show no obvious trend in grain size or bed thickness.

These units may show sharp to transitional bases from fine-grained sandstones or channel units and may be also transitional to heterolithic and fine-grained sandstones towards the top, when they are not eroded by channel units. They range from millimetre-thick intercalations between channel units up to 5 m thick successions that are more frequent and thicker towards the top of the unit studied.

Mudstones are usually dark grey to dark green and mainly massive, although a subtle horizontal lamination may be present. They also show occasional root casts and abundant desiccation cracks (Fig. 12). Sandstone layers range from 5 to 30 cm thick and are very fine- to medium-grained with sparse rip-up clasts up to 2 cm in diameter. Current ripples, climbing ripple cross-lamination and horizontal lamination are common but sandstones can be also massive. Bioturbation, as well as isolated (*Skolithos*) and paired (*Arenicolites*) vertical burrows, is also present. Palynological studies carried out in mudstone clasts of this unit (Prámparo & Volkheimer, 1999) suggest the proliferation of a complex flora, including moss, hepaticas and different specimens of ferns, related to a humid environment.

These deposits are interpreted as the accumulation of mud by suspension fallout in a subaqueous environment with episodic unidirectional flows responsible for deposition of the sandstone beds. The vertical and lateral relationship to fluvial channels suggests that these deposits might represent accumulation in a floodplain environment under a relatively humid climate, where temporary water bodies or small ponds were developed. Sandstone layers were probably the result of overbank flows associated with flood events, but they might also be related to density underflows where small distributary channels debouched into temporary ponds (Plint & Browne, 1994).

Fining upward tabular sandstones

Tabular sandstone bodies with a clear fining upward trend are closely associated with floodplain/lacustrine deposits. These bodies are 0.2–0.6 m thick with sharp and horizontal lower boundaries, sometimes with evidence of local erosion. They are composed of medium- to fine-grained sandstones in a fining upward succession and are usually transitional to floodplain mudstones. In some cases, several fining upward units group together vertically, defining up to 1.5 m successions, with no evident vertical trend. The lower boundary of these units may show flutes. Internally, they may be entirely massive or may show small-scale cross-lamination, asymmetrical subcritical climbing ripples and horizontal lamination.

Erosive-based tabular sandstone bodies with fining upward trend, high-regime structures and climbing ripples are usually related to sheet-flood deposits (or sand sheets) accumulated from unconfined floods in floodplain environments (Bridge, 2003). These deposits show a much greater thickness than individual sandstone beds in floodplain deposits, and are therefore related to more important floods than the ones produced by overbank splay near the margins of active channels. They might be related to exceptionally large-scale seasonal floods.

Sandstone lobes

These deposits are characterized by an external lenticular geometry with a flat and sharp basal surface and a convex-up top. The thickness of these lens-shaped units ranges between 0.2 and 0.8 m and their lateral extent is in the order of tens of metres. They are exclusively comprised of fine- to medium-grained sandstones with scarce isolated rip-up clasts up to a few millimetres in diameter. These sandstone bodies are usually massive but they can also show current ripples, cross-lamination and convolute lamination in the lower portions.

These units are interpreted as sandstone terminal lobes or small-scale splay deposits developed in a floodplain environment due to overbank flooding events. They are characterized by the relatively simple internal organization and the absence of mudstones intercalated with the sandstones. This implies that they might be related to single flooding events in contrast to the more complex nature of normal crevasse splay deposits that are related to periodic sheet flooding (Miall, 1996)

Small-scale bars (crevasse splays/lacustrine mouth bars/levees)

These units are characterized by medium- to very fine-grained sandstones and mudstones. They usually define coarsening and thickening upward successions up to 1.5 m thick in contrast with background sedimentation (Fig. 12). Each sandstone layer may be massive or show cross-lamination, climbing ripples or horizontal lamination, and may reach up to 0.3 m thick. Mudstones are usually massive or may show a subtle horizontal lamination and reach up to 0.5 m thick.

Coarsening and thickening upward heterolithic successions closely related to floodplain deposits can be interpreted as the result of the progradation of levees (Bridge, 2003) or crevasse splays, related to overbank flows close to main fluvial channel margins (Smith *et al.*, 1989; Clemente & Pérez-Arlucea, 1993). However, if relatively permanent water bodies were developed in a floodplain environment, these successions may be related to the development of mouth bars in the distal portions of crevasse channels (Smith & Pérez Arlucea, 1994; Perez-Arlucea & Smith, 1999), or minor distributary channels diverting from main feeders (Tye & Coleman, 1989). However, lacustrine deltaic sediments are difficult to distinguish (based on facies association) from crevasse splay deposits (Miall, 1996), and they are common in fluvially dominated sedimentary basins recording periods of rapid basin alluviation (Tye & Coleman, 1989).

Large-scale bars (lacustrine bars)

These deposits are not very common in the Avilé Sandstone but, where present, they comprise large-scale sandstone bodies up to 5 m thick, composed mainly of well-sorted, medium- to coarse-grained sandstones showing horizontal to low-angle cross-stratification, with isolated mudclasts. They are characterized by an irregular and horizontal lower boundary with evidence of erosion and by an external wedge shape with an inclined upper boundary (Fig. 13). Internally some erosion surfaces are present but these units are composed almost entirely of a single set of low-angle cross-stratified sandstones dipping to the west and northwest.

The large scale of these units precludes their interpretation as in-channel bars, as their thickness and lateral extent exceed by far the dimension of the documented fluvial channels which would have to contain them. The preferred interpretation is, therefore, one of subaqueous bars associated either to exceptionally large flooding events, or to relatively deep water bodies with important and continuous sediment influx. Under these circumstances, a combination of inertia- and friction-dominated hyperpycnal flows may develop in the river mouth, inhibiting the development of large inclined foresets. These bars may be equivalent to mouth bars associated with distributary channel complexes and developed during relatively long-lived lacustrine environments. The scale of these deposits distinguishes them from the small-scale lacustrine mouth bars, and they are thought to be related to the distal terminations of the main channels of the system. The accumulation of sandstone units up to 5 m thick also implies a continuous sand supply that could be promoted by important flood events in the upstream end of the fluvial system.

Aeolian dunes and sandsheets

Aeolian deposits characterize most of the Avilé Member in the eastern part of the basin (Veiga *et al.*, 2002). However, wind-laid accumulations are not common in the study area. They are present in the uppermost stratigraphic section in the Pichi Neuquén and Chos Malal (west) areas (Figs 3 & 4) and they may be present sporadically elsewhere, as small intercalations between fluvial deposits. These deposits are characterized by tabular bodies, up to 3 m thick, of well sorted, fine- to medium-grained sandstones and constitute the only sandstone bodies that do not contain rip-up clasts. They usually show a very sharp and horizontal lower boundary above overbank

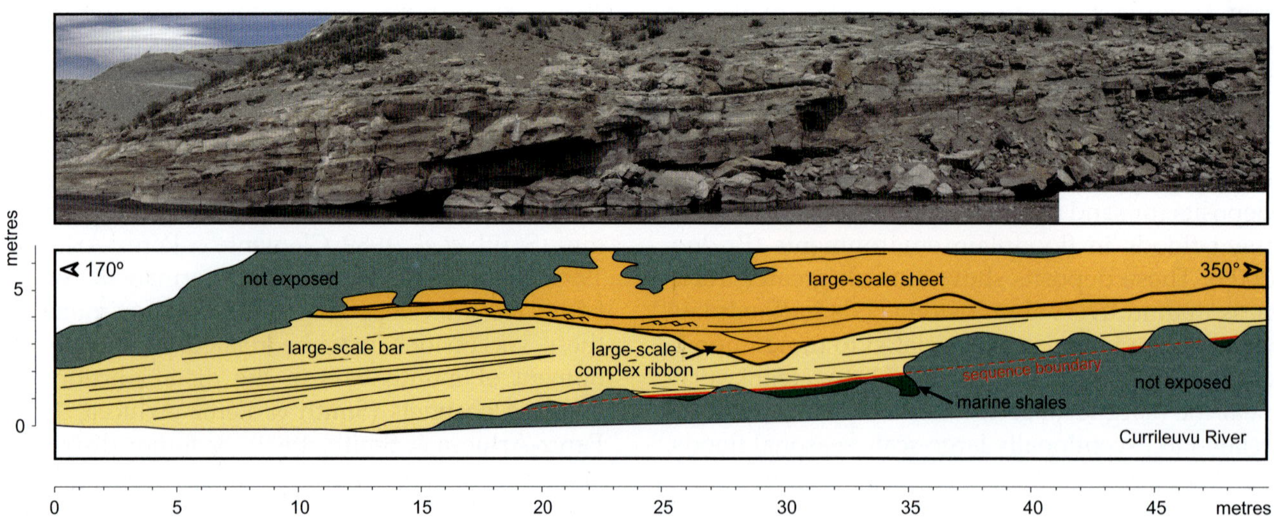

Fig. 13 Architectural element analysis panel showing large-scale bar deposits at the base of the Avilé Member at the Currileuvu Creek (CLV) locality. Note the vertical relationship with complex ribbons and the sharp basal contact with the marine shales of the Lower Member of the Agrio Formation.

mudstones with abundant desiccation cracks or they may be vertically related to fluvial deposits. They are characterized by bimodally sorted, horizontally laminated sandstone or by large-scale, tangential cross-stratified sets up to 2 m thick with internal bounding surfaces that separate subsets that show small changes in foreset orientation. In the bottom portion of steeply inclined foresets, the alternation of wedge-shaped grainflow and structureless grainfall laminations is common. The presence of preserved isolated, low-amplitude ripples is also common at the bottom of inclined sets or intercalated between horizontally stratified deposits. Horizontally stratified units are usually transitional from fluvial channel deposits. In places they can be transitional from rip-up-clast conglomerates within complex sheets but their lateral extent is difficult to establish due to subsequent fluvial erosion.

The presence of well-sorted sands in large-scale cross-stratified sets, with bimodal lamination interpreted as wind-ripple lamination (Hunter, 1977) and the alternation of grainfall and grainflow deposits in the bottom portion of high-angle foresets, suggests the development of slipfaced aeolian dunes. Internal bounding surfaces of these sets may represent reactivation surfaces associated with small local changes in wind speed and orientation. These deposits are relatively simple and thin, and could have formed under a reduced rate of accumulation (Kocurek, 1996). In these circumstances dunes did not climb on top of the previously developed bedforms (or climb at a very reduced angle), giving rise to a rather simple internal organization. Thick packages of horizontally laminated sandstones with bimodal grain-size sorting may represent the accumulation of aeolian sand sheets, under limited sand supply (Kocurek & Nielson, 1986). The fact that these deposits overlie fluvial channel facies suggests that they may be the result of wind reworking of fluvial sands, probably with a high water table that locally reduced the sand availability and prevented the development of slipfaced aeolian dunes.

DEPOSITIONAL SETTING OF THE AVILÉ MEMBER

Since the early studies of the Mesozoic succession of the Neuquén Basin, the Avilé Member of the Agrio Formation has been defined as a non-marine unit dominated by the interaction of fluvial and aeolian processes (Weaver, 1931; Groeber, 1946). However, perhaps due to the limited thickness of this unit at several localities, no further characterization of the fluvial environments represented has been carried out, despite the great variability in channel and non-channel units.

One of the most outstanding characteristics of the Avilé Sandstone that arises from this study is that in the northern sector of the study area, the non-marine interval reaches up to 140 m in thickness, and although thickness decreases southward, this change is not uniform and important thickness variations are recorded between closely spaced localities (Fig. 4). Palaeocurrent indicators within the main channels of the system suggest a SSE–NNW trend for the fluvial system (Fig. 4) with no major changes throughout the sequence; therefore the thickness increases approximately down the depositional dip.

Up-dip sector

The Avilé Sandstone in the southernmost sector of the study area is less than 5 m to 30 m thick and dominated by sandstone, with few intercalations of fine-grained muddy facies (Fig. 4). The most extreme case is the outcrop just south of the Salado River (RSA locality, Fig. 3), where the unit is only 5 m thick and composed exclusively of medium- to coarse-grained sandstone.

Internally, the up-dip Avilé Member (especially in RSA and APJ localities, Fig. 3) shows the vertical amalgamation of complex channel sheets. This reflects high sand supply and relatively low accommodation. Under these conditions a braided fluvial network was developed, with a complex pattern of unstable shallow channels that migrated across a sandy fluvial plain. There is a gradual decrease in grain-size towards the top of the Avilé Member at some localities (PIN and CHC, Fig. 3), with the development of aeolian deposits.

Thickness increases obliquely down-dip to the northwest, suggesting that this unit is not a uniform wedge that increases its thickness progressively towards the north. It is possible that some localized incision of the Agrio ramp occurred and that individual channel networks might have a SE–NW flow towards the northwest.

Only in these more 'distal' areas of the up-dip sector (such as PSS and CHC, Fig. 3) do fine-grained deposits, up to 6 m thick, intercalate within sandstone sheets. These fine-grained facies are interpreted as floodplain deposits in which small-scale bars and simple small-scale sandy ribbons intercalate. They are also laterally related to large-scale simple ribbons that may show large-scale inclined surfaces. An important amount of erosive relief (up to 2 m) is seen on top of these floodplain deposits and complex sandstone sheets overlie these fine-grained intervals.

The superimposition of complex sheets and fine-grained floodplain deposits, together with the development of high-sinuosity channels closely related to floodplain facies, suggests contrasting conditions during the accumulation of the Avilé Member in the upstream sector. Episodes of increased runoff and sand supply (under low accommodation creation) might have been suitable for the development of complex sandstone sheets in a sandy braided fluvial environment. The abundance of soft sediment deformation in the cross-bedded sandstone complexes also suggests a high sedimentation rate. During periods of reduced sand supply or even exceptionally high accommodation creation, a complete change in fluvial style is recorded, with the development of high-sinuosity channels laterally associated with persistently wet (flooded) floodplains.

Down-dip sector

The northern part of the area studied is characterized by a greater thickness of the Avilé Member, which reaches up to 140 m in the northernmost locality, and, especially, by an increase in the proportion of fine-grained deposits. These fine-grained facies may constitute up to 10-m-thick packages, which are more frequent towards the top of the unit.

This area is characterized by a wide variety of channel and non-channel units interbedded throughout the unit. Although not as common as in the up-dip sector, complex sheets are also present, mainly in the lower half of the unit. In some northern localities (TRI, CLV, Fig. 4), the basal portion of the Avilé Member is defined by the amalgamation of complex sheet units. This sandy lower interval can reach up to 20 m (TRI); however, the amalgamation of sheet units is not complete and thin intercalations (up to 50 cm) of fine-grained deposits are present between them (Fig. 10). In addition, sandy intervals up to 10 m thick, resulting from the accumulation of complex sandstone sheets, are present in the northern sector throughout the lower portion of the unit studied. These intervals show an important basal relief and are sharply overlain by thick packages of fine-grained deposits. As for the upstream sector, these sandy intervals may reflect the recurrent development of a bedload-dominated fluvial system with a complex pattern of multiple channels and with no preservation of floodplain deposits.

Fine-grained intervals are more abundant in this downstream sector throughout the entire Avilé Member but particularly towards the top. They are mainly composed of background floodplain/lacustrine sediments in which small-scale bars and tabular sandstone sheets intercalate. Bioturbation of fine-grained deposits is also more common in this area. Channel units associated with these fine-grained intervals include large-scale simple ribbons with well-defined lateral accretion surfaces and small-scale sandy and heterolithic channels. A gradual increase in the proportion of floodplain facies is recorded towards the top of the Avilé, with channel units almost absent. If present in the upper half of the unit, channel deposits are characterized by isolated large-scale simple ribbons. Another aspect that characterizes the down-dip sector of the Avilé Member is the development of large-scale subaqueous bars and large-scale complex ribbons, closely associated either with fine-grained intervals or complex sheets. These fine-grained intervals represent the accumulation of a mixed-load fluvial system, with important aggradation of the alluvial plain and significant accumulation of floodplain deposits, associated with important and maybe more permanent water bodies. The development of large-scale subaqueous bars implies the progradation of distributary channel-mouth bar complexes into important water bodies. Therefore, the alternation of intervals that depict contrasting fluvial styles, although suggested for the up-dip sector, is more obvious in this northern area and present in every study locality.

CHANGES IN FLUVIAL STYLE: DISCUSSION

High-frequency sequences within the Avilé Member

One important character of the Avilé Member in the study area is the development of packages 20–40 m thick in which contrasting fluvial styles are recorded (Fig. 14). Short-term intercalation of sandy intervals associated with the development of a braided fluvial system and thick packages of fine-grained deposits with isolated high-sinuosity ribbons suggests contrasting accumulation conditions, especially in the relationship between clastic supply and the rate of accommodation creation. During periods characterized by a lower accommodation/sediment supply ratio, multiple unstable channels migrated significantly across the fluvial plain, removing any contemporaneous fine-grained overbank material. This generated the complex sandy sheets. During periods in which accommodation was created at a faster rate than sediment was supplied, there was a higher rate of aggradation of the fluvial plain and fluvial channels became isolated within floodplain deposits (Wright & Marriott, 1993, Marriott, 1999), which in turn were preserved as packages up to

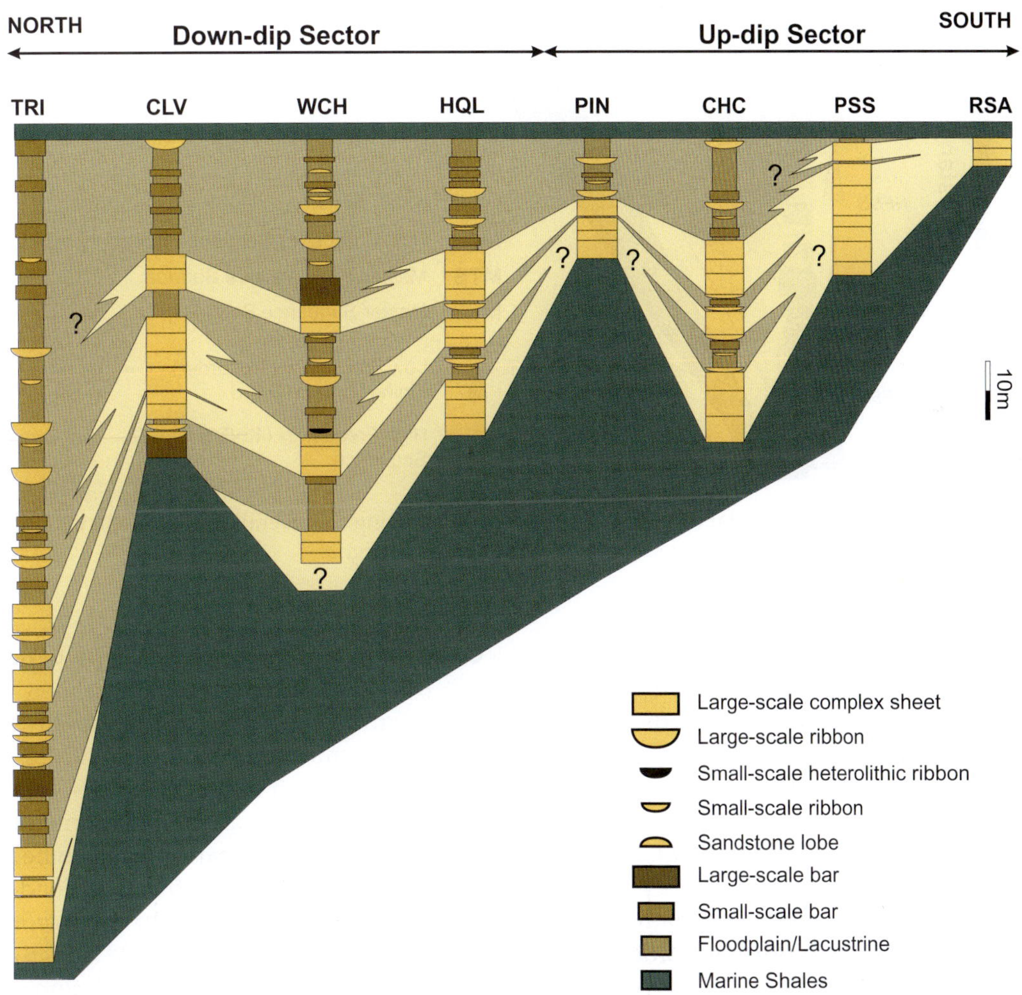

Fig. 14 Distribution of sedimentary units identified for the Avilé Member in the study area. Pale yellow areas depict bedload-dominated fluvial systems, darker yellow areas mixed-load fluvial systems, and dark grey areas deep-marine accumulation. Note the overall wedge shape of the Avilé deposits. No horizontal scale.

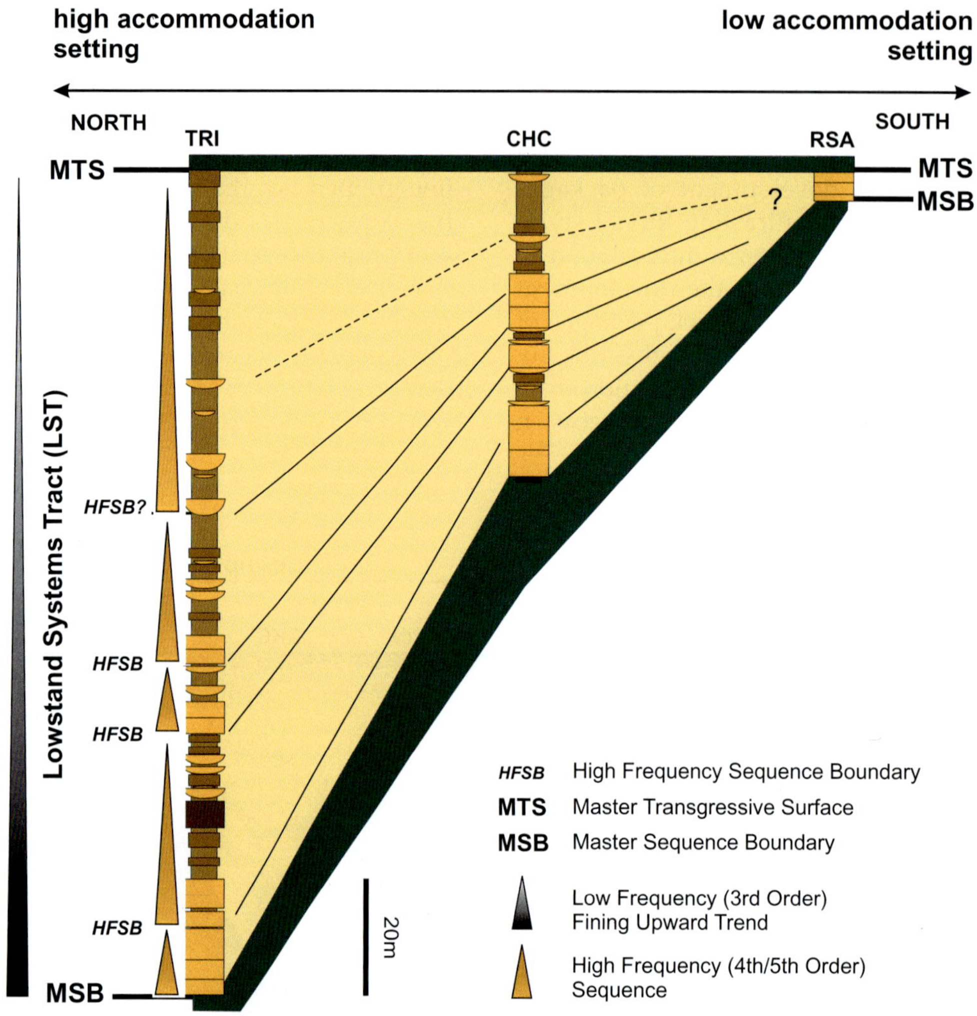

Fig. 15 Sequence stratigraphic interpretation of the Avilé Member showing the overall low-frequency fining upward trend and the development of high-frequency sequences. Localities as in Fig. 3.

5 m thick and closely associated with small-scale floodplain channels and minor bars. The fact that these fine-grained intervals reflect periods of high accommodation is also suggested by the development of more permanent water bodies associated with the floodplains, where lacustrine bars and more complex distributary channel/sandy mouth bar (large-scale bars) systems, up to 5 m thick, developed. These small-scale cycles within the Avilé Member may represent, therefore, high-order sequences (4th to 5th order) developed in response to changes in external factors that modified the relationship between accommodation and sediment supply within this low-order lowstand succession (Fig. 15).

Controls on high-frequency sequence development

The relationship between sediment supply and rate of accommodation creation can be altered and modified through different factors. Local increases in subsidence rate can create anomalous episodes of accommodation creation and can be responsible for the upward passage from bedload-dominated fluvial systems to lacustrine environments (Plint & Browne, 1994). No evidence of active tectonics is recorded during the accumulation of the Avilé Member. The Lower Cretaceous in the Neuquén Basin was characterized by roughly uniform subsidence within a post-rift stage and no major episodes of tectonic inversion have been identified for the Hauterivian (Legarreta & Uliana, 1993; Vergani *et al.*, 1995).

The fact that there is evidence of erosion at the base of the sandy (bedload-dominated) intervals suggests that rejuvenation of the fluvial system occurred together with the change in fluvial style. This could be associated either with a drastic hydrological change or with short-term fall in the base level to which the fluvial system graded.

It is possible that some of the changes observed in fluvial style may have been caused by climate-driven changes in clastic sediment supply. It has been proposed that during wetter periods, there is an increase in vegetation cover that can reduce considerably the clastic sediment supply (Cecil, 1990). Evidence for abundant plant growth can be found associated with fine-grained intervals within the Avilé Member, including *in situ* and reworked rhizoliths and abundant plant debris in the large-scale bar deposits. During dryer periods, plant cover reduces and sand availability and clastic supply increase considerably, reaching a maximum. Also during dryer periods, low base levels can be linked to a low water table and locally aeolian deposits can be developed. Within the Avilé Member, abundant aeolian reworking was recorded associated with episodes of increased clastic supply (aeolian reworking of complex sheets). However, aeolian deposits are also present in the Avilé Member associated with fine-grained deposits in the uppermost portions. This may reflect that, in this part of the basin, the development of aeolian deposits may be related more to local conditions (of low water table) than to a general vertical pattern, as in other parts of the basin (cf. Veiga *et al.*, 2002). Evidence within the Avilé Member, such as abundant desiccation cracks in mudstones and aeolian reworking of fluvial deposited material in the sandy intervals, suggests a strong seasonal climate, but no important changes in climate throughout the studied interval were recorded.

Changes in fluvial style and rejuvenation of the fluvial system at the base of the bedload-dominated fluvial intervals also can be associated with changes in the base level of the fluvial system, which in turn can also be strongly related to climate. If these systems graded to a local lacustrine base level and the basin was completely disconnected from the Pacific Ocean, as suggested by some authors (Rossi, 2001, Legarreta, 2002), during wetter periods lake-level rise would be accompanied by increasing accommodation and decreasing clastic sediment supply. During dryer periods, base level could have lowered enough to enhance the rejuvenation of the fluvial system, producing important incision at the bases of these intervals. Evidence for short-term changes in climate associated with orbital processes has been recorded in the black shales of the Agrio Formation with which this unit is intercalated (Sagasti, 2000, 2005). Cycles ranging from 20 up to 100 kyr were recorded in the deep-marine deposits of the Agrio Formation, although most of this cyclicity was related to the 19–21 kyr precessional signal.

If, in turn, the Neuquén Basin was partially connected to the Pacific Ocean during the accumulation of the Avilé Member, and as eustatic sea level could have been affected by these orbital processes, high-frequency eustatic oscillation could have controlled changes in the base level to which river systems graded. Therefore, a eustatic control can be proposed for accumulation of the whole Agrio Formation, including the non-marine deposits of the Avilé Member in the down-dip part of the study area. Although it is not possible to precisely date these high-frequency oscillations in these non-marine deposits, in this context the basal erosional surfaces of the bedload-dominated intervals can be regarded as the sequence boundaries of these high-order sequences developed within the Avilé Member (Fig. 15).

Low-order lowstand–transgressive systems tract

The Avilé Member shows an overall, low-frequency, upward change from bedload-dominated fluvial systems to fine-grained deposits related to high-sinuosity mixed-load systems (Figs 14 & 15). This is interpreted as a function of reducing fluvial gradient/increasing non-marine accommodation related to a general, long-term base-level rise. Similar trends have been interpreted from depositional systems of all ages around the world (Shanley & McCabe, 1994; Aitken & Flint, 1995; Olsen *et al.*, 1995; Marriott, 1999). Considering that these non-marine deposits were accumulated in the central part of the basin following a major relative sea-level fall (lowstand), the amalgamation of thick, coarse-grained packages associated with bedload braided fluvial systems is interpreted as a response to low accommodation conditions generated by the relative sea-level fall. The upper limit of the Avilé Member is a major transgressive surface, above which deep-marine deposition was re-established across the whole basin, suggesting a relative sea-level rise of at least the same magnitude as the relative fall that produced the accumulation of this non-marine unit. No detailed bathymetric data are available for the black shales of the Agrio Formation. However, considering that these deposits accumulated below storm wave-base, a relative sea-level fall (and subsequent rise) of at least 50 m can be estimated.

The black shales that cap the Avilé Sandstone contain ammonites with boreal affinities, suggesting that this relative sea-level rise can be correlated worldwide and therefore can be associated with a low-frequency eustatic rise (Aguirre-Urreta & Rawson, 1997). If these fluvial systems were at least in part related to marine base level, the internal organization of the Avilé Member, especially in the upper portion, can be associated with an overall transgressive trend that produced a gradual increase in the rate at which accommodation was created in the downstream end of the fluvial system.

An integrated model

The data from the Avilé Member and its basin context lead to a model in which the system was probably marine-connected and thus influenced by marine sea-level changes in the north. This marine connection is based on the presence of shallow-marine deposits in the northern part of the basin (Mendoza Province) that could be coeval with the Avilé Sandstone (Sagasti, 2002).

The tectonic quiescence during this part of the Cretaceous argues for little or no local tectonic control, with the high-frequency sequence development being dominated by the eustatic component. The effect of sea-level fall on the Avilé system was to cut high-frequency sequence boundaries (with a subsequent gradient steepening), followed by deposition of bedload-dominated sandy intervals with high sediment supply and low base level. During the high-frequency rise in relative sea level and linked warming/wetting, the fluvial system evolved to a mixed-load high sinuosity fluvial/lacustrine/high water table situation. The rising base level created new non-marine accommodation, reduced sediment supply and resulted in increased sinuosity of the rivers. If during the history of the Avilé Member the Neuquén Basin was completely disconnected from the Pacific Ocean, then the oscillations in base level could have been produced by climatic/lake-level fluctuations but a similar stratigraphic response would be expected (Keighley *et al.*, 2003).

The estimated duration of 0.5 Myr for the Avilé Member suggests that it represents a 3rd-order lowstand systems tract. However, it is common that these low-order systems tracts may in fact contain high-order sequences. In this context, and considering that high-frequency changes in fluvial style within the Avilé Member may represent high-frequency sequences (4th or 5th order), it can be regarded as a composite sequence or a lowstand sequence set (*sensu* Mitchum & Van Wagoner, 1991). The high-frequency sequences at the top of this sequence set are dominated by an overall increase in accommodation produced by a low-order base-level rise and the transition from the low-order lowstand to transgressive systems tract. Therefore their identification becomes more difficult (Fig. 15).

Up-dip to down-dip changes in Avilé Member architecture

In fluvial systems, base-level oscillations can control the rate and magnitude of accommodation creation, but this effect can be restricted to the downstream part of the fluvial system (Blum, 1993). The fact that several localities in the up-dip sector of the Avilé are completely dominated by the accumulation of bedload fluvial systems implies that this area behaved as a low accommodation setting throughout the whole history of the Avilé Member (Figs 14 & 15). This can be seen as evidence that more accommodation was being created in the down-dip sector of the fluvial system, associated with fluctuations in base level. Despite the overall long-term rising base level (as indicated in the down-dip areas by the upward change to a mixed-load, fine-grained fluvial system dominated by thick floodplain/lacustrine deposits) up-dip areas continued behaving as low accommodation zones (Fig. 15). It is only in the uppermost portion of the Avilé Member in the up-dip area that fine-grained accumulation is recorded, suggesting that only in the final stages of development of this lowstand wedge, and prior to the main transgressive event at the top, was enough accommodation created in the upstream sector to change depositional style. It is also possible that accommodation was not created in the up-dip sector until the latest stages of development of this lowstand. Therefore, an important part of the Avilé Sandstone in the down-dip sector could be time-equivalent to the basal sequence boundary in the up-dip area. The high-frequency sequences are therefore difficult to delineate in the accommodation-limited southern area.

CONCLUSIONS

1 The Avilé Member was studied in 12 different localities, where detailed sedimentological observations and architectural element analysis allowed the identification of 11 channel and non-channel sedimentary units.

2 The up-dip (southern) sector of the study area is characterized by the superposition of large-scale complex fluvial sheets, interpreted as a bedload fluvial system under low accommodation conditions. Only in the uppermost portion of the unit or in the more distal areas are fine-grained floodplain intercalations associated with high-sinuosity channels present, here associated with the development of a mixed-load, high sinuosity fluvial system.

3 The down-dip area is also dominated by the intercalation of sandy intervals associated with bedload-dominated fluvial systems and fine-grained intervals accumulated in a mixed-load fluvial system. Fine-grained intervals are thicker in this area and evidence for more permanent water bodies, and the development of large-scale lacustrine bars, is also present. Also, an overall upward increase in the proportion of fine-grained deposits is more clearly recorded in this area.

4 Small-scale, high-frequency changes in fluvial style are interpreted as high-frequency sequences developed within this low-order lowstand systems tract as a result of short-term oscillations in base level. Base-level fluctuations could have been associated with climatic changes and temporary lake levels or eustatic sea-level changes associated with orbital processes. The presence of shallow-marine deposits down dip to the north that are probably coeval with the Avilé Sandstone supports the eustatic control on the development of these sequences.

5 A low-frequency change in fluvial style is also recorded by the gradual change from bedload-dominated fluvial systems at the base, to fine-grained, mixed-load fluvial systems towards the top of the Avilé. This change can be regarded as an overall increase in accommodation associated with a low-frequency (possibly global) transgressive trend developed after the major relative sea-level fall that produced the accumulation of these deposits.

6 High- and low-frequency accommodation changes are better identified in the down-dip sector of the study area. In the up-dip sector, the unit has a reduced thickness and is completely dominated by bedload sandy braided deposits. This suggests that this area behaved as a low accommodation setting during the accumulation of this unit and that accommodation was being created especially in the downstream part of the fluvial system.

7 The analysis of the lateral and vertical variations in fluvial style within the Avilé Sandstone shows that high-resolution regional correlations within non-marine deposits is extremely difficult, especially if different accommodation settings are present, and if the influence of external controls is partly out of phase in proximal and distal parts of the fluvial system.

ACKNOWLEDGEMENTS

This research was funded by the Ministerio de Educación, Ciencia y Tecnología of Argentina through Research Grant IM40 #34 and the Agencia Nacional de Promoción Científica y Tecnológica of Argentina (ANPCYT) Research Grant PICT 07-08451. We thank E. Schwarz, F. Colombo Piñol and J. Franzese for field assistance and fruitful discussions. S. Ballent (Museo de La Plata, Argentina) is thanked for micropalaeontological analysis. Sue Marriot and Piret Plink-Björklund are thanked for their essential and constructive reviews.

REFERENCES

Aguirre Urreta, M.B. and Rawson, P.F. (1997) The ammonite sequence in the Agrio formation (Lower Cretaceous), Neuquén Basin, Argentina. *Geol. Mag.*, **134**, 449–458.

Aguirre-Urreta, M.B., Concheyro, A., Lorenzo, M., Ottone, E.G. and Rawson, P.F. (1999) Advances in biostratigraphy of the Agrio Formation (Lower

Cretaceous) of the Neuquén Basin, Argentina: ammonites, palynomorphs, and calcareous nannofossils. *Palaeogeogr. Palaeoclimatol. Palaeoecol.*, **150**, 33–47.

Aitken, J.F. and Flint, S.S. (1995) The application of high-resolution sequence stratigraphy to fluvial systems: a case study from the Upper Carboniferous Breathitt Group, eastern Kentucky, USA. *Sedimentology*, **42**, 3–30.

Blum, M. (1993) Genesis and architecture of incised valley fill sequences: a Late Quaternary example from the Colorado River, Gulf coastal plain of Texas. In: *Siliciclastic Sequence Stratigraphy: Recent Developments and Applications* (Eds P. Weimer and H. Posamentier), pp. 259–283. Memoir 58, American Association of Petroleum Geologists, Tulsa, OK.

Bridge, J.S. (1993) Description and interpretation of fluvial deposits: a critical perspective. *Sedimentology*, **40**, 801–810.

Bridge, J.S. (2003) *Rivers and Floodplains: Forms, Processes, and Sedimentary Record*. Blackwell Publishing, Oxford, 491 pp.

Cecil, C.B. (1990) Paleoclimate controls on stratigraphic repetition of chemical and siliciclastic rocks. *Geology*, **18**, 533–536.

Clemente, P. and Pérez-Arlucea, M. (1993) Depositional architecture of the Cuerda del Pozo Formation, Lower Cretaceous of the extensional Cameros Basin, North-Central Spain. *J. Sediment. Petrol.*, **63**, 437–452.

Friend, P.F., Slater, M.J. and Williams, R.C. (1979) Vertical and lateral building of river sandstone bodies, Ebro Basin, Spain. *J. Geol. Soc. London*, **136**, 39–46.

Groeber, P. (1946) Observaciones geológicas a lo largo del meridiano 70. Hoja Chos Malal. *Rev. Soc. Geol. Arg.*, **1**, 177–208.

Hunter, R.E. (1977) Basic types of stratification in small eolian dunes. *Sedimentology*, **24**, 361–387.

Keighley, D., Flint, S., Howell, J. and Moscariello, A. (2003) Sequence stratigraphy in lacustrine basins: a model for part of the Green River Formation (Eocene), southwestern Uinta basin, Utah, USA. *J. Sediment. Res.*, **73**, 987–1006.

Kocurek, G. (1996) Desert aeolian systems. In: *Sedimentary Environments: Processes, Facies and Stratigraphy* (Ed H.G. Reading), pp. 125–153. Blackwell Science, Oxford.

Kocurek, G. and Nielson, J. (1986) Conditions favourable for the formation of warm-climate aeolian sand sheets. *Sedimentology*, **33**, 795–816.

Legarreta, L. (2002) Eventos de desecación en la Cuenca Neuquina: depósitos continentales y distribución de hidrocarburos. In: *V Congreso de Exploración y Desarrollo de Hidrocarburos*, Mar del Plata, Argentina. Proceedings on CDRom.

Legarreta, L. and Gulisano, C.A. (1989) Análisis estratigráfico secuencial de la Cuenca Neuquina (Triásico Superior-Terciario inferior). In: *Cuencas Sedimentarias Argentinas* (Eds G. Chebli and L. Spalletti), pp. 221–243. Serie Correlación Geológica 6, S. M. de Tucumán, Universidad Nacional de Tucumán.

Legarreta, L. and Uliana, M.A. (1991) Jurassic-Cretaceous marine oscillations and geometry of a back-arc basin fill, central Argentine Andes. In: *Sedimentation, Tectonics and Eustacy. Sea level Changes at Active Margins* (Ed D.I.M. MacDonald), pp. 429–450. Special Publication 12, International Association of Sedimentologists. Blackwell Scientific Publications, Oxford.

Marriott, S.B. (1999) The use of models in the interpretation of the effects of base-level change on alluvial architecture. In: *Fluvial Sedimentology VI* (Eds N.D. Smith and J. Rogers), pp. 271–281. Special Publication 28, International Association of Sedimentologists. Blackwell Science, Oxford.

Miall, A.D. (1996) *The Geology of Fluvial Deposits: Sedimentary Facies, Basin Analysis and Petroleum Geology*. Springer-Verlag, Berlin, 582 pp.

Mitchum, R.M. and Van Wagoner, J.C. (1991) High frequency sequences and their stacking patterns: sequence stratigraphic evidence of high frequency eustatic cycles. *Sediment. Geol.*, **70**: 135–144.

Mjøs, R., Walderhaug, O. and Prestholm, E. (1993) Crevasse splay sandstone geometries in the Middle Jurassic Ravenscar Group of Yorkshire, UK. In: *Alluvial Sedimentation* (Eds M. Marzo and C. Puigdefábregas), pp. 167–184. Special Publication 17, International Association of Sedimentologists. Blackwell Scientific Publications, Oxford.

Olsen, T., Steel, R.J., Hogseth, K., Skar, T. and Roe, S.L. (1995) Sequential architecture in a fluvial succession: sequence stratigraphy in the Upper Cretaceous Mesaverde Group, Price Canyon, Utah. *J. Sediment. Res.*, **B65**, 265–280.

Pérez-Arlucea, M. and Smith, N.D. (1999) Depositional patterns following the 1870s avulsion of the Saskatchewan river (Cumberland Marshes, Saskatchewan, Canada). *J. Sediment. Res.*, **69**, 62–73.

Plint, A.G. and Browne, G.H. (1994) Tectonic event stratigraphy in a fluvio/lacustrine, strike-slip setting: the Boss Point Formation (Westphalia A), Cumberland Basin, Maritime Canada. *J. Sediment. Res.*, **B64**, 341–364.

Prámparo, M.B. and Volkheimer, W. (1999) Palinología del Miembro Avilé (Formación Agrio, Cretácico Inferior) en el cerro de la Parva, Neuquén. *Ameghiniana*, **36**, 217–227.

Ramos, V.A. (1978) Estructura. In: *Septimo Congreso Geológico Argentino. Relatorio*, pp. 99–118. Asociación Geológica Argentina, Buenos Aires.

Rossi, G. (2001) *Arenisca Avilé: facies, ambientes sedimentarios y estratigrafía de una regresión forzada del Hauteriviano Inferior de la Cuenca Neuquina.* Unpublished PhD thesis, Universidad Nacional de La Plata, La Plata.

Sagasti, G. (2000) La sucesión rítmica de la Formación Agrio (Cretácico inferior) en el sur de la provincia de Mendoza, y su posible vinculación con Ciclos de Milankovitch. *Rev. Asoc. Arg. Sediment.*, **7**, 1–22.

Sagasti, G. (2002) *Estudio Sedimentológico y de Estratigrafía Secuencial de las sedimentitas carbonáticas de la Formación Agrio (Cretácico Inferior) en el sector surmendocino de la Cuenca Neuquina, República Argentina.* Unpublished PhD thesis, Universidad Nacional de La Plata, La Plata, 280 pp.

Sagasti, G. (2005) Hemipelagic record of orbitally-induced dilution cycles in Lower Cretaceous sediments of the Neuquén Basin. In: *The Neuquén Basin, Argentina: a Case Study in Sequence Stratigraphy and Basin Dynamics* (Eds G.D. Veiga, L.A. Spalletti, J. Howell and E. Schwarz), pp. 231–250. Special Publication 252, Geological Society Publishing House, Bath.

Shanley, K.W. and McCabe, P.J. (1994) Perspectives on the sequence stratigraphy of continental strata. *Am. Assoc. Petrol. Geol. Bull.*, **78**, 544–568.

Smith, N.D. and Pérez-Arlucea, M. (1994) Fine-grained deposition in the avulsion belt of the lowers Saskatchewan river, Canada. *J. Sediment. Res.*, **B64**, 159–168.

Smith, N.D., Cross, T.A., Dufficy, J.P. and Clough, S.R. (1989) Anatomy of an avulsion. *Sedimentology*, **36**, 1–23.

Thomas, R.G., Smith, D.G., Wood, J.M., Visser, J., Calverley-Range, E.A. and Koster, E.H. (1987) Inclined heterolithic stratification – terminology, description, interpretation and significance. *Sediment. Geol.*, **53**, 123–179.

Tye, R.S. and Colemam, J.M. (1989) Depositional processes and stratigraphy of fluvially dominated lacustrine deltas: Mississippi delta plain. *J. Sediment. Petrol.*, **59**: 973–996.

Veiga, G.D., Spalletti, L.A. and Flint, S. (2001) Anatomy of the Cretaceous Avilé Sandstone Lowstand Wedge in Central Neuquén Basin (Argentina). In: *2001 American Association of Petroleum Geologists Annual Convention*, Denver, p. A207.

Veiga, G.D., Spalletti, L.A. and Flint, S. (2002) Aeolian/fluvial interactions and high resolution sequence stratigraphy of a non-marine lowstand wedge: the Avilé Member of the Agrio Formation (Lower Cretaceous) in central Neuquén Basin, Argentina. *Sedimentology*, **49**, 1001–1019.

Veiga, R. and Vergani, G.D. (1993) Depósitos de nivel bajo: nuevo enfoque sedimentológico y estratigráfico del Miembro Avilé en el Norte del Neuquén. Argentina. In: *XII Congreso Geológico Argentino y II Congreso de Exploración de Hidrocarburos*, Mendoza Actas I, pp. 55–65.

Vergani, G.D., Tankard, A.J., Belotti, H.J. and Welsink, H.J. (1995) Tectonic evolution and paleogeography of the Neuquén Basin, Argentina. In: *Petroleum Basins of South America* (Eds A.J. Tankard, R. Suarez Soruco and H.J. Welsink), pp. 383–402. Memoir 62, American Association of Petroleum Geologists, Tulsa, OK.

Weaver, C.E. (1931) Paleontology of the Jurassic and Cretaceous of West Central Argentina. *Univ. Washington Mem.*, **I**, 1–496.

Wright, V.P. and Marriott, S.B. (1993) The sequence stratigraphy of fluvial depositional systems: the role of floodplain sediment storage. *Sediment. Geol.*, **86**, 302–210.

Anatomy of a transgressive systems tract revealed by integrated sedimentological and palaeoecological study: the Barcellona Pozzo di Gotto Basin, northeastern Sicily, Italy

CARLO MESSINA*[1], MARIA ANTONIETTA ROSSO*, FRANCESCO SCIUTO*, ITALO DI GERONIMO*, WOJCIEK NEMEC†, TATIANA DI DIO*, RAFFAELLA DI GERONIMO*, ROSANNA MANISCALCO* *and* ROSSANA SANFILIPPO*

**Dipartimento di Scienze Geologiche, Università di Catania, 95129 Catania, Italy*
†Department of Earth Science, University of Bergen, 5007 Bergen, Norway

ABSTRACT

The Barcellona Pozzo di Gotto Basin of northeastern Sicily, central Mediterranean, is a Plio-Pleistocene peri-Tyrrhenian shelf embayment that formed by the collapse and marine inundation of bedrock fault-blocks in response to regional tectonic extension. The study focuses on the well-developed transgressive systems tract of the lower bay-fill sequence. This succession of middle Pliocene to Lower Pleistocene marine deposits has a mixed siliciclastic to bioclastic composition and is ~ 73 m thick in mid-bay outcrop section. The deposits are sandy to silty facies indicating a wave-dominated bay rich in suspended sediment and influenced by storms and tidal currents. Facies associations represent upper and lower shoreface, offshore-transition and mid-bay offshore zones. The abundance of silty to sandy suspension is attributed to the entrapment of fine sediment entrained by storms and tides and possibly derived from nearby streams. The supply of sediment from the bay's shoreline zone probably combined with fine-grained sediment drift from offshore areas, as is also suggested by admixtures of outer circalittoral benthic microfauna. Facies-based estimates indicate a water depth of ≤ 25 m for the mid-bay area, with a mean depth of ~ 10 m for fairweather wave base and ~ 15–16 m for storm wave base. The shallow bay hosted circalittoral benthic fauna typical of deeper water Mediterranean shelves, which can be attributed to the high turbidity of the bay water (limited light penetration).

The stratigraphic organization of the facies associations and their fauna assemblages reveals that the succession consists of six parasequences, or transgressive–regressive cycles, bounded by marine flooding surfaces and showing an overall deepening upward trend. The parasequences are 4–17 m thick, and some include well-developed transgressive deposits and also a relatively thick mid-cycle condensation zone. The latter indicates a prolonged balance between the rates of accommodation development and its filling by slow aggradation. Palaeoecological and taphonomic criteria defining a condensation maximum allow the maximum flooding surface to be identified, typically in the upper part of the mid-cycle condensation zone. The parasequences have time spans of ~ 300 kyr and correlate with the 4th-order regional sequences recognized in the central Mediterranean. Accordingly, they are inferred to be the local equivalents of these high-frequency sequences, owing their facies architecture to a relatively high rate of tectonic subsidence in the peri-Tyrrhenian coastal region of northern Sicily. These would thus be type 2 sequences involving little or no fall in relative sea level and hence developed as parasequences. The integration of sedimentological, biostratigraphic, palaeoecological and taphonomic data proves to be a powerful method for high-resolution sequence stratigraphy and palaeoenvironment reconstruction, including sediment dynamics, palaeogeography and bathymetric changes.

Keywords Shelf embayment, facies analysis, taphonomy, palaeoenvironment, sequence stratigraphy, benthic fauna, Tyrrhenian Sea.

[1]Present address: Statoil Research Centre, Rotvoll, 7005 Trondheim, Norway (Email: cmes@statoil.com).

INTRODUCTION

Sequence stratigraphy is a rapidly developing frontier of conceptual models, striving for verification through rock-record studies and for high-resolution methods allowing ancient environmental changes to be deciphered from sedimentary successions. Important applications of sequence stratigraphy span a wide range of research areas, from reconstruction of sea-level history and basin analysis, to petroleum exploration and reservoir modelling. Although depositional sequences are considered to be the principal stratigraphic units (Mitchum, 1977; Vail *et al.*, 1977; Jervey, 1988; Posamentier & Vail, 1988; Posamentier *et al.*, 1988), parasequences are now widely regarded as the more fundamental component blocks of stratigraphic successions (Van Wagoner *et al.*, 1990; Swift *et al.*, 1991; Posamentier & Allen, 1993; Helland-Hansen & Martinsen, 1996; Coe, 2003; Storms & Hampson, 2005). Parasequences of local extent typically reflect sediment supply, whereas the more extensive ones reflect regional tectonism and/or eustasy. The recognition of parasequences and an understanding of their facies architecture are thus crucial to stratigraphic analysis.

The present study from a peri-Tyrrhenian shelf embayment of northeastern Sicily focuses on a set of parasequences that constitute a relatively thick transgressive systems tract in the lower part of the shallow-marine bay-fill succession. This series of transgressive–regressive cycles spans the middle Pliocene to Early Pleistocene time, and a good quality mid-bay outcrop section allows the internal facies architecture of the individual parasequences to be analysed in detail. Their unusual features include well-developed deposits of the cycle's transgressive phase and a relatively thick condensation zone of mid-cycle turnabout. The multidisciplinary study demonstrates that an integration of sedimentological, biostratigraphic, palaeoecological and taphonomic data can be a powerful method for high-resolution sequence-stratigraphic analysis and palaeoenvironmental reconstruction.

GEOLOGICAL SETTING

The Barcellona Pozzo di Gotto Basin of northeastern Sicily, central Mediterranean (Fig. 1A), is a Plio-Pleistocene peri-Tyrrhenian shelf embayment comprising two adjoining palaeobays (Fig. 1B). The embayment formed in middle Pliocene time by rapid marine drowning of the triangular-shaped coastal depressions, referred to as the Castroreale and Furnari palaeobays, each several kilometres across and 10–12 km in length, surrounded by a high-relief landscape. Relict deposits of similar palaeobays occur also farther to the east and to the west (Messina, 2003). Bedrock belongs to the Kabilo–Calabride massif (Fig. 1A), which consists of ophiolites and Variscan metamorphic rocks with a cover of Tertiary deposits, including Messinian evaporites and Early Pliocene chalks. The basin formed after a pulse of regional uplift, by the collapse and marine inundation of bedrock fault-blocks in response to regional tectonic extension (Ghisetti, 1981; Di Geronimo *et al.*, 1997; Messina, 2003).

The Tyrrhenian Sea is a late Neogene backarc basin that opened due to the westward subduction of the Ionian Sea plate under the Calabrian Arc (Fig. 1A) and involved eastward shifts of the axis of crustal separation and a similar migration of the subduction arc (Amodio Morelli *et al.*, 1976; Malinverno & Ryan, 1986; Boccaletti *et al.*, 1990; Patacca *et al.*, 1990; Knott & Turco, 1991; Catalano *et al.*, 1995; Robertson & Grasso, 1995; Monaco *et al.*, 1996; Lentini *et al.*, 2000; Bonardi *et al.*, 2001). The backarc tectonic extension led to a structural foundering of the Kabilo-Calabride massif, involving listric detachments, high-angle normal faults and strike-slip rotation of crustal blocks (Speranza *et al.*, 2000; Van Dijk *et al.*, 2000; Bonardi *et al.*, 2001; Guarnieri & Carbone, 2003).

Progressive unconformities and buried normal faults in the Barcellona Pozzo di Gotto Basin indicate syndepositional tectonic extension (Messina, 2003), and the basin-fill stratigraphy itself bears a high-resolution local record of peri-Tyrrhenian palaeogeography and relative sea-level changes controlled by tectonic subsidence.

METHODS AND TERMINOLOGY

The data for the present study have been acquired by detailed sedimentological logging of the mid-bay Castroreale section (Fig. 1B), combined with biostratigraphic, palaeoecological, ichnological and taphonomic analyses. Macrofauna (hand-picked

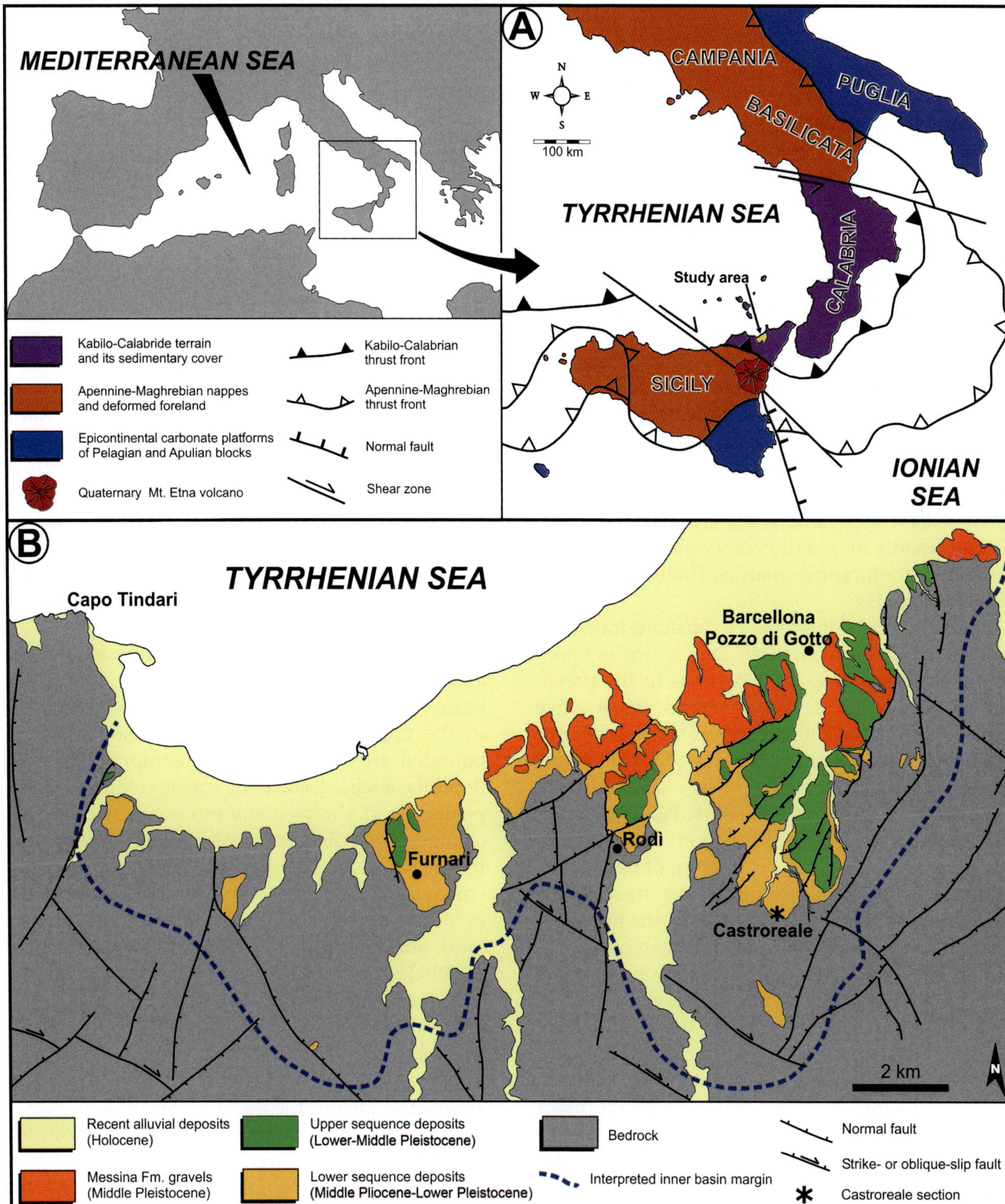

Fig. 1 (A) Locality map of the study area in northeast Sicily, southern Italy, showing the tectonic framework of the eastern Tyrrhenian Sea region. (B) Simplified geological map of the Barcellona Pozzo di Gotto Basin (modified from Messina, 2003). The basin comprises the adjoining Furnari (western) and Castroreale (eastern) palaeobays, and the deposits studied form the bulk of the lower bay-fill sequence. The interpreted inner basin margin shows the basin's minimum extent during the deposition of the lower sequence; the embayment was much larger during the deposition of the second, upper sequence. Note the location of the Castroreale outcrop section.

shells) and microfauna (bulk sediment samples) were sampled systematically, and some of the data were quantified as semi-continuous plots. Palaeoenvironmental and palaeobathymetric inferences involved all taxonomic groups, but were based particularly on bryozoans, molluscs and ostracods, which are consistently the most abundant throughout the outcrop section.

The descriptive sedimentological terminology follows Collinson & Thompson (1982) and Harms *et al.* (1982). The term 'sedimentary facies' refers to the basic types of sedimentary deposit, distinguished on a macroscopic basis and attributed to different modes of sediment deposition (Harms *et al.*, 1975; Walker, 1984). The term 'facies association' denotes an assemblage of spatially and genetically related facies, considered to represent a particular depositional environment (or 'system' in the parlance of sequence stratigraphy; Posamentier *et al.*, 1988). Facies associations are the basic architectural elements of a sedimentary succession in its sequence-stratigraphic analysis (Emery & Myers, 1996; Coe, 2003).

The distinction of shoreface, offshore-transition and offshore zones in an ancient record is based on sedimentary facies and pertains to the prevalent depths of the fairweather and storm wave bases (Reading & Collinson, 1996, fig. 6.6). Benthic biocoenoses are classified according to such factors as the substrate type, water energy and sedimentation rate (Pérès & Picard, 1964; Pérès, 1982; Di Geronimo, 1985). The distinction of ecological zones pertains chiefly to the amount of light transmitted by the water column. The mesolittoral zone is intertidal and corresponds to the foreshore environment. The infralittoral zone extends to the water depth where light penetration becomes insufficient for phanerogam photosynthesis; this critical depth varies from 35 to 70 m on Mediterranean shelves (according to the distribution of *Posidonia oceanica*), and the infralittoral zone corresponds to the shoreface and at least the inner part of the offshore-transition environment. The circalittoral zone extends further to the water depth where light penetration becomes insufficient for algal photosynthesis, which normally means the outermost shelf and a depth of 120–130 m in the Mediterranean; this zone corresponds to the offshore environment and commonly includes also the outer part of the offshore transition. The term epibathyal pertains to the uppermost bathyal zone, typically the shelf break and upper slope environment.

Ecological control, however, may also include other factors, and because the penetration depth of light itself depends strongly on water turbidity, the palaeobathymetric estimates based on biocoenoses may differ from those based on sedimentary facies. Such discrepancies are meaningful, and the two types of criteria supplement and verify each other.

BASIN-FILL STRATIGRAPHY

The sedimentary succession of the Barcellona Pozzo di Gotto Basin has previously been studied mainly from the point of view of regional backarc tectonics (Catalano & Cinque, 1995; Catalano & Di Stefano, 1997; Lentini *et al.*, 2000). There have been a few palaeoecological studies and tentative palaeoenvironmental reconstructions (Barrier, 1987; Kezirian, 1993; Di Geronimo *et al.*, 2002, 2005; Messina, 2003), but no detailed sedimentary facies analysis and sequence-stratigraphic model.

The Plio-Pleistocene basin-fill succession (Fig. 2) exceeds 200 m in thickness and crops out in many parts of the basin (Fig. 1B), but is best preserved in the Castroreale palaeobay. The succession is bounded by unconformities, comprises marine deposits of siliciclastic to bioclastic calcareous composition and shows many environmental (facies) changes, including an erosional unconformity in the middle part. The deposits recorded the shelf embayment's palaeogeographical history, which involved relative sea-level changes due to both local and regional subsidence and to episodic regional uplift (Kezirian, 1993; Messina, 2003). The overlying fluvio-deltaic and alluvial Middle Pleistocene–Holocene deposits (Fig. 1B) are not limited to the Barcellona Pozzo di Gotto Basin in their lateral extent and hence are not regarded as an integral part of the basin-fill succession.

Messina (2003) divided the marine basin-fill succession into two type-1 sequences (*sensu* Van Wagoner *et al.*, 1990), which are bounded by unconformities (Fig. 2) and have been mapped (Fig. 1B). The first, lower sequence consists of a thick transgressive systems tract (TST_1) and a relict highstand systems tract (HST_1), whereas the upper sequence is thinner and consists of a lowstand

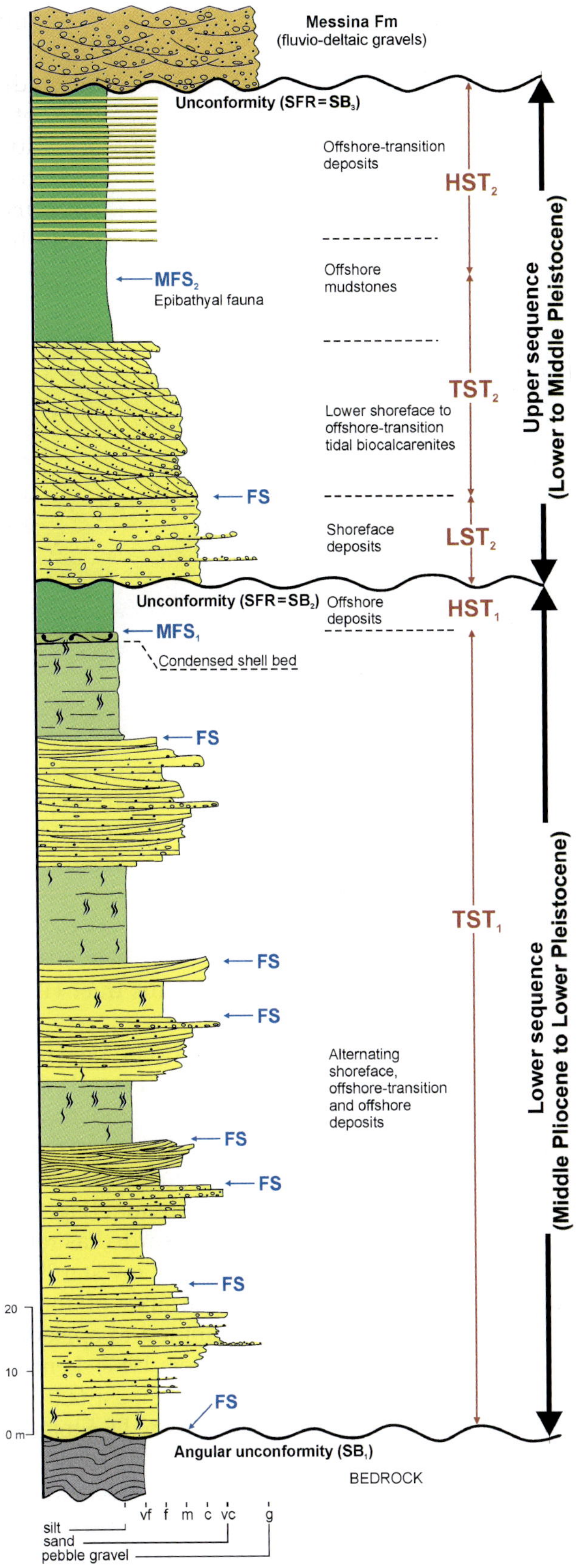

systems tract (LST$_2$) overlain by a transgressive (TST$_2$) and a highstand systems tract (HST$_2$).

The lower sequence comprises mixed siliciclastic–bioclastic deposits of middle Pliocene (Piacenzian) to Early Pleistocene age, overlying unconformably the deformed bedrock that includes karstified Messinian limestones (Fig. 3A). This basal surface of subaerial exposure is regarded as a sequence boundary (SB$_1$). The TST$_1$ consists of alternating littoral to neritic deposits that are up to 125 m thick and show an overall upward deepening. As discussed further in the paper, this transgressive succession bears a record of higher-frequency transgressive–regressive cycles bounded by marine-flooding surfaces (Fig. 2), which indicates that the relative sea-level rise was incremental, punctuated by normal regressions.

The overlying HST$_1$ consists of laminated siltstones and silty sandstones, up to 9 m thick, whose fauna indicates middle to lower circalittoral zone and hence a water depth of possibly 80–100 m (Messina, 2003). The boundary of TST$_1$ and HST$_1$ (Fig. 2) is a condensed stratigraphic horizon indicating sediment-starved seafloor conditions and is regarded as the maximum flooding surface (MFS$_1$). The HST$_1$ is relatively thin, lacks a shallowing-upward facies signature and is overlain sharply by coarse-grained littoral deposits, which implies that its upper part was eroded prior to the deposition of the upper sequence (Fig. 2). This erosional unconformity is attributed to a forced regression and considered to be the upper sequence boundary (SB$_2$).

The upper sequence comprises deposits of late Early to Middle Pleistocene age. The SB$_2$ is overlain by littoral deposits (Fig. 2), ~ 13 m thick, representing a lowstand systems tract (LST$_2$). The LST$_2$ indicates aggradation combined with seaward sediment bypass and its fauna suggests lowest

Fig. 2 (*left*) Interpreted stratigraphy of the Barcellona Pozzo di Gotto Basin. Letter symbols: FS, marine flooding surface; HST, highstand systems tract; LST, lowstand systems tract; MFS, maximum flooding surface; SB, sequence boundary; SFR, surface of forced regression; TST, transgressive systems tract. The letter symbols in mean grain-size scale indicate very fine (vf), fine (f), medium (m), coarse (c) and very coarse (vc) sand and pebble gravel (g).

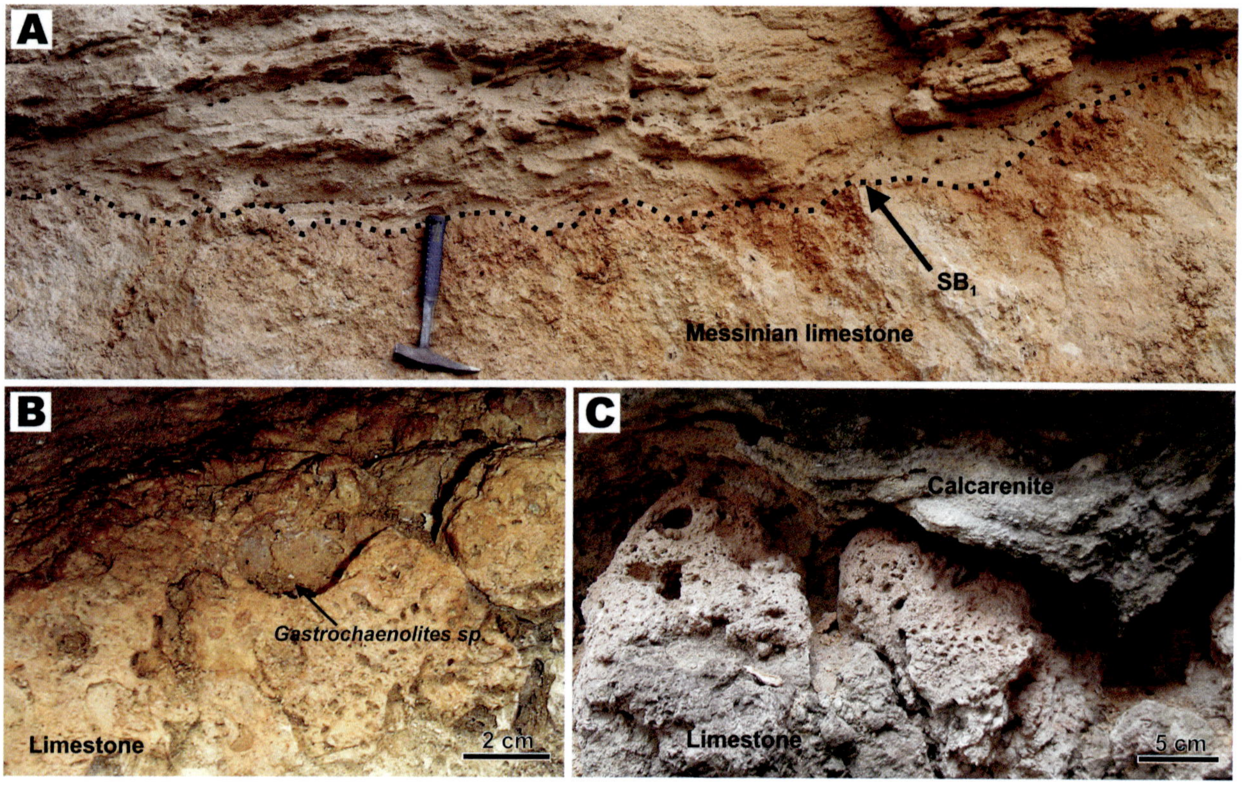

Fig. 3 Base of the bay-fill succession. (A) Outcrop detail of the basal boundary (surface SB_1 in Fig. 2) in the Castroreale section; the lower-shoreface calcarenites of facies association B overlie here the weathered, karstified and wave-swept surface of Messinian evaporitic limestone. (B & C) Close-up details of the limestone at the boundary show boring traces of *Gastrochaenolites* sp. (B) and *Lithophaga lithophaga* (C).

infralittoral zone, with a water depth of possibly 30–40 m (Messina, 2003). The overlying, fining upward package of cross-stratified tidal biocalcarenites (Fig. 2), up to 25 m thick, indicates water deepening and is regarded as a transgressive systems tract (TST_2). Its fossil fauna includes brachiopods, bryozoans and pectinid bivalves indicative of a lower circalittoral zone. The tidal dune foresets show variable palaeocurrents, but mainly towards the north-northeast (Messina, 2003). The TST_2 culminates in laminated mudstones (Fig. 2), with the fossil fauna indicating rapid deepening to epibathyal conditions. A maximum flooding surface (MFS_2) is inferred at this level, for the mudstones above are increasingly intercalated with calcarenitic tempestite sheets, which indicates shallowing to an offshore-transition environment. This muddy to heterolithic succession is up to 40 m thick and the MFS_2 separates the TST_2 from the overlying highstand systems tract HST_2, ~ 25 m thick.

The overlying Middle Pleistocene gravelly deposits of the Messina Formation (Figs 1B & 2) represent raised Gilbert-type deltas and associated alluvium. They are underlain by an erosional unconformity (SB_3) and are erosionally covered (SB_4) by the deeply incised recent alluvium (Fig. 1B). These two youngest units can be regarded as relict sequences, each comprising a late lowstand to highstand systems tract. The surfaces of forced regressions SB_3 and SB_4 are attributed to episodes of regional tectonic uplift (Barrier, 1987; Lentini *et al.*, 2000), with a mean rate for the past 600 kyr estimated at 1.1 mm yr^{-1} (Catalano & Di Stefano, 1997).

The present study focuses on the thick transgressive systems tract of the lower sequence (TST_1 in Fig. 2), which is well-developed and has isolated outcrops at several localities in the basin, but is best preserved and well-exposed in the Castroreale section (Fig. 1B).

THE CASTROREALE SECTION

The transgressive systems tract of the lower sequence (TST_1 in Fig. 2) is particularly well-developed, extensive (Fig. 1B) and relatively thick (~ 73 m). This succession consists of fossiliferous marine deposits that vary from mainly siliciclastic to predominantly calcareous, bioclastic, and include siltstones, sandstones and subordinate granule conglomerates. The stratigraphic data reviewed in the present section are summarized in Fig. 4. For the sake of an easy reference to its particular intervals, the stratigraphic succession has been divided into consecutive transgressive (T), mid-cycle (MC) and regressive (R) facies zones (Fig. 4). These facies zones, much like the related flooding surfaces (FS), are time-stratigraphic portions of the succession that can be traced and correlated across the basin.

Biostratigraphy

Both macro- and microfauna indicate a middle to Late Pliocene age for the lower part of the succession and an Early Pleistocene age for its upper part (Fig. 4). The basal part bears abundant *in situ* bryozoans and a diverse ostracod assemblage, including *Aurila* gr. *punctata*, *A. pigadiana*, *A.* gr. *convexa*, *A. intrepretis*, *A. latisolea*, *Mutilus elegantulus*, *Ruggeria tetraptera* and *Quadracythere salebrosa*. Bryozoans in the lower part include *Cheiloporina campanulata* (present in the basal zone), *Hippopleurifera sedgwichi* (at the base of facies zone MC_2), *H. surgens*, *Celleporaria palmata*, *Smittina canavari* (in facies zone MC_2 to the lower mid-part of zone MC_4) and *Umbonula monoceros* (in the uppermost zone MC_4), which all are species known only from the pre-Pleistocene, or specifically Pliocene, record (Poluzzi, 1975; Pouyet, 1976; Barrier *et al.*, 1987; Moissette, 1988; Pouyet & Moissette, 1992; Spjeldnaes & Moissette, 1997). The same pertains to the barnacle *Archaeobalanus stellaris* (Menesini, 1984; Menesini & Casella, 1988), found in facies zone MC_2.

Planktonic foraminifers throughout the section include a high percentage of reworked early and middle Pliocene species, such as *Globorotalia margaritae* and *G. puncticulata*, which occur up to facies zone T_5 (Fig. 4B). The lowest part of the succession, from its base to the upper part of zone R_2, contains rich and diverse planktonic foraminifers, with common *G. crassaformis* accompanied by rare and juvenile other species. The juvenile forms can be identified as *G. bononiensis* and/or *G. inflata*, because the two are linked by an evolutionary trend and share morphological features (Colalongo & Sartoni, 1967; Stainforth *et al.*, 1975; Brolsma, 1978; Bossio *et al.*, 1997). The lowest part of the succession would then represent either biozone MPL6 (if the juvenile forms are *Globorotalia inflata*) or the lower part of biozone MPL5a (if the juvenile ones are assigned to *Globorotalia bononiensis*) (Cita, 1975). The notion of biozone MPL5a is supported by the occurrence of the bivalve *Pecten benedictus* at the base of facies zone T_1 and the ostracod *M. elegantulus* from the basal part to facies zone MC_2 (Fig. 4A). This latter species is not known from the stratigraphic record after the earliest Late Pliocene (*sensu* Bonaduce *et al.*, 1987) or the middle Pliocene in modern terms (Gradstein *et al.*, 2004), and the former species is known only from the Mediterranean Pliocene Molluscan Unit 1 of Monegatti & Raffi (2001), no younger than 3.0 Ma.

Biozone MPL6 is well recognizable from the middle part of facies zone T_4 upwards, where well-preserved adult forms of *G. inflata* occur. The top of biozone MPL6, or the Pliocene–Pleistocene boundary, is in the middle part of facies zone R_4 (Fig. 4B), where the first common occurrence (FCO) of left-coiled *Neogloboquadrina pachyderma* has been recognized. The left-coiled forms of *N. pachyderma* are rare in the Plio-Pleistocene of the Mediterranean (Sprovieri *et al.*, 1998) and their FCO is defined as the first relative increase in abundance, which amounts to about 10% of the bulk *N. pachyderma* population in the present case.

The topmost part of the succession bears the ostracod *Triebelina raripila* and also contains the first occurrence of the benthic foraminifer *Hyalinea baltica* in the upper part of facies zone R_5, which indicate an Emilian age (Ruggieri, 1980; Pasini & Colalongo, 1994). An Early Pleistocene age is indicated also by macrofauna, particularly the bivalve *Arctica islandica*. The top part of facies zone MC_5 (Fig. 4B) bears the serpulid *Pomatoceros triqueter*, the 'giant' form of which is typical of the region's cool Pleistocene waters (Di Geronimo *et al.*, 2000).

A

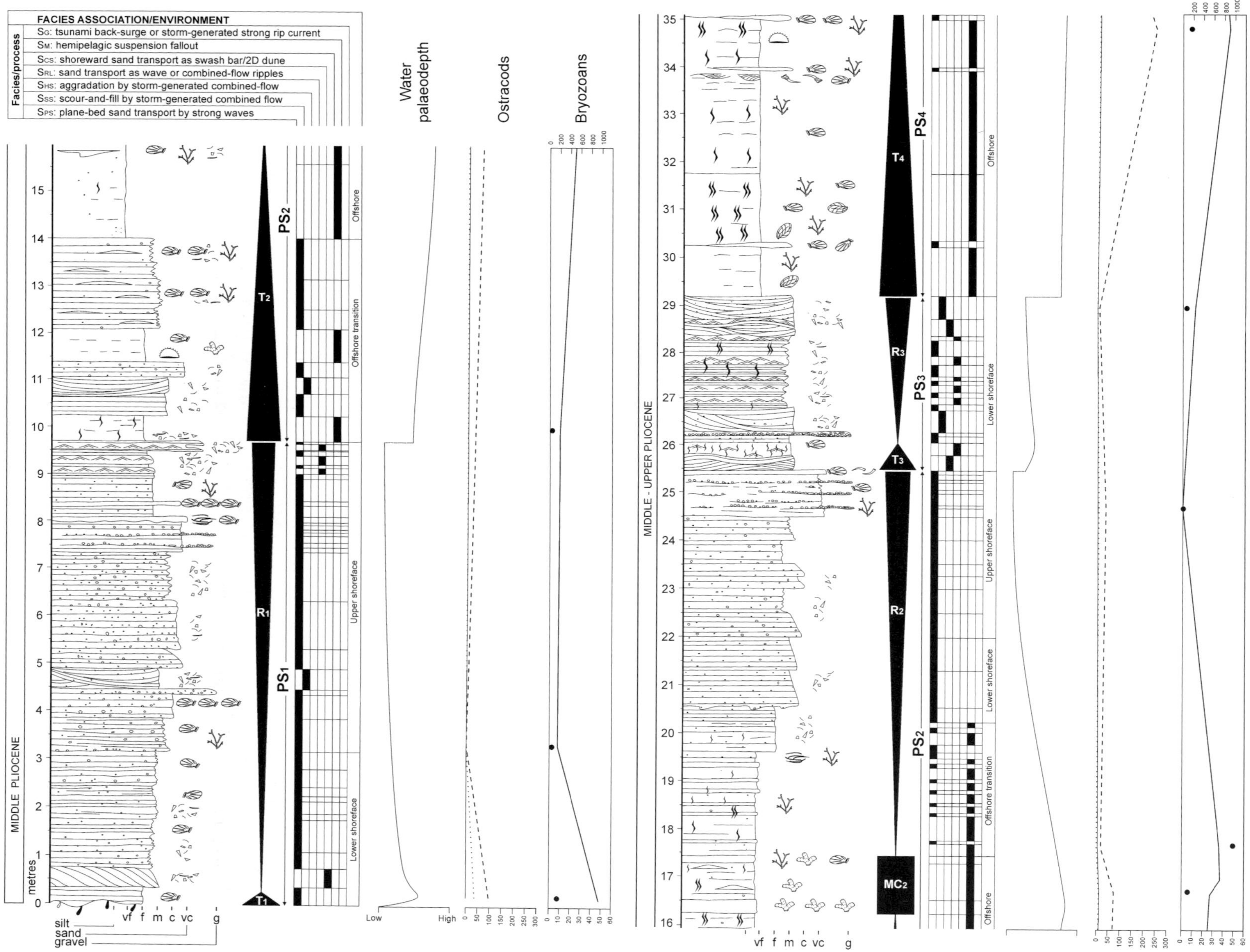

Facies/process
FACIES ASSOCIATION/ENVIRONMENT
SG: tsunami back-surge or storm-generated strong rip current
SM: hemipelagic suspension fallout
SCS: shoreward sand transport as swash bar/2D dune
SRL: sand transport as wave or combined-flow ripples
SHS: aggradation by storm-generated combined-flow
SSS: scour-and-fill by storm-generated combined flow
SPS: plane-bed sand transport by strong waves
Water palaeodepth
Ostracods
Bryozoans
MIDDLE PLIOCENE
MIDDLE - UPPER PLIOCENE
metres
silt
sand
gravel
vf f m c vc g
Low
High
T1
R1
T2
T3
R2
R3
T4
MC2
PS1
PS2
PS3
PS4
Lower shoreface
Upper shoreface
Offshore transition
Offshore

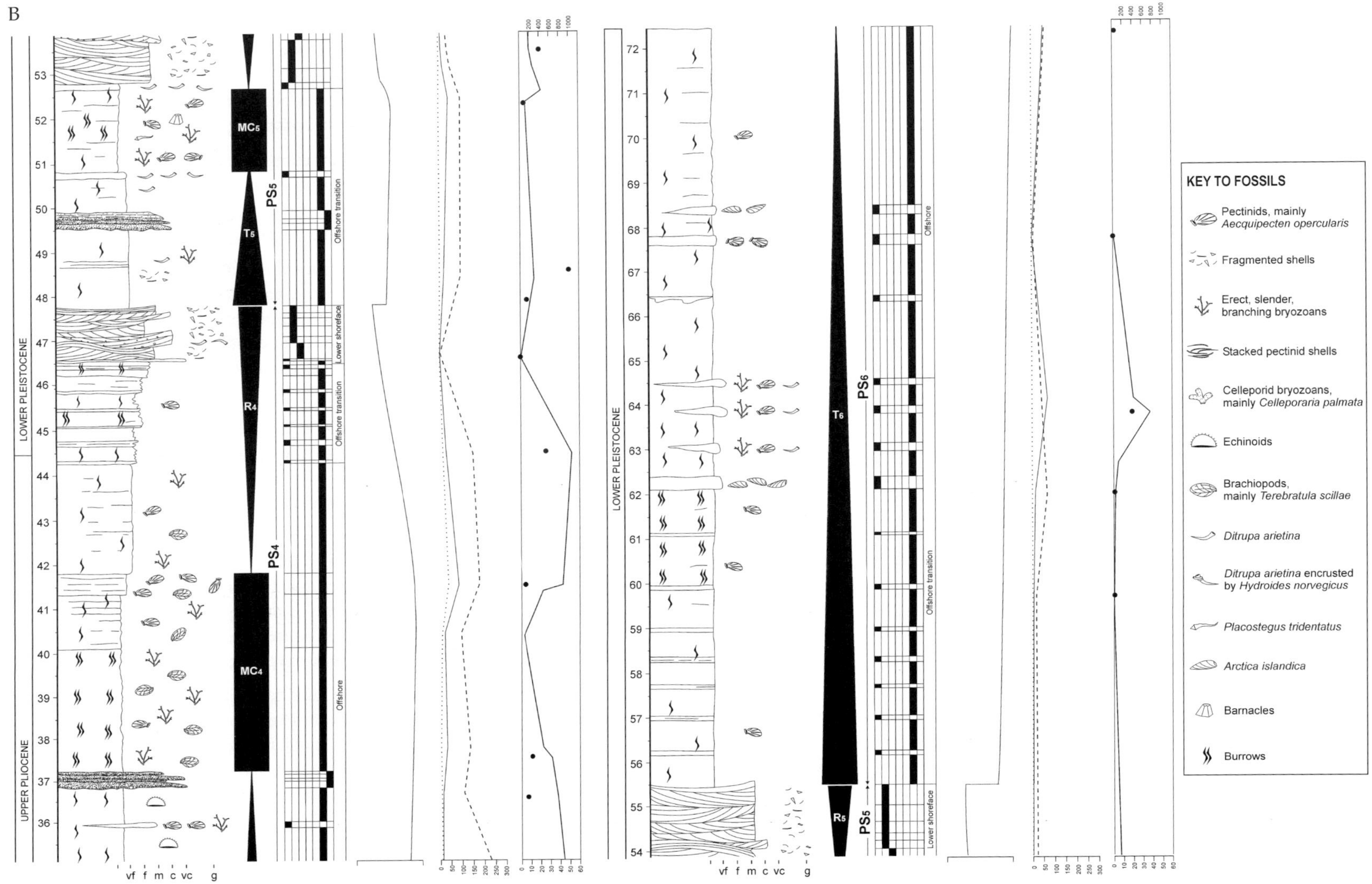

Fig. 4 Sedimentological log of the Plio-Pleistocene succession in the Castroreale outcrop section (see location in Fig. 1B). (A) Middle–Upper Pliocene. (B) Upper Pliocene to Lower Pleistocene. Facies are as defined in Table 1, their associations are as described in the text and the legend for grain-size scale is in the caption to Fig. 2. The consecutive parasequences (PS) are divided into transgressive (T), mid-cycle (MC) and regressive (R) facies zones. The plot of water palaeodepth is qualitative, indicating relative changes, with the local bathymetric maximum at the log top. The ostracod plot indicates the abundance (specimen number per sample) of infralittoral species (dotted line), infra-circalittoral species (dashed line) and circalittoral species (solid line). The bryozoan plot similarly indicates the abundance of specimens (solid line, with scale at the top) and the number of intact and apparently *in situ* species (dots, with scale at the bottom).

Table 1 Sedimentary facies of the Castroreale bay-fill succession. The stratigraphic distribution of facies is shown in Fig. 4

Facies code		Facies name and description	Genetic interpretation
S_{PS}	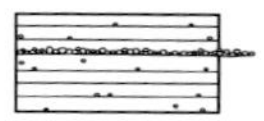	**Sandstones with planar parallel stratification** — Fine/medium to very coarse and granule-bearing arenitic sandstones, with common stringers/lenses of granule gravel and scattered small pebbles. Planar parallel strata form sets 3–8 cm thick, separated by subhorizontal or gently undulating erosional surfaces lined patchily with fine gravel. Facies units are 3 to 102 cm thick, averaging 23 cm.	Tractional deposition of sand by relatively strong waves, with 'supercritical' orbital velocities and planar seafloor configuration (Harms *et al.*, 1982, their Fig. 2–14). The deposition of coarser, erosional and gravel-bearing beds probably involved also storm-generated seaward currents (Arnott, 1993).
S_{SS}	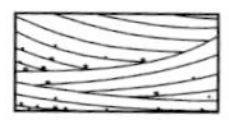	**Sandstones with swaley cross-stratification** — Fine/medium to coarse arenitic sandstones, commonly with scattered granules in the lower parts of swales. Cross-strata sets are 10–30 cm thick, with concave-upward, pod-shaped erosional bases and widths of 0.8–4 m. Facies units are 15 to 52 cm thick, mainly ~ 30 cm.	Tractional deposition of sand by scour-and-fill action of storm-generated combined-flow currents (Bourgeois, 1980; Dott & Bourgeois, 1982; Arnott & Southard, 1990).
S_{HS}		**Sandstones with hummocky cross-stratification** — Fine/medium to coarse arenitic sandstones with granules scattered along the bases. Dome-shaped strata sets are 10–30 cm thick, with concave-upward or nearly planar erosional bases and wavelengths of 1–5 m. Some hummocks are draped with symmetrical ripples. Facies units are 25 to 38 cm thick, averaging 33 cm.	Tractional deposition of sand by mainly aggradational action of storm-generated combined-flow currents (Bourgeois, 1980; Dott & Bourgeois, 1982; Harms *et al.*, 1982; Duke, 1985; Arnott & Southard, 1990; Duke *et al.*, 1991).
S_{RL}	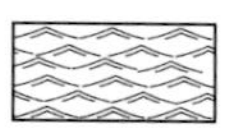	**Sandstones with ripple cross-lamination** — Fine/medium to coarse sandstones, mainly arenitic, with cross-lamination indicating bedforms ranging from symmetrical or asymmetrical 2D (long-crested) ripples to dome-shaped 3D ripples ('micro-hummocks' of Kreisa, 1981). Facies units are 3 to 5 cm thick.	Tractional deposition of sand as 2D to 3D vortex ripples by oscillatory waves with 'subcritical' orbital velocities, or as 3D ripples by storm-generated combined-flow currents (Harms *et al.*, 1982; Clifton & Dingler, 1984; Yokokawa *et al.*, 1995).
S_{CS}		**Sandstones with planar cross-stratification** — Fine- to medium-grained arenitic sandstones forming solitary sets of planar cross-strata, sigmoidal in shape. Such sporadic, isolated cross-sets are ≤ 40 cm thick, encased in facies S_{PS}. Foreset dip directions are towards the palaeoshoreline.	Shoreward transport of sand as 2D dune or swash bar by shoaling wave-generated currents during a post-storm phase of shoreface recovery (Clifton *et al.*, 1971; Hobday & Banks, 1971; Clifton, 1976; Clifton & Dingler, 1984).
S_M	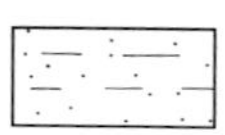	**Massive or faintly laminated siltstones and silty sandstones** — Siltstones alternating with very fine-grained silty sandstones, homogeneous or weakly parallel laminated, with ≤ 11 wt.% mud matrix; rich fauna and abundant burrows. Facies units are 8 to 330 cm thick.	Deposition of silt and very fine sand by fallout from suspension in quiet-water conditions; seafloor influenced by tidal currents and subject to extensive bioturbation (Reineck & Singh, 1975).
S_G	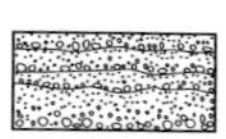	**Sandstones with graded bedding** — Very coarse to medium/fine arenitic sandstones, forming isolated fining-upward packages of 3 or 4 beds, massive and normally graded, 5–16 cm thick and superimposed upon one another. These rare units have sharp, erosional bases and occur embedded in facies S_M.	Non-tractional deposition of sand by rapid fallout from turbulent suspension, attributed to exceptionally strong rip currents or possibly tsunami backwash surges (Gruszczyński *et al.*, 1993; Myrow & Southard, 1996; Cantalamessa & Di Celma, 2005).

Sedimentary facies

The following seven sedimentary facies have been distinguished as the basic building blocks of the marine bay-fill succession:

1 sandstones with planar parallel stratification (facies S_{PS});
2 sandstones with swaley cross-stratification (facies S_{SS});
3 sandstones with hummocky cross-stratification (facies S_{HS});
4 sandstones with wave-ripple cross-lamination (facies S_{RL});
5 sandstones with planar cross-stratification (facies S_{CS});
6 homogeneous siltstones alternating with silty sandstones (facies S_M);
7 massive sandstones with graded bedding (facies S_G).

Their descriptive characteristics and genetic interpretation are summarized in Table 1. Sand varies from predominantly siliciclastic to bioclastic, bearing up to 70–80% by volume of skeletal detritus in many layers. Admixtures of non-skeletal gravel consist of quartz and other bedrock fragments. Mud is similarly of mixed, calcareous–siliciclastic composition, and its admixture in sandstones varies from < 1 to ~ 11 wt.%. The spectrum of facies indicates a wave-dominated environment influenced by storms and tides, and rich in suspension comprising mud, silt and very fine sand. As discussed further in this paper, the abundance of fine-grained sediment suspension is attributed to its entrapment by waves in the coastal embayment.

SEDIMENTARY FACIES ASSOCIATIONS

The sedimentary facies (Table 1) have been recognized to form five major facies associations, which are indicated in Fig. 4 and are described and interpreted in the present section. These facies assemblages represent depositional environments ranging from the upper shoreface to offshore zones, and it is their vertical stacking pattern that allows parasequences, or transgressive–regressive cycles, to be identified in the bay-fill succession (Figs 2 & 4).

Shoreline record

Description

No palaeoshoreline deposits are preserved at the palaeobay's landward fringe, because the whole inner periphery of the coastal embayment has been eroded by the Middle Pleistocene and Holocene episodes of subaerial denudation (Fig. 1B). However, an early shoreline record is preserved at the base of the bay-fill succession (Fig. 4), where the transgressive marine deposits overlie the unconformity surface of marine flooding (SB_1). The underlying evaporitic limestones are karstified and pinkish-grey to reddish-brown in colour, with a surface relief of several decimetres (Fig. 3A). Animal boring traces include *Gastrochaenolites* sp. and *Lithophaga lithophaga* (Fig. 3B, C), representing the *Trypanites* ichnofacies. The weathered limestone lacks regolith mantle, and also most of the borings are incompletely preserved, with their inlet parts eroded.

Interpretation

The basal unconformity (SB_1) is clearly a ravinement surface of marine transgression, swept by waves and recording the erosive encroachment of the shoreline across a drowning coastal embayment. The karstified, reddish-coloured limestone indicates fersiallitic weathering of terra-rossa type (Duchaufour, 1977). The truncated borings and a lack of regolith cover indicate net erosion under a fluctuating wave-energy level. The *Trypanites* ichnofacies is typical of a hard substrate in a shoreline environment with high water energy (Pemberton *et al.*, 1990).

Facies association A: upper shoreface deposits

Description

This facies assemblage occurs only in the lower part of the succession studied (see PS_1 and PS_2 in Fig. 4A) and constitutes ~ 14% of its total thickness. The deposits are sandstones and subordinate granule conglomerates. The bulk (~ 93% thickness) of this assemblage consists of facies S_{PS} (Fig. 5A & Table 1), comprising fine/medium to very coarse, parallel-stratified sandstones with scattered granules and

Fig. 5 Outcrop details of shoreface deposits (facies associations A and B). (A) Thinly bedded, planar parallel-stratified sandstones of facies S_{PS}; the flat and slightly inclined erosional surfaces E are attributed to storm events. (B) Cross-laminated sandstones of facies S_{RL}, showing combined-flow ripple forms and storm erosion surfaces. (C) Swaley cross-stratified sandstones of facies S_{SS}. (D) Shell bed in facies S_{PS} sandstones, rich in disarticulated and broken pectinid shells; note that the majority of shells are resting parallel to the planar strata, with convex sides upwards.

small pebbles and with discrete lenses (broad patches) of granule gravel, up to 9 cm thick. Gravel clasts are subrounded to well-rounded and up to 2.5 cm in size. This facies is locally intercalated with the cross-laminated, medium to coarse sandstones of facies S_{RL} (Fig. 5B & Table 1), the thin (3–5 cm) interbeds of which constitute merely 2.3% of the association thickness. An interbed of fine–medium, swaley cross-stratified and granule-bearing sandstone of facies S_{SS}, 44 cm thick (Fig. 5C), makes up the remainder (4.4%) of the association thickness in parasequence PS_1.

Broken shells occur scattered as highly abraded pieces up to a few centimetres in length, but are locally concentrated into thin lenses (patches) or form layers up to 5–10 cm thick, rich in bivalves (*Aequipecten opercularis* and subordinate *Pecten jacobaeus*). Broken branches of bryozoan *Myriapora truncata* are scattered throughout. Shells are invariably fragmented, disarticulated, and rest horizontally with their convex sides upwards (Fig. 5D), although locally in more haphazard positions, with either convex or concave side upwards. No microfauna and small macrofossils have been found, and there are also no trace fossils.

Statistical analysis of the orientation of pectinid shells, concentrated with their convex sides upwards in facies S_{PS}, indicates a preferential alignment of their longest axes, with a mean direction of 350° ± 19°. The Kuiper, Watson and Rayleigh tests have all rejected the null hypothesis of distribution uniformity with ≥ 95% confidence, which indicates a circular-normal, Von Mises-type frequency distribution (Nemec, 2005).

Interpretation

This assemblage of relatively coarse-grained, wave-worked and storm-modified facies (Table 1) indicates deposition well above the fairweather wave base, which means an upper shoreface environment. The lack of muddy or silty interlayers implies perennial wave action, with predominantly high orbital wave velocities and punctuated by storm events (Clifton *et al.*, 1971; Clifton, 1976; Bourgeois, 1980; Dott & Bourgeois, 1982; Harms *et al.*, 1982; Clifton & Dingler, 1984; Arnott & Southard, 1990; Duke *et al.*, 1991). The gravel component was apparently derived by storms from a contemporaneous adjacent beach zone (Clifton, 1973; Leithold & Bourgeois, 1984; Leckie, 1988; Arnott, 1993). The lack of recognizable breaker bars indicates a reflective shoreline (Komar, 1976; Wright *et al.*, 1979).

The disarticulated and abraded shells support the notion of a persistent and generally strong wave action. The concentrations of aligned shells with convex sides upwards can be attributed to their accumulation by short-lived unidirectional currents (Kidwell *et al.*, 1986; Kidwell & Bosence, 1991), which could be storm-generated and tidal. Weak waves and tidal currents probably dispersed shells that are more haphazardly oriented.

Facies association B: lower shoreface deposits

Description

This facies association occurs in parasequences PS_1 to PS_5 (Fig. 4) and consists of fine- to coarse-grained sandstones with scattered granules. Its units are between 1.4 and 3.7 m thick and constitute ~ 18% of the total thickness of the succession studied. The assemblage includes facies S_{PS}, S_{SS}, S_{HS}, S_{RL} and S_{CS} (Table 1).

Facies S_{PS} has similar characteristics as in the previous association, except that the gravel component here is sparser, limited to granules and forming local lenses (patches) no thicker than 2 cm. Shells are also sparser, represented mainly by abraded fragments of pectinids and *Ditrupa arietina*. A horizon of concentrated *Aequipecten opercularis* shells has been found in parasequence PS_1 (Fig. 4A), with valves in convex-upward positions and slightly encrusted. Fine-grained sandstone layers contain highly fragmented and undeterminable shells, as well as diverse ostracods. The units of facies S_{PS} are 4–76 cm thick, averaging 31 cm, and constitute up to 52% of the association thickness.

Facies S_{SS} sandstones occur as interbeds 15–52 cm thick and constitute 20–35% of the association thickness, but occasionally form amalgamated units 100–230 cm thick and predominate (see parasequence PS_5 in Fig. 4B). Granules are common in the basal parts of swaley cross-sets, where also many strata contain granule- to pebble-sized shell fragments. Otherwise, both shells and microfossils are sparse, and their little-abraded specimens indicate transport in suspension followed by rapid burial. However, the same facies at the top of parasequence PS_3 (Fig. 4A) contains abraded specimens of bryozoans and foraminifers, filled with pelitic sediment and indicating considerable displacement and/or reworking.

Facies S_{SS} is commonly underlain by facies S_{HS} (Fig. 6), which consists of medium to fine sandstones with minor granules scattered along the hummock bases. Facies S_{HS} units are 25–38 cm thick, averaging 33 cm, and constitute little more than 10% of the association thickness. Hummocks are 7–20 cm in amplitude and 1–4 m in wavelength, locally draped with veneers of bryozoan colonies. Some bedding surfaces show horizontal burrows (Fig. 6, lower part), but shelly macrofauna is nearly absent, except for sporadic pieces of pectinid and *D. arietina* shells and rare internodes of articulated bryozoans, relatively well preserved.

Facies S_{RL} sandstones are fine- to medium-grained and form interbeds 3.5–5 cm thick. These layers are moderately bioturbated, contain scattered shell fragments and constitute ≤ 2% of the association thickness. Cross-lamina sets indicate symmetrical to asymmetrical two-dimensional ripples and oval dome-shaped three-dimensional ripples (Harms *et al.*, 1982; 'micro-hummocks' *sensu* Kreisa, 1981). Many asymmetrical ripples have rounded crests and relatively narrow troughs (Fig. 5B), similar to those reported from laboratory experiments by Yokokawa *et al.* (1995).

Facies S_{CS} occurs as an isolated set of cross-strata, 40 cm thick, in the lower part of facies association B at the base of the succession (see parasequence PS_1 in Fig. 4A). This sandstone interbed is medium- to fine-grained and contains only sparse, small fragments of pectinid shells. Foreset strata indicate transport direction towards the palaeoshoreline.

Interpretation

Facies association B shows evidence of a perennial wave action punctuated by storm-generated currents, which implies deposition above the prevalent

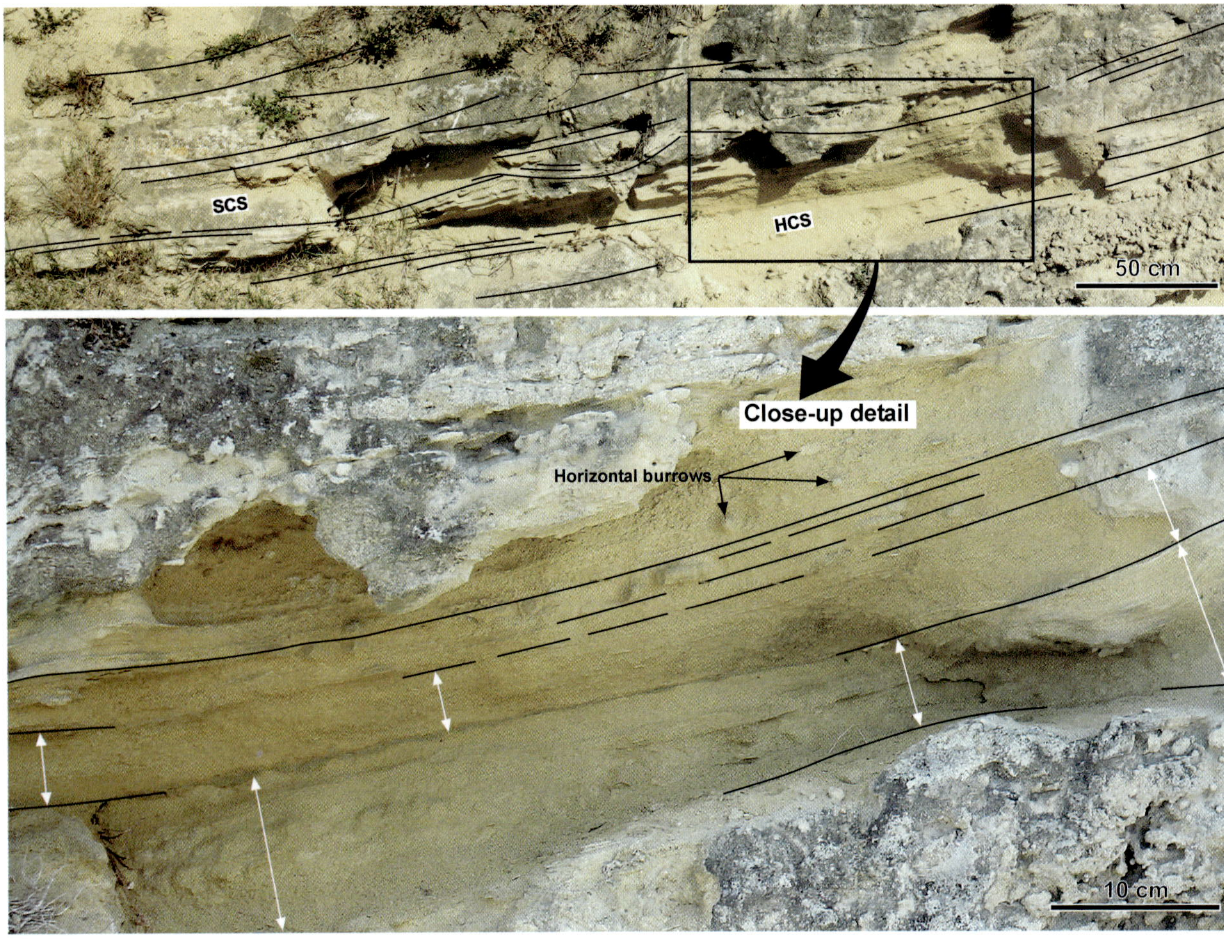

Fig. 6 Sandstones of facies S_{HS}, showing dome-shaped sets of convex-upward strata (HCS) with a wavelength of 2–2.5 m, overlain by sandstones of facies S_{SS} with swaley cross-stratification (SCS).

fairweather wave base (Clifton *et al.*, 1971; Clifton, 1976, 1981; Kumar & Sanders, 1976; Bourgeois, 1980; Leckie & Walker, 1982; Clifton & Dingler, 1984; DeCelles & Cavazza, 1992). However, this facies association differs from the previous one, because the range of facies here is wider, the sandstones are generally finer grained and gravel is sparser, and also shells are sparse, whereas microfauna abounds and burrows are common. The evidence indicates a lower-energy nearshore zone, interpreted to be a lower shoreface environment. This interpretation is supported further by the fact that facies association B underlies directly association A in the regressive parts of parasequences (see PS_1 and PS_2 in Fig. 4A).

Hummocky and swaley cross-stratifications (facies S_{CS} and S_{SS}) are widely attributed to storm-generated, combined-flow currents (Hamblin & Walker, 1979; Dott & Bourgeois, 1982; Duke, 1985; Tillman, 1985; Myrow & Southard, 1996), and the former is considered to be typical of a lower shoreface zone (Bourgeois, 1980; Brenchley, 1985). The interbeds of facies S_{RL} represent periods of relatively weak wave action, whereas the occasional layers with 'micro-hummocky' (three-dimensional ripple) cross-lamination are products of weak, storm-generated combined-flow currents dominated by oscillatory waves (Harms, 1969; Dott & Bourgeois, 1982; Yokokawa *et al.*, 1995; Myrow & Southard, 1996). The isolated set of planar cross-strata (facies S_{CS}) in parasequence PS_1 is probably a rare two-dimensional dune, or swash bar, formed by the shoreward sweep of sand during a post-storm phase of shoreface recovery (Hobday & Banks, 1971; Fitzgerald *et al.*, 1984; Massari & Parea, 1988).

Quasi-perennial wave action and frequent storms would inevitably remove the record of quiet-water sedimentation corresponding to the brief rises of fairweather wave base, and hence the lack of muddy or silty interlayers. The horizons with burrows are probably a relict record of such conditions, representing periods of stable floor, lowest sedimentation rate and incipient colonization by benthic macrofauna. Significant phases of sediment winnowing are indicated by horizons of concentrated pectinid shells in convex-upward positions.

Facies association C: offshore-transition deposits

Description

These deposits are silty, very fine-grained to medium/coarse sandstones, occur in parasequences PS_2, PS_4, PS_5 and PS_6 (Fig. 4), and constitute ~ 32% of the total thickness of the succession studied. Their units are 2.5–9.4 m thick, characteristically underlie facies association B (Fig. 4) and consist of facies S_M, S_{PS}, S_{SS} and S_G (Table 1).

Facies S_M predominates and its units are 8–185 cm thick, constituting ~ 73% of the association thickness. These sandy siltstones alternate with very fine-grained, silty sandstones and are homogeneous to faintly laminated. They are rich in shells (Fig. 7A) and show strong bioturbation. Macrofauna is typically dispersed and includes mainly *Aequipecten opercularis* valves (sporadically articulated) and large fragments of convolute laminar and erect branched bryozoan colonies, among which phidoloporids predominate. *In situ* echinoids are rare, including spatangids and *Echinocardium* sp. with articulated spines. In zone MC_5 (Fig. 4B), this facies assemblage contains also a discontinuous layer of tightly packed pectinid coquinas (Fig. 7B), predominantly *A. opercularis* and subordinate *Pecten jacobaeus*, moderately encrusted by barnacles (mainly *Balanus mylensis* and *B. amphitrite*), serpulid tubes (mainly *Pomatoceros triqueter*) and bryozoans.

Finer-grained sediment layers consist mainly of a skeletal hash (≤ 70 vol.%), comprising species that are ecologically incompatible and have apparently been mixed by cross-zonal drift. For example, ostracods include mainly infralittoral to upper mid-circalittoral species (such as *Aurila* gr. *punctata*, *A. latisolea*, *Loxoconcha tumida* and *Paracytheridea* gr. *depressa*), mixed with broader circalittoral species, such as *Bythocythere turgida*, *Celtia quadridentata* and *Monoceratina mediterranea*. Bryozoans include internodes of bushy, rhizoid-bearing, erect articulated forms, such as *Crisia* spp., *Caberea boryi*, *Scrupocellaria* spp. and *Cellaria* spp. (mainly *C. fistulosa*).

Units of facies S_{PS} are 3.5–80 cm thick (Fig. 4) and constitute ~ 24% of the association thickness. This sandstone facies here is only fine- to medium-grained and lacks a gravelly component. Skeletal debris abounds, including scattered large fragments of the bivalves *P. jacobaeus* and *A. opercularis* and branches of the bryozoan *M. truncata* and small celleporids, accompanied by finely ground shell detritus derived from the shallower water zone.

Locally present are lenticular (patchy) concentrations of *A. opercularis* valves resting in convex-upward position, accompanied by celleporids, mainly *Celleporina mangnevillana*, and the serpulid *Ditrupa arietina*. Some of the pectinid valves are encrusted by bryozoans, including *Onychocella marioni*, *Crassimarginatella manzonii*, *Calyptotheca* sp. and *Thalamoporella 'neogenica'*. Fragments of *D. arietina* tubes are particularly abundant in the upper part of parasequence PS_5 (facies zone T_5 in Fig. 4B), where they locally form nearly monotypic concentrations (Fig. 7C).

A solitary interbed of facies S_{SS} occurs in this association in the lower part of parasequence PS_2 (see zone T_2 in Fig. 4B). This swaley cross-stratified sandstone is medium-grained and 35 cm thick, separating beds of facies S_{PS}. Similarly rare is facies S_G, which occurs in the lower part of parasequence PS_5 (see zone T_5 in Fig. 4B) as a fining-upward package of three superimposed normal-graded beds, 8–15 cm thick, composed of coarse to medium/fine sand. This composite, sheet-like unit has an erosional, undulatory base (Fig. 7D) and forms a striking split in facies S_M.

Interpretation

The abundance of the fine-grained, bioturbated 'background' facies S_M indicates deposition below the fairweather wave base in an environment with perennially abundant sediment suspension rich in silt and very fine skeletal sand. The sandstone interbeds of facies S_{PS} and rare facies S_{SS} are

Fig. 7 Offshore-transition deposits of facies association C. (A) Offshore-transition deposits overlain sharply by the lower shoreface deposits of facies association B at the transition of zones MC_5 and R_5 in Fig. 4; the latter association is dominated by facies S_{SS}, whereas the former is dominated by faintly laminated, bioturbated siltstones of facies S_M, rich in shells in the lower part. (B) Close-up detail of a shell bed in facies S_M in the middle of zone MC_5 in Fig. 4; the shells are densely packed, disarticulated pectinid valves, mainly *Aequipecten opercularis*, with a disorderly orientation and common encrustations. (C) Disorderly concentration of serpulid *Ditrupa arietina* tubes in a sandstone bed of facies S_{PS}; detail from the uppermost zone T_5 in Fig. 4. (D) Normal-graded sandstone bed of facies S_G, overlying bioturbated facies S_M with an undulatory erosional contact; detail from zone T_5 in Fig. 4.

tempestites attributed to storm events, implying an offshore-transition environment. The isolated thin unit of coarse-grained facies S_G indicates the abrupt incursion of an exceptionally powerful and pulsating current with high sediment concentration, depositing sand by rapid dumping directly from turbulent suspension (Lowe, 1988; Vrolijk & Southard, 1997). This could be a strong, density-enhanced rip current or backwash surge generated by a rare extreme storm (Gruszczyński *et al.*, 1993; Myrow & Southard, 1996) or possibly a tsunami (Cantalamessa & Di Celma, 2005).

The variable proportion of facies S_M and S_{PS} in the successive occurrences of this association (Fig. 4) may reflect changes in the shelf wave climate, but more likely represents somewhat different bathymetric parts of an offshore-transition zone. This interpretation seems to be supported by the corresponding slight differences in fauna assemblages.

The notion of episodic currents and/or wave action is supported by the occurrence of pectinid shell beds (lags), with shells in convex-upward position, indicating sediment winnowing. The shell lags in parasequence PS_5 (Fig. 4B) consist almost exclusively of juvenile and diagenetically fragmented *D. arietina*. The abundance of this species is consistent with the notion of high water turbidity (Sanfilippo, 1999), and the juvenile forms suggest that the optimal conditions for *Ditrupa* populations were relatively short-lived, interrupted by frequent storms or excessive suspension fallout.

Quiet water conditions and a generally high sedimentation rate are indicated by facies S_M, which bears abundant burrows and locally contains *in situ* macro- and microfauna, including well-preserved bryozoan colonies and articulated echinoids and bivalves. The encrustation of *A. opercularis* shells is limited to a few thin layers on the shell outer surface, apparently formed during the pectinid life time, and this sparse development of epifauna is consistent with a high rate of burial. The assemblages of *in situ* micro- and macrofauna indicate an upper circalittoral palaeoenvironment.

The range and preservation of *in situ* fauna in facies S_M indicate an upper circalittoral zone. Ostracods include infra- to circalittoral species. The fossil content of facies S_{PS} tempestites suggests derivation from a coastal detritic (DC) biocoenosis, as indicated by the bivalve *P. jacobaeus* (Pérès & Picard, 1964) and the bryozoans *Frondipora verrucosa*, *Smittina cervicornis* and *Turbicellepora coronopus* (Harmelin, 1976; Rosso, 1996). In the middle part of parasequence PS_2 (Fig. 4A), facies S_{PS} contains abundant *C. mangnevillana*, whose small, stoutly branched celleporiform colonies are typical of its relatively deep-water occurrences in the lowermost infralittoral zone (Gautier, 1962). Overall, the palaeoecological and facies evidence is consistent with an offshore-transition environment subject to sediment incursions from a shallower zone.

Facies association D: offshore deposits

Description

The three occurrences of these deposits in parasequences PS_2, PS_4 and PS_6 (Fig. 4) are 3.4 to nearly 15 m thick and constitute ~ 36% of the succession's total thickness. This assemblage comprises chiefly facies S_M (~ 95% thickness), with units that are 15–330 cm thick, consist of bioturbated, massive to faintly laminated siltstones interlayered with very fined-grained silty sandstones (Fig. 8A) and are similar to those in the previous association. The sediment is mainly skeletal (≤ 80 vol.%), and the diffuse laminaea are up to 0.7 cm thick, verging on thin layering. Noticeable are sporadic occurrences of gutter casts, which are filled with cross-laminated, very fine skeletal sand rich in adult specimens of *D. arietina*, commonly colonized by the serpulid *Hydroides norvegicus* (Fig. 8B, C).

Facies S_{PS} sandstones form isolated interbeds 5–25 cm thick, mainly sheet-like, but laterally discontinuous and commonly consisting of broad lenses. The skeletal sand varies from very fine- to medium-grained, and the beds have sharp, slightly erosional bases. This facies constitutes < 4% of the association thickness and occurs mainly in the transgressive parts of parasequences (Fig. 4).

Facies S_G occurs as an isolated, fining-upward unit of four superimposed beds, which are normal-graded and 5–16 cm thick, composed of coarse or very coarse to medium sand (see the top of facies zone T_4 in Fig. 4B). This solitary unit of facies S_G, splitting the silty facies S_M, is strikingly similar to its isolated occurrence in the previous facies association.

Shelly fauna remains are mainly *in situ* or only slightly transported, and occur dispersed or are locally concentrated in pockets, small flat lenses or shell-rich beds. There are also some significant differences in faunal assemblages in the successive stratigraphic occurrences of this facies association.

In the mid-cycle zone MC_2 of parasequence PS_2 (Fig. 4A), facies S_M includes layers, up to 25 cm thick, that are remarkably rich in large fossils. Digitate-branched celleporiform bryozoan colonies predominate, accompanied by disarticulated valves of *A. opercularis* in mainly concave-upward position and locally stacked upon one another (Fig. 8D). The bryozoans are bushy, densely branched colonies of *Celleporaria palmata*, large (≤ 20 cm high and ≤ 25 cm in diameter) and mainly well-preserved, accompanied by subordinate smaller and less ramified colonies of *C. mangnevillana*. Sporadically found are the more delicate and slender colonies of *Smittina canavari* and rare *Diporula verrucosa*. The pectinid shells and many celleporiforms have been colonized by the serpulid *Pomatoceros triqueter*, the barnacle *A. stellaris* and various encrusting bryozoans, including *Calpensia nobilis*, *Onychocella marioni*, *Thalamoporella 'neogenica'* and *Crassimarginatella manzonii*. Cemented sessile specimens of the bivalves *Pododesmus* (*P. aculeatus*, *P. patelliformis*, *P. squamula*) are also common, whereas local *Entobia* borings indicate bioerosion by clionid sponges. The surrounding skeletal sediment abounds in non-abraded bryozoan fragments, echinoid spines and ostracod tests. The latter include several *Aurila* species (*A.* gr. *convexa*, *A.* gr. *punctata* and *A. cymbaeformis*), but are less diversified upwards

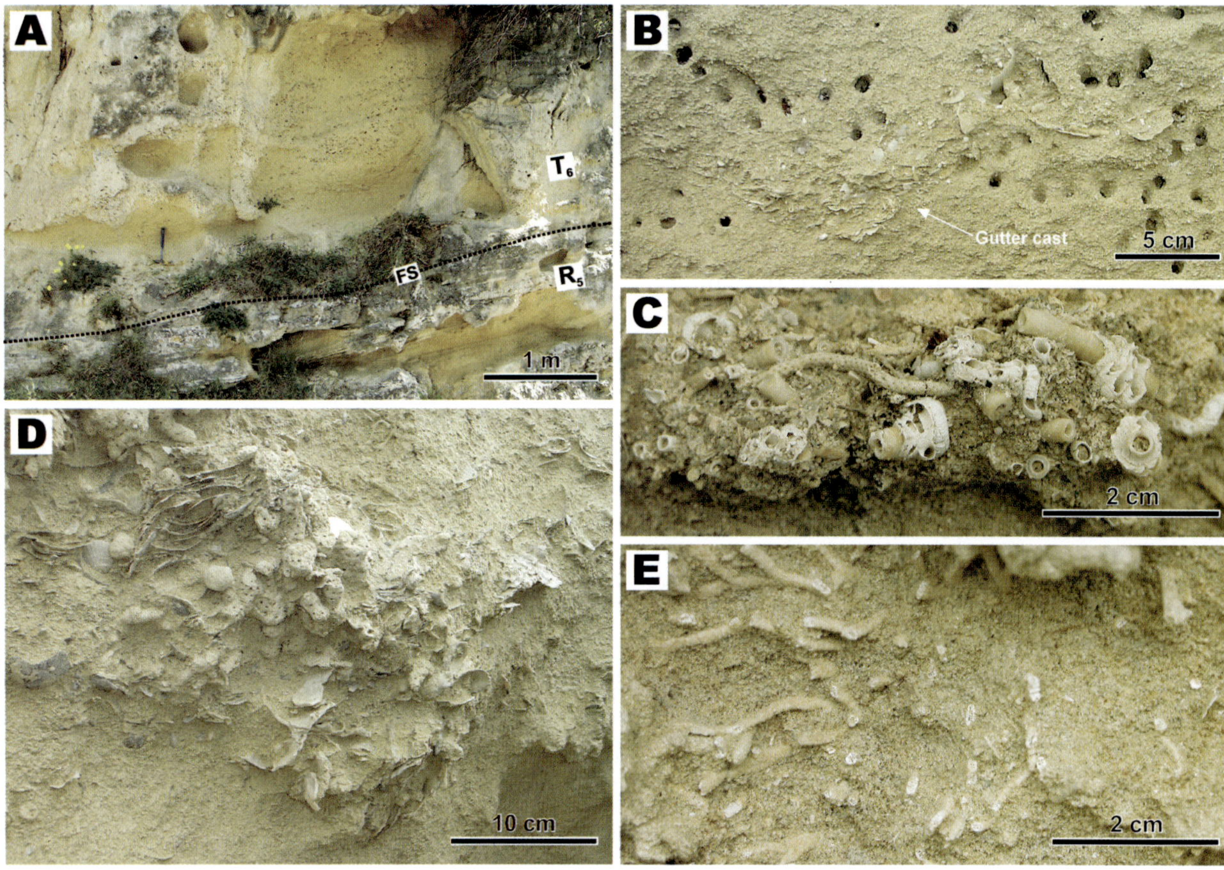

Fig. 8 Offshore deposits of facies association D. (A) Lower shoreface deposits dominated by medium-grained sandstones of facies S_{SS}, overlain sharply by offshore deposits composed of shell-bearing facies S_M; outcrop detail of the contact between log zones R_5 and T_6, separated by a flooding surface FS (cf. Fig. 4). (B) Sand-filled, fossil-rich gutter cast in facies S_M siltstone. (C) Fossils concentrated in a gutter cast, including serpulid *Ditrupa arietina* (with apertures encrusted by coiled *Hydroides norvegicus*) and rare slender branches of bryozoan *Diporula verrucosa*. (D) Concentrations of fossils in facies S_M in the log zone MC_2 (cf. Fig. 4), including celleporiform colonies of bryozoan *Celleporaria palmata* in upside-down position and haphazardly packed valves of *Aequipecten opercularis*. (E) Dichotomous, slender branches of adeonelliform bryozoan *Smittina cervicornis* in subprimary positions in facies S_M.

in the MC_2 zone, where circalittoral, deeper water species (*Celtia quadridentata* and *Monoceratina mediterranea*) predominate.

A similar faunal assemblage characterizes this facies association in zone T_4 of parasequence PS_4 (Fig. 4A), where the degree of bioturbation varies from low to high and facies S_M locally bears large bryozoan colonies, up to 20 cm in height, accompanied by bivalves, echinoids, serpulids and rare brachiopods (*Terebratula scillae*). Bryozoans are mainly *Smittina canavarii*, *S. landsborovii* and *Biflustra savartii*, which form slender, highly brittle colonies with convolute laminar structure and are accompanied by erect, rigid adeonelliform colonies of ribbon-like and dichotomously branched *S. cervicornis* (Fig. 8E). The colonies are in life position, occasionally slightly tilted or broken, but with the fragments largely in place. Similarly contiguous are fragments of broken serpulid tubes. Present also are relatively large and well-preserved fragments of erect rigid bryozoans *Diporula verrucosa* and *Hornera frondiculata*, internodes of *Cellaria sinuosa* and rare spatangid echinoids with interconnected spines. Some layers contain sparse specimens of the byssate bivalve *Limatula subauriculata*, with shells diagenetically decalcified, but in near-life position. Ostracods abound and are highly diversified, dominated by *Cytheretta judaea*, *Aurila cymbaeformis*,

A. gr. *punctata*, *A. latisolea*, *A. convexa*, *Pseudocytherura calcarata*, *Celtia quadridentata* and *Monoceratina mediterranea*. The sporadic interbeds of facies S_{PS} bear thin concentrations of disarticulated pectinid shells in convex-upward or occasionally haphazard position, along with rare bryozoan fragments and exhumed spatangids encrusted by serpulids.

Fossils are increasingly more abundant in the overlying facies zone MC_4 (Fig. 4B), where they occur concentrated and often densely packed in layers (Fig. 9A). The uppermost layer consists almost exclusively of disarticulated *A. opercularis* valves (Fig. 9B), some encrusted by bryozoan colonies, mainly on the outer surfaces. The brachiopod *Terebratula scillae*, which occurs scattered in the lower part of zone T_4, here becomes abundant and its articulated valves lack preferential orientation. Scattered anomiids (mainly *Pododesmus squamula*), pectinids (*P. jacobaeus*) and echinoids (*Cidaris* sp.) occur. Bryozoan colonies are sparse, represented mainly by celleporiform branches of *C. palmata* and *Turbicellepora tubigera*, articles of *C. sinuosa* and relatively large fragments of *B. savartii*. The abundance and diversity of ostracods increase progressively upwards in zone MC_4, reaching 50 species (Fig. 4B). This trend is paralleled by an increase in typical circalittoral species, such as *Bythocythere turgida*, *C. quadridentata*, *M. mediterranea* and *Bosquetina carinella*, which persist in the lower part of the overlying zone R_4, although fossils are less abundant therein.

In zone T_6 of parasequence PS_6 (Fig. 4B), the faintly laminated deposits of this facies association (Fig. 8A) contain less abundant fossils. Only the valves of *A. opercularis* are scattered throughout, disarticulated or rarely with both valves, usually

Fig. 9 Offshore deposits of facies association D. (A) Fossil-rich layers in facies S_M, containing almost exclusively pectinid shells in the upper part; portion of zone MC_4 in Fig. 4B. (B) Close-up detail of the previous outcrop (see arrow), showing concentrations of *Aequipecten opercularis* shells, mainly disarticulated. (C) Well-preserved *A. opercularis* bivalves in facies S_M, with valves only slightly displaced relative to each other; detail from zone T_6 in Fig. 4B. (D) A well-cemented layer of facies S_{PS} sandstone, 25 cm thick, rich in *Arctica islandica* moulds; the disorderly concentration of disarticulated valves is attributed to an event of live infauna exhumation by storm-generated current. (E) Close-up view of the *A. islandica* moulds in the same layer in log zone T_6 (Fig. 4B).

displaced slightly relative to each other (Fig. 9C). Patches of concentrated and nearly *in situ* fossils occur in the middle part of zone T_6. They include mainly circalittoral bryozoans, represented by convolute reteporiform colonies of phydoloporids (*Reteporella couchi couchi*, *R. mediterranea*), subordinate celleporiforms (*C. mangnevillana*, *T. tubigera*), bundles of *Cellaria* spp. (mainly *C. sinuosa*) and fragments of *D. arietina*. Bryozoan internodes commonly occur in patches and are aligned.

Facies zone T_6 also bears fragments of small, erect brittle bryozoans, such as *Tessaradoma boreale*, *D. verrucosa*, *Omalosecosa ramulosa* and *Buskea dichotoma*. Molluscs are subordinate, including bivalves (*Chlamys multistriata*, *C. varia*, *Pseudamussium clavatum*, *Hyalopecten similis*, *Palliolum incomparabile*, *P. squamula*, *P. patelliformis* and *Limatula subauriculata*) and minor gastropods (*Charonia nodifera* and *Epitonium* sp.). Rare carapaces of the small crab *Ebalia* sp. and fragments of the gorgonacean *Funiculina* sp. benthic ostracods abound, including several infra-circalittoral species, such as *Aurila cymbaeformis*, *A.* gr. *punctata*, *A.* gr. *convexa*, *Cytheretta judaea* and *Pterigocythereis jonesii*, and also exclusively circalittoral species, such as *C. quadridentata* and *B. carinella* (Bonaduce *et al.*, 1975; Montenegro *et al.*, 1998). The latter predominate in some layers. All of these are species that generally thrive on fine sandy to silty bottoms (Neale, 1964; Sciuto & Rosso, 2002).

The two thickest (25–30 cm) sandstone interbeds of facies S_{PS} in zone T_6 (Fig. 4B) abound in moulds of the bivalve *A. islandica* (Fig. 9D & E), with only few shells partly preserved, apparently due to their encrustation with bryozoan sheets and minor serpulid tubes. The moulds indicate chaotic deposition and dense packing of valves, mainly in convex-upward position. Local moulds of *Entobia* spp., with variable sizes and distribution of globular chambers, are evidence of clionid sponge borings in the *Arctica* shells. The dissolution of aragonitic *A. islandica* valves commonly left their calcitic encrustations preserved, which indicates selective diagenesis.

Interpretation

The predominance of fine-grained and bioturbated facies S_M indicates deposition below the average storm wave base in an offshore environment. The diffuse and often thick lamination can be due to a rhythmic shedding of sediment from suspension (Kerr, 1991; Nemec, 1995, fig. 40) and/or cyclic action of tidal currents (Reineck & Singh, 1975). The abundance of silt and very fine sand in a mid-bay offshore zone is attributed to the persistent entrapment of storm- and tidally-entrained sediment suspension in the coastal embayment (see subsequent discussion). The notion of weak currents is supported by the patches of aligned bryozoan internodes, slight rearrangement of shells and the contamination of faunal *in situ* assemblages with skeletal remains derived from shallower water biotopes.

The interbeds of sandstone facies S_{PS} are tempestites representing the sporadic distal incursions of sand from the strongest storms, with magnitudes above the average. The isolated unit of facies S_G resembles closely its occurrence in the previous facies association and is similarly attributed to a rare event, probably a tsunami or a hurricane storm.

Fossil assemblages consist of shallow circalittoral fauna, slightly deeper in parasequence PS_6 (Fig. 4B), with an admixture of transported and variously abraded skeletal remains derived from a shallower-water zone. Fauna *in situ* is akin to the coastal detritic (DC) and/or muddy detritic (DE) biocoenosis. The assemblage of *C. palmata* and *A. opercularis* in zone MC_2 is comparable to the peculiar build-up communities known from the Mediterranean modern circalittoral soft bottoms and referred to as the shelf coralligenous biocoenosis (Pérès & Picard, 1964; Pérès, 1982).

The bivalve *A. opercularis* is a tolerant species, well adapted to silty/muddy substrates (Gamulin-Brida, 1974). The species *C. palmata* is now extinct, but its durable colonies are known to have been capable of forming thickets in suitable conditions, typically at water depths around 30–50 m (Spjeldnæs & Moissette, 1997). The large *Celleporaria* colonies apparently grew on non-fossilizable organisms and/or directly on bottom sediment, with skeletal remains of other bryozoans found locally beneath the bases of such colonies (Moissette & Pouyet, 1991). Modern *Celleporaria* species are known to form extensive thickets in low-energy or sub-swell wave-base settings, preferably on muddy to silty bottoms with moderate sedimentation rates (Hageman *et al.*, 2003). Taphonomic features of the

C. palmata–A. opercularis assemblages include significant encrustation and bioerosion, which implies relatively long exposure at the seafloor and hence phases of non-deposition, probably due to sediment bypass or winnowing. The predominantly quiet seafloor conditions apparently also allowed the bryozoan *S. canavarii* (now extinct) to thrive on the bay floor and form hollow slender branches.

The large *in situ* colonies of *S. cervicornis, S. landsborovii, S. canavarii* and *B. savartii* in facies zone T_4, up to 20–25 cm high and accompanied by the smaller colonies of *D. verrucosa* and *H. frondiculata*, were likely derived from an original DC biocoenosis. These species occur as erect, arborescent and subordinate fan-shaped colonies, some of which (notably *S. landsborovii, S. canavarii* and *B. savartii*) formed convolute uni- or bilaminar sheets directly on the soft substrate. Comparable assemblages, although involving different species, are known from 40 to 50 m deep circalittoral areas of northeast Adriatic (McKinney & Jaklin, 1993, 2001), where a transitional DC–DE biocoenosis has been recognized by Gamulin-Brida (1974). Similar build-up assemblages, although laterally wider and slightly more diversified, have been reported from shallower water Pleistocene environments of Sicily (Rosso, 1987).

The decline of bryozoans and increased amount of the brachiopod *T. scillae* in facies zone MC_4, accompanied by an increase in both abundance and diversity of ostracods, indicate a DE biocoenosis (cf. Gaetani & Saccà, 1983; Taddei Ruggiero, 1994), which is consistent with the notion of a sand-starved bottom dominated by suspension fallout.

The bryozoan-rich patches of *in situ* fauna in facies zone T_6 are comparable to the modern multi-species clumps formed by epibenthic filter-feeding organisms that exploit soft-bodied, unpreservable other fauna in an environment where sedimentary substrate is unsuitable for colonization (Zuschin *et al.*, 1999). *Cellaria* colonies, particularly *C. sinuosa*, are commonly intergrown with erect rigid phidoloporids and other bryozoans. As documented from the Mediterranean (Mckinney & Jaklin, 2000, 2001) and other regions (Henrich *et al.*, 1995; Bader, 2001), these organisms are able to colonize the substrate directly by means of rootlet bundles and can withstand diverse environmental conditions. The coexistence of *Cellaria* and erect brittle bryozoans suggests relatively tranquil water, possibly 50–60 m deep, as might be expected in an open-shelf setting for *C. sinuosa* and some of the other bryozoan species (*D. verrucosa, O. ramulosa, B. dichothoma* and *T. boreale*), and also for some of the associated molluscs, ostracods and the pennatulacean *Funiculina*.

There is no evidence of coralline algae, probably because the water turbidity was high and the rate of suspension fallout fluctuated. Phases of intense fallout are indicated by sediment layers with the communities of *D. arietina, A. islandica* and *A. opercularis*, which imply temporal seafloor colonization by biocoenoses DE, transitional DC–DE and incipient heterogeneous PE1. High turbidity allowed nearly monotypic communities of this first species to flourish (Di Geronimo & Robba, 1989), albeit briefly, for most of these occurrences involve juvenile forms (Sanfilippo, 1999). The second species is known to thrive as infauna on relatively stable fine-grained bottoms smothered with mud (Malatesta & Zarlenga, 1986), and also the last species represents mobile epifauna adapted to similar conditions. Layers rich in *A. islandica* and/or *D. arietina* with moderate to heavy bryozoan encrustations and traces of bioerosion indicate exhumation of dead and/or alive infauna by sediment winnowing over periods of several years, which implies persistent weak currents, most probably tidal.

THE PALAEOBAY ENVIRONMENT

The coastal embayment was probably fringed with a reflective sand–gravelly shoreline (presently non-preserved) and had a sandy, gravel-strewn shoreface dominated by waves, from where sand was episodically transported by storms to the offshore-transition zone and sporadically spread farther offshore. The surface temperature of the Tyrrhenian Sea fluctuated (Thunnel *et al.*, 1990), but the bay's water salinity was normal and fauna productivity was high, providing abundant skeletal sediment. Influence of tidal currents is recognizable in the offshore-transition and mid-bay offshore deposits, but not obvious in the shoreface zone, where the tidal record was probably obliterated by persistent waves.

The region was tectonically active, and at least two tsunami events are inferred to have affected the bay. Alternatively, the rare occurrences of facies S_G may be a record of strong rip currents. Reflective shorelines generally lack rip currents, but these occasionally can be generated by extreme storms, when coastal swell causes strandplain inundation and may result in edge waves (Bowen & Guza, 1978; Gruszczyński *et al.*, 1993).

The microtidal range in the western Mediterranean is little more than 30 cm, but the local straits and many bays are known to amplify tidal currents (see review by Longhitano & Nemec, 2005). Strong tidal influence is recorded by a thick succession of bioclastic two-dimensional dunes in the overlying, Pleistocene bay-fill sequence (Fig. 2). The lack of any similar record in the lower sequence suggests that the embayment at its earlier stage failed to intensify water-mass tidal oscillations. As discussed by Pugh (1987), tidal currents move as long-period internal waves with a wavelength $L = T\sqrt{gd}$, where d is the water depth, g is the gravity constant and the period T is 12.42 h for M2 tides. These waves come into resonance in a bay if its length is $L_B = (2n+1)\frac{1}{4}L$ (for $n = 0, 1, 2, 3$, etc.), which means when $L_B = \frac{1}{4}L, \frac{3}{4}L, \frac{5}{4}L, \frac{7}{4}L$, etc. The embayment became considerably larger during the deposition of the upper sequence, where the maximum-flooding zone (Fig. 2) bears epibathyal fauna, and hence it is possible that the bay's previous size and water depth simply did not match a resonance condition.

The thicknesses of regressive shoreface deposits (facies associations A and B) are consistently in the range of 9–10 m, which allows the average depth of fairweather wave base to be estimated as no more than 10–11 m. The estimate takes into account the succession burial depth of < 500 m and ~ 7.5 vol.% compaction (Baldwin & Butler, 1985), but assumes that no significant rise in relative sea level occurred during the normal regressions, which means that the wave base in reality might have been even shallower. A depth of ~ 10 m can thus be taken as an approximate average. The thicknesses of regressive offshore-transition deposits (facies association C) are in the range of 3–5 m, which suggests that the prevalent storm-wave base was only a few metres deeper (~ 15–16 m can be assumed, with a correction for sediment compaction). The water depth in the Castroreale area thus probably did not exceed 25 m, although the bay could have been somewhat deeper in its outer part. The relatively shallow storm-wave base can be attributed to the fact that the coastal embayment was perched above the general level of the adjoining peri-Tyrrhenian shelf (cf. Fig. 1B), such that the incoming storm waves were probably attenuated by the seafloor topography.

Accordingly, the presence of upper to middle circalittoral fauna in facies association B and its predominance in association C indicate ecological zones significantly shallower than on the western Mediterranean shelves, where similar faunal assemblages are documented to occur at water depths generally greater than 35–70 m. The striking shallowness of the boundary of infralittoral and circalittoral zones in the Barcellona Pozzo di Gotto Basin is attributed to the high turbidity of the bay water, which would reduce the penetration depth of light and affect the bathymetric distribution of local biocoenoses. The notion of high turbidity is strongly supported by the abundance of silt and very fine sand in the 'background' facies S_M.

Coastal embayments commonly act as sediment traps, where both bedload and plumes of suspension become arrested by waves and tidal processes (Fig. 10; Avoine & Larsonneur, 1987; Kirkby, 1987; Gao *et al.*, 1990; Ke *et al.*, 1996; Plater *et al.*, 2000). Large volumes of sediment suspension entrained by storms and tides can be entrapped in bays with no significant fluvial discharges, and may be subject to circulation by remnant turbulence and residual currents. The Wash embayment of eastern England, for example, is dominated by sediment suspension derived from the adjacent marine areas and coastal erosion, with supply rates two orders of magnitude higher than the net landward transport of bedload by tides and waves (Evans & Collins, 1987; Ke *et al.*, 1996). The admixtures of relatively deep-water circalittoral microfauna in the mid-bay facies association D in the present case support the notion of fine-grained sediment import from offshore areas.

Due to the lower density and cohesionless platy nature of its particles, fine skeletal sediment is more easily entrained and suspended than similar siliciclastic sediment (Southard *et al.*, 1971; Mantz, 1977). Likewise, skeletal carbonate sand requires lower bottom-shear stresses for transport initiation (Young & Southard, 1978; Young & Mann, 1985; Prager *et al.*, 1996), which can explain extensive winnowing by weak tidal currents.

The palaeobay and its neighbourhood bear no record of contemporaneous fluvial deposits, but

Fig. 10 Sediment suspension entrapped in bays. (A) Plumes of stream-derived and tidally entrained suspension entrapped by a weak and reversing sea breeze in a coastal embayment, ~ 3.5 km across; Longyearbyen, Spitsbergen. (B) Plumes of suspension from small bedload streams, drifted alongshore into adjacent embayment; Kapp Ekholm, Spitsbergen. (C) River-derived suspension entrapped in a coastal embayment; Heron Bay, Lake Superior shore.

the regional climate was mainly cool and humid (Bertoldi *et al.*, 1989), and it is likely that the coastal zone hosted some small deltas of bedload streams, presently not preserved. Bedload streams can supply large volumes of mud to the nearshore zone (see Nemec, 1995, pp. 32–34) and contribute suspension to coastal embayments (Fig. 10). At least some of the siliciclastic mud admixture in facies S_M could be derived from such a coastal source.

STRATIGRAPHIC ORGANIZATION

The stratigraphic organization of the facies associations and their faunal assemblages (Fig. 4) indicates that the thick transgressive systems tract of the lower bay-fill sequence consists of six parasequences, or transgressive–regressive cycles, with an overall deepening-upward bathymetric trend. The parasequences indicate repetitive episodes of relative sea-level rise and landward shoreline shift, followed by a gradual readvance of the shoreline (Fig. 11A). The succession as a whole can be regarded as a back-stepping (transgressive) parasequence set. The landward shoreline shifts were probably limited by the high coastal relief of the bay, which would render the transgressive shoreline trajectory relatively steep.

The evidence from sedimentary facies is consistent with that from palaeoecological and taphonomic analyses, and the two data types importantly supplement each other. For example, facies associations allow the prevalent depths of fairweather and storm wave bases to be estimated, which fauna does

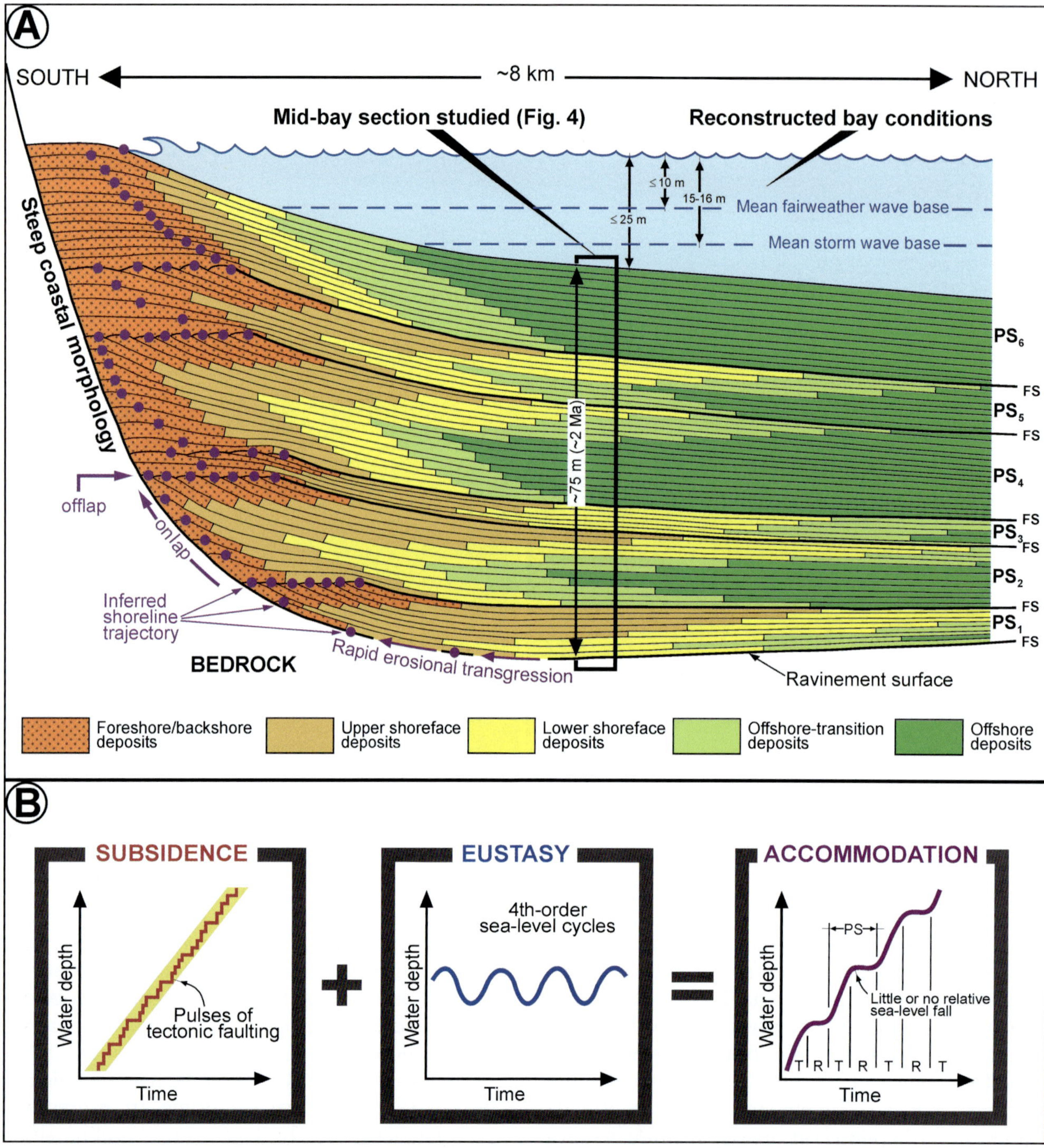

Fig. 11 Models for the marine bay-fill succession in the Barcellona Pozzo di Gotto Basin. (A) Stratigraphic model for the marine bay-fill succession studied in the Barcellona Pozzo di Gotto Basin (cf. Fig. 4). Letter symbols: FS, marine flooding surface; PS, parasequence. (B) A generic model for the succession's component parasequences, explaining their origin by changes in accommodation resulting from the combination of tectonic subsidence and high-frequency eustatic sea-level changes (based on Jervey, 1988).

not permit, whereas the taphonomic data allow sediment winnowing to be recognized and the turnabout zone of a transgressive–regressive cycle not only to be identified, but its thickness to be estimated and the surface of maximum condensation to be recognized. No similar detailed information can be derived from sedimentary analysis alone, and maximum flooding surfaces are often placed arbitrarily in the middle parts of turnabout zones.

The parasequence thicknesses in the Castroreale section range from ~4 m to nearly 19 m. Each sequence commences with a transgressive component, which varies from negligibly thin (see zones T_1 and T_3 in Fig. 4A) to impressively thick (see zones T_4 and T_6 in Fig. 4A & B) and leads to the mid-cycle turnabout zone, which itself may similarly be thin (as in parasequences PS_1 and PS_3) or up to a few metres in thickness (see zones MC_4 and MC_5 in Fig. 4B). The thicknesses of the overlying regressive components vary from ~3 m in parasequences PS_3 and PS_5 to 9.5 m in parasequence PS_1. The regressive components bear the characteristic shallowing- upward facies signature of shoreline progradation, although the shoreline itself has never advanced far enough to reach the mid-bay area studied (cf. Fig. 11A).

Both sedimentary facies and fauna indicate that the successive transgressive zones T_1–T_6 have an overall deepening-upward trend. A similar indication comes from the increasing thicknesses of mid-cycle zones and from the facies of corresponding regressive zones, since the R_1 and R_2 zones culminate in upper shoreface deposits (facies association A), whereas the subsequent zones R_3–R_5 culminate in lower shoreface deposits (facies association B). The varied thicknesses of the parasequences and their transgressive zones imply large temporal differences in the rate of accommodation development. The relatively thick mid-cycle zones in parasequences PS_2, PS_4 and PS_5 imply periods when the sedimentation rate, although itself at a minimum, apparently kept pace with the rate of subsidence and relative sea-level rise, which means a roughly constant accommodation space.

Palaeoecological and taphonomic evidence indicates that the mid-cycle zones are hiatal (*sensu* Kidwell, 1991), with common concentration of shells due to sediment winnowing, probably by tidal currents. These zones are similar to the 'mid-sequence shell beds' of Banerjee & Kidwell (1991) and the 'mid-cycle shell beds' of Abbott & Carter (1994) and Carter *et al.* (1998), but the thicker ones are more compound, formed by numerous episodes of incremental deposition alternating with erosion (Messina, 2003; Messina & Rosso, 2005). For example, the concentration of shells in the upper part of zone MC_4 clearly increases (Fig. 9B), which indicates a net decline of sedimentation rate and greater condensation (cf. type I shell bed of Kidwell, 1986) and thus bears directly on the positioning of the maximum flooding surface (Loutit *et al.*, 1988). The latter would correspond to the horizon of maximum condensation, which may not necessarily coincide with a surface of stratal downlap and hence is more reliable (Abbott, 1997a,b; Abbot & Carter, 1997).

DISCUSSION

Parasequence internal architecture

As pointed out by Arnott (1995), the existing definition of a parasequence gives little provision for significant 'transgressive' deposition during the phase of water deepening. Parasequences are expected to be extremely asymmetrical successions of shallowing-upward facies associations, with the maximum flooding surface corresponding roughly to the parasequence boundary (e.g. Posamentier *et al.*, 1988; Van Wagoner *et al.*, 1990). It is worth noting, therefore, that the episodes of water deepening in the present case involved significant accumulation of sediment, often quite thick (see zones T_2–T_6 in Fig. 4), and these transgressive deposits amount to as much as 49% of the succession total thickness. Parasequences with substantial transgressive deposits have been reported by other authors (e.g. Penland *et al.*, 1988; Posamentier & Allen, 1993; Arnott, 1995; Naish & Kamp, 1997; Ilgar & Nemec, 2005; Longhitano & Nemec, 2005), demonstrating that transgressions in some settings may be highly depositional. These deposits bear valuable information on the transgression rate and sediment dynamics, and should thus not be disregarded in stratigraphic models.

Two factors could enhance transgressive deposition in the present case. First, the relative sea-level rises were probably rapid, because the surrounding high-relief topography would cause the transgressive shoreline to climb up, rather than shift abruptly landwards (Fig. 11A). Second, the availability of sediment in the bay, unlike in a deltaic, estuarine or barrier setting, would not decline with the relative sea-level rise and could actually increase as sediment was swept by waves from the inundated areas.

The marine flooding surface underlying the transgressive deposits is generally easy to recognize by a sharp facies change indicating abrupt deepening (see the FS surfaces in Fig. 2 and the corresponding R–T boundaries in Fig. 4). In contrast, their upper boundaries, or the surfaces of maximum flooding (MFS), are easy to recognize in some cases, but difficult to pinpoint in parasequences with relatively thick turnabout zones of mid-cycle condensation (see zones MC_2, MC_4 and MC_5 in Fig. 4). These zones represent periods when the rate of seafloor aggradation stayed in balance with the rate of accommodation development. The taphonomic and palaeoecological evidence allows the condensation maximum to be recognized and indicates that the common practice of placing the MFS arbitrarily in the middle of a mid-cycle zone may be incorrect and misleading, obscuring the actual rate of relative sea-level change and sedimentation dynamics.

The sedimentary succession is considered to have recorded changes in the accommodation space controlled by relative sea-level changes and sediment supply. The latter was generally high, and the episodes of relative sea-level rise can be attributed to the regional and local tectonic subsidence. The magnitude and rate of accommodation change, whether an increase or a decrease, would control the temporal and spatial evolution of the bay-filling depositional systems and generate parasequences. These former parameters are reflected in the vertical spacing of the successive FSs and MFSs and in the corresponding lateral shifts of facies association belts (Fig. 11A).

Parasequence time span

The succession spans the middle Pliocene to Early Pleistocene time, and hence ~ 2 Myr. The parasequences are by no means identical in terms of their component facies associations (Fig. 4), but if the thickness proportionality alone is considered, their time spans would appear to range between 100 kyr (PS_3) and 540 kyr (PS_4). However, such simplistic estimates are unlikely to be realistic, since the parasequences contain numerous small hiatuses, consist of facies with widely different sedimentation rates and also differ in the thickness proportion of particular facies (Fig. 4). It is thus not surprising that, for example, the tentative estimates of the time span of parasequences PS_4 and PS_5 (540 kyr and 210 kyr, respectively) differ considerably from those based on biostratigraphic data (see below).

The available biostratigraphic evidence indicates that:

1 the lowermost part of the succession corresponds to biozone MPL5a and the basal deposits, containing *Pecten benedictus*, are probably older than 3.0 Ma;
2 the succession interval from the middle part of facies zone T_4 to the middle part of zone R_4 represents biozone MPL6 (defined by the appearance of *G. inflata* ~ 2.13 Ma and the first common occurrence of left-coiled *Neogloboquadrina pachyderma* ~ 1.79 Ma, at the base of Pleistocene; Sprovieri *et al.*, 1998);
3 the upper part of facies zone R_5 is of Emilian age, as indicated by *Hyalinea baltica* with its first occurrence ~ 1.50 Ma (Sprovieri *et al.*, 1998) – although the benthic foraminifer *H. baltica* is facies-dependent, its lack in similar facies in the underlying part of the succession indicates that the species was absent earlier.

The interval from mid-zone T_4 to mid-zone R_4 (i.e. the bulk of parasequence PS_4) would thus appear to have a time span of 340 kyr and the subsequent interval to the uppermost zone R_5 (i.e. the bulk of parasequence PS_5) to have a time span of 290 kyr. If an average time span of these parasequences (~ 315 kyr) is assumed for each of the underlying ones (PS_1–PS_3), the base of the succession would appear to be 3.07 Ma in age, which matches biozone MPL5a and corresponds quite well with a regional unconformity dated to 3.05 Ma in the central Mediterranean (Catalano *et al.*, 1998).

Interestingly, the number and ages of the parasequences appear to match the regional Plio-Pleistocene chronostratigraphy of the central Mediterranean (Catalano *et al.*, 1993, 1998), with the six parasequences correlating reasonably well with the 4th-order sequences P.4.a (3.05–2.7 Ma), P.4.b (2.7–2.53 Ma), P.4.c (2.53–2.1 Ma), P.5 (2.1–1.85 Ma), Q.1 (1.85–1.58 Ma) and Q.2 (1.58–1.4 Ma) of Catalano *et al.* (1998). This correspondence may not be coincidental, which raises the question as to whether the parasequences may in reality be high-frequency sequences recording eustatic sea-level changes.

Parasequence interpretation

The distinction between parasequences and high-frequency 'true' sequences, although itself important, is not always an easy task (Jervey, 1988; Swift *et al.*, 1991; Posamentier & James, 1993). The key criterion used is the nature of their bounding surfaces. Sequences are supposed to be bounded by unconformities, with an unconformity defined as 'a surface separating older from younger strata along which there is evidence of subaerial erosion or subaerial exposure with a significant hiatus indicated' (Van Wagoner *et al.*, 1988). Parasequences are bounded by marine-flooding surfaces, which mark an abrupt increase in water depth recorded as a change in facies and/or ecological conditions; these facies discontinuities may involve 'minor marine erosion, but no subaerial erosion or basinward shift in facies' (Van Wagoner *et al.*, 1988). In short, a sequence bears the record of a relative sea-level fall and rise, whereas a parasequence represents an episode of accommodation space increase and filling, with the transgressive and regressive phases of deposition reflecting little more than changes in the filling rate.

The sedimentary succession in the present case is fully marine, lacks internal evidence of subaerial erosion and matches well these latter circumstances (cf. Fig. 4). On the other hand, the transgressive–regressive cycles here involve relatively thick transgressive deposits and mid-cycle condensation zones, are in the Milankovitch frequency band and appear to correlate with the central Mediterranean 4th-order sequences, attributed to eustatic sea-level changes driven by astronomical eccentricity cycles (Catalano *et al.*, 1998). It is possible, therefore, that the parasequences are a local expression of the 4th-order eustatic cycles, specific to the peri-Tyrrhenian coastal zone of northern Sicily, where the local rates of tectonic subsidence could be sufficiently high to minimize or virtually eliminate the signal of a sea-level fall. In other words, these would be sequences of type 2 (*sensu* Posamentier *et al.*, 1988) that involved little or no effective fall in relative sea level and hence developed as parasequences (see Jervey, 1988; Van Wagoner *et al.*, 1990; Helland-Hansen & Martinsen, 1996).

The combination of low-amplitude eustatic cycles and pronounced tectonic subsidence would result in a highly asymmetrical curve of relative sea-level changes, with the rises being potentially magnified and accelerated and the falls being negligible or absent (Fig. 11B). The subsidence would probably be incremental and occur in pulses (see 'seismic cycles' of McCalpin, 1996; Sieh, 2000), which could render the accommodation changes irregular and might thus explain the varied thicknesses and facies architecture of the parasequences.

Even if modest falls in relative sea level did occur, the record of forced regressions might be unrecognizable in the present case, not least because the outcrop section provides very limited lateral control. First, a high-relief coastal topography would cause the shoreline to climb down, rather than shift seawards across the bay. Second, the erosional effect of a moderate fall in relative sea level would be limited to the shoreface zone (Storms & Hampson, 2005), where a composite discontinuity with multiple erosional surfaces would form ('regressive surface of marine erosion' *sensu* Plint & Nummedal, 2000). This discontinuity surface would practically be hidden in the shoreface deposits, impossible to distinguish from the numerous erosional surfaces produced by regular storms and tidal winnowing.

CONCLUSIONS

The Barcellona Pozzo di Gotto Basin of northeast Sicily is a peri-Tyrrhenian shelf embayment of middle Pliocene to Pleistocene age. The study has focused on the transgressive systems tract of the lower bay-fill sequence, which is a marine, mixed siliciclastic–bioclastic succession ~ 75 m thick in mid-bay outcrop section. The deposits are a wide range of sandy to silty facies indicating a wave-dominated embayment influenced by storms and tidal currents, hosting an abundant sediment suspension and probably affected by rare tsunami events. Facies associations represent the upper and lower shoreface, offshore-transition and offshore zones. The abundance of silty to sandy suspension in the bay is attributed to a perennial entrapment of fine sediment entrained by storms and tides and possibly also derived from nearby streams. The high turbidity of the water would reduce the penetration depth of light, which might explain the shallowness of the infra- and circalittoral ecological zones in the bay.

The evidence from sedimentary facies analysis is consistent with that from palaeontological observations, and the two data types both supplement and verify each other. The integration of sedimentological, biostratigraphic, palaeoecological and taphonomic data has proved to be a powerful method for high-resolution sequence-stratigraphic analysis and palaeoenvironmental reconstruction, including sediment dynamics, palaeogeography and bathymetric changes. The multidisciplinary study has provided new insight in the anatomy of a transgressive systems tract, far more detailed than can be acquired from a conventional sedimentary analysis in sequence-stratigraphic studies.

The stratigraphic organization of the facies associations and their faunal assemblages indicates that the succession consists of six parasequences, bounded by marine-flooding surfaces and showing an overall deepening-upward trend. The parasequences are several metres thick and have well-developed transgressive and regressive deposits. Some parasequences also have a relatively thick condensation zone of mid-cycle turnabout, which indicates a prolonged balance between the rate of accommodation development and the rate of its filling by slow seafloor aggradation. Palaeoecological and taphonomic evidence of maximum condensation allows the maximum flooding surface to be identified, typically in the upper part of the zone. The thick transgressive deposits and mid-cycle condensation zones thus bear valuable information on the transgression rate and sediment dynamics.

The parasequences have time spans of the order of 300 kyr and appear to correlate with the high-frequency regional sequences recognized in the central Mediterranean and attributed to the 4th-order eustatic cycles (Catalano *et al.*, 1998). Therefore, the parasequences are inferred to be local equivalents of the 4th-order sequences, owing their specific facies architecture to a relatively high rate of tectonic subsidence in the coastal zone of northern Sicily. These would thus be sequences of type 2 (Jervey, 1988) involving little or no relative fall in sea level and are hence masquerading as parasequences.

ACKNOWLEDGEMENTS

This study is an extension of Carlo Messina's earlier doctoral project, with fieldwork supported by MIUR grants to Antonietta Rosso and a Statoil travel grant to W. Nemec. The authors thank Alfio Viola for grain-size analysis of sediment samples, and William Helland-Hansen, John Howell and Ayhan Ilgar for helpful discussions. The manuscript was critically reviewed by Andre Freiwald, Massimiliano Ghinassi, Francesco Massari and Gary Nichols, whose constructive comments are much appreciated by the authors.

REFERENCES

Abbot, S.T. (1997a) Mid-cycle condensed shellbeds from mid-Pleistocene cyclothems, New Zealand: implications for sequence architecture. *Sedimentology*, **44**, 805–824.

Abbot, S.T. (1997b) Foraminiferal paleobathymetry and mid-cycle architecture of mid-Pleistocene depositional sequences, Wanganui Basin, New Zealand. *Palaios*, **12**, 267–281.

Abbott, S.T. and Carter, R.M. (1994) The sequence architecture of mid-Pleistocene (0.35–0.95 Ma) cyclothems from New Zealand: facies development during a period of known orbital control on sea-level cyclicity. In: *Orbital Forcing and Cyclic Sequences* (Eds P.L. De Boer and D.G. Smith), pp. 367–394. Special Publication 19, International Association of Sedimentologists. Blackwell Scientific Publications, Oxford.

Abbott, S.T. and Carter, R.M. (1997) Macrofossil associations from shoreline-shelf depositional sequences in the Castlecliff section (mid-Pleistocene, Wanganui Basin, New Zealand). *Palaios*, **12**, 188–210.

Amodio Morelli, L., Bonardi, G., Colonna, V., *et al.* (1976) L'Arco Calabro-Peloritano nell'Orogene Appenninico-Maghrebide. *Soc. Geol. Ital. Mem.*, **17**, 1–6.

Arnott, R.W.C. (1993) Quasi-planar laminated sandstone beds of the Lower Cretaceous Bootlegger Member, north-central Montana: evidence of combined-flow sedimentation. *J. Sediment. Petrol.*, **63**, 488–494.

Arnott, R.W.C. (1995) The parasequence definition – Are transgressive deposits inadequately addressed? *J. Sediment. Res.*, **B65**, 1–6.

Arnott, R.W.C. and Southard, J.B. (1990) Exploratory flow-duct experiments on combined-flow bed configurations, and some implications for interpreting storm-event stratification. *J. Sediment. Petrol.*, **60**, 211–219.

Avoine, J. and Larsonneur, C. (1987) Dynamics and behaviour of suspended sediment in macrotidal estuaries along the south coast of the English Channel. *Cont. Shelf Res.*, **7**, 1301–1305.

Bader, B. (2001) Modern bryomol sediments in a cool-water, high energy setting: the inner shelf off northern Brittany. *Facies*, **44**, 81–104.

Baldwin, B. and Butler, C.O. (1985) Compaction curves. *Am. Assoc. Petrol. Geol. Bull.*, **69**, 622–626.

Banerjee, I. and Kidwell, S.M. (1991) Significance of molluscan shell beds in sequence stratigraphy: an example from the Lower Cretaceous Mannville Group of Canada. *Sedimentology*, **38**, 913–934.

Barrier, P. (1987) Stratigraphie des dépôts pliocènes et quaternaires du Détroit de Messine. *Doc. Trav. IGAL (Paris)*, **11**, 59–81.

Barrier, P., Casale, V., Costa, B., Di Geronimo, I., Oliveri, O. and Rosso, A. (1987) La sezione plio-pleistocenica di Pavigliana (Reggio Calabria). *Boll. Soc. Paleont. Ital.*, **25**, 107–144.

Bertoldi R., Rio D. and Thunnel R. (1989) Pliocene–Pleistocene vegetational and climatic evolution of the south-central Mediterranean. *Palaeogeogr. Palaeoclimatol. Palaeoecol.*, **72**, 263–275.

Boccaletti, M., Ciaranfi, N., Casentino, D., *et al.* (1990). Palinspastic restoration and paleogeographic reconstruction of the peri-Tyrrhenian area during the Neogene. *Palaeogeogr. Palaeoclimatol. Palaeoecol.*, **77**, 41–50.

Bonaduce, G., Ciampo, G. and Masoli, M. (1975) Distribution of Ostracoda in the Adriatic Sea. *Pubbl. Staz. Zool. Napoli, Suppl.* **40**, 304 pp.

Bonaduce, G., Ruggieri, G. and Russo, A. (1987) The ostracod genus *Mutilus* and so-called *Mutilus* from the Mediterranean Miocene–Pleistocene. *Boll. Soc. Paleont. Ital.*, **26**, 251–268.

Bonardi, G., Cavazza, W., Perrone, V. and Rossi, S. (2001) Calabria-Peloritani terrane and northern Ionian Sea. In: *Anatomy of an Orogen: the Apennines and Adjacent Mediterranean Basin* (Eds G.B. Vai and I.P. Martini), pp. 287–306. Kluwer Academic Publishers, Dordrecht.

Bossio, A., Foresi, L.M., Mazzanti, R., Mazzei, R. and Salvatorini, G. (1997) Note micropaleontologiche sulla successione miocenica del Torrente Morra e su quella pliocenica del Bacino dei Fiumi Tora e Fine (Provincie di Livorno e Pisa). *Atti Soc. Tosc. Sci. Nat., Mem. Ser. A*, **104**, 85–134.

Bourgeois, J. (1980) A transgressive shelf sequence exhibiting hummocky stratification: the Cape Sebastian Sandstone (Upper Cretaceous), southwestern Oregon. *J. Sediment. Petrol.*, **50**, 681–702.

Bowen, A.J. and Guza, R.T. (1978) Edge waves and surf beat. *J. Geophys. Res.*, **83**, 1913–1920.

Brenchley, P.J. (1985) Storm-influenced sandstone beds. *Modern Geol.*, **9**, 369–396.

Brolsma, M.J. (1978) Quantitative foraminiferal analysis and environmental interpretation of the Pliocene and topmost Miocene of the south coast of Sicily. *Utrecht Micropaleontol. Bull.*, **18**, 1–159.

Carter, R.M., Fulthorpe, C.S. and Naish, T.R. (1998) Sequence concepts at seismic and outcrop scale: the distinction between physical and conceptual stratigraphic surfaces. *Sediment. Geol.*, **122**, 165–179.

Cantalamessa, G. and Di Celma, C. (2005) Sedimentary features of tsunami backwash deposits in a shallow marine Miocene setting, Mejillones Peninsula, northern Chile. *Sediment. Geol.*, **178**, 259–273.

Catalano, R., Di Stefano, E., Lo Cicero, G., Infuso, S., Vail, P.R. and Vitale, F.P. (1993) Basin analysis and sequence stratigraphy of the Plio-Pleistocene of Sicily. In: *Geological Development of the Sicilian-Tunisian Platform* (Eds M.D. Max and P. Colantoni). *UNESCO Rep. Mar. Sci.*, **58**, 99–104.

Catalano, R., Infuso, S. and Sulli, A. (1995) Tectonic history of the submerged Maghrebian Chain from the southern Tyrrhenian Sea to the Pelagian Foreland. *Terra Nova*, **7**, 179–188.

Catalano, R., Di Stefano, E., Sulli, A., Vitale, F.P., Infuso, S. and Vail, P.R. (1998) Sequences and systems tracts calibrated by high-resolution bio-chronostratigraphy: the Central Mediterranean Plio-Pleistocene record. In: *Mesozoic and Cenozoic Sequence Stratigraphy of European Basins* (Eds P.C. De Graciansky, J. Hardenbol, T. Jacquin and P.R. Vail), pp. 155–177. Special Publication 60, Society of Economic Paleontologists and Mineralogists, Tulsa, OK.

Catalano, S. and Cinque, A. (1995) L'evoluzione neotettonica dei Peloritani settentrionali (Sicilia nord-orientale): il contributo di un'analisi geomorfologica preliminare. *Studi Geol. Camerti Spec. Vol.*, **2**, 113–123.

Catalano, S. and Di Stefano, A. (1997) Sollevamenti e tettogenesi pleistocenica lungo il margine tirrenico dei Monti Peloritani: integrazione di dati geomorfologici, strutturali e biostratigrafici. *Il Quaternario*, **10**, 337–342.

Cita, M.B. (1975) Studi sul Pliocene e sugli strati di passaggio dal Miocene al Pliocene. VIII Planktonic foraminiferal biozonation of the Mediterranean Pliocene deep sea record. A revision. *Riv. Ital. Paleontol.*, **81**, 527–544.

Clifton, H.E. (1973) Pebble segregation and bed lenticularity in wave-worked versus alluvial gravel. *Sediment. Geol.*, **20**, 173–187.

Clifton, H.E. (1976) Wave-formed sedimentary structures – a conceptual model. In: *Beach and Nearshore Sedimentation* (Eds R.A. Davis, Jr. and R.L. Ethington), pp. 126–148. Special Publication **24**, Society of Economic Paleontologists and Mineralogists, Tulsa, OK.

Clifton, H.E. (1981) Progradational sequences in Miocene shoreline deposits, southeastern Caliente Range, California. *J. Sediment. Petrol.*, **51**, 165–184.

Clifton, H.E. and Dingler, J.R. (1984) Wave-formed structures and paleoenvironmental reconstruction. *Mar. Geol.*, **60**, 165–198.

Clifton, H.E., Hunter, R.E. and Phillips, R.L. (1971) Depositional structures and processes in the non-barred high-energy nearshore. *J. Sediment. Petrol.*, **44**, 651–670.

Coe, A.L. (Ed.) (2003) *The Sedimentary Record of Sea-Level Change*. Cambridge University Press, Cambridge, 288 pp.

Colalongo, M.L. and Sartoni, S. (1967) *Globorotalia hirsuta* aemiliana nuova sottospecie cronologica del Pliocene in Italia. *Giorn. Geol. Ser. 2*, **34**, 265–284.

Collinson, J.D. and Thompson, D.B. (1982) *Sedimentary Structures*. Allen and Unwin, London, 207 pp.

DeCelles, P.G. and Cavazza, W. (1992) Constraints on the formation of Pliocene hummocky cross-stratification in Calabria (southern Italy) from consideration of hydraulic and dispersive equivalence, grain-flow theory, and suspended-load fallout rate. *J. Sediment. Petrol.*, **62**, 555–568.

Di Geronimo, I. (1985) La bionomie benthique appliquèe à l'étude des peuplements fossiles de la Méditerranée: contribution des chercheurs italiens. *Tethys*, **11**, 243–248.

Di Geronimo, I. and Robba, E. (1989) The structure of benthic communities in relation to basin stability. *Atti Acc. Lincei*, **80**, 341–352.

Di Geronimo, I., D'Atri, A., La Perna, R., Rosso, A., Sanfilippo, R. and Violanti, D. (1997) The Pleistocene bathyal section of Archi (Southern Italy). *Boll. Soc. Paleont. Ital.*, **36**, 189–212.

Di Geronimo, I., Di Geronimo, R., La Perna, R., Rosso, A. and Sanfilippo, R. (2000) Cooling evidences from Pleistocene shelf assemblages in SE Sicily. In: *Climates: Past and Present* (Ed. H.B. Hart), pp. 113–120. Special Publication 181, Geological Society Publishing House, Bath.

Di Geronimo, I., Messina, C., Rosso, A. and Sanfilippo, R. (2002) Taphonomic data in palaeoenvironmental reconstruction of shelly concentrations in a dune system. In: *Current Topics on Taphonomy and Fossilization* (Eds M. De Renzi, M.V. Pardo Alonso, M. Belinchòn, E. Penalver, P. Montoya and A. Màrquez-Aliaga), pp. 187–191. Ajuntament de Valencia, Valencia.

Di Geronimo, I., Messina, C., Rosso, A., Sanfilippo, R., Sciuto, F. and Vertino, A. (2005) Enhanced biodiversity in the deep: Early Pleistocene coral communities from southern Italy. In: *Cold-Water Coral Ecosystems* (Eds A. Freiwald and J.M. Roberts), pp. 61–86. Springer-Verlag, Heidelberg.

Duchaufour, P. (1977) *Pedology*. Allen & Unwin, London, 448 pp.

Dott, R.H., Jr. and Bourgeois, J. (1982) Hummocky stratification: significance of its variable bedding sequences. *Geol. Soc. Am. Bull.*, **93**, 663–680.

Duke, W.L. (1985) Hummocky cross-stratification, tropical hurricanes, and intense winter storms. *Sedimentology*, **32**, 167–194.

Duke, W.L., Arnott, R.W. and Cheel, R.J. (1991) Shelf sandstones and hummocky cross stratification; new insights on a stormy debate. *Geology*, **19**, 625–628.

Emery, D. and Myers, K.J. (1996) *Sequence Stratigraphy*. Blackwell Science, Oxford, 291 pp.

Evans, G. and Collins, M.B. (1987) Sediment supply and deposition in The Wash. In: *Nearshore Sediment Dynamics and Sedimentation* (Eds J. Hails and A. Carr), pp. 237–306. John Wiley, Chichester.

Fitzgerald, D.M., Penland, S. and Nummedal, D. (1984) Control of barrier island shape by inlet sediment bypassing: East Frisian Islands, West Germany. *Mar. Geol.*, **60**, 355–376.

Gaetani, M. and Saccà, D. (1983) Brachiopodi batiali nel Pliocene e Pleistocene di Sicilia e Calabria. *Riv. Ital. Paleontol. Stratigr.*, **90**, 407–458.

Gamulin-Brida, H. (1974) Biocoenoses benthiques de la mer Adriatique. *Acta Adriat.*, **15**, 1–102.

Gao, S., Xie, Q. and Feng, Y. (1990) Fine-grained sediment transport and sorting by tidal exchange in Xiangshan Bay, Zhejiang, China. *Estuar. Coast. Shelf. Sci.*, **31**, 397–409.

Gautier, Y.V. (1962) Recherches écologiques sur les bryozoaires chilostomes en Méditerranée occidentale. *Rec. Trav. Mar. Endoume*, **24**, 1–434.

Ghisetti, F. (1981) L'evoluzione strutturale del bacino plio-pleistocenico di Reggio Calabria nel quadro geodinamico dell'Arco Calabro. *Boll. Soc. Geol. Ital.*, **100**, 433–466.

Gradstein, F.M., Ogg, J.G., Smith, A.G., Bleeker, W. and Lourens, L.J. (2004) A new Geologic Time Scale, with special reference to Precambrian and Neogene. *Episodes*, **27**, 83–100.

Gruszczyński, M., Rudowski, S., Semil, J., Skomiński, J. and Zrobek, J. (1993) Rip currents as a geological tool. *Sedimentology*, **40**, 217–236.

Guarnieri, P. and Carbone, S. (2003) Assetti geologico e lineamenti morfostrutturali dei bacini plio-quaternari del Tirreno meridionale. *Boll. Soc. Geol. Ital.*, **122**, 377–386.

Hageman, S.J., Lukasik, J., McGowran, B. and Bone, Y. (2003) Paleoenvironmental significance of Celleporaria (Bryozoa) from modern and Tertiary cool-water carbonates of Southern Australia. *Palaios*, **16**, 510–527.

Hamblin, A.P. and Walker, R.B. (1979) Storm-generated shallow marine deposits: the Fernie-Kootenay (Jurassic) transition, southern Rocky Mountains. *Can. J. Earth Sci.*, **16**, 1673–1690.

Harmelin, J.-G. (1976) Le sous-ordre des Tubuliporina (Bryozoaires Cyclostomes) en Méditerranée. *Inst. Océanogr. Monaco Mém.*, **10**, 1–326.

Harms, J.C. (1969) Hydraulic significance of some sand ripples. *Geol. Soc. Am. Bull.*, **80**, 363–396.

Harms, J.C., Southard, J.B., Spearing, D.R. and Walker, R.G. (1975) *Depositional Environments as Interpreted from Primary Sedimentary Structures and Stratification Sequences*. SEPM Short Course No. 2 Lecture Notes, Society of Economic Paleontologists and Mineralogists, Dallas, 161 pp.

Harms, J.C., Southard, J.B. and Walker, R.G. (1982) *Structures and Sequences in Clastic Rocks*. SEPM Short Course No. 9 Lecture Notes, Society of Economic Paleontologists and Mineralogists, Tulsa, 250 pp.

Helland-Hansen, W. and Martinsen, O.J. (1996) Shoreline trajectories and sequences: description of variable depositional-dip scenarios. *J. Sediment. Res.*, **66**, 670–688.

Henrich, R., Freiwald, A., Betzler, C., *et al.* (1995) Controls on modern carbonate sedimentation on warm-temperate to Arctic coasts, shelves and seamounts in the northern hemisphere: implications for fossil counterparts. *Facies*, **32**, 71–108.

Hobday, D.K. and Banks, N.L. (1971) A coarse grained pocket beach complex, Tanafjord (Norway). *Sedimentology*, **16**, 129–134.

Ilgar, A. and Nemec, W. (2005) Early Miocene lacustrine deposits and sequence stratigraphy of the Ermenek Basin, Central Taurides, Turkey. *Sediment. Geol.*, **173**, 233–275.

Jervey, M.T. (1988) Quantitative geological modeling of siliciclastic rock sequences and their seismic expression. In: *Sea-level Changes – an Integrated Approach* (Eds C.K. Wilgus, B.S. Hastings, H.W. Posamentier, J.C. Van Wagoner, C.A. Ross and C.G.St.C. Kendall), pp. 47–69. Special Publication 42, Society of Economic Paleontologists and Mineralogists, Tulsa, OK.

Ke, X., Evans, G. and Collins, M.B. (1996) Hydrodynamics and sediment dynamics of The Wash embayment, eastern England. *Sedimentology*, **43**, 157–174.

Kerr, R.C. (1991) Erosion of a stable density gradient by sedimentation-driven convection. *Nature*, **353**, 423–425.

Kezirian, F. (1993) Evolution tectono-sédimentaire post-nappes des Monts Péloritains (Sicile Nord-Orientale, Italie). *Mém. IGAL (Paris)*, **49**, 1–260.

Kidwell, S.M. (1986) Models for fossil concentrations: paleobiologic implications. *Paleobiology*, **12**, 6–24.

Kidwell, S.M. (1991) The stratigraphy of shell concentrations. In: *Taphonomy: Releasing the Data Locked in the Fossil Record* (Eds P.A. Allison and D.E.G. Briggs), pp. 211–290. Plenum Press, New York.

Kidwell S.M. and Bosence, D.W. (1991) Taphonomy and time-averaging of marine shelly faunas. In: *Taphonomy: Releasing the Data Locked in the Fossil Record* (Eds P.A. Allison and D.E.G. Briggs), pp. 115–209. Plenum Press, New York.

Kidwell, S.M., Fürsich, F.T. and Aigner, T. (1986) Conceptual framework for the analysis and classification of fossil concentrations. *Palaios*, **1**, 228–238.

Kirkby, R. (1987) Introduction: sediment exchanges across the coastal margins of NW Europe. *J. Geol. Soc. London*, **144**, 121–126.

Knott, S.D. and Turco, E. (1991) Late Cenozoic kinematics of the Calabrian Arc, southern Italy. *Tectonics*, **10**, 1164–1172

Komar, P.D. (1976) *Beach Processes and Sedimentation*. Prentice-Hall, Englewood Cliffs, NJ, 417 pp.

Kreisa, R.D. (1981) Storm-generated sedimentary structures in subtidal marine facies with examples from the Middle and Upper Ordovician of southwestern Virginia. *J. Sediment. Petrol.*, **51**, 823–848.

Kumar, N. and Sanders, J.E. (1976) Characteristics of shoreface storm deposits: modern and ancient examples. *J. Sediment. Petrol.*, **46**, 145–162.

Leckie, D.A. (1988) Wave-formed, coarse-grained ripples and their relationship to hummocky cross-stratification. *J. Sediment. Petrol.*, **58**, 607–622.

Leckie, D.A. and Walker, R.G. (1982) Storm- and tide-dominated shorelines in Cretaceous Moosebar–Lower Gates interval – outcrop equivalents of Deep Basin gas trap in Western Canada. *Am. Assoc. Petrol. Geol. Bull.*, **66**, 138–157.

Leithold, E.L. and Bourgeois, J. (1984) Characteristics of coarse-grained sequences deposited in nearshore, wave-dominated environments – examples from the Miocene of south-west Oregon. *Sedimentology*, **31**, 749–775.

Lentini, F., Carbone, S. and Catalano, S. (2000) *Carta Geologica della Provincia di Messina*. Provincia Regionale di Messina Assessorato Territorio, Servizio Geologico, Selca, Florence, 70 pp.

Longhitano, S.G. and Nemec, W. (2005) Statistical analysis of bed-thickness variation in a Tortonian succession of biocalcarenitic tidal dunes, Amantea Basin, Calabria, southern Italy. *Sediment. Geol.*, **179**, 195–224.

Loutit, T., Hardenbol, J., Vail., P.R. and Baum, G.R. (1988) Condensed sections: the key to age determination and correlation of continental margin sequences. In: *Sea-level Changes – an Integrated Approach* (Eds C.K. Wilgus, B.S. Hastings, H.W. Posamentier, J.C. Van Wagoner, C.A. Ross and C.G.St.C. Kendall), pp. 183–213. Special Publication 42, Society of Economic Paleontologists and Mineralogists, Tulsa, OK.

Lowe, D.R. (1988) Suspended-load fallout rate as an independent variable in the analysis of current structures. *Sedimentology*, **35**, 765–776.

Malatesta, A. and Zarlenga, F. (1986) Northern guests in the Pleistocene Mediterranean Sea. *Geol. Romana*, **25**, 91–154.

Malinverno, A. and Ryan, W.B.F. (1986) Extension in Tyrrhenian Sea and shortening in the Apennines as a result of arc migration driven by sinking of the lithosphere. *Tectonics*, **5**, 227–254.

Mantz, P.A. (1977) Incipient transport of fine grains and flakes by fluids – extended Shields diagram. *Proc. Am. Soc. Civ. Eng., J. Hydraul. Div.*, **103**, 601–615.

Massari, F. and Parea, G.C. (1988) Progradational gravel beach sequences in a moderate- to high-energy microtidal marine environment. *Sedimentology*, **35**, 881–913.

McCalpin, J.P. (1996) Application of paleoseismic data to seismic hazard assessment and neotectonic research. In: *Introduction to Paleoseismology* (Ed. J.P. McCalpin), pp. 439–493. Academic Press, San Diego.

McKinney, F. and Jaklin, A. (1993) Living populations of free-lying bryozoans: implications for post-Paleozoic decline of the growth habit. *Lethaia*, **26**, 171–179.

McKinney, F. and Jaklin, A. (2000) Spatial niche partitioning in the Cellaria meadow epibiont association, northern Adriatic Sea. *Cah. Biol. Mar.*, **41**, 1–17.

McKinney, F. and Jaklin, A. (2001) Sediment accumulation in a shallow-water meadow carpeted by a small erect bryozoan. *Sediment. Geol.*, **145**, 397–410.

Menesini, E. (1984) Distribution of some Mediterranean species of Balanomorpha (Cirripedia, Thoracica) from the Tertiary to the Actual. *Atti Soc. Tosc. Sci. Nat., Mem. Ser. A*, **94**, 291–303.

Menesini, E. and Casella, C. (1988) Balanidi Pliocenici della provincia di Almeria (Andalusia orientale – Spagna). Studio sistematico. *Atti Soc. Tosc. Sci. Nat., Mem. Ser. A*, **95**, 231–269.

Messina, C. (2003) *Il bacino plio-quaternario di Barcellona P.G. (Sicilia nord-orientale): analisi di facies ed evoluzione stratigrafico-sequenziale*. Unpublished PhD thesis, University of Catania, 120 pp.

Messina, C. and Rosso, A. (2005) Taphonomy of shell concentrations in backstepping parasequence sets: the case study of Barcellona P.G. basin (NE Sicily, Italy). In: *Proceedings 2nd International Meeting Taphos 2005*, pp. 63–64. University of Barcelona.

Mitchum, R.M. (1977) Seismic stratigraphy and global changes of sea level. Part I: Glossary of terms used in seismic stratigraphy. In: *Seismic Stratigraphy – Applications to Hydrocarbon Exploration* (Ed. C.E. Payton), pp. 49–212. Memoir 26, American Association of Petroleum Geologists, Tulsa, OK.

Moissette, P. (1988) Faunes de bryozoaires du Messinien d'Algerie occidentale. *Doc. Lab. Géol. Lyon*, **102**, 1–351.

Moissette, P. and Pouyet, S. (1991) Bryozoan masses in the Miocene–Pliocene and Holocene of France, North Africa and the Mediterranean. In: *Bryozoa Living and Fossil* (Ed. F.P. Bigey). *Bull. Soc. Sci. Nat. Ouest France*, **1**, 271–279.

Monaco, C., Tortorici, L., Nicolich, R., Cernobori, L. and Costa, M. (1996) From collisional to rifted basins: an example from the southern Calabrian arc (Italy). *Tectonophysics*, **266**, 233–249.

Monegatti P. and Raffi R. (2001) Taxonomic diversity and stratigraphic distribution of Mediterranean Pliocene bivalves. *Palaeogeogr. Palaeoclimatol. Palaeoecol.*, **165**, 171–193.

Montenegro, M.E., Pugliese, N. and Bonaduce, G. (1998) Shelf ostracods distribution in the Italian seas. *Bull. Centres Rech. Explor.-Prod. Elf Aquitaine*, **20**, 91–101.

Myrow, P.M. and Southard, J.B. (1996) Tempestite deposition. *J. Sediment. Res.*, **66**, 875–887.

Naish, T. and Kamp, P.J.J. (1997) Sequence stratigraphy of sixth-order (41 k.y.) Pliocene–Pleistocene cyclothems, Wanganui Basin, New Zealand; a case for the regressive systems tract. *Geol. Soc. Am. Bull.*, **109**, 978–999.

Neale, J.W. (1964) Factors influencing distribution of recent British Ostracoda. *Publ. Staz. Zool. Napoli, Suppl.*, **33**, 247–307.

Nemec, W. (1995) The dynamics of deltaic suspension plumes. In: *The Geology of Deltas* (Eds M.N. Oti and G. Postma), pp. 31–93. Balkema, Rotterdam.

Nemec, W. (2005) *Geostatistics*. Course GEOL368 Lecture Notes, University of Bergen, 172 pp.

Pasini, G. and Colalongo, M.L. (1994) Proposal for the erection of the Santernian/Emilian boundary-stratotype (lower Pleistocene) and new data on the Pliocene–Pleistocene boundary-stratotype. *Boll. Soc. Paleont. Ital.*, **33**, 101–120.

Patacca, E., Sartori, R. and Scandone, P. (1990) Tyrrhenian basin and Apenninic arcs: kinematic relations since late Tortonian times. *Soc. Geol. Ital. Mem.*, **45**, 425–451.

Pemberton, S.G., Frey, R.W. and Saunders, T.D. (1990) Trace fossils. In: *Palaeobiology – a Synthesis* (Eds D.E.G. Briggs and P.R. Crowther), pp. 335–362. Blackwell Scientific Publications, Oxford.

Penland, S., Boyd, R. and Suter, J.R. (1988) Transgressive depositional systems of the Mississippi River delta plain: a model for barrier shoreline and shelf development. *J. Sediment. Petrol.*, **58**, 932–949.

Pérès, J.M. (1982) Major benthic assemblages. In: *Marine Ecology – a Comprehensive Integrated Treatise on Life in Oceans and Coastal Waters*, Vol. V, Part 1: *Ocean Management* (Ed. O. Kinne), pp. 373–522. John Wiley & Sons, Chichester.

Pérès, J.M. and Picard, J. (1964) Nouveau manuel de bionomie bentique de la Mer Méditerranée. *Rec. Trav. Stat. Mar. Endoume*, **47**, 5–137.

Plater, A.J., Ridgway, J., Rayner, B., *et al.* (2000) Sediment provenance and flux in the Tees Estuary: the record from the Late Devensian to the present. In: *Holocene Land–Ocean Interaction and Environmental Change around the North Sea* (Eds I. Shennan and J. Andrews), pp. 171–195. Special Publication **166**, Geological Society Publishing House, Bath.

Plint, A.G. and Nummedal, D. (2000) The falling systems tract: recognition and importance in sequence stratigraphic analysis. In: *Sedimentary Responses to Forced Regressions* (Eds D. Hunt and R.L. Gawthorpe), pp. 1–17. Special Publication 172, Geological Society Publishing House, Bath.

Poluzzi, A. (1975) I Briozoi cheilostomi del Pliocene della Val d'Arda (Piacenza, Italia). *Soc. Ital. Sci. Nat. Mem., Mus. Civ. Staz. Nat. Milano*, **21**, 37–77.

Posamentier, H.W. and Allen, G.P. (1993) Variability of the sequence stratigraphic model: effects of local basin factors. *Sediment. Geol.*, **86**, 91–109.

Posamentier, H.W. and James D.P. (1993) An overview of sequence-stratigraphic concepts: uses and abuses. In: *Sequence Stratigraphy and Facies Associations* (Eds H.W. Posamentier, C.P. Summerhayes, B.U. Haq and G.P. Allen), pp. 3–18. Special Publication 18, International Association of Sedimentologists. Blackwell Scientific Publications, Oxford.

Posamentier, H.W. and Vail P.R. (1988) Eustatic controls on clastic deposition II – sequence and systems tract models. In: *Sea-level Changes – an Integrated Approach* (Eds C.K. Wilgus, B.S. Hastings, H.W. Posamentier, J.C. Van Wagoner, C.A. Ross and C.G.St.C. Kendall), pp. 125–154. Special Publication 42, Society of Economic Paleontologists and Mineralogists, Tulsa, OK.

Posamentier, H.W., Jervey, M.T. and Vail, P.R. (1988) Eustatic controls on clastic deposition I – Conceptual framework. In: *Sea-level Changes – an Integrated Approach* (Eds C.K. Wilgus, B.S. Hastings, H.W. Posamentier, J.C. Van Wagoner, C.A. Ross and C.G.St.C. Kendall), pp. 109–124. Special Publication 42, Society of Economic Paleontologists and Mineralogists, Tulsa, OK.

Pouyet, S. (1976) Bryozoires cheilostomes du Pliocène d'Aguilas (Espagne Méridionale). *Nouv. Arch. Mus. Hist. Nat. Lyon*, **14**, 53–82.

Pouyet, S. and Moissette, P. (1992) Bryozaires du Pliocene d'Altavilla (Sicilie, Italie): revision de la collection Cipolla, nouvelles donnees, paleoecologie. *Palaeontographica*, **A223**, 19–101.

Prager, E.J., Southard, J.B. and Vivoni-Gallart, E.R. (1996) Experiments on the entrainment threshold of well-sorted and poorly sorted carbonate sands. *Sedimentology*, **43**, 33–40.

Pugh, D.T. (1987) *Tides, Surges and Mean Sea-Level*. John Wiley & Sons, Chichester, 472 pp.

Reading, H.G. and Collinson, J.D. (1996) Clastic coasts. In: *Sedimentary Environments: Processes, Facies and Stratigraphy* (Ed. H.G. Reading), pp. 154–231. Blackwell Science, Oxford.

Reineck, H.-E. and Singh, I.B. (1975) *Depositional Sedimentary Environments*, 2nd edn. Springer-Verlag, Berlin, 439 pp.

Robertson, A.H.F. and Grasso, M. (1995) Overview of the Late Tertiary–Recent tectonic and palaeoenvironmental development of the Mediterranean region. *Terra Nova*, **7**, 114–127.

Rosso, A. (1987) Popolamenti a Briozoi nel Pleistocene di Monte dell'Apa (Sicilia sud-orientale). *Atti Acc. Gioenia Sci. Nat. Catania Ser. 20*, **331**, 167–197.

Rosso, A. (1996) Popolamenti e tanatocenosi a Briozoi di fondi mobili circalitorali del Golfo di Noto (Sicilia SE). *Naturalista Siciliano Ser. 4*, **20**, 189–225.

Ruggieri, G. (1980) Sulla distribuzione stratigrafica di alcuni ostracodi nel Pleistocene italiano. *Boll. Soc. Paleont. Ital.*, **19**, 127–135.

Sanfilippo, R. (1999) *Ditrupa brevis* n.sp., a new serpulid from the Mediterranean Neogene with comments on the ecology of the genus. *Riv. Ital. Paleont. Stratigr.*, **105**, 455–464.

Sciuto, F. and Rosso, A. (2002) Contributo alla conoscenza di tanatocenosi ad ostracodi di fondi circalitorali al largo di Aci Trezza (CT, Sicilia orientale). *Boll. Acc. Gioenia Sci. Nat. Catania Ser. 35*, **361**, 293–309.

Sieh, K. (2000) The repetition of large-earthquake ruptures. In: *Active Fault Research for the New Millenium* (Eds K. Okumura, H. Takada and H. Goto), pp. 465–468. Proceedings of the International Symposium School on Active Faulting, Hokudan, Hokudan.

Southard, J.B., Young, R.A. and Hollister, C.D. (1971) Experimental erosion of calcareous ooze. *J. Geophys. Res.*, **76**, 5903–5909.

Speranza, F., Mattei, M., Sagnotti, L. and Grasso, F. (2000) Palaeomagnetism of upper Miocene sediments from the Amantea basin (Calabria, Italy): rotational differences between the northern and southern Tyrrhenian domains. *J. Geol. Soc. London*, **157**, 327–334.

Spjeldnæs, N. and Moissette, P. (1997) Celleporid (bryozoans) thickets from the Upper Pliocene of the island of Rhodes, Greece. In: *Cool-water Carbonates* (Eds N.P. James and J.A.D. Clarke), pp. 263–270. Special Publication 56, Society of Economic Paleontologists and Mineralogists, Tulsa, OK.

Sprovieri R., Di Stefano, E., Howell, M., Sakamoto, T., Di Stefano, A. and Marino, M. (1998) Integrated calcareous plankton biostratigraphy and cyclostratigraphy at Site 964. *Proc. ODP Sci. Results*, **160**, 155–165.

Stainforth, R.M., Lamb, J.L., Luterbacher, H., Beard, J.H. and Jeffords, R.M. (1975) Cenozoic planktonic foraminiferal zonation and characteristics of index forms. *Univ. Kansas Paleontol. Contrib.*, **62**, 1–425.

Storms, J.E.P. and Hampson, G.J. (2005) Mechanisms for forming discontinuity surfaces within shoreface-shelf parasequences: sea level, sediment supply, or wave regime? *J. Sediment. Res.*, **75**, 67–81.

Swift, D.J.P., Phillips, S. and Thorne, J.A. (1991) Sedimentation on continental margins, V: Parasequences. In: *Shelf Sand and Sandstone Bodies: Geometry, Facies and Sequence Stratigraphy* (Eds D.J.P. Swift, G.F. Oertel, R.W. Tillman and J.A. Thorne), pp. 153–188. Special Publication **14**, International Association of Sedimentologists. Blackwell Scientific, Oxford.

Taddei Ruggiero, E. (1994) Neogene Salento brachiopod palaeocommunities. *Boll. Soc. Paleont. Ital.*, **33**, 197–213.

Thunnel, R., Williams, D., Tappa, E., Rio, D. and Raffi, I. (1990) Pliocene–Pleistocene stable isotope record for ocean drilling program site 653, Tyrrhenian Basin: implications for the paleoenvironmental history of the Mediterranean Sea. *Proc. ODP Sci. Results*, **107**, 387–399.

Tillman, R.W. (1985) A spectrum of shelf sands and sandstones. In: *Shelf Sands and Sandstone Reservoirs* (Eds R.W. Tillman, D.J.P. Swift and R.G. Walker), pp. 1–46. SEPM Short Course No. 13 Lecture Notes, Society of Economic Paleontologists and Mineralogists, Tulsa.

Vail, P.R., Mitchum, R.M., Jr. and Thomson, S., III (1977) Seismic stratigraphy and global changes of sea level, part 3: relative changes of sea level from coastal onlap. In: *Seismic Stratigraphy – Applications to Hydrocarbon Exploration* (Ed. C.E. Payton), pp. 63–81. Memoir 26, American Association of Petroleum Geologists, Tulsa, OK.

Van Dijk, J.P., Bello, M., Brancaleoni, G.P., *et al.* (2000) A regional structural model for the northern sector of the Calabrian Arc (southern Italy). *Tectonophysics*, **324**, 267–320.

Van Wagoner, J.C., Posamentier, H.W., Mitchum, R.M., Jr., *et al.* (1988) An overview of the fundamentals of sequence stratigraphy and key definitions. In: *Sea-level Changes – an Integrated Approach* (Eds C.K. Wilgus, B.S. Hastings, H.W. Posamentier, J.C. Van Wagoner, C.A. Ross and C.G.St.C. Kendall), pp. 39–45. Special Publication 42, Society of Economic Paleontologists and Mineralogists, Tulsa, OK.

Van Wagoner, J.C., Mitchum, R.M., Jr., Campion, K.M. and Rahmanian, V.D. (1990) *Siliciclastic Sequence Stratigraphy in Well Logs, Cores and Outcrops*. Methods in Exploration 7, American Association of Petroleum Geologists, Tulsa, OK, 55 pp.

Vrolijk, P.J. and Southard, J.B. (1997) Experiments on rapid deposition of sand from high-velocity flows. *Geosci. Can.*, **24**, 45–54.

Walker, R.G. (1984) General introduction: facies, facies sequences and facies models. In: *Facies Models* (Ed. R.G. Walker), 2nd edn. *Geosci. Can. Reprint Ser.*, **1**, 1–9.

Wright, L.D., Chappell, J., Thom, B.G., Bradshaw, M.P. and Cowell, P. (1979) Morphodynamics of reflective and dissipative beach and inshore systems, south-eastern Australia. *Mar. Geol.*, **32**, 105–140.

Yokokawa, M., Masuda, F. and Endo, N. (1995) Sand particle movement on migrating combined-flow ripples. *J. Sediment. Res.*, **A65**, 40–44.

Young, R.A. and Mann, R. (1985) Erosion velocities of skeletal carbonate sands, St. Thomas, Virgin Islands. *Mar. Geol.*, **69**, 171–185.

Young, R.A. and Southard, J.B. (1978) Erosion of fine-grained marine sediments: sea-floor and laboratory experiments. *Geol. Soc. Am. Bull.*, **89**, 663–672.

Zuschin, M., Stachowitsch, M., Pervesler, P. and Kollmann, H. (1999) Structural features and taphonomic pathways of a high-biomass epifauna in the northern Gulf of Trieste, Adriatic Sea. *Lethaia*, **32**, 299–317.

Late Cretaceous to early Eocene sedimentation in the Sinop–Boyabat Basin, north-central Turkey: a deep-water turbiditic system evolving into littoral carbonate platform

BEATE L.S. LEREN*[1], NILS E. JANBU*[2], WOJCIEK NEMEC*, EDIZ KIRMAN† *and* AYHAN ILGAR‡

**Department of Earth Science, University of Bergen, 5007 Bergen, Norway*
†Department of Geological Engineering, University of Ankara, 06100 Ankara, Turkey
‡General Directorate of Mineral Research and Exploration (MTA), 06520 Ankara, Turkey

ABSTRACT

The Sinop–Boyabat Basin is a southeast-trending elongate basin in the Central Pontides, northern Anatolia, filled with a succession of Lower Cretaceous to middle Eocene deposits, nearly 7 km thick. The basin evolved from a backarc rift related to the Western Black Sea crustal extension into a retroarc foreland basin of the Central Pontides, and was eventually inverted by tectonic compression in the late Eocene. The present sedimentological study, supplemented with petrographical, micropalaeontological and ichnological data, is focused on the Upper Cretaceous to lowest Eocene part of the basin-fill succession, which is ~ 2 km thick, comprises the Gürsökü, Akveren and Atbaşı formations and corresponds to the basin's transformation from a failed rift into an orogenic foreland. The succession's facies associations reveal a deep-marine turbiditic system that underwent prolonged aggradation and evolved into a wave-dominated littoral carbonate platform, to be drowned again due to a eustatic sea-level rise and rapid tectonic subsidence in late Paleocene time.

The upper Campanian to lower Maastrichtian Gürsökü Formation consists of alternating siliciclastic sandstones, calcareous mudstones and subordinate marlstones. The deposits represent a basin-floor turbiditic system directed towards the east, supplied with epiclastic volcanic detritus and increasingly more abundant bioclastic sediment from the basin's southwestern margin. The bioclastic admixture indicates development of a reefal platform at the basin margin. The northeastern margin was submerged and insignificant as a sediment source. The sheet-like turbidites indicate non-channelized currents of low to high density, and the succession represents transition from the medial to distal part of the system. At least one isolated palaeochannel occurs in the lowermost, thicker bedded part of the succession. The system was supplied with sediment from a storm-dominated littoral ramp perched on the basin margin and was subject to gradual retreat (back-stepping) by onlapping the margin.

The upper Maastrichtian–Paleocene Akveren Formation consists of sheet-like calcarenitic turbidites interbedded with marlstones and calcareous mudstones. Its uppermost part is dominated by tempestites, with wave-worked shoreface calcarenites and a reefal limestone unit at the top. Eastward sediment dispersal persisted, and the basin-floor turbiditic system was supplied with sediment from a distally steepened ramp with bypass chutes. As the turbiditic system aggraded, the ramp became homoclinal and the ignition of turbidity currents declined, giving way to

[1]Present address: Statoil ASA, PO Box 273, 7501 Stjordal, Norway (Email: beler@statoil.com).
[2]Present address: Statoil Research Centre, Rotvoll, 7005 Trondheim, Norway.

tempestitic sedimentation. The rate of sediment accumulation outpaced the declining subsidence rate, and this imbalance culminated in the rapid shallowing of the basin recorded by the uppermost part of the formation.

The uppermost Paleocene to lowest Eocene Atbaşı Formation recorded a dramatic rise in relative sea level, which could initially be mainly eustatic, with brief normal regressions, but then was greatly accelerated by foreland subsidence due to the crustal loading by Central Pontide thrust sheets. Basal transgressive shoreface and offshore-transition deposits are overlain by deep-water, variegated calcareous mudstones interspersed with thin calcarenitic turbidites. The muddy deposits indicate a sand-starved basin with a very low sedimentation rate and widespread seafloor oxidation.

Keywords Turbidites, tempestites, carbonate ramp, shoreface, backarc rift, retroarc foreland, Central Pontides.

INTRODUCTION

The Sinop–Boyabat Basin of north-central Anatolia, Turkey, is located within the central segment of the Pontide orogenic belt (Fig. 1), a Neotethyan suture zone formed by collision of the Kırşehir Massif with the volcanic arc and rifted margin of Eurasia. The elongate basin originated in the Early Cretaceous as a 'failed' backarc rift adjacent to the 'successful' and rapidly subsiding Western Black Sea Rift to the north, but evolved into a retroarc foreland basin of the Central Pontides and was tectonically inverted by the end of the Eocene. The basin-fill succession of Cretaceous to Eocene clastic deposits has a combined stratigraphic thickness of nearly 7000 m and bears an important record of the region's tectonic and palaeogeographical history. Thrusts have elevated large parts of the basin to ≥ 1000 m above sea level, with the coastal cliffs, river canyons, road cuts and quarries resulting in good exposure.

The basin has previously been mapped and its stratigraphy, tectonic structure and regional plate-tectonic setting have been discussed by many authors (Badgley, 1959; Göksu *et al.*, 1974; Aydın *et al.*, 1986, 1995a; Sonel *et al.*, 1989; Tüysüz, 1990, 1993, 1999; Robinson *et al.*, 1995; Tüysüz *et al.*, 1995; Görür, 1997; Görür & Tüysüz, 1997; Okay & Şahintürk, 1997; Ustaömer & Robertson, 1997; Okay & Tüysüz, 1999; Gürer & Aldanmaz, 2002; Meredith & Egan, 2002). However, few sedimentological studies are available, none of which is detailed and all have been published in local Turkish journals (Ketin & Gümüş, 1963; Aydın *et al.*, 1982, 1995b; Gedik & Korkmaz, 1984; Gedik *et al.*, 1984). Most of this previous research was intended to assess the hydrocarbon potential of the Turkish part of the Black Sea region, as the Ukrainian northern counterpart has been a significant petroleum province (Aydın *et al.*, 1982; Robinson *et al.*, 1996; Ziegler & Roure, 1999).

The present paper focuses on the Campanian to lowest Eocene part of the basin-fill succession, ~ 2000 m thick, which recorded the basin's gradual transformation from a rift into an orogenic foreland. The change in tectonic regime was accompanied by a gradual replacement of the original sources of siliciclastic sediment by a contemporaneous bioclastic carbonate source. The study documents facies assemblages of a deep-marine turbiditic system that underwent prolonged aggradation, accompanied by a change in sediment source from siliciclastic into bioclastic. The system eventually became dominated by tempestites and turned into a wave-dominated shoreface, which allowed brief basinward expansion of a carbonate platform, terminated by a late Paleocene dramatic rise in relative sea level. Only one of the basin margins acted as a sediment-supplying narrow shelf, with no evidence of contemporaneous fluvio-deltaic deposits, while the other one was submerged and dormant. Few similar cases of a continuous facies transition from bathyal to non-deltaic littoral environment have been described in the literature.

The sedimentological data have been acquired by detailed lithostratigraphic logging and petrographic analysis of three component formations of the succession (Gürsökü, Akveren and Atbaşı formations), which are well-exposed in roadcut sections and isolated outcrops over an area of ~ 4500 km^2. The spatial distribution of facies assemblages,

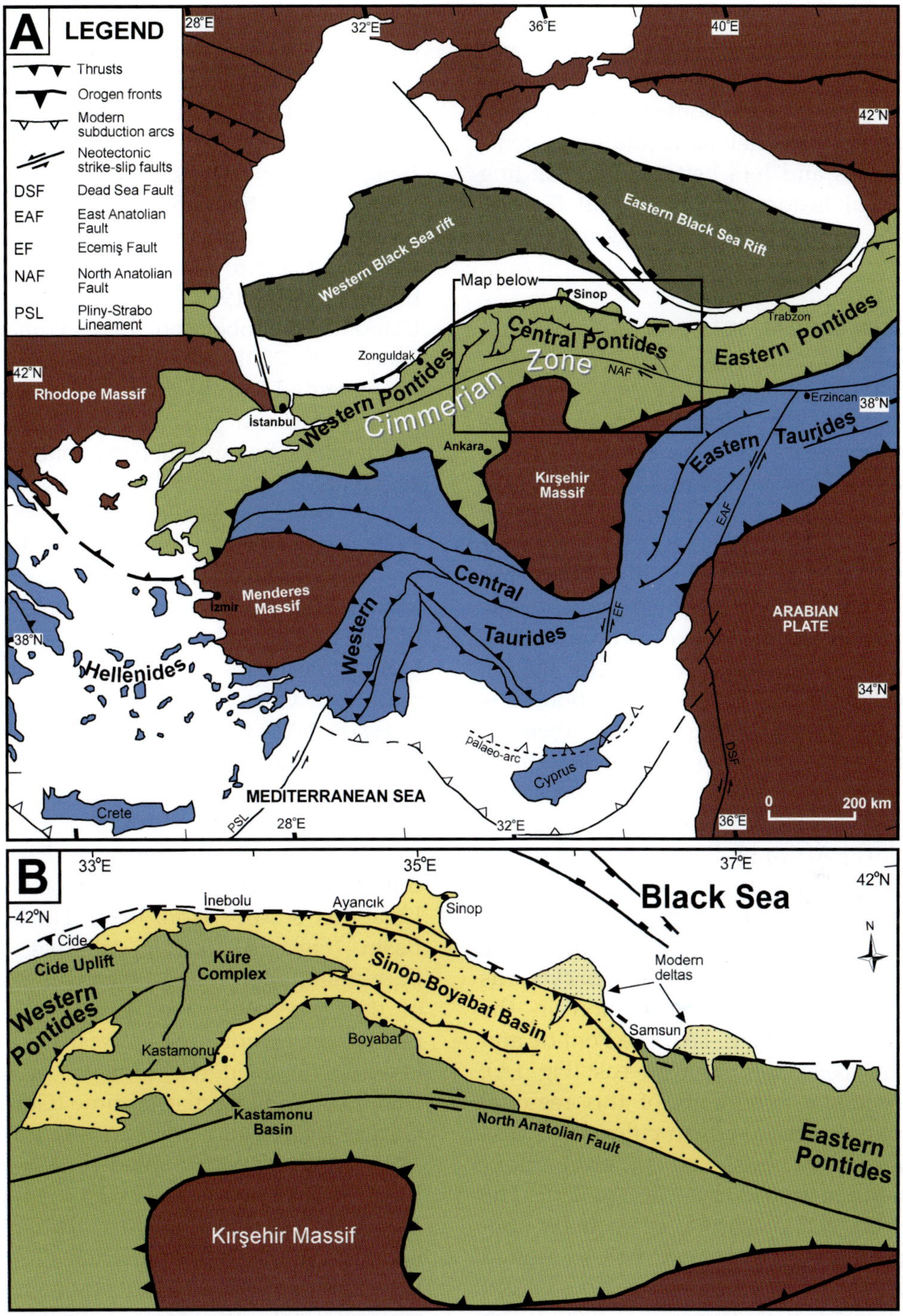

Fig. 1 Regional setting of the study area. (A) Tectonic map of Anatolia and surrounding areas, showing the Pontide and Tauride orogenic belts enveloping the Kırşehir Massif. (B) A simplified map of the Central Pontides, showing the location of the Sinop–Boyabat Basin. Maps compiled with modifications from Robinson *et al.* (1996), Tüysüz (1999), Okay *et al.* (2001) and Nikishin *et al.* (2003).

combined with biostratigraphic data and palaeocurrent directions, allowed the palaeogeography and stratigraphic architecture of the evolving basin to be reconstructed. The study contributes to a general knowledge on non-deltaic turbiditic systems in foreland basins and to a better understanding of the geological history of the Central Pontide orogen and the southern Black Sea region.

REGIONAL GEOLOGICAL SETTING

The compound Anatolian craton (Fig. 1A) was assembled through the Alpine orogeny in the Eastern Mediterranean, by a progressive closure of Neotethyan oceanic branches and suturing of Africa-derived crustal blocks (Şengör, 1987; Dilek & Moores, 1990; Dilek & Rowland, 1993; Okay & Tüysüz, 1999; Görür & Tüysüz, 2001). The earlier suturing of Cimmerian microcontinents to the southern margin of Eurasia in Jurassic time marked the closure of the Palaeotethys ocean in the region, with the development of a new subduction zone south of the Cimmeride orogenic zone (Şengör, 1984). The subsequent northward subduction of Neotethyan crust led to backarc extension and a stepwise accretion of other cratonic blocks, torn away from North Africa by Permo-Triassic rifting and the opening of Neotethys branches.

As the successive microcontinents collided with the accretionary Cimmerian margin of Eurasia, the subduction zone was shifting further backwards until reaching its present-day position in the Cyprian and Cretan arcs (Fig. 1A). The Pontide and Tauride orogenic belts, trending west–east (Fig. 1A), represent two main increments of this regional process of plate accretion, which culminated in a direct collision of Africa's Arabian promontory with the Eurasian margin in mid-Miocene time. The accretion process was diachronous, spatially non-uniform, and the Pontide orogenic belt continued to evolve during the development of the adjacent Tauride belt (Özgül, 1976; Fayon *et al.*, 2001). The Pontide orogeny commenced in Late Cretaceous time and culminated at the end of the Eocene (Okay, 1989; Okay & Şahintürk, 1997; Ustaömer & Robertson, 1997; Yılmaz *et al.*, 1997; Okay & Tüysüz, 1999), whereas the Tauride orogeny began near the end of Cretaceous time and proceeded until the middle Oligocene in the central part of Anatolia (Andrew & Robertson, 2002), and until the late Miocene in its western (Hayward, 1984; Collins & Robertson, 1998, 1999) and eastern part (Michard *et al.*, 1984; Aktaş & Robertson, 1990; Dilek & Moores, 1990; Yılmaz, 1993; Yılmaz *et al.*, 1993; Sunal & Tüysüz, 2002).

The subduction of Neotethys under Eurasia's Cimmerian margin was accompanied by volcanism and backarc rifting from at least Barremian time (Tüysüz, 1990, 1999; Robinson *et al.*, 1996; Ustaömer & Robertson, 1997; Yılmaz *et al.*, 1997; Okay *et al.*, 2001; Nikishin *et al.*, 2003). A volcanic arc extended from Georgia in the east to Bulgaria in the west, and parts of it were probably associated with the subduction zone's intra-oceanic segments, rather than hosted by the continental margin itself (Peccerillo & Taylor, 1975; Eğin *et al.*, 1979; Akıncı, 1984; Tüysüz *et al.*, 1995), although the area of volcanic activity became considerably widened by the backarc rifting and crustal break-up (Baş, 1986; Bektaş & Gedik, 1986). The calcalkaline volcanism in the Central Pontides occurred mainly in Coniacian to middle Campanian time (Tüysüz, 1999; Okay *et al.*, 2001), with minor activity in the late Eocene (Güven, 1977).

The backarc tectonic extension led to the development of the Black Sea rift system along an intra-Cimmerian suture in Early Cretaceous time (Fig. 1A; Okay *et al.*, 1994; Robinson *et al.*, 1996; Ustaömer & Robertson, 1997). The Sinop–Boyabat Basin (Fig. 1B) formed at that time, in the Barremian, as a 'failed (abortive)' southern sister of the 'successful' Western Black Sea Rift. The crustal separation in the Western Black Sea Rift is widely considered to have occurred in late Cenomanian to Coniacian time (Görür *et al.*, 1984; Görür, 1988; Okay *et al.*, 1994; Robinson *et al.*, 1995, 1996; Okay & Şahintürk, 1997; Meredith & Egan, 2002; Rangin *et al.*, 2002; Cloetingh *et al.*, 2003; Nikishin *et al.*, 2003), whereas the timing of crustal break-up in the Eastern Black Sea Rift is more controversial and is thought to have occurred at approximately the same time (Nikishin *et al.*, 2003) or possibly later, in the Maastrichtian (Okay & Şahintürk, 1997), or even Paleocene (Robinson *et al.*, 1995, 1996).

The collisional Pontide orogeny subsequently converted the failed intracontinental rift into a foreland basin, which became progressively deformed by northward thrusting and was eventually inverted by uplift near the end of the Eocene (Janbu *et al.*, 2003; Janbu, 2004).

DYNAMIC STRATIGRAPHY OF THE SINOP–BOYABAT BASIN

The Sinop–Boyabat Basin (Fig. 1B) formed as an extensional graben, 'hanging' structurally between the strongly subsiding Western Black Sea Rift to the north and the Central Pontide accretionary zone to the south. The basin is estimated to have been ~ 80 km wide and at least 200 km long (Janbu, 2004) before becoming subject to orogenic contraction and progressive tectonic inversion. The basin's southeastern end is unpreserved, eroded as a result of the strong uplift of the Eastern Pontides. The northwestern part extends offshore, where it has not been explored, whereas the southern, Boyabat part passes to the southwest into the narrow adjacent Kastamonu Basin (Güven, 1977; Aydın *et al.*, 1986; Koçyiğit, 1986; Şengün *et al.*, 1990).

The stratigraphy of the basin-fill succession is summarized in Fig. 2, based on the present study (Leren *et al.*, 2002; Janbu *et al.*, 2003, this volume, pp. 457–517; Leren, 2003; Janbu, 2004) and previous publications (Ketin & Gümüş, 1963; Gedik & Korkmaz, 1984; Aydın *et al.*, 1986, 1995b; Tüysüz, 1990, 1999; Görür & Tüysüz, 1997). The pre-rift 'bedrock' unit comprises thick platform carbonates, Late Jurassic to Early Cretaceous in age. The onset of rifting was recorded by the Çağlayan Formation of Barremian–Albian age, which consists of calcareous turbidites intercalated with olistostromal breccias and large slide blocks of resedimented bedrock limestones. These deposits are locally up to 2000 m thick and their varied thickness reflects a horst-and-graben submarine topography of the early-stage rift basin. The sediment was derived from both margins of the rift, with the turbidity currents filling in the basin-floor relief and flowing mainly westwards along the basin axis. The sediment supply to the basin declined in Turonian to earliest Coniacian time, when the Kapanboğazı Formation was deposited in a sand-starved deep-water environment. This formation is ≤ 40 m thick and consists of variegated, reddish-grey mudstones intercalated with pelagic marlstone layers. The cessation of sediment supply was probably due to a post-rift phase of broader thermal subsidence

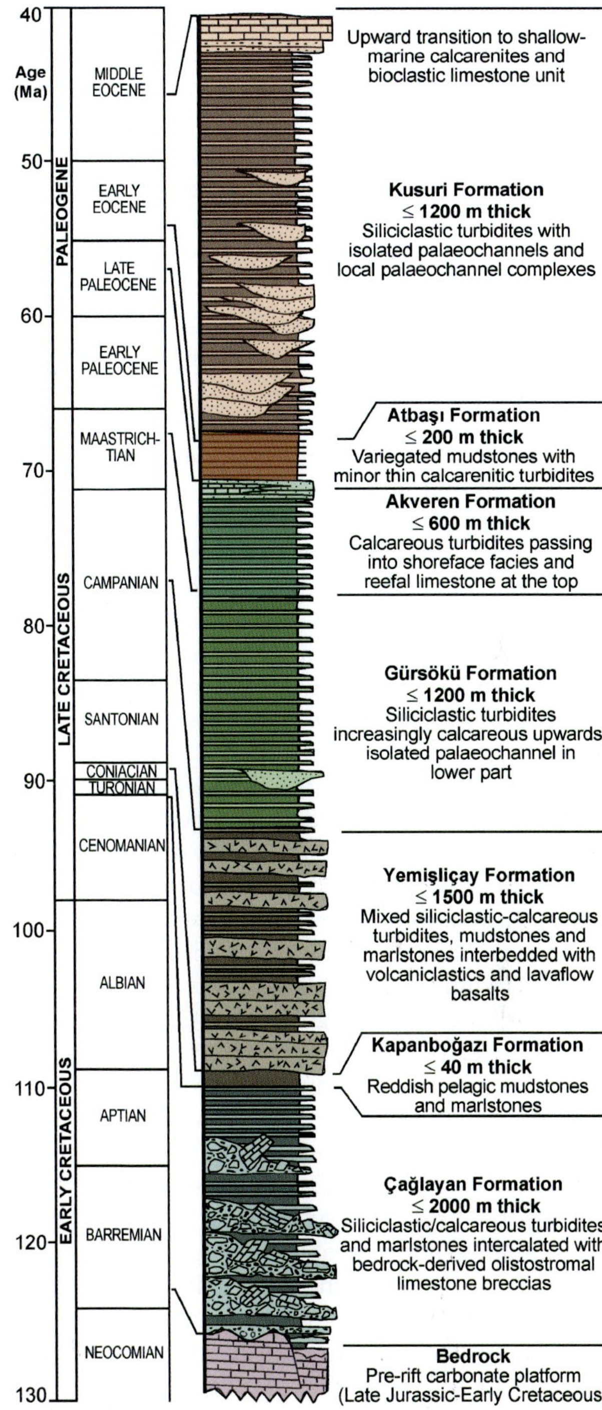

Fig. 2 Stratigraphy of the Sinop–Boyabat Basin (modified from Ketin & Gümüş, 1963; Gedik & Korkmaz, 1984; Aydın *et al.*, 1982, 1995b). The youngest Eocene formation in the profile pertains to the northern half of the basin (the Sinop trough in Fig. 3), but a coeval succession of turbidites and shallow-marine to fluvio-deltaic deposits occurs in the adjacent Boyabat trough.

that caused the contemporaneous shorelines to shift landwards, away from the rift margins.

Another strong rifting pulse was recorded by the overlying Yemişliçay Formation (Coniacian–Campanian), which is ≤ 1500 m thick and consists of turbidites with a mixed calcareous–siliciclastic composition, interbedded with abundant volcaniclastic deposits and lavaflow basalts. The sediment was still derived from both rift margins, but the northern margin soon became submerged below wave base and remained practically inactive as a clastic source (Aydın *et al.*, 1995b; Tüysüz, 1999). This asymmetrical development of the basin is attributed to crustal break-up and margin fault-block collapse in the adjacent Western Black Sea Rift (Janbu, 2004).

The aborted Sinop–Boyabat rift then became progressively affected by orogenic compression from the south, which converted it into a retroarc foreland basin of the growing Central Pontides and eventually inverted it by structural closure (Tüysüz, 1999; Janbu *et al.*, 2003; Janbu, 2004). The present study indicates that the compressional tectonic deformation in the Sinop–Boyabat Basin commenced in Late Cretaceous time, concurrently with the deposition of the Gürsökü Formation (Campanian–Maastrichtian). This turbiditic succession, ≤ 1200 m thick, consists of mixed siliciclastic–calcareous sediment that was supplied mainly from the west/southwest and spread eastwards along the basin axis (Leren *et al.*, 2002; Leren, 2003).

The easterly sediment transport and cessation of volcanism are attributed to the collision of the Kırşehir Massif with the Cimmerian margin (Fig. 1A), which commenced in the transition area of the Western/Central Pontides in Late Cretaceous time (Tüysüz *et al.*, 1995; Okay & Tüysüz, 1999) and probably caused the subduction zone to roll back to the rear side of this large 'indentor' block. As the accreted massif was pushed further to the north and underwent counter-clockwise rotation (Sanver & Ponat, 1981; Görür *et al.*, 1984; Kaymakcı *et al.*, 2003), the Central Pontide nappes began to be emplaced northwards and to affect the foreland basin, with a carbonate platform developing along the basin's southwestern margin and supplying abundant sediment. The Maastrichtian–Paleocene Akveren Formation (Fig. 2) is a succession of calcareous turbidites, ≤ 600 m thick, with mainly eastward palaeocurrents and evidence of rapid shallowing in the uppermost part.

The overlying Atbaşı Formation (upper Paleocene to lowest Eocene) consists of deep-water variegated mudstones, ~ 200 m thick, intercalated with thin calcareous turbidites. The rapid deepening of water and sand-starved basin conditions are attributed to a broad subsidence of the foreland due to crustal loading by nappes (Nikishin *et al.*, 2003), which coincided with the Thanetian eustatic sea-level rise (Haq *et al.*, 1988). The present paper focuses on the Campanian to earliest Eocene sedimentation history of the Sinop–Boyabat Basin, when its transformation from a failed rift into the Central Pontide foreland basin occurred.

The younger basin-fill deposits recorded the late Paleocene culmination of orogeny in the Eastern Pontides and further contraction of the Central Pontide foreland, with reversal of pre-existing normal faults and active thin-skinned thrust tectonics (Aydın *et al.*, 1995b). In the early Eocene, the Erikli thrust coupled with the antithetic Ekinveren back-thrust (Fig. 3) to form a structural pop-up ridge that split the Sinop–Boyabat Basin longitudinally into two subparallel troughs (Janbu *et al.*, 2003; Janbu, 2004): a northern foredeep trough referred to as the Sinop Basin and a southern wedge-top ('piggyback') trough referred to as the Boyabat Basin.

The lower–middle Eocene Kusuri Formation in the Sinop Basin (Fig. 2) is a siliciclastic turbiditic succession, ≤ 1200 m thick, which recorded an abundant sediment supply from the east, through a fluvio-deltaic system draining the adjacent, emerged Eastern Pontide foreland (Janbu *et al.*, this volume, pp. 457–517). The turbidites are increasingly calcareous in the formation's shale-rich uppermost part, where they give way to tempestites with a rapid upward transition into littoral bioclastic limestones. This part of the formation also shows structural evidence of the basin's progressive closure between the Erikli thrust and the younger Balıfakı thrust to the north (Fig. 3; Janbu, 2004). The coeval siliciclastic succession in the Boyabat Basin is ~ 900 m thick and shows an upward transition from turbiditic to shallow-marine, deltaic and fluvial sedimentation, with the sediment supplied similarly from the east. As the structural contraction continued until the final stages of the Tauride orogeny (Özgül, 1976; Okay & Tüysüz, 1999), the two basins were gradually inverted by tectonic

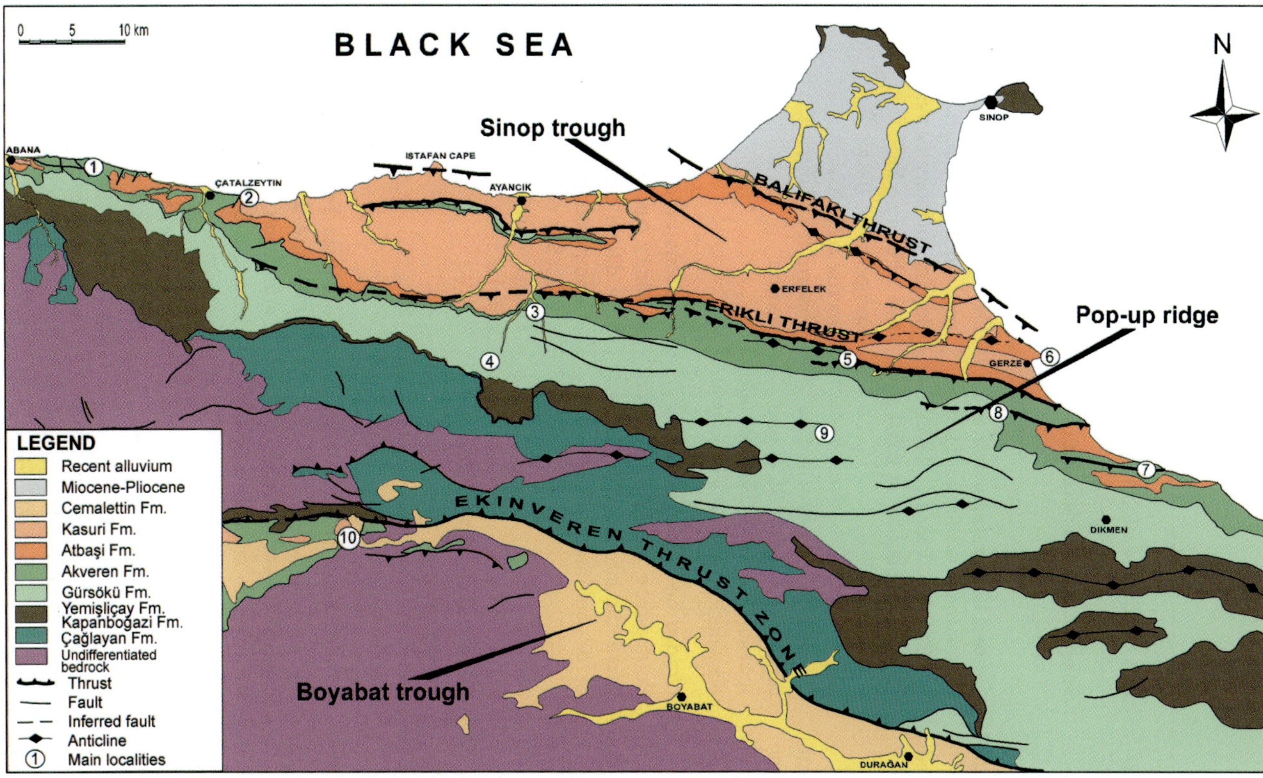

Fig. 3 Geological map of the Sinop–Boyabat Basin, showing the areal distribution of the basin-fill formations. Modified from Gedik & Korkmaz (1984), Barka *et al.* (1985) and Aydın *et al.* (1995b). Note that the northward Erikli thrust and the southward Ekinveren back-thrust turned the whole axial part of the basin into a pop-up ridge, which effectively split the original basin into a narrow northern trough (Sinop Basin) and a similarly narrow southern trough (Boyabat Basin) in early Eocene time. The younger Balıfakı thrust to the north, together with the Erikli thrust, resulted in subsequent tectonic closure and uplift of the Sinop Basin.

uplift in late Eocene to early Miocene times (Okay & Şahintürk, 1997). Paratethyan shallow-marine deposits of Miocene age occur in the Sinop peninsula area (Fig. 3), outside the Sinop Basin, where they overlie a major unconformity and are dominated by tidal calcarenites and bioclastic limestones (Görür *et al.*, 2000).

The late Miocene also witnessed the onset of neotectonics, with a westward 'tectonic escape' (strike-slip expulsion) of the Anatolian craton bounded by the sinistral East Anatolian Fault and the dextral North Anatolian Fault (Fig. 1A; Şengör *et al.*, 1985; Flerit *et al.*, 2004).

SEDIMENTARY FACIES ASSOCIATIONS

The Campanian to lowest Eocene sedimentary succession has been studied and sampled in outcrop sections all over the basin, and its various parts have been logged in detail at 22 localities. The logs have a cumulative stratigraphic thickness of ~ 840 m and include five logs from the Gürsökü Formation, 13 logs from the Akveren Formation and four logs from the Atbaşı Formation and its transitional basal part (Fig. 2). Only selected portions of a few representative logs are shown in the present paper. The descriptive sedimentological terminology used is mainly after Harms *et al.* (1975) and Collinson & Thompson (1982). The division of a neritic to littoral environment into offshore, offshore-transition, shoreface and foreshore zones is according to Reading & Collinson (1996, fig. 6.6).

The succession studied consists of a wide range of sedimentary facies (Table 1 and Figs 4 & 5), including calcareous mudstones and marlstones, several varieties of sandstone, subordinate gravelstones and a prominent bioclastic limestone unit.

Table 1 Sedimentary facies of the Campanian–Ypresian succession in the Sinop–Boyabat Basin (see also Figs 4 and 5)

Facies		Subfacies		Bed geometry and thickness	Internal bed characteristics	Interpretation
A	Thick- to medium-bedded sandstones with planar parallel stratification and/or current-ripple cross-lamination	A1	Very coarse- to very fine-grained siliciclastic sandstones with calcareous admixture	Sheet-like beds ≤ 75 cm thick, some thinning laterally; wedge-shaped and ≤ 145 cm thick in palaeochannel only	Graded beds categorized as Bouma-type turbidites T*bcd* (~ 88%) and T*abcd* (~ 12%), some with pebbles in basal part	Deposits of low-density and subordinate high-density turbidity currents; the difference in composition (facies A1 versus A2) reflects a change in sediment source
		A2	Fine- to very fine-grained and silt-rich calcarenites	Sheet-like beds ≤ 78 cm thick; some thinning or pinching out laterally	Graded beds categorized as turbidites Tbcd (~ 95%) and sporadic T*abcd* (~ 5%); common diagenetic chert concretions	
B	Thin-bedded sandstones and siltstones with current-ripple cross-lamination	B1	Coarse- to very fine-grained siliciclastic sandstones and siltstones with calcareous admixture	Sheet-like tabular to uneven beds ≤ 10 cm thick, some thinning or pinching out laterally	Graded beds categorized as Bouma-type turbidites T*cd* and T(*c*)*d*, the latter with only local or discontinuous division *c*	Deposits of low-density turbidity currents, some very dilute; the difference in composition (facies B1 versus B2) reflects a change in sediment source
		B2	Very fine-grained to silty calcarenites	Sheet-like tabular to uneven beds ≤ 10 cm thick, some thinning or pinching out laterally	Graded beds categorized as Bouma-type turbidites T*cd* and T(*c*)*d*; common chert concretions	
C	Massive and top-stratified, very coarse- to very fine-grained calcarenites			Sheet-like or lenticular, scour-confined beds ≤ 44 cm thick, commonly amalgamated	Graded beds with scattered carbonate pebbles, multiple internal scours and parallel-stratified to cross-laminated top	Turbidites T*ab* and composite beds T*a*(*b*)*a*(*b*)*ab* . . . *abc* deposited by pulsating high-density currents
D	Stratified calcarenites, very coarse- to very fine-grained, with wave-formed and combined-flow structures			Tabular to wedge-shaped beds ≤ 80 cm thick, isolated or amalgamated, with erosional bases and often undulatory tops	Various types of wave-ripple cross-lamination (with two- and three-dimensional vortex ripples), planar parallel stratification and hummocky or swaley cross-stratification	Wave-worked shoreface deposits (amalgamated) and offshore-transition tempestites (isolated)

E	Clast-supported gravelstones			Beds 15–70 cm thick, lenticular and isolated or wedge-shaped, stacked upon one another and gently inclined (< 10°)	Submature to mature, granule to cobble gravel made of limestone, marlstone and sporadic volcanic clasts, locally rich in granule-armoured mudclasts; sand-filled framework, massive or parallel stratified	Beachface (foreshore) deposits; patchy transgressive lag; and beach-derived gravelly component of isolated tempestites and possible tsunamites
F	Bioclastic limestones	F1	Bedded limestones	Uneven beds ≤ 50 cm thick, commonly lenticular, separated by thin marlstone layers or amalgamated	Medium- to coarse-grained bioclastic grainstones, internally massive, with a wide range of bioclasts, including bryozoan corals and coralline red algae	Shallow-marine reefal platform with a varying degree of sediment reworking by waves and tidal currents (facies F1 versus F2)
		F2	Non-bedded limestone	Homogeneous units 2–10 m thick		
G	Marlstones			Sheet-like and mainly tabular beds ≤ 130 cm thick, whitish-grey in colour, some with a slightly undulatory base and/or top	Massive, some with a silty, faintly parallel-laminated and/or cross-laminated basal part (turbidites T*de* and T*cde*), commonly grading upwards into grey mudstone	Fallout of 'background' pelagic suspension; hemipelagic cappings of calcarenitic turbidites and tempestites; also thin deposits of highly dilute turbidity currents or storm-generated suspension surges
H	Calcareous mudstones	H1	Grey to greenish-grey mudstones	Sheet-like layers ≤ 44 cm thick, capping sandstone, siltstone or marlstone beds	Massive, bioturbated and mainly grey, occasionally whitish to greenish, pinkish or blackish, with sporadic coaly plant detritus	Hemipelagic 'background' deposits; facies H2 indicates seafloor oxidation in sediment-starved conditions
		H2	Variegated mudstones	Monotonous unit ≤ 40 m thick, interspersed with thin isolated calcarenite sheets of facies B2	Massive, bioturbated, with irregular bands (≤ 95 cm thick) of olive green to brownish- or purple-red colouration	

Fig. 4 Sedimentary facies of the Gürsökü and Akveren formations. Facies are defined in Table 1. (A) Siliciclastic turbidites of facies A1 and B1, capped with mudstones of facies H1. (B) Calcarenitic turbidites of facies A2 and B2, capped with marlstones of facies G and/or mudstones of facies H1. (C) Turbidite T*bcd* of facies B1, with a thin planar-stratified division overlain by climbing-ripple cross-lamination and silty marlstone capping. (D) Turbidite T*bcd* of facies B2, separated by marlstone with sand-filled burrows. (E) Turbidite T*bcd* of facies A1. (F) Flute casts on the sole of facies A2 turbidite. (G) Turbidite T*bcd* of facies A2, capped with a whitish-grey marlstone grading upwards into grey calcareous mudstone. (H) Erosional packages of amalgamated turbidites of facies C in chutes within a succession dominated by sheet-like turbidites of facies C, G and H1. Photograph (A) is from facies subassociation 1b; (B, C, E, F and G) are from subassociation 2c; (D) is from subassociation 1a; and (H) from subassociation 2a.

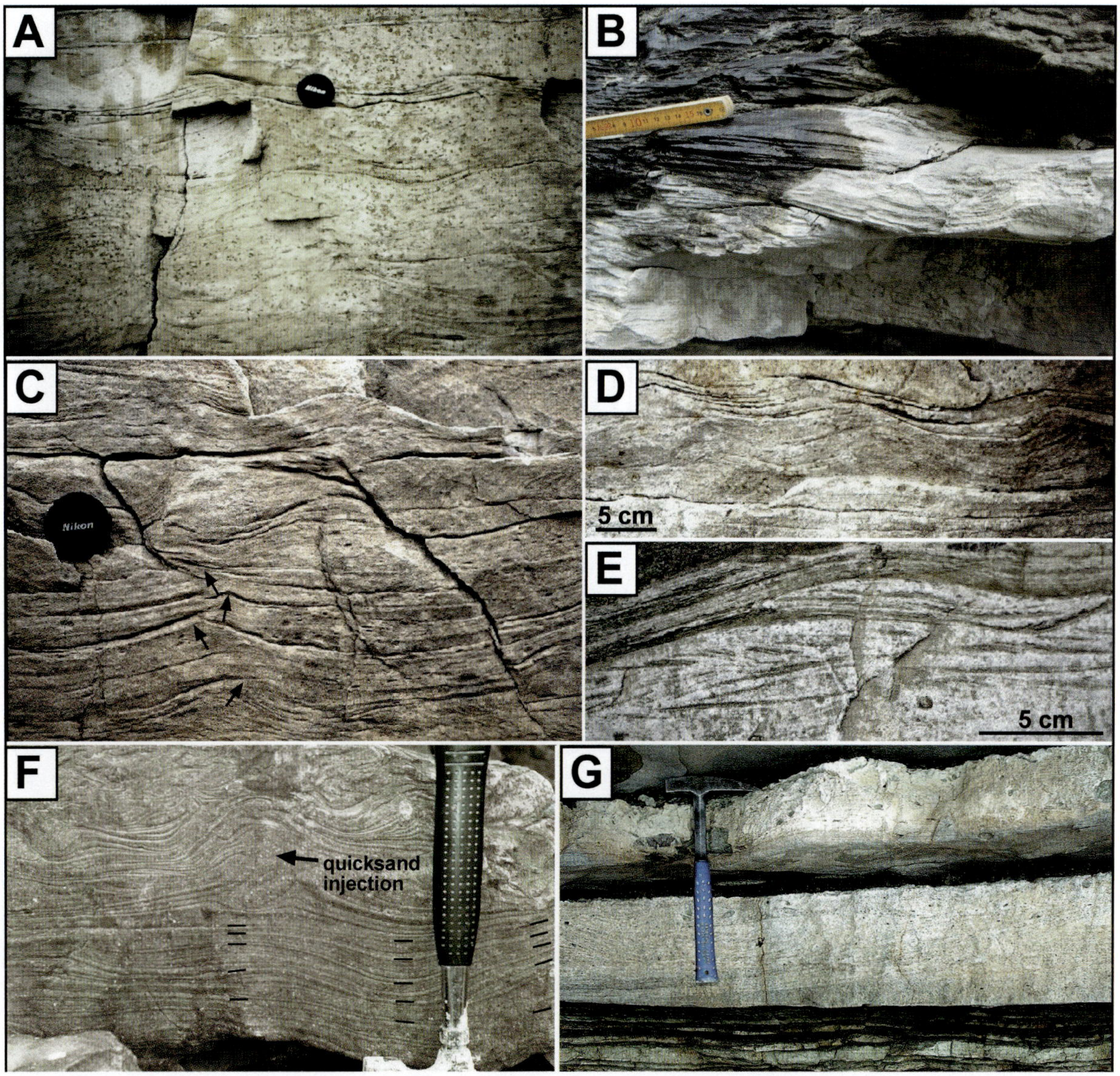

Fig. 5 Sedimentary facies of the uppermost Akveren and Atbaşı formations. Facies are defined in Table 1. (A) Calcarenite of facies D with hummocky stratification passing upwards into cross-lamination representing symmetrical to asymmetrical three-dimensional vortex ripples, covered by planar parallel strata. (B) Hummocky stratification in calcarenite of facies D. (C) Cross-laminated calcarenite of facies D showing reversing-crest two-dimensional ripples (arrows) passing upwards into three-dimensional vortex ripples. (D) Calcarenite of facies D showing planar parallel stratification overlain by cross-lamination representing slightly asymmetrical, highly aggradational (climbing) two-dimensional ripples. (E) Calcarenite of facies D showing wave-ripple cross-lamination with bundled upbuilding, draped by three-dimensional vortex ripple laminae. (F) Calcarenitic tempestite of facies D with a slightly undulatory base, parallel stratification overlain by 'micro-hummocky' lamination (three-dimensional vortex ripples) and a convoluted upper part with liquefaction features. (G) Thin to moderately thick tempestites of facies D, with sharp bases and tops, separated by calcareous mudstone layers of facies H1; the thin calcarenites are mainly parallel-stratified and/or cross-laminated, commonly with rippled tops, whereas the tabular thicker ones contain mudclasts and show also swaley and/or hummocky stratification.

Fig. 5 (*cont'd*) Sedimentary facies of the uppermost Akveren and Atbaşı formations. Facies are defined in Table 1. (H) Calcarenite of facies D showing planar parallel stratification overlain by hummocky and swaley stratification. (I) Calcarenite of facies D showing planar parallel and swaley stratification. (J) Detail from facies E, showing a coarse gravelstone bed composed of rounded, flat-lying marlstone and limestone cobbles mixed with granule-armoured mudstone pebbles and cobbles (weathering makes the mudclasts look like matrix); the bed is underlain and overlain by granule gravelstones, the upper one rich in small pebbles. (K) Facies E showing disconformable packages of gently inclined, parallel-stratified bed sets composed of granule gravel and cobble-bearing pebble gravel. (L) Fractured, alternating limestones of facies F1 and F2, underlain by parallel-stratified calcarenites of facies D. (M) Marlstones of facies G, forming graded beds with silty basal parts and cappings of facies H1 mudstone. (N) Mudstone of facies H2, with a vertical *Ophiomorpha annulata* burrow (arrow) and weak bedding marked by silty interlayers. Photographs (A–E) are from facies subassociation 3b; (F & G) from subassociation 3c; (H & I) from subassociation 4a; (J & K) from subassociation 3b; (L) is from subassociation 3a; (M) from subassociation 3c; and (N) from subassociation 4c.

The component facies and subfacies, distinguished on the basis of descriptive sedimentological criteria, are considered to be the basic 'building blocks' of the sedimentary succession (Harms *et al.*, 1975; Walker, 1984a). They are indicated in the outcrop logs and have been the basis for an interpretation of the depositional processes involved.

Based on the spatial grouping and stratigraphic distribution of sedimentary facies, four major facies assemblages, or associations, have been recognized. These have been divided further into two or more subassociations. These 'building megablocks' are described and interpreted in the present section, with reference to depositional processes inferred from their component facies (Table 1) along with micropalaeontological (Table 2) and ichnological data (Leren, 2003; Uchman *et al.*, 2004). The facies associations are discussed in their

Table 2 A summary list of planktonic (P) and small benthic foraminifers (B) and nanoplankton species (N) found in the Gürsökü (GÜ, 2 samples), Akveren (AK, 8 samples) and Atbaşı formations (AT, 2 samples)

Microfossil taxa	Type	Formation		
		GÜ	AK	AT
Ahmuellerella octaradiata (Gorka)	N		x	
Archaeglobigerina sp.	P		x	
Arkhangelskiella cymbiformis Vekshina	N		x	
Bathysiphon sp.	B		x	
B. vitta Nauss	B		x	
Biscutum constans (Gorka)	N		x	
Blackites creber (Deflandre)	N			x
Braarudosphaera bigelowii (Gran & Braarud)	N		x	
Ceratolithoides kamptneri Bramlette & Martini	N		x	
Chiasmolithus grandis (Bramlette & Riedel)	N			x
Chiastozygus amphipons (Bramlette & Martini)	N		x	
C. danicus (Brotzen)	N		x	
Coccolithus eopelagicus (Bramlette & Riedel)	N		x	
Coccolithus pelagicus (Wallich)	N		x	
Coronocyclus prionion (Deflandre & Fert)	N		x	
Cribrosphaerella ehrenbergii (Arkhangelsky)	N		x	
Cyclagelosphaera deflandrei (Manivit)	N		x	
Cylindiralithus cf. *oweinae* Perch-Nielsen	N		x	
C. serratus Bramlette & Martini	N		x	
Discoaster binodosus Martini	N			x
Eiffellithus parallelus Perch-Nielsen	N		x	
E. turriseiffelii (Deflandre)	N		x	
Ellipsolithus bollii Perch-Nielsen	N		x	
Ericsonia cava (Hay & Mohler)	N		x	x
E. formosa (Kamptner)	N			x
E. ovalis Black	N		x	x
E. subpertusa Hay & Mohler	N		x	
Fasciculithus janii Perch-Nielsen	N		x	
F. tympaniformis Hay & Mohler	N		x	
Globigerina sp.	P			x
G. inaequispira Subbotina	P			x
Globotruncana sp.	P	x	x	
G. cf. *arca* (Cushman)	P	x	x	
G. cf. *lapparenti* Brotzen	P	x		

Table 2 *(cont'd)*

Microfossil taxa	Type	Formation		
		GÜ	AK	AT
Globotruncanita conica (White)	P		x	
G. stuarti (de Lapparent)	P		x	
Helicosphaera seminulum Bramlette & Sullivan	N			x
Kathina? sp.	B		x	
Lagenidae	B		x	
Lithraphidites carniolensis Deflandre	N		x	
L. quadratus Bramlette & Martini	N		x	
Marginotruncana cf. *pseudolinneiana* Pessagno	P	x		
Markalius apertus Perch-Nielsen	N		x	
Micrantholithus sp.	N		x	
Microrhabdulus attenuatus (Deflandre)	N		x	
M. undosus Perch-Nielsen	N		x	
Micula concava (Stradner)	N		x	
M. decussata Vekshina	N		x	
M. staurophora (Gardet)	N		x	
Miscellanea ?primitiva Rahaghi	B		x	
Morozovella sp.	P			x
M. cf. *aragonensis* (Nuttall)	P			x
Neochiastozygus chiastus (Bramlette & Sullivan)	N		x	
N. concinnus (Martini)	N		x	
N. perfectus Perch-Nielsen	N		x	
Neococcolithes protenus (Bramlette & Sullivan)	N		x	
Nodosaria latejugata Gumbel	B		x	
Parhabdolithus embergeri (Noël)	N		x	
Placozygus fubiliformis (Reinhardt)	N		x	
Pontosphaera sp.	N		x	
P. plana (Bramlette & Sullivan)	N			x
Prediscosphaera cretacea (Arkhangelsky)	N		x	
P. spinosa (Bramlette & Martini)	N		x	
Prinsius bisulcus (Stradner)	N		x	
P. dimorphosus (Perch-Nielsen)	N		x	
Pseudocuvillierina sireli (Inan)	B		x	
Pseudotextularia sp.	B		x	
Rhabdolekiskus parallelus Wind & Ćepek	N		x	
Rhagodiscus splendens (Deflandre)	N		x	
Rhomboaster cuspis Bramlette & Sullivan	N		x	
Sphenolithus anarrhopus Bukry & Bramlette	N		x	
S. editus Perch-Nielsen	N			x
S. moriformis (Brönnimann & Stradner)	N			x
S. primus Perch-Nielsen	N		x	
S. radians Deflandre	N			x
Stradneria crenulata (Bramlette & Martini)	N		x	
Thoracosphaera saxea Stradner	N		x	
Tranolithus exiguus Stover	N		x	
Vagalapilla matalosa (Stover)	N		x	
Watznaueria barnesae (Black)	N		x	
Zygodiscus bramlettei Perch-Nielsen	N		x	

stratigraphic order and considered to represent different sedimentary environments, whose physical nature and spatial relationships provide information on the basin's palaeogeographical evolution (discussed in a subsequent section).

Facies association I: siliciclastic turbidites

This facies association constitutes the Gürsökü Formation (Fig. 2) and consists of the predominantly siliciclastic turbidites of facies A1 and B1 (Table 1), which are normally graded and capped with calcareous mudstone (facies H1) and/or marlstone layers (facies G). The formation as a whole shows a gradual upward increase in the content of calcareous bioclasts and a slight overall fining of the sand fraction (Leren, 2003). Its contact with the underlying Yemişliçay Formation (Fig. 2) is conformable and gradational, as is also the upward transition to the calcareous Akveren Formation.

The turbidites are moderately sorted, carbonate-cemented litharenites (Fig. 6) composed of angular to subrounded quartz grains (7–50 vol.%); various rock fragments (30–90 vol.%), including volcanic glass and finely crystalline igneous rocks; plagioclase and microcline grains (3–5 vol.%); sporadic mica flakes and common dark/opaque mineral grains (≥ 5 vol.%) of mainly volcanic-rock provenance (Leren, 2003). The sediment contains also an admixture of calcareous bioclasts (foraminifers and fragments of bryozoans, brachiopods and echinoderms), which are quantitatively negligible in the lower part, but increasingly more abundant upwards in the formation. Foraminifers in the mudstone layers indicate a deep-water environment, and the deposits bear a typical *Nereites* ichnofacies, with trace fossils including *Chondrites targionii*, *C. intricatus*, *Ophiomorpha rudis*, *Trichichnus linearis*, *Arthrophycus tenius*, *Thalassinoides*, *Nereites*, *Ubina* and *Palaeodictyon majus*. The siliciclastic turbidites form

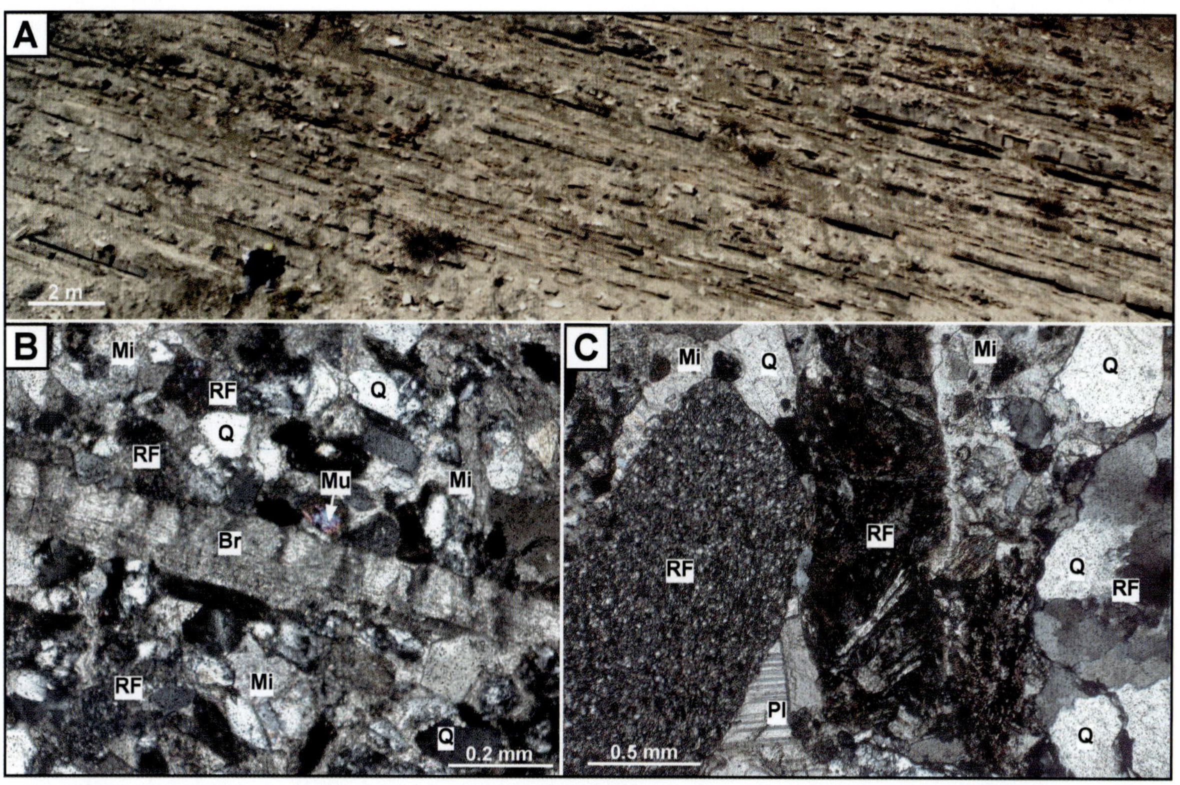

Fig. 6 Gürsökü Formation turbidites. (A) Outcrop of the middle part of the formation in the central part of the basin, showing the mudstone-capped siliciclastic turbidites of facies subassociation 1a; roadcut section ~ 0.4 km south of Çakıldak, in the area indicated as locality 9 in Fig. 3. (B) Photomicrograph (thin-section XPL view) of typical fine-grained turbiditic sandstone of facies A1 in the lower–middle part of Gürsökü Formation; letter symbols: Br, brachiopod fragment; Mi, micritic cement; Mu, muscovite; Q, quartz; RF, volcanic rock fragment. (C) Photomicrograph (thin-section XPL view) of the granule-bearing, very coarse-grained basal part of a turbiditic sandstone of facies A1 in the same part of the Gürsökü Formation; letter symbols as above and Pl, plagioclase.

two distinct subassociations, differing markedly in their depositional architecture: (i) the main assemblage of sheet-like, mudstone- and/or marlstone-capped turbidites that constitute the bulk of the Gürsökü Formation; and (ii) a minor assemblage of amalgamated turbidites that form the infill of an isolated palaeochannel in the lower part of the formation. Their detailed characteristics are summarized and interpreted below (with reference to facies in Table 1).

Facies subassociation 1a

These sheet-like, Bouma-type turbidites (Figs 7 & 8) range from the fine-grained beds T*cd* and T(*c*)*d* of facies B1, mainly < 10 cm thick and with a laterally continuous or discontinuous cross-laminated *c*-division, to the coarser-grained beds T*bcd* and T*abcd* of facies A1, mainly 10–50 cm thick, sporadically including also a trough cross-stratified division. The beds are capped with calcareous mudstone (facies H1) and/or marlstone layers (facies G). Some beds pinch out laterally, as is more commonly observed in the lower part of the Gürsökü Formation. Palaeocurrent measurements indicate predominantly eastward sediment dispersal with subordinate northward directions (see the rose diagram in Fig. 7). The lower portion of facies subassociation 1a, notably in the western to central part of the basin, abounds in relatively thick turbidite beds (facies A1) alternating with thin ones (facies B1). The corresponding value of the bed-thickness coefficient of variation (CV) is 1.53, indicating a bimodal or polymodal bed population with a clustering tendency of thinner and thicker beds (Ball *et al.*, 1997). This pattern contrasts with that in the middle to upper portion of the formation, where facies B1 predominates and the coefficients of variation CV ≈ 1 indicate a disorderly bed-thickness succession (see plots 1 and 2 in Fig. 8). The CV values are all > 0.5, indicating a strongly skewed and apparently non-normal bed-thickness frequency distribution.

Facies subassociation 1b

These deposits (Figs 9A & 10) occur as an isolated succession, ~ 22 m thick, in the lower part of the Gürsökü Formation, near the village of Yenikonak in the central part of the basin (locality 3 in Fig. 3). The assemblage comprises turbidite beds of facies A1, ≤ 145 cm thick, with highly uneven erosional bases and mainly amalgamated, intercalated with minor packages of the thin, mudstone- or marlstone-capped turbidites of facies B1 (Fig. 10). A heterolithic package of alternating facies B1 and H1 beds occurs at the top.

The depositional architecture consists of two superimposed bed sets, the lower one inclined gently to the southeast and the upper to the northwest, separated by a gravel-bearing zone with abundant irregular scours (Figs 9B & 10). The inclined bed sets indicate lateral accretion (LAPs *sensu* Abreu *et al.*, 2003), and their superposition resembles the stacking of point bars accreted to the opposite banks of a sinuous submarine channel (Deptuck *et al.*, 2003; Janbu *et al.*, this volume, pp. 457–517). Flute marks and trough-shaped scours indicate local palaeocurrent azimuths ranging between the northeast and southeast, which is consistent with the notion of a sinuous thalweg and northeasterly palaeochannel trend. This localized turbidite succession differs markedly from the surrounding sheet-like deposits of subassociation 1a and is considered to be a palaeochannel, ~ 22 m thick and much wider than the outcrop limits. Channel margins are not exposed, but the lateral-accretion architecture itself provides compelling evidence of channel-fill deposits. The covering package of the alternating thin beds of facies B1 and H1 (Fig. 10) indicates channel abandonment.

Interpretation

The deposition of facies subassociation 1a is attributed to non-channelized turbidity currents of high to low density (*sensu* Lowe, 1982), derived from a siliciclastic source area dominated by volcanic rocks, but with an increasing contribution of calciclastic sediment. As described in the next section, the calcareous component was derived from a contemporaneous, shallow-water bioclastic source. Palaeocurrent directions indicate sediment supply from the basin's western part and southwestern margin, with a predominantly eastward dispersal along the basin axis. The succession of sheet-like, non-amalgamated turbidites with local evidence of lateral thinning or pinch-out of beds suggests a basin-floor turbiditic system, possibly smoothing out subtle seafloor depressions (cf.

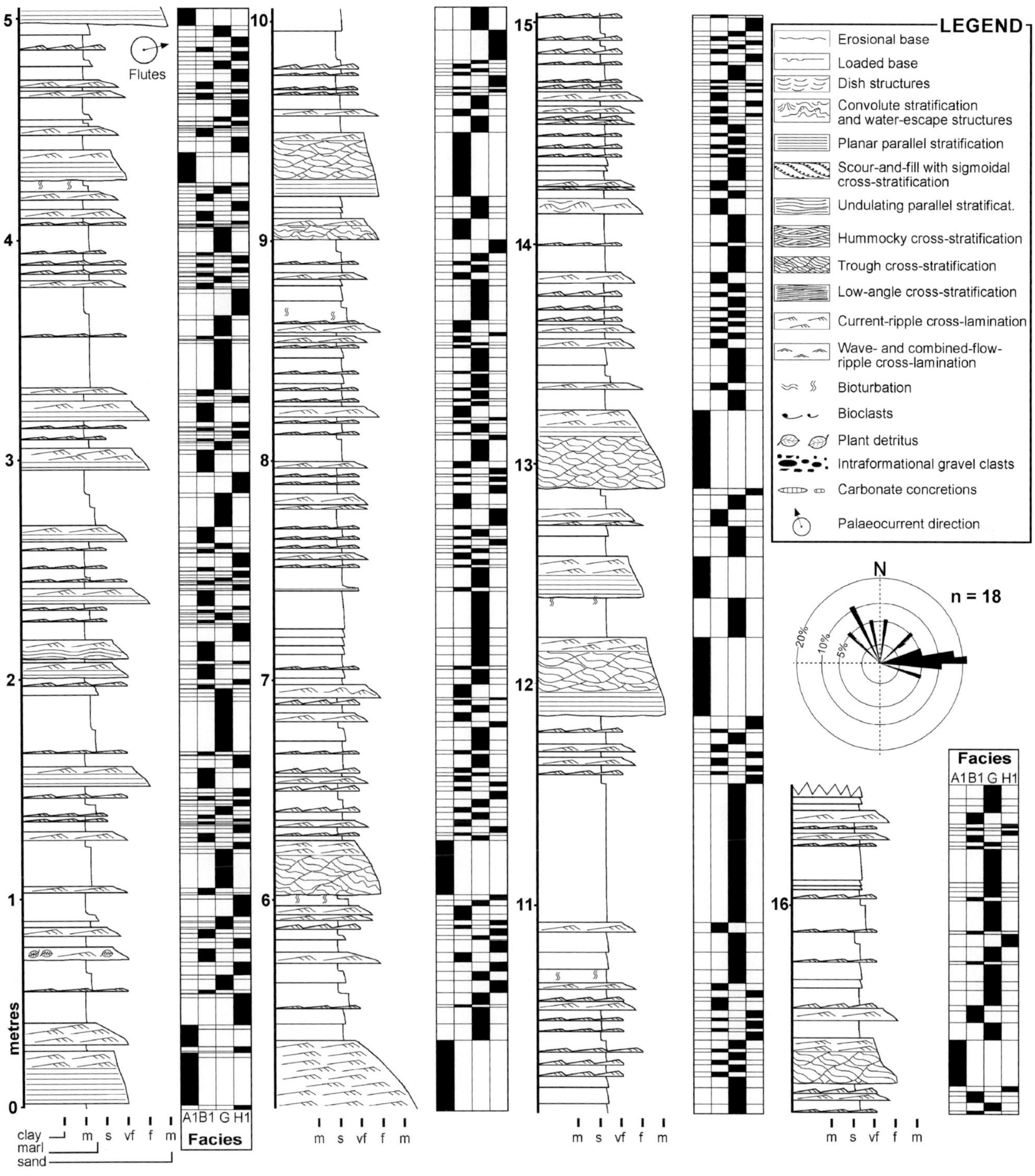

Fig. 7 Example log from the middle part of Gürsökü Formation in the east-central part of the basin, showing facies subassociation 1a. Roadcut section ~ 0.3 km north of Çakıldak, in the area indicated as locality 9 in Fig. 3. Facies symbols are as in Table 1. The rose diagram summarizes palaeocurrent data from subassociation 1a, based mainly on flutes (n = number of data). The legend (inset) pertains to all sedimentary logs in this paper.

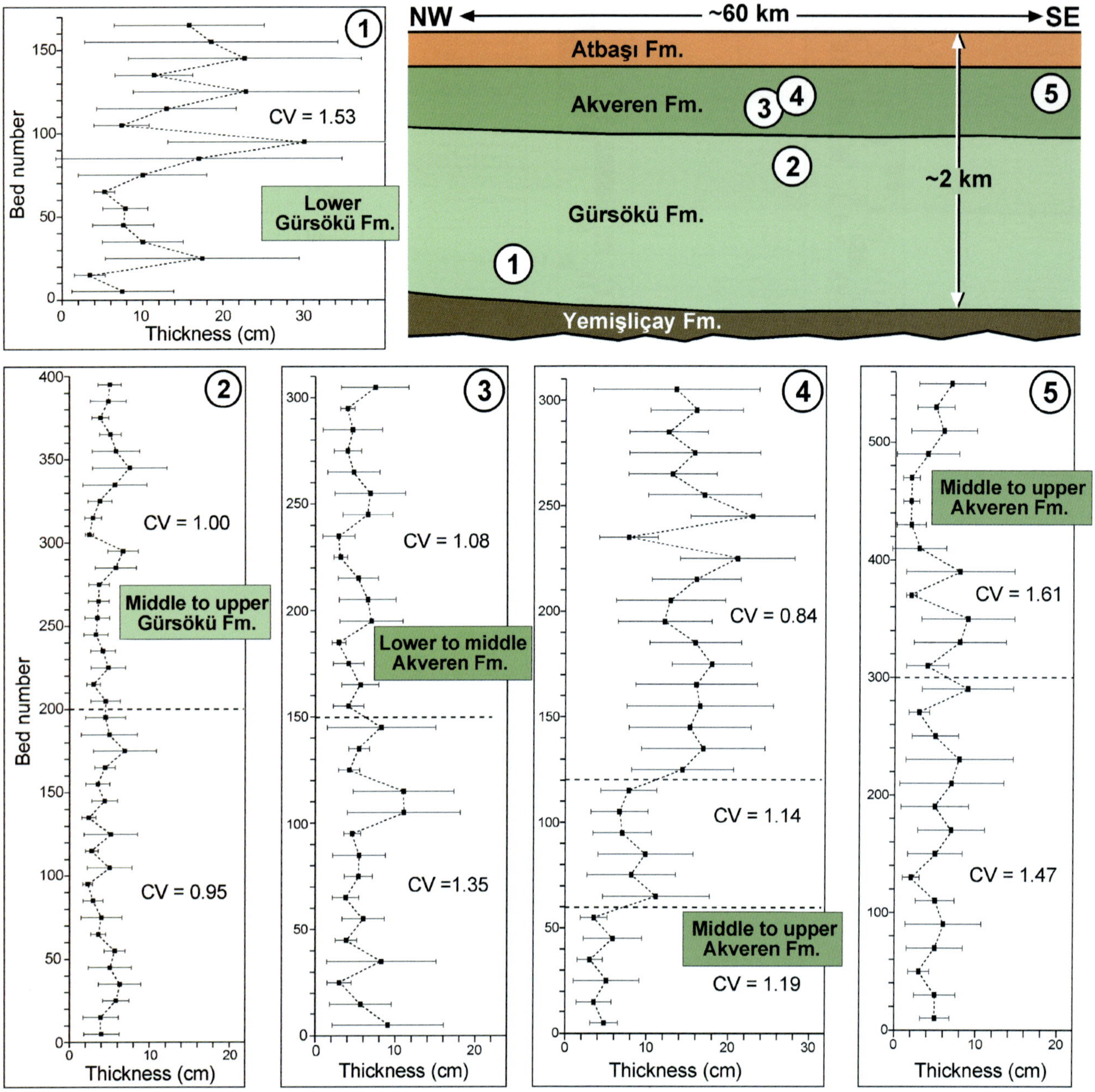

Fig. 8 Statistical summary of selected logs representing different portions of the Gürsökü and Akveren formations (see inset stratigraphic cross-section). The plots show the mean thickness (± standard deviation) of every 10 consecutive sandstone beds, or every 20 consecutive beds in plot 5; the coefficient of variation (CV) is the ratio of standard deviation and arithmetic mean, used as a dimensionless measure of bed-thickness variability expressed as a fraction of the mean (Ball *et al.*, 1997).

Haughton, 2000; Satur *et al.*, 2000; Johnson *et al.*, 2001; Grecula *et al.*, 2003) or forming poorly defined depositional lobes (cf. Pickering & Hiscott, 1985; Mutti, 1992; Lien *et al.*, 2003).

Facies subassociation 1b is interpreted as the infill of an east-trending submarine channel with a sinuous thalweg and lateral-accretion bars. Such point bars are considered to be a result of the turbidity currents eroding the cut-bank side of a sinuous channel and depositing sediment against the opposite bank (e.g. Stetling *et al.*, 1985; Kolla *et al.*, 2001; Abreu *et al.*, 2003). In the outcrop section (Fig. 9B), the channel's right-hand point bar (RPB) was superimposed upon the left-hand bar (LPB) as the

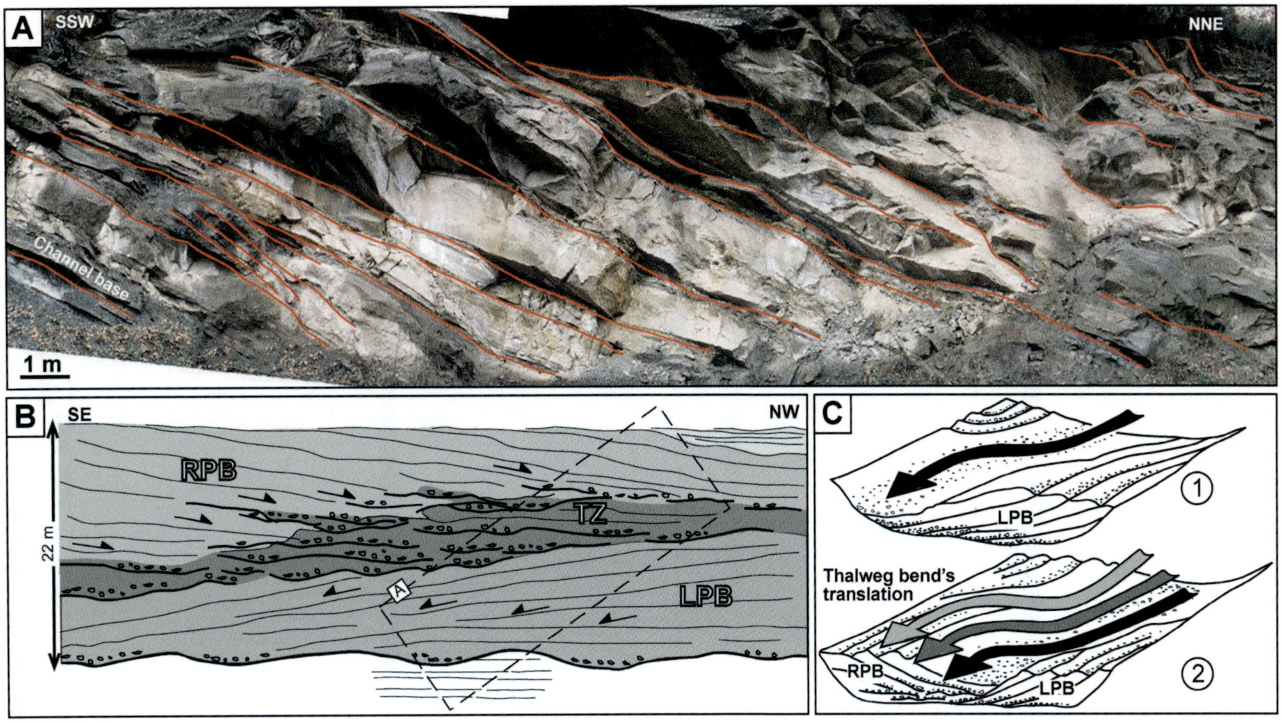

Fig. 9 Facies subassociation 1b (channel-fill turbidites). (A) Outcrop section in the lower part of Gürsökü Formation in the central part of the basin. Dirt-road section 10 km south of Ayancık, in the area indicated as locality 4 in Fig. 3. The palaeochannel direction is to the northeast, obliquely towards the viewer. (B) Interpretation of this tectonically tilted channel-fill succession (section transverse to channel axis; exposed part indicated by the dash-line frame), showing the channel's right-bank point bar, RPB, superimposed on a left-bank point bar, LPB, with an intervening, erosional and gravel-bearing thalweg/riffle zone, TZ. (C) Schematic diagrams explaining the erosional stacking of point bars as a result of the downflow translation of the channel's sinuous thalweg.

erosional, gravel-bearing riffle zone (TZ) shifted gradually onto the latter with multiple scouring. The stacking of one point bar upon and against another indicates a marked downflow translation of thalweg loop (*sensu* Jackson, 1976) combined with a lateral migration and channel-floor aggradation (Fig. 9C), similar to that described from mid-fan submarine channels (Stetling *et al.*, 1985; Kastens & Shor, 1986; Peakall *et al.*, 2000; Posamentier & Kolla, 2003). As pointed out by Peakall *et al.* (2000) and Kolla *et al.* (2001), pronounced aggradation distinguishes sinuous turbiditic channels from fluvial ones. The channel-fill architecture here would be comparable to the three-stage model of Peakall *et al.* (2000), except that the lateral accretion apparently accompanied aggradation, rather than preceded it, before the channel was abandoned.

The bed-set inclinations have been measured by disregarding the tectonic tilt and assuming that the underlying and overlying sheet-like turbidites were horizontal. The mean primary inclination of the left-bank bed set (see LPB in Fig. 9B) is ~ 5° and that of the right-bank bed set (RPB in Fig. 9B) is ~ 10°, which suggests channel half-widths of 250 m and 125 m, respectively, on account of the channel-fill thickness of ~ 22 m (Fig. 10). The riffle zone of point-bar superposition (see TZ in Fig. 9B) is ~ 75 m wide, and the bulk channel width is thus estimated at ~ 300 m. The channel-fill thickness, when corrected for ~ 3000 m burial depth (Fig. 2) and ~ 25 vol.% compaction (Baldwin & Butler, 1985), suggests a bulk channel depth of ~ 27.5 m. The depth/width ratio of 1/11 would appear to be only slightly lower than the average aspect ratio of 1/10 indicated by a worldwide dataset compiled by Clark & Pickering (1996), but much lower than, for example, the aspect ratio of 1/7.5 reported by Abreu *et al.* (2003) from the offshore Angola channels. Similar 'extra-wide', low aspect-ratio palaeochannels with superimposed point bars occur in the Eocene Kusuri Formation in the basin (Janbu *et al.*, this volume, pp. 457–517).

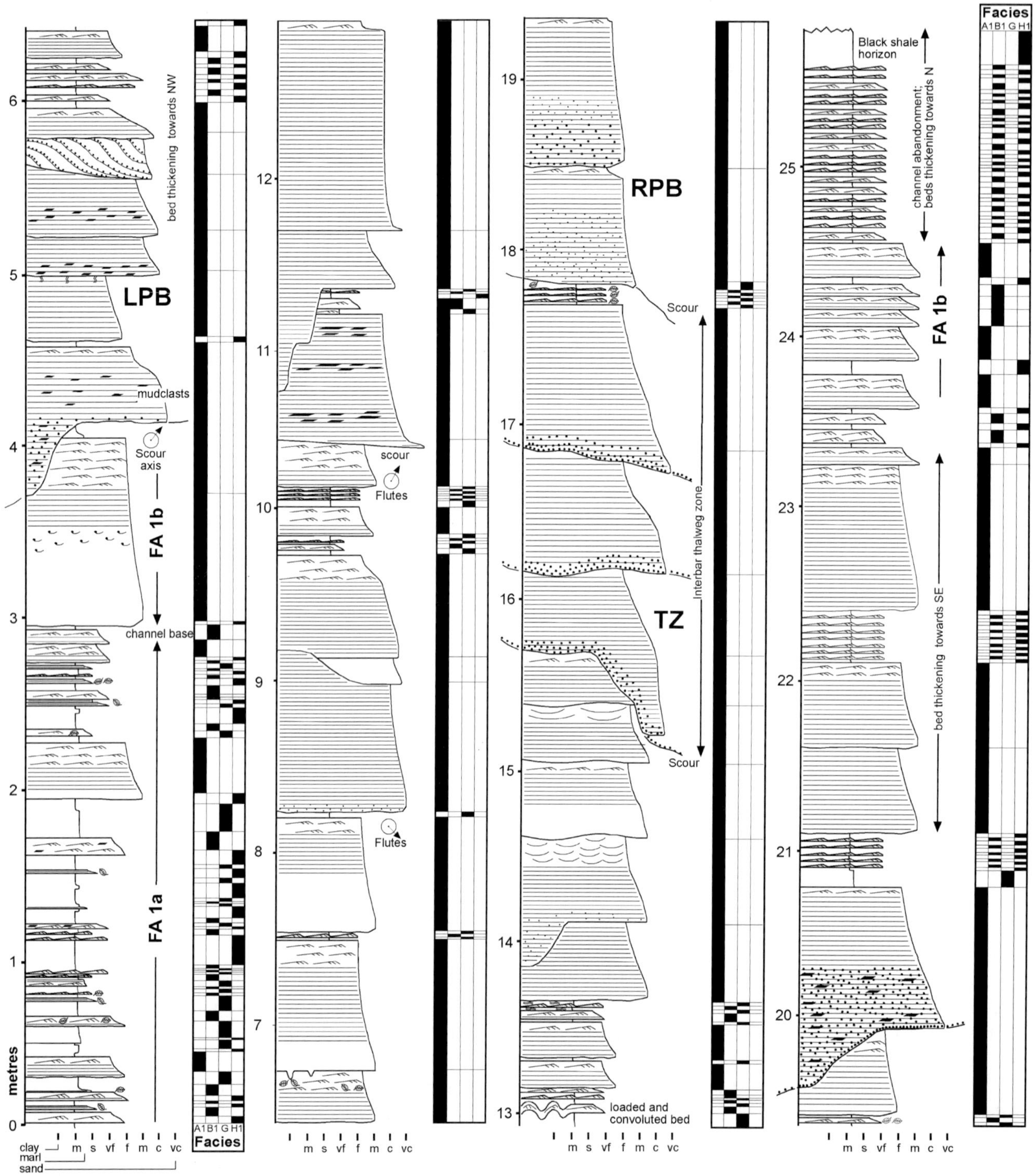

Fig. 10 Log through the central part of the channel-fill turbiditic succession (subassociation 1b) in the outcrop section shown in Fig. 9A. For log legend, see Fig. 7; facies symbols are as in Table 1 and the letter symbols LPB, RPB and TZ are as in the caption to Fig. 9B.

Facies association 1 as a whole is thought to represent the medial to distal part of a deep-water turbiditic system supplied with sediment from a narrow shelf zone at the basin's western part and southwestern margin. The shelf zone was accumulating epiclastic sediment derived from erosion of volcanic rocks, but hosted also an expanding carbonate platform. Storms probably transported abundant sediment to the shelf edge, causing its gravitational failures, and possibly also generated some sediment-laden, gravity-driven surges of seaward-returning water (Hamblin & Walker, 1979; Walker, 1984b; Snedden *et al.*, 1988; Myrow & Southard, 1996) and/or shelf-crossing rip currents (Bowen & Inman, 1969; Dalrymple, 1975; Cacchione *et al.*, 1984, 1994; Gruszczyński *et al.*, 1993). The notion of storm-generated currents is supported by the tempestites of facies subassociation 3c in the shallower part of the evolving depositional system (see subsequent section). When plunging below the effective wave base, these flows could ignite into turbidity currents (Parker, 1982; Walker, 1984b; Fukushima *et al.*, 1985; Myrow & Southard, 1996; Mulder *et al.*, 2001). Accordingly, the turbiditic system of the Gürsökü Formation can be regarded as a line-source ramp (*sensu* Reading & Richards, 1994) of siliciclastic to mixed silici-calciclastic composition, with the shelf edge-derived and/or storm-triggered turbidity currents coalescing and turning eastwards along the basin axis.

The basin-floor turbiditic system was subject to aggradation, rather than progradation, as it was probably backlapping the basin margin's submarine slope. The notion of back-stepping and prolonged aggradation is supported by the general upward thinning of sandstone beds (cf. plots 1 and 2 in Fig. 8), with bed-thickness clustering (CV = 1.53) giving way to a disorderly pattern (CV ≈ 1.00). The axial sinuous channel formed at an early stage, when the siliciclastic sediment supply was high, the basin margin's submarine relief was at a maximum and the basin-floor topography was probably uneven, affected by blind thrusts related to the compressional uplift to the west. These factors could jointly promote channelized turbidity currents. The solitary submarine channel and generally thicker bedding (plot 1 in Fig. 8) are considered to signify a medial part of the system, comparable to a 'mid-fan' setting, and it is likely that similar channels were more common in the unexposed proximal part of the turbiditic system (Fig. 11; cf. Braga *et al.*, 2001; Savary & Ferry, 2004).

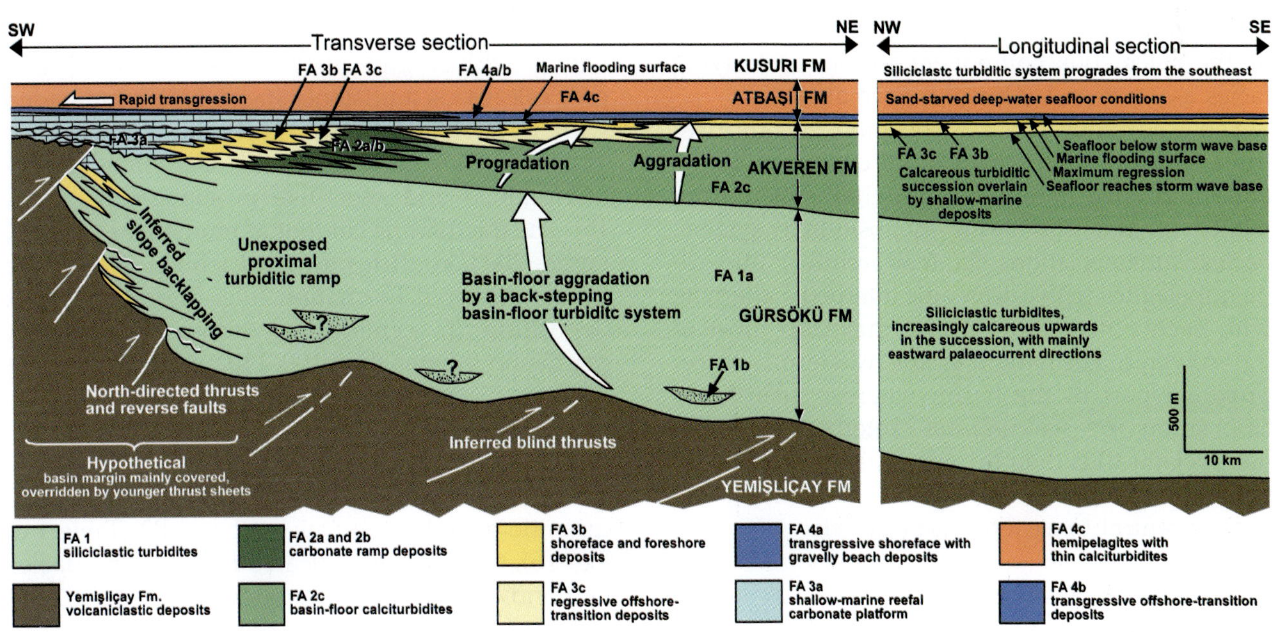

Fig. 11 Schematic transverse and longitudinal cross-sections through the Campanian–Ypresian part of the basin-fill succession, showing the interpreted spatial relationships among facies associations.

Facies association 2: calcareous turbidites

This turbiditic assemblage consists of calcarenites, marlstones and calcareous mudstones, and constitutes the main, lower to middle part of the Akveren Formation (Fig. 2). The latter formation overlies the siliciclastic Gürsökü Formation with a transitional contact, which renders their lithostratigraphic boundary somewhat arbitrary, especially since it is seldom well-exposed. Facies association 2 is little more than 100 m thick in the western part of the basin, but thickens to ~ 550 m along the basin axis towards the east (Fig. 11). *Nereites* ichnofauna and the microfauna content of basinal mudstones indicate a deep-water environment. The calcarenites are mainly packstones (*sensu* Dunham, 1962), with a micritic calcite cement and common chert concretions. The sand consists of well-sorted, submature to mature bioclasts with a minor (< 5 vol.%) admixture of mainly subangular to subrounded grains of quartz, mica, plagioclase and various rock fragments, including volcanic glass and related detritus (Leren, 2003). Bioclasts have thick micritic envelopes and are fragments of foraminifers, bryozoans, brachiopods, bivalves, other molluscs, echinoderms and coralline red algae, predominantly crustose *Corallinaceae melobesieae* (Fig. 12). The detritus represents a foramol carbonate facies (Lees & Buller, 1972) and was derived from a contemporaneous reefal platform, similarly to that of the calcareous component in the underlying facies association 1 (Gürsökü Formation).

The calcareous facies association 2 comprises three subassociations, which are described below and interpreted to represent different subenvironments of the submarine depositional system. Facies subassociations 2a (calcarenites) and 2b (hemipelagites with calcarenitic interbeds) alternate with each other (Fig. 13) and make up most of the Akveren Formation in the western, source-proximal part of the basin, whereas facies subassociation 2c (calcarenites interbedded with hemipelagites) is their 'distal' equivalent and constitutes most of the Akveren Formation in the basin's central to eastern part (Fig. 11).

Facies subassociation 2a

This assemblage is dominated by the calcarenitic turbidites of facies C, interbedded with thin marlstones of facies G (Table 1) and forming packages 2–4 m thick (Fig. 13). The beds of facies C are mainly turbidite T*a*(*b*), ≤ 41 cm thick and roughly tabular, with the planar-stratified *b*-division not always present. They are commonly stacked erosionally upon one another as the infill of broad and shallow scours, 1–2 m in relief and > 200 m wide (Figs 4H & 14). Most of these calcarenite beds show only weak grading, but some are coarse- or very coarse-grained and granule-bearing in their lower parts. Many beds seem to contain multiple *a*-divisions, which are separated by scours and merge laterally into a single massive division. Some beds contain scattered clasts of marlstone, calcareous mudstone and/or limestone (packstone), mainly subrounded and ≤ 5 cm in length. The scour-based sets of amalgamated beds T*a* and T*ab* occasionally pass upwards into beds T(*a*)*bc*, with laterally discontinuous *a*-divisions and with or without marlstone cappings. Palaeocurrent directions are mainly towards the north-northeast and northeast. In outcrop sections approximately parallel to the palaeocurrent direction, the bed sets commonly show internal downlapping architecture (Fig. 13), which suggests deposition on a gently sloping substrate.

Facies subassociation 2b

This assemblage (Fig. 13) consists mainly of alternating whitish-grey marlstones (facies G) and grey calcareous mudstones (facies H1), interbedded with isolated calcarenitic turbidites of facies C. The calcarenite beds are mainly thin, fine- to very fine-grained and sheet-like in shape (Figs 14 & 15). The microfauna content of mudstones (Table 2; Leren, 2003) confirms a Maastrichtian age for the lowest Akveren Formation (Fig. 2) and indicates deposition in a neritic environment, at water depths no greater than 100–150 m.

Facies subassociation 2c

This assemblage consists of the sheet-like calcarenitic turbidites of facies A2 and B2 (Table 1 & Fig. 16), capped with calcareous mudstone (facies H1) and/or marlstone (facies G). The entire succession is evenly bedded, dominated by thin to moderately thick, fine- to very fine-grained turbidites T*bcd*, T*cd* and T(*c*)*d*, composed of bioclastic

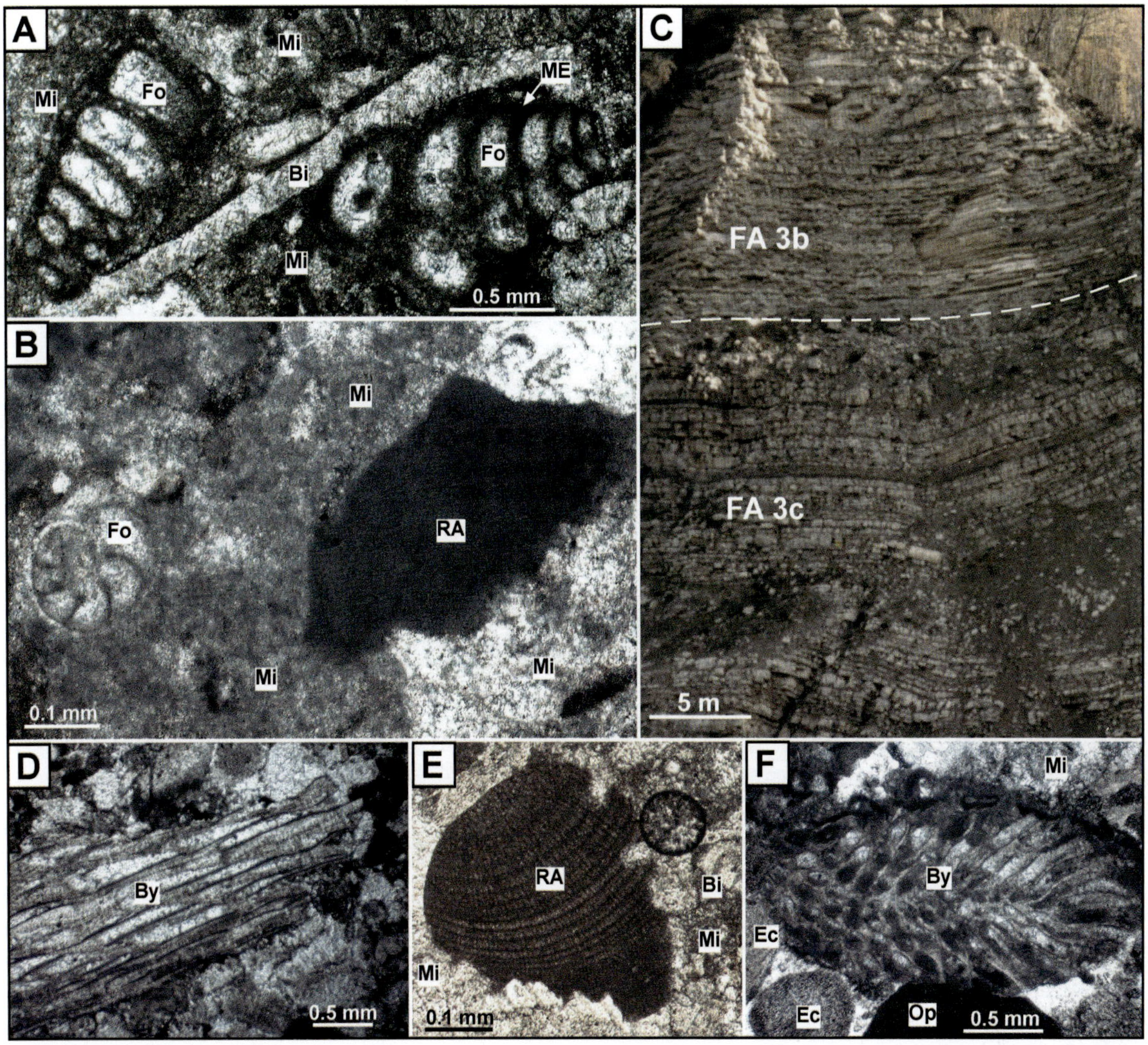

Fig. 12 Facies of the middle to upper Akveren Formation. Thin-section views show: (A) calcarenite of facies A2 from facies association 2c; (B) calcarenite of facies C from facies association 4b. (C) Outcrop photograph showing facies subassociation 3c overlain by subassociation 3b in the uppermost part of the formation. Photomicrographs: (D & E) massive limestone of facies F2 from facies association 2a; (F) calcarenite of facies D from facies association 3c. Letter symbols: Bi, bivalve fragment; By, bryozoan fragment; Ec, echinoderm fragment; Fo, foraminifer test; ME, micritic envelope; Mi, micritic cement; Op, opaque grain; RA, fragment of coralline red algae.

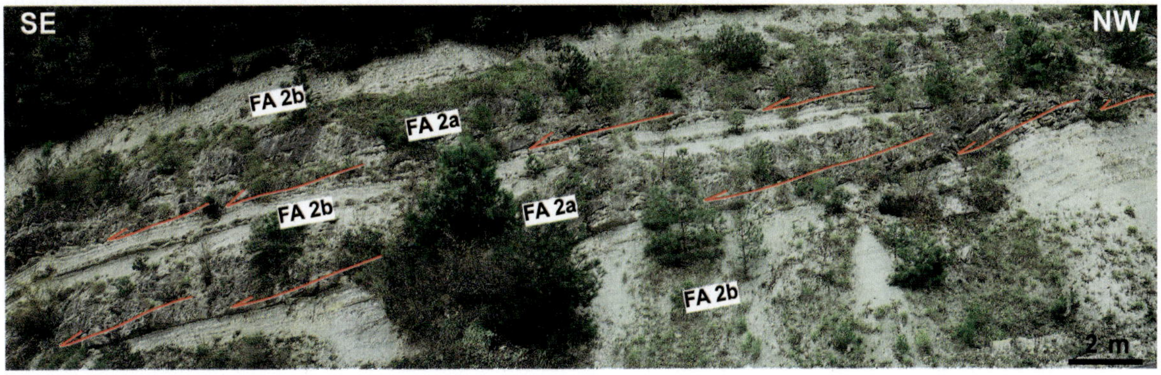

Fig. 13 Alternating turbiditic packages of facies subassociations 2a and 2b in the middle portion of the Akveren Formation in the western part of the basin (locality 1, Fig. 3). Note the downlapping pattern of gently inclined bedding (arrows) in the sand-rich units of subassociation 2a.

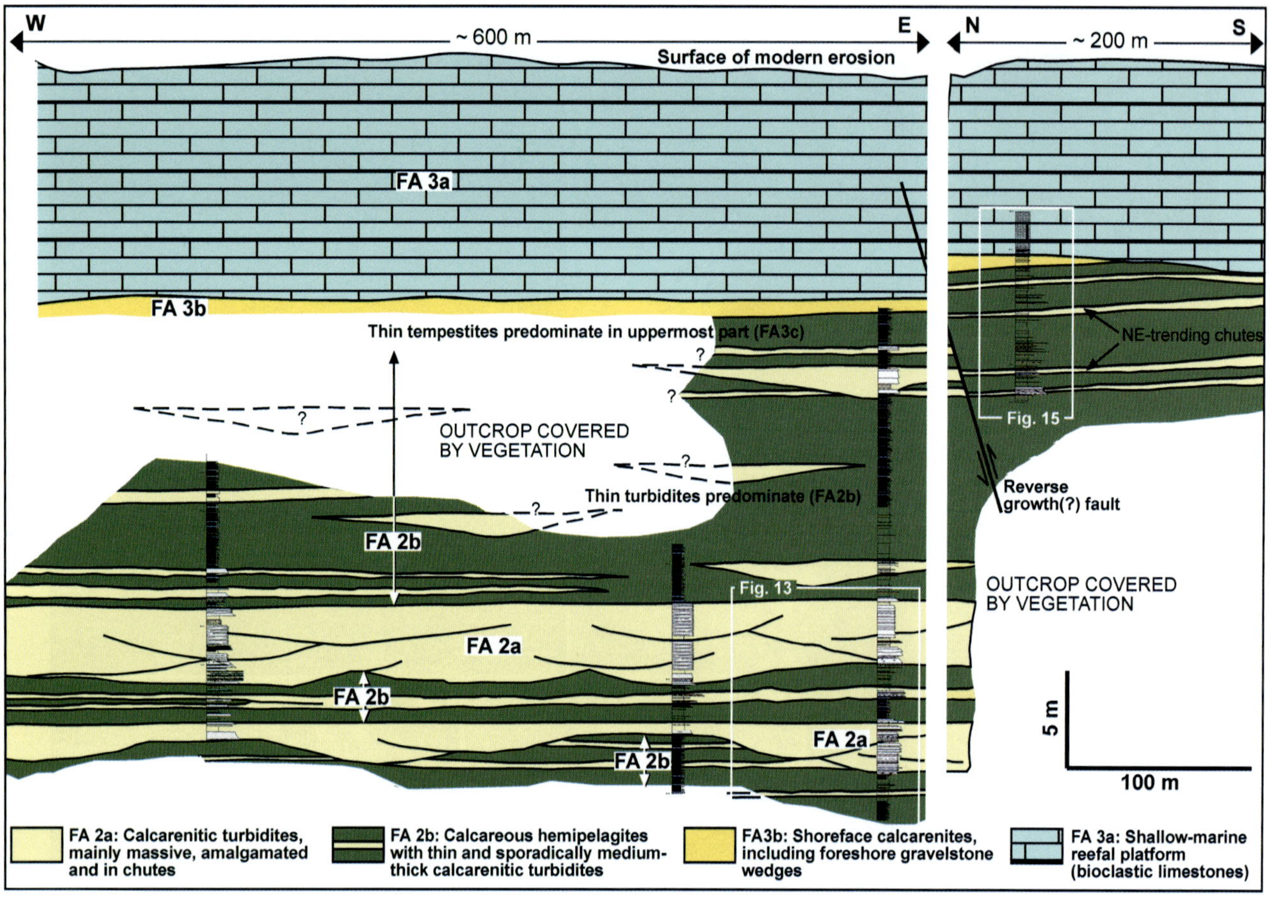

Fig. 14 Interpretative log-correlation panel based on an outcrop section near Kuğuköy (locality 1, Fig. 3), showing the upward transition of facies subassociations 2a and 2b into subassociations 3a and 3b; see outcrop details in Figs 13 & 15.

sediment with minor siliciclastic admixture. Bed thicknesses vary on an outcrop scale (Fig. 17) and show a thickening-upward trend in the upper portion of their succession in the basin's western to central part (plot 4 in Fig. 8), whereas no systematic upward change is recognizable in the lower to middle portion and in the basin's eastern part (plots 3 and 5 in Fig. 8). The corresponding coefficients of variation (Fig. 8) indicate strongly skewed, non-normal bed-thickness frequency distributions. Turbidite thicknesses tend to be clustered, bimodal or polymodal (CV > 1) in the lower portion of the succession (plot 3) and also in its whole middle to upper portion in the basin's eastern part (plot 5), but are random (CV ≈ 1) to anticlustered, with regularly spaced values (CV < 1), in the middle to upper portion in the basin's central and western part (plots 3 and 4). Flute marks and ripple trough axes indicate mainly eastward (east/southeast) palaeocurrents, with sporadic flows towards the northeast and south (see the rose diagram in Fig. 16A). Maastrichtian foraminifers indicate a warm-water subneritic environment, with a deep-water *Nereites* ichnofacies including *Chondrites intricatus*, *C. targionii*, *Cosmorhaphe lobata*, *Gyrolithes*, *Halimedites annulata*, *Megagrapton irregulare*, *Ophiomorpha annulata*, *Palaeodictyon latum*, *P. majus*, *Phycosiphon incertum*, *Phymatoderma*, *Pilichnus dichotomus*, *Planolites*, *Scolicia strozzii*, *Talassinoides suevicus*, *Trichichnus linearis* and *Zoophycos* (Uchman *et al.*, 2004).

Interpretation

Facies subassociation 2c is thought to represent a basin-floor turbiditic system supplied with sediment

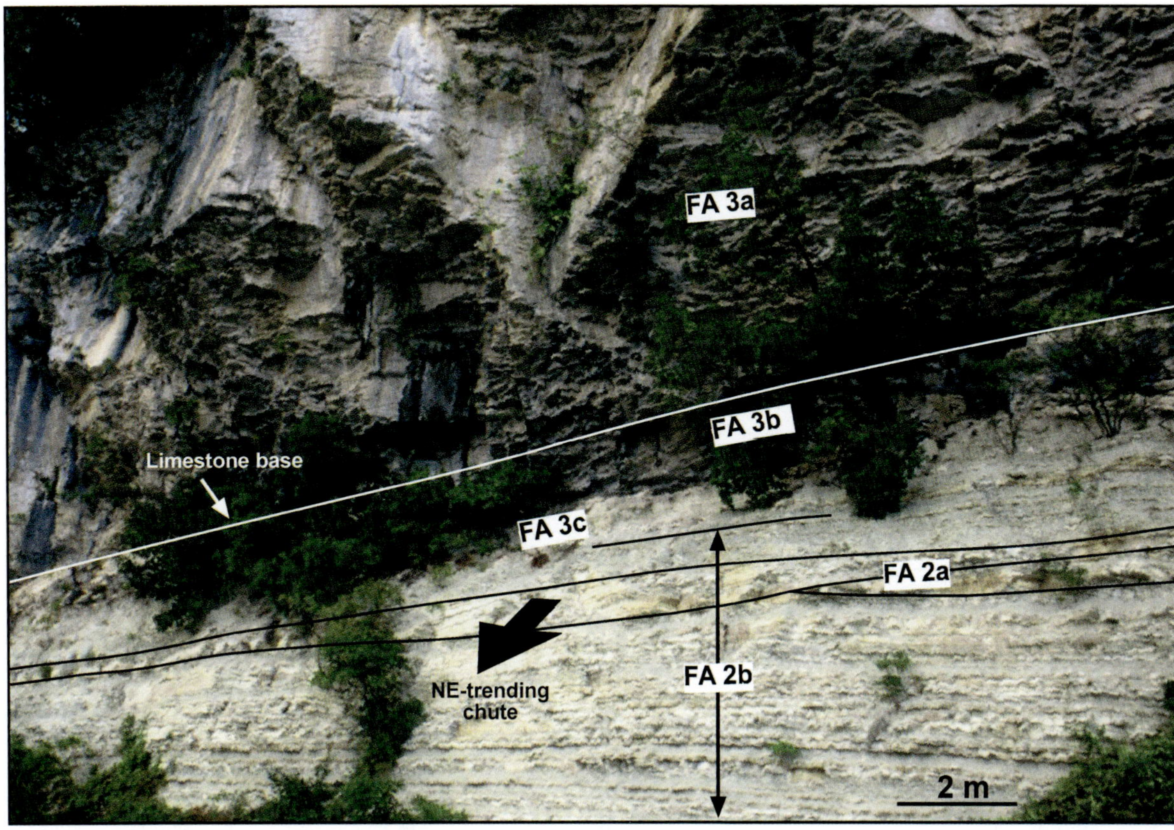

Fig. 15 The upward transition of facies association 2 to association 3 in the upper part of the Akveren Formation in the western part of the basin; a northeast-facing cliff near Kuğuköy (locality 1, Fig. 3).

Fig. 16 Akveren Formation calcareous turbidites. (A) Outcrop of facies subassociation 2c (lower Akveren Formation) near Tangal in the central part of the basin (locality 5, Fig. 3). The rose diagram (inset) summarizes palaeocurrent data from this subassociation (n = number of data). (B) Outcrop of facies subassociation 2c (middle Akveren Formation) in the east-central part of the basin (locality 6, Fig. 3). (C) Close-up detail of the latter outcrop, showing thin to moderately thick calcarenite sheets (facies A2 and B2) capped with marlstone (facies G) and commonly also calcareous mudstone (facies H1); note the flute casts on the overhanging sole surface.

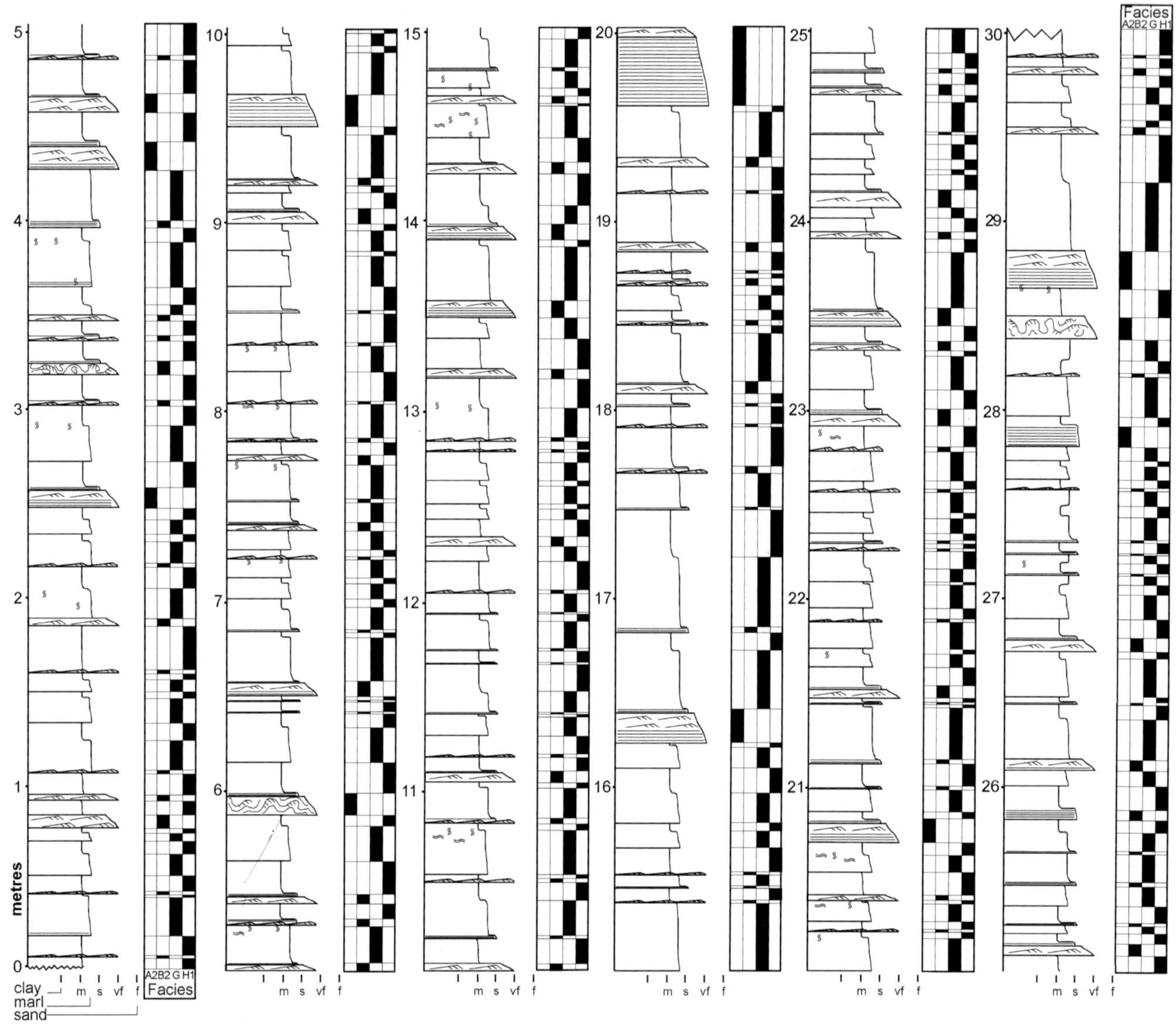

Fig. 17 Log of facies subassociation 2c in the middle to upper part of Akveren Formation in the east-central part of the basin (locality 7, Fig. 3). See also outcrop details in Fig. 16B & C and statistical summary plot 5 in Fig. 8. For log legend, see Fig. 7.

from a carbonate ramp sourced by a contemporaneous basin-margin reefal platform (Fig. 18). Somewhat similar carbonate-ramp apron systems have been described by Mullins & Cook (1986), Burchette & Wright (1992), Coniglio & Dix (1992) and Harris (1994) among others. Calciturbiditic successions deposited by non-channelized and mainly low-density currents have been widely reported from the lower slope, base-of-slope and basin-plain environments related to shelf-edge carbonate platforms and ramps (e.g. Colacicchi & Baldanza, 1986; Mullins & Cook, 1986; Eberli, 1987; Hazlett & Warme, 1988; Tucker, 1990; Braga *et al.*, 2001; Drzewiecki & Simó, 2002; Savary & Ferry, 2004). The origin of chert concretions was discussed by Bustillo & Ruiz-Ortiz (1987). The observed pattern of turbidite thickness variation (plots 3–5 in Fig. 8) is consistent with the notion

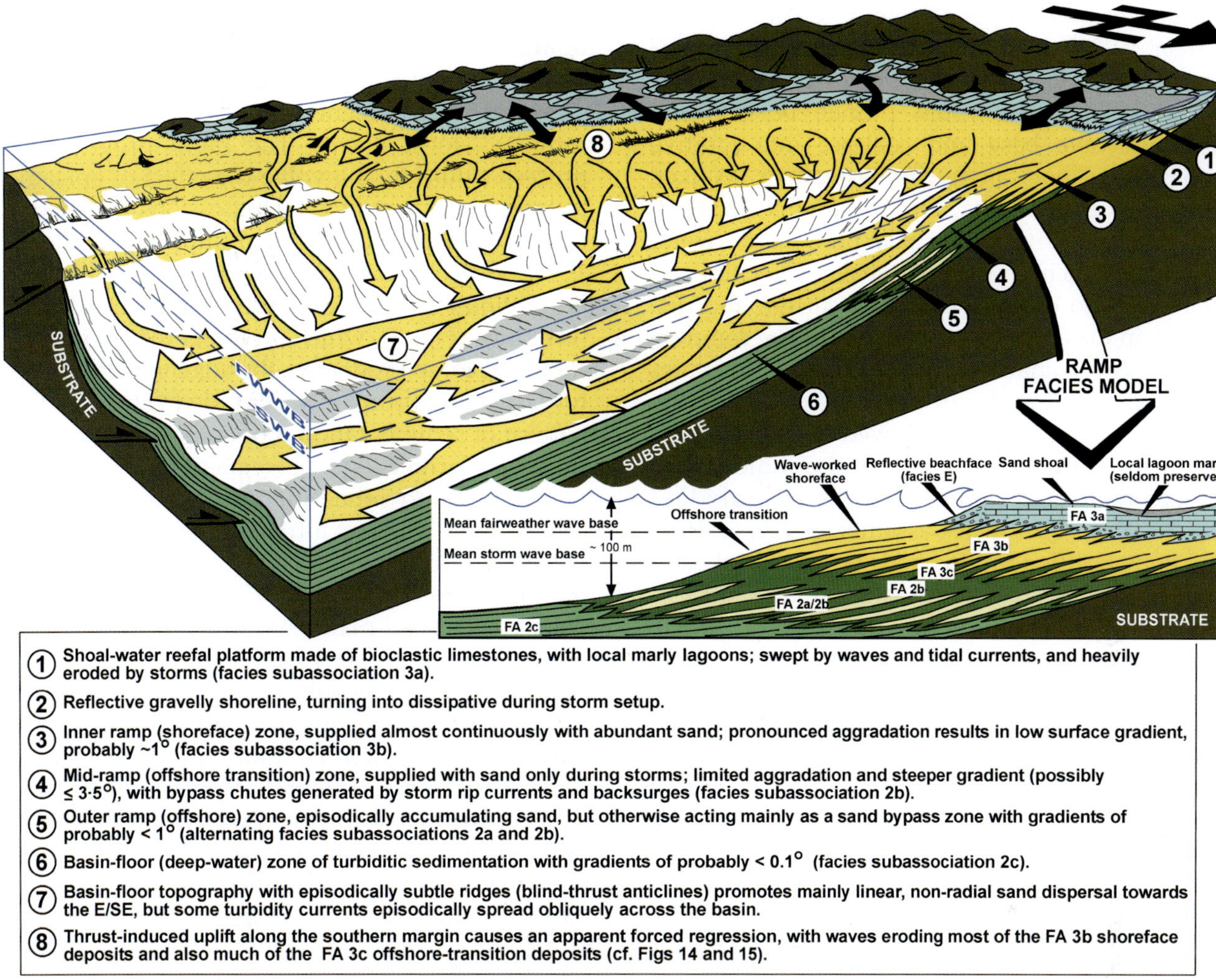

Fig. 18 Depositional model for the ramp-sourced, calciclastic basin-floor turbiditic system of the Akveren Formation. Generalized palaeogeographical scenario and sediment dispersal pattern for late Maastrichtian time (schematic, not to scale).

of an aggrading and prograding depositional system, yet lacking distributary channels. The tendency for bimodal clustering of bed thicknesses in the lower and distal eastern part of the succession may reflect a combination of turbidity currents generated by storms and shed by gravitational failures from a skeletal sand-prone apron perched on the basin margin. The thicker bedding and bed-thickness anticlustering (CV < 1.00) in the proximal upper part of the succession may represent chiefly this former triggering factor, reflecting basin-floor shallowing and ramp advance (cf. Fig. 11).

The reefal carbonate platform apparently formed in the western part of the basin and extended eastwards along the basin's southwestern margin, as the latter became subject to uplift by tectonic thrusting and formed a narrow shelf (Fig. 18). As discussed in the next section, the reefal platform was a sand-prone shoal fringed with a reflective, wave-dominated gravelly shoreline and relatively narrow, sandy shoreface zone. Abundant sediment could be transferred from the shoreface zone to the deep-water environment by storm-generated rip currents (Bowen & Inman, 1969; Dalrymple, 1975; Cacchione *et al.*, 1984, 1994; Gruszczyński *et al.*, 1993) and geostrophic surges (Hamblin & Walker, 1979; Walker, 1984b; Snedden *et al.*, 1988; Myrow & Southard, 1996). The sediment-laden, gravity-enhanced currents would plunge beneath the effective wave base and coalesce into ignitive turbidity currents along the basin floor (Walker, 1984b; Fukushima *et al.*, 1985; Mulder *et al.*, 2001). The term 'effective' wave base here means water depth at which the orbital wave velocities are insufficient to dilute the density current and disperse it in the ambient water column. The common occurrence of shallow chutes (Fig. 14) indicates frequent scour-and-fill phenomena, implying localized sand bypass followed by deposition. Bypassing turbidity currents are also indicated by the common marl-capped 'top-absent' turbidites T*a*, T*ab* and T*b* in facies assemblages 2a and 2b. The non-stratified graded beds with multiple and laterally impersistent *a*-divisions resemble some of the so-called fluxoturbidites (Leszczyński, 1989), attributed to pulsating, highly non-uniform turbidity currents subject to internal 'density fluxes'.

Reflective shorelines generally lack rip currents, but these erosive, pulsating and topographically controlled jets of seaward-returning water could form episodically during storms, when dissipative shoreline conditions occurred on the carbonate shoal, with edge waves produced from the breaking (Bowen & Guza, 1978), reflected (Guza & Davis, 1974; Guza & Inman, 1975) or swash-excited waves (Huntley & Bowen, 1975). The bottom velocities of rip currents generated by moderate storms are commonly up to 2–3 m s^{-1}, whereas megarips can attain speeds of 5–10 m s^{-1} and reach offshore distances of a few tens of kilometres (Gruszczyński *et al.*, 1993). Rip currents flowing at 1.5 m s^{-1} have been reported by the latter authors to carry sand and small pebbles in turbulent suspension and coarser gravel in bedload traction.

The alternating turbidite packages of facies subassociations 2a and 2b are gently inclined (≤ 1°) clinothems representing deposition in the lower ramp zone, which acted as a sand-bypass area dominated by hemipelagic sedimentation, but episodically accumulated abundant sand, commonly coarse-grained and bearing intraformational gravel clasts derived from the carbonate platform. The calcarenitic packages downlapping the lower ramp slope (Fig. 13) imply a marked increase in basinward sand flux and are attributed to episodic uplift of the basin margin, which would enhance density currents and cause an extensive erosional sweeping of sediment from the carbonate platform by storm waves. The ramp progradation process would thus involve the effect of episodic, tectonically induced forced regressions recorded as pulses of increased sediment supply (cf. Tucker, 1990; Eberli, 1991). Facies subassociation 2c shows no obvious upward coarsening or thickening of turbidites (Fig. 8), which suggests that, while the carbonate ramp prograded, the adjoining basin-floor turbiditic system mainly aggraded (Fig. 11). This pattern of sedimentation may indicate ponding of turbidity currents (see subsequent discussion).

The neritic to subneritic lower ramp is considered to have acted as a transitional zone between the littoral to sublittoral fringe of the carbonate platform (see facies association 3 below) and the deep-water basin-plain environment, where sediment aggradation kept pace with and eventually exceeded the rate of subsidence (cf. Figs 11 & 14). The basin was elongate, but still wide (possibly ≤ 80 km) in Maastrichtian time, and the non-radial pattern of easterly sediment dispersal might reflect subtle ridge-and-swale topography of a basin floor affected by blind thrusts (Fig. 18; see also Janbu *et al.*, this volume, pp. 457–517). The axial turbiditic system had a very low gradient (probably < 0.1°; cf. Reading & Richards, 1994; Betzler *et al.*, 1999), which was reduced further by aggradation and allowed some of the turbidity currents to spread transversely across the basin. The axial turbidity currents were probably subject to distal ponding in the eastern, 'dead-end' part of the basin, where the margins remained largely submerged and hemipelagic sedimentation predominated until the earliest Eocene (Janbu *et al.*, this volume, pp. 457–517).

The predominance of facies subassociation 2b in the upper part of the progradational ramp succession, prior to its rapid upward transition to the shallow-marine facies association 3 (Figs 14 & 15), is thought to represent a mid-ramp zone that was undersupplied with sand and hence relatively steep (possibly ≤ 3–5°), traversed by sand-transferring chutes during storm events (Fig. 18). The gradual upward transition to subassociation 3c (Fig. 14), described further below, involves an alternation of isolated calcarenitic turbidites and tempestites.

This 'mixed' mode of offshore sand transport reflects a progressive loss of self-ignition capacity (Parker, 1982; Fukushima *et al.*, 1985) by the storm-generated currents on an aggrading subhorizontal seafloor, with increasingly more of the currents spreading seawards as typical combined-flow surges, rather than turbidity currents (cf. Myrow & Southard, 1996).

Facies association 3: carbonate reefal platform and ramp deposits

This calcareous facies association constitutes the uppermost part of the Akveren Formation (Fig. 11) and consists of reefal limestones (facies subassociation 3a) underlain by and passing basinwards into well-stratified, wave-worked calcarenites (subassociation 3b), which themselves are underlain by assemblages of silty marlstones and calcareous mudstones interspersed with sheet-like calcarenitic tempestites (subassociation 3c). These subassociations are described and interpreted below. Facies association 3 as a whole is little more than 15 m thick in the western part of the basin, where the limestone unit is most prominent (Figs 14 & 15), but it thickens to ~ 65 m towards the southeast, where the limestone unit pinches out (Fig. 12C). This facies association has a transitional boundary with the underlying facies association 2 and a conformable, transitional to sharp boundary with the overlying facies association 4 (Fig. 11).

Facies subassociation 3a

In its outcrops in the basin interior, this facies assemblage is ≤ 15 m thick and at least a few tens of kilometres in basinward extent (Fig. 11), consisting of massive (non-stratified), weakly bedded to non-bedded bioclastic limestones (facies F2 and subordinate facies F1, Table 1). The limestones are whitish- to light pinkish-grey in colour, strongly cemented and densely fractured, which generally obscures their primary bedding (Fig. 15). The homogeneous, microcrystalline limestone is rich in bioclasts, mainly ≤ 0.5 cm in size, with scattered fragments of greenish-grey volcanic rocks (≤ 0.1 cm in size) and numerous intraformational clasts of marlstone, mudstone and fine-grained calcarenite (≤ 1 cm in size). Siliciclastic detritus includes grains of quartz, microcline, plagioclase, opaque minerals and volcanic glass, apparently derived from eroded tephra. This non-bioclastic detritus is subordinate (≤ 2 vol.%), and the majority of clasts bear thin micritic coatings. Bioclasts range from angular to subrounded and include fragments of brachiopod shells, colonies of coralline red algae (mainly crustose *Corallinaceae melobesieae*), cyclostome bryozoans, echinoderm plates and foraminifer tests (Fig. 12D & E). Cement is micro- to macrocrystalline calcitic sparite, locally micrite, with common evidence of syntaxial overgrowths around coarse bioclasts. The homogeneous limestones can be classified as packstones with crystalline zones (Dunham, 1962) and are considered to represent a foramol carbonate facies (Lees & Buller, 1972).

At the basin's southwestern margin, the corresponding limestone unit is ≤ 150 m thick (Fig. 19A), includes lenses of calcareous lagoonal mudstones and overlies a thin siliciclastic shelfal succession covering lavaflow basalts and volcaniclastic rocks. The limestone consists of algal biomicrite/biosparite and bryozoan biosparite, biomicrite and boundstone varieties, with common corals and brachiopods. Bedding is inclined steeply basinwards (Fig. 19B), and these progradational clinothems include massflow and slump deposits (Fig. 19C). Nanoplankton species include *Discoaster megastypus*, *Toweius tovae*, *Ericsonia subpertusa*, *Fasciculithus tympaniformis* and *Throacosphaera* sp., indicating a late Paleocene (Thanetian) age. Planktonic foraminifers include *Morozovella aequa*, *Planorotalites* sp. and *Globigerina* sp., consistent with a late Paleocene age, and also the benthic foraminifers are a mixture of Paleocene and reworked Maastrichtian species. No species younger than Paleocene have been found in samples from the lower and middle part of this unit, but its uppermost part is reportedly of early to middle Eocene age (Tunoğlu, 1994), coeval with the carbonate platform that covered the Sinop Basin prior to its complete inversion (Janbu, 2004).

Facies subassociation 3b

This facies assemblage (Figs 12C & 20) reaches a thickness of ~ 20 m in the east-central part of the basin and consists of the calcarenite beds of facies

Fig. 19 Carbonate platform deposits at the southwestern margin of the Sinop–Boyabat Basin (locality 10, Fig. 3). (A) Broad view towards the north, showing cliff section roughly parallel to the basin margin. (B) Progradational clinothems sloping northwards, towards the basin. (C) Slump deposit within the northward-sloping clinothems.

D (Table 1), which are mainly amalgamated, although commonly separated by thin and laterally discontinuous layers of silty marlstone (facies G) or calcareous mudstone (facies H1). In wide (> 100 m) outcrops, some of the packages of amalgamated beds appear to be thinning laterally, either filling some broad and shallow scours or forming subtle topographic mounds. The calcarenites are of whitish- to light yellowish-grey colour and vary from fine- to coarse-grained. Their beds are mainly 5–65 cm thick, commonly slightly graded (Fig. 20) and generally tabular, with planar or slightly undulatory erosional bases. Some of the coarse- and very coarse-grained calcarenite beds contain pebbles and/or granules, mainly subrounded to rounded and of intraformational provenance, but occasionally including volcanic-rock clasts. Gravel also occurs sporadically as scour-based lenses, ≤ 10 cm thick and a few metres in lateral extent, with a clast-supported, sand-filled pebbly framework containing shell fragments and flat-lying marlstone or limestone cobbles, some ≤ 37 cm in length.

Internal sedimentary structures include planar parallel stratification; undulatory parallel stratification resembling hummocky and/or swaley structures, with wavelengths of 50–220 cm (Fig. 5A & B); wave-ripple cross-lamination attributed to two-dimensional oscillatory ripples (Fig. 5C–E) and to symmetrical or asymmetrical three-dimensional vortex ripples (Fig. 5A, C & F), as described by Harms *et al.* (1982); and cross-lamination attributed to asymmetrical combined-flow ripples, with broad rounded crests, narrow troughs and unidirectional foresets similar to those reported from laboratory experiments by Yokokawa *et al.* (1995). Ripple indices are summarized in Fig. 21. Some beds have massive, graded basal parts, and the scale of cross-stratification commonly decreases upwards within a bed. Small-scale hummocky cross-sets (three-dimensional vortex ripples) at the bed top have wavelengths of 10–25 cm and locally show dewatering features, including convolute lamination (Fig. 5F). Erosionally bounded sets of planar parallel strata, where superimposed upon one another, often differ slightly in inclination and may imitate low-angle cross-stratification. Intraformational mudclasts, ≤ 4 cm in size, are locally scattered along the planar strata. The calcarenites bear chert concretions, which locally form semi-continuous

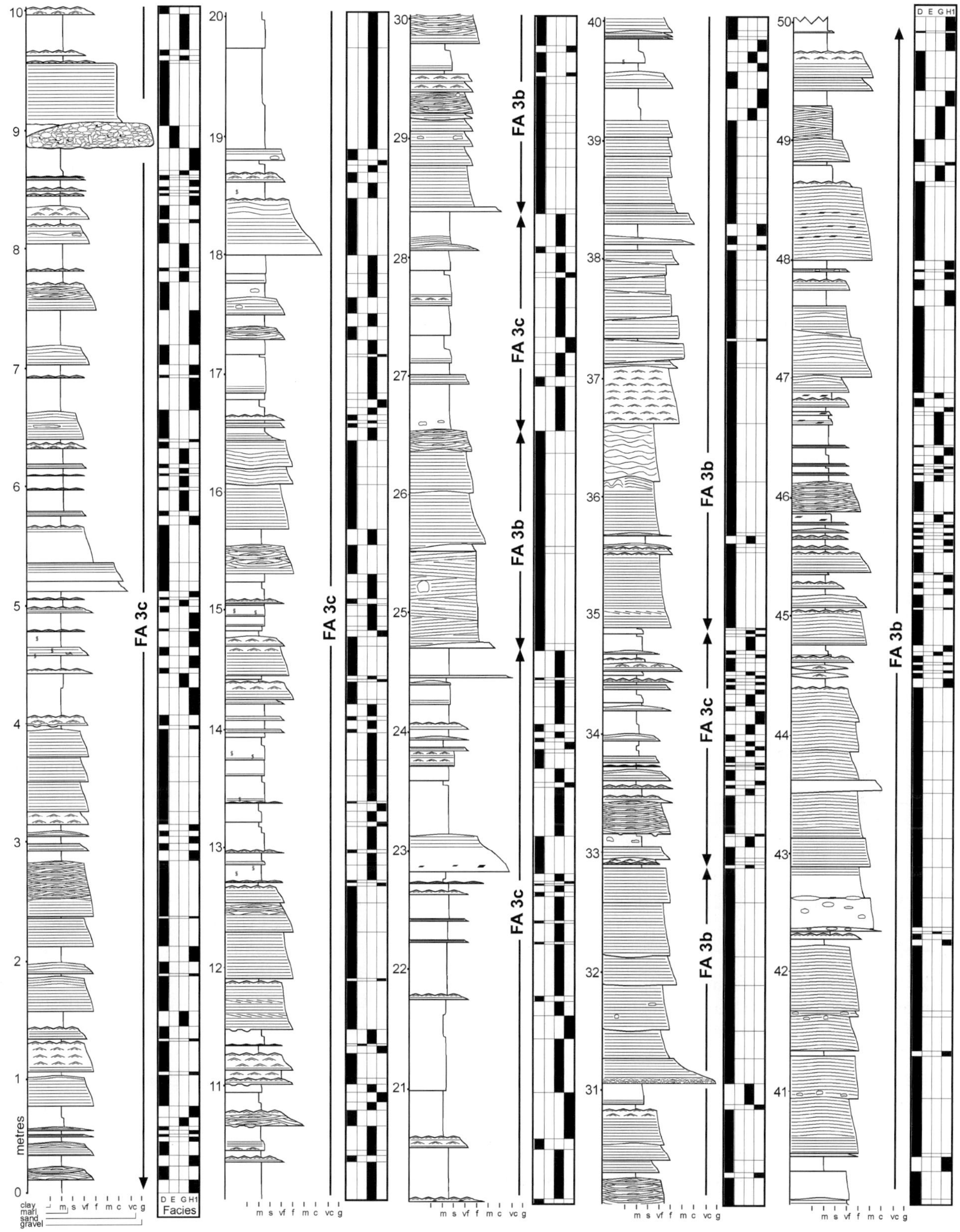

Fig. 20 Log of the upward transition from facies subassociation 3c to subassociation 3b in the upper part of the Akveren Formation, near Yenikonak in the central part of the basin (locality 3, Fig. 3). For log legend, see Fig. 7.

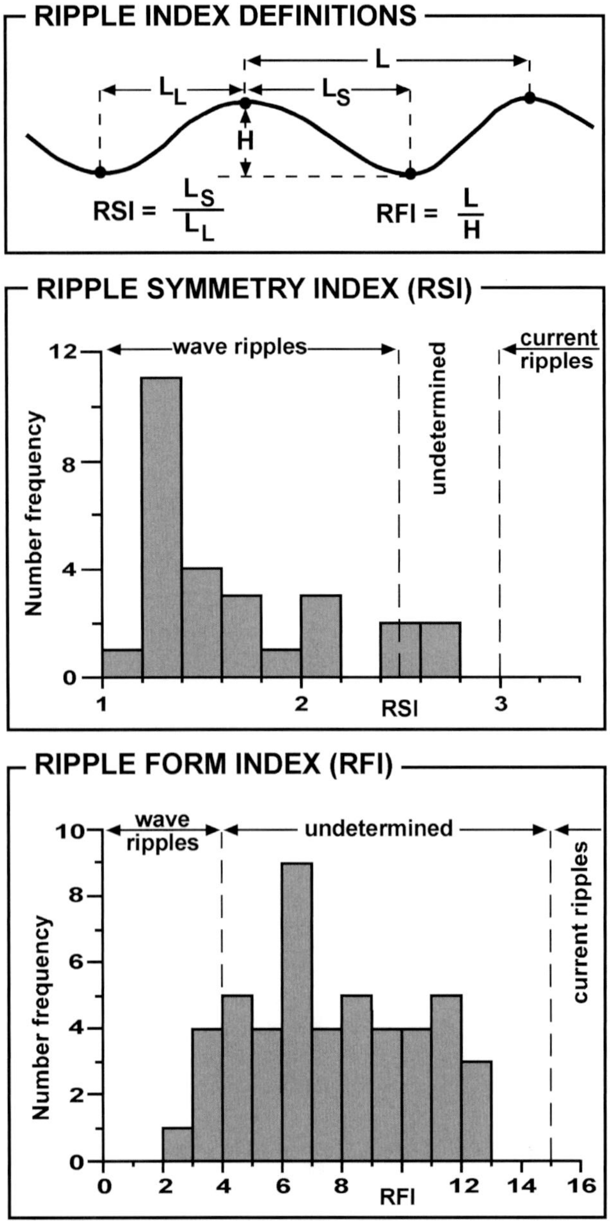

Fig. 21 Definition of ripple indices (upper diagram; after Collinson & Thompson, 1982) and a summary of the RSI and RFI values derived from the calcarenites of facies D in subassociations 3b and 3c. Data from outcrop section in Gerze (locality 6, Fig. 3).

bands ≤ 9 cm thick, roughly parallel to the bed boundaries. Bed bases sporadically show vertical and horizontal burrows, in the form of sand-filled pipes 0.5–1 cm in diameter, and at least one trace of *Ophiomorpha ?nodosa* has been identified at the base of a package of amalgamated sandstone beds (Leren, 2003).

Facies subassociation 3c

This facies assemblage is merely 2–5 m thick in the western part of the basin (Fig. 15), but thickens to > 45 m in the east-central part (Fig. 12C). It consists of whitish-grey marlstones (facies G) intercalated with light grey to greenish-grey calcareous mudstones (facies H1) and densely interspersed with sheet-like, graded calcarenite beds, mainly very fine- to fine-grained and 2–25 cm thick, but occasionally medium- to very coarse-grained and ≤ 80 cm in thickness (Fig. 22). Calcarenites are packstones composed chiefly of skeletal sand (Fig. 12F), as in facies subassociation 3b, similar to the foramol facies of Lees & Buller (1972). Bioclasts represent foraminifers, bryozoans, brachiopods, other molluscs, coralline red algal colonies and echinoderm plates (Fig. 12F). Cement is mainly micritic calcite (< 10 vol.%), with sparite locally present as syntaxial crystal overgrowths around bioclasts. Siliciclastic admixture (< 1 vol.%) includes quartz, volcanic rock fragments and opaque grains.

The calcarenite beds are predominantly tempestites (facies D), similar to those described by many others (e.g. Brenchley *et al.*, 1979; Handford, 1986; Arnott, 1993; Molina *et al.*, 1997; Vera & Molina, 1998). They have sharp bases and also fairly sharp tops, and show planar parallel stratification overlain by cross-lamination (Fig. 5F) representing 'micro-hummocks' (*sensu* Kreisa, 1981), or symmetrical to asymmetrical three-dimensional vortex ripples (Harms *et al.*, 1982). Some beds contain a massive basal division and/or hummocky to swaley stratification (Fig. 5G; Dott & Bourgeois, 1982; Harms *et al.*, 1982), or show planar or hummocky stratification overlain by unidirectional foresets of asymmetrical combined-flow ripples (Hamblin & Walker, 1979; Yokokawa *et al.*, 1995; Myrow & Southard, 1996). Ripple indices support the notion of bedforms produced by waves or combined flows (Fig. 21). Palaeocurrent indices suggest sand transport towards the east (see the rose diagram in Fig. 22). Some of the calcarenite beds contain chert concretions and many bear scattered intraformational clasts of mudstone, marlstone or limestone, locally concentrated in the bed's basal massive division (Fig. 22). A few thickest beds are

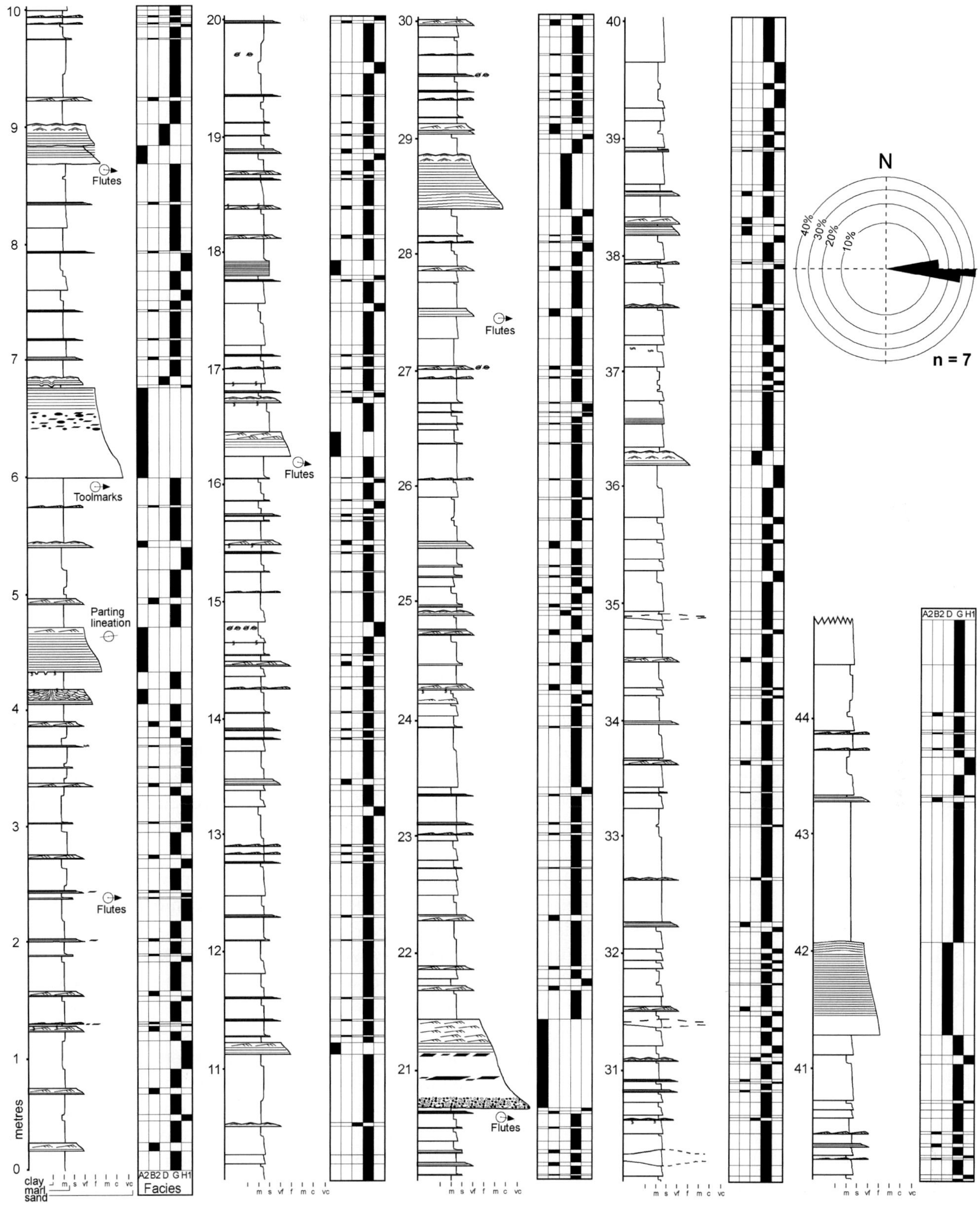

Fig. 22 Log of the upper part of the Akveren Formation in the east-central part of the basin (locality 7, Fig. 3). Facies subassociation 2c passes upwards into subassociation 3c; the first tempestites (facies D) occur at the log height of 6.8 m. For log legend, see Fig. 7. The rose diagram summarizes the corresponding palaeocurrent data.

underlain by wide lenses of facies E (gravelstone patches), which are massive, weakly graded, slightly mounded and occasionally ≤ 20 cm thick, composed of subrounded marlstone and limestone clasts (≤ 16 cm in length in the thickest lens; Fig. 20, top left).

Many calcarenite beds in the lower part of facies subassociation 3c (Fig. 22) lack evidence of combined-flow features, have gradational tops and resemble closely turbidites T*abcd* and T*bcd* of the underlying facies subassociation 2c. Accordingly, these beds are classified as deposits of facies B2 and subordinate facies A2 (Table 1). The lower part of facies subassociation 3c would thus appear to contain calcarenite beds of both facies D and facies B2/A2, and represent an upward transition from facies subassociation 2c. The proportion of calcarenite beds increases towards the top of facies subassociation 3c, where only the tempestites of facies D occur, increasingly amalgamated, and where this heterolithic assemblage is intercalated with the calcarenitic facies subassociation 3b (Fig. 20).

The microfauna content of mudstones indicates a late Paleocene age and basin bathymetry from ≤ 100 m to < 40 m (Leren, 2003), which is consistent with the presence of tempestites and suggests a sublittoral, shallow neritic environment. On the other hand, the tempestitic facies subassociation 3c lacks trace fossils typical of a neritic *Cruziana ichnofacies* and, instead, appears to contain a mixture of ichnotaxa characteristic of *Zoophycos* and *Nereites* ichnofacies, such as *Chondrites targionii*, *C. intricatus*, *Ophiomorpha annulata*, *Planolites*, *Thalassinoides*, *Trichichnus* and *Zoophycos*. This trace-fossil assemblage is rather unusual, especially since the underlying turbidites of facies association 2 still bear a typical *Nereites* ichnofacies, including agrichnial graphoglyptids (Uchman *et al.*, 2004), even though the uppermost turbidites of this facies association were apparently deposited in a relatively shallow environment, probably no deeper than 150–200 m.

Interpretation

The deposition of facies association 3 is attributed to a basin-floor turbiditic system supplied with sediment from a gently inclined, prograding and shallowing-upward carbonate ramp characterized by a wave-dominated shoreline and sandy forereef shoreface zone (Fig. 18; cf. Burchette & Wright, 1992). As discussed further below, the ramp was probably of a distally steepened type, involving mid-ramp chutes and evolving into a homoclinal type (Read, 1982, 1985). There is no evidence of a forereef talus apron, and the sand-prone shoreface is thought to have formed a gentle forereef slope in topographic continuity with the reef platform, separated from the latter by a beach zone of wave breaking (e.g. Handford, 1986; Wright & Bruchette, 1996). The foreshore to shoreface transition was initially steep enough to generate slumping (Fig. 19), but decreased in gradient as the platform margin advanced in the basin.

The foramol bioclastic limestones of facies subassociation 3a indicate a littoral environment in temperate climatic conditions, with water temperatures of < 20°C, normal salinity and depth of < 25 m (Lees & Buller, 1972; Wray, 1978; Wright & Burchette, 1996). The reefal platform was dominated by coralline red algae and bryozoan corals, and its crestal areas were wave-worked sand shoals, probably influenced by tidal currents and swept by storms. Bryozoans and coralline algae commonly dominate in Cenozoic reefs and related environments (Adams *et al.*, 1984; Tucker, 2001), which is attributed to the scarcity of reef-building large corals after the episode of mass extinction at the K–T boundary (Burchette & Wright, 1992; James & Bourque, 1992; Wright & Burchette, 1996). The calcareous deposits of facies subassociations 3b and 3c were all derived from this broad basin-margin platform (Fig. 18), which apparently acted as a highly efficient, line-type sediment source (*sensu* Reading & Richards, 1994).

The bryozoan fauna, the lack of stratification in the bioclastic limestones and the occasional abundance of marlstone and calcareous mudstone clasts in the beachface and beach-derived gravel of facies E (see facies subassociation 3b, Figs 5J & 25) suggest that the subtidal to intertidal platform included lagoonal areas, only episodically eroded by storms (Fig. 18). The abundance of the delicately branched cyclostome bryozoans seems to preclude a high-energy environment. Bryozoans are small colonial organisms with little tolerance to strong waves, living in shallow to moderately deep waters, commonly in excess of 70 m (James & Bone, 1991; Tucker, 2001). Coralline red algae are encrusting and binding organisms which prefer clear, generally shallow (< 25 m) and low-turbidity waters, and that thrive in high-energy reefal shoals and bank-edge environments (Wray, 1978; Adams *et al.*, 1984; Adams & MacKenzie, 1998). It is possible that bryozoan colonies, when strengthened by coralline algae, became more tolerant of moderate waves and were able to grow in littoral waters sheltered by a beach breaker zone. The fact that the platform deposits are packstones, with no preserved intact coral colonies, indicates piecemeal grinding by littoral waves and episodic storm erosion, probably combined with reef fragmentation by bioerosional processes.

A criterion most commonly used to distinguish between reefal and resedimented bioclastic deposits, such as shelf tidal sand banks and ridges, is whether the detritus was formed and accumulated *in situ* or as a result of major transport (Tucker, 2001). In the present case, the fragmented skeletal material varies from angular to subrounded, but is mainly subangular and poorly sorted, texturally submature, and lacks well-developed micritic envelopes, which indicates limited abrasion and relatively little degradation by endolithic bacteria. This evidence and the lack of distinct stratification support the notion of a reefal platform with moderate levels of hydraulic energy and limited transport of skeletal sand by waves and tidal currents, but with a strong erosional impact of frequent storms (cf. James & Bone, 1991; Jones & Desrochers, 1992; Wright & Bruchette, 1996).

The protective edge of the carbonate platform was a reflective shoreline with a steep foreshore slope dominated by breaking waves (Howard & Reineck, 1981; Komar, 1998), and the platform topography with extensive sand shoals and lagoonal depressions would further dissipate wave energy and reduce sediment abrasion. However, the impact of waves would increase greatly during storms, when coastal setup could render the shoreline dissipative and sand-laden rip currents would be generated (Komar, 1998), turning into turbidity currents by plunging beneath the effective wave base (Walker, 1984b; Myrow & Southard, 1996). Foreshore slumping and storm-generated currents are thought to have been the main agents for the origin of turbidity currents that deposited the basinward facies association 2 (Fig. 11). Tectonic uplift of the basin margin (Fig. 18) would intensify erosion and force a rapid basinward advance of the carbonate platform and associated ramp (Fig. 11).

The characteristics of facies subassociation 3b indicate nearly perennial action of waves with a fluctuating and often high energy, including frequent combined-flow currents, which implies deposition above the mean fairweather wave base in a shoreface environment (Clifton *et al.*, 1971; Clifton, 1976, 1981; Kumar & Sanders, 1976; Bourgeois, 1980; Leckie & Walker, 1982; Clifton & Dingler, 1984). Plane-bed transport occurs when the wave orbital velocities and resulting bottom shear stresses exceed the stability limit for oscillatory ripples (Harms *et al.*, 1982). The abundant amalgamation surfaces reflect erosional storm events, when the shoreface sand was episodically swept seawards by currents (Leithold & Bourgeois, 1984; Duke *et al.*, 1991; Hequette & Hill, 1993, 1995). Many of the scour-based beds with planar parallel stratification can be attributed to the tractional transport by unidirectional currents in the upper flow regime (Harms *et al.*, 1975), and the massive bed divisions may represent rapid, non-tractional dumping of sand from dense turbulent suspension (Dott & Bourgeois, 1982; Lowe, 1988; DeCelles & Cavazza, 1992; Myrow & Southard, 1996; Vrolijk & Southard, 1997). The packages of amalgamated, erosional calcarenite beds of facies D indicate an upper shoreface zone, whereas the bed packages with interlayers of silty marlstone (facies G) indicate lower shoreface. Their alternation in the stratigraphic succession (Fig. 20, log part above 35 m height) implies changes in the basin's wave climate and/or relative sea level (Clifton, 1981; Simpson & Eriksson, 1990; Walker & Bergman, 1993). The prograding system was clearly affected by relative sea-level changes, as is indicated by the stratigraphic alternation of facies subassociations 3b and 3c (Fig. 20, log interval 22–35 m).

The characteristics of facies subassociation 3c (Fig. 22), with the tempestites (facies D) and subordinate turbidites (facies B2 and A2) separated by marlstone (facies G) and commonly also mudstone layers (facies H1), indicate deposition in an offshore-transition zone, between the prevalent fairweather and storm wave bases, where hemipelagic sedimentation was punctuated by the episodic incursions of storm-derived sand. Violent storms would likely produce abundant intraclasts in the platform area and could spread some of the gravel seawards (Mount & Kidder, 1993; Seguret *et al.*, 2001; see also Forbes & Boyd, 1987; Leckie, 1988). Sheet-like sandstone beds with planar parallel stratification, hummocky or swaley stratification and wave-ripple cross-lamination are widely attributed to combined-flow currents (Dott & Bourgeois, 1982; Leckie & Walker, 1982; Nøttvedt & Kreisa, 1987; Arnott & Southard, 1990). When heavily laden with sediment, storm-generated currents can be boosted by density and result in deposits resembling turbidites (Hamblin & Walker, 1979; Handford, 1986; Myrow & Southard, 1996), or may plunge beneath the wave base and ignite into true turbidity currents (Walker, 1969, 1984b; Myrow & Southard, 1996). These phenomena might explain the occurrence of both facies D and facies A2/B2 in the lower part of the progradational succession of facies subassociation 3c (Fig. 22), which is thought to have recorded a decrease in depositional slope from ~ 1–2°, steep enough for turbidity current ignition (Parker, 1982; Fukushima *et al.*, 1985), to probably ≤ 0.1° (cf. Handford, 1986; Wright & Bruchette, 1996).

It is worth noting that some of the isolated tempestites of facies D in outcrops a few tens of kilometres away from the palaeoshoreline (Figs 20 & 22) are much thicker and coarser grained than the offshore deposits of great modern storms (e.g. Gagan *et al.*, 1988; Snedden & Nummedal, 1991; Hubbard, 1992; Keen & Slingerland, 1993). For example, the September 1961 hurricane Carla and the August 1980 hurricane Allen on the Texas shelf spread sand to offshore distances of ~ 50 km, but the resulting tempestite in each case was fine-grained and ≤ 6 cm thick (Snedden *et al.*, 1988). The thicker beds in the present case have gently undulating tops and were probably deposited either by some very rare, extreme storms, unwitnessed in modern times, or by the combined-flow relaxation surges of tsunami events (Dott & Bourgeois, 1982; Saito, 1989; Myrow & Southard, 1996; Massari & D'Alessandro, 2000; Rossetti *et al.*, 2000). The basin was tectonically active and earthquake-generated tsunamis almost certainly occurred (cf. Altınok & Ersoy, 2000; Bryant, 2001; Hebenstreit, 2001), and these could episodically spread abundant sand and beach-derived gravel to the offshore zone (cf. Saito, 1989; Yamazaki *et al.*, 1989; Young & Bryant, 1992; Yeh *et al.*, 1994; Cantalamessa & Di Celma, 2005).

The reefal carbonate platform (facies subassociation 3a) expanded basinwards over a prograding inner ramp composed of facies subassociations 3b and 3c (Fig. 11), with the latter passing seawards into and underlain by the alternating turbiditic facies subassociations 2a and 2b (Figs 11 & 14). The sandy shoreface zone was subject to perennial wave action and had a relatively high aggradation rate compared with the offshore zone, which was receiving sand only episodically during the strongest storms. The differential aggradation is thought to have resulted in a distally steepened inner ramp (Fig. 18, inset; cf. Read, 1985), with a mid-ramp slope of possibly up to 3–5° ('shelf break' *sensu* Plink-Björklund *et al.*, 2001), favouring the formation of bypass chutes (Fig. 14) by the most powerful turbidity currents. The succession as a whole indicates dramatic shallowing of the basin, similar to that described by several others (e.g. Handford, 1986; Monaco, 1992; Johnson & Baldwin, 1996; Vera & Molina, 1998; Bádenas & Aurell, 2001). The thickness of facies subassociations 3b and 3c decreases markedly towards the basin margin (cf. Figs 14 & 20), which reflects progressive erosion of the inner ramp by waves as a result of forced regression attributed to the basin margin uplift (Figs 11 & 18). This gradual cannibalization of high-productivity carbonate platform and sand-prone inner ramp would explain the large amount of sand transferred to the basin-floor turbiditic system, causing its strong aggradation (Fig. 11). The scarcity of ichnofauna in the shoreface deposits of facies subassociation 3b, with only one trace of *Ophiomorpha ?nodosa* found, is consistent with the notion of rapid shallowing and a substrate intensely reworked by waves.

The mixed assemblage of *Zoophycos* and *Nereites* ichnofauna in facies subassociation 3c and the occurrence of typical *Nereites* ichnofacies in the underlying, shallow subneritic turbidites are rather unusual facts deserving special attention. A similar relationship in the basin is repeated at the top of the Kusuri Formation (Fig. 2; Uchman *et al.*, 2004; Janbu *et al.*, this volume, pp. 457–517). A somewhat analogous case was also reported from a Late Cretaceous open-marine African shelf (Gierlowski-Kordesch & Ernst, 1987; Ernst & Zander, 1993), where the shallow occurrence of *Nereites* ichnofauna was attributed to the existence of protected, deep intra-shelf troughs only sporadically affected by the strongest storms.

It has been suggested (Pemberton *et al.*, 1992; Pervesler & Uchman, 2004) that the main factors controlling ichnofauna ecology are not so much the bathymetry or distance from shoreline, but rather the substrate type, near-bottom hydraulic energy, sediment deposition rate, water turbidity, oxygen and salinity levels, and the quality and quantity of nutrient supply. In this respect, a storm-punctuated offshore-transition environment could resemble a turbiditic one, whereby some of the *Nereites* trace-makers might survive when the latter environment evolved rapidly into the former. A critical change in ecological conditions would be expected with the onset of littoral sedimentation, as is indeed indicated by the sparse *Skolithos* ichnofacies in the overlying shoreface deposits of facies subassociation 3b. The deep-water basin lacked stable, well-developed shelf habitats, and hence no expansion of a *Cruziana* ichnofacies occurred when the seafloor reached neritic depths. The extensive carbonate platform sheltered the basin from land sources of plant detritus, which may explain the moderate diversity of ichnofauna and relative scarcity of *Ophiomorpha annulata* (Uchman *et al.*, 2004). The present evidence indicates that the bathymetric upper limit for *Nereites* ichnofacies in a rapidly shallowing basin may be a neritic environment little more than 100 m deep.

Facies association 4: transgressive platform cover

This facies association constitutes the Atbaşı Formation (Fig. 2), including its transition from the underlying Akveren Formation. The succession is ~ 200 m thick and has a roughly layer-cake stratigraphy comprising three facies assemblages (Fig. 11): littoral calcarenites with subordinate basal calcirudites (subassociation 4a, Fig. 23A); neritic to

Fig. 23 Calcarenitic facies subassociation 4a. (A) Facies subassociation 4a erosionally overlying the reefal platform limestones of subassociation 3a. (B) Lower part of facies subassociation 4b, composed of densely alternating calcarenite sheets and calcareous hemipelagic deposits. Outcrops near Kuğuköy in the western part of the basin (locality 1, Fig. 3).

subneritic, calcareous mudstones intercalated with thin marlstones and sheet-like calcarenite beds (subassociation 4b, Fig. 23B); and deep-water, variegated calcareous mudstones interspersed with marlstone layers and sporadic thin calcarenite sheets (subassociation 4c). This deepening upward, transgressive succession clearly recorded a rapid, dramatic rise in relative sea level. The first two facies assemblages form the transitional basal part of the Atbaşı Formation, which is 20–25 m thick in the central part of the basin, but thinner towards the east/southeast, where subassociation 4a pinches out, and also towards the west/southwest – where this subassociation is reduced to local gravelstone patches and where subassociation 4b is intercalated with bioclastic limestones of retreating reefal platform (Figs 11 & 24). Microfauna in the mudstones of subassociation 4c indicates an early Eocene age for the Atbaşı Formation and a palaeowater depth of > 200 m, which is consistent with an impoverished *Nereites* ichnofacies including *Planolites*, *Ophiomorpha annulata*, *Phycodes*, *Planolites*, *Nereites* and *Scolicia strozzii* (Uchman *et al.*, 2004).

Accordingly, the boundary between the Akveren and Atbaşı formations is considered to be a marine flooding surface. It is taken at the erosional top of the drowned carbonate platform in the western to central part of the basin (Figs 22A & 24), but is more arbitrary in the eastern part, where the transgressive subassociations 4a and 4b overlie similar regressive deposits of subassociations 3c and 3b (Fig. 11). A visual difference recognizable in the field is that the calcareous mudstone interbeds in subassociation 4b vary in colour from dark grey to pinkish or reddish grey, in contrast to the predominantly grey mudstone interlayers in the underlying subassociation 3b. Furthermore, the calcarenites of subassociations 4a and 4b are somewhat richer in volcanic-rock detritus, including rounded granules. Bioclasts represent the same foramol-type reefal source and include fragments of bryozoans, echinoderms, bivalves, nummulite foraminifers and coralline red-algal colonies.

Facies subassociation 4a

This lowest assemblage (Fig. 23A) consists of the calcarenitic facies D interlayered with minor marlstones (facies G), thinly bedded limestones (facies F1) and calcareous mudstones (facies H1), and is locally underlain by or intercalated with the gravelstone facies E (Table 1). Calcarenite beds are mainly amalgamated (Fig. 25, lower part), with planar erosional boundaries and common burrows. The sporadic marlstone and calcareous mudstone interlayers are thin, silty and bioturbated. Trace fossils include *Chondrites*, *Thalassinoides*, *Helminthopsis* and *Ophiomorpha nodosa*. The calcarenites are coarse- to very fine-grained, generally well-sorted and also well-stratified, showing planar parallel stratification, hummocky and swaley stratification (Fig. 5H & I) and a range of wave-ripple cross-lamination types (similar to those in Fig. 5C–F). The underlying gravelstone facies E occurs as broad lenses (patches), mainly non-stratified and < 20 cm thick, resting on the uneven erosional top of the platform limestone unit.

Where intercalated with calcarenites (Fig. 25, lower part), the gravelstone facies E forms one or more wedges, ~ 1–2 m thick, which are composed of amalgamated beds 15–70 cm thick, gently inclined (< 10°) basinwards (Fig. 5K) and slightly undulating in a direction parallel to the inferred palaeoshoreline. The beds are mainly planar parallel-stratified, have planar or slightly inclined erosional boundaries and consist of a submature to mature gravel of granule to cobble grade, with predominantly flat fabric and a clast-supported framework filled with sand. Gravel clasts are rounded fragments of limestone (fine packstone), marlstone and minor volcanic rock. Many beds consist solely of sand-filled, well-rounded to subrounded granule gravel (Fig. 5K), and some of the cobbly beds are rich in mudstone intraclasts armoured with granules (Fig. 5J).

Facies subassociation 4b

This overlying subassociation (Fig. 23B) consists of alternating calcareous mudstone (facies H1) and marlstone layers (facies G), densely interspersed with sheet-like calcarenite beds. The latter are mainly isolated, coarse- to very fine-grained and 2–20 cm thick, but occasionally amalgamated into thicker beds and containing scattered calcareous intraclasts (Fig. 25, upper part). Marlstones are mainly whitish-grey, whereas the colour of mudstones varies from grey and dark grey to pinkish- or light brownish-grey. The calcarenite beds in the

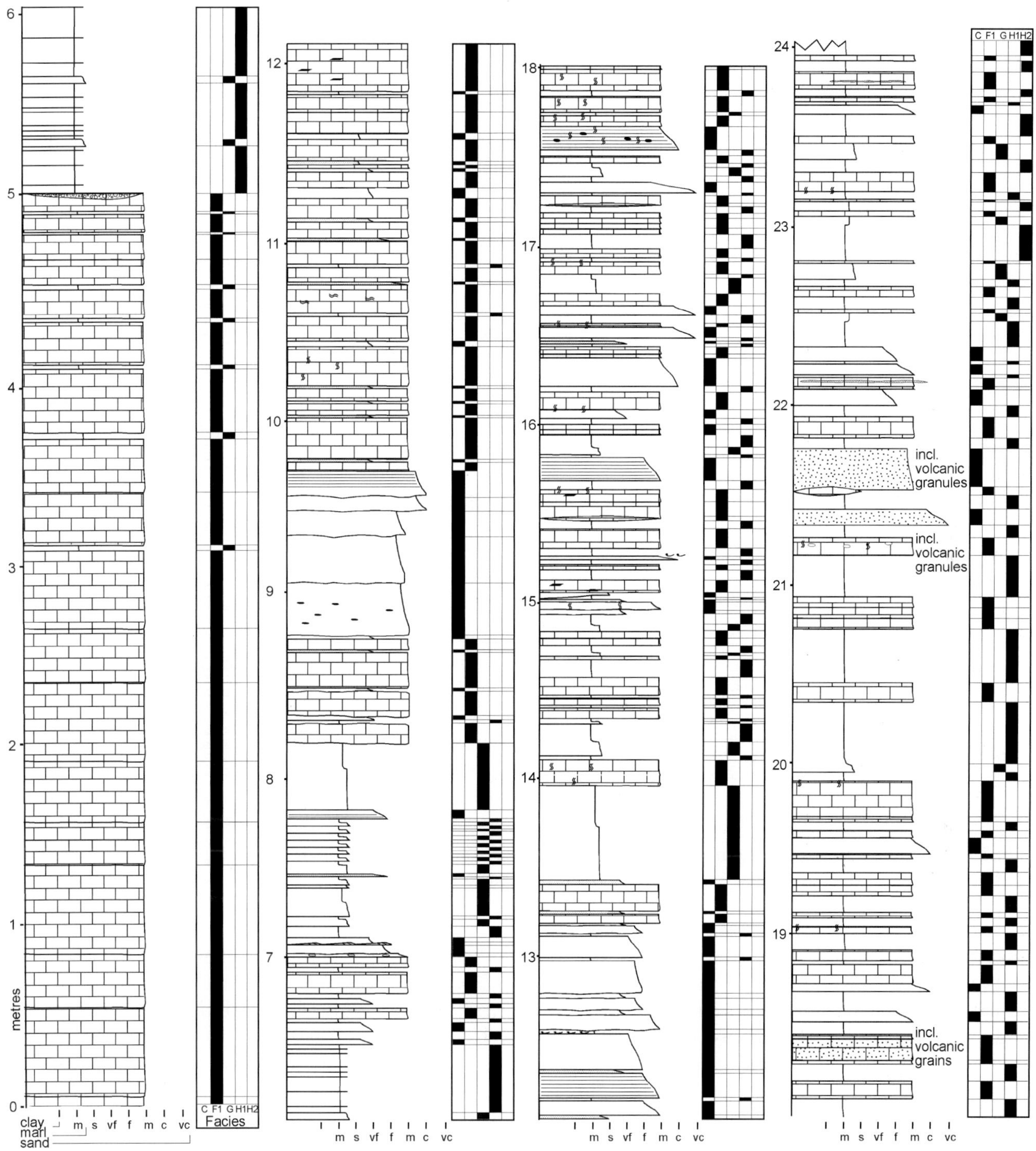

Fig. 24 Log of the lowermost part of the Atbaşı Formation. Alternating deposits of facies subassociations 4a and 4b overlie the reefal unit of subassociation 3a and contain numerous limestone interbeds. Outcrop section near Kuğuköy in the western part of the basin (locality 1, Fig. 3). For log legend, see Fig. 7.

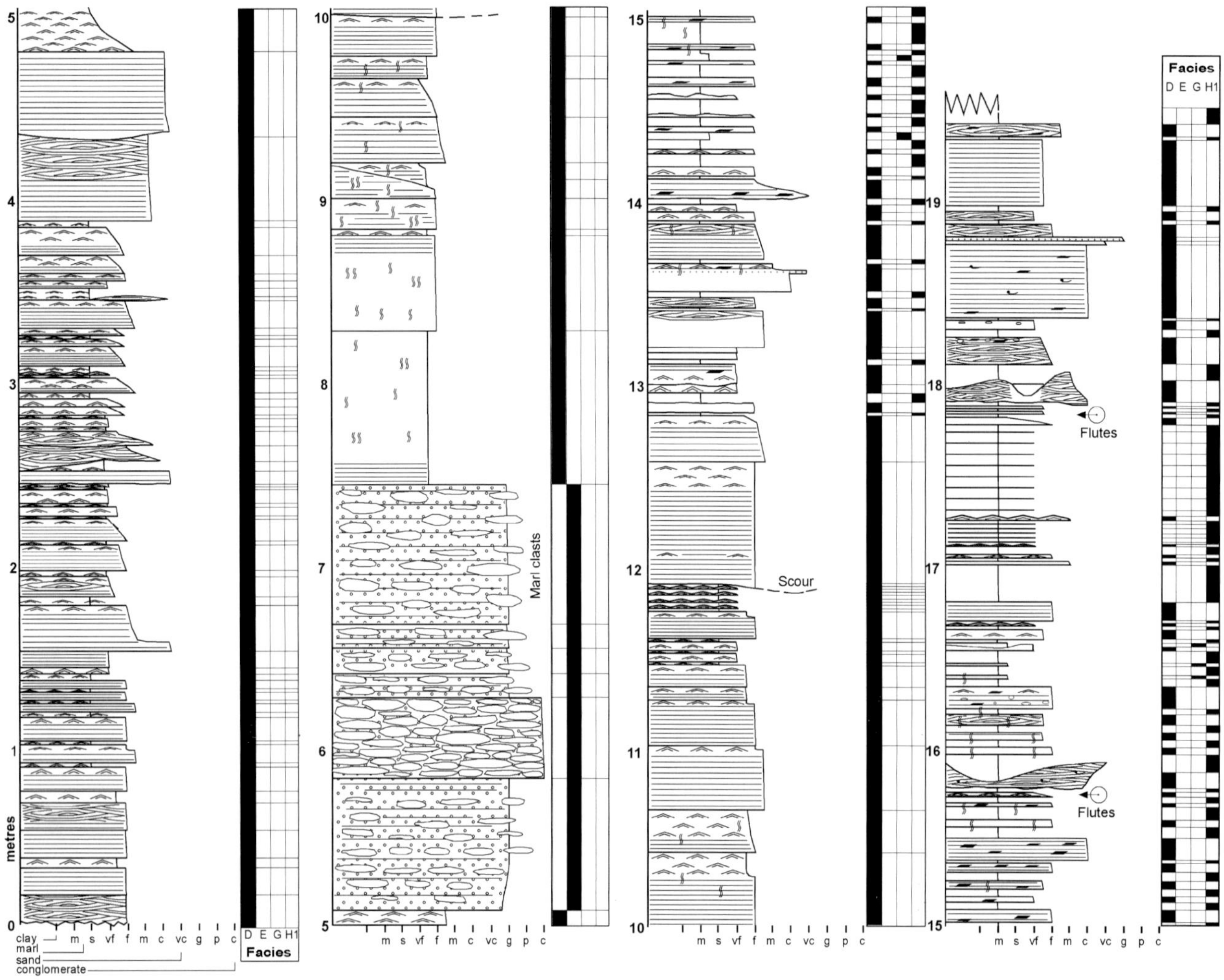

Fig. 25 Log of the upward transition from facies subassociation 4a (with a gravelly wedge of facies E) to subassociation 4b. Coastal outcrop west of Türkeli in the western part of the basin (locality 2, Fig. 3). For log legend, see Fig. 7.

lower part of the succession show features similar to the tempestites of facies D in subassociation 3c (see earlier text), including sharp boundaries and a planar parallel stratification commonly passing upwards into hummocky stratification and/or wave-ripple cross-lamination (mainly three-dimensional vortex ripples). In the upper part of the succession, most of the calcarenite beds lack features attributable to a combined flow or wave action and, instead, resemble closely the thin to moderately thick turbidites T*cd*, T*bcd* and T*abcd* of facies B2/A2 (described earlier in subassociation 2c). Translatory ripple cross-lamination and flute casts on bed soles indicate sediment transport towards the northeast and east. Trace fossils include *Ophiomorpha annulata*, *Phycodes*, *Belorhaphe*, *Chondrites*, *Thalassinoides* and *Zoophycos*, which indicate an impoverished *Nereites* ichnofacies (Uchman *et al.*, 2004).

This succession as a whole has a thinning and fining upward trend, as the calcarenite beds become thinner and finer grained and the net thickness proportion of hemipelagic deposits increases (Leren, 2003). In the western part of the basin, where subassociation 4b is ~ 25 m thick and only locally separated from the underlying reefal unit by gravelly deposits, the succession also includes numerous discrete interlayers of micritic limestone (massive packstones of facies F1), mainly 5–20 cm thick (Fig. 24).

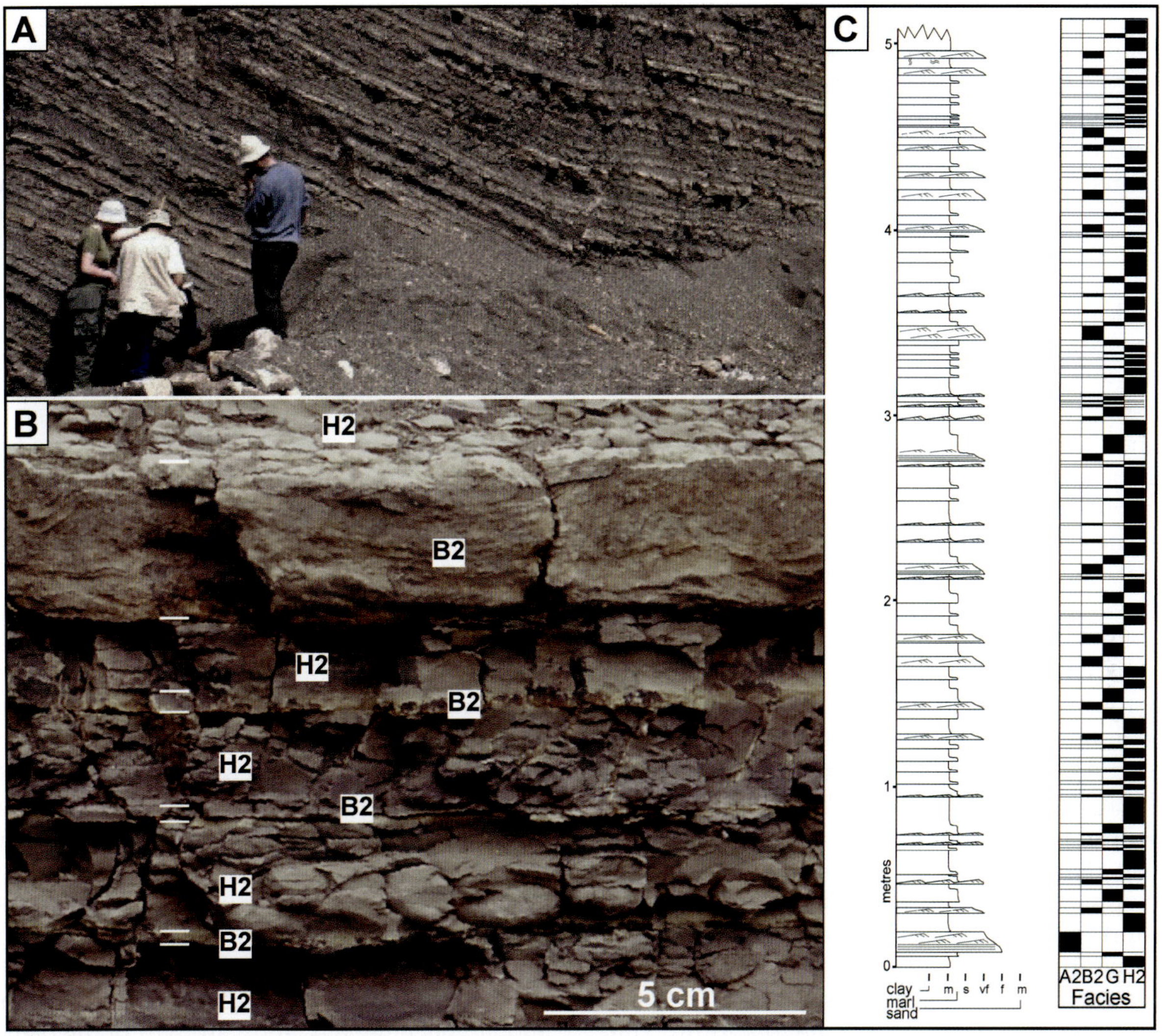

Fig. 26 Facies subassociation 4c. (A) Outcrop at Tangal (locality 5, Fig. 3). (B) Close-up detail of the variegated mudstones of facies H2 with thin interbeds of facies B2 calcarenites. (C) Log of subassociation 4c from the same outcrop section; for log legend, see Fig. 7.

Facies subassociation 4c

This upper facies assemblage (Fig. 26A & C) is ~ 175 m thick, has a transitional boundary with the underlying subassociation 4b and constitutes the main, higher part of the Atbaşı Formation. The deposits are predominantly variegated calcareous mudstones (facies H2) intercalated with thin marlstone layers (facies G) and isolated calcarenite beds (Fig. 26B). The calcarenite beds are mainly 1–10 cm thick, normally graded, fine- to very fine-grained and commonly silty. They lack features attributable to oscillatory waves or combined flow, and are categorized as turbidites T*cd* and sporadic T*bcd* or T*bd* (facies B2, Table 1). The turbidite spectrum includes common thin beds T(*c*)*d*, showing laterally discontinuous *c*-divisions and composed of a graded silty marlstone capped with mudstone. The mudstones are massive, burrowed (Fig. 5N) and mainly purple-red in colour, with irregular pinkish-grey and olive-green bands roughly parallel to bedding, but are brownish-grey to grey in the uppermost part of the succession, where they pass gradually into the dark grey non-calcareous mudstones of the overlying Kusuri Formation (Fig. 2; Janbu *et al.*, this volume, pp. 457–517). Identifiable trace fossils include *Planolites*, *Ophiomorpha annulata* and *Scolicia strozzii*, indicating an impoverished *Nereites* ichnofacies (Uchman *et al.*, 2004). Relatively few measurable palaeocurrent indices have been found, all showing southeastward transport directions (Leren, 2003).

Interpretation

The calcarenites of facies subassociation 4a show features similar to those of subassociation 3b (see earlier text) and are interpreted to be shoreface deposits, accumulated above the average fair-weather wave base. The sporadic marlstone and mudstone interlayers are relics of sedimentation during seasons when the wave base stayed above its mean level. The gravelstone facies E occurs as a basal transgressive lag, deposited as a horizon of gravel patches in substrate depressions, and also forms progradational foreshore wedges, with varying texture, inclined stratification and numerous internal truncations that indicate a wave-dominated reflective shoreline episodically scoured by storms (Bluck, 1967, 1999; Orford, 1977; Clifton, 1981; Massari & Parea, 1988; Komar, 1998). The undulation of strata in a direction parallel to depositional strike probably reflects the development of cusps on a beachface affected by storm waves (Sallenger, 1979; Caldwell & Williams, 1985; Sherman *et al.*, 1993; Bluck, 1999). The coarse debris was apparently derived by erosion of the inundated carbonate platform. The abundance of marlstone and granule-armoured mudstone clasts supports the notion that the platform originally included some protected lagoonal areas where mud was deposited (see Fig. 18 and earlier discussion of facies subassociation 3a).

The pinkish- to brownish-grey colour of some of the interlayers of platform-derived calcareous mud suggests that the landward parts of the carbonate platform were probably emerged during the preceding regression and subject to fersiallitic weathering. When swept by waves under transgression, these areas would episodically yield reddened mud. Fersiallitic weathering of limestones and development of red soils occurred in Anatolia through most of Cenozoic time, signifying warm, temperate to subtropical climatic conditions with alternating humid and dry seasons (Duchaufour, 1977).

The overlying succession of facies subassociation 4b, with the tempestites giving way to turbidites, is a reversed stratigraphic mirror-image of subassociation 3c (see earlier text). Also here the deep-marine ichnofauna appears to be incompatible with the neritic depositional environment and tempestitic facies of the lower part of the succession, although it corresponds well with its turbiditic upper part (Fig. 26). Facies subassociation 4b clearly recorded a deepening of neritic to subneritic environment, with frequent incursions of sand spread by storm-generated combined-flow and turbidity currents. The relative sea-level rise and landward shift of the basin shoreline might explain the gradual upward replacement of tempestites by turbidites, because the storm wave base would rise and the margin-onlapping shoreface would increase its gradient, promoting the ignition of turbidity currents (Parker, 1982; Fukushima *et al.*, 1985).

The foreshore gravel wedges in subassociation 4a and the basinward-thinning limestone sheets in subassociation 4b in the western part of the basin indicate that the initial drowning of the carbonate platform occurred in a stepwise manner, with brief transgressions followed by immediate shoreline readvances due to high sediment supply (normal regressions). The rising sea level and wave action must have reactivated the abandoned and largely emerged landward part of the platform as a sediment source, shedding abundant gravel and sand during storms. The sea-level rise then markedly accelerated, which brought about deep-water conditions and the deposition of facies subassociation 4c. The shoreline shifted landwards and the basinward flux of calcareous sand declined, whereby hemipelagic sedimentation prevailed in the basin. The dramatic decrease in sediment supply led to seafloor oxidation, reflected in the purple-red colouration of the variegated mudstones of subassociation 4c (cf. Franke & Paul, 1980; Görür *et al.*, 1993; Dreyer *et al.*, 1999; Eren & Kadir, 1999). The isolated, thin calcarenite sheets represent sporadic low-density turbidity currents, triggered probably by some of the strongest storms and/or by the shedding of sediment from underwater basin-margin slopes by earthquakes.

DISCUSSION

The origin and spatial organization of the facies associations (Figs 11 & 18) and the depositional setting of the three formations interpreted above shed new light on the history of the Sinop–Boyabat Basin (Fig. 18). The interpreted tectono-palaeogeographical evolution of the basin during Campanian to Ypresian time is depicted in Fig. 27 and discussed further in this section, with references to the regional information reviewed at the beginning of the paper.

The early rifting phases

The first phase of rifting, which formed the Sinop–Boyabat Basin, occurred in Barremian–Albian time and was recorded by the lower to middle part of the Çağlayan Formation (Fig. 2). The formation's upper part indicates cessation of olistostromal massflow processes and a gradual decline of coarse sediment supply, which culminated in the sand-starved depositional conditions of the overlying Kapanboğazı Formation (Fig. 2). This post-rift phase of sedimentation (Cenomanian–Coniacian) implies that the shorelines shifted away from the graben, with main sediment accumulation outside the rift proper.

The second phase of rifting, recorded by the Yemişliçay Formation (Fig. 2), occurred in Santonian to early Campanian time and was accompanied by strong volcanism. The sedimentation involved mixed volcaniclastic and calcareous turbidity currents, pyroclastic currents and lava flows derived from both basin margins (Fig. 27A), perhaps more from the northern side (Aydın *et al.*, 1995b). The dispersal of abundant sediment was controlled by an uneven seafloor topography created by fault blocks, burying them gradually and smoothing out the basin floor. With nearly 1500 m of sediment and lavaflow basalts deposited in little more than 10 Myr and no recognizable shallowing, the rate of basin-floor subsidence must have kept pace with the rate of sediment accumulation and was considerably higher than during the first rifting phase, when a similar sediment thickness was deposited in ~ 20 Myr. The Sinop–Boyabat Basin formed as a southern sister-branch of the Western Black Sea Rift, extending towards the southeast (Fig. 1A), and the tectonic subsidence driven by crustal stretching probably combined with the effect of sediment compaction and the increasing sedimentary load on the crust – similar to that in the adjacent main rift (Cloetingh *et al.*, 2003; Nikishin *et al.*, 2003).

Onset of the compressional foreland regime

The uppermost part of the Yemişliçay Formation (Fig. 2) recorded cessation of volcanism in the Central Pontides (Okay *et al.*, 2001) and consists of mudstones intercalated with mainly thin turbidites, which indicates the onset of a post-rift phase in the Sinop–Boyabat Basin. Volcanic activity persisted until the end of the Campanian in the Western Pontides (Robinson *et al.*, 1995; Tüysüz, 1999; Sunal & Tüysüz, 2002) and until the Early Paleocene in the Eastern Pontides (Okay & Şahintürk, 1997; Okay & Tüysüz, 1999). Only minor volcanism re-occurred briefly in the adjacent Kastamonu Basin in Eocene time (Güven, 1977).

The post-rift phase was interrupted by a pulse of compression that affected the western part of the basin (Fig. 27B), activating a southwestern source of siliciclastic (chiefly epiclastic volcanic) sediment and leading to the deposition of the Gürsökü Formation in late Campanian to Maastrichtian time. The tectonic compression is attributed to the collision of the Kırşehir Massif with the volcanic arc of the southern margin of the Cimmerian zone (Fig. 1A), beginning with its northward indention in the transitional area of Western and Central Pontides. The bedrock Küre Complex and adjacent Cide Uplift (Fig. 1A) were elevated in the late Campanian, to be gradually submerged again in Maastrichtian time (Aydın *et al.*, 1995a; Tüysüz, 1999).

The Gürsökü Formation is ≤ 1200 m thick, thickening towards the southeast, and spans a period of < 10 Myr, which indicates a new phase of high sediment supply. The microfossils and ichnofauna indicate deep water and there is no facies evidence of basin shallowing, which means that the subsidence rate generally kept pace with the rate of sediment accumulation. The formation consists of siliciclastic turbidites that are increasingly richer in bioclastic admixture upwards in the succession, and the evidence of a foramol-type

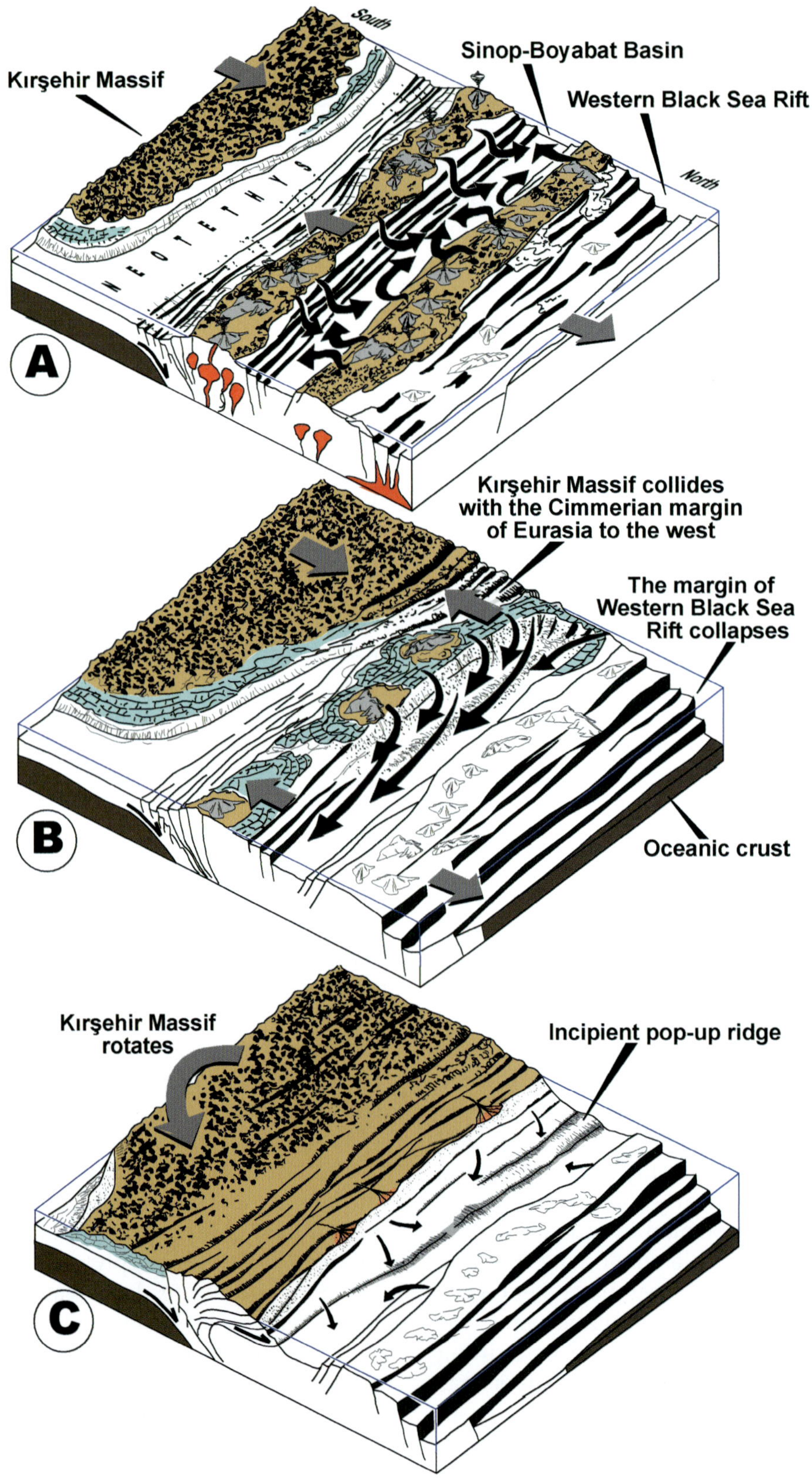

Fig. 27 Interpreted tectono-palaeogeographical development and sediment dispersal pattern in the Sinop–Boyabat Basin. (A) Early Campanian time. (B) Maastrichtian time (cf. Fig. 18). (C) Paleocene–Eocene transition. Schematic reconstruction, discussed in the text; for subsequent stages of this cartoon, see Janbu *et al.* (this volume, pp. 457–517, fig. 27).

source implies development of a contemporaneous reefal platform. The carbonate platform probably formed in the western part of the basin and gradually expanded eastwards along the basin's rising southwestern margin (Fig. 27B). Palaeocurrent directions are mainly to the northeast, turning towards the east and southeast in the central to eastern part of the basin. There is no evidence of significant sediment supply from the basin's northern margin, which was probably submerged below wave base, pulled down with the collapsing flank of the adjacent Western Black Sea Rift (Fig. 27B).

The Gürsökü Formation was deposited as the medial to distal part of a basin-floor turbiditic system. There is no evidence of deltaic deposits beneath the basin-margin carbonate platform (Fig. 19A), and the epiclastic sediment of volcanic provenance, rich in glass shards, does not support the notion of a major fluvial supply, from which a wider range of 'exotic' detritus might be expected. Apart from their bioclastic admixture, the successive turbidites lack any obvious differences in mineral composition, which also does not lend support to the notion of a 'multi-point source' (*sensu* Reading & Richards, 1994). The turbiditic system is thought to have been fed by a 'line-source' littoral ramp, which was probably narrow, perched on the steep basin margin, and was cannibalized by the margin uplift caused by thrusting.

The depositional system was dominated by non-channelized turbidity currents of low to high density (facies subassociation 1a), which probably originated from the ramp slumping and from the ignition of sediment-laden, storm-generated currents that plunged beneath the effective wave base on the steep underwater slope of the basin margin. Both storms and earthquakes would probably generate wide, non-channelized flows, which would coalesce further on the southeast-inclined floor of the elongate basin. A similar depositional setting has been suggested, for example, for some of the Paleocene–Eocene turbiditic systems in the Viking Graben, North Sea (Rochow, 1981; Lovell, 1990; Bowman, 1998), and the Eocene turbidites in the Tyee Basin of western Oregon (Chan & Dott, 1983).

It is possible that the dispersal system included channels in its proximal part, for at least one isolated and apparently sinuous palaeochannel (facies subassociation 1b) occurs in the lowermost part of the formation, which itself is thicker bedded and thought to represent the system's medial part. The bed-thickness upward trend (plots 1 and 2 in Fig. 8) and the lack of palaeochannels at higher stratigraphic levels suggest an overall back-stepping of the turbiditic system, which itself was apparently dominated by aggradation, rather than progradation. This notion is supported by the impressive thickness and monotonous character of the whole middle to upper part of the succession (cf. plot 2 in Fig. 8), with no recognizable upward change other than a gradual predominance of increasingly calcareous and low-density turbidity currents (Leren, 2003). The pronounced aggradation can be attributed to the ponding of turbidity currents in a 'blind-end' basin, where the distal eastern part was topographically closed. As a result, the head zone of the confined turbiditic system would retreat by backlapping the proximal basin-margin slope (McCaffrey & Kneller, 2001; Smith & Joseph, 2004).

The proximal part of the turbiditic systems might have one or more 'perennial' channels, which could be shifting by avulsion and occasionally extending to the medial zone; or might involve only ephemeral channels, formed by series of unusually large and robust currents generated by the ramp's rapid retrogressive slumping in response to earthquakes. Alternatively, ephemeral channels might have formed in the system's medial part only, by turbidity currents that were locally confined and boosted by seafloor topography, as discussed further below.

The Sinop–Boyabat Basin at this stage was ~ 80 km wide and at least 200 km long, which allowed the vast majority of currents to spread widely and deposit sheet-like turbidites. No distinct depositional pattern of channels and lobes is recognizable, but it cannot be precluded that the thick succession resulted from the vertical stacking of wide and poorly defined depositional lobes (cf. Stow *et al.*, 1996; McCaffrey & Kneller, 2001). It is possible that the largest turbidity currents were basin-wide (cf. Ricci Lucchi & Valmori, 1980; Chan & Dott, 1983; Smith & Joseph, 2004), but the eastward palaeocurrent directions and lack of radial dispersal can be attributed to the eastward basin-floor inclination and possible development of gentle blind-thrust anticlines (Fig. 18), rather than to the basin confinement as such. A gentle synclinal swale, before being buried, could briefly confine and boost turbidity currents, which might also explain the origin of the isolated palaeochannel in the present case. The recognition of subtle palaeotopographic features in a tectonically deformed and discontinuously exposed sedimentary succession is a formidable task, but there is evidence of local synsedimentary folding (intrabasinal slumps) and lateral pinch-out of beds (Leren, 2003) that supports the notion of a seafloor subject to mild syndepositional deformation.

Cessation of rifting and basin shallowing

The upward transition of the Gürsökü Formation to the Akveren Formation (Fig. 2) marks a Late Maastrichtian predominance of sediment supply from the basin-margin carbonate platform. The Akveren Formation is no more than 600 m thick and spans a period of ~ 10 Myr, which indicates a decrease in sedimentation rate; and the shallowing of facies in the uppermost part of the formation (Fig. 11) implies a marked decline in subsidence rate. The basin at this stage was probably decoupled from the extensional regime of the Western Black Sea Rift (Fig. 27B), while the latter broke up and underwent seafloor spreading (Okay & Şahintürk, 1997; Nikishin *et al.*, 2003). The 'failed' Sinop–Boyabat rift thus turned into the Central Pontides' retroarc foreland basin (*sensu* Dickinson, 1974), and gradually filled up with sediments by middle Late Paleocene time. As the reefal littoral platform expanded along the basin's southwestern margin, the basin-floor turbiditic system began to be supplied with sediment from an extensive carbonate ramp (Fig. 18), which gradually advanced in the basin. Eastward sediment dispersal and ponding of currents persisted, and the basin's accommodation began to be exhausted. A marked shallowing of the basin is evidenced by the deep-water turbiditic association 2 overlain by the regressive, neritic to littoral facies association 3, topped with the reefal limestones (Fig. 11).

The foramol-type reefal platform included large sand shoals worked by waves and some protected lagoonal areas, locally ≥ 20 m deep, where bryozoans thrived and calcareous mud was deposited (Fig. 18). The platform had a wave-dominated reflective shoreline, turning into a dissipative shoreline during storms, and was linked to the basin-floor turbiditic system by a distally steepened ramp. The inner ramp was a wave-dominated, sand-prone shoreface extending out from a gravelly beach zone (facies subassociation 3b) and passing basinwards into a storm-dominated offshore-transition zone (facies subassociation 3c). The turbidity currents are thought to have been generated by storms and triggered by earthquakes, and largely bypassed the neritic outer ramp by forming transient chutes (facies subassociation 2b, Fig. 14). Periodic accumulation of abundant sand in the offshore zone, involving multiple chutes (facies subassociation 2a), is attributed to episodes of basin-margin uplift by orogenic thrusting (Fig. 11), which would probably cause cannibalization of the carbonate platform and associated shoreface by wave erosion.

As the rate of basin-floor aggradation outpaced the rate of subsidence, the carbonate ramp became homoclinal (Read, 1985), the ignition of turbidity currents declined and the seafloor became increasingly influenced by storm-generated combined-flow currents. This gradual change is evidenced by the upward transition of the turbiditic facies association 2 into the tempestitic facies subassociation 3c overlain by the shoreface calcarenites of subassociation 3b (Figs 11 & 14). The rapid shallowing allowed basinward expansion of the carbonate platform over a few tens of kilometres, while the platform's landward part was probably emerged by tectonic uplift and subject to fersiallitic weathering and denudation.

Rapid subsidence and sea-level rise

The subsequent marine transgression, leading to the deposition of the Atbaşı Formation in latest Paleocene to earliest Eocene time, indicates a dramatic rise in relative sea level. The rise was initially punctuated by brief normal regressions (facies subassociations 4a and 4b; Fig. 11) and could be eustatic, but was greatly accelerated as a result of the foreland subsidence due to the crustal loading by Central Pontide thrust sheets (Fig. 27C). These basal facies assemblages form a retrogradational, back-stepping parasequence set of a transgressive systems tract, separated by a type 2 sequence boundary (*sensu* Posamentier *et al.*, 1988) from the underlying highstand systems tract.

Global sea level is known to have risen dramatically in Thanetian time (see the eustatic supercycle TA2 of Haq *et al.*, 1988), which must have affected the basin. Effects of this eustatic change are recognizable also in other basins of the Pontides (e.g. Aydın *et al.*, 1995a; Tüysüz, 1999). The Kırşehir Massif had meanwhile rotated counter-clockwise and moved further northwards (Fig. 27C), adjusting its position with respect to the Cimmerian margin in response to the onset of the Tauride orogeny to the south (cf. Fig. 1A). The resulting oroclinal indentation led to emplacement of the Central Pontide nappes, which loaded the Cimmerian crust and caused flexural subsidence of the foreland zone (cf. Beaumont, 1981; DeCelles & Giles, 1996).

The Atbaşı Formation, deposited in ~ 5 Myr, is little more than 200 m thick and its whole middle to upper part consists of deep-water mudstones (facies subassociation 4c), which implies a rapid increase in basin accommodation and a marked decline in sediment supply. The variegated mudstones (facies H2) indicate a sand-starved basin with a very low sedimentation rate and extensive seafloor oxidation.

Subsequent development

The youngest Kusuri Formation (Lower–Middle Eocene; Fig. 2) is a siliciclastic turbiditic succession sourced from the east and capped with calciclastic littoral deposits. The development of this formation and the corresponding, latest part of the tectonic history of the Sinop–Boyabat Basin are discussed by Janbu *et al.* (this volume, pp. 457–517). The reversal of transport direction and change in sediment provenance are attributed to the uplift and fluvial denudation of the adjacent foreland zone of the Eastern Pontides, coeval with an active tectonic subsidence of the Central Pontide foreland. The tectonic inversion of the Central Pontide foreland basin near the end of Eocene time is attributed to the climax of the Tauride orogeny, which pushed the Kırşehir Massif further to the north and closed the basin by forming a series of thrusts led by the Balıfakı thrust (Fig. 3).

Eocene deltaic and alluvial deposits occur only in the wedge-top Boyabat trough, where an axial fluvio-deltaic system prograded along the basin. The deltaic feeder of the turbiditic Kusuri Formation in the foredeep Sinop trough was switched off prior to the basin shallowing and is not preserved, because of the excessive uplift of the basin's easternmost part. Repetitive tectonic uplift by orogenic thrusting and accompanying erosion may also explain the lack of pre-Eocene terrestrial deposits along the southwestern margin of the Sinop–Boyabat Basin.

CONCLUSIONS

The study has reconstructed the depositional history of a deep-marine, basin-floor turbiditic system that recorded the tectonic transformation of the Sinop–Boyabat Basin from a failed backarc rift into a retroarc foreland basin of the Central Pontides. The sediment source changed gradually from epiclastic volcanic into reefal bioclastic, while the turbiditic system underwent remarkable aggradation (~ 1800 m rock thickness), enhanced by the ponding of turbidity currents in a 'dead-end' basin. The basin floor eventually reached littoral bathymetry, which allowed basinward expansion of a carbonate platform, terminated by a dramatic rise in relative sea level.

The upper Campanian to lower Maastrichtian Gürsökü Formation represents a siliciclastic turbiditic system directed towards the east (NE–ESE), supplied with recycled volcanic detritus and increasingly more abundant bioclastic sediment from the basin's southwestern margin. The northern margin was submerged below wave base and supplied little sediment. The sheet-like turbidites indicate basin-wide currents of low to high density, and the succession represents transition from the medial to distal part of a back-stepping turbiditic system. At least one isolated sinuous palaeochannel occurs in the lowermost, thicker bedded part of the succession. This solitary channel could have formed due to a temporal confinement of currents by blind-thrust anticlines or could be a conduit extended from the proximal and possibly channelized part of the system. The sediment was supplied from a storm-dominated littoral ramp perched on the basin margin, and the ponded, aggrading basin-floor system was subject to a gradual retreat by backlapping the margin. Bioclastic admixture indicates development of a reefal platform at the basin margin.

The overlying, upper Maastrichtian to Paleocene Akveren Formation recorded the predominance of a reefal carbonate source along the basin margin. This calcareous succession of sheet-like turbidites represents a non-channelized, aggrading basin-floor system. The sediment was supplied from a distally steepened ramp with bypass chutes, as the eastward dispersal direction and flow ponding persisted. As the turbiditic system aggraded, the ramp became homoclinal and the ignition of turbidity currents declined, giving way to sublittoral tempestitic sedimentation. The rate of sediment accumulation thus outpaced the subsidence rate, and this imbalance culminated in a rapid shallowing of the basin. The uppermost part of the formation is dominated by tempestites, with shoreface calcarenites and a reefal limestone unit at the top.

The uppermost Paleocene to lowest Eocene Atbaşı Formation recorded a dramatic rise in relative sea level, which initially involved shoreline readvances, but was greatly accelerated when rapid subsidence occurred due to the crustal loading by the Central Pontide nappes. Transgressive basal shoreface and offshore-transition deposits are overlain by deep-water, variegated calcareous mudstones interspersed with thin calcareous turbidites. The deposits indicate a sand-starved basin with a very low sedimentation rate and widespread seafloor oxidation.

The overlying Eocene Kusuri Formation represents a channelized turbiditic system directed towards the west (west-northwest), which recorded further subsidence combined with an abundant supply of siliciclastic sediment from the east, from the uplifted foreland of the adjacent Eastern Pontides (Janbu *et al.*, this volume, pp. 457–517). The topmost calciclastic deposits of this formation recorded shallowing of the basin and the final stages of its tectonic closure (Janbu, 2004).

The Upper Cretaceous to lower Eocene sedimentary succession thus provides a legible record of the tectonic evolution and palaeogeographical history in the Sinop–Boyabat Basin. The first pulse of compression, which affected the basin's western part, occurred in the late Campanian. In late Maastrichtian time, the basin was apparently decoupled from the extensional regime of the Western Black Sea Rift and became fully controlled by the compressional regime of the Pontide orogeny, with flexural subsidence due to the crustal loading by thrust sheets and with the basin floor increasingly affected by thrusts. The latter eventually inverted the basin in late Eocene time, during the climax of the Tauride orogeny to the south. The study points to a differential development of the foreland zones of the Central and Eastern Pontides and sheds more light on the geological history of the Central Pontides and the southern Black Sea region.

ACKNOWLEDGEMENTS

The study was sponsored by the Norsk Hydro Research Centre under a research programme co-ordinated by Ole J. Martinsen, whose kind interest, support and stimulating discussions are greatly appreciated. We also thank Mehmet C. Alçiçek, Łukasz Gągała and Volkan Özaksoy for field assistance; Alfred Uchman for identification of trace fossils in the field; and Birkan Alan, Emin Erkan, Enis K. Sagular and Ercüment Sirel for micropalaeontological analyses of our sediment samples. The manuscript was critically reviewed by Lawrence Amy, John Howell, Gilbert Kelling, Allard Martinius, Carlo Messina and Gary Nichols, whose constructive comments are much appreciated by the authors.

REFERENCES

Abreu, V., Sullivan, M., Pirmez, C. and Mohrig, D. (2003) Lateral accretion packages (LAPs): an important reservoir element in deep water sinuous channels. *Mar. Petrol. Geol.*, **20**, 631–648.

Adams, A.E. and MacKenzie, W.S. (1998) *A Colour Atlas of Carbonate Sediments and Rocks under the Microscope*. Manson Publishing, London, 108 pp.

Adams, A.E., MacKenzie, W.S. and Guilford, C. (1984) *Atlas of Sedimentary Rocks under the Microscope*. Longman, Harlow, 104 pp.

Akıncı, Ö. (1984) The Eastern Pontide volcano-sedimentary belt and associated massive sulphide deposits. In: *The Geological Evolution of the Eastern Mediterranean* (Eds J.E. Dixon and A.H.F. Robertson), pp. 415–428. Special Publication 17, Geological Society, London.

Aktaş, G. and Robertson, A.H.F. (1990) Tectonic evolution of the Tethys suture zone in SE Turkey:

evidence from the petrology and geochemistry of Late Cretaceous and Middle Eocene extrusives. In: *Ophiolites – Oceanic Crustal Analogues* (Eds J. Malpas, E.M. Moores, A. Panayiotou and C. Xenophontos), pp. 311–328. Proceedings of the International Symposium, Troodos 1987, Cyprus Geological Survey, Nicosia.

Altınok, Y. and Ersoy, Ş. (2000) Tsunami observed on and near the Turkish coasts. *Nat. Hazards*, **21**, 185–205.

Andrew, T. and Robertson, A.H.F. (2002) The Beyşehir–Hoyran–Hadım nappes: genesis and emplacement of Mesozoic marginal and oceanic units of the northern Neotethys in southern Turkey. *J. Geol. Soc. London*, **159**, 529–543.

Arnott, R.W.C. (1993) Quasi-planar laminated sandstone beds of the Lower Cretaceous Bootlegger Member, north-central Montana: evidence of combined-flow sedimentation. *J. Sediment. Petrol.*, **63**, 488–494.

Arnott, R.W.C. and Southard, J.B. (1990) Exploratory flow-duct experiments on combined-flow bed configurations, and some implications for interpreting storm-event stratification. *J. Sediment. Petrol.*, **60**, 211–219.

Aydın, M., Serdar, H.S. and Şahintürk, Ö. (1982) Orta Karadeniz bölgesi jeolojisi ve petrol olanakları. In: *Proceedings of the 6th Petroleum Congress of Turkey*, pp. 63–71. Turkish Petroleum Company (TPAO), Ankara.

Aydın, M., Şahintürk, Ö., Serdar, H.S., *et al.* (1986) The geology of the area between Ballıdağ and Çangaldağı (Kastamonu). *Bull. Geol. Soc. Turk.*, **29**, 1–16.

Aydın, M., Demir, O., Özçelik, Y., Terzioğlu, N. and Satır, M. (1995a) A geological revision of İnebolu, Devrekani, Ağlı and Küre areas: new observations on Paleotethys-Neotethys sedimentary successions. In: *Geology of the Black Sea Region* (Eds A. Erler, E. Tuncay, E. Bingöl and S. Örçen), pp. 33–38. General Directorate of Mineral Research and Exploration (MTA), Ankara.

Aydın, M., Demir, O., Serdar, H.S., Özaydın, S. and Harput, B. (1995b) Tectono-sedimentary evolution and hydrocarbon potential of the Sinop–Boyabat Basin, North Turkey. In: *Geology of the Black Sea Region* (Eds A. Erler, E. Tuncay, E. Bingöl and S. Örçen), pp. 254–263. General Directorate of Mineral Research and Exploration (MTA), Ankara.

Bádenas, B. and Aurell, M. (2001) Proximal-distal facies relationships and sedimentary processes in a storm dominated carbonate ramp (Kimmeridgian, northwest of the Iberian Ranges, Spain). *Sediment. Geol.*, **139**, 319–340.

Badgley, P.C. (1959) *Stratigraphy and Petroleum Possibilities of the Sinop Region*. Tidewater Oil Co. Report, Petrol İşleri Genel Müdürlüğü Arşivi, Ankara, 38 pp.

Baldwin, B. and Butler, C.O. (1985) Compaction curves. *Am. Assoc. Petrol. Geol. Bull.*, **69**, 622–626.

Ball, L.D., Corbertt, P.W.M., Jensen, J.L. and Lewis, J.J.M. (1997) The role of geology in the behavior and choice of permeability predictors. *SPE Form. Eval.*, **12**, 32–39.

Barka, A., Sütçü, Y.F., Gedik, A., Tekin, T.F., Arel, E., Özdemir, M. and Erkal, T. (1985) *Final Report on the Geological Investigation for the Sinop Nuclear Power Plant*. Report No. 7963, General Directorate of Mineral Research and Exploration (MTA), Ankara.

Baş, H. (1986) Petrology and geochemistry of the Sinop volcanics. *Bull. Geol. Soc. Turk.*, **29**, 143–156.

Beaumont, C. (1981) Foreland basins. *Geophys. J. Roy. Astron. Soc.*, **65**, 291–329.

Bektaş, O. and Gedik, İ. (1986) Sinop volkanitlerinin ved jeokimyası Tartışma. *Bull. Geol. Soc. Turk.*, **29**, 73–74.

Betzler, C., Reijmer, J.J.G., Barnet, K., Eberli, G.P. and Anselmetti, F.S. (1999) Sedimentary patterns and geometries of the Bahamian outer carbonate ramp (Miocene-lower Pliocene Great Bahama Bank). *Sedimentology*, **46**, 1127–1143.

Bluck, B.J. (1967) Sedimentation of beach gravels: examples from South Wales. *J. Sediment. Petrol.*, **37**, 128–157.

Bluck, B.J. (1999) Clast assembling, bed-forms and structure in gravel beaches. *Trans. Roy. Soc. Edinb. Earth Sci.*, **89**, 291–332.

Bourgeois, J. (1980) A transgressive shelf sequence exhibiting hummocky stratification: the Cape Sebastian Sandstone (Upper Cretaceous), southwestern Oregon. *J. Sediment. Petrol.*, **50**, 681–702.

Bowen, A.J. and Guza, R.T. (1978) Edge waves and surf beat. *J. Geophys. Res.*, **83**, 1913–1920.

Bowen, A.J. and Inman, D.L. (1969) Rip currents, 2: laboratory and field evidence. *J. Geophys. Res.*, **74**, 5479–5490.

Bowman, M.B.J. (1998) Cenozoic. In: *Petroleum Geology of the North Sea: Basic Concepts and Recent Advances* (Ed. K.W. Glennie,), 4th edn, pp. 350–375. Blackwell Science, Oxford.

Braga, J.C., Martin, J.M. and Wood, J.L. (2001) Submarine lobes and feeder channels of redeposited, temperate carbonate and mixed siliciclastic-carbonate platform deposits (Vera Basin, Almeria, southern Spain). *Sedimentology*, **48**, 99–116.

Brenchley, P.J., Newall, G. and Stanistreet, I.G. (1979) A storm surge origin for sandstone beds in an epicontinental platform sequence, Ordovician, Norway. *Sediment. Geol.*, **22**, 185–217.

Bryant, E. (2001) *Tsunami: the Underrated Hazard*. Cambridge Univer. Press, Cambridge, 320 pp.

Burchette, T.P. and Wright, V.P. (1992) Carbonate ramp depositional systems. *Sediment. Geol.*, **79**, 3–57.

Bustillo, M.A. and Ruiz-Ortiz, P.A. (1987) Chert occurrences in carbonate turbidites: examples from the Upper Jurassic of the Betic Mountains (southern Spain). *Sedimentology*, **34**, 611–621.

Cacchione, D.A., Drake, D.E., Grant, W.D. and Tate, G.B. (1984) Rippled scour depressions on the inner continental shelf off central California. *J. Sediment. Petrol.*, **54**, 1280–1291.

Cacchione, D.A., Drake, D.E., Ferreira, J.T. and Tate, G.B. (1994) Bottom stress estimates and sand transport on northern California inner continental shelf. *Cont. Shelf Res.*, **14**, 1273–1289.

Caldwell, N.E. and Williams, A.T. (1985) The role of beach profile configuration in the discrimination between differing depositional environments affecting coarse clastic beaches. *J. Coastal Res.*, **1**, 129–139.

Cantalamessa, G. and Di Celma, C. (2005) Sedimentary features of tsunami backwash deposits in a shallow marine Miocene setting, Mejillones Peninsula, northern Chile. *Sediment. Geol.*, **178**, 259–273.

Chan, M.A. and Dott, R.H., Jr. (1983) Shelf and deep-sea sedimentation in Eocene forearc basin, western Oregon – fan or non-fan? *Am. Assoc. Petrol. Geol. Bull.*, **67**, 2100–2116.

Clark, J.D. and Pickering, K.T. (1996) Architectural elements and growth patterns of submarine channels: application to hydrocarbon exploration. *Am. Assoc. Petrol. Geol. Bull.*, **80**, 194–221.

Clifton, H.E. (1976) Wave-formed sedimentary structures – a conceptual model. In: *Beach and Nearshore Sedimentation* (Eds R.A. Davis, Jr. and R.L. Ethington), pp. 126–148. Special Publication 24, Society of Economic Paleontologists and Mineralogists, Tulsa, OK.

Clifton, H.E. (1981) Progradational sequences in Miocene shoreline deposits, southeastern Caliente Range, California. *J. Sediment. Petrol.*, **51**, 165–184.

Clifton, H.E. and Dingler, J.R. (1984) Wave-formed structures and paleoenvironmental reconstruction. *Mar. Geol.*, **60**, 165–198.

Clifton, H.E., Hunter, R.E. and Phillips, R.L. (1971) Depositional structures and processes in the non-barred high-energy nearshore. *J. Sediment. Petrol.*, **44**, 651–670.

Cloetingh, S., Spadini, G., Van Wees, J.D. & Beekman, F. (2003) Thermo-mechanical modelling of Black Sea Basin (de)formation. *Sediment. Geol.*, **156**, 169–184.

Colacicchi, R. and Baldanza, A. (1986) Carbonate turbidites in a Mesozoic pelagic basin: Scaglia Formation, Apennines; comparison with siliciclastic depositional models. *Sediment. Geol.*, **48**, 81–105.

Collins, A.C. and Robertson, A.H.F. (1998) Processes of Late Cretaceous to Late Miocene episodic thrust sheet translation in the Lycian Taurides, SW Turkey. *J. Geol. Soc. London*, **155**, 759–772.

Collins, A.C. & Robertson, A.H.F. (1999) Evolution of the Lycian allochthon, western Turkey, as a north-facing late Palaeozoic to Mesozoic rift and passive continental margin. *Geol. Jour.*, **34**, 107–138.

Collinson, J.D. and Thompson, D.B. (1982) *Sedimentary Structures*. Allen and Unwin, London, 207 pp.

Coniglio, M. and Dix, G.R. (1992) Carbonate slopes. In: *Facies Models: Response to Sea Level Change* (Eds R.G. Walker and N.P. James), pp. 349–373. Geological Association of Canada, St. John's.

Dalrymple, R.A. (1975) A mechanism for rip current generation on an open coast. *J. Geophys. Res.*, **80**, 3485–3487.

DeCelles, P.G. and Cavazza, W. (1992) Constraints on the formation of Pliocene hummocky cross stratification in Calabria (Southern Italy) from consideration of hydraulic and dispersive equivalence, grain-flow theory, and suspended-load fallout rate. *J. Sediment. Petrol.*, **62**, 555–568.

DeCelles, P.G. and Giles, K.A. (1996) Foreland basin systems. *Basin Res.*, **8**, 105–123.

Deptuck, M.E., Steffens, G.S., Barton, M. and Pirmez, C. (2003) Architecture and evolution of upper fan channel-belts on the Niger Delta slope and in the Arabian Sea. *Mar. Petrol. Geol.*, **20**, 649–676.

Dickinson, W.R. (1974) Plate tectonics and sedimentation. In: *Tectonics and Sedimentation* (Ed. W.R. Dickinson), pp. 1–27. Special Publication 22, Society of Economic Paleontologists and Mineralogists, Tulsa, OK.

Dilek, Y. and Moores, E.M. (1990) Regional tectonics of the eastern Mediterranean ophiolites. In: *Ophiolites – Oceanic Crustal Analogues* (Eds J. Malpas, E.M. Moores, A. Panayiotou and C. Xenophontos), Proceedings of the International Symposium, Troodos 1987, pp. 295–309. Cyprus Geological Survey, Nicosia.

Dilek, Y. and Rowland, J.C. (1993) Evolution of a conjugate passive margin pair in Mesozoic southern Turkey. *Tectonics*, **12**, 954–970.

Dott, R.H., Jr. and Bourgeois, J. (1982) Hummocky stratification: significance of its variable bedding sequences. *Geol. Soc. Am. Bull.*, **93**, 663–680.

Dreyer, T., Corregidor, J., Arbues, P. and Puigdefábregas, C. (1999) Architecture of the tectonically influenced Sobrarbe deltaic complex in the Ainsa Basin, northern Spain. *Sediment. Geol.*, **127**, 127–169.

Drzewiecki, P.A. and Simó, J.A. (2002) Depositional processes, triggering mechanisms and sediment composition of carbonate gravity flow deposits: examples from the Late Cretaceous of the south-central Pyrenees, Spain. *Sediment. Geol.*, **146**, 155–189.

Duchaufour, P. (1977) *Pedology*. Allen & Unwin, London, 448 pp.

Duke, W.L., Arnott, R.W. and Cheel, R.J. (1991) Shelf sandstones and hummocky cross stratification; new insights on a stormy debate. *Geology*, **19**, 625–628.

Dunham, R.J. (1962) Classification of carbonate rocks according to depositional textures. In: *Classification of Carbonate Rocks* (Ed. W.E. Ham), pp. 108–121. Memoir 1, American Association of Petroleum Geologists, Tulsa, OK.

Eberli, G.P. (1987) Carbonate turbidite sequences deposited in rift-basins of the Jurassic Tethys ocean (Eastern Alps, Switzerland). *Sedimentology*, **34**, 363–388.

Eberli, G.P. (1991) Calcareous turbidites and their relationship to sea-level fluctuations and tectonism. In: *Cycles and Events in Stratigraphy* (Eds G. Einsele, W. Ricken and A. Seilacher), pp. 340–359. Springer-Verlag, Berlin.

Eğin, D., Hirst, D.M. and Phillips, R. (1979) The petrology and geochemistry of volcanic rocks from the northern Harşit River area, Pontide volcanic province, northeastern Turkey. *J. Volcanol. Geoth. Res.*, **6**, 105–123.

Eren, M. and Kadir, S. (1999) Colour origin of Upper Cretaceous pelagic red sediments within the Eastern Pontides, northeast Turkey. *Int. J. Earth Sci.*, **88**, 593–595.

Ernst, G. and Zander, J. (1993) Stratigraphy, facies development, and trace fossils of the Upper Cretaceous of Southern Tanzania (Kilwa District). In: *Geology and Mineral Resources of Somalia and Surrounding Regions* (Eds E. Abbate, M. Sagri and F.P. Sassi), pp. 259–278. Istituto Agronomico per l'Oltremare, Florence.

Fayon, A.K., Whitney, D.L., Teyssier, C., Garver, J.I. and Dilek, Y. (2001) Effects of plate convergence obliquity on timing and mechanisms of exhumation of a mid-crustal terrain, the Central Anatolian Crystalline Complex. *Earth Planet. Sci. Lett.*, **192**, 191–205.

Flerit, F., Armijo, R., King, G. and Meyer, B. (2004) The mechanical interaction between the propagating North Anatolian Fault and the back-arc extension in the Aegean. *Earth Planet. Sci. Lett.*, **224**, 347–362.

Forbes, D.L. and Boyd, E. (1987) Gravel ripples on the inner Scotian Shelf. *J. Sediment. Petrol.*, **57**, 46–54.

Franke, W. and Paul, J. (1980) Pelagic redbeds in the Devonian of Germany: deposition and diagenesis. *Sediment. Geol.*, **25**, 231–256.

Fukushima, Y., Parker, G. and Pantin, H.M. (1985) Prediction of ignitive turbidity currents in Scripps submarine canyon. *Mar. Geol.*, **67**, 55–81.

Gagan, M.K., Johnson, D.P. and Carter, R.M. (1988) The cyclone Winifred storm bed, central Great Barrier Reef shelf, Australia. *J. Sediment. Petrol.*, **58**, 845–861.

Gedik, A. and Korkmaz, S. (1984) Sinop havzasının jeolojisi ve petrol olanakları. *Jeol. Mühend. Derg.*, **19**, 53–79.

Gedik, A., Ercan, T. and Korkmaz, S. (1984) Orta Karadeniz (Samsun-Sinop) havzasının jeolojisi ve volkanik kayaçların petrolojisi. *Bull. Miner. Res. Explor. Inst. Turk.*, **99**, 33–50.

Gierlowski-Kordesh, E. and Ernst, G. (1987) A flysch trace fossil assemblage from the Upper Cretaceous shelf of Tanzania. In: *Current Research in African Earth Sciences* (Eds G. Mathesis and H. Schandelmeier), pp. 217–221. A.A. Balkema, Rotterdam.

Göksu, E., Pamir, H.N. and Erentöz, C. (1974) *Samsun Map Sheet 1:500000*. General Directorate of Mineral Research and Exploration (MTA), Ankara.

Görür, N. (1988) Timing of opening of the Black Sea basin. *Tectonophysics*, **147**, 247–262.

Görür, N. (1997) Cretaceous syn- to postrift sedimentation on the southern continental margin of the Western Black Sea Basin. In: *Regional and Petroleum Geology of the Black Sea and Surrounding Regions* (Ed. A.G. Robinson), pp. 227–240. Memoir 68, American Association of Petroleum Geologists, Tulsa, OK.

Görür, N. and Tüysüz, O. (1997) Petroleum geology of the southern continental margin of the Black Sea. In: *Regional and Petroleum Geology of the Black Sea and Surrounding Regions* (Ed. A.G. Robinson), pp. 241–254. Memoir 68, American Association of Petroleum Geologists, Tulsa, OK.

Görür, N. and Tüysüz, O. (2001) Cretaceous to Miocene palaeogeographic evolution of Turkey: implications for hydrocarbon potential. *J. Petrol. Geol.*, **24**, 119–146.

Görür, N., Oktay, F.Y., Seymen, İ. and Şengör, A.M.C. (1984) Palaeotectonic evolution of the Tuzgölü basin complex, Central Turkey: sedimentary record of a Neotethyan closure. In: *The Geological Evolution of the Eastern Mediterranean* (Eds J.E. Dixon and A.H.F. Robertson), pp. 467–482. Special Publication 17, Geological Society, London.

Görür, N., Tüysüz, O., Aykol, A., Sakınç, M., Yiğitbaş, E. and Akkök, R. (1993) Cretaceous red pelagic carbonates of northern Turkey: their place in the opening history of the Black Sea. *Eclogae Geol. Helv.*, **86**, 819–838.

Görür, N., Çağatay, N., Sakınç, M., Akkök, R., Tchapalyga, A. and Natalin, B. (2000) Neogene Paratethyan succession in Turkey and its implications for the palaeogeography of the Eastern Paratethys. In: *Tectonics and Magmatism in Turkey and the Surrounding Areas* (Eds E. Bozkurt, J.A. Winchester and J.D.A. Piper), pp. 251–269. Special Publication 173, Geological Society Publishing House, Bath.

Grecula, M., Flint, S., Potts, G., Wickens, D. and Johnson, S. (2003) Partial ponding of turbidite systems in a basin with subtle growth-fold topography: Laingsburg-Karoo, South Africa. *J. Sediment. Res.*, **73**, 603–620.

Gruszczyński, M., Rudowski, S., Semil, J., Skomiński, J. and Zrobek, J. (1993) Rip currents as a geological tool. *Sedimentology*, **40**, 217–236.

Gürer, Ö.F. and Aldanmaz, E. (2002) Origin of the Upper Cretaceous-Tertiary sedimentary basins within the Tauride-Anatolide platform in Turkey. *Geol. Mag.*, **139**, 191–197.

Güven, A. (1977) *Stratigraphy and sedimentology of Eocene formations, Karabük Area, Turkey*. Unpublished PhD thesis, University College, Swansea, 307 pp.

Guza, R.T. and Davis, R.E. (1974) Excitation of edge waves by wave incident on the beach. *J. Geophys. Res.*, **79**, 1285–1291.

Guza, R.T. and Inman, D.L. (1975) Edge waves and beach cusps. *J. Geophys. Res.*, **80**, 2997–3012.

Hamblin, A. and Walker, R.G. (1979) Storm-dominated shallow marine deposits: the Fernie-Kootenay (Jurassic) transition, southern Rocky Mountains. *Can. J. Earth Sci.*, **16**, 1673–1690.

Handford, C.R. (1986) Facies and bedding sequences in shelf-storm-deposited carbonates – Fayetteville Shale and Pitkin Limestone (Mississipian), Arkansas. *J. Sediment. Petrol.*, **56**, 123–137.

Haq, B.U., Hardenbol, J. and Vail, P.R. (1988) Mesozoic and Cenozoic chronostratigraphy and eustatic cycles. In: *Sea-Level Changes – An Integrated Approach* (Eds C.K. Wilgus, B.S. Hastings, H.W. Posamentier, J.C. Van Wagoner, C.A. Ross and C.G.St.C. Kendall), pp. 71–108. Special Publication 42, Society of Economic Paleontologists and Mineralogists, Tulsa, OK.

Harms, J.C., Southard, J.B., Spearing, D.R. and Walker, R.G. (1975) *Depositional Environments as Interpreted from Primary Sedimentary Structures and Stratification Sequences*. SEPM Short Course No. 2 Lecture Notes, Society of Economic Paleontologists and Mineralogists, Dallas, TX, 161 pp.

Harms, J.C., Southard, J.B. and Walker, R.G. (1982) *Structures and Sequences in Clastic Rocks*. SEPM Short Course No. 9 Lecture Notes, Society of Economic Paleontologists and Mineralogists, Calgary, 250 pp.

Harris, M.T. (1994) The foreslope and toe-of-slope facies of the Middle Triassic Latemar buildup (Dolomites, northern Italy). *J. Sediment. Res.*, **B64**, 132–145.

Haughton, P.D.W. (2000) Evolving turbidite systems on a deforming basin floor, Tabernas, SE Spain. *Sedimentology*, **47**, 497–518.

Hayward, A.B. (1984) Sedimentation and basin formation related to ophiolite nappe emplacement, Miocene, SW Turkey. *Sediment. Geol.*, **40**, 105–129.

Hazlett, B.H. and Warme, J.E. (1988) Shelf-to-basin resedimented carbonates of the southern margin of the Jurassic central High Atlas trough, Morocco. *Am. Assoc. Petrol. Geol. Bull.*, **72**, 1006.

Hebenstreit, G.T. (Ed.) (2001) *Tsunami Research at the End of a Critical Decade*. Kluwer Academic Publishers, Dordrecht, 282 pp.

Hequette, A. and Hill, P.R. (1993) Storm-generated currents and offshore transport on a sandy shoreface, Tibjack Beach, Canadian Beaufort Sea. *Mar. Geol.*, **113**, 282–304.

Hequette, A. and Hill, P.R. (1995) Response of the seabed to storm-generated combined flows on a sandy arctic shoreface, Canadian Beaufort Sea. *J. Sediment. Res.*, **A65**, 461–471.

Howard, J.D. and Reineck, H.-E. (1981) Depositional facies of high-energy beach-to-offshore sequence, comparison with low energy sequence. *Am. Assoc. Petrol. Geol. Bull.*, **65**, 807–830.

Hubbard, D.K. (1992) Hurricane-induced sediment transport in open-shelf tropical systems: an example from St. Croix, U.S. Virgin Islands. *J. Sediment. Petrol.*, **62**, 946–960.

Huntley, D.A. and Bowen, A.J. (1975) Field observations of edge waves and their effect on beach material. *J. Geol. Soc. London*, **131**, 69–81.

Jackson, R.G., II (1976) Depositional model of point bars in the lower Wabash River. *J. Sediment. Petrol.*, **46**, 579–594.

James, N.P. and Bone, Y. (1991) Origin of a cool-water, Oligo-Miocene deep shelf limestone, Eucla Platform, southern Australia. *Sedimentology*, **38**, 323–341.

James, N.P. and Bourque, P.-A. (1992) Reefs and mounds. In: *Facies Models: Response to Sea Level Change* (Eds R.G. Walker and N.P. James), pp. 323–347. Geological Association of Canada, St. John's.

Janbu, N.E. (2004) *Tectonic control on turbiditic sedimentation: the Late Cretaceous–Eocene successions in the Sinop–Boyabat Basin of north-central Turkey*. Unpublished Dr. Scient. Dissertation, University of Bergen, 334 pp.

Janbu, N.E., Leren, B.L.S., Kırman, E. and Nemec, W. (2003) The Late Cretaceous-Eocene turbiditic sedimentation in the Sinop Basin, north-central Turkey: response to tectonic changes in basin morphology. In: *Abstracts AAPG-SEPM Annual Convention*, Part A, p. 84. American Association of Petroleum Geologists, Salt Lake City.

Johnson, H.D. and Baldwin, C.T. (1996) Shallow clastic seas. In: *Sedimentary Environments: Processes, Facies and Stratigraphy* (Ed. H.G. Reading), pp. 232–280. Blackwell Science, Oxford.

Johnson, S.D., Flint, S., Hinds, D. and Wickens, H. DeV. (2001) Anatomy, geometry and sequence stratigraphy of basin floor to slope turbidite systems,

Tanqua Karoo, South Africa. *Sedimentology*, **48**, 987–1023.

Jones, B. and Desrochers, A. (1992) Shallow platform carbonates. In: *Facies Models: Response to Sea Level Change* (Eds R.G. Walker and N.P. James), pp. 277–301. Geological Association of Canada, St. John's.

Kastens, K.A. and Shor, A.N. (1985) Depositional processes of a meandering channel on Mississippi Fan. *Am. Assoc. Petrol. Geol. Bull.*, **69**, 190–202.

Kaymakcı, N., Duermeijer, C.E., Langereis, C., White, S.H. and Van Dijk, P.M. (2003) Palaeomagnetic evolution of the Çankırı Basin (central Anatolia, Turkey): implications for oroclinal bending due to indentation. *Geol. Mag.*, **140**, 343–355.

Keen, T.R. and Slingerland, R.L. (1993) Four storm-event beds and the tropical cyclones that produced them: a numerical hindcast. *J. Sediment. Petrol.*, **63**, 218–232.

Ketin, İ. and Gümüş, Ö. (1963) *Sinop-Ayancık arasındaki, III. Bölgeye dahil sahaların jeolojisi hakkında rapor, 2. Kısım, Jura ve Kretase formasyonlarının etüdü*. Report No. 213–288, Turkish Petroleum Company (TPAO), Ankara, 118 pp.

Koçyiğit, A. (1986) Stratigraphy and nature of the northern margin of the Karabük-Safranbolu Tertiary basin. *Bull. Geol. Soc. Turk.*, **30**, 61–69.

Kolla, V., Bourges, Ph., Urruty, J.-M. and Safa, P. (2001) Evolution of deep-water Tertiary sinuous channels offshore Angola (west Africa) and implications for reservoir architecture. *Am. Assoc. Petrol. Geol. Bull.*, **85**, 1373–1405.

Komar, P.D. (1998) *Beach Processes and Sedimentation*, 2nd edn. Prentice-Hall, Englewood Cliffs (N.J.), 544 pp.

Kreisa, R.D. (1981) Storm-generated sedimentary structures in subtidal marine facies with examples from the Middle and Upper Ordovician of southwestern Virginia. *J. Sediment. Petrol.*, **51**, 823–848.

Kumar, N. and Sanders, J.E. (1976) Characteristics of shoreface storm deposits: modern and ancient examples. *J. Sediment. Petrol.*, **46**, 145–162.

Leckie, D.A. (1988) Wave-formed coarse-grained ripples and their relationship to hummocky cross stratification. *J. Sediment. Petrol.*, **58**, 607–622.

Leckie, D.A. and Walker, R.G. (1982) Storm- and tide-dominated shorelines in Cretaceous Moosebar-Lower Gates interval – outcrop equivalents of Deep Basin gas trap in Western Canada. *Am. Assoc. Petrol. Geol. Bull.*, **66**, 138–157.

Lees, A. and Buller, A.T. (1972) Modern temperate-water and warm-water shelf carbonate sediments contrasted. *Mar. Geol.*, **13**, 67–73.

Leithold, E.L. and Bourgeois, J. (1984) Characteristics of coarse-grained sequences deposited in nearshore, wave-dominated environments – examples from the Miocene of south-west Oregon. *Sedimentology*, **31**, 749–775.

Leren, B.L.S. (2003) *Late Cretaceous to Early Eocene Sedimentation in the Sinop–Boyabat Basin, North-Central Turkey: Facies Analysis of Turbiditic to Shallow-Marine Deposits*. Unpublished Cand. Scient. thesis, University of Bergen, 140 pp.

Leren, B.L.S., Janbu, N.E., Kırman, E. and Nemec, W. (2002) The Late Cretaceous-Eocene turbiditic succession in the Sinop Basin, Turkey: Sedimentation in an evolving continental-margin rift closed by inversion. In: *Abstracts 16th International Sedimentological Congress* (Eds M. Knoper and B. Cairncross), pp. 221–222. Rand Afrikaans University, Johannesburg.

Leszczyński, S. (1989) Characteristics and origin of fluxoturbidites from the Carpathian Flysch (Cretaceous-Palaeogene), south Poland. *Ann. Soc. Geol. Pol.*, **59**, 351–390.

Lien, T., Walker, R.G. and Martinsen, O.J. (2003) Turbidites in the Upper Carboniferous Ross Formation, western Ireland: reconstruction of a channel and spillover system. *Sedimentology*, **50**, 113–148.

Lovell, J.P.B. (1990) Cenozoic. In: *Introduction to the Petroleum Geology of the North Sea*, 3rd Edn (Ed. K.W. Glennie), pp. 273–293. Blackwell Scientific Publications, Oxford.

Lowe, D.R. (1982) Sediment gravity flows, II. Depositional models with special reference to the deposits of high-density turbidity currents. *J. Sediment. Petrol.*, **52**, 279–297.

Lowe, D.R. (1988) Suspended-load fallout rate as an independent variable in the analysis of current structures. *Sedimentology*, **35**, 765–776.

Massari, F. and D'Alessandro, A. (2000) Tsunami-related scour-and-drape undulations in Middle Pliocene restricted-bay carbonate deposits (Salento, south Italy). *Sediment. Geol.*, **135**, 265–281.

Massari, F. and Parea, G.C. (1988) Progradational gravel beach sequences in a moderate- to high-energy, microtidal marine environment. *Sedimentology*, **35**, 881–913.

McCaffrey, W.D. and Kneller, B.C. (2001) Process controls on the development of stratigraphic trap potential on the margins of confined turbidite systems, and aids to reservoir evaluation. *Am. Assoc. Petrol. Geol. Bull.*, **85**, 971–988.

Meredith, D.J. and Egan, S.S. (2002) The geological and geodynamic evolution of the eastern Black Sea basin: insights from 2-D and 3-D tectonic modelling. *Tectonophysics*, **350**, 157–179.

Michard, A., Whitechurch, H., Ricou, L.E., Montigny, R. and Yazgan, E. (1984) Tauric subduction (Malatya-Elazığ provinces) and its bearing on tectonics of the Tethyan realm in Turkey. In: *The Geological Evolution of the Eastern Mediterranean* (Eds J.E. Dixon and

A.H.F. Robertson), pp. 361–373. Special Publication **17**, Geological Society, London.

Molina, J.M., Ruiz-Ortiz, P.A. and Vera, J.A. (1997) Calcareous tempestites in pelagic facies (Jurassic, Betic Cordillera, southern Spain). *Sediment. Geol.*, **109**, 95–109.

Monaco, P. (1992) Hummocky cross-stratified deposits and turbidites in some sequences of the Umbria-Marche area (central Italy) during the Toarcian. *Sediment. Geol.*, **77**, 123–142.

Mount, J.F. and Kidder, D. (1993) Combined flow origin of edgewise intraclasts conglomerates: Sellick Hill Formation (Lower Cambrian), South Australia. *Sedimentology*, **40**, 315–329.

Mulder, T., Weber, O., Anschutz, P., Jorissen, F.J. and Jouanneau, J.-M. (2001) A few month-old storm-generated turbidite deposited in the Capbreton Canyon (Bay of Biscay, SW France). *Geo-Mar. Lett.*, **21**, 149–156.

Mullins, H.T. and Cook, H.E. (1986) Carbonate apron models: alternatives to the submarine fan model for paleoenvironmental analysis and hydrocarbon exploration. *Sediment. Geol.*, **48**, 37–79.

Mutti, E. (1992) *Turbidite Sandstones*. Agip, San Donato Milanese, 275 pp.

Myrow, P.M. and Southard, J.B., 1996. Tempestite deposition. *J. Sediment. Res.*, **66**, 875–887.

Nikishin, A.M., Korotaev, M.V., Ershov, A.V. and Brunet, M.-F. (2003) The Black Sea basin: tectonic history and Neogene-Quaternary rapid subsidence modelling. *Sediment. Geol.*, **156**, 149–168.

Nøttvedt, A. and Kreisa, R.D. (1987) Model for the combined flow origin of hummocky cross-stratification. *Geology*, **15**, 357–361.

Okay, A.I. (1989) Tectonic units and sutures in the Pontides, northern Turkey. In: *Tectonic Evolution of the Tethyan Region* (Ed. A.M.C. Şengör), pp. 109–115. Kluwer Academic Publishers, Dordrecht.

Okay, A.I. and Şahintürk, Ö. (1997) Geology of the Eastern Pontides. In: *Regional and Petroleum Geology of the Black Sea and Surrounding Regions* (Ed. A.G. Robinson), pp. 291–311. Memoir 68, American Association of Petroleum Geologists, Tulsa, OK.

Okay, A.I. and Tüysüz, O. (1999) Tethyan sutures of northern Turkey. In: *The Mediterranean Basins: Tertiary Extension within the Alpine Orogen* (Eds B. Durand, L. Jolivet, F. Horváth and M. Séranne), pp. 475–515. Special Publication 156, Geological Society Publishing House, Bath.

Okay, A.I., Şengör, A.M.C. and Görür, N. (1994) Kinematic history of the opening of the Black Sea and its effect on the surrounding regions. *Geology*, **22**, 267–270.

Okay, A.I., Tansel, İ. and Tüysüz, O. (2001) Obduction, subduction and collision as reflected in the Upper Cretaceous-Lower Eocene sedimentary record of western Turkey. *Geol. Mag.*, **138**, 117–142.

Orford, J.D. (1977) A proposed mechanism for beach sedimentation. *Earth Surf. Proc. Land.*, **2**, 381–400.

Özgül, N. (1976) Torosların bazı temel jeoloji özellikleri. *Bull. Geol. Soc. Turk.*, **19**, 65–78.

Parker, G. (1982) Conditions for the ignition of catastrophically erosive turbidity currents. *Mar. Geol.*, **46**, 307–327.

Peakall, J., McCaffrey, B. and Kneller, B. (2000) A process model for the evolution, morphology, and architecture of sinuous submarine channels. *J. Sediment. Res.*, **70**, 434–448.

Peccerillo, A. and Taylor, S.R. (1975) Geochemistry of Upper Cretaceous volcanic rocks from the Pontide chain, northern Turkey. *Bull. Volcanol.*, **39**, 1–13.

Pemberton, S.G., MacEachern, J.A. and Frey, R.W. (1992) Trace fossil facies models: environmental and allostratigraphic significance. In: *Facies Models: Response to Sea Level Change* (Eds R.G. Walker and N.P. James), pp. 47–72. Geological Association of Canada, St. John's.

Pervesler, P. and Uchman, A. (2004) Ichnofossils from the type area of the Ground Formation (Miocene, Lower Badenian) in northern Lower Austria (Molasse Basin). *Geol. Carpath.*, **55**, 103–110.

Pickering, K.T. and Hiscott, R.N. (1985) Contained (reflected) turbidity currents from the Middle Ordovician Cloridorme Formation, Quebec, Canada: an alternative to the antidune hypothesis. *Sedimentology*, **32**, 373–394.

Plink-Björklund, P., Mellere, D. and Steel, R.J. (2001) Turbidite variability and architecture on sand-prone deepwater slopes, Eocene clinoforms in the Central Tertiary Basin, Spitsbergen. *J. Sediment. Res.*, **71**, 895–912.

Posamentier, H.W. and Kolla, V. (2003) Seismic geomorphology and stratigraphy of depositional elements in deep-water settings. *J. Sediment. Res.*, **73**, 367–388.

Posamentier, H.W., Jervey, M.T. and Vail, P.R. (1988) Eustatic controls on clastic deposition I – Conceptual framework. In: *Sea-level Changes – an Integrated Approach* (Eds C.K. Wilgus, B.S. Hastings, H.W. Posamentier, J.C. Van Wagoner, C.A. Ross and C.G.St.C. Kendall), pp. 109–124. Special Publication 42, Society of Economic Paleontologists and Mineralogists, Tulsa, OK.

Rangin, C., Bader, A.G., Pascal, G., Ecevitoğlu, B. and Görür, N. (2002) Deep structure of the Mid- Black Sea High (offshore Turkey) imaged by multi-channel seismic survey (BLACKSIS cruise). *Mar. Geol.*, **182**, 265–278.

Read, J.F. (1982) Carbonate platforms of passive (extensional) continental margins: types, characteristics and evolution. *Tectonophysics*, **81**, 195–212.

Read, J.F. (1985) Carbonate ramp facies models. *Am. Assoc. Petrol. Geol. Bull.*, **69**, 1–21.

Reading, H.G. and Collinson, J.D. (1996) Clastic coasts. In: *Sedimentary Environments: Processes, Facies and Stratigraphy* (Ed. H.G. Reading), pp. 154–231. Blackwell Science, Oxford.

Reading, H.G. and Richards, M. (1994) Turbidite systems in deep-water basin margins classified by grain size and feeder system. *Am. Assoc. Petrol. Geol. Bull.*, **78**, 792–822.

Ricci Lucchi, F. and Valmori, E. (1980) Basinwide turbidites in a Miocene, oversupplied deep-sea plain: a geometrical analysis. *Sedimentology*, **27**, 241–270.

Robinson, A.G., Banks, C.J., Rutherford, M.M. and Hirst, J.P.P. (1995) Stratigraphic and structural development of the Eastern Pontides, Turkey. *J. Geol. Soc. London*, **152**, 861–872.

Robinson, A.G., Rudat, J.H., Banks, C.J. and Wiles, R.L.F. (1996) Petroleum geology of the Black Sea. *Mar. Petrol. Geol.*, **13**, 195–223.

Rochow, K.A. (1981) Seismic stratigraphy of the North Sea 'Palaeocene' deposits. In: *Petroleum Geology of the Continental Shelf of North-West Europe* (Eds L.V. Illing and G.D. Hobson), pp. 255–266. Heyden & Son, London.

Rossetti, D. de F., Góes, A.M., Truckenbrodt, W. and Anaisse, J., Jr. (2000) Tsunami-induced large-scale scour-and-fill structures in Late Albian to Cenomanian deposits of the Grajaú Basin, northern Brazil. *Sedimentology*, **47**, 309–323.

Saito, Y. (1989) Modern storm deposits in the inner shelf and their recurrence intervals, Sendai Bay, northeast Japan. In: *Sedimentary Facies in the Ancient Plate Margin* (Eds A. Taira and F. Masuda), pp. 331–344. Terra Scientific, Tokyo.

Sallenger, A.H., Jr. (1979) Beach-cusp formation. *Mar. Geol.*, **29**, 23–37.

Sanver, M. and Ponat, E. (1981) Kırşehir ve dolaylarına ilişkin paleomagnetik bulgular: Kırşehir Masifinin rotasyonu. *İstanbul Yerbil.*, **2**, 2–8.

Satur, N., Hurst, A., Cronin, B.T., Kelling, G. and Gürbüz, K. (2000) Sand body geometry in a sand-rich, deep-water clastic system, Miocene Cingöz Formation of southern Turkey. *Mar. Petrol. Geol.*, **17**, 239–252.

Savary, B. and Ferry, S. (2004) Geometry and petrophysical parameters of a calcarenitic turbidite lobe (Barremian-Aptian, Pas-de-la-Cluse, France). *Sediment. Geol.*, **168**, 281–304.

Seguret, M., Moussine-Pouchkine, A., Raja Gabaglia, G. and Bouchette, F. (2001) Storm deposits and storm-generated coarse carbonate breccias on a pelagic outer shelf (South-East Basin, France). *Sedimentology*, **48**, 231–254.

Şengör, A.M.C. (1984) The Cimmeride Orogenic System and the Tectonics of Eurasia. *Geol. Soc. Am. Spec. Pap.*, **195**, 82 pp.

Şengör, A.M.C. (1987) Tectonics of the Tethysides: orogenic collage development in a collisional setting. *Ann. Rev. Earth Planet. Sci.*, **15**, 213–244.

Şengör, A.M.C., Görür, N. and Şaroğlu, F. (1985) Strike-slip faulting and related basin formation in zones of tectonic escape: Turkey as a case study. In: *Strike-Slip Deformation, Basin Formation, and Sedimentation* (Eds K.D. Biddle and N. Christie-Blick), pp. 227–264. Special Publication 17, Society of Economic Paleontologists and Mineralogists, Tulsa, OK.

Şengün, M., Keskin, H., Akçören, F., *et al.* (1990) Geology of the Kastamonu region and geological constraints for the evolution of the Paleotethyan domain. *Geol. Bull. Turk.*, **33**, 1–16.

Sherman, D.J., Orford, J.D. and Carter, R.W.G. (1993) Development of cusp-related, gravel size and shape facies at Malin Head, Ireland. *Sedimentology*, **40**, 1139–1152.

Simpson, E.L. and Eriksson, K.A. (1990) Early Cambrian progradational and transgressive sedimentation patterns in Virginia: an example of the early history of a passive margin. *J. Sediment. Petrol.*, **60**, 84–100.

Smith, R. and Joseph, P. (2004) Onlap stratal architectures in the Grès d'Annot: geometrical models and controlling factors. In: *Deep-Water Sedimentation in the Alpine Foreland Basin of SE France: New Perspectives on the Grès d'Annot and Related Systems* (Eds P. Joseph and S.A. Lomas), pp. 389–399. Special Publication 221, Geological Society Publishing House, Bath.

Snedden, J.W. and Nummedal, D. (1991) Origin and geometry of storm-deposited sand beds in modern sediments of the Texas continental shelf. In: *Shelf Sand and Sandstone Bodies: Geometry, Facies and Sequence Stratigraphy* (Eds D.J.P. Swift, G.F. Oertel, R.W. Tillman and J.A. Thorne), pp. 283–308. Special Publication 14, International Association of Sedimentologists. Blackwell Scientific Publications, Oxford.

Snedden, J.W., Nummedal, D. and Amos, A.F. (1988) Storm- and fair-weather combined flow on the central Texas continental shelf. *J. Sediment. Petrol.*, **58**, 580–595.

Sonel, N., Sarı, A., Coşkun, B. and Tozlu, E. (1989) Boyabat (Sinop) havzası Ekinveren Fayının petrol aramalarındaki önemi. *Geol. Bull. Turk.*, **32**, 39–49.

Stetling, C.E. and DSDP Leg 96 Shipboard Scientists (1985) Migratory characteristics of a mid-fan meander belt, Mississippi Fan. In: *Submarine Fans and Related Turbidite Systems* (Eds A.H. Bouma, W.R. Normark and N.E. Barnes), pp. 283–290. Springer-Verlag, New York.

Stow, D.A.V., Reading, H.G. and Collinson, J.D. (1996) Deep seas. In: *Sedimentary Environments: Processes, Facies and Stratigraphy* (Ed. H.G. Reading), pp. 395–453. Blackwell Science, Oxford.

Sunal, G. and Tüysüz, O. (2002) Palaeostress analysis of Tertiary post-collisional structures in the Western Pontides, northern Turkey. *Geol. Mag.*, **139**, 343–359.

Tucker, M.E. (1990) Carbonate depositional systems: II. Deeper-water facies of pelagic and resedimented limestones. In: *Carbonate Sedimentology* (Eds M.E. Tucker, V.P. Wright and J.A.D. Dickson), pp. 228–283. Blackwell Science, Oxford.

Tucker, M.E. (2001) *Sedimentary Petrology – an Introduction to the Origin of Sedimentary Rocks*. Blackwell Science, Oxford, 262 pp.

Tunoğlu, C. (1994) Microfacies analysis of the Upper Paleocene-Middle Eocene carbonate sequence in Devrekani Basin (northern Kastamonu). *Geol. Bull. Turk.*, **37**, 43–51.

Tüysüz, O. (1990) Tectonic evolution of a part of the Tethyside orogenic collage: the Kargı Massif, northern Turkey. *Tectonics*, **9**, 141–160.

Tüysüz, O. (1993) Karadeniz'den Orta Anadolu'ya bir jeotravers: Kuzey Neo-Tetisin tektonik evrimi. *Türk. Petrol. Jeol. Der. Bül.*, **5**, 1–33.

Tüysüz, O. (1999) Geology of the Cretaceous sedimentary basins of the Western Pontides. *Geol. J.*, **34**, 75–93.

Tüysüz, O., Dellaloğlu, A.A. and Terzioğlu, N. (1995) A magmatic belt within the Neo-Tethyan suture zone and its role in the tectonic evolution of northern Turkey. *Tectonophysics*, **243**, 173–191.

Uchman, A., Janbu, N.E. and Nemec, W. (2004) Trace fossils in the Late Cretaceous-Eocene flysch of the Sinop–Boyabat Basin, Central Pontides, Turkey. *Ann. Soc. Geol. Pol.*, **74**, 197–235.

Ustaömer, T. and Robertson, A. (1997) Tectonic-sedimentary evolution of the North Tethyian margin in the Central Pontides of northern Turkey. In: *Regional and Petroleum Geology of the Black Sea and Surrounding Regions* (Ed. A.G. Robinson), pp. 255–290. Memoir 68, American Association of Petroleum Geologists, Tulsa, OK.

Vera, J.A. and Molina, J.M. (1998) Shallowing-upward cycles in pelagic troughs. *Sediment. Geol.*, **119**, 103–121.

Vrolijk, P.J. and Southard, J.B. (1997) Experiments on rapid deposition of sand from high-velocity flows. *Geosci. Can.*, **24**, 45–54.

Walker, R.G. (1969) The juxtaposition of turbidites and shallow-water sediments: study of a regressive sequence in the Pennsylvanian of North Devon, England. *J. Geol.*, **77**, 125–142.

Walker, R.G. (1984a) General introduction: facies, facies sequences and facies models. In: *Facies Models*, 2nd edn (Ed. R.G. Walker). *Geosci. Can. Reprint Ser.*, **1**, 1–9.

Walker, R.G. (1984b) Shelf and shallow marine sands. In: *Facies Models*, 2nd edn (Ed. R.G. Walker). *Geosci. Can. Reprint Ser.*, **1**, 141–170.

Walker, R.G. and Bergman, K.M. (1993) Shannon Sandstone in Wyoming: a shelf-ridge complex reinterpreted as lowstand shoreface deposits. *J. Sediment. Petrol.*, **63**, 839–851.

Wray, J.L. (1978) Calcareous algae. In: *Introduction to Marine Micropaleontology* (Eds B.U. Haq and A. Boersma), pp. 171–187. Elsevier, Amsterdam.

Wright, V.P. and Burchette, T.P. (1996) Shallow-water carbonate environments. In: *Sedimentary Environments: Processes, Facies and Stratigraphy* (Ed. H.G. Reading), pp. 325–394. Blackwell Science, Oxford.

Yamazaki, T., Yamaoka, M. and Shiki, T. (1989) Miocene offshore tractive current-worked conglomerates – Tsubutegura, Chita Peninsula, central Japan. In: *Sedimentary Facies in the Ancient Plate Margin* (Eds A. Taira and F. Masuda), pp. 483–494. Terra Scientific, Tokyo.

Yeh, H., Liu, P.L.-F., Briggs, M. and Synolakis, C.E. (1994) Propagation and amplification of tsunamis at coastal boundaries. *Nature*, **372**, 353–355.

Yılmaz, Y. (1993) New evidence and model on the evolution of the southeast Anatolian orogen. *Geol. Soc. Am. Bull.*, **105**, 251–271.

Yılmaz, Y., Yiğitbaş, E. and Genç, Ş.C. (1993) Ophiolitic and metamorphic assemblages of southeast Anatolia and their significance in the geological evolution of the orogenic belt. *Tectonics*, **12**, 1280–1297.

Yılmaz, Y., Tüysüz, O., Yiğitbaş, E., Genç, Ş.C. and Şengör, A.M.C. (1997) Geology and tectonic evolution of the Pontides. In: *Regional and Petroleum Geology of the Black Sea and Surrounding Regions* (Ed. A.G. Robinson), pp. 183–226. Memoir 68, American Association of Petroleum Geologists, Tulsa, OK.

Yokokawa, M., Masuda, F. and Endo, N. (1995) Sand particle movement on migrating combined-flow ripples. *J. Sediment. Res.*, **A65**, 40–44.

Young, R.W. and Bryant, E.A. (1992) Catastrophic wave erosion on the southeastern coast of Australia: impact of the Lanai tsunamis ca. 105 ka? *Geology*, **20**, 199–202.

Ziegler, P.A. and Roure, F. (1999) Petroleum systems of Alpine-Mediterranean foldbelts and basins. In: *The Mediterranean Basins: Tertiary Extension within the Alpine Orogen* (Eds B. Durand, L. Jolivet, F. Horváth and M. Séranne), pp. 517–540. Special Publication 156, Geological Society Publishing House, Bath.

Facies anatomy of a sand-rich channelized turbiditic system: the Eocene Kusuri Formation in the Sinop Basin, north-central Turkey

NILS E. JANBU*[1], WOJCIEK NEMEC*, EDIZ KIRMAN† *and* VOLKAN ÖZAKSOY‡

**Department of Earth Science, University of Bergen, 5007 Bergen, Norway*
†Department of Geological Engineering, Ankara University, 06100 Ankara, Turkey
‡General Directorate of Mineral Research and Exploration (MTA), 06520 Ankara, Turkey

ABSTRACT

This study focuses on a basin-floor turbiditic system in an Eocene foredeep basin, using facies analysis supplemented with micropalaeontological and ichnological data. Sediment dispersal processes are interpreted from sedimentary facies, and the morphogenesis, spatial relationships and stratigraphic distribution of facies associations are used to reconstruct the behaviour and morphodynamic evolution of the turbiditic system. The case study sheds more light on the development of submarine channels and related patterns of overbank sedimentation in narrow foreland basins, and contributes to a better understanding of the geological history of the Central Pontides.

The lower to middle Eocene Kusuri Formation in the Sinop Basin, north-central Anatolia, is a succession of siliciclastic turbidites ~ 1200 m thick, well-exposed on the Turkish Black Sea coast. The deposition occurred in a west-trending foredeep trough of the Central Pontides, ~ 30 km wide and > 150 km long, and involved a deep-water axial dispersal system supplied with coarse sediment by a fluvio-deltaic feeder draining the emerged adjacent foreland of the Eastern Pontides. Sedimentary facies include hemipelagic 'background' mudstones, thin muddy turbidites and 'classic' Bouma-type turbidites, a wide range of non-classic turbidites attributed to low- and high-density sustained currents, and gravelly debrisflow deposits. These facies form four major associations:

1 mudstones interspersed with thin turbidite sheets;
2 broad depositional lobes with thickening-upward bedding trends;
3 poorly defined wide palaeochannels, solitary sinuous palaeochannels and multistorey palaeochannel complexes;
4 packages of thin overbank turbidites with tabular, wedge-shaped or sigmoidal bedding.

The first assemblage forms the lowermost and uppermost part of the Kusuri Formation, whereas the others occur in its middle main part. The poorly defined palaeochannels are 20–25 m thick, typically overlie the depositional lobes and are themselves overlain by the sinuous palaeochannels, 20–30 m thick and ≤ 400–500 m wide, which suggests that the former channels tended to evolve into the latter. The sinuous channels show lateral accretion (point bars) indicating meander-bend expansion combined with a marked downstream translation, and their depth/width aspect ratios are much lower than those of many modern submarine channels. The multistorey complexes of sinuous palaeochannels are 100–160 m thick and estimated to be ≤ 3–5 km wide. The vertical stacking of multistorey channels is attributed to the growth of syndepositional blind-thrust anticlines on the basin floor. The overbank facies assemblages indicate basin-wide flows (tabular bed packages),

[1]Present address: Statoil Research Centre, Rotvoll, 7005 Trondheim, Norway (Email: nilj@statoil.com).

small and highly depletive spill-over flows forming minor levées (wedge-shaped bed packages), and overbank flows deflected by local topography (sigmoidal packages of laterally plastered turbidites).

The narrowness of the basin, the rate of sediment supply (turbidity current discharges) and the temporal topographic confinement provided by syndepositional seafloor deformation are considered to have been the main factors controlling the behaviour and morphodynamic evolution of the turbiditic system. The basin was subject to orogenic compression, with the pulses of tectonic contraction causing both seafloor deformation and an increased sediment supply. As the Sinop foredeep was eventually converted by thrusting into a wedge-top ('piggyback') basin, the fluvial feeder was diverted away from the latter and contributed to a rapid shallowing and deltaic system advance in the adjacent Boyabat Basin. Both basins were gradually inverted by compressional tectonics in late Eocene to early Miocene time, during the climax and final stages of the Tauride orogeny to the south.

Keywords Siliciclastic turbidites, sinuous channels, channel-fill facies, overbank facies, foredeep basin, Central Pontides.

INTRODUCTION

Deep-water turbiditic systems, although formed basically by recurrence of the same episodic process of sediment gravity flow, appear to vary enormously in their morphodynamic characteristics. The great variation of turbiditic systems is reflected in the discrepancies among published models and the difficulty in reconciling features of modern and ancient systems (Barnes & Normark, 1985; Shanmugam & Moiola, 1985, 1988; Mutti & Normark, 1987). Notably, relatively few modern submarine fans are actually fan-shaped; most show channels extending almost throughout the system, which contrasts with the relative paucity of such features in many ancient turbiditic successions; and many turbiditic systems cannot be described in terms of the conventional upper, middle and lower morphometric divisions, the distinction of which is itself disputed. From their attempt to categorize modern turbiditic systems, Stow *et al.* (1985, p. 19) concluded that 'hybrid' system types, combining morphodynamic features of a wide range of idealized models, 'are probably the norm rather than the exception'. This reality is highlighted by the more recent classification scheme of Reading & Richards (1994), which shows several classes to be underrepresented or virtually empty, and has led the authors to observe that many, if not most, of the natural systems actually may be 'intermediate' in character. This notion has been confirmed by numerous case studies published in the past decade (see references below).

It must be concluded that the existing knowledge of the plethora of deep-marine turbiditic systems is simply too limited (Normark *et al.*, 1993; Stow & Mayall, 2000), which also bears directly on our understanding of subsurface turbiditic successions containing hydrocarbon resources, where controversies abound (e.g. Downie & Stedman, 1993; Jensen *et al.*, 1993; Timbrell, 1993; Shanmugam *et al.*, 1994, 1995; Anderton, 1995; Hiscott *et al.*, 1997; Shanmugam, 1997, 2000; Spaak *et al.*, 1999). A wide spectrum of natural hybrids, to be satisfactorily recognized and understood, obviously requires wide exploration, and hence the importance of all new case studies, including potential analogues for petroleum reservoirs. The present case study is such a contribution, although not drawing any particular analogies. As postulated by Normark *et al.* (1993, p. 112): 'Detailed stratigraphic and facies analysis carried out without preconceived models in mind [is] the most effective tool for the study of ancient turbiditic systems'.

One of the main difficulties in understanding turbiditic systems is that the existing knowledge of their responses to principal controlling factors is too limited. The great diversity of turbiditic systems was suggested by Stow *et al.* (1985) to result chiefly from the variation in sediment type and style of supply, or feeder type, a notion elaborated upon further by Reading & Richards (1994). As a first-order control, eustatic sea-level changes have been considered to be of primary importance, by modulating the supply. Tectonics have been thought to play an important secondary role: by creating the deep-water setting for turbiditic sedimentation; by driving the basin-floor subsidence; and by affecting the source area, and hence also the feeder system (Shanmugam *et al.*, 1985; Stow *et al.*, 1985; Haq, 1991; Vail *et al.*, 1991; Bouma, 2000).

However, researchers have come to realize that the overall control in reality is highly complex, with hybrid conditions resulting in hybrid systems. For example, it has been noted that:

1 different feeder types can produce similar turbiditic systems (Alonso & Ercilla, 2003);
2 the response of deep-water systems to sea-level cycles may be quite inconsistent, whereas morphodynamic self-regulation may play an important role (Kolla & Macurda, 1988; Ross *et al.*, 1994; Burgess & Hovius, 1998; Cronin *et al.*, 1998, 2000a; Galloway, 1998; Pirmez *et al.*, 2000; Wynn *et al.*, 2002);
3 the impact of climatic changes may be more pronounced than that of sea-level fluctuations (Postma *et al.*, 1993; Weltje & De Boer, 1993; Winkler, 1993; Winkler & Gawenda, 1999; Beaudouin *et al.*, 2004);
4 the concepts of sequence stratigraphy as a comparative basis may readily apply to some systems (e.g. Pujalte *et al.*, 1998; Johnson *et al.*, 2001a), but not necessarily to others (Saito & Ito, 2002; Helle, 2003);
5 unusual flow geometries and sediment partitioning patterns can result where currents encounter obstacles (Alexander & Morris, 1994; Chikita *et al.*, 1996; Bursik & Woods, 2000; Morris & Alexander, 2003) or fail grossly to scale with pre-existing topography (Normark *et al.*, 1980; Baines, 1984; Clark *et al.*, 1992; Nakajima *et al.*, 1998; Kneller & Buckee, 2000; Migeon *et al.*, 2001; Wynn *et al.*, 2002; Habgood *et al.*, 2003; Sinclair & Cowie, 2003);
6 the controlling role of tectonics and seafloor morphology is of primary importance in many basins (Thornburg *et al.*, 1990; Mutti, 1992; Sinclair, 1992; Agirrezabala & García-Mondéjar, 1994; Haughton, 1994, 2000; Cronin, 1995; Kneller & McCaffrey, 1995; Cronin *et al.*, 2000b; Satur *et al.*, 2000; Felletti, 2002; McCaffrey *et al.*, 2002; Sinclair & Tomasso, 2002; Friès & Parize, 2003; Grecula *et al.*, 2003a,b; Lomas & Joseph, 2004).

As pointed out by Haughton (2000), there has been a wide recognition of how tectonics can affect the feeder systems, but relatively few studies dealing with the effects of syndepositional basin-floor deformation and direct tectonic forcing on turbiditic system behaviour. In an active-margin foreland setting, for example, tectonics can boost sediment supply by uplifting the basin margin and forcing rapid local regression, or by inverting one segment of the foreland and turning it into a sand-prone catchment, while causing flexural subsidence and possibly major water deepening in an adjacent segment (Van Vliet, 1978; Labaume *et al.*, 1985; Puigdefàbregas *et al.*, 1992; DeCelles & Giles, 1996; Sinclair, 1997, 2000; Bryn, 1998; Avramidis *et al.*, 2000). The differential, tectonically forced changes in relative sea level, and the impact of tectonics on basin-floor configuration and sediment flux, may effectively determine the behaviour and morphodynamic character of a turbiditic system.

These issues are addressed by the present study of an Eocene turbiditic system in the retroarc foreland basin of the Central Pontides, northern Anatolia, Turkey. The siliciclastic turbiditic succession, referred to as the Kusuri Formation, is ~ 1200 m thick and well-exposed, but has previously been little studied. The Eocene Sinop Basin was a narrow foredeep trough, gradually converted into a 'piggyback' (wedge-top) basin and inverted by tectonic compression. The basin confinement and syndepositional deformation had a major impact on the axial, sand-rich turbiditic system, inferred to have been fed by a fluvio-deltaic system draining the uplifted adjacent foreland of the Eastern Pontides. The study focuses on the sedimentary facies assemblages, morphodynamic evolution and depositional history of this channelized, basin-floor turbiditic system.

GEOLOGICAL SETTING

The Pontide and Tauride orogenic belts of Anatolia (Fig. 1A) resulted from the suturing of Africa-derived microcratons that were successively accreted to the Cimmerian margin of Eurasia during the Alpine orogeny (Şengör, 1987; Okay & Tüysüz, 1999; Görür & Tüysüz, 2001). The Jurassic accretion of Cimmerian microcontinents marked the closure of the Palaeotethys ocean in the region (Şengör, 1984) and was followed, in Late Cretaceous to Paleogene time, by the accretion of the Kırşehir and Menderes massifs and then the group of Tauric blocks (Yazgan, 1984; Dilek & Moores, 1990; Dilek & Rowland, 1993). The latter two-step accretion resulted in the Pontide and Tauride orogenic belts,

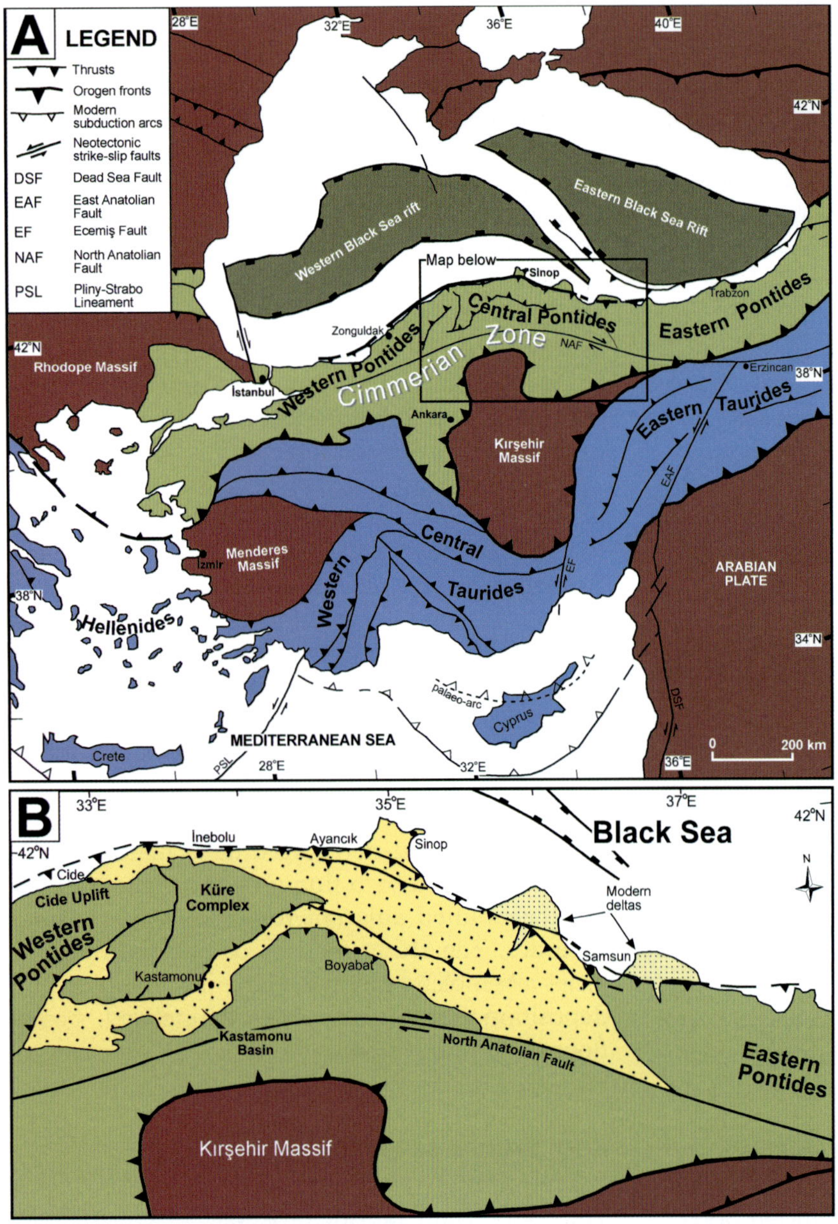

Fig. 1 Regional setting of the study area. (A) Tectonic map of Anatolia and surrounding areas, showing the Pontide and Tauride orogenic belts enveloping the Kırşehir Massif. (B) Simplified map of the Central Pontides, showing the location of the Sinop–Boyabat Basin. Maps compiled with modifications from Robinson *et al.* (1996), Tüysüz (1999), Okay *et al.* (2001) and Nikishin *et al.* (2003).

respectively, and the orogeny ended when Africa's Arabian promontory collided with the Eurasian margin to the east (Fig. 1A). The accretion process was driven by a progressive northward subduction of the Neotethyan oceanic slivers separating the microcontinents, with the subduction zone stepping backwards and eventually shifting, in the early Neogene, to its present-day position in the Cyprian and Cretan arcs (Fig. 1A).

The accretion process was diachronous and spatially non-uniform, as the successive microcratons collided with the Cimmerian margin and themselves, undergoing further adjustments. The large Kırşehir Massif (Fig. 1A) indented the margin and underwent counter-clockwise rotation (Sanver & Ponat, 1981; Görür *et al.*, 1984; Kaymakcı *et al.*, 2003), which caused northward emplacement of the Central Pontide nappes. The subsequent accretion of the Tauric blocks caused further tectonic deformation, whereby the Pontide orogen continued to evolve during the development of the adjacent Tauride orogen to the south. The Pontide orogeny commenced in Late Cretaceous time and culminated at the end of the Eocene (Okay, 1989; Aydın *et al.*, 1995a,b; Okay & Şahintürk, 1997; Ustaömer & Robertson, 1997; Yılmaz *et al.*, 1997; Okay & Tüysüz, 1999), whereas the Tauride orogeny began near the end of Cretaceous time and lasted until the Middle Oligocene in the central part (Andrew & Robertson, 2002) and until the Late Miocene in the western (Hayward, 1984; Collins & Robertson, 1998, 1999) and eastern part of Anatolia (Michard *et al.*, 1984; Aktaş & Robertson, 1990; Dilek & Moores, 1990; Yılmaz, 1993; Yılmaz *et al.*, 1993; Sunal & Tüysüz, 2002).

The northward subduction of Neotethys under the Cimmerian margin was accompanied by backarc extension that led to the formation of the Black Sea rift system (Fig. 1A) along a former intra-Cimmerian suture in Early Cretaceous time (Tüysüz, 1990, 1993; Okay *et al.*, 1994, 2001; Robinson *et al.*, 1996; Ustaömer & Robertson, 1997; Yılmaz *et al.*, 1997; Nikishin *et al.*, 2003). A volcanic arc initially extended from Georgia in the east to Bulgaria in the west (Peccerillo & Taylor, 1975; Eğin *et al.*, 1979; Akıncı, 1984; Tüysüz *et al.*, 1995; Yılmaz *et al.*, 2000), and the zone of volcanic activity was broadened by the backarc rifting and subsequent crustal break-up. The calcalkaline volcanism in the Central Pontides occurred in Coniacian to mid-Campanian times (Tüysüz, 1993; Göncüoğlu *et al.*, 2000; Okay *et al.*, 2001), with minor activity in the Eocene (Güven, 1977). The crustal break-up in the Western Black Sea Rift occurred in late Cenomanian–Coniacian time (Görür, 1988; Okay *et al.*, 1994; Robinson *et al.*, 1995, 1996; Okay & Şahintürk, 1997; Meredith & Egan, 2002; Rangin *et al.*, 2002; Cloetingh *et al.*, 2003; Nikishin *et al.*, 2003), whereas the timing of crustal separation in the Eastern Black Sea Rift is uncertain; considered to have occurred at approximately the same time (Görür, 1988; Nikishin *et al.*, 2003), or possibly in the Maastrichtian (Okay & Şahintürk, 1997), or even Paleocene (Robinson *et al.*, 1995, 1996).

The Sinop–Boyabat Basin (Fig. 1B) formed in Barremian time as a 'failed (abortive)' southern sister of the 'successful' Western Black Sea Rift, failing to achieve crustal separation. The southeast-trending deep-water graben was ~ 80 km wide and ≥ 200 km long, 'hanging' structurally between the strongly subsiding Western Black Sea Rift to the north and the Central Pontide accretionary zone to the south. The basin underwent two main rifting phases before becoming subject to orogenic compression in the late Campanian and being decoupled from the extensional Black Sea regime in late Maastrichtian time (Leren *et al.*, this volume, pp. 401–456). In the earliest Eocene, the basin was split axially into two subparallel troughs by a structural pop-up ridge formed by the northward Erikli thrust and the antithetic, southward Ekinveren back-thrust (Fig. 2). The southern wedge-top ('piggyback') trough, referred to as the Boyabat Basin, was initially subneritic but underwent rapid shallowing, whereas the northern foredeep trough, referred to as the Sinop Basin, remained bathyal and hosted the Kusuri turbiditic system (foreland terminology after DeCelles & Giles, 1996; and also Ori & Friend, 1984). The Sinop Basin underwent contraction and was affected by blind thrusts, until the northernmost Balıfakı thrust (Fig. 2) turned the foredeep into another wedge-top basin in the early middle Eocene. Both basins were then tectonically inverted in the late Eocene, although alluvial sedimentation in the Boyabat Basin probably persisted into the Oligocene (Aydın *et al.*, 1995b). Large parts of the basins were uplifted to ≥ 1000 m above sea level, resulting in good overall exposure. Coastal cliffs, river canyons, road-cut sections and abandoned quarries afford excellent outcrops in the Sinop Basin.

The foredeep Sinop Basin (Fig. 2) is estimated to have been ~ 30 km wide and ≥ 150 km long before its conversion into a 'piggyback' trough and tectonic uplift. The easternmost part of the basin is not preserved, being eroded due to the strong uplift of the Eastern Pontides. The northwestern part extends offshore, where it has not been explored. The Boyabat Basin was smaller, ≤ 20 km wide, passing to the southwest into the narrow Kastamonu Basin (Fig. 1B; Güven, 1977; Aydın *et al.*, 1986; Koçyiğit, 1986; Şengün *et al.*, 1990).

BASIN HISTORY

The Sinop–Boyabat Basin has been mapped and its stratigraphy, tectonic structure and regional plate-tectonic setting have been discussed by many (Badgley, 1959; Göksu *et al.*, 1974; Aydın *et al.*, 1982, 1986; Sonel *et al.*, 1989; Tüysüz, 1990, 1993, 1999; Robinson *et al.*, 1995; Tüysüz *et al.*, 1995; Görür, 1997; Görür & Tüysüz, 1997; Okay &

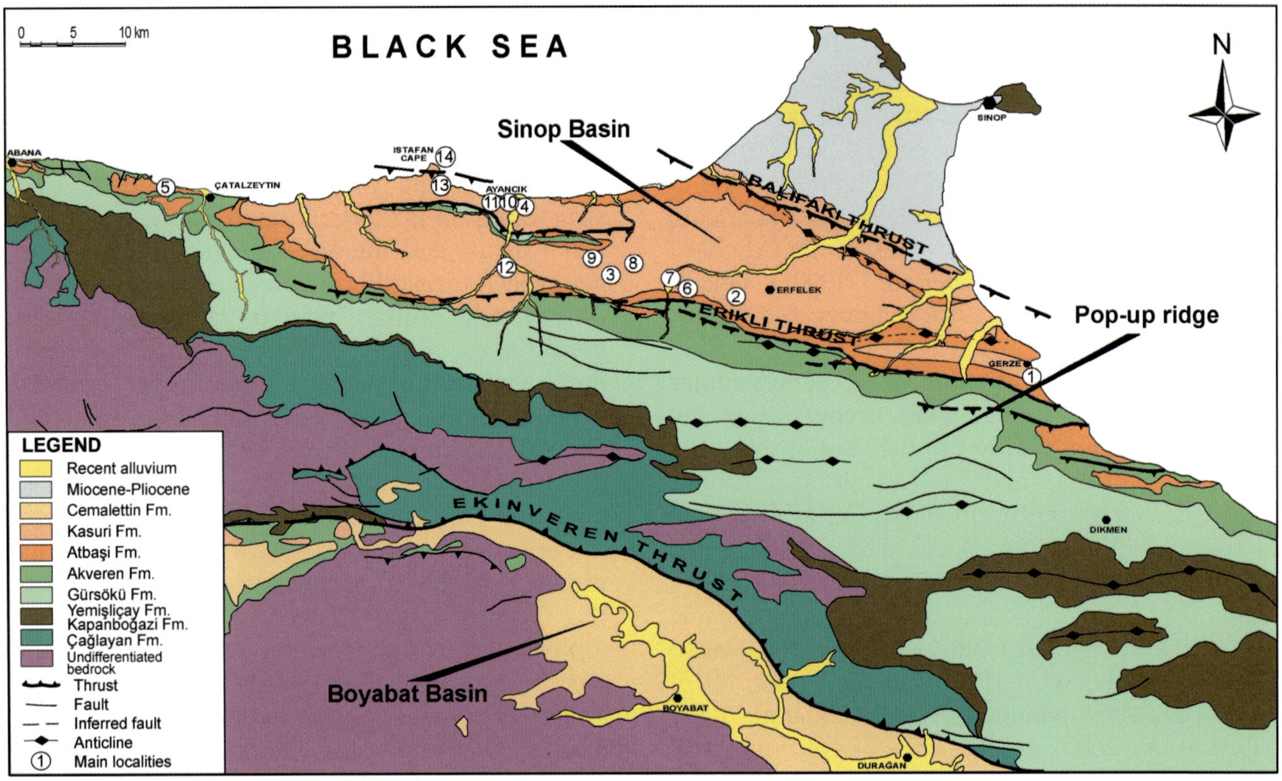

Fig. 2 Geological map of the Sinop–Boyabat Basin, showing the areal distribution of the basin-fill formations. Modified from Gedik & Korkmaz (1984), Barka *et al.* (1985) and Aydın *et al.* (1995b). Note that the northward Erikli thrust and the southward Ekinveren back-thrust turned the axial part of the basin into a pop-up ridge, which split the original basin into a southern wedge-top trough (Boyabat Basin) and a northern foredeep trough (Sinop Basin). The younger Balıfakı thrust to the north eventually converted the Sinop Basin into a wedge-top trough.

Şahintürk, 1997; Ustaömer & Robertson, 1997; Okay & Tüysüz, 1999), but with no detailed research and all sedimentological studies published in local Turkish journals (Ketin & Gümüş, 1963; Gedik & Korkmaz, 1984; Gedik *et al.*, 1984; Aydın *et al.*, 1995a,b). Most of this previous research was intended to assess the hydrocarbon potential of the Turkish part of the Black Sea region, where the Ukrainian northern part has been a significant petroleum province (Aydın *et al.*, 1982; Robinson *et al.*, 1996; Ziegler & Roure, 1999). A few wells were drilled onshore (Aydın *et al.*, 1995b), but these data are not accessible and outcrops have served as the primary basis for an interpretation of offshore seismic sections (Robinson *et al.*, 1996; Meredith & Egan, 2002; Rangin *et al.*, 2002; Cloetingh *et al.*, 2003; Nikishin *et al.*, 2003).

The basin-fill succession of Early Cretaceous to middle Eocene deposits (Fig. 3) has a combined stratigraphic thickness of ~ 7000 m and provides an important record of the region's tectonic development and palaeogeographical history. The ensuing review of the basin's dynamic stratigraphy compiles the results of previous studies (Ketin & Gümüş, 1963; Gedik & Korkmaz, 1984; Aydın *et al.*, 1986, 1995b; Tüysüz, 1990, 1993; Görür *et al.*, 1993; Görür & Tüysüz, 1997) and the present research (see also Leren, 2003; Janbu, 2004; Uchman *et al.*, 2004; Leren *et al.*, this volume, pp. 401–456).

Pre-Eocene development

The pre-rift 'bedrock' consists mainly of thick platform carbonates, Late Jurassic to Early Cretaceous in age. The onset of rifting and establishment of a deep-water graben are recorded by the Barremian–Albian Çağlayan Formation (Fig. 3), which consists of calcareous and subordinate siliciclastic turbidites intercalated with olistostromal breccias and large slide blocks of bedrock limestones. These deposits

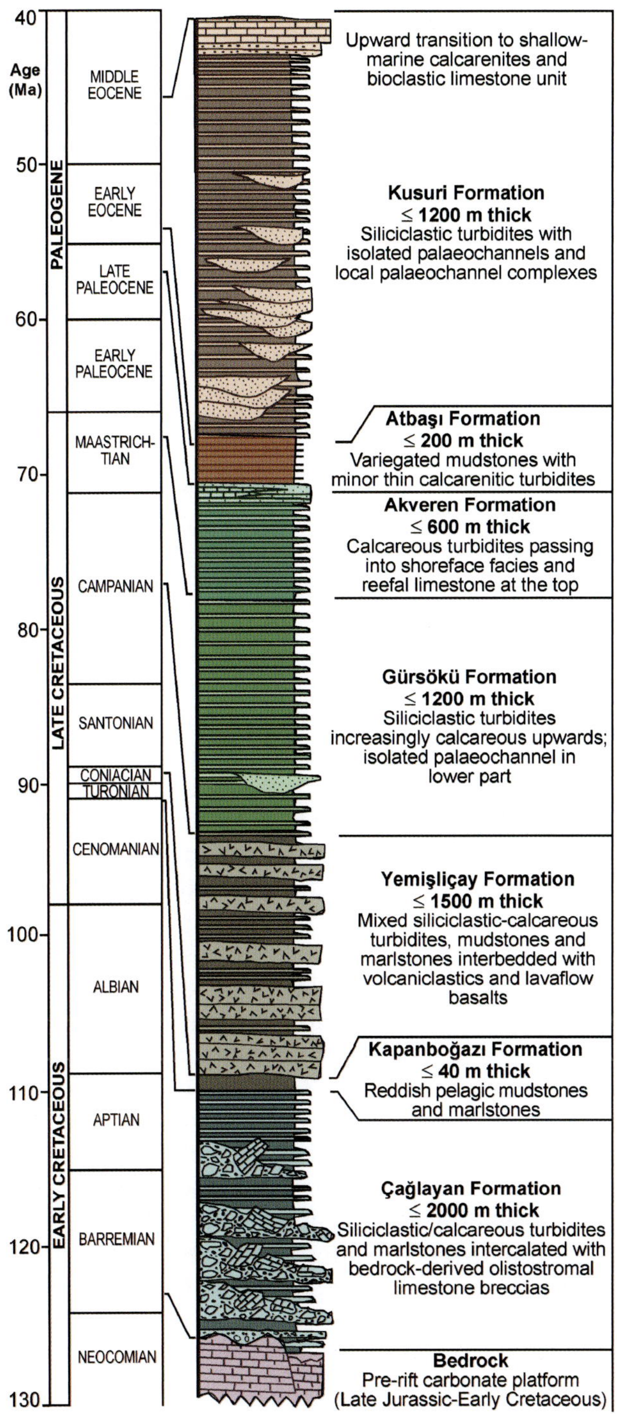

Fig. 3 Stratigraphy of the Sinop–Boyabat Basin (modified from Ketin & Gümüş, 1963; Gedik & Korkmaz, 1984; Aydın *et al.*, 1982, 1995b). The Eocene (post-Atbaşı Formation) part of the profile pertains to the Sinop Basin (cf. Fig. 2).

are locally ≤ 2000 m thick and their varied thickness reflects rugged fault-block topography of the early stage rift. The sediment was derived from both margins of the graben, with the turbidity currents filling in the seafloor relief and flowing mainly westwards (northwest) along the basin axis. The sediment supply declined in Turonian to earliest Coniacian time, when the Kapanboğazı Formation (Fig. 3) was deposited in a sand-starved deep-water environment. This formation is ≤ 40 m thick and consists of reddish-grey, variegated mudstones interspersed with pelagic marls. The cessation of sediment supply indicates a post-rift phase of broader thermal subsidence that caused the contemporaneous shorelines to shift away from the graben.

Another phase of rifting is recorded by the overlying, Coniacian–Campanian Yemişliçay Formation, which is ≤ 1500 m thick and consists of turbidites with a mixed calcareous–siliciclastic composition, interbedded with abundant volcaniclastic deposits and lavaflow basalts (Fig. 3). The sediment was still derived from both basin margins and possibly more from the northern one (Aydın *et al.*, 1995b; Tüysüz, 1999), but the latter was subsequently submerged below wave base and became insignificant as a clastic source. This asymmetrical development of the basin is attributed to the crustal break-up and fault-block margin collapse in the adjacent Western Black Sea Rift (Leren *et al.*, this volume, pp. 401–456, their fig. 27B).

The aborted Sinop–Boyabat rift then became increasingly affected by orogenic thrust tectonics from the south, which converted it into a retroarc foreland basin of the Central Pontides (Janbu, 2004). The conversion commenced in Late Cretaceous time with the deposition of the Campanian–Maastrichtian Gürsökü Formation (Fig. 3). This turbiditic succession is ≤ 1200 m thick and consists of siliciclastic (mainly epiclastic volcanic) sediment increasingly richer in calcareous bioclastic admixture, supplied mainly from the west/southwest and spread eastwards along the basin axis (Leren, 2003; Leren *et al.*, this volume, pp. 401–456). The formation shows little evidence of channelized currents, except for an isolated palaeochannel in its lowermost part, and the sediment was supplied from a sublittoral ramp perched on the basin-margin slope, with the turbidity currents generated mainly by storms and earthquakes. High subsidence rates prevented seafloor shallowing, while the aggrading, ponded turbiditic system tended to retreat by onlapping the sourcing basin-margin slope (Leren *et al.*, this volume, pp. 401–456).

The supply of sediment from the west/southwest and cessation of volcanism are attributed to the collision of the Kırşehir Massif with the Cimmerian margin in the transition area of the Western and Central Pontides in the Late Cretaceous (Tüysüz *et al.*, 1995; Okay & Tüysüz, 1999), when the subduction process probably shifted to the southern side of the accreted massif. As this large 'indentor' block was pushed northwards and rotated counterclockwise (Kaymakcı *et al.*, 2003), the Central Pontide nappes began to be emplaced northwards and affect the foreland, whereby a reefal platform expanded along the basin's southwestern margin. The overlying Maastrichtian–Paleocene Akveren Formation (Fig. 3) is a succession of calciclastic sheet-like turbidites, ≤ 600 m thick, with mainly eastward palaeocurrent directions and evidence of rapid shallowing in the uppermost part. The sediment was derived from a distally steepened carbonate ramp, which eventually became homoclinal and advanced across the shallowing basin, as the ignition of turbidity currents declined and the basin floor became influenced by storms (Leren *et al.*, this volume, pp. 401–456). The marked decrease in subsidence rate suggests that the basin by this time was largely decoupled from the extensional Black Sea regime.

The overlying, upper Paleocene to lowest Eocene Atbaşı Formation (Fig. 3) consists of deep-water variegated mudstones, ~ 200 m thick, interspersed with thin calciclastic turbidites. The rapid deepening of water and sediment-starved basin conditions are attributed to foreland flexural subsidence due to crustal loading by the Central Pontide nappes (Nikishin *et al.*, 2003; Janbu, 2004), coeval with the Thanetian eustatic sea-level rise (Haq *et al.*, 1988). Late Cretaceous to earliest Eocene sedimentation in the basin is discussed in detail by Leren *et al.* (this volume, pp. 401–456).

Eocene development

The continental collision then culminated in the Eastern Pontides, where the accretion of a suprasubduction volcanic arc to the Cimmerian margin in Late Paleocene time marked the local climax of the Pontide orogeny (Yazgan, 1984; Şengör, 1987; Dilek & Rowland, 1993; Okay & Tüysüz, 1999; Gürer & Aldanmaz, 2002). The subsequent accretion of Tauric blocks to the Kırşehir Massif was followed by the extension of the Tauride orogeny to the east (Dilek & Moores, 1990). The full-scale onset of the Tauride orogeny caused further contraction in the Central Pontide foreland, with reversal of pre-existing normal faults and thin-skinned thrust tectonics (Aydın *et al.*, 1995b). In the early Eocene, the axial pop-up ridge split the original Sinop–Boyabat Basin into the 'piggyback' Boyabat Basin and the foredeep Sinop Basin (Fig. 2; Janbu, 2004).

The lower to middle Eocene Kusuri Formation in the Sinop Basin (Fig. 3) is a turbiditic succession that recorded an abundant supply of siliciclastic sediment from the east, attributed to a fluvio-deltaic system draining the adjacent, emerged Eastern Pontide foreland. This succession is ≤ 1200 m thick, with a mud-rich lower part dominated by sheet-like turbidites and a sand-rich middle part containing solitary and multistorey palaeochannels. The mud-rich upper part of the succession contains sheet-like turbidites that are increasingly calcareous, with a rapid upward transition to neritic calciclastic tempestites and littoral bioclastic limestones at the top. This latest part, deposited in connection with northward thrusting led by the younger Balıfakı thrust (Fig. 2), recorded the structural closure of an inverted basin (Janbu, 2004).

The coeval Eocene succession in the Boyabat Basin, referred to as the Cemalettin Formation, is ~ 900 m thick and similarly siliciclastic, and rich in sand and gravel derived from the east. Its lower part consists of turbidites, but the thick middle part is shallow marine and mainly fluvio-deltaic, including incised valleys and recording major fluctuations in relative sea level. The upper part consists of conglomeratic braided-river alluvium. A detailed study of the Cemalettin Formation is under way.

Miocene development

The two basins were gradually inverted by tectonic contraction in late Eocene to Early Miocene times, during the climax and final stages of the Tauride orogeny (Okay & Şahintürk, 1997; Okay & Tüysüz, 1999). Paratethyan shallow-marine deposits of

Miocene age occur only at the northern rim of the Sinop Basin (see the Sinop peninsula area in Fig. 2), where they unconformably overlie a deformed bedrock. These deposits are dominated by calcareous tidal facies and shell-rich bioclastic limestones (Görür *et al.*, 2000).

The Late Miocene also witnessed the onset of the neotectonic regime, with a westward tectonic escape (strike-slip expulsion) of the compound Anatolian craton between the sinistral East Anatolian Fault and the dextral North Anatolian Fault (Fig. 1A; Şengör *et al.*, 1985). In the Central Pontides, this latter fault corresponds roughly to the Neotethyan suture zone and the northern margin of the Kırşehir Massif.

Syndepositional basin-floor deformation

Eocene turbiditic sedimentation in the Sinop Basin occurred in a compressional tectonic regime, after the Erikli thrust formed and several minor thrusts extended further to the north, until the northernmost Balıfakı thrust took the basin 'piggyback' into inversion (Fig. 2). The outward propagation of thrusting, as is typical of foreland basins (Ori & Friend, 1984; Allen & Homewood, 1986; Puigdefábregas *et al.*, 1992; DeCelles & Giles, 1996; Mascle & Puigdefàbregas, 1998), inevitably affected the basin floor. The Eocene succession in the Sinop Basin abounds in asymmetrical folds and minor thrusts, some of which are mappable (Fig. 2). The north-verging anticlines commonly have a steep thrust at the axial plane and/or pass laterally into thrusts, suggesting blind-thrust folds. At least some of these features apparently represent syndepositional deformation of the basin floor.

The main evidence of syndepositional deformation invoked and documented further in the paper includes: (i) intrabasinal slump deposits and local angular unconformities within mudstone-rich facies successions, attributed to sediment failure and formation of slump scars due to localized upwarping of the seafloor; (ii) buried synsedimentary faults; (iii) local growth anticlines, with multiple 'progressive unconformities' due to erosion by currents and sporadic gravitational failure; and (iv) synclinal downwarping of substrate deposits beneath palaeochannel complexes, attributed to tectonic folding combined with localized compaction.

THE KUSURI FORMATION

Lithostratigraphic definition

The lithostratigraphic nomenclature for the Sinop–Boyabat Basin has a history of numerous changes, partly because it was based on an inconsistent combination of lithostratigraphic mapping and biostratigraphic correlations, rather than an understanding of the spatial facies relationships. The lower to middle Eocene Kusuri Formation in the Sinop Basin (Figs 2 & 3) was first mapped and described by Ketin & Gümüş (1963), who referred to this unit as the 'Ayancık and Kusuri Formation'. Gedik & Korkmaz (1984) and Gedik *et al.* (1984) subsequently changed this name to 'Yenikonak Formation' and referred to its sandstone-dominated and mudstone-rich parts as the Ayancık and Kusuri members, respectively, although these could not be mapped at conventional scales. Sonel *et al.* (1989), Aydın *et al.* (1995b) and Görür & Tüysüz (1997) used the name 'Kusuri Formation' for the whole unit in their maps of the basin. However, Görür & Tüysüz (1997) also used the name 'Ayancık Sandstone Member' for the thick sandstone bodies within the formation, thus following partly the earlier lithostratigraphic notions of Ketin & Gümüş (1963) and Gedik *et al.* (1984). This division is abandoned in the present study, because the isolated sandstone bodies (solitary and multistorey palaeochannels) form discontinuous outcrops, are not mappable at conventional scales (≤ 1:25,000) and do not constitute a coherent part of the formation. Although the sandstone bodies occur chiefly in the middle part of the formation, they are widely scattered at various stratigraphic levels and surrounded by mudstone-rich heterolithic deposits.

The name 'Kusuri Formation' has also been used for the coeval turbiditic deposits in the adjacent Boyabat Basin (Sonel *et al.*, 1989; Aydın *et al.*, 1995b; Görür & Tüysüz, 1997; Tüysüz, 1999), primarily because of their similar siliciclastic composition and content of early Eocene microfossils, but despite their deposition in a separate and tectonically different ('piggyback') basin. The name 'Cemalettin Formation' was used for the overlying shallow-marine and fluvio-deltaic deposits. The lower part of the Eocene succession in the Boyabat Basin was also referred to as the Gökırmak Formation and the upper part as the Sakızdağ Formation by Gedik &

Table 1 Sedimentary facies of the turbiditic succession of the Kusuri Formation, Sinop Basin (for outcrop examples, see Figs 4–8)

Facies	Subfacies	Bed characteristics	Interpretation
A	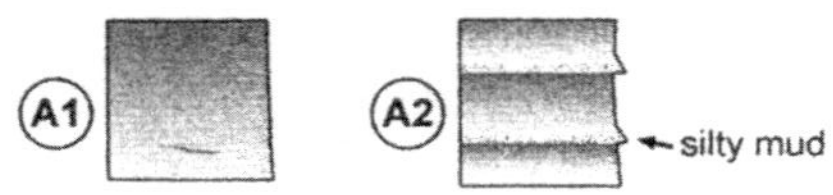	**Facies A1**: Grey, massive mudstone beds, 2–42 cm thick, separating turbidites; generally darker than the silt-rich turbiditic e-divisions and commonly bioturbated.	Products of hemipelagic 'background' sedimentation (**A1**), with episodic incursions of silt-rich suspension derived from turbidity currents and possibly also shed by deltaic hypopycnal plumes (**A2**).
		Facies A2: Composite mudstone units, up to 60 cm thick, split by graded silty interlayers into beds 2–10 cm thick; the thin interlayers are mainly sharp-based turbidites T*de* and Te.	
B	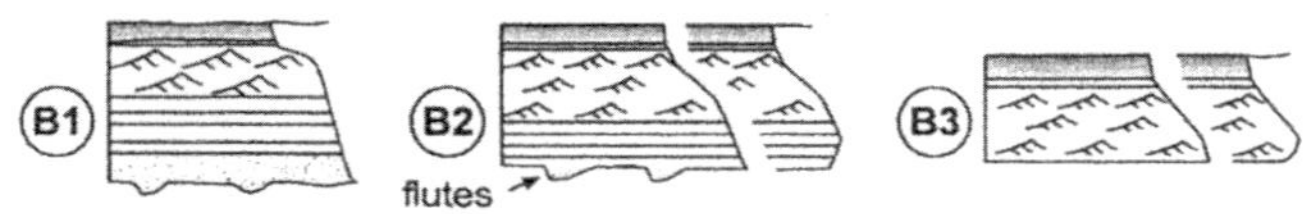	Bouma-type and predominantly sheet-like turbidites. **Facies B1**: Turbidites T*abcde* composed of coarse/very coarse to fine sand and mainly 20–80 cm thick, but ranging from 6 to 314 cm. **Facies B2:** Turbidites T*bcde* made of medium to fine sand and mainly 10–30 cm thick, ranging from 2 to 270 cm. **Facies B3:** Turbidites T*cde* composed of fine to very fine sand and mainly 2–5 cm thick, ranging from < 1 cm to 19 cm.	Classic turbidites deposited by turbidity currents of high (**B1**) to low density (**B2** & **B3**), mainly non-channelized, spread in overbank and terminal-lobe areas, but occasionally flowing through channel or other local topographic confinement.
C	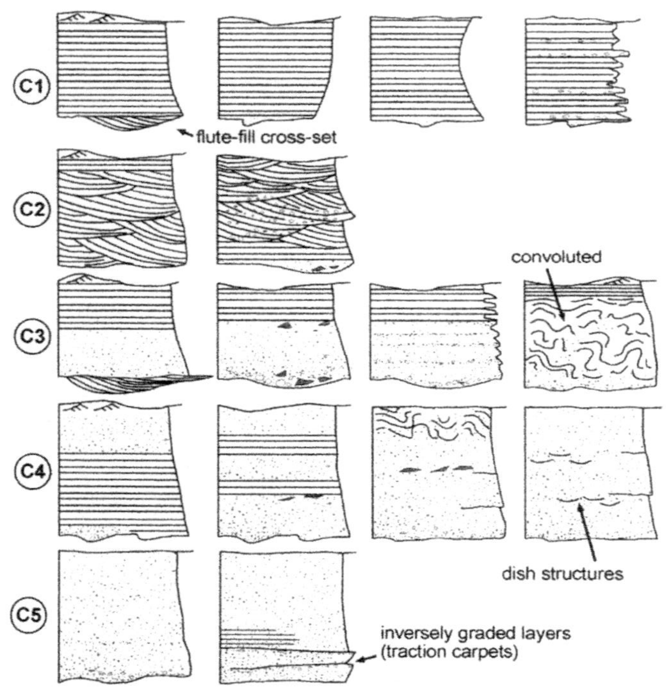	**Facies C1**: Channel-fill turbidites T*b(c)* composed of coarse/very coarse to medium sand and mainly 40–60 cm thick, but ranging from 10 to 260 cm; the relatively thick *b*-division shows normal and/or inverse grading, often multiple, or an irregular grain-size changes with granule- or pebble-rich strata in the last case. Some of the thickest beds show large basal flutes, up to 35 cm deep and 500 cm long, filled with sigmoidal 'microdelta-type' cross-stratification.	Non-classic turbidites deposited by low-density currents flowing through channels; the relatively thick *b*-division and variable grain-size trend suggest sustained flows with waning, waxing, waning/waxing or more fluctuating behaviour. Flute-fill cross-sets indicate an initial bypass phase.
		Facies C2: Channel-fill turbidites composed of coarse to very coarse sand, locally pebbly and mainly 70–120 cm thick, but ranging from 15 to 300 cm; beds show mainly trough cross-stratification, with simple or multiple fining-upward trend, locally underlain by massive and/or parallel-stratified division and occasionally overlain by ripple cross-lamination.	Non-classic turbidites deposited by low-density currents flowing through channels and forming 3-D dunes; the stratification and variable grain-size trend suggest sustained and quasi-steady flows.
		Facies C3: Channel-fill turbidites T*ab(c)* composed of very coarse (locally gravelly) to medium sand and mainly 60–110 cm thick, but ranging from 10 to 305 cm; some beds show varied grain-size trend, or strong convolution and partial homogenization by dewatering. Some of the thickest beds show large basal flutes, up to 60 cm deep and 950 cm long, filled with sigmoidal 'microdelta-type' cross-stratification.	'Top-absent' classic turbidites deposited by high-density turbidity currents flowing through channels, with a rapid dumping of suspension load followed by plane-bed traction. Flute-fill cross-sets indicate an initial phase of tractional sand bypass.
		Facies C4: Channel-fill turbidites T*(a)ba(c)*, T*ababa(b)* and T*aaa(b)*, composed of very coarse to medium/fine sand and mainly 50–100 cm thick, but ranging from 25 to 346 cm; poorly graded beds, with local evidence of dish structures, convolute parallel stratification, alternating *a*- and *b*-divisions or multiple *a*-divisions, and scattered mudclasts.	Non-classic turbidites deposited by high-density turbidity currents flowing through channels; the repetitive massive and stratified divisions suggest sustained, but highly fluctuating currents.
		Facies C5: Channel-fill turbidites T*a*, or T*(b)a* with one or more basal traction-carpet layers; composed of coarse/very coarse to medium sand, occasionally with small pebbles and/or granules at the base; mainly 50–80 cm thick, but ranging from 17 to 205 cm. The traction-carpet layers are up to 10 cm thick, laterally discontinuous, with two or three superimposed layers locally merging into one.	Non-classic turbidites deposited by high-density turbidity currents flowing through channels and dumping suspended load; the multiple carpets and sparse planar stratification indicate tractional bypass of sand.
D	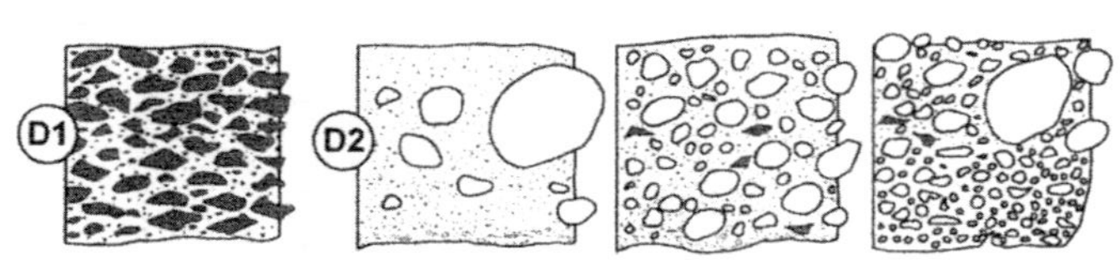	**Facies D1**: Lenticular, non-graded massive beds, up to 50–70 cm thick, composed of pebble- to cobble-sized mudclasts and sand matrix.	Deposits of in-channel debrisflows; some cohesive, generated within the channel by bank collapse and/or by slurrying of a turbidity current's 'moving bed' (**D1**); others cohesionless, derived from the feeder system and emplaced in the channel thalweg zone (**D2**).
		Facies D2: Lenticular, mounded massive beds of gravelly sandstone or gravelstone, 15–460 cm thick, composed of subrounded to rounded extraformational clasts, up to boulder size; the gravelstone beds are rich to poor in sand matrix and some of these latter have a pebbly and slightly inversely-graded basal part.	

Korkmaz (1984). In the present paper, the name Cemalettin Formation is used for the whole succession, as an equivalent to the Kusuri Formation in the Sinop Basin.

Sedimentary facies

The bulk of the Eocene Kusuri Formation in the Sinop Basin (Fig. 3) consists of siliciclastic turbidites and associated mudstones. The succession has been studied in all accessible outcrop sections, and its various parts have been logged in detail at 14 localities (Fig. 2). The logs have a cumulative stratigraphic thickness of 1060 m, with five logs from the formation's lower part (localities 1–5), nine logs from the middle part, including palaeochannels (localities 4–12), and two logs from the upper part (localities 13 and 14). Only selected portions of a few representative logs are shown in the present paper. The calcareous, shallow-marine topmost part of the Kusuri Formation, which recorded the basin's tectonic closure, is described in detail elsewhere (Janbu, 2004).

The turbiditic succession consists of a wide range of sedimentary facies, which have been distinguished on the basis of macroscopic sedimentological criteria and are described briefly and interpreted in Table 1. The descriptive sedimentological nomenclature used is after Bouma (1962), Harms *et al.* (1975, 1982) and Collinson & Thompson (1982), and the terminology for turbidity currents is according to Lowe (1982) and Kneller & Buckee (2000). The sedimentary facies include (Table 1):

A Hemipelagic 'background' mudstones (facies A1) and thin muddy turbidites (facies A2), generally homogeneous and bioturbated (Fig. 4A).
B A typical range of 'classic', Bouma-type turbidites with solemarks and silty mudstone cappings (facies B1–B3; Fig. 4B–F).
C A wide spectrum of non-classic turbidites, mainly amalgamated, attributed to low-density (facies C1 and C2) and high-density currents (facies C3–C5; Figs 5–7), with common evidence of sustained, long-duration flows. This evidence includes: thick beds with monotonous planar-parallel stratification and fluctuating grain sizes, indicating multiple waxing or waning and waxing of the flows (Figs 5 & 6G); cosets of trough cross-stratification (Fig. 6A-E), indicating migration of three-dimensional dunes under fairly steady flow; multiple traction-carpet layers (Figs 5C, D & 7G); and large, scoop-shaped basal flutes (1–2 m wide, 3–5 m long and 0.1–0.3 m in scour relief) filled with massive sediment (Figs 6G & 7G) or with a foreset of sigmoidal strata (Figs 5D, E & 6C, F) similar to the 'microdelta' cross-stratification related to local hydraulic jump (Jopling, 1965, fig. 9).
D Subordinate gravelstone and gravelly sandstone beds, massive and bearing extra- and/or intraformational clasts (facies D1 and D2; Fig. 8), interpreted as deposits of in-channel debrisflows derived from the feeder system or local bank collapse.

The sediment is siliciclastic. The gravel consists of quartz and various igneous and metamorphic rock clasts, but commonly also bears calcarenite and marlstone debris derived from the underlying Akveren Formation (Fig. 8). The sand is submature, poorly rounded and moderately sorted, dominated by quartz, feldspar and various rock fragments of mainly ophiolitic and epiclastic volcanic provenance; mineral detritus also includes muscovite, pyroxene/hornblende, garnet and opaque grains. Many sandstone beds contain coalified plant detritus and sporadic larger wood fragments. The mudstones are commonly rich in mica flakes and plant detritus. Some of the coarser-grained sandstone beds are rich in granules and small pebbles (Figs 5 & 6), or are gravelly in their lower parts (Fig. 7). The palaeocurrent directions measured from flutes and cross-strata are generally towards the west, which supports the notion of an axial basin-floor turbiditic system.

The assemblage of associated trace fossils (Table 2) represents a highly diverse *Nereites* ichnofacies, indicating a deep-sea environment with well-nourished, oxygenated bottom waters with plant detritus (for details, see Uchman *et al.*, 2004). The microfossil content of mudstone beds (Table 3) confirms bathyal palaeobathymetry and an early to early middle Eocene age of the turbiditic succession. Unidentified fish teeth and a rich admixture of redeposited late Maastrichtian and/or Paleocene foraminifers have also been found in the mudstone samples.

FACIES ASSOCIATIONS

The sedimentary facies (Table 1) are considered to be the basic 'building blocks' of the sedimentary

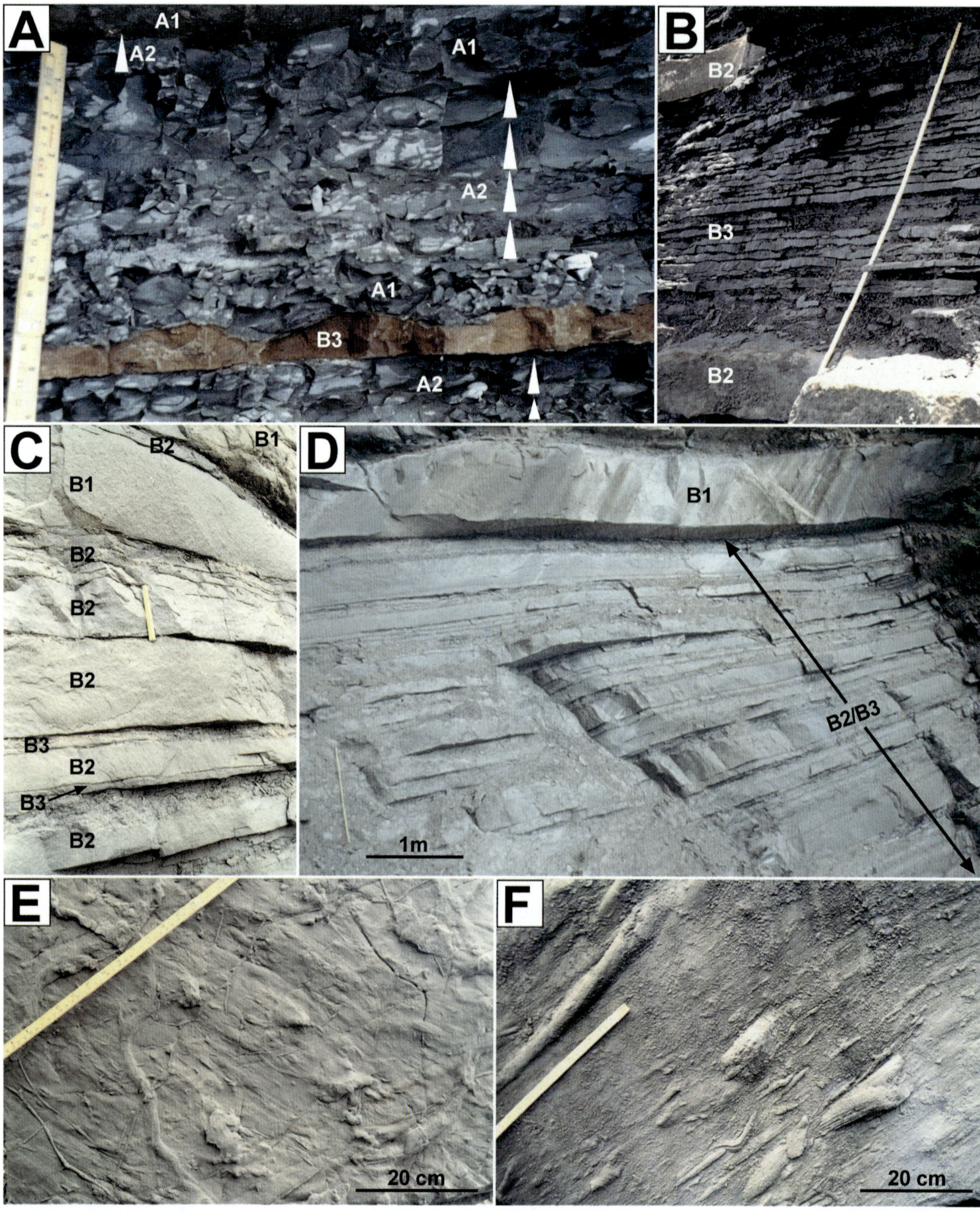

Fig. 4 Outcrop details of sedimentary facies A and B. (A) Succession of alternating facies A1 and A2 beds, with an isolated thin turbidite of facies B3; from FA 1 at locality 1. (B) Thin turbidites of facies B3, interspersed with thicker beds of facies B2; from 'distal' (lower) FA 2 at locality 4. (C) Turbidites of facies B2 interspersed with thicker beds of facies B3; from 'proximal' (upper) FA 2 at locality 3. (D) Alternating beds of facies B2 and B3, overlain by a facies B1 bed; portion of a thickening upward FA 2 at locality 6. (E) Flute marks and *Ophiomorpha annulata* traces on the sole of facies B2 bed. (F) Flute and groove marks on the sole of facies B1 bed. The locality numbers are as in Fig. 2 and facies association (FA) code as in Fig. 9. See Table 1 for facies definitions.

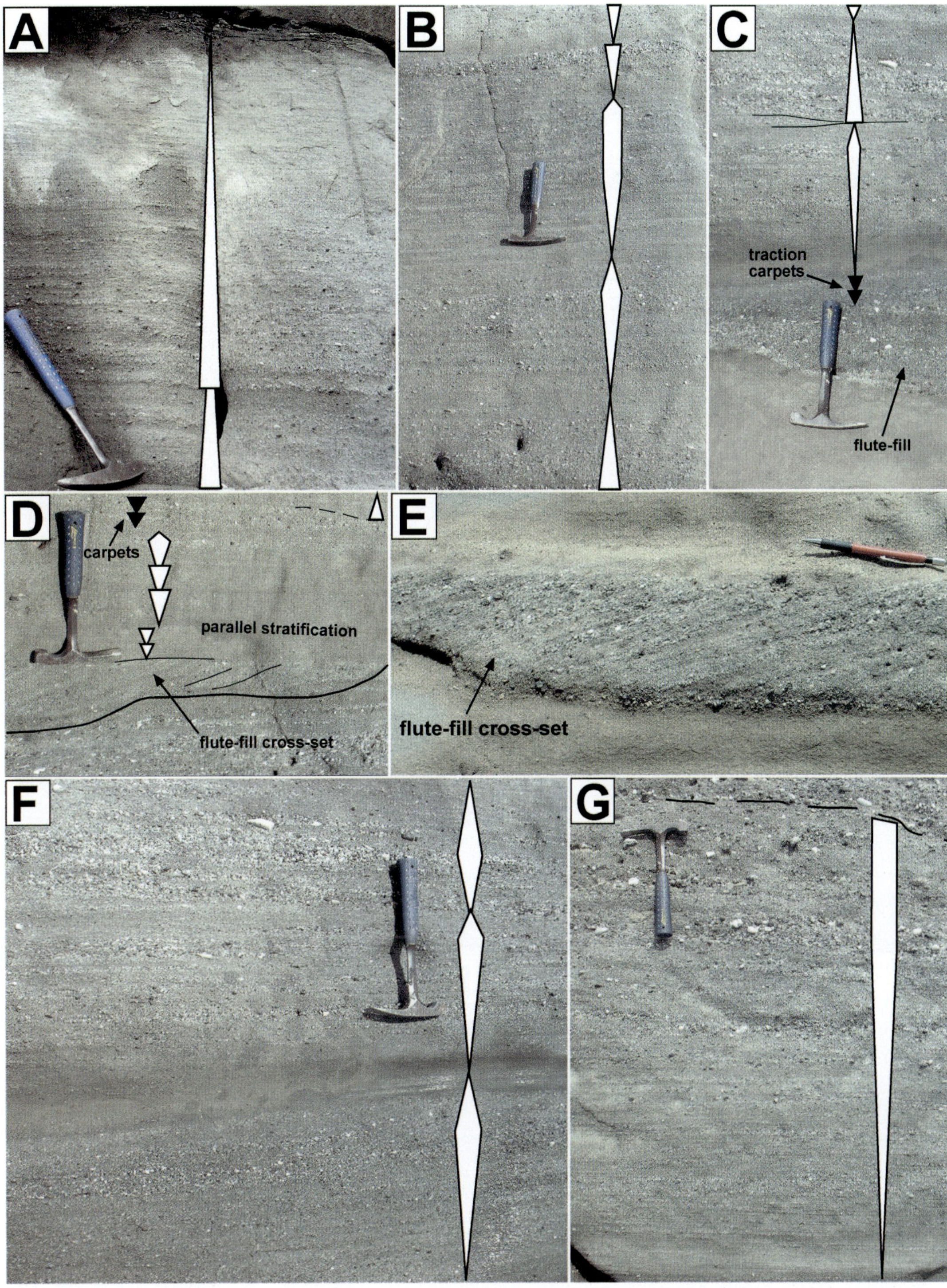

Fig. 5 Outcrop details of sedimentary facies C1, attributed to sustained turbidity currents. (A) Planar parallel-stratified turbidites with multiple normal grading. (B) Alternating inverse to normal grading; the hammer is 35 cm. (C) Similar thick beds with basal flute-fills overlain by traction-carpet layers or by (D) inversely graded planar strata sets covered with analogous layers. (E) Close-up view of a flute-fill cross-strata set at turbidite base; palaeoflow direction to the left; the pen is 15 cm. (F) Portion of a stratified thick bed with repetitive inverse-to-normal grading motif and a markedly finer grained horizon. (G) Stratified thick bed with an overall inverse grading. All examples are from FA 3C at locality 4 (Fig. 2). See Table 1 for facies definitions.

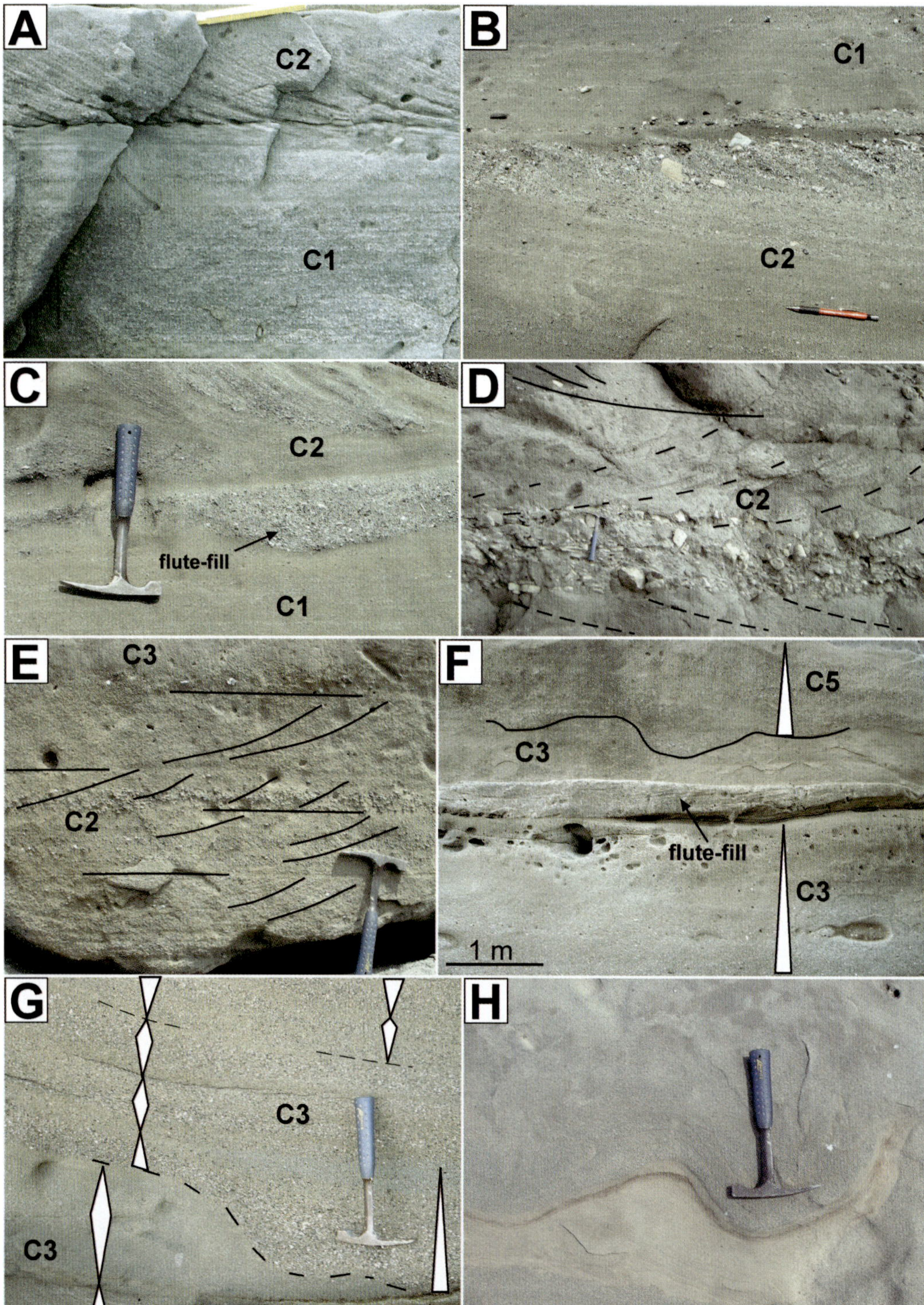

Fig. 6 Outcrop details of sedimentary facies C2 and C3, attributed mainly to sustained turbidity currents. (A) Planar parallel-stratified and normally graded bed of facies C1 overlain by a cross-stratified bed of facies C2; the ruler is 25 cm. (B) Trough cross-stratified bed of facies C2 overlain by planar-stratified bed of facies C1; the pen is 15 cm. (C) Flute-fill at the base of trough cross-stratified facies C2 bed; the hammer is 35 cm. (D) Gravel-rich cross-set within a facies C2 bed. (E) Trough cross-stratified bed of facies C2 overlain by facies C3 bed. (F) Amalgamated beds of facies C3, with an intervening flute-fill cross-set (palaeoflow direction to the left), overlain by a bed of facies C5 with uneven erosional base. (G) Amalgamated and deformed beds of facies C3. (H) Loaded contact between facies C3 beds, the upper one strongly convoluted. All examples are from FA 3C at localities 4, 6 and 10 (Fig. 2). See Table 1 for facies definitions.

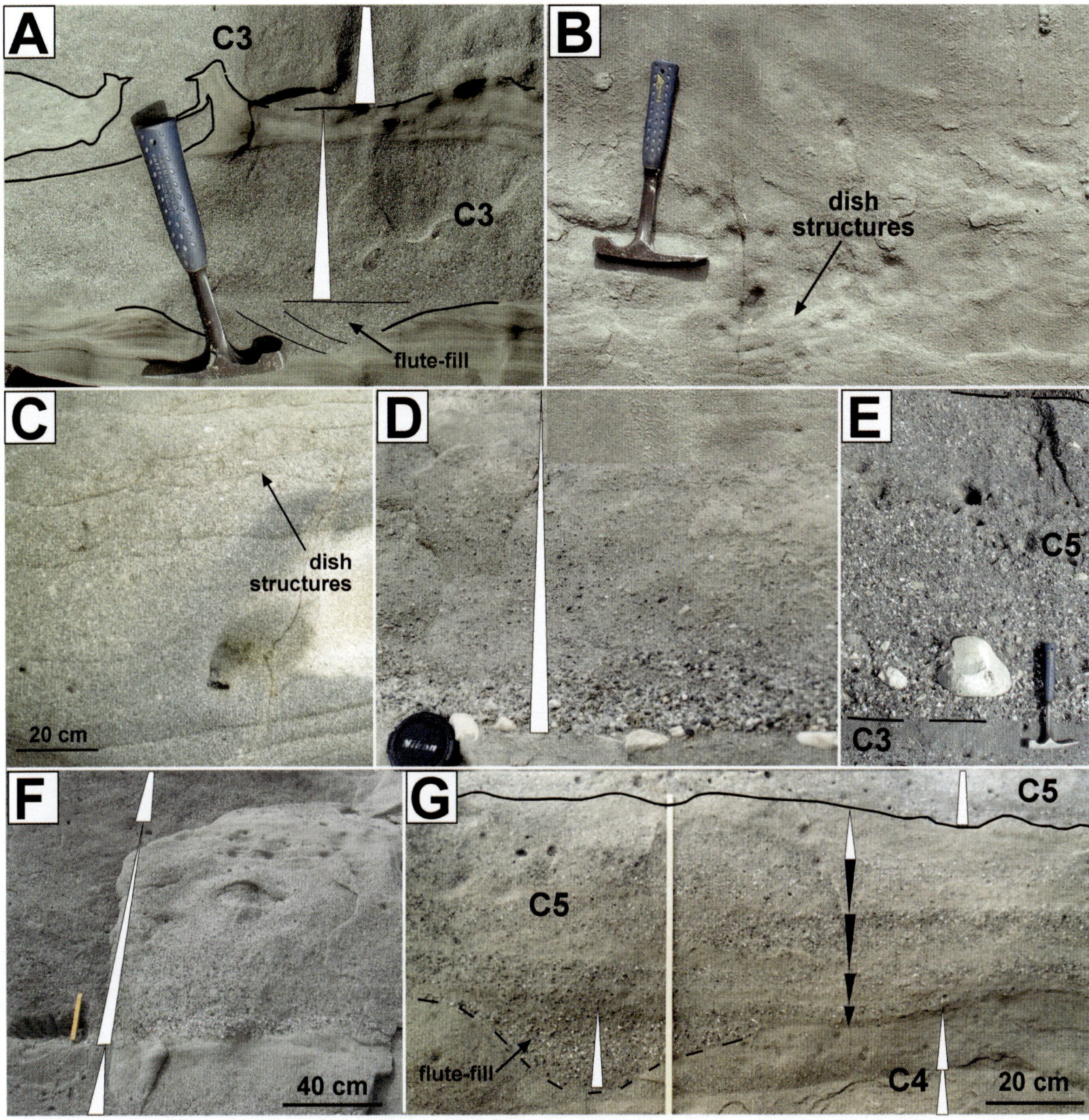

Fig. 7 Outcrop details of sedimentary facies C3–C5, attributed to high-density turbidity currents. (A) Amalgamated beds of facies C3 with a strongly loaded, deformed contact; the hammer is 35 cm. (B & C) Dewatering structures in homogenized beds of facies C4. (D & E) Massive, graded beds of facies C5 with outsized gravel clasts scattered along the base or within the basal part; the lens cap is 5 cm. (F) Amalgamated massive beds of facies C5. (G) Facies C4 bed with multiple weak normal grading, overlain by a facies C5 bed with a basal flute-fill, multiple traction-carpet layers (inversely graded) and erosionally truncated, normally graded upper part. All examples are from FA 3C at localities 4 and 10 (Fig. 2). See Table 1 for facies definitions.

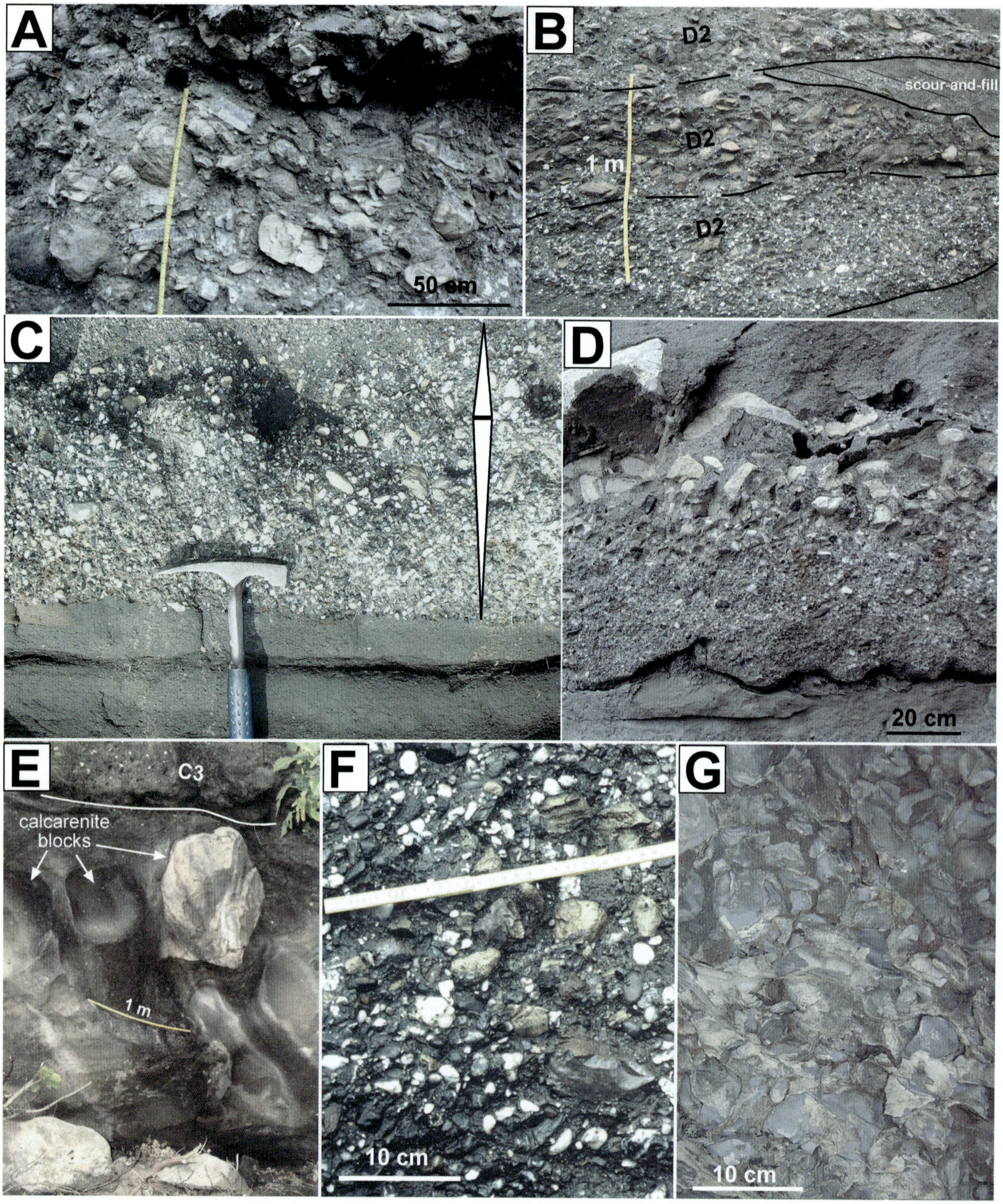

Fig. 8 Outcrop details of sedimentary facies D, attributed to in-channel debrisflows. (A) Chaotic bouldery gravelstone of facies D2, with a sand-supported texture and subangular to rounded calcarenite and marlstone clasts. (B) Amalgamated gravelstone beds of facies D2, with an intervening, trough-shaped scour-and-fill feature. (C) Inversely graded bed of facies D2 overlain, with a diffuse contact, by a normally graded turbidite of facies C3; also the underlying thinner beds belong to the latter facies. (D) Gravelstone of facies D2, with a pebbly, clast-supported, inversely graded lower part (note the loaded base) and sand-supported upper part, including mudclasts and rafted calcarenite boulders at the top. (E) Massive pebbly sandstone of facies D2, with floating calcarenite boulders (some of which have fallen off the outcrop wall, leaving moulds). (F) Massive, matrix-supported pebbly gravelstone of facies D2, with floating calcarenite cobbles and abundant well-rounded pebbles of vein quartz and ophiolitic rocks. (G) Intraformational gravelstone of facies D1, composed of mudclasts and a muddy sand matrix. All examples are from FA 3C at localities 4 and 6 (Fig. 2). See Table 1 for facies definitions.

Table 2 Trace fossils recognized in the turbiditic deposits of the Kusuri Formation (locality numbers as in Fig. 2); for systematic description and discussion, see Uchman *et al.* (2004)

Ichnotaxa	Localities						
	1	2	4	6	11	12	13
Chondrites intricatus	x	x	x		x	x	
Chondrites targionii	x						
Halopoa isp.			x			x	
Planolites isp.	x	x	x			x	
Ophiomorpha rudis		x	x	x	x	x	x
Ophiomorpha annulata	x	x	x	x	x	x	x
Thalassinoides suevicus		x					
Phymatoderma isp.						x	
Lorenzinia ?apenninica					x		
Lorenzinia isp.					x		
cf. *Cosmorhaphe* isp.	x						
Gordia isp.		x	x		x	x	
?Gordia isp.					x	x	
Helminthopsis isp.			x				
Helminthorhaphe flexuosa						x	x
Helminthorhaphe japonica							x
Nereites irregularis							x
Scolicia vertebralis			x		x		
Scolicia prisca		x	x		x		
Scolicia strozzii	x	x	x	x	x	x	x
Spirorhaphe involuta		x	x		x	x	
?Acanthorhaphe isp.						x	
Belocosmorhaphe aculeata		x					
Belorhaphe zigzag					x		
Desmograpton dertonensis		x	x				
Helicolithus sampelayoi		x	x		x		x
Protopaleodictyon incompositum					x		
Megagrapton submontanum		x	x		x	x	x
Megagrapton irregulare			x				
Paleodictyon strozzii		x	x		x	x	x
Paleodictyon cf. *maximum*	x						
Paleodictyon majus		x				x	
Squamodictyon tectiforme					x		

succession (*sensu* Harms *et al.*, 1975; Walker, 1984). They are indicated in the outcrop logs shown in the paper and have been the basis for an interpretation of the various modes of sediment deposition (Table 1). Based on their spatial grouping and depositional architecture, the sedimentary facies have been recognized to form four main assemblages, or facies associations (Fig. 9), including a number of varieties (subassociations). A facies association is defined as an assemblage of spatially and genetically related facies representing a particular morphodynamic style of turbiditic sedimentation, involving specific facies, bed geometries and depositional architecture. These 'building megablocks' are described and interpreted in the present section. They are regarded as the main architectural elements of basin-fill succession (cf. Mutti & Normark, 1987; Miall, 1989; Clark & Pickering,

Table 3 A summary list of planktonic (P) and small benthic foraminifers (B) and nanoplankton species (N) found in the Kusuri Formation (14 samples) and the underlying Atbaşı Formation (2 samples). The samples from the Kusuri Formation also contain an admixture of redeposited late Maastrichtian and/or Paleocene species, locally up to 30–50%

Microfossil taxa	Type	Formation	
		Atbaşı	Kusuri
Acarinina cf. *bullbrooki* (Bolli)	P		x
Bathysiphon sp.	B		x
B. vitta Nauss	B		x
Blackites creber (Deflandre)	N	x	
Chiasmolithus consuetus (Bramlette & Sullivan)	N		x
C. grandis (Bramlette & Riedel)	N	x	
Clausicoccus fenestratus (Deflandre & Fert)	N		x
Coccolithus pelagicus (Wallich)	N		x
Coronocyclus prionion (Deflandre & Fert)	N		x
Discoaster barbadiensis Tan	N		x
D. bifax Bukry	N		x
D. binodosus Martini	N	x	x
D. diastypus Bramlette & Sullivan	N		x
Ericsonia cava (Hay & Mohler)	N	x	
E. formosa (Kamptner)	N	x	x
E. ovalis Black	N	x	x
E. robusta (Bramlette & Sullivan)	N		x
Globigerina sp.	P	x	x
G. inaequispira Subbotina	P	x	x
Helicosphaera lophota Bramlette & Sullivan	N		x
H. seminulum Bramlette & Sullivan	N	x	x
Lagenidae	B		x
Morozovella sp.	P	x	x
M. cf. *aragonensis* (Nuttall)	P	x	x
M. cf. *caucasica* (Glassner)	P		x
Pontosphaera plana (Bramlette & Sullivan)	N	x	
P. scissura (Perch-Nielsen)	N		x
Reticulofenestra hampdanensis Edwards	N		x
R. coenura (Reinhardt)	N		x
R. dictyoda (Deflandre)	N		x
Rhabdosphaera pinguis Deflandre	N		x
R. truncata Bramlette & Sullivan	N		x
Sphenolithus editus Perch-Nielsen	N	x	x
S. moriformis (Brönnimann & Stradner)	N	x	x
S. primus Perch-Nielsen	N		x
S. pseudoradians Bramlette & Wilcoxon	N		x
S. radians Deflandre	N	x	x
Toweius ?gammation (Bramlette & Sullivan)	N		x
T. occultatus (Locker)	N		x
T. pertusus (Sullivan)	N		x
Transversopontis pulcher (Deflandre)	N		x
Tribrachiatus orthostylus Shamrai	N		x
Zygrhablithus bijugatus (Deflandre)	N		x

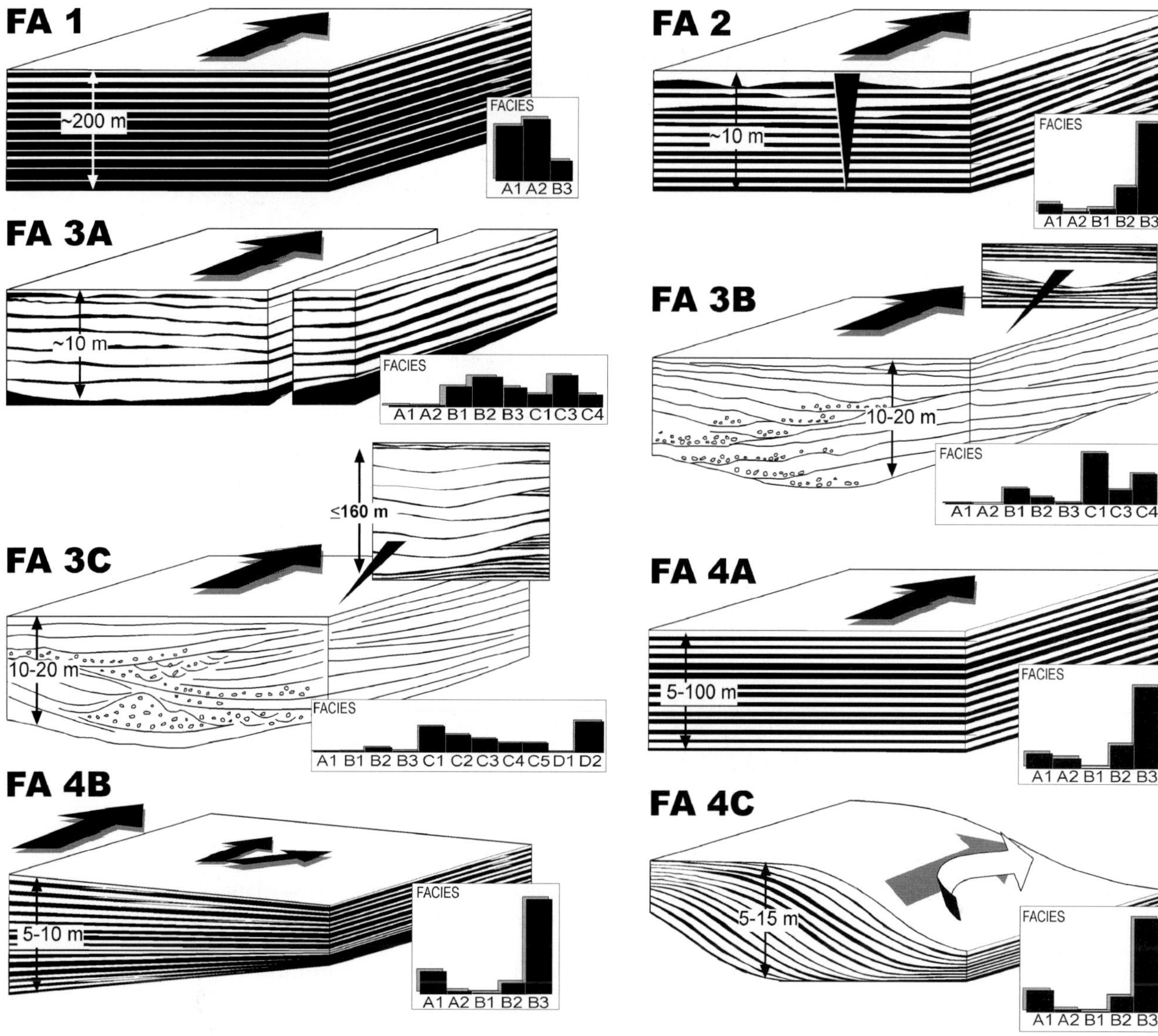

Fig. 9 Turbidite facies associations of the Kusuri Formation. For description and interpretation, see text. The frequency histograms indicate the relative volumetric proportion of component sedimentary facies (code as in Table 1), calculated as thickness percentages.

1996), because their morphogenesis, spatial relationships and stratigraphic distribution reveal the morphodynamic evolution of the basin's turbiditic system.

Facies association 1: mudstones with thin sheet-like turbidites

Description

This facies assemblage is ≤ 200 m thick and constitutes the lowermost part of the Kusuri Formation (Fig. 10), transitional with the underlying Atbaşı Formation (Fig. 3), which itself is dominated by variegated mudstones and represents deposition in a sand-starved deep-water environment (Leren *et al.*, this volume, pp. 401–456). The basal contact is difficult to pinpoint in any single small outcrop, because the stratigraphic transition is gradual, with the variegated mudstones becoming increasingly grey and non-calcareous, and with the calcarenitic interbeds giving way to sandstone sheets of mixed to siliciclastic composition. Both the Atbaşı Formation and the overlying unit of facies association

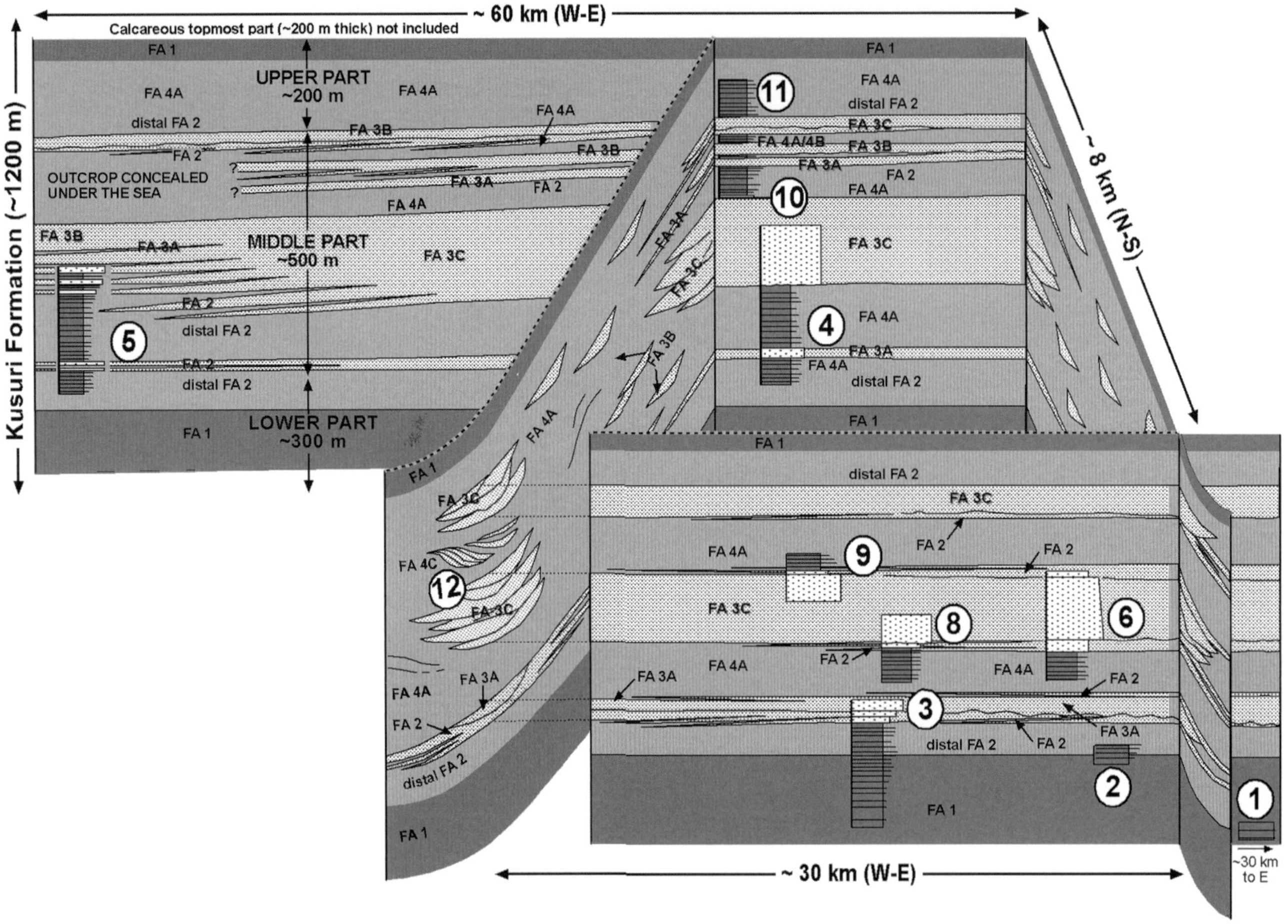

Fig. 10 Schematic log-correlation panels showing the stratigraphic distribution of facies associations in the Kusuri Formation. The locality numbers are as in Fig. 2 and the facies association (FA) code is as in Fig. 9.

1 thin westwards, where the latter decreases its thickness to ~ 100 m over a distance of ~ 70 km.

The deposits of facies association 1 (Fig. 11A & B) are grey mudstones of facies A1 (~ 40 vol.%) interbedded with thin muddy turbidites of facies A2 (~ 45 vol.%) and the fine-grained, thin, sheet-like sandstone turbidites of facies B3 (~ 15 vol.%). Although predominantly siliciclastic in composition, the mudstones are commonly calcareous and also most of the sandstone beds bear a calcareous admixture, similar to the reefal bioclastic component in the underlying formations (see Leren *et al.*, this volume, pp. 401–456). The relative thickness proportion of sandstone interbeds increases upwards in the succession, from < 10 to > 20 vol.%, but there is no obvious lateral change. The sandstone sheets in the easternmost coastal outcrop in Gerze harbour (locality 1 in Fig. 2) are ≤ 5.5 cm thick and average 2.2 cm (Fig. 11A), similar to the westernmost outcrop (locality 5), where they are only slightly less common (11 vol.%) than at the former locality (14 vol.%). The associated beds of facies A1 and A2 are ≤ 17 cm thick, averaging 3.5 cm, and their microfossil content points to an early Eocene (Ypresian) age. Both benthic microfauna and ichnofauna (Uchman *et al.*, 2004) indicate a deep-marine environment. The mudstone facies A1 and A2 predominate and thin sandstone beds of facies B3 amount to ~ 5 vol.% in a small sliver of this facies assemblage at the basin's southern margin, near the eastern onshore terminus of the Erikli thrust (an outlier outcrop not mappable at the scale of Fig. 2).

The outcrop section at locality 1 shows a marked erosional unconformity (Fig. 11C) that truncates the underlying beds at an angle of ~ 20°. The structural attitude of bedding implies pre-erosional tectonic tilting. The steep unconformity, with no facies change across the truncation surface and no facies evidence of strong currents, suggests an

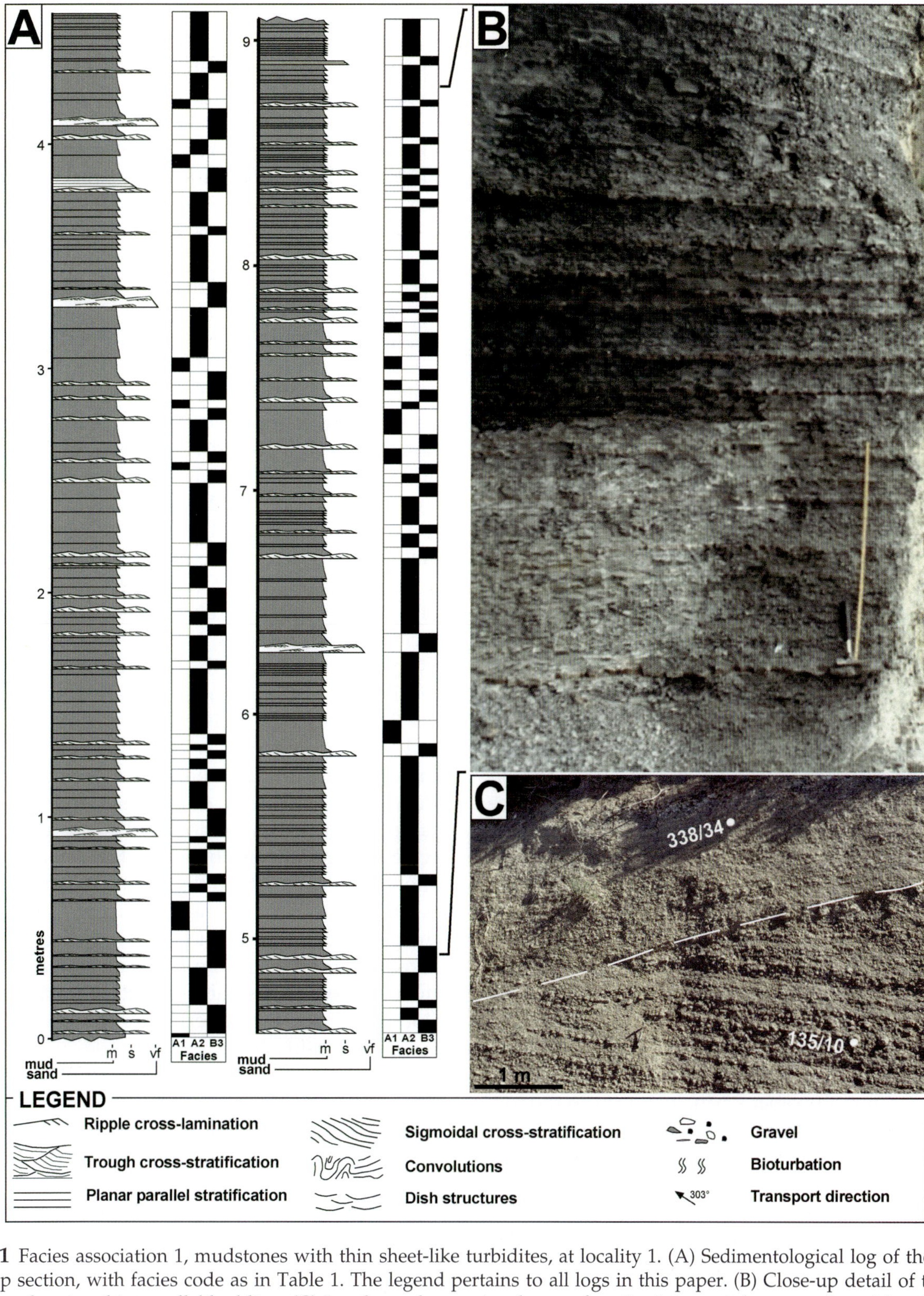

Fig. 11 Facies association 1, mudstones with thin sheet-like turbidites, at locality 1. (A) Sedimentological log of the outcrop section, with facies code as in Table 1. The legend pertains to all logs in this paper. (B) Close-up detail of the outcrop, showing thin parallel bedding. (C) Local angular erosional unconformity, interpreted as a syndepositional slump scar, within a muddy heterolithic succession that lacks facies evidence of strong currents; the whole sedimentary succession is tectonically tilted, and the measurements (dip azimuth and angle) indicate present-day bedding attitude.

intra-basinal slump scar. The outcrop is small and shows no associated slump deposit, but such features occur in the thinly bedded heterolithic deposits elsewhere in the succession (see facies association 4A later in the text), which supports the notion of syndepositional slumping due to basin-floor deformation.

The fine-grained, thinly bedded deposits of facies association 1 crop out in topographically low areas and tend to be strongly weathered, and relatively few turbidites show clear palaeocurrent indicators. Flute casts and ripple cross-lamination, where measurable, indicate transport directions towards the northwest and west.

Interpretation

Facies association 1 represents sporadic incursions of low-density turbidity currents in a deep-water environment dominated by muddy hemipelagic sedimentation. Many of the currents were volumetrically small and dilute, carrying sediment no coarser than silt (facies A2; see the fine-grained turbidites of Stow & Bowen, 1980; Stow & Shanmugam, 1980). The sediment composition indicates derivation from a siliciclastic source area with remnants of an eroded carbonate platform, apparently similar to the foramol-type reefal source of turbidites in the underlying formations (Leren *et al.*, this volume, pp. 401–456). The sediment provenance and palaeocurrent directions indicate supply from the southern margin and westward dispersal along the basin floor. In contrast to the higher part of the formation (Fig. 10), there is no evidence of a major supply of sand from the eastern end of the basin, where muddy facies A1 and A2 seem to predominate at this stratigraphic level (Janbu, 2004).

The turbidity currents were probably derived from the thrust-deformed southern margin of the original Sinop–Boyabat Basin, just prior to the formation of the intrabasinal pop-up ridge (cf. Fig. 2) and the activation of sand-prone sources at the southeastern ends of the resulting sub-basins. The westward thinning of facies association 1 and the underlying Atbaşı Formation is thought to reflect the basin-floor topography inherited from the Late Cretaceous uplift in the western part of the Sinop–Boyabat Basin and the resulting northeastward advance of a carbonate ramp (Leren *et al.*, this volume, pp. 401–456). These early turbidity currents were volumetrically small, and the seafloor topography could render them subject to ponding (Sinclair, 2000; Sinclair & Tomasso, 2002; Lomas & Joseph, 2004). As the foreland foundered under the load of the Central Pontide thrust sheets in Late Paleocene time, the muddy deposits smoothed out the pre-existing relief of a rapidly subsiding basin floor and the basin re-opened to the west.

Facies association 2: depositional lobes

Description

This facies association (Fig. 9) overlies directly the previous one in the lower part of the Kusuri Formation, but occurs also at higher stratigraphic levels and virtually predominates again at the transition to the upper part (Fig. 10), where mudstones begin to be increasingly calcareous and also the associated sandstone beds are mainly calcarenitic.

Facies association 2 (Figs 12A & 13) consists of interbedded, sheet-like sandstone turbidites of facies B3 (60–80 vol.%), B2 (10–30 vol.%) and B1 ($\leq$ 5 vol.%). Most of these turbidites have silty mudstone cappings and some are also separated by the mudstone beds of facies A1 and/or A2 ($\leq$ 10 vol.%), generally $<$ 5 cm thick. Palaeocurrent directions are towards the west, with mean vectors in an azimuthal range of 270–290° and dispersion of $\geq$ 45°, although seldom $>$ 35° on an outcrop scale. The sandstone beds are tabular and rarely show significant lateral thinning within an outcrop width of $\leq$ 100 m, whether parallel or transverse to the palaeocurrent direction. There is also no recognizable difference in the facies and palaeocurrent directions of this assemblage in its western outcrops (locality 5) and the eastern, more source-proximal sections (locality 3).

The relative thickness proportion of sandstones and mudstones varies. Facies association 2 shows one or more coarsening upward (CU) motifs, 8–25 m thick, recognizable from a net increase in the thickness, frequency and/or grain size of sandstone beds (Fig. 12A and the lower parts of logs in Figs 13 & 14). The thinner CU successions typically consist of the tabular, mudstone-capped beds of facies B3 (75–90 vol.%) and subordinate facies B2 (5–15 vol.%), very fine- to fine-grained and $\leq$ 30–35 cm thick, but mainly $\leq$ 6 cm (Fig. 13F). The

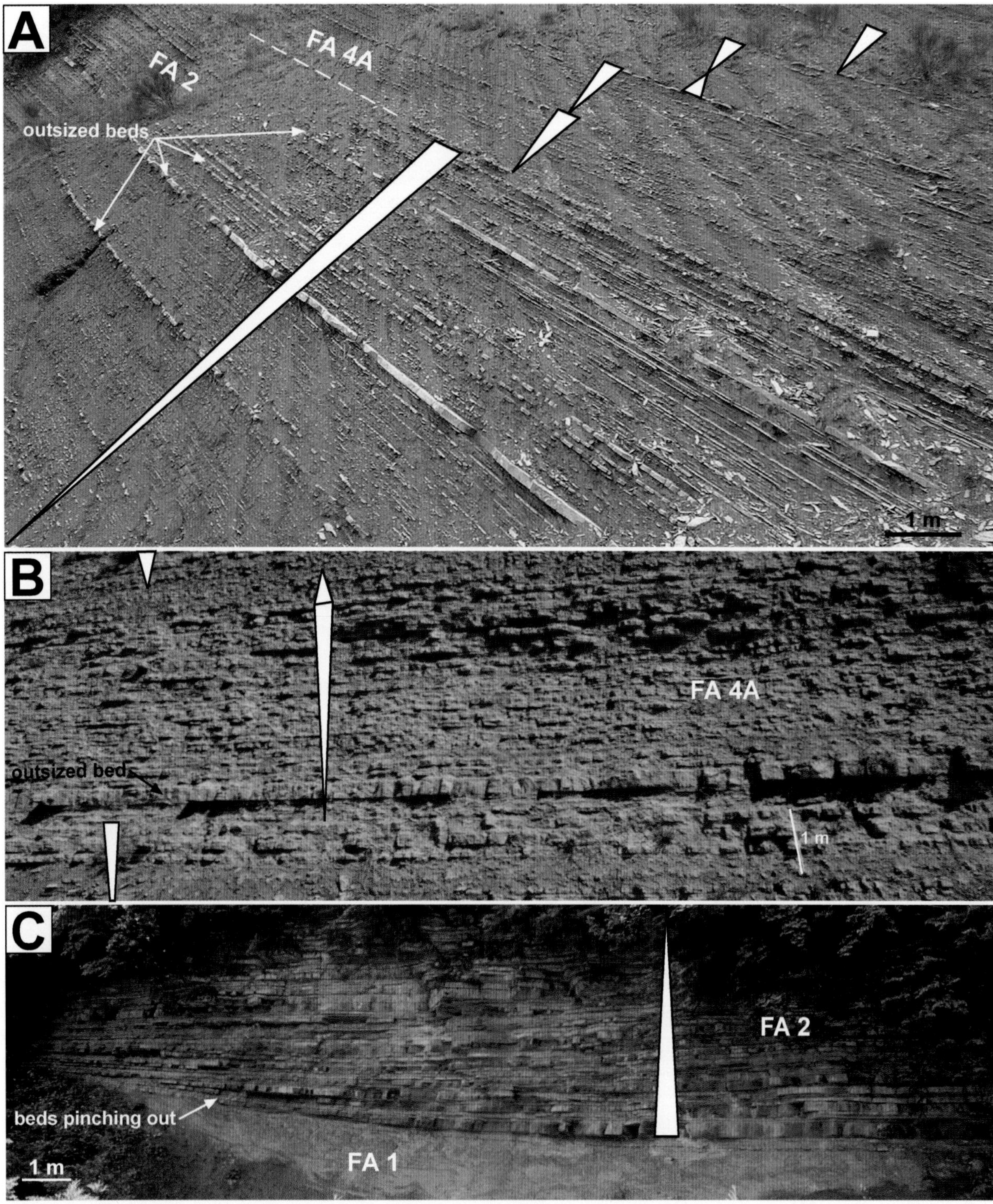

Fig. 12 Facies association 2, depositional lobes, in roadcut section between localities 4 and 10 (A) and localities 2 and 6 (B & C); for localities see Fig. 2. (A) FA 2 succession of thin, alternating sandstone and mudstone sheets with an overall thickening upward trend and sporadic 'outsized' beds. The overlying FA 4A consists of several minor thickening upward bed packages, 1–3 m thick. (B) Close-up detail of the latter bed packages. (C) A succession of FA 2 deposits with an overall thinning upward trend, onlapping laterally the relief of a gentle syncline formed in the underlying deposits of FA 1.

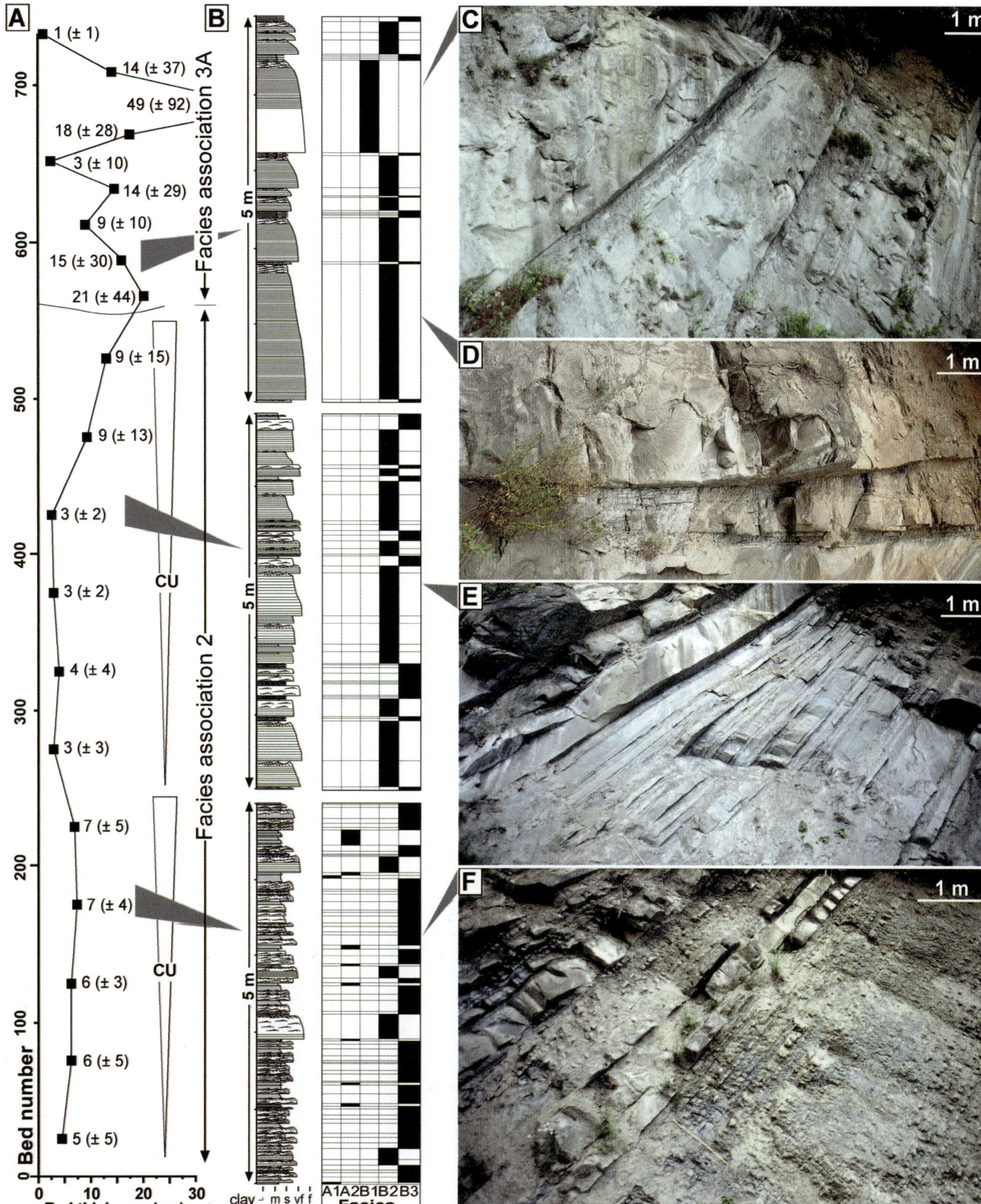

Fig. 13 Facies association 2 (depositional lobes) overlain by facies association 3A (poorly defined palaeochannels) at locality 3. (A) Stratigraphic plot of mean bed thickness (± standard deviation) calculated for every 50 consecutive turbidite beds, and for every 20 beds in the more varied upper part, in a measured outcrop succession 78 m thick. (B) Three portions of the corresponding detailed log; facies code as in Table 1. (C & D) Outcrop photographs showing details of facies association 3A. (E & F) The underlying facies association 2. See Figs 2 & 10 for locality details.

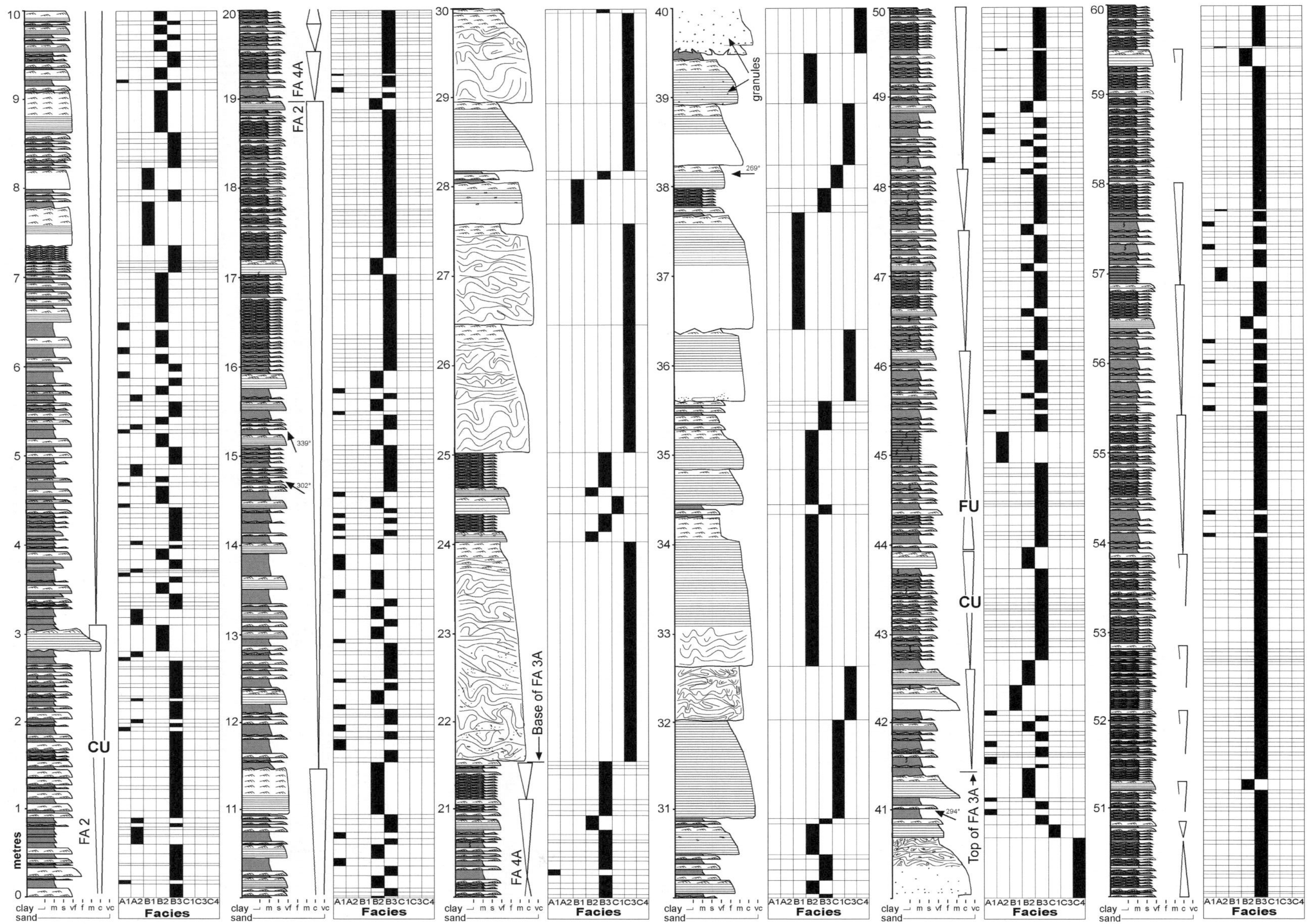

Fig. 14 Detailed log from locality 4, showing facies association 2 (distal lobe deposits) overlain by facies associations 4A (sheet-like overbank deposits) and 3A (poorly defined palaeochannel), with another unit of facies association 4A at the top. Facies code as in Table 1 and log legend in Fig. 11. Note the large-scale coarsening upward (CU) trends in FA 2 and the smaller-scale CU and minor fining upward (FU) trends in FA 4A, recognizable from the upward changes in sandstone bed frequency, thickness and/or grain size. See Figs 2 & 10 for locality details. For a corresponding outcrop photograph, see Fig. 15A.

intervening mudstone beds of facies A1 and A2 (5–10 vol.%) are < 10 cm thick, only occasionally ≤ 15 cm. In a thicker CU succession, this 'lower' facies package passes upwards into a package of the tabular beds of facies B2 and B3 (~ 90–95 vol.%) and subordinate facies B1 (2–5 vol.%), very fine- to medium-grained and mainly < 20 cm thick, but occasionally coarse-grained and ≤ 66 cm (Fig. 13E). These sandstone beds tend to be amalgamated, and the thicker ones have uneven bases with a scour relief of ≤ 10 cm. Interbeds of facies A1 and A2 are rare (1–2 vol.%), mainly < 5 cm thick and commonly discontinuous due to erosion. Such 'upper' packages are less common in the CU successions at the western localities.

For example, the succession of facies association 2 at locality 5 (Figs 2 & 10) is nearly 200 m thick and consists of vertically stacked CU 'lower' packages, 8–10 m thick. Only the uppermost part of the succession abounds in medium- to coarse-grained, amalgamated sandstone beds ≤ 86 cm thick. The succession also includes minor fining upward (FU) bed packages, 2–5 m thick.

The CU successions commonly contain scattered 'outsized' beds (relatively coarse-grained and/or thick), which occur isolated or sporadically as couplets or triplets (see Figs 12A, 13B & 14). These are typically turbidites of facies B1 or B2, which apparently punctuated the succession quite randomly, without disturbing its overall CU trend.

Facies association 2 shows an atypical, pronounced FU trend at locality 2 (Fig. 2), where the exposed turbidite package (Fig. 12C) is > 10 m thick and dominated by sandstone beds (78 vol.%). The turbidites are thin to moderately thick (≤ 30 cm) beds of facies B3 and B2, fine- to medium-grained, and include sporadic beds of facies B1, ≤ 56 cm thick, in the lower part. The sandstone beds have silty mudstone cappings and show little amalgamation. The lowest turbidites pinch out laterally over a distance of 30 m, by onlapping the relief of a gentle syncline formed in the underlying deposits of facies association 1 (Fig. 12C). Apart from their synclinal bedding, the underlying deposits show no erosional truncation.

Interpretation

The extensive tabular bedding, relatively high palaeocurrent dispersion and CU trend suggest depositional lobes (Mutti & Normark, 1987; Shanmugam & Moiola, 1988; Pickering *et al.*, 1989) formed by low-density (facies B2 and B3) and subordinate high-density turbidity currents (facies B1), with a minor contribution of hemipelagic sedimentation (facies A1) and highly dilute, muddy currents (facies A2). The CU trend is thought to indicate progradation, and the common occurrence of palaeochannels (facies association 3) at the top and/or directly upstream to the east (Fig. 10) supports the notion of channel-related terminal lobes. Accordingly, the 'lower' and 'upper' facies packages of the CU successions are considered to represent the relatively distal and proximal parts of depositional lobes, respectively (see FA 2 and 'distal' FA 2 in Fig. 10). The multiple thinner CU motifs may represent pulses of a single lobe advance. The subordinate FU motifs may reflect retrogradation or lateral migration of the channel's terminal depocentre. The random 'outsized' beds are probably seismites, deposited by turbidity currents triggered by strong seismic tremors.

Cases such as the FU succession at locality 2 (Fig. 12C) are attributed to the semi-confinement of a developing lobe by local seafloor synclines. The localized, intrabasinal angular unconformities of this kind (see also Fig. 11C) probably represent the syndepositional development and burial of blind-thrust folds, which would affect turbidity currents by semi-confining their bedload and limiting their capacity for broader sand dispersal.

Facies association 3A: poorly defined palaeochannels

Description

Facies association 3A (Fig. 9) occurs as isolated units at various stratigraphic levels in the middle part of the Kusuri Formation (Fig. 10), where these deposits overlie and/or pass westwards into the depositional lobes of facies association 2.

Facies association 3A consists of medium to thick and mainly amalgamated sandstone turbidites of facies B and C, which are stacked vertically in an alternating manner, forming isolated packages 20–30 m thick (Fig. 13; log interval 21.5–41.5 m in Fig. 14; Fig. 15). The sandstone beds are stacked upon one another with little erosion and are tabular on an outcrop scale, but some have uneven lower boundaries and a few of the thickest ones show concave-upward bases with a scour relief of

Fig. 15 Facies association 3A, poorly defined palaeochannels. (A) Tectonically tilted deposits at locality 4, showing a thinning upward bedding trend. The palaeocurrent direction is towards the viewer, obliquely to the left; the corresponding log is in Fig. 14. (B) Facies association 3A at locality 11. Note that the lowermost few beds of the sandstone turbidite package are heavily loaded and locally turned into pillows; the palaeocurrent direction is to the right, at ~ 10° into the outcrop section, and the corresponding log is shown in Fig. 16 (see lower 13 m). The locality numbers are as in Figs 2 & 10.

≤ 90 cm over a lateral distance of 10–15 m. The facies composition varies (Figs 14 & 16), including turbidites of facies B1 (5–10 vol.%), B2 (10–20 vol.%), B3 (15–30 vol.%), C1 (10–20 vol.%), C3 (10–45 vol.%) and C4 (5–15 vol.%), with minor interbeds of facies A1 and A2 (1–3 vol.%). The sandstone turbidites are mainly ≤ 50 cm thick, but some exceed 100 cm and exceptionally reach 233 cm (facies C3) or even 314 cm (facies C4), without internal evidence of amalgamation. Most of the thinner beds are medium- to fine-grained, but the thicker ones, or their basal divisions, consist of coarse to very coarse sand, commonly with granules and scattered mudclasts. Some of the thicker (> 50 cm) beds of facies B1 contain flat-lying marlstone clasts, ≤ 35 cm in length, scattered along the base. Flute casts and ripple cross-lamination show low dispersion (< 30°) and indicate palaeocurrent directions consistently towards the west.

These turbiditic packages have sharp bases and also fairly sharp tops, typically overlain by the mudstone-rich facies association 4A (Figs 14–16 & 17A & B). In an outcrop section transverse to the palaeocurrent direction, the base of the turbiditic package is broadly concave upwards (Fig. 15A) as a result of scour combined with substrate loading. The basal sandstone beds commonly show evidence of rapid dewatering (dish structures) and hydroplastic deformation (Figs 15B & 16), including extensive convolutions (Fig. 14), and the package as a whole is characterized by a thinning upward bedding trend (Fig. 15A).

The thick units of facies association 3A logged at localities 3 and 4 (Figs 13, 14 & 15A) differ slightly from the thinner ones measured at localities 5 and 11 to the west (Fig. 15B and log interval 2.5–13 m in Fig. 16). The latter turbiditic packages have similarly sharp bases and tops, and show a similar aggradational pattern of bed stacking, but are only 11–13.5 m thick, lack a distinct thinning upward trend and also comprise somewhat different facies. They consist mainly of facies B1 (30–40 vol.%), B2 (20–30 vol.%), B3 (2–5 vol.%), C1 (< 5 vol.%), C3 (5–30 vol.%) and C4 (< 10 vol.%), with a variable amount of facies A1 and A2 interlayers (3–5 vol.%). The mudstone interlayers are ≤ 15 cm thick and laterally more persistent, and the sandstone beds show less amalgamation. Despite

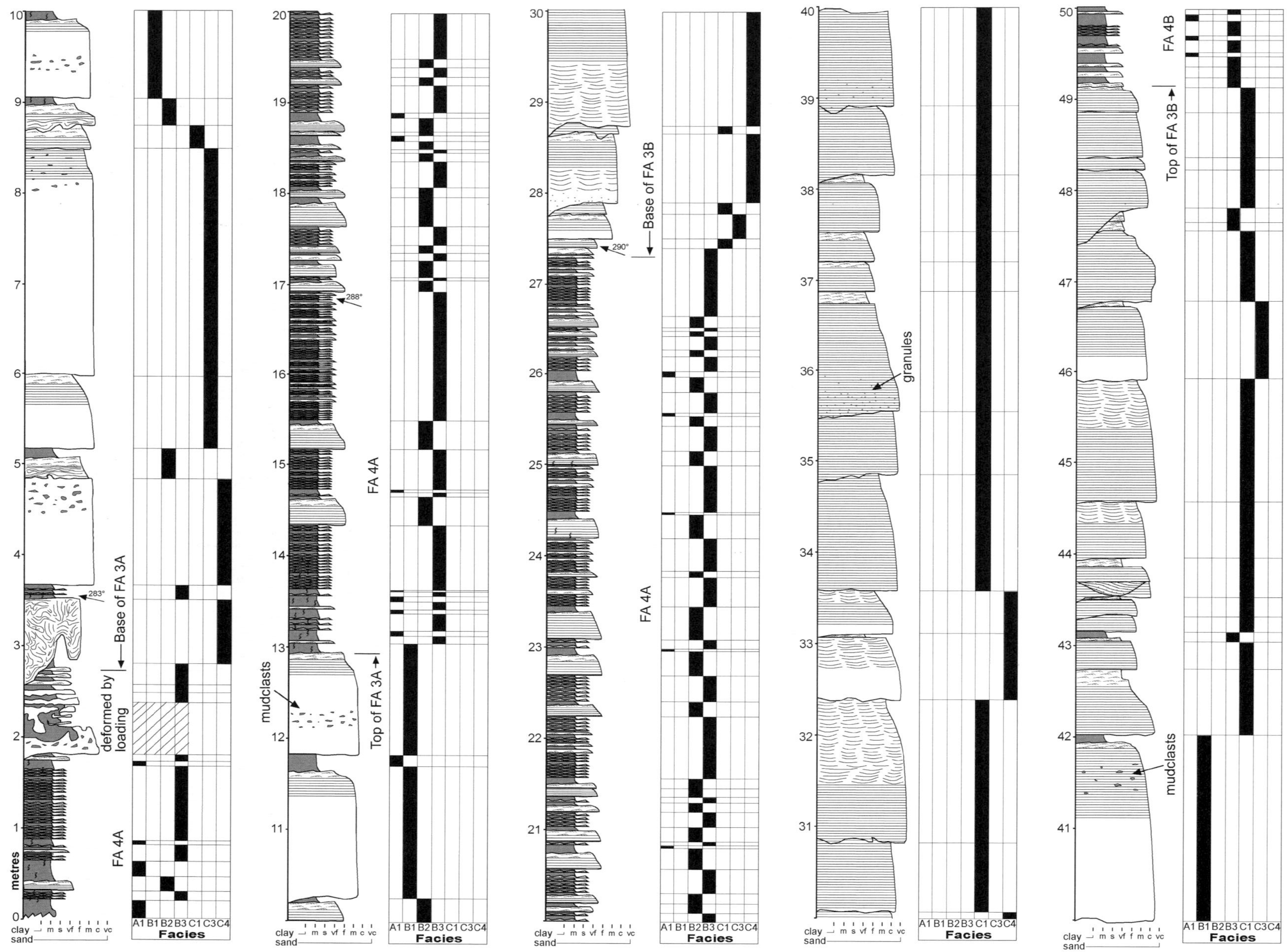

Fig. 16 Detailed log from locality 11, comprising a sandstone body of facies association 3A underlain and overlain by facies association 4A, with a thicker sandstone body of facies association 3B above. Facies code as in Table 1 and log legend in Fig. 11. See Figs 10 & 15B for further details.

Fig. 17 Facies association 3B, solitary sinuous palaeochannels. (A) Sandstone bodies of facies associations 3A and 3B at locality 11. The modal palaeocurrent direction is to the right (westwards), at ~ 10° into the outcrop section for unit FA 3A and at ~ 15° out of the section for unit FA 3B. (B) Sandstone bodies of facies associations 3A and 3B separated by facies association 4A, at the same locality; note the tabular bedding in unit FA 3A and the broadly convex-upward bed sets (PB) in the overlying unit FA 3B. (C) Close-up detail of unit FA 3B, showing two superimposed, broadly convex-upward bed sets, one dipping into the outcrop and the other out of the outcrop relative to the palaeochannel basal surface. (D) Close-up detail of the sharp top of the same unit FA 3B, overlain by facies association 4A. Facies association (FA) code as in Fig. 9. See Figs 2 & 10 for locality details.

the greater proportion of classic turbidites (facies B) and mudstone interlayers (facies A), the sandstone beds themselves are commonly coarse- to medium-grained and mainly > 50 cm thick, occasionally 120–240 cm. Many beds contain scattered mudclasts. The flute casts and ripple cross-lamination here have a directional dispersion of ≤ 40° on an outcrop scale, but indicate consistent palaeocurrents towards the west or northwest.

Interpretation

The isolated units of facies association 3A are interpreted to be poorly defined palaeochannels, because: (i) they are spatially linked with the depositional lobes of facies association 2 (Fig. 10); (ii) they show a thinning upward aggradational bedding pattern (Figs 15A & 16) and low-dispersion westward palaeocurrents; and (iii) their sharp bases are broadly concave upwards in flow-transverse outcrop sections (Fig. 15A) and roughly planar in flow-parallel sections (Figs 15B & 17A, B). The interpretation is supported by the sparseness of mudstone facies and the predominance of relatively thick and coarse-grained sandstone beds, which are a mixture of classic (facies B) and non-classic turbidites (facies C). The low-relief bases are sharp, but show limited erosion and the channel-fill deposits themselves show no evidence of lateral accretion, which suggests deposition by poorly confined turbidity currents in wide, non-sinuous conduits. These aggradational channels were apparently responsible for the deposition of the terminal turbiditic lobes of facies association 2, over which they often extended in the downstream direction, towards the west (Fig. 10).

The channel-fill deposits are products of low-density (facies B2, B3, C1 and C2) and high-density turbidity currents (facies B1, C3 and C4), many of which were apparently voluminous and of sustained (long-duration) type, depositing relatively thick beds in quasi-steady flow conditions (Table 1). The thinner units of facies association 3A, with a greater proportion of classic turbidites and mudstone interlayers, may represent channel flanks (cf. FA 3A in Fig. 9) or lower reaches (localities 5 and 11 in Fig. 10).

The basin floor was subject to tectonic deformation, and the formation of these channels could have been instigated by the development of gentle synclines on the seafloor. A similar cause for the formation of poorly defined turbiditic channels was inferred by Grecula *et al.* (2003b), although the channel-fill facies in that case were different.

Facies association 3B: solitary sinuous palaeochannels

Description

This facies assemblage (Fig. 9) occurs as isolated sandstone bodies, 18–28 m thick, in the middle part of the Kusuri Formation, where they overlie facies associations 2, 3A or 4A, and are in turn overlain by facies associations 2, 4A or 4B (Figs 10, 16 & 17A, B). The local palaeocurrent directions measured from flute casts and ripple cross-lamination have dispersion commonly > 45–50°, but the vector means are consistently towards the west, within a range of < 30°. The sandstone bodies have sharp, erosional bases (Figs 16 & 17C) and sharp, flat tops (Fig. 17D). The basal relief is low (< 30–50 cm) in outcrop sections parallel to the palaeocurrent direction and also in small (20–50 m wide) transverse sections. However, the widest (> 300 m) outcrop sections transverse or oblique to the palaeocurrent direction show the bases of these sandstone bodies to be broadly concave upwards, with a relief of several metres over a flow-transverse distance of ~ 100–150 m. For example, the apparent westward thickening of the flat-topped sandstone body at locality 11 (FA 3B in Fig. 17A) indicates that its concave-upward base deepens by 7 m within a lateral distance that corresponds to little more than 100 m in a direction transverse to the mean palaeocurrent vector; the outcrop section is at ~ 15° to the local mean vector (which is to the right, obliquely out of the section), and the thicker part of the sandstone body is its inner, more axial part. The sandstone body reaches a thickness of ≥ 28 m farther to the west and is estimated to be ~ 500 m wide.

The sandstone bodies of facies association 3B consist of thick (50–150 cm) and highly amalgamated turbidites (Fig. 17B and the corresponding log interval 27.5–49 m in Fig. 16), mainly coarse- to very coarse-grained and commonly bearing granules. The majority of beds have uneven, erosional bases. The subordinate thinner (< 30 cm) beds are medium-grained, and are commonly tabular where stacked upon one another. The deposits are alternating turbidites of facies C1 (30–40 vol.%), C4 (15–25 vol.%), C3 (10–15 vol.%), B1 (5–15 vol.%) and B2 (< 7 vol.%), with minor thin interbeds of facies B3

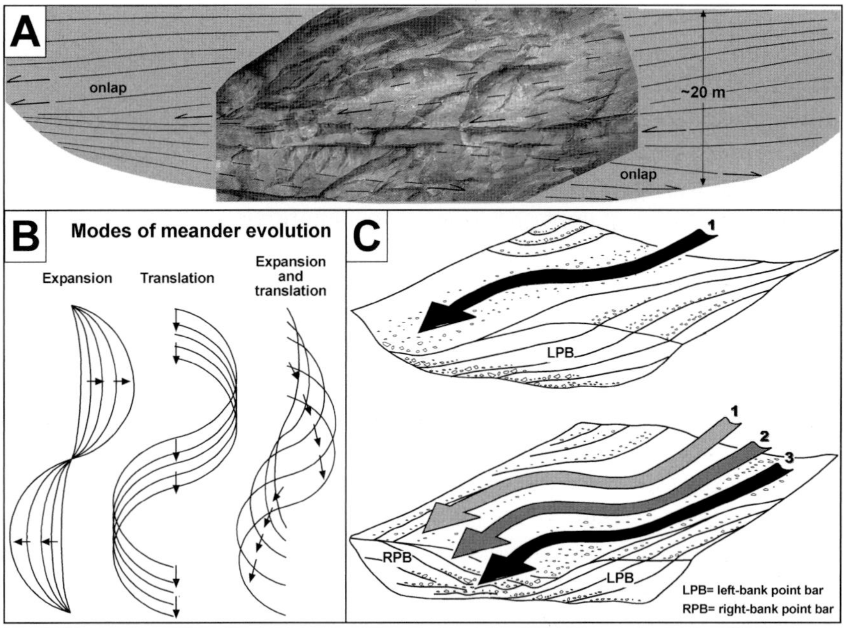

Fig. 18 Facies association 3B, solitary sinuous palaeochannels. (A) Idealized transverse cross-section, with an outcrop photograph, showing the stacking pattern of inclined bed sets in a sandstone body of facies association 3B; the modal palaeocurrent direction is towards the viewer. (B) Meander-bend evolution by lateral expansion, downstream translation, and combined expansion and translation. (C) Schematic diagrams explaining the stacking of one point bar against and upon another as a result of channel-bend expansion and translation.

(< 3 vol.%) and facies A1/A2 (< 2 vol.%). Many beds have loaded bases and show dewatering structures.

In outcrop sections perpendicular or highly oblique to the mean local palaeocurrent direction, these deposits occur as a package of beds that are gently inclined (5–10°) in one flow-transverse direction and commonly overlain by a similar bed set inclined in the opposite direction (Figs 17C & 18A; cf. FA 3B in Fig. 9). The inclined bed sets indicate lateral accretion of turbidites (epsilon cross-bedding *sensu* Allen, 1963; or amalgamated LAPs *sensu* Abreu *et al.*, 2003). In sections parallel or only slightly oblique to the palaeocurrent direction, these superimposed packages are recognizable as broad mounds of gently convex-upward beds, 1.7–2 km in downstream extent and offset relative to each other (see the bed sets PB in Fig. 17B), with the beds showing bi-directional downlap or onlap.

Interpretation

The geometry and internal bedding architecture of these sandstone bodies indicate sinuous, well-defined channels filled with deposits of large, low- to high-density turbidity currents, commonly sustained and quasi-steady (see facies interpretation in Table 1). The laterally inclined bed sets, broadly convex upwards in longitudinal section, are interpreted to be point bars related to the channel meanders. The preferential deposition on the inner bank of a channel bend is attributed to the local deceleration of currents due to flow expansion where the channel widens slightly, with the concurrent erosion of the outer bank adding sediment (mass) to the current (Abreu *et al.*, 2003). The meandering channels and the character of their infill facies would be consistent with the notion of predominantly sustained currents, such as the hyperpycnal flows associated with many river deltas (Elliott, 2000).

Based on the geometrical estimates and the channel-fill thicknesses corrected for ≤ 1000 m burial and ~ 15 vol.% compaction (Baldwin & Butler, 1985), the channels would appear to have been 22–34 m deep and ~ 400–500 m wide, with meander wavelengths in the range of 1.7–2 km. The channels would thus be in the middle range of sinuous channel dimensions reported from modern and ancient submarine fans (e.g. Flood & Damuth, 1987; Clark *et al.*, 1992; Cronin, 1995; Damuth *et al.*, 1995, 1998; Cronin *et al.*, 2000b; Kolla *et al.*, 2001; Abreu *et al.*, 2003; Posamentier & Kolla, 2003; Lomas & Joseph, 2004). However, the channel depth/width ratios of 1/15 to 1/19 would appear to be lower than the approximate average of 1/10 indicated by a worldwide dataset compiled by Clark & Pickering (1996) and much lower than, for example, the aspect ratio of 1/7.5 reported by Abreu *et al.* (2003) from the offshore Angola channels. The bulk meander-belt widths are difficult to estimate in the present case, because of the limited exposure, but may probably be ~ 2–3 km.

The stacking of point bars against and upon each other (Fig. 18A) implies a meander-bend evolution by lateral expansion combined with a pronounced downstream translation (Fig. 18B & C). This style of channel recurving would cause its marked widening, which might explain the relatively low aspect ratios. A similar pattern of submarine channel-belt evolution has been reported by Deptuck *et al.* (2003) and Posamentier & Kolla (2003). This evidence contradicts the notion that sinuous turbiditic channels, in contrast to fluvial ones, evolve solely by meander-bend expansion (Peakall *et al.*, 2000a, b). Furthermore, the channel-fill architecture in the present case does not seem to support the three-stage model for submarine channel evolution postulated by Peakall *et al.* (2000a, b), with the lateral accretion due to meander-belt widening followed by an equilibrium stage of pure aggradation and the final stage of abandonment. Instead, the channel-fill architecture (Fig. 18) indicates lateral accretion accompanied by aggradation and downcurrent sweep, leading to the stacking of adjacent point bars upon each other and hence to a complete filling of the widened channel and its eventual abandonment. The sharp tops of the palaeochannels (Fig. 17D) imply an abrupt abandonment by avulsion.

The formation of these channels was probably facilitated by a semi-confinement of large, powerful turbidity currents by basin-floor folds related to blind thrusts (see evidence shown in the next section). In the basin-fill succession, these palaeochannels overlie the terminal-lobe deposits of facies association 2, the poorly defined palaeochannels of facies association 3A (themselves also associated with depositional lobes) or the heterolithic overbank deposits of facies association 4B. This latter relationship is consistent with the notion of channel shifting by avulsion, whereas the two former relationships suggest that the sinuous channels evolved from the poorly defined, straight and wider ones, probably as a result of the flow interaction with the conduit in response to greater turbidity-current discharges (Pirmez & Flood, 1995; Imran *et al.*, 1999; Abreu *et al.*, 2003; Deptuck *et al.*, 2003). The channel transformation from straight to sinuous would appear to have occurred prior to or after significant aggradation, depending on the timing of the increase in flow confinement and discharge. The transformation from a wide channel might partly explain the low depth/width aspect ratio of the resulting sinuous channel.

Facies association 3C: multistorey palaeochannel complexes

Description

This facies assemblage (Fig. 9) forms at least four thick and gravel-rich sandstone bodies in the middle part of the Kusuri Formation. Based on their estimated stratigraphic order in the basin-fill succession, these units of facies association 3C are herein referred to as sandstone bodies 1 to 4. Sandstone body 1 is ~ 160 m thick and crops out at localities 6–9, and at locality 12 (Fig. 10) near the southern basin margin, where also the highest sandstone body 4 is partly exposed, separated from the former body by a thick succession comprising facies associations 4A and 4C. Sandstone body 2 is ~ 130 m thick and crops out at localities 4 and 10 in the basin's mid-northern part (Fig. 10), where also sandstone body 3, ~ 80 m thick, is exposed in the poorly accessible, high coastal cliff at locality 11 and over a few kilometres farther to the west. These two sandstone bodies are separated by a muddy succession of facies association 4A intercalated with facies associations 3A, 3B and 4B, and they both pass westwards into sandy successions involving facies associations 2, 3A and 3B (Fig. 10).

The widths of these sandstone bodies are unknown, but have been estimated from the present-day topography of outcrop ridges to be ≤ 5 km. The sandstone bodies have a similar internal architecture, but their facies composition shows differences and their gravel content is generally higher towards the east.

Sandstone body 2 is exposed at locality 4, in an old quarry in Ayancık (Fig. 19A), where it sharply overlies facies association 4A (Fig. 19B) and consists of highly amalgamated sandstone and gravelstone beds (Fig. 20). Only the lower to middle part (~ 110 m thickness) of this succession is accessible, showing an alternation of facies C1 (32 vol.%), C3 (29 vol.%) and D2 (16 vol.%), with subordinate beds of facies C5 (8.5 vol.%), C2 and C4 (4 vol.%) and facies B1–B3 (9 vol.%). Interbeds of facies A1 are minor (~ 1 vol.%), thin and laterally discontinuous, as are also sporadic lenticular beds of facies D1 (< 1 vol.%), which occur as isolated erosional

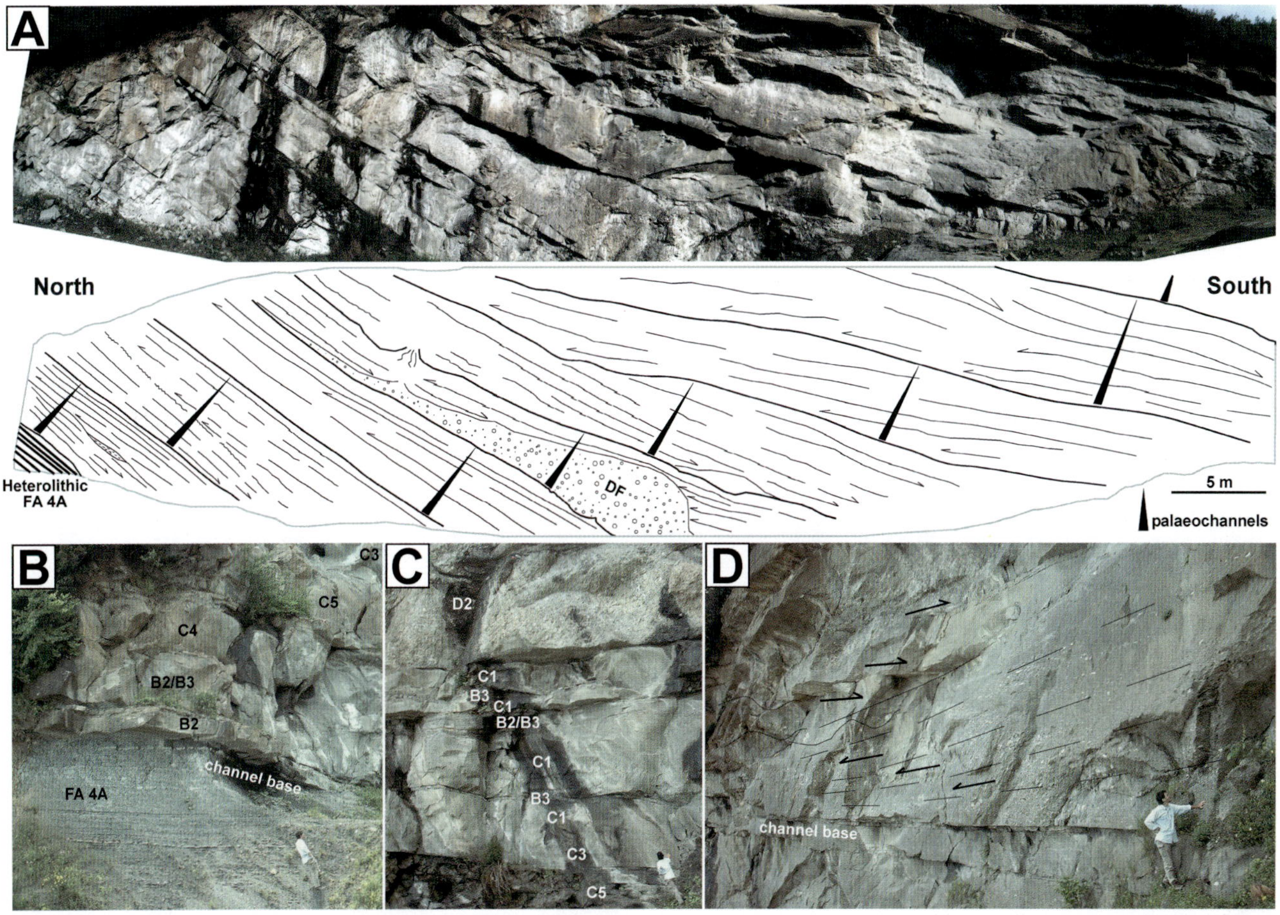

Fig. 19 Facies association 3C, multistorey palaeochannel complexes. (A) Outcrop section and corresponding overlay drawing of a tectonically tilted sandstone body of facies association 3C at locality 4. The modal palaeocurrent direction is towards the viewer and DF is a mound of gravelly debrisflow deposits (facies D2). Note the vertical stacking of laterally accreted bed sets. The corresponding log is shown in Fig. 20. (B & C) Facies details from the basal and upper mid-part of the same sandstone succession. Facies code as in Table 1. (D) Laterally accreted bed sets dipping in opposite directions within a palaeochannel in the lower mid-part of the same succession. Note the gravel-rich zone with multiple scours separating the lower bed set from the onlapping upper one, interpreted as a riffle zone of the channel thalweg. The uppermost part of this FA 3C sandstone complex crops out at the adjacent locality 10 (see Figs 21 & 22). See Figs 2 & 10 for locality details.

remnants ≤ 53 cm thick. Most of the sandstone beds are thicker than 50 cm and many beds of facies C1–C3, C5 and B2 are > 150 cm in thickness (Figs 19C & 20), although bed thicknesses show considerable lateral variation due to erosion. Their grain size varies from fine/medium to very coarse sand, and many beds abound in granules and pebbles, including intraformational mudclasts.

The associated gravelstone beds of facies D2 are mound-shaped, truncated by the overlying turbidites and locally ≤ 4 m thick (Fig. 19A). The gravel ranges from pebbles to cobbles, including subangular to subrounded clasts of marlstone and calcarenite; rounded clasts of vein quartz and subordinate ophiolitic and metamorphic rocks; and subangular to rounded mudclasts. The lenticular mounds of facies D2 occur in several places within the sandstone body, but are particularly abundant (84 vol.%) in a 9-m-thick interval ~ 24 m above its base (Fig. 20), where they are amalgamated or intercalated with minor interbeds of facies C5. The higher part of the outcrop section, 77 m thick, is dominated by amalgamated sandstone beds of facies C1 and C3, which are mainly 50–150 cm thick and commonly rich in granules and small pebbles (Fig. 20). They occasionally contain isolated,

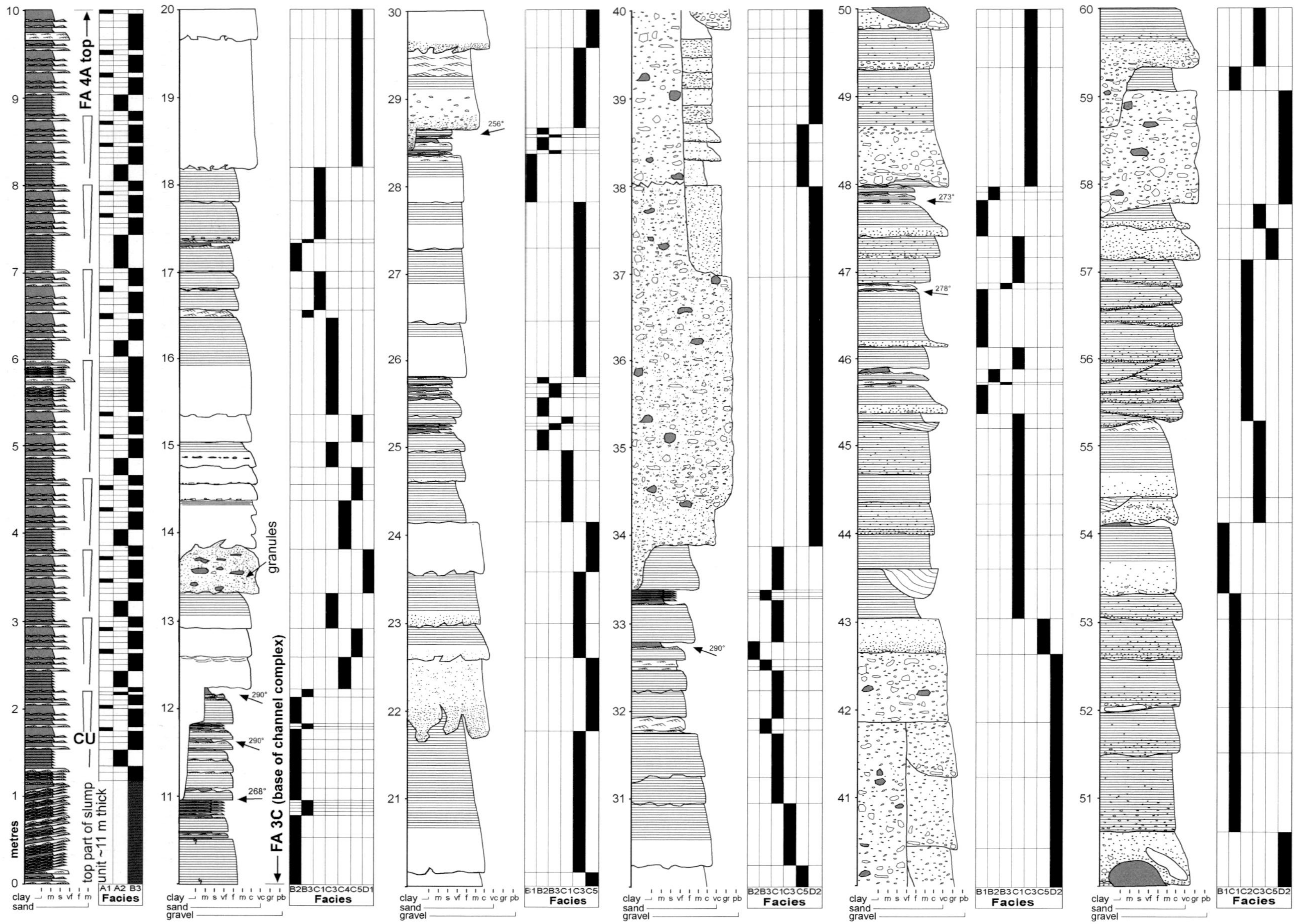

Fig. 20 Detailed log from the lower part of the sandstone body of facies association 3C exposed at locality 4 (Fig. 19A), including the underlying facies association 4A. Note the topmost portion of an intraformational slump deposit (~ 11 m thick) at the log base. Facies code as in Table 3 and log legend as in Fig. 11.

rounded cobbles or boulders of calcarenite and marlstone, derived from the underlying Akveren Formation. A couple of thick, massive sandstone beds of facies D2 contain floating calcarenite boulders > 1 m in length (Fig. 8E).

The palaeocurrent directions measured from flute casts and large trough-shaped scours have a dispersion of ~ 60° and westward mode. The beds have uneven, erosional bases and form packages that are gently inclined northwards or southwards with respect to the basal surface of the sandstone body (Fig. 19A, D), which suggests lateral accretion (cf. Fig. 18). The oppositely inclined bed sets, where stacked against and upon each other, are separated in their toe parts by a gravel-rich zone with multiple high-relief scours (Fig. 19D) and debris-flow deposits of facies D1 and/or D2 (Fig. 19A).

The upper part of the same sandstone body is exposed at the nearby coastal locality 10 in Ayancık, < 1.5 km to the west (Fig. 2), where it is overlain sharply by facies association 4A (Figs 21 & 22). The sandstone succession here is dominated by amalgamated beds of facies C1 and C3 (75 vol.%), mainly 50–120 cm thick and occasionally ≤ 220 cm. They are intercalated with subordinate beds of facies C2, C5, B2 and B3. The sandstone beds are medium- to very coarse-grained and commonly rich in quartz granules. Gravelstone beds are lacking, as are also mudstone interbeds, except for a couple of thin (< 2.5 cm) and discontinuous layers. The sandstone beds have uneven, erosional bases with common load features and sporadic large flutes. The topmost turbidites of facies C1 and B1–B3 are relatively thin and capped with silty mudstones (Fig. 22), but the transition to the overlying mudstones of facies association 4A is abrupt (Fig. 21B).

Sandstone body 1 has been studied at localities 6–9 and 12 west of Erfelek (Figs 2 & 10), where an extensive outcrop section is afforded by an adandoned quarry and adjoining roadcut escarpments at an earth-dam construction site. The outcrop section has a north–south trend, transverse to the westward mean palaeocurrent direction determined from flutes, trough-shaped scours and the axes of trough cross-strata sets. The sandstone body here is ~ 160 m thick (Fig. 23A), and is underlain and overlain by deposits of facies association 2 (Fig. 10). The exposed portion of the underlying deposits, ~ 10 m thick (Fig. 23A), consists of thin to moderately thick sandstone beds, mainly amalgamated and some nearly 100 cm thick, occasionally showing cross-stratification (facies C2). The lack of facies A1/A2 and the abundance of facies C1 and C2 (~ 30 vol.%), in addition to facies B1–B3, render this FU succession similar to facies association 3A, passing upwards into a CU succession of facies association 2 (see log 1 in Fig. 23B).

The overlying facies association 3C has a highly uneven, erosional base and its lowermost part consists of amalgamated gravelstone beds of facies D2, ≤ 290 cm thick, commonly cobbly and boulder-bearing, intercalated with subordinate pebbly sandstone beds of facies C3 (Fig. 23B, log 1). This gravelly basal part of the succession is ≥ 17 m thick. The beds are thinning laterally and many pinch out towards the south, and they are broadly concave-upwards in shape, apparently due to mild deformation. The gravel includes subrounded clasts of vein quartz and ophiolitic rocks, ≤ 5 cm in size; angular to subrounded marlstone clasts, ranging from < 1 cm to > 50 cm in length; angular to subrounded calcarenite clasts, ≤ 100 cm in length, commonly silicified and elongate; and scattered intraformational clasts of mudstone and sandstone, the latter oblate, with margins defused by the surrounding sandy matrix. Marlstone and calcarenite clasts commonly predominate.

The middle part of the succession, ~ 70 m thick (Fig. 23A), consists of amalgamated, thick sandstone beds, predominantly facies C1–C5 (65 vol.%). They are mainly medium- to coarse-grained and commonly gravel-bearing, intercalated with the gravelstone beds of facies D2 (25 vol.%). The turbidites are alternating beds of facies C1 (21 vol.%), C2 (16 vol.%), C3 (12 vol.%), C4 (8 vol.%), C5 (8 vol.%), B1 (3 vol.%), B2 (5 vol.%) and B3 (1.5 vol.%). The majority of beds have uneven, erosional bases and are > 50 cm thick (Fig. 23B, log 2), some reaching 460 cm (facies C4). Sporadic interbeds of facies A1 (< 1 vol.%) are thin and laterally discontinuous. The upper part of the succession, 60 m thick (Fig. 23A), is dominated by medium-grained sandstones and consists of thick beds of facies C1, C2 and B1 (~ 75 vol.%), with subordinate beds of facies C5 and B2. Minor packages of thin facies B3 beds occur as intercalations near the top (Fig. 23B, log 3).

In the flow-transverse outcrop section, the turbidites form recognizable bed sets, 5–15 m thick, which show crude upward fining (Fig. 23B, log 2) and are gently inclined (5–10°) towards the north

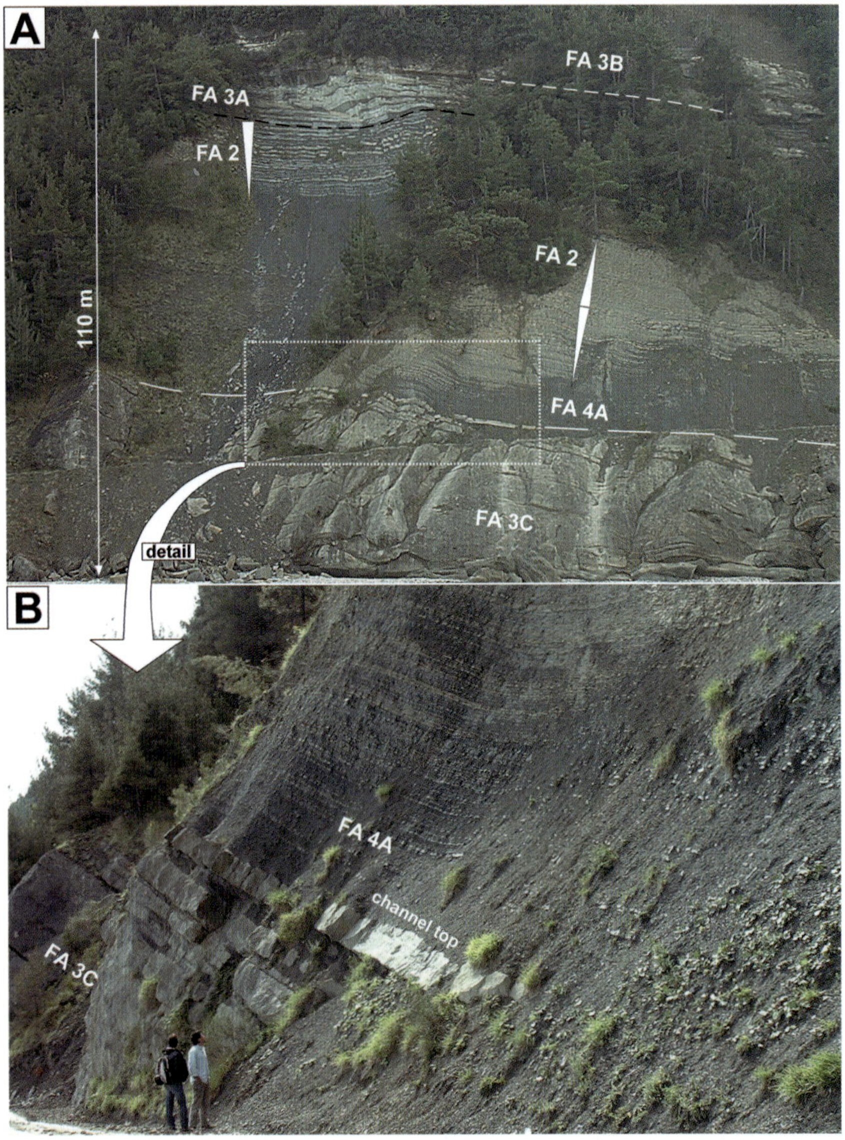

Fig. 21 Facies association 3C, multistorey palaeochannel complexes. (A) The upper part of the sandstone body of facies association 3C (same as in Figs 19A & 20) separated by facies associations 4A and 2 from the overlying sandstone bodies of facies associations 3A and 3B, locality 10 (Figs 2 & 10); general palaeocurrent direction to the right (westwards). (B) Close-up view of the sharp top of the sandstone body (FA 3C), overlain by heterolithic FA 4A.

or south, transverse to the general palaeoflow direction and hence indicating lateral accretion. The set boundaries are highly uneven, with local erosional relief of ≤ 1 m and a general relief of several metres. These bed sets are apparently erosional remnants of point bars, similar to those seen at other localities (Figs 18 & 19A, D). Except for its uppermost 30–40 m, the turbiditic succession is gently concave upwards, apparently bent by syndepositional deformation (syncline growth) combined with the effect of loading and substrate compaction.

The same sandstone body is exposed near the village of Aksu to the west, where a winding roadcut section at localities 8 and 9 (Fig. 2) shows the succession's lower and upper part, respectively (Fig. 10). The lower part sharply overlies the deposits of facies association 2, which are > 15 m thick and consist of facies B2 and B3 (88 vol.%) intercalated with facies A1/A2 (12 vol.%). The sandstone beds here are mainly fine- to medium-grained and ≤ 55 cm thick, commonly amalgamated, occasionally coarse-grained and ≤ 95 cm thick. The overlying facies association 3C (~ 75 m thickness exposed) consists of amalgamated, fine- to coarse-grained sandstone beds intercalated with gravelstone beds. The range and relative proportion of turbiditic facies are similar to those at localities 6

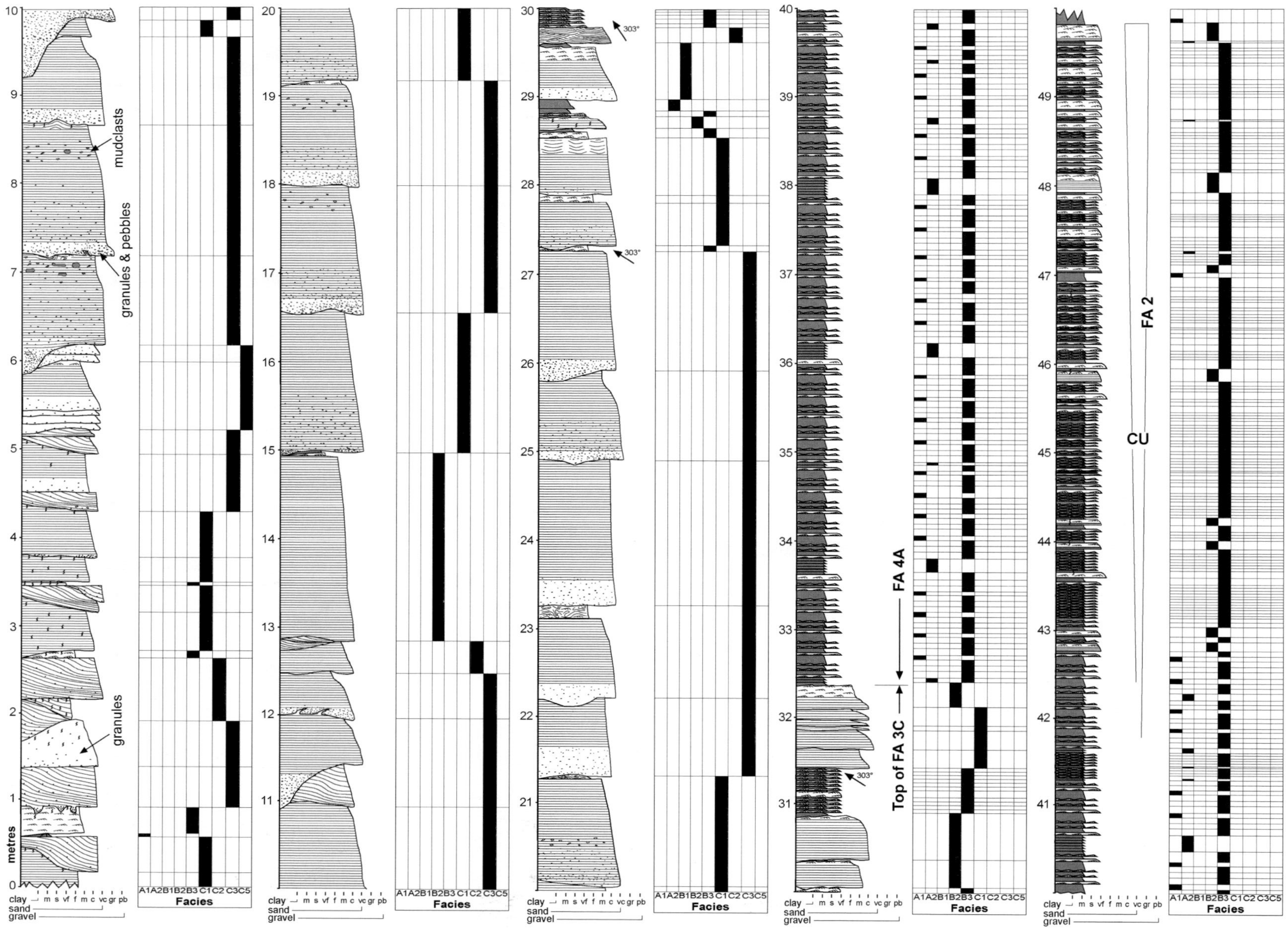

Fig. 22 Detailed log of the upper part of the sandstone body of facies association 3C, overlain by facies associations 4A and 2, as exposed at locality 10 (Fig. 21). Facies code as in Table 1 and log legend in Fig. 11. The lower to middle part of this palaeochannel complex crops out at the adjacent locality 4 (see Figs 19A & 20).

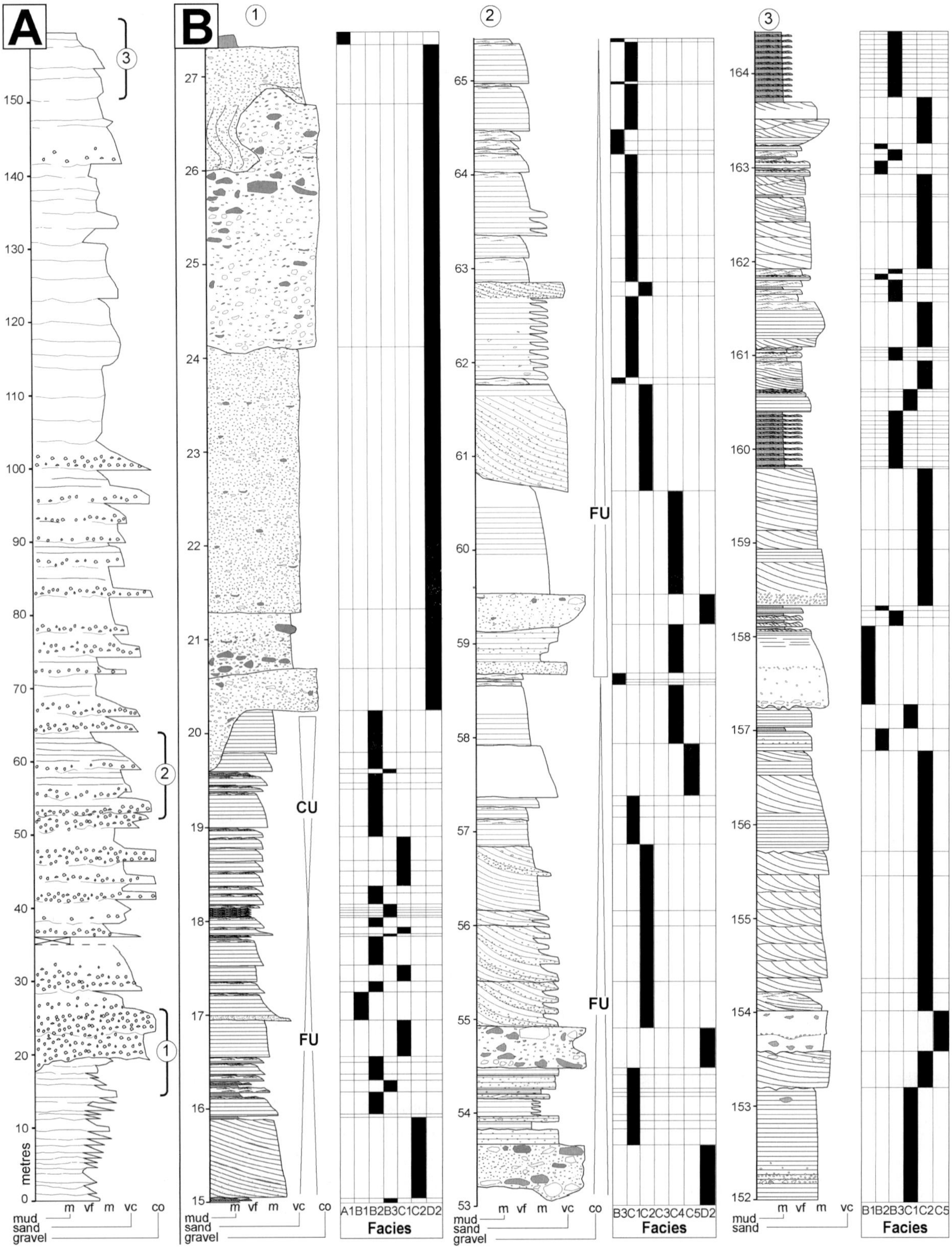

Fig. 23 Facies association 3C, multistorey palaeochannel complexes. (A) Simplified log of the whole sandstone body of facies association 3C exposed at locality 6. (B) Selected portions of the corresponding detailed log; facies code as in Table 1 and log legend in Fig. 11. See Figs 2 & 10 for locality details.

and 7 described above, but the gravelstone beds of facies D2 here are subordinate (12 vol.%). Flute casts, large trough-shaped scours and ripple foresets indicate palaeocurrent directions generally towards the west. The lowest part (41 m) of the succession is dominated by facies B1, B2 and C1–C5, with many beds ≤ 220 cm thick (facies C3) and some beds showing cross-stratification (facies C2). The remaining part (34 m) of the outcrop section shows amalgamated turbidites of facies C1 and C3–C5, with several mound-shaped beds of facies D2 (23 vol.%) in a narrow stratigraphic interval of ~ 8 m. The thickest sandstone beds are those of facies C2 (≤ 220 cm) and C3 (≤ 270 cm). The gravelstone beds are cobbly and commonly contain small boulders, but consist chiefly of pebbles supported by a medium to coarse sand matrix. Clast composition is similar to that at locality 6, including debris derived from the Akveren Formation. Turbidite bed sets, 5–15 m thick, are commonly inclined at 5–8° in directions transverse or oblique to the mean palaeocurrent trend and onlapping one another, which indicates a similar pattern of lateral accretion as observed in facies association 3B (Fig. 18) and elsewhere in facies association 3C (Fig. 19A & D).

The upper part of the sandstone body, ~ 40 m thick, is exposed at locality 9 (Fig. 10) ~ 1.5 km to the west, where it is overlain by facies association 2. The sandstone succession here consists of facies C1 (47 vol.%), C2 (26 vol.%) and C3 (25 vol.%). Gravelstone beds of facies D2 are rare (< 1 vol.%), as are also muddy interbeds of facies A1/A2 (~ 1 vol.%). The beds of facies C1 are mainly < 50 cm thick, but some are ≤ 190 cm, and their grain size ranges from very coarse to very fine sand. Several beds contain scattered granules and small pebbles, as well as sporadic mudclasts.

The sandstone body at localities 8 and 9 thus shows somewhat different facies composition and grain-size range than at the upstream localities 6 and 7 to the east (Fig. 2), although the depositional architecture of turbidites seems to be much the same.

The same sandstone body and the stratigraphically higher sandstone body 4 are exposed in a winding roadcut section at locality 12 south of Ayancık (Fig. 2), where they are separated by a succession of facies associations 4A and 4C, ~ 80 m thick (Fig. 10). The deposits are tectonically tilted and the outcrop shows < 30 m of their lateral extent, but the high local topography suggests that each of these sandstone bodies may be a few kilometres wide. The local palaeocurrent directions vary, but are generally to the west. The lower sandstone body has its upper ~ 37 m exposed and shows similar facies characteristics as the above-described sandstone succession at locality 9, including an upward transition to the thinner bedded, finer grained turbidites of facies association 2. Only the lower 18 m of sandstone body 4 are exposed, but its total thickness is estimated at ~ 70 m from the local topography. It consists of amalgamated, coarse-grained sandstone beds of facies B1 and cobble-rich facies D2, mainly 100–210 cm thick, with uneven erosional bases (scour relief ≤ 95 cm), overlain by similarly amalgamated, thick beds of facies C1 and C3, some with a basal relief of ≤ 80 cm. The sandstones are medium- to very coarse-grained and commonly contain granules, pebbles and scattered mudclasts, some up to cobble size. Many beds pinch out laterally due to erosion.

Interpretation

Similar thick-bedded, coarse-grained sandstone and gravelstone facies have been described from the feeder canyons and large 'proximal' channels of submarine fans (e.g. Stanley *et al.*, 1978; Winn & Dott, 1979; Kelling *et al.*, 1987; Cronin, 1995). The internal bedding architecture of the sandstone bodies of facies association 3C resembles closely that shown by the palaeochannels of facies association 3B (Fig. 18), except that the palaeochannels in the present case are stacked erosionally upon one another and are considerably richer in gravel. The gravel-rich zones separating the sets of oppositely inclined beds, superimposed point bars (Fig. 19A), are riffle zones of the channel thalweg. The thickest, southern sandstone body 1 (localities 6–9 and 12 in Fig. 10) consists of up to 12 superimposed palaeochannels, 10 to 20 m thick, incised into one another, whereas the number of palaeochannels stacked in the younger sandstone body 4 at locality 12 is uncertain, estimated to be four or five. The northern sandstone body 2 at localities 4 and 10 consists of at least eight superimposed palaeochannels, and the number of palaeochannels in the higher-lying sandstone body 3 at locality 11 is estimated to be five or six. The individual channels apparently had dimensions comparable to those of the solitary channels of facies association 3B, but conveyed stronger and coarser-grained turbidity currents, accompanied by debrisflows. The vertical stacking of channels is attributed to the flow confinement by intrabasinal growth-fold synclines, probably related to blind thrusts (see later text). Similar nested channels, although not necessarily filled with similar facies, have been described by Cronin *et al.* (2000b), Kolla *et al.* (2001), McCaffrey *et al.* (2002) and Grecula *et al.* (2003b).

These multistorey palaeochannel complexes have sharp bases, but their lowermost part is neither the coarsest-grained nor the thickest-bedded (see Figs 20 & 23A). It is the gravel-rich middle part of a palaeochannel complex that indicates the highest sediment fluxes and strongest currents. This evidence suggests that the vertical stacking of channels involved increasingly stronger and better-confined currents, with a reverse trend towards the top of the multistorey complex (Figs 22 & 23A).

The lowest palaeochannel complex (sandstone body 1) overlies erosionally a coarsening upward relict portion of a precursory depositional lobe (Fig. 23B, log 1) and is also covered by a sandstone-rich, fining upward facies association 2 (Fig. 23B, log 3), which suggests that the latest channel here underwent back-filling and was covered by its retrograding terminal lobe (see Saito & Ito, 2002). The three other palaeochannel complexes overlie overbank deposits, which suggests channel nesting initiated by an abrupt shift due to avulsion. The second palaeochannel complex (sandstone body 2) is overlain sharply by mudstone-rich facies association 4A (Fig. 21B), which suggests an abrupt abandonment of the latest channel by back-filling and upstream avulsion (Saito & Ito, 2002), similar to the case of the solitary palaeochannels of facies association 3B (Figs 17B, D & 24A). The top parts of the higher-lying palaeochannel complexes are either inaccessible (sandstone body 3 at locality 11) or not exposed (sandstone body 4 at locality 12), but the overlying deposits are, respectively, facies association 4A and facies association 2 covered by association 4A, and hence consistent with the abandonment pattern postulated above.

Facies association 4A: tabular overbank turbidites

Description

This heterolithic facies association forms laterally extensive units at various stratigraphic levels in the whole middle part of the Kusuri Formation, enveloping the sandstone bodies of facies associations 3A–C (Fig. 10). The deposits are thin, alternating sheet-like beds of sandstone and mudstone, predominantly facies B2 and B3 (Table 1).

One of the thickest units of facies association 4A (145 m) is exposed at locality 4 (Figs 14 & 15A), where it separates stratigraphically the sandstone bodies of facies associations 3A and 3C (Fig. 10). The succession is monotonous, with a nearly equal proportion of sandstones and mudstones, but shows numerous coarsening upward and occasional fining upward motifs, mainly 1–3 m thick, recognizable from an upward change in the frequency, thickness and/or grain size of sandstone beds (Figs 14 & 20). The sandstone beds are fine- to very fine-grained, tabular and mainly < 10 cm in thickness, but sporadically 30–35 cm thick. Most of the deposits are mudstone-capped turbidites of facies B3 (70–90 vol.%), commonly separated by mudstone layers of facies A1 and/or thin packages of facies A2 sheets (Fig. 20). The succession is also interspersed with beds of facies B2 and rare facies B1. Sporadic flute casts and measurable ripple foresets indicate currents flowing generally towards the west, but with azimuth dispersion of ≤ 45–50° and locally directed to the northwest. A large slump unit of similar deposits, 11 m thick, is found in the uppermost part of the succession (see the slump top near the log base in Fig. 20), with asymmetrical hydroplastic folds and thrust-like listric shears indicating local displacement direction towards the south.

A similar thick succession of thin tabular turbidites with thickening upward bedding motifs and sporadic 'outsized' beds (Fig. 12A & B) separates units of facies association 2 at localities 6 and 8 (Fig. 10). The more common thinner units of facies association 4A, 5–15 m in thickness, have been studied at locality 11, where they separate the palaeochannels (Figs 16 & 17B); at locality 10, directly above the palaeochannel complex (Figs 21 & 22); and at locality 4, between facies association 2 and the overlying palaeochannel (log interval 19–21.5 m in Fig. 14). These heterolithic successions are dominated by the mudstone-capped turbidites of facies B2 and B3 or only the latter, mainly < 5 cm thick, with intervening thin packages of the muddy turbidites of facies A2 (see Fig. 22, log interval 32.3–41.7 m). The local palaeocurrent directions

Fig. 24 Facies associations 4A and 4B, tabular and wedge-shaped overbank turbidites. (A) Sharp top of facies association 3B (the uppermost sandstone body in Fig. 21A, photographed at locality 11 to the west) overlain by facies associations 4A and 4B. The modal palaeocurrent direction is to the right, at ~ 35° into the outcrop. (B) Broad view of facies association 4B at the same locality. (C) Close-up detail of the upper portion of the latter deposits. See Figs 2 & 10 for locality details.

measured from flute casts and ripple crests are broadly towards the west, but deviate by up to ± 45° from the basin axis and the trend of adjacent palaeochannels.

The deposits of facies association 4A resemble those of facies association 2 (Fig. 9), but are distinguishable by: (i) the presence of small-scale CU and minor FU bedding motifs; (ii) the lack of a large-scale upward coarsening; (iii) the lack of an upstream transition into a sandstone-richer 'proximal' assemblage; and (iv) a direct spatial association with the levée deposits of facies assemblage 4B (e.g. Fig. 24A).

Interpretation

The characteristics of facies association 4A and its stratigraphic distribution among the palaeochannel bodies indicate overbank deposition by widespread, low-density turbidity currents overflowing an active channel. The varying proportion of sandstones and mudstones from outcrop to outcrop may reflect the relative distances from contemporaneous channels, whereas the small-scale coarsening upward and minor fining upward trends probably indicate short-term fluctuations in the volumes of successive turbidity currents.

The westward palaeocurrent directions, roughly parallel to the basin axis and palaeochannels, suggest that the low-density overbank currents were probably basin-wide or were directed westwards by subtle synclinal depressions of the seafloor. Many sand-rich submarine channel systems in relatively narrow basins are reported to have sheet-like overbank deposits, instead of levée ridges (e.g. Saito & Ito, 2002; Takano *et al.*, 2005). For example, basin-wide overbank flows characterize the deep-sea Toyama Trough, the Sea of Japan, which is 30–40 km wide and > 100 km long (Nakajima, 1996).

The intrabasinal slump unit at locality 4 (log base in Fig. 20) indicates gravitational instability of the seafloor, attributed to syndepositional tectonic deformation. This notion is supported by the overlying palaeochannel complex (FA 3C in Figs 19 & 20), the nesting of which is ascribed to the growth of an intrabasinal syncline, and by the occurrence of slump scars elsewhere in the sedimentary succession (e.g. Fig. 11C).

Facies association 4B: wedge-shaped overbank turbidites

Description

This facies assemblage (Fig. 9) occurs only locally in the middle part of the Kusuri Formation, forming isolated, wedge-shaped packages of sheet-like turbidites sandwiched between the units of facies association 4A in the neighbourhood of palaeochannels (Fig. 24A). These packages have thicknesses of up to 5–10 m and consist of sandstone turbidites that are visibly thinning and commonly pinching out in one direction, over a lateral distance of 200–300 m (Fig. 24B). The sandstone beds are mainly thinner than 10 cm and seldom thicker than 25 cm (Fig. 24C), have flat bases and show little or no amalgamation. These are almost exclusively fine- to very fine-grained turbidites of facies B3 (70–90 vol.%) and facies B2 (10–15 vol.%), capped with silty mudstones and also intercalated with thin mudstone layers of facies A1 and A2 (5–10 vol.%). Some beds show climbing-ripple cross-lamination and/or convolutions, but these features are not particularly common. Intraformational mudclasts are generally rare, though present in some beds. The upward bed-thickness trend varies from a gradual thickening to more abrupt changes, in some cases irregular or involving upward thinning. However, these bedding trends are not necessarily followed by the relative thickness proportion of sandstone and mudstone, such that a thinner-bedded interval may be dominated by sandstones and a thicker-bedded one by mudstones in a local vertical profile.

These turbidite packages thin either northwards or southwards, in directions roughly perpendicular to the westward trend of palaeochannels, although the palaeocurrent directions measured from flute casts seldom deviate by more than 45° from the latter trend.

Interpretation

The geometry, stratigraphic position and internal characteristics of facies association 4B indicate levée deposits lateral to palaeochannels. Their occurrence between the broader packages of the tabular overbank turbidites of facies association 4A suggests that the levées were formed only episodically, during periods when the overflowing low-density currents were relatively small and highly depletive, depositing most of their sand load within a short distance from the channel margin. The pattern of palaeocurrent directions suggests that the spill-over currents tended to flow obliquely away from the channel, rather than orthogonally (cf. Nakajima *et al.*, 1998), which may reflect the relatively low relief of the levée and the narrowness of the basin (see preceding discussion of facies association 4A).

It is worth noting that these deposits are not quite similar to the so-called 'CCC turbidites' (with climbing-ripple lamination, convolutions and clasts of rip-up mud), which are widely considered to be characteristic of levées (Walker, 1985). However, the diagnostic value of the CCC turbidite facies has recently been questioned (Cronin *et al.*, 2000b).

This kind of small levée would form only locally, at the outer cut-bank margins of channel bends, and could be limited to the broadest meanders, which might explain the scarcity of facies association 4B in outcrop sections. However, it cannot be precluded that the sedimentary succession contains also some larger, broader levée deposits, which would be difficult to distinguish from facies association 4A and might have been lumped with the latter in the outcrop studies. The distinction of levées, as geomorphological ridges, obviously relies on the outcrop width and orientation, but only a few outcrops of facies association 4A can be regarded as being suitable for this purpose.

Facies association 4C: sigmoidal overbank turbidites

Description

This facies assemblage is rather unusual and occurs only at locality 12 (Figs 2 & 10), in a roadcut section ~ 5 km south of Ayancık, where it forms four packages, 5–15 m thick, of sigmoidal-shaped turbidite beds (Fig. 25). These bed sets, separated by thinner units of facies association 4A, are stacked upon one another and comprise turbidites that have been accreted laterally with respect to their modal palaeocurrent direction (which is here towards the west-northwest). The individual beds are lenticular in cross-section, thinning and predominantly pinching out in the downdip direction and also thinning, flattening and occasionally pinching out in the updip direction (Fig. 25A).

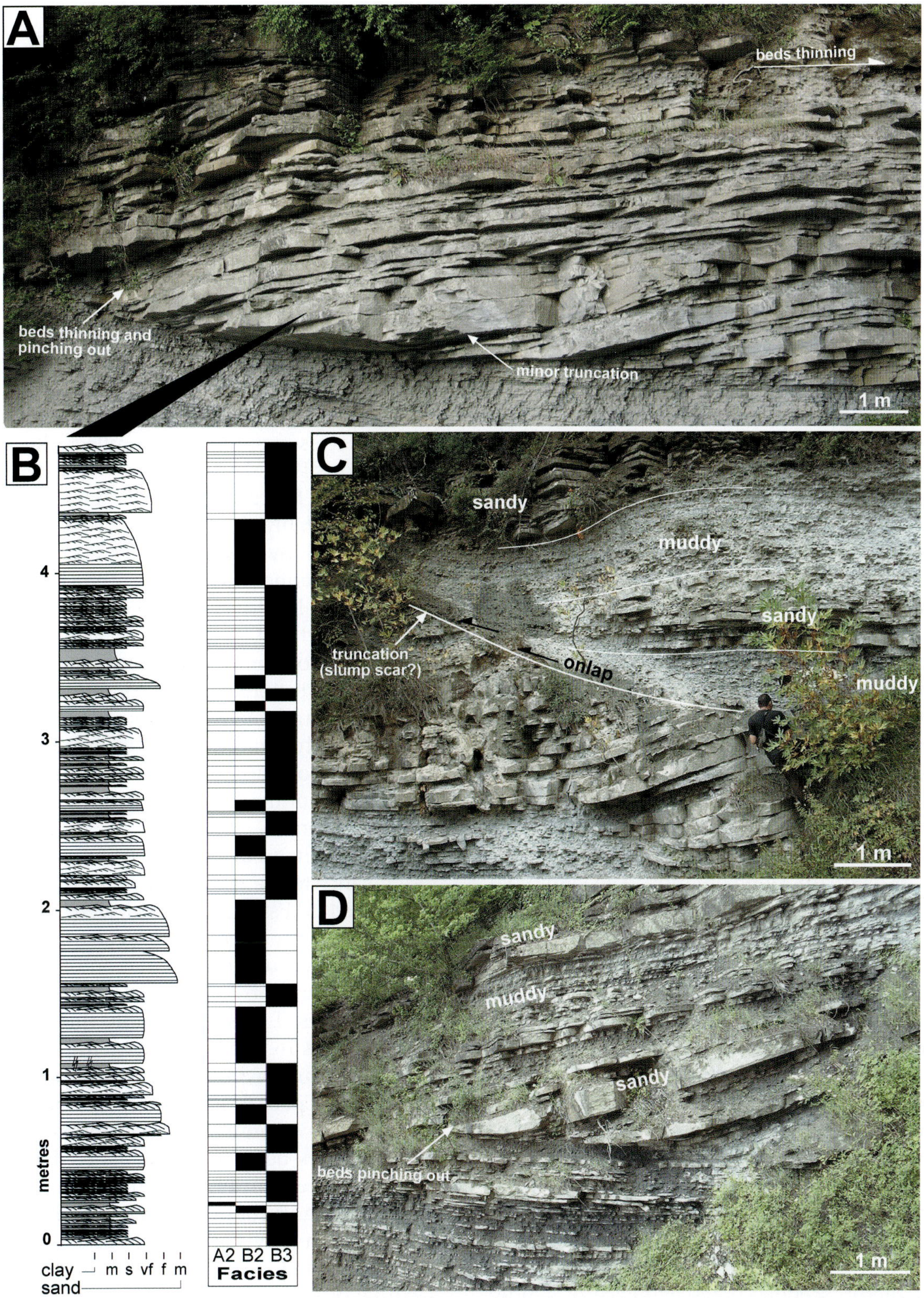

Fig. 25 Facies association 4C, sigmoidal overbank turbidites, at locality 12. (A) A sigmoidal set of laterally accreted lenticular turbidites. (B) A corresponding detailed log (facies code as in Table 1). (C) Similar sigmoidal bed sets in the lower part of the outcrop section; note the separating truncation surface, interpreted as a slump scar, and the alternating sand-rich and mud-rich bed packages in the upper set. (D) An analogous sigmoidal bed set, underlain by facies association 4A, in the higher part of the outcrop section (cf. Fig. 26A). The palaeocurrent direction (west-northwest) is out of the outcrop section, slightly to the left. See Figs 2 & 10 for locality details.

The angle of bed inclination relative to the bed-set base varies from < 5° to 25°. The beds are fine-grained sandstones of facies B (Table 1), mainly 5–20 cm thick and rarely ≤ 38 cm, but the proportion of turbidite subfacies varies (Fig. 25B). Some portions of a bed set consist of facies B2 and B3 (Fig. 25A); others are dominated by facies B3 (> 70 vol.%), with subordinate beds of facies B2 (10–15 vol.%) and B1 (2–3 vol.%) and common mudstone interlayers of facies A1 (Fig. 25D); and yet other portions are very thinly bedded and distinctly muddy, composed solely of facies B3 and A1/A2 (Fig. 25C).

A concave-upward erosional surface separates two of the sigmoidal bed sets (Fig. 25C), although there is no associated sedimentary evidence of an exceptionally powerful current. Intraset erosional truncations and bed amalgamation are rare and negligible (Fig. 25A).

Interpretation

The deposits of facies association 4C are rather puzzling. Contour currents can be precluded in the narrow basin, closed to the east, and also tidal currents are unlikely to have operated here at bathyal depths. The deposits themselves resemble closely the overbank facies assemblages 4A and 4B, and the underlying, intervening and directly overlying facies association 4A (Figs 25D & 26A) supports the notion of overbank sedimentation. The sigmoidal bedding architecture indicates lateral accretion, a depositional style typifying sinuous channels (Abreu *et al.*, 2003), but the turbidite facies and accretion pattern here differ from those of channel-fill point bars (see earlier text). The successive beds are thinning tangentially and pinching out in downdip direction, against the bed-set basal surface, which indicates pure lateral accretion, with little or no aggradation and negligible degree of intervening erosion. The facies and depositional pattern suggest relatively weak, low-density currents, with the turbidites 'plastered' on the outer side of a sharp bend in the flow course and thus mimicking a point-bar architecture.

The outcrop section shows evidence of syndepositional tectonic deformation (Fig. 26A), and the overbank flows in this area could have been semi-confined and/or sharply deflected by the growth of a local intrabasinal anticline. The deflection would have decelerated the successive flows and rendered them strongly dissipative, causing localized plastering of turbidites against the topographic obstacle (Alexander & Morris, 1994; Kneller & McCaffrey, 1999; McCaffrey & Kneller, 2001; Morris & Alexander, 2003). Tectonic deformation of the seafloor is capable of deflecting channelized flow courses (Cronin, 1995), and can thus have a similar effect on overbank flows, particularly where these are directed against an obstacle by the basinal confinement or seafloor relief. The multiple stacking of sigmoidal bed sets (Fig. 26A) is attributed to a tectonic rejuvenation of the structural obstacle (anticline growth). The variation in the component facies would simply reflect the varying magnitude of overbank flows, as recorded also by the overbank facies associations 4A and 4B (see earlier text). Notably, the notion of a local growth fold is supported by the preceding and subsequent nesting of channels in this part of the basin (see FA 3C bodies at locality 12 in Fig. 10).

As pointed out earlier in the text, the Kusuri Formation shows much compelling evidence of syndepositional tectonic deformation (Fig. 26), which probably stimulated the development of channels and caused their multistory stacking. The deposits of facies association 4C would then represent a specific case of the tectonic influence on overbank sedimentation.

The deposits of facies association 4C bear some resemblance to the 'non-amalgamated, suspension-dominated LAPs' of Abreu *et al.* (2003, fig. 18), distinguished with reference to field examples from the Carboniferous Ross Formation of Ireland and Jackfork Group of Arkansas. These lateral accretion packages (LAPs) are mainly thicker bedded, inclined at ≤ 16° and found in the lower parts of palaeochannels, but – as pointed out by Abreu *et al.* (2003, p. 643) – are not obviously related to the main channel-fill succession of coarser-grained, thick-bedded amalgamated LAPs. Their origin has been attributed to the fine-grained tails of bypassing turbidity currents, although it is unclear as to why the tail deposit of one current was not eroded by the powerful head of the consecutive one, since the currents would have been extremely 'efficient' (*sensu* Mutti, 1992) and presumably shaping the original channel.

In the light of the present examples, it is tempting to speculate that, alternatively, the channel in these rare cases may have been accidentally

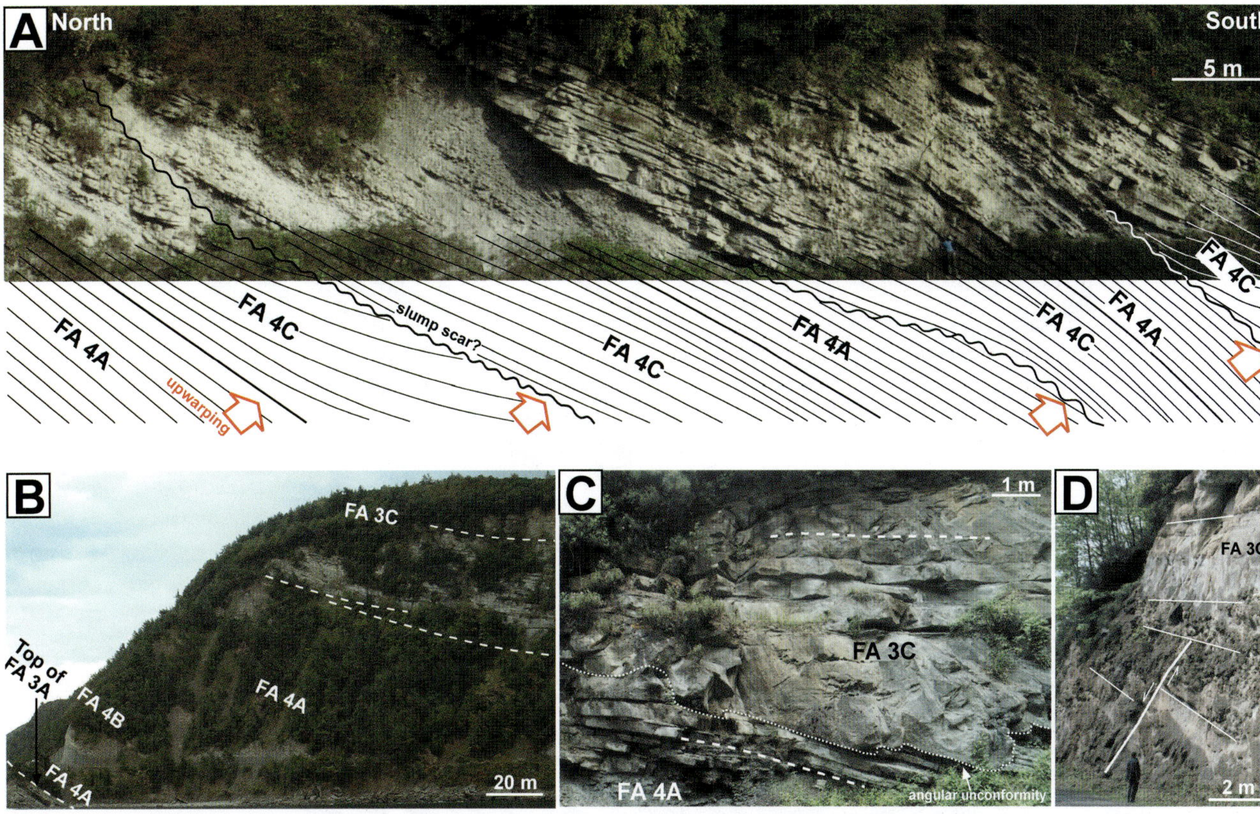

Fig. 26 Evidence of syndepositional tectonic deformation. (A) Progressive unconformities (wavy lines) related to tectonic upwarping in the outcrop section at locality 12. The growth of a local intrabasinal anticline is thought to have been responsible for the lateral plastering of the overbank turbidites of facies association 4C (see also Fig. 25). (B) Synclinal bending of deposits beneath a palaeochannel complex (sandstone body 3 of FA 3C at the top of coastal cliff west of locality 11) attributed to syndepositional seafloor deformation. Note the upward-decreasing inclination of bedding. (C) Close-up detail of an angular erosional unconformity beneath a palaeochannel complex. (D) A buried syndepositional fault in FA 3C deposits at locality 8. Localities as in Fig. 2.

abandoned at its early stage and conveyed relatively small currents, or overbank flows from a coeval active conduit, before being reactivated and filled up with the coarse-grained 'main' deposits (see Abreu *et al.*, 2003). In short, the actual significance of such sporadic in-channel heterolithic LAPs is by no means established (see also Martinsen *et al.*, 2000; Lien *et al.*, 2003).

DISCUSSION

The ensuing discussion of the Kusuri Formation refers to the preceding sedimentological documentation and to the regional tectonic framework of the foreland basin reviewed with references at the beginning of the paper.

Depositional setting

The deep-water sedimentation of the Kusuri Formation commenced in a sand-starved basin in Ypresian time, after the Sinop–Boyabat retroarc foreland was submerged under the load of the Central Pontide thrust sheets and the pre-existing reefal source began to be replaced by siliciclastic sediment supply (Fig. 27A and the basal unit FA 1 in Fig. 10). Tectonic inversion of the Eastern Pontide foreland then led to an abundant supply of coarse siliciclastic sediment from the east, while the contraction also split the Central Pontide foreland into the 'piggy-back' Boyabat Basin and the foredeep Sinop Basin (Fig. 27B). The development of the northward Erikli thrust and the antithetic Ekinveren back-thrust, which formed the axial pop-up ridge (Fig. 27B), was

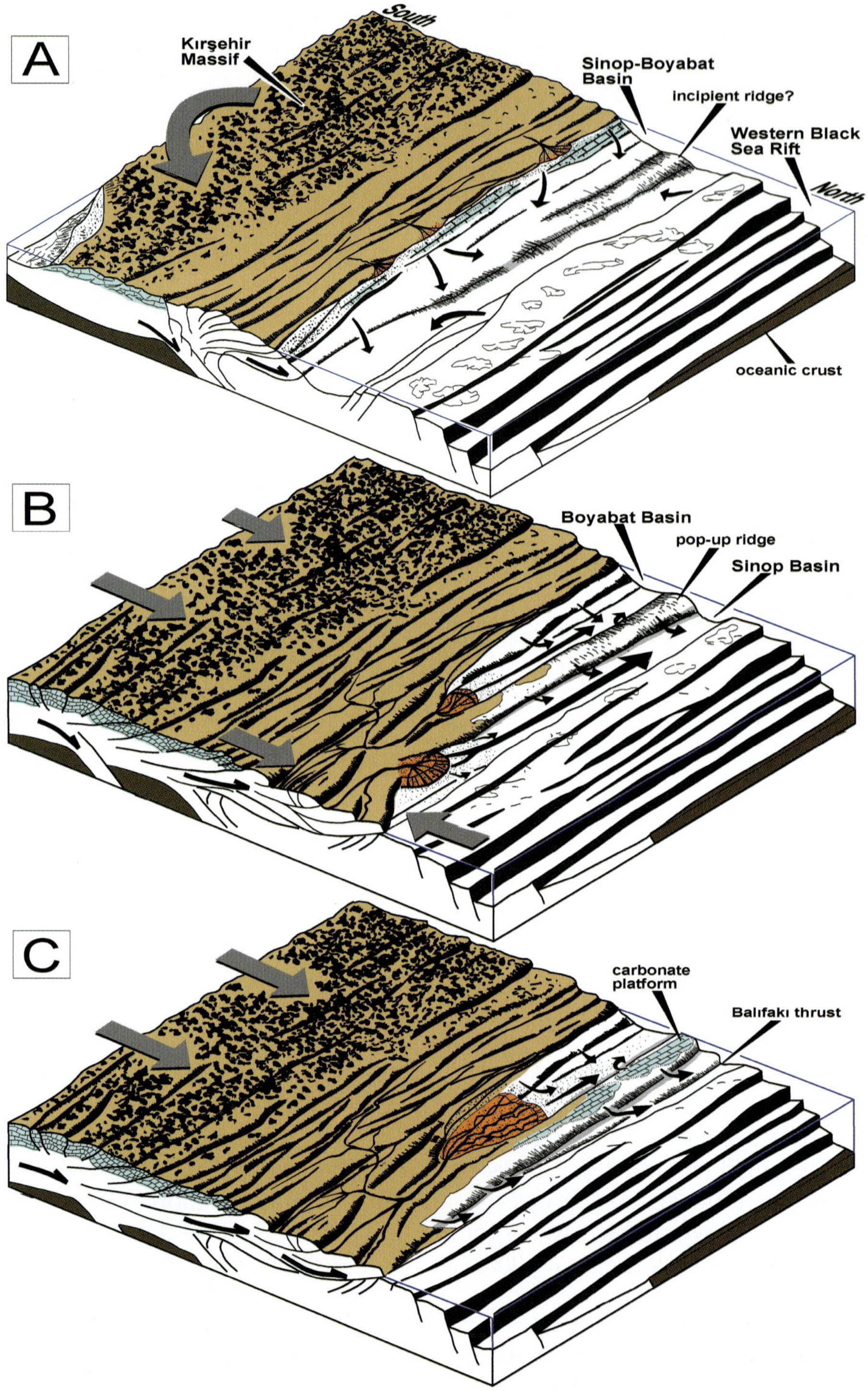

Fig. 27 Interpreted tectono-palaeogeographical setting of the Eocene sedimentation in the Sinop–Boyabat Basin. (A) Late Paleocene to earliest Eocene scenario for the deposition of the Atbaşı and lowermost Kusuri formations in a sand-starved basin (Leren *et al.*, this volume, pp. 401–456). (B) Early Eocene reconstruction for the deposition of the turbiditic Kusuri Formation in the Sinop Basin and the coeval turbiditic part of the Cemalettin Formation in the adjacent Boyabat Basin. (C) Middle Eocene model for the cessation of fluvial sediment supply to the Sinop Basin and an increased supply to the shallowing Boyabat Basin. For the preceding episodes of this basinal reconstruction, see Leren *et al.* (this volume, pp. 401–456, fig. 27).

followed by a northward propagation of thrusting, until the northermost Balıfakı thrust formed as a major basin-sole detachment (Fig. 2). The bulk of the Kusuri Formation was deposited between the early Eocene development of the pop-up ridge and the early middle Eocene conversion of the Sinop foredeep into a 'piggyback' basin, leading to its tectonic uplift and structural closure (Fig. 27C).

The turbiditic sedimentation occurred in a west-trending deep-water basin that was ~ 30 km wide and ≥ 150 km long, supplied with coarse siliciclastic sediment from the east and subject to active compressional deformation. The Kusuri Formation has a time span of ~ 6 Myr and its total thickness is nearly 1200 m, which might suggest a mean sedimentation rate of ~ 20 cm kyr^{-1}. However, the sedimentation rate of the sand-rich turbiditic succession was probably at least twice as high, because the lower 300 m and the upper 200 m of the formation are rich in hemipelagic mudstones that were deposited slowly in a sediment-starved environment over a considerable amount of time.

The depositional setting of the Kusuri Formation resembles to some extent the settings of many other turbiditic systems, fed mainly or solely by river deltas (Van Vliet, 1978, 1982; Labaume *et al.*, 1985; Sinclair, 1992, 1997, 2000; Bryn, 1998; Nakajima *et al.*, 1998; Dreyer *et al.*, 1999; Winkler & Gawenda, 1999; Avramidis *et al.*, 2000; Cronin *et al.*, 2000b; Haughton, 2000; Saito & Ito, 2002; Sinclair & Tomasso, 2002; Grecula *et al.*, 2003a,b; Lien *et al.*, 2003; Takano *et al.*, 2005). However, the Kusuri turbiditic system differs from most of these systems, which also themselves show great morphodynamic variation. The nature of the turbiditic system in the Sinop Basin thus deserves special consideration.

Morphodynamics of the turbiditic system

This basin-floor turbiditic system involved axial channels (FA 3A–C) with terminal depositional lobes (FA 2), which were formed by volumetrically large turbidity currents of high to low density (*sensu* Lowe, 1982) and commonly extended throughout the exposed length of the basin (> 70 km). The palaeochannels trend westwards, and the channel-fill deposits are thicker bedded and coarser grained in the eastern, source-proximal reaches. Gravelly debris-flow deposits are common in many palaeochannels as far as 10 km from the system's head zone and occur at downflow distances of ≤ 25–30 km in some cases. The channels ranged from poorly defined conduits, which were wide (≥ 500 m) and filled up by aggradation (FA 3A), to well-defined sinuous conduits that were filled by lateral accretion and subject to meander-bend expansion combined with marked downstream translation (FA 3B). The sinuous channels are estimated to have been 23–34 m deep and about 400–500 m wide, with meander wavelengths of 1.7–2 km and bulk meander-belt widths of possibly 2–3 km. The belt width to channel width ratios are fully comparable to those reported from other submarine channels (e.g. Kolla *et al.*, 2001), and much lower than those of meandering fluvial systems. However, the channel depth/width ratios of 1/15 to 1/19 would appear to be much lower than the aspect ratios of the majority of other submarine channels (cf. Clark & Pickering, 1996; Kolla *et al.*, 2001; Abreu *et al.*, 2003; Deptuck *et al.*, 2003). The development of long and sinuous submarine channels is consistent with the evidence of sustained turbidity currents (see facies C earlier in Table 1), which, in the context of a deltaic feeder might support the notion of river-generated hyperpycnal flows (Nakajima *et al.*, 1998; Elliott, 2000; Martinsen *et al.*, 2000; Sinclair, 2000; Johnson *et al.*, 2001b).

The palaeochannels occur in the middle part of the formation and are scattered across the basin width, which suggests channel shifting by avulsion. It is likely that essentially one channel was active at any particular time of the turbiditic sedimentation (cf. Nakajima *et al.*, 1998). The development of successive channels is considered to have been instigated by high sediment supply combined with the formation of seafloor synclines (cf. Mutti *et al.*, 1988; Takano *et al.*, 2005), probably related to blind-thrust folds. The sinuous palaeochannels (FA 3B) occur encased in overbank deposits (FA 4A), or overlie directly the poorly defined straight palaeochannels (FA 3A) and/or depositional lobes (FA 2). This latter relationship suggests that a straight channel tended to evolve into a sinuous one, by the flow interaction with the conduit (Pirmez & Flood, 1995; Imran *et al.*, 1999), prior to or after significant aggradation. The lateral shifting of channels was probably due to backfilling, which led to their abrupt abandonment by upstream avulsion or to a shortening of the conduit accompanied by terminal lobe retrogradation (cf. Saito & Ito, 2002). The synclinal confinement, where more pronounced by fold growth, caused localized nesting of sinuous channels, resulting in multistorey palaeochannel complexes 80–160 m thick and possibly ≤ 5 km wide (FA 3C). These multistorey palaeochannels contain gravelly facies, including local debrisflow mounds in the thalweg zone, and are stacked upon one another with a considerable degree of erosion. Some palaeochannels occur as remnants. A similar style of channel nesting, attributed to topographic confinement, has been described from modern (Kolla *et al.*, 2001; Posamentier & Kolla, 2003) and ancient turbiditic systems (Cronin *et al.*, 2000b; Grecula *et al.*, 2003b).

The overbank sedimentation (FA 4A) involved wide, low-density currents flowing roughly parallel to the basin axis, rather than at high angles to the channels, which may suggest flows that were mainly basin-wide or were directed westwards by subtle seafloor depressions. Basin-wide overbank flows often predominate in narrow basins (Baines, 1984; Nakajima, 1996; Nakajima *et al.*, 1998; Bursik & Woods, 2000; Gorsline *et al.*, 2000; Saito & Ito, 2002; Wynn *et al.*, 2002; Lien *et al.*, 2003; Sinclair & Cowie, 2003; Takano *et al.*, 2005). The large, sand-laden turbidity currents probably developed strong density layering and, when not scaling with the conduit, were spilling out a major part of their fine-grained suspension load as a voluminous, low-density overbank flow (Baines, 1984; Kneller & McCaffrey, 1999) constrained by the basin width. Recognizable levées are minor features in the present case (FA 4B), isolated and ≤ 10 m in thickness. They apparently formed during periods when relatively small turbidity currents predominated, with highly depletive spill-over flows. The local occurrence of the sigmoidal-shaped packages of laterally plastered thin turbidites (FA 4C) is attributed to a sharp deflection of overbank flows by the topography of an intrabasinal growth fold. These heterolithic LAPs are considered to be a striking example of the effect of syndepositional seafloor deformation on overbank sedimentation.

Inferences on deltaic feeder

The head part of the turbiditic system is not preserved, as the easternmost part of the basin had been uplifted and eroded (Fig. 2). The turbiditic system is inferred to have been supplied with sediment from a large delta of a bedload-dominated fluvial system draining the adjacent, uplifted foreland of the Eastern Pontides (Fig. 27B). The basin had a negligibly narrow shelfal rim, and a shelf-edge delta would be the most likely feeder to supply large volumes of coarse siliciclastic sediment, including gravel up to boulder grade, to a deep-water basin over a relatively long time (Reading & Richards, 1994; Burgess & Hovius, 1998; Nakajima *et al.*, 1998; Talling, 1998; Dreyer *et al.*, 1999; Sinclair, 2000; Lønne *et al.*, 2001; Muto & Steel, 2002; Alonso & Ercilla, 2003; Porębski & Steel, 2003; Posamentier & Kolla, 2003).

The delta probably formed as a Gilbert-type system, for it never advanced far into the basin and caused no significant shallowing in the basin's eastern part. A deltaic system can be arrested at the edge of a deep-water basin and keep shedding abundant sediment into it (Lønne *et al.*, 2001; Muto & Steel, 2002; Porębski & Steel, 2003). If excessive water depth does not allow major progradation, the arrested delta may have a negligible preservation potential. In fact, such margin-attached deltas in deep-water foreland basins are seldom preserved, being erased by erosion during the basin uplift. For example, a 'ghost' large deltaic feeder draining the eastern part of the North Pyrenean foreland has been widely invoked for the sand-rich Eocene turbiditic systems in the adjacent Basque–Cantabrian Foreland Basin to the west (Kruit *et al.*, 1972; Crimes, 1976; Van Vliet, 1978, 1982; Bryn, 1998).

The notion of a deltaic feeder in the present case is supported by the coarseness, siliciclastic composition and submature character of the sediment, which includes 'exotic' components and plant detritus, corresponds well with a fluvial provenance and contrasts with the calcareous sediment of the underlying formations, sourced from the basin's southwestern margin (Leren *et al.*, this volume, pp. 401–456). The northern basin margin remained submerged below wave base, and also the pop-up ridge to the south was still mainly underwater during the turbiditic sedimentation (Fig. 27B). Even when fully emerged prior to the basin inversion, this narrow (< 40 km) ridge failed to provide any major river catchments.

The notion of a deltaic feeder is also consistent with the coeval development of a large, coarse-grained delta at the eastern end of the adjacent Boyabat Basin (Fig. 27C), where subsequent shallowing allowed the deltaic feeder to advance westwards over the whole basin length.

With a Gilbert-type delta acting as a feeder, some of the large turbidity currents would probably have been generated by river floods as hyperpycnal flows (Prior *et al.*, 1986; Wright *et al.*, 1986; Nemec, 1990; Carlson *et al.*, 1992; Chikita *et al.*, 1996; Nemec *et al.*, 1999; Johnson, Paull *et al.*, 2001; Kassem & Imran, 2001; Parsons *et al.*, 2001), which is consistent with the facies indications of common sustained, long-duration currents. In the basin's active tectonic setting, voluminous sediment-gravity flows from a delta slope could have also been generated by earthquakes (Gorsline *et al.*, 2000) or by more spontaneous events of retrogressive slumping (Mastbergen & Van den Berg, 2003). Coarse-grained deltas of bedload-dominated rivers are also extremely efficient suppliers of hemipelagic mud (Nemec, 1995, pp. 32–34), which would explain the abundance of siliciclastic mudstones in the basin.

Stratigraphic evolution of the turbiditic system

The channelized turbiditic system formed after a prolonged phase of low and gradually increasing sand supply (see the basal unit FA 1 in Fig. 10), which probably heralded the development of a large, sand-prone fluvial catchment leading to delta formation. The deposition of a few turbiditic lobes (FA 2), formed by poorly defined and laterally shifting channels (FA 3A), was followed by an apparent decline in sand dispersal, which is attributed to the nesting of channel in the southern, thrust-loaded part of the basin (see the lower palaeochannel complex FA 3C at localities 6–9 and 12 in Fig. 10). The distributary channel was subsequently shifting laterally, mainly in the basin axial zone, until it became nested in the basin's northern part (see the lower palaeochannel complex FA 3C at localities 4 and 10 in Fig. 10), probably due to the northward propagation of blind thrusts that elevated the seafloor in the basin's southern part.

The subsequent wide shifting of a solitary channel, poorly to well-defined (FA 3A and 3B in Fig. 10) and associated with depositional lobes (FA 2), suggests a temporal decline in seafloor deformation and a reduction of structural confinement, until the channel was briefly nested again in the basin's northern part (see the upper palaeochannel complex FA 3C at locality 11 in Fig. 10). The channel then continued to shift laterally, before being nested in the basin's southern part (see the upper palaeochannel complex FA 3C at locality 12 in Fig. 10). This event is thought to have recorded the conversion of the Sinop foredeep into a 'piggyback' basin, with the Balıfakı sole thrust tilting the basin floor southwards and the Erikli thrust loading it along the southern margin (Fig. 2). The supply of coarse-grained siliciclastic sediment then declined (see the uppermost unit FA 2, ~ 50 m thick, in Fig. 10), as the onset of tectonic inversion diverted the fluvial feeder system away from the Sinop Basin, while boosting delta advance in the adjacent Boyabat Basin (Fig. 27C) and causing its shallowing. The Sinop Basin became dominated by muddy sedimentation, increasingly punctuated by non-channelized calcareous turbidity currents derived from a reefal platform that meanwhile formed along the southern pop-up ridge (Fig. 27C) and eventually prograded across the rapidly shallowing basin (Janbu, 2004).

Controlling factors

In broad terms, the most crucial factor responsible for the development of the Kusuri turbiditic system was the subsidence of the Central Pontide foreland coeval with the tectonic inversion of its adjacent Eastern Pontide counterpart. The pulses of differential tectonics would necessarily affect the deltaic feeder, through sediment supply and relative sea-level changes (Burgess & Hovius, 1998; Talling, 1998; Dreyer *et al.*, 1999; Porębski & Steel, 2003; Posamentier & Kolla, 2003), and could have caused slope readjustments (Ross *et al.*, 1994; Cronin *et al.*, 2000a). The basin floor itself was subject to deformation, and this may have affected the turbiditic system's base level and equilibrium profile (Pirmez *et al.*, 2000). The signal of eustatic sea-level changes may have been muted in the active tectonic setting, but the marked increase in coarse sediment supply recorded in the middle part of the formation might reflect also the late Ypresian fall in global sea-level (see the boundary of supercycles

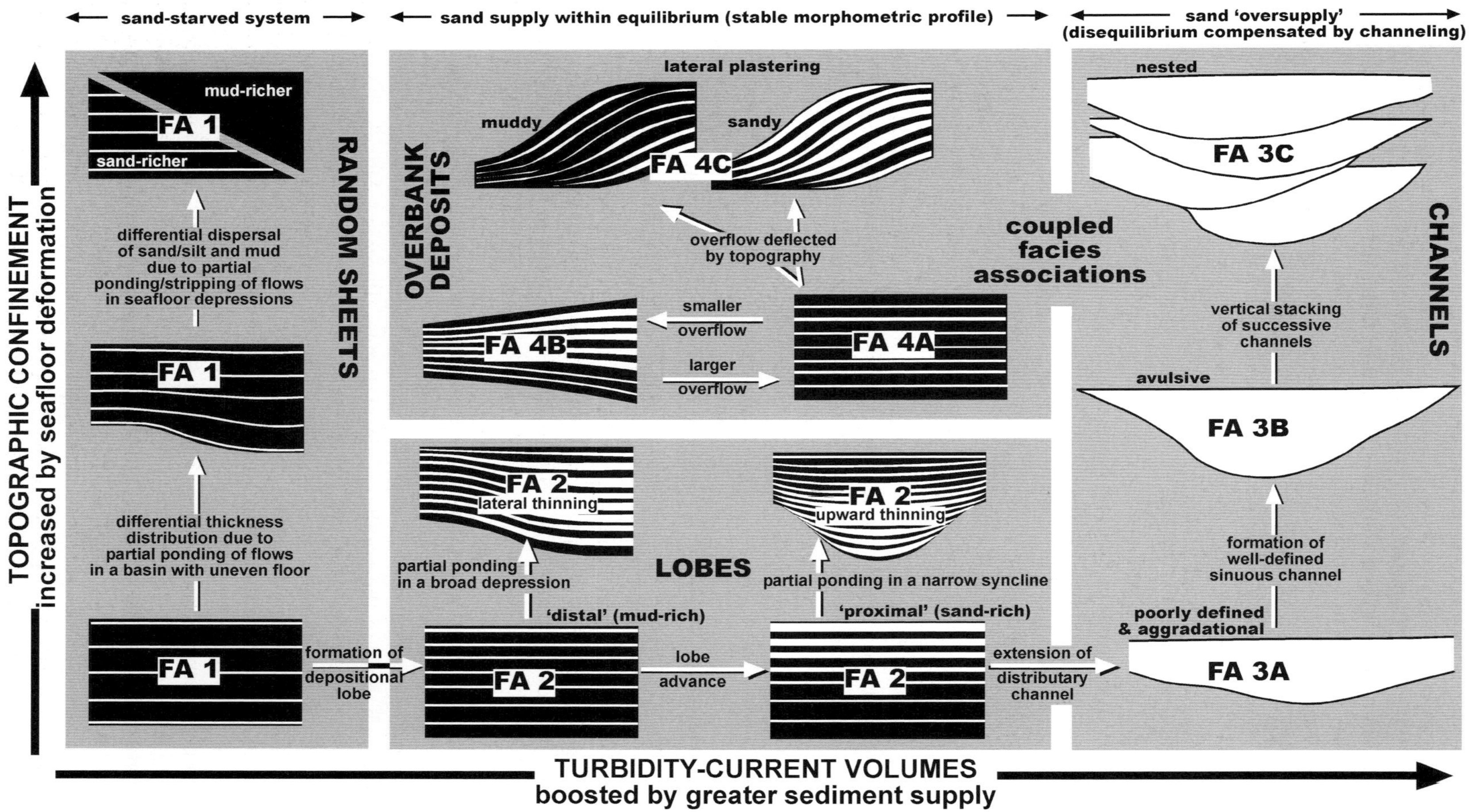

Fig. 28 Conceptual generic model for the spectrum of facies associations recognized in the Kusuri Formation. The deposition of particular facies associations (see Fig. 9) is considered to have been a function of the local turbidity-current discharges and the available topographic confinement (for discussion, see text). The evolutionary trends indicated by the arrows pertain to changes in local conditions, rather than to spatial relationships, although the model obviously bears some spatial implications. The model is meant to be an explanatory guide to, and predictive tool for, the origin and stratigraphy of the facies associations (Fig. 10).

TA2 and TA3 in Haq *et al.*, 1988). No supporting evidence of coastal onlaps is recognizable, because the southern margin's deposits have been eroded or concealed by the Erikli thrust (Fig. 2).

The Eocene regional climate was mainly moderate-humid, as testified by plant detritus and contemporaneous fluvial successions, but short-term climatic fluctuations cannot be excluded. The feeder delta's catchment would be sensitive to climatic changes, and these could also be reflected in the associated turbiditic system (Postma *et al.*, 1993; Weltje & De Boer, 1993; Beaudouin *et al.*, 2004). The short-term changes in turbidity current discharges recorded by the overbank facies successions (FA 4A–C), including small-scale trends of upward coarsening and fining, may possibly reflect climatic fluctuations.

The morphodynamics of the turbiditic system is considered to have been determined by the basin narrowness and controlled by a combination of two main factors (Fig. 28):

1 the magnitude of sediment supply, including its effect on local turbidity current discharges;
2 the availability of topographic confinement on the seafloor.

These factors are basically different modes of tectonic control on sedimentation, including its forcing of relative sea-level changes, combined further with the impact of climatic conditions and eustatic fluctuations. Seafloor topography is widely regarded as a major factor in the development of submarine channels (e.g. Kolla *et al.*, 2001; Posamentier & Kolla, 2003). Pulses of tectonic contraction would uplift the source area and boost sediment supply, while producing thrusts and causing basin-floor deformation. The narrow basin and seafloor deformation effectively allowed coarse sediment to be transported by turbidity currents over distances of > 100 km, despite the apparent lack of channels with well-developed, high levées (cf. Damuth *et al.*, 1998; Kolla *et al.*, 2001; Deptuck *et al.*, 2003).

In a conceptual model (Fig. 28), the volumetrically modest, low-density turbidity currents would spread widely on a flat basin floor, depositing thin and planar 'random sheets' (FA 1). A gentle seafloor relief would result in differential cumulative thickness of such deposits, whereas a more pronounced relief might cause partial ponding of flows and spatial partitioning of sediment, with sand-richer deposits in topographic depressions and mainly muddy deposits in higher-lying areas (Fig. 28, left-hand part). Larger, higher-density currents would be more depletive and form depositional lobes on a flat basin floor (FA 2), with a thickening upward bedding trend in the lobe 'proximal' part, where progradation would be more prounced (Fig. 28, lower mid-part). When subject to partial ponding due to local seafloor relief, these currents would deposit laterally thinning bed packages which, if insufficiently exposed, might be misinterpreted as levées. A depositional lobe filling in a synclinal depression might show a thinning upward bedding trend, as the flows would increasingly spill out and be more uniform; the resulting succession might thus resemble a broad channel-fill (Fig. 28, lower mid-centre part).

The largest currents, to balance their discharges with an equilibrium profile, would tend to produce channels (Fig. 28, right-hand part). These might form by default as poorly defined conduits, filled by aggradation (FA 3A), but would evolve into sinuous conduits dominated by lateral accretion (FA 3B) once the sustained flows became sufficiently channelized or were semi-confined by mild seafloor deformation. The development of growth folds, combined with sediment-supply maxima and substrate compaction, would cause channel nesting and the formation of multistorey channel complexes (FA 3C). The overbank flows associated with channels could be voluminous, forming basin-wide packages of thin heterolithic sheets (FA 4A), or might be periodically smaller and highly depletive, forming spill-over levées (FA 4B) (Fig. 28, upper mid-centre part). In response to a local topographic relief, the overbank flows could be sharply deflected and form packages of laterally plastered beds (FA 4C), which would vary between sand-rich and mud-rich varieties, depending on the prevalent flow magnitude (Fig. 28, top mid-centre part).

CONCLUSIONS

This case study contributes to the existing knowledge on deep-marine turbiditic systems in narrow, elongate foreland basins, particularly on the development of turbiditic channels and the associated styles of overbank sedimentation, with special emphasis on the impact of tectonic activity. The study contributes also to a better understanding of the geological history of the Central Pontides and of the southern Black Sea region.

The Eocene Sinop Basin evolved as a foredeep trough of the Central Pontides, ~ 30 km wide and > 150 km long, trending towards the west-northwest. The Kusuri Formation studied is ≤ 1200 m thick and comprises siliciclastic deposits of an axial, westward-directed turbiditic system supplied with coarse sediment by a fluvio-deltaic feeder draining the adjacent uplifted foreland of the Eastern Pontides. The sedimentological study of outcrop sections, supplemented by micropalaeontological and ichnological data, indicates deposition in a deep-sea environment.

The sedimentary facies of the Kusuri Formation include: hemipelagic 'background' mudstones interspersed with thin muddy turbidites; a range of 'classic' Bouma-type turbidites; a wide spectrum of non-classic turbidites deposited by channelized, low- to high-density currents, with common amalgamation of deposits and evidence of sustained (long-duration) flows; and subordinate massive gravelstones and gravelly sandstones attributed to in-channel debrisflows. Based on their spatial grouping, depositional architecture and stratigraphic distribution, the sedimentary facies are recognized to form four main facies associations:

1 basin-wide mudstones interspersed with thin sheet-like turbidites;
2 broad depositional lobes with thickening upward bedding trends;
3 poorly defined wide palaeochannels, solitary sinuous palaeochannels and multistorey palaeochannel complexes;
4 packages of overbank turbidites with tabular, wedge-shaped or sigmoidal bedding.

These facies assemblages are considered to be the principal architectural elements of the turbiditic succession. The first assemblage forms the lowermost and the uppermost part of the Kusuri Formation, whereas the others occur in its middle main part.

The poorly defined palaeochannels are 20–25 m thick, typically overlie the depositional lobes and are themselves overlain by the sinuous palaeochannels, 20–30 m thick and ≤ 400–500 m wide, which suggests that the former channels tended to evolve into the latter, prior to or after significant aggradation. The sinuous palaeochannels have sharp tops, commonly occur also encased in overbank deposits and some are overlain by depositional lobes in the distal to medial reaches. The lateral shifting of channels was thus probably due to their backfilling, which led to an abrupt abandonment by upstream avulsion or to a rapid shortening of the conduit accompanied by terminal lobe deposition. The channel-fill architecture of lateral accretion and point-bar stacking indicates meander-bend expansion combined with a marked downstream translation. The channel depth/width aspect ratios are much lower than those of many modern and ancient counterparts. The multistorey complexes of sinuous palaeochannels are 100–160 m thick and estimated to be ≤ 3–5 km wide. The vertical stacking of channels is attributed to the growth of syndepositional blind-thrust anticlines on the basin floor, combined with localized substrate compaction. The overbank facies assemblages indicate basin-wide flows (tabular bed packages), small and highly depletive spill-over flows forming minor levées (wedge-shaped bed packages), and overbank flows deflected by local topography (sigmoidal packages of laterally plastered turbidites).

The principal factors responsible for the behaviour and morphodynamic evolution of the turbiditic system are considered to have been:

1 the structural narrowness of the basin;
2 the topographic confinements provided by syndepositional seafloor deformation;
3 the rate of sediment supply, or turbidity current discharges.

Pulses of tectonic contraction are thought to have formed blind-thrust growth folds and/or increased sediment supply, and the coupling of these factors, at their maxima, resulted in multistorey channel stacking.

The siliciclastic supply was terminated by the foredeep transformation into a 'piggyback' basin, with the abandoned turbiditic system becoming mud-rich and increasingly calcareous, and turning gradually into a storm-influenced, shallowing neritic environment. The fluvial feeder system was diverted to the adjacent Boyabat Basin, contributing to its rapid shallowing and causing a fluvio-deltaic system to advance along its axis. Both basins were subsequently inverted by tectonic contraction in late Eocene to Early Miocene times, during the climax and final stages of the Tauride orogeny to the south.

ACKNOWLEDGEMENTS

The field project was sponsored by the Norsk Hydro Research Centre under a deep-marine research programme co-ordinated by Ole J. Martinsen, whose kind support and stimulating discussions we greatly appreciate. We thank also Alfred Uchman for field identification of trace fossils; Ercüment Sirel and Enis K. Sagular for micropalaeontological analyses of sediment samples; and Mehmet C. Alçiçek, Łukasz Gągała, Ayhan Ilgar, Beate L.S. Leren and Takeshi Nakajima for field assistance. The manuscript was critically reviewed by J.H. Baas, B.T. Cronin, S.O. Johnson, G. Kelling, S.J. Lippard, A.W. Martinius, G. Nichols and S.J. Porębski, whose helpful comments are much appreciated by the authors.

REFERENCES

Abreu, V., Sullivan, M., Pirmez, C. and Mohrig, D. (2003) Lateral accretion packages (LAPs): an important reservoir element in deep water sinuous channels. *Mar. Petrol. Geol.*, **20**, 631–648.

Agirrezabala, L.M. and García-Mondéjar, J. (1994) A coarse grained turbidite system with morphotectonic control (Middle Albian, Ondarroa, northern Iberia). *Sedimentology*, **41**, 383–407.

Akıncı, Ö. (1984) The Eastern Pontide volcano-sedimentary belt and associated massive sulphide deposits. In: *The Geological Evolution of the Eastern Mediterranean* (Eds J.E. Dixon and A.H.F. Robertson), pp. 415–428. Special Publication 17, Geological Society, London.

Aktaş, G. and Robertson, A.H.F. (1990) Tectonic evolution of the Tethys suture zone in SE Turkey: evidence from the petrology and geochemistry of Late Cretaceous and Middle Eocene extrusives. In: *Ophiolites – Oceanic Crustal Analogues* (Eds J. Malpas, E.M. Moores, A. Panayiotou and C. Xenophontos), Proceedings of the International Symposium, Troodos 1987, pp. 311–328. Cyprus Geological Survey, Nicosia.

Alexander, J. and Morris, S.A. (1994) Observations on experimental non-channelized turbidites: thickness variations around obstacles. *J. Sediment. Petrol.*, **64**, 899–909.

Allen, J.R.L. (1963) The classification of cross-stratified units, with notes on their origin. *Sedimentology*, **2**, 93–114.

Allen, P.A. and Homewood, P. (Eds) (1986) *Foreland Basins*. Special Publication 8, International Association of Sedimentologists, Blackwell Scientific Publications, Oxford, 462 pp.

Alonso, B. and Ercilla, G. (2003) Small turbidite systems in a complex tectonic setting (SW Mediterranean Sea): morphology and growth patterns. *Mar. Petrol. Geol.*, **19**, 1225–1240.

Anderton, R. (1995) Sequences, cycles and other nonsense: Are submarine fan models any use in reservoir geology? In: *Characterization of Deep Marine Clastic Systems* (Eds A.J. Hartley and D.J. Prosser), pp. 5–11. Special Publication 94, Geological Society Publishing House, Bath.

Andrew, T. and Robertson, A.H.F. (2002) The Beyşehir–Hoyran–Hadım nappes: genesis and emplacement of Mesozoic marginal and oceanic units of the northern Neotethys in southern Turkey. *J. Geol. Soc. London*, **159**, 529–543.

Avramidis, P., Zelilidis, A. and Kontopoulos, N. (2000) Thrust dissection control of deep-water clastic dispersal patterns in the Klematia–Paramythia foreland basin, western Greece. *Geol. Mag.*, **137**, 667–685.

Aydın, M., Serdar, H.S. and Şahintürk, Ö. (1982) Orta Karadeniz bölgesi jeolojisi ve petrol olanakları. In: *Proceedings of the 6th Petroleum Congress of Turkey*, pp. 63–71. Turkish Petroleum Company (TPAO), Ankara.

Aydın, M., Şahintürk, Ö., Serdar, H.S., *et al.* (1986) The geology of the area between Ballıdağ and Çangaldağ (Kastamonu). *Bull. Geol. Soc. Turk.*, **29**, 1–16.

Aydın, M., Demir, O., Özçelik, Y., Terzioğlu, N. and Satır, M. (1995a) A geological revision of Inebolu, Devrekani, Ağlı and Küre areas: new observations on Paleotethys–Neotethys sedimentary successions. In: *Geology of the Black Sea Region* (Eds A. Erler, E. Tuncay, E. Bingöl and S. Örçen), pp. 33–38. General Directorate of Mineral Research and Exploration (MTA), Ankara.

Aydın, M., Demir, O., Serdar, H.S., Özaydin, S. and Harput, B. (1995b) Tectono-sedimentary evolution and hydrocarbon potential of the Sinop–Boyabat Basin, North Tűrkiye. In: *Geology of the Black Sea Region* (Eds A. Erler, E. Tuncay, E. Bingöl and S. Örçen), pp. 254–263. General Directorate of Mineral Research and Exploration (MTA), Ankara.

Badgley, P.C. (1959) *Stratigraphy and Petroleum Possibilities of the Sinop Region*. Tidewater Oil Co. Report, Petrol Işleri Genel Müdürlüğü Arşivi, Ankara, 38 pp.

Baines, P.G. (1984) A unified description of two-layer flow over topography. *J. Fluid Mech.*, **146**, 127–167.

Baldwin, B. and Butler, C.O. (1985) Compaction curves. *Am. Assoc. Petrol. Geol. Bull.*, **69**, 622–626.

Barka, A., Sütçü, Y.F., Gedik, A., *et al.* (1985) *Final Report on the Geological Investigation for the Sinop Nuclear Power Plant*. Report No. 7963, General Directorate of Mineral Research and Exploration (MTA), Ankara.

Barnes, N.E. and Normark, W.R. (1985) Diagnostic parameters for comparing modern submarine fans and ancient turbiditic systems. In: *Submarine Fans and Related Turbidite Systems* (Eds A.H. Bouma, W.R. Normark and N.E. Barnes), pp. 13–14. Springer-Verlag, New York.

Beaudouin, C., Dennierlou, B., Melki, T., *et al.* (2004) The Late-Quaternary climatic signal recorded in a deep-sea turbiditic levee (Rhône Neofan, Gulf of Lions, NW Mediterranean): palynological constraints. *Sediment. Geol.*, 172, 85–97.

Bouma, A.H. (1962) *Sedimentology of Some Flysch Deposits: a Graphic Approach to Facies Interpretation*. Elsevier, Amsterdam, 168 pp.

Bouma, A.H. (2000) Coarse-grained and fine-grained turbidite systems as end member models: applicability and dangers. *Mar. Petrol. Geol.*, **17**, 137–143.

Bryn, B.K.L. (1998) Sedimentology of Monte Jaizkibel Formation: an Early Eocene Turbiditic Fan in the Basque-Cantabrian Basin. Unpublished Cand. Scient. thesis, University of Bergen, 163 pp.

Burgess, P.M. and Hovius, N. (1998) Rates of delta progradation during highstands: consequences for timing of deposition in deep-marine systems. *J. Geol. Soc. London*, **155**, 217–222.

Bursik, M.I. and Woods, A.W. (2000) The effects of topography on sedimentation from particle-laden turbulent density currents. *J. Sediment. Res.*, **70**, 53–63.

Carlson, P.R., Powell, R.D. and Phillips, A.C. (1992) Submarine sedimentary features on a fjord delta front, Queen Inlet, Glacier Bay, Alaska. *Can. J. Earth Sci.*, **29**, 565–573.

Chikita, K.A., Smith, N.D., Yonemitsu, N. and Perez-Arlucea, M. (1996) Dynamics of sediment-laden underflows passing over a subaqueous sill: glacier-fed Peyto Lake, Alberta, Canada. *Sedimentology*, **43**, 865–875.

Clark, J.D. and Pickering, K.T. (1996) Architectural elements and growth patterns of submarine channels: application to hydrocarbon exploration. *Am. Assoc. Petrol. Geol. Bull.*, **80**, 194–221.

Clark, J.D., Kenyon, N.H. and Pickering, K.T. (1992) Quantitative analysis of the geometry of submarine channels: implications for the classification of submarine fans. *Geology*, **20**, 633–636.

Cloetingh, S., Spadini, G., Van Wees, J.D. and Beekman, F. (2003) Thermo-mechanical modelling of Black Sea Basin (de)formation. *Sediment. Geol.*, **156**, 169–184.

Collins, A.C. and Robertson, A.H.F. (1998) Processes of Late Cretaceous to Late Miocene episodic thrust sheet translation in the Lycian Taurides, SW Turkey. *J. Geol. Soc. London*, **155**, 759–772.

Collins, A.C. & Robertson, A.H.F. (1999) Evolution of the Lycian allochthon, western Turkey, as a north-facing late Palaeozoic to Mesozoic rift and passive continental margin. *Geol. J.*, **34**, 107–138.

Collinson, J.D. and Thompson, D.B. (1982) *Sedimentary Structures*. Allen and Unwin, London, 207 pp.

Crimes, T.P. (1976) Sand fans, turbidites, slumps and the origin of the Bay of Biscay: a facies analysis of the Guipuzcoan flysch. *Palaeogeogr. Palaeoclimatol. Palaeoecol.*, **19**, 1–15.

Cronin, B.T. (1995) Structurally-controlled deep sea channel courses: examples from the Miocene of southeast Spain and the Alboran Sea, southwest Mediterranean. In: *Characterization of Deep Marine Clastic Systems* (Eds A.J. Hartley and D.J. Prosser), pp. 115–135. Special Publication 94, Geological Society Publishing House, Bath.

Cronin, B.T., Owen, D., Hartley, A.J. and Kneller, B. (1998) Slumps, debris flows and sandy deep-water channel systems: implications for the application of sequence stratigraphy to deep water clastic systems. *J. Geol. Soc. London*, **155**, 429–432.

Cronin, B.T., Hartley, A.J., Celik, H., Hurst, A., Türkmen, I. and Kerey, E. (2000a) Equilibrium profile development in graded deep-water slopes: Eocene, Eastern Turkey. *J. Geol. Soc. London*, **157**, 943–955.

Cronin, B.T., Hurst, A., Celik, H. and Türkmen, I. (2000b) Superb exposure of a channel, levee and overbank complex in an ancient deep-water slope environment. *Sediment. Geol.*, **132**, 205–216.

Damuth, J.E., Flood, R.D., Pirmez, C. and Manley, P.L. (1995) Architectural elements and depositional processes of Amazon deep sea fan imaged by long-range side-scan sonar (GLORIA), bathymetric swath mapping (Sea Beam), high-resolution seismic and piston-core data. In: *Atlas of Deep-water Environments: Architectural Style in Turbidite Systems* (Eds K.T. Pickering, R.N. Hiscott, N.H. Kenyon, F. Ricci-Lucchi and R.D. Smith), pp. 105–121. Chapman and Hall, London.

Damuth, J.E., Flood, R.D., Kowsmann, R.O., Belderson, R.H. and Gorini, M.A. (1998) Anatomy and growth pattern of Amazon deep-sea fan as revealed by long-range side-scan sonar (GLORIA) and high-resolution seismic studies. *Am. Assoc. Petrol. Geol. Bull.*, **72**, 885–911.

DeCelles, P.G. and Giles, K.A. (1996) Foreland basin systems. *Basin Res.*, **8**, 105–123.

Deptuck, M.E., Steffens, G.S., Barton, M. and Pirmez, C. (2003) Architecture and evolution of upper fan channel-belts on the Niger Delta slope and in the Arabian Sea. *Mar. Petrol. Geol.*, **20**, 649–676.

Dilek, Y. and Moores, E.M. (1990) Regional tectonics of the eastern Mediterranean ophiolites. In: *Ophiolites – Oceanic Crustal Analogues* (Eds J. Malpas, E.M. Moores, A. Panayiotou and C. Xenophontos), Proceedings of the International Symposium, Troodos 1987, pp. 295–309. Cyprus Geological Survey, Nicosia.

Dilek, Y. and Rowland, J.C. (1993) Evolution of a conjugate passive margin pair in Mesozoic southern Turkey. *Tectonics*, **12**, 954–970.

Downie, R.A. and Stedman, C.I. (1993) Complex deformation and fluidization structures in Aptian sediment gravity flow deposits of the Outer Moray Firth. In: *Petroleum Geology of Northwest Europe: Proceedings of the 4th International Conference* (Ed. J.R. Parker), pp. 185–188. Geological Society, London.

Dreyer, T., Corregidor, J., Arbues, P. and Puigdefábregas, C. (1999) Architecture of the tectonically influenced Sobrarbe deltaic complex in the Ainsa Basin, northern Spain. *Sediment. Geol.*, **127**, 127–169.

Eğin, D., Hirst, D.M. and Phillips, R. (1979) The petrology and geochemistry of volcanic rocks from the northern Harşit River area, Pontide volcanic province, northeastern Turkey. *J. Volcanol. Geoth. Res.*, **6**, 105–123.

Elliott, T. (2000) Depositional architecture of a sand-rich, channelized turbidite system: the Upper Carboniferous Ross Sandstone Formation, western Ireland. In: *Deep-Water Reservoirs of the World, Proceedings of the 20th Annual Research Conference* (Eds P. Weimar, R.M. Slatt, A.H. Bouma and D.T. Lawrence), pp. 342–373. Gulf Coast Section Foundation, Society of Economic Paleontologists and Mineralogists, Austin, TX.

Felletti, F. (2002) Complex bedding geometries and facies associations of the turbiditic fill of a confined basin in a transpressive setting (Castagnola Fm., Tertiary Piedmont Basin, NW Italy). *Sedimentology*, **49**, 645–667.

Flood, R.D. and Damuth, J.E. (1987) Quantitative characteristics of sinuous distributary channels on the Amazon deep-sea fan. *Geol. Soc. Am. Bull.*, **98**, 728–738.

Friès, G. and Parize, O. (2003) Anatomy of ancient passive margin slope systems: Aptian gravity-driven deposition on the Vocontian palaeomargin, western Alps, south-east France. *Sedimentology*, **50**, 1231–1270.

Galloway, W.E. (1998) Siliciclastic slope and base-of-slope depositional systems: component facies, stratigraphic architecture and classification. *Am. Assoc. Petrol. Geol. Bull.*, **82**, 569–595.

Gedik, A. and Korkmaz, S. (1984) Sinop havzasının jeolojisi ve petrol olanakları. *Jeol. Mühend. Derg.*, **19**, 53–79.

Gedik, A., Ercan, T. and Korkmaz, S. (1984) Orta Karadeniz (Samsun-Sinop) havzasının jeolojisi ve volkanik kayaçlarin petrolojisi. *Bull. Mineral Res. Explor. Inst. Turk.*, **99/100**, 33–50.

Gorsline, D.S., De Diego, T. and Nava-Sanchez, E.H. (2000) Seismically triggered turbidites in small margin basins: Alfonso Basin, Western Gulf of California and Santa Monica Basin, California Borderland. *Sediment. Geol.*, **135**, 21–35.

Göksu, E., Pamir, H.N. and Erentöz, C. (1974) *Samsun Map Sheet 1:500 000*. General Directorate of Mineral Research and Exploration (MTA), Ankara.

Göncüoğlu, M.C., Turhan, N., Şentürk, K., *et al.* (2000) A geotraverse across northwestern Turkey: tectonic units of the Central Sakarya region and their tectonic evolution. In: *Tectonics and Magmatism in Turkey and the Surrounding Areas* (Eds E. Bozkurt, J.A. Winchester and J.D.A. Piper), pp. 139–161. Special Publication 173, Geological Society Publishing House, Bath.

Görür, N. (1988) Timing of opening of the Black Sea basin. *Tectonophysics*, **147**, 247–262.

Görür, N. (1997) Cretaceous syn- to postrift sedimentation on the southern continental margin of the Western Black Sea Basin. In: *Regional and Petroleum Geology of the Black Sea and Surrounding Region* (Ed. A.G. Robinson), pp. 227–240. Memoir 68, American Association of Petroleum Geologists, Tulsa, OK.

Görür, N. and Tüysüz, O. (1997) Petroleum geology of the southern continental margin of the Black Sea. In: *Regional and Petroleum Geology of the Black Sea and Surrounding Region* (Ed. A.G. Robinson), pp. 241–254. Memoir 68, American Association of Petroleum Geologists, Tulsa, OK.

Görür, N. and Tüysüz, O. (2001) Cretaceous to Miocene palaeogeographic evolution of Turkey: implications for hydrocarbon potential. *J. Petrol. Geol.*, **24**, 119–146.

Görür, N., Oktay, F.Y., Seymen, I. and Şengör, A.M.C. (1984) Palaeotectonic evolution of the Tuzgölü basin complex, Central Turkey: sedimentary record of a Neotethyan closure. In: *The Geological Evolution of the Eastern Mediterranean* (Eds J.E. Dixon and A.H.F. Robertson), pp. 467–482. Special Publication 17, Geological Society, London.

Görür, N., Tüysüz, O., Aykol, A., Sakınç, M., Yiğitbaş, E. and Akkök, R. (1993) Cretaceous red pelagic carbonates of northern Turkey: their place in the opening history of the Black Sea. *Eclogae Geol. Helv.*, **86**, 819–838.

Görür, N., Çağatay, N., Sakınç, M., Akkök, R., Tchapalyga, A. and Natalin, B. (2000) Neogene Paratethyan succession in Turkey and its implications for the palaeogeography of the Eastern Paratethys. In: *Tectonics and Magmatism in Turkey and the Surrounding Areas* (Eds E. Bozkurt, J.A. Winchester and J.D.A. Piper), pp. 251–269. Special Publication 173, Geological Society Publishing House, Bath.

Grecula, M., Flint, S., Potts, G., Wickens, D. and Johnson, S. (2003a) Partial ponding of turbidite systems in a basin with subtle growth-fold topography: Laingsburg-Karoo, South Africa. *J. Sediment. Res.*, **73**, 603–620.

Grecula, M., Flint, S., Wickens, D. and Johnson, S. (2003b) Upward-thickening patterns and lateral continuity of Permian sand-rich turbidite channel fills, Laingsburg Karoo, South Africa. *Sedimentology*, **50**, 831–853.

Gürer, Ö.F. and Aldanmaz, E. (2002) Origin of the Upper Cretaceous-Tertiary sedimentary basins within the Tauride-Anatolide platform in Turkey. *Geol. Mag.*, **139**, 191–197.

Güven, A. (1977) *Stratigraphy and sedimentology of Eocene formations, Karabük area, Turkey.* Unpublished PhD thesis, University College, Swansea, 307 pp.

Habgood, E.L., Kenyon, N.H., Masson, D.G., *et al.* (2003) Deep-water sediment wave fields, bottom current sand channels and gravity flow channel-lobe systems: Gulf of Cadiz, NE Atlantic. *Sedimentology*, **50**, 483–510.

Haq, B.U. (1991) Sequence stratigraphy, sea-level change, and significance for the deep sea. In: *Sedimentation, Tectonics and Eustasy: Sea-Level Changes at Active Margins* (Ed. D.I.M. Macdonald), pp. 3–40. Special Publication 12, International Association of Sedimentologists. Blackwell Science, Oxford.

Haq, B.U., Hardenbol, J. and Vail, P.R. (1988) Mesozoic and Cenozoic chronostratigraphy and eustatic cycles. In: *Sea-level Changes – an Integrated Approach* (Eds C.K. Wilgus, B.S. Hastings, H.W. Posamentier, J.C. Van Wagoner, C.A. Ross and C.G.St.C. Kendall), pp. 71–108. Special Publication 42, Society of Economic Paleontologists and Mineralogists, Tulsa, OK.

Harms, J.C., Southard, J.B., Spearing, D.R. and Walker, R.G. (1975) *Depositional Environments as Interpreted from Primary Sedimentary Structures and Stratification Sequences.* SEPM Short Course No. 2 Lecture Notes, Society of Economic Paleontologists and Mineralogists, Dallas, 161 pp.

Harms, J.C., Southard, J.B. and Walker, R.G. (1982) *Structures and Sequences in Clastic Rocks.* SEPM Short Course No. 9 Lecture Notes, Society of Economic Paleontologists and Mineralogists, Calgary, 250 pp.

Haughton, P.D.W. (1994) Deposits of deflected and ponded turbidity currents, Sorbas Basin, southeastern Spain. *J. Sediment. Res.*, **A47**, 223–246.

Haughton, P.D.W. (2000) Evolving turbidite systems on a deforming basin floor, Tabernas, SE Spain. *Sedimentology*, **47**, 497–518.

Hayward, A.B. (1984) Sedimentation and basin formation related to ophiolite nappe emplacement, Miocene, SW Turkey. *Sediment. Geol.*, **40**, 105–129.

Helle, K. (2003) *Anatomy and allostratigraphy of deep-marine Mount Messenger Formation (Miocene), eastern-margin Taranaki Basin, New Zealand.* Unpublished Cand. Scient. thesis, University of Bergen, 124 pp.

Hiscott, R.N., Pickering, K.T., Bouma, A.H., *et al.* (1997) Basin-floor fans in the North Sea: sequence stratigraphic models vs. sedimentary facies: Discussion. *Am. Assoc. Petrol. Geol. Bull.*, **81**, 662–665.

Imran, J., Parker, G. and Pirmez, C. (1999) A numerical model of flow in meandering submarine and subaerial channels. *J. Fluid Mech.*, **400**, 295–331.

Janbu, N.E. (2004) *Tectonic control on turbiditic sedimentation: the Late Cretaceous–Eocene succession in the Sinop–Boyabat Basin of north-central Turkey.* Unpublished Dr. Scient. Dissertation, University of Bergen, 334 pp.

Jensen, A.I., Bergslien, D., Rye-Larsen, M. and Lindholm, R.M. (1993) Origin of complex mound geometry of Paleocene submarine-fan sandstone reservoirs, Balder Field, Norway. In: *Petroleum Geology of Northwest Europe: Proceedings of the 4th International Conference* (Ed. J.R. Parker), pp. 135–143. Geological Society, London.

Johnson, K.S., Paull, C.K., Barry, J.P. and Chavez, F.P. (2001a) A decadal record of underflows from a coastal river into the deep sea. *Geology*, **29**, 1019–1022.

Johnson, S.D., Flint, S., Hinds, D. and Wickens, H. DeV. (2001b) Anatomy, geometry and sequence stratigraphy of basin floor to slope turbidite systems, Tanqua Karoo, South Africa. *Sedimentology*, **48**, 987–1023.

Jopling, A.V. (1965) Hydraulic factors controlling the shape of laminae in laboratory deltas. *J. Sediment. Petrol.*, **35**, 777–791.

Kassem, A. and Imran, J. (2001) Simulation of turbid underflows generated by the plunging of a river. *Geology*, **29**, 655–658.

Kaymakcı, N., Duermeijer, C.E., Langereis, C., White, S.H. and Van Dijk, P.M. (2003) Palaeomagnetic evolution of the Çankırı Basin (central Anatolia, Turkey): implications for oroclinal bending due to indentation. *Geol. Mag.*, **140**, 343–355.

Kelling, G., Davies, P. and Holroyd, J. (1987) Style, scale and significance of sand bodies in the Northern and Central Belts, southwest Southern Uplands. *J. Geol. Soc. London*, **144**, 787–805.

Ketin, I. and Gümüş, Ö. (1963) *Sinop-Ayancık arasindaki, III. Bölgeye dahil sahaların jeolojisi hakkında rapor; 2. Kısım, Jura ve Kretase formasyonlarının etüdü*. Report No. 213–288, Turkish Petroleum Company (TPAO), Ankara, 118 pp.

Kneller, B.C. and Buckee, C. (2000) The structure and fluid mechanics of turbidity currents: a review of some recent studies and their geological implications. *Sedimentology*, **47**, 62–94.

Kneller, B.C. and McCaffrey, W.D. (1995) Modelling the effects of salt-induced topography on deposition from turbidity currents. In: *Salt, Sediment and Hydrocarbons* (Eds C.J. Travis, H. Harrison, M.R. Hudeac, B.C. Vendeville, F.J. Peel and R.F. Perkins), pp. 137–145. Gulf Coast Section, Society of Economic Paleontologists and Mineralogists, Houston, TX.

Kneller, B.C. and McCaffrey, W.D. (1999) Depositional effects of flow non-uniformity and stratification within turbidity currents approaching a bounding slope: deflection, reflection, and facies variation. *J. Sediment. Res.*, **69**, 980–991.

Koçyiğit, A. (1986) Stratigraphy and nature of the northern margin of the Karabük-Safranbolu Tertiary basin. *Bull. Geol. Soc. Turk.*, **30**, 61–69.

Kolla, V. and Macurda, D.B., Jr. (1988) Sea-level changes and timing of turbidity-current events in deep-sea fan systems. In: *Sea-level Changes – an Integrated Approach* (Eds C.K. Wilgus, B.S. Hastings, H.W. Posamentier, J.C. Van Wagoner, C.A. Ross and C.G.St.C. Kendall), pp. 381–392. Special Publication 42, Society of Economic Paleontologists and Mineralogists, Tulsa, OK.

Kolla, V., Bourges, Ph., Urruty, J.-M. and Safa, P. (2001) Evolution of deep-water Tertiary sinuous channels offshore Angola (west Africa) and implications for reservoir architecture. *Am. Assoc. Petrol. Geol. Bull.*, **85**, 1373–1405.

Kruit, C., Brouwer, J. and Ealey, P. (1972) A deep-water sand fan in the Eocene Bay of Biscay. *Nature Phys. Sci.*, **240**, 59–61.

Labaume, P., Séguret, M. and Seyve, C. (1985) Evolution of a turbiditic foreland basin and analog with an accretionary prism: example of the Eocene South-Pyrenean basin. *Tectonics*, **4**, 661–685.

Leren, B.L.S. (2003) *Late Cretaceous to Early Eocene sedimentation in the Sinop–Boyabat Basin, north-central Turkey: Facies analysis of turbiditic to shallow-marine deposits*. Unpubl. Cand. Scient. Thesis, Univer. of Bergen, 140 pp.

Lien, T., Walker, R.G., Martinsen, O.J. (2003) Turbidites in the Upper Carboniferous Ross Formation, western Ireland: reconstruction of a channel and spillover system. *Sedimentology*, **50**, 113–148.

Lomas, S.A. and Joseph, P. (Eds) (2004) *Confined Turbidite Systems*. Special Publication 222, Geological Society Publishing House, Bath, 328 pp.

Lønne, I., Nemec, W., Blikra, L.H. and Lauritsen, T. (2001) Sedimentary architecture and dynamic stratigraphy of a marine ice-contact system. *J. Sediment. Res.*, **B71**, 922–943.

Lowe, D.R. (1982) Sediment gravity flows, II. Depositional models with special reference to the deposits of high-density turbidity currents. *J. Sediment. Petrol.*, **52**, 279–297.

Martinsen, O.J., Lien, T. and Walker, R.G. (2000) Upper Carboniferous deep water sediments: analogues for passive margin turbidite plays. In: *Deep-Water Reservoirs of the World, Proceedings of the 20th Annual Research Conference* (Eds P. Weimar, R.M. Slatt, A.H. Bouma and D.T. Lawrence), pp. 533–555. Gulf Coast Section Foundation, Society of Economic Paleontogists and Mineralogists, Austin, TX.

Mascle, A. and Puigdefàbregas, C. (1998) Tectonics and sedimentation in foreland basins: results from the Integrated Basin Studies project. In: *Cenozoic Foreland Basins of Western Europe* (Eds A. Mascle, C. Puigdefábregas, H.P. Luterbacher and M. Fernàndez), pp. 1–28. Special Publication 134, Geological Society Publishing House, Bath.

Mastbergen, D.R. and Van den Berg, J.H. (2003) Breaching in fine sands and the generation of sustained turbidity currents in submarine canyons. *Sedimentology*, **50**, 625–637.

McCaffrey, W.D. and Kneller, B.C. (2001) Process controls on the development of stratigraphic trap potential on the margins of confined turbidite systems, and aids to reservoir evaluation. *Am. Assoc. Petrol. Geol. Bull.*, **85**, 971–988.

McCaffrey, W.D., Gupta, S. and Brunt, R. (2002) Repeated cycles of submarine channel incision, infill and transition to sheet sandstone development in the Alpine Foreland Basin, SE France. *Sedimentology*, **49**, 623–635.

Meredith, D.J. and Egan, S.S. (2002) The geological and geodynamic evolution of the eastern Black Sea basin: insights from 2-D and 3-D tectonic modelling. *Tectonophysics*, **350**, 157–179.

Miall, A.D. (1989) Architectural elements and bounding surfaces in channelized clastic deposits: notes on comparisons between fluvial and turbiditic systems. In: *Sedimentary Facies in the Active Plate Margins* (Eds A. Taira and F. Masuda), pp. 3–15. Terra Scientific, Tokyo.

Michard, A., Whitechurch, H., Ricou, L.E., Montigny, R. and Yazgan, E. (1984) Tauric subduction (Malatya-

Elazığ provinces) and its bearing on tectonics of the Tethyan realm in Turkey. In: *The Geological Evolution of the Eastern Mediterranean* (Eds J.E. Dixon and A.H.F. Robertson), pp. 361–373. Special Publication 17, Geological Society, London.

Migeon, S., Savoye, B., Zanella, E., Mulder, T., Faugères, J.-C. and Weber, O. (2001) Detailed seismic-reflection and sedimentary study of turbidite sediment waves on the Var Sedimentary Ridge (SE France): significance for sediment transport and deposition for the mechanisms of sediment-wave construction. *Mar. Petrol. Geol.*, **18**, 179–208.

Morris, S.A. and Alexander, J. (2003) Changes in flow direction at a point caused by obstacles during passage of a density current. *J. Sediment. Res.*, **73**, 621–629.

Muto, T. and Steel, R.J. (2002) In defense of shelf-edge delta development during falling and lowstand of relative sea level. *J. Geol.*, **110**, 421–436.

Mutti, E. (1992) *Turbidite Sandstones*. Agip, San Donato Milanese, 275 pp.

Mutti, E. and Normark, W.R. (1987) Comparing examples of modern and ancient turbidite systems: problems and concepts. In: *Deep Water Clastic Deposits: Models and Case Histories* (Eds J.K. Leggett and G.G. Zuffa), pp. 1–38. Graham and Trotman, London.

Mutti, E., Seguret, M. and Sgavetti, M. (1988) *Sedimentation and Deformation in the Tertiary Sequences of the Southern Pyrenees*. AAPG Mediterranean Basin Conference, Field Trip 7, University of Parma Special Publication, 153 pp.

Nakajima, T. (1996) *Turbidite sedimentation along the Toyama deep sea channel in the Japan Sea*. Unpublished Doctoral Thesis, Kyoto University, 108 pp.

Nakajima, T., Satoh, M. and Okamura, Y. (1998) Channel-levee complexes, terminal deep-sea fans and sediment wave fields associated with the Toyama Deep-Sea Channel system in the Japan Sea. *Mar. Geol.*, **147**, 25–41.

Nemec, W. (1990) Aspects of sediment movement on steep delta slopes. In: *Coarse-Grained Deltas* (Eds A. Colella and D.B. Prior), pp. 29–73. Special Publication 10, International Association of Sedimentologists. Blackwell Scientific Publications, Oxford.

Nemec, W. (1995) The dynamics of deltaic suspension plumes. In: *Geology of Deltas* (Eds M.N. Oti and G. Postma), pp. 31–93. A.A. Balkema, Rotterdam.

Nemec, W., Lønne, I. and Blikra, L.H. (1999) The Kregnes moraine in Gauldalen, west-central Norway: anatomy of a Younger Dryas proglacial delta in a palaeofjord basin. *Boreas*, **28**, 454–476.

Nikishin, A.M., Korotaev, M.V., Ershov, A.V. and Brunet, M.-F. (2003) The Black Sea basin: tectonic history and Neogene-Quaternary rapid subsidence modelling. *Sediment. Geol.*, **156**, 149–168.

Normark, W.R., Hess, G., Stow, D.A.V. and Bowen, A.J. (1980) Sediment waves on the Monterey Fan levee: a preliminary physical interpretation. *Mar. Geol.*, **37**, 1–18.

Normark, W.R., Posamentier, H. and Mutti, E. (1993) Turbidite systems: state of the art and future directions. *Rev. Geophys.*, **31**, 91–116.

Okay, A.I. (1989) Tectonic units and sutures in the Pontides, northern Turkey. In: *Tectonic Evolution of the Tethyan Region* (Ed. A.M.C. Şengör), pp. 109–115. Kluwer Academic Publishers, Dordrecht.

Okay, A.I. and Şahintürk, Ö. (1997) Geology of the Eastern Pontides. In: *Regional and Petroleum Geology of the Black Sea and Surrounding Region* (Ed. A.G. Robinson), pp. 291–311. Memoir 68, American Association of Petroleum Geologists, Tulsa, OK.

Okay, A.I. and Tüysüz, O. (1999) Tethyan sutures of northern Turkey. In: *The Mediterranean Basins: Tertiary Extension within the Alpine Orogen* (Eds B. Durand, L. Jolivet, F. Horváth and M. Séranne), pp. 475–515. Special Publication 156, Geological Society Publishing House, Bath.

Okay, A.I., Şengör, A.M.I. and Görür, N. (1994) Kinematic history of the opening of the Black Sea and its effect on the surrounding regions. *Geology*, **22**, 267–270.

Okay, A.I., Tansel, İ. and Tüysüz, O. (2001) Obduction, subduction and collision as reflected in the Upper Cretaceous-Lower Eocene sedimentary record of western Turkey. *Geol. Mag.*, **138**, 117–142.

Ori, G.G. and Friend, P.G. (1984) Sedimentary basins, formed and carried piggyback on active thrust sheets. *Geology*, **12**, 475–478.

Parsons, J.D., Bush, J.W.M. and Syvitski, J.P.M. (2001) Hyperpycnal plume formation from riverine outflows with small sediment concentrations. *Sedimentology*, **48**, 465–478.

Peakall, J., McCaffrey, B. and Kneller, B. (2000a) A process model for the evolution, morphology, and architecture of sinuous submarine channels. *J. Sediment. Res.*, **70**, 434–448.

Peakall, J., McCaffrey, B., Kneller, B., Stelting, C.E., McHargue, T.R. and Schweller, W.J. (2000b) A process model for the evolution of submarine fan channels: implications for sedimentary architecture. In: *Fine-Grained Turbidite Systems* (Eds A.H. Bouma and C.G. Stone), 73–88. Memoir 72, American Association of Petroleum Geologists, Tulsa, OK.

Peccerillo, A. and Taylor, S.R. (1975) Geochemistry of Upper Cretaceous volcanic rocks from the Pontide chain, northern Turkey. *Bull. Volcanol.*, **39**, 1–13.

Pickering, K.T., Hiscott, R.N. and Hein, F.J. (1989) *Deep-marine Environments: Clastic Sedimentation and Tectonics*. Unwin Hyman, London, 41 pp.

Pirmez, C. and Flood, R.D. (1995) Morphology and structure of Amazon channel. *Proc. ODP Initial Rep.*, **155**, pp. 23–45.

Pirmez, C., Beaubouef, R.T., Friedmann, S.J. and Mohrig, D.C. (2000) Equilibrium profile and base-level in submarine channels: examples from late Pleistocene systems and implications for the architecture of deep-water reservoirs. In: *Deep-water Reservoirs of the World* (Ed. P. Weimer), Proceedings of the 20th Annual Bob Perkins Research Conference, pp. 782–804. Gulf Coast Section, Society of Economic Paleontologists and Mineralogists, Houston, TX.

Porębski, S.J. and Steel, R.J. (2003) Shelf-margin deltas: their stratigraphic significance and relation to deepwater sands. *Earth-Sci. Rev.*, **62**, 283–326.

Posamentier, H.W. and Kolla, V. (2003) Seismic geomorphology and stratigraphy of depositional elements in deep-water settings. *J. Sediment. Res.*, **73**, 367–388.

Postma, G., Hilden, F.J. and Zachariasse, W.J. (1993) Procession-punctuated growth of a late Miocene submarine-fan lobe on Gavdos (Greece). *Terra Nova*, **5**, 438–444.

Prior, D.B., Bornhold, B.D. and Johns, M.W. (1986) Active sand transport along a fjord-bottom channel, Bute Inlet, British Columbia. *Geology*, **14**, 581–584.

Puigdefàbregas, C., Muñoz, J.A. and Vergés, J. (1992) Thrusting and foreland basin evolution in the Southern Pyrenees. In: *Thrust Tectonics* (Ed. K.R. McClay), pp. 247–254. Chapman & Hall, London.

Pujalte, V., Baceta, J.I., Orue-Etxebarria, X. and Payros, A. (1998) Paleocene strata of the Basque Country, Western Pyrenees, Northern Spain: facies and sequence development in a deep-water starved basin. In: *Mesozoic and Cenozoic Sequence Stratigraphy of European Basins* (Eds P.-C. de Graciansky, J. Hardenbol, T. Jacquin and P.R. Vail), pp. 311–325. Special Publication 60, Society of Economic Paleontologists and Mineralogists, Tulsa, OK.

Rangin, C., Bader, A.G., Pascal, G., Ecevitoğlu, B. and Görür, N. (2002) Deep structure of the Mid-Black Sea High (offshore Turkey) imaged by multi-channel seismic survey (BLACKSIS cruise). *Mar. Geol.*, **182**, 265–278.

Reading, H.G. and Richards, M. (1994) Turbidite systems in deep-water basin margins classified by grain size and feeder system. *Am. Assoc. Petrol. Geol. Bull.*, **78**, 792–822.

Robinson, A.G., Banks, C.J., Rutherford, M.M. and Hirst, J.P.P. (1995) Stratigraphic and structural development of the Eastern Pontides, Turkey. *J. Geol. Soc. London*, **152**, 861–872.

Robinson, A.G., Rudat, J.H., Banks, C.J. and Wiles, R.L.F. (1996) Petroleum geology of the Black Sea. *Mar. Petrol. Geol.*, **13**, 195–223.

Ross, W.C., Halliwell, B.A., May, J.A., Watts, D.E. and Syvitski, G.P.M. (1994) Slope readjustment: a model for the development of submarine fans and aprons. *Geology*, **22**, 511–514.

Saito, T. and Ito, M. (2002) Deposition of sheet-like turbidite packets and migration of channel-overbank systems on a sandy submarine fan: an example from the Late Miocene–Early Pliocene forearc basin, Boso Peninsula, Japan. *Sediment. Geol.*, **149**, 265–277.

Sanver, M. and Ponat, E. (1981) Kırşehir ve dolaylarına ilişkin paleomagnetik bulgular: Kırşehir Masifinin rotasyonu. *Istanbul Yerbil.*, **2**, 2–8.

Satur, N., Hurst, A., Cronin, B.T., Kelling, G. and Gürbüz, K. (2000) Sand body geometry in a sand-rich, deep-water clastic system, Miocene Cingöz Formation of southern Turkey. *Mar. Petrol. Geol.*, **17**, 239–252.

Şengör, A.M.C. (1984) *The Cimmeride Orogenic System and the Tectonics of Eurasia*. Special Paper 195, Geological Society of America, Boulder, CO, 82 pp.

Şengör, A.M.C. (1987) Tectonics of the Tethysides: orogenic collage development in a collisional setting. *Ann. Rev. Earth Planet. Sci.*, **15**, 213–244.

Şengör, A.M.C., Görür, N. and Saroğlu, F. (1985) Strike-slip faulting and related basin formation in zones of tectonic escape: Turkey as a case study. In: *Strike-Slip Deformation, Basin Formation, and Sedimentation* (Eds K.D. Biddle and N. Christie-Blick), pp. 227–264. Special Publication 17, Society of Economic Paleontologists and Mineralogists, Tulsa, OK.

Şengün, M., Keskin, H., Akçören, F., *et al.* (1990) Geology of the Kastamonu region and geological constraints for the evolution of the Paleotethyan domain. *Geol. Bull. Turk.*, **33**, 1–16.

Shanmugam, G. (1997) Basin-floor fans in the North Sea: sequence stratigraphic models vs. sedimentary facies: Reply. *Am. Assoc. Petrol. Geol. Bull.*, **81**, 666–672.

Shanmugam, G. (2000) 50 years of the turbidite paradigm (1950s–1990s): deep-water processes and facies models – a critical perspective. *Mar. Petrol. Geol.*, **17**, 285–342.

Shanmugam, G. and Moiola, R.J. (1985) Submarine fan models: problems and solutions. In: *Submarine Fans and Related Turbidite Systems* (Eds A.H. Bouma, W.R. Normark and N.E. Barnes), pp. 29–34. Springer-Verlag, New York.

Shanmugam, G. and Moiola, R.J. (1988) Submarine fans: characteristics, models, classification, and reservoir potential. *Earth-Sci. Rev.*, **24**, 383–428.

Shanmugam, G., Moiola, R.J. and Damuth, J.E. (1985) Eustatic control of submarine fan development. In: *Submarine Fans and Related Turbidite Systems* (Eds A.H. Bouma, W.R. Normark and N.E. Barnes), pp. 23–28. Springer-Verlag, New York.

Shanmugam, G., Lehtonen, L.R., Straume, T., Syvertsen, S.E., Hodgkinson, R.J. and Skibeli, M. (1994) Slump and debris-flow dominated upper slope facies in the Cretaceous of the Norwegian and northern North Sea (61–67°N): implications for sand distribution. *Am. Assoc. Petrol. Geol. Bull.*, **78**, 910–937.

Shanmugam, G., Bloch, R.B., Mitchell, S.M., Beamish, W.J., Hodgkinson, R.J., Damuth, J.E., Straume, T., Syvertsen, S.E. and Shields, K.E. (1995) Basin-floor fans in the North Sea: sequence stratigraphic models vs. sedimentary facies. *Am. Assoc. Petrol. Geol. Bull.*, **79**, 477–512.

Sinclair, H.D. (1992) Turbidite sedimentation during Alpine thrusting: the Taveyannaz sandstones of eastern Switzerland. *Sedimentology*, **39**, 837–856.

Sinclair, H.D. (1997) Tectono-stratigraphic model for underfilled peripheral foreland basin: an Alpine perspective. *Geol. Soc. Am. Bull.*, **109**, 324–346.

Sinclair, H.D. (2000) Delta-fed turbidites infilling topographically complex basins: a new depositional model for the Annot Sandstones, SE France. *J. Sediment. Res.*, **70**, 504–519.

Sinclair, H.D. and Cowie, P.A. (2003) Basin-floor topography and the scaling of turbidites. *J. Geol.*, **111**, 277–299.

Sinclair, H.D. and Tomasso, M. (2002) Depositional evolution of confined turbidite basins. *J. Sediment. Res.*, **72**, 451–456.

Sonel, N., Sarı, A., Coşkun, B. and Tozlu, E. (1989) Boyabat (Sinop) havzası Ekinveren Fayının petrol aramalarındaki önemi. *Geol. Bull. Turk.*, **32**, 39–49.

Spaak, P., Almond, J., Salahudin, S., Mohd Salleh, Z. and Tosun, O. (1999) Fulmar: a mature field revisited. In: *Petroleum Geology of Northwest Europe: Proceedings of the 5th Conference* (Eds A.J. Fleet and S.A.R. Boldy), pp. 1089–1100. Geological Society, London.

Stanley, D.J., Palmer, H.D. and Dill, R.F. (1978) Coarse sediment transport by mass flow and turbidity current processes and downslope transformations in Annot Sandstone canyon-fan valley systems. In: *Sedimentation in Submarine Canyons, Fans and Trenches* (Eds D.J. Stanley and G. Kelling), pp. 85–115. Dowden, Hutchinson and Ross, Stroudsburg, PA.

Stow, D.A.V. and Bowen, A.J. (1980) A physical model for the transport and sorting of fine-grained sediments by turbidity currents. *Sedimentology*, **27**, 31–46.

Stow, D.A.V. and Mayall, M. (2000) Deep-water sedimentary systems: new models for the 21st century. *Mar. Petrol. Geol.*, **17**, 125–135.

Stow, D.A.V. and Shanmugam, G. (1980) Sequence of structures in fine-grained turbidites; comparison of recent deep-sea and ancient flysch sediments. *Sediment. Geol.*, **25**, 23–42.

Stow, D.A.V., Howell, D.G. and Nelson, C.H. (1985) Sediment, tectonics, and sea-level changes. In: *Submarine Fans and Related Turbidite Systems* (Eds A.H. Bouma, W.R. Normark and N.E. Barnes), pp. 15–22. Springer-Verlag, New York.

Sunal, G. and Tüysüz, O. (2002) Palaeostress analysis of Tertiary post-collisional structures in the Western Pontides, northern Turkey. *Geol. Mag.*, **139**, 343–359.

Takano, O., Tateishi, M. and Endo, M. (2005) Tectonic controls of a backarc trough-fill turbiditic system: the Pliocene Tamugigawa Formation in the Niigata-Shin'etsu inverted rift basin, Northern Fossa Magna, central Japan. *Sediment. Geol.*, **176**, 247–279.

Talling, P.J. (1998) How and where do incised valleys form if sea level remains above the shelf edge? *Geology*, **26**, 87–90.

Thornburg, T.M., Kulm, L.D. and Hussong, D.M. (1990) Submarine-fan development in the southern Chile Trench: a dynamic interplay of tectonics and sedimentation. *Geol. Soc. Am. Bull.*, **102**, 1658–1680.

Timbrell, G. (1993) Sandstone architecture of the Balder Formation depositional system, UK Quadrant 9 and adjacent areas. In: *Petroleum Geology of Northwest Europe: Proceedings of the 4th International Conference* (Ed. J.R. Parker), pp. 107–121. Geological Society, London.

Tüysüz, O. (1990) Tectonic evolution of a part of the Tethyside orogenic collage: the Kargı Massif, northern Turkey. *Tectonics*, **9**, 141–160.

Tüysüz, O. (1993) Karadeniz'den Orta Anadolu'ya bir jeotravers: kuzey Neo-Tetisin tektonik evrimi. *Türk. Petrol. Jeol. Der. Bül.*, **5**, 1–33.

Tüysüz, O. (1999) Geology of the Cretaceous sedimentary basins of the Western Pontides. *Geol. J.*, **34**, 75–93.

Tüysüz, O., Dellaloğlu, A.A. and Terzioğlu, N. (1995) A magmatic belt within the Neo-Tethyan suture zone and its role in the tectonic evolution of northern Turkey. *Tectonophysics*, **243**, 173–191.

Uchman, A., Janbu, N.E. and Nemec, W. (2004) Trace fossils in the Cretaceous-Eocene flysch of the Sinop–Boyabat Basin, Central Pontides, Turkey. *Ann. Soc. Geol. Pol.*, **74**, 197–235.

Ustaömer, T. and Robertson, A.H.F. (1997) Tectonic-sedimentary evolution of the North Tethyan margin in the Central Pontides of Northern Turkey. In: *Regional and Petroleum Geology of the Black Sea and Surrounding Region* (Ed. A.G. Robinson), pp. 255–290. Memoir 68, American Association of Petroleum Geologists, Tulsa, OK.

Vail, P.R., Audemard, F., Bowman, S.A., Eisner, P.N. and Pirez-Cruz, G. (1991) The stratigraphic signatures of tectonics, eustasy and sedimentology – an overview.

In: *Cycles and Events in Stratigraphy* (Eds G. Einsele, W. Ricken and A. Seilacher), pp. 617–659. Springer-Verlag, Berlin.

Van Vliet, A. (1978) Early Tertiary deep-water fans of Guipuzcoa, northern Spain. In: *Sedimentation in Submarine Canyons, Fans and Trenches* (Eds D.J. Stanley and G. Kelling), pp. 190–209. Dowden, Hutchinson and Ross, Stroudsburg, PA.

Van Vliet, A. (1982) *Submarine Fans and Associated Deposits in the Lower Tertiary of Guipuzcoa (Northern Spain)*. Doctoral Dissertation, Leiden University, 45 pp.

Walker, R.G. (1984) General introduction: facies, facies sequences and facies models. In: *Facies Models*, 2nd edn (Ed. R.G. Walker). *Geosci. Can. Reprint Ser.*, **1**, 1–9.

Walker, R.G. (1985) Mudstones and thin-bedded turbidites associated with the Upper Cretaceous Wheeler Gorge conglomerates, California: a possible channel-levee complex. *J. Sediment. Petrol.*, **55**, 279–290.

Weltje, G. and De Boer, P.L. (1993) Astronomically induced paleoclimatic oscillations reflected in Pliocene turbidite deposits of Corfu (Greece): implications for the interpretation of higher order cyclicity in fossil turbidite systems. *Geology*, **21**, 307–310.

Winkler, W. (1993) Control factors on turbidite sedimentation in a deep-sea trench setting. The example of the Schlieren Flysch (Upper Maastrichtian-Lower Eocene, Central Switzerland). *Geodin. Acta*, **6**, 81–102.

Winkler, W. and Gawenda, P. (1999) Distinguishing climatic and tectonic forcing of turbidite sedimentation, and the bearing on turbidite bed scaling: Palaeocene–Eocene of northern Spain. *J. Geol. Soc. London*, **156**, 791–800.

Winn, R.D., Jr. and Dott, R.H., Jr. (1979) Deep-water fan-channel conglomerates of Late Cretaceous age, southern Chile. *Sedimentology*, **26**, 203–228.

Wright, L.D., Yang, Z.-S., Bornhold, B.D., Keller, G.H., Prior, D.B. and Wiseman, W.J., Jr. (1986) Hyperpycnal plumes and plume fronts over the Huanghe (Yellow River) delta front. *Geo-Mar. Lett.*, **6**, 97–105.

Wynn, R.B., Weaver, P.P.E., Masson, D.G. and Stow, D.A.V. (2002) Turbidite depositional architecture across three interconnected deep-water basins on the north-west African margin. *Sedimentology*, **49**, 669–695.

Yazgan, E. (1984) Geodynamic evolution of the Eastern Taurus region. In: *International Symposium on the Geology of the Taurus Belt* (Eds O. Tekeli and M.C. Göncüoğlu), pp. 199–208. Institute of Mineral Research and Exploration (MTA), Ankara.

Yılmaz, A., Adamia, S., Chabukiani, A., Chkhotua, T., Erdoğan, K., Tuzcu, S. and Karabıyıkoğlu, M. (2000) Structural correlation of the southern Transcaucasus (Georgia)-eastern Pontides (Turkey). In: *Tectonics and Magmatism in Turkey and the Surrounding Area* (Eds E. Bozkurt, J.A. Winchester and J.D.A. Piper), pp. 171–182. Special Publication 173, Geological Society Publishing House, Bath.

Yılmaz, Y. (1993) New evidence and model on the evolution of the southeast Anatolian orogen. *Geol. Soc. Am. Bull.*, **105**, 251–271.

Yılmaz, Y., Yiğitbaş, E. and Genç, Ş.C. (1993) Ophiolitic and metamorphic assemblages of southeast Anatolia and their significance in the geological evolution of the orogenic belt. *Tectonics*, **12**, 1280–1297.

Yılmaz, Y., Tüysüz, O., Yiğitbaş, E., Genç, Ş.C. and Şengör, A.M.C. (1997) Geology and tectonic evolution of the Pontides. In: *Regional and Petroleum Geology of the Black Sea and Surrounding Region* (Ed. A.G. Robinson), pp. 183–226. Memoir 68, American Association of Petroleum Geologists, Tulsa, OK.

Ziegler, P.A. and Roure, F. (1999) Petroleum systems of Alpine-Mediterranean foldbelts and basins. In: *The Mediterranean Basins: Tertiary Extension within the Alpine Orogen* (Eds B. Durand, L. Jolivet, F. Horváth and M. Séranne), pp. 517–540. Special Publication 156, Geological Society Publishing House, Bath.

River morphologies and palaeodrainages of western Africa (Sahara and Sahel) during humid climatic conditions

GIAN GABRIELE ORI[1], GAETANO DIACHILLE, GORO KOMATSU, LUCIA MARINANGELI *and* ANGELO PIO ROSSI[2]

International Research School of Planetary Sciences, Universita' d'Annunzio, Viale Pindaro 42, 65127 Pescara, Italy (Email: ggori@irsps.unich.it)

ABSTRACT

River systems in desert regions are basically of two types: ephemeral streams and exotic (or allogenic) rivers. In contrast to ephemeral streams, exotic rivers are perennial and survive hydrological crisis. Unique features of exotic rivers are their inland deltas, where they form intricate patterns of small channels and lose a large part of their water. Extensive exotic rivers flowed in the Sahara during wet climatic periods, and they have left a large number of dry streams and palaeovalleys. Most of these ancient courses are at the present time ephemeral streams or they are now totally inactive. During wet climatic periods the Niger River, for example, was split into two unconnected reaches: the upper reach flowed to the north, forming a large inland delta in the area of Azouad (north of Timbuktu); the lower reach flowed to the south, from the Adrar des Iforhas and Air. The upper part of this southward flowing system is now dry, and corresponds to the palaeovalley of Azaouak and adjacent palaeovalleys. During wetter climatic periods, the Sahara was covered by lakes and swamps in the deepest parts of the inland basins. These lakes were supplied by exotic rivers forming deltas at their mouths. The rivers formed palaeovalleys that are now inactive or are occupied by ephemeral streams with episodic floods, and terminal fans have replaced the original deltaic systems.

Keywords Palaeodrainage, ephemeral streams, exotic rivers, Sahara, western Africa.

INTRODUCTION

The entire Sahara has been affected in the geological past by a number of climatic variations that changed its environment from the fluvial- to aeolian-dominated and vice versa. During high-precipitation climatic periods a large network of river systems criss-crossed the entire area of the Sahara (e.g. Nichol, 1988; Goudie, 2002, 2005). The large-scale pattern of the drainage systems can be identified and the main river courses and basins have been already described (Petit-Maire & Risier, 1981; Petit-Marie, 1991; Gasse, 2001; Goudie, 2002; Griffin, 2002). In addition, a large amount of work has been carried out on lacustrine basins (Fontes & Gasse, 1991). The Sahara and Sahel area of western Africa was split into several basins that displayed different kinds of river systems and depocentre deposits.

Field investigations of the alluvial systems have been inhibited by the difficulties in recognizing the sedimentary and geomorphological features on the ground. The analysis has been improved by the availability of satellite imagery and by high-precision GPS measurements (Smith, 1963; Mainguet & Callot, 1978; Breed & Grow, 1979; Breed *et al.*, 1979; El Baz & Maxwell, 1982). However, until recently major drawbacks were the limited availability of images at about 15 m pixel^{-1} and high-resolution elevation models. Two

[1]Also at: Ibn Battuta Centre, Faculté des Sciences Semlalia, Université Cadi Ayyad, Marrakech, Morocco.
[2]Present address: ESTEC, European Space Agency, Noordwijk, The Netherlands.

recent Earth observation space missions contributed to filling these gaps and allow detailed remote sensing analyses: (i) the ASTER instrument onboard the satellite TERRA; (ii) the SRTM onboard a Space Shuttle mission. The ASTER instrument is a multispectral imager built by Japan's Ministry of International Trade and Industry (MITI) and is sending images with a resolution ranging from 15 to 30 m pixel^{-1}. The SRTM experiment was carried out by NASA, the Agenzia Spaziale Italiana (ASI) and the German space agency DLR. The SRTM instrument provides altimetry data of Earth's surface with a horizontal resolution of 90 m pixel^{-1} and a vertical accuracy of $\pm$ 6 m.

The aim of this paper is to describe the palaeoriver systems of the western Sahara and link them with the landforms preserved today. The approach is descriptive and the analysis deals with fluvial processes without stratigraphical reconstructions. It is intended to reveal basic regional patterns of fluvial landforms in desert areas and not the time-stratigraphic sequences of the fluvial system evolution in the Sahara. For this reason, this paper does not include any data and discussion regarding the climatic history of Sahara and adjoining areas that have been the subject of a host of publications (see below). The data sets used for this analysis are mainly the ASTER images and the SRTM data from the Terra satellite and the Space Shuttle respectively. Other data sets used in this analysis are LANDSAT images (NASA), ERS 1 and 2 radar data (ESA), and declassified CORONA photos (US DoD). Fieldwork has been carried out at several locations.

THE PALAEORIVER SYSTEMS

The river systems in desert areas show a variety of patterns. Their variability and complexity depend on a large number of variables that affect the river systems and their evolution through geological time. This large variability is described in several papers (see reviews by Tooth, 2000; Goudie, 2002) that clearly show the difficulty of comprehensively classifying the river systems in drylands. Nevertheless, it is possible to define a scheme for river systems that is applicable to specific cases. In order to describe the river systems of the drylands of western Africa, the scheme proposed by Ori (1988) will be used. It comprises two different river types: (i) ephemeral streams that undergo episodic flooding and long 'normal' dry periods; (ii) exotic (or allogenic) rivers that are perennial but affected by remarkable fluctuations in water level along their courses (Figs 1 & 2). This classification is not exhaustive of the river system of Sahara, as many details are not taken into account, and it is used only for the study areas described in this paper.

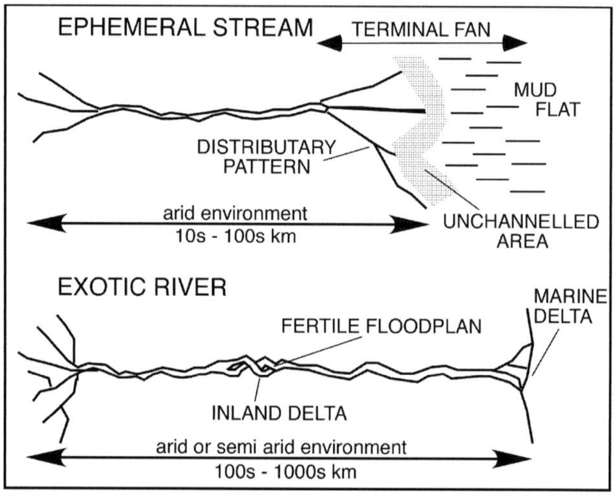

Fig. 1 Outline of the main features of exotic (allogenic) rivers and ephemeral streams occurring in drylands. (After Ori, 1988.)

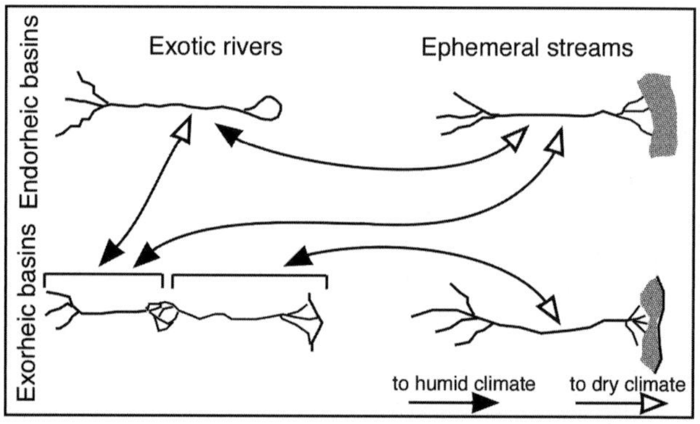

Fig. 2 Generalized and simplified changes in river types according to climatic changes. Exotic rivers tend to become ephemeral streams under drier conditions (and vice versa). Exorheic systems may also become disconnected from the sea and form endorheic basins. Exotic rivers may be split into two or more reaches under drier conditions and form ephemeral streams.

The ephemeral streams are usually short headed with relatively small drainage basins. Their major characteristic is the episodic occurrence of large floods with high peak discharge and upper-flow regime currents. Aeolian sands accumulate in the riverbeds during the dry interflood periods. These rivers occur in hyperarid to arid zones all over the Sahara. Evaporation and infiltration play a major role in their discharge behaviour, causing a massive loss of water. Consequently, the rivers lose capacity to transport and the channels diverge and become shallower and broader. Farther downstream, the channels fade out by spreading their water and sediments over large unchannelized flat areas. These distributary systems were identified for the first time in the arid zones of India (Mukerij, 1976), and their importance in modern and ancient fluvial deposits was recognized by Friend (1978). Actually, even these terminal fans show a large variability in patterns and facies (Tooth, 1999). In the western Sahara two basic types of terminal fans are recognized: (i) those merging into a mud flat and (ii) those merging into a sabkha. The former type of terminal fan is similar to that originally described by Mukerij (1976), and consists of channels that disappear in open areas without feeding into a standing body of water. An example of this type of terminal fan is the one associated with Oued Saoura (western Algeria, Fig. 3; Masini *et al.*, 1988).

In cases where an ephemeral stream debouches onto a sabkha, it is probable that the sabkha will not be dry, but will be inundated, forming a shallow body of water. The terminal fan formed in this environment is more similar to deltaic deposits, with a number of relatively well-defined channels forming a fan-shaped distributary pattern. The channelled area passes into the unchannelized zone, which is constituted by the salt flat of the sabkha, with the formation of tear-shaped bars. A good example of this type of fan occurs in Sabkha Aridal, near Cap Boujoudur along the Atlantic coast (Fig. 4). Of course, terminal fans display a continuum of facies and patterns spanning between these two end members. A clear example of this variability is the terminal fan formed at the termini of Oued el Mellah in the Chott el Rharsa (southern Tunisia, Ori *et al.*, 2001).

The perennial rivers that cross deserts are called exotic (Czaya, 1981), allogenic (Tooth, 2000), or allochthonous (Deodhar & Kale, 1999) (Fig. 1). These types of river flow from rainy areas across arid lands, maintaining an active watercourse throughout the year and along the entire fluvial reach. Typical North African examples are the Nile, the Niger and the Senegal rivers. Their

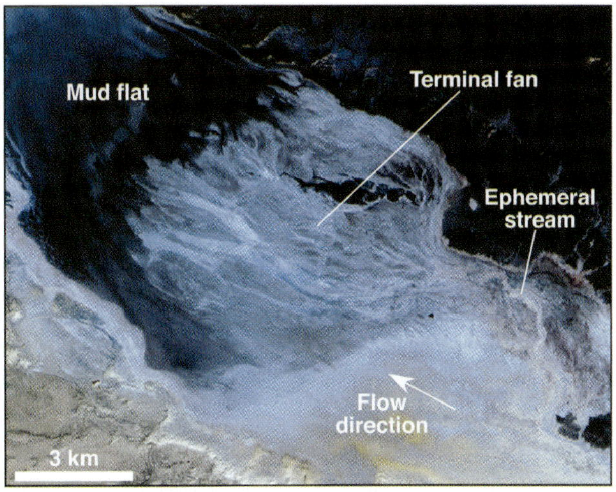

Fig. 3 An ASTER image (false colour) of the terminal fan of Sabkha el Mellah formed by the northward flowing branch of Oued Saoura (see location in Fig. 5). This terminal fan is debouching onto a large mud flat. The flat area is about 5 km wide and the sheet floods expand without forming a shallow water body near the fan.

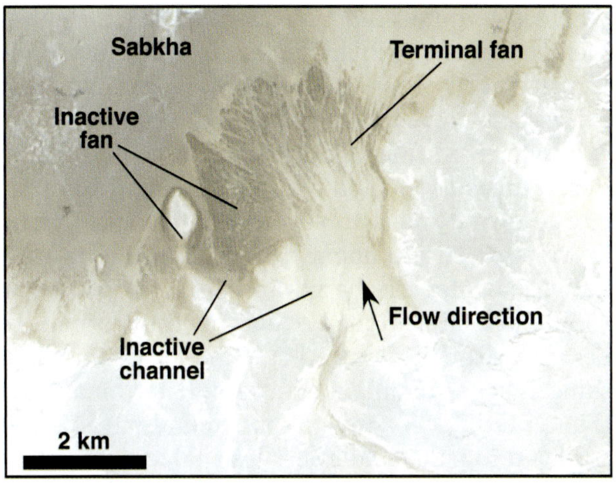

Fig. 4 An ASTER image (false colour) of the terminal fan debouching in Sabkha Aridal near Cap Boujoudur (see location in Fig. 5). This terminal fan is flowing directly in a shallow water body of the sabkha, forming tear-shaped bars.

common characteristic is the loss of water along their course without, however, vanishing entirely. In several instances, the greatest hydrological crisis, with a massive loss of water, occurs in the inland deltas; these are areas where the fluvial channel diverges and forms an intricate pattern of channels, swamps, marshes and lakes. In Africa there are several inland deltas (McCarthy, 1993), such as the one occurring in western Africa formed by the Niger River south of Timbuktu. However, several others occur in the continent, including the Sudd along the Nile River (Howell *et al.*, 1988) and the Okavango delta (McCarthy *et al.*, 1991) in Botswana. The inland deltas share some features with the terminal fans because both tend to form distributary systems in areas where the rivers tend to lose their singularity. Nevertheless, the differences are plentiful: (i) inland deltas show large numbers of well-developed channels, whereas terminal fans are commonly dominated by unchannelized sheet floods; (ii) inland deltas are basically wetland, whereas terminal fans are desert features; (iii) inland deltas are associated with exotic rivers, whereas terminal fans are associated with ephemeral streams; (iv) inland deltas are associated with lacustrine environments, whereas terminal fans are associated with mud flats and sabkhas.

Exotic rivers may occur in arid and hyperarid desert areas as well as in wetter zones. The capacity of exotic rivers to survive hydrological crisis in the arid environment depends on the balance between water discharge and the loss of water. Climatic changes may cause a transformation between exotic rivers and ephemeral streams. The same is true for inland deltas and terminal fans (Fig. 2; Ori, 1988). Several ephemeral streams are confined to palaeovalleys. These palaeovalleys are much larger than the ephemeral channels, which are active just a few times each decade. As a consequence, the valleys are probably the products of much higher discharge permanent rivers. This inference is also supported by the meandering nature of several valley reaches and the complex river terraces (Czaya, 1981, Masini *et al.*, 1988). An exotic river undergoing transformation into an ephemeral stream may split into several reaches, each of them containing independent ephemeral streams and associated terminal fans (see below and Fig. 2).

The desert environment during pluvial periods and the related deposits are a much more complicated issue and are beyond the scope of this paper. Most palaeoclimatic data are based on the analysis of lacustrine deposits spread all over the Sahara. These deposits have to be supplied by rivers with large discharges, suggesting that the ages of the lacustrine deposits correspond to the ages of the large-scale fluvial landforms. The large number of papers published (e.g. Fontes *et al.*, 1985; Petit-Maire, 1986; COHMAP, 1988; Fabre & Petit-Maire, 1988; Gasse, 2000) on the palaeoclimatic record of Sahara concur to define the last climatic optimum as occurring between 10.5 and 5.7 ka, with regional differences on the order of 2–3 kyr (Gasse, 2001). However, the age of the fluvial landforms pre-dates the Holocene hydrological optimum. Goudie (2005) identified several river systems in Sahara that have been active since the Cretaceous over the entire African continent. It is very probable that parts of the river systems that are described in this paper have been active through the Cenozoic, and some of the basins were already established in the Messinian (Griffin, 2002).

A large variety of terminal systems in endorheic basins can be recognized if other large drylands, such as the Australian and Central Asia deserts, are considered. Several exotic rivers and ephemeral streams characterize these zones, but they display a number of different fluvial and lacustrine patterns. In the warm and cold desert of central Asia (Ambolt & Norin, 1982) the most representative fluvial systems with inland deltas are the Tarim River, with its terminal lake Lop Nur, and the Amu Darya and Syr Darya rivers, which are connected to the Aral Sea. However, smaller systems, such as the River Tsakya Tsangpo and the associated Lake Serling on the Tibetan Plateau, show different deltaic patterns. The same is true for Lake Eyre, the largest endorheic basin of Australia (Tooth, 1999, 2000). Distributary channel systems are complicated geological features in terms of facies, channel patterns and system hierarchy (Nichols & Fisher, 2007), and include a large number of different channel patterns, floodplain deposits and terminal bodies.

Alluvial basins

The western Sahara is dominated, like the rest of Africa, by uplifts and sags. The uplifts act as source areas for the drainage basins and the sags

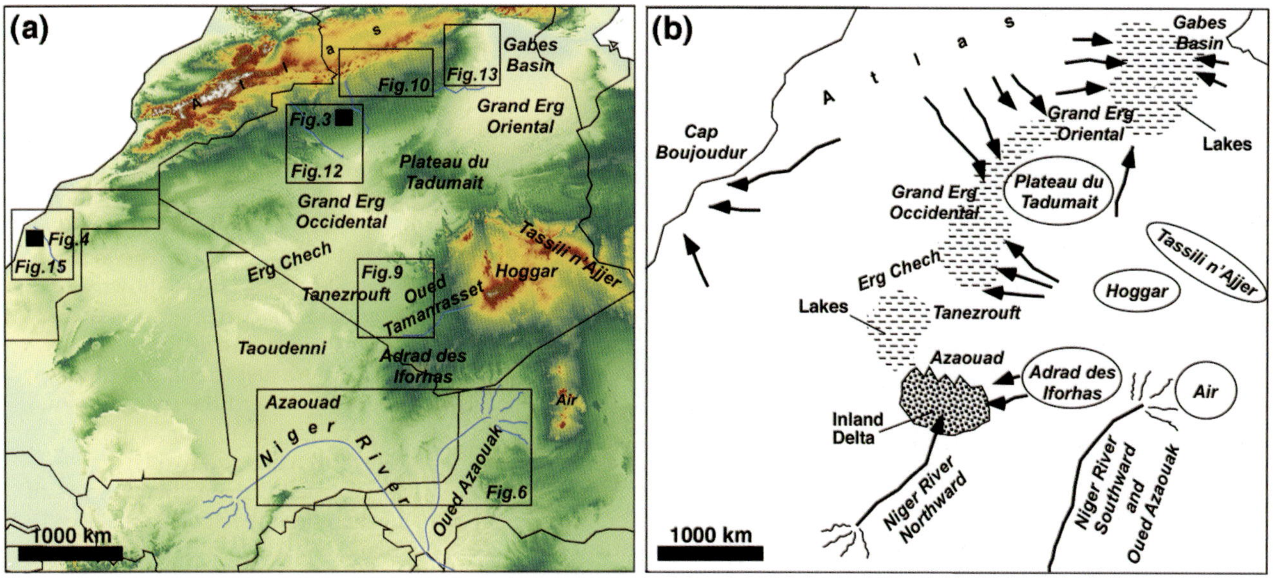

Fig. 5 The study area. (a) Elevation model of western Africa based on SRTM data. The major streams and localities cited in the text are shown as well as the location of several figures. (b) Sketch of the major fluvial features activated during wetter climatic periods. The exotic rivers are marked by lines and their terminal depositional systems by arrows.

as interior endorheic alluvial basins. The digital elevation model (DEM) obtained by the SRTM data shows three main inland drainage basins (Fig. 5): the Azaouad basin, directly north of the Niger River, the Erg Chech basin and the Gabes basin. The Azouad basin is the furthest south and is bordered by the Sahel at its southern limit, which corresponds to the course of the Niger River. The deepest part of the basin is at present an area of aeolian deposits and deflation surfaces. To the north, and separated from the southern Azaouad basin by a low relief sill, there is the Erg Chech basin, which comprises an erg in the area with low elevation and the Tanezrouft plateau to the east. This basin extends northward into the Grand Erg Occidental. The Gabes basin is well defined in the elevation model as it is surrounded by topographic relief. It was called the Gabes basin by Griffin (2002), but it extends considerably out of the Gabes area into the Grand Erg Oriental to the south.

Another kind of drainage pattern to be considered is the one that flows into the Atlantic Ocean (Fig. 5). This open system is formed by a number of rivers flowing into the ocean and originates in the interiors of Mauritania and Morocco. This system is in someway different from the former ones because it is an exorheic (open to the sea) basin and its deepest area is not covered by ergs or thick aeolian deposits.

The inland basins identified show common characteristics. They are surrounded by highlands that in some cases can be only a few tens of metres higher than the lowlands in the central part of the basins. The highlands are dissected by palaeovalleys that debouch in the central flat part of the basin, which is covered by sand seas (Breed *et al.*, 1979).

The Niger system and the Azaouad

The present-day Niger River is a remarkable fluvial system with a number of unique features (Fig. 5). The river flows to the north into a large inland delta (Makaske, 1998) where the river loses up to 70% of its water (NEDECO, 1959). Downstream, the river bends to the east and then flows to the south up to the Gulf of Guinea forming its large delta. There are, however, a number of lines of evidence indicating that the upper northern reach formerly flowed to the north forming a large inland delta in the Azaouad area, north of Timbuctu (Urvoy, 1942; Czaya, 1981; Petit-Maire, 2002; Goudie, 2005) (Figs 6 & 7). This inland delta was large enough to reach the area of lacustrine deposits farther north in the Taoudenni area (see below). The remnants of the northern course of the Niger

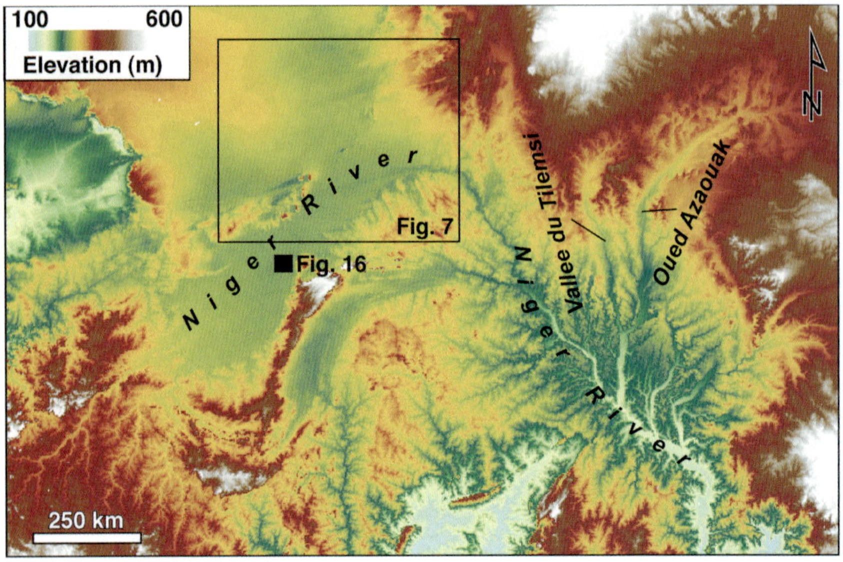

Fig. 6 Elevation model of the Niger delta and the Azaouak basin. The locations of Figs 7 & 16 are shown. Figure 7 is located where the palaeo-inland delta extended. The location of Lake Debo (Fig. 16) is shown. Compare the current Azaouak and Tilemsi river course, when the streams are just tributaries of the Niger River, and the river pattern during humid climatic conditions (Fig. 5b), when the two rivers represented the proximal reaches of the southern reach of the palaeo-Niger. See location in Fig. 5.

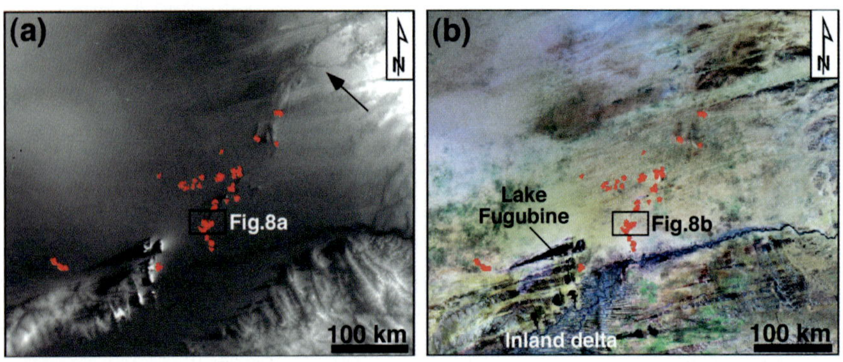

Fig. 7 (a) Elevation model and (b) LANDSAT image (false colour) of the present inland delta and of the palaeo-inland delta. The palaeochannels of the latter are shown in red and indicate that the inland delta extended farther north during humid climatic conditions. See location in Fig. 6.

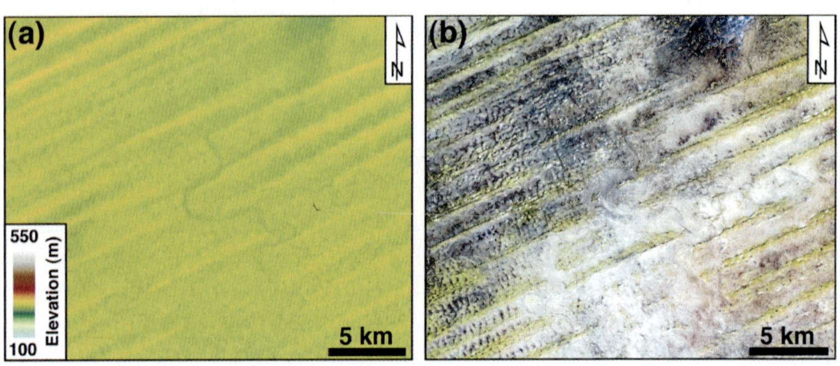

Fig. 8 A close-up of one of the channels in the palaeo-inland delta as observed in (a) the elevation model and (b) the LANDSAT image (false colour). The palaeochannels wandered between dune fields forming isolated meanders or meander belts. See location in Fig. 7.

River are still observable as meandering bodies and belts surrounded by linear dunes in the Azaouad basin (Fig. 8), which was also supplied by water from the Adrar des Iforhas to the east (Fig. 5). Urvoy (1942) was the first to recognize the possibility that the Niger River used to flow northward during pluvial periods. However, he suggested that the river was flowing directly into a lacustrine basin, whereas the recent satellite imagery clearly shows the presence of a past inland delta (Fig. 7; Petit-Maire, 2002).

A large ephemeral stream, called Oued Tamanrasset and extending today from the Hoggar up to the Tanezrouft (Fig. 5), has been proposed as a supplier of water to the Azaouad basin during wetter climatic phases (Chorowicz & Fabre, 1977,

Czaya, 1981). However, the analysis of the altimetry from SRTM does not support this hypothesis. The lower (southward) reach of the river was probably disconnected from the upper Niger River reach and most of its water was supplied from the Hoggar and the Air by the Azaouak exotic river. Currently, the Azaouak is just an epehemeral stream (Oued Azaouak), but its valley and associated drainage pattern clearly show that they were shaped by much larger discharges than the present ones.

It is impossible to define, with the current data, the relationships between the upper and lower reach of the Niger River and it is not known when and how the two reaches were connected (Jacobberger, 1981; McIntosh, 1983). As Urvoy (1942) pointed out, the change in the Niger River due to different climatic conditions is marked simply by a modification of the river course, and not by a change in river pattern from exotic river to ephemeral stream. This is due to the position of the Niger River at the border of the Sahara. The increase in aridity produced by the climatic change was not so dramatic as to transform the Niger River into a smaller desert stream, but simply changed its course and increased the rate of loss of water into the inland delta. Actually, the current extension of the inland delta is probably less than half the inland delta during the hydrological optimum.

The Erg Chech and the Tanezrouft

North of the Azaouad there is another depression, which has the Erg Chech at its centre (Fig. 5). This depression, along with its northern extension into the area of the Grand Erg Occidental, is called here the Erg Chech basin. In its southern portion a large number of Holocene lacustrine deposits have been identified (Petit-Maire, 1991). This basin was supplied by a large river system located in the Tanezrouft (Chorowicz & Fabre, 1997). The origin of this fluvial system was in the Hoggar and the rivers cross a flat plateau gently dipping toward the northwest (Fig. 9). On the surface the channels are unrecognizable because the channels are only 2–5 m deep and are 2–5 km wide, with broad banks and flat bottoms. The stream slope is near horizontal with an elevation change of a few metres from the

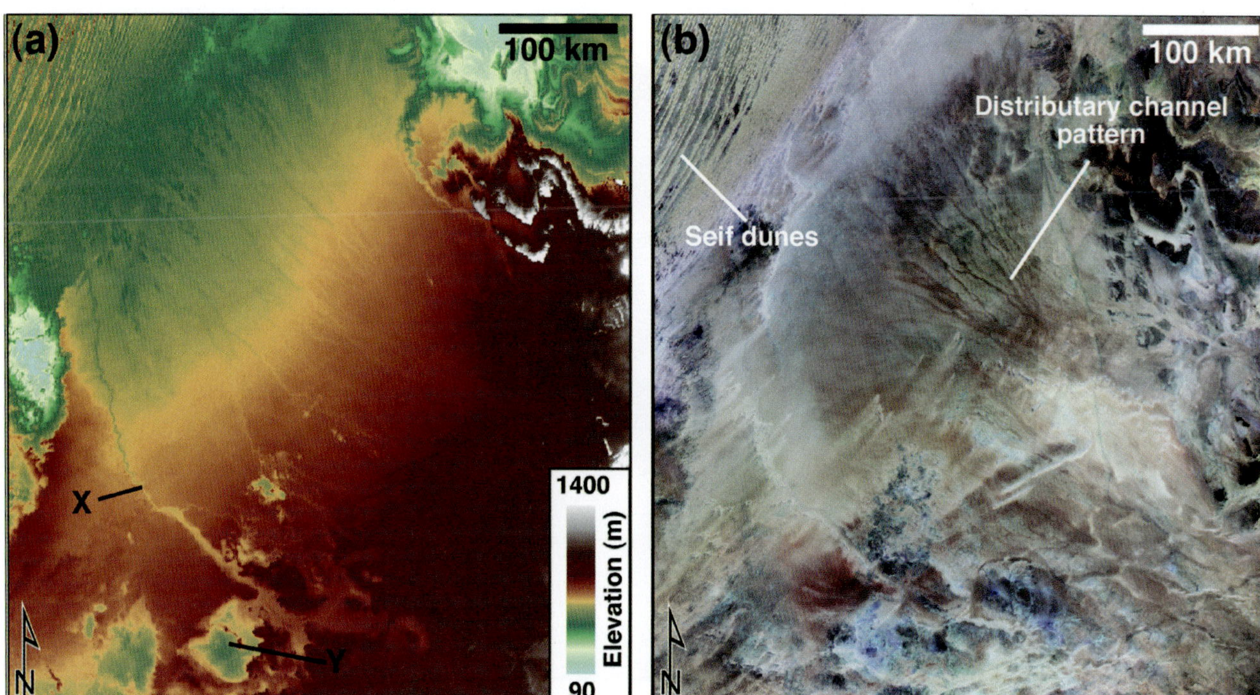

Fig. 9 The distributary system of the Tanezrouft with straight to slightly braided streams. At the base of the slope the channels diverge and form fan-shaped features. The well-defined channel (X) that originated as a sapping depression (Y) occurs in the southwest of the image. (a) Elevation model and (b) LANDSAT image (false colours). See location in Fig. 5.

proximal part of the plain to the basin. The channel pattern is variable, from braided to low-sinuosity single channels, and forms a large distributary pattern partially resembling a large megacone like the Kosi River in India (Wells & Dorr, 1987; Fig. 9). The term 'distributary pattern' is used in this paper as a descriptive term identifying just the geometry, in plan view, of a divergence of channels. At present the system seems to be entirely inactive because there is no evidence from the sediments of even episodic flash floods, nor any record in the oral communication and tradition of local nomadic populations. The stream channels fade out in a distal, almost flat area and are covered by sands of the Erg Chech. The channels form at their mouths small distributary systems that resemble terminal fans. These features are observable in the elevation models but are not well defined in ASTER and LANDSAT images (Fig. 9). No ground observation has been carried out at these specific sites.

South of this system, a palaeovalley 1–20 m deep and about 5 km wide is present. The valley borders the southern margin of the Tanezrouft and originates in a set of depressions to the south (Fig. 9). The depressions seem to be a product of sapping, that is, the slow seepage of underground water emerging at a free face or slope (Nash, 1997). The channel is distinctively different from those of the adjacent distributary system because it is larger and deeper.

The Erg Chech basin has been an area of lacustrine sedimentation in its southern part (in the Taoudenni area) but there is no indication of lacustrine facies below the aeolian erg. However, scattered lacustrine deposits occur in the broad area (Petit-Maire & Kropelin, 1991). Farther north, the Erg Chech basin occupies an area of the Grand Erg Occidental (Fig. 5). This part of the basin is bordered to the north by the Atlasic Hamada, which, in turn, borders the Atlas mountains. The Hamada is dissected by a number of palaeovalleys flowing south (Fig. 10). These palaeovalleys are about 5 m deep, their width ranges from 15 to 5 km and they have an extremely gentle slope. The valleys are straight, but internally they show medium to high sinuosity rivers that form, in places, point bars and other features typical of meandering streams. Some of these streams reach the deepest part of the basin: Oued Saoura, which is next to the western margin of the Hamada, and

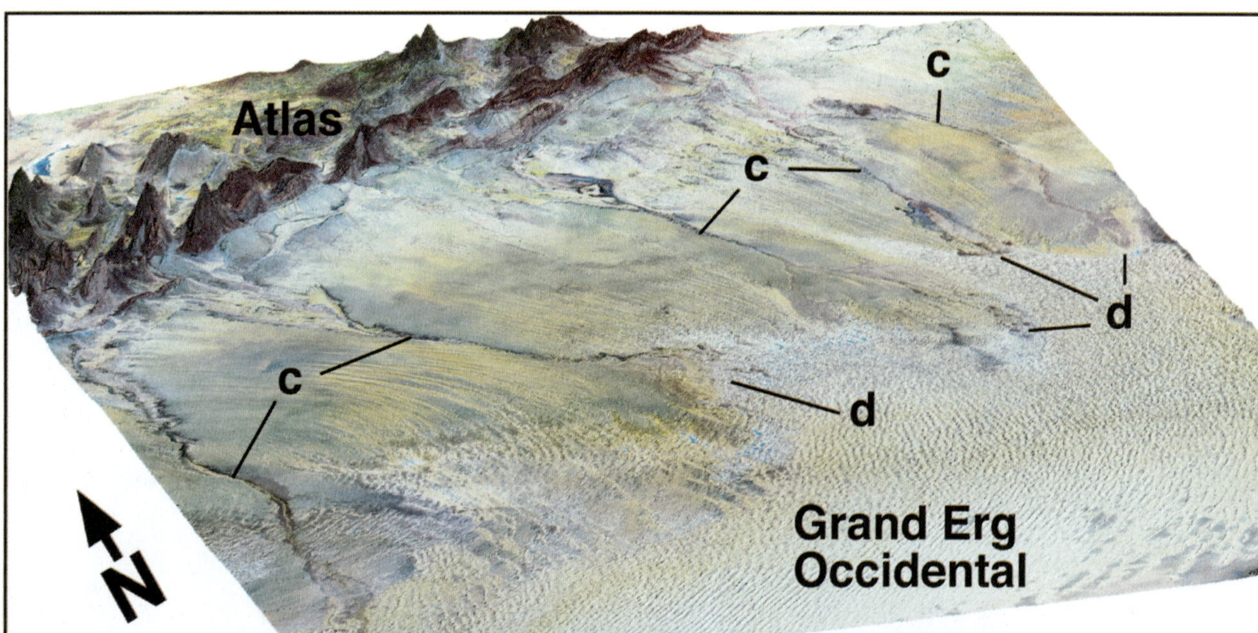

Fig. 10 Oblique view of the Atlasic Hamada with the palaeovalleys (c) flowing from the Atlas Mountains to the Grand Erg Occidental. Oued Namous is to the left. The palaeochannels are clearly visible and their distributary channel systems have their termini at the border of the erg. (d) LANDSAT images (false colour) draped on a SRTM elevation model. See location in Fig. 5.

Fig. 11 An ASTER image (false colour) showing a plan view of distributary patterns of three palaeochannels. These are the palaeochannels at the right-hand side of Fig. 10.

Oued Namous, which is the westernmost palaeovalley of the Hamada itself (Fig. 5). Other palaeovalleys debouch in the area of the Grand Erg Occidental, in a wing of the basin that has an elevation of 200–300 m above the depocentre. These palaeovalleys display distributary patterns at their mouths (Fig. 11) similar to the ones present at the mouths of the channels in the Tanezrouft, but they are larger and have better defined channels.

At present, the palaeovalleys contain ephemeral streams that, due to their energy and the flash flood nature of their activity, cannot be responsible for the erosion of large valleys and the formation of meandering river courses. The valleys were probably active during the wetter climatic periods and contained exotic rivers flowing from the Atlas Mountains. It is possible that Oued Saoura and Oued Namous flowed directly to lakes in areas corresponding to the current Erg Chech, forming lacustrine deltas (Fig. 12). Alternatively, the northern valleys could have been flowing into inland deltas in swampy areas that now lie in the Grand Erg Oriental. Even if no sediment confirming the presence of palustrine deposits below the aeolian cover has been observed in the field, the presence of delta-like features at the mouths of the valleys supports the palaeopalustrine interpretation. From these features it is clear that the channels were diverging and pouring water into a large and unchannelled area. The presence of lakes is unlikely, due to the fact that this is at a higher altitude than the base level represented by the lakes farther south.

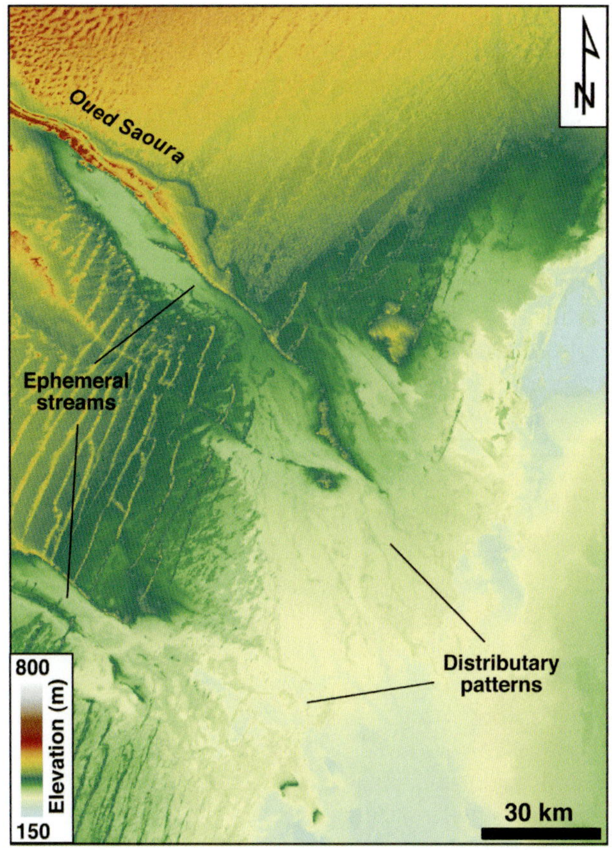

Fig. 12 Elevation model of the distributary channel system at the termini of Oued Saoura (see location in Fig. 5). The channels (in the distributary fans) are slightly sinuous and are a few metres higher than the interdistributary plains due to the embedded levee complexes.

The Gabes basin

This basin extends from the Chott el Jerid in southern Tunisia to the entire Grand Erg Oriental to the south (Fig. 5). The basin is surrounded by higher relief areas that are crossed by a number of palaeovalleys (Ori *et al.*, 2001). The most remarkable examples of these palaeovalleys are the ones flowing from the west in the area of M'zab (Figs 13 & 14). This plateau is gently dipping toward the basin at about 0.0035 m km^{-1}. The palaeovalleys on this side of the basin display a number of features that cannot be observed in the palaeovalleys of the other margins. They resemble the palaeovalleys of the Atlasic Hamada, but the sinuosity of the internal channels is higher. The valleys are about 40 m deep and 5 km wide. Delta-like

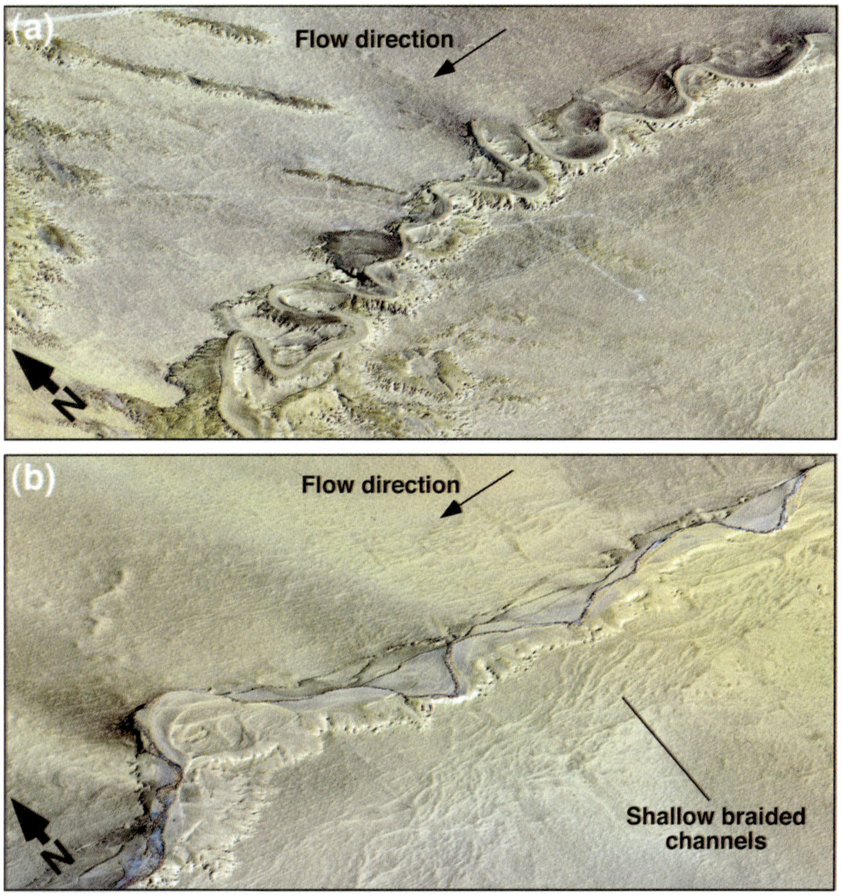

Fig. 13 Two examples of palaeovalleys on the western margin of the Gabes basin. (a) Oued M'zab shows a meandering pattern (see also Fig. 14). (b) A palaeovalley adjacent to Oued M'zab shows a straighter pattern. The plateau is crossed by remnants of braided streams. See location in Fig. 5.

Fig. 14 The meandering pattern of the palaeovalley of Oued M'zab as seen from the plateau.

features similar to those previously described are observable. Lacustrine deposits have been reported in the northern part of the basin (Petit-Maire & Kropelin, 1991), but most of the central area of the basin remains covered by a dune field and no data are available.

The Atlantic slope

A number of ephemeral streams cross drylands adjacent to the Atlantic coast and flow into it (Fig. 5). The largest one is the Draa River that crosses a substantial part of southern Morocco and can be considered an exotic river that is supplied from the highest part of the Atlas chain. However, the other streams are generated inside the lowland arid zone and are currently ephemeral. These ephemeral streams were, during wetter periods, exotic rivers analogous to the other Sahara rivers described above. These river systems display poorly defined palaeovalleys due to the fact that they mainly crossed lowlands, whereas the palaeovalleys in the northern Sahara are well defined because they cross highlands or plateaux, and are deeply incised into them. One of the best examples is the system terminating in the Aridal sabkha near Cap Boujoudur (Figs 5 & 15). At the present day, a 100 km long ephemeral stream debouches into a sabkha, forming a terminal fan. However,

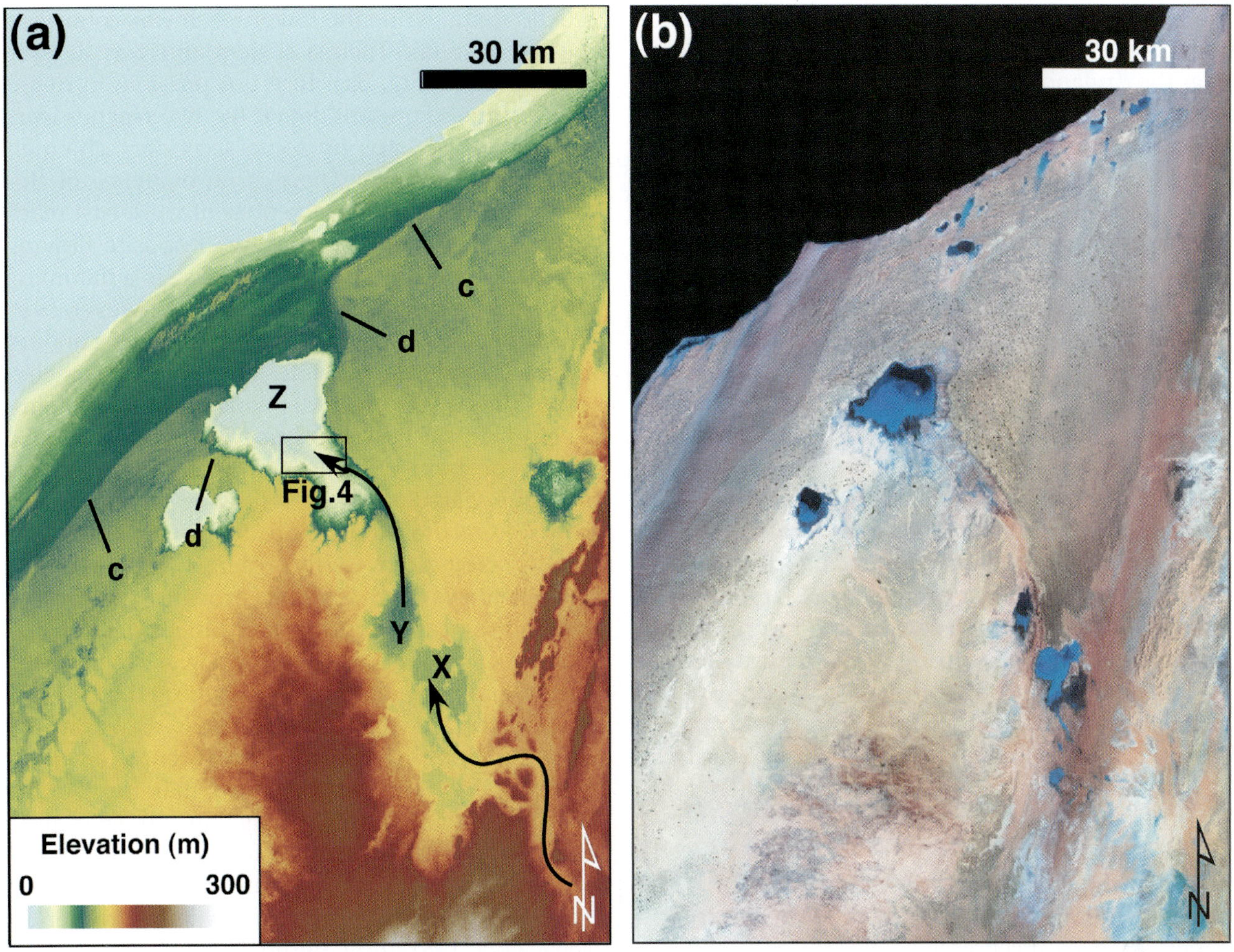

Fig. 15 The palaeovalley of Sabkha Aridal (see location in Fig. 5). (a) Elevation model and (b) LANDSAT images (false colour). Two ephemeral streams (marked by the lines and the arrows are their terminal fans) are contained in the same palaeovalley; these are the watercourse of an exotic river during the wetter climatic conditions. The sabkha (X) represents the terminal area of the southern ephemeral stream. The other northern ephemeral stream originates in the depression (Y) and terminates in the Sabkha Aridal (Z) where the terminal fan of Fig. 4 is formed. The margins of its palaeodelta are marked (d) and correspond to an old coastline (c). This is one of the few cases where an exotic river developed during hydrological and climatic optimum conditions is split into several ephemeral streams during a climatic change toward drier climatic conditions (see Fig 2).

the stream is contained in a valley 20–25 m deep and about 10 km wide. This palaeovalley can be followed to the west, and along its course two other ephemeral streams with terminal fans and a related sabkha are present. It seems probable that the entire palaeovalley was, during humid climatic conditions, a single exotic river that reached the Atlantic Ocean. From the elevation data it is possible to see that in the area of the most distal terminal fan and sabkha, the palaeovalley enlarges and flows in correspondence with the most internal coastline system, defined by linear remnants of beach ridges. As far as has been observed, this is the best example in western Sahara of the splitting of an exotic river into several ephemeral streams as a consequence of climatic changes.

DEPOSITIONAL SYSTEMS

The large number of palaeovalleys and large palaeorivers in the western Sahara are the remnants of the river systems produced under humid climatic conditions. This type of pattern is observed all over the drylands of the world (Tooth, 2000). These palaeovalleys and palaeorivers occur as morphological features or coarse-grained deposits that cannot be satisfactorily dated. The climatic fluctuations are recognized in fine-grained depocentre deposits, which are usually lacustrine or palustrine. It is clear that the observable morphological units are the result of a number of environmental changes, but it is not easy to discern different events or to date them. The high-energy processes (both erosional and depositional) and the coarse grain-size sediments both mask stratigraphic relations. This is particularly the case in western Sahara, where the fluvial landforms are the result of a complex history initiated in the early Cenozoic, and possibly also in the Mesozoic (Griffin, 2002; Goudie, 2005). The last events of the large perennial river activity can be dated using the youngest lacustrine deposits occurring in the central part of basins, and in the southern part of the Erg Chech basin near Taoueddeni (Petit-Maire, 1991); these show a climatic optimum from about 10 to 6 ka. This age is confirmed in other areas around Sahara and Sahel (Gasse, 2000).

Recent tectonic movements may have played a role in the changes of the river patterns, overprinting their effects on climatic changes. River pattern changes are observed to be largely controlled by tectonic movements in the Atlas foothills, but a few examples have been reported in the southern Sahara. The river pattern changes in the Tanezrouft area have been proposed to be controlled by recent tectonism. Chorowicz & Fabre (1977), on the basis of Landsat images, suggested that Oued Tammanraset and the adjacent channels underwent modifications induced by changes in the large-scale fault and fracture patterns. Analysis of the DEM of the area does not support this interpretation but does not rule out some tectonic forcing. On the other hand, from a large-scale, basin-wide point of view, the basin depocentres and the source areas were not affected by major tectonic changes in the Quaternary (Griffin, 2002; Goudie, 2005).

The southern portion of western Sahara is dominated by the presence of the Niger River (Fig. 5). During climatically wetter conditions the upper reach was probably flowing into a large northern inland delta, while the lower reach was connected to the Azaouak–Tilemsi system and was flowing to the south (Fig. 5b). It is not possible to determine with the present data if the two reaches were connected, perhaps by some secondary channels of the inland delta. There is no evidence of this palaeoconnection, and the present west–east reach of the Niger that links the two opposite flowing reaches of the Niger seems to be due to a damming effect of dune fields (Goudie, 2005). However, even if a connection existed it was probably secondary because evidence indicates that the two reaches could have been two independent fluvial systems. The upper reach was forming a large inland delta (Fig. 7) of about 140,000 km^2 in area that could have accommodated the water flowing in the watercourse. On the other hand, the lower reach was clearly connected to the Tilemsi and Azaouak palaeovalleys (Fig. 5b) and, probably, with other rivers from the west. The dimensions of the palaeovalleys fit well with the lower reach of the Niger, and they were clearly connected.

As noted above, the inland basins, as well as the Atlantic slope of Sahara, were occupied during wetter periods by exotic rivers that formed the palaeovalleys observable today (Fig. 5b). The nature of the palaeovalleys and the features of the palaeowatercourses clearly indicate that their origin was shaped by exotic perennial rivers (for references see Czaya, 1981 and Goudie, 2005). A distributary pattern similar to the Kosi megacone in the Himalaya is observed with the Tanezrouft (Fig. 9) and it provides evidence for wetter periods. The dimensions of this distributary pattern and of the examples associated with other African palaeovallyes are smaller than the greatest megacones that formed by flows from the Himalaya (Wells & Dorr, 1987). However, there is at least another megacone in Sudan larger than the one in the Tanezrouft, and this flows into the Nile River valley (Ori, unpublished observations).

The courses of the former exotic rivers are well defined because they correspond to marked palaeovalleys or incised channels that can be detected by satellite imagery or in high-precision DEMs. However, a key point with analysis of depositional systems of western Sahara is to identify the nature of the deposits at the termini of the rivers. Fluvial distributary patterns have been described in a number of papers (Nichols and

Fig. 16 The shallow-water delta in Lake Debo produced by the main channel of the Niger River in the Inland Delta. Compare the shape and dimension of the delta with the terminal distributary patterns of the exotic rivers (palaeovalleys) of the Atlasic Plateau (Fig. 4), and with the one of Oued Saoura (Fig. 12). See location in Fig. 6. LANDSAT image (false colour).

Fisher, 2007) and display a variety of facies and patterns. A common feature at the termini of the channels of the exotic rivers is a distributary pattern of channels confined by levees a few metres higher than the channel floors. These terminal systems are larger than the terminal fans of ephemeral streams and show channelling and levees, which are rather uncommon for simple terminal fan. In plan view, they resemble the delta produced by the main course of the Niger River where it flows into Lake Debo (Fig. 16), a large water body of the Inland Delta. The delta in Lake Debo does not show mouth bars, but a number of diverging channels are bordered by levees. The absence of mouth bars is typical of rivers flowing into shallow-water lakes, whereas the presence of prominent levees is typical of low-energy rivers carrying suspended load and flowing into a vertical accretionary basin, such as in the case of anastomosed channels (Smith and Smith, 1980). Therefore, the distributary patterns at the termini of the exotic palaeorivers are interpreted as components of inland deltaic systems and the rivers were flowing into shallow-water lakes, marshes and swamps.

CONCLUSIONS

1 In western Sahara during climatically wetter periods exotic rivers were flowing into inland basins and on the Atlantic slope.
2 The inland basins correspond roughly to the present-day large ergs and were occupied by lakes and palustrine environments during wetter periods.
3 The Niger River was split into an upper reach flowing to the north into a large inland delta, and a lower reach forming a southward-flowing system along with the palaeo-Azaouak.
4 The exotic rivers formed shallow-water deltas where they entered marshes and lakes.
5 The systems of exotic rivers, inland deltas and ephemeral streams, and terminal fans occurred intermittently in western Sahara in response to wetter or drier climatic conditions respectively.

ACKNOWLEDGEMENTS

This work has been carried out in past decades with a number of expeditions in the Sahara and Sahel. The principal participants in these expeditions provided fundamental support: Luca Masini, Gioia Domeniconi, Riccardo Sabbadini, Stefano Dalla, Claudio Vinsentin, Patrizia Bianconi, and several others. Paolo Sammartino processed part of the data and supported the interpretation of some areas. Ian Reid, Gary Nichols and Victor Baker provided extremely useful comments. The work is part of a project on the analysis of terrestrial analogues for Mars. The financial support for field and laboratory work was provided by Agenzia Spaziale Italiana and Ministero dell' Universita' e Ricerca.

REFERENCES

Ambolt, N. and Norin, E. (1982) *Sven Hedin central Asia Atlas*. Memoir on Maps 3, Sven Hedin Foundation, Stockholm, 61 pp.

Breed, C. and Grow, T. (1979) Morphology and distribution of dunes in sand seas observed by remote sensing. In: *A Study of Global Sand Seas* (Ed. E.D. McKee). *U.S. Geol. Surv. Prof. Pap.*, **1052**, 253–303.

Breed, C.S., Fryberger, S.G., Andrews, S., *et al.* (1979) Regional studies of sand seas using LANDSAT (ERTS) imagery. In: *A Study of Global Sand Seas* (Ed. E.D. McKee). *U.S. Geol. Surv. Prof. Pap.*, **1052**, 305–397

Chorowicz, J. and Fabre, J. (1997) Organization of drainage networks from space imagery in the Tanezrouft plateau (western Sahara): implications for recent intracratonic deformations. *Geomorphology*, **21**, 139–151.

COHMAP (1988) Climatic changes of the last 18,000 years: observations and model simulations. *Science*, **241**, 1043–1052.

Czaya, E. (1981) *Rivers of the World*. Cambridge University Press, Cambridge, 352 pp.

Deodhar, L.A. and Kale, V.S. (1999) Downstream adjustments in allochthonous rivers: western deccan Trap upland region, India. In: *Varieties of Fluvial Forms* (Eds. A.J. Miller and A. Gupta), pp. 295–315, Wiley, Chichester

El Baz, F. and Maxwell, T.A. (Eds) (1982) *Desert Landforms of Southwest Egypt: a Basis for Comparison with Mars*. CR-3611, NASA, Washington, DC, 372 pp.

Fabre, J. and Petit-Maire, N. (1988) Holocene climatic evolution at 22–23°N from two paleolakes in the Taoudenni area (northern Mali). *Palaeogeogr. Palaeoclimatol. Palaeoecol.*, **65**, 133–148.

Fontes, J.C. and Gasse, F. (1991) PALHYDAF (Palaeohydrology in Africa program: objectives, methods, results. *Palaeogeogr. Palaeoclimatol. Palaeoecol.*, **84**, 191–215.

Fontes, J.C., Gasse, E., Callot, Y., *et al.* (1985) Freshwater to marine-like environments from Holocene lakes in northern Sahara. *Nature*, **317**, 608–610.

Friend, P.F. (1978) Distinctive featuers of some ancient river systems. In: *Fluvial Sedimentology* (Ed. A.D. Miall). *Can. Soc. Petrol. Geol. Mem.*, **5**, 531–542.

Gasse, F. (2000) Hydrological changes in the African tropics since the last glacial maximum. *Quatern. Sci. Rev.*, **19**, 189–211.

Gasse, F. (2001) Diatom-inferred salinity and carbobate oxygen isotopes in Holocene waterbodies of western Sahara and sahel (Africa). *Quatern Sci. Rev.*, **21**, 737–767.

Goudie, A.S. (2002) *Great Warm Desert of the World: Landscapes and Evolution*. Oxford University Press, Oxford, 444 pp.

Goudie, A.S. (2005) The drainage of Africa since Cretaceous. *Geomorphology*, **67**, 437–456.

Griffin, D.L. (2002) Aridity and humidity: two aspects of the late Miocene climate of North Africa and the Mediterranean. *Palaeogeogr. Palaeoclimatol. Palaeoecol.*, **182**, 65–91.

Howell, P., Lock, M. and Cobb, S. (Eds) (1988) *The Jongley Canal*. Cambridge University Press, Cambridge, 535 pp.

Jacobberger, P.A. (1981) Geomorphology of the upper inland Niger delta. *J. Arid Environ.*, **13**, 95–112.

Mainguet, M. and Callot, Y. (1978) L'Erg de Fachi–Bilma (Tchad–Niger). *CNRS Mem. Doc.*, **18**, 184 pp.

Masini, L., Dalla, S., Ori, G.G., Bianconi, P. and Visentin, C. (1988) Oued Saoura: un fiume effimero nel Sahara nord-occidentale. *Giorn. Geol.*, **50**, 177–184.

McCarthy, T.S. (1993) The great inland deltas of Africa. *J. Afr. Earth Sci.*, **17**, 275–291.

McCarthy, T.S., Stainstreet, I.G. and Cairncross, B. (1991) The sedimentary dynamics of active fluvial channels on the Okavango fan, Botswana, *Sedimentology*, **38**, 471–487.

McIntosh, R.J. (1983) Floodpalin geomorphology and human occupation of the upper inland delta of the Niger. *Geogr. J.*, **149**, 182–201.

Makaske, B. (1998) *Anastomosing Rivers: Forms, Processes and Sediments*. Nederlandse Geografische Studies, 249, Utrecht, 287 pp.

Mukerij, A.B. (1976) Terminal fans on inland streams in Sutlej-Yamuna Plain, India. *Z. Geomorphol*, **20**, 190–204.

Nash, D.J. (1997) Groundwater as a geomorphological agend in dryland, In: *Arid Zone Geomorphology: Process, Form and Change in Drylands* (Ed. D.S.G. Thomas), pp. 321–348. Wiley, Chichester

NEDECO (1959) *River Studies and Recommendations on Improvement of Niger and Benue*. North-Holland., Amsterdam, 450 pp.

Nichol, J.E. (1988) Quaternary climate and landscape development in west Africa: evidence from satellite images. *Z. Geomorphol.*, **42**, 229–347.

Nichols, G.J. and Fisher, J.A. (2007) Processes, facies and architecture of fluvial distributary system deposits: a review. *Sediment. Geol.* **195**, 75–90.

Ori, G.G. (1988) River systems in arid and semi-arid climatic conditions: examples from Sahara and adjacent areas. In: *Deserts: Evolution Passé et Future* (Ed. N. Petit-Maire), pp. 179–181. Proceedings Fuerteventura Meeting IGCP 252, Marseille.

Ori, G.G., Komatsu, G. and Marinangeli, L. (Eds) (2001) *Exploring Mars surface and its terrestrial analogues: Field Trip Guidebook to Chott el Gharsa and Chott el Jerid*. Alenia Spazio, Torino, 91 pp.

Petit-Maire, N. (1986) Paleoclimates in the Sahara of Mali: a multidisciplinary study. *Episodes*, **9**, 7–16.

Petit-Maire, N. (Ed.) (1991) *Paleoenvironnements du Sahara*. CNRS Editions, Paris, 237 pp.

Petit-Maire, N. (2002) *Sahara, sous le Sanle des Lacs*. CNRS Editions, Paris, 127 pp.

Petit-Maire, N. and Kropelin, S. (1991) Les climats Holocenes du Sahara le long du tropique du cancer. In: *Paleoenvironnements du Sahara* (Ed. N. Petit-Maire), pp. 205 –210. CNRS, Paris.

Petit-Maire, N. and Risier, J. (1981) Holocene lake deposits and palaeoenvironments in northeastern Mali, *Palaeogeogr. Palaeoclimatol. Palaeoecol.*, **35**, 45–61.

Smith, H.T.U. (1963) *Eolian geomorphology, wind direction, and climatic change in North Africa*, Air Force Cambridge Research Laboratories, Cambridge, 63–443, 48 pp.

Smith, D.G. and Smith, N.D. (1980) Sedimentation in anastomosed river systems: examples from alluvial valleys near Banff, Alberta, *J. Sediment. Petrol.*, **50**, 157–164.

Tooth, S. (1999) Floodouts in Central Australia. In: *Varieties of Fluvial Forms* (Eds A.J. Miller and A. Gupta), pp. 219–247. Wiley, Chichester.

Tooth, S. (2000) Process, form and change in dryland rivers: a review of recent approach. *Earth-Sci. Rev.*, **51**, 67–107.

Urvoy, Y. (1942) *Les Basins du Niger*. Memoir 4, l'Institut Francais d'Afrique Noir, Paris, 139 pp.

Wells, N.A. and Dorr, J.A. (1987) A reconnaissance of sedimentation in the Kosi alluvial fan in India. In: *Recent Developments in Fluvial Sedimentology* (Eds F.G. Ethridge, R.M. Flores and M.D. Harvey), pp. 1–40. Special Publication 17, Society of Economic Paleontogists and Mineralogists, Tulsa, OK.

Floodplain sediments of the Tagus River, Portugal: assessing avulsion, channel migration and human impact

T.M. AZEVÊDO*, A. RAMOS PEREIRA†, C. RAMOS†, E. NUNES*, M.C. FREITAS*, C. ANDRADE* *and* D.I. PEREIRA‡

**Departamento de Geologia and Centro de Geologia, Universidade de Lisboa, Ed. C6, 4° Piso, Campo Grande, 1749-016 Lisboa, Portugal (Email: terazeve@fc.ul.pt)*

†Centro de Estudos Geográficos, Alameda da Universidade, Faculdade de Letras, 1600-214 Lisboa, Portugal

‡Departamento de Ciências da Terra, Universidade do Minho, Campus de Gualtar, 4710-057 Braga, Portugal

ABSTRACT

A study of the Tagus River floodplain (Portugal) has been carried out using a variety of methods including sedimentological, geochemical and geochronological analyses, as well as geomorphological and hydrological studies, performed in order to characterize the flood sediments and the dynamics of the river during the Holocene. Until the 19th century, the Tagus was an anastomosed river, with multiple channels separated by large areas of floodplain; today, it is a single channel river with alternate bars, mainly as a result of anthropogenic modification. In order to study its behaviour during the Holocene, four cores ranging in length from 3.70 to 8.04 m were obtained from the floodplain and 232 samples were analysed. Detailed textural analysis was necessary owing to the lack of preservation of sedimentary structures in the cores. The sediments of the present-day geomorphological elements of the floodplain (bars, natural levees, crevasse-splay deposits, flood basin and abandoned channels) were also studied in order to compare their textural characteristics with those of the cored samples. Both the present analogues and core sediments were well discriminated using mean diameter versus standard deviation and average mean diameter versus average mud percentage graphical correlations. The textural parameters defined (sand/mud ratio, mean, standard deviation, skewness) and particularly the interparameter correlations, together with 12 ^{14}C numerical ages of organic matter obtained, allowed the evaluation of: (i) sedimentological changes in the floodplain (channel migration, avulsion and crevasse-splay development); and (ii) the chronological evolution of the different energetic environments of the floodplain for the past 4 kyr. These approaches permitted the determination of sedimentation rates for the different alluvial plain environments. The highest sedimentation rates occurred in the flood basin and channel-fill domains, with values ranging from 2.2 mm yr^{-1} to 4.7 mm yr^{-1} and the lowest in the channel (0.3 mm yr^{-1}). Values from 0.8 to 1.6 mm yr^{-1} were recorded in sedimentary environments proximal to the channel, where several crevasse-splay episodes have been recognized. In the period common to the four cores, i.e. the past 4000 yr, the sedimentation rates decreased towards the present. In spite of increasing human intervention in the hydrographic basin during this time, the increasing aridity of the climate is considered to have outweighed the sediment availability, which resulted in a decreasing sedimentation rate.

Keywords Tagus, floodplain, channel changes, chronology, sedimentation rates.

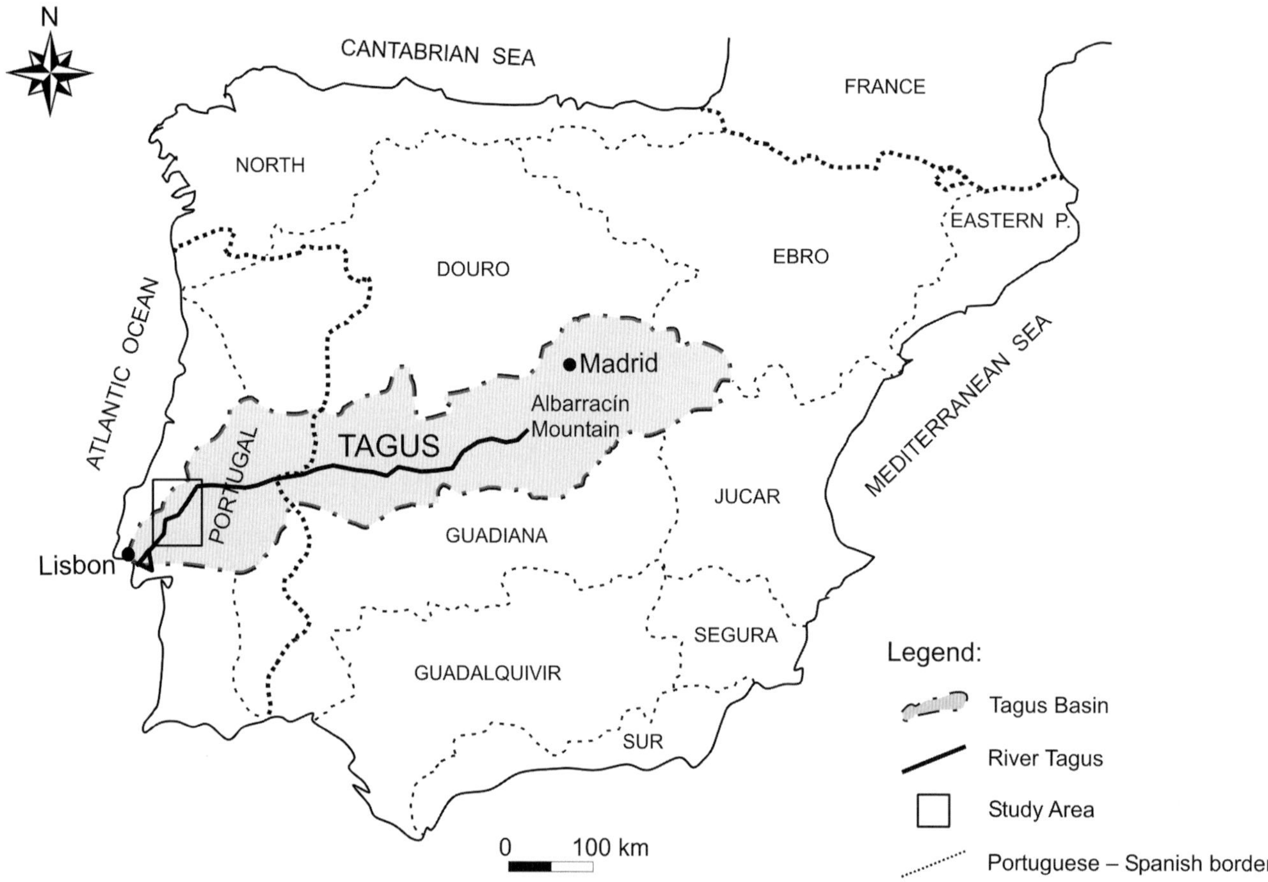

Fig. 1 Position of the Tagus River in the Iberian Peninsula and the study area.

INTRODUCTION

From the Albaracin Mountains in Spain, until it reaches the sea in the Lisbon estuary, the Tagus River has a total length of 1110 km and a drainage basin area of 80,630 km^2. About one-third of the catchment lies in Portugal (Fig. 1). This relation between drainage area and length is consistent with Potter's (1978) plot of drainage basin area versus length for the world's 50 largest rivers. The Tagus River cuts rocks with ages ranging from Pre-Cambrian to Quaternary; their lithological variety has resulted in a diverse fluvial landscape. In the 19th century the Portuguese Tagus was divided into three reaches to simplify its study and administration (Fig. 2). The High Tagus, between the Portuguese–Spanish border and Tancos, deeply incises old resistant igneous and metamorphic rocks; the Middle Tagus, between Tancos and Vila Franca de Xira, incises the sediments of its Tertiary Basin, where the channel changes direction and the valley widens rapidly leading to sediment deposition on a 2 to 13 km wide alluvial plain, bounded by Pleistocene terrace sediments (Fig. 2). The Lower Tagus, between Vila Franca de Xira and its mouth west of Lisbon, occupies a complex tectonic depression filled by a few hundred to 2000 m of Cenozoic sediments.

The study area (Fig. 2), located 50–100 km northeast of Lisbon in the Middle Tagus, is an agro-industrial area with a population of approximately 500,000 inhabitants, and is the reach most affected by flooding. Since the Arab occupation, the alluvial plain of the Tagus River has been considered comparable to the Nile in its high fertility, and comprises the best Portuguese agricultural soils. These soils owe their fertility to centimetre-thick sheets of fine sediment, locally referred to as 'nateiros', which settle after each flood, which in turn has led to the accumulation of a thick sequence of organic-rich muds alternating with fine sands. In this region, the main channel of the

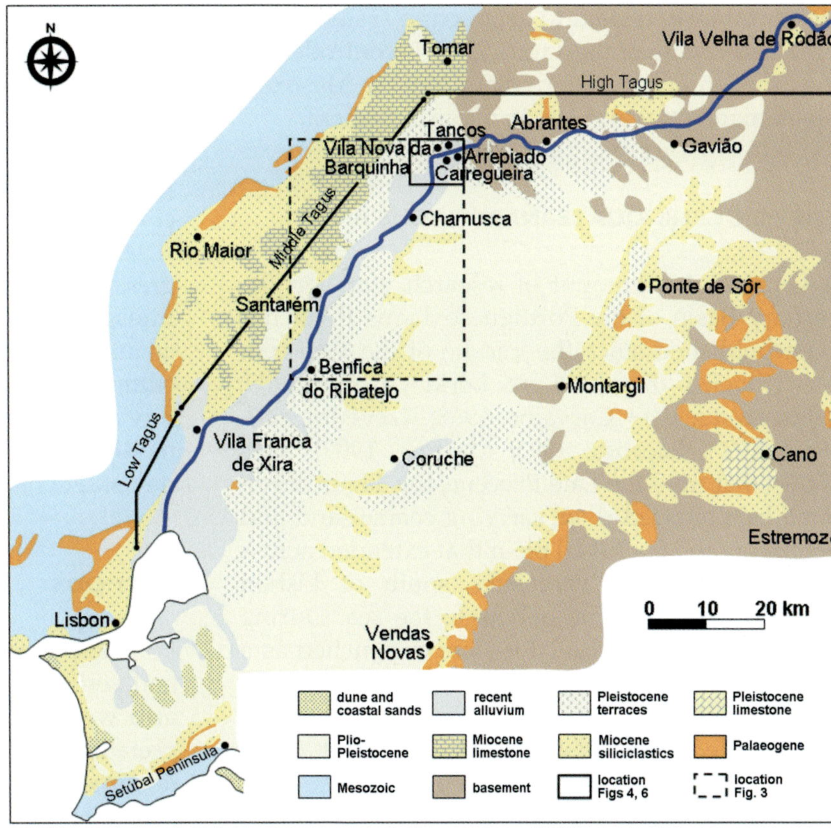

Fig. 2 Simplified geology of the Portuguese Tagus basin (modified after Daveau, 1970). Miocene siliciclastics include Paleogene age sediments in the eastern part of the map. Mesozoic sediments mark the eastern margin of the Lusitanian Basin.

Tagus has changed its course during the Holocene and in historical times, as a result of both natural processes (avulsion and lateral migration) and human intervention.

This study constitutes part of a research project that aims to identify present and past dynamic features of this alluvial plain with emphasis on low-frequency flood events, which constitute an element of hazard and risk to this region. Floods, channel variations and anthropogenic interference can be identified using sedimentological and geomorphological methods. The main goal of this paper is to present results of textural analysis and interpretation of four sections through the Tagus River floodplain that represent part of its Holocene alluvial stratigraphic record. This paper also provides quantitative information on sediment accumulation rates and timing of channel avulsions relevant to work on modern and ancient floodplain sediments (e.g. Bridge & Leeder, 1979; Bridge, 1984; Kraus & Bown, 1993; Weerts & Bierkens, 1993; Stouthamer, 2001).

GEOLOGICAL SETTING

During the Eocene, the collision between the African and Eurasian plates, with an approximate NNE–SSW convergence vector, was followed by the extensional reactivation of NNE–SSW-striking faults. As a consequence, an elongate depression was opened orthogonally to this direction: the Cenozoic Tagus Basin. The fault system of the Lower Tagus Valley has been interpreted to reach the Moho (Victor *et al.*, 1980; Hirn *et al.*, 1981). In the Miocene, following the collision between Africa and the Iberian micro-plate, the tectonic inversion of the Mesozoic Lusitanian Basin occurred through the rotation of the stress field towards the NW–SE (Ribeiro *et al.*, 1979, 1990; Rasmussen *et al.*, 1998). The tectonic control of this segment of the Lower Tagus Valley was recently established, through the analysis of seismic reflection data (Rassmussen *et al.*, 1998; Cabral *et al.*, 2003). As noted by Potter (1978), the location of major river systems on cratons largely follows structural lows related to geofracture systems. The principal course of the Tagus River is controlled by the junction of the eastern margin of the Mesozoic Lusitanian Basin and the Variscan basement.

THE PLIOCENE–QUATERNARY TAGUS RIVER

A considerable amount of research work on the Tertiary Basin of the Portuguese Tagus River has been published since the middle of the 19th century (Ribeiro, 1866; Dolfus & Berkley-Cotter, 1909; Zbyszewski, 1949; Carvalho, 1968; Azevêdo, 1983, 1987, 1997; Barbosa, 1995; Martins, 1999). This shows that until the Late Pliocene, the Tagus developed as a braided river, carrying coarse sand and pebbles with an alluvial plain that extended across the entire Setúbal Peninsula, south of Lisbon (Fig. 2), with multiple outlets to the sea. During the Pleistocene the fluvial system entrenched as a response to uplift controlled by a NNE–SSW-trending regional structure. The uplift generated four stepped terrace levels: Q1 and Q2 at 80–60 m, Q3 at 50 m and Q4 at 20–25 m (Cabral, 1995; Martins, 1999). The Middle Terrace (Q3) contains archaeological evidence of human occupation, the banks of the Tagus having been occupied since the Late Palaeolithic (Mozzi *et al.*, 2000).

Holocene sedimentary sequences up to 70 m thick, revealed in several borehole cores made for groundwater exploitation (Mendonça, 1990), overlie an erosively based, very coarse quartzitic pebble-rich stratigraphical unit, up to 40 m in thickness, that indicates strong erosion of the catchment prior to deposition of the Holocene sequence. This unit is assumed by different authors to represent the maximum glacial episode, during which the local relative sea level stood about 120 m below that of the present day. Fluvio-glacial erosion, seasonal ice melting and strong spring and autumn rainfall are thought to have resulted in high fluvial discharges, with consequent transport of very coarse material (Dias, 1987). The present-day distal parts of the rivers corresponded then to deeply incised valleys, confirmed by numerous palaeoenvironmental reconstructions of the post-Late-glacial infill of several Portuguese fluvial valleys and lagoons (e.g. Cearreta *et al.*, 2002; Freitas *et al.*, 2002, 2003; Freitas & Ferreira, 2004; Alday *et al.*, 2006). According to Daveau (1980), 18,000 yr ago the rainy seasons were much longer than today, and a large contrast existed between the wet environment of the Atlantic border and the dryer and hotter climate of the remainder of the Peninsula. The distinctiveness of the Portuguese Tagus basin relies on its geographical position, between these two contrasting climatic environments.

Throughout the Holocene the deep valley progressively filled-up, in tune with the Holocene transgression: first by coarse materials, related to an inherited high slope, and later by finer sediments, culminating in the build-up of an anastomosed fluvial system. Finally, anthropogenic activities modified the natural features, imposing the present-day single channel.

METHODS

For the purposes of this study, geomorphological and sedimentological investigations of the study area were undertaken. The description and interpretation of the floodplain morphology relied upon documentary evidence, which included written accounts since the 10th century, historical maps dated from 1560 onwards, both ground and aerial photographs and film footage (including cinema and television sources since the 1950s), and recent 1:20,000 scale maps. A digital elevation model (DEM) of the floodplain in the studied area was constructed (Figs 3 & 4) with ArcView, using topographic data from 1:25,000 maps and georeferenced with ArcGis.

Fieldwork was carried out to retrieve 27 surface sediment samples (at 10–20 cm depth) of the present-day morphosedimentary units. These were collected from channel bars, former and present-day natural levees, abandoned channels and flood basins (Fig. 3). These samples were used as analogues to interpret the cored sediments.

Following a preliminary regional sedimentological and stratigraphic survey, four sites were selected for coring: Santarém Entre Valas (SEV), Quinta da Boavista (QB), Fonte Bela (FB) and Goucharia (G); the first three located on the right bank of the Tagus River and the last on the left bank (Fig. 3). The SEV core site, 6 km north of Santarém and 5 km from the main channel, was selected due to its distal geomorphological position on the floodplain and its frequency of flooding, i.e. it is

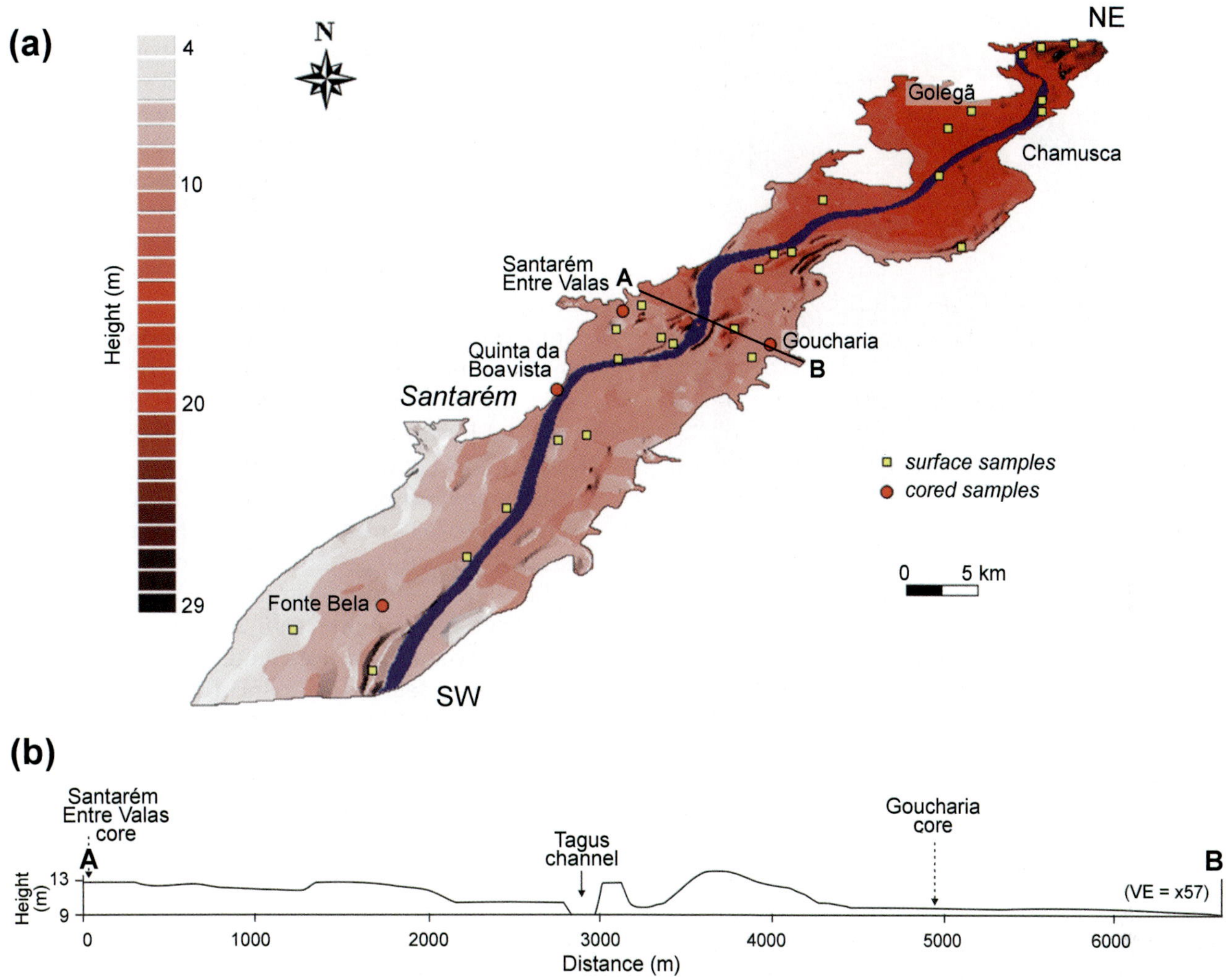

Fig. 3 Location of the study cores. (a) Digital elevation model (DEM) of the floodplain and the morphological position of cores and surface samples (see Fig. 2 for regional location; Pereira *et al.*, 2004). (b) Transverse section of the floodplain close to the Santarém Entre Valas–Goucharia cores.

inundated mainly during major floods. Core QB was located in the lowest part of the embankment in the main channel. Core site FB was selected due to its position in the medial zone of the floodplain. The core, totalling 7.4 m, is composed of two different parts: an upper section 2.5 m thick corresponding to an existing ditch in agricultural land; and a lower 4.9 m section below the base of the ditch. Core G was selected due to its position on the left bank, opposite the SEV core site; the location is 5 km from the main channel and is frequently flooded. Coring sites were coordinated and connected to the national UTM grid and vertical Portuguese Datum (mean sea level).

The SEV core was retrieved using 75 mm diameter and 1 m long Shelby samplers driven by hydraulic pressure and operated inside a cased borehole, down to 19.35 m below surface. At this depth, a coarse gravel unit, interpreted as Pleistocene in age, was encountered and coring was stopped. In this paper, only the top 8.04 m of the SEV core are considered, in order to make comparison with the three shorter cores. The G, QB and FB cores were obtained using steel, hand-operated gauge augers (35–50 mm in diameter and 0.5 m long) driven to 3.80, 3.70 and 7.40 m below surface, respectively. In spite of maximum care to achieve continuous coring, retrieval

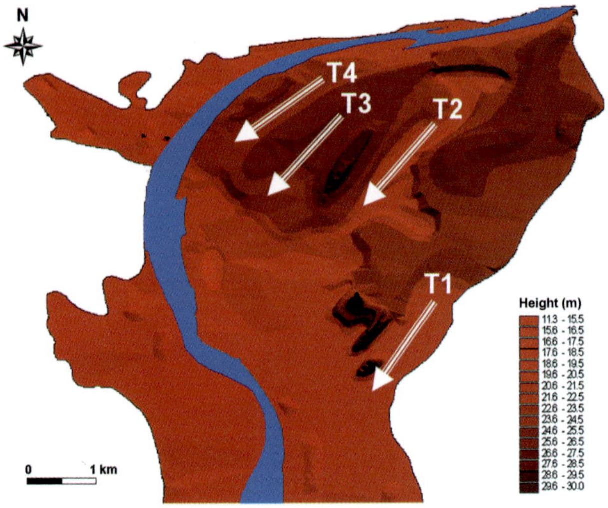

Fig. 4 Digital elevation model (DEM) of the Tagus alluvial plain in the Tancos area. See Fig. 2 for location. T1, the original channel; T2, artificial channel; T3 and T4, former channel positions prior to the present-day channel (blue). (After Pereira *et al.*, 2004.)

problems occurred, limiting the continuity of cores.

Interpretation of the cores neglected the effects of post-depositional compaction. This follows the work of Baldwin (1971) and Perrier & Quiblier (1974), the latter suggesting that compaction effects need to be considered only after the first 10 m of burial.

In the laboratory, the cores were cut lengthwise, described and subsampled at 10-cm intervals. Each subsample consisted of a 1-cm-thick slice of sediment that was oven-dried at 60°C and processed for grain size using a series of standard sieves (from −3 to 4 ϕ; 8.0 to 0.0625 mm), a SEDIGRAPH instrument (from 4 to 11 ϕ; 0.0625 to 0.0005 mm), and the software SEDPC for the determination of textural parameters (Henriques, 1998, 2003, 2004). The phi (ϕ) numerical system, based on the logarithmic transformation of the Wentworth (1922) sediment grain-size scale, was used as it is a convenient method for the calculation of grain-size parameters. Sediment samples followed a log-normal size distribution, and statistical parameters on central tendency, sorting and symmetry were computed using the moment's method. Textural classification and sorting and symmetry parameters followed Fleming (2000) and Friedman (1961), respectively. A number of exploratory samples of the cored sediments were used for heavy mineral and clay mineral analyses. The heavy mineral contents of the 3 and 4 ϕ (very fine sand) size fractions were separated using bromoform, and grains were mounted on glass slides for microscopic observation and identification. The < 0.0625 mm fraction was studied for clay mineralogy by X-ray diffraction using a Phillips X'Pert Diffractometer and the data output was processed using the Phillips Profile Fit software.

Organic matter content was determined in samples from the SEV core only, using 1 g of dried sediment, by oxidation with potassium dichromate, followed by titration using iron sulfate (Standard E-201; LNEC, 1967).

Sediment samples collected for isotopic dating consisted of 1-cm-thick slices of bulk organic material (including fragments of roots, wood and coal). Twelve samples (four from SEV, four from FB, three from QB and one from G) were dated by ^{14}C radiometric standard accelerated mass spectrometry (AMS) at Beta Analytic Inc. (USA). Calibration of radiocarbon dates was performed using the 'Fairbanks0805' calibration curve method (Reimer *et al.*, 2004; Fairbanks *et al.*, 2005).

CHANNEL AND FLOODPLAIN CHARACTERISTICS

In order to place in context the sediments of the floodplain described below, the principal characteristics of the Tagus River channel belt in the study reach are given. The middle reach of the present Tagus River is defined as a single bedload channel with alternate bars, according to the classification of Miall (1996). The channel has a low sinuosity ($\rho = 1.05$), but in this reach it shows a large variability: just downstream of Tancos (Figs 2 & 4) it describes a large meander ($\rho = 1.5$) evolving to an almost straight reach before the next bend (Fig. 3). The channel width (W) varies between 270 and 590 m, its depth (D) ranges between 3.58 and 6.3 m and the mean W/D ratio is 97. However, because this reach of the Tagus shows (i) highly variable sinuosity, (ii) highly variable W/D ratio, (iii) a very low channel slope (0.0006) and (iv) a sand-dominant bed-load, then according to Rosgen's classification of natural rivers (Rosgen, 1994, 1996) it is more typical of, and can be classified as, an anastomosing river (D_A5 in Rosgen's classification; 1994, 1996).

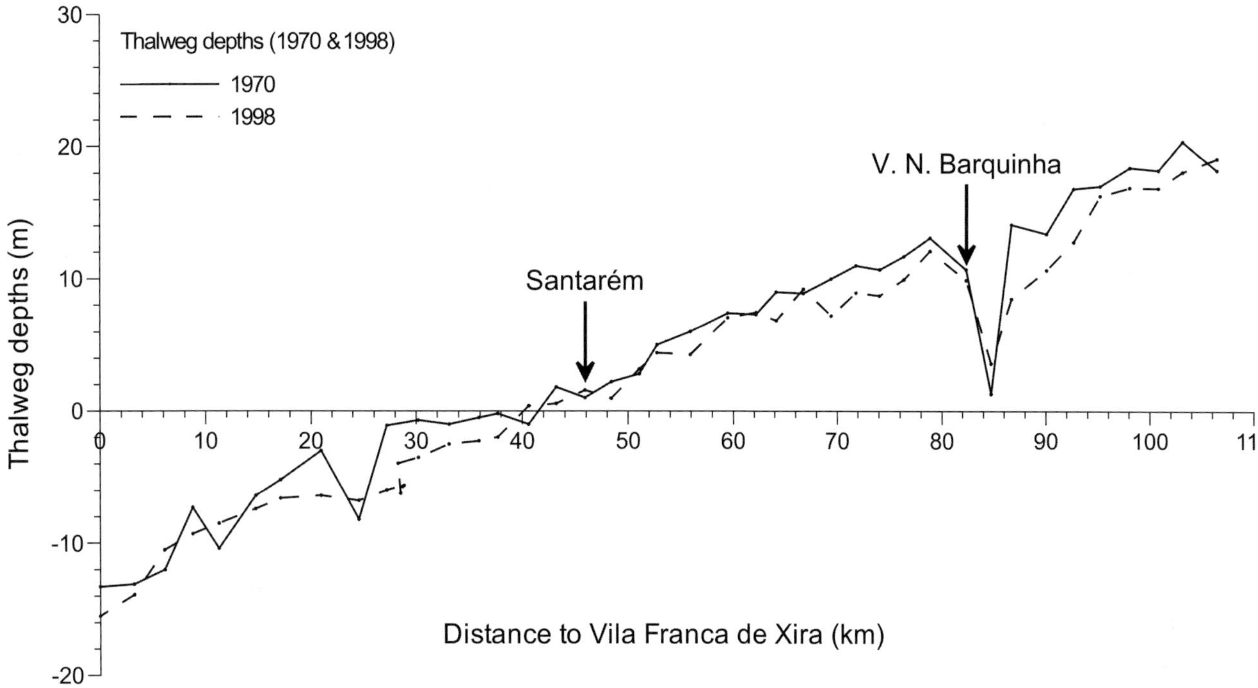

Fig. 5 The channel incision between 1970 and 1998 (data from PNA, 2001).

The present situation (a single channel), however, is the result of multiple human interventions, known to have happened since the Roman occupation (1st century BC to 4th century AD; Custódio, 1992–93).

The grain size of the river's main bedload is medium to coarse sand but it can carry large pebbles during floods. The main in-channel bars are longitudinal alternate bars close to the banks or located in the middle of the channel, reaching 1 km in length and 200 m in width. At the Ómnias gauging station (the only one on the floodplain), close to Santarém (Fig. 2), the discharge records (from 1972 onwards) show that the average annual discharge reaches 360 $m^3 s^{-1}$, but displays large monthly variations between the dry (summer) and wet seasons. In the summer, the daily mean discharge can be as low as 8 $m^3 s^{-1}$, whereas the peak flood discharge can surpass 10,000 $m^3 s^{-1}$.

Comparing the river regime before (natural regime) and after 1950 (artificial, due to dam construction in the Spanish and Portuguese Tagus basins), it has been shown that there has been a decrease in both flood frequency and peak flood discharges, which diminished fivefold (Ramos & Reis, 2002; Ramos *et al.*, 2002). As a consequence, the total load transported by the river has fallen by 82%, increasing the erosive capacity of the channel and leading to its strong incision. Between 1970 and 1998 its maximum bottom incision reached 5.62 m, upstream of Vila Nova da Barquinha (Figs 2 & 5; PNA, 2001). Another important feature is that, in spite of regulation of the river regime since the 1950s, the suspended load has been much more important than the bedload transport (six to eight times higher). Nowadays, the suspended load reaches a volume of 1887.8 $dam^3 yr^{-1}$ and the bedload is 300.1 $dam^3 yr^{-1}$ (PNA, 2001).

The Tagus River floodplain between Tancos and Benfica do Ribatejo (Fig. 2), a down-valley distance of 47 km, varies in height from 22 m upstream to 7 m downstream, with a 0.31 m km^{-1} slope (approximately half the channel slope). The floodplain shows large lateral variations in width of between 2 and 13 km.

In the floodplain, topographic highs (former and present-day natural levees and crevasse splays) and depressions (flood basins and abandoned channels) are found (Fig. 3). The former group reaches 1–4 m above the average elevation of the floodplain and the latter group 1–5 m below that mark (Pereira *et al.*, 2002). The spatial distribution of these elements over the floodplain can be used to define the former positions of the Tagus channels.

RECENT AVULSION HISTORY OF THE MIDDLE TAGUS

Several changes in channel position have been due to human intervention, such as those observed in the Tancos area (Fig. 2) where, in the past 450 yr, the river has shifted laterally by 2 km towards the northwest. Initially, the river flowed near the southeastern limit of the floodplain, close to the Arrepiado and Carregueira terraces (Fig. 6), adapting its course to the NNE–SSW lower Tagus fault (Martins, 1999). However, because flooding damaged the agricultural land of King João III's brother, in 1550 he asked the king to shift the river channel by 1 km to the northwest, in a 10 km long segment between Lagoa Fedorenta (Tancos) and Chamusca (Figs 2 & 6). In one month, 30,000 workers altered the course of the Tagus and engineered a straight channel in this area (Fig. 6; Alves Dias, 1984; Azevêdo, 2001, 2004; Azevêdo *et al.*, 2004). The river, however, was not stable in this new location and began migrating to the northwest, adopting two new courses before stabilizing in the present one, as shown in Fig. 4. All three successive avulsions (to the west-northwest) occurred very rapidly, as demonstrated by historical research; in fact, in 1565, the monks of Quinta da Cardiga asked the king to shift the channel back to its original position, as the river was strongly eroding their agricultural land. These avulsions transferred the flow from one channel into the other, building a major point bar where abandoned channels and natural levees can be recognized today (Fig. 4). These shifts occurred during floods, once the flow migrated to the lowest parts of the floodplain (flood basin). The present course of the river is at a height of 19 m, whereas the artificial channel was built at 22 m.

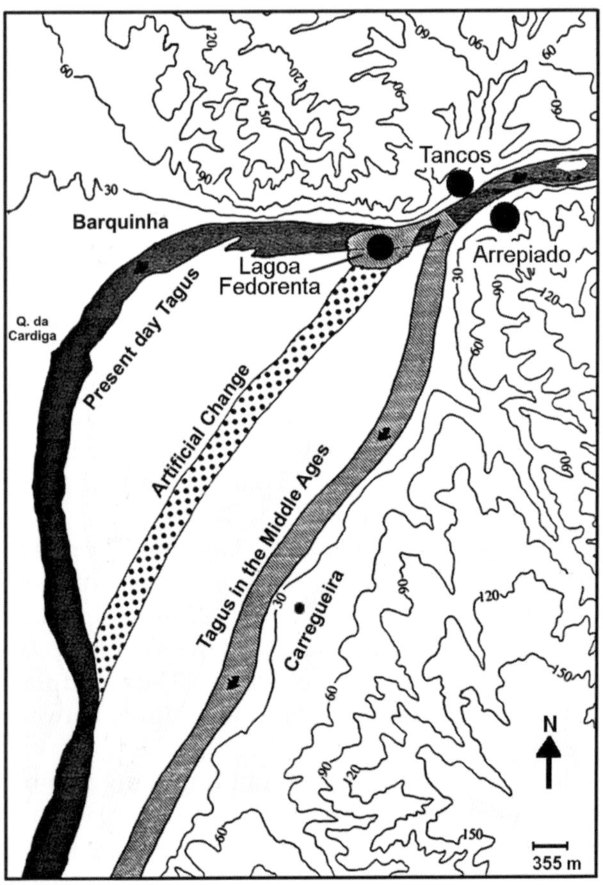

Fig. 6 River Tagus channel change in the 16th century. See Fig. 2 for location. Contours are in metres. (After Alves Dias, 1984.).

Downstream, works of different types progressively modified the channel pattern of the Tagus. In the originally anastomosed reach of the river, local landowners increased the area of their properties by transforming the old channels into agriculture fields. This practice was forbidden by law only in the 18th century. Two more engineered avulsions, carried out in the 18th century, transformed the anastomosing Tagus into a single-channel river.

TEXTURAL ANALYSIS

According to Bridge (1984, p. 583), thick modern floodplain sequences ($\geq$ 10 m) are difficult to study because of problems of exposure, and they are difficult to interpret due to rapid climatic and relative sea-level changes during the Quaternary. Because of this, subsurface sedimentary facies may be unrelated to modern flow and sedimentation conditions and, as a result, postulated floodplain origins of ancient alluvial sequences have not always been based on close comparisons with modern facies. However, given that this work considers only the second half of the Holocene, it is reasonable to assume that climatic conditions contemporaneous with the first 8 m of the sedimentary sequence of the Tagus floodplain were not substantially different from the present day, and that changes in sea-level were negligible (Dias, 1987). Attention is focused on textural analysis and, to some extent, composition because sedimentary structures (including directional structures), which could have yielded complementary information about the sedimentary environments, are not preserved.

Present analogues

The results from the study of the present-day geomorphological elements of the alluvial plain, such as bars, natural levees, flood basins and abandoned channels, were used to compare their textural parameters with those of the cored samples. This methodology allowed the identification, in the subsurface sequences, of episodes of channel avulsion and progressive migration. Textural results of 27 samples taken from and characterizing present-day morphological elements of the alluvial plain in the study area are presented in Fig. 7.

The plot of mean grain size (Mz) versus standard deviation (SD) shows that samples cluster in three distinct fields (Fig. 7). Samples representative of the flood basins, infilled abandoned channels and alluvial plain sediments are essentially coarse to fine silt (4–7 ϕ) and are extremely poorly sorted (SD > 2.5 ϕ). Although it was not possible to sample swamps located in the deepest parts of the flood-basins, field knowledge indicates that sediment from these environments would extend the floodplain domain, as indicated in Fig. 7, towards finer mean grain sizes. Deposits closer to the channel,

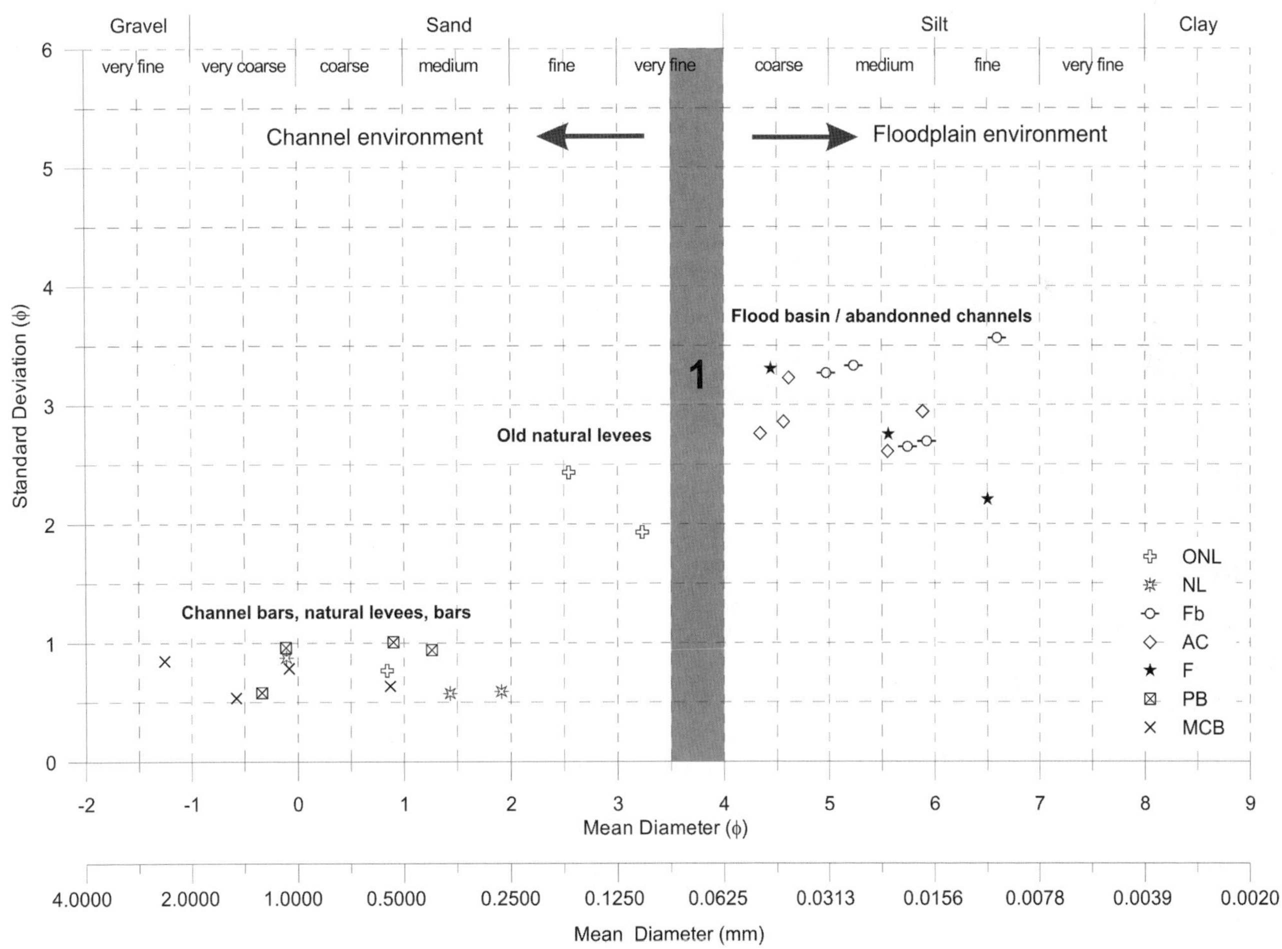

Fig. 7 Mean grain size versus standard deviation for the surface samples of the main morphological features of the Tagus floodplain. AC, abandoned channels; F, floodplain; Fb, flood basin; MCB, marginal channel bar; NL, natural levees; ONL, old natural levees; PB, point bars; 1, transition between channel and floodplain environments.

natural levees and present-day marginal channel bars and point bars, are sandy and occasionally gravelly (especially channel bars), with a mean grain size varying between –2 and 2 ϕ (very fine gravel to medium sand) and with a SD of 0.5–1.5 ϕ (moderately well sorted to moderately sorted). The abandoned natural levees consist of finer and less sorted sands than active levees close to the present Tagus channel. This suggests post-depositional dimensional and fabric rearrangement in relation to either pedogenic processes or illuviation of finer particles following high-turbidity floods. In fact, historical and hydrological data do not point to a change in flood magnitudes through time and/or a decrease of flow capacity in the past.

These results agree with the location of sampling points in relation to the main channel and with the textural attributes typical of the same morphological elements of the fluvial system described by others (Friedman, 1961; Middleton, 1976; Haner, 1984).

Core samples (Goucharia, Quinta da Boavista, Fonte Bela and Santarém Entre Valas)

Based on textural parameters each of the four cores shows three main units (I, II, III). Given that the SEV core is much longer than the others, both its unit I (19.35–13.50 m) and the lower part of unit II (13.50–8.04 m) were not used in this study; unit III of this core was further divided in two subunits IIIa and IIIb, for comparative analysis (Fig. 8).

Quinta da Boavista core (QB)

The results for the QB core demonstrate a clear coarsening upward trend (Table 1). The basal unit (unit I; 370–250 cm) consists essentially of muddy sand and sandy mud, the sediments corresponding to extremely poorly sorted and positively skewed very fine sand and very coarse silt. The intermediate unit (unit II; 250–70 cm) consists of slightly muddy sand to muddy sand. The sediment is coarser than in the lower unit and the mean grain size is variable; it consists of fine and very fine sand and occasionally very coarse silt and coarse to medium sand, in general, extremely poorly sorted and strongly positively skewed. The upper unit (unit III; 0–70 cm) consists of clean sand, which is coarse to medium grade, moderately to moderately well sorted (better sorted than units I and II) and strongly positively skewed. In general, sedimentation in the QB core becomes coarser, better sorted and more positively skewed.

The lower unit (I) is interpreted as the lowest energy environmental conditions of the floodplain (Fig. 9a). Further up-core (in the intermediate unit), a change to higher energy levels occurred, either progressively or abruptly, to (more proximal) floodplain facies, until the present-day channel conditions were eventually established (unit III; Fig. 9a).

Goucharia core (G)

The Goucharia core (Table 1, Fig. 8) comprises three main units. Unit I (380–300 cm) is dominantly slightly muddy sand to muddy sand, corresponding to very poorly to extremely poorly sorted and strongly positively skewed fine sand to coarse silt. Unit II (300–80 cm) is a monotonous sequence of mud to sandy mud (88.5% average content of mud), very poorly to extremely poorly sorted and near-symmetrical to strongly negatively skewed. Unit III (0–80 cm) is composed of extremely poorly sorted sandy mud (56% average of mud content) that shows positive to strongly positive skewness.

The entire sequence is thought to represent a change in energy level of deposition within the floodplain. The basal unit was deposited under the highest energy conditions, which decreased continuously through time. The intermediate unit is interpreted to represent a floodplain channel-fill with quiet and monotonous settling of fine sediment from suspension. This was replaced abruptly by the present-day environment (top unit), i.e. the floodplain (Fig. 9b).

Santarém Entre Valas core (SEV)

The lower section of the SEV core examined in this study starts at 804 cm in unit II (804–297 cm; Table 1 & Fig. 8), which is a very uniform, monotonous accumulation of slightly sandy mud. The modal sediment is medium and fine silt (mean diameter: 6.3–8.0 ϕ), poorly to extremely poorly sorted with a near symmetrical to strongly negative skewness. The mud content ranges between 79 and 95%, averaging 90% for the entire unit.

At 2.97 m, the mud content drops drastically to 21% and the sign of the skewness changes from

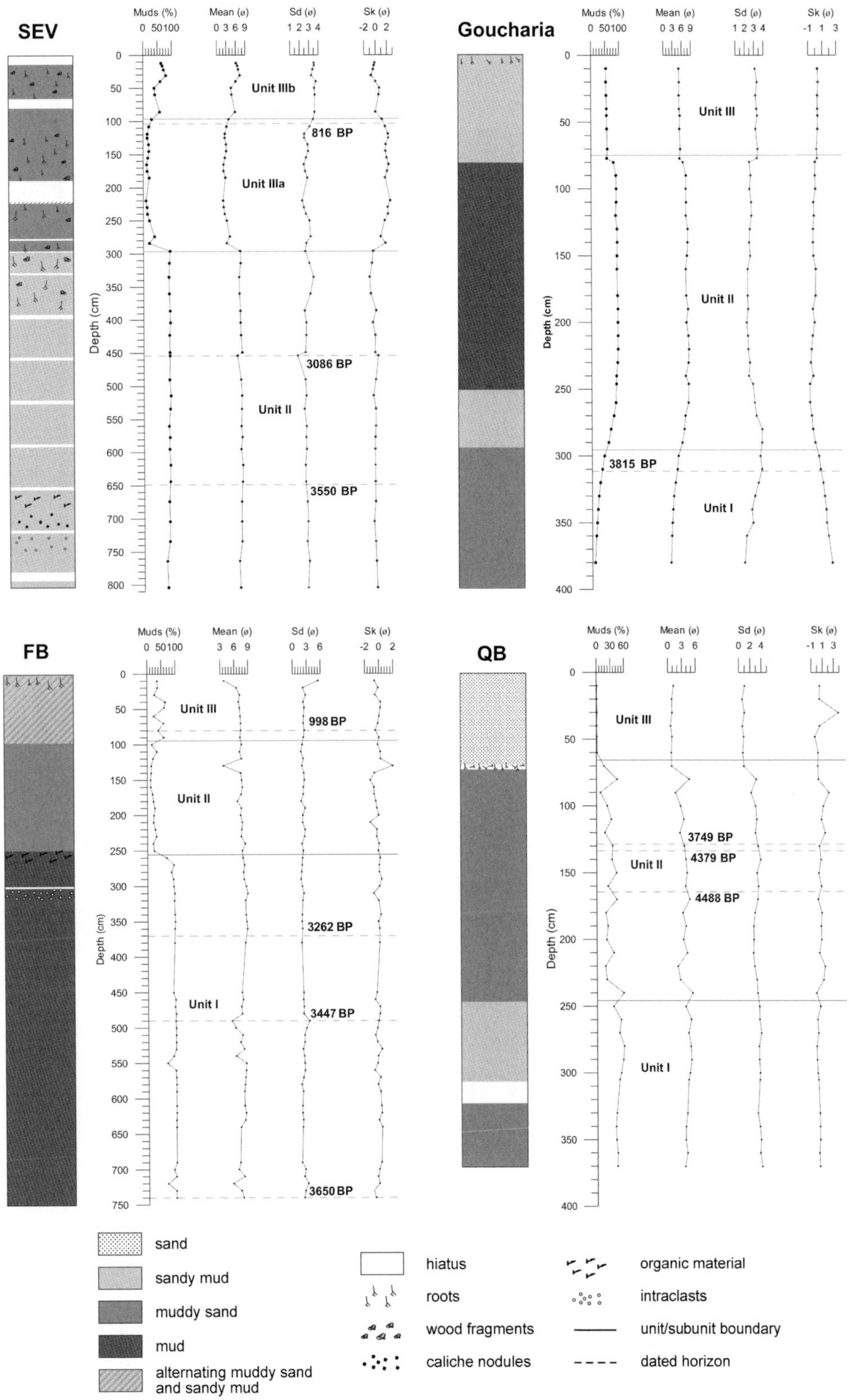

Fig. 8 Lithological columns and mud content, textural parameters, unit division and dated horizons of the Santarém Entre Valas (SEV), Goucharia (G), Fonte Bela (FB) and Quinta da Boavista (QB) cores.

Table 1 Mud content and textural parameters of the gross stratigraphical units identified in the Quinta da Boa Vista (QB), Goucharia (G), Santarém Entre Valas (SEV) and Fonte Bela (FB) cores

Unit		Quinta da Boavista (QB) ($n = 36$)					Unit		Goucharia (G) ($n = 37$)				
		Depth (cm)	Mud (%)	Mean (ϕ)	SD (ϕ)	Sk (ϕ)			Depth (cm)	Mud (%)	Mean (ϕ)	SD (ϕ)	Sk (ϕ)
III	Range	70–0	1.0–1.5	0.6–1.2	0.6–1.0	−0.3–(+3.9)	III	Range	80–0	53.7–58.8	5.1–5.6	3.1–3.4	0.2–0.4
	Mean		1.3	0.9	0.8	0.8		Mean		56.0	5.3	3.3	0.3
II	Range	250–70	8.5–45.1	0.9–4.7	0.9–3.9	0.2–2.2	II	Range	300–80	61.0–96.0	6.0–8.3	2.2–3.8	−0.8–0.1
	Mean		27.2	3.2	2.9	0.8		Mean		88.5	7.5	2.7	−0.3
I	Range	370–250	35.7–58.5	3.6–5.3	3.3–4.1	−0.1–(+0.6)	I	Range	380–300	9.6–45.1	2.2–4.7	2.0–3.8	0.5–2.4
	Mean		47.5	4.3	3.7	0.3		Mean		24.8	3.3	3.0	1.4

Unit/ Subunit		Santarém Entre Valas (SEV) ($n = 46$)					Unit		Fonte Bela (FB) ($n = 61$)				
		Depth (cm)	Mud (%)	Mean (ϕ)	SD (ϕ)	Sk (ϕ)			Depth (cm)	Mud (%)	Mean (ϕ)	SD (ϕ)	Sk (ϕ)
IIIb	Range	97–0	39.1–80.3	4.6–7.4	3.2–3.7	−0.7–(+0.7)	III	Range	100–0	23.3–64.3	3.8–7.4	2.2–5.5	−0.6–(+0.2)
	Mean		60.2	6.0	3.5	0.0		Mean		44.9	4.0	2.6	0.7
IIIa	Range	297–97	9.4–39.0	1.9–4.1	2.3–3.4	0.9–2.6	II	Range	260–100	11.5–34.5	3.7–8.3	1.8–2.6	−1.3–2.0
	Mean		18.7	2.7	2.7	1.9		Mean		21.1	3.1	1.9	0.5
II	Range	804–297	79.2–94.7	6.3–8.0	1.7–3.4	−1.1–(+0.4)	I	Range	740–260	67.9–99.7	5.3–8.8	1.7–3.4	−0.9–(+0.4)
	Mean		90.1	7.4	2.7	−0.2		Mean		94.1	7.6	2.1	−0.1

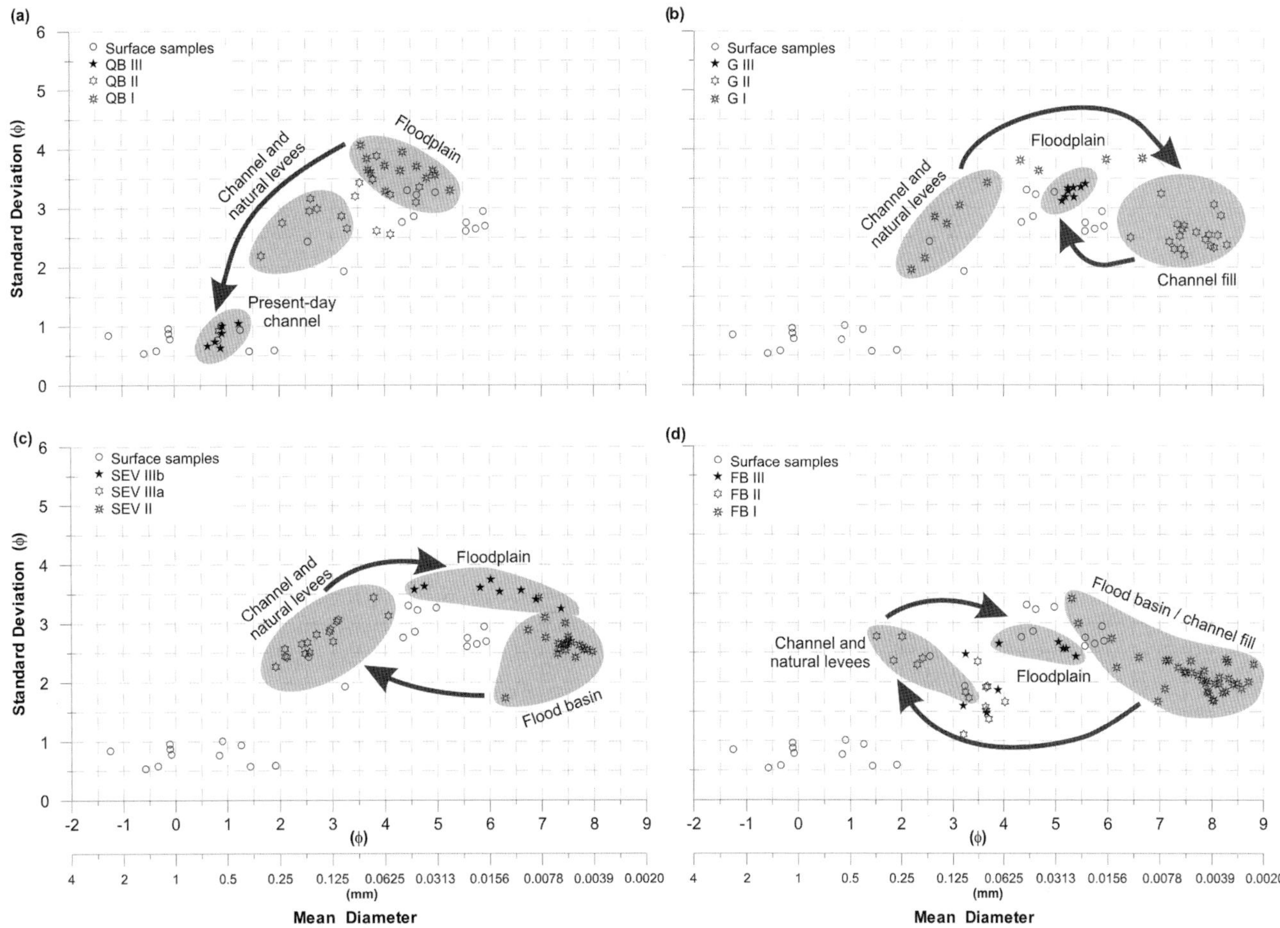

Fig. 9 Plot of the mean grain size versus standard deviation for the surface and core samples: (a) Quinta da Boavista, (b) Goucharia, (c) Santarém Entre Valas and (d) Fonte Bela. See Fig. 7 for details of surface samples. Core units are described in Fig. 8. The arrows show the chronological sequence of the fluvial environments.

negative to positive, showing clear evidence of a change in the sedimentological environment in the transition from unit II to unit IIIa. This unit (297–97 cm; Table 1 & Fig. 8) consists essentially of fine to very fine, slightly muddy sand to muddy sand (mean = 2.7 ϕ), with a low mud content, averaging 19% (9–39%). The organic matter content is very low (< 1%). The sediment is very poorly to extremely poorly sorted and strongly positively skewed. It shows textural similarity with unit I of the Goucharia core. Unit IIIb (0–97 cm), the uppermost unit, shows strong oscillations in mud content (39–80%) with an average of 60%, corresponding to slightly sandy mud to muddy sand. The sediment is extremely poorly sorted very coarse to fine silt (mean diameter: 4.6–7.4 ϕ); the sign of the skewness changes from strongly positive or near-symmetrical in the lower half of the unit to negative and strongly negative in the upper half.

The environmental changes interpreted in this core are: a very low energy (floodbasin) environment represented by the lower unit, dominated by suspension deposition (Fig. 9c). This was followed by the approach of a channel and associated natural levees illustrated by the intermediate unit and finally the present-day floodplain environment (Fig. 9c).

Fonte Bela core (FB)

The basal unit of this core (740–260 cm; Table 1 & Fig. 8) consists essentially of mud and slightly sandy mud (average mud content 94%). The sediment is poorly sorted to very poorly sorted fine to very fine silt, with a strongly negative to near symmetrical skewness. The topmost sample of this unit is a sandy mud, and it is unclear if this sediment reflects the present-day dynamics of flow in the cored ditch or the transition to the coarser sediments of unit II. The intermediate unit (unit II, 260–100 cm; Table 1 & Fig. 8) consists of slightly muddy sand and occasionally, especially in the lower half, muddy sand, corresponding to moderately to extremely poorly sorted, medium to very fine sand. The transition to the upper unit (0–100 cm) is again marked by an abrupt increase in the mud content. Unit III consists of alternating muddy sand and sandy mud. The sediment consists of coarse silt to very fine sand, with the exception for the top layer where coarse sand occurs; the sorting is poor to extremely poor.

The location of the coring site close to an artificial ditch, which is breached occasionally during higher floods, may have influenced the large amplitude oscillations found in the mud/sand content: breaching of the ditch induces coarser sedimentation while progressive flooding favours the accumulation of finer sediments. The plot of grain size versus standard deviation of samples from FB shows that the cluster of unit I is representative of the lower part of the floodplain (flood basin or channel fill) (Fig. 9d). The transition between units I and II is interpreted to correspond to an avulsion, with the establishment of a channel environment. Unit III represents the present-day floodplain with several crevasse splays, which no longer occur due to the building of an artificial ditch.

HEAVY MINERALS AND CLAY MINERALOGY

Study of the heavy mineral suite revealed two populations. The most abundant transparent mineral is andalusite, followed by tourmaline, garnet and zircon; secondary in abundance were the minerals fibrolite, sillimanite, staurolite, anatase, rutile, sphene, brookite, disthene, epidote and hornblende. This assemblage was found to be virtually constant in all the cores and corresponds to 'assemblage B' defined by Oliveira (1967), which is typical of Palaeozoic magmatic and metamorphic rocks. The assemblage is thought to have had a polycyclic origin, supplied by the Tertiary detrital sediments that crop out in both banks of the river. The vertically invariant heavy mineral assemblage illustrates that provenance throughout the Holocene remained the same.

Analysis of the clay mineralogy in the cores revealed a very consistent assemblage in both space and time. Smectite, kaolinite and illite were found to occur in different percentages. Instead of making a core by core characterization, reference to the clay minerals will only be made when these complement other information on the sedimentary environment.

Comparative analysis of cores and analogue environments

The plot of mean diameter (Mz) versus standard deviation (SD) of present-day near-surface samples (Fig. 7) enabled two main environmental domains to be identified, corresponding to (i) fluvial channel and natural levees and (ii) floodplain, separated at a grain size of 3.5–4 ϕ (very fine sand). Around this boundary should plot samples of crevasse splay sediments. The first domain can be divided in two fields.

1 The present-day fluvial channel (marginal channel bars, point bars, natural levees), consisting of very fine gravels to medium sands ($-1.5\ \phi < Mz < 2\ \phi$) with a standard deviation between 0.5 ϕ and 1.0 ϕ. This assemblage is not represented in the SEV, FB and G cores (Fig. 9), but only in QBIII, corresponding to the present-day Tagus channel.
2 The second field is bounded by $2.5\ \phi < Mz < 3.5\ \phi$ and $1.9\ \phi < SD < 2.5\ \phi$ where two samples of former natural levees (surface samples) cluster. The subsurface counterpart of this environment was found in SEV IIIa, GI and QB II (Fig. 9), although the limits of grain size and sorting bracketing the equivalent field in the cored sediments were larger. This suggests for these core sediments a sedimentary environment similar to the late Holocene Tagus River channel.

The second environmental domain includes samples from the present-day floodplain, defined by $4\ \phi < Mz < 7\ \phi$ and $2.25\ \phi < SD < 3.6\ \phi$ (Fig. 7). Samples from SEV II, SEV IIIb, G II, G III, QB I, FB I, FB II and FB III plot in this field (Fig. 9), but correspond to mean grain sizes between 3.5 ϕ and 9 ϕ and SDs between 1.5 ϕ and 4.5 ϕ. This is also greater

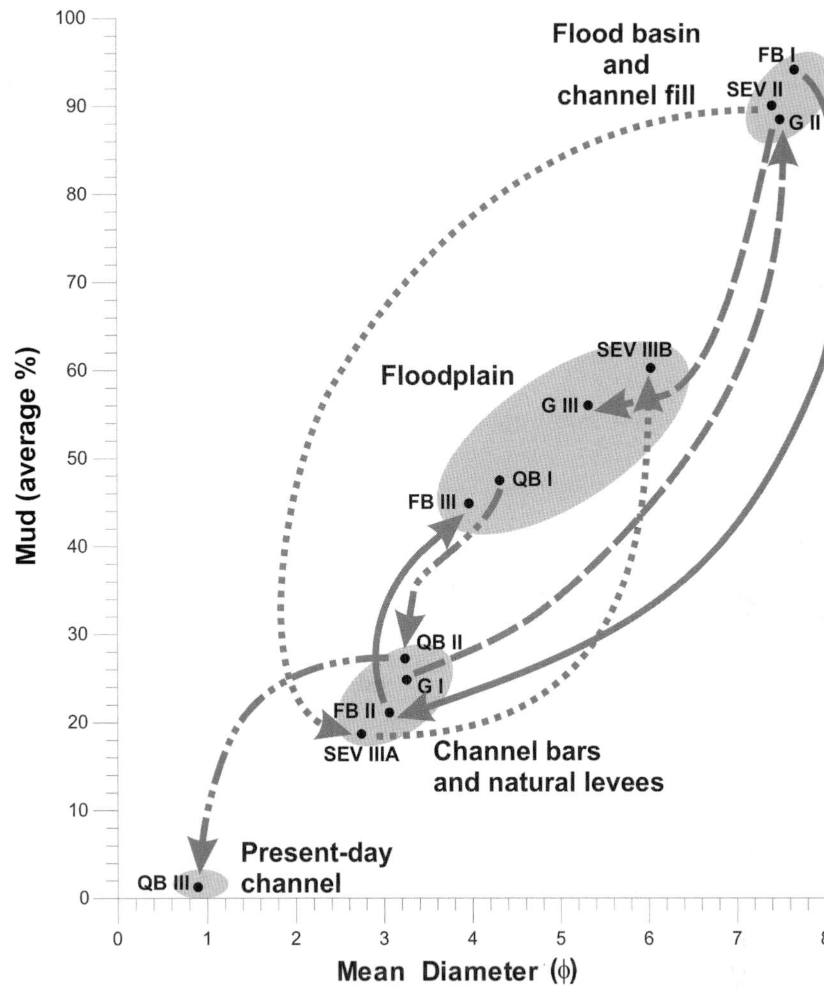

Fig. 10 Plot of the mean diameter versus the average of mud (silt + clay) percentage, showing the position of the sedimentological units identified in the cores in terms of the interpreted environmental domains.

than the range of variation of the present-day analogues.

The floodplain sedimentary units can be organized into two different floodplain domains when the average of the mean diameter versus the average of mud percentage in each unit is correlated (Fig. 10). Units SEV II, G II and FB I represent very fine sediments (7 ϕ > Mz > 8 ϕ) with a high mud content (> 80%) which, again, have no analogues on the present-day Tagus floodplain. As these are the finest sediments, they are most likely to have settled in the lowest topographical features of the floodplain system, i.e. they may represent the infill of abandoned floodplain drainage or crevasse channels (Bridge, 1984), or more distal areas from the main channel (backswamp or floodbasin deposits). This is suggested by the presence of smectite, which only forms in poorly drained areas, where cations can remain available for incorporation in the lattice of clay minerals.

The changing environments revealed by the textural parameters (Fig. 8) permit three different kinds of changes of the Tagus River to be proposed, as follows.

1 *Avulsions*: from unit II to unit III in QB; from unit II to unit III in G; from unit II to unit IIIa in SEV; from unit I to unit II in FB.
2 *Channel migration*: from unit I to unit II in G. The textural parameters show a progressive deviation of the channel environment (Fig. 8).
3 *Crevasse splays*: in the topmost section of unit I and along unit II of QB and in unit III of FB. The QB core (Fig. 8) reveals at least six crevasse splays prior to avulsion of the channel, represented by unit III. In the upper unit of core FB, four crevasse splays are recognized (Fig. 8).

Table 2 Radiometric (^{14}C) ages from organic matter horizons in the Quinta da Boavista (QB), Goucharia (G), Santarém Entre Valas (SEV) and Fonte Bela (FB) cores

Laboratory code	Bank	Depth below surface (m)	$\delta^{13}C$ (‰)	Date (^{14}C yr BP ± 1σ)	Calibrated age* (2σ calibrated results) (yr BP)
Beta-150354 QB6	Right	1.28–1.29	−25.2	3480 ± 40	3749 (3850–3640)
Beta-150356 QB4(B)	Right	1.32–1.33	−25.5	3920 ± 40	4379 (4440–4240)
Beta-150355 QB4(T)	Right	1.62–1.63	−24.8	4020 ± 50	4488 (4780–4780 and 4600–3400)
Beta-184658 G-310	Left	3.10–3.11	−24.0	3530 ± 40	3815 (3900–3700)
Beta-174116 SEV82	Right	1.03–1.04	−24.4	900 ± 40	816 (920–720)
Beta-174117 SEV454	Right	4.54–4.55	−25.9	2930 ± 40	3086 (3220–2950)
Beta-184659 SEV649	Right	6.49–6.50	−25.1	3320 ± 40	3550 (3640–3460)
Beta-184660 SEV1074	Right	10.74–10.75	−25.1	6090 ± 40	6960 (7020–6850 and 6840–6800)
Beta-138920 FB2	Right	0.69–0.70	−25.0	1090 ± 70	998 (1170–910)
Beta-184657 FB-370	Right	3.69–3.70	−24.9	3040 ± 40	3262 (3350–3140)
Beta-150351 FB(−240)	Right	4.89–4.90	−26.1	3230 ± 40	3447 (3550–3370)
Beta-150352 FB(−495)	Right	7.44–7.45	−26.0	3400 ± 40	3650 (3720–3560)

*See Fig. 8 for stratigraphic location.

DATING, SEDIMENTATION RATES AND CHANNEL CHANGES

Numerical ^{14}C dates were used to evaluate sedimentation rates and the timing of channel avulsions, channel migration and emplacement of crevasse splays. Table 2 contains results of the ^{14}C dating of sediment samples taken from all of the cores. The ^{14}C dating of sediments studied in this paper indicates a middle to late Holocene age (Table 2 & Fig. 8).

Based on the dating results (Table 2), a mean sedimentation rate during the past 4000 yr at each core location was integrated, using the radiocarbon date nearest that age and the present day; this method yielded 1.8 mm yr^{-1} for SEV on the right bank, roughly double the value of 0.8 mm yr^{-1} found on the left bank (G). The FB core site, in a medial position on the floodplain, showed the highest value, 2.0 mm yr^{-1}, while at QB it was only 0.3 mm yr^{-1}. This low value is probably related to the present-day position and the erosion of the channel. The calculated sedimentation rates at SEV and FB were higher in the lower part of the cores (before circa 3200 cal. yr BP) and show overall uniformity in the upper part of the two cores (after circa 3200 cal. yr BP). This may be explained by a decrease in erosion in the drainage basin or a decrease in the flood frequency related to climatic events. However, only future research will clarify this issue.

In order to interpolate dates for the main sedimentological changes recognized, linear or polynomial fits to the ^{14}C age and depth data were performed;

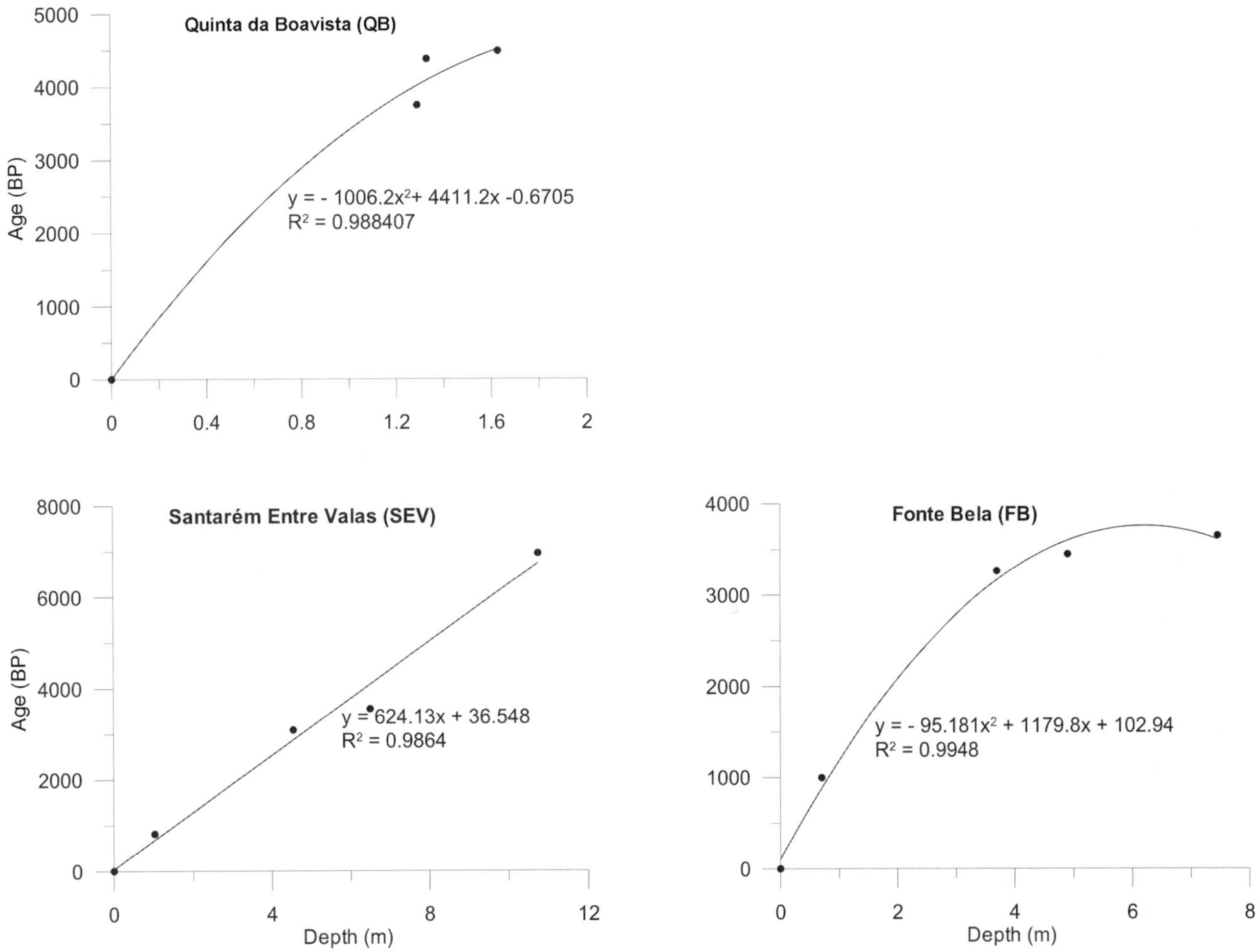

Fig. 11 Cross-plots showing the correlation between ^{14}C numerical ages and depth in three cores.

correlation coefficients (R^2) were greater than 0.98 (Fig. 11). In Quinta da Boavista (QB), a more distal floodplain environment prevailed until *c.* 4700 cal. yr BP, followed by more proximal floodplain environments where several crevasse splay episodes occurred, before the establishment of channel-belt sedimentation at *c.* 2590 cal. yr BP. Goucharia (not shown in Fig. 11) reveals a channel environment before *c.* 3570 cal. yr BP, which was filled-in until *c.* 950 cal. yr BP, when the present-day floodplain environment was established. On the opposite bank, Santarém Entre Valas (SEV) changed from a floodplain to a channel environment at *c.* 1770 cal. yr BP; the present-day situation, between ditches, was reached *c.* 540 cal. yr BP. Fonte Bela (FB) seems to have been a flood basin environment until the occurrence of an avulsion between *c.* 2530 and *c.* 2460 cal. yr BP. The channel environment prevailed until *c.* 1100 cal. yr BP when the present-day floodplain environment was established, with several crevasse-splay episodes.

CONCLUSIONS

The Tagus River is presently a sand bedload, single-channel river with alternate bars, although historical and cartographic documents indicate a former anastomosed river. These documents have revealed that since Roman times, and especially from the 16th to the 19th century, significant anthropogenic modification of the river and its floodplain took place. In the floodplain, former and present natural levees, crevasse splays as well as flood basins and abandoned channels have been recognized. The spatial distribution of these morphological elements over the floodplain has been used to define the historical position of the Tagus channels.

A sedimentological approach enabled the comparison of textural parameters of present-day geomorphological elements with those of the samples collected in the cores. This comparison allowed four different domains to be defined: (i) the present-day channel, with a low content of mud (< 5%) and the coarsest material (< 1 ϕ, medium sand); (ii) channel bars and natural levees from previous channel-belts, with mud contents from 15 to 30% and a mean grain-size range from 2.5 to 3.5 ϕ (fine to very fine sand); (iii) the floodplain itself, with 40–60% mud (3.5–5.5 ϕ, very fine sand to coarse silt); (iv) the more distal environment (flood basin and floodplain channel fill) with the highest content of mud (> 85%) and a mean grain size ranging from 6.5 to 7.5 ϕ (fine to very fine silt).

The correlation of mud percentage versus mean diameter and the grain-size statistics of sediment show the dynamics of this complex alluvial plain, where several avulsions and crevasse splays, as well as channel migration events, have been recorded during the last 4000 years. The results demonstrate that textural analysis is still a valuable tool in characterizing and evaluating sedimentary environments, and in identifying environmental changes in alluvial plain sequences.

After analysing the different sedimentary units of the cores, and ascribing them bounding ages obtained from linear and polynomial fits of ^{14}C dates, it is possible to affirm that:

1 the highest sedimentation rates occurred in the flood basin and floodplain channel fill domains, with values of 2.2 mm yr^{-1} in SEV II and 4.72 mm yr^{-1} in FB I;
2 lowest sedimentation rates were recorded in the channel-belt (0.3 mm yr^{-1} in QB III);
3 intermediate environments were characterized by sedimentation rates between 0.81 and 1.60 mm yr^{-1}, corresponding to sedimentary environments in the proximal floodplain, with several crevasse-splay episodes.

In the Goucharia core, it was not possible to make a similar distinction as there is only one numerical date available, giving an average rate of 0.8 mm yr^{-1} in the past *c.* 3815 cal. yr BP.

It is also possible to conclude that in the past 4000 yr (the period common to the four cores), sedimentation rates become lower when approaching the present. Increasing human intervention in the drainage basin during this period is expressed by a drastic reduction in the forested area and by the expansion of pasture and agricultural land, especially from 3000 yr ago. This situation is thought to favour an increasing supply of sediment. However, a simultaneous trend towards the decrease of long rainy periods was observed. This trend is due to the progressive mediterraneanization of the climate, expressed by increasing aridity. This xerification seems to have outweighed the sediment availability as a factor, which resulted in a decreasing sedimentation rate. In the interval studied, however, it is nevertheless possible to identify an episode of high sedimentation rate, circa 3500–3000 cal. yr BP, which reached 6.5 mm yr^{-1} at Fonte Bela; the reasons for this are not yet clear.

ACKNOWLEDGEMENTS

This work is part of the project POCTI/CTA/39427/2001-C.C.2079, approved by the Fundação para a Ciência e a Tecnologia (FCT) and co-founded by FEDER. It has also been supported by the Centro de Geologia da Universidade de Lisboa, the Centro de Estudos Geográficos da Universidade de Lisboa and the Centro de Ciências da Terra da Universidade do Minho. We thank our colleagues Eng. João Rocha, João Fernandes and Nuno Charneca, from the National Laboratory of Civil Engineering, for the data provided. Finally, the authors would also like to thank Edward Williams, Esther Stouthamer and Ruth Robinson for their critical and constructive reviews.

REFERENCES

Alday, M., Cearreta, A., Cachão, M., Freitas, M.C., Andrade, C. and Gama, C. (2006) Micropaleontological record of Holocene estuarine and marine stages in the Corgo do Porto rivulet (Mira River, SW Portugal). *Estuar. Coast. Shelf Sci.*, **66**, 532–543.

Alves Dias, J.J. (1984) Uma grande obra de engenharia em meados do século XVI: A mudança do curso do rio Tejo. *Nova Hist. Sec. XVI*, **1**, 66–82.

Azevêdo, T.M. (1983) *O Sinclinal de Albufeira – evolução pós-miocénica e reconstituição paleogeográfica.* Unpublished PhD thesis. C.E.G. da F.C.U.L., Lisboa, 321 pp.

Azevêdo, T.M. (1987) Reconstituição paleogeográfica do Tejo no Plio-quaternário. Que Tejo, que futuro? *Assoc. Amigos Tejo*, **2**, 27–31.

Azevêdo, T.M. (1997) Arquitectura deposicional del pré-Tajo en el Plio-Quaternario. Abstracts, *III Cong. Grupo Esp. del Terciário*, Cuenca, pp. 49–52.

Azevêdo, T.M. (2001) A utilização dos dados históricos no estudo das cheias do Tejo. *Estud. Quatern.*, **4**, 69–77.

Azevêdo, T.M. (2004) As mudanças de percurso do Tejo nos tempos modernos. Causas naturais e antrópicas. In: *Evolução Geohistórica do Litoral Português e Fenómenos Correlativos. Geologia, História, Arqueologia e Climatologia* (Eds A.A. Tavares, M. José, F. Tavares and J.L. Cardoso), pp. 517–567. Universidade Aberta, Lisbon.

Azevêdo, T.M., Nunes, E. and Ramos, C. (2004) Some morphological aspects and hydrological characterization of the Tagus floods in the Santarém region, Portugal. *Nat. Hazards*, **31**, 587–601.

Baldwin, B. (1971) Ways of deciphering compact sediments. *J. Sediment. Petrol.*, **41**, 293–301.

Barbosa, B.P. (1995) *Alostratigrafia e Litostratigrafia das Unidades Continentais da Bacia Terciária do Baixo Tejo. Relações com o eustatismo e a tectónica.* Unpublished PhD thesis, University of Lisbon, 253 pp.

Bridge, J.S. (1984) Large-scale facies sequences in alluvial overbank environments. *J. Sediment. Petrol.*, **54**, 583–588.

Bridge, J.S. and Leeder, M.R. (1979) A simulation model of alluvial stratigraphy. *Sedimentology*, **26**, 617–644.

Cabral, J. (1995) Neotectónica em Portugal Continental. *Mem. Inst. Geol. Min.*, **31**, 265 pp.

Cabral, J., Moniz, C., Ribeiro, P., Terrinha, P. and Matias, L. (2003) Analysis of seismic reflection data as a tool for the seismotectonic assessment of a low activity intraplate basin – the Lower Tagus Valley (Portugal). *J. Seismol.*, **7**, 431–447.

Carvalho, A.M.G. (1968) Contribuição para o Conhecimento Geológico da Bacia Terciária do Tejo. *Mem. Serv. Geol. Port.*, **15**(Nova Série), 210 pp.

Cearreta, A., Alday, M., Freitas, M.C., Andrade, C. and Cruces, A. (2002) Modern and Holocene foraminiferal record of alternating open and restricted environmental conditions in the Santo André lagoon, SW Portugal. *Hydrobiologia*, **475/476**, 21–27.

Custódio, J. (1992–93) Alpiarça, o lugar, a freguesia e o concelho no distrito de Santarém. *PDM de Alpiarça.* Alpiarça.

Daveau, S. (1970) Le bassin tertiaire du Tage: problèmes d'interpretation géomorphologique, *Finisterra*, **10**, 291–300.

Daveau, S. (1980) Espaço e tempo. Evolução do ambiente geográfico de Portugal ao longo dos tempos pré-históricos. *Clio*, **2**, 13–37.

Dias, J.M.A. (1987) *Dinâmica sedimentary e evolução recente da plataforma continental portuguesa setentrional.* Unpublished PhD thesis, University of Lisbon, 384 pp.

Dolfus, G.F. and Berkley-Cotter, J.C. (1909) Mollusques tertiaires du Portugal. Le Pliocène au Nord du Tage. *Mem. Com. Serv. Geol. Port.*, Lisboa, 103 pp.

Fairbanks, R.G., Mortlock, R.A., Chiu, T.C., *et al.* (2005) Marine radiocarbon calibration curve spanning 10,000 to 50,000 years B.P. based on paired ^{230}Th/^{234}U/^{238}U and ^{14}C dates on pristine corals. *Quatern. Sci. Rev.*, **24**, 1781–1796.

Fleming, B.W. (2000) A revised textural classification of gravel-free muddy sediments on the basis of ternary diagrams. *Cont. Shelf Res.*, **20**, 1125–1137.

Freitas, M.C. and Ferreira, T. (2004) *A Lagoa de Albufeira. Geologia.* Instituto da Conservação da Natureza/Centro de Zonas Húmidas, pp. 11–52.

Freitas, M.C., Andrade, C. and Cruces, A. (2002) The geological record of environmental changes in southwestern Portuguese coastal lagoons since the Late glacial. *Quatern. Int.*, **93–94**, 161–170.

Freitas, M.C., Andrade, C., Rocha, F., *et al.* (2003) Late glacial and Holocene environmental changes in Portuguese coastal lagoons: 1. The sedimentological and geochemical records of the Santo André coastal area (SW Portugal). *The Holocene*, **13**, 433–446.

Friedman, G.M. (1961) Distinction between dune, beach and river sands from their textural characteristics. *J. Sediment. Petrol.*, **31**, 514–529.

Haner, B.E. (1984) Santa Ana River: an example of a sandy braided floodplain system showing sediment source area imprintation and selective sediment modification. *Sediment. Geol.*, **38**, 247–261.

Henriques, R. (1998) *Propostas Metodológicas para a Monitorização das Zonas Costeiras – Aspectos Sedimentológicos.* Unpublished MSc Thesis, Faculdade de Ciências da Universidade do Porto, 185 pp.

Henriques, R. (2003) SEDMAC/SEDPC: programa informático de apoio à análise dimensional de populações detríticas. *Ciências da Terra* – Volume Especial, VI Congresso Nacional de Geologia, Faculdade de Ciências e Tecnologia da Universidade Nova de Lisboa, 40.

Henriques, R.F. (2004) SEDMAC/SEDPC: An application to support particle size analysis of unconsolidated sediments. *32nd International Geological Congress*, Abstracts Vol., pt. 1, abs. **154–6**, 726.

Hirn, A., Senos, L. and Caetano, H. (1981) *Variações da profundidade Moho na região da grande falha do Alentejo*. Inst. Nac. Met. Geof. (INMG), Lisboa.

Kraus, M.J. and Bown, T.M. (1993) Short-term sediment accumulation rates determined from Eocene alluvial palaeosols. *Geology*, **21**, 743–746.

LNEC (1967) Especificação E-201–1967. *Solos – Determinação do Teor em Matéria Orgânica*, documentação normativa, 3.

Martins, A.A. (1999) *Caracterização morfotectónica e morfosedimentar da Bacia do Baixo Tejo (Pliocénico e Quaternário)*. Unpublished PhD thesis, University of Évora, 500 pp.

Mendonça, J.J.L. (1990) *Sistema aquífero aluvionar do Vale do Tejo (V.N. da Barquinha a Alverca). Características e funcionamento hidráulico*. Unpublished PhD thesis, University of Coimbra, 343 pp.

Miall, A.D. (1996) *The Geology of Fluvial Deposits. Sedimentary Facies, Basin Analysis and Petroleum Geology*. Springer-Verlag, New York, 582 pp.

Middleton, G.V. (1976) Hydraulic interpretation of sand size distributions. *J. Geol.*, **84**, 405–426.

Mozzi, P., Azevêdo, M.T., Nunes, E. and Raposo, L. (2000) The Middle Terrace deposits of the Tagus River in Alpiarça, Portugal, in relation to early human occupation. *Quatern. Res.*, **54**, 359–371.

Oliveira, R. (1967) Contribuição para o Estudo do Estuário do Tejo (sedimentologia). *LNEC Mem.*, **296**, 61 pp.

Pereira, A.R., Ramos, C., Reis, E., *et al.* (2002) A dinâmica da planície aluvial do Baixo Tejo no Holocénico recente: aplicação de métodos de análise geomorfológica e sedimentológica. *Publ. Assoc. Port. Geomorphol.*, **I**, 67–76.

Pereira, A.R., Ramos, C., Azevêdo, M.T. and Nunes, E. (2004) Geomorphological and textural analysis as a tool to evaluate the migration of Tagus river channel during the late Holocene (Portugal). *Geophys. Res. Abstr.*, **6**, 04749, European Geosciences Union, Nice.

Perrier, R. and Quiblier, J. (1974) Thickness changes in sedimentary layers during compaction history; methods for quantitative evaluation. *Am. Assoc. Petrol. Geol. Bull.*, **58**, 507–520.

PNA (2001) *Plano Nacional da Água*, 2 vols. Instituto da Água, Ministério do Ambiente e Ordenamento do Território, Lisboa.

Potter, P.E. (1978) Significance and origin of big rivers. *J. Geol.*, **86**, 13–33.

Ramos, C. and Reis, E. (2002) Floods in southern Portugal: their physical and human causes, impacts and human response. *Mitig. Adapt. Strat. Global Change*, **7**, 267–284.

Ramos, C., Reis, E., Pereira, A.R., *et al.* (2002) Late Holocene evolution of the Lower Tagus alluvial plain and heavy metals content: preliminary results. *Environmental Change and Water Sustainability* (Eds J.M. García-Ruiz, J.A.A. Jones and J. Arnáez), pp. 167–182. Ins. Pirenaico de Ecología, Zaragoza.

Rassmussen, E.S., Lomholt, S.; Andersen, C. and Vejbæk, O.V. (1998) Aspects of structural evolution of the Lusitanian Basin in Portugal and the shelf and slope area offshore Portugal. *Tectonophysics*, **300**, 199–225.

Reimer, P.J., Baillie, M.G.L., Bard, E., *et al.* (2004) Intcal04 Tterrestrial radiocarbon age calibration, 26–0 ka BP. *Radiocarbon*, **46**, 1029–1058.

Ribeiro, A., Antunes, M.T., Ferreira, M.P., *et al.* (1979) *Introduction a la Géologie Général du Portugal*. Serv. Geol. de Portugal, Lisboa, 114 pp.

Ribeiro, A., Kullberg, M.C., Kullberg, J.C., Manuppella, G. and Phipps, S. (1990) A review of Alpine tectonics in Portugal: foreland detachment in basement and cover rocks. *Tectonophysics*, **184**, 357–366.

Ribeiro, C. (1866) Descripção do solo quaternário das bacias hydrográphicas do Tejo e do Sado. *Com. Geol. Portugal*, Lisboa, **6**, 164 pp.

Rosgen, D.L. (1994) A classification of natural rivers. *Catena*, **22**, 169–199.

Rosgen, D.L. (1996) *Applied River Morphology*. Wildland Hydrology, Pagosa Springs, CO, 390 pp.

Stouthamer, E. (2001) Sedimentary products of avulsions in the Holocene Reimer-Meuse Delta, The Netherlands. *Sediment. Geol.*, **145**, 73–92.

Victor, L.A.M., Hirn, A. and Veinante, J.L. (1980) A seismic section across the Tagus valley, Portugal: possible evolution of the crust. *Am. Geophys.*, **36**, 469–476.

Weerts, H.J.T. and Bierkens, M.F.P. (1993) Geostatistical analysis of overbank deposits of anastomosing and meandering fluvial systems; Rhine-Meuse delta, The Netherlands. *Sediment. Geol.*, **85**, 221–232.

Wentworth, C.K. (1922) A scale of grade and class terms for clastic sediments. *J. Geol.*, **30**, 377–392.

Zbyszewski, G. (1949) Contribution à la connaissance du Pliocène Portugais. *Com. Ser. Geol. Port.*, **30**, Lisboa, 59 pp.

Creation and preservation of channel-form sand bodies in an experimental alluvial system

BENJAMIN A. SHEETS[1], CHRIS PAOLA *and* J. MICHAEL KELBERER

National Center for Earthsurface Dynamics / Saint Anthony Falls Laboratory, Department of Geology and Geophysics, University of Minnesota, Minneapolis MN 55414, USA (Email: benjamin.a.sheets@exxonmobil.com)

ABSTRACT

An experiment is described that was designed to investigate the relationship between channel-form sand bodies and scour-and-fill processes in an alluvial system. Despite the experiment's simplified conditions (e.g. restricted grain-size distribution, constant discharge) channel sand bodies preserved in the deposits record a remarkably complex sequence of erosion and deposition. The life span of a channel can be summarized in three phases:

1 initial incision produced by scour associated with convergent flow;
2 a period of multiple episodes of abandonment and reoccupation, which lead to multiple storeys of deposit;
3 burial and preservation by an expanding depositional flow, which creates convex-up topography, and prevents flow reoccupation.

The resultant channel bodies are larger (depth, width) than the cross-sectional geometry of the fluvial channels that created them, although their aspect ratio (width/depth) is comparable to that of scouring flow. The sequence of events leading to the formation and preservation of channel bodies is consistent throughout the experiment, and leads to the preservation of multiple storeys and lateral accretion – sedimentary structures commonly produced by more complex natural systems, and common in the stratigraphic record. Convex topography associated with channel filling and abandonment causes regional avulsion and reorganization of the channel system, which leads to cyclic compensational filling of the experimental basin.

Keywords Alluvial stratigraphy, fluvial stratigraphy, channel body formation, experimental stratigraphy, physical experiments, channel fill.

INTRODUCTION

It is intuitively obvious that low aspect ratio (width:depth < 15:1) ribbon sand bodies (Friend *et al.*, 1979) are related to fluvial channel scour-and-fill, as discussed in fundamental studies by Bersier (1958), Fisk (1944, 1947) and Schumm (1960), as well as the reviews by Allen (1965) and Potter (1967), and numerous subsequent studies. Despite recent advances in quantitative modelling of channel and bar deposition (Bridge, 2003), there remains no quantitative understanding of how instantaneous channel shapes are related to the geometries of preserved channel-form deposits. This is generally due to the fact that scour, fill and preservation of channel sand bodies occur on time-scales somewhat longer than can usually be observed in modern systems. Even where recent studies have been able to relate modern fluvial geometries to shallow subsurface data, the complex

[1]Present address: ExxonMobil Upstream Research Company, PO Box 2189, Houston TX 77252-2189, USA.

sequence of fluvial events that are recorded in the deposits cannot be observed directly (e.g. Best *et al.*, 2003; Lunt *et al.*, 2004).

The present study is an investigation of the relationship between scour-and-fill processes and their stratigraphic products in an experimental alluvial fan-delta setting. The experimental system provides nearly continuous records of topographic evolution and flow pattern that can be used to detail the sequence of events leading to the creation of preserved sand bodies (Ashworth *et al.*, 1999; Moreton *et al.*, 2002; Sheets *et al.*, 2002). In particular, previous experimental studies have shown a clear relationship between erosively based, low aspect ratio channel bodies and convergent flow or confluence scour, as well as between high aspect ratio depositional sheets and overbank sheet-flows (Cazanacli *et al.*, 2002; Sheets *et al.*, 2002). The sheet-flows are expansional, and represent the major mechanism of deposition in the experimental systems.

HIGH-RESOLUTION TOPOGRAPHY EXPERIMENT

The motivation for the experiment (DB 03-1) on which this paper is based was to obtain detailed records of fluvial processes, topographic evolution and stratigraphy, with sufficient spatial and temporal resolution to observe and quantify the deposition of individual ribbon and sheet sand bodies. The experiment was conducted in a 5 m by 5 m experimental facility designed to form physical stratigraphy through sustained net deposition (Fig. 1). Sediment and water were mixed in a funnel and fed into the basin at one corner, producing a radially symmetrical fan-delta, which averaged 2.50 m from source to shoreline. The sediment supply was a mixture comprising 70% 0.120 mm quartz and 30% bimodal (0.190 and 0.460 mm) anthracite. This relatively restricted grain-size distribution has been used effectively in previous experiments (Heller *et al.*, 2001; Paola *et al.*, 2001; Cazanacli *et al.*, 2002; Sheets *et al.*, 2002; Hickson *et al.*, 2005; Strong *et al.*, 2005) due to the optical contrast between quartz and anthracite, and the fact that the anthracite behaves as an analogue for fine-grained sediments due to its relatively low density (1.7 g cm^{-3} versus 2.65 g cm^{-3} for quartz).

Water was withdrawn from the flume through a siphon, attached to a motorized weir, allowing precise (± 0.1 mm) control of base level during an experiment. Subsidence was simulated in the delta basin via a gradual (5 mm h^{-1}, Table 1) rise in the relative base level, at a rate equal to the total sediment discharge (Q_s) into the experiment divided by the desired fluvial system plan-view area (analogous to dropping the basin floor 5 mm h^{-1} while holding the water level constant). This is equivalent to a spatially uniform (piston) subsidence pattern (cf. Moreton *et al.*, 2002).

The experiment began with an initial phase during which there was no subsidence and the fan-delta prograded into standing water until the desired system length (2.50 m) was attained. This was followed by a 30-h equilibrium (no transgression or regression of the shoreline) aggradation phase during which subsidence was active. An average of 150 mm of deposition occurred during this phase. The duration of the experiment was chosen to be long enough to preserve several scour-depths (~ 20 mm) worth of sedimentation.

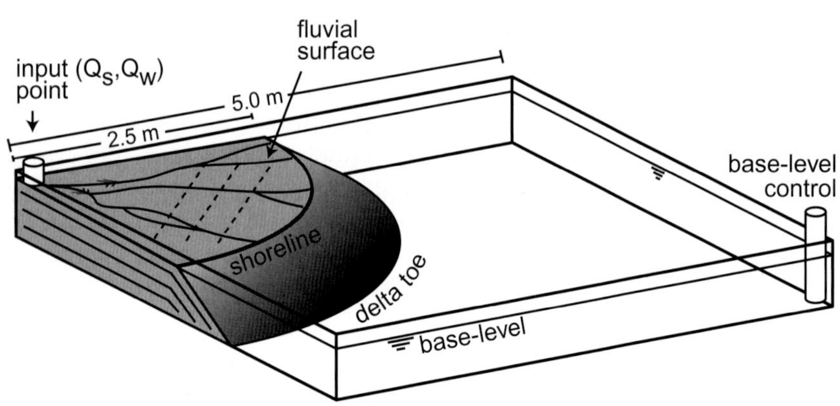

Fig. 1 Schematic diagram of the Delta Basin experimental facility. Positions of the topographic transects are indicated by dashed lines on the fluvial surface. Note that the base level control drain is in the opposite corner of the flume from the experiment. Motorized weir discussed in text is not shown.

Table 1 DB 03-1 experimental parameters

Q_w (L s^{-1})	Q_s (L s^{-1})	Q_w:Q_s	Average aggradation rate (mm h^{-1})	Average slope	Average *Fr**
0.4	0.01	40:1	5.0	0.05	1.1

Q_w, total volumetric water discharge; Q_s, total volumetric sediment discharge.
*A slightly supercritical Froude measurement; see discussion in text.

The topography along three flow-perpendicular transects, located 1.50 m, 1.75 m, and 2.00 m from the infeed point (Fig. 1), was measured at 2-min intervals through the course of the experiment (900 total measurements). These measurements were made from oblique digital images of lines cast by vertical laser sheets from which the true topography can be calculated. Overhead digital images of the experiment were recorded at 15-s intervals, allowing for time-lapse movies of fan-delta evolution.

After the experiment, the deposit was sectioned and imaged at each of the topographic strike-transects. Care was taken to ensure that images of the stratigraphic panels were recorded with the same camera used for the topographic measurements, so that the laser lines could be directly superimposed on the deposits.

It is stressed that the experimental fan-delta was not designed to reproduce all of the detailed behaviour of natural systems, and as such was not scaled to, or intended to simulate, any particular natural system. Specific geometric data measured in the experiment cannot in general be simply 'scaled up' to field scales. It is instead intended to be a self-organized, distributary depositional system in which many of the processes characteristic of its larger relatives ('similarity of process' of Hooke, 1968; Paola, 2000) could be observed with a level of detail impossible to obtain in the field. Thus, the focus is on the mechanisms by which surface topography and kinematics combine with net deposition to produce channel-form stratigraphic bodies, and what this indicates about how preserved channel-form sand bodies are and are not related to the channels that produce them. The gross morphology and processes operating in the experimental system are also heuristically valuable (cf. Moreton *et al.*, 2002), as they suggest phenomena that may be important, although difficult to quantify, at natural spatial and temporal scales.

EXPERIMENTAL RESULTS

The experimental conditions (Table 1) produced a remarkably active fan-delta with relatively high slopes and high wetted fraction. An important consequence of the relatively high sediment discharge to water discharge ratio was a relatively high fluvial slope. These parameters led to an average Froude number that was slightly supercritical, a phenomenon with important consequences for scour behaviour, as discussed below.

Topographic measurements indicate that erosion on the fluvial surface was associated with convergent flow and scour (Fig. 2). Scour points tended to migrate upstream with time, in a manner analogous to headward knickpoint migration, eventually dissipating near the infeed point. The tendency toward upstream migration was largely a consequence of the slightly supercritical flow. Lateral migration of persistent channels is rare in this experiment. Deposition in this system is typically associated with flow expansion, often immediately downstream of scour points (Fig. 2).

Convergent and expanding flow structures are reflected in the deposits, which comprise both low aspect ratio (width/depth) channel bodies and high aspect ratio sheet deposits (Fig. 3). Preserved channel bodies, characterized by their concave-up erosive bases, have a mean aspect ratio of 6, comparable to, but slightly larger than, the aspect

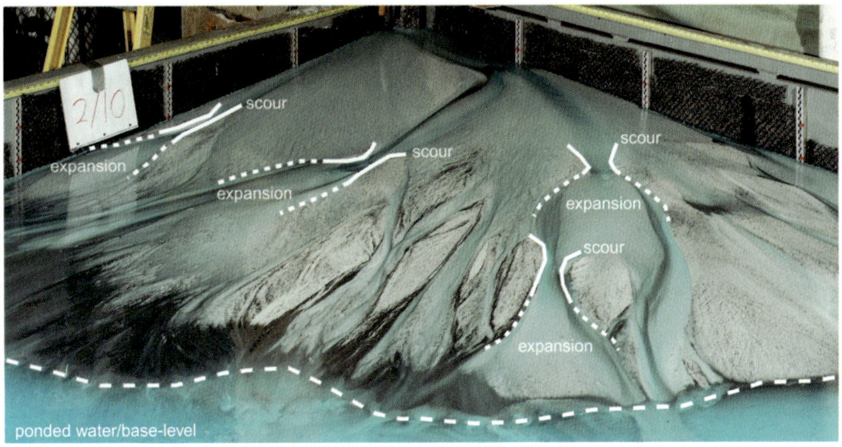

Fig. 2 Photograph taken approximately 1600 min into the DB 03-1 experiment. Major scour and flow-expansion features active at this time are indicated. System is approximately 2.5 m in length from source (back centre) to shoreline. Scour points not occupied by flow are not annotated.

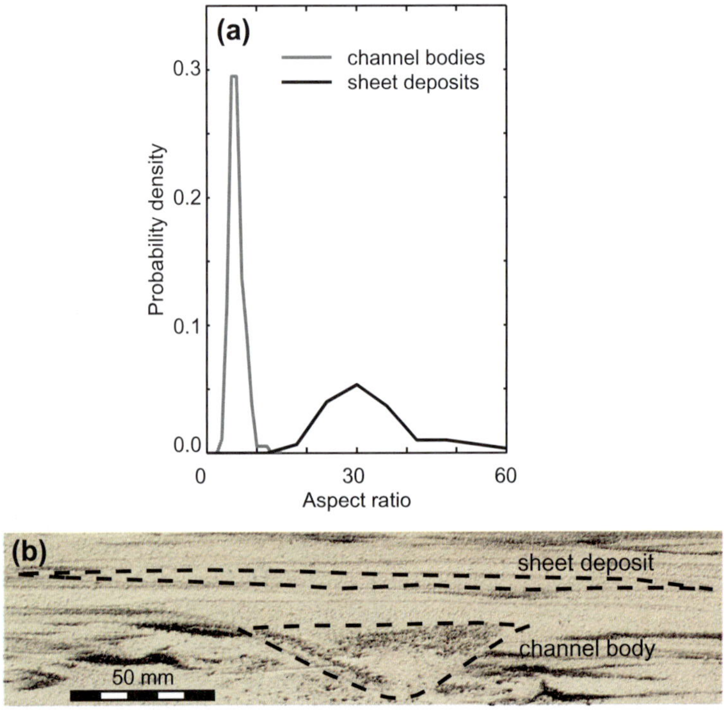

Fig. 3 Channel bodies and sheet deposition. (a) Probability density distribution of channel body and sheet deposit aspect ratios (based on 186 channel measurements; 50 sheets). (b) Examples of a ribbon channel body and a sheet deposit. Note that the channel body is bounded by an erosional surface that cuts older sheet deposits on the right, but is amalgamated with older channel bodies to the left. Sheet deposit is conformable.

ratio of active scours (4). Relatively low channel body aspect ratios are a consequence of the fact that lateral migration of persistent channels was rare under these experimental conditions. Therefore, these channel bodies are analogous to the ribbon sandstone bodies of Friend *et al.* (1979), and the 'fixed channel' classification of Friend (1983), as well as depositional niche F described by Ashworth *et al.* (1999). The 'corps central' (Bersier, 1958) of the majority of the channel bodies in the experimental strata are laterally connected with thinly bedded, depositionally based 'wings' (Bersier, 1958; Friend *et al.*, 1979). The channel bodies often contain multiple storeys of sedimentation, although these can be difficult to identify in the experimental strata, a point that will be returned to below.

Topographic evolution

Aggradation at 2-min intervals is remarkably discontinuous, both temporally and spatially. At each of 27 points (nine per transect), successive elevations were differenced, and the net change compared with a chosen threshold (2.5 mm in the

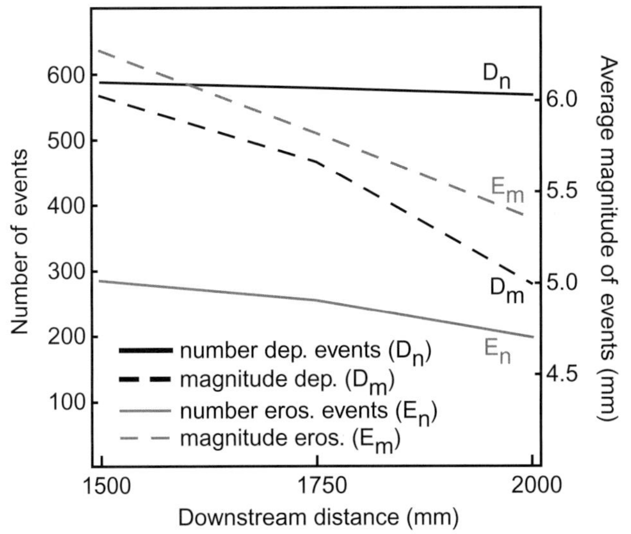

Fig. 4 Statistics of depositional (dep.) and erosional (eros.) 'events' relative to downstream position over course of entire experiment. Numbers of events given by solid lines and magnitude (depth) given in dashed lines. Black indicates deposition and grey erosion. The average number is roughly constant with downstream position, but there are twice as many depositional events.

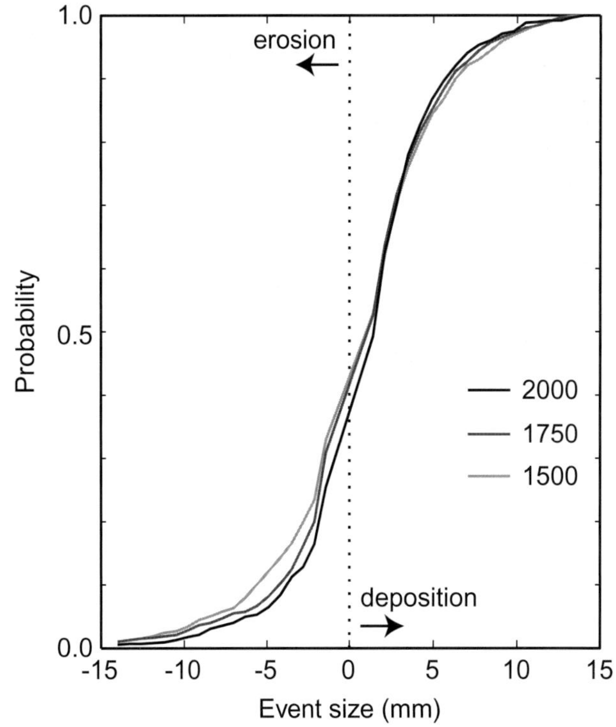

Fig. 5 Cumulative distribution functions of depositional and erosional event sizes (negative values indicate erosion) for all three topographic transects. Note that the distributions become slightly steeper with downstream distance, indicating a decrease in variability.

data presented here; large enough to omit noise in the measurements, but small enough to capture the smallest events). When this threshold is exceeded, the vertical size and time of the event are recorded.

Figure 4 shows the average magnitude, in vertical distance, and average number of both depositional and erosional events. Depositional events are, on average, 5% smaller than, and twice as frequent as, erosional events. Therefore, the total depth of erosion is approximately 50% of the total depth of aggradation. In other words, this system took one step down for every two up. Comparison between topographic transects shows a tendency for the average magnitude of the events to decrease downstream, although the absolute number remains relatively constant with downstream distance. The cumulative distribution functions shown in Fig. 5 indicate decreased variability in the size of depositional and erosional events distally.

Spatial sedimentation patterns

Time–space visualizations (Wheeler, 1958; Wheeler, 1964) of DB 03-1 sedimentation illustrate several interesting aspects of the experimental alluvial system. Figure 6 shows strike-oriented time–space plots of sedimentation at each of the topographic transects, with periods of erosion in white tones, periods of deposition in black tones, and static topography in grey. The bimodal nature of the stratigraphy is reflected in these plots at each of the topographic transects. Relatively large erosional and depositional events are generally paired, apparent as bright white immediately below dark black. This is the record of scour and subsequent filling, which leads to low aspect ratio channel bodies. Sheet deposition is apparent as comparatively lighter (thinner deposits) and wider black regions in the plots, which are typically not associated with an erosional episode. Furthermore, the distal decrease in variability and size of depositional and erosional events discussed above (Figs 4 & 5) is also apparent in Fig. 6 as a general decrease in the intensity of white or black with downstream distance.

The average strike position of depositional or erosional activity is indicated on the centre ($x = 1.75$ m)

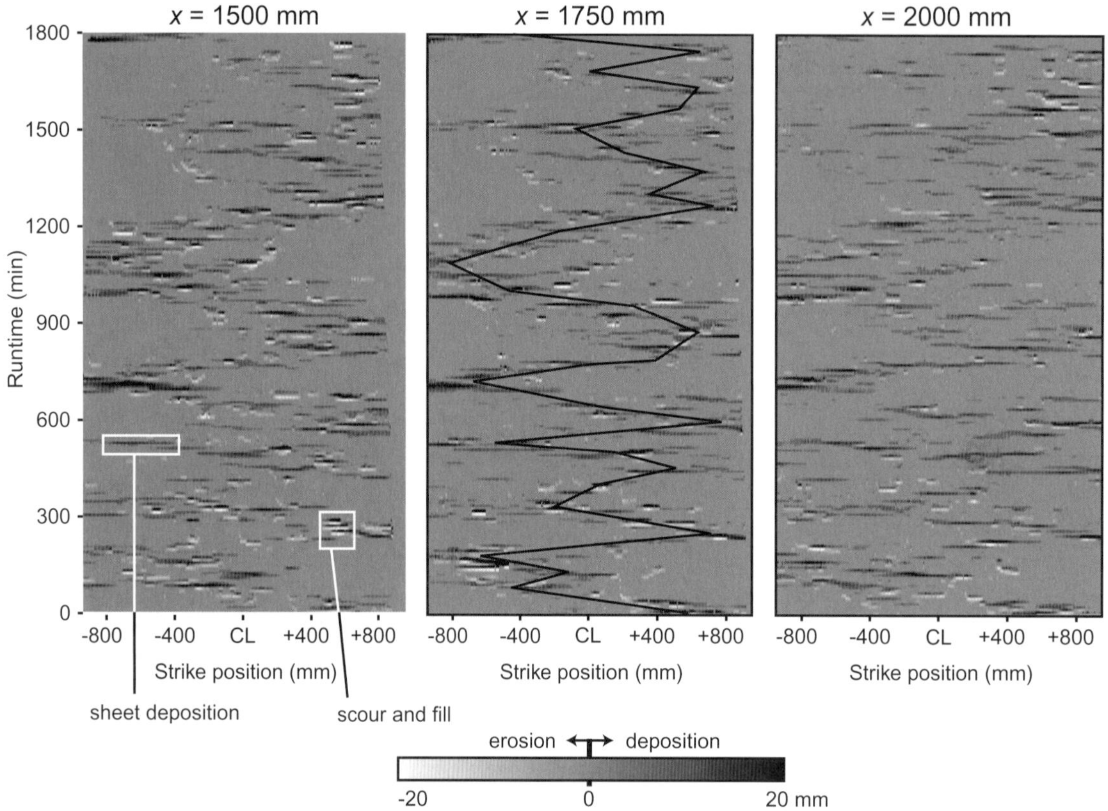

Fig. 6 Time–space plots of DB 03-1 sedimentation at all three topographic transects. Greyscale values and average activity trend (see text for explanation) indicated in $x = 1750$ mm plot. Strike position is given relative to the centreline (CL) of the deposit, which is perpendicular to the topographic transects, and bisects the right-angle walls of the flume (Fig. 1). Representative scour-and-fill event and sheet-deposition event annotated (see text for discussion). Note that there is no indication of lateral migration of persistent channels.

plot in Fig. 6. This is calculated as the along-strike average position of topographic change over the time interval, weighted by magnitude of change. Note that the trend indicated by this line is present at the 2.0 m and 1.5 m transects, but is not shown as it obscures the erosional and depositional data. The trend indicates a pronounced lateral cyclicity to DB 03-1 basin sedimentation. This cyclicity was produced as the fluvial system, which has a finite width over which it can deposit sediment, migrated laterally in order to fill available accommodation. The period between major occupations of particular portions of the floodplain in this experiment is, on average, 4 h.

Relating surface morphology and stratigraphy

As the same camera was used for the surface, topographic and stratigraphic images, the three can be directly compared. This simplifies analysis of the events leading to the creation and preservation of multistorey channel bodies. Figure 7a–d shows the evolution of four representative channel bodies from initial incision (or occupation by flow) to final filling (and flow abandonment). In each case, the sequence of events that led to the ultimate preserved channel body form will be detailed.

Figure 7a shows channel body preservation due to a relatively long-duration sequence. The initial incision and formation of channel topography at this location were due to flow occupation and scour 1619 min into the experiment. The flow subsequently wanes, and the channel widens with sediment depositing on the bed. A second flow occupation and scour occurs 6 min later (1625 min); this time erosion is deeper than before, leading to a maximum topographic (and stratigraphic) depth of approximately 20 mm. In this case, the channel

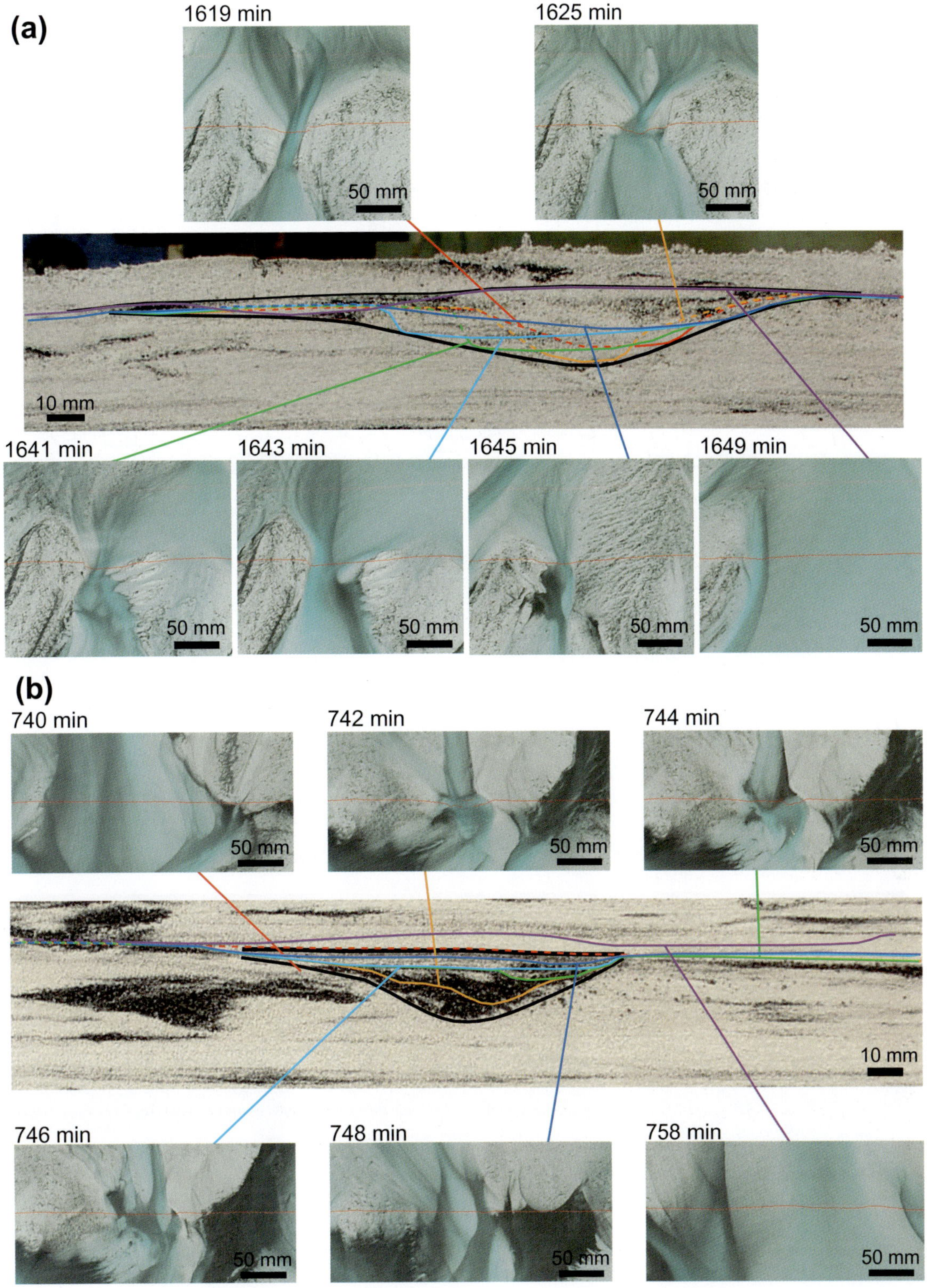

Fig. 7 Sequence of events leading to channel body preservation. (a–d) Evolution of four representative channel bodies. See text for interpretations. Central images show stratigraphic detail of channel bodies with timelines superimposed in colour. Timelines dashed where unambiguously eroded by subsequent topographic scans. Interpreted channel body bounding surfaces indicated by black lines. Surrounding images show fluvial morphology at run time indicated with topographic data superimposed in red. Flow is out of the page. Note that scales vary among examples, particularly in (d). In each example, the initial scour is due to a regional avulsion into this portion of the floodplain. The adjacent, older deposits are unambiguously associated with erosion and deposition from a previous occupation, filling and abandonment of this area (i.e. the older deposits were not produced by scour or bar migration associated with this sequence of flows).

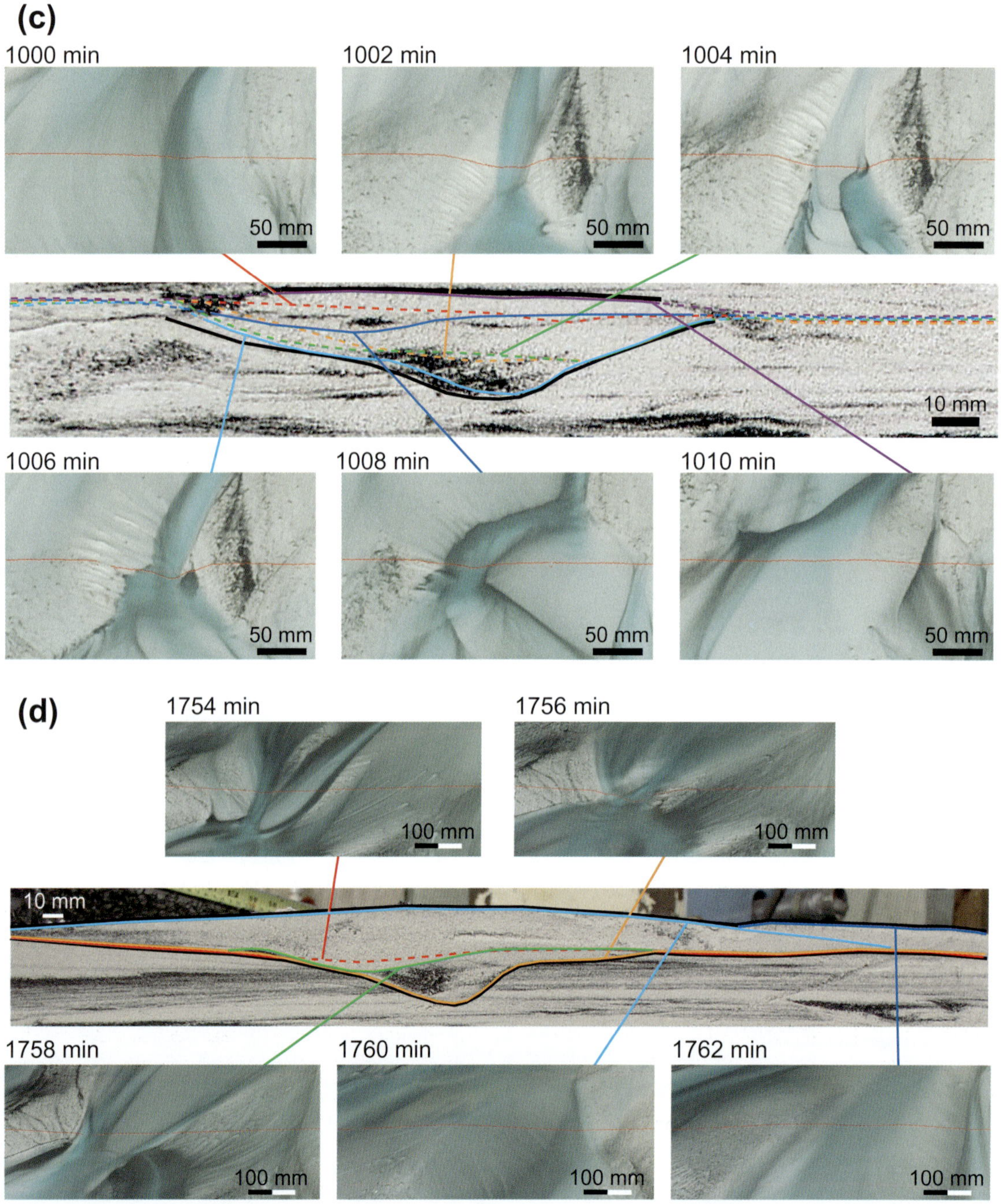
(c)
1000 min
50 mm
1002 min
50 mm
1004 min
50 mm
10 mm
1006 min
50 mm
1008 min
50 mm
1010 min
50 mm
(d)
1754 min
100 mm
1756 min
100 mm
10 mm
1758 min
100 mm
1760 min
100 mm
1762 min
100 mm

Fig. 7 *(cont'd)*

topography is filled slightly before being abandoned (not shown). There is little topographical change at this location until 32 minutes after the initial scour (1641 min). This time, flow returns to the channel from a slightly different direction, which leads to lateral migration (to the left in this perspective). Continuous lateral migration continues for nearly 4 min at a rate of approximately 5 mm min^{-1}. As flow subsequently wanes, the channel bed aggrades and widens, preserving coal along the left channel margin (1645 min). Ultimately, the channel topography is filled by a rapid (< 2 min) 'dump' of sediment produced by an expanding overbank flow associated with an adjacent channel. This expanding flow deposition creates a convex (domed) deposit, preventing future reoccupation of the channel. After this topography-capping event, flow is diverted to other portions of the basin.

The stratigraphic detail of the deposits associated with this 30-min sequence of events shows that the form of the preserved channel body reflects an integration of the various formative events. In particular, the significant confluence scours at 1619 min and 1625 min led to the deepest, most concave portion of the bounding surface. The lateral migration and scour from 1641 to 1643 min led to the more gradual inclination of the bounding surface to the left. Coal that was apparent at the margin of the fluvial channel at 1645 min and 1649 min is preserved along the left margin of the channel body. Note that the preserved size and shape of the channel body are not the same as the cross-sectional geometry of any of the fluvial channels that sculpted it. Rather, the bounding surface and fill of this channel body are composite, multistorey features, owing their geometry to several reoccupations of persistent channel topography.

Figure 7b details a somewhat shorter duration sequence of events leading to the preservation of a channel body. In this case, the channel body bounding surface is created sometime between 740 min and 742 min, rapidly enough to escape direct topographic measurement. After this initial scour event flow wanes, eventually dying out after 748 min. As in Fig. 7a, this channel topography is filled and capped by rapid sedimentation associated with an expanding overbank flow associated with an adjacent channel, again preventing fluvial reoccupation. The deposits associated with this sequence of events are relatively coal-rich, contrasting with the channel body of Fig. 7a. Lateral accretion (from left to right) occurred as the flow waned between 742 and 748 min. The comparatively high coal content of this fill is apparently due to several factors: the relatively long period over which flow waned (6+ min); the relatively small size (width, depth, discharge less competent for sand transport) of the flows (compare with Fig. 7a, c & d); as well as the substrate, which was apparently relatively coal-rich at this location and time.

Figure 7c shows a sequence of events of similar duration to Fig. 7b, although the channel fill is sand-rich. The entire sequence of events lasts only 8 min, from an initial scour at 1002 min to a final filling at 1010 min. The final filling (1010 min) is again accomplished by an expanding flow, but this time originating from within this channel, as the confluence that led to the initial scour migrates upstream out of the field of view. Evidence of lateral accretion is preserved in the deposit, associated with sedimentation forcing the flow up and to the left. The comparatively large flows associated with this sequence led to a larger and more sand-rich channel body than in Fig. 7b. Note that at the time of deepest scour (1002 min), the flow is not bank-full, occupying a narrow region at the base of the channel topography. As such, the preserved channel body cross-sectional shape is, again, not representative of any particular flow that sculpted it. The right and left corners of this channel body have been removed by subsequent erosion during the remaining ~ 800 min of experiment.

Finally, Fig. 7d shows unique preservation of a relatively large, convex capping deposit, as this channel body was preserved during the final hour of the experiment. The time between initial scour to final fill is 4 min (1756–1760 min), the briefest of the four examples presented here. As the initial scour migrates upstream from this point, its associated flow expansion rapidly fills the channel topography. As in each of the cases in Fig. 7, expanding flow overfills the channel depression, leading to convex topography and preventing flow reoccupation. Note that the ratio of depth of incision below the initial surface to total depth of aggradation before abandonment is approximately 1:2, similar to that measured from topographic scans (Fig. 4). Preservation of the full thickness of convex deposits is rare in this experiment, however, due to post-depositional reworking. As in Fig. 7c, the flow occupies only the lowest portion of the channel topography at the time of maximum scour (1756 min), reinforcing the notion that the channel bodies are typically larger than the flows that sculpted them.

While the four sequences of events detailed in Fig. 7 contrast in duration, grain size, and overall size, they are similar in several important respects. The aspect ratios of all channel bodies are comparable. Except in the case of Fig. 7d, where, due to its timing in the experiment, a wide capping deposit has been preserved, the aspect ratios are in the range 6–7, comparable to the aspect ratios of fluvial scours. Therefore, the aspect ratio of fluvial scour is faithfully represented in the deposits, and the depth and width of fluvial scour are comparably exaggerated in the deposits. These ratios contrast with those of sheet deposits (at 1649 min in Fig. 7a or 758 min in Fig. 7b), which are deposited by expanding, short-lived, unchannelized flows.

The mechanisms of channel creation and final filling are similar among these channel bodies. Each sequence begins with scour, which, in this relatively non-cohesive sediment, typically leads to the formation of channel topography that is deeper and wider than the fluvial channels themselves. Whether that channel topography persists depends on whether an expanding flow event occurs, and associated convex topography is formed. In each of these cases, the filling and final abandonment of channel topography are caused by rapid sedimentation associated with expanding flow. In Fig. 7a, the channel topography persists and continues to attract flow for 30 min before the final deposition. In the three other examples, the capping event occurs sooner after the initial incision, leading to shorter overall durations.

The sedimentary character of the channel bodies imperfectly reflects the sequence of fluvial flows responsible for their creation. There is some indication of the multiple stages of abandonment and reoccupation in Fig. 7a, such as the coal channel margins preserved to the left, and the asymmetric form. Bounding surfaces associated with multiple occupations, however, are preserved as sand on sand contacts, which would be difficult to interpret without topographic information. Although the grain-size contrast between Figs 7b & 7c is indicative of the size and competency of the fluvial channels, this is also controlled by preexisting substrate.

DISCUSSION

Channel body depositional sequence

The sequence of events that leads to the formation and preservation of the experimental channel bodies in the subsurface comprises three major phases, each of which tend to be preserved to some extent in the stratigraphic record:

1 *Initial incision occurs due to autogenic scour in the fluvial system.* This phase occurs via avulsion to, and occupation of, a portion of the floodplain where there is no pre-existing channel topography. Following the terminology of Slingerland & Smith (2004), these events can be termed regional avulsions. Channel topography is formed by upstream migration of a scour point associated with converging flow, as in the incisional avulsion model of Mohrig *et al.* (2000). In this experiment, the scour points typically nucleate over subtle breaks in the fluvial slope, such as those associated with either pre-existing depositional topography, transition from the sand-dominated to coal-dominated region of the system, or near the shoreline.

2 *Abandonment and reoccupation of the channel topography.* During this phase, the depth and width of the initial scour are often increased via lateral migration and scour. The shape of the highest-order channel bounding surface represents an integration of the various occupations (as in Fig. 7a). Deposition on the bed of the channel often leads to complex, multistorey internal architecture, including lateral accretion packages. Each reoccupation of the channel topography during this phase is analogous to a regional avulsion by annexation (Slingerland & Smith, 2004), a phenomenon common in natural systems (e.g. Aslan & Blum, 1999; Mohrig *et al.*, 2000; Morozova & Smith, 2000).

3 *Burial and preservation of the channel body.* This is accomplished by an unchannelized expanding flow, which overfills the channel topography, and creates a convex deposit. These events lead to topography that diverts flow laterally, preventing reoccupation of that particular channel course. In contrast to deposition during the second phase, which is relatively gradual, and associated with confined flow, these

burial events occur rapidly, and involve expanding flows which overtop the channel banks. In this experiment the expanding flows may be associated with scour in the active (and ultimately preserved) channel, or expansion in adjacent channels. In either case, the final event in the abandonment of the channel and preservation of the channel body is caused by a flow of a very different character than the confined flows of the second phase.

It is proposed that this may be a common sequence of events in relatively non-cohesive, coarse-grained channelized flow systems. The first two phases, initial occupation and reoccupation, are common in natural systems, as discussed above. The third phase, however, is less commonly recognized. The convex deposit of the third phase is a result of flow expansion that was caused by sedimentation immediately downstream from a confluence scour, as in the choking avulsion of Leddy *et al.* (1993). Similarly, Schumm & Hadley (1957) and Schumm (1961, 1968) described processes that lead to alluviation along reaches of a stream where the sediment load increases faster than the water discharge due to headward knickpoint migration and local sediment production, leading to sediment plugging the channel. Although this is presented as a mechanism to produce discontinuous gullies via subsequent erosion, the notion that sediment load might increase more rapidly than water discharge along a particular reach is applicable to the experiment described here.

The experimental convex deposits associated with this third phase might be analogous to the channel wings of Friend *et al.* (1979). These have been interpreted as channel levees (e.g. Friend *et al.*, 1979; Allen *et al.*, 1983; Marzo *et al.*, 1988; Bridge *et al.*, 2000; Mohrig *et al.*, 2000), but the experiment suggests that in some cases they may not represent gradual flood deposition throughout the life of the channel. Rather, it is possible that some of these features are produced during a relatively rapid depositional phase associated with expanding flow that occurs as channel topography is filled.

Processes leading to channel plugging by expansive flow have also been interpreted in deep-water channelized systems. The build-cut-fill-spill model of Gardner & Borer (2000) invokes processes similar to those operating in the present experiment. It is suggested that due to the fact that submarine channels generally backfill, 'a transition from erosion and bypass, to confined aggradation, to focused, unconfined deposition' (Gardner & Borer, p. 195) will be recorded in the deposits at a particular downstream position. Indeed, erosion and bypass correlate to the first phase, confined aggradation to the second phase, and unconfined deposition to the final capping phase. The experimental alluvial deposits described here, however, do not show substantial longitudinal variation in the degree to which the various channel body phases are preserved, as observed for instance in the Brushy Canyon Formation deep-water deposits (Gardner & Borer, 2000). Indeed, the non-cohesiveness of the sediment mixture used in this fluvial experiment may be more similar to the bed-load fraction of a deep-water system such as the Brushy Canyon than to fines-rich fluvial systems.

The character of the events leading to channel body formation and preservation in this experiment is relatively insensitive to downstream position in the experiment, although systematic changes in preservation (or reworking) of various phases in a larger natural system might be expected, where accommodation, sediment flux and associated fluvial characteristics (depth, width) varied more substantially longitudinally. The only marked longitudinal trend in the character of the experimental deposits is a distal decrease in the average thickness of channel bodies (19%). This is comparable to the distal decrease in the size (depth) of both depositional and erosional events (Fig. 4) and their resultant deposits (16%). The numbers of depositional and erosional events, however, are comparable in all three topographic transects. Whether there is a comparable trend in the absolute numbers of channel bodies is difficult to evaluate, however, due to post-depositional reworking and amalgamation in the deposits.

Cyclicity and flow diversion

Convex deposition and subsequent diversion of flow can be quantitatively related to the cyclicity evident in the time–space visualization of DB 03-1 sedimentation (Fig. 6). The period of the lateral cyclicity is, on average, 4 h, during which an average of 20 mm of sedimentation occurs with these experimental conditions (subsidence rate of 5 mm h^{-1}; Table 1). This is, in turn, comparable to the mean thickness of a channel body measured in the DB 03-1 strata. Therefore, the period of the lateral cycles corresponds with approximately one channel-depth worth of aggradation. This has an appealing physical explanation: in order for the fluvial system to free itself from a particular flow path it must fill in, more or less completely, the existing channel topography.

CONCLUSIONS

1 The creation and preservation of channel sand bodies is remarkably complex, involving multiple storeys of deposition even under simplified experimental conditions. The creation of channel bodies comprises three major phases:

(a) initial incision via a regional avulsion and subsequent upstream migrating confluence scour;

(b) abandonment and reoccupation by avulsion into existent channel topography;

(c) filling and preservation by convex deposits associated with expanding and depositional flows.

Each of these phases is preserved, to some extent, in all channel bodies in the experimental system. It is proposed that this sequence of events might be a characteristic mode for the creation and preservation of channel-form bodies in relatively non-cohesive natural channelized systems, including submarine systems.

2 The convex topography associated with the deposits of the third phase described above diverts flow, and leads to lateral cyclicity in sedimentation. The time-scale of this cyclicity is set by the time necessary for the deposition of one channel-depth worth of sediment at the average sedimentation rate. This suggests that in order for a reorganization of the fluvial system to occur, a particular arrangement of channels must be filled completely, eliminating the possibility of reoccupation.

ACKNOWLEDGEMENTS

The authors appreciate the thorough reviews of Greg Sambrook Smith and John Bridge, as well as the comments of Gary Nichols, which significantly improved the focus of this work. The DB 03-1 experiment would not have happened without the help of Craig Hill. This work was supported by the National Center for Earth-surface Dynamics (NCED), a Science and Technology Center funded by the Office of Integrative Activities of the National Science Foundation (under agreement number EAR-0120914), and by the St. Anthony Falls Laboratory Industrial Consortium (ExxonMobil, Chevron, ConocoPhillips, Anadarko, and JOGMEC).

REFERENCES

Allen, J.R.L. (1965) A review of the origin and characteristics of recent alluvial sediments. *Sedimentology*, **5**, 89–191.

Allen, P.A., Cabrera, L., Colombo, F. and Matter, A. (1983) Variations in fluvial style on the Eocene-Oligocene alluvial fan of the Scala Dei Group, SE Ebro Basin, Spain. *J. Geol Soc. London*, **140**, 133–146.

Ashworth, P.J., Best, J.L., Peakall, J. and Lorsong, J.A. (1999) The influence of aggradation rate on braided alluvial architecture: field study and physical scale-modelling of the Ashburton River gravels, Canterbury Plains, New Zealand. In: *Fluvial Sedimentology VI* (Eds N.D. Smith and J. Rogers), pp. 333–346. Special Publication 28, International Association of Sedimentologists. Blackwell Science, Oxford.

Aslan, A. and Blum, M.D. (1999) Contrasting styles of Holocene avulsion, Texas Gulf Coastal Plain, USA. In: *Fluvial Sedimentology VI* (Eds N.D. Smith and J. Rogers), pp. 193–209. Special Publication 28, International Association of Sedimentologists. Blackwell Science, Oxford.

Bersier, A. (1958) Sequences detritiques et divigations fluviales. *Eclogae Geol. Helv.*, **51**, 854–893.

Best, J.L., Ashworth, P.J., Bristow, C.S. and Roden, J. (2003) Three-dimensional sedimentary architecture of a large, mid-channel sand braid bar, Jamuna River, Bangladesh. *J. Sediment. Res.*, **73**, 516–530.

Bridge, J.S. (2003) *Rivers and Floodplains: Forms, Processes, and Sedimentary Record*. Blackwell Science, Oxford, 491 pp.

Bridge, J.S., Jalfin, G.A. and Georgieff, S.M. (2000) Geometry, lithofacies, and spatial distribution of Cretaceous fluvial sandstone bodies, San Jorge Basin, Argentina: outcrop analog for the hydrocarbon-bearing Chubut Group. *J. Sediment. Res.*, **70**, 341–359.

Cazanacli, D.A., Paola, C. and Parker, G. (2002) Experimental steep, braided flow: application to flooding risk on fans. *J. Hydraul. Eng.*, **128**, 322–330.

Fisk, H.N. (1944) *Geological investigation of the alluvial valley of the lower Mississippi River*. Mississippi River Commission, Vicksburg, MI, 78 pp.

Fisk, H.N. (1947) *Fine-grained Alluvial Deposits and their Effects on Mississippi River Activity*. Mississippi River Division, U.S. Army Corps of Engineers, 82 pp.

Friend, P.F. (1983) Towards the field classification of alluvial architecture or sequence. In: *Modern and Ancient Fluvial Systems* (Eds J.D. Collinson and J. Lewin), pp. 531–542. Special Publication 6, International Association of Sedimentologists. Blackwell Science, Oxford.

Friend, P.F., Slater, M.J. and Williams, R.C. (1979) Vertical and lateral building of river sandstone bodies, Ebro Basin, Spain. *J. Geol. Soc. Lond.*, **136**, 39–46.

Gardner, M.H. and Borer, J.M. (2000) Submarine channel architecture along a slope to basin profile, Brushy Canyon Formation, West Texas. In: *Fine-grained Turbidite Systems* (Eds A.H. Bouma and C.G. Stone), pp. 195–214. Memoir 72, American Association of Petroleum Geologists/Special Publication 68, Society of Economic Paleontogists and Mineralogists, Tulsa, OK.

Heller, P.L., Paola, C., Hwang, I.-G., John, B. and Steel, R. (2001) Geomorphology and sequence stratigraphy due to slow and rapid base-level changes in an experimental subsiding basin (XES 96-1). *Am. Assoc. Petrol. Geol. Bull.*, **85**, 817–838.

Hickson, T.A., Sheets, B.A., Paola, C. and Kelberer, J.M. (2005) Experimental test of tectonic controls on three-dimensional alluvial facies architecture. *J. Sediment. Res.*, **75**, 710–722.

Hooke, R.L. (1968) Model geology: prototype and laboratory streams: discussion. *Geol. Soc. Am. Bull.*, **79**, 391–394.

Leddy, J.O., Ashworth, P.J. and Best, J.L. (1993) Mechanisms of anabranch avulsion within gravel-bed braided rivers: observations from a scaled physical model. In: *Braided Rivers* (Eds J.L. Best and C.S. Bristow), pp. 119–127. Special Publication 75, Geological Society Publishing House, Bath.

Lunt, I.A., Bridge, J.S. and Tye, R.S. (2004) A quantitative, three-dimensional depositional model of gravelly braided rivers. *Sedimentology*, **51**, 377–414.

Marzo, M., Nijman, W. and Puigdefabregas, C. (1988) Architecture of the Castissent fluvial sheet sandstones, Eocene, South Pyrenees, Spain. *Sedimentology*, **35**, 719–738.

Mohrig, D.C., Heller, P.L., Paola, C. and Lyons, W.J. (2000) Interpreting avulsion process from ancient alluvial sequences: Guadalope-Matarranya system (northern Spain) and Wasatch Formation (western Colorado). *Geol. Soc. Am. Bull.*, **112**, 1787–1803.

Moreton, D.J., Ashworth, P. and Best, J. (2002) The physical scale modelling of braided alluvial architecture and estimation of subsurface permeability. *Basin Res.*, **14**, 265–285.

Morozova, G.S. and Smith, N.D. (2000) Holocene avulsion styles and sedimentation patterns of the Saskatchewan River, Cumberland Marshes, Canada. *Sediment. Geol.*, **130**, 81–105.

Paola, C. (2000) Quantitative models of sedimentary basin filling. *Sedimentology*, **47** (supplement 1), 121–178.

Paola, C., Mullin, J., Ellis, C., *et al.* (2001) Experimental stratigraphy. *GSA Today*, **11**, 4–9.

Potter, P.E. (1967) Sand bodies and sedimentary environments – a review. *Am. Assoc. Petrol. Geol. Bull.*, **51**, 337–365.

Schumm, S.A. (1960) The effect of sediment type on the shape and stratification of some modern fluvial deposits. *Am. J. Sci.*, **258**, 177–184.

Schumm, S.A. (1961) Effect of sediment characteristics on erosion and deposition in ephemeral stream channels. *U.S. Geol. Surv. Prof. Pap.*, **352C**, 31–70 pp.

Schumm, S.A. (1968) River adjustment to altered hydrologic regimen–Murrumbidgee River and paleochannels, Australia. *U.S. Geol. Surv. Prof. Pap.*, **598**, 65 pp.

Schumm, S.A. and Hadley, R.F. (1957) Arroyos and the semiarid cycle of erosion. *Am. J. Sci.*, **255**, 161–174.

Sheets, B., Hickson, T. and Paola, C. (2002) Assembling the stratigraphic record: depositional patterns and time-scales in an experimental alluvial basin. *Basin Res.*, **14**, 287–301.

Slingerland, R. and Smith, N.D. (2004) River avulsions and their deposits. *Ann. Rev. Earth Planet. Sci.*, **32**, 257–285.

Strong, N., Sheets, B.A., Hickson, T.A. and Paola, C. (2005) A mass-balance framework for quantifying downstream changes in fluvial architecture. In: *Fluvial Sedimentology VII* (Eds M.D. Blum, S.B. Marriott and S. Leclair), pp. 243–253. Special Publication 28, International Association of Sedimentologists. Blackwell Science, Oxford.

Wheeler, H.E. (1958) Time-stratigraphy. *Am. Assoc. Petrol. Geol. Bull.*, **42**, 1047–1063.

Wheeler, H.E. (1964) Baselevel, lithosphere surface, and time-stratigraphy. *Geol. Soc. Am. Bull.*, **75**, 599–610.

Fluvial systems in desiccating endorheic basins

GARY NICHOLS*

Department of Geology, Royal Holloway, University of London, Egham, Surrey, TW20 0EX, UK (Email g.nichols@gl.rhul.ac.uk)

ABSTRACT

Endorheic basins are basins of internal drainage with no direct hydrological connection to the marine environment. In relatively humid settings, the depositional environments of the basin will be dominated by a basin centre lake, but if the climate is more arid, fluvial systems will be important depositional mechanisms, along with ephemeral lake, alluvial plain and aeolian reworking. Rivers in desiccating basins show a decrease in discharge down-flow because the loss of water by evaporation and soak-away exceeds the input to the system. One of the features of internal drainage is that all sediment supplied is deposited in the basin. Therefore, base level will be determined by the balance between sediment supply and basin subsidence. If sediment supply exceeds subsidence, the river channels will not deeply incise into the alluvial plain of the medial and distal parts of the system, and overbank flow in the distal areas will result in a high proportion of thin sheets of sand and mud deposited by unconfined flow. The depositional gradient will be very low (and may effectively be horizontal over much of the fluvial depositional tract), and a rising base level may also cause the rivers to back-fill the feeder valleys in the proximal areas. Avulsion and lateral migration of the channels across the alluvial plain result in a fan-shaped body of sediment being built up as the rivers distribute sediment, a geomorphological form referred to as a fluvial distributary system. However, the conditions for forming a fluvial distributary system are sensitive to climate, and with an increase in water supply, the basin-centre lake may become perennial: as the rivers will be feeding into a standing body of water, the distal part of the fluvial depositional system is therefore a delta. Lake deltas and fluvial distributary systems can hence be considered as members of a spectrum of depositional settings determined by climate. This continuum of processes and environments can be extended to include aeolian facies, which will dominate if conditions in an endorheic basin are too arid for a perennial fluvial system to form.

Keywords Endorheic basins, fluvial distributary systems, base level, palaeoclimate, ephemeral rivers, lakes.

INTRODUCTION

Basins of internal drainage (endorheic basins) can form in many different tectonic settings and range in size over several orders of magnitude. Present-day examples include the large Lake Eyre Basin, Australia (Kotwicki & Isdale, 1991), the Caspian Sea (Kroonenberg *et al.*, 1997, 2000) and much smaller basins such as Death Valley, California, the East African Rift Valley lake basins and the Dead Sea (Jordan). They can be important sites of sediment accumulation at high altitudes, such as the Tarim Basin in China, the Tibetan Plateau and the Puna-Altiplano in the Central Andes (Sobel *et al.*, 2003). Endorheic basins occupy about 20% of the Earth's land surface, but are found mostly in arid regions and collect only 2% of the global river runoff (Garcia-Castellanos *et al.*, 2003). Many ancient sedimentary basins have been interpreted as endorheic, including late Cenozoic of the South Caspian Sea (Hinds *et al.*, 2004), the Triassic Newark Basin in North America (Faill, 1973), Devonian basins in the

*Also at: UNIS, University Centre on Svalbard, PO Box 156, N-9171 Longyearbyen, Norway.

North Atlantic area (Friend *et al.*, 2000) and the Oligocene to late Miocene Ebro foreland basin in Spain (Nichols, 2004). The fill of these basins may consist of hundreds to thousands of metres thick successions of continental strata deposited in lakes, by rivers, as accumulations of aeolian facies or extensive evaporite deposits.

The tectonic and climatic controls on lacustrine systems have been summarized in Carroll & Bohacs (1999): a basin may be considered to be overfilled, balanced-fill or underfilled with respect to the balance between accommodation and water plus sediment supply. An underfilled basin has no hydrological pathway for water to flow out of the basin because the base level is well below the basin sill (Carroll & Bohacs, 1999), and therefore the basin is endorheic. The proportions of fluvial and evaporitic facies of the alluvial plain and ephemeral lakes will be determined by the balance between water and sediment supply from the hinterland and the desiccation of the basin by evaporation.

In this paper, the characteristics of fluvial depositional systems in endorheic basins are considered. Under climatic regimes where loss of water through evaporation in the basin exceeds supply, lakes are ephemeral or are restricted to the basin centre. There may, however, be sufficient supply of water from nearby hinterland areas for rivers to be active, semi-permanent features in the basin, which are responsible for a high proportion of the sedimentary succession. The deposits of these fluvial systems display features that are a consequence of their land-locked setting; these include a tendency for the rivers to form a radial pattern of deposits, an absence of incised valleys and, in the distal area, a high proportion of deposits resulting from unconfined flow on the alluvial plain. The effects of changing climate in endorheic basins are also explored, and the links between fluvial systems, lake deltas and arid basins dominated by aeolian facies are considered in terms of a spectrum of depositional models.

Examples are drawn from two main areas. In the Miocene of the northern part of the Ebro Basin in northern Spain (Fig. 1) there is a very well exposed succession of fluvial deposits which have

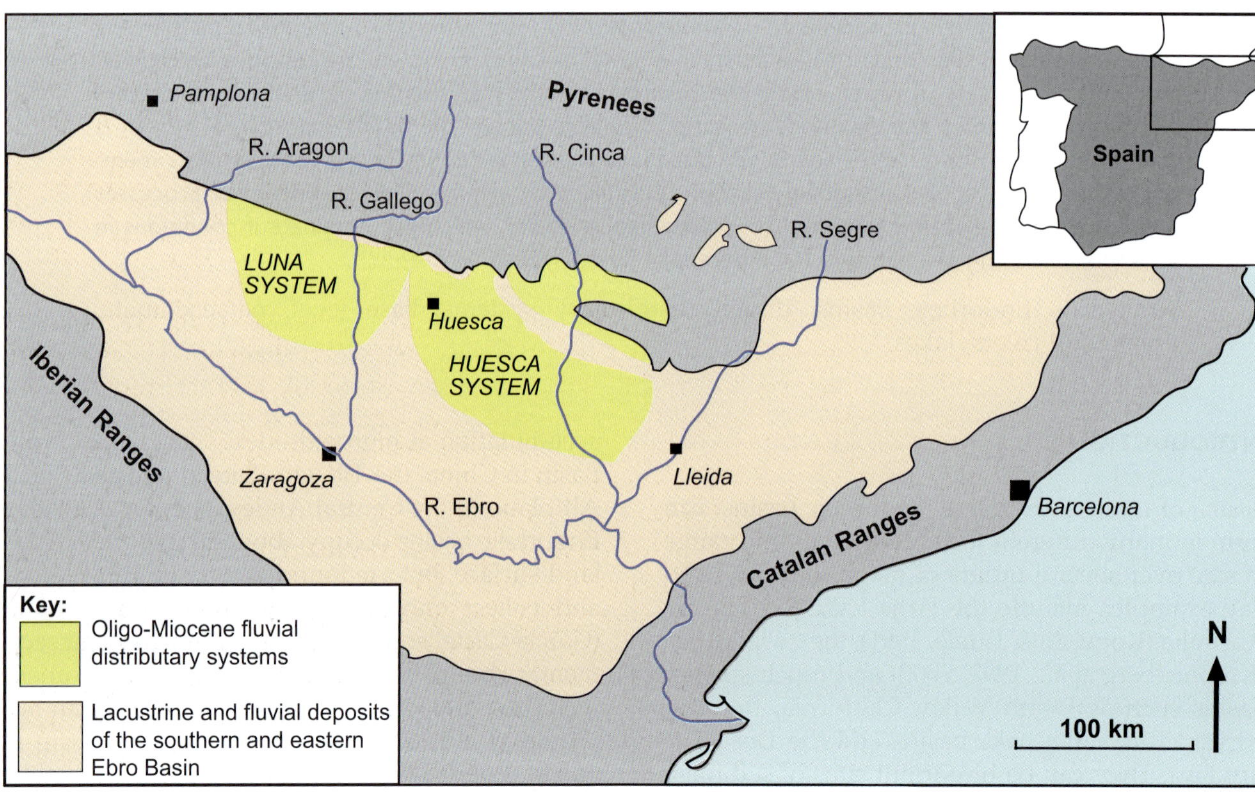

Fig. 1 The Ebro Basin is the southern foredeep of the Pyrenean orogenic belt: in the Oligocene and early Miocene it was a basin of internal drainage and was the site of fluvial and lacustrine sedimentation. Deposits of the Huesca and Luna fluvial distributary systems are preserved in the north-central part of the basin.

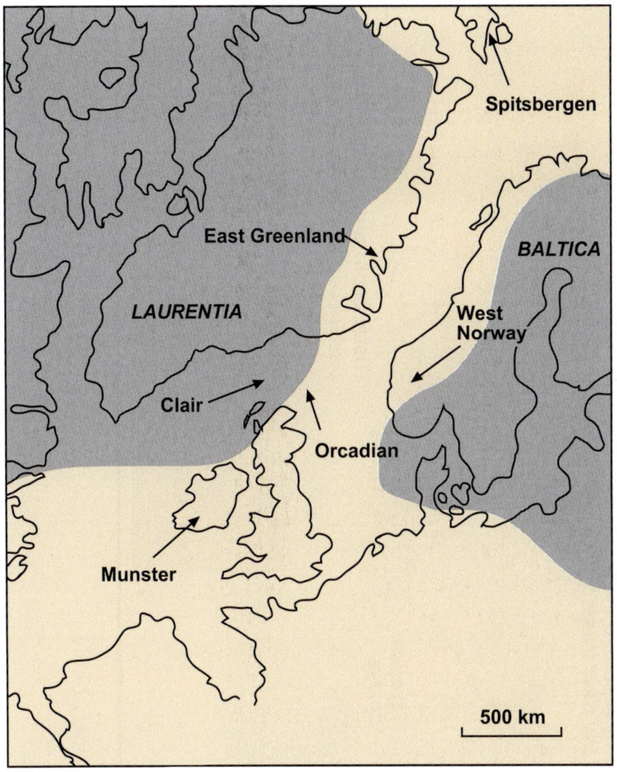

Fig. 2 Old Red Sandstone basins in the North Atlantic area. (From Friend *et al.*, 2000.)

been interpreted as having been formed by *fluvial distributary systems* (Hirst & Nichols, 1986). These provide details of the depositional processes and the architecture of fluvial facies in an endorheic basin. The second main group of examples are from Devonian strata in the North Atlantic area (Fig. 2), where fluvial, lacustrine and aeolian facies have been deposited in endorheic basins under different climatic conditions: southern Ireland (Graham, 1983; Williams *et al.*, 1989; MacCarthy, 1990; Sadler & Kelly, 1993; Richmond & Willliams, 2000; Williams, 2000), east and northeast Greenland (Friend *et al.*, 1983; Kelly & Olsen, 1993) and Spitsbergen (Friend & Moody-Stuart, 1972). A further example of an endorheic basin is taken from the Lower Jurassic of the Hartford Basin, New England (Demicco & Kordesch, 1986).

FLUVIAL SYSTEMS IN ENDORHEIC BASINS

Friend (1978) recognized that there are deposits of ancient river systems which appear to be characterized by a loss of discharge downstream by evaporation and soak-away such that the river channels tend to become smaller and shallower distally. As channels of these systems migrate laterally and avulse they generate a broad, low-angle fan-shaped body of sediment which may be referred to as a fluvial distributary system (Fig. 3; Nichols, 1987; Nichols & Fisher, 2007). The distal portions of fluvial distributary systems may form a fan-shaped body of channel and overbank deposits, which some authors refer to as a 'terminal fan' (Kelly & Olsen, 1993); see Nichols & Fisher (2007) for discussion of the terminology associated with fluvial deposits of this style. Unconfined flow on the floodplain is a feature of the more distal parts of these systems if the discharge is not contained within the river channels and the interfluve areas on the alluvial plain are close to horizontal (Kelly & Olsen, 1993).

Small terminal fans have been documented from modern examples in the southern foothills of the Himalayas (Parkash *et al.*, 1983) and Sudan (Abdullatif, 1989); see discussion in Kelly & Olsen (1993). However, the models for larger fluvial distributary systems are largely derived from the interpretation of fluvial deposits in the stratigraphic record. Nichols (1987) and Hirst & Nichols (1986) documented the Luna and Huesca Systems in Oligo-Miocene fluvial deposits in the northern part of the Ebro Basin, Spain, and considered that hundreds of metres of channel and overbank deposits were the products of fluvial distributary systems that were 40 km and 60 km in radius, respectively. Deposits of similar character and dimensions have also been recognized from the Devonian of Britain and Ireland, for example the Munster Basin (Graham, 1983; Williams *et al.*, 1989; MacCarthy, 1990; Sadler & Kelly, 1993; Williams, 2000).

Examples of fluvial distributary systems

The Ebro Basin in northeastern Spain (Fig. 1) is the Oligo-Miocene foreland basin formed by flexural loading of the Iberian plate on the southern side of the Pyrenean orogenic belt (Choukroune *et al.*, 1989; Muñoz, 1992). It became endorheic in the late Eocene when connection to the Atlantic was blocked as part of the N–S shortening between Iberia and Europe (Coney *et al.*, 1996). Sedimentation in the basin from the Oligocene through to the middle

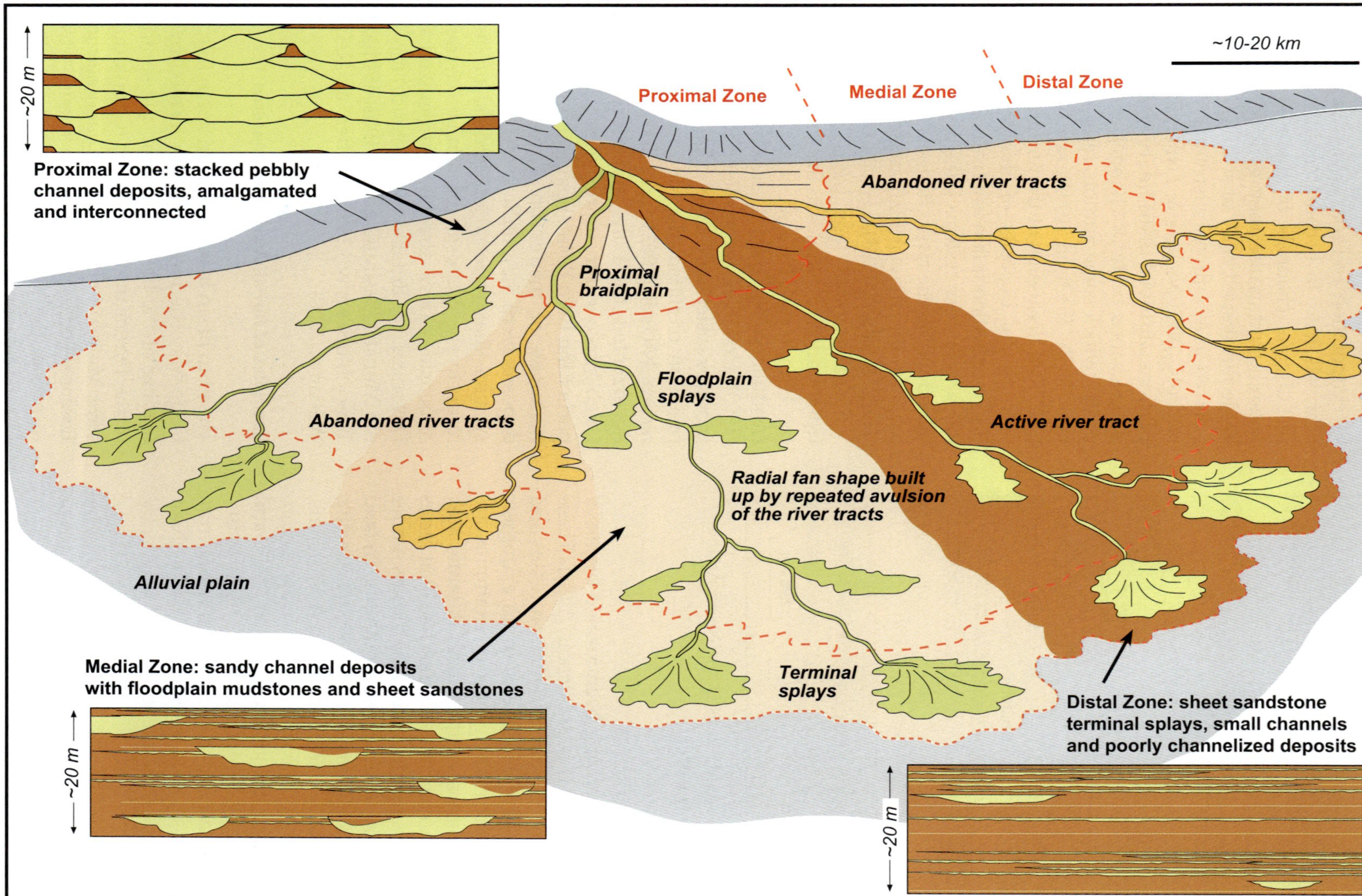

Fig. 3 Characteristics of a fluvial distributary system. (Modified from Nichols & Fisher, 2007.)

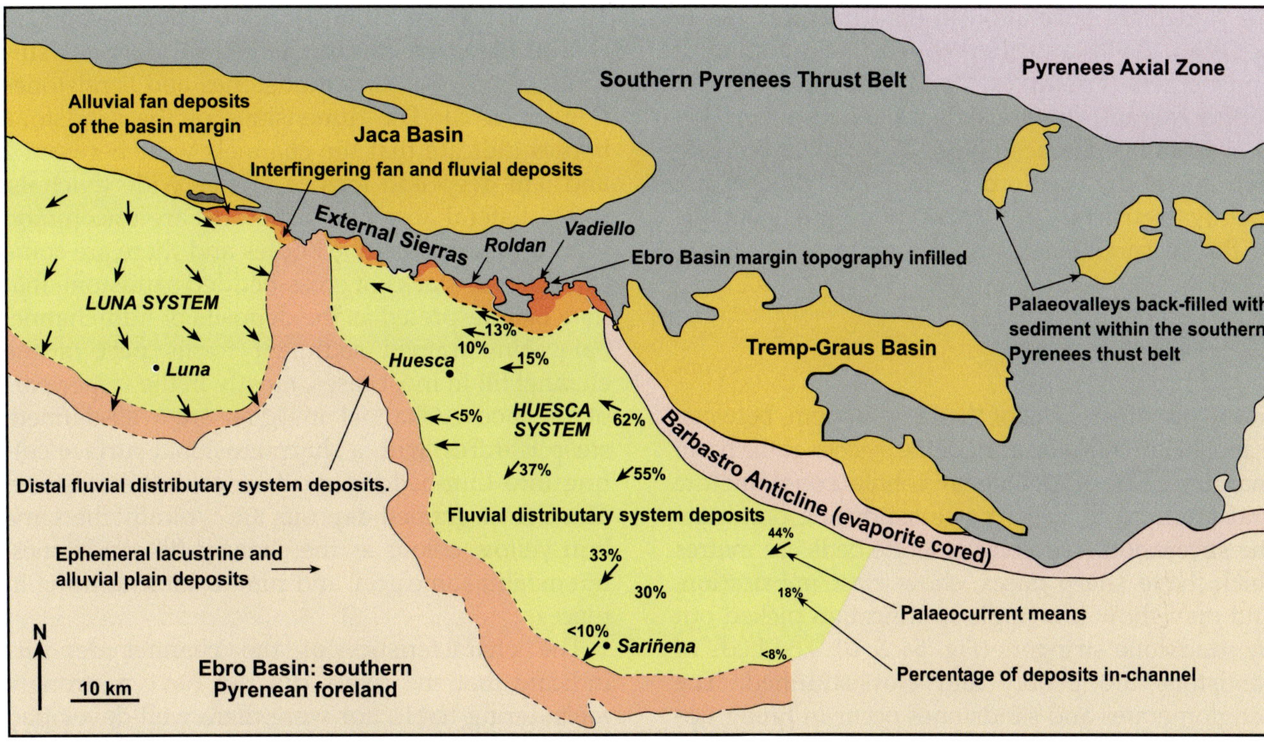

Fig. 4 The Luna and Huesca fluvial distributary systems in the Oligo-Miocene deposits of the northern Ebro Basin (from Hirst & Nichols, 1986; Nichols & Hirst, 1998). Palaeocurrent data and variations in the proportions of channel and overbank facies in the Huesca System are from Nichols (1987) and Hirst (1991).

Miocene was entirely continental and without any connection to the marine realm until the Río Ebro started to drain the area into the Mediterranean in the late Miocene (Evans & Arche, 2002). In the northern part of the basin, in the area that is now around the city of Huesca (Fig. 4), rivers draining the central part of the Pyrenees supplied water and sediment to two large fluvial distributary systems in the Miocene (Hirst & Nichols, 1986). Alluvial-fan deposits at the basin margin have been shown to be distinct and separate from the fluvial systems that provided the bulk of the sediment supply to this part of the basin (Hirst & Nichols, 1986; Nichols & Hirst, 1998). The deposits of the fluvial systems, the Luna System in the west and the Huesca System in the east (Fig. 4), comprise hundreds of metres of strata that are largely undeformed and it is possible to correlate horizons over large areas by tracing them out across the landscape (Arenas *et al.*, 2001). The Luna System is the more completely exposed of the two, and in a transect through coeval strata out from the basin margin, downstream changes in the character of the fluvial channel and overbank deposits can be recognized (Nichols, 1987, 1989; Arenas *et al.*, 2001). The deposits of the fluvial system can be divided into three concentric zones, proximal, medial and distal, which are bordered distally by alluvial plain and lacustrine facies.

The following description of facies and architectural relationships within fluvial deposits of an endorheic basin is based largely on the Miocene Luna and Huesca systems of the Ebro Basin because they exhibit remarkably complete exposure across the basin. Reference and comparison is made to other examples described in the literature, mainly Devonian rocks of the North Atlantic area. The Old Red Sandstone facies of the Munster Basin indicate that deposition was by fluvial systems which exhibited a downstream decrease in grain size and channel magnitude, but they were somewhat larger systems, with a radius of 90–110 km, almost twice the size of the Huesca and Luna systems (Graham, 1983; Williams *et al.*, 1989; MacCarthy, 1990; Sadler & Kelly, 1993). The Devonian Snehvide Formation in east Greenland (Friend *et al.*, 1983), the Rødebjerg Formation in the same area (Kelly & Olsen, 1993), and the Wood Bay Formation of Spitsbergen (Friend & Moody-Stuart, 1972) are further examples of fluvial depositional systems that show similar characteristics. The Lower Jurassic deposits of the Hartford Basin, New England, are mainly lacustrine mudrocks with intercalations of fluvial sandstone units (Demicco & Kordesch, 1986).

Proximal fluvial facies

In the proximal parts of the Luna System, between 5 and 10 km from the apex of the system at the basin margin, beds of pebble to cobble conglomerate and medium to very coarse sandstone dominate the succession. The conglomerate beds are metres thick, have sharp bases, show clast imbrication, and may show low-angle stratification picked out by sandstone stringers (Fig. 5a & b). The beds of sandstone are pebbly and cross-stratified. The conglomerates and sandstones occur in fining upward units at least 7 m thick with deeply scoured bases (Nichols, 1987). These deposits are interpreted as the products of coarse sandy and pebbly braided rivers (Nichols, 1987). Individual channel-fill units (*sensu* Bridge, 2003) are difficult to recognize due to amalgamation of channel deposits, but where individual channel-fill units can be identified they indicate channel dimensions up to 7 m deep and 50–100 m wide (Nichols, 1987). There is poor preservation of overbank facies within this part of the succession.

In the Devonian Munster Basin the conglomeratic fluvial deposits occur up to 40 km from the basin margin and the channel-fill successions are up to 10 m thick (MacCarthy, 1990; Williams, 2000). They are also interpreted as the deposits of pebbly braided rivers (Graham, 1983; MacCarthy, 1990; Sadler & Kelly, 1993) which formed laterally extensive conglomerate and sandstone bodies. Proximal facies in the Devonian of east Greenland are a succession of trough cross-bedded pebbly sandstones and fine pebble conglomerates deposited by braided rivers (Friend *et al.*, 1983).

Medial fluvial facies

The medial parts of the Luna and Huesca systems are well exposed and cover the area between about 10 km and 40 km from the apex. The deposits are characterized by sandy channel-fill deposits surrounded by overbank mudstones and sandstones (Fig. 5c & d). The dimensions of the sandstone bodies indicate that the channels were between 2 and 5 m deep and tens of metres wide (Nichols, 1987). Lateral accretion structures are uncommon within these sandstone bodies and there are some examples of stacks of cross-bedded sandstone that may be interpreted as the deposits of mid-channel bars. Fine-grained sediment forms part of the channel fill in most cases, mostly in the upper part of the body. Channel margins are well-defined, steep features, with a sharp erosional surface cutting into thin-bedded mudstones and sandstone. These finer grained deposits are typically the same buff-yellow colour as the channel-fill sandstones, but exhibit some grey and pink colour mottling in places.

The characteristics of the channel deposits indicate that the rivers did not have a strongly meandering habit, nor were there well-developed mid-channel bars: a straight to sinuous simple form of channel (Schumm, 1981) was probably most common. The final stages of the fill of the channels would have been by waning flow of the river, as clay plugs are commonly observed, and this probably occurred as part of a process of avulsion. New channels formed by scour into the sandstone sheets and mudstone of the floodplain. The colour mottling in the overbank facies represents palaeosol development. Analysis of data collected from 363 sandstone bodies across the Luna System (Nichols, 1987) indicates there is a proximal to distal decrease in the channel body thickness and a decrease in maximum channel-fill grain size.

Channel facies in the medial parts of the fluvial systems in the Munster Basin are commonly trough cross-bedded sandstones and pebbly sandstones interpreted as the deposits of laterally mobile bedload streams (Graham, 1983; Sadler & Kelly, 1993). A consistent decrease in the proportion of 'in-channel' deposits has been qualitatively established in the Devonian examples from Ireland (Graham, 1983; MacCarthy, 1990; Sadler & Kelly, 1993), Spitsbergen (Friend & Moody-Stuart, 1972) and east Greenland (Friend *et al.*, 1983), and quantitatively measured in the Huesca System in the Ebro Basin by Hirst (1991). In the medial parts of the system, 25 km from the calculated apex,

Fig. 5 Ebro Basin lithofacies. (a & b) Conglomerate and pebbly sandstone deposited by braided rivers in the proximal parts of the Luna System. The pen in (a) is 10 mm wide; outcrop in (b) is 3 m high. (c & d) Channel-fill bodies of sandstone scoured into overbank facies in the medial part of the Luna System. Car for scale in (c) is 1.5 m high; outcrop in (d) is 8 m high. (e & f) Distal facies of the Luna System: thin, sometimes irregular, sheets of sandstone deposited by unconfined overbank flows interbedded with floodplain mudstone. Person for scale in (e) is 1.65 m tall; height of outcrop in (f) is 2 m.

62% of the deposits occupied channels, and this decreased to 30 to 40% between 40 and 45 km from the apex.

Distal fluvial facies

In the outer parts of the Luna and Huesca systems, exposures of channel-fill sandstone are less common and the succession is dominated by mudstone and thin sheets of sandstone (Fig. 5e & f). The patchy nature of the exposure in the Luna System precludes any quantitative analysis of the changes in the proportions of in-channel and overbank facies, but in the Huesca System Hirst (1991) calculated that around 10% of the distal part of the system was deposited within channels (Fig. 4). The sandstone sheets are fine-grained, with horizontal (parallel) lamination or current ripple cross-lamination preserved in places, although commonly they are structureless. They are centimetres to tens of centimetres thick and they may have sharp or erosional basal surfaces with pronounced localized scours. There is a spectrum of geometries from beds that are broad sheets, tens to hundreds of metres across, to beds only a few metres wide with deeply scoured bases. They are interpreted as the products of unconfined flow on the floodplain, occurring when the flow in the rivers was not contained within the channels (Fisher *et al.*, 2007). Similar unconfined flows at the distal ends of river systems are referred to as *floodouts* by Tooth (1999a, b) or *terminal splays* by Lang *et al.* (2004). The high proportion of floodplain sandstone sheets in the distal areas suggests that unconfined flow was more frequent in these areas compared with the medial and proximal parts of the distributary systems.

The distal zone of the fluvial deposits in the Munster Basin also contains fewer and smaller channels than the proximal and medial zones (Graham, 1983; Sadler & Kelly, 1993) and many of the channels are shallow and rather poorly defined. In this basin the thin sheet sandstones of the distal zone have sharp, sometimes clearly erosive bases that suggest local channelization of flow (Graham, 1983). The channel-fill sandstone units in the distal zone of the fluvial system that formed the Rødebjerg Formation, Greenland, are documented as less than a metre thick and tens of metres wide (Kelly & Olsen, 1993).

Basin centre deposits: alluvial plain, lacustrine and aeolian facies

The more distal deposits of the Luna and Huesca systems interfinger with sheets of thin-bedded pale grey to pale brown sandstone and mudstone. The sandstone beds have variable internal character and may be structureless, parallel-laminated or wave-ripple cross-laminated. Many of the beds are calcareous and both nodular and vein gypsum is locally common. They are interpreted as lacustrine facies formed during periods of relatively high lake level, when the most distal fringes of the distributary systems were flooded (Hirst & Nichols, 1986; Nichols, 1987). The gypsum deposits indicate that the lake and its margins were sites of evaporation at times. More extensive lacustrine facies occur farther south, beyond the southern fringes of the Luna and Huesca systems, but these outcrops are stratigraphically younger than the exposed parts of the fluvial distributary systems (Arenas *et al.*, 2001).

Lacustrine facies in the Hartford Basin, New England, can be divided into two main types (Demicco & Kordesch, 1986): dark laminated mudstones which are interpreted as the deposits of perennial lakes; and paler green/red/grey mudstones that show disruption and desiccation features, which are considered to be ephemeral lake deposits. Beds of cross-bedded and planar stratified sandstone that occur interbedded with the mudrocks are interpreted as fluvial channel and overbank facies. The succession shows a cyclicity of facies that represents alternations between perennial lake conditions under a more humid climate, and ephemeral lakes with sandy alluvial plain facies that formed under more arid conditions (Demicco & Kordesch, 1986).

Mudrocks showing evidence of desiccation and colour mottling associated with palaeosol development occur extensively interbedded with the distal fluvial facies in the Munster Basin (MacCarthy, 1990; Sadler & Kelly, 1993) and in the Devonian deposits of Spitsbergen (Friend & Moody-Stuart, 1972). The basinal zones in these areas are considered to be extensive alluvial mudflats that periodically received water from the distal parts of the fluvial systems, but evaporation and soak-away into the dry alluvial plain prevented the formation of any lakes. Sediments deposited in basins in an arid climatic regime will be subject to aeolian reworking. The Devonian Rødebjerg Formation in eastern Greenland (Kelly & Olsen, 1993) includes aeolian deposits which range from small dune deposits intercalated with distal fluvial facies to extensive dune facies with cross-bed sets up to 15 m thick. Aeolian facies are also reported from the Devonian of the Dingle Basin, southwest Ireland (Richmond & Williams, 2000), where aeolian sands and distal fluvial facies are intercalated.

Patterns of channel and floodplain deposition

The trends in channel and overbank facies across the distributary systems indicate that there was a change from the proximal area, where large braided rivers deposited coarse material in channels which migrated across the alluvial plain with little preservation of overbank facies, to the distal parts of the system, which were characterized by smaller channels and a higher proportion of preserved floodplain deposits. This pattern is partly attributable to a decrease in discharge in the rivers downstream, but the low depositional slope in the distal areas is also likely to have been a factor. As rivers flowed across the flat alluvial plain, the lack of gradient into a body of water (a lake or sea) inhibited incision of channels. With a limited capacity within these distal channels, water flow would have spread out onto the floodplain, depositing both bedload and suspended load on the overbank areas. The sands were deposited by flows that locally scoured into the floodplain forming erosively based sheets. The high proportion of overbank sandstone sheets in the outer parts of the distributary systems suggests that unconfined and poorly channelized flow was a significant process in the distal zone.

Intercalation between shallow lake facies and distal fluvial deposits indicates that the boundary between the floodplain and the lake varied with time and that the two environments were adjacent. In this region, overbank flows may have merged into the lake at its margin, making it difficult to draw a boundary between the floodplain environment and the shallow lake margin setting at times of flood and high lake level. Similarly, during drier periods the floodplain would have extended further basinwards as the lake level fell and the shoreline receded. The distinction between the floodplain environment and an ephemeral lake margin may therefore be difficult to determine, either in the modern environment or in the deposits of these settings.

ALLUVIAL ARCHITECTURE IN ENDORHEIC BASINS

In basins connected to the oceans, the sea level defines the 'base level' of all depositional systems connected to the sea, including the 'equilibrium profile' of rivers which flow into it (Shanley & McCabe, 1994). Endorheic basins in relatively humid settings have a lake in the basin depocentre, and any changes in the lake level will influence depositional systems in much the same way as sea-level fluctuations. However, where an endorheic basin is in a more arid setting and there is no permanent lake, the alluvial plain in the centre acts as the 'base level' for rivers feeding the basin.

The lack of connection to the ocean means that everything brought in by water from the surrounding erosion areas, as bedload, suspended load or as ions in solution, is deposited within the basin. Only wind-blown material can escape, and this is likely to be aeolian dust. Consequently, the base level in the basin will be determined by the interplay of tectonic subsidence and sediment supply. If sediment supply is greater than subsidence the base level will rise through time, resulting in an aggradational pattern, which Bohacs *et al.* (2000) considered to be the most common situation for underfilled, evaporitic lake basins. In the examples considered here, there is evidence of aggradation, and a rise in base level will reduce the overall gradient of rivers feeding the basin; cases where there is evidence of increasing fluvial gradient as a consequence of a high subsidence rate compared with sediment supply have not been recognized.

River valley incision

When there is a relative fall in base level, rivers cut down to form valleys, which incise into the previous depositional surface (Shanley & McCabe, 1994) if the exposed former lake floor slope has a higher gradient than the distal floodplain (Vincent *et al.*, 1998). During lowstand, the channel deposits

Fig. 6 Outcrop of medial Huesca System deposits at La Serreta, near the village of Piraces, 15 km southeast of Huesca city. Channel sandstone bodies are only incised a few metres and there is no evidence of confinement of channels to a more deeply incised valley (see also Hirst, 1991). View is towards the east, approximately up the palaeoflow; height of the cliff is 80 m.

will be concentrated into these incised valleys, to the sides of which lie areas that are exposed for long periods, subject to extensive pedogenic alteration and the formation of mature palaeosols. Friend (1978) noted the absence of large-scale alluvial incision within river systems of Devonian age from Spitsbergen and East Greenland.

Within the Luna and Huesca systems in the Ebro Basin, neither incised valleys filled with sandstone bodies nor well-developed palaeosols have been recognized. In all places where there is sufficient exposure to assess the architectural relationships of the channel sandstone bodies, the stacking arrangement reflects an aggradational pattern. This is most clearly demonstrated in the medial part of the Huesca System where there is an exposure of channel and overbank deposits over 100 m thick and over 2 km long in a canyon beside the hill 'La Serreta' near the village of Piraces (Hirst, 1991; Fig. 6). Channel-fill sandstone bodies are on average 3.8 m thick, and have lateral extents of hundreds of metres (Hirst, 1991), but none of them show deep incision. Other exposures throughout the Huesca and Luna systems are less complete, but no examples of valley incision within the medial and distal parts of the systems have been recognized in the outcrops along the valleys of the major rivers, roads or irrigation channels that cut through the Oligo-Miocene deposits.

Back-filling of feeder valleys

If sediment supply exceeds subsidence and there is a rise in relative base level within an endorheic basin, the transfer valleys that feed a fluvial distributary system from a mountain hinterland (Vincent & Elliott, 1997) may become sites of accumulation of sediment as they become back-filled. The most proximal parts of both the Luna and Huesca systems are not exposed in the Ebro Basin as a result of deformation and erosion at the basin margin, but there are exposures of Oligo-Miocene conglomerates and sandstones farther to the north within the fold and thrust belt of the southern Pyrenees. These beds are largely undeformed and lie unconformably on older strata. The basal contacts of these deposits define N–S palaeovalleys, indicating that they were components of drainage systems that fed water and sediment into the Ebro Basin (Vincent & Elliott, 1997; Vincent, 2001; Jones, 2004; Luzón, 2005). These palaeovalley-fill conglomerates may have fed the Huesca System (Coney *et al.*, 1996; Jones, 2004), and this may indicate that the topography within the southern Pyrenees was back-filled by a rising base level in the Ebro Basin (Coney *et al.*, 1996). Other authors (González *et al.*, 1997) have disputed this assertion, arguing that the External Sierras, which form the northern margin of the Ebro Basin, were not blanketed by fluvial deposits in the early Miocene. However, evidence for submergence of the basin-margin topography has been presented by Nichols (2004), who estimated that there was a rise in base level of approximately 1000 m within the Ebro Basin during the late Oligocene and early Miocene.

A second, less direct line of evidence of aggradation of the continental succession in the Ebro Basin in the early Miocene comes from the deposits

Fig. 7 Northern margin of the Ebro Basin north of Huesca. (a) Fluvial deposits with palaeoflow to the west (left) in the foreground interfinger with alluvial fan deposits that form the 400 m high pinnacles at Salto de Roldan (centre back of the photograph with limestone beds of the External Sierras behind). (b) A vertical unconformity (picked out by a line) at the basin margin, with deformed limestone beds to the left and on the right 400 m thickness of alluvial fan conglomerates deposited against the palaeotopography.

of alluvial fans at the northern margin of the basin, which are coeval with the fluvial deposits (Hirst & Nichols, 1986; Nichols, 2005a). To the north of the city of Huesca, the fluvial sediments of the Huesca System interfinger with sandstones and conglomerates that were the deposits of one of these alluvial fans (the Roldán fan, Fig 7a; Nichols & Hirst, 1998). The fan shows an aggradational pattern, with no evidence of incision by channels within the proximal and medial parts of the fan succession, indicating that the fan was deposited during a period of constantly rising base level (Nichols, 2004, 2005a). The contact between the Roldán fan deposits and the deformed strata of the thrust front is striking because it is a near-vertical unconformity, formed as the fan deposits banked up against very steep basin margin topography (Fig. 7b), with over 400 m of vertical aggradation (Nichols, 2004). To the east of Roldán, another coeval conglomeratic fan body at Vadiello shows a clear on-lapping relationship to the thrust-front strata and fills a palaeovalley (Friend *et al.*, 1989). These relationships demonstrate that the base level in the Ebro Basin rose during the Miocene and that topography at the basin margin was onlapped and infilled by marginal facies, and provides indirect evidence that the proximal parts of the coeval fluvial systems, which are no longer exposed, must have also filled in topography at the basin margin.

CLIMATIC CONTROLS ON ENDORHEIC BASIN SEDIMENTATION

The association of fluvial and lacustrine facies in the Ebro Basin in the Miocene was formed under a climatic regime in which there was a sufficient supply of water from the Pyrenees to develop the fluvial system, but with a high enough rate of evaporation in the basin to exclude the formation of a large, deep, basin-centre lake. Fluvial channel deposits include complexes of stacked, cross-bedded sandstones which suggest that river flow was reasonably steady for long enough for well-developed bar forms to accrete. On the other hand, desiccation cracks in mudstone layers within channel-fill successions (Nichols, 1987) provide equivocal evidence of ephemeral flow. Within the overbank facies, palaeosols are common: organic matter is rarely preserved, suggesting oxidizing, relatively dry conditions, but calcretes are absent from within these successions, despite the abundance of available calcium carbonate provided by the limestone lithoclasts in the sands. This suggests that the climate was not sufficiently arid to produce the rates of surface evaporation required to generate calcretes.

The fluvial distributary systems of the Ebro Basin required a particular climatic regime. Under a cooler climate with less evaporation and more rain in the basin, a large basin-centre lake could have formed and the whole depositional system could have been dominated by lacustrine facies and with marginal fluvial deposits. Conversely, if the climate in the hinterland had been drier, and less water had been supplied, the rivers would have played a smaller role in distributing sediment in the basin. Under more arid conditions in the basin, aeolian processes would have become more important as sediment brought in by the rivers would have been reworked by wind. The distribution of depositional environments within an endorheic basin is therefore determined by both the climatic conditions within the basin and the climate of the region the rivers are sourced from.

A spectrum of depositional environments can therefore be envisaged (Fig. 8). Under humid conditions a lacustrine basin fed by rivers that form lake deltas would exist (Fig. 8a). With increasing aridity in the basin, the lake margin would retreat and the water body would become largely ephemeral (Fig. 8b). The area of subaerial deposition would expand and at this stage the rivers feeding the lake delta can start to be considered a fluvial distributary system as the river channel and overbank deposition start to become the dominant process of sedimentation. When considered in terms of this climatically determined continuum of processes and environments, the relationship between fluvial distributary systems and lake deltas becomes more apparent and the fluvial systems can be considered to be ephemeral lake deltas ('ephemeral-lacustrine floodplain delta' of Blair & McPherson, 1994). In the absence of any substantial lake (Fig. 8c), the outer fringes of the fluvial distributary systems become areas where flow spreads out onto terminal splays. Under more arid conditions, the sands deposited in river channels and as overbank splays may be reworked by the wind (Fig. 8d) in the absence of a complete vegetation cover, and in arid basins with a low fluvial input, ephemeral streams are restricted to the basin margin and aeolian processes dominate (Fig. 8e).

The fluvial depositional systems in the basins described above can be considered in terms of this spectrum. The more humid end of the spectrum is represented by the Lower Jurassic deposits of the Hartford Basin, which show alternations between perennial lake conditions (Fig. 8a) and periods when the lake was ephemeral, allowing fluvial systems to prograde across the basin (Fig. 8b). The northern part of the Ebro Basin in the early Miocene was predominantly an area of fluvial deposition in channels with mud and sand deposited as splays in the overbank area. There must have been sufficient water from the Pyrenean orogenic belt to supply the rivers but evaporation was high enough to keep the basin-centre lakes shallow and ephemeral: there is no evidence of aeolian reworking and the depositional model would therefore be equivalent to Fig. 8b & c. Devonian deposits in the Munster Basin (MacCarthy, 1990) appear to have been largely fluvial with ephemeral lacustrine deposits but no aeolian facies reported from the distal parts of the alluvial plain (Fig. 8c). The Wood Bay Formation in

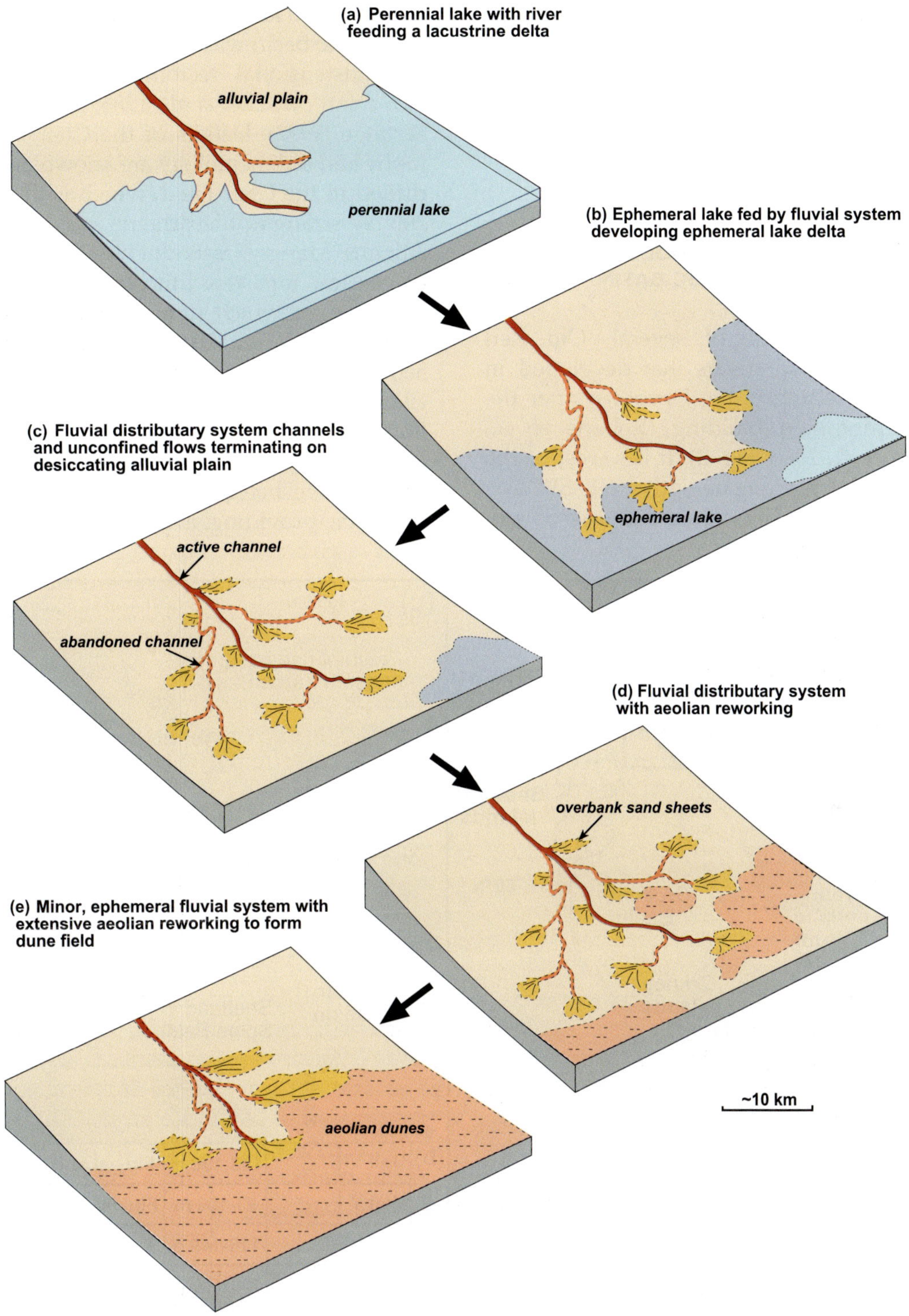

Fig. 8 Conceptual models for deposition in an endorheic basin. (a) Lake-dominated system formed in a humid climate. (b) Fluvially dominated delta system feeding into a lake, which may be ephemeral. (c) Fluvial distributary system with terminal splays bordering a desiccating alluvial plain. (d) Fluvial distributary system, which may be ephemeral, with aeolian reworking of deposits. (e) Arid, aeolian-dominated environment with a minor, ephemeral fluvial system. Fluvial distributary systems fall within this spectrum, forming where the climate in the basin is too dry for a permanent lake body, but water supply from the hinterland establishes a well-developed river system across a very low gradient alluvial plain.

Spitsbergen has similar characteristics and would fall within the same part of the spectrum. More arid conditions have been determined for the Rødebjerg Formation in Greenland (Kelly & Olsen, 1993), where aeolian sands are important in the distal parts of the depositional system (Fig. 8d & e).

CASE STUDY: CLIMATIC CONTROL ON DEPOSITION IN AN ENDORHEIC BASIN

The Clair Basin is one of several 'Old Red Sandstone' continental basins that developed in a post-orogenic setting in the Devonian after the Caledonian mountain building (Roberts *et al.*, 1999; Friend *et al.*, 2000; Fig. 2). It lies west of the Shetland Isles and is elongate with a NE–SW axis (Fig. 9a): it is about 20 km wide and at least 55 km, possibly 130 km, long (Duindam & Van Hoorn, 1987). The basin was formed under an extensional to transtensional tectonic regime in the mid-Devonian (McClay *et al.*, 1986; Norton *et al.*, 1987; Seranne, 1992). Details of the Clair Basin stratigraphy and sedimentology are known only from core drilled in the Clair Field, which is in UKCS Block 206. A stratigraphic scheme was established by Allen & Mange-Rajetzky (1992), who divided the succession into ten lithostratigraphic units, the lower six of which (Units I–VI) form the Lower Clair Group (Fig. 9b). The Lower Clair Group is 500–550 m thick, is the main reservoir for the Clair Field and is Givetian (Middle Devonian) to Frasnian (Late Devonian) in age (Allen & Mange-Rajetzky, 1992). The units of the Lower Clair Group have been drilled and extensively cored (Fig. 9c), providing a database of wireline logs

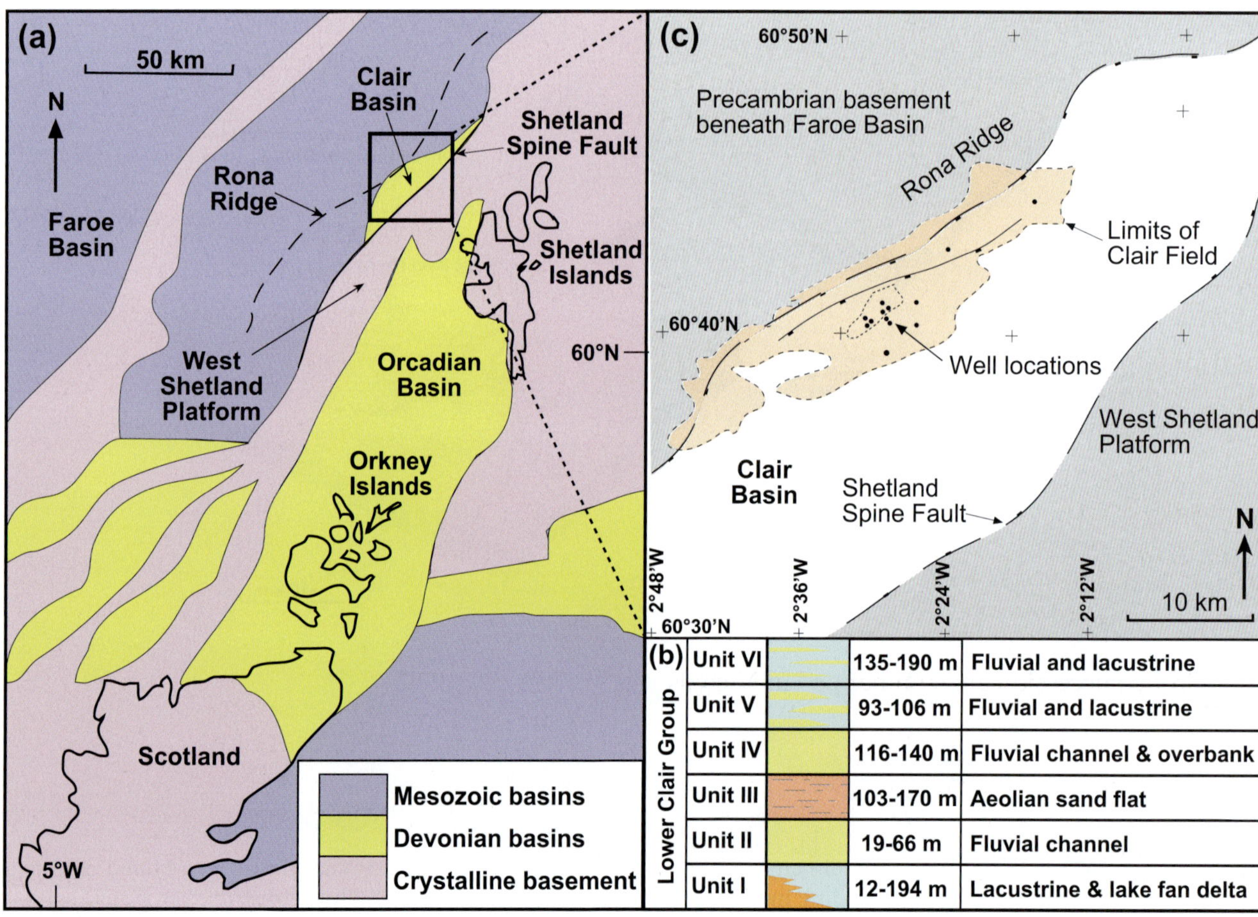

Fig. 9 Location and depositional setting of the Clair Basin. (a) The Clair Basin lies in the subsurface offshore of the Shetland Islands, northwest of the Orcadian Basin. (b) Stratigraphy of the Lower Clair Group, Devonian, west of Shetland. (Modified from Allen & Mange-Rajetzky, 1992.) (c) Location of boreholes in the Clair Basin. (Modified from Nichols, 2005b.)

and core material. Published sedimentological studies have been by Allen & Mange-Rajetzky (1992), who carried out a palaeogeographical synthesis and an analysis of the heavy mineral suites in the sandstone units, McKie & Garden (1996), who recognized climatically controlled stratigraphic cycles, and Nichols (2005b), who presented a model for the Lower Clair Group in terms of an evolving depositional system.

Endorheic basin setting

An endorheic setting for the Clair Basin was proposed by Nichols (2005b) because it lay within a large continental area where all the surrounding basins show limited evidence of connection to the marine environment. To the north the nearest areas of Old Red Sandstone deposition were internal basins in eastern Greenland (Friend *et al.*, 1983) and western Norway (Nilsen & McLaughlin, 1985), whilst to the south there were several endorheic basins in southern Ireland (Graham, 1983; Williams *et al.*, 1989; MacCarthy, 1990). To the east, the Orcadian Basin was mainly internal, but shows evidence of periodic marine influence during the Middle Devonian (Mykura, 1991; Bluck *et al.*, 1992; Friend *et al.*, 2000). The first signs of marine influence in the Clair Basin were in uppermost Devonian to lower Carboniferous strata (Allen & Mange-Rajetzky, 1992). The close proximity of the Orcadian Basin to the east raises the possibility of a connection between the two, particularly at times of high lake level, but correlation between the two basins is difficult because of poor biostratigraphic control on the age of the Lower Clair Group.

Facies and environments

The principal depositional facies in the Lower Clair Group succession are summarized in Nichols (2005b). In brief they are:

1 coarse pebble to granule conglomerate beds, with a clast-supported fabric and crude stratification, and interpreted as subaqueous gravity flow deposits of a lacustrine fan delta (cf. Nemec, 1990; Wescott & Ethridge, 1990);
2 sharp-based, fining upward successions 2–4 m thick comprising clast-supported, pebble to granule conglomerate, moderately well stratified, cross-bedded very coarse to very fine sandstone with common mud clasts, which are considered to be the deposits of sandy and pebbly braided rivers;
3 thin (less than a metre thick) very coarse to very fine sandstone beds with horizontal (parallel) lamination and pedogenic features in the upper parts of the beds, interpreted as sheet deposits formed by overbank flow or at the terminations of channels where the current became unconfined;
4 siltstone and mudstone in beds centimetres to tens of centimetres thick that show evidence of pedogenic alteration, interpreted as deposits of dry floodplains or palustrine lake margin settings;
5 very well laminated, very fine sandstone, siltstone and mudstone interpreted as the products of deposition in lakes by gravity flows and from suspension (Sturm & Matter, 1978);
6 sandstone beds that show climbing ripple cross-lamination, flaser lamination, wave ripples and soft sediment deformation features, characteristics that indicate deposition by currents carrying sand and flowing into a standing body of water at a lake margin (cf. Dam & Surlyk, 1993; Talbot & Allen, 1996).
7 sandstone beds interpreted as the products of aeolian processes are well-sorted, very fine to medium (rarely coarse) grained sands that do not contain mud clasts or granules; cross-bedding is uncommon and the beds are typically horizontally stratified with a 'crinkly lamination', thought to have formed on a periodically wet aeolian sand flat (cf. Goodall *et al.*, 2000).

The six units of the Lower Clair Group (Fig. 9b) are defined on their lithological characteristics and these reflect changes in the processes and environment of deposition (Allen & Mange-Rajetzky, 1992; Nichols, 2005b). The basal unit (I) comprises conglomerate and pebbly sandstone deposited on a coarse-grained fan delta and laminated sandstone and siltstone interpreted as open lake facies. Unit II is interpreted as the deposits of a fluvial system that built out into the basin (Fig. 10a); the deposits were almost exclusively braided river sands and gravels (Nichols, 2005b). The fluvial deposits in the Lower Clair Group are considered to be the products of a distributary system (Nichols, 2005b).

In unit III (Fig. 10b), the deposits are mainly aeolian sands with some coarser river deposits, representing a period of drier conditions in the basin. A return to a more humid climate led to the formation of a unit (IV), dominated by fluvial

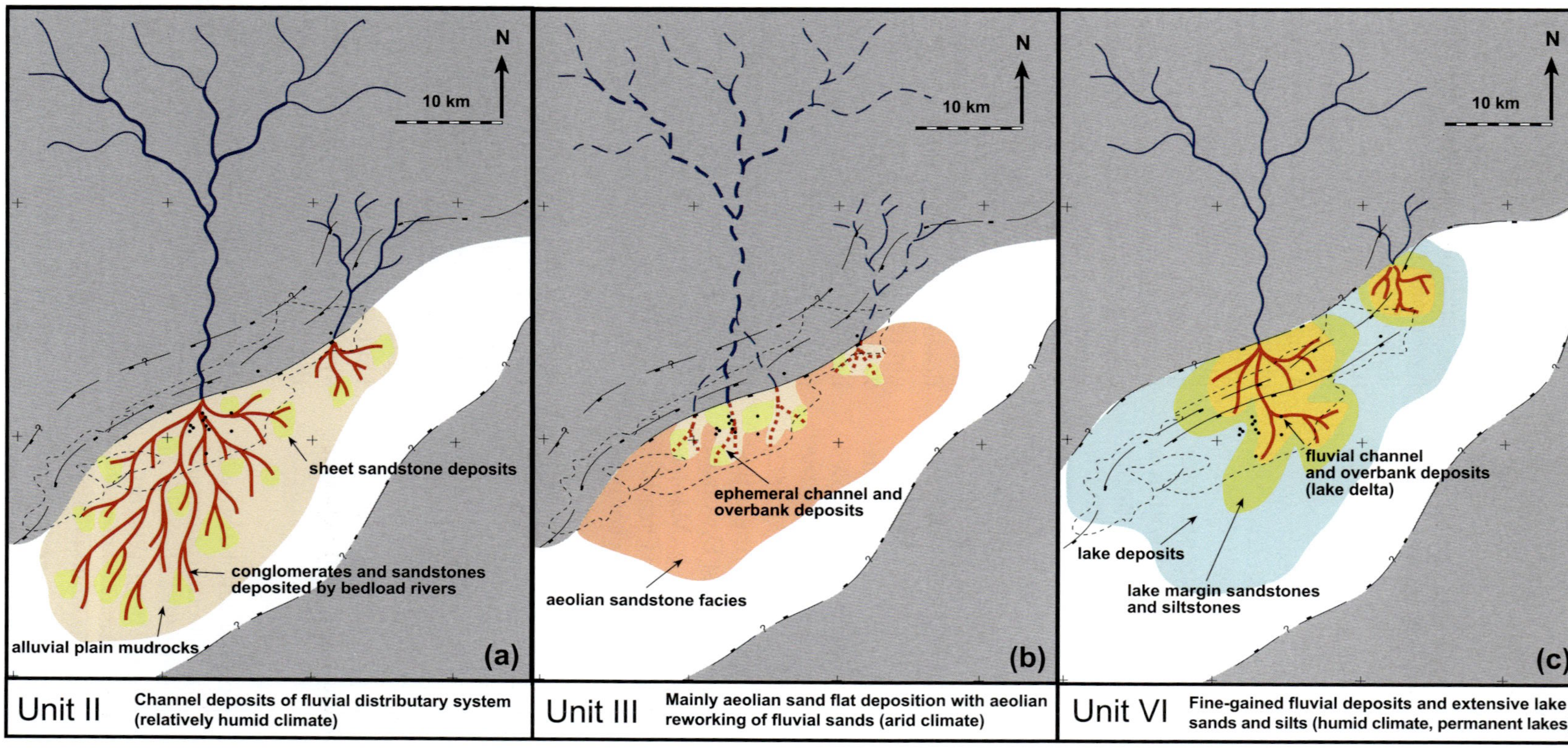

Fig. 10 Palaeogeographical models for units II, III and VI, Lower Clair Group. (Modified from Nichols, 2005b.) (a) Dominance of fluvial distributary system (unit II). (b) Dominance of aeolian conditions (unit III). (c) Dominance of lacustrine conditions (unit IV). These represent different stages in the Middle to Late Devonian history of the Clair Basin.

channel and overbank facies, with some aeolian reworking of the waterlain deposits (McKie & Garden, 1996). Unit V is gradational with the unit below, but the fluvial channel-fill successions are thinner (2.0–2.5 m thick) and generally composed of finer grained sediment; much of the deposition may have been from unconfined flows (Nichols, 2005b). In unit VI (Fig. 10c), open lake, lake margin and lacustrine delta facies are common, interbedded with the deposits of unconfined flows on the alluvial plain; pedogenic horizons are common in the lake margin and alluvial plain facies (Nichols, 2005b).

Climatic controls on deposition

The primary control on facies in the Clair Basin was climate (McKie & Garden, 1996; Nichols, 2005b). During periods when the water supply was at its maximum most of the basin was covered by a lake (units I and VI), and at the other extreme there were times (unit III) when the basin was relatively dry and aeolian processes were dominant. In between these two end members there were periods when river deposition was extensive (unit II) or there was a mixture of aeolian and fluvial processes (unit IV) or lacustrine and fluvial conditions (unit V). Depositional environments in the Clair Basin can therefore be considered in terms of the climatically controlled spectrum present in Fig. 8. During 'wet' phases a drainage network on the west side of the basin provided water and sediment to a basin-wide lake, where lake deltas were formed. With decreasing water supply the lake-shore receded and sedimentation occurred in rivers and on the floodplain of a fluvial distributary system. Under the driest conditions, the fluvial system probably became ephemeral and sediment was reworked by aeolian processes (Nichols, 2005b). The basin was therefore subject to the influence of both the climate in the catchment area, which determined the water supply to the river systems, and the intrabasinal climate which affected evaporation, and hence the extent of lacustrine and aeolian facies.

Other features of endorheic basin sedimentation

It is difficult to establish from the relatively widely spaced cores across the Clair Field whether the river channels are confined within incised valleys, so this aspect of fluvial architecture in the basin cannot be tested easily. However, in the upper parts of the succession (unit V) there is a high proportion of sandstone units tens of centimetres thick, which are interpreted as the deposits of unconfined splays on the alluvial plain (Nichols, 2005b), a feature of the distal portions of fluvial distributary systems. There is also some limited evidence for back-filling of valleys in the Clair Basin: in one borehole on the western flank of the basin the lower part of the succession (units I to III) is absent and unit IV lies directly on the basement (Nichols, 2005b), indicating that the area of deposition expanded as the basin filled.

CONCLUSIONS

The depositional systems in endorheic basins are strongly controlled by climate. A spectrum of depositional environments can be envisaged, ranging from lake basins that form in relatively humid settings, to sandy deserts under arid climates. Under conditions where there is a relatively high water supply from the hinterland and a basin where the rate of evaporation is greater than the rate of water supply, fluvial depositional systems will dominate. The rivers will show a decrease in discharge down-flow, resulting in a reduction of channel dimensions distally, and the rivers will terminate either in ephemeral lakes or form terminal splays on the alluvial plain. Avulsion and lateral migration of the channels results in the formation of a radial pattern of distribution of sediment through time, and the fan-shaped sediment body will have the form of a fluvial distributary system.

The documented examples of fluvial distributary systems appear to have formed under conditions where the rate of sediment supply exceeded the subsidence rate. There is no evidence of relative base-level fall within these successions, and hence no development of incised valleys; incision was limited to scouring to form new channels following avulsion, and this decreased down-system to the distal areas where the flow was largely dechannelized. Aggradation of the fluvial system also resulted in the back-filling of the main feeder valleys, and infilled basin margin topography.

An increase in water supply to the basin will result in a fluvial distributary system becoming a lake delta. Conversely, a decrease in water supply may lead to aeolian conditions becoming dominant, as sediment brought in by ephemeral streams is reworked by wind. Climate in both the hinterland and the basin will be an important control on the facies, as it will determine the water supply by rivers and the loss of water by evaporation on lakes and alluvial plains.

ACKNOWLEDGEMENTS

Thanks are particularly due to Doug Boyd, formerly of BP Aberdeen, who gave me the opportunity to look at the Clair Field data and BP, Dyce, for access to these data. This manuscript was greatly improved by the thoughtful comments and suggestions provided by Steve Vincent, Colin North, John Graham and Ed Williams. Most of all, the author would like to thank Peter Friend for introducing him to the world of fluvial sedimentology, for his friendship over a period of 25 years, and for providing the insight that although geology itself is fun, it is the people you meet whilst doing it that provide the greatest pleasure.

REFERENCES

Abdullatif, O.M. (1989) Channel-fill and sheet-flood facies sequences in the ephemeral terminal River Gash, Kassala, Sudan. *Sediment. Geol.*, **63**, 171–184.

Allen, P.A. and Mange-Rajetzky, M.A. (1992) Devonian–Carboniferous sedimentary evolution of the Clair area, offshore north-western UK: impact of changing provenance. *Mar. Petrol. Geol.*, **9**, 29–52.

Arenas, C., Millán, H., Pardo, G. and Pocoví, A. (2001) Ebro Basin continental sedimentation associated with late compressional Pyrenean tectonics (north-eastern Iberia): controls on basin margin fans and fluvial systems. *Basin Res.*, **13**, 65–89.

Blair, T.C. and McPherson, J.G. (1994) Alluvial fans and their natural distinction from rivers based on morphology, hydraulic processes, sedimentary processes and facies assemblages. *J. Sediment. Res.*, **A64**, 450–589.

Bluck, B.J., Cope, J.C.W., and Scrutton, C.T. (1992) Devonian. In: *Atlas of Palaeogeography and Lithofacies* (Eds J.C.W. Cope, J.K. Ingham, and P.F. Rawson), pp. 57–66. Memoir 13, Geological Society Publishing House, Bath.

Bohacs, K.M., Carroll, A.R., Neal, J.E., and Mankiewicz, P.J. (2000) Lake-basin type, source potential, and hydrocarbon character: an integrated sequence stratigraphic–geochemical framework. In: *Lake Basins through Space and Time* (Eds E. Gierlowski-Kordesch and K. Kelts), pp. 3–37. Studies in Geology 46, American Association of Petroleum Geologists, Tulsa, OK.

Bridge, J.S. (2003) *Rivers and Floodplains: Forms, Processes, and Sedimentary Record.* Blackwell Science, Oxford, 491 pp.

Carroll, A.R. and Bohacs, K.M. (1999) Stratigraphic classification of ancient lakes: Balancing tectonic and climatic controls. *Geology*, **27**, 99–102.

Choukroune, P. and ECORS-Pyrenees Team (1989) The ECORS Pyrenean deep seismic profile reflection data and the overall structure of an orogenic belt. *Tectonics*, **8**, 23–39.

Coney, P.J., Muñoz, J.A., McClay, K.R. and Evenchick, C.A. (1996) Syntectonic burial and post-tectonic exhumation of the southern Pyrenees foreland fold-thrust belt. *J. Geol. Soc. London*, **153**, 9–16.

Dam, G. and Surlyk, F. (1993) Cyclic sedimentation in a large wave- and storm-dominated anoxic lake; Kap Stewart Formation (Rhaetian-Sinemurian), Jameson Land, East Greenland. In: *Sequence Stratigraphy and Facies Associations* (Eds H.W. Posamentier, C.P. Summerhayes, B.U. Haq and G.P. Allen), pp. 419–448. Special Publication 18, International Association of Sedimentologists. Blackwell Scientific Publications, Oxford.

Demicco, R.V. and Kordesch, E.G. (1986) Facies sequences of a semi-arid closed basin: the Lower Jurassic East Berlin Formation of the Hartford Basin, New England, USA. *Sedimentology*, **33**, 107–118.

Duindam, P. and Van Hoorn, B. (1987) Structural evolution of the West Shetland Continental margin. In: *Petroleum Geology of Northwest Europe* (Eds J. Brooks and K. Glennie), pp. 711–723. Graham and Trotman, London.

Evans, G. and Arche, A. (2002) The flux of siliciclastic sediment from the Iberian Peninsula with particular reference to the Ebro. In: *Sediment Flux to Basins: Causes, Controls and Consequences* (Eds S.J. Jones and L.E. Frostick), pp. 199–208. Special Publication 191, Geological Society Publishing House, Bath.

Faill, R.T. (1973) Tectonic development of the Triassic Newark-Gettysburg Basin in Pennsylvania. *Geol. Soc. Am. Bull.*, **84**, 725–740.

Fisher, J.A., Nichols, G.J. and Waltham, D.A. (2007) Unconfined flow deposits at the distal sectors of fluvial distributary systems: examples from the Luna and Huesca Systems, northern Spain. *Sediment. Geol.*, **195**, 55–73.

Friend, P.F. (1978) Distinctive features of some ancient river systems. In: *Fluvial Sedimentology* (Ed. A.D. Miall), pp. 531–542. Memoir 5, Canadian Society of Petroleum Geologists, Calgary.

Friend, P.F. and Moody-Stuart, M. (1972) Sedimentation of the Wood Bay Formation (Devonian) of Spitsbergen: regional analysis of a late orogenic basin. *Nor. Polarinst. Skr.*, **157**, 1–77.

Friend, P.F., Alexander-Marrack, P.D., Allen, K.C., Nicholson, J. and Yeats, A.K. (1983) Devonian sediments of East Greenland VI: Review of Results. *Medd. Gronl. Geosci.*, **206**(6), 96 pp.

Friend, P.F., Hirst, J.P.P., Hogan, P.J., *et al.* (1989) *Pyrenean Tectonic Control of Oligo-Miocene River Systems, Huesca, Aragon, Spain.* Excursion Guidebook 4, 4th International Conference on Fluvial Sedimentology. Publicacions del Servei Geològic de Catalunya, 142 pp.

Friend, P.F., Williams, B.P.J., Ford, M. and Williams, E.A. (2000) Kinematics and dynamics of Old Red Sandstone basins. In: *New perspectives on the Old Red Sandstone* (Eds P.F. Friend and B.P.J. Williams), pp. 29–60. Special Publication 180, Geological Society Publishing House, Bath.

Garcia-Castellanos, D., Vergés, J., Gaspar-Escribano, J. and Cloetingh, S. (2003) Interplay between tectonics, climate, and fluvial transport during the Cenozoic evolution of the Ebro Basin (NE Iberia). *J. Geophys. Res.*, **108**, B7, 2347.

González, A., Arenas, C. and Pardo, G. (1997) Discussion on syntectonic burial and post-tectonic exhumation of the southern Pyrenees foreland fold-thrust belt. *J. Geol. Soc. London*, **154**, 361–365.

Goodall, T.M., North, C.P. and Glennie, K.W. (2000) Surface and sub-surface sedimentary structures produced by salt crusts. *Sedimentology*, **47**, 99–118.

Graham, J.R. (1983) Analysis of the Upper Devonian Munster Basin, an example of a fluvial distributary system. In: *Modern and Ancient Fluvial Systems* (Eds J.D. Collinson and J. Lewin), pp. 473–484. Special Publication 6, International Association of Sedimentologists. Blackwell Scientific Publications, Oxford.

Hinds, D.J., Aliyeva, E., Allen, M.B., Davies, C.E. Kroonenberg, S.B., Simmons, M.D. and Vincent, S.J. (2004) Sedimentation in a discharge dominated fluvial–lacustrine system: the Neogene Productive Series of the South Caspian Basin, Azerbaijan. *Mar. Petrol. Geol.*, **21**, 613–638.

Hirst, J.P.P. (1991) Variations in alluvial architecture across the Oligo-Miocene Huesca fluvial system, Ebro Basin, Spain. In: *The Three-dimensional Facies Architecture of Terrigenous Clastic Sediments and its Implications for Hydrocarbon Discovery and Recovery* (Eds A.D. Miall and N. Tyler), pp. 111–121. Concepts in Sedimentology and Palaeontology 3. Society of Economic Paleontogists and Mineralogists, Tulsa, OK.

Hirst, J.P.P. and Nichols, G.J. (1986) Thrust tectonic controls on alluvial sedimentation patterns, southern Pyrenees. In: *Foreland Basins* (Eds P.A. Allen and P. Homewood), pp. 153–164. Special Publication 8, International Association of Sedimentologists. Blackwell Scientific Publications, Oxford.

Jones, S.J. (2004) Tectonic controls on drainage evolution and development of terminal alluvial fans, southern Pyrenees, Spain. *Terra Nova*, **16**, 121–127.

Kelly, S.B. and Olsen, H. (1993) Terminal fans – a review with reference to Devonian examples. *Sediment. Geol.*, **85**, 339–374.

Kotwicki, V. and Isdale, P. (1991) Hydrology of Lake Eyre, Australia: El Nino link. *Palaeogeog., Palaeoclimat., Palaeoecol.*, **84**, 87–98.

Kroonenberg, S.B., Rusakov, G.V. and Svitoch, A.A. (1997) The wandering of the Volga delta: a response to rapid Caspian sea-level change. *Sediment. Geol.*, **107**, 189–209

Kroonenberg, S.B., Badyukovab, E.N., Stormsa, J.E.A., Ignatovb, E.I. and Kasimovb, N.S. (2000) A full sea-level cycle in 65 years: barrier dynamics along Caspian shores. *Sediment. Geol.*, **134**, 257–274.

Lang, S.C., Payenberg, T.H.D., Reilly, M.R.W., Hicks, T., Benson, J. and Kassan, J. (2004) Modern analogues for dryland sandy fluvial-lacustrine deltas and terminal splay reservoirs. *Aust. Petrol. Prod. Explor. Assoc.*, **44**, 329–356.

Luzón, A. (2005) Oligo-Miocene alluvial sedimentation in the northern Ebro Basin, NE Spain: tectonics controls and palaeogeographical evolution. *Sediment. Geol.*, **177**, 19–39.

MacCarthy, I.A.J. (1990) Alluvial sedimentation patterns in the Munster Basin, Ireland. *Sedimentology*, **37**, 685–712.

McClay, K.R., Norton, M.G., Coney, P. and Davies, G.H. (1986) Collapse of the Caledonian orogen and the Old Red Sandstone. *Nature*, **525**, 147–149.

McKie, T. and Garden, I.R. (1996) Hierarchical stratigraphic cycles in the non-marine Clair Group (Devonian) UKCS. In: *High Resolution Sequence Stratigraphy: Innovations and Application* (Eds J.A. Howell and J.F. Aitken), pp. 139–157. Special Publication 104, Geological Society Publishing House, Bath.

Muñoz, J.A. (1992) Evolution of a continental collision belt: ECORS-Pyrenees crustal balanced cross-section. In: *Thrust Tectonics* (Ed. K.R. McClay), pp. 235–246. Chapman and Hall, London.

Mykura, W. (1991) Old Red Sandstone. In: *The Geology of Scotland* (Ed. G.Y. Craig), pp. 205–251. Geological Society, London.

Nemec, W. (1990) Aspects of sediment movement on steep delta slopes. In: *Coarse-Grained Deltas* (Eds A. Colella and D.B. Prior), International Association of Sedimentologists. Blackwell Science, Oxford Special Publication 10, 29–74

Nichols, G.J. (1987) Structural controls on fluvial distributary systems – the Luna System, Northern Spain. In: *Recent Developments in Fluvial Sedimentology* (Eds F.G. Ethridge, R.M. Flores and M.D. Harvey), pp. 269–277. Special Publication 39, Society of Economic Paleontogists and Mineralogists, Tulsa, OK.

Nichols, G.J. (1989) Structural and sedimentological evolution of part of the west central Spanish Pyrenees in the Late Tertiary. *J. Geol. Soc. London*, **146**, 851–857.

Nichols, G.J. (2004) Sedimentation and base level controls in an endorheic basin: the Tertiary of the Ebro Basin, Spain. *Bol. Inst. Geol. Min. Esp.*, **115**, 427–438.

Nichols, G.J. (2005a) Tertiary alluvial fans at the northern margin of the Ebro Basin. In: *Alluvial Fans: Geomorphology, Sedimentology, Dynamics* (Eds A.M. Harvey, A.E. Mather and M. Stokes), Geological Society Publishing House, Bath. Special Publication 251, 187–206.

Nichols, G.J. (2005b) Sedimentary evolution of the Lower Clair Group, Devonian, west of Shetland: climate and sediment supply controls on fluvial, aeolian and lacustrine deposition. In: *Petroleum Geology: North West Europe and Global Perspectives – Proceedings of the 6th Petroleum Geology Conference* (Eds A.G. Doré and B.A. Vining), pp. 957–967. Geological Society, London.

Nichols, G.J. and Fisher, J.A. 2007. Processes, facies and architecture of fluvial distributary system deposits: a review. *Sediment. Geol.* **195**, 75–90.

Nichols, G.J. and Hirst, J.P.P. (1998) Alluvial fans and fluvial distributary systems, Oligo-Miocene, northern Spain: contrasting processes and products. *J. Sediment. Res.*, **68**, 879–889.

Nilsen, T.H. and McLaughlin, R.J. (1985) Comparison of tectonic framework and depositional patterns of the Hornelen strike-slip basin of Norway and the Ridge and Little Sulphur Creek strike-slip basins of California. In: *Strike-Slip Deformation, Basin Formation and Sedimentation* (Eds K.T. Biddle and N. Christie-Blick), pp. 79–103. Special Publication 37, Society of Economic Paleontogists and Mineralogists, Tulsa, OK.

Norton, M.G., McClay, K.R. and Way, N.A. (1987) Tectonic evolution of Devonian basins in northern Scotland and southern Norway. *Nor. Geogr. Tidsskr.*, **67**, 323–338.

Parkash, B., Awasthi, A.K. and Gohain, K. (1983) Lithofacies of the Markanda terminal fan, Kurukshetra district, Haryana, India. In: *Modern and Ancient Fluvial Systems* (Eds J.D. Collinson and J. Lewin), pp. 337–344. Special Publication 6, International Association of Sedimentologists. Blackwell Science, Oxford.

Richmond, L.K. and Williams, B.P.J. (2000) A new terrane in the Old Red Sandstone of the Dingle Peninsula, SW Ireland. In: *New Perspectives on the Old Red Sandstone* (Eds P.F. Friend and B.P.J. Williams), pp. 147–184. Special Publication 180, Geological Society Publishing House, Bath.

Roberts, D.G., Thompson, M., Mitchener, B., Hossack, J., Carmichael, S. and Bjørnseth, H-M. (1999) Palaeozoic to Tertiary rift and basin dynamics: mid-Norway to the Bay of Biscay – a new context for hydrocarbon prospectivity in the deep water frontier. In: *Petroleum Geology of Northwest Europe: Proceedings of the 5th Conference* (Eds A.J. Fleet and S.A.R. Boldy), pp. 7–40. Geological Society, London.

Sadler, S.P. and Kelly, S.B. (1993) Fluvial processes and cyclicity in terminal fan deposits: an example from the Late Devonian of southwest Ireland. *Sediment. Geol.*, **85**, 375–386.

Schumm, S.A. (1981) Evolution and response of the fluvial system; sedimentologic implications. In: *Recent Developments in Fluvial Sedimentology* (Eds F.G. Ethridge, R.M. Flores and M.D. Harvey), pp. 19–29. Special Publication 39, Society of Economic Paleontogists and Mineralogists, Tulsa, OK.

Seranne, M. (1992) Devonian extensional tectonics versus Carboniferous inversion in the northern Orcadian Basin. *J. Geol. Soc. London*, **149**, 27–37.

Shanley, K.W. and McCabe, P.J. (1994) Perspectives on the sequence stratigraphy of continental strata. *Am. Assoc. Petrol. Geol. Bull.*, **78**, 544–568.

Sobel, E.R., Hilley, G.E. and Strecker, M.R. (2003) Formation of internally-drained contractional basins by aridity-limited bedrock incision. *J. Geophys. Res*, **108**(B7), 2344.

Sturm, M. and Matter, A. (1978) Turbidites and varves in Lake Brienz (Switzerland): deposition of clastic detritus by density currents. In: *Modern and Ancient Lake Sediments* (Eds A. Matter and M.E. Tucker), pp. 147–168. Special Publication 2, International Association of Sedimentologists. Blackwell Science, Oxford.

Talbot, M.R. and Allen, P.A. (1996) Lakes. In: *Sedimentary Environments: Processes, Facies and Stratigraphy* (Ed. H.G. Reading), pp. 83–124. Blackwell Science, Oxford.

Tooth, S. (1999a) Floodouts in Central Australia. In: *Varieties of Fluvial Form* (Eds A.J. Miller and A. Gupta), pp. 219–247. Wiley and Sons, London.

Tooth, S. (1999b) Downstream changes in floodplain character on the Northern Plains of arid central Australia. In: *Fluvial Sedimentology VI* (Eds N.D. Smith and J. Rogers), International Association of

Sedimentologists. Blackwell Science, Oxford Special Publication 28, 93–112.

Vincent, S.J. (2001) The Sis palaeovalley: a record of proximal fluvial sedimentation and drainage basin development in response to Pyrenean mountain building. *Sedimentology*, **48**, 1235–1276.

Vincent, S.J. and Elliott, T. (1997) Long-lived transfer-zone paleovalleys in mountain belts: an example from the Tertiary of the Spanish Pyrenees. *J. Sediment. Res.*, **67**, 303–310.

Vincent, S.J., Macdonald, D.I.M. and Gutteridge, P. (1998) Sequence Stratigraphy. In: *Unlocking the Stratigraphical Record: Advances in Modern Stratigraphy* (Eds P. Doyle and M.R. Bennett), pp. 299–350. Wiley and Sons, London.

Wescott, W.A. and Ethridge, F.G. (1990) Fan deltas – alluvial fans in coastal settings. In: *Alluvial Fans: a Field Approach* (Eds A.H. Rachocki and M. Church), pp. 195–212. Wiley and Sons, Chichester.

Williams, E.A. (2000) Flexural cantilever models of extensional subsidence in the Munster Basin (SW Ireland) and Old Red Sandstone fluvial dispersal systems. In: *New Perspectives on the Old Red Sandstone* (Eds P.F. Friend and B.P.J. Williams), pp. 239–268. Special Publication 180, Geological Society Publishing House, Bath.

Williams, E.A., Bamford, M.L.F., Cooper, M.A., *et al.* (1989) Tectonic controls and sedimentary response in the Devonian–Carboniferous Munster and South Munster basins, south-west Ireland. In: *The Role of Tectonics in Devonian and Carboniferous Sedimentation in the British Isles* (Eds R.S. Arthurton, P. Gutteridge and S.C. Nolan), pp. 123–141. Occasional Publication 6, Yorkshire Geological Society.

Anatomy and architecture of ephemeral, ribbon-like channel-fill deposits of the Caspe Formation (Upper Oligocene to Lower Miocene of the Ebro Basin, Spain)

JOSÉ LUIS CUEVAS MARTÍNEZ, PAU ARBUÉS CAZO, LLUIS CABRERA PÉREZ *and* MARIANO MARZO CARPIO

Departament d'Estratigrafia, Paleontología i Geociències Marines, Universitat de Barcelona, Martí i Franqués s/n. 08028, Barcelona, Spain (Email: jcuevas@ub.edu)

ABSTRACT

The Caspe Formation comprises the fluvial facies of the Guadalope–Matarranya alluvial system, which encompasses a transition from alluvial to fluvial and fluvio-lacustrine facies assemblages over 60 km from south to north. This formation crops out over an area of 900 km^2 and is characterized by the presence in the landscape of ribbon-like sandstone ridges that are interpreted to be the infills of mostly laterally stable fluvial palaeochannels. The palaeoclimatic setting was semi-arid. The internal structure of the channel sandstones and their sedimentary architecture show some characteristic features.

1 The main sedimentary units that compose the channel fills are downstream accretion macroforms, with well-developed topsets and foresets. Channel-fill sequences record the migration of these macroforms rather than reflecting a gradual decrease in channel activity typical of channel fills in many other ancient systems.
2 A strongly episodic regime can be inferred from the presence of drapes of pedogenically altered mudstones, separating the accretion units, and by the presence of ant nests in the channel sandstones.
3 The presence of frontal lobes at the distal end of some channels.
4 In some cases, the geometrical arrangement of channel units suggests that avulsion processes were conditioned by the depositional topography of the antecedent channel.

From these observations, it is concluded that the facies of the Caspe Formation were deposited by a fluvial system with markedly episodic flow in time and, possibly, space. The resulting depositional architecture consists of discontinuous, low-sinuosity ribbon sandstone bodies with a relatively poor degree of interconnection.

Keywords Stream channels, fluvial sediments, sedimentary deposits, Oligo-Miocene.

INTRODUCTION

The Caspe Formation represents an exceptional example of fluvial ribbon-like, ephemeral channel deposits (Riba *et al.*, 1967; Williams, 1975; Friend *et al.*, 1979) that developed under semi-arid conditions (Cabrera & Sáez, 1987; Agustí *et al.*, 1988). The exceptional outcrop spans an area of about 900 km^2 in which the channel fills crop out extensively as elongated sandstone ridges, often traceable for some hundreds of metres up to several kilometres. On closer scrutiny, the Caspe ribbons also display other unusual features that make the interpretation of the Caspe Formation facies problematic in terms of classic fluvial models: channel-fill sequences that sometimes show coarsening and thickening upwards sections; channel cross-sections that seldom display abandonment facies; and the presence of facies that can be interpreted as frontal-lobe deposits. The importance of this work is that the Caspe Formation can be considered as an example of the deposits of a fluvial system developed in an arid to semi-arid palaeoclimatic context. Whereas the processes and products developed in this kind of environment have been dealt with in a number of previous, essentially geomorphological works, the number of ancient examples in the sedimentological literature remains relatively low.

This work focuses on the following features of the Caspe Formation:

1 characterization of the channel-fill structures and inferred depositional processes;
2 interpretation of the style of fluvial channel evolution and its relation to the floodplain facies;
3 the architecture of the channel deposits and their relation to avulsion and reoccupation processes.

The special nature of the Caspe outcrops allows for an unusual approach to the study of the internal anatomy of the channel deposits. In addition to the analysis of two-dimensional cross-sections, as typically seen along road cuts and cliffs, the ribbon sandstone outcrops allow for the analysis of the structure of longitudinal sections, along the palaeoflow, and thus reveal the three-dimensional relationship between channel and floodplain facies.

GEOLOGICAL SETTING

The study area is located in the southeast of the Ebro Basin (Fig. 1), and is part of the Ebro foreland basin. The tectonic loading of the Pyrenean orogen caused lithospheric flexure and gave rise to subsidence in its peripheral southern foreland zones, with the maximum subsidence (from 3 to 5 km) located near the main orogen. Late Oligocene frontal thrusting and uplift of the surrounding ranges (the Pyrenees, Iberian and Catalan Ranges)

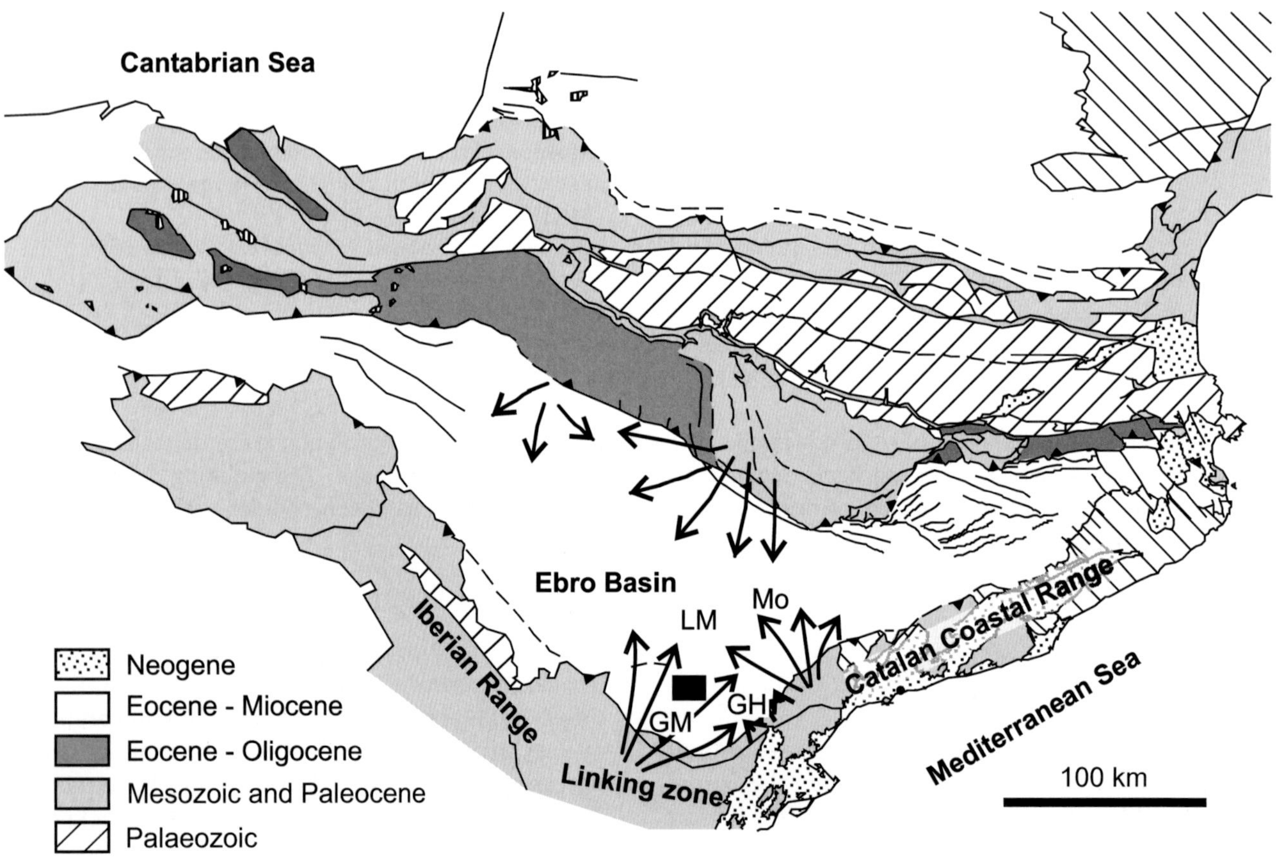

Fig. 1 Tectonic map of the northeast Iberian peninsula showing the Pyrenees, the Catalan Coastal Ranges and the Iberian Range surrounding the Ebro foreland basin. Location of the studied area (black box) and the related late Oligocene to Miocene depositional systems: GH, Gandesa-Horta; GM, Guadalope-Matarranya; LM, Los Monegros; Mo, Montsant.

resulted in the generation of the late Ebro foreland basin. The Oligo-Miocene evolution of this foreland basin was largely controlled by the Pyrenean Range (Fig. 1). Nevertheless, sedimentation in the basin was also influenced by tectonics along the Iberian and Catalan Coastal Ranges, which constitute its southern boundaries. Thrusting and folding within these ranges resulted in structural accommodation in the basin, whereas in the inner range zones it led to the development of high relief and extensive source areas (Guimerà, 1984; Anadón *et al.*, 1985, 1986, 1989).

During the Oligocene and Early–Middle Miocene the whole Ebro Basin was closed and progressively infilled with alluvial and lacustrine deposits. Terrigenous contributions into the basin resulted in the generation of synorogenic alluvial fans and fluvial megafans, which interfingered with central basin lacustrine systems, where terrigenous, carbonate and evaporite sedimentation took place (Cabrera & Sáez, 1987; Cabrera *et al.*, 2002). The Huesca and Luna fluvial distributary systems spread 40–60 km radially southwards from the Pyrenees (Hirst & Nichols, 1986; Nichols & Hirst, 1998, Arenas *et al.*, 2001, Nichols, this volume, pp. 567–587). The Montsant and Guadalope–Matarranya megafans also attained a wide radial spread (up to 40–60 km) northwards from the Iberian and the Catalan Coastal Ranges (Anadón *et al.*, 1986, 1989; González & Guimerà, 1997; Cabrera *et al.*, 2002). Radially more restricted alluvial fan systems (i.e. Horta–Gandesa system) were coeval with the larger fluvial fans (Jones *et al.*, 2004).

Palinspastic plate reconstructions (Smith, 1996) and palaeomagnetic data (Barberà *et al.*, 2001; Pérez Rivares *et al.*, 2004) show that during Oligocene to Early Miocene times the Iberian Peninsula was located slightly south of its present latitude, resulting in warmer and drier conditions. This is corroborated by the sedimentary and palaeobiological records of the sequences studied (Cabrera, 1983; Cabrera & Sáez, 1987; Cabrera *et al.*, 2002).

THE CASPE FORMATION

General characteristics

In the study area (Figs 1 & 2) the Caspe Formation comprises middle to distal fluvial facies (Williams,

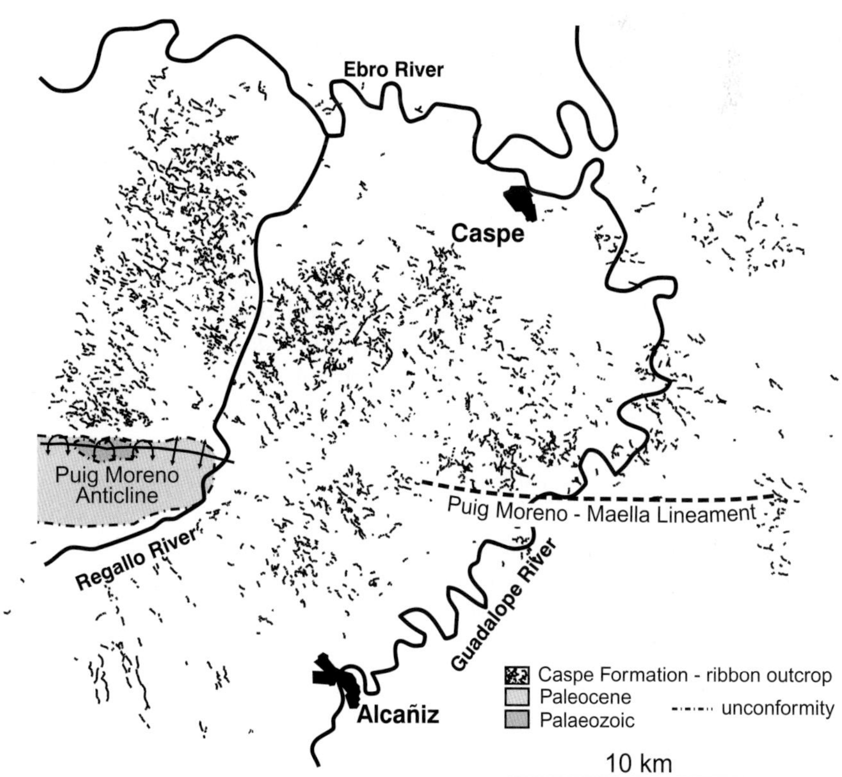

Fig. 2 Detailed map of the study area. Small segments represent sandstone ribbon axes.

1975) deposited on the Guadalope–Matarranya fluvial fan system. This system was in the southernmost part of the Ebro Basin and distributed sediment northward from late Oligocene to Early Miocene times (Fig. 1). It encompasses, over 60 km, a full northward transition from proximal (conglomerate-dominated) to distal alluvial (terminal mud flats) and fluvio-lacustrine facies assemblages (Cabrera *et al.*, 2002).

The fluvial successions are folded only in the region of the WSW–ENE trending Puig Moreno anticline, which probably is related to a blind thrust (Klimovitz, 1992). All along the Puig Moreno–Maella lineament, which corresponds to the eastward extension of the Puig Moreno anticline (Fig. 2), the fluvial beds are tilted and often display nearly vertical dips. The total thickness of the Caspe Formation is uncertain, given the lack of detailed subsurface data, but it is estimated to be around 400 m, taking into account the sections with the most complete outcrops and the available regional measurements.

Fluvial facies assemblages

The Caspe Formation is made up mainly of two major facies assemblages: the channel-infill assemblages and the floodplain assemblages (Williams, 1975). Although ribbon-like (Riba *et al.*, 1967; Friend *et al.*, 1979, 1986), low- to high-sinuosity channel-fill sandstone bodies are by far the most frequent type, tabular sandstone bodies that correspond to point-bar deposits also have been reported at some localities (Williams, 1975; Anadón *et al.*, 1989). Point-bar deposits are more frequent in the distal zones of the Caspe Formation, beyond the area studied (Cabrera *et al.*, 1985; Anadón *et al.*, 1989). The palaeocurrents measured from the channel trends are mainly directed to the north and northwest (Figs 1 & 2).

The channel sandstones are composed of medium to fine sands. It is only in the southern parts of the area studied that centimetre-scale pebbles, mostly composed of Mesozoic limestones also appear as a minor component of the sandstone bodies (up to 5–10%). The size and amount of these pebbles diminish northward, so that 10 km north of Alcañiz they are very scarce or absent. The sandstone grains are limestone lithoclasts (up to 50–55%), quartz (35–45%) and other minor lithologies (up to 5%) (Williams, 1975).

The floodplain facies assemblages are mostly composed of reddish and ochre mudstones with interbeds of fine to very fine, thin sandstone and siltstone layers. Poorly defined horizons of millimetre- to centimetre-scale alabastrine gypsum nodules occur commonly in the mudstones. Incipient palaeosols are also common, usually marked by horizons of rootlets and pedotubules and, less commonly, by nodular caliche horizons. Laterally extensive (up to several hundred metres), decimetre-thick micritic limestone layers also occur interbedded in the floodplain facies assemblages. These limestone beds become more frequent and thicker northward, in the lateral transition to the lacustrine, carbonate-dominated facies.

Outcrop conditions and morphology of the sandstone ribbons

The ribbon-like sandstone bodies (Figs 3–5) crop out extensively in the flat region between Caspe and Alcañiz and the rivers Regallo and Guadalope, an area some 20 km (N–S) by 35 km (E–W) (Fig. 2). The areal distribution of the ribbon sandstones within the area studied is shown in Fig. 2. Most of the ribbons' axes show a SE–NW and S–N trend. The areal distribution of the ribbons shows a loosely distributive pattern, although there is no evidence of a hierarchical drainage arrangement (Fig. 2).

These exceptional outcrop conditions in the Caspe area resulted from three factors:

1 cementation of the sandstones, which makes them less susceptible to erosion than the fines;
2 a very low regional dip of the fluvial sequences (near to zero);
3 a relatively flat landscape, with a general gentle slope oriented to the north that coincides in most of the area with the regional dip.

Factors which may contribute to bias the preferential preservation of the sandbodies are as follows.

1 Some very gentle changes of the low-angle regional dip, which can lead to selective preservation of the sandbodies with axes that are parallel to the strike. This factor seems to be of very local influence, and only considered to be important in areas close to the tectonic structures (Puig Moreno anticline and Puig Moreno–Maella lineament).

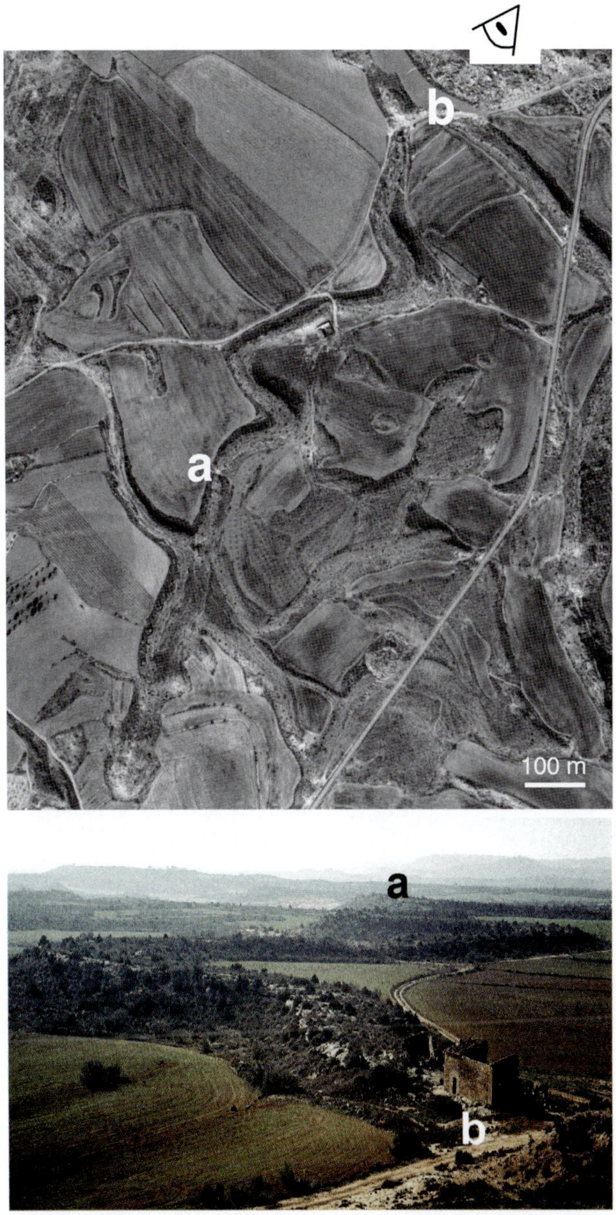

Fig. 3 Aerial photograph of a ribbon and oblique view of the same ribbon. UTM: latitude 4559929, longitude 748031 (datum European 1950, zone 30N).

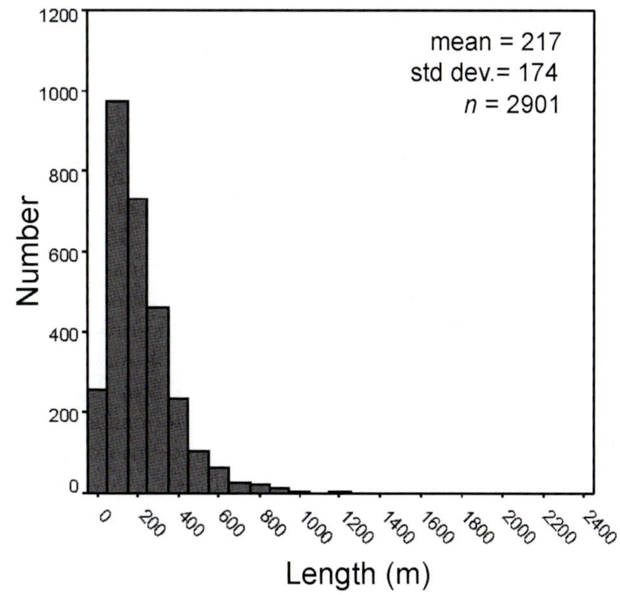

Fig. 4 Length histogram from a sample of 2901 ribbons. Measurements were taken from SPOT satellite images of the area shown in Fig. 2.

2 The existence of a sparse network of joints can influence the shape of the preserved sandbodies, so that parts of the ribbon sandstones that are almost perpendicular to the joints' strike will be more densely fractured and, hence, more readily eroded.
3 The topographic gradient, which potentially is the most important factor. The Caspe ribbons can crop out extensively only in areas with a gentle topographic gradient. When they crop out in steep areas, only the ribbons with axes that are parallel to the contour levels crop out extensively.

Studies of the planimetric morphology of the ribbons (Williams, 1975) indicate that the sinuosity is lower than 1.1 for 66% of the preserved length of a sample of 749 ribbons; values up to 6 have been recorded in some extreme cases. The preserved lengths range from a few tens of metres up to 3 km, following a log-normal distribution, with a mean of 217 m and a standard deviation of 174 m (Fig. 4).

The maximum thickness of the ribbons rarely exceeds 15 m. Their widths are in the order of a few tens of metres. The width/thickness ratio is difficult to determine for most ribbon sandstones for the following reasons.

1 Width/thickness ratio measured from ribbons preserved in the present-day topography will typically be underestimated, as preserved width will be a fraction of the original width owing to landscape erosion.
2 Width/thickness ratio measured from cross-sections is closer to the real value, although it must be corrected because the cross-section width is a function of the angle between the ribbon axis and the cross-section plane. A mean width:thickness ratio of 7.2, with a standard deviation of 3.1, has been determined from nine measurements (Williams, 1975).

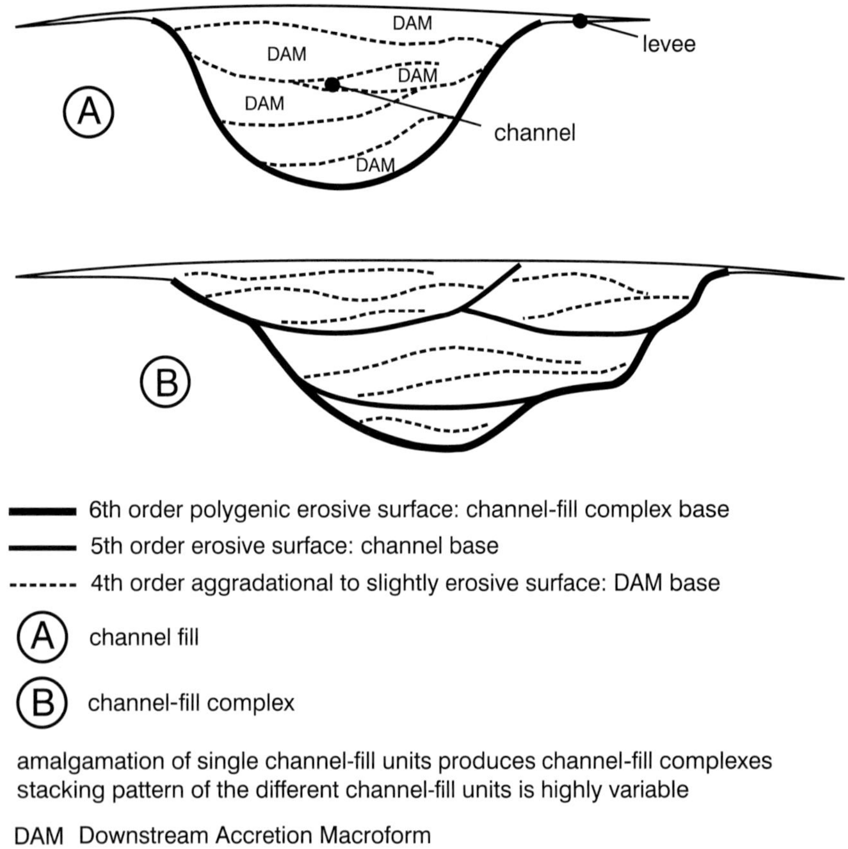

Fig. 5 Hierarchy and nomenclature of the channel-fill elements.

TERMINOLOGY

In order to describe the architecture of the Caspe Formation, a hierarchy of elements has been defined (Fig. 5). The order classification of Miall (1988, 1996) has been applied to describe the bounding surfaces. It must be stressed that the term ribbon, as applied in this work, is a strictly geomorphological term that corresponds to the sandstone outcrops formed by channel-fill complexes and single channel fills. On the other hand, the terms storey and multistorey, which have been frequently applied to the structure of the channel facies (Williams, 1975; Anadón *et al.*, 1989; Mohrig *et al.*, 2000) are considered to be of ambiguous genetic meaning and, hence, they will be avoided; such terms refer to the segmentation of the channel deposits into different architectural units (e.g. channel fills and accretion macroforms) that are the product of different sedimentary processes operating at different scales of time and space.

From higher to lower order, the following architectural elements have been considered.

1 Channel-fill complexes: sandstone bodies composed of several stacked channel fills. The bases of these units are bounded by 6th-order surfaces. The stacking pattern of the individual channel fill units is highly variable, resulting in highly variable, often lenticular cross-sectional geometries.
2 Channel fills (Fig. 5A): lenticular sandstone bodies bounded by 5th-order surfaces. Bases of this type of unit are denoted by strongly erosive, concave-upwards surfaces.
3 Accretion units: prismatic sandstone bodies limited by 4th-order surfaces that comprise the infilling of the channels. Two kinds of these units have been distinguished: downstream accretion macroforms (DAM, *sensu* Miall, 1988, 1996) and lateral accretion macroforms (point-bar deposits). These latter units are very rarely found in the study area.
4 Cosets of cross-stratified sandstones bounded by 3rd-order surfaces within the accretion units.
5 Sets of cross-bedded and cross-laminated sandstones bounded by lower order surfaces.

Higher order bounding surfaces contain, and can coincide with, lower order surfaces. Lower order surfaces are totally contained within higher order surfaces.

CHANNEL FILLS

Structure

Channel fills are the basic building blocks of the Caspe Formation architecture. Representative channel fills are illustrated in Figs 5–11. The cross-sectional shape (Figs 5 & 6) is generally lenticular, with two distinct elements (Williams, 1975).

1 A central, lenticular shaped 'body' composed of trough and tabular cross-bedded sandstones in sets up to some decimetres thick. Sandstone bases are markedly concave-upwards, 5th-order surfaces that correspond to a single channel scour. Thickness ranges from a few metres up to some 15 m. Width is in the order of a few tens of metres. Tops are often slightly convex-upwards, marked by a sharp surface

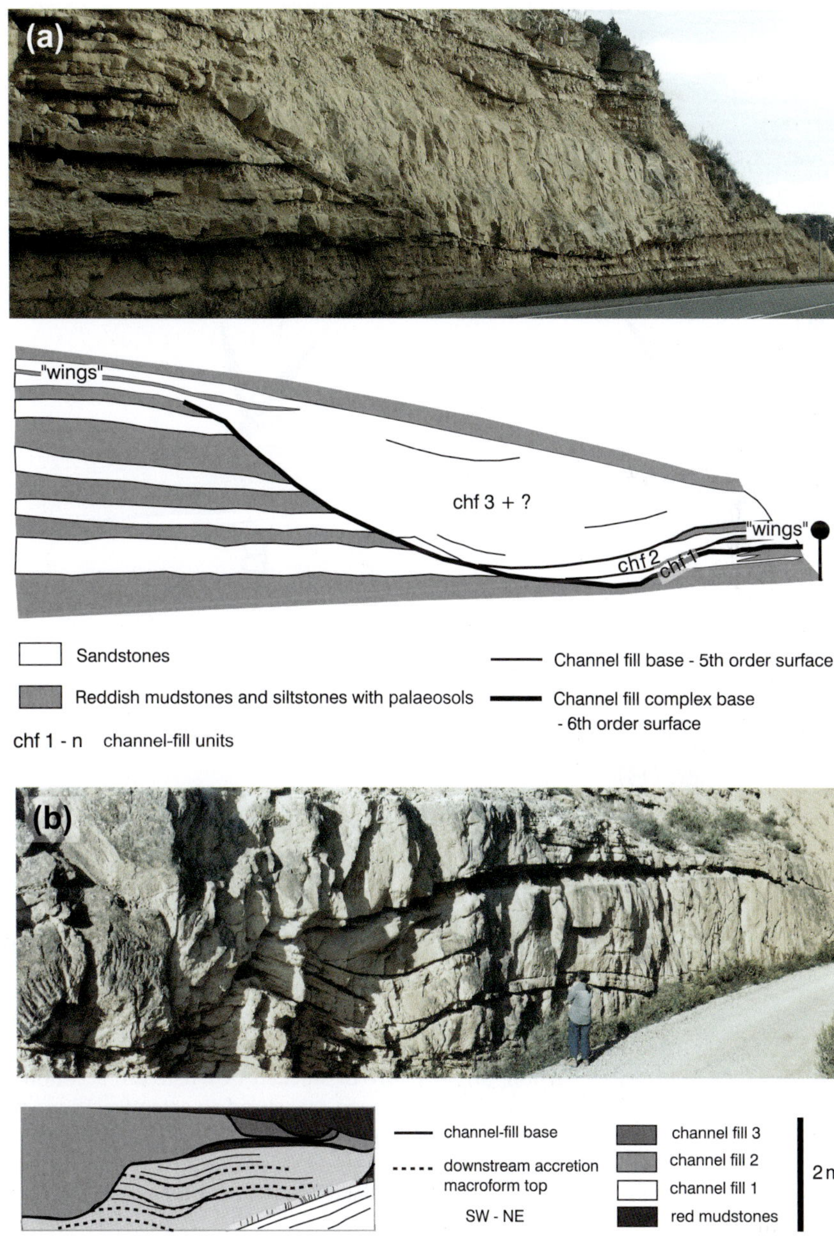

Fig. 6 Cross-sectional views of channel-fill complexes. (a) Each of the individual channel fills that compose this complex show a restricted lateral shifting to the northwest (left of the image). Height of the road sign (right) is 175 cm. UTM: latitude 4567241, longitude 744653. (b) Undulating depositional surfaces at channel-fill 1 are interpreted as downstream accetion macroform (DAM) tops. Note the convex top of channel-fill 1 as well as the preservation of drapes of red mudstones at the DAM tops, and at the contact between channel-fill units 1 and 2. UTM: latitude 4546710, longitude 747170.

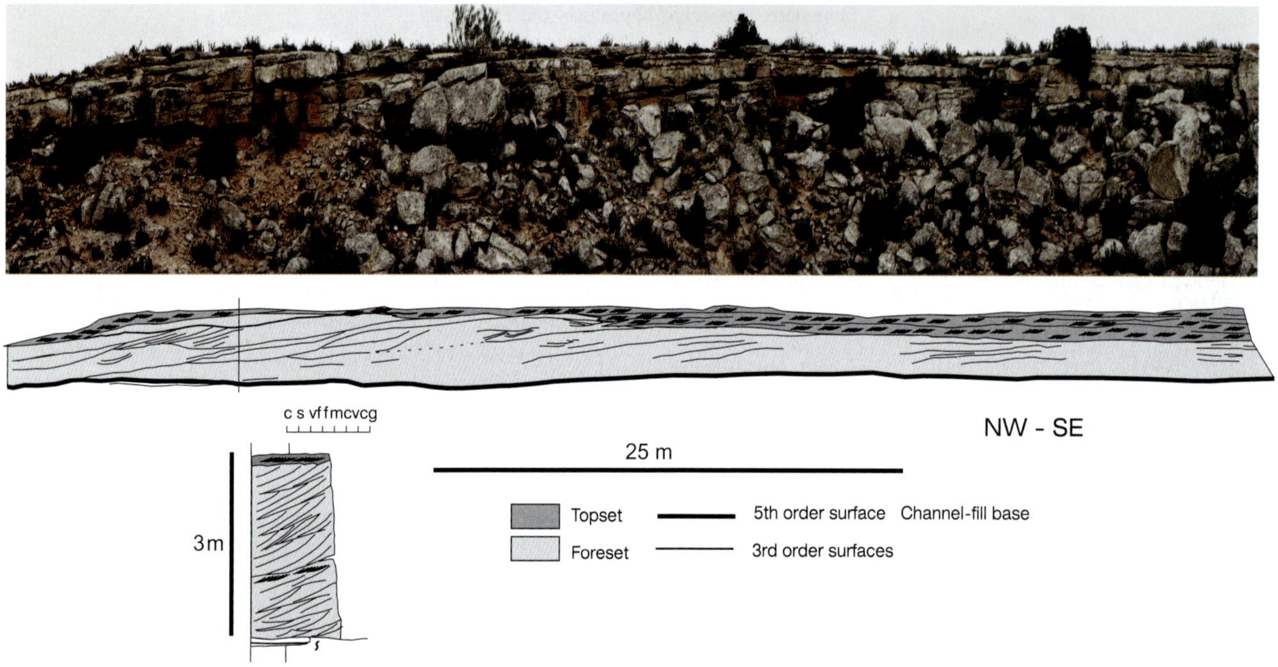

Fig. 7 Hoya del Guallar outcrop. Longitudinal section of a ribbon sandstone showing the structure of a downstream accetion macroform. UTM: latitude 4560454, longitude 748267.

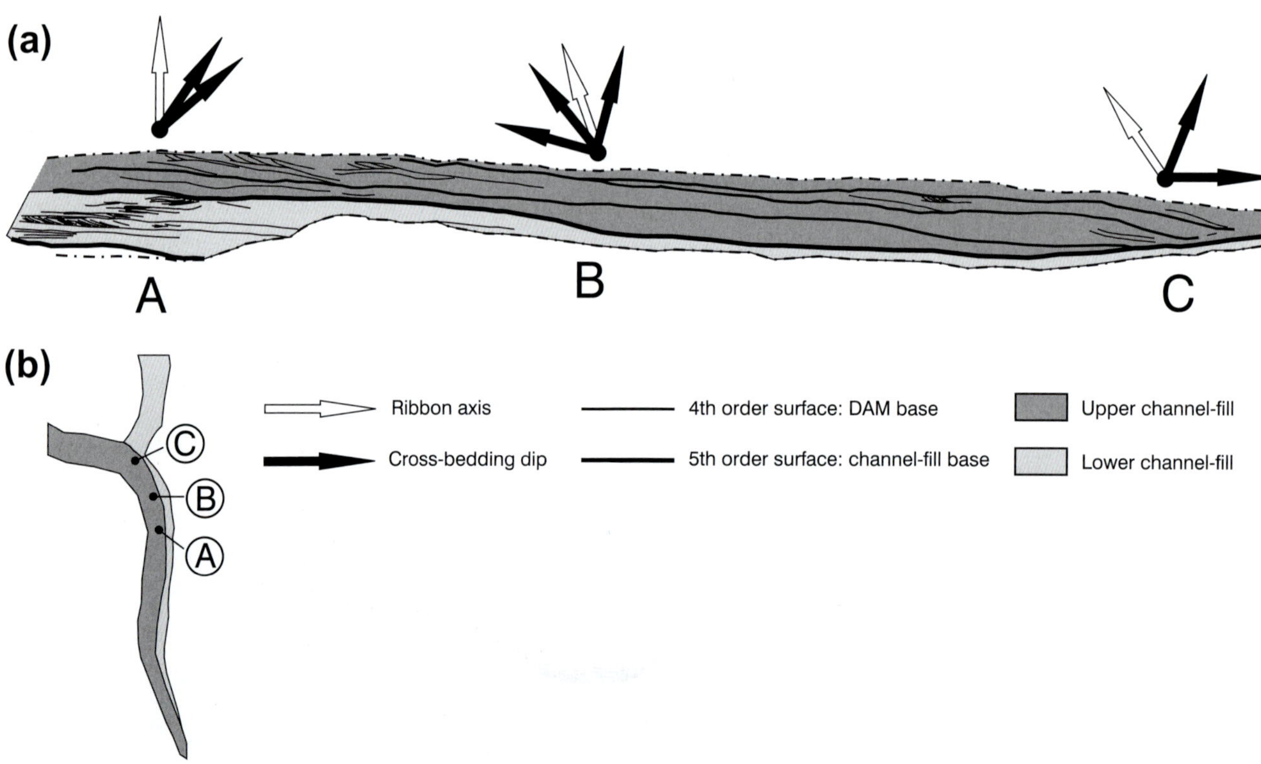

Fig. 8 Hoya del Guallar section. (a) Structure of stacked downstream accetion macroforms (DAMs) in a section parallel to the channel-fill axis. Palaeocurrents have been measured in sets of planar and trough cross-bedded sandstones. Set thickness ranges from 25 to 40 cm. Note the gently downcurrent dip of the DAMs, and the trend of the cross-bedding with respect to the ribbon axis: oblique at the straight reach and almost perpendicular, oriented to the outer margin at the bend. (b) Plan view of the channel fills. Distance A–C is approximately 100 m. UTM: latitude 4559761, longitude 747862.

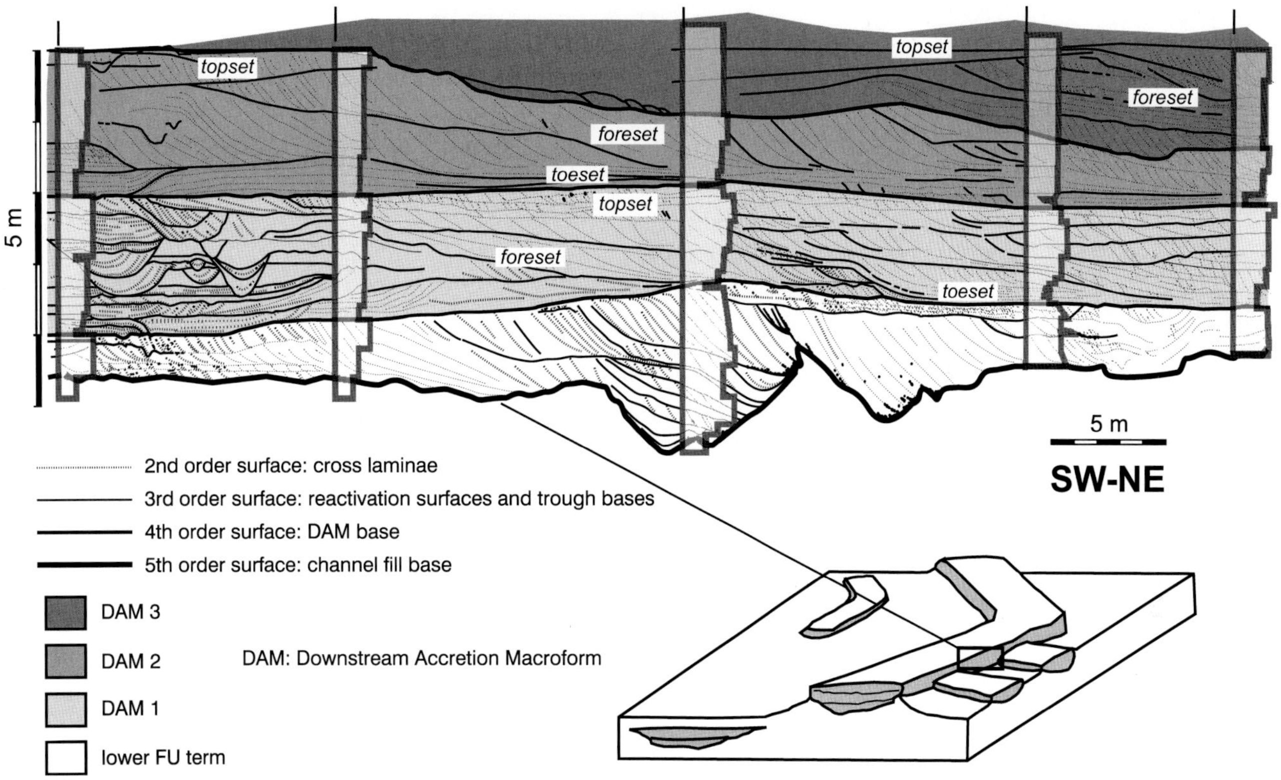

Fig. 9 Mocatero outcrop. Internal structure of downstream accretion units in longitudinal section. Note the coarsening upwards trend of these depositional units. UTM: latitude 4570718, longitude 734147.

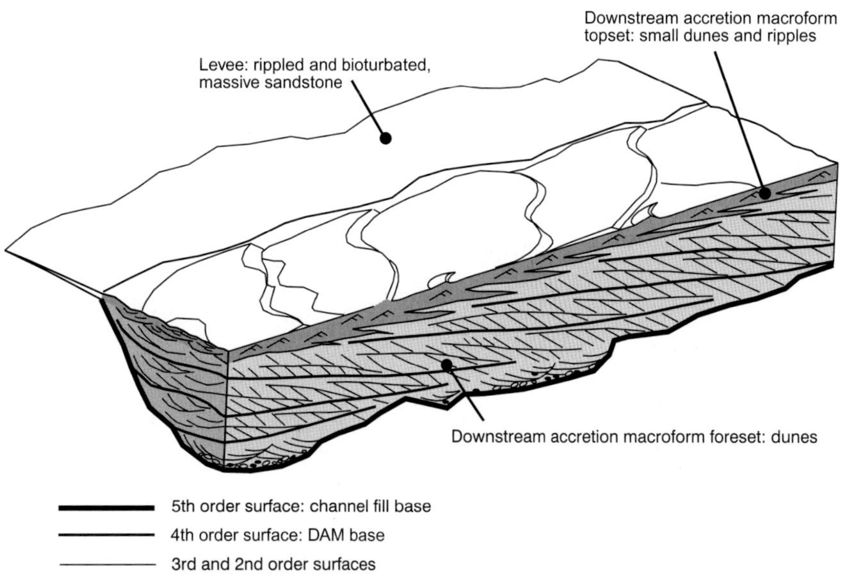

Fig. 10 Three-dimensional structure of a channel fill. Not to scale.

or a rapid transition to the fine-grained floodplain materials, and are composed of rippled sandstones, with occasionally preserved ripples and small dunes up to 20 cm thick. Mottling and bioturbation in the form of pedotubules are frequent at the channel sandstone tops. Clay plugs, in the form of dark grey mudstones overlying the channel sandstone tops, are rarely observed.

Fig. 11 Ant nest in channel sandstones. Zaragoceta outcrop. The hammer is 30 cm long. UTM: latitude 4561319, longitude 752727.

2 'Wings': in the more complete sections, tabular bodies composed of medium to very fine sandstones and siltstones extend laterally from the channel sandstone tops (Figs 5 & 6a). Sometimes they can be seen pinching out into the floodplain mudstones. These bodies show an apparent extent away from the palaeochannel ranging from a few tens of metres to a few hundreds of metres, and their thickness ranges from a few decimetres to a few metres. Internally they are organized into centimetre- to decimetre-scale layers and form poorly defined sequences. Sedimentary structures are dominantly formed by ripples and small-scale dunes up to 15 cm thick and frequent erosive scars, although bioturbation has often obliterated the depositional texture. The most common biogenic structures are pedotubules and rootlets. Alabastrine gypsum nodules often appear inside the pedotubules. The character of the tabular sandstone bases can laterally vary from slightly erosive to aggradational. Such bodies are interpreted as levee deposits.

The longitudinal structure of the channel-fill units is well observed in the Mocatero ribbon outcrop (Fig. 9). This outcrop illustrates the structure and sequential trends of a channel-fill unit that is dominated by sandstone bodies up to 2 m thick by 20 m wide and are probably more than several hundreds of metres in length. The 4th-order surfaces that bound these bodies are overall aggradational and gently inclined, although they can be locally erosive. Each of these units displays a topset–foreset–toeset structure. The topsets are generally thinner than 20 cm and consist of sets of trough/planar cross-strata composed of medium to very coarse-grained sandstone with granules. The foresets are mostly characterized by sets of tabular cross-bedded sandstones up to 1 m thick, sloping as much as 40°, and they display a down-set decrease in grain size from medium- to fine-grained sandstone. Some of the sets of tabular cross-stratification include smaller scale cross-stratification corresponding to minor superimposed bedforms. The toesets are generally thinner than 10 cm and consist of reddish, parallel-laminated, very fine-grained sandstones. The downstream migration of successive topset–foreset–toeset cosets resulted in the Mocatero ribbon channel fill being dominated by stacked coarsening upwards units.

Figure 10 summarizes the most frequent three-dimensional structure of the channel fills. In general, the longitudinal structure of the individual channel-fill units is often characterized by the presence of large, downcurrent-dipping prismatic bodies (Figs 7 & 8). The downstream extent ranges from a few tens of metres up to a few hundreds of metres. Thickness ranges from a few decimetres up to a few metres. Their bases are marked by 4th-order surfaces dipping from nearly flat up to 5° downcurrent; the tops are sometimes draped by reddish mudstones with pedogenic structures. In some cases (Fig. 9), these 4th-order surfaces show a downcurrent transition from erosive to aggradational. The most prominent feature of these bodies is the presence of three distinct elements that include an upper topset, an intermediate foreset and a lower toeset. The topset element is composed of medium- to small-scale trough and planar cross-bedded sandstones in sets up to 20 cm thick. The foreset is made up mostly by tabular cross-bedded sandstones in sets up to some decimetres thick. Along the straight segments of the ribbon sandstones, foreset laminae dips are at angles up to 30° to the ribbon axis, whereas at ribbon bends dips are oriented towards the outer margin, reaching almost 90° to the ribbon axis (Fig. 8). These sets are limited by downcurrent-dipping surfaces, and show frequent internal scars and reactivation surfaces (Fig. 9) sometimes draped by reddish mottled mudstones. Sets of trough cross-bedded sandstones also appear, although they are a minor component. The toeset consists of parallel-laminated medium- to fine-grained sandstones, which developed at the base of the foreset. The thickness of this toeset ranges from some centimetres to a few decimetres. These bodies show complex sequences, sometimes coarsening upwards, in which some of the elements, frequently the toeset, can be absent.

Sequences

The idealized, most complete sequence of the Caspe channel fills shows an overall, poorly defined, fining upwards trend, and can be punctuated by coarsening upwards segments that correspond to the stacking of downstream accretion macroform (DAM) elements. A lower, coarse-grained interval is sometimes present, with abundant intraformational clasts and dominated by trough cross-bedded, coarse to medium sandstones. Centimetre pebbles are sometimes present in these lower units, which usually display a fining upwards trend. On top of this lower element or directly overlying the channel basal scour, an interval dominated by tabular and trough cross-bedded, coarse to medium sandstones is typical. This element may be organized into poorly defined or coarsening upwards sequences that record the migration of the DAM, sometimes truncated by films of reddish mudstone interbedded with the DAM sandstone. Bioturbation and pedogenic structures appear in three parts of the sequence:

1 at the sandstone top, in the form of rootlets and pedotubules – mottling is also frequent, and the thickness of this bioturbated unit is of the order of a few decimetres;
2 within the mudstone films at the DAM (Fig 6b);
3 in the lower parts of the sequence, *Polychresichnia* ichnofossils (Hasiotis, 2003) are interpreted as ant nests (Fig. 11; Hasiotis, pers. comm., 1999), although they can also appear (less frequently) in the middle and upper intervals.

Interpretation

The channel-fill sequences record a process of channel incision, episodic infill of the channel, including phases of overbank deposition, and rapid abandonment. The lower coarse-grained intervals are interpreted as the deposits of channel infilling, almost contemporary with the initial phase of channel incision. Intraformational clasts, frequent in these intervals, correspond to the development of the erosive scour of the channel base. Fining upwards trends probably reflect the decrease in energy due to the waning of the initial flows. Following this phase of channel incision and initial infilling, the DAM architectural elements record episodic phases of channel infilling by large macroforms that occupy a large part of the initial channel scour width (Fig. 6b). The overall disposition of the cross-stratification described above indicates the dominant downcurrent migration of such macroforms, with limited lateral accretion at channel bends. These macroforms can be interpreted as the deposits of mid-stream and bank-attached unit bars (Bridge *et al.*, 1998; Bridge, 2003) that migrated downcurrent in channels that had little capability for lateral migration. The Platt-type macroforms described by Crowley (1983) show important analogies with the Caspe bar deposits:

1 they are characterized by an internal topset–foreset–toeset structure;
2 the sequences are coarsening upwards, as it has been observed in some cases in the Caspe bar facies;
3 the dimensions of the cross-strata sets are similar in both cases.

Nevertheless, there are two important differences.

1 The topset interval of the Platte macroforms is separated from the foreset by a scour surface, whereas the topset of the Caspe macroforms usually grades laterally and downwards into the foreset (Figs 7 & 9). This suggests that the formation of the topset of the Caspe macroforms was contemporary with the development of the foreset, whereas the topset of the Platte macroforms probably was not directly related to the development of the foreset.
2 The Platte macroforms do not contain layers of pedogenically altered mudstones that could indicate long periods of subaerial exposure.

Formation of the toeset has been interpreted by Crowley (1983) as formed by grain-fall deposition of sediments taken into suspension from the stoss side of the macroform. Preservation of the toeset and topset elements in the Caspe macroforms could be related to a high sediment transport rate.

The approximate palaeoflow depth, measured from the height of cosets of cross-strata corresponding to individual bars, averages 1.38 m with a maximum of 3.3 m (see Mohrig *et al.*, 2000). It must be stressed that the sedimentation of these macroforms was essentially a discontinuous process, as evidenced by the presence of drapes of reddish mudstones that show processes of initial palaeosol development, interbedded with the DAM sandstones (Fig. 6b), indicating that these channels were intermittently inactive during periods long enough to allow the development of incipient palaeosols. Moreover, the presence of ant nests, more frequent at the base of the channel sequences, indicates the total abandonment of the channel sands by phreatic water.

During the later stages of channel activity, as the accommodation space was reduced, overbank flow led to deposition of levees before the channel was fully plugged by sand. In this way, the deposits reflect the structure of the last active bars, instead of a gradual waning of the last channel flows. This reduction of the bedform size is rather abrupt and it is interpreted as corresponding to the vertical transition from the foreset to the topset element of the DAM. Following this last phase of channel activity, reddish floodplain mudstones directly overlie the top of the channel fill, indicating the total abandonment of the channel activity and the development of pedogenic processes. In a few cases, dark grey mudstones infill the upper parts of the palaeochannels, but these clay plugs are rare.

CHANNEL-FILL COMPLEXES

Channel-fill complexes are sandstone bodies resulting from the stacking of several channel-fill units. Sixth-order surfaces that bound these complexes are polygenetic, resulting from the amalgamation of the 5th-order channel scours that bound the individual channel fills. Axes of the individual channel-fill units are typically nearly parallel. However, in some cases the individual channel fills either diverge or converge.

The Hoya del Guallar outcrop illustrates a divergent situation (Fig. 12). In this outcrop, a channel-fill complex formed by two channel fills traceable along some 500 m splits downflow into two separated channel-fill sandstone bodies. A bioturbated, 20–30 cm thick horizon at the top of the lower unit is present all along the contact between the two channel fills, suggesting that incision by the upper channel was relatively small, and that the upper unit mostly followed the remnant depression of the lower channel fill. The point of divergence may have been controlled by the topography of the infill of the lower unit as it is located just before an elevation of the lower channel-fill top. No evidence of crevasse-splay deposits has been found at the divergence area. This particular spatial arrangement of channel fill units can be interpreted as the product of the following sequence of events:

1 active channel sands plugged some reaches almost up to the bankfull depth;
2 the channel was abandoned and pedogenic processes developed at the top of the channel-fill sands – a more or less irregular topography, at least in part of depositional origin, was preserved along the channel axis;
3 the abandoned channel, preserved on the floodplain locally as a topographic low, was reactivated by a later channel which partially followed the trace of the earlier one, up to a point where an obstacle, such as a bar, caused a deflection in the later channel course.

The above sequence of events may be related to the ephemeral nature of this type of channel;

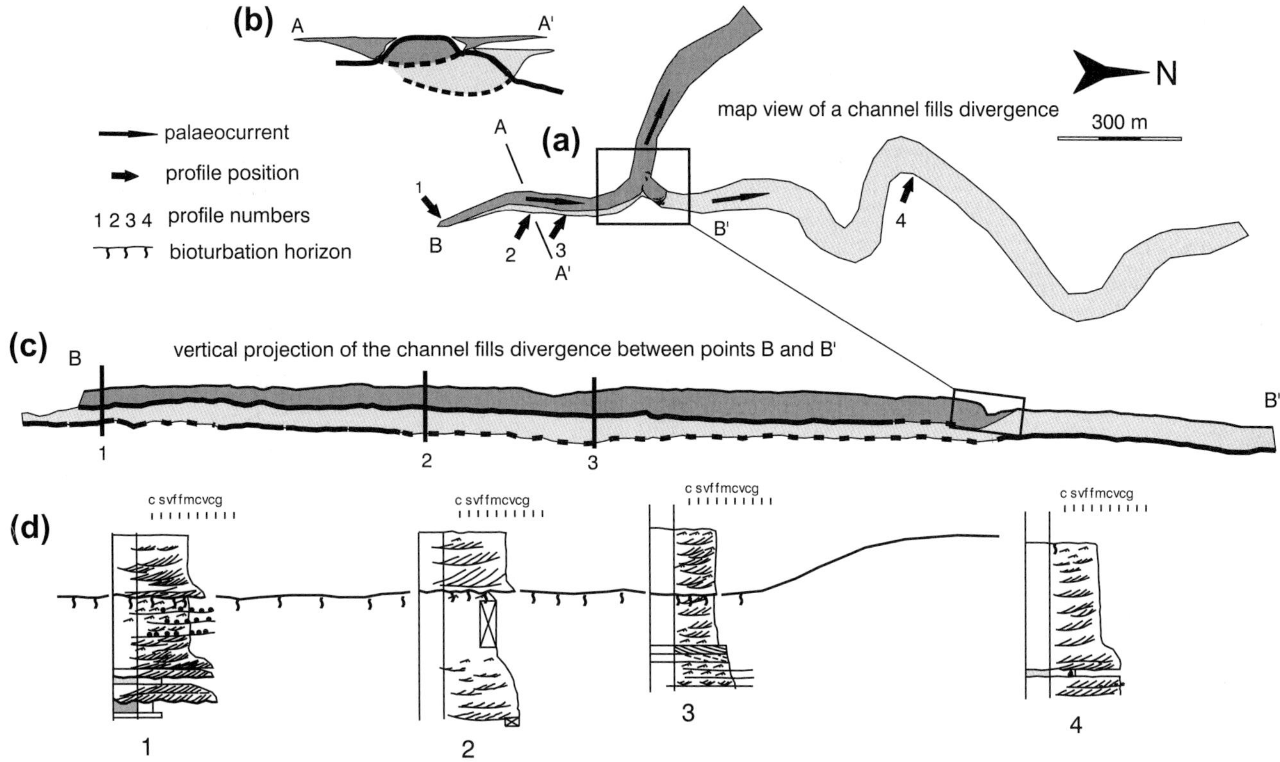

Fig. 12 Hoya del Guallar outcrop. Example of divergent channel-fill units. (a) Plan view. (b) Cross-section. (c) Longitudinal section. (d) Log correlation. UTM: latitude 4559761, longitude 797862.

abandoned channels were partially reactivated, up to reaches where obstructions such as sediment accumulations acted as barriers to the successor channel, resulting in a bifurcation of channel courses. This type of avulsion would belong to group 3 of Jones & Schumm (1999), in which avulsions are the result of a decrease in the channel's ability to carry sediments, instead of progressive channel destabilization due to the upbuilding of alluvial ridges.

The inverse case, in which two channel-fill sandstone units merge down-palaeoflow to form a single channel-fill complex is observed at Mas de Ciuzón, some 15 km southwest of Caspe (Fig. 13). In this case, an upper channel-fill sandstone that follows a SSW–NNE direction merges with a lower, previous channel-fill sandstone that follows a SE–NW direction. Downflow from the convergence point, the upper unit follows the trace of the lower unit, partially eroding its top. This example suggests that former, partially infilled channels could capture newly developing channels (Mohrig *et al.*, 2000). The inferred avulsion process described above, the lack of splay facies and the observation that newly formed, possibly avulsed channels adapted their course to previous channels suggest that avulsion by annexation (Slingerland & Smith, 2004) took place in the Caspe fluvial network.

Downstream evolution of the channel-fill complexes

Channel-fill complexes commonly erosively overlie relatively thin (some centimetres to a few decimetres), crudely coarsening and thickening upwards sequences consisting of very fine bioturbated sandstones. In some cases, as described below, these fine-grained thickening and coarsening upwards sequences can be traced into channel-fill facies. Figure 14 shows a longitudinal section of a channel-fill complex in the Zaragoceta area. Three channel fills (A, B and C in Fig. 14) make up this complex. Palaeocurrents, measured on DAM foresets, are subparallel to the channels' axes and are directed mostly to the northwest and west-northwest. The lowermost channel fill (C), which is composed of cross-bedded, medium to coarse sandstone with

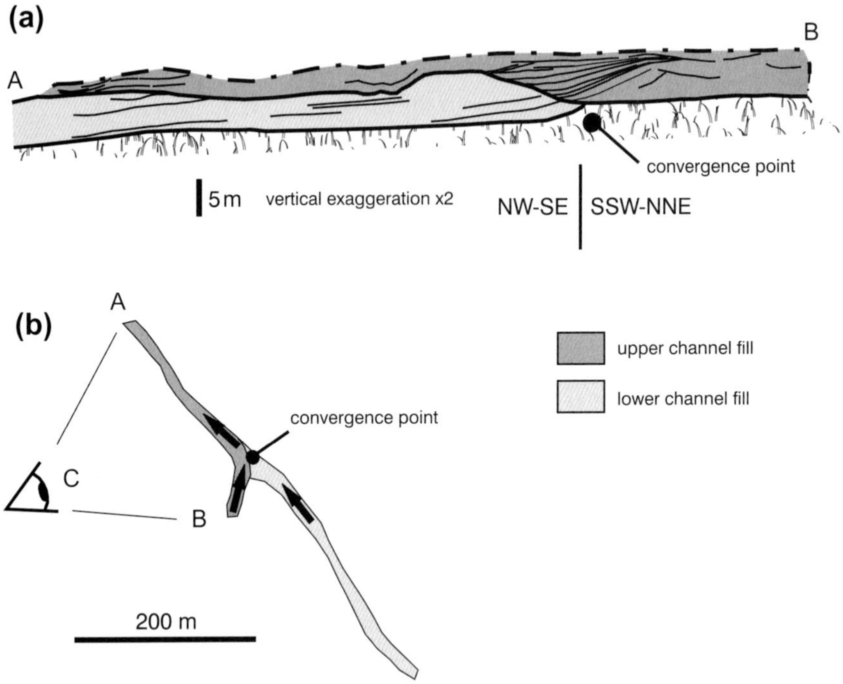

Fig. 13 Mas de Ciuzón outcrop. (a) Perspective view from point C. (b) Plan view of two converging channel-fill units. UTM: latitude 4563149, longitude 744168.

pebble-sized intraclasts (profile 3, Fig. 14), grades downcurrent into fine-grained, bioturbated layers of sandstone organized into coarsening upwards sequences that belong to unit D (profiles 1 and 2, Fig. 14). Bioturbation also increases in the same direction. Cross-lamination is still recognizable or only slightly obliterated by bioturbation in unit D at profile 2, whereas the same unit is composed of almost structureless sandstones and siltstones at profile 1. In addition, unit D in profile 1 has a relatively high carbonate content in the lower part, suggesting deposition in a poorly drained, possibly ponded area.

Channel-fill C, as well as part of unit D, is eroded by channel-fill B. In turn, channel-fill A is incised into the top of channel-fill B. Sandstones of channel-fills A and B are cross-bedded, organized into DAM units. Channel-fill B shows bioturbation by ant nests (*Polychresichnia*) close to the base in profile 1 (Fig. 11). Some key features are to be stressed on this outcrop: the whole thickness of channel-fill C grades into unit D, showing a full downcurrent transition from channel to finer grained, strongly bioturbated facies; palaeocurrents in channel-fills A, B and C, measured on sets of tabular cross-beds in sets up to 40 cm thick, are nearly parallel, with minor flow divergences between the deposition of channel-fills A and B. Each of these channel fills partly erodes the top of the preceding one.

These features indicate that this outcrop was the result of a process of frontal lobe development, followed by downcurrent propagation and incision of channels. In this way, unit D is interpreted as the deposits of a frontal lobe related to the downstream termination of channel-fill C, whereas channel-fills B and A were developed during later phases of channel reactivation and downcurrent incision. Frontal-lobe deposits have been interpreted from ephemeral stream deposits in the Paleocene of the Tremp-Graus Basin (southern Pyrenees, Spain) by Dreyer (1993), although this case is in a higher energy alluvial facies. Rundle (1985) described a process of lobe formation at the mouth of braided channels, where flow from the channel expands into a relatively shallow, quieter mass of water that has some analogies with the lobe formation process inferred in this work.

Frontal-lobe facies are arranged into coarsening and thickening upwards decimetre-scale sequences. Two facies associations have been distinguished: proximal and distal lobe. Key criteria to distinguish between these two associations are the intensity of bioturbation and grain size. The proximal-lobe

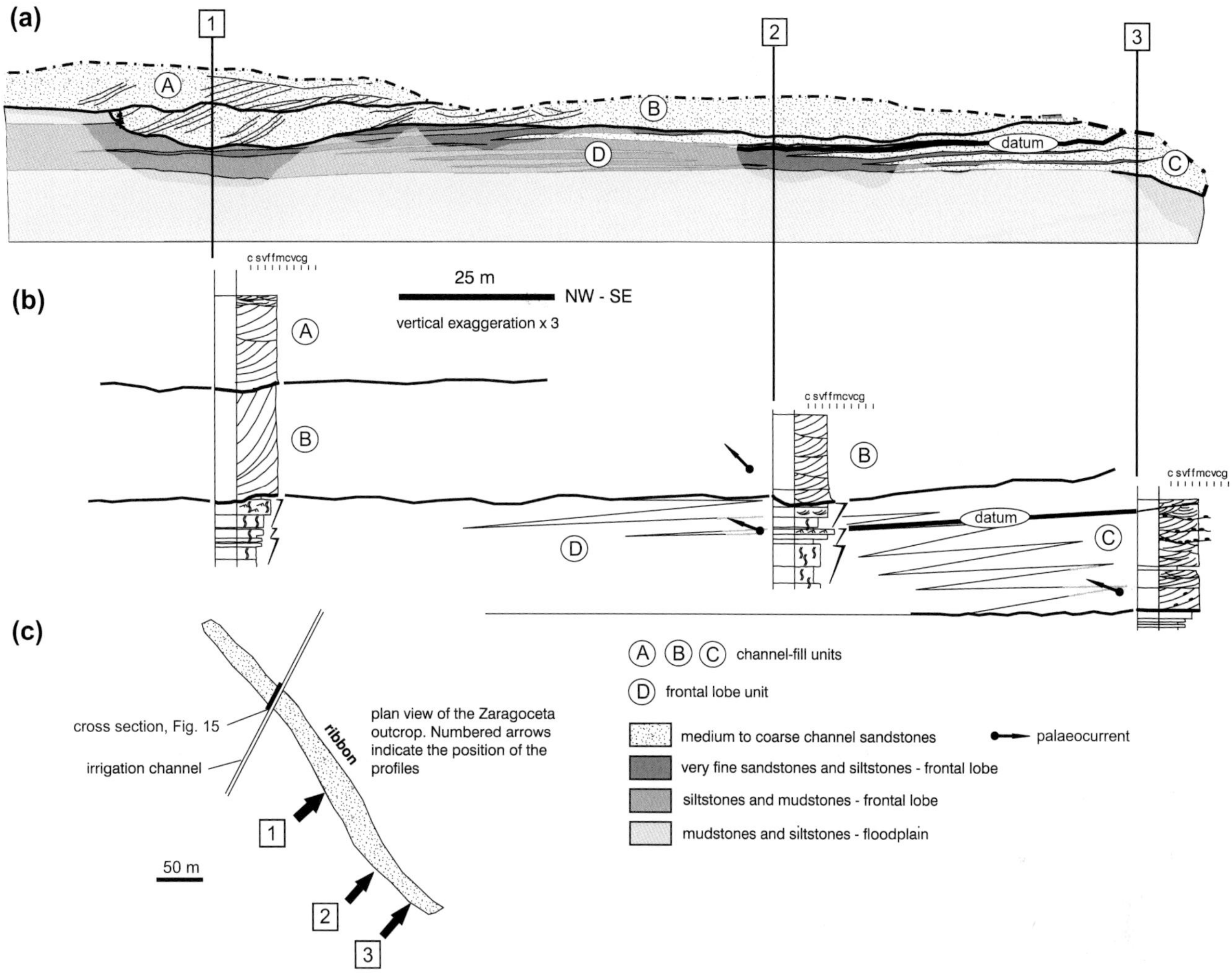

Fig. 14 Zaragoceta outcrop. (a) Longitudinal cross-section of a channel frontal-lobe complex. (b) Sediment logs. (c) Plan view of sandstone ribbon showing log locations. UTM: latitude 4561319, longitude 752727.

facies consists of fine to very fine sandstones with cross-laminated sandstone layers, whereas cross-stratification is almost totally obliterated by bioturbation in the very fine sandstones and siltstones of the distal-lobe facies.

About 50 m northwest of profile 1, a cross-section perpendicular to the palaeocurrent direction shows that channel-fills A and B appear incised on decimetre-scale tabular layers of sandstones organized into coarsening and thickening upwards sequences (Fig. 15), which are equivalent to the upper parts of the frontal-lobe facies of unit D. This fact suggests that, at least at the position of the cross-section of Fig. 15, the lobe facies of unit D were deposited in an unconfined environment. This type of sandstone body, laterally extensive and organized into coarsening and thickening upwards sequences, forms the bulk of the floodplain sandstone bodies in the area studied. Although it can be difficult to differentiate them from levee sandstones, some criteria, such as their organization into clearly defined thickening and coarsening upwards sequences and their wide lateral continuity, with no evidence of lateral transitions to ribbon channel sandstones, suggest that part of the floodplain facies of the Caspe Formation is in fact dominated by frontal-lobe deposits.

Downstream channel incision, instead of upstream channel erosion and head-cut migration, as suggested from the analysis of the Zaragoceta outcrop, was probably the dominant mechanism for channel propagation. Spatial relationships between

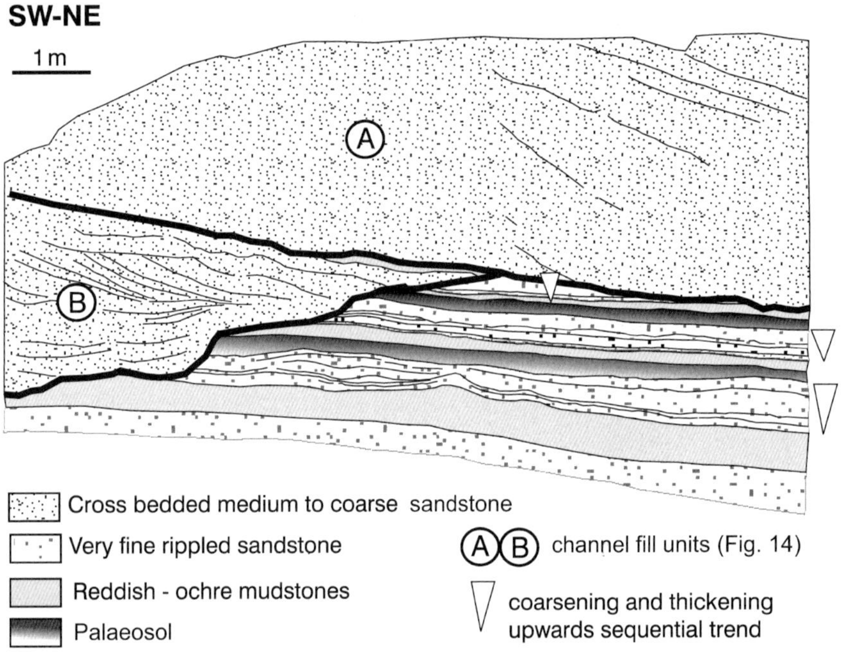

Fig. 15 Cross-section of the Zaragoceta channel-fill complex. UTM: latitude 4561367, longitude 753695.

the different channel fills and the frontal lobe of the Zaragoceta outcrop suggest that the Caspe channels evolved in an episodic manner: phases in which individual channels terminated and developed frontal lobes were followed by phases of channel reactivation and downstream incision (Fig. 16c).

THE NATURE OF THE CASPE FLUVIAL SYSTEM: DISCUSSION

Analysis of the Caspe Formation facies shows a fluvial succession dominated by ribbon-like channel deposits with some distinctive features.

1 An ephemeral regime, which has been inferred from two observations. First the presence of layers of pedogenically modified mudstones and, second, ant nests in the channel fills. Other features, such as the association with aeolian deposits (Miall, 1996), the presence of high-regime plane beds and flash-flood deposits, commonly described in modern and ancient ephemeral streams (Stear, 1983, 1985; Abdullatif, 1989; Reid & Frostick, 1997) have not been recognized in the Caspe Formation, although there is not a conclusive, defined set of features diagnostic of ephemeral streams (Tooth, 2000).
2 Infilling of the channels was by downstream and vertical accretion, leading in most cases to the plugging by sand up to bankfull level.
3 Some of the Caspe channels terminated in the middle of the floodplain, developing frontal lobes, without reaching the distal terminal alluvial zones.
4 Some of the avulsions can be shown to have been controlled by the inherited topography of the reactivated channels.
5 The position of the channels on the floodplain was in some cases controlled by the presence of former channels, which captured the newly forming channels.
6 The areal distribution of ribbons shows no evidence of hierarchical structuring.
7 The form and position of the initial channel scours remained almost unmodified by the channel processes.

The fluvial system is interpreted to have been formed by a network of ephemeral channels, which rarely show signs of lateral migration (point-bar development); some of the channels terminated at frontal lobes. Immaturity of the Caspe fluvial network is inferred from the following observations.

1 Some channel fills show an uneven distribution of sediments longitudinally, with local accumulations (Fig. 12) and irregular profiles. The ephemeral regime may have contributed to the development of ungraded channel profiles.
2 The apparent lack of hierarchy, which suggests that the Caspe channels were short lived.

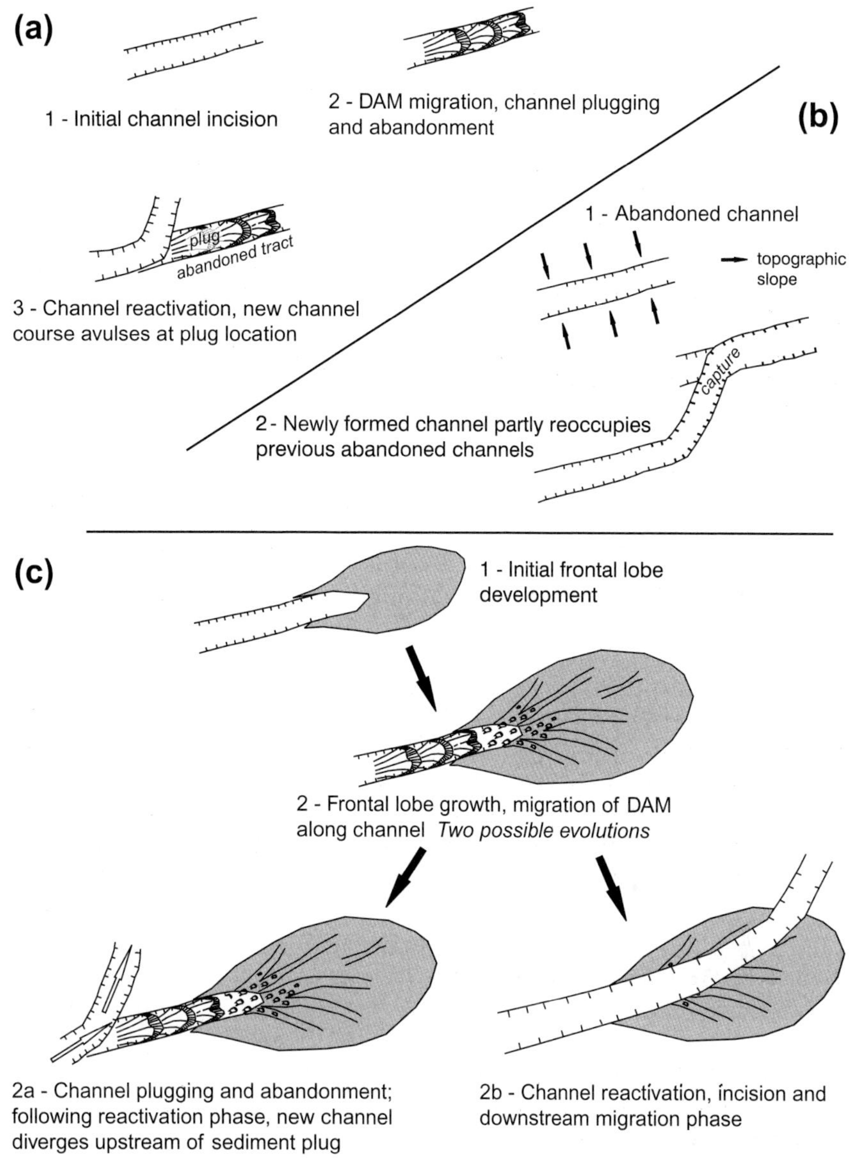

Fig. 16 Evolution of a channel-fill complex. (a) Avulsion process. (b) Reoccupation of abandoned channels. (c) Evolution of channel frontal-lobe complexes.

3 Scarcity of evidence of lateral migration of the channels has been interpreted by Williams (1975) and Friend *et al.* (1979) as probably related to the low erodibility of the banks. On the other hand, such scarcity could be another indicator of the short life time of these channels; channel activity was too short to migrate laterally in an extensive way.

Avulsion, reoccupation and frontal-lobe development were important elements of the fluvial network. Vertical and downstream channel infill led to channel plugging and avulsion in an essentially episodic process (Fig. 16a). The position and planimetric shape of the newly formed and avulsed channels were partially conditioned by the position of antecedent abandoned channels, which in some cases formed topographic lows (Fig. 16b) favouring the capture of the incipient channels. Depressions and probably slope breaks on the floodplain acted as sediment traps that conditioned the development of frontal lobes (Fig. 16c).

The Caspe fluvial system shows some analogies with the Cooper Creek anastomosed, multichannel ephemeral fluvial system (Rust, 1981; Rust & Legun, 1983). The more remarkable similarities are:

1 The Caspe ribbons indicate deposition in a network of low w/t ratio, laterally stable low-sinuosity channels.

Table 1 Comparison of Cooper Creek and Caspe Formation fluvial channels

Parameter	Study area	
	Cooper Creek (Gibling *et al.*, 1998)	Caspe Formation (Williams, 1975)
Width/thickness	< 10:1	7.2
Sinuosity	1.7	1.1
Width (maximum)	100 m	Tens of metres
Channel depth	Typically 3–5 m	3.3 m (mean palaeoflow depth, Mohrig *et al.*, 2000)

Values of w/t ratio, sinuosity and channel dimensions are comparable in both systems (Table 1). Channel lateral stability and restricted lateral migration also has been observed in the Cooper Creek channel, and it has been related to bank cohesivity, and stabilization by plants (Rust, 1981; Gibling *et al.*, 1998; Fagan & Nanson, 2003;) (see Table 1).
2 Cooper Creek channels are ephemeral, as has been inferred for the Caspe channels.
3 Floodplain deposits are fine grained, composed mostly of mud and silt and layers of evaporites in both systems.
4 Climatic conditions in the Cooper Creek area are comparable to the inferred palaeoclimatic setting of the Caspe Formation.
5 Terminal splays are present at some downstream ends of Cooper Creek channels (Gibling *et al.*, 1998).
6 Anabranch formation processes in the anastomosed channels of Cooper Creek have been related to channel constriction by local sediment accumulations (Gibling *et al.*, 1998). The inferred avulsion process described in this paper could be considered, to some extent, as a comparable mechanism of channel-capacity reduction.

Nevertheless, significant differences are observed in both systems. The most prominent are:

1 Cooper Creek channels usually have mud aggregates as an important component of the channel-fill sediments. Fine-grained sediments are less frequent in the Caspe channel fills.
2 The Cooper Creek floodplain evolves into a system of braided channels during high-water stages (Fagan & Nanson, 2004). This type of floodplain environment has not been recognized in the Caspe Formation.
3 Gilgai and vertisol structures, as well as aeolian deposits, are frequent in the Cooper Creek floodplain but have not been recognized in the Caspe Formation.
4 The tectonic context is different in both systems: the Cooper Creek fluvial system drains an intracratonic area, with minor tectonic activity related to fold development (Rust, 1981), whereas the Caspe Formation corresponds to synorogenic sediments related to the development of a fold and thrust belt. Accretion rates are much lower for Cooper Creek (0.04 mm yr^{-1}, Gibling *et al.*, 1998) than for Caspe (1.1 mm yr^{-1} in the distal areas, Barberà *et al.*, 1994).

Taking into account the above-mentioned similarities, together with the morphological features of the ribbon outcrops, two hypotheses concerning the morphology of the Caspe palaeochannels network can be formulated.

A The ribbon sandstone outcrops are the preserved fragments of much longer, highly connected channel-fill sandstones deposited in a network of anastomosed channels.
B The ribbons correspond to the almost complete preservation of essentially discontinuous, disconnected sandstone channel fills. Since temporal and spatial discontinuity in the activity of fluvial systems is a characteristic of arid to semi-arid zones (Tooth, 2000), there is a question as to what extent the present-day outcrops of the Caspe ribbons reflect the true architecture of the channel deposits.

If hypothesis **B** is correct, the ribbons are either the product of a process of discontinuous sedimentation in a network of essentially continuous, possibly anastomosed channels, or they correspond to the infilling of a network of essentially discontinuous channels. Networks of discontinuous ephemeral channels have been widely described in the geomorphological literature (Bull, 1997; Bourke & Pickup, 1999; Tooth, 1999, 2000). Although all these examples have been described in different geomorphological contexts, they share a common feature: a range of semi-arid to arid climatic environments that encompass the palaeoclimatic conditions of the Caspe Formation.

Hypotheses **A** and **B** have very different implications for sandstone connectivity in this type of fluvial deposit: the degree of interconnection of the channel-fill sandstone bodies would be much lower if hypothesis **B** is the case. Sedimentological criteria presented in this work seem to favour hypothesis **B**, although more conclusive observations may arise from geophysical studies of the subsurface structure.

CONCLUDING REMARKS

From the analysis of the facies, a number of conclusions can be drawn:

1 The most frequent channel-fill sequences record the episodic migration of downstream accreting macroforms. Well-developed topset structures overlie the larger foreset cross-beds. This upwards reduction in size of the structures is rather abrupt and corresponds to the downstream accretion arrangement.

2 The infill of the Caspe Formation channels indicates a markedly discontinuous process, as indicated by the presence of pedogenic structures at the upper surfaces bounding successive downstream accretion macroforms. Periods of inactivity long enough to allow the development of such incipient palaeosols were probably longer than seasonal. The presence of ant nests, which in some cases cross the entire thickness of the channel sandstones, is another indicator of strong ephemerality in the hydraulic regime of the channels, and at the very least indicates significant falls in water table.

3 The Caspe Formation channels could terminate, at least during part of their evolution, in frontal lobes. After a phase of lobe construction, phases of channel reactivation and downstream incision followed. The typical sequence of a channel-lobe complex is a composite one (Fig. 13, profiles 1 and 2), with a coarsening and thickening upwards lower interval that corresponds to the lobe construction and an upper interval, corresponding to the fill of a later down-current propagating channel, with an often poorly defined profile.

4 Reoccupation of channels was frequent, as indicated by the stacking of channel fills in the channel-fill complexes. New channels partially followed the trace of the preceding ones, conditioned by the irregular depositional topography of the preserved channel beds.

5 The Caspe facies correspond to the deposits of an immature fluvial system dominated by ephemeral channels. The ribbon sandstone outcrop could be representative of the original architecture of discontinuous channel fills developed in a network of ephemeral, poorly graded channels.

ACKNOWLEDGEMENTS

This research was funded by the research projects of the Spanish Ministerio de Educación y Ciencia CARES (BTE2001-3650) and MARES (CGL2004-05816-C02-02/BTE). The authors thank J.P.P. Hirst, P. Ashworth and G.J. Nichols for their comments that greatly helped to improve the clarity and quality of this paper.

REFERENCES

Abdullatif, O.M. (1989) Channel-fill and sheet-flood facies sequences in the ephemeral River Gash, Kassala, Sudan. *Sediment. Geol.*, **63**, 171–184.

Agustí, J., Cabrera, Ll., Anadón, P. and Arbiol, S. (1988) A late Oligocene–Early Miocene rodent biozonation from the SE Ebro Basin (NE Spain): a potential mammal stratotype. *Newsl. Stratigr.*, **18**(2), 81–97.

Anadón, P., Cabrera, Ll., Guimerà, J. and Santanach, P. (1985) Paleogene strike-slip deformation and sedimentation along the southeastern margin of the Ebro Basin. In: *Strike-slip Deformation, Basin Deformation and Sedimentation* (Eds K. Biddle and N. Christie-Blick), pp. 303–318. Special Publication 37, Society of Economic Paleontogists and Mineralogists, Tulsa, OK.

Anadón, P., Cabrera, Ll., Colombo, F., Marzo, M. and Riba, O. (1986) Syntectonic intraformational unconformities in alluvial fan deposits, eastern Ebro Basin margins (NE Spain). In: *Foreland Basins* (Eds P.A. Allen and P. Homewood), pp. 259–271. Special Publication 8, International Association of Sedimentologists. Blackwell Scientific Publications, Oxford.

Anadón, P., Cabrera, Ll., Colldeforns, B., Colombo, F., Cuevas, J.L. and Marzo, M. (1989) Alluvial fan evolution in the SE Ebro Basin: response to tectonics and lacustrine base level changes. *Excursion Guidebook. 4th International Conference on Fluvial Sedimentology*. Publicacions del Servei Geològic de Catalunya. Barcelona.

Arenas, C., Millán, H., Pardo, G. and Pocovi, A. (2001) Ebro Basin continental sedimentation associated with late compressional Pyrenean tectonics (northeastern Iberia): controls on basin margin fans and fluvial systems. *Basin Res.*, **13**, 65–89.

Barberà, X, Parés, J.M., Cabrera, L. and Anadón, P. (1994) High resolution magnetic stratigraphy across the Oligocene–Miocene boundary in an alluvial-lacustrine succession (Ebro Basin, NE Spain). *Phys. Earth Planet. Inter.*, **85**, 181–193.

Barberà, X., Cabrera, L., Marzo, M., Parés, J.M. and Agustí, J. (2001) A complete terrestrial Oligocene magnetobiostratigraphy from the Ebro Basin, Spain. *Earth Planet. Sci. Lett.*, **187**, 1–16.

Bourke, M.C. and Pickup, G. (1999) Fluvial form variability in arid Central Australia. In: *Varieties of Fluvial Form* (Eds A.J. Miller and A. Gupta), pp. 249–271. John Wiley and Sons, Chichester.

Bridge, J. (2003) *Rivers and Floodplains*. Blackwell Scientific, Oxford, 504 pp.

Bridge, J., Collier, R. and Alexander, J. (1998) Large-scale structure of Calamus River deposits (Nebraska, USA) revealed using ground-penetrating radar. *Sedimentology*, **45**, 977–986.

Bull, W.B. (1997) Discontinuous ephemeral streams. *Geomorphology*, **19**, 227–276.

Cabrera, Ll. (1983) *Estratigrafía y Sedimentología de las formaciones lacustres del tránsito Oligoceno-Mioceno del SE de la Cuenca del Ebro*. Unpublished PhD thesis, University of Barcelona, 443 pp.

Cabrera, Ll. and Sáez, A. (1987) Coal deposition in carbonate-rich shallow lacustrine systems: the Calaf and Mequinenza sequences (Oligocene, Eastern Ebro Basin, NE Spain). *J. Geol. Soc. London*, **144**, 451–461.

Cabrera, L., Colombo, F. and Robles, S. (1985) Sedimentation and tectonics interrelationships in the Paleogene marginal alluvial systems of the SE Ebro Basin. Transition from alluvial to shallow lacustrine environments. In: *Excursion Guide Book of the 6th European Regional IAS Meeting* (Eds M. Mila and J. Rosell), Excursion 10, pp. 393–492. Llieda.

Cabrera, L., Cabrera, M., Gorchs, R. and de las Heras, F.X.C. (2002) Lacustrine basin dynamics and organosulphur compound origin in a carbonate-rich lacustrine system (Late Oligocene Mequinenza Formation, SE Ebro Basin, NE Spain). *Sediment. Geol.*, **148**, 289–317.

Crowley, K.D. (1983) Large-scale bed configurations (macroforms), Platter River Basin, Colorado and Nebraska: primary structures and formative processes. *Geol. Soc. Am. Bull.*, **94**, 117–133.

Dreyer, T. (1993) Quantified fluvial architecture in ephemeral stream deposits of the Esplugafreda Formation (Paleocene), Tremp-Graus Basin, northern Spain. In: *Alluvial Sedimentation* (Eds M. Marzo and C. Puigdefàbregas), pp. 37–49. Special Publication 17, International Association of Sedimentologists. Blackwell Scientific Publications, Oxford.

Fagan, S.D. and Nanson, G.C. (2004) The morphology and formation of floodplain-surface channels, Cooper Creek, Australia. *Geomorphology*, **60**, 107–126.

Friend, P.F., Slater, M.J. and Williams, R.C. (1979) Vertical and lateral building of river sandstone bodies, Ebro Basin, Spain. *J. Geol. Soc. London*, **136**, 39–46.

Friend, P.F., Hirst, J.P.P. and Nichols, G.J. (1986) Sandstone-body structure and river process in the Ebro Basin of Aragon, Spain. *Cuad. Geol. Ibérica*, **10**, 9–30.

Gibling, M.R., Nanson, G.C. and Maroulis, J.C. (1998) Anastomosing river sedimentation in the Channel Country of Central Australia. *Sedimentology*, **45**, 595–619.

González, A. and Guimerà, J. (1997) Marco estructural de la sedimentación durante el Mioceno inferior en el extremo meridional de la cuenca del Ebro. In: *Avances en el conocimiento del Terciario Ibérico* (Eds J.P. Calvo and J. Morales), pp. 97–100. Museo Nacional de Ciencias Naturales. Madrid.

Guimerà, J. (1984) Paleogene evolution of deformation in north eastern Iberian Peninsula. *Geol. Mag.*, **121**, 413–420.

Hasiotis, S.T. (2003) Complex ichnofossils of solitary and social organisms: understanding their evolution and roles in terrestrial paleoecosystems. *Palaeogeogr, Palaeoclimatol. Palaeoecol.*, **192**, 259–320.

Hirst, J.P.P. and Nichols, G.J. (1986) Thrust tectonic controls on Miocene alluvial distribution patterns, southern Pyrenees. In: *Foreland Basins* (Eds P.A. Allen and P. Homewood), pp. 247–258. Special Publication 8, International Association of Sedimentologists. Blackwell Scientific Publications, Oxford.

Jones, L.S. and Schumm, S.A. (1999) Causes of avulsion: an overview. In*: Fluvial Sedimentology VI* (Eds N.D. Smith and J. Rodgers), pp. 171–178. Special Publication 28, International Association of Sedimentologists. Blackwell Science, Oxford.

Jones, M.A., Séller, P.L., Roca, E., Garcés, M. and Cabrera, Ll. (2004) Time lag of syntectonic sedimentation across an alluvial basin: theory and example from the Ebro Basin, Spain. *Basin Res.*, **16**, 467–488.

Klimowitz, J. (1992) Estratigrafía y disposición estructural del Terciario Inferior en el subsuelo del Sector

Central de la Cuenca del Ebro. *Acta Geol. Hisp.*, **27**, 117–125.

Miall, A.D. (1988) Reservoir heterogeneities in fluvial sandstones: lessons from outcrop studies. *Am. Assoc. Petrol. Geol. Bull.*, **72**, 682–697.

Miall, A.D. (1996) *The Geology of Fluvial Deposits*. Springer-Verlag, Berlin, 504 pp.

Mohrig, D., Heller, P., Paola, C. and Lyons, W.J. (2000) Interpreting avulsion processes from ancient alluvial sequences: the Guadalope-Matarranya system (northern Spain) and Wasatch Formation (western Colorado). *Geol. Soc. Am. Bull.*, **112**, 1787–1803.

Nichols, G.J. and Hirst, J.P.P. (1998) Alluvial fans and fluvial distributary systems, Oligo-Miocene, northern Spain: contrasting processes and products. *J. Sediment. Res.*, **68**, 879–889.

Pérez Rivares, F.J., Garcés, M., Arenas, C. and Pardo, G. (2004) Magnetostratigraphy of the Miocene continental deposits of the Montes de Castejón (central Ebro Basin, Spain): geochronological and paleoenvironmental implications. *Geol. Acta*, **2**, 221–234.

Reid, I. and Frostick, L.E. (1997) Channel form, flows and sediments in deserts. In: *Arid Zone Geomorphology: Process, Form and Change in Drylands* (Ed. D.S.G. Thomas), pp. 205–229. Wiley and Sons, Chichester.

Riba, O., Villena, J. and Quirantes, J. (1967) Nota preliminar sobre la sedimentación en paleocanales terciarios de la zona Caspe-Chiprana (Provincia de Zaragoza). *Anal. Edafol. Agrobiol.*, **XXVI**, 1–4. Madrid.

Rundle, A. (1985) Braid morphology and the formation of multiple channels. The Rakaia, New Zealand. *Z. Geomorphol.*, **55**, 15–37.

Rust, B.R. (1981) Sedimentation in an arid-zone anastomosing fluvial system: Cooper's Creek, Central Australia. *J. Sediment. Petrol.*, **51**, 745–755.

Rust, B.R. and Legun, A.S. (1983) Modern anastomosing fluvial deposits in arid Central Australia, and a Carboniferous analogue in New Brunswick, Canada. In: *Modern and Ancient Fluvial Systems* (Eds J.D. Collinson and J. Lewin), pp. 385–392. Special Publication 6, International Association of Sedimentologists. Blackwell Science, Oxford.

Slingerland, R. and Smith, N.D. (2004) River avulsions and their deposits. *Ann. Rev. Earth. Planet. Sci.*, **32**, 257–285.

Smith, A.G. (1996) Cenozoic latitudes, positions and topography of the Iberian Peninsula. In: *Tertiary Basins of Spain* (Eds P.F. Friend and C.J. Dabrio), pp. 6–9. Cambridge University Press.

Stear, W.M. (1983) Morphological characteristics of ephemeral stream channel and overbank splay sandstone bodies in the Permian Lower Beaufort Group, Karoo Basin, South Africa. In: *Modern and Ancient Fluvial Systems* (Eds J.D. Collinson and J. Lewin), pp. 405–420. Special Publication 6, International Association of Sedimentologists. Blackwell Science, Oxford.

Stear, W.M. (1985) Comparison of the bedform distribution and dynamics of modern and ancient sandy ephemeral flood deposits in the southwestern Karoo region, South Africa. *Sediment. Geol.*, **45**, 209–230.

Tooth, S. (1999) Floodouts in Central Australia. In: *Varieties of Fluvial Form. (Eds A.J. Miller and A. Gupta)*, pp. 217–249. John Wiley and Sons, Chichester.

Tooth, S. (2000) Process, form and change in dryland rivers: a review of recent research. *Earth-Sci. Rev.*, **51**, 67–107.

Williams, R.C. (1975) *Fluvial deposits of Oligo-Miocene of Southern Ebro Basin, Spain*. Unpublished PhD thesis, University of Cambridge, 220 pp.

Index

Note: page numbers in *italics* refer to figures, those in **bold** refer to tables